Zurich Lectures in Advanced Mathematics

Edited by

Mathematics in Zurich has a long and distinguished tradition, in which the writing of lecture notes volumes and research monographs plays a prominent part. The *Zurich Lectures in Advanced Mathematics* series aims to make some of these publications better known to a wider audience. The series has three main constituents: lecture notes on advanced topics given by internationally renowned experts, in particular lecture notes of "Nachdiplomvorlesungen", organized jointly by the Department of Mathematics and the Institute for Research in Mathematics (FIM) at ETH, graduate text books designed for the joint graduate program in Mathematics of the ETH and the University of Zürich, as well as contributions from researchers in residence. Moderately priced, concise and lively in style, the volumes of this series will appeal to researchers and students alike, who seek an informed introduction to important areas of current research.

Previously published in this series:

Yakov B. Pesin, *Lectures on partial hyperbolicity and stable ergodicity*
Sun-Yung Alice Chang, *Non-linear Elliptic Equations in Conformal Geometry*
Sergei B. Kuksin, *Randomly forced nonlinear PDEs and statistical hydrodynamics in 2 space dimensions*
Pavel Etingof, *Calogero-Moser systems and representation theory*
Guus Balkema and Paul Embrechts, *High Risk Scenarios and Extremes – A geometric approach*
Demetrios Christodoulou, *Mathematical Problems of General Relativity I*
Camillo De Lellis, *Rectifiable Sets, Densities and Tangent Measures*
Paul Seidel, *Fukaya Categories and Picard–Lefschetz Theory*
Alexander H.W. Schmitt, *Geometric Invariant Theory and Decorated Principal Bundles*
Michael Farber, *Invitation to Topological Robotics*
Alexander Barvinok, *Integer Points in Polyhedra*
Christian Lubich, *From Quantum to Classical Molecular Dynamics: Reduced Models and Numerical Analysis*
Shmuel Onn, *Nonlinear Discrete Optimization – An Algorithmic Theory*
Kenji Nakanishi and Wilhelm Schlag, *Invariant Manifolds and Dispersive Hamiltonian Evolution Equations*
Erwan Faou, *Geometric Numerical Integration and Schrödinger Equations*
Alain-Sol Sznitman, *Topics in Occupation Times and Gaussian Free Fields*
François Labourie, *Lectures on Representations of Surface Groups*
Isabelle Gallagher, Laure Saint-Raymond and Benjamin Texier, *From Newton to Boltzmann: Hard Spheres and Short-range Potentials*
Robert J. Marsh, *Lecture Notes on Cluster Algebras*
Emmanuel Hebey, *Compactness and Stability for Nonlinear Elliptic Equations*
Sylvia Serfaty, *Coulomb Gases and Ginzburg–Landau Vortices*
Alessio Figalli, *The Monge–Ampère Equation and Its Applications*
Walter Schachermayer, *Asymptotic Theory of Transaction Costs*
Anne Thomas, *Geometric and Topological Aspects of Coxeter Groups and Buildings*

Published with the support of the Huber-Kudlich-Stiftung, Zürich

Todd Fisher
Boris Hasselblatt

Hyperbolic Flows

European Mathematical Society

Authors:

Todd Fisher
Department of Mathematics
Brigham Young University
Provo, UT 84602
USA

E-mail: tfisher@math.byu.edu

Boris Hasselblatt
Department of Mathematics
Tufts University
503 Boston Avenue
Medford, MA 02155
USA

E-mail: boris.hasselblatt@tufts.edu

2010 Mathematics Subject Classification (primary; secondary): 37D40, 37D20; 37A30, 37A35

Key words: hyperbolic, hyperbolicity, flow, ergodic theory, topological dynamics, rigidity, expansiveness, shadowing, specification, geodesic flow, Anosov flow, Axiom A, entropy, equilibrium states, stable manifold, topological pressure, symbolic flows, Markov partitions

ISBN 978-3-03719-200-9

The Swiss National Library lists this publication in The Swiss Book, the Swiss national bibliography, and the detailed bibliographic data are available on the Internet at http://www.helveticat.ch.

Contact address:

European Mathematical Society Publishing House
TU Berlin, Mathematikgebäude
Strasse des 17. Juni 136
10623 Berlin
Germany

Email: info@ems-ph.org
Homepage: ems-ph.org

Typeset using the author's TEX files: Alison Durham, Manchester, UK
Printing and binding: Beltz Bad Langensalza GmbH, Bad Langensalza, Germany
∞ Printed on acid free paper
9 8 7 6 5 4 3 2 1

Anatole Katok
in memoriam

(Author: Konrad Jacobs. Source: Photo collection of the Mathematical Research Institute of Oberwolfach, Germany.)

Acknowledgments

On Thu, 15 May 2014, Todd Fisher wrote,

```
Boris,
I hope things are going well for you.  I have an undergraduate student working
on a problem about hyperbolic flows.  I am having a hard time coming up with a
survey or reference for the background material.  Mostly, I find books and
surveys on hyperbolic maps and then in the papers they apply these to flows
without specific references.  It would help his understanding if there was such
a reference.  I was hoping you might know of one.  If you do can you send me
some places to look.
Best, Todd
```

On Thu, May 15, 2014 and Fri, 16 May 2014 Boris Hasselblatt wrote,

```
Dear Todd,
Greetings from Somerville.  I hope that you are well.
I am here for a few days between a 6-month stint in Marseille and a month in
Tokyo.  3 cheers for sabbaticals.
...
“Introduction to the Modern Theory of Dynamical Systems” is largely guilty of
the sins you mention, but Tolya and I have long been working on a successor, and
some of the intent is to have fewer transgressions of this sort (also in ergodic
theory).  I could send the current PDF...
...
While we do more with flows (with respect to hyperbolicity but also ergodic
theory) than in “Introduction to the Modern Theory of Dynamical Systems,” there
may be a place for a survey that makes a point of it.  It might be interesting
to write one, and I am game.
...
Best regards
Boris
```

Thus emerged the modest germ of a plan to collect in one place the core of hyperbolic dynamics with continuous time. That summer, Hasselblatt became Associate Provost at Tufts, which significantly impacted this project. Modest progress occurred until Fisher was able to spend a significant amount of time on the book in the first half of 2016 during a sabbatical. In 2016, Hasselblatt's assignment of Associate Provost at Tufts ended with the commitment by the Office of the Provost to fund a semester leave. Soon after, Hasselblatt was also informed that a previously deferred sabbatical was due to expire if not taken by the spring semester of 2018. Thus, a spring 2017 topics course at Tufts, spring 2018 Nachdiplom lectures in Zürich, and fall 2018 lectures in Tokyo hugely boosted this project, and it grew well beyond its initial scope by embracing a far broader view of "core," including some new results, and by encompassing a concise and lively introduction to important areas of current research, variously with proofs, proof outlines, proof ideas, or references to proofs.

There is much to acknowledge that has significantly helped us write this book. We owe a debt to our respective institutions for their support, and to the Simons Foundation for providing research support for the book.[1] We also want to thank

---

[1] Simons Collaborative Grants # 585027 and # 587001.

students from Brigham Young University, Tufts University, Brandeis University, the University of Tokyo, and the ETH Zürich for their forbearance, support, and criticism;[2] colleagues and students who commented helpfully on book drafts from afar; Manfred Einsiedler and Michael Struwe for arranging the Nachdiplom lectures at the ETH in the spring of 2018; their colleagues and Grete Einsiedler for making Zürich a home; Takashi Tsuboi and Masahiko Kanai for arranging lectures on hyperbolic flows at the Graduate School of Mathematical Sciences of the University of Tokyo in the fall of 2018; their colleagues in Tokyo, Kyoto, Nagoya, and beyond for their interest and warm hospitality; Thomas Kappeler for shepherding this project into publication; and Thomas Hintermann from the European Mathematical Society for his enthusiasm, wisdom, and support in all stages of the publication process over more than a year, right up to the start of the publication process; and Vera Spillner who, with Sylvia Fellmann and Alison Durham, took this project to completion from there with exemplary attention to quality. It seems highly appropriate and satisfying that during the last stages of writing, Boris Hasselblatt was a department colleague of Masahiko Kanai, whose work was foundational for substantial parts of the rigidity theory described near the end of the book, as well as of Shuhei Hayashi, who with his proof of the stability theorems for hyperbolic flows placed one of the crowning glories atop hyperbolic dynamics in the 20th century. Toshitake Kohno, Dean of the Graduate School of Mathematical Sciences, not only provided a most conducive environment but also access to the model collection and permission to make the photograph in Figure 5.2.6. Boris Hasselblatt is grateful for and will long remember the generous hospitality and outstanding working conditions of both the Graduate School of Mathematical Sciences at the University of Tokyo and the Forschungsinstitut für Mathematik at the ETH Zürich. Both of us thank our respective home institutions for the leaves which made this project possible. We are also grateful to the faculty writing group in the Department of Mathematics at Brigham Young University for the many suggestions and improvements they provided.

Some of this book is owed to earlier books and research articles by one or the other of us, which included text we deemed—in more or less adapted form—to be an excellent fit for this work. This implies a debt to our respective coauthors of such prior works, Anatole Katok foremost among them. Indeed, some of this text is adapted from [213] and from unfinished projects of Katok and Hasselblatt. In some cases, original research papers by others still remain the best exposition of ideas we could not omit from this book, so it will be apparent and often explicit where we followed their ideas; Bowen foremost comes to mind. And occasionally, unpublished lecture notes (such as by Lanford at the ETH) provided the most elegant proofs we know of a needed fact.

[2]In the Talmud, R. Chanina remarked, "I have learned much from my teachers, more from my colleagues, and the most from my students" (Ta'anis 7a).

In that category we particularly appreciate having the permission of Flavio Abdenur and Marcelo Viana to reproduce their proof of absolute continuity of the invariant foliations in the generality of partially hyperbolic dynamical systems (Section B.7.b). We are also grateful to the countless colleagues who generously answered questions, read drafts, and commented on the text, Clark Butler, Vaughn Climenhaga, and Daniel Thompson foremost among them: Ethan Akin, Joseph Auslander, Lennard Bakker, Aaron Brown, Keith Burns, Manfred Einsiedler, Sergio Fenley, Andrey Gogolev, Shuhei Hayashi, François Ledrappier, and Davi Obata. Several of them also encouraged us greatly by pointing out just how valuable a reference this text is, even for researchers.

Above all, we are grateful to our families for the support they provided for our obsession and absences. Mary Fisher was supportive of her husband, while he traveled often, so he could work on the book. Kathleen Hasselblatt deserves particular gratitude and recognition for dealing with the adversities of an old house and the various challenges of New England seasons while her husband lived on other continents for most of 2018—during which year some 500 pages of this book were written. We could not have done it without them.

Provo and Medford, May 2019
Todd Fisher, Boris Hasselblatt

Contents

* = optional

0

Introduction

0.1 About this book

This book presents the theory of flows, that is, continuous-time dynamical systems from the topological, smooth, and measurable points of view, with an emphasis on the theory of (uniformly) hyperbolic dynamics. It includes both an introduction and an exposition of recent developments in uniformly hyperbolic dynamics, and it can be used as both a textbook and a reference for students and researchers.

Books on dynamics tend to focus on discrete time, largely leaving it to the reader (or unaddressed) to transfer those insights to flows, where the origins of the theory actually lie.[1] It is thus often implicit that "things work analogously for flows," or that "this is different for flows," and aside from geodesic flows, many theorems about flows have had little visibility beyond the research literature. Although much about flows can indeed be found in the research literature, doing so usually involves a combination of diligence and consultation with experts. We fill this gap in the expository literature by giving a deep "flows-first" presentation of dynamical systems and focusing on continuous-time systems, rather than treating these as afterthoughts or exceptions to methods and theory developed for discrete-time systems.

We point to a few additional features of interest and to some new results in this book:

- Chapter 5 is to our knowledge unique in the literature for the extent to which it implements the Anosov–Katok–Bowen program of developing the dynamical features of hyperbolic sets for flows from shadowing alone.[2]
- Section 5.2 and Chapter 8 provide an exceptional range of examples of (uniformly) hyperbolic flows. To our knowledge, these are more examples than have appeared in any one other source, in no small part because several of them are quite recent discoveries.
- Chapter 5 may be the first account to provide a proper, natural definition of a

[1]This might in some part be because there are simpler examples available in discrete time, and longitudinal issues do not obscure the main effects of hyperbolicity—however, these longitudinal effects are quite interesting and indeed define the forefront of some research areas in dynamical systems.

[2]Specifically, the Shadowing Lemma and the Shadowing Theorem, which include uniqueness, so in terms of customary usage one should say that shadowing and expansivity produce the insights in Chapter 5.

(uniformly) hyperbolic flow (Definition 5.3.50) based on the equivalence of the three popular notions (Theorem 5.3.47 on page 305). Although this equivalence is not new, it does not seem to be broadly known.

- Section 6.6 gives a stronger theorem about existence of Markov sections than anywhere else in the literature.
- Other new results are that discreteness of centralizers is a *topological* fact (Theorem 9.1.3) and our results on trivial centralizers in Section 9.1.
- In addition to topological and smooth dynamics, we cover the ergodic theory of flows to a considerable extent, and this as well may be singular in the literature—while most of what we present can be found *somewhere* in the (often original research) literature, the ergodic theory of flows is not common textbook material.
- We also call attention to a proof (by Abdenur and Viana) of absolute continuity of the invariant foliations in the generality of partially hyperbolic dynamical systems (Section B.7). This exceeds what we need, but seemed like a desirable addition to the literature.
- Chapters 8 and 9 include a range of advanced topics mainly from the theory of Anosov flows (such as their topology and dynamics, as well as rigidity phenomena); some of these are from recent research and have not previously appeared in any expository literature.
- At the end of the main chapters there are a number of exercises.
- Thorough indexing facilitates the use of this book as a reference.

Chapters 8 and 9 are no less accessible than the introductory subjects, but here we take even more opportunities to augment the results we prove with complements whose proofs we do not include, or to outline proofs rather than giving full proofs. This is meant to provide a substantial introduction to these subjects with proofs.

As complements to this book the reader can choose from an abundance of books on ergodic theory and dynamical systems [221, 204, 289, 348, 317, 97, 98, 274, 329, 55, 70, 112, 32, 283, 251, 242, 263, 264, 315, 301, 334, 279, 268, 281, 257, 113, 114, 154, 176, 178, 133, 71]. Here we single out just a brisk introduction in a similar spirit that focuses on discrete time [174], the rather larger books [213] and [314], and the more example-driven text [177].

The text is divided into two parts. The first of these develops the general theory of flows (that is, not assuming hyperbolicity, but with a bias toward those aspects of the theory that are most pertinent to hyperbolic flows), both in the topological and measurable realm. The second part is about hyperbolicity and includes an introduction, advanced material, and a panorama of current topics. The book is self-contained in the technical sense, that is, it includes definitions of all dynamics concepts with which we work, but without any pretense to being comprehensive with introductory material.

We intend this book to be useful for courses, directed study, self-study, and as a reference. For the latter, the broad and deep coverage combined with thorough indexing should be helpful. It has been written in a way that it can be adapted to a course (or independent study) in a number of different ways, depending on the purpose of the course. Starred chapters and sections are optional. They are not necessarily "harder," but the material is not needed for further sections except for an occasional result that can be used as a black box. Much of this material is hard to find in the literature except for original sources.

The core chapters are Chapters 1, 5, and 6. If one wants to emphasize ergodic properties of flows then one could include Chapters 3, 4, and 7, or at least portions of them. For a more topological or geometric course one would instead include Chapter 2, and portions from Chapters 8 or 9 (several sections of the latter invoke some ergodic theory, however). A topics course, especially to an audience with some prior knowledge, could more extensively cover those last chapters. The core chapters include exercises.

The appendices contain background material on discrete-time dynamical systems, some of which is invoked on a few occasions in the main text. Those already familiar with it can omit it, refer to it as needed, or review it quickly. For those not familiar with the discrete-time theory, the appendices should provide sufficient background to understand either the material on ergodic theory in Chapter 3 or the material on invariant foliations in Chapter 6.

0.2 Continuous and discrete time

To give our selection of flows versus discrete-time systems some context, we describe a few connections between these. Historically, dynamical systems came about in the form of flows, such as those that arise from differential equations that describe a mechanical system. Poincaré is widely regarded as the founder of the discipline of dynamical systems as we know it, and among the wealth of notions he created is that of a local section, known also as a Poincaré section. This is natural when using periodic orbits (trajectories) of a continuous-time dynamical system as anchors to study other motions in the system. Such a nearby motion will track the periodic motion for possibly considerable amounts of time, and it is often of less interest whether it lags or leads a little than how it moves closer to or further from the periodic orbit. To focus on these transverse phenomena Poincaré considered a small hypersurface perpendicular to the periodic orbit on which he could track successive "hits" by a nearby motion. This defines a map on this disk, called the Poincaré (first) return map; see Figure 0.2.1. This is an early way in which discrete-time dynamical systems arose.

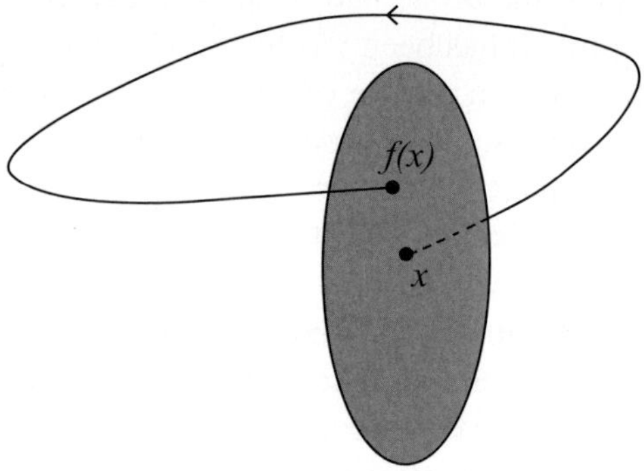

Figure 0.2.1. Poincaré section and map. [Reprinted from [213] (© Cambridge University Press, all rights reserved) with permission.]

Coming from a different direction, billiard systems illustrate how a similar approach works both naturally and globally. A mathematical billiard system idealizes physical billiards by ignoring the spin and rolling of the balls: a point particle moves along straight lines and is reflected in the boundary with incoming angle equal to outgoing angle. This makes them more like air hockey or a description of light in a mirrored room, and tables of shapes other than rectangular are of considerable interest. These

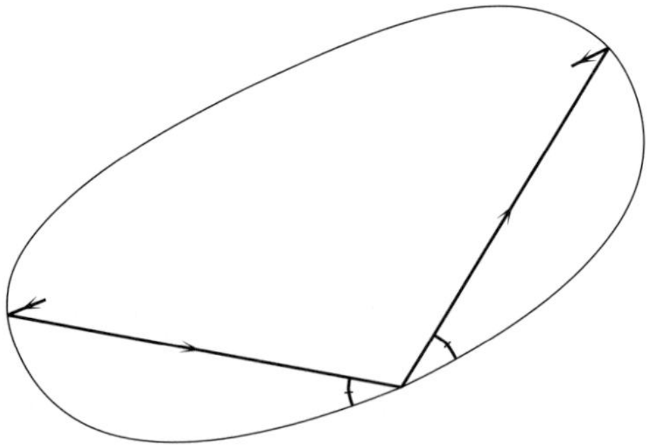

Figure 0.2.2. Billiard. [Reprinted from [213] (© Cambridge University Press, all rights reserved) with permission.]

are naturally continuous-time systems, but they come with natural discrete moments in time: the moments in which collisions occur. Indeed, all information about the evolution of such a system is contained in the locations and velocities of all balls at the moment of a collision, because this determines the motion until the next collision and the positions and velocities at that subsequent moment. Therefore, the dynamics can be described as a map on the "collision space" that sends each collision configuration to the next one. Once again, a discrete-time system describes the dynamics of a continuous-time system.[3]

This latter process can be reversed. Given the discrete-time system, one more piece of information reconstructs the flow entirely: the "return time" from one collision to the next. We call this assembly of a map and a return-time function a suspension if the return time is constant (Definition 1.2.8), and a special flow or flow under a function otherwise (Definition 1.2.11).

We digress to note that some discrete-time dynamical systems arise directly in scientific problems, such as the population biology of species with nonoverlapping generations (cicadas, for instance). While the preceding process could be used to embed this in continuous-time dynamics, this is neither helpful nor meaningful.

There are also aspects of dynamics in which pronounced differences between flows and discrete-time systems are manifested. On one hand, this occurs when "longitudinal" effects matter, that is, when time changes make a difference. In the case of a special flow this amounts to properties that are affected by the choice of "roof" or return-time function versus those that are not. For instance, the existence of a dense orbit is unaffected by the choice of roof function, but whether all periodic orbits are commensurate (their periods are various multiples of one positive number) clearly does depend on return times. Another notable feature of flows is that they permit surgery constructions to construct new flows. Accordingly, such a construction establishes that Anosov flows need not have a dense orbit (Section 8.3), but it is a long-open and exceedingly difficult problem to decide whether Anosov *diffeomorphisms* always have a dense orbit. In fact, it is not even known whether every Anosov diffeomorphism has a fixed point.

The theory of continuous-time dynamical systems does not directly reduce to that of discrete-time dynamical systems in the most obvious way: few diffeomorphisms arise as time-t maps of flows (Definition 1.1.1) since (every time-t map of) every flow is isotopic to the identity.[4] Also, time-t maps of flows have "roots" of all orders, being the nth iterate of the time-t/n map. But one might say that a full continuous-time theory yields a full discrete-time theory because every diffeomorphism can be represented as a Poincaré section for some flow via the suspension/special-flow construction—provided one has a comprehensive understanding of the dynamics of

[3]This goes back to Birkhoff; see the discussion leading up to Theorem 5.2.49.

[4]One point of view from which flows produce a "sparse" set of maps of a given manifold is related to the mapping-class group. For a manifold M the *mapping-class group* is the set of isotopy-classes of homeomorphisms (or diffeomorphisms) of M. Flows are contained in the trivial equivalence class of the mapping-class group.

a section in terms of that of the flow. This does not work in reverse because that construction is not unique, and many flows generate a given diffeomorphism, with confounding "longitudinal" effects as above.

More to the point, for the study of hyperbolic flows (Chapter 5) it may be useful to know all about hyperbolic maps, but that theory does not apply directly to time-1 maps of flows unless the periodic points for the flow are all hyperbolic equilibria. More specifically, the time-t map of a hyperbolic flow satisfies a weaker condition called partial hyperbolicity due to the flow direction, in which neither contraction nor expansion occur. Thus, this "*flows-first*" book complements the existing literature emphasizing discrete-time systems.

Once more, beyond the general theory, our emphasis is on uniformly hyperbolic dynamics. Neither partial nor nonuniform hyperbolicity are themselves subjects in this book. (The sole exception being the proof of absolute continuity of the invariant foliations for partially hyperbolic diffeomorphisms: while it is provided here to be applied to uniformly hyperbolic flows via time-1 maps, the proof covers partially hyperbolic diffeomorphisms in full generality.)

In short, discrete-time dynamics and continuous-time dynamics have closely related toolkits and close interactions, but the discrete-time focus of the existing literature leaves room for an explicit presentation of continuous-time dynamics.[5]

0.3 Historical sketch

We now outline some of the developments that brought about the theory of hyperbolic flows.[6] There are several intertwined strands of the history of hyperbolic dynamics, including geodesic flows and statistical mechanics on one hand, and hyperbolic phenomena ultimately traceable to some application of dynamical systems. Geodesic flows were studied, for example, by Hadamard, Hedlund, Hopf (primarily either on surfaces or in the case of constant curvature) and Anosov–Sinai (negatively curved surfaces and higher-dimensional manifolds). Other hyperbolic phenomena appear in the work of Poincaré (homoclinic tangles in celestial mechanics [295]), Perron (differential equations [285]), Cartwright, Littlewood (relaxation oscillations in radio circuits [87, 88, 243]), Levinson (the van der Pol equation [241]) and Smale (horseshoes [338, 337]), to name a few.

0.3.a Homoclinic tangles and negative curvature. The advent of complicated dynamics took place in the context of Newtonian mechanics, according to which

[5] To be clear, the *research* literature does not omit the continuous-time theory altogether; it is among *books* that this work occupies a unique place.

[6] An expanded version can be found in [174].

simple underlying rules governed the evolution of the world in clockwork fashion. The successes of classical and especially celestial mechanics in the 18th and 19th centuries were seemingly unlimited and Pierre Simon de Laplace felt justified in saying (in the opening passage he added to [231, p. 2]),

> Nous devons donc envisager l'état présent de l'univers, comme l'effet de son état antérieur, et comme la cause de celui qui va suivre. Une intelligence qui pour un instant donné, connaîtrait toutes les forces dont la nature est animée, et la situation respective des êtres qui la composent, si d'ailleurs elle était assez vaste pour soumettre ces données à l'analyse, embrasserait dans la même formule les mouvemens des plus grands corps de l'univers et ceux du plus léger atome: rien ne serait incertain pour elle, et l'avenir comme le passé, serait présent à ses yeux.[7]

The enthusiasm in this passage is understandable and its forceful description of (theoretical) determinism is a good anchor for an understanding of one of the basic aspects of dynamical systems. Moreover, the titanic life's work of Laplace in celestial mechanics earned him the right to make such bold pronouncements. Another bold pronouncement of his, that the solar system is stable, came under renewed scrutiny later in the 19th century, and Henri Poincaré was expected to win a competition to finally establish this fact. However, Poincaré came upon hyperbolic phenomena in revising his prize memoir [295] on the three-body problem. He found that a phenomenon now called homoclinic tangles (Figure 6.5.1) (which he had initially overlooked) caused great difficulty and necessitated essentially a reversal of the main thrust of that memoir [34]. He perceived that there is a highly intricate web of invariant curves and that this situation produces dynamics of unprecedented complexity:

> Que l'on cherche à se représenter la figure formée par ces deux courbes et leurs intersections en nombre infini dont chacune correspond à une solution doublement asymptotique, ces intersections forment une sorte de treillis, de tissu, de réseau à mailles infiniment serrées; chacune des deux courbes ne doit jamais se recouper elle-même, mais elle doit se replier sur elle-même d'une manière trés complexe pour venir recouper une infinité de fois toutes les mailles du réseau. On sera frappé de la complexité de cette figure, que je ne cherche même pas à tracer.[8]

[7] We ought then to consider the present state of the universe as the effects of its previous state and as the cause of that which is to follow. An intelligence that, at a given instant, could comprehend all the forces by which nature is animated and the respective situation of the beings that make it up, if moreover it were vast enough to submit these data to analysis, would encompass in the same formula the movements of the greatest bodies of the universe and those of the lightest atoms. For such an intelligence nothing would be uncertain, and the future, like the past,

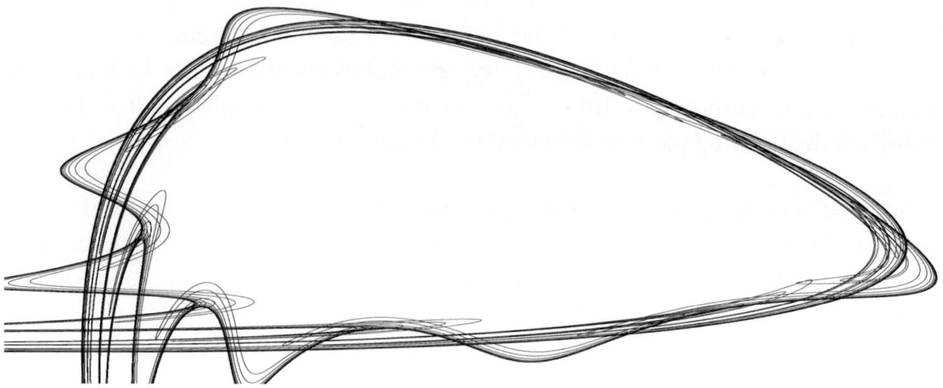

Figure 0.3.1. Homoclinic tangles. [Reprinted from [213] (© Cambridge University Press, all rights reserved) with permission.]

This is often viewed as the moment chaotic dynamics was first noticed. He concluded that in all likelihood the prize problem could not be solved as posed; which was to find series expansions for the motions of the bodies in the solar system that converge uniformly for all time. Indeed, when Birkhoff picked up the study of this situation in his prize memoir [51] for the Papal Academy of Sciences, he noted that and described how this implies complicated dynamics [51, p. 184] (Theorem 6.5.2).

0.3.b Geodesic flows. A major class of mathematical examples motivating the development of hyperbolic dynamics is that of geodesic flows (that is, free-particle motion) of Riemannian manifolds of negative sectional curvature. Hadamard considered (noncompact) surfaces in $\mathbb{R}^3$ of negative curvature [166] and found, with apparent delight, that if the unbounded parts are "large" (do not pinch to arbitrarily small diameter as you go outward along them) then at any point the initial directions of bounded geodesics form a Cantor set. Since only countably many directions give geodesics that are periodic or asymptotic to a periodic one, this also proves the existence of more complicated bounded geodesics. Hadamard was fully aware of the connection to Cantor's work and similar sets discovered by Poincaré, and he appreciated the relation between the complicated dynamics in the two contexts. Hadamard also showed that each homotopy class (except for the "waists" of cusps) contains a unique geodesic.

---

would be open to its eyes.

[8]If one tries to imagine the figure formed by these two curves with an infinite number of intersections, each corresponding to a doubly asymptotic solution, these intersections form a kind of trellis, a fabric, a network of infinitely tight mesh; each of the two curves must not cross itself but it must fold on itself in a complicated way to intersect all of the meshes of the fabric infinitely many times. One will be struck by the complexity of this picture, which I will not even attempt to draw

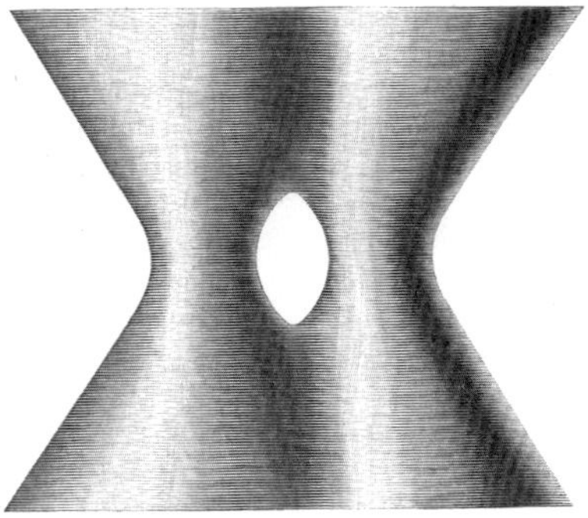

Figure 0.3.2. Negatively curved surface. [Reproduced from Hadamard [166] (© 1898 Elsevier Masson SAS, all rights reserved) with permission.]

Duhem [123] seized upon this to describe the dynamics of a geodesic flow in terms of what might now be called deterministic chaos—the system is completely determined (no randomness), but one would need infinite precision for long-term predictability.

Several authors trace the introduction of symbolic dynamics to the work of Hadamard on geodesic flows. Birkhoff is among them. Indeed, in his proof of the Birkhoff–Smale Theorem (see Theorem 6.5.2) symbolic sequences appear (as well as a picture that resonates with Figure 6.5.2). It appears, however, that only in 1944 did symbol spaces begin to be seen as dynamical systems, rather than as a coding device [99].

0.3.c Boltzmann's Fundamental Postulate. Well before Poincaré's work, James Clerk Maxwell (1831–1879) and Ludwig Boltzmann (1844–1906) had aimed to give a rigorous formulation of the kinetic theory of gases and statistical mechanics. A central ingredient was Boltzmann's Fundamental Postulate, which says that the time and space (phase or ensemble) averages of an observable (a function on the phase space) agree. Apparently because of a misstatement by Maxwell,[9] one often ascribes to him the so-called ergodic hypothesis:

> *The trajectory of the point representing the state of the system in phase space passes through every point on the constant-energy hypersurface of the phase space.*

Poincaré and many physicists doubted its validity since no example satisfying it had been exhibited [296]. Accordingly, in 1912 Paul and Tatiana Ehrenfest [127] proposed the alternative quasi-ergodic hypothesis:

[9]"The system, if left to itself in its actual state of motion, will, sooner or later, pass through every phase."

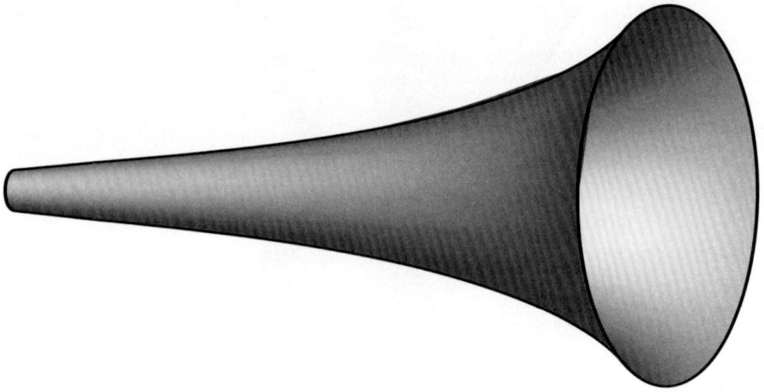

Figure 0.3.3. The pseudosphere. [Reprinted from [213] (© Cambridge University Press, all rights reserved) with permission.]

The trajectory of the point representing the state of the system in phase space is dense on the constant energy hypersurface of the phase space.

Indeed, within a year proofs (by Rosenthal and Plancherel) appeared that the ergodic hypothesis fails [291, 316]. (This is obvious today because a trajectory has measure 0 in an energy surface.) These difficulties led to the search for *any* mechanical systems with this second property. The motion of a single free particle (that is, the geodesic flow) in a negatively curved space (beginning with the pseudosphere, Figure 0.3.3) emerged as the first and for a long time sole class of examples with this property. Within a decade, the understanding of the problem led to the pertinent contemporary notion, and this turned out to be probabilistic in nature.[10] The 1931 Birkhoff Ergodic Theorem (Theorem 3.2.15) ("time averages exist a.e.")[11] laid the foundation for the definition of ergodicity now in use, which is that "any invariant set has zero measure or full measure."[12] If this is the case, then time averages agree with space averages—Boltzmann's Fundamental Postulate. Furthermore, almost every orbit is dense in the support of the measure.

The 1930s saw a flurry of work in which Artin's 1924 work on the modular surface was duly extended to other manifolds of *constant* negative curvature. For constant curvature, finite volume, and finitely generated fundamental groups the

[10]This serves to point out that the earlier quote by Laplace about determinism comes from his "Philosophical essay on probabilities," where he goes on to say that we often do not have sufficiently detailed initial data, and must hence resort to a probabilistic approach. The motion of a molecule of air was a prominent instance he mentioned in that context.

[11]This was proved after the von Neumann Ergodic Theorem (Theorem 3.2.4) but published earlier [358]—and the true foundational paper of ergodic theory is much more likely [265].

[12]These two combine to give the Strong Law of Large Numbers.

geodesic flow was shown to be topologically transitive [225, 249], topologically mixing [191], ergodic [196], and mixing [192, 197]. (In the case of an infinitely generated fundamental group the geodesic flow may be topologically mixing without being ergodic [327].) If the curvature is allowed to vary between two negative constants then finite volume implies topological mixing [155] (see also [158, p. 183]). But as Hedlund noted in an address delivered before the New York meeting of the American Mathematical Society on October 27, 1938,

> Outstanding problems remain unsolved, a notable one being the problem of metric transitivity [ergodicity] of the geodesic flow on a closed analytic surface of *variable* negative curvature.

It so happens that Eberhard Hopf was just then working on this problem [197]. He considered compact surfaces of nonconstant (predominantly) negative curvature and was able to show ergodicity of the Liouville measure (phase volume).

From Hopf's work there was no progress in the direction of ergodicity of geodesic flows for almost 30 years. Hopf's argument had shown roughly that Birkhoff averages of a continuous function must be constant on almost every leaf of the horocycle foliation, and, since these foliations are C^1, the averages are constant a.e. He realized that much of the argument was independent of the dimension of the manifold (indeed, he carried much of the work out in arbitrary dimension), but could not verify the C^1 condition in higher dimension. Dmitri Anosov [10] axiomatized Hopf's instability, defining Anosov flows, and he showed that differentiability may indeed fail in higher dimension, but that the Hopf argument can still be used because the invariant laminations have an absolute continuity property [10, 12, 303, 23, 70, 31]. This extension is interesting because, despite the ergodicity paradigm central to statistical mechanics, Boltzmann's Fundamental Postulate, there was a dearth of examples of ergodic Hamiltonian systems. The quintessential model for the Fundamental Postulate, the gas of hard spheres, resisted sustained attempts to prove ergodicity for half a century [332, 330, 331].[13]

The Hopf argument remains a main method for establishing ergodicity of volume in hyperbolic dynamical systems without an algebraic structure (the alternative tool being the theory of equilibrium states; see [213, Theorem 20.4.1]).

0.3.d Picking up from Poincaré. Like Hadamard, several mathematicians had begun to pick up some of Poincaré's work during his lifetime; Birkhoff did so soon after Poincaré's death. He addressed issues that arose from the mathematical development of mechanics and celestial mechanics such as Poincaré's Last Geometric Theorem and the complex dynamics necessitated by homoclinic tangles [49, Section 9].

[13] Half a century because Sinai convinced physicists that he had solved this problem in 1963 [232].

He was also important in the development of ergodic theory,[14] notably by proving the Pointwise Ergodic Theorem (Theorem 3.2.15).

The work of Cartwright and Littlewood during World War II on relaxation oscillations in radar circuits [88, 87, 243] consciously built on Poincaré's work. Further study of the van der Pol equation by Levinson [241] contained the first example of a structurally stable diffeomorphism with infinitely many periodic points. Structural stability had originated in 1937 with Andronov and Pontryagin [9] (necessary and sufficient conditions on singularities and periodic orbits for structural stability of vector fields on a disk) but began to flourish only 20 years later—thanks in no small part to Pontryagin's favorite student, Anosov. Inspired by Peixoto's work, which generalized [9] to any orientable closed surface [284], Smale had been after a program of studying diffeomorphisms with a view to classification [339], and he proved that Morse–Smale systems (finitely many periodic points with stable and unstable sets in general position) are structurally stable. The Cartwright–Littlewood example was brought to his attention by Levinson just as he conjectured that Morse–Smale systems are the only structurally stable ones [336]. He eventually extracted from Levinson's work the horseshoe [338, 337]. Independently, Thom (unpublished) studied hyperbolic toral automorphisms (Example 1.5.26) and their structural stability.

Smale in turn was in contact with the Russian school, where Anosov systems (then C- or U-systems) had been shown to be structurally stable, and their ergodic properties were studied by way of further development of the study of geodesic flows in negative curvature.

This book focuses on uniformly hyperbolic flows, and even in this realm there are plenty of new developments. Section 5.2 gives instances of uniformly hyperbolic flows of which several are quite new, and Chapter 8 includes various further constructions of such (notably in Sections 8.2 and 8.3). Our presentation of these includes results in a range of directions that still await publication.

The initial development of the theory of hyperbolic systems in the 1960s was followed by the founding of the theory of *nonuniformly* hyperbolic dynamical systems in the 1970s, mostly by Pesin [273, 286, 32] (during which time the hyperbolic theory continued its development). One of the high points in the development of smooth dynamics is the proof by Robbin, Robinson, Mañé, and Hayashi [189] that structural stability indeed characterizes hyperbolic dynamical systems. For diffeomorphisms this was achieved in the 1980s, for flows in the 1990s. Starting in the 1980s the field of geometric and smooth rigidity came into being and is flourishing now (Chapter 9). At the same time topological and stochastic properties of attractors began to be better understood with techniques that nowadays blend ideas from hyperbolic and 1-dimensional dynamics. Meanwhile, the theory of *partially* hyperbolic dynamical

[14] The Poincaré Recurrence Theorem (Theorem 3.2.1) is proved in Poincaré's prize memoir [295].

systems, which goes back to seminal works of Brin and Pesin in the 1970s, has seen explosive development since the last years of the 20th century [288], which in turn has entailed renewed interest in the methods of uniformly hyperbolic dynamical systems and their possible extensions to this new realm.

Of course, insights into complicated dynamics have penetrated well beyond pure mathematics. In the sciences, these ideas have fundamentally changed the appreciation of nonlinear behavior and that complex data may arise from simple models; they have also provided terminology for describing complexity [152]. Celestial mechanics is the realm where applications have most clearly gone beyond the descriptive; since the 1980s the design of trajectories for space probes has irreversibly moved beyond perturbing the 2-body problem in ways that make entirely new mission designs feasible and economical in astonishing ways [38]. This can also be said to have added to the very foundation of how evidence is used to build science [351].

Part I

Flows

Introduction to Part I

This book begins with an introduction to flows—as opposed to hyperbolic flows, to which the second part is dedicated. While this serves as a self-contained introduction to the basic concepts in dynamics, readers familiar with dynamics in discrete time will see this as a parallel development in which the distinctions between dynamics in continuous time and discrete time begin to reveal themselves. And while we treat flows in rather great generality here, the selection of notions and phenomena we include is informed by those we will later be able to observe in and apply to hyperbolic flows.

Chapter 1 introduces flows from a topological point of view and after introducing the most basic notions explores recurrence ideas all the way to Conley's Fundamental Theorem of Dynamical Systems, the existence of dense orbits and (topological) mixing properties, concluding with a strong notion of sensitive dependence on initial conditions (expansivity) and symbolic flows, both of which are central to hyperbolic dynamical systems. While this chapter is replete with examples, the optional Chapter 2 introduces a foundational example in hyperbolic dynamics, the geodesic flow of a surface of constant negative curvature. This treatment is elementary and explicit but foreshadows features, and to some extent methods, that in Part II will be seen as characteristic of hyperbolic flows. Section 5.2 can be seen as a direct follow-up. More broadly, Chapters 5 and 6 build on these chapters.

Chapter 3 is a probabilistic counterpart to Chapter 1. It introduces measurable and measure-preserving flows and the pertinent notions and results from ergodic theory. This development and that in the next chapter are supported by Appendix A which introduces the basic notions regarding probability and Lebesgue spaces as well as the discrete-time counterparts to the theory developed here. Chapter 4 continues and connects the developments in both Chapters 1 and 3 by presenting the core of entropy theory, which, being centered on notions of exponential orbit complexity, is a natural and important tool in hyperbolic dynamics. Indeed, the notions from these chapters will be applied in Chapter 7.

1

Topological dynamics

This chapter provides a foundation for the remainder of the book and includes a number of examples to refer to in this and subsequent chapters; they are chosen to illustrate notions and phenomena that will be encountered throughout the book. We begin with the basic notions of dynamical behaviors from a topological point of view, such as definition and basic properties of flows, properties of individual orbits, techniques for varying speed or time, notions of equivalence for flows, and the interplay between continuous and discrete time.

We then explore the orbit structure of flows by first defining various notions of recurrence (including periodicity) and sensitive dependence, and then turn to a more global approach. The chain decomposition and the Conley Theorem can be viewed as pinnacles of organizing recurrent behavior in a global context, and they later turn out to be basic for hyperbolic flows.

With a view to hyperbolic flows, we also introduce properties of topological flows that involve even tighter entanglement of orbits: transitivity, mixing, and expansivity. Lastly, we describe symbolic flows, which will provide finitary models for hyperbolic flows.

1.1 Basic properties

We begin by introducing flows, the central concept of this book. The notion of a flow arose from studying solutions to differential equations. Over time, mathematicians realized that the notion of a flow could be generalized to the definition we give below.[1] We relate flows to solutions of a differential equation in Section 1.1.b.

Definition 1.1.1 (Flow). A *flow* Φ on a set X is a mapping $\Phi\colon X \times \mathbb{R} \to X$ such that

- $\Phi(x, 0) = x$, and
- $\Phi(\Phi(x, t), s) = \Phi(x, s + t)$.

[1]The step from differential equations to Definition 1.1.1 is not only a matter of generalization but of adopting a global point of view by advancing from thinking about individual functions of time (solutions of a differential equation) to a comprehensive consideration of all time evolutions together.

In other words, Φ defines an $\mathbb{R}$-action $t \mapsto \varphi^t$ of morphisms of X. Here, X is variously referred to as the *phase space* or *state space* of the flow, and the morphisms φ^t can variously be homeomorphisms, diffeomorphisms, or measurable isomorphisms—and sometimes additionally preserve a group, symplectic, contact, or other structure. A flow is C^r for $0 \le r \le \infty$ if Φ is C^r. When we use the term *smooth flow* we will mean a flow that is at least C^1. A smooth flow has a *generating vector field* given by

$$V(x) := \dot{\varphi}(x) := \varphi'(x) := \frac{d}{dt}\varphi^t(x)_{|_{t=0}}.$$

We usually assume that either X is a topological (or metric) space and Φ is continuous or that X is a measure space and each φ^t is measure preserving.

This definition describes the *subject* of this book and of (continuous-time) dynamics. The *object* is to describe long-term behavior. On one hand, this is possible because time is parametrized by the set of real numbers, which is not compact and hence has a notion of going to infinity, and on the other hand, it is interesting because we tend to focus on spaces or subsets thereof that are compact (in the topological or differentiable context) or have finite measure (in the measurable context), which forces some accumulation or recurrence. We will define various forms of recurrence in Section 1.5.

1.1.a Time-*t* maps and orbits. It is illuminating in different ways to fix one input variable at a time. Fixing t yields self-maps of the space X, where a *self-map* is a map from X to X. For $t \in \mathbb{R}$ the *time-t* map is $\varphi^t := \Phi(\cdot, t) \colon X \to X$.[2] We will typically refer to a flow by $\Phi = \{\varphi^t\}_{t\in\mathbb{R}}$ to avoid confusion with φ^t.

Claim 1.1.2. *If Φ is a flow and $t \in \mathbb{R}$, then the time-t map φ^t of the flow is a bijection with inverse φ^{-t}.*

Proof. Taking $s = -t$ in Definition 1.1.1 gives $\varphi^t \circ \varphi^{-t} = \mathrm{Id} = \varphi^{-t} \circ \varphi^t$. □

Thus, $\varphi^0 = \mathrm{Id}$ and $\varphi^s \circ \varphi^t = \varphi^{s+t}$ for all $s, t \in \mathbb{R}$. Hence, a flow is a group action of the real numbers. One can also study actions of groups other than $\mathbb{R}$, but for the most part we will restrict to group actions by $\mathbb{R}$.

Definition 1.1.3. The *inverse flow* of a flow $t \mapsto \varphi^t$ is the flow $t \mapsto \varphi^{-t}$.

Remark 1.1.4. Note that if $a < b$ where $a, b \in \mathbb{R}$, then the flow is completely determined by the mapping $\Phi \colon X \times (a,b) \to X$. (By inversion and the group law, this determines the flow for t in the interval $I := (a,b) - (a,b) \ni 0$, and by iteration, this determines the flow for $t \in \mathbb{Z}I = \mathbb{R}$.)

[2]Our notation ":=," "=:," ":⇔," and "⇔:" defines the quantity/property on the side of the ":."

We now provide a number of simple examples of flows.

Example 1.1.5 (Translations). If $v \in \mathbb{R}$ and $\varphi^t(x) := x + tv$, then $\varphi^0(x) = x$ and if $s, t \in \mathbb{R}$, then

$$\varphi^{s+t}(x) = x + (s + t)v = (x + sv) + tv = \varphi^t(\varphi^s(x)).$$

So $\Phi = \{\varphi^t\}_{t \in \mathbb{R}}$ is a flow on $\mathbb{R}$.

Note the contrast to discrete-time dynamical systems: a translation and its iterates constitute an action of $\mathbb{Z}$ (or $\mathbb{N}$), and here we have a family of translations parametrized by a continuous parameter. In fact, it contains all translations of $\mathbb{R}$.

Example 1.1.6 (Rotations). By considering $\mathbb{R} \pmod 1 = \mathbb{R}/\mathbb{Z}$ in Example 1.1.5, one projects the flow from the previous example to a flow on a circle—which can also be represented as $(z, t) \mapsto e^{2\pi i t} z$ for $|z| = 1$ in $\mathbb{C}$.

This illustrates that the gap between continuous and discrete time is greater than suggested by the previous example: while any two translations of $\mathbb{R}$ are dynamically the same, circle rotations as maps have quite disparate behaviors. Rotations by a rational number are periodic, while rotations by an irrational number exhibit rather nontrivial dynamics. By contrast, this circle flow is about as simple as a flow can be. There is just a little more complexity in the next example.

Example 1.1.7 (Scaling). If $a \in \mathbb{R}$ and $\varphi^t(x) := x \cdot e^{at}$, then $\varphi^0(x) = x$ and

$$\varphi^{s+t}(x) = x \cdot e^{a(s+t)} = (x \cdot e^{as}) \cdot e^{at} = \varphi^t(\varphi^s(x)) \quad \text{for } s, t \in \mathbb{R}.$$

So $\Phi = \{\varphi^t\}_{t \in \mathbb{R}}$ is a flow on $\mathbb{R}$.

More generally, flows in dimension 1 are dynamically quite simple, so they do not take up much of this book. We now proceed to higher-dimensional examples, where more interesting dynamics can be present.

Example 1.1.8 ("Asteroids"—toral translations). The linear flow Φ_v on the n-torus $\mathbb{T}^n$ in the direction $v \in \mathbb{R}^n$ is defined by $\varphi^t(x) = \varphi_v^t(x) = (x + tv) \bmod 1$. As in Example 1.1.5, this defines a flow (which generalizes the one in Example 1.1.6). Geometrically, a point moves with constant speed along a straight line and (like in old video games such as Asteroids) reemerges from one side of a fundamental domain after encountering the opposite side (Figure 1.1.1).

This example is not quite as new as it first seems. Taking n copies of Example 1.1.6 with possibly different speeds produces Example 1.1.8 as their cartesian product, which is described more generally as follows.

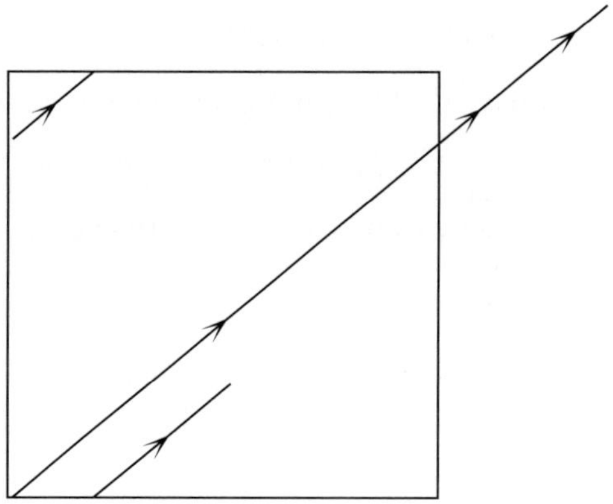

Figure 1.1.1. Linear flow on $\mathbb{T}^2$. [Reprinted from [213] (© Cambridge University Press, all rights reserved) with permission.]

Example 1.1.9 (Product flows). If Φ and Ψ are flows on X and Y, respectively, then their cartesian product $\Phi \times \Psi$ on $X \times Y$ defined by $(\varphi \times \psi)^t(x, y) = (\varphi^t, \psi^t)(x, y) := (\varphi^t(x), \psi^t(y))$ is a flow on the product space $X \times Y$.

Having explored flows as families of maps, we now take the complementary approach of fixing $x \in X$ and letting t vary in order to focus attention on the time evolutions of individual initial conditions, that is, curves in X.

Definition 1.1.10 (Orbits). The *orbit* of $x \in X$ under the flow Φ on X is

$$\mathcal{O}(x) := \varphi^{\mathbb{R}}(x) := \{\varphi^t(x) \mid t \in \mathbb{R}\} \quad \text{or} \quad \mathcal{O}(x) := \varphi\restriction_{\{x\}\times\mathbb{R}} : t \mapsto \varphi^t(x),$$

depending on whether we wish to keep track of the time parameter or not.[3] Similarly, the *forward (semi-)orbit of* x is

$$\mathcal{O}^+(x) := \varphi^{[0,\infty)}(x) = \{\varphi^t(x) \mid t \geq 0\},$$

and the *backward (semi-)orbit of* x is

$$\mathcal{O}^-(x) := \varphi^{(-\infty,0]}(x) = \{\varphi^t(x) \mid t \leq 0\}.$$

We say that x is a *fixed point* (or *equilibrium* or *singularity*) of Φ if $\mathcal{O}(x) = \{x\}$ (so $\varphi^t(x) = x$ for all $t \in \mathbb{R}$).

A point $x \in X$ is *periodic* for a flow Φ if there exists some $t > 0$ such that $\varphi^t(x) = x$ and x is not a fixed point. The point x is t-periodic, and its orbit is said to be *closed*.

[3] Here, the orbit of x and $\varphi^s(x)$ are the same even in the former case, so parametrized orbits are identified if they differ only by precomposition with a translation of $\mathbb{R}$.

The set of periodic points is denoted by Per(Φ). Both fixed points and periodic orbits are referred to as compact orbits. The (*least* or *prime*) period of x is the infimum of all t such that x is t-periodic.

As a fixed point does not have a flow direction, fixed points and periodic points for flows require somewhat different analysis.

Remark 1.1.11. Example 1.1.5 has only a single orbit, which is $\mathbb{R}$ parametrized with speed $|v|$. Even though the parametrizations differ for different initial points, they differ only by a constant offset of time, so we do not consider these to be different orbits even if there is an intent to pay attention to the parametrization. Example 1.1.7 has three orbits: the origin and two half-lines. Unless $n = 1$ (in which case there is a single orbit), Example 1.1.8 has uncountably many orbits, each of which is the projection of a line to $\mathbb{T}^n$; these lines are all parallel. If $v = (1, 0, \ldots, 0)$ then they all project to circles. If $v = (1, \sqrt{2}, 0, \ldots, 0)$ then they all lie in the projection of the x–y plane and fill it densely.

In Example 1.1.8 the existence of a periodic orbit implies that there is a $T \neq 0$ for which $Tv \in \mathbb{Z}^n$, in which case *every* orbit is periodic. (This is the case, for example, if $v \in \mathbb{Q}^n$.) Fixed points occur only if $v = 0$, in which case the flow is trivial. Summarized in slightly different terms, Φ_v has periodic orbits (and is itself periodic) if and only if $v \in \mathbb{R}\mathbb{Z}^n$.

Near a fixed point a flow is going to be "slow," and it is plausible that the absence of fixed points implies positive minimum speed by a compactness argument. This is indeed easy when "speed" makes sense, as it does when the flow is generated by a nonzero vector field (Definition 1.1.1). But continuity of the flow is sufficient:

Proposition 1.1.12 ("Minimum speed"). *If Φ is a continuous flow without fixed points on a compact space, then there is a $T_0 > 0$ such that for any $t \in (0, T_0)$ there is a $\gamma_t > 0$ with $d(\varphi^t(x), x) \geq \gamma_t$ for all x.*

Proof. If t is such that for all $n \in \mathbb{N}$ there is an x_n with $d(\varphi^t(x_n), x_n) < 1/n$, then an accumulation point x of the x_n satisfies $d(\varphi^t(x), x) = 0$. If there is no periodic point, let $T_0 := 1$, otherwise $T_0 := \inf\{t > 0 \mid \exists x \in X : \varphi^t(x) = x\} > 0$. □

Definition 1.1.13 (Flow box). Let Φ be a flow on a topological space X. A *flow-box* neighborhood of a point $x \in X$ is a neighborhood U of x with a continuous embedding $h\colon U \to \mathbb{R}^{n+1}$ such that $h \circ \Phi = \Psi \circ h$, where $\psi^t : (x, s) \mapsto (x, s + t)$; see Figure 1.1.2.

A C^r ϵ-flow box (for $r \in \mathbb{N}_0$) is $\varphi^{[-\epsilon,\epsilon]}(D)$ where D is a C^r-embedded disk and ϵ is such that $\Phi\restriction_{[-\epsilon,\epsilon]\times D}$ is a C^r-embedding.

Proposition 1.1.14 (Flow box). *If Φ is a C^r-flow, then any point where the generating vector field is nonzero has a C^r-flow-box neighborhood.*

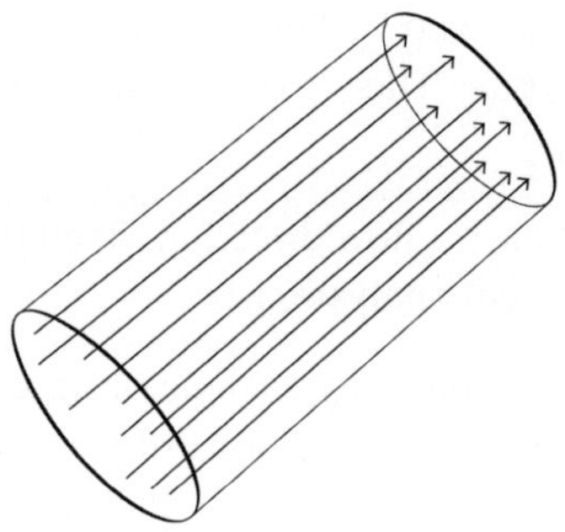

Figure 1.1.2. A flow box.

Proof. By the Inverse-Function Theorem there is an $\epsilon > 0$ such that if B is an ϵ-ball transverse to the vector field at the point in question, then

$$B \times [-\epsilon, \epsilon] \to M, \quad (x,t) \mapsto \varphi^t(x)$$

is a C^r-embedding. □

Figure 1.1.4 illustrates the obvious fact that this only works away from fixed points. In the purely measurable context, however, something much like this is indeed possible not just locally, but *globally* by the Ambrose–Kakutani–Rokhlin Special-Flow Representation Theorem (Theorem 3.6.2). In the topological setting there are some flows with a global flow box. This is the notion of a suspension flow (Definition 1.2.8).

1.1.b Differential equations. The study of flows originated in the field of ordinary differential equations, and smooth flows always arise in this way. We now make this connection explicit. This is meant to serve readers coming to this subject from a background in differential equations, but we can reassure others that knowledge of differential equations is not required.

Proposition 1.1.15. *If* $f\colon \mathbb{R}^n \to \mathbb{R}^n$ *and for all* $\xi \in \mathbb{R}^n$ *the initial-value problem*

$$\begin{cases} \dfrac{dx}{dt} = f(x), \\ x(0) = \xi, \end{cases}$$

has a unique solution $x_\xi(t)$ *defined for all* $t \in \mathbb{R}$, *then* $\Phi\colon (\xi,t) \mapsto x_\xi(t)$ *is a flow.*

Proof. Given $s \in \mathbb{R}$ and $y\colon \mathbb{R} \to \mathbb{R}^n$ defined by $y(t) = x_\xi(t+s)$ we write $x' := \frac{dx}{dt}$ and $y' := \frac{dy}{dt}$ and have $y'(t) = x'_\xi(t+s) = f(x(t+s,\xi)) = f(y(t))$. So y is a solution of $\frac{dx}{dt} = f(x)$ with $y(0) = x_\xi(s)$. Since a solution is unique we have

$$\varphi^{t+s}(\xi) = x_\xi(t+s) = y(t) = x_{y(0)}(t) = x_{x_\xi(s)}(t) = x_{\varphi^s(\xi)}(t) = (\varphi^t \circ \varphi^s)(\xi),$$

as well as $\varphi^0(\xi) = x_\xi(0) = \xi$. □

Remark 1.1.16. Example 1.1.5 arises in this way from $\frac{dx}{dt} = v$ and Example 1.1.7 from $\frac{dx}{dt} = ax$. In both cases, and in general, $f(x) = \frac{d}{dt}\big|_{t=0}\varphi^t(x) =: \dot{\varphi}(x)$ will serve. Note that the solution x_ξ of the differential equation is the orbit of ξ under the flow.

We now consider another class of flows that are defined by a differential equation but for which we can find explicit formulas. A flow Φ on a vector space X is *linear* if all φ^t are linear, that is, if for all $x, y \in X$, $t \in \mathbb{R}$ and scalars α and β we have

$$\varphi^t(\alpha x + \beta y) = \alpha\varphi^t(x) + \beta\varphi^t(y).$$

An example of a linear flow is the flow generated by the differential equation $x' = Ax$ where A is an $n \times n$ real-valued matrix ($A \in \mathcal{M}_n(\mathbb{R})$). If $n = 1$, then solutions of $x' = ax$ are easily seen to be of the form $x(t) = Ce^{at}$ (Example 1.1.7). We will see that solutions to $x' = Ax$ have a similar form.

Definition 1.1.17. If $A \in \mathcal{M}_n(\mathbb{R})$, then the *exponential* of A is $e^A = \sum_{k=0}^{\infty} \frac{A^k}{k!}$.

Proposition 1.1.18. *The exponential e^A is well defined.*

Proof. The series converges since $\|AB\| \le \|A\|\|B\|$ and hence $\|A^n\| \le \|A\|^n$: if $M < N$ then

$$\Big\|\sum_{k=0}^{N} \frac{A^k}{k!} - \sum_{k=0}^{M} \frac{A^k}{k!}\Big\| = \Big\|\sum_{k=M+1}^{N} \frac{A^k}{k!}\Big\| \le \sum_{k=M+1}^{N} \frac{\|A^k\|}{k!} \le \sum_{k=M+1}^{N} \frac{\|A\|^k}{k!}.$$

Since $\sum_{k=0}^{\infty} \frac{\|A\|^k}{k!}$ is convergent, hence Cauchy (in $\mathbb{R}$), this shows that $\sum_{k=0}^{m} \frac{A^k}{k!}$ is Cauchy, hence convergent in $\mathcal{M}_n(\mathbb{R})$. □

Analogously to the power series representation of the exponential function on $\mathbb{R}$, this gives $e^{At} = \sum_{k=0}^{\infty} \frac{(At)^k}{k!}$, $e^{A0} = \mathrm{Id}$, and (by differentiating termwise)

$$\frac{d}{dt}e^{At} = \sum_{k=1}^{\infty} \frac{k(At)^{k-1}}{k!} \cdot A = \sum_{k=0}^{\infty} A\frac{(At)^k}{k!} = Ae^{At}.$$

So for each $x \in \mathbb{R}^n$ the function $e^{At}x$ is the solution to $y' = Ay$ with initial condition $y(0) = x$. The flow $\varphi^t(x) = e^{At}x$ is a linear flow.

If $A \in \mathcal{M}_n(\mathbb{R})$ has n linearly independent eigenvectors, then we can diagonalize A and explicitly compute e^{At} from the power series:

$$e^{\begin{pmatrix} \lambda_1 & & & \\ & \lambda_2 & & \\ & & \ddots & \\ & & & \lambda_n \end{pmatrix}} = \begin{pmatrix} e^{\lambda_1} & & & \\ & e^{\lambda_2} & & \\ & & \ddots & \\ & & & e^{\lambda_n} \end{pmatrix}.$$

Otherwise, we can use the decomposition into generalized eigenspaces (Theorem B.2.5). Specifically, since $e^{\lambda It} = e^{\lambda t}I$ and $(A - \lambda I)$ commutes with λI, we can (as such, vacuously) write $e^{At} = e^{(A-\lambda I)t}e^{\lambda It} = e^{\lambda t}e^{(A-\lambda I)t}$ and then find a basis of vectors for each of which the matrix exponential on the right collapses to a polynomial. If λ is an eigenvalue of A, let $M(\lambda)$ be the generalized eigenspace (Section B.2) and $r(\lambda)$ the natural number at which the nullspace of $(A - \lambda I)^k$ stabilizes. If $v \in M(\lambda)$, then $(A - \lambda I)^{r(\lambda)}v = 0$ and the series for $e^{(A-\lambda I)t}v$ terminates:

$$e^{(A-\lambda I)t}v = \lim_{n\to\infty} \sum_{k=0}^{n} \frac{t^k}{k!}(A - \lambda I)^k v = \sum_{k=0}^{r(\lambda)-1} \frac{t^k}{k!}(A - \lambda I)^k v,$$

so

$$e^{At}v = e^{\lambda t}e^{(A-\lambda I)t}v = e^{\lambda t} \sum_{k=0}^{r(\lambda)-1} \frac{t^k}{k!}(A - \lambda I)^k v.$$

We have proved the following theorem.

Theorem 1.1.19. *Let $\lambda_1, \dots, \lambda_p$ be eigenvalues with $M(\lambda_1), \dots, M(\lambda_p)$ the corresponding generalized eigenspaces (Theorem* B.2.5*). If $\zeta = v_1 + \cdots + v_p$ for $v_j \in M(\lambda_j)$ and $1 \le j \le p$, then $x' = Ax$ and $\zeta = x(0)$ has the solution*

$$x(t) = \sum_{j=1}^{p} e^{\lambda_j t} \left[\sum_{k=0}^{r(\lambda_j)-1} (A - \lambda_j I)^k \frac{t^k}{k!} \right] v_j.$$

Remark 1.1.20. This could be written as $x(t) = \sum_{j=1}^{p} e^{\lambda_j t}e^{(A-\lambda_j I)t}v_j$, but the point is that the matrix exponential in the middle is effectively a polynomial rather than an exponential, that is, up to polynomial error we have $x(t) \approx \sum_{j=1}^{p} e^{\lambda_j t}v_j$.

It is rare for a flow to be given in terms of explicit formulas, just like it is rare for a discrete-time dynamical system to admit simple closed-form expressions of arbitrary iterates. Indeed, Poincaré's insight that founded the discipline of dynamical systems was that one can study the dynamics without expressing solutions, that is, orbits, explicitly in closed form or in terms of power series.

In what follows, the following regularity notion is particularly suitable.

Definition 1.1.21. A map $f\colon X \to Y$ between metric spaces is said to be *Lipschitz continuous* if there is an $L \in \mathbb{R}$ such that $d(f(x), f(y)) \le Ld(x, y)$ for all $x, y \in X$.

For a smooth manifold M and $0 \le r \le \infty$, let $\mathcal{X}^r = \mathcal{X}^r(M)$ be the space of C^r vector fields on M and $\mathcal{X}^{\mathrm{Lip}}(M)$ the space of Lipschitz-continuous vector fields. This implies that if $V \in \mathcal{X}^{\mathrm{Lip}}(M)$, then for each $x \in M$ we have $V(x) \in T_xM$ and the map $x \mapsto V(x)$ is a section of the tangent bundle $TM = \bigcup_{x\in M} T_xM$, which varies in a Lipschitz-continuous manner. In a local coordinate chart we can use an existence-and-uniqueness theorem by Picard[4] and Proposition 1.1.15 to show that the flow exists for small values of t. If M is compact without boundary (that is, "closed"), then for arbitrary values of t we can use compositions of maps defined in local coordinates to define the flow on M for all t. Then

$$\frac{dx}{dt} = V(x) \tag{1.1.1}$$

generates a flow (Proposition 1.1.15), which we will denote by Φ_V. Conversely, if Φ is a C^1-flow on a smooth manifold, then the generating vector field

$$V(x) := \dot{\varphi}(x) := \varphi'(x) := \frac{d}{dt}\varphi^t(x)_{|t=0} \tag{1.1.2}$$

defines a continuous vector field on the manifold.

Remark 1.1.22. Equation (1.1.1) extends to *nonautonomous* or time-dependent differential equations, that is, differential equations of the form

$$\begin{cases} \dfrac{dx}{dt} = V(x,t), \\ x(0) = \xi, \end{cases} \tag{1.1.3}$$

when V is continuous and $x \mapsto V(x,t)$ is Lipschitz continuous; the aforementioned theorem of Picard gives (local) existence and uniqueness of solutions, which then extend to globally defined ones as in (1.1.1). The resulting maps $t \mapsto \varphi^t(\cdot)$ may not satisfy $\varphi^s \circ \varphi^t = \varphi^{s+t}$, however, so we may not obtain a flow. This corresponds to having a time-dependent vector field in (1.1.1), and some of the results and techniques presented in this book can be adapted to this context. It is possible, of course to make a nonautonomous differential equation autonomous by treating t as an additional independent variable. The price is that the resulting differential equation on $M \times \mathbb{R}$ no longer "lives" on a compact space. Furthermore, one may lose structural information;

[4]The proof due to Picard considers a Banach space of candidates for solutions of the differential equation and constructs an operator that is a contraction mapping and whose fixed points are solutions of the differential equation. By the Contraction Mapping Theorem (Proposition B.1.3) there is a unique fixed point, hence a unique solution to the differential equation, and this depends smoothly on initial values and the vector field.

for instance, a nonautonomous linear differential equation may in this way become nonlinear because of nonlinearities in the time dependence.

We now give a classical example of a nonlinear ordinary differential equation.

Example 1.1.23 (The pendulum). Consider a pendulum consisting of a point mass in the plane attached by a rod to a fixed joint. If we take $2\pi x$ to be the angle of deviation from the vertical then (with a suitable choice of units) the pendulum is described by the differential equation

$$\frac{d^2x}{dt^2} + \sin 2\pi x = 0.$$

Writing $v = \frac{dx}{dt}$ (velocity) we obtain the system of first-order differential equations

$$\begin{cases} \dfrac{dx}{dt} = v, \\ \dfrac{dv}{dt} = -\sin 2\pi x, \end{cases}$$

for $x \in S^1$, $v \in \mathbb{R}$. The total energy of the system is the kinetic energy plus the

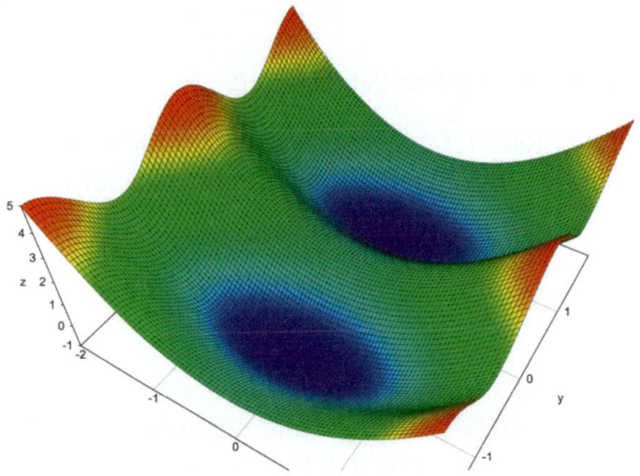

Figure 1.1.3. Total energy of the pendulum. [https://academo.org/demos/3d-surface-plotter/.]

potential energy. It is not hard to show that in this case the total energy is given by $H(x, v) = \frac{1}{2}v^2 - \frac{1}{2\pi}\cos 2\pi x$ (see Figure 1.1.3). As this equation is for the undamped (frictionless) pendulum, energy is conserved for a solution, and so the function H on the cylinder $S^1 \times \mathbb{R}$ is invariant under the flow (constant along orbits):

$$\frac{d}{dt}H(x, v) = v\frac{dv}{dt} + \frac{dx}{dt}\sin 2\pi x = 0.$$

We say that H is a *constant of motion* (sometimes also referred to as a (first) integral), and this means that the orbits are on level curves H = const. as shown in Figure 1.1.4. This means that without solving the system of differential equations, we can describe

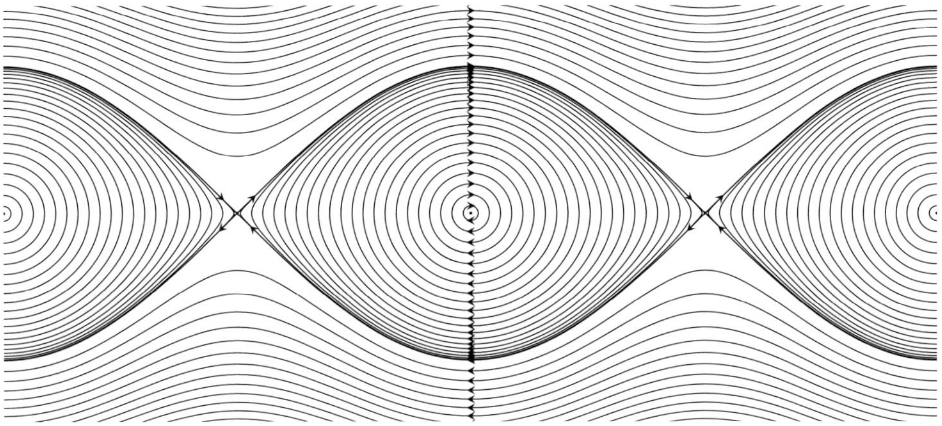

Figure 1.1.4. Energy levels of the pendulum rolled out to the plane.

the solution curves precisely. (It helps that $v = \frac{dx}{dt}$ tells us that in the upper half of the picture the direction of motion is to the right and to the left in the lower half.)

Because of its utility, we formalize the notion of constant of motion:

Definition 1.1.24. A *constant of motion* or *first integral* for a flow is a continuous invariant function. Here, a function $f\colon X \to \mathbb{R}$ is said to be *invariant* for a flow Φ if $f \circ \varphi^t = f$ for all $t \in \mathbb{R}$.

Because a constant function is always trivially a constant of motion, we abuse semantics (omitting "nonconstant") and follow established terminology by saying that *a flow has no constants of motion* if each continuous invariant function is constant.

For $-1/2\pi < H < 1/2\pi$ each energy level consists of a single closed curve corresponding to oscillations around the stable equilibrium $(x, v) = (0, 0)$; these are periodic orbits, and the period increases monotonically to $+\infty$ as a function of $H \in (-1/2\pi, 1/2\pi)$.

Those closed periodic orbits are separated from higher-energy orbits corresponding to rotation around the joint by a *homoclinic loop* (an orbit that joins an equilibrium to itself) with $H = 1/2\pi$ containing the equilibrium $(x, v) = (1/2, 0)$. For $H > 1/2\pi$ each energy level consists of two orbits corresponding to rotation in opposite directions.

This is a good moment to note the character of the fixed points that are joined by these homoclinic loops. They satisfy the definition given below.

Definition 1.1.25. A fixed point p of a flow Φ is said to be *hyperbolic* if for any $t \neq 0$ the differential $D_p\varphi^t$ of the time-t map φ^t at p has no eigenvalues on the unit circle (or, more generally, if $D_p\varphi^t$ is a hyperbolic linear map as in Definition B.4.1).

Figure 1.1.4 is a little less confusing than the picture on the actual phase cylinder. It shows two hyperbolic "saddle" points connected by two arcs; the upper one consists of points tending to the right saddle in positive time and to the left one in negative time, and points on the lower one do likewise in reverse. Such arcs are called *saddle connections*. Likewise, these can be seen as unstable sets for the respective other saddle. The full stable set of the saddle on the right consists of the upper of these arcs as well as the corresponding lower arc to its right, of which only half is shown. So its stable set is an arc with the saddle in its interior, and likewise for the unstable set. (We formally define these later, in Definition 1.3.26.) The points in any of the arcs in Figure 1.1.4 are *heteroclinic*, positively asymptotic to one saddle and negatively asymptotic to another. On the phase cylinder they are instead *homoclinic* because they are positively and negatively asymptotic to the same saddle, which thus has two *homoclinic loops*.

1.1.c Geodesic flows. We now describe flows that arise naturally from differential geometry. As these will be important in later chapters we have separated them into a separate subsection and revisit them again in Chapter 2.

In the case of a complete Riemannian manifold, flows arise naturally from the geodesics on the manifold. We will restrict ourselves at present to closed connected orientable surfaces. Any such surface is homeomorphic to one of the following: a sphere, a torus, or a higher-genus surface. By the Uniformization Theorem, each of these surfaces admits a metric of constant Gauss curvature, which is either positive, zero, or negative respectively. We point this out to note that not much knowledge of differential geometry is required in the sequel.

We first describe the concept of geodesic flow, and then study the geodesic flow for the torus and sphere. We will wait to discuss the geodesic flow on surfaces of negative curvature until Chapter 2 and show in Theorem 5.2.4 that these provide the classical example of a hyperbolic flow. In the case of negative curvature we will see that the geodesic flow is dynamically far more complicated.

The flow in Example 1.1.8 describes a point moving in a fixed direction with constant speed; this is the motion of a particle that is not subject to any external forces. This flow is connected to acceleration via $F = ma$ and the absence of an external force implies zero acceleration and hence constant velocity.[5]

The motion of a free particle is described as in Example 1.1.8, except that any velocity vector v is allowed. Thus, a state of this system is given by a location and a

[5]Here, F denotes force, m the mass, and a the acceleration.

velocity, that is, by a point on the torus and a tangent vector. The geodesics of $\mathbb{R}^n$ are exactly the straight lines, and so the geodesics on the (flat) torus $\mathbb{T}^n = \mathbb{R}^n/\mathbb{Z}^n$ are exactly the projections of straight lines. Therefore the description of the motion of a free particle on $\mathbb{T}^n$, also known as the *geodesic flow* on $\mathbb{T}^n$, is given as follows. On the tangent bundle $\mathbb{T}^n \times \mathbb{R}^n$ of $\mathbb{T}^n$ the geodesic flow is defined by

$$g^t(x, v) := (x + tv \pmod 1, v).$$

The geodesic flow on $\mathbb{T}^n$ is completely integrable, that is, it decomposes into invariant tori carrying linear flows—with frequency vector ω on the invariant torus $\{(x, v) \mid x \in \mathbb{T}^n,\ v = \omega\}$.

In like manner, the motion of a free particle on any Riemannian manifold can be described as motion along geodesics, the "straight lines" for the manifold, except that there usually are no formulas as explicit as in the formula for the torus given above to describe the time evolution. Indeed, the geodesic equation in differential geometry describes geodesics as having zero acceleration, and for each vector v at a point x of a manifold there is a unique geodesic $\gamma_{(x,v)}$ such that $\gamma_{(x,v)}(0) = x$ and $\dot{\gamma}_{(x,v)}(0) = v$, where $\dot{\gamma}$ denotes the t-derivative or tangent vector (that is, velocity vector). The geodesic flow is defined as follows.

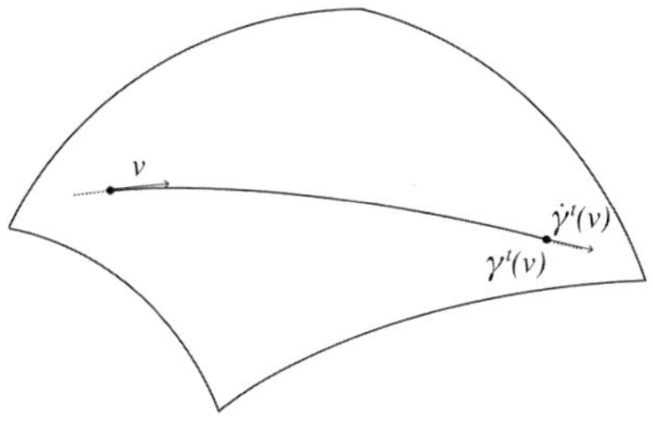

Figure 1.1.5. Geodesic flow.

Definition 1.1.26 (Geodesic flow). The *geodesic flow* g^t of a Riemannian manifold M is defined on the tangent bundle TM of M by

$$g^t(x, v) = (\gamma_{(x,v)}(t), \dot{\gamma}_{(x,v)}(t)),$$

that is, the tangent vector v at x is sent to the tangent vector of the geodesic $\gamma_{(x,v)}$ at time t (that is, at the point $\gamma_{(x,v)}(t)$).

Although we try to visualize the geodesic flow, as in Figure 1.1.5, on its *configuration space* M, it is important to remember that it is a flow *not* on M but on TM. We will see in Chapter 2 that this can lead to misunderstandings if one is not careful: horocycles give pairs of curves in TM that are not tangent to each other, but whose projections to M are; and so the inattentive reader may be confused. Proper use of prepositions can keep this distinction clear: the geodesic flow *of* a Riemannian manifold is a flow *on* the tangent bundle.

Remark 1.1.27. The fact that the geodesic flow on the flat torus decomposes naturally into linear flows as in Example 1.1.8 is specific to the torus, but the fact that the speed $\|v\|$ is preserved holds generally.[6] Moreover, restricting attention to vectors of a given norm produces a flow that is much like the flow obtained by restricting to vectors of any other norm, except for being uniformly faster or slower. Therefore we will normally (and often implicitly) restrict the geodesic flow to unit vectors, that is, to the *unit tangent bundle*. Note that this is a fixed-point-free flow on a *compact space*.

Example 1.1.28. The sphere with constant positive curvature has a particularly simple geodesic flow: it involves motion along great circles with unit speed and is hence periodic, that is, the time-2π map is the identity.

Example 1.1.29 (Magnetic flows). That the geodesics are so simple in the case of the sphere makes it easy to explore a slight variation on the theme of free-particle motion. A deformation of this geodesic flow is obtained by modeling a constant magnetic field perpendicular to the sphere. For a charged particle this produces a constant deflection from a constant force/acceleration perpendicular to the velocity, which means that instead of moving along great circles, such a particle moves along curves with constant and nonzero geodesic curvature, and these happen to be circles of latitude when viewed as a perturbation of the equator. Each vector is tangent to two such circles, one of which "bends right" and the other of which "bends left" according to the sign of the geodesic curvature or the orientation of the magnetic field. This is called the *magnetic flow* or *twisted geodesic flow*.[7] In all cases, we retain the feature that all orbits are periodic with the same period.

Remark 1.1.30 (Reversibility and the flip map). Magnetic flows call attention to a symmetry of geodesic flows that the magnetic flows lack. A geodesic flow can be reversed by reversing vectors, that is, $-\varphi^t(-v) = \varphi^t(v)$. Flows with this property are said to be *reversible*, and because of its importance in this respect, the map $v \mapsto -v$ is called the *flip map*. Magnetic flows lack this symmetry; the closest match would be to flip vectors as well as the magnetic field at the same time. While we concentrate on

[6]This is conservation of kinetic energy.

[7]In the limit of infinitely strong magnetic field one obtains motion along circles with infinite geodesic curvature, that is, circles with radius 0. This means that the motion reduces to tangent vectors spinning over a fixed basepoint.

Riemannian metrics when we study geodesic flows, Finsler metrics (where instead of an inner product, each tangent space is given a norm) give rise to additional examples of flows, and those often also lack reversibility.

Beyond surfaces, there are, of course, Riemannian manifolds of higher dimension. To see what makes the geodesic flows of the torus and the sphere tractable it will be useful to put them into a framework that will enable us to study geodesic flows in other cases when a Riemannian manifold possesses a lot of isometries and "symmetries," and therefore, the geodesic flow can be described without explicitly solving the geodesic equation. Chapter 2 does this for constant negative curvature.

1.1.d Smooth flows: The elliptic–parabolic–hyperbolic paradigm. While this chapter considers topological dynamics, the examples so far and those to come are mainly smooth flows, often because they can be seen as arising from differential equations. This is the context in which one can explain the focus of the book. Linear maps and flows can be divided into elliptic, parabolic, and hyperbolic ones, and in a less precise fashion one can view flows as elliptic, parabolic, and hyperbolic, respectively. Rotations provide the iconic exemplars of linear elliptic maps and flows, and the periodic part of Figure 1.1.4 is a classical nonlinear counterpart, as are linear flows on tori and billiard flows and maps (Figure 0.2.2) on an elliptic table and the geodesic flow of a round sphere. In terms of properties that we will study, these kinds of flows are characterized by (essentially) no orbit growth and by having a fairly regular structure such as a foliation (Definition 7.1.14) into pieces on which the flow acts isometrically.

Planar linear maps and flows are exemplified by $\begin{pmatrix}1&1\\0&1\end{pmatrix}$ and $t \mapsto \begin{pmatrix}1&t\\0&1\end{pmatrix}$, respectively. Parabolic flows have polynomial orbit growth and locally look like this example, exhibiting "twist" or "shear." The geodesic flow on the flat torus is an example, and a particularly important one arises in the next chapter, the horocycle flow. Another source of such flows is polygonal billiards. In the context of ergodic theory, these in turn are related to interval-exchange transformations.

Hyperbolic flows being the core subject of this book, we have an abundance of examples throughout, so there is little need to list any here. Nonetheless, we mention $\begin{pmatrix}2&0\\0&1/2\end{pmatrix}$ and $\begin{pmatrix}2&1\\1&1\end{pmatrix}$ as planar linear such maps and the flows in Section 1.4.a as continuous-time counterparts. The suspension of a hyperbolic toral automorphism (Example 1.5.26) and geodesic flows in negative curvature (Chapter 2) are nonlinear archetypes. They are characterized by exponential behavior both in the relative behavior of orbits and in orbit-growth properties.

1.2 Time change, flow under a function, and sections

In this section we study phenomena specific to continuous-time dynamics and without a true counterpart in the discrete-time case. Specifically, we investigate reparametrizations of a flow (Remark 1.1.27 is suggestive of this) and look at connections between flows and maps by use of suspensions and sections as first mentioned around page 4.

Definition 1.2.1 (Time change). A flow Ψ on M is a *time change* of another flow Φ if for each $x \in M$ the orbits $\mathcal{O}_\Phi(x) = \varphi^{\mathbb{R}}(x)$ and $\mathcal{O}_\Psi(x) = \psi^{\mathbb{R}}(x)$ coincide and the orientations given by the change of t in the positive direction are the same.[8]

If Φ and Ψ are generated by vector fields V and W, respectively, then equivalently, Ψ is a time change of Φ if $W = \rho V$ for some continuous $\rho\colon M \longrightarrow [0,\infty)$ with $\rho \neq 0$ away from fixed points. Usually we (implicitly) assume that ρ (or α below) is as smooth as Φ (in order for Ψ to be equally smooth).

Proposition 1.2.2. *If Ψ is a time change of Φ then their fixed points coincide, and $\psi^t(x) = \varphi^{\alpha(t,x)}(x)$ for every $x \in M$, where*

$$\alpha(t+s,x) = \alpha(t,x) + \alpha(s,\psi^t(x)) \tag{1.2.1}$$

and

$$\alpha(t,x) \geq 0 \quad \text{if } t \geq 0. \tag{1.2.2}$$

Indeed, either $\psi^t(x) = x$ for all $t \in \mathbb{R}$, or $\alpha(t,x) > 0$ if $t > 0$.

Proof. The "group" property $\psi^{t+s} = \psi^s \circ \psi^t$ gives (1.2.1), while (1.2.2) reflects preservation of orientation. The factor ρ in Definition 1.2.1 generates α as follows: $\alpha(t,x) = \int_0^t \rho(\psi^s(x))\,ds$; this gives the last claim. □

Remark 1.2.3. If Φ and Ψ are C^r in Proposition 1.2.2 and x is not a fixed point, then $\alpha(t,x)$ is C^r in both variables by the Implicit Function Theorem (though at fixed points, α might not even be continuous in x), while $\rho = V\alpha = \frac{\partial}{\partial t}\alpha(t,x)_{|_{t=0}}$ is C^{r-1}.

Sometimes the term "time change" is used for the flow generated by a scalar multiple of the vector field V even if it vanishes at some points where $V \neq 0$. Interesting examples of time changes arise in connection with constructions that will be seen to amount to a reversal of finding Poincaré sections (Figure 0.2.1).

The *cocycle equation* (1.2.1) is important beyond time changes and defines a notion one can consider in greater generality.

[8]The regularity of a time change depends on context.

Definition 1.2.4. A *cocycle* over a flow Ψ on X is a group-valued function $\alpha\colon \mathbb{R}\times X \to G$ such that the *cocycle equation* (1.2.1) holds (using additive notation).

Note that (1.2.1) with $t = 0$ gives $\alpha(0, x) = 0$. The cocycle equation can be motivated and remembered as saying that to go for time $t + s$, go for time t and then go further for time s *from that point.*[9]

Example 1.2.5. The differential of a flow is a cocycle, that is, $\alpha(t,x) = D\psi^t(x)$, where the cocycle equation is the chain rule with composition of linear maps as the group operation (so instead of "+" we have composition).

Example 1.2.6. A function $a\colon X \to \mathbb{R}$ generates a cocycle by setting $\alpha(t,x) := \int_0^t a(\psi^s(x))\,ds$; this is how the cocycles in time changes come about.

Remark 1.2.7. The most pertinent cocycle here is a real-valued cocycle that defines a time change—the cocycle equation ensures that the time-changed map is a flow (Proposition 1.2.2). Note that expressing the new time through the old time gives rise to a cocycle over the "new" flow Ψ. To recapitulate, suppose a flow Φ is generated by V and a flow Ψ is generated by aV for some positive function a. If $b := 1/a$, then $\beta(x,t) := \int_0^t b(\varphi^s(x))\,ds$ is a cocycle over Φ and a cocycle α over Ψ can be defined equivalently by $\alpha(x,t) := \int_0^t a(\varphi^s(x))\,ds$ or by $\beta(x,\alpha(x,t)) \equiv t \equiv \alpha(x,\beta(x,t))$ for all x to give $\psi^t(x) = \varphi^{\alpha(x,t)}(x)$ and $\varphi^t(x) = \psi^{\beta(x,t)}(x)$ for all x, t.

We now study how some flows (such as Example 1.5.26 and Definition 6.5.4) arise naturally from a map.

Definition 1.2.8 (Suspension). For a homeomorphism $f\colon M \to M$ of a topological space we define the *suspension flow* $f_\circ$ as the "vertical" flow generated by the vector field $\frac{\partial}{\partial t}$ on the *suspension manifold* (or *mapping torus*) $M_f := (M \times \mathbb{R})/\sim$, where $(x,s) \sim \alpha^n(x,s)$ for all $n \in \mathbb{Z}$ with $\alpha(x,s) := (f(x), s-1)$. (This is well defined because the vertical flow commutes with α.)

The notion of a suspension flow is related to the solution of differential equations with periodic coefficients.

Example 1.2.9. On $M = S^1 = \{z \in \mathbb{C} \mid |z| = 1\}$ consider the following situations:

(1) If $f(z) = e^{2\pi i\alpha}z$, then the suspension manifold M_f is homeomorphic to the 2-torus and $f_\circ$ is linear. All orbits are periodic if α is rational, and all orbits are dense if α is irrational; see Example 1.6.2 below.

(2) If $f(z) = \bar{z}$, then M_f is the Klein bottle and $f_\circ$ has two orbits of period 1, and all others have period 2.

[9]The word "cocycle" rightly hints at a cohomology theory (Definition 1.3.18).

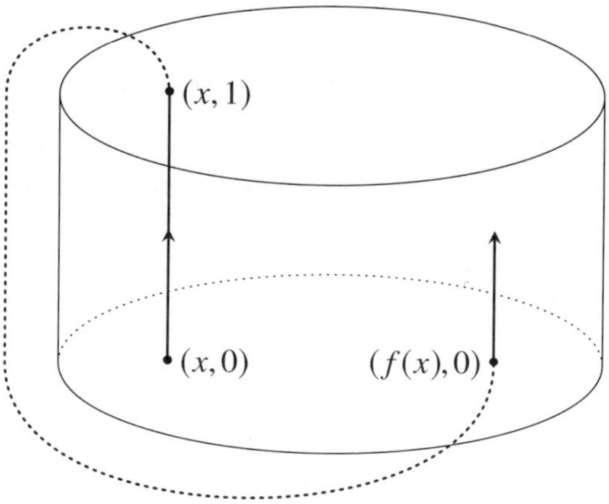

Figure 1.2.1. Suspension.

Remark 1.2.10 (Metric for a suspension manifold). That Definition 1.2.8 produces a topological space is not surprising, but at times a suitable distance function on M_f is needed if M is a metric space, and the one induced from the suspension construction is not well defined. To this end it is convenient to think of M_f as $M \times [0,1]$ with $(x,1) \sim (f(x),0)$. Let ρ be a metric (that is, distance function) on M and assume (up to scaling, hence without loss of generality) that the ρ-diameter of M is at most 1. Then

$$\begin{aligned}\rho_t((y,t),(z,t)) &:= (1-t)\rho(y,z) + t\rho(f(y),f(z))\\ &\geq \min(\rho(y,z),\rho(f(y),f(z))) =: \rho'(y,z)\end{aligned}$$

defines a metric on $M \times \{t\} \subset M \times [0,1]$. To define the distance between arbitrary $x_1, x_2 \in M \times [0,1]$ consider finite "paths" $x_1 = w_0, w_1, \dots, w_n = x_2$ such that for each i either $w_i, w_{i+1} \in M \times \{t\}$ for some t (in which case we call the pair a horizontal segment of length $\rho_t(w_i, w_{i+1})$) or $w_i = (\alpha, t_1)$ and $w_{i+1} = (\alpha, t_2)$ for some $\alpha \in M$ (in which case we call the pair a vertical segment of length $|t_1 - t_2|$). The length of such a path is the sum of the lengths of its segments, and $d(x_1, x_2)$ is the infimum of such path lengths. This is nondegenerate (since $d((y,t),(z,s)) \geq \rho'(y,z) + |t-s|$) and symmetric, and it satisfies the triangle inequality and induces the given topology.

The next construction is closely related to the idea of a Poincaré section (Figure 0.2.2) where the return time to the section is not necessarily a constant function. (We will make the connection precise after the definition.) In this case, the return time varies continuously with the basepoint on the section, and this gives a generalization of a suspension flow as defined below.

Definition 1.2.11 (Special flow, flow under a function). Starting with a map $f\colon M \to M$ define the *special flow* or *flow under a function* $r\colon M \to (0,\infty)$ as the flow $\Phi_r = \Phi_{f,r}$ generated by the vector field $\frac{\partial}{\partial t}$ on

$$M_{f,r} := M \times \mathbb{R}/\sim\,, \quad \text{where } (x,s) \sim \alpha^n(x,s) \text{ for all } n \in \mathbb{Z},$$

with $\alpha(x,s) := (f(x), s - r(x))$. (This is well defined because the vertical flow commutes with α.) The function r is also called a *roof function*.

Topologically the flows on M_f and $M_{f,r}$ are related by a time change (scale the vector field $\frac{\partial}{\partial t}$ on $M_{f,r}$ to $r(x)\frac{\partial}{\partial t}$). Equivalently, consider the manifold $M_{f,r}$ obtained from $M_r := \{(x,t) \mid x \in M,\ t \in \mathbb{R},\ 0 \le t \le r(x)\}$ by identifying pairs $(x, r(x))$ and $(f(x), 0)$.

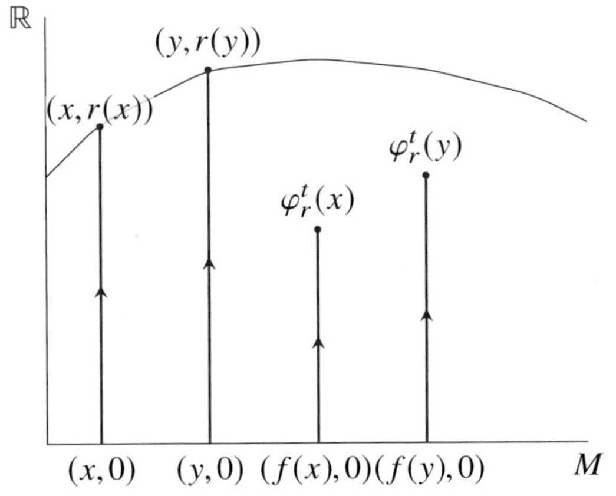

Figure 1.2.2. Special flow.

Remark 1.2.12. Special flows will arise in Example 1.2.13, Definition 1.9.5, Example 1.5.26, Definition 6.5.4, and Section 6.6. Indeed, in ergodic theory, this is the universal model (Theorem 3.6.2).

From a flow under a function one can recover the original map f in the construction as follows. Identifying X with the projection S of $X \times \{0\}$ (or the graph of any function on X) to the identification space, the desired map sends each point of this (global) *section* to the point of first return. Formally, if Φ denotes the flow, this *first-return map* on S is given by $x \mapsto \varphi^{\min\{t>0 \mid \varphi^t(x) \in S\}}(x) \in S$.

Locally, one can often use this first-return idea to define a map that reflects the local transverse character of a flow near a reference orbit. This is most naturally a way to study a periodic orbit via a small transversal, an idea that goes back to Poincaré's approach of taking periodic orbits as anchors for studying the overall dynamics by working outward from behavior near them. With a little care this is still possible and useful if a point x and $\varphi^t(x)$ are so close to each other for some t that there is a small hypersurface S through x transverse to $\dot{\varphi}$ (as in Remark 1.1.16 and Figure 1.2.3), called

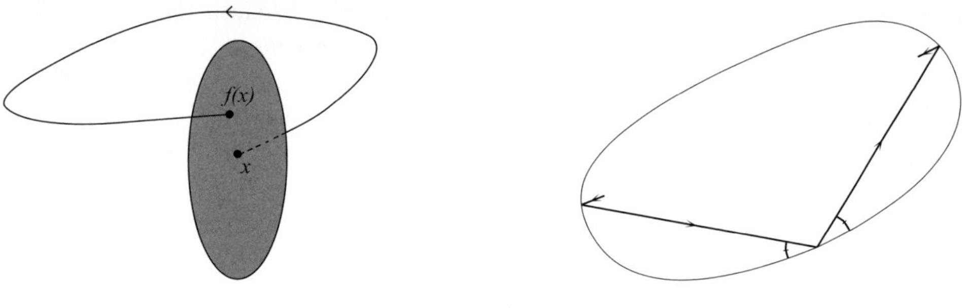

Figure 1.2.3. Poincaré section and map, and a billiard. [Reprinted from [213] (© Cambridge University Press, all rights reserved) with permission.]

a *local section* or *Poincaré section*. If φ and S are smooth then by the Implicit Function Theorem so is the *first-return time* $T(x) := \min \{t > 0 \mid \varphi^t(x) \in S\}$ on a neighborhood of x in S, and on a possibly smaller neighborhood the map $f(x) := \varphi^{T(x)}(x)$ is well defined and continuous (hence smooth if φ and S are). Example 1.1.8 illustrates this in a global way: either of the circles $S^1 \times \{0\}$ or $\{0\} \times S^1$ is a section that meets every orbit.

Example 1.2.13 (Billiard flow). Consider a strictly convex region $R \subset \mathbb{R}^2$ with smooth boundary Γ and define a flow on unit vectors as follows. For a vector v at a point $x \in R \smallsetminus \Gamma$ follow the line through x in the direction v with unit speed until it encounters Γ (geodesic flow as in Definition 1.1.26). If $x \in \Gamma$, and v points outside R, reflect v in Γ according to "angle in = angle out" ("optical" or "specular" reflection) and follow this inward direction as before (see Figure 1.2.3). In this case, there is a *global section* or *global transversal* given by inward-pointing vectors at points of Γ. The induced map on this section is called the *billiard* map, and the billiard flow is the flow over the billiard map under the function given by the free path length (until the next encounter with the boundary).

In this section we have seen differences as well as natural connections between discrete-time and continuous-time dynamics. In summary, there is no discrete-time counterpart to time changes (other than passing to an iterate), suspensions produce flows from maps, and sections produce (usually local) maps from a flow or part thereof. The flexibility of flows in time adds challenges and a richness to this subject compared to discrete-time dynamics. Sections can provide a useful tool for the local study of flows, and suspensions (as well as special flows) are on one hand topologically special but on the other hand a source of topologically interesting dynamical systems.

1.3 Conjugacy and orbit equivalence

Our ambition is to study classes of flows, and it helps to have effective means to relate or identify different flows with substantially the same or similar features. They might be naturally equivalent in that they differ only by a global change of coordinates, or one may be a subsystem of another. It turns out that for flows there are several notions of equivalence that correspond to a single notion for maps. This is related to the fact that the notion of conjugacy, which is the main notion of similarity used for maps, is of less use for flows due to longitudinal effects (effects in the flow direction). We now make some of these notions precise.

1.3.a Conjugacy and semiconjugacy. The first notion we define is topological conjugacy, an equivalence relation (Exercise 1.21) that preserves all topological properties of a flow. The related notion of a semiconjugacy is not an equivalence relation but carries enough dynamical information to be useful. The purpose of this section is to explore these using examples that also have independent interest.

Definition 1.3.1 (Factor, conjugacy). We say that the flow Φ on M is a *lift* or *extension* of Ψ on N, and Ψ is a *factor* of Φ, if there exists a continuous surjection $h\colon M \to N$ such that $h \circ \varphi(t,x) = \psi(t, h(x))$ for all $x \in M$ and all t. In that case we say that Φ and Ψ are *semiconjugate*.[10] If h is a homeomorphism, then we say that the flows are *topologically conjugate* (or *flow equivalent*). If, furthermore, h is C^r, then we say that the flows are C^r-conjugate or C^r-flow equivalent.

Remark 1.3.2. Thus, topological conjugacy preserves the entire orbit structure of a flow, and the orbit structure of a factor is naturally "included" in its extension.

This notion was central to Smale's idea of classifying dynamical systems; it provides an equivalence relation for which there is hope to understand the equivalence classes. (Smooth conjugacy is much rarer outside of elliptic dynamics.)

Proposition 1.3.3. *The circle flow (Example 1.1.6) is a factor of any suspension.*

Proof. The factor map is the projection to the fiber direction. □

[10] A drawback to this terminology is that it sounds more symmetric than it is by being unclear about which flow is a factor of which.

Example 1.3.4. The cartesian product of two flows has either flow as a factor by the projection.

Example 1.3.5. The flow $(x_1, x_2) \mapsto (x_1, x_2 + tx_1)$ on $\mathbb{T}^2$ has the identity on $S^1 = \mathbb{T}^1$ as a factor via $h\colon (x_1, x_2) \mapsto x_1$.

Remark 1.3.6. The factor map in the preceding examples collapses a lot; in Proposition 1.3.3, the preimage of any point is a copy of the base. This is useful nonetheless, but we will also encounter factors that are definitely not homeomorphisms, but close to it. This is well illustrated by the surjective continuous map $\sum_{i\in\mathbb{N}} \frac{2a_i}{3^i} \mapsto \sum_{i\in\mathbb{N}} \frac{a_i}{2^i}$ from the ternary Cantor set $\left\{ \sum_{i\in\mathbb{N}} \frac{2a_i}{3^i} \mid a_i \in \{0,1\}\right\}$ (Figure 1.3.1) to $[0,1]$, which is injective on points that are not endpoints of complementary intervals of the Cantor set, and two-to-one on those endpoints, a countable set. The semiconjugacies we obtain from "coding" later on are sometimes conjugacies (Example 1.9.18) and at least also this close to being conjugacies (Example 1.9.20). In those situations, the Cantor model has a useful combinatorial structure.

Figure 1.3.1. The ternary Cantor set.

For future reference, the *Cantor function* $\mathfrak{c}\colon [0,1] \to [0,1]$ (Figure 1.3.2) is defined by linearly interpolating this map across complementary intervals—which makes it constant on those.

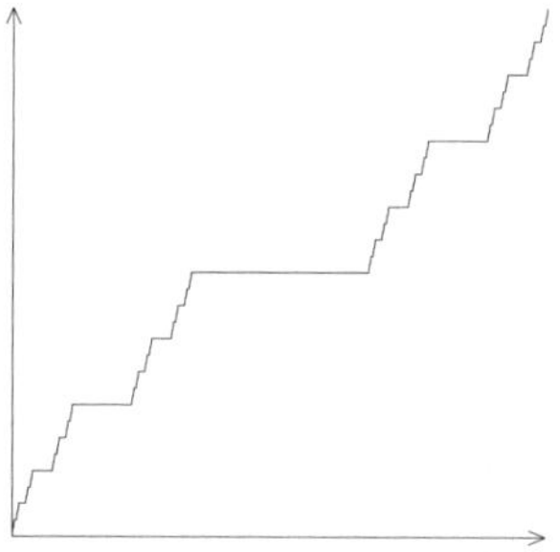

Figure 1.3.2. The Cantor function $\mathfrak{c}$. [Reprinted from [177] (© Cambridge University Press, all rights reserved) with permission.]

Example 1.3.7. Consider an interval $I = (a,b) \subset \mathbb{R}$ and a flow Φ on it defined by $\frac{dx}{dt} = f(x)$ with $f(x) > 0$ on I and $f \xrightarrow[x\to\{a,b\}]{} 0$ (see Proposition 1.1.15). Then Φ

is topologically conjugate to the flow $\psi^t\colon x \mapsto x + t$ on $\mathbb{R}$ (see Example 1.1.5) via $h_{\Phi,c}\colon \mathbb{R} \to I$, $s \mapsto \varphi^s(c)$ for any fixed $c \in (a,b)$ because

$$h_{\Phi,c}(\Psi^t(s)) = h_{\Phi,c}(s+t) = \varphi^{s+t}(c) = \varphi^t(\varphi^s(c)) = \varphi^t(h_{\Phi,c}(s)).$$

It is rare to have an explicit formula for a topological conjugacy; in Example 1.3.7 it helps that the dynamical system in question consists of a single orbit. On the other hand, the conjugacy is not unique, the choices being parametrized by $c \in (a,b)$. This corresponds to the fact that $h_{\Phi,c'}^{-1} \circ h_{\Phi,c}$ is a self-conjugacy by a constant time shift.

Example 1.3.8. The definition of φ in Example 1.3.7 naturally extends to $[a,b]$ by taking a, b to be fixed points, that is, $\varphi^t(a) = a$ and $\varphi^t(b) = b$ for all t (see Definition 1.1.10).

Example 1.3.9 (North–south dynamics). Let $S^2 = \{(x,y,z) \mid x^2+y^2+z^2 = 1\}$ be the standard unit sphere in $\mathbb{R}^3$. We consider the flow that moves every point downward (or "southward," if we think of S^2 as the surface of the globe and take the earth's axis to be vertical) along a great circle (meridian) connecting the point $(0,0,1)$ (the "north pole") and $(0,0,-1)$ (the "south pole"). The speed of the motion is equal to the derivative of the vertical coordinate along the meridian. In other words, our flow is generated by integrating the following vector field v on the sphere:

$$v(x,y,z) = (xz, yz, -x^2 - y^2).$$

To see this note that the downward unit vector tangent to the sphere at (x,y,z) is given by $(xz, yz, -(x^2+y^2))/\sqrt{x^2+y^2}$. The absolute value of its z-coordinate $\sqrt{x^2+y^2}$ gives the norm of the gradient vector. The two poles are the only zeros of v and hence the only fixed points for the flow. Every point except for the north pole asymptotically approaches the south pole as time goes to $+\infty$. In fact, this convergence is exponential. Similarly, every point except for the south pole exponentially approaches the north pole as time goes to $-\infty$. This example can be extended to gradient flow on any n-sphere for $n \ge 1$ and will have similar dynamics.

Interestingly, the lowest-dimensional case of the preceding example turns out to be an ingredient in the study of an important class of hyperbolic flows (Figure 2.3.2). We revisit gradient flows in Example 1.4.14.

Example 1.3.10 (Uniqueness and smoothness of conjugacies). In the context of Proposition 1.1.15 with $n = 1$, consider a flow on $\mathbb{R}$ generated by $\frac{dx}{dt} = f(x)$ with $x \cdot f(x) < 0$ for $x \neq 0$ and f continuous on $\mathbb{R}$. Then $\varphi^t(x) \xrightarrow[t\to\infty]{} 0$ monotonically for any $x \in \mathbb{R}$, that is, Φ is contracting, and Φ is conjugate to a linear flow as follows. Example 1.3.7 with $a = 0$, $b = \infty$ shows that $\Phi\restriction_{\mathbb{R}^+}$ is conjugate to the flow

$(t,x) \mapsto x - t$ on $\mathbb{R}$, hence to the flow $(t,y) \mapsto ye^{-t}$ on $\mathbb{R}^+$. It is similar on $\mathbb{R}^-$, and setting $h(0) = 0$ gives a conjugacy to the flow $(t,y) \mapsto ye^{-t}$ on $\mathbb{R}$. This further implies that any two such flows are topologically conjugate. As noted in Example 1.3.7, the conjugacy is not unique (we freely chose the image of 0), and in this context we can make independent choices on $\mathbb{R}^+$ and $-\mathbb{R}^+$.

That 0 is the fixed point is not central here because the homeomorphism $x \mapsto x + c$ is a conjugacy to a contracting flow that fixes $c \in \mathbb{R}$. Thus, *all contracting flows on* $\mathbb{R}$ *are pairwise topologically conjugate.*

Remark 1.3.11. One can also show that the choice of conjugacy we have described is all there is. The first indication is given by linear flows: if h conjugates the linear contracting flow $(t,x) \mapsto \lambda^t x$ on $\mathbb{R}^+$ to itself (that is, it commutes with the linear flow), then $h(\lambda^t x) = \lambda^t(x)$ for all $t \in \mathbb{R}$, so h is linear (on $\mathbb{R}^+$) and hence determined by choosing the image of a single point. Since all contracting flows are conjugate to a linear flow, this gives the complete story: if h_1, h_2 both conjugate Φ to Ψ and h conjugates Φ to a linear flow, then $hh_2h_1h^{-1}$ conjugates the linear flow to itself and is hence (linear and) unique up to a choice of scale factor.

Remark 1.3.12. Lastly, the topological conjugacy between two differentiable contracting flows we have obtained here is differentiable, except possibly (indeed, probably) at 0: if it is differentiable at 0, then differentiating the conjugacy relation at 0 shows that $(\varphi^t)'(0) = (\psi^t)'(0)$ for all $t \in \mathbb{R}$.[11] In this case, uniqueness up to a scale factor shows that a conjugacy that is differentiable at 0 is unique once we prescribe 1 as the derivative at 0.

The presence of this obstruction shows that using differentiable conjugacy to define equivalence of flows creates "structural fragility" in the sense that a flow would not be equivalent to a typical perturbation. This is an important reason for preferring continuous conjugacy as the natural notion of equivalence. Indeed, with respect to this notion we will find the opposite of structural "fragility" for hyperbolic flows (Corollary 5.4.7) (though with a weakening of "conjugacy" to "orbit equivalence"; see Section 1.3.b). Looking even further ahead we note that the rarity of smooth conjugacy (or indeed, orbit equivalence) can in this context make it intensely interesting in some rather particular respects; this is central to rigidity theory (Chapter 9, for example, Lemma 9.1.7).

The following is a close cousin to Examples 1.3.7 and 1.3.8.

Example 1.3.13 (South–south dynamics). Identifying a and b in Example 1.3.8 gives a flow on the circle with a single fixed point (Figure 1.3.3). (Note that this is

[11] This is sufficient for the existence of a conjugacy that is differentiable at 0, but the argument is not elementary. Theorem 6.8.10 implies this, and it might be interesting to simplify its proof for the present situation.

also included in Figure 1.5.4.) With more specific choices one can describe this as generated by the differential equation $\frac{dx}{dt} = f(x)$ on $[0, 1]$ mod 1 with $f(0) = 0$ and $f(x) > 0$ otherwise.

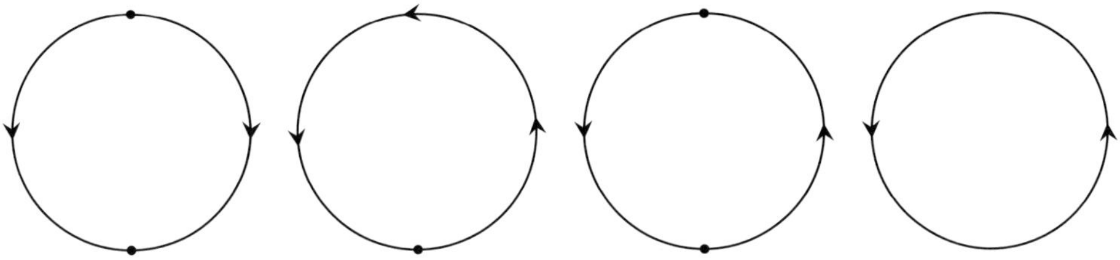

Figure 1.3.3. North–south (Example 1.3.9), south–south (Example 1.3.13), south–north–south (Example 1.3.14) dynamics, rotation (Example 1.1.6).

Although they are closely related, no two of the flows in Examples 1.3.7, 1.3.8, and 1.3.13 are topologically conjugate because the spaces on which they are defined are not homeomorphic. However, Examples 1.3.8 and 1.3.13 are related by a semiconjugacy, as are Examples 1.3.9 on S^1 and 1.3.8.

Example 1.3.14 (South–north–south dynamics). A companion example to the preceding circle flows reverses one arrow in the north–south dynamics (Example 1.3.9) on S^1, giving two fixed points with one orbit each connecting them in one direction versus the other (Figure 1.3.3). Note that this dynamics also appears as part of Figure 1.1.4 (Figure 1.3.4).

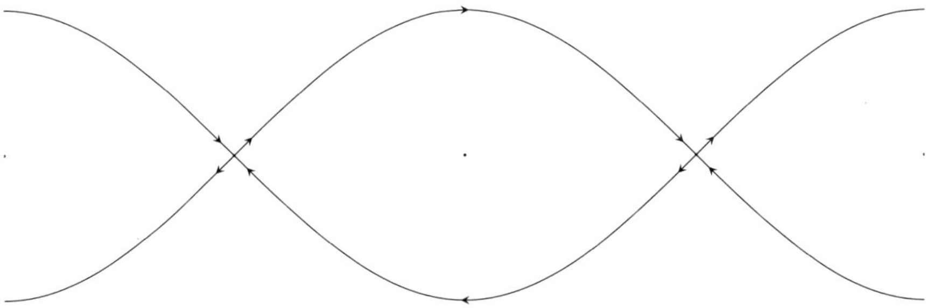

Figure 1.3.4. The south–north–south dynamics in the context of Figure 1.1.4.

Clearly, no two flows in Figure 1.3.3 are topologically conjugate.

Example 1.3.15. More generally than Example 1.3.8, consider an interval $I = [a,b] \subset \mathbb{R}$ and a flow φ^t on I defined by $\frac{dx}{dt} = f(x)$ with f continuous on I and $f(a) = f(b) = 0$ (see Proposition 1.1.15), that is, we do not assume $f > 0$. The zeros of f are the fixed points of this flow. It is illuminating to prove that two such flows are conjugate if (and clearly only if) there is an increasing homeomorphism that identifies the respective sets where f is zero, positive, and negative.

Example 1.3.16 (Akin). Example 1.3.15 describes a class of flows which could likewise be defined on $S^1 = [0,1]/\mathbb{Z}$, and we specialize this to a pair of examples on $[0,1]$ and S^1. Choose $f\colon [0,1] \to [0,1]$ continuous such that $f^{-1}(\{0\}) = C$, the ternary Cantor set (Remark 1.3.6), and define the Akin flow $A = (\alpha^t)_{t\in\mathbb{R}}$ on $[0,1]$ by $\frac{dx}{dt} = f$, $A_\circ$ its projection to S^1. Note that here we do not only specialize the fixed-point set but also unidirectional motion.

Note that the Cantor function $\mathfrak{c}$ (Remark 1.3.6) is a constant of motion for A, and $\mathfrak{c}(1-\mathfrak{c})$ is a constant of motion for both A and $A_\circ$.

Example 1.3.17. Let $v = (v_1, \dots, v_{n-1}, 1)$. The linear flow Φ_v (Example 1.1.8) on the n-torus is C^∞-conjugate to the suspension of the translation $f_\gamma\colon x \mapsto x + \gamma$ on the $(n-1)$-torus, where $\gamma := (v_1, \dots, v_{n-1})$: Consider the map H from the suspension manifold $M = \mathbb{T}^{n-1}_{T_\gamma}$ to the torus $\mathbb{T}^n$ given by

$$H(x_1, \dots, x_{n-1}, t) = (x_1 + v_1 t, x_2 + v_2 t, \dots, x_{n-1} + v_{n-1} t, x_n + t).$$

It is differentiable for $t \neq 0$. Differentiability at $t = 0$ follows from the definition of the smooth structure on the suspension manifold. The differential of H carries the upward vector field $\frac{\partial}{\partial t}$ to the vector field $v_1 \frac{\partial}{\partial x_1} + v_2 \frac{\partial}{\partial x_2} + \cdots + v_n \frac{\partial}{\partial x_n}$ and hence conjugates the flows generated by those vector fields, which are exactly the suspension flow and the linear flow, respectively.

Special flows admit a straightforward sufficient criterion for conjugacy that is useful, even though this is viewed as a trivial example of conjugacy.

Definition 1.3.18. For an invertible map $f\colon X \to X$ two functions $r_1, r_2\colon X \to (0,\infty)$ are *cohomologous* via a *transfer function* $g\colon X \to \mathbb{R}$ if $r_1(x) = r_2(x) + g(f(x)) - g(x)$ for all $x \in X$.

For a flow Φ on X generated by a vector field V, two functions $r_1, r_2\colon X \to \mathbb{R}$ are *cohomologous* via a *transfer function* $g\colon X \to \mathbb{R}$ if $r_1 = r_2 + Vg$, where Vg is the derivative along the flow. If $r_2 \equiv 0$ then r_1 is *null cohomologous*.

Proposition 1.3.19. *Let $f\colon X \to X$ be an invertible map and $r_1, r_2\colon X \to (0,\infty)$ be cohomologous via a transfer function g. Then Φ_{r_1} and Φ_{r_2} are conjugate via a conjugacy with the same regularity as g.*

Proof. The function $\overline{h}\colon X \times \mathbb{R} \to X \times \mathbb{R}$ defined by $(x, s) \mapsto (x, s + g(x))$ is as regular as g and commutes with the vertical flow, while by assumption

$$\overline{h} \circ \alpha_1(x, s) = (f(x), s - r_1(x) + g(f(x))) = (f(x), s + g(x) - r_2(x)) = \alpha_2 \circ \overline{h}(x, s). \quad \square$$

Example 1.3.20 (Trivial time change). Let Φ be a smooth flow with generating vector field V. Let $h(x) = \varphi^{b(x)}(x)$, where b is a differentiable function with

$$(Vb)(x) = db(V)(x) = \frac{db(\varphi^t(x))}{dt}\Big|_{t=0} > 0 \quad \text{when } V(x) \neq 0,$$

that is, the derivative in the flow direction is positive if $V(x) \neq 0$. Then

$$(h \circ \varphi^t \circ h^{-1})(hx) = h(\varphi^t(x)) = \varphi^{b(\varphi^t(x))}(\varphi^t(x)) = \varphi^{t+b(\varphi^t(x))}(x) = \varphi^{t+b(\varphi^t x)-b(x)}(hx),$$

and

$$\beta(t, x) := t + b(\varphi^t x) - b(x)$$

satisfies (1.2.2). This kind of time change is said to be *trivial* or *canonical*. An equivalent way to describe these is as follows.

Proposition 1.3.21 (Trivial time changes). *Consider a flow Φ generated by the vector field V and a smooth $f\colon M \to \mathbb{R}$ such that $1 + df(V) > 0$. Then $h\colon x \mapsto \varphi^{f(x)}(x)$ conjugates the flow generated by the vector field $V_f := \frac{V}{1+df(V)}$ to Φ.*

Proof. Smoothness of f and $1 + df(V) > 0$ ensure that h is a diffeomorphism. Now we write $x_t = \varphi^t(x)$ and use the chain rule to compute

$$\begin{aligned} dh(V(x)) &= \frac{dh}{dt} = \frac{d}{dt}\varphi^{f(x_t)}(x_t)\Big|_{t=0} = \frac{d\varphi}{dt}\Big|_{t=0} df(V)(x) + V(\varphi^{f(x)}(x)) \\ &= V(\varphi^{f(x)}(x)) \cdot df(V)(x) + V(\varphi^{f(x)}(x)) = (1 + df(V)(x))V(\varphi^{f(x)}(x)), \end{aligned}$$

which gives $dh(V_f) = V$ upon division by $1 + df(V)(x)$. $\square$

1.3.b Orbit equivalence. Example 1.3.10 showed that differentiable conjugacy is more restrictive than we want because even small perturbations of a flow can remove it from a given smooth conjugacy class. However, even topological conjugacy is restrictive because the slightest time changes may render topological conjugacy impossible, that is, the equivalence classes of topological conjugacy are often too small to be helpful for classifying flows. The notion of an orbit equivalence is often a more natural equivalence relation for flows although, unlike topological conjugacy, an orbit equivalence fails to preserve some important topological properties (such as mixing) and quantities (such as entropy).

Definition 1.3.22 (Orbit equivalence). A flow Ψ on N is said to be an *orbit factor* of a flow Φ on M if there exists a continuous surjection $h\colon M \longrightarrow N$ that sends orbits of Φ to orbits of Ψ. We also say that Ψ and Φ are *semiequivalent*. If h is a homeomorphism, then the flows are *orbit equivalent*. Two orbit equivalences $h, k\colon M \longrightarrow N$ are said to be *flow related* if there is a continuous $\alpha\colon M \longrightarrow \mathbb{R}$ such that $h(\varphi^{\alpha(t)}(x) = k(x)$ for all x.[12]

Remark 1.3.23. Orbit equivalence occurs more commonly for flows than conjugacy and therefore tends to be the more prominent concept of these two. However, it does not preserve some topological properties sensitive to "longitudinal" effects, notably topological mixing (Definition 1.6.31) and topological entropy (Definition 4.2.2; see equation (4.3.6)). This is a reason we do not refer to this as topological equivalence. It is, of course, an equivalence relation (Exercise 1.24).

In some simple contexts (Example 1.3.15) there is little distinction between orbit equivalence and topological conjugacy.

Remark 1.3.24. If a flow Φ without fixed points is topologically orbit equivalent to a flow Ψ via h (that is, h is a homeomorphism that maps orbits of Φ to orbits of Ψ), then $h^{-1} \circ \psi^t \circ h$ is a flow with the same orbits as Φ, and the reparametrization is homeomorphic. If $x \in X$ is not periodic, then $\sigma_x\colon \mathbb{R} \longrightarrow \mathbb{R}$ defined by $h^{-1}(\psi^t(h(x))) = \varphi^{\sigma_x(t)}(x)$ is a homeomorphism with $\sigma_x(0) = 0$. If $x \in X$ is periodic for Φ with least period ν and μ is its least period for $h^{-1} \circ \psi^t \circ h$, then $h^{-1}(\psi^t(h(x))) = \varphi^{\sigma_x(t)}(x)$ defines a strictly monotone continuous map σ_x on $[0, \mu]$ whose range is an interval of length ν with 0 as an endpoint. Extending naturally to $[n\mu, (n+1)\mu]$ gives a homeomorphism of $\mathbb{R}$.

Remark 1.3.25. No two of the flows in Examples 1.3.7, 1.3.8, and 1.3.13 are orbit equivalent because the spaces on which they are defined are not homeomorphic.

We now begin to study the *relative* behavior of orbits, an important concern for topological dynamics.

Definition 1.3.26. For a flow Φ and point $x \in X$ define the (strong) *stable* and *unstable sets* of x by

$$\begin{aligned} W^s(x) &:= \{y \in X \mid d(\varphi^t(x), \varphi^t(y) \xrightarrow[t\to+\infty]{} 0\}, \\ W^u(x) &:= \{y \in X \mid d(\varphi^t(x), \varphi^t(y) \xrightarrow[t\to-\infty]{} 0\}. \end{aligned} \tag{1.3.1}$$

The sets

$$W^{cs}(x) := \bigcup_{t\in\mathbb{R}} \varphi^t(W^s(x)) \quad \text{and} \quad W^{cu}(x) := \bigcup_{t\in\mathbb{R}} \varphi^t(W^u(x))$$

are called the *center-stable* and *center-unstable* sets of x.

[12] This last notion is a hint at questions of uniqueness: any orbit equivalence can be altered to a flow-related one, so uniqueness questions have to be about the existence of another orbit equivalence that is *not* flow related.

Remark 1.3.27. In the literature one also sees the notation W^{ss} or W^- for the stable set and W^{uu}, W^{su}, or W^+ for the (strong) unstable set. And at this time these are indeed merely (un)stable *sets*—rather than manifolds, as will be the case for hyperbolic flows later.

The triangle inequality gives (Exercise 1.18) the following proposition:

Proposition 1.3.28. *If Φ is a flow and $x, y \in X$, then*

- $W^s(x) \cap W^s(y) \neq \varnothing$ *implies* $W^s(x) = W^s(y)$;
- $W^{cs}(x) \cap W^{cs}(y) \neq \varnothing$ *implies* $W^{cs}(x) = W^{cs}(y)$;
- $W^u(x) \cap W^u(y) \neq \varnothing$ *implies* $W^u(x) = W^u(y)$; *and*
- $W^{cu}(x) \cap W^{cu}(y) \neq \varnothing$ *implies* $W^{cu}(x) = W^{cu}(y)$.

By uniform continuity, conjugacies and orbit equivalences preserve these sets as follows:

Proposition 1.3.29. *If Φ and Ψ are flows conjugate by h, then $h(W^s(x)) = W^s(h(x))$ and $h(W^u(x)) = W^u(h(x))$ for all x, hence likewise for center-stable and center-unstable sets. For an orbit equivalence a similar statement holds for the center-stable and center-unstable sets (only).*

Proposition 1.3.30 (Longitudinal regularity of orbit equivalence). *For $r \in \mathbb{N}$ an orbit equivalence between fixed-point-free C^r-flows can be chosen to depend C^r on time.*

Proof ([147]). If h maps orbits of Φ homeomorphically to orbits of Ψ, then (as in Proposition 1.2.2) there is a continuous cocycle α over Ψ such that

$$h(\varphi^t(x)) = \psi^{\alpha(t,x)}(h(x)),$$

that is, $\alpha(t+s, x) = \alpha(t, \varphi^s(x)) + \alpha(s, x)$. For some $T > 0$ set $k(x) := \psi^{\frac{1}{T}\int_0^T \alpha(\tau,x)\,d\tau}(h(x))$ to get (reverting to the original notation $\Psi(t, x) = \psi^t(x)$)

$$\begin{aligned}
k(\varphi^t(x)) &= \Psi\Big(\frac{1}{T}\int_0^T \underbrace{\alpha(\tau, \varphi^t(x))}_{=\alpha(t+\tau,x)-\alpha(t,x)}\,d\tau, \underbrace{h(\varphi^t(x))}_{=\psi^{\alpha(t,x)}(h(x))}\Big) \\
&= \underbrace{\Psi\Big(\underbrace{\frac{1}{T}\int_0^T \alpha(t+\tau, x) - \alpha(t,x)\,d\tau}_{=\frac{1}{T}\int_0^T \alpha(t+\tau,x)\,d\tau = \frac{1}{T}\int_t^{T+t}\alpha(\tau,x)\,d\tau} + \alpha(t,x), \underbrace{h(x)}\Big)}_{=\Psi(-\frac{1}{T}\int_0^T \alpha(\tau,x)\,d\tau, k(x))} \\
&= \Psi\Big(\underbrace{\frac{1}{T}\int_t^{T+t}\alpha(\tau,x)\,d\tau - \frac{1}{T}\int_0^T \alpha(\tau,x)\,d\tau}_{=:\beta(t,x)=\frac{1}{T}\int_0^t \alpha(T+\tau,x)-\alpha(\tau,x)\,d\tau}, k(x)\Big).
\end{aligned}$$

Since Ψ is smooth, differentiability of $t \mapsto k(\varphi^t(x)) = \psi^{\beta(t,x)}(k(x))$ follows because $\frac{d}{dt}\beta(t,x) = \frac{1}{T}(\alpha(T+t,x) - \alpha(t,x))$ by the Fundamental Theorem of Calculus, so $t \mapsto \beta(t,x)$ is C^1.

Now, if $t \mapsto \alpha(t,x)$ is C^1, then this construction leads to a β such that $t \mapsto \beta(t,x)$ is C^2, that is, recursively, we can make the orbit equivalence C^r in the time direction so long as the flows themselves are C^r. □

Naturally, orbit equivalence fails to distinguish between flows under a function (Definition 1.2.11) and suspensions (Definition 1.2.8).

Proposition 1.3.31. *Let M be a compact differentiable manifold, $f\colon M \to M$ a C^m-diffeomorphism, and $r\colon M \to \mathbb{R}_+$ a C^m-function on M. Then the special flow on the manifold M_f^r is C^m orbit equivalent to the suspension flow on M_f.*

Proof. Consider a C^∞-function $g\colon [0,1] \times [\min r, \max r] \to \mathbb{R}$ such that

(1) $g(t,s) = t$ for $t \in [0, 1/4]$,

(2) $g(t,s) = t + s - 1$ for $t \in [3/4, 1]$,

(3) $\frac{\partial}{\partial t} g(t,s) > 0$.

Then the map $(x,t) \mapsto (x, g(t, r(x)))$ is a diffeomorphism between M_f and M_f^r which takes the vertical vector field $\frac{\partial}{\partial t}$ on M_f to a vertical vector field on M_f^r, and hence conjugates the suspension flow to a time change of the special flow under r. □

Both orbit equivalence and conjugacy preserve periodic orbits, although only conjugacy preserves their periods. In the next sections we investigate the structure and relative behavior of orbits for a flow with a focus on stability and recurrence. These are topological notions, so much of this provides further invariants of conjugacy and orbit equivalence.

1.4 Attractors and repellers

We now describe certain sets, attractors and repellers, that are important anchors for describing the dynamics of a flow. We begin with the notion of attracting and repelling fixed points before moving to arbitrary attracting and repelling sets for a flow. The notion of an attracting fixed point is related to the notion of having a steady state in an engineered system that is stable in the sense that the system returns to it if it is ever jolted away. In terms of differential equations, a "stable" solution is such that "nearby" solutions either don't go far away from the solution or converge to the stable solution in some sense. The desirability of this is also related to the fact that in practice we don't have infinite precision in starting a time evolution, so stability can provide the mechanism for settling into the desired state. This notion is of particular interest with respect to fixed or periodic points, but Poincaré focused attention on using these as anchors for the understanding of other orbits.

Definition 1.4.1 (Attracting fixed points). A fixed point p of a flow Φ is *asymptotically stable* or *attracting*[13] if it is in the interior of its stable set (Definition 1.3.26), that is, given $\epsilon > 0$ there is a $\delta > 0$ such that $d(x,p) < \delta \Rightarrow d(\varphi^t(x),p) < \epsilon$ for all $t \geq 0$ and there is a $\gamma > 0$ such that $d(x,p) < \gamma \Rightarrow \varphi^t(x) \xrightarrow[t\to\infty]{} p$. In other words, the *basin of attraction* is open. A fixed point is *repelling* if it is attracting for φ^{-t}.

A periodic point p is stable if for $\epsilon > 0$ there is a $\delta > 0$ such that $d(x,p) < \delta \Rightarrow$ there is a continuous increasing $s\colon (0,\infty) \to (0,\infty)$ such that $d(\varphi^{s(t)}(x),\varphi^t(p)) < \epsilon$ for all $t > 0$. It is *asymptotically stable* or *attracting* if it is stable and there exists some $\gamma > 0$ such that $d(x,p) < \gamma \Rightarrow \exists\,\delta\colon d(\varphi^t(x),\varphi^{t+\delta}(p)) \xrightarrow[t\to\infty]{} 0$. A periodic point is *repelling* if it is attracting for φ^{-t}.

In Example 1.3.8 the fixed point b is attracting but a is not. In Example 1.1.7 the origin is attracting if and only if $a < 0$. In Example 1.1.23 the origin is not attracting because all orbits nearby are periodic and hence not asymptotic to the origin. (This changes with damping; see Figure 1.4.1.) In Example 1.3.8 the fixed point b is attracting. The fixed point in Example 1.3.13 is neither attracting nor repelling.

Example 1.4.2. The orbit of $(0,0)$ is periodic and asymptotically stable for the flow $(x,y) \mapsto (x+t \pmod 1, ye^{-t})$ on $S^1 \times \mathbb{R}$. Likewise, the orbit of $(1,0)$ is periodic and asymptotically stable for the flow on $\mathbb{R}^2$ described in polar coordinates by the differential equations $\left\{\begin{smallmatrix} dr/dt=r(1-r), \\ d\theta/dt=1. \end{smallmatrix}\right.$

Attracting fixed points have a neighborhood, each point of which limits on the fixed point as time approaches $+\infty$. Similarly, for repelling fixed points the same happens as time approaches $-\infty$.

1.4.a Linear flows.

We now investigate stability and conjugacy for the classical example of a linear flow arising from a matrix. Let $A \in \mathcal{M}_n(\mathbb{R})$ and e^{At} be the flow on $\mathbb{R}^n$ generated by A for the equation $x' = Ax$. To investigate the stability of the origin for these linear constant coefficient flows we will use the closed-form solution from Theorem 1.1.19, which is particularly amenable to discerning asymptotic growth and decay. There is an easy criterion for asymptotic stability of 0 for $x' = Ax$.

Theorem 1.4.3. *For $A \in \mathcal{M}_n(\mathbb{R})$ there are $K,\alpha > 0$ with $\|e^{At}\| \leq Ke^{-\alpha t}$ if and only if all eigenvalues have negative real part.*

[13] In the theory of differential equations this is called an asymptotically stable fixed point.

Remark 1.4.4. The proof gives a sharper version: if $-\alpha \in (\max_j \operatorname{Re}(\lambda_j), 0)$, then there is a K such that $\|e^{At}\| \le Ke^{-\alpha t}$ for all t.

Proof. Let $\zeta \in \mathbb{C}^d$ and $x(t)$ be the solution of $x' = Ax$ with $x(0) = \zeta$. Let $\lambda_1, \dots, \lambda_p$ be the eigenvalues of A. For $\lambda_j = \alpha_j + i\beta_j$ and $\zeta = v_1 + \dots + v_p$ where $v_j \in M(\lambda_j)$, for each $1 \le j \le p$ we have

$$x(t) = \sum_{j=1}^{p} e^{\lambda_j t} \sum_{k=0}^{r(\lambda_j)-1} (A - \lambda_j I)^k \frac{t^k}{k!} v_j.$$

Let $M = \max \{ \|A - \lambda_j I\|^k \mid 1 \le j \le p,\ 0 \le k \le r(\lambda_j) - 1 \}$. Then

$$\|x(t)\| \le \sum_{j=1}^{p} e^{\alpha_j t} \sum_{k=0}^{r(\lambda_j)-1} M \frac{|t|^k}{k!} \|v_j\|.$$

We now define a norm $\|\cdot\|_A$ on $\mathbb{C}^d$ by

$$\|\zeta\|_A = \|v_1\| + \dots + \|v_p\|.$$

Then there exists some $C > 0$ such that $\|v_i\| \le \|\zeta\|_A \le C\|\zeta\|$, and

$$\|x(t)\| \le \left[\sum_{j=1}^{p} e^{\alpha_j t} \sum_{k=0}^{r(\lambda_j)-1} \frac{|t|^k}{k!} \right] MC\|\zeta\|.$$

If all $\alpha_j < 0$, then choose $0 > \alpha > \max_{1 \le j \le p} \alpha_j$. So

$$\lim_{t \to \infty} \frac{e^{\alpha_j t} t^k}{e^{\alpha t}} = 0,$$

and there is a $K_0 > 0$ such that

$$\frac{e^{\alpha_j t} t^k}{e^{\alpha t}} \le K_0$$

for all $t \ge 0$, $1 \le j \le p$, and $0 \le k \le r(\lambda_j) - 1$. This implies that $e^{\alpha_j t} t^k \le K_0 e^{\alpha t}$. Define $K = (r(\lambda_1) + r(\lambda_2) + \dots + r(\lambda_p))K_0$.

The converse is easy (consider eigenvectors). □

Theorem 1.4.3 helps us establish topological conjugacy between linear flows with attracting origin.

Proposition 1.4.5. *If all eigenvalues of $A, B \in M_n(\mathbb{R})$ have negative real part, then the flows e^{At} and e^{Bt} are topologically conjugate.*

Remark 1.4.6. By reversing the flow, the conclusion also holds when all eigenvalues have positive real part.

Proof. By Theorem 1.4.3 there are $C \geq 1$, $a_0 > 0$ such that $\|e^{At}x\| \leq Ce^{-a_0 t}\|x\|$ for all $x \in \mathbb{R}^d$ and $t \geq 0$. For $a \in (0, a_0)$ there is a T such that $\|e^{At}x\| \leq e^{-at}\|x\|$ for all $t \geq T$. Define a new norm $\|x\|_A = \int_0^T e^{as}\|e^{As}x\|ds$. Then

$$\|e^{At}x\|_A = \int_0^T e^{as}\|e^{As}e^{At}x\|ds.$$

Write $t = nT + \tau$ with $0 \leq \tau < T$. Then

$$\begin{aligned}\|e^{At}x\|_A &= \int_0^{T-\tau} e^{as}\|e^{AnT}e^{A(\tau+s)}x\|ds + \int_{T-\tau}^T e^{as}\|e^{A(n+1)T}e^{A(\tau-T+s)}x\|ds \\ &\leq \int_\tau^T e^{a(u-\tau-nT)}\|e^{Au}x\|du + \int_0^\tau e^{a(u+T-\tau-(n+1)T)}\|e^{Au}x\|du \\ &= e^{-at}\int_0^T e^{au}\|e^{Au}x\|du = e^{-at}\|x\|_A.\end{aligned}$$

Such a norm $\|\cdot\|_A$ is called an *adapted norm* or *Lyapunov norm* and always exists in the hyperbolic case (Propositions 5.1.5 and B.3.8). We likewise define an adapted norm $\|\cdot\|_B$ for B. Now let

$$S_A = \{x \mid \|x\|_A = 1\} \quad \text{and} \quad S_B = \{x \mid \|x\|_B = 1\}.$$

Each nonzero orbit for A crosses S_A exactly once, and likewise for B, so we call these *fundamental domains* of the flows,[14] and

$$h_0\colon S_A \to S_B, \quad x \mapsto \frac{x}{\|x\|_B}$$

is a homeomorphism with $h_0^{-1}(y) = \frac{y}{\|y\|_A}$. Extend h_0 to $\mathbb{R}^n$ by using that for $x \in \mathbb{R}^n \smallsetminus \{0\}$ there is a unique $\tau(x)$ such that $e^{A\tau}x \in S_A$, and $\tau(e^{At}x) = \tau(x) - t$ for all $t \in \mathbb{R}$:

$$h(x) := \begin{cases} e^{-B\tau(x)}h_0(e^{A\tau(x)}x) & \text{if } x \neq 0, \\ 0 & \text{if } x = 0, \end{cases}$$

is a bijection of $\mathbb{R}^n$, continuous on $\mathbb{R}^n \smallsetminus \{0\}$, and satisfies

$$\begin{aligned} h(e^{At}x) &= e^{-B\tau(e^{At}x)}h_0(e^{A\tau(e^{At}x)}e^{At}x) \\ &= e^{-B(\tau(x)-t)}h_0(e^{A(\tau(x)-t)}e^{At}x) \\ &= e^{Bt}e^{-B\tau(x)}h_0(e^{A\tau(x)}x) \\ &= e^{Bt}h(x). \end{aligned}$$

[14] The terminology reflects the fact that the images under the flow disjointly fill the space (except for the origin).

To check that h is continuous at 0, let $x_j \xrightarrow[j\to\infty]{} 0$. Then $\tau_j = \tau(x_j) \xrightarrow[j\to\infty]{} -\infty$. Let $y_j = h_0(e^{A\tau_j}x_j)$. Since $\|y_j\|_B = 1$ for all j we have

$$\|h(x_j)\|_B = \|e^{-B\tau_j}y_j\|_B \le e^{-b|\tau_j|} \xrightarrow[j\to 0]{} 0,$$

so h is continuous at the origin. □

In the present context, we say that a matrix A is *hyperbolic* if none of the eigenvalues have zero real part. The eigenvalues are roots of the characteristic polynomial, which vary continuously with the coefficients of the characteristic polynomial, and these coefficients vary continuously with the matrix. Thus, the set of hyperbolic matrices is open (among square matrices of the same size), and it can be shown to be dense. The more general definition of hyperbolicity (Definition 5.3.50) is similarly shown to be an open property (Theorem 5.4.5). However, in this generality the set of hyperbolic flows is not dense among all smooth flows.

There is a related notion of hyperbolicity for maps (Definition 1.1.25) that can be seen by taking the time-t or time-1 map of the flow. In the case of the time-1 map we are looking at e^A and so we take the exponential of the eigenvalues for the flow. The corresponding notion is that the derivative of the map has no eigenvalues with modulus 1 (corresponding to strictly imaginary eigenvalues for the flow).

For a hyperbolic matrix A the *stable space* for A is E^s_A, or E^s when there is no ambiguity, and consists of all vectors that are in the span of the generalized eigenspaces corresponding to eigenvalues with negative real part. Similarly, the *unstable space* for A is E^u_A, or E^u when there is no ambiguity, and consists of all vectors that are in the span of the generalized eigenspaces corresponding to eigenvalues with positive real part. Then $\mathbb{R}^n = E^s \oplus E^u$. Furthermore, Theorem 1.1.19 shows that if $x \in E^\sigma$ for $\sigma = u$ or s, then $x(t) \in E^\sigma$ for all $t \in \mathbb{R}$. So these subspaces are invariant under the flow. Remark 1.4.4 shows that there exist K and $\alpha > 0$ such that

$$\|e^{At}x\| \le Ke^{-\alpha t} \quad \text{for all } x \in E^s,\ t \ge 0$$

and

$$\|e^{At}x\| \le Ke^{\alpha t} \quad \text{for all } x \in E^u,\ t \le 0.$$

For the linear flow $x' = Ax$ the stable space is then the stable set and the unstable space is then the unstable set from Definition 1.3.26. Beyond the linear context, these rates of exponential contraction in forward or backward time will be the defining feature of hyperbolicity (Definition 5.3.50).

The next result shows that there are topological conjugacies between linear flows with constant coefficients if they are hyperbolic and the dimensions of the splittings into stable and unstable subspaces are equal.

Theorem 1.4.7. *If $A, B \in \mathcal{M}_n(\mathbb{R})$ are hyperbolic with stable/unstable splittings of the same dimension, then their flows e^{At} and e^{Bt} are topologically conjugate.*

Proof. Let $h_\sigma \colon E^\sigma_A \to E^\sigma_B$ for $\sigma = u$ or s where h is the conjugacy from Proposition 1.4.5. Let π_σ be the projection from $\mathbb{R}^d$ to E^σ for $\sigma = u$ or s. Then $x = \pi_u(x) + \pi_s(x)$. It is now a straightforward calculation to show that $h(x) = h_u(\pi_u(x)) + h_s(\pi_s(x))$ defines the desired conjugacy. □

The Hartman–Grobman Theorem (Theorem 5.5.1) states that if Φ is a flow with fixed point p such that the linear approximation of Φ at p is given by a hyperbolic matrix, then locally the nonlinear flow is topologically conjugate to the linearized flow.

For nonhyperbolic matrices one does not expect, in general, that there is a conjugacy between the nonlinear flow and the linearized flow near a fixed point. In fact, here is an example of an attracting fixed point where the linearized flow is not asymptotically stable at the origin.

Example 1.4.8. The linearization of the system $\begin{cases} dx/dt = y - x^3, \\ dy/dt = -x - y^3, \end{cases}$ at 0 is $\vec{x}' = \begin{pmatrix} 0 & 1 \\ -1 & 0 \end{pmatrix} \vec{x}$, which has eigenvalues $\pm i$ and hence gives a rotational flow for which all orbits are periodic. Since $\frac{d}{dt}(x^2 + y^2) = 2x\frac{dx}{dt} + 2y\frac{dy}{dt} = 2x(y - x^3) + 2y(-x - y^3) = -2(x^4 + y^4) < 0$ along solutions, except at 0, the origin is an attractor.

1.4.b Lyapunov functions and attractors. Until the 1950s, local analysis (that is, the study of asymptotic stability and hyperbolicity) was largely limited to fixed points and periodic points. Attention focused on fixed points whose linearization is hyperbolic and periodic orbits whose return map is hyperbolic as described above. From the late 1950s onward, more complicated invariant sets came into view as attractors, that is, possessing an open set of points that asymptotically limit on these sets. Such a set is called an *attractor* if this happens as time approaches ∞ or a *repeller* if this happens as time approaches $-\infty$ (Definition 1.4.18).

Lyapunov developed a method to determine the basin of attraction for ordinary differential equations that does not require solving the equation, but instead uses something called a *Lyapunov function*—a continuous function that decreases along orbits, such as $x^2 + y^2$ in Example 1.4.8 (Definition 1.4.10). The difficulty with this method is to find a Lyapunov function. In certain physical situations there are ways to do this; for instance, energy will be decreasing in mechanical systems with friction. Differential equations that admit Lyapunov functions sometimes allow heuristic approaches to guessing a Lyapunov function.

Example 1.4.9. If we modify the pendulum in Example 1.1.23 to account for friction, then a possible model is given for some $c > 0$ by the differential equation

$$\frac{d^2x}{dt^2} + c\frac{dx}{dt} + \sin 2\pi x = 0.$$

With $v := \frac{dx}{dt}$ we obtain the system of first-order differential equations

$$\begin{cases} \dfrac{dx}{dt} = v, \\ \dfrac{dv}{dt} = -\sin 2\pi x - cv, \end{cases}$$

for $x \in S^1$, $v \in \mathbb{R}$. Hence the total energy given by $H(x,v) = \frac{1}{2}v^2 - \frac{1}{2\pi}\cos 2\pi x$ (Figure 1.1.3) on the cylinder $S^1 \times \mathbb{R}$ decreases along orbits of the flow:

$$\frac{d}{dt}H(x,v) = v\frac{dv}{dt} + \frac{dx}{dt}\sin 2\pi x = -cv^2 \le 0,$$

with strict inequality when $v \neq 0$. Therefore, energy is now a Lyapunov function rather than a constant of motion, so orbits no longer lie on the energy level sets in Figure 1.1.4 but cross them "downward" at all times, which gives the phase portrait in

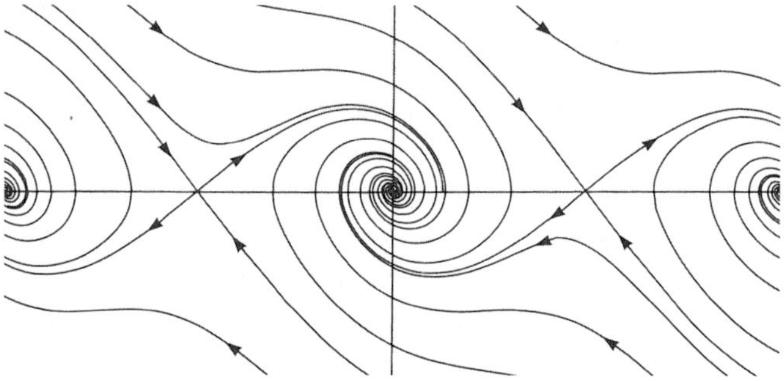

Figure 1.4.1. Phase portrait of the damped pendulum. [Picture by James Schlesinger, from [163] (© Sonia Guterman, all rights reserved) with permission.]

Figure 1.4.1. Friction thus changes the character of the stable equilibria: they are still stable but now furthermore asymptotically stable.

Definition 1.4.10 (Lyapunov function). For a flow Φ on a space X a continuous function $L\colon X \to \mathbb{R}$ is a *Lyapunov function* if $L(\varphi^t(x)) \le L(x)$ for all $x \in X$ and all $t \ge 0$.

Remark 1.4.11. Note that constant functions are therefore always Lyapunov functions. It is thus tempting to require strict inequality when $t > 0$ and x is not fixed, as is the case in Example 1.4.9. However natural that might be for situations like the damped pendulum, there are important applications in which it is crucial to avoid this restriction (see Example 1.4.12, for instance), and the accepted notion of a *strict* Lyapunov function is more realistic (Remark 1.5.45).

On the other hand, if f is a Lyapunov function, then so are $\arctan f$, $f + c$ for any constant c, and cf for any positive constant c, so one can always assume without loss of generality that a Lyapunov function takes values in a prescribed closed bounded interval (of positive length).

Example 1.4.12. Using polar coordinates, $(r-1)^2$ is a Lyapunov function for the system $\left\{\begin{smallmatrix} dr/dt=r(1-r), \\ d\theta/dt=1, \end{smallmatrix}\right.$ of differential equations. Its global minimum value is attained on the periodic orbit $r = 1$, a *limit cycle*. The origin is a maximum and an isolated equilibrium, and this function strictly decreases along all other orbits.

Example 1.4.13. The expression $x \mapsto \mathfrak{c}(x) - x$ (Figure 1.3.2) is a Lyapunov function, in fact a strict Lyapunov function (Remark 1.5.45), for the Akin flow (Example 1.3.16) on $[0, 1]$.

Aside from Example 1.4.9, other previous instances of flows admit somewhat obvious Lyapunov functions. In Example 1.3.9 the height (that is, the z-coordinate) will do, and in Example 1.3.8 $L(x) = -x$ clearly works because (see Example 1.3.7) $\frac{d}{dt}L(x) = f(x) < 0$ away from the endpoints. In Example 1.4.2, $|y|$ is a Lyapunov function whose absolute minimum is attained on the attracting periodic orbit.

Example 1.4.14 (Gradient flows). Example 1.3.9 is a specific manifestation of a class of flows that by design have a Lyapunov function that one can think of as a "height." Consider a Riemannian metric on a compact smooth manifold M and a real-valued function F on M. At each point $x \in M$ that is not a critical point for F one can define the unique direction of fastest increase for F, that is, the unit tangent vector $\zeta(x) \in T_xM$ such that $\mathcal{L}_{\zeta(x)}F = \max_{\eta \in T_xM} \mathcal{L}_\eta F/\|\eta\|$, where $\mathcal{L}_\eta F$ denotes the Lie (directional) derivative of the function F along the vector η.

We define the gradient vector field ∇F by

$$\nabla F(x) = \begin{cases} \mathcal{L}_{\zeta(x)}F \cdot \zeta(x) & \text{if } x \text{ is noncritical,} \\ 0 & \text{if } x \text{ is critical.} \end{cases}$$

Suppose that in local coordinates $(x_1, \dots, x_n)$ the Riemannian metric has the form $ds^2 = \sum g_{ij}(x_1, \dots, x_n)\, dx_i\, dx_j$. Then

$$\nabla F(x_1, \dots, x_n) = G^{-1}(x)\left(\frac{\partial F}{\partial x_1}, \dots, \frac{\partial F}{\partial x_n}\right),$$

where $G(x) = \{g_{ij}(x)\}$ and G^{-1} is the inverse matrix, so it is a smooth vector field on M. The flow generated by the gradient vector field ∇F is called the *gradient flow* of F.

From calculus we know that the gradient defined in coordinates is orthogonal to level sets of the function. This is still true in this setting because the direction of the gradient vector field is that of the fastest increase of the function F.

Example 1.3.9 is the gradient flow for the function $F(x, y, z) = -z$ on the 2-sphere provided with the Riemannian metric induced from the standard Euclidean metric in $\mathbb{R}^3$.

Example 1.4.15 (Toral gradient flows). To consider a less simple instance than Example 1.3.9, let $M \approx \mathbb{T}^2$ embedded in $\mathbb{R}^3$ as a doughnut standing up as in Figure 1.4.2, and as before, $F(x, y, z) = -z$, the negative of the height function. The function F has four critical points on the torus: a maximum A, two saddles B and C, and a minimum D. All orbits of the gradient flow other than those fixed points and six special orbits described below approach the minimum D as $t \to +\infty$ and the maximum A as $t \to -\infty$. Two special orbits connect A with B, two more connect B with C, and the last two connect C with D.

Now tilt this torus slightly, that is, change the embedding but keep the function F the same. Equivalently, consider instead the function $F = -z + \epsilon x$ for small $\epsilon > 0$. Four critical points remain, as well as the special orbits connecting the maximum with the upper saddle and the lower saddle with the minimum. However, the orbits connecting the two saddles disappear. Instead of these two orbits we have four: two connecting the maximum with the lower saddle and two connecting the upper saddle with the minimum; see Figure 1.4.2.

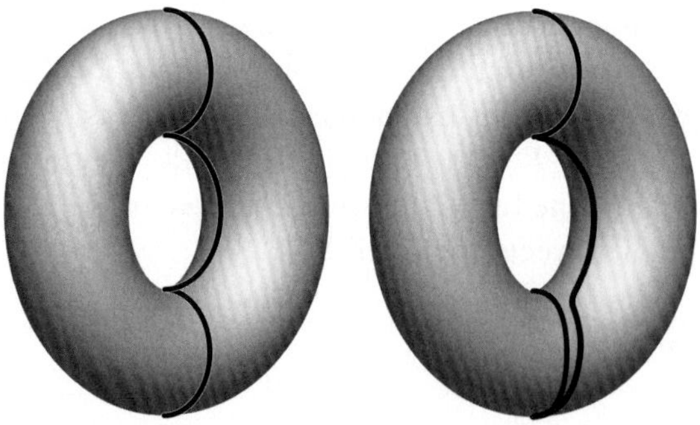

Figure 1.4.2. Gradient flows on the torus. [Reprinted from [213] (© Cambridge University Press, all rights reserved) with permission.]

Example 1.4.16 (Hot vinyl). Orbits of a gradient flow need not be asymptotic to a single fixed point. Consider an old-fashioned vinyl record suspended flat from its rim but sagging toward the center. The music is encoded by a groove that spirals toward a circular groove around the center. Consider such a grooved "bowl" but with an infinite spiral toward a circle. The gradient flow then has the bottom of the spiral groove as an orbit that is asymptotic to that entire circle—with ever-diminishing speed.

Lyapunov functions impose a gradient-like structure on the dynamics, but without the requirement that critical sets consist of fixed points. We will see that this makes them a universal tool for disentangling transient and recurrent dynamics. If, as in the case of gradient flows, a Lyapunov function is *strictly* decreasing along nonconstant orbits, then there is no nontrivial recurrence, and as in Example 1.4.9 this has long been a tool for establishing (asymptotic) stability in differential equations. Note that $L \equiv 0$ is always a Lyapunov function, so it is usually understood that a Lyapunov function is not meant to be constant. Functions that are merely nonincreasing along orbits can provide deep insights into the interplay between different invariant pieces of a dynamical system, as we now begin to demonstrate.

Definition 1.4.17. Let Φ be a flow on a metric space X. A set $\varnothing \neq U \subsetneq X$ is a *trapping region* if $\varphi^t(U) \subset U$ for all $t \geq 0$ and there exists a $T > 0$ such that $\varphi^T(\overline{U}) \subset \operatorname{int}(U)$.[15]

Since φ^t is a homeomorphism for each t, if U is a trapping region, then $X \smallsetminus \overline{U}$ is a trapping region for φ^{-t}. We need this fact in the next definition.

Definition 1.4.18. If $U \subset X$ is open, write $A_U := \bigcap_{t \geq 0} \varphi^t(U)$ and $R_U := \bigcap_{t \leq 0} \varphi^t(X \smallsetminus \overline{U})$. A set $A \subset X$ is an *attracting set* or *attractor* for the flow φ^t provided there exists a trapping region U such that $A = A_U$. We say that U is a trapping region for A. Similarly, the *repelling set* or *repeller* associated with U is R_U. For a given trapping region U the pair (A_U, R_U) of attracting and repelling sets for U is called an *attractor–repeller pair* for U. We denote the set of attractor–repeller pairs by $\mathcal{AR}(\Phi)$. An orbit is called a *sink* if it is an attractor and a *source* if it is a repeller.

It is illuminating to explore these notions in the examples of flows that have appeared so far (Examples 1.1.5, 1.1.7, 1.1.8, 1.3.7, 1.3.8, 1.3.13, 1.3.15, 1.3.16, 1.3.17, and 1.4.16), and a more subtle context is provided by Example 1.5.16 below.

The set of attractor–repeller pairs is not as large as one might suspect:

Lemma 1.4.19. *The set of attractor–repeller pairs* $\mathcal{AR}(\Phi)$ (*Definition* 1.4.18) *is countable.*

[15]Note the improvement in Corollary 1.4.22.

Proof. For an attractor–repeller pair (A_U, R_U) and U, T as in Definition 1.4.18, there is a finite union $V \subset U$ of elements of a countable base for the topology such that $\varphi^T(\overline{U}) \subset V$ and hence $(A_U, R_U) = (A_V, R_V)$. Thus $V \mapsto (A_V, R_V)$ maps onto $\mathcal{AR}(\Phi)$ and has countable domain. □

A set $\Lambda \subset X$ is *invariant* under a flow Φ if $\varphi^t(\Lambda) = \Lambda$ for each $t \in \mathbb{R}$. Invariant sets make it possible to take a reductionist approach by restricting the flow to such sets, and hyperbolic invariant sets are a mainstay of the second part of this book. Here we note that attractors and repellers are instances.

Lemma 1.4.20. *If Φ is a flow on a compact metric space, then any attracting or repelling set is nonempty, closed, and invariant.*

Proof. If U is a trapping region and $T > 0$ is such that $\varphi^T(\overline{U}) \subset \operatorname{int}(U)$, then

$$\bigcap_{t\geq 0} \varphi^{t+T}(\overline{U}) \subset \bigcap_{t\geq 0} \varphi^t(U) = A_U \subset \bigcap_{t\geq 0} \varphi^t(\overline{U}) \subset \bigcap_{t\geq 0} \varphi^{t+T}(\overline{U}).$$

So A_U is an intersection of nonempty nested compact sets, hence nonempty and closed, and

$$\begin{aligned}\varphi^s(A_U) = \varphi^s\Bigl(\bigcap_{t\geq 0}\varphi^t(U)\Bigr) &= \bigcap_{t\geq 0}\varphi^{s+t}(U)\\ &= \begin{cases}\displaystyle\bigcap_{t\geq s}\varphi^t(U) \subset \bigcap_{t\geq 0}\varphi^t(U) = A_U & \text{if } s \leq 0,\\ \displaystyle\bigcap_{t\geq 0}\varphi^t(\underbrace{\varphi^s(U)}_{\subset U}) \subset \bigcap_{t\geq 0}\varphi^t(U) = A_U & \text{if } s \geq 0.\end{cases}\end{aligned}$$

The proofs for repelling sets are similar. □

Attractor–repeller pairs are separated by Lyapunov functions:

Proposition 1.4.21. *For $(A, R) \in \mathcal{AR}$ there is a Lyapunov function $L\colon X \to [0, 1]$ such that $L^{-1}(\{0\}) = A$, $L^{-1}(\{1\}) = R$, and L is strictly decreasing along orbits of points outside $A \cup R$.*

Proof. The sets A and R are disjoint and compact, so

$$V(x) := \frac{d(x, A)}{d(x, A) + d(x, R)}$$

is continuous with $V^{-1}(\{0\}) = A$ and $V^{-1}(\{1\}) = R$. From this we will obtain a like function that is strictly decreasing off $A \cup R$.

Since every orbit outside $A \cup R$ converges to A as $t \to \infty$ and to R as $t \to -\infty$, the supremum $\overline{V}(x) := \sup V(\varphi^{[0,\infty)}(x)) = \max V(\varphi^{[0,t_x]}(x))$ is attained and hence continuous by compactness, continuity of V, and equicontinuity of the flow on $[0, t_x]$, where t_x is such that $V(\varphi^t(x)) < \overline{V}(x)/2$ for $t \geq t_x$. Also, $\overline{V}(\varphi^t(x)) \leq \overline{V}(x)$ for all $x \in X$ and $t \geq 0$.

To make $\overline{V}$ strictly decreasing off $A \cup R$, let

$$L(x) := \int_0^\infty e^{-s}\overline{V}(\varphi^s(x))\, ds$$

be the weighted average of $\overline{V}$ along the forward orbit. Since $\overline{V}$ is continuous and nonincreasing, so is L: if $t \geq 0$, then

$$L(\varphi^t(x)) = \int_0^\infty e^{-s}\overline{V}(\varphi^{s+t}(x))ds \leq \int_0^\infty e^{-s}\overline{V}(\varphi^s(x))ds = L(x). \tag{1.4.1}$$

Now suppose $x \notin R$ is such that $L(\varphi^t(x)) = L(x)$ for some $t > 0$. Then on one hand $\varphi^t(x) \to A$ as $t \to \infty$ and on the other hand $\overline{V}(\varphi^{s+t}(x)) = \overline{V}(\varphi^s(x))$ for all $s > 0$ by (1.4.1), so $\overline{V}(x) = \overline{V}(\varphi^t(x)) \to 0$, and $x \in A$. □

This result allows us to "improve" trapping regions over Definition 1.4.18:

Corollary 1.4.22. *For any attracting set A there exists a trapping region U for A such that $\varphi^t(\overline{U}) \subset U$ for all $t > 0$.*

Proof. Let L be as in Proposition 1.4.21 and $U := L^{-1}([0, \epsilon))$ for $\epsilon > 0$ sufficiently small. Then $\overline{U} \subset L^{-1}([0, \epsilon])$ and $\varphi^t(\overline{U}) \subset L^{-1}([0, \epsilon))$ for all $t > 0$. □

The above results presage a remarkable general structural result: any flow admits a Lyapunov function, so the dynamics flows "downward" except for indecomposable dynamics on level sets of the Lyapunov function (Theorem 1.5.44). The next sections develop these indecomposability notions.

Since our use of Lyapunov functions centers on a global view of a flow, we here briefly mention some of the "pedestrian" uses of Lyapunov functions for ordinary differential equations; this is variously called the *second method of Lyapunov* or the *direct method of Lyapunov* [168, 103]. Examples 1.4.8, 1.4.9, and 1.4.12 illustrate their use for stability of compact orbits. The following detects ω-limit sets.

Theorem 1.4.23. *If L is a differentiable Lyapunov function for a flow Φ on $\mathbb{R}^n$ and $\varphi^{(0,\infty)}(x)$ is bounded, then $\omega(x) \subset M := \bigcap_{t\geq 0} \varphi^t(S)$, the largest invariant subset of $S := \{x \in \mathbb{R}^n \mid \frac{dL}{dt} = 0\}$.*

Corollary 1.4.24. *If here $G := L^{-1}((-\infty, R])$ is bounded, then $\omega(x) \subset M$ for all $x \in G$.*

Corollary 1.4.25. *If* $L(x) \xrightarrow[\|x\|\to\infty]{} +\infty$ *in Theorem* 1.4.23, *then* $\omega(x) \subset M$ *for all* $x \in \mathbb{R}^n$; *in particular, if* $M = \{0\}$, *then* 0 *is a global attractor.*

1.5 Recurrence properties and chain decomposition

Our study of dynamical behaviors has so far been limited to single orbits, and simple ones at that. We mainly considered fixed points, periodic orbits, and points that approach these orbits as time approaches ∞ or $-\infty$ (asymptotic behavior). For example, orbits near an asymptotically stable fixed point have rather simple asymptotic behavior themselves; they converge to the fixed point. In particular, they are *transient* in the sense that there is a neighborhood to which they never return.

1.5.a Recurrent points. We now develop terminology to describe more complicated asymptotic behavior.

Definition 1.5.1 (Limit set). The *ω-limit set* of $x \in X$ for a flow Φ is the (closed) set

$$\omega(x) := \bigcap_{t\geq 0} \overline{\varphi^{[t,\infty)}(x)} \tag{1.5.1}$$

of accumulation points of the positive semiorbit. The α-limit set is defined similarly for negative time or as the ω-limit set for the inverse flow.

The closure $\mathcal{L}(\Phi)$ of the union of all ω-limit sets and all α-limit sets is the *limit set* of Φ.

Remark 1.5.2. For instance, if x is fixed or periodic, then $\alpha(x) = \omega(x) = \mathcal{O}(x)$. It is a good exercise to determine these sets in the context of Examples 1.1.5, 1.1.7, 1.1.8, 1.3.7, 1.3.8, 1.3.13, 1.3.15, 1.3.16, 1.3.17, 1.4.16, 1.5.16, and 1.6.2.

Note that $\omega(x)$ may be empty (but rarely so in this book; see Proposition 1.5.7). In the context of Definition 1.4.1, the fixed point is the ω-limit set for all orbits that ever come close enough. This motivates the following.

Proposition 1.5.3. $q \in \omega(x) \Leftrightarrow$ *there is a sequence* $t_k \xrightarrow[k\to\infty]{} \infty$ *with* $\varphi^{t_k}(x) \xrightarrow[k\to\infty]{} q$.

Proof. For $q \in \omega(x)$ and $k \in \mathbb{N}$ there exist $t_k \geq k$ such that $d(\varphi^{t_k}(x), q) < 1/k$. Conversely, $q = \lim_{k\to\infty} \varphi^{t_k}(x) \in \overline{\{\varphi^t(x) \mid t \geq m\}}$ for all $m \geq 0$. □

Starting earlier or later does not affect the asymptotics:

Proposition 1.5.4. *The set* $\omega(x)$ *is* φ^t*-invariant*: *if* $s \in \mathbb{R}$, *then* $\varphi^s(\omega(x)) = \omega(x) = \omega(\varphi^s(x))$.

Proof. Since φ^s is a homeomorphism,

$$\bigcap_{T=0}^{\infty} \overbrace{\underbrace{\overline{\varphi^s(\varphi^{[T,\infty)}(x))}}_{=\varphi^{[T+s,\infty)}(x)}}^{=\varphi^s(\overline{\varphi^{[T,\infty)}(x)})} = \begin{cases} \varphi^s\Big(\bigcap_{T=0}^{\infty} \overline{\varphi^{[T,\infty)}(x)}\Big) = \varphi^s(\omega(x)), \\ \bigcap_{T=0}^{\infty} \overline{\varphi^{[T,\infty)}(\varphi^s(x))} = \omega(\varphi^s(x)), \\ \bigcap_{T=s}^{\infty} \overline{\varphi^{[T,\infty)}(x)} = \bigcap_{T=0}^{\infty} \overline{\varphi^{[T,\infty)}(x)} = \omega(x). \end{cases}$$ □

Definition 1.5.5. If Λ is an invariant set for a flow Φ on X, define its *basin* of attraction or *stable set* and basin of repulsion or *unstable set* by

$$W^s(\Lambda) := \{x \in X \mid \varnothing \neq \omega(x) \subset \Lambda\},$$
$$W^u(\Lambda) := \{x \in X \mid \varnothing \neq \alpha(x) \subset \Lambda\}.$$

Remark 1.5.6. Compare with Definition 1.4.1. Examples 1.3.9 and 1.3.13 provide quite complementary simple instances of these sets, and Figure 1.5.4 shows a rather more interesting situation.

Proposition 1.5.7. *If* $O^+(x) := \varphi^{[0,\infty)}(x) \subset K$ *with* $K \subset X$ *compact, then*

(1) $\varnothing \neq \omega(x) \subset K$,

(2) $\omega(x)$ *is compact,*

(3) $d(\varphi^t(y), \omega(y)) \xrightarrow[t\to\infty]{} 0$: *if* $\omega(x) \subset O$ *open, then* $\exists\, T \in \mathbb{R}$ *with* $\varphi^{[T,\infty)}(x) \subset O$,

(4) $\omega(x)$ *is connected, so it is either a single point or infinite.*

Proof. (1) and (2) follow directly from (1.5.1) since the closed nested $\overline{\varphi^{[t,\infty)}(x)}$ are in K.

(3): Otherwise there are $t_i \to +\infty$ with $\varphi^{t_i}(x) \in K \smallsetminus O$, and these points accumulate in the compact set $K \smallsetminus O$, contrary to $\omega(x) \subset O$.

(4): We show that if $p, q \in \omega(x)$ have disjoint neighborhoods O_p, O_q, then $\omega(x) \not\subset O_p \cup O_q$. Pick $\tau_n \to \infty$, $t_n \geq 0$ such that $p_n := \varphi^{\tau_n}(x) \to p$ in O_p and $q_n := \varphi^{t_n}(p_n) \to q$ in O_q, and let

$$s_n := \max\big\{t \in [0, t_n] \mid \varphi^{[0,t)}(p_n) \subset O_p\big\}.$$

Then $\varphi^{s_n}(p_n) = \varphi^{\tau_n + s_n}(x) \in K \cap \partial O_p$ has an accumulation point which is in ∂O_p, hence outside $O_p \cup O_q$ and on the other hand in $\omega(x)$ since $\tau_n \to \infty$. □

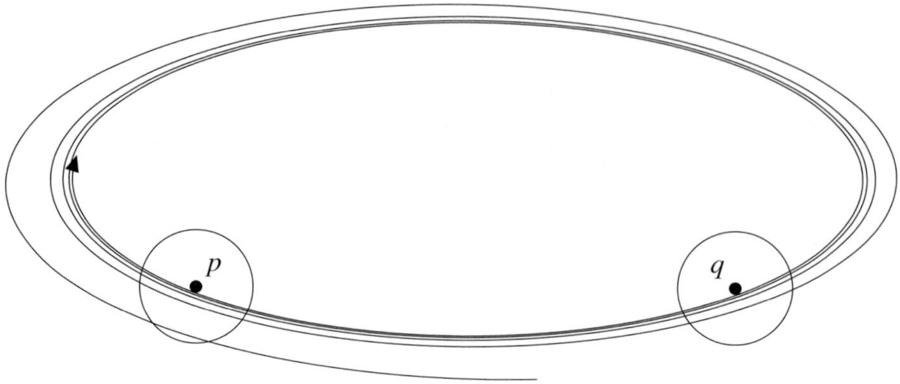

Figure 1.5.1. Proof of Proposition 1.5.7(4).

Proposition 1.5.8. *If h is a constant of motion (Definition 1.1.24) for a flow Φ on a* compact *space X, then $h(X) = h(\mathcal{L}(\Phi))$ (Definition 1.5.1).*

Proof. If $x \in X$, then $h(\{x\}) = h(\overline{\varphi^{\mathbb{R}}(x)}) = h(\omega(x)) \subset h(\mathcal{L}(\Phi))$. □

Remark 1.5.9. Together with Figure 1.3.3, this brings to mind a notion inspired by symmetrizing and extending the prolongational limit set from Remark 1.5.14 below: any constant of motion is clearly constant on the *elongation* $\mathcal{E}(x) \supset \mathcal{O}(x)$ of any point x, where

$$\mathcal{E}(A) := \big\{ \lim_{i\to\infty} \varphi^{t_i}(x_i) \mid \lim_{i\to\infty} x_i \in A,\ t_i \in \mathbb{R} \big\} = \bigcap \big\{ \overline{\varphi^{\mathbb{R}}(O)} \mid A \subset O \text{ open} \big\},$$

and recursively on the *elongational limit set*

$$\overline{\mathcal{E}}(x) := \overline{\bigcup_{n\in\mathbb{N}} \underbrace{\mathcal{E}(\mathcal{E}(\dots\mathcal{E}}_{n \text{ times}}(x)\dots))}$$

of x, and indeed on the *elongational hull* of a point x, the smallest set containing x that is invariant under closure and under application of $\mathcal{L}$ (and hence Φ).

Proposition 1.5.10. *A flow Φ on a connected space has no constant of motion if $\mathcal{L}(\Phi)$ (Definition 1.5.1) is contained in at most countably many elongational hulls, in particular, if each connected component of $\mathcal{L}(\Phi)$ is contained in an elongational hull.*

Proof. A constant of motion h is constant on elongational hulls and takes all its values on $\mathcal{L}(\Phi)$, so $h(X) = h(\mathcal{L}(\Phi)) \subset \mathbb{R}$ is countable and connected, hence a point. □

Definition 1.5.11 (Recurrence). A point x is *ω-recurrent* or *positively recurrent* if $x \in \omega(x)$, *α-recurrent* or *negatively recurrent* if $x \in \alpha(x)$, and *recurrent* (or *Poisson stable*) if $x \in \alpha(x) \cap \omega(x)$. We denote the closure of the set of recurrent points by $\mathcal{B}(\Phi)$—for "Birkhoff center" (Remark 1.5.40).

Remark 1.5.12. $\overline{\mathrm{Per}(\Phi)} \subset \mathcal{B}(\Phi) \subset \mathcal{L}(\Phi)$ (see Definition 1.1.10).

1.5.b Nonwandering. The next generalization of recurrence we study is that a point x may not come back close to x, but a different point arbitrarily close to x comes back close to x.

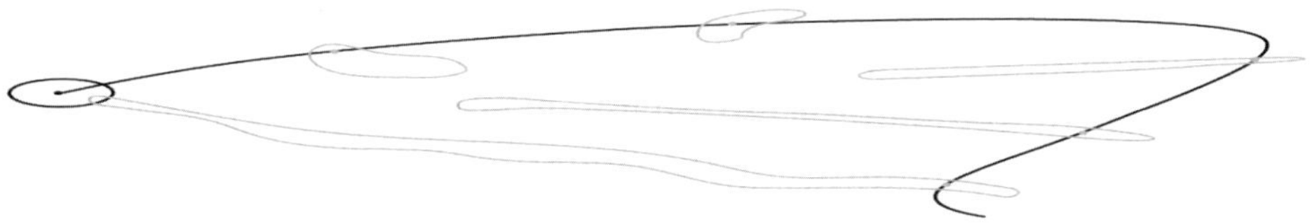

Figure 1.5.2. A nonwandering point.

Definition 1.5.13 (Nonwandering). A point $x \in X$ is *nonwandering* for a flow Φ on X if for any neighborhood U of x and $T_0 > 0$ there is a $t > T_0$ with $\varphi^t(U) \cap U \neq \varnothing$;[16] otherwise x is said to be *wandering*. The set of nonwandering points is denoted by $\mathrm{NW}(\Phi)$. We say that Φ is *regionally recurrent* if $\mathrm{NW}(\Phi) = X$.[17]

Remark 1.5.14 (Auslander). Analogously to defining recurrence of x as $x \in \omega(x)$, x being nonwandering is equivalent to the existence of $x_i \xrightarrow[i\to\infty]{} x$, $t_i \xrightarrow[i\to\infty]{} +\infty$ such that $\lim_{i\to\infty} \varphi^{t_i}(x_i)$ and hence to

$$x \in \mathrm{PL}(x) := \big\{ \lim_{i\to\infty} \varphi^{t_i}(x_i) \mid x_i \xrightarrow[i\to\infty]{} x,\ t_i \xrightarrow[i\to\infty]{} +\infty \big\} = \bigcap_{t\in\mathbb{R}} \bigcap_{\epsilon>0} \overline{\varphi^{(t,\infty)}(B(x,\epsilon))},$$

the first *prolongational limit set* of x. Figure 1.5.3 shows that this can be nonempty even when there is no recurrence at all.

The definition of the nonwandering set looks asymmetric in time, but this is only apparent since $\varphi^t(U) \cap U \neq \varnothing \Leftrightarrow \varnothing \neq \varphi^{-t}(\varphi^t(U) \cap U) = \varphi^{-t}(U) \cap U$:

Proposition 1.5.15. *Note* $\mathrm{NW}(\varphi^t) = \mathrm{NW}(\varphi^{-t})$: *$x \in X$ is nonwandering if and only if for all neighborhoods U of x and all $T_0 < 0$ there is a $t < T_0$ with $\varphi^t(U) \cap U \neq \varnothing$.*

[16]The following from [194, p. 22] may be helpful: "A better choice of words, suggested to us by K. Sigmund, is that a point is called *nostalgic* iff its neighborhoods U keep returning as in the definition of [nonwandering]. The point itself may or may not return near by, but its thoughts (nearby points) always do."

[17]In earlier days a flow was said to be regionally recurrent *at* x if $x \in \mathrm{NW}(\Phi)$.

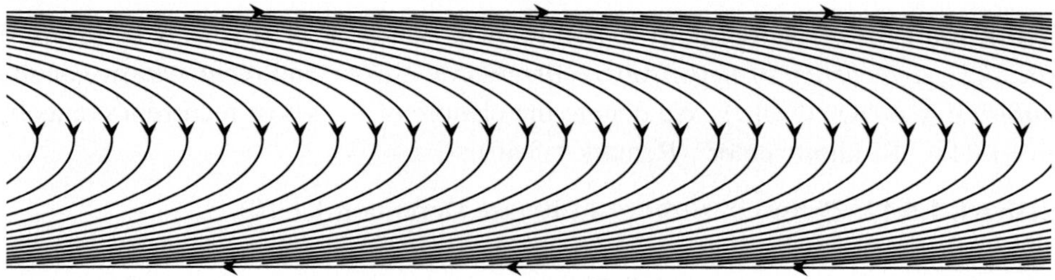

Figure 1.5.3. A Reeb component: the first prolongational limit set of each point on the top line is the bottom line.

Example 1.5.16. From the examples so far it is not apparent that being nonwandering is a strictly weaker notion than recurrence, so Figure 1.5.4 shows a planar flow[18] with nonwandering nonrecurrent points: only the three fixed points are recurrent, but the nonwandering set includes the entire "∞" curve. Furthermore, the flow *restricted* to this nonwandering set has only fixed nonwandering points, that is, $\mathrm{NW}(\Phi\restriction_{\mathrm{NW}(\Phi)}) \subsetneq \mathrm{NW}(\Phi)$.

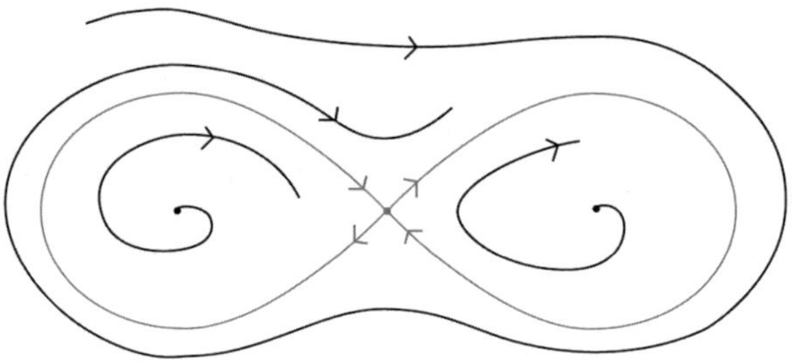

Figure 1.5.4. The Bowen–Katok "figure eight attractor" [210, p. 140].

Proposition 1.5.17. $\mathrm{NW}(\Phi)$ *is closed and* $\overline{\mathrm{Per}(\Phi)} \subset \mathcal{B}(\Phi) \subset \mathcal{L}(\Phi) \subset \mathrm{NW}(\Phi)$.

Proof. A wandering point $x \in X$ has a neighborhood U and a $T > 0$ with $\varphi^t(U) \cap U = \varnothing$ for all $t > T$. Then every point in U is wandering, so the set of wandering points is open. If $x \in X$, $y \in \omega(x)$ and $y \in O$ open, $T_0 > 0$, take $t_1 > 0$ and $t > T_0$ such that $\varphi^{t_1}(x) \in O$ and $\varphi^{t_1+t}(x) \in O$ (since $y \in \omega(x)$), hence $\varphi^{t_1+t}(x) \in \varphi^t(O) \cap O$. Thus, $\forall\, x\colon \omega(x) \subset \mathrm{NW}(\Phi)$, so $\mathcal{L}(\Phi) \subset \mathrm{NW}(\Phi)$ (Proposition 1.5.15). The rest follows from Remark 1.5.12 and Proposition 1.5.17. □

[18]By including ∞ as a repelling fixed point, this becomes an example on the 2-sphere with four fixed points.

Remark 1.5.18. Examples show that each of the inclusions in Proposition 1.5.17 can be strict, but a deep and important result, the proof of which is well beyond our scope, says that C^1-generically, they are not (Theorem 1.5.22).

Here we used the following definition.

Definition 1.5.19. In a topological space the intersection of a countable collection of open sets is called a G_δ-set. A property of elements of a topological space is said to be *generic* if it holds for each member of a dense G_δ-set.[19] (A G_δ-set is a countable intersection of open sets.)

Definition 1.5.20. For $k \geq 0$ the C^k distance between two C^k-flows on a C^k-manifold M is the usual (uniform) C^k distance between their restrictions to $[0,1] \times M$.

Theorem 1.5.21 (Pugh Closing Lemma [309, 308, 307, 14, 189]). *For a nonwandering point of a vector field there is an arbitrarily C^1-close vector field for which this point is periodic.*

Proof strategy. The basic task seems rather obvious: consider a tube around an orbit segment that starts and ends near enough the nonwandering point p and make a perturbation of the vector field inside the tube that moves p onto this orbit at the start of it and at the end. The difficulty arises from the fact that we are aiming for the reverse of a usual perturbation result: those usually involve arbitrarily small modifications, but here we must, for a given length of nearby orbit, change the dynamics by a definite amount. For instance, a tube as described might well have to necessarily self-intersect because the nearby orbit is long and tangled. Thinning the tube might avert the problem but make it difficult to perturb p enough and moreover, localizing perturbations more requires larger derivatives, which counters the desired C^1-smallness of the perturbation. Instead, choosing many flow boxes along parts of that orbit that aren't too close to others is a better strategy. This balancing act makes for a formidable proof in which the gentlest possible deformations are just barely made to accumulate enough total change over the length of the orbit. Counterexamples to C^2 versions of this underscore the delicacy of what is required. (On the other hand, there are astonishing results in low dimensions [201, 18].) □

Together with general genericity results and Theorem 6.1.6, this implies the following theorem:

Theorem 1.5.22 (Pugh General Density Theorem [308]). *Generically among C^1-flows,* $\overline{\mathrm{Per}(\Phi)} = \mathrm{NW}(\Phi)$.

[19] While this notion can be defined in this generality, it is usually applied in complete metric spaces where the Baire Theorem can be used.

Recurrence other than periodicity is usually referred to as *nontrivial recurrence*. Since smooth curves locally separate the plane, flows on simply connected surfaces have only trivial recurrence:

Theorem 1.5.23 (Poincaré–Bendixson Theorem). *Let Φ be a C^1-flow on an open subset of the sphere S^2. Then all positively or negatively recurrent orbits are periodic. Furthermore, if the ω-limit set of a point contains no fixed points, then it consists of a single periodic orbit.*

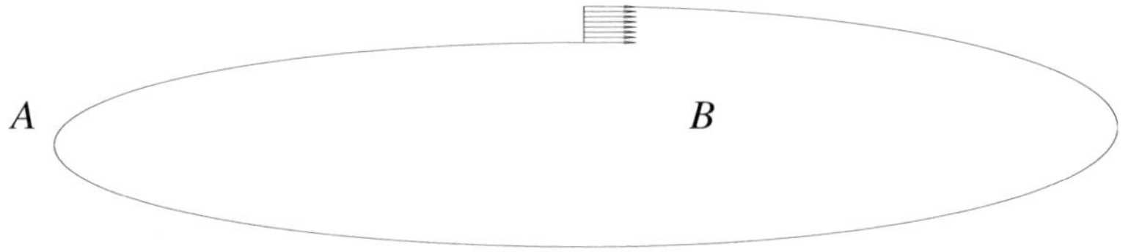

Figure 1.5.5. A pretransversal.

Proof. Suppose p is positively recurrent and neither fixed nor periodic. Take a short transversal γ at p and let t be the smallest positive number for which $\varphi^t(p) \in \gamma$. Then the union of the orbit segment $\{\varphi^s(p)\}_{0 \le s \le t}$ and the piece of γ between p and $\varphi^t(p)$ is a simple closed curve C called a pretransversal. By the Jordan Curve Theorem the complement of C consists of two disjoint open sets A and B. We may label them such that near γ the flow goes from A to B. This implies that the positive semiorbit of $\varphi^t(p)$, hence the ω-limit set $\omega(p)$ of p, is in B. Since p is recurrent we have $A \ni \varphi^{-\epsilon}(p) \in \mathcal{O}(p) \subset \omega(p) \subset B$, a contradiction.

Now assume that $W := \omega(p)$ contains no fixed points. By Remark 1.6.29 below there are recurrent points in W. By the preceding, these are periodic. Thus let $q \in W$ be periodic. Consider a small transverse segment γ containing q. By continuity the return map to this segment is defined on a neighborhood of q in γ. Take a one-sided neighborhood I of q small enough so that the first point $\varphi^t(p)$ in γ is not in I, but infinitely many of these returns are. Parametrizing this neighborhood by $[0, \delta)$ gives a continuous map f from an interval $[0, \delta)$ to an interval $(0, \delta')$ that fixes 0. The orbit of p provides infinitely many $x \in (0, \delta)$ for which $f(x) < x$, so either $f(x) < x$ for all $x \in [0, \delta)$ or $[0, \delta)$ contains a fixed point y. The latter case is impossible, since the interval $[0, y]$ would be invariant under f and hence there would be an invariant annulus for the flow that separates the orbit of q from that of p, so $q \notin \omega(p)$. But if $f(x) < x$ then all $x \in (0, \delta)$ are positively and monotonically asymptotic to 0. Since the return times to I are bounded this means that the orbit segments of p between successive returns converge to the orbit of q, so $\omega(p)$ coincides with the orbit of q. □

By contrast, higher-dimensional flows can be rather more complex; it is a good exercise to also explore the various recurrence notions in the next examples. Examples 1.5.24 and 1.5.26 will also serve as standard examples of hyperbolic flows, which we describe later.

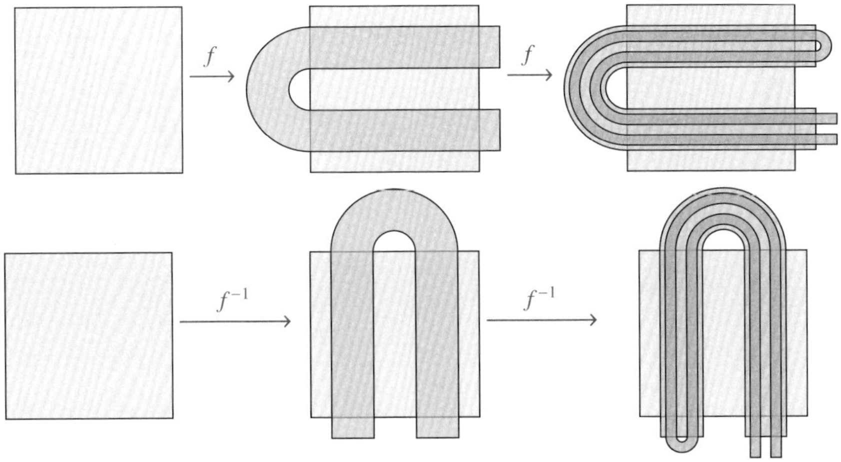

Figure 1.5.6. Horseshoe.

Example 1.5.24 (Smale Horseshoe). The flow we introduce here is a suspension or special flow over a map of R^2 or S^2 (or of any surface) which arises naturally in a Poincaré section. This map f squeezes a rectangle Δ vertically, stretches it horizontally and folds it over the original rectangle (Figure 1.5.6) in a horseshoe shape. Specifically, let us assume that the map is linear on the two halves, so the contraction factor is a constant $\lambda < 1/2$ and the expansion is by a factor $\mu > 2$ (to ensure that there are gaps between the branches) and that there are two complete strips that are folded back over Δ. The set $\Lambda := \bigcap_{n\in\mathbb{Z}} f^n(\Delta)$ of points whose orbits are in Δ is then a Cantor set with vertical contracting direction and horizontal expanding direction (Figure 1.5.7). A horseshoe flow is the time-1 suspension. Note that we have partially or implicitly defined a smooth flow but will focus on the continuous flow obtained by restricting to the suspension of Λ. Variants on this original construction allow for more crossings in Δ as well as nonlinearly expanding and contracting directions.

Example 1.5.25 (Linked horseshoes). More generally, several rectangles might be mapped across each other in a like fashion. Figure 1.5.8 shows an instance that involves two rectangles with horizontal stretching: the left (black) rectangle is mapped across both rectangles plus across itself a second time, while the other rectangle is mapped once across both rectangles.

Figure 1.5.7. The invariant set of the Smale Horseshoe.

Example 1.5.26 (Suspending toral automorphism). Consider the suspension (Definition 1.2.8) of the following map of the 2-torus. The linear map of $\mathbb{R}^2$ given by the matrix $A = \left(\begin{smallmatrix} 2 & 1 \\ 1 & 1 \end{smallmatrix}\right)$ has integer entries and hence induces a well-defined map F_A of the 2-torus $\mathbb{T}^2 = \mathbb{R}^2/\mathbb{Z}^2$, called the Cat Map or Arnold's Cat Map. Since it has unit determinant, the same goes for the inverse, which means that it defines a diffeomorphism (indeed, algebraic automorphism) of $\mathbb{T}^2$—which, furthermore, preserves area. The eigenvalues are

$$\lambda_1 = \frac{3+\sqrt{5}}{2} > 1 \quad \text{and} \quad \lambda_1^{-1} = \lambda_2 = \frac{3-\sqrt{5}}{2} < 1.$$

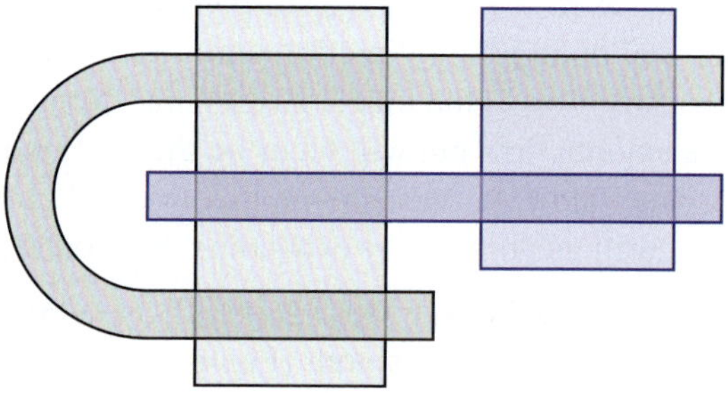

Figure 1.5.8. Linked horseshoes (Example 1.5.25).

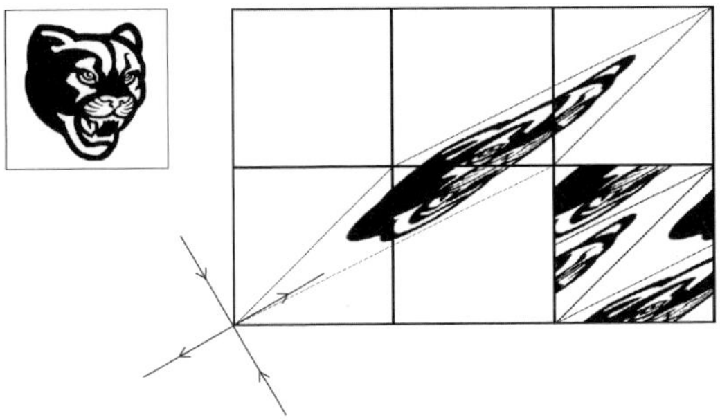

Figure 1.5.9. Example 1.5.26, Cat (cougar) Map.

The eigenvectors for the first eigenvalue are on the line $y = \frac{\sqrt{5}-1}{2}x$. The family of lines parallel to it is invariant, and distances on those lines are expanded by a factor λ_1. Similarly, there is an invariant family of contracting lines $y = \frac{-\sqrt{5}-1}{2}x + \text{const.}$

The tangent space spanned by the direction of expansion and contraction together with the flow direction define hyperbolicity (formally so in Definition 5.1.1). Note that the stable and unstable sets (Definition 1.3.26) for any point in the suspension flow are given by translations of the contracting and expanding eigendirections, respectively.

It is an interesting exercise to show that the collection of periodic points for the map F_A is exactly the set of points with rational coordinates, so the periodic orbits in the suspension flow are dense. But we will see later that there are also dense orbits for both F_A and the suspension, indeed, almost every orbit is dense.

Example 1.5.27. More generally, any $A \in GL(m, \mathbb{Z})$ induces an automorphism F_A of $\mathbb{T}^m$ that preserves Lebesgue measure.[20] We say that it is hyperbolic if A has no eigenvalues on the unit circle. We will see that the suspension is then hyperbolic on the whole suspension manifold.

Remark 1.5.28. In contrast with a phenomenon we will see later (Theorem 8.5.1), the universal cover $\mathbb{R}^2 \times \mathbb{R}$ of the suspension manifold has a (dynamical) global product structure as follows: each contracting plane $\left\{y = \frac{-\sqrt{5}-1}{2}x + \text{const.}\right\} \times \mathbb{R}$ meets each expanding plane $\left\{y = \frac{\sqrt{5}-1}{2}x + \text{const.}\right\} \times \mathbb{R}$ in an orbit (which is necessarily unique).

1.5.c Chain recurrence. The notion of a nonwandering point involves nearby orbits; another variant of recurrence behavior is expressed in terms of objects that are nearly orbits.

[20]Here, $GL(m, \mathbb{Z})$ consists of the integer matrices that are invertible *among integer matrices*, which requires that they have determinant ± 1.

Definition 1.5.29 (Pseudo-orbit, chain). An ϵ*-pseudo-orbit* or ϵ*-chain* for a flow Φ on a space X is a map $g\colon I \to X$ on a nontrivial interval $I \subset \mathbb{R}$ such that

$$d(g(t+\tau), \varphi^\tau(g(t))) < \epsilon \quad \text{for } t, t+\tau \in I \text{ and } |\tau| < 1.$$

It is a pseudo-orbit from x to y of length T if $0, T \in I$ and $g(0) = x$, $g(T) = y$.

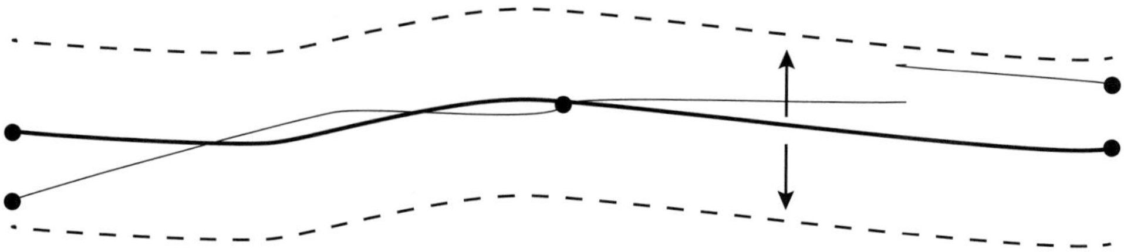

Figure 1.5.10. ϵ-pseudo-orbit.

Note that g need not be continuous; see Figure 1.5.10.

This important notion is a little more involved for flows than for diffeomorphisms. Specifically, an alternate definition of an ϵ-pseudo-orbit is that there is a sequence of points $\{x = x_0, \dots, y = x_n\}$ and times $t_j \geq 1$ with $t_1 + \dots + t_n = T$ and $d(\varphi^{t_j}(x_{j-1}), x_j) < \epsilon$ for all $1 \leq j \leq n$. These variants are related as follows.

Proposition 1.5.30. *Let X be a metric space, Φ a flow on X, $\epsilon > 0$, and $\delta > 0$ such that $d(x,y) < \delta \Rightarrow d(\varphi^t(x), \varphi^t(y)) < \epsilon$ for $0 \leq t \leq 2$. Then we have the following:*

(1) *If there exist points $\{x = x_0, \dots, x_n = y\}$ and times $t_j \geq 1$ with $t_1 + \dots + t_n = T$ and $d(\varphi^{t_j}(x_{j-1}), x_j) < \delta$ for all $1 \leq j \leq n$, then there is an ϵ-pseudo-orbit of length T from x to y.*

(2) *If there is a δ-pseudo-orbit of length $T > 1$ from x to y, then there are points $\{x = x_0, \dots, y = x_n\}$ and times $t_j \geq 1$ with $t_1 + \dots + t_n = T$ and $d(\varphi^{t_j}(x_{j-1}), x_j) < \epsilon$ for all $1 \leq j \leq n$.*

Proof. (1): For $t \in [0,T]$ there is a unique $j \in \{1, \dots, n\}$ with $t_1 + \dots + t_j < t \leq t_1 + \dots + t_j + t_{j+1}$. Define $g^t(x) = \varphi^{t-(t_1+\dots+t_j)}(x_{j-1})$ and check that $g\colon [0,T] \to X$ is an ϵ-pseudo-orbit of length T from x to y.

(2): Let $g\colon [0,T] \to X$ be a δ-pseudo-orbit from x to y. Set $x_0 = x$, $x_n = y$, $n = \lceil T \rceil - 1 \in (\frac{T}{2}, T]$, $t_j = \frac{T}{n} \in [1,2)$ for $j \in \{1, \dots, n\}$, and $x_j = g(j\frac{T}{n})$ for $j \in \{1, \dots, n-1\}$. Then the choice of δ gives $d(\varphi^{t_j}(x_{j-1}), x_j) < \epsilon$ when $1 \leq j \leq n$. □

By Proposition 1.5.30 the two ways of defining a pseudo-orbit can be used interchangeably, and so we will do that.

Remark 1.5.31. Pseudo-orbits arise in different ways. A pseudo-orbit might consist of orbit segments with jumps, that is, it is given by a sequence of points $x_k \in M$ and times $t_k \in \mathbb{R}^+$ such that $\inf t_k > 0$, $\sup t_k < \infty$, and $d(\varphi(t_k, x_k), x_{k+1}) < \delta$. The term "chain" seems particularly apt in this case. It might "drift" if it is the orbit of a perturbation of the given vector field; an orbit for the new vector field will be a pseudo-orbit for the old vector field. In this case there are no jumps (discontinuities) but there can be a "drift" from a true orbit. In full generality one may combine jumps and drift.

Moreover, the arguments in the proof of Proposition 1.5.30 combined with interpolation show that on a topological manifold one can without loss of generality take a pseudo-orbit to be continuous, and on a smooth manifold one can take it to be smooth. In that case one can, with additional work, furthermore arrange for the tangent vectors to the pseudo-orbit to be close to the vector field that generates Φ.

Chains produce trapping regions (Definition 1.4.17) in a way that plays an important role in understanding the global structure of a flow.

Proposition 1.5.32. *The set $\mathcal{R}_\epsilon(x)$ of endpoints of ϵ-pseudo-orbits that start at x*

(1) *is open,*

(2) *satisfies $\varphi^{(0,\infty)}(x) \subset \mathcal{R}_\epsilon(x)$ (an orbit segment is an ϵ-chain),*

(3) *satisfies $y \in \mathcal{R}_\epsilon(x) \Rightarrow \mathcal{R}_\epsilon(y) \subset \mathcal{R}_\epsilon(x)$ (by concatenation of chains),*

(4) *satisfies $\varphi^t(\mathcal{R}_\epsilon(x)) \subset \mathcal{R}_\epsilon(x)$ for $t \geq 0$ (by concatenation of chains), and*

(5) *is a trapping region.*

Proof. We only need to prove (1) and (5) as the other facts are clear.
(1): If $y \in \mathcal{R}_\epsilon(x)$, then $B_\delta(y) \subset \mathcal{R}_\epsilon(x)$ for sufficiently small δ by modifying the connecting ϵ-chain.
(5): If $y \in \overline{\mathcal{R}_\epsilon(x)}$ take $\mathcal{R}_\epsilon(x) \ni y_n \xrightarrow[n\to\infty]{} y$, so $\varphi^1(y_n) \xrightarrow[n\to\infty]{} \varphi^1(y)$ by continuity. Then there is an $N \in \mathbb{N}$ such that $d(\varphi^1(y_n), \varphi^1(y)) < \epsilon/2$ for all $n \geq N$, so $\varphi^1(y) \in \mathcal{R}_\epsilon(y_n) \subset \mathcal{R}_\epsilon(x)$. We have shown that $\varphi^1(\overline{\mathcal{R}_\epsilon(x)}) \subset \mathcal{R}_\epsilon(x)$. □

This helps to understand the global structure of a flow via the following important recurrence notion.

Definition 1.5.33 (Chain recurrence, equivalence, components, decomposition). Let Φ be a continuous flow on a metric space X. A point x is *chain recurrent* if $x \in \bigcap_{\epsilon>0} \mathcal{R}_\epsilon(x)$, that is, for all $\epsilon > 0$ there is an ϵ-pseudo-orbit from x to x. In other words, x lies on a closed ϵ-chain for any $\epsilon > 0$. The set $\mathcal{R}(\Phi)$ of chain recurrent points is the *chain recurrent set* of Φ.

For points $x, y \in \mathcal{R}(\Phi)$ we say $x \sim y$ or x, y are *chain equivalent* or *chainable* if $x \in \bigcap_{\epsilon>0} \mathcal{R}_\epsilon(y)$ and $y \in \bigcap_{\epsilon>0} \mathcal{R}_\epsilon(x)$, that is, for all $\epsilon > 0$ there is an ϵ-pseudo-orbit from x to y and an ϵ-pseudo-orbit from y to x. In other words, x, y lie on a common closed ϵ-chain for any $\epsilon > 0$.

The equivalence classes of $\sim$ define the *chain decomposition* into the *chain (-transitive) components*, *chain recurrent classes*, *chain-equivalence classes*, or (in hyperbolic flows) *homoclinic classes* of $\mathcal{R}(\Phi)$.

Then Φ is said to be *chain transitive* if $\mathcal{R}(\Phi) = X$ and there is only one chain component.

Remark 1.5.34. Again, while these notions are surprisingly effective for hyperbolic flows, they do not in themselves imply any complexity—see the Conley example (Figure 1.5.11) or, for that matter, the constant flow $(t, x) \mapsto x$, or (more naturally) Figure 1.1.4 (on the cylinder $S^1 \times \mathbb{R}$; this example also illustrates that a constant of motion need not be constant on $\mathcal{R}(\Phi)$), or the geodesic flow on the torus $\mathbb{T}^n$.

The following justifies the term "chain equivalence":

Proposition 1.5.35. *Chain equivalence is an equivalence relation on* $\mathcal{R}(\Phi)$.

Proof. Symmetry is clear, and reflexivity follows by definition of $\mathcal{R}(\Phi)$. Transitivity: $x \sim y \sim z \Rightarrow x \in \mathcal{R}_\epsilon(y) \subset \mathcal{R}_\epsilon(z)$ because $y \in \mathcal{R}_\epsilon(z)$ (Proposition 1.5.32(3)). □

We note that "small" changes can make a big difference. The chain-recurrent set is very different for the south–south dynamics (Example 1.3.13) on the circle (chain transitive) versus its interval counterpart (only the two fixed points are chain recurrent). Likewise, the chain-recurrent set of the Akin flow on the interval (Example 1.3.16) is the ternary Cantor set, while the projection $A_\circ$ to the circle is chain transitive.

Remark 1.5.36. $\mathrm{NW}(\Phi) \subset \mathcal{R}(\Phi) = \overline{\mathcal{R}(\Phi)}$ (Exercise 1.38). As before, it is good to examine this notion in the context of our examples so far, for instance by identifying the recurrent, nonwandering, and chain-recurrent sets in Figures 1.3.3, 1.5.4 and Example 1.3.8, as well as in Conley's example of a continuous vector field that is zero on the boundary of a rectangle and nonzero pointing downward inside (Figure 1.5.11). This is a somewhat "pathological" situation, which should induce skepticism about the notion of chain recurrence, and Figure 1.1.4 shows a natural chain-transitive example where a meaningful analysis would produce much finer information than chain transitivity alone. However, the value of the chain decomposition in understanding the global structure of a continuous flow justifies the notion, particularly in the context of hyperbolicity, which precludes the occurrence of such pathology (Corollary 5.3.15(1)).

To summarize, Proposition 1.5.17 and Remark 1.5.36 give the following proposition:

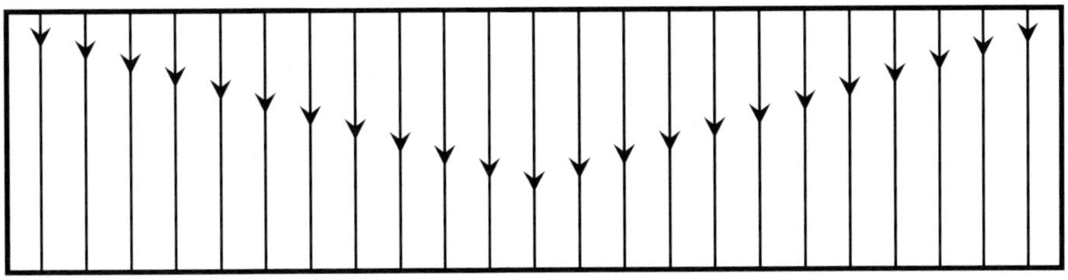

Figure 1.5.11. The Conley example.

Proposition 1.5.37. $\overline{\mathrm{Per}(\Phi)} \subset \mathcal{B}(\Phi) \subset \mathcal{L}(\Phi) \subset \mathrm{NW}(\Phi) = \overline{\mathrm{NW}(\Phi)} \subset \mathcal{R}(\Phi) = \overline{\mathcal{R}(\Phi)}$.

Remark 1.5.38. One should not expect a strengthening of Proposition 1.5.37—each of these inclusions can be strict (Exercise 1.34). This does not mean that we do not often have equality. These levels of recurrence are all conflated when there is no chain recurrence at all, and also C^1-generically (Theorem 1.5.22). This is the case in our simplest examples. More importantly for us, however, these sets tend to coincide in hyperbolic flows not despite but because of the complexity of the dynamics. This is the content of Proposition 5.3.32, and "semilocal" counterparts follow from Theorem 5.3.37.

Proposition 1.5.35 suggests studying a continuous flow through the strategy of restricting to chain components, so we pause to note that this is not a recursive process, that is, that chain components are themselves chain recurrent:

Theorem 1.5.39 (Restriction property). *Let Φ be a flow on a compact metric space X. Then $\mathcal{R}(\Phi\restriction_{\mathcal{R}(\Phi)}) = \mathcal{R}(\Phi)$ and has the same chain decomposition.*

Proof. The "$\subset$" is obvious: $\mathcal{R}(\Phi\restriction_A) \subset A$ for any A, and if x, y lie on a common periodic ϵ-chain in $\mathcal{R}(\Phi)$ then these trivially are periodic ϵ-chains in X. Conversely, let $x \in \mathcal{R}(\Phi)$ and $g_n\colon \mathbb{R} \to X$ be a periodic $1/n$-pseudo-orbit for $n \in \mathbb{N}$ with $g_n(0) = x$ (and $g_n(t) = y$ for some t to prove heredity of chain equivalence). Note first that it suffices to show that for any neighborhood U of $\mathcal{R}(\Phi)$ there is an $N \in \mathbb{N}$ with $g_n(\mathbb{R}) \subset U$ for $n \geq N$. To show this, suppose (by compactness) to the contrary that there are a $z \notin \mathcal{R}(\Phi)$ and sequences $n_k \xrightarrow[k\to\infty]{} +\infty$ and t_k with $g_{n_k}(t_k) \xrightarrow[k\to\infty]{} z$. But then the periodic pseudo-orbits

$$\bar{g}_k(t) := \begin{cases} g_{n_k}(t + t_k) & \text{if } i \notin p_k\mathbb{Z}, \\ z & \text{if } i \in p_k\mathbb{Z}, \end{cases}$$

where p_k is the period of g_{n_k}, show that $z \in \mathcal{R}(\Phi)$. □

Remark 1.5.40. In contrast to this heredity of chain recurrence, $\mathrm{NW}(\Phi\restriction_{\mathrm{NW}(\Phi)}) \neq \mathrm{NW}(\Phi)$ in general; see Example 1.5.16. The *Birkhoff center* of a flow is defined by recursively restricting to the nonwandering set, that is, from $\mathrm{NW}(\Phi)$ pass to $\mathrm{NW}(\Phi\restriction_{\mathrm{NW}(\Phi)})$, etc.[21] The ultimate intersection is called the Birkhoff center and can be characterized as the maximal set C such that $C = \mathrm{NW}(\Phi\restriction_C)$. It is a closed set that contains the recurrent points, and for flows on complete metric spaces it coincides with their closure. This explains the notation $\mathcal{B}$ in Definition 1.5.11.

Remark 1.5.41. The heredity of chain recurrence gives it a somewhat intrinsic nature, and this makes it natural to note that ω-limit sets are characterized by being connected (Proposition 1.5.7) and chain recurrent (Proposition 1.5.17): if Φ is a continuous flow on a connected space X and $\mathcal{R}(\Phi) = X$, then Φ is topologically conjugate to the restriction of some continuous flow to the ω-limit set of some point [139].

Lyapunov functions (Definition 1.4.10) and Proposition 1.5.32 make it possible to connect the notion of chain recurrence with stability as represented by attractor–repeller pairs (see Definition 1.4.18), including a characterization of chain equivalence in terms of attractor–repeller pairs:

Theorem 1.5.42 (Attractor–repeller pairs). *Let Φ be a flow on a compact metric space. Then*

$$\mathcal{R}(\Phi) = \bigcap_{(A,R)\in\mathcal{AR}} A \cup R.$$

If $x, y \in \mathcal{R}(\Phi)$, then $x \sim y$ if and only if for each $(A,R) \in \mathcal{AR}$ (see Definitions 1.4.18 *and* 1.5.33*), x and y are either both in A or both in R.*

Remark 1.5.43. In Example 1.3.15 one can describe attractor–repeller pairs explicitly. A connected component of a trapping region is an interval $[a,c)$ or $(c,b]$ with $f(c) \neq 0$ or (α,β) with $f(\alpha) > 0 > f(\beta)$. For example, if the trapping region is an interval $[a,c)$ or $(c,b]$ with $f(c) \neq 0$, then the corresponding attractor–repeller pair consists of $[a,c_1]$ and $[c_2,b]$ (which is the attractor and which is the repeller depends on the sign of $f(c)$), where $f(c_1) = 0 = f(c_2)$ and $f \neq 0$ on (c_1,c_2). An illustrative special case is $f^{-1}(\{0\}) = \{a,b,c\}$ with $f \geq 0$ (or $f \leq 0$), when each $A \cup R$ is either $[a,c] \cup \{b\}$ or $\{a\} \cup [c,b]$. In either case one member of the pair contains points that are not chain recurrent, so the intersection over $\mathcal{AR}$ is essential in Theorem 1.5.42. (If f takes both positive and negative values, this is a little different.)

[21] More precisely, for each ordinal α set $N_\alpha = X$ if $\alpha = 0$, $N_\alpha = \mathrm{NW}(\Phi\restriction_{N_\beta})$ when α is the successor of β, and $N_\alpha = \bigcap_{\beta<\alpha} N_\beta$ if α is a limit ordinal; this terminates after at most countably many steps because each closed set in this sequence is characterized by the elements in a countable base for the topology from which it is disjoint.

Proof. "$\supset$": If $x \notin \mathcal{R}(\Phi)$, then there is an $\epsilon > 0$ with $x \notin \mathcal{R}_\epsilon(x) \supset A_{\mathcal{R}_\epsilon(x)}$. On the other hand, $\varphi^t(x) \in \mathcal{R}_\epsilon(x)$ for $t > 0$ (Proposition 1.5.32), so $x \notin R_{\mathcal{R}_\epsilon(x)}$ since $R_{\mathcal{R}_\epsilon(x)}$ is invariant. Therefore, $x \notin A_{\mathcal{R}_\epsilon(x)} \cup R_{\mathcal{R}_\epsilon(x)}$, and we have shown that $x \notin \bigcap_{(A,R)\in\mathcal{AR}} A \cup R$.

Furthermore, if x and y are in different chain components, then there is no ϵ-chain from x to y for sufficiently small ϵ, so $y \notin \mathcal{R}_\epsilon(x)$ (see Proposition 1.5.32), and $x \in A_{\mathcal{R}_\epsilon(x)}$. Hence, $y \in R_{\mathcal{R}_\epsilon(x)}$.

"$\subset$": Let $x \notin A \cup R$ for some attractor–repeller pair. Proposition 1.4.21 yields a Lyapunov function L that is strictly decreasing off $A \cup R$. If $c_0 := L(x)$ and $c_1 := L(\varphi^1(x))$, then L is strictly decreasing on $L^{-1}([c_1, c_0])$. By compactness there is a $\delta \in (0, (c_0 - c_1)/2)$ such that if $y \in L^{-1}([c_1, c_1 + \delta])$, then $\varphi^1(y) \in L([0, c_1])$ and there is an $\epsilon > 0$ such that $L(y') < c_1 + \delta$ for all $y \in L^{-1}([0, c_1])$ and $y' \in B_\epsilon(y)$. Even with pseudo-orbits we "can't go back up," that is, ϵ-chains starting at x cannot be closed: To see this we use the sequence definition of ϵ-chains (Proposition 1.5.30). Let $\{x = x_0, \dots, x_n; t_0, \dots, t_{n-1}\}$ be an ϵ-chain with $t_k \ge 1$ for each $0 \le k \le n-1$; then $L(\varphi^{t_0}(x_0)) \le L(\varphi^1(x_0)) = c_1$. Also, $L(x_1) \le c_1 + \delta$ by the choice of ϵ and $L(\varphi^{t_1}(x_1)) \le c_1$. Inductively, we have $L(x_k) \le c_1 + \delta$ for $1 \le k \le n$. Hence, there is no ϵ-chain from x to x, and $x \notin \mathcal{R}(\Phi)$. Thus, $\mathcal{R}(\Phi) \subset \bigcap_{(A,R)\in\mathcal{AR}} A \cup R$ by contraposition.

If $x \sim y$, then we see from this that x and y are in the same piece of any attractor–repeller pair (A, R). □

Theorem 1.5.42 suggests that a continuous flow should be studied by analyzing the dynamics on the chain components, which can be done by restriction because they are compact invariant sets, and to then augment this analysis by determining the transient dynamics between them. We are indeed ready to give a complete global description of the dynamics: a Lyapunov function will disentangle transient and recurrent behavior systematically.

Theorem 1.5.44 (Conley's Fundamental Theorem of Dynamical Systems). *Let Φ be a flow on a compact metric space X. Then there is a Lyapunov function $L\colon X \to [0,1]$ such that $L(\mathcal{R}(\Phi))$ is nowhere dense, $x \notin \mathcal{R}(\Phi) \Rightarrow L(\varphi^t(x)) < L(x)$ for all $t > 0$, and if $x, y \in \mathcal{R}(\Phi)$, then $L(x) = L(y) \Leftrightarrow x \sim y$ (see Proposition* 1.5.35).

Remark 1.5.45. A Lyapunov function with these properties is called a *complete Lyapunov function*. A Lyapunov function is said to be *strict* if $x \notin \mathcal{R}(\Phi) \Rightarrow L(\varphi^t(x)) < L(x)$ for all $t > 0$.

Example 1.3.16 exhibits uncountably many chain components.

Proof. By Lemma 1.4.19 write $\mathcal{AR}(\Phi) = \{(A_j, R_j)\}_{j=1}^M$ with $M \in \mathbb{N} \cup \{\infty\}$. Proposition 1.4.21 gives Lyapunov functions $L_j\colon X \to [0,1]$ that strictly decrease along orbits off $A_j \cup R_j$. The continuous function defined by the uniformly convergent series

$$L(x) := 2\sum_{j=1}^{M} 3^{-j} L_j(x) \in [0,1]$$

is nondecreasing along orbits since the summands are, and $L(\mathcal{R}(\Phi))$ is contained in the ternary Cantor set.

If $x \notin \mathcal{R}(\Phi)$, then $x \notin A_j \cup R_j$ for some j, and $L_j(\varphi^t(x)) < L_j(x)$ for $t > 0$. Also, $L_k(\varphi^t(x)) \le L_k(x)$ for all k, so $L(\varphi^t(x)) < L(x)$ and L is strictly decreasing off $\mathcal{R}(\Phi)$.

Theorem 1.5.42 shows that $x, y \in \mathcal{R}(\Phi)$ are chain equivalent if and only if for each attractor–repeller pair (A_j, R_j) they are in the same component. Hence, $L(x) = L(y)$. Conversely, if $x \nsim y$, then there is a minimal $j < \infty$ with $x \in A_j$, $y \in R_j$ after possibly relabeling, so $L_j(x) = 0$ and $L_j(y) = 1$. Then

$$L(y) - L(x) \ge \frac{2}{3^j} - 2\sum_{k=j+1}^{M} \frac{1}{3^k} \ge \frac{2}{3^j} - \frac{2}{3^{j+1}}\left(\frac{1}{1-\frac{1}{3}}\right) = \frac{1}{3^j} > 0. \qquad \square$$

With the chain decomposition, the phase space or an essential part of it splits into a well-behaved union of closed invariant subsets, and the dynamics on these may be studied separately. This is highly effective, especially for hyperbolic flows. Therefore, our next agenda is to concentrate on such pieces, and we will investigate ways in which the recurrence on them can be stronger than just chain recurrence. Before doing so, we make a few digressions.

In the context of smooth flows on manifolds, it is natural to expect that Lyapunov functions can be taken smooth. Unfortunately, in full generality this is not even true for flows on a circle [129, Example 2.11, p. 1197]. Nonetheless, we have the following theorem:

Theorem 1.5.46 ([130, Corollary 2.3]). *If Φ is a continuous flow on a compact C^∞-manifold M which is generated by a continuous uniquely integrable vector field X, then there is a C^∞ Lyapunov function $L\colon M \to \mathbb{R}$ such that*

- $N(L) := \{x \in M \mid \exists t > 0 : L(\varphi^t(x)) = L(x)\} = \mathcal{R}(\Phi) = \mathrm{Crit}(L)$, *the critical set of L,*
- $XL < 0$ *off $\mathcal{R}(\Phi)$,*
- *L is constant on each chain component, and*
- *L takes distinct values on different chain components.*

Note that smoothness of Φ is not required. By Sard's Theorem, $L(\mathcal{R}(\Phi)) \subset \mathbb{R}$ is then a Lebesgue-null set.

Remark 1.5.47 (Generalized recurrent set). The finest decomposition of the space X by Lyapunov functions for the flow Φ is given by the (closed invariant) generalized recurrent set $\mathrm{GR}(\Phi)$ of points along whose orbits *any* Lyapunov function for the flow is constant. Then $\mathrm{NW}(\Phi) \subset \mathrm{GR}(\Phi) \subset \mathcal{R}(\Phi)$. Each of these inclusions can be strict (see Exercise 1.39 and Figure 1.3.3), so in light of this and Proposition 1.5.37,

$$\overline{\mathrm{Per}(\Phi)} \subset \mathcal{B}(\Phi) = \overline{\mathcal{B}(\Phi)} \subset \mathcal{L}(\Phi) \subset \mathrm{NW}(\Phi) = \overline{\mathrm{NW}(\Phi)} \subset \mathrm{GR}(\Phi) \subset \mathcal{R}(\Phi) = \overline{\mathcal{R}(\Phi)},$$

with each set closed and each inclusion strict in some of our examples (Exercises 1.34, 1.39, and 1.40).

Analogously to the proof of Conley's Theorem one shows the following theorem:

Theorem 1.5.48 (Auslander [19, Theorem 2]). *There is a Lyapunov function L such that $x \in \mathrm{GR}(\Phi)$ iff L is constant on $\mathcal{O}(x)$, and $x \notin \mathrm{GR}(\Phi) \Rightarrow L(\varphi^t(x)) < L(x)$ for $t > 0$.*

Proof. The space $\mathcal{L}$ of Lyapunov functions $L\colon X \to [-1,1]$ (with the topology of uniform convergence on compact sets) is separable, so there is a dense subset $\{L_k\}_{k\in\mathbb{N}}$, and $x \in \mathrm{GR}(\Phi)$ if and only if $L_k(\varphi^t(x)) = L_k(x)$ for all $t \in \mathbb{R}$ and $k \in \mathbb{N}$. Then $F := \sum_{k\in\mathbb{N}} \frac{L_k}{2^k} \in \mathcal{L}$, and $L(x) := \int_0^\infty \frac{F(\varphi^s(x))}{s^2+1}\,ds$ is as desired: If $F(\varphi^t(x)) = F(x)$ for all $t > 0$, then $L_k(\varphi^t(x)) = L_k(x)$ for all $t > 0$ and $k \in \mathbb{N}$, so $x \in \mathrm{GR}(\Phi)$. Conversely, if $x \notin \mathrm{GR}(\Phi)$, there are $t_n \to +\infty$ with $F(\varphi^{t_{n+1}}(x)) < F(\varphi^{t_n}(x))$ for all $n \in \mathbb{N}$, hence the claim. □

The level sets of the Lyapunov function in Theorem 1.5.44 dynamically decompose the manifold in a way that is coherent with the chain components. The dynamics still can be (and for continuous discrete-time systems generically (Definition 1.5.19) is; [5]) rather complicated, but for hyperbolic flows (Definition 5.3.50) the chain components are open and hence finite in number (Corollary 5.3.36). Then this decomposition by level sets can even more effectively describe the overall dynamics.

Definition 1.5.49. Let Φ be a flow on a compact manifold M. A *filtration* $\mathbf{M}$[22] for Φ is a nested sequence $\varnothing = M_0 \subsetneq M_1 \subsetneq \cdots \subsetneq M_k = M$ of compact sets such that $\varphi^t(M_i) \subset \mathrm{int}(M_i)$ for any $t > 0$ and any $i \in \{1, \dots, k\}$.

Remark 1.5.50. This notion is not obviously hereditary: the existence of a filtration for $\Phi\restriction_\Lambda$ does not imply the existence of a filtration for Φ.

So (the set of interiors of the members of) a filtration is a nested sequence of trapping regions. Note that $K_i^\Phi(\mathbf{M}) := \bigcap_{t\in\mathbb{R}} \varphi^t(M_i \smallsetminus M_{i-1})$ is compact and the maximally Φ-invariant subset in $M_i \smallsetminus M_{i-1}$ for $i \in \{1, \dots, k\}$. We let $K^\Phi(\mathbf{M}) := \bigcup_{i=1}^k K_i^\Phi(\mathbf{M})$.

[22] More generally, a filtration of M is a decomposition $I \to 2^M$ that includes M, with I a totally ordered set and such that if $i \le j$ in I then $M_i \subset M_j$.

Theorem 1.5.51 (Filtration). *Let Φ be a continuous flow on X with finite chain decomposition $\Lambda_1, \dots, \Lambda_k$. Then there is a filtration $\mathbf{M}$ of X composed of $M_0 \subset M_1 \subset \cdots \subset M_k$ such that $\Lambda_i = K_i^\Phi(\mathbf{M})$ for each $i \in \{1, \dots, k\}$.*

Proof. Theorem 1.5.44 gives a Lyapunov function $L\colon M \to \mathbb{R}$ for Φ with $L(\Lambda_k) > L(\Lambda_{k-1}) > \cdots > L(\Lambda_2) > L(\Lambda_1)$ after possibly relabeling. Fix $a_1, \dots, a_k \in \mathbb{R}$ such that

$$a_k > L(\Lambda_k) > a_{k-1} > L(\Lambda_{k-1}) > \cdots > a_2 > L(\Lambda_2) > a_1 > L(\Lambda_1).$$

The $M_i := L^{-1}(-\infty, a_i]$ define a filtration with $\Lambda_i \subset M_i \smallsetminus M_{i-1}$. If $x \in K_i^\Phi(\mathbf{M})$, then $\omega(x) \subset \mathcal{R}(\Phi) \cap K_i^\Phi(\mathbf{M}) \subset \Lambda_i$. Similarly, $\alpha(x) \subset \Lambda_i$, so $x \in \Lambda_i$. □

While the Conley Theorem is quite general, we are motivated by hyperbolic flows. Therefore we mention that there are quite complementary uses of Lyapunov functions for describing dynamics far from hyperbolicity [142].

While Lyapunov functions and the chain decomposition are effective in narrowing in on recurrent dynamics and organizing it to some extent, we saw that constants of motion can do so to some extent (Proposition 1.5.8) but also previously pointed to Figure 1.1.4 viewed on the cylinder $S^1 \times \mathbb{R}$ as an illustration that a constant of motion need not be constant on $\mathcal{R}(\Phi)$. Indeed, in this chain-transitive example the level sets of energy provide a far better disaggregation of the dynamics: except for the energy level of the saddle, each level set here is an orbit. This motivates studying finer decompositions as well as stronger dynamical entanglements. This is the goal of the next section.

1.6 Transitivity, minimality, and topological mixing

As we mentioned after the proof of Theorem 1.5.44, the chain decomposition splits a flow into chain-transitive pieces. We now investigate ways in which orbits in a given chain component might be more tightly entangled than chain recurrence alone implies. This is our task for the present section.

Definition 1.6.1 (Topological transitivity). We say that a flow on a metric space X is *topologically transitive* if there is a point $x \in X$ such that $\overline{\mathcal{O}^+(x)} = X$. An invariant subset Λ of X is said to be (topologically) transitive if it is the forward orbit closure of a point in Λ.

This is a recurrence property in two ways: on one hand the point x is recurrent, and on the other hand, this property implies that every point is nonwandering. Transitivity will also play a major role in studying hyperbolicity for flows. One of the fundamental notions in hyperbolicity is the idea of a basic set (Definition 5.3.16), that is a transitive subset of the flow satisfying some additional hypotheses.

Example 1.6.2. Example 1.1.6 is a trivial example of a topologically transitive system; it consists of a single periodic orbit. If $n = 2$ and $0 \neq v_1 = \alpha v_2$ in Example 1.1.8 with irrational α, then the corresponding linear flow is topologically transitive (see also Remark 1.1.11). This can be shown by adapting the observation in Example 1.3.17 to reduce this to studying the rotation $x \mapsto x + \alpha$, whose orbits are $x_0 + \alpha\mathbb{Z}$ mod 1, hence dense. This shows that indeed *every* (semi-) orbit is dense (Definition 1.6.21). By contrast, all orbits are periodic if $\alpha \in \mathbb{Q}$ (Remark 1.1.11); moreover, if $pv_1 = qv_2$, write $q = Tv_1, p = Tv_2$ to get $T(v_1, v_2) = (q, p)$ and $\varphi^T = \mathrm{Id}$.

Remark 1.6.3. A similarly homogeneous example arises below in a geometric context (Example 2.1.16), and it is profoundly different in terms of longitudinal behavior: while toral translations (Example 1.1.8) are suspensions (Example 1.3.17), those flows are not (Theorem 3.4.44).

The notion of transitivity proves useful immediately.

Proposition 1.6.4. *A topologically transitive flow has no constant of motion* (*Definition* 1.1.24).

Proof. A constant of motion is constant on the closure of the dense orbit. □

As noted in Proposition 1.5.10, we can easily amplify this in the context of the chain decomposition.

Proposition 1.6.5. *A flow on a connected space X whose chain components are transitive and at most countable in number has no constant of motion.*

Proof. A constant of motion h is constant on orbit closures, hence

- h is constant on each chain component, and
- for any $x \in X$, $h(x) = h(\overline{\varphi^{\mathbb{R}}(x)}) = h(\omega(x))$, where $\omega(x) := \bigcap_{t \in \mathbb{R}} \overline{\varphi^{[t,\infty)}(x)}$ is contained in a chain component.

Thus, $h(X)$ is at most countable and connected. □

Remark 1.6.6. This is a good moment to look back at Figure 1.3.3. None of those flows have a constant of motion, but not all have a dense orbit. Only one is topologically transitive, and the chain-recurrent set of the north–south dynamics consists of the fixed points, whereas it is the circle in the other examples—which shows that Proposition 1.6.5 is not sharp. For the first three flows, the nonwandering set consists of the fixed points, but when the south–north–south dynamics is included in Figure 1.1.4 rather than viewed in isolation, then all its points are nonwandering. This recalls Remark 1.5.9.

Proposition 1.6.7. *A flow is transitive if and only if $\omega(x) = X$ for some $x \in X$.*

Proof. Since $\omega(x) \subset \overline{O^+(x)}$, a flow is transitive if there exists a point $x \in X$ such that $\omega(x) = X$. Conversely, suppose $X = \overline{O^+(x)}$. Unless x is periodic and hence $X = O^+(x) = O(x)$, we have $\varphi^{-1}(x) \in X \smallsetminus O^+(x) = \overline{O^+(x)} \smallsetminus O^+(x) \subset \omega(x)$, so (since $\omega(x)$ is closed and invariant) $X = \overline{O(x)} \subset \omega(x)$. □

It is common to define topological transitivity as the existence of a dense orbit, rather than a forward dense orbit. While there are flows that satisfy the first of these and not the latter (Example 1.3.8 or 1.3.13), this is a 1-dimensional phenomenon. This suggests a natural terminology in analogy to discrete-time dynamics, where the various definitions of topological transitivity agree on a *perfect set*, that is, a compact set without isolated points.[23]

Definition 1.6.8. A compact set is said to be *flow perfect* if it has no isolated segments, that is, no open subset is homeomorphic to an interval.

Proposition 1.6.9 (Transitivity). *For a continuous flow Φ on a flow-perfect metric space X, the following four conditions are equivalent:*[24]

(1) *Φ has a dense positive semiorbit (topological transitivity, Definition* 1.6.1).

(2) *Φ has a dense orbit.*

(3) *If $\varnothing \neq U, V \subset X$ are open, then there exists a $t \in \mathbb{R}$ such that $\varphi^t(U) \cap V \neq \varnothing$.*

(4) *If $\varnothing \neq U, V \subset X$ are open, then there exists a $t \geq 0$ such that $\varphi^t(U) \cap V \neq \varnothing$.*

Remark 1.6.10. Implications (4) ⇒ (3) and (1) ⇒ (2) are clear. We prove (2) ⇒ (3) ⇒ (4) ⇒ (1). Note that (1) ⇒ (2) ⇒ (3) (and (4) ⇒ (3)) use no assumptions on the topology of X. Considering Example 1.3.8 or 1.3.13 in light of these four statements may help clarify Proposition 1.6.9 and its proof.

Remark 1.6.11. Item (3) can be strengthened. Since $\{B(x, \epsilon/2) \times B(y, \epsilon/2) \mid x, y \in X\}$ has a finite subcover by compactness of $X \times X$,

$$\forall \epsilon > 0\ \exists T \in \mathbb{R}\ \forall x, y \in X\ \exists t \in [0, T]\colon \varphi^t(B(x, \epsilon)) \cap B(y, \epsilon) \neq \varnothing.$$

[23] Isolated points are not a problem for flows because if there is one, then the four assertions in Proposition 1.6.9 are all clearly equivalent to it being the whole space.

[24] See also Exercise 1.20.

Proof of Proposition 1.6.9. (2) $\Rightarrow$ (3): If $\varnothing \neq U, V \subset X$ are open and $\overline{O(x)} = X$, then there are $t, s \in \mathbb{R}$ with $\varphi^t(x) \in U$ and $V \ni \varphi^s(x) = \varphi^{t-s}(\varphi^t(x)) \in \varphi^{t-s}(U)$, so $\varphi^{t-s}(U) \cap V \neq \varnothing$.

(3) $\Rightarrow$ (4): This is the "uphill" step. Here we "symmetrize time." To that end, first "symmetrize space" by considering the case $U = V =: W$ in (4). We show that

$$\text{given } \varnothing \neq W \subset X \text{ open and } T > 0 \text{ there is a } t \geq T \text{ with } \varphi^t(W) \cap W \neq \varnothing. \quad (1.6.1)$$

It is important here that t can be taken arbitrarily large.

Claim 1.6.12. *For $\varnothing \neq W \subset X$ open there are $t \geq 1$ and $\varnothing \neq W' \subset W$ open with $\varphi^t(W') \subset W$.*

This implies (1.6.1) because applying it to W' recursively, we find that given $\varnothing \neq W \subset X$ and $T > 0$ there are an open $W' \subset W$ and $t \geq T$ such that $\varphi^t(W') \subset W$. We use the following notation several times.

Definition 1.6.13. For a topological space X and $x \in A \subset X$ we denote by $\mathbb{C}(A, x)$ the connected component of A containing x.

Proof of Claim 1.6.12. If W consists of fixed points, then so does its closure, and by (3) the closure is X, hence again by (3), X is a point, in which case (4) holds (trivially, as do the other three statements). Otherwise, pick a point $x \in W$ that is not fixed, and let $I := \mathbb{C}\big(\{t \in (-2,2) \mid \varphi^t(x) \in W\}, 0\big) \subset \mathbb{R}$. Then $\varphi^{[-1,1] \smallsetminus I}(x)$ is compact, and we can replace W by $W \smallsetminus \varphi^{[-1,1] \smallsetminus I}(x)$. Since W is not homeomorphic to an interval, there is a $y \in W \smallsetminus \varphi^I(x)$, and there are disjoint neighborhoods W_1 of $\varphi^I(x)$ and W_2 of y.[25] By (3) (and the choice of I) there is an $s \in \mathbb{R} \smallsetminus [-1,1]$ with $\varphi^s(W_1) \cap W_2 \neq \varnothing$. Let $t := |s| \geq 1$.

- If $s < 0$ set $W' := \varphi^s(W_1) \cap W_2 \subset W_2 \subset W$ to get

$$\varphi^t(W') = \varphi^{-s}(f^s(Z_1) \cap Z_2) \subset Z_1 \subset W.$$

- If $s > 0$ set $W' := W_1 \cap \varphi^{-s}(W_2) \subset W_1 \subset W$ to get

$$\varphi^t(W') = \varphi^s(W_1 \cap \varphi^{-s}(W_2)) \subset W_2 \subset W. \qquad \square$$

We now return to the proof of Proposition 1.6.9. Statement (1.6.1) implies (3) $\Rightarrow$ (4) in Proposition 1.6.9: If $\varnothing \neq U, V \subset X$ are open, then there exists an $s \in \mathbb{R}$ such

[25] This uses that X is a metric space, and "regular Hausdorff" would suffice.

that $W := \varphi^s(U) \cap V \neq \varnothing$. If $s > 0$ we are done. Otherwise, (1.6.1) gives a $t > -s$ with

$$\varnothing \neq \varphi^t(W) \cap W = \varphi^t(\varphi^s(U) \cap V) \cap \varphi^s(U) \cap V \subset \varphi^{t+s}(U) \cap V.$$

Since $t + s > 0$, this proves (4).

(4) $\Rightarrow$ (1). Since X is second countable, let $U_1, U_2, \dots$ be a base for the topology. We inductively construct a semiorbit that intersects every U_n and is hence dense.

As the first step, take an open $W_1 \neq \varnothing$ with $\overline{W_1} \subset U_1 =: W_0$ compact and $t_1 = 0$.

Suppose for $1 \leq j \leq n$ there are $t_j \geq 0$ and open $\varnothing \neq W_j \subset \overline{W_j} \subset W_{j-1}$ with $\varphi^{t_j}(x) \in U_j$ for all $x \in W_j$. Item (4) then gives $t_{n+1} > 0$ with $\varphi^{t_{n+1}}(W_n) \cap U_{n+1} \neq \varnothing$. Since Φ is continuous, $W_n' := W_n \cap \varphi^{-t_{n+1}}(U_{n+1}) \neq \varnothing$ is open, and there is a nonempty open $W_{n+1} \subset \overline{W_{n+1}} \subset W_n'$.

Then $\varnothing \neq K := \bigcap_{j \in \mathbb{N}} \overline{W_j} \subset \bigcap_{j \in \mathbb{N}} W_{j-1}$ and $x \in K, j \in \mathbb{N} \Rightarrow \varphi^{t_j}(x) \in \varphi^{t_j}(W_j) \subset U_j$. □

Remark 1.6.14. Examples 1.3.8 and 1.3.13 are not the only ones showing the need for the assumption on X in Proposition 1.6.9. More generally, if a point $x \in X$ has a dense positive semiorbit for a flow Φ, consider the cartesian product of Φ and the flow in Example 1.3.8 or 1.3.13. If y is a nonfixed point in the latter factor, then Φ restricted to the orbit closure of (x, y) has a dense orbit by definition, but no dense semiorbit.

Example 1.1.8 in dimension higher than 2 does not yield a transitive-versus-periodic dichotomy as in Example 1.6.2, but Proposition 1.6.9 gives a convenient criterion for transitivity.

Proposition 1.6.15. *A linear flow $x \mapsto x + tv$ on $\mathbb{T}^n$ is topologically transitive if and only if the components of v are rationally independent (that is, if $k \in \mathbb{Z}^n$ and $\langle k, v\rangle = 0$, then $k = 0$).*

We prove this via a converse to Proposition 1.6.4:

Lemma 1.6.16. *If Φ is a continuous flow on $\mathbb{T}^n$ and every bounded Lebesgue measurable Φ-invariant function is constant, then Φ is topologically transitive.*

Proof. If O is an open Φ-invariant set then χ_O is Φ-invariant, hence constant almost everywhere, so O has Lebesgue measure 0 or 1. Thus, there are no disjoint nonempty open Φ-invariant sets. If now $U, V \subset X$ are open then the Φ-invariant open sets $\varphi^{\mathbb{R}}(U)$ and $\varphi^{\mathbb{R}}(V)$ are therefore not disjoint, so $\varphi^t(U) \cap \varphi^s(V) \neq \varnothing$ for some $t, s \in \mathbb{R}$, and $\varphi^{t-s}(U) \cap V \neq \varnothing$. □

Proof of Proposition 1.6.15. We show both implications by contraposition. If there is a $k \in \mathbb{Z}^n \setminus \{0\}$ with $\langle k, v\rangle = 0$, then $\sin\big(2\pi\langle k, x\rangle\big)$ is a nontrivial constant of motion, and Φ is not transitive by Proposition 1.6.4.

Conversely, suppose f is a nonconstant bounded Lebesgue measurable (hence L^2) invariant function and use the Fourier expansion

$$\begin{aligned}\sum_{k\in\mathbb{Z}^n} f_k e^{2\pi i\langle k,x\rangle} &= f(x) = f(x+tv)\\ &= \sum_{k\in\mathbb{Z}^n} f_k e^{2\pi i\langle k,x+tv\rangle}\\ &= \sum_{k\in\mathbb{Z}^n} f_k e^{2\pi it\langle k,v\rangle} e^{2\pi i\langle k,x\rangle}.\end{aligned}$$

Since f is not constant, there is a $k \neq 0$ with $f_k \neq 0$, so the uniqueness of this expansion implies $e^{2\pi it\langle k,v\rangle} = 1$ for *all* $t \in \mathbb{R}$, hence $\langle k,v\rangle = 0$. □

The criteria in Lemmas 1.6.4 and 1.6.16 are not meant to be optimal, but they are well suited for the purpose at hand and also yield Proposition 3.3.6 below.

Remark 1.6.17. Proposition 1.6.15 gives a clean connection between a dynamical property and a parameter of the flow. This makes it natural to discuss this whole family of linear flows as such rather than viewing each in isolation. Among this family, flows with rather different kinds of orbit structures are tightly interspersed. A rational vector v gives rise to a flow all of whose orbits are closed, but arbitrarily near v there are rationally independent vectors, and they define flows with dense orbits; conversely each of these in turn is arbitrarily close to a rational vector and, on $\mathbb{T}^n$ with $n \geq 3$, also to "intermediate" flows with neither periodic nor dense orbits but orbit closures that form tori of smaller dimension. In particular, such distinct flows are definitely not orbit equivalent. This indicates a great deal of structural "fragility" of these flows.

From this perspective we revisit some earlier examples. Example 1.4.15 has similarities but also a pronounced difference. We noted that the gradient flow on a "stand-up" torus undergoes a qualitative change when the torus is tilted slightly; this is akin to the "fragility" for toral flows. On the other hand, the description in Example 1.4.15 of the dynamics after this slight tilt did not depend on the amount of the tilt, so structurally all these perturbations of the initial gradient flow look rather the same. One might conjecture that they are pairwise orbit equivalent. In a rather similar vein, Example 1.4.9 was obtained from the undamped pendulum and behaves quite differently—but as we change the amount of damping, Figure 1.4.1 changes geometrically (the spirals will approach the stable equilibrium more quickly) but not topologically, so here as well, we have a whole range of parameters with structurally "constant" behavior. The dynamics here is simple enough that one can try to slightly refine the ideas in the proof of Proposition 1.4.5 to show that any two of these damped-pendulum flows are topologically conjugate.

Let us clarify as well that these features of the various families of flows are not an artifact of the parametrization; in a natural sense, these are continuous parametrizations.

Definition 1.6.18 (C^r-closeness). Two flows Φ and Ψ on M are said to be C^r-close if $\Phi\restriction_{[0,1]\times M}$ and $\Psi\restriction_{[0,1]\times M}$ are uniformly C^r-close.

Remark 1.6.19. This is the compact-open C^r-topology for maps on $\mathbb{R}\times M$. Since a flow Φ is determined by the mapping $\Phi\restriction_{[\alpha,\beta]\times M}$ for any $\alpha < \beta$ (Remark 1.1.4), this definition incorporates all information about the flows without unrealistically imposing bounds that are uniform in time.

For toral flows, changing v slightly produces slight changes in this C^r sense for any r, and similarly for the angle of tilting the torus with the gradient flow or for increasing the damping parameter.

Remark 1.6.20. Returning to the study of a linear flow by itself, we note that any two orbits of a linear flow on $\mathbb{T}^n$ are isometric (by a translation), so whenever such a linear flow is topologically transitive, every orbit is dense. The latter feature is a natural indecomposability condition for topological dynamical systems, a property stronger than topological transitivity and, after periodicity, the next case of strong and uniform recurrence.

Definition 1.6.21. A flow is said to be *minimal* if every orbit is dense or, equivalently, if every closed invariant set is empty or the whole space. A Φ-invariant set A is said to be minimal if $\Phi\restriction_A$ is minimal (or A has no proper closed invariant subset).

Remark 1.6.20 gives the next result.

Proposition 1.6.22. *A linear flow $x \mapsto x + tv$ on $\mathbb{T}^n$ is minimal iff the components of v are rationally independent (meaning that if $k \in \mathbb{Z}^n$ and $\langle k, v\rangle = 0$, then $k = 0$).*

Example 1.6.23. A topologically transitive flow that is not minimal is easy to construct from a minimal linear flow on a torus (which is generated by the constant vector field v) by considering the flow generated by the vector field fv with $f\colon \mathbb{T}^n \to [0,\infty)$ such that $f^{-1}(0)$ (the set of fixed points) is nonempty and finite. Also, the south–north–south flow (Example 1.3.14) has a dense orbit and a nondense one.

Theorem 1.6.24. *If a flow is minimal then so are its time-τ maps for all but countably many $\tau \in \mathbb{R}$.*

Proof outline. If φ^τ is not minimal, then there is a proper minimal set A_τ for the map φ^τ. Note first that no orbit stays in A_τ for an interval $[0,\epsilon)$ of time because by minimality of $\varphi^\tau\restriction_{A_\tau}$, and by an approximation argument, every point of A_τ would stay in A_τ for all $s \in [0, \epsilon)$ and hence forever, so A_τ is a proper invariant set for the flow φ^t, contrary to minimality. Thus every point of A_τ has a positive first-return time, which again by minimality of $\varphi^\tau \restriction_{A_\tau}$ and an approximation argument is a constant τ_1 on A_τ, and then $\tau \in \tau_1 \mathbb{Z}$. We define a continuous nonconstant eigenfunction f_τ for φ^t by taking $f_\tau(x) = 1$ for $x \in A_\tau$ and imposing $f(\varphi^s(x)) = e^{2\pi i s/\tau_1} f_\tau(x)$ (then $|f_\tau| = 1$). Note that the f_τ are distinct for different first-return times, so by separability of $C(M)$ there are hence only countably many τ for which φ^τ is not minimal. □

Proposition 1.6.25. *A continuous flow Φ on a compact metric space X is minimal if and only if for every $\epsilon > 0$ there is a $T > 0$ such that $\varphi^{[0,T]}(x)$ is ϵ-dense in X for each $x \in X$.*

Proof. The latter condition clearly implies minimality. On the other hand, if it fails then $\exists \epsilon > 0\ \forall n \in \mathbb{N}\ \exists c_n, x_n \in X\colon \varphi^{[-T,T]}(x_n) \cap B(c_n, \epsilon) = \varnothing$. By compactness there are accumulation points x of $(x_n)_{n\in\mathbb{N}}$ and c of $(c_n)_{n\in\mathbb{N}}$, and we claim that the orbit of x misses $B(c, \epsilon/3)$. To that end, take $N \in \mathbb{N}$ and choose $n \geq N$ such that

- $c_n \in B(c, \epsilon/3)$,
- $\varphi^t(x_n) \in B(\varphi^t(x), \epsilon/3)$ for $|t| \leq N$.

Then for $|t| \leq N$ we have

$$d(\varphi^t(x), c) \geq d(\varphi^t(x_n), c_n) - d(\varphi^t(x_n), \varphi^t(x)) - d(c_n, c) \geq \epsilon - \epsilon/3 - \epsilon/3 = \epsilon/3.$$

Since N was arbitrary, this proves the claim. □

Remark 1.6.26. For the linear flows in Proposition 1.6.15 the exceptional values of τ are those of the form $l/\langle k, v\rangle$ with $l \in \mathbb{Z}$, $k \in \mathbb{Z}^n \setminus \{0\}$ because for such τ we have $t\langle k, v\rangle = l$, so $\sin\big(2\pi\langle k, x\rangle\big)$, say, is φ^τ-invariant. This is illuminating even for $n = 1$.

The next result can be proved using Zorn's Lemma, but we will provide a different proof.

Proposition 1.6.27. *A continuous flow on a compact metric space has a nonempty minimal subset.*

Lemma 1.6.28. *The set of closed invariant sets of a flow Φ on a metric space is closed with respect to the Hausdorff metric.*

Proof. The flow Φ acts homeomorphically on the collection of closed subsets with the Hausdorff metric, and invariant sets are the fixed points, so the set of these is closed. □

Proof of Proposition 1.6.27. For a closed and Φ-invariant set B, let $m(B) := \max\{d(A,B) \mid A \subset B \text{ closed invariant}\}$ and M be such that $m(M) = m_0 := \min m$. Then M has no proper closed invariant subsets: Otherwise $m_0 > 0$. Take a closed invariant $M_1 \subset M$ such that $d(M_1, M) = m_0$. By assumption M_1 is not minimal and contains M_2 such that $d(M_2, M_1) \geq m_0$ and hence $d(M_2, M) > m_0$. We continue this process to obtain a sequence M_i such that $d(M_i, M_j) \geq m_0$ contradicting compactness with respect to the Hausdorff metric. □

Remark 1.6.29. For a continuous flow Φ on a compact metric space denote the closure of the union of all invariant minimal sets by $M(\Phi)$. Then $\mathcal{B}(\Phi) \supset M(\Phi) \neq \varnothing$ since every point of a minimal set is recurrent.

The following proposition is an obvious and useful observation.

Proposition 1.6.30. *Each of topological transitivity, minimality, and density of periodic orbits is invariant under time changes and holds for a special flow if and only if its discrete-time counterpart* (*defined in the obvious way*) *holds for the base.*

While minimality is a strengthening of topological transitivity as defined by density of an orbit, a strengthening of transitivity as defined by open sets gives a criterion for much greater dynamical complexity: images of an open set persistently overlap with another given open set.

Definition 1.6.31 (Topological mixing). A flow φ^t on a topological space X is said to be *topologically mixing* if for any two open sets U and V there exists a $T > 0$ such that $\varphi^t(U) \cap V \neq \varnothing$ for all $t \geq T$.

Remark 1.6.32. Figure 1.6.1 shows this in the context of Example 1.5.26.

Analogously to Remark 1.6.11 this implies a uniform property if X is compact: $\forall\, \epsilon > 0\; \exists\, T \in \mathbb{R}\; \forall\, x, y \in X,\ t \geq T : \varphi^t(B(x,\epsilon)) \cap B(y,\epsilon) \neq \varnothing$.

This can be seen as an extreme form of unpredictability: if ϵ is taken to be the size of observational accuracy, then this statement says that after time T, an initial state can evolve to literally any state whatsoever, within ϵ, that is, no prediction at all is possible beyond this time T.

Proposition 1.6.9(4) immediately gives the following corollary:

Corollary 1.6.33. *Topologically mixing flows are topologically transitive.*

In contrast with Proposition 1.6.30, topological mixing depends on longitudinal effects, that is, the time parametrization matters. The clearest illustration is given by suspensions:

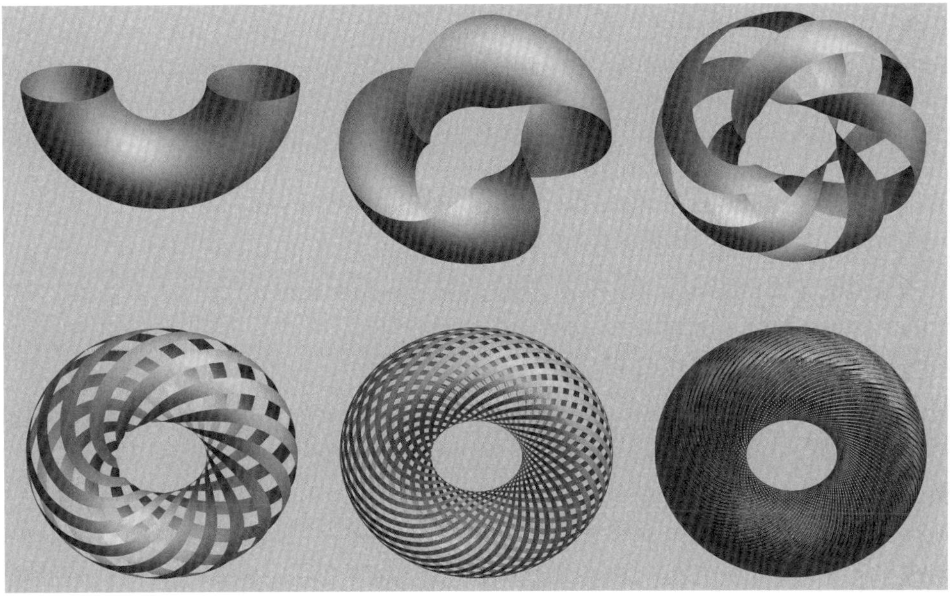

Figure 1.6.1. Mixing in Example 1.5.26 (David Carraher, after [156]).

Example 1.6.34 (Suspensions are not mixing). A suspension flow Φ over a homeomorphism of a space X is not mixing: if $U := X \times (0, 1/2)$ and $V := X \times (1/2, 1)$, then $\varphi^n(U) \cap V = \varnothing$ for all $n \in \mathbb{Z}$.

This is more generally true for special flows whose roof function is cohomologous to a constant: if Φ_f and $\Phi_{\hat{f}}$ are special flows over X, and f and $\hat{f}$ are cohomologous, that is,

$$\hat{f}(x) = f(x) + v(x) - v(\sigma x)$$

for some continuous function v, then by Proposition 1.3.19 the two flows are topologically conjugate via $\pi(x, t) = (x, t + v(x))$.

In particular, a flow under a function that is cohomologous to a constant is topologically conjugate to a suspension and hence not mixing.

For discrete-time dynamical systems we define topological mixing in the same way (but with integers t, T). Example 1.6.34 shows that unlike with topological transitivity, a special flow over a topologically mixing homeomorphism need not be topologically mixing. On the other hand, Example 1.6.35 below is a special flow with mixing base that is mixing. Thus, topological mixing (Definition 1.6.31) is sensitive to time changes and hence to the choice of roof function for special flows.

Example 1.6.35. Consider Φ_{r_c}, a mixing special flow over the map F_A (Example 1.5.26) when the roof function is $r_c\colon \mathbb{T}^2 \to \mathbb{R}^+$, $p \mapsto 1 + c\beta(d(p,0))$ with $c \notin \mathbb{Q}$ and $\beta\colon \mathbb{R} \to [0,1]$ smooth, even, decreasing on $[1/4, 1/2]$, and such that $\beta(x) = 1$ for $|x| < 1/4$, $\beta(x) = 0$ for $|x| > 1/2$: r_c is irrational at the fixed point associated with the origin, but 1 at the period 3-point associated with the orbit $(1/2, 1/2)$, $(0, 1/2)$, and $(0, 1/2)$, so the periods of these two points for the special flow are incommensurate, and Φ_{r_c} is mixing by Proposition 6.2.18 below.

The details for the next example are given in the following chapter, but we mention it now as another important instance of a mixing flow. Furthermore, we will see that this is the foundational example of a hyperbolic flow.

Example 1.6.36. The geodesic flow on a compact factor of the hyperbolic plane (Section 2.4) is topologically mixing (Remark 7.1.13 and Corollary 8.1.6).

Remark 1.6.37. We pick up once more from Remark 1.6.20. Whether or not a toral translation is minimal, the orbit closures provide a natural decomposition of the torus: each orbit closure is a translate (or coset) of the closure of the orbit of 0, which is itself an embedded (sub-)torus. This is, however, not an instance of a general phenomenon but rather a reflection of the homogeneity of toral translations, specifically the fact that any two orbits differ by a translation. (The orbit closure of 0 is a compact subgroup; other orbit closures are its cosets.) When this is not the case, a decomposition into orbit closures does not usually go well. The next section provides abundant examples of this. Recall, though, that while orbit closures do not usually partition the space neatly, Proposition 1.5.35 and Theorem 1.5.44 provide a natural and effective decomposition in great generality, which is also particularly well suited to hyperbolic flows, where there is a finite partition by transitive pieces (Theorem 5.3.37).

1.7 Expansive flows

While Section 1.5 explored recurrence fully and Section 1.6 introduced ways in which orbits are entangled, we now take a global view of the stability notions we studied early on in connection with fixed and periodic points: the relative behavior of orbits. Hyperbolic flows are characterized by universal orbit instability, and we capture this here with the concept of expansivity, a property that, together with compactness of the space, provides a mechanism for complicated dynamical phenomena.

This instability can be paraphrased as saying that any two orbits will separate over time, either in the past or in the future (and, in fact, usually both). We first make this explicit in discrete time both as a warm-up and because this more clearly calls attention to the "longitudinal" aspects we need to allow for flows.

Definition 1.7.1 (Expansivity for maps). A homeomorphism $f\colon X \to X$ is said to be *expansive* if there exists a constant $\delta > 0$ such that if $d(f^n(x), f^n(y)) < \delta$ for all $n \in \mathbb{Z}$ then $x = y$.

The adaptation of expansivity to flows is subtler because of the flow direction and the possibility of reparametrization. For any two orbits of a flow one expects to be able to reparametrize one of them in such a way that at some time the orbits are substantially separated. Expansivity says that this will happen for *any* reparametrization, or conversely, that *no* reparametrization can make two orbits stay close forever. This definition has proven to have the desired properties, and we now formalize it and then study some of its consequences as well as equivalent formulations.[26]

Definition 1.7.2 (Expansivity). A flow Φ on a compact metric space X is *expansive* if for all $\epsilon > 0$ there is a $\delta > 0$, called an *expansivity constant* (for ϵ), such that

$$\textbf{if } x, y \in X,\ s\colon \mathbb{R} \to \mathbb{R} \text{ continuous, } s(0) = 0, \text{ and } d(\varphi^t(x), \varphi^{s(t)}(y)) < \delta\ \forall\, t \in \mathbb{R},$$

$$\textbf{then } y = \varphi^t(x) \text{ for some } t \in (-\epsilon, \epsilon).$$

Remark 1.7.3. By contraposition this says that any two orbits will separate by δ at some time, no matter how you reparametrize (one of) them.

For flows a few notable features of expansivity contrast with the discrete time:

- As in the discrete-time context, expansivity implies that points on different orbits separate by δ in the future or the past. In particular, no orbit is stable for both the flow and the reversed flow. (In discrete time, this characterizes expansivity.)
- Expansivity is independent of the metric and preserved by orbit equivalence (Theorem 1.7.8), time changes (Corollary 1.7.9), and the forming of cartesian products. (Likewise in discrete time for topological conjugacy and products.)
- A suspension is expansive if and only if the base is (Proposition 1.7.10).
- Expansivity implies that fixed points of Φ are isolated points of X (Lemma 1.7.4), so one can omit these ($X \smallsetminus \{\text{fixed points}\}$ is compact) and thereby in most contexts assume without loss of generality that an expansive flow has no fixed points. In particular, an expansive flow on a connected space X has no fixed points unless X is a point. For flows without fixed points, expansivity can be easier to check; see Theorem 1.7.5.
- We do not instead use the natural-looking simpler variant

$$\text{“}\exists\, \delta > 0\ \big(d(\varphi^t(x), \varphi^t(y)) < \delta\ \forall\, t \in \mathbb{R}\big) \Rightarrow x = y\text{”}$$

[26] Our presentation follows that by Bowen and Walters [64].

because it does not hold for any nontrivial flow: $\forall\,\delta > 0\,\exists\,\eta > 0$ such that $\big(y = \varphi^s(x)$ with $|s| < \eta\big) \Rightarrow \big(d(\varphi^t(x), \varphi^t(y)) < \delta\ \forall\, t \in \mathbb{R}\big)$.

It is illuminating to show directly that the Smale Horseshoe (Example 1.5.24) is an expansive flow; this also follows from Example 1.9.18 below. Likewise, Example 1.5.26 (the suspension of $\left(\begin{smallmatrix}2&1\\1&1\end{smallmatrix}\right)$) also provides an instance of an expansive flow. This is a consequence of Proposition 1.7.10 below but also not hard to see directly. Indeed, the orbits of two points x, y will separate (exponentially) for positive time unless y is in the local center-stable set of x, in which case such separation occurs in negative time. Hence, the only points that remain close are on the same orbit. It might be interesting to consider this argument in the case of a special flow over $\left(\begin{smallmatrix}2&1\\1&1\end{smallmatrix}\right)$, or one can reduce this to the suspension by invoking Proposition 1.3.31 and Theorem 1.7.8 or Proposition 1.7.11.

We noted that expansivity is a central feature of hyperbolicity, and Chapter 5 shows that combined with shadowing it yields many of the topological features of hyperbolic dynamics. We add here that on 3-manifolds, expansivity of a flow alone implies the existence of (topological) stable and unstable sets that actually form foliations (Definition 7.1.14), which are central to the even finer analysis of hyperbolic dynamics in Chapter 6 [282]. This is the reason for the term "stable manifolds."

Lemma 1.7.4. *Each fixed point of an expansive flow on a compact metric space (X, d) is an isolated point of X.*

Proof. For an expansive flow Φ, let $\epsilon > 0$ and $\delta > 0$ be an expansive constant corresponding to ϵ. If $x \in X$ is a fixed point and $y \in X$ is such that $d(x, y) < \delta$, take $s(t) \equiv 0$ for all $t \in \mathbb{R}$ to get $d(\varphi^t(x), \varphi^{s(t)}(y)) = d(x, y) < \delta$ for all t and hence $y = \varphi^t(x) = x$ (for some t). □

Theorem 1.7.5. *Expansivity of a fixed-point-free flow Φ is equivalent to each of*

(1) $\forall\,\epsilon > 0\,\exists\,\alpha > 0$ *such that if* $x, y \in X$, $h\colon \mathbb{R} \to \mathbb{R}$ *is an increasing homeomorphism,* $h(0) = 0$, $d(\varphi^t(x), \varphi^{h(t)}(y)) < \alpha\ \forall\, t \in \mathbb{R}$, *then* $y = \varphi^t(x)$ *for some* $|t| < \epsilon$ *(Komuro-expansivity).*[27]

(2) $\forall\,\eta > 0\,\exists\,\delta > 0$ *such that if* $x, y \in X$, $s\colon \mathbb{R} \to \mathbb{R}$ *is continuous,* $s(0) = 0$, *and* $d(\varphi^t(x), \varphi^{s(t)}(y)) < \delta\ \forall\, t \in \mathbb{R}$, *then* $y \in \mathcal{O}(x)$ *and the orbit segment from x to y lies in the ball* $B_\eta(x)$.

(3) $\forall\,\epsilon > 0\,\exists\,\alpha > 0$ *as follows: if* $t_{\pm i} \xrightarrow[i \to +\infty]{} \pm\infty$, $0 < t_{i+1} - t_i \le \alpha$, $|u_{i+1} - u_i| \le \alpha$, $u_0 = t_0 = 0$, $d(\varphi^{t_i}(x), \varphi^{u_i}(y)) \le \alpha$ *for all* $i \in \mathbb{Z}$, *then* $\exists\,|t| < \epsilon\colon y = \varphi^t(x)$.[28]

[27] Despite its name, this notion is due to Bowen and Walters [64, Theorem 3]. It has been used in the study of Lorenz attractors, where fixed points are not isolated points.

[28] This last characterization is particularly useful for Proposition 4.2.23 and hence Theorem 4.2.24.

Proof. That expansivity implies (1) and (2) is clear (for (2) take $\epsilon > 0$ such that $\varphi^t(x) \in B_\eta(x)$ for $|t| < \epsilon$). That (2) implies expansivity is also easy: For $\epsilon \in (0, T_0)$ take $\eta > 0$ such that $d(x, \varphi^\epsilon(x)) > \eta$ for all $x \in X$ by Proposition 1.1.12. Then the orbit segment from x to y lying in $B_\eta(x)$ implies $y = \varphi^t(x)$ with $|t| < \epsilon$.

Showing that (1) implies expansivity involves deforming a continuous $s(\cdot)$ in the definition of expansivity to a homeomorphism. As a first step we show that in a coarse way s is uniformly increasing.

Claim 1.7.6. *If T_0 is as in Proposition* 1.1.12*, $T \in (0, T_0/3)$, then there is a τ_T such that if $x, y \in X$, $s\colon \mathbb{R} \to \mathbb{R}$ continuous, $s(0) = 0$, $d(\varphi^t(x), \varphi^{s(t)}(y)) < \delta_T := \gamma_T/3$ (where γ_T is as in Proposition* 1.1.12*) for all $t \in \mathbb{R}$, then $s(t+T) - s(t) \geq \tau_T$ for all $t \in \mathbb{R}$.*

Proof. Proposition 1.1.12 gives

$$\begin{aligned} &d(\varphi^{s(t)}(y), \varphi^{s(t+T)}(y)) \\ &\quad \geq d(\varphi^t(x), \varphi^{t+T}(x)) - d(\varphi^t(x), \varphi^{s(t)}(y)) - d(\varphi^{s(t+T)}(y), \varphi^{t+T}(x)) \\ &\quad \geq \gamma_T - 2\delta_T > 0, \end{aligned}$$

so continuity of Φ yields a $\tau_T > 0$ such that $|s(t+T) - s(T)| \geq \tau_T$ for all $t \in \mathbb{R}$.

We still need to "remove the absolute value," that is, to check that $s(t+T) \geq s(t)$ for all t, and it suffices to do so for $t = 0$. Suppose to the contrary that there is a $T \in (0, T_0/3)$ such that for all $n \in \mathbb{N}$ there are $x_n, y_n \in X$ and continuous $c_n\colon \mathbb{R} \to \mathbb{R}$ with $s_n(0) = 0$ for which $d(\varphi^t(x_n), \varphi^{s_n(t)}(y_n)) < 1/n$ for all $t \in \mathbb{R}$ but $s_n(T) < 0$ and (by passing to a subsequence) that $x_n \to x$ and hence $y_n \to x$. We will see that this produces a periodic orbit of period less than T_0, contrary to the choice of T_0.

If $s_n(T) \geq -T$ for infinitely many n, then $s_{n_i}(T) \to -L \in [-T, 0]$ for a subsequence, so $d(\varphi^T(x), \varphi^{-L}(x)) = 0$, and x is periodic with period $L + T < T_0$, a contradiction. Otherwise, $s_n(T) < -T$ for all large n, so $s_n(t_n) = -T$ for some $t_n \in [0, T]$ and $t_{n_i} \to t$, hence likewise $x = \varphi^{T+t}(x)$, a contradiction. □

That (1) implies expansivity follows from Claim 1.7.6 because it shows that if $d(\varphi^t(x), \varphi^{s(t)}(y)) < \delta_T$, then the desired increasing homeomorphism h_T of $\mathbb{R}$ is obtained from s by taking $h_T(nT) = s(nT)$ for $n \in \mathbb{Z}$ and linear in between. Moreover, for $t \in [nT, (n+1)T]$ there is a $t' \in [nT, (n+1)T]$ such that $h_T(t) = s(t')$ and thus

$$d(\varphi^t(x), \varphi^{h_T(t)}(y)) = d(\varphi^t(x), \varphi^{s(t')}(y)) \leq \underbrace{d(\varphi^t(x), \varphi^{t'}(x))}_{\leq \sup_{x \in X, u \in [0,T]} d(x, \varphi^u(x))} + d(\varphi^{t'}(x), \varphi^{s(t')}(y)).$$

From this, we obtain expansivity as follows. For $\epsilon > 0$ and α as in (1) choose $T \in (0, T_0/3)$ such that $\sup_{x \in X, u \in [0,T]} d(x, \varphi^u(x)) < \alpha/2$. Then

$$d(\varphi^t(x), \varphi^{s(t)}(y)) < \delta < \min(\delta_T, \alpha/2) \quad \text{for all } t \in \mathbb{R}$$

implies $d(\varphi^t(x), \varphi^{h_T(t)}(y)) < \alpha$ for all $t \in \mathbb{R}$, so $y = \varphi^t(x)$ for some $t \in (-\epsilon, \epsilon)$.

Finally, we prove that (3) is equivalent to expansivity. If Φ is expansive, $\epsilon > 0$, δ as in Definition 1.7.2, $\alpha > 0$ such that $\alpha + 2\sup\{d(z, \varphi^t(z)) \mid z \in X, |t| \le \alpha\} < \delta$, t_i, u_i, x, y as in (3), $s(t_i) := u_i$, then interpolate linearly to $s(t)$ for $t \in [t_i, t_{i+1}]$ to get

$$d(\varphi^t(x), \varphi^{s(t)}(y)) \le \underbrace{d(\varphi^t(x), \varphi^{t_i}(x)) + d(\varphi^{t_i}(x), \varphi^{u_i}(y)) + d(\varphi^{u_i}(y), \varphi^{s(t)}(y))}_{\le \alpha + 2\sup\{d(z,\varphi^t(z)) \mid z \in X,\ |t| \le \alpha\}} < \delta,$$

so $y = \varphi^t(x)$ for some $|t| < \epsilon$ by choice of δ.

Conversely, choose $\epsilon > 0$ and α as in (3). If $d(\varphi^t(x), \varphi^{h(t)}(y)) < \alpha$ for all $t \in \mathbb{R}$ and an increasing homeomorphism $h\colon \mathbb{R} \to \mathbb{R}$ with $h(0) = 0$ let $t_0 = 0$ and $t_{\pm i} \xrightarrow[i\to+\infty]{} \pm\infty$ such that $0 < t_{i+1} - t_i \le \alpha$ and $0 < h(t_{i+1}) - h(t_i) \le \alpha$. Then (3) with $u_i := h(t_i)$ gives $y = \varphi^t(x)$ with $|t| < \epsilon$. □

Remark 1.7.7. One can improve a homeomorphism h as in Theorem 1.7.5(1) in the case of smooth flows: h can be taken smooth with derivative close to 1 for small enough α. By the Whitney Embedding Theorem assume that the flow is on a manifold embedded in some $\mathbb{R}^n$, and endow it with the induced Riemannian metric. At each $\varphi^t(x)$ consider the α-disk $D(\varphi^t(x), \alpha)$ (of dimension $n - 1$) centered at $\varphi^t(x)$ and orthogonal to the generating vector field at that point. There is a unique $\tau(t)$ near $h(t)$ such that $\varphi^{\tau(t)}(y) \subset D(\varphi^t(x), \alpha)$, and by the Implicit Function Theorem, the derivative of $t \mapsto \tau(t)$ is near 1. More explicitly, for every $\delta > 0$ there is an α such that the preceding construction results in a τ such that the derivative of $\tau - \mathrm{Id}$ is less than δ in absolute value.

Proposition 7.3.12 illustrates the utility of the characterization (3) in Theorem 1.7.5.

Theorem 1.7.8. *Expansivity is preserved by orbit equivalence.*

Proof. If h is a homeomorphism that maps orbits of an expansive flow Φ on X to orbits of a flow Ψ on Y, then the fixed points of both flows are isolated and can hence be omitted (Remark 1.7.3). For $\eta' > 0$ choose $\eta > 0$ such that $h(B_\eta(x)) \subset B_{\eta'}(h(x))$ for all $x \in X$ and δ as in Theorem 1.7.5(2), as well as $\delta' > 0$ such that $d_Y(y_1, y_2) < \delta' \Rightarrow d_X(h^{-1}(y_1), h^{-1}(y_2)) < \delta$.

Suppose now that $x_1, x_2 \in X$ are such that there is a continuous $s\colon \mathbb{R} \to \mathbb{R}$ with $s(0) = 0$ and $d_Y(\psi^t(h(x_1)), \psi^{s(t)}(h(x_2))) < \delta'$ for all $t \in \mathbb{R}$. Then Remark 1.3.24 and the choice of δ' give

$$d_X(\varphi^{\sigma_{x_1}(t)}(x_1), \varphi^{\sigma_{x_2}(t)}(x_2)) < \delta, \quad \text{that is, } d_X(\varphi^t(x_1), \varphi^{\sigma_{x_2}(s(\sigma_{x_1}^{-1}(u)))}(x_2)) < \delta$$

for all $t \in \mathbb{R}$. Thus, by Theorem 1.7.5(2), $x_2 \in \mathcal{O}(x_1)$, and the Φ-orbit segment from x_1 to x_2 is in $B_\eta(x_1)$, so $h(x_2) \in \mathcal{O}(h(x_1))$ and the Ψ-orbit segment from $h(x_1)$ to $h(x_2)$ is in $B_{\eta'}(h(x_1))$. □

Corollary 1.7.9. *A time change of an expansive flow is expansive.*

Here is a counterpart of Proposition 1.6.30 (with Proposition 1.7.11 providing a broader one); Example 1.5.26 illustrates this.

Proposition 1.7.10. *A suspension is expansive if and only if the base is.*

Proof. With the conventions of Remark 1.2.10 suppose the suspension flow Φ is expansive. For $\epsilon \in (0, 1/2)$ take $\delta > 0$ as in Definition 1.7.1 and suppose $y_1, y_2 \in M$ are such that $d(f^n(y_1), f^n(y_2)) < \delta$ for all $n \in \mathbb{Z}$. Then

$$\begin{aligned} d(\varphi^t((y_1,0),\varphi^t(y_2,0)) &\le \underbrace{\rho_{t-\lfloor t\rfloor}(\varphi^{\lfloor t\rfloor}(y_1),\varphi^{\lfloor t\rfloor}(y_1))}_{=(1-t+\lfloor t\rfloor)\rho(\varphi^{\lfloor t\rfloor}(y_1),\varphi^{\lfloor t\rfloor}(y_1))+(t-\lfloor t\rfloor)\rho(\varphi^{\lfloor t\rfloor+1}(y_1),\varphi^{\lfloor t\rfloor+1}(y_1))} \\ &< (1-t+\lfloor t\rfloor)\delta + (t-\lfloor t\rfloor)\delta = \delta. \end{aligned}$$

Thus, $(y_2, 0) = \varphi^t(y_1, 0)$ with $|t| < \epsilon < 1/2$, so $y_1 = y_2$, and f is expansive.

Conversely, if f is expansive and $\epsilon > 0$ take $\delta < \min(1/4, \epsilon)$ less than the expansivity constant of f with respect to ρ' and $x_1, x_2 \in M_f$ such that $d(\varphi^t(x_1), \varphi^{s(t)}(x_2)) < \delta$ for all $t \in \mathbb{R}$ and some continuous $s\colon \mathbb{R} \to \mathbb{R}$ with $s(0) = 0$.

We will later reduce to the case where $x_1 \sim (y_1, 1/2) \in M \times [0, 1]$, and with $x_2 \sim (y_2, t_2)$ we then get $\rho'(y_1, y_2) \le d(x_1, x_2) < \delta < 1/4$. Since $\varphi^1(x_1) \sim (f(y_1), 1/2)$ and $d(\varphi^t(x_1), \varphi^{s(t)}(x_2)) < \delta$ for all $t \in [0, 1]$, we have $\varphi^{s(1)}(x_2) \sim (f(y_2), s)$, and therefore $\rho'(f(y_1), f(y_2)) \le d(\varphi^1(x_1), \varphi^{s(1)}(x_2)) < \delta$. Continuing this gives $\rho'(f^n(y_1), f^n(y_2)) < \delta$ for all $n \in \mathbb{Z}$ and hence $y_1 = y_2$, which also gives $x_2 = \varphi^t(x_1)$ for some $|t| < \delta < \epsilon$.

For arbitrary x_1 find $r \in [-1/2, 1/2]$ with $x_1' := \varphi^r(x_1) \sim (y_1, 1/2)$. With $x_s' := \varphi^{s(r)}(x_2)$ this gives $d(\varphi^t(x_1'), \varphi^{s(t+r)-s(r)}(x_2')) < \delta$ for all $t \in \mathbb{R}$, so the foregoing implies $x_2' = \varphi^t(x_1')$ for some $|t| < \delta$, hence $x_2 = \varphi^{t+r-s(r)}(x_1)$ with $|t + r - s(r)| = d(x_1, x_2) < \delta < \epsilon$. Thus, Φ is expansive. □

With Proposition 1.3.31 and Theorem 1.7.8 this also implies the following result:

Proposition 1.7.11. *A special flow is expansive if and only if the base is.*

Remark 1.7.12 (Topological dynamics of hyperbolic flows). Although we have not introduced hyperbolic flows as such, this chapter has placed an emphasis on exploring topological features that are associated with hyperbolicity. We resisted the temptation to go much further, but we will mention several parts of our later exposition that we could have chosen to include here with little change—and which therefore a reader is at this point prepared to study. One of these is the topological dynamics of hyperbolic sets in Section 5.3. While explicitly carried out for hyperbolic sets, we note there that it is based on the shadowing property (Definition 5.3.1), which is of a topological nature, combined with expansivity. Combining the shadowing property with topological transitivity or mixing implies the specification property, and Section 7.3 shows how this (together with expansivity) leads directly to the central concerns of the ergodic theory of hyperbolic sets—which, of course could not be included here because the basic pertinent notions will only appear in Chapter 3. However, parts of another subject we mainly bring up later also fit well right here: Proposition 1.8.7 and Theorem 9.1.3 show that (for instance) no transitive expansive flow commutes with any other flow. The proof in the purely topological context is brief and elementary enough to include in the next section, but the smooth context belongs with deeper explorations of dynamics commuting with hyperbolic flows. We emphasize that the topological results are of an entirely new character; we are not aware of any statement in the literature that establishes such restrictions on the centralizer of a *continuous* flow.

Finally, Section 1.9 below is highly pertinent to hyperbolic dynamics as well (Section 6.6), but beyond definitions it is mostly illustrative.

1.8 Weakening expansivity*

Expansivity plays a central role in our investigations of hyperbolic dynamics, but for a variety of reasons weaker notions play a part in describing or understanding complicated topological dynamics. We describe several of them here, and some of these have distinct technical uses.

The first weakening plays no role in our book, but it describes an essential feature of unpredictability in deterministic systems in a less stringent way than expansivity: some (but not necessarily all) nearby orbits separate in time—so *some* microscopic deviations in initial conditions can lead to macroscopic differences in the orbits. This is sometimes takes to be a defining ingredient for "chaos."

Definition 1.8.1 (Sensitive dependence). Suppose Φ is a flow on a metric space X. A point $x \in X$ is said to exhibit *sensitive dependence on initial conditions* if there is an $\epsilon > 0$ as follows: for all $\delta > 0$ there is a $y \in X$ such that $d(y, x) < \delta$ and for any continuous map $s\colon \mathbb{R} \to \mathbb{R}$ there is a $t \in \mathbb{R}$ with $d(\varphi^t(x), \varphi^{s(t)}(y)) \geq \epsilon$. If this is the case for all $x \in X$, then we say that Φ has sensitive dependence on initial conditions.

Together with compactness of the metric space this reflects chaotic dynamics. There are two reasons we do not use the notion of sensitive dependence further. One is that hyperbolic flows are expansive (the stronger notion). The other is that it is redundant as follows:

Proposition 1.8.2 ([24]). *Topological transitivity and density of periodic orbits imply sensitive dependence (unless the space is a circle).*

Proof. Pick two periodic orbits and choose $\epsilon > 0$ such that the distance between them is 8ϵ. Note that any $p \in X$ is at least 4ϵ from one of these periodic orbits. Thus, given $x \in X$, a neighborhood N of x and a T-periodic point $p \in U := N \cap B(x, \epsilon)$, there is a periodic q such that $d(p, \mathcal{O}(q)) \geq 4\epsilon$. Then q is an interior point of $V := \bigcap_{0 \leq t \leq T} \varphi^{-t}(B(\varphi^t(q), \epsilon))$, so transitivity gives a $y \in U$ and $\tau > 0$ such that $\varphi^\tau(y) \in V$. Then $j := \lfloor \frac{k}{n} \rfloor + 1$ satisfies $1 \leq jT - \tau \leq T$, and

$$\varphi^{jT}(y) = \varphi^{jT-\tau}(\varphi^\tau(y)) \in \varphi^{jT-\tau}(V) \subset B(\varphi^{jT-\tau}(q), \epsilon),$$

so

$$\underbrace{d(\varphi^{jT}(p), \varphi^{jT}(y))}_{=p} \geq \underbrace{d(x, \varphi^{jT}(q))}_{\geq 4\epsilon} - \underbrace{d(\varphi^{jT}(q), \varphi^{jT}(y))}_{<\epsilon} - \underbrace{d(p, x)}_{<\epsilon} > 2\epsilon.$$

Thus, either $d(\varphi^{jT}(x), \varphi^{jT}(y)) > \epsilon$ or $d(\varphi^{jT}(x), \varphi^{jT}(p)) > \epsilon$. □

Two other weakened notions of expansivity turn out to be natural, for instance, for limiting how many symmetries a flow admits.

Definition 1.8.3 (Kinematic expansivity, separation). A flow Φ on a compact metric space X is said to be *kinematically expansive* if for all $\epsilon > 0$ there is a $\delta > 0$, called a *separation constant* such that

$$\forall \epsilon > 0 \, \exists \delta > 0\colon \big(d(\varphi^t(x), \varphi^t(y)) < \delta \, \forall t \in \mathbb{R}\big) \Rightarrow y \in \varphi^{(-\epsilon,\epsilon)}(x).$$

It is *separating* if there is a $\delta > 0$ such that if $x, y \in X$ and $d(\varphi^t(x), \varphi^t(y)) < \delta \, \forall t \in \mathbb{R}$, then $y \in \mathcal{O}(x)$.

Compactness and contraposition give what is usually stated as a consequence of expansivity:

Proposition 1.8.4. *If Φ is a kinematically expansive flow on a compact metric space X and δ a separation constant for $\epsilon > 0$ (Definition 1.8.3), then for any $\rho > 0$ there is a $T > 0$ with*

$$d(\varphi^t(x), \varphi^t(y)) < \delta \textit{ for all } t \in [-T, T] \Rightarrow d(y, \varphi^t(x)) < \rho \textit{ for some } t \in [-\epsilon, \epsilon].$$

Proof. Otherwise, take $x_n, y_n \in X$ such that $d(y_n, \varphi^t(x_n)) > \rho$ for all $t \in [-\epsilon, \epsilon]$ and $d(\varphi^t(x_n), \varphi^t(y_n)) < \eta$ for all $t \in [-n, n]$, and (without loss of generality) $x_n \longrightarrow x$ and $y_n \longrightarrow y$. Then on one hand, $\varphi^t(x) \neq y$ when $|t| \leq \epsilon$, while on the other hand for any $r \in \mathbb{R}$ we have $d(\varphi^r(x_n), \varphi^r(y_n)) < \eta$ for all $n \geq K := |r|$, so $d(\varphi^r(x), \varphi^r(y)) < \eta$, so, since r was arbitrary, $y = \varphi^t(x)$ for some $t \in [-\epsilon, \epsilon]$, a contradiction. □

Lemma 1.7.4 extends as follows:

Lemma 1.8.5. *If Φ is a separating flow, then fixed points are isolated, hence finite in number, and Φ does not have arbitrarily small positive periods.*

Proof. If Φ is a separating flow, then for δ as in the definition let $\eta > 0$ be such that $d_{C^0}(\varphi^t, \mathrm{id}) < \delta$ for $|t| < \eta$. If Φ has arbitrarily small periods or infinitely many fixed points, then there are points x_n with periods $p_n \to 0$ which, without loss of generality, converge to a fixed point x, which then has a δ-neighborhood that contains a point y with positive period or a fixed point $y \neq x$. Since $d(\varphi^t(x), \varphi^t(y) < \delta$ for all $t \in \mathbb{R}$, this contradicts being separating. □

Remark 1.8.6.
- Unlike expansivity, kinematic expansivity is not invariant under orbit equivalence: it holds for the "twist" flow $(x, y) \mapsto (x + ty \pmod 1), y)$ on $S^1 \times [1, 2]$ or equivalently, rotation of the annulus $1 \le r \le 2$ in $\mathbb{R}^2$ with constant *linear* speed, even though this can be time-changed to a rigid rotation, and no two orbits ($y = \text{const.}$) separate. We say that a flow is *strongly* kinematically expansive or strongly separating if all orbit-equivalent flows are kinematically expansive or separating, respectively. The twist example is far from hyperbolic in that it has a continuum of closed orbits. Hence the preference for allowing arbitrary continuous reparametrizations in the hypothesis. Orbits separate "kinematically" but not geometrically. However, we will find this notion quite natural in some contexts. It implies that fixed points are isolated, and so does being separating.

- The proof of Theorem 8.6.14 shows that the horocycle flow (Definition 2.1.13) of a negatively curved surface is strongly separating (but not expansive) [162].

- In light of recent developments we point out that kinematic expansivity is sufficient for existence (Exercise 4.25) and uniqueness of equilibrium states (Remark 7.3.21), which is a central motivation for the notion of expansivity.[29] That is to say, with respect to our principal purpose for this notion, this would serve. We prefer our choice because of the utility of Theorem 1.7.5 and because it reflects that no orbit can closely track another even if we are flexible with the timing, that is, the parametrization.

As promised, we now apply these notions to symmetries of flows. A symmetry of a flow is a map that sends the space to itself in a flow-equivariant way, that is, a homeomorphism commuting with the flow. It is clear that these form a group under composition, and these various expansivity notions imply that this is a discrete group.

[29]This can also be found in [94, Section 2.5 (definition), Theorems A & 2.9 (application)] but that context is far outside our uniformly hyperbolic setting.

We note a reason for which this is particularly interesting in the topological context. A symmetry is a (time-preserving) conjugacy of the flow with itself (in particular, it maps orbits to orbits), so it also reflects the extent to which a conjugacy with another flow may fail to be unique. Specifically, if $\varphi^t \circ h = h \circ \psi^t$ and $\varphi^t \circ k = k \circ \psi^t$ for all $t \in \mathbb{R}$ and $f := h \circ k^{-1}$, then $\varphi^t \circ f = h \circ \psi^t \circ k^{-1} = h \circ k^{-1} \circ \varphi^t = f \circ \varphi^t$, and similarly in reverse.

Proposition 1.8.7. *Suppose Φ is a continuous flow on X and f a homeomorphism of X that commutes with Φ, that is, $f \circ \varphi^t = \varphi^t \circ f$ for all $t \in \mathbb{R}$.*

(1) *If Φ is kinematically expansive* (*Definition* 1.8.3), *$\epsilon > 0$, $\delta > 0$ a separation constant for ϵ, $d_{C^0}(f, \mathrm{Id}) < \delta$, then $f(x) \in \varphi^{(-\epsilon,\epsilon)}(x)$ for all $x \in X$.*

(2) *If Φ is separating, δ as in Definition* 1.8.3, *$d_{C^0}(f, \mathrm{Id}) < \delta$, then $f(x) \in \mathcal{O}(x)$ for all $x \in X$.*

(3) $f(x) \in \mathcal{O}(x) \Rightarrow \exists\, \tau = \tau(\mathcal{O}(x)) \in \mathbb{R}\colon f\restriction_{\overline{\mathcal{O}(x)}} = \varphi^\tau\restriction_{\overline{\mathcal{O}(x)}}$.

(4) (*Discrete symmetry group I*). *If Φ is transitive and separating, δ as in Definition* 1.8.3, *f a homeomorphism that commutes with Φ, $d_{C^0}(f, \mathrm{Id}) < \delta$, then $f = \varphi^\tau$ for some τ.* (*See also Exercise* 1.50.)

(5) *If Φ is kinematically expansive* (*Definition* 1.7.2), *$\epsilon > 0$, $\delta > 0$ a separation constant for ϵ, $d_{C^0}(f, \mathrm{Id}) < \delta$, then $x \mapsto \tau(\mathcal{O}(x)) \in (-\epsilon, \epsilon)$ from* (3) *is continuous.*

(6) *If Ψ is a continuous flow, then $\forall\, \epsilon > 0\ \exists\, \delta_0 > 0\colon |\delta| < \delta_0 \Rightarrow d_{C^0}(\psi^\delta, \mathrm{Id}) < \epsilon$.*

(7) *If two flows are the same* set *of maps, then they are the same* group *of maps: if Φ, Ψ are continuous flows, $\{\psi^s \mid s \in \mathbb{R}\} = \{\varphi^t \mid t \in \mathbb{R}\}$, and $\forall\, \epsilon > 0\ \exists\, \delta > 0\colon d_{C^0}(\varphi^t, \mathrm{Id}) < \delta \Rightarrow |t| < \epsilon$, then $\exists\, c \in \mathbb{R}\ \forall\, t \in \mathbb{R}\ \psi^t = \varphi^{ct}$.*

(8) (*Discrete symmetry group II*). *If a homeomorphism f of a connected space commutes with a kinematically expansive flow Φ with countably many chain components, each topologically transitive, and ϵ is a separation constant, then $d_{C^0}(f, \mathrm{Id}) < \epsilon \Rightarrow f = \varphi^\tau$ for some τ.*

(9) (*Discrete symmetry group III*). *If a homeomorphism f of a connected space commutes with a kinematically expansive flow Φ whose limit set is contained in at most countably many elongational hulls* (*Remark* 1.5.9), *and ϵ is a separation constant, then $d_{C^0}(f, \mathrm{Id}) < \epsilon \Rightarrow f = \varphi^\tau$ for some τ.*

(10) (*Discrete symmetry group IV*). *If a homeomorphism f of a connected space commutes with a kinematically expansive flow Φ, each connected component of whose limit set is contained in an elongational hull, and ϵ is a separation constant, then $d_{C^0}(f, \mathrm{Id}) < \epsilon \Rightarrow f = \varphi^\tau$ for some τ.*

Proof. (1) and (2): Otherwise, $d_{C^0}(f, \mathrm{Id}) \geq \delta$ because there are $x \in X$ and $t \in \mathbb{R}$ with $\delta \leq d(\varphi^t(f(x)), \varphi^t(x)) = d(f(\varphi^t(x)), \varphi^t(x))$.

(3): Writing $f(x) = \varphi^\tau(x)$ gives $f(\varphi^t(x)) = \varphi^t(f(x)) = \varphi^{t+\tau}(x) = \varphi^\tau(\varphi^t(x))$, so $f\restriction_{\mathcal{O}(x)} = \varphi^\tau\restriction_{\mathcal{O}(x)}$, and the claim follows by continuity of f and φ^τ.

(4): If Φ is also topologically transitive, then applying (1) and (3) to a dense orbit shows that $f = \varphi^\tau$ for some $\tau \in \mathbb{R}$.

(5): At fixed points x, (5) is vacuous because they are isolated points of X (and finite in number) by expansivity (Remark 1.7.3). Elsewhere, for $\epsilon < T_0/3$ in (1) with $T_0 > 0$ as in the proof of Proposition 1.1.12, $\tau \in (-\epsilon, \epsilon)$ is uniquely determined for each $x \in X$. If $x_n \to x$, then $n \mapsto \tau(x_n)$ (being bounded) has an accumulation point $\tau_0 = \lim_{k\to\infty} \tau(x_{n_k})$, and continuity of Φ and f gives $\varphi^{\tau_0}(x) = \lim_{k\to\infty} \varphi^{\tau(x_{n_k})}(x_{n_k}) = \lim_{k\to\infty} f(x_{n_k}) = f(x)$, so $\tau(x) = \tau_0 = \lim_{n\to\infty} \tau(x_n)$.

(6) is uniform continuity (in t) of $\Psi\restriction_{[0,1]\times X}$.

((7)): The assumptions on Φ imply that $t \mapsto \tau(t)$ is well defined on $\mathbb{R}$ by $\psi^t = \varphi^{\tau(t)}$ (since $t \mapsto \varphi^t$ is injective) and continuous at 0. Then $\varphi^{\tau(t+s)} = \psi^{t+s} = \psi^t \circ \psi^s = \varphi^{\tau(t)} \circ \varphi^{\tau(s)} = \varphi^{\tau(t)+\tau(s)}$ gives $\tau(t+s) = \tau(t) + \tau(s)$ for all $s, t \in \mathbb{R}$. This implies that τ is continuous and then that $\tau(t) \equiv ct$ for some $c \in \mathbb{R}$ by [1, §2.1].

(8): τ from (3) is constant on orbits and continuous by (5), hence a constant of motion, and thus constant (Proposition 1.6.5).

(9) and (10): Argue likewise using Proposition 1.5.10. □

Example 1.8.8. The north–south flow on S^2 (Example 1.3.9) commutes with the rotational flow that fixes the poles.

Remark 1.8.9. Proposition 1.8.7(4), (8), and (9) say that the *centralizer* of the flow in question is discrete in the following sense: the (closed) group of homeomorphisms f that commute with Φ always includes $\{\varphi^t \mid t \in \mathbb{R}\}$ as a normal subgroup, and the factor group modulo this closed normal subgroup is discrete in the context of (4), (8), and (9). While this usage of "centralizer" is the correct one in the sense of group theory (applied to the homeomorphism group of the space), later chapters will focus on commuting *flows*; with that usage, Proposition 1.8.7(4), (8), or (9) combine with (7) to give a trivial rather than discrete centralizer in the sense that no other *flow* commutes with the given one (Theorem 9.1.3).

1.9 Symbolic flows, coding

We now describe a class of topological flows that provide the standard model for representing hyperbolic flows, in a way that we will later make explicit. A finite coding can help investigate the dynamics of deterministic systems that are so complex as to appear random. The flows that arise from coding a system are symbolic flows, and we will show that they are expansive. Chapter 2 will introduce the paradigmatic case of smooth flows for which these notions are pertinent.

Symbolic flows are particularly amenable to careful study of the orbit structure, as well as, later, statistical features. They are constructed as special flows over finite-state systems, that is, over a system that is described in terms of allowed sequences of symbols from a finite alphabet. Symbolic flows can also exhibit recurrence properties from among those listed in the previous sections and thus also provide new examples of systems with such features.

The symbolic examples with which we do this are central to the study of hyperbolic dynamical systems. In fact, we will show later that hyperbolic flows have a lift to a symbolic system that is uniformly finite-to-one and so will preserve many of the important properties of the hyperbolic flow (Section 6.6).

Definition 1.9.1. Let $\mathcal{A}_n$ be a finite set with the discrete topology (the "alphabet," whose members are called the symbols), where $n = \#\mathcal{A}_n$, and $\Sigma_n := \mathcal{A}_n^{\mathbb{Z}}$ with the product topology, which makes it a compact metric space. A point $\mathbf{t} = \{t_i\}_{i\in\mathbb{Z}} \in \Sigma_n$ is a bi-infinite sequence with each $t_i \in \mathcal{A}_n$. The product topology is induced by a product metric: for $a > 1$ (and usually $a = 2$) we define a metric on Σ_n by $d_a(\mathbf{s},\mathbf{t}) = a^{-N}$ where $N := \max\{k \in \mathbb{N} \mid s_i = t_i \text{ for } |i| < k\}$.

A basis for the topology on Σ_n is given by the *cylinder sets*

$$C^{n_1,\dots,n_k}_{i_1,\dots,i_k} = \{\mathbf{s} \in \Sigma_n \mid s_{n_j} = i_j \text{ for all } 1 \le j \le k\}, \quad \text{where } n_j \in \mathbb{Z},\ i_j \in \mathcal{A}_n, \tag{1.9.1}$$

consisting of the sequences with prescribed symbols in a finite set of locations. Since the complement of a cylinder is a union of cylinders, hence open, cylinders are both open and closed.

The (left) *shift map* is the homeomorphism $\sigma\colon \Sigma_n \to \Sigma_n$ such that $\sigma(s)_i = s_{i+1}$. The space (Σ_n, σ) is the *full shift on n-symbols*. A closed σ-invariant set $\Lambda \subset \Sigma_n$ together with the shift map is a *subshift*.

A subshift Λ is said to be of *finite type* if there is a finite $A \subset \mathcal{A}_n^{k+1}$ for some $k \in \mathbb{N}$ ("allowed words") such that

$$\forall (\xi_0,\dots,\xi_k) \in A\ \exists\, \alpha,\omega \in \mathcal{A}_n\colon (\alpha,\xi_1,\dots,\xi_k) \in A,\ (\xi_0,\dots,\xi_{k-1},\omega) \in A,$$

and

$$\xi \in \Lambda \Leftrightarrow (\xi_i,\dots,\xi_{k+i}) \in A \text{ for all } i \in \mathbb{Z};$$

in this case, Λ is also called a k-step topological Markov chain. The case $k = 1$ is worth singling out: if $A\colon \mathcal{A}_n \times \mathcal{A}_n \to \{0,1\}$ is a function (that is, an $n \times n$ matrix) such that

$$\forall\, i \in \mathcal{A}_n\ \exists\, j \in \mathcal{A}_n\colon A(i,j) = 1 \quad \text{and} \quad \forall\, j \in \mathcal{A}_n\ \exists\, i \in \mathcal{A}_n\colon A(i,j) = 1,$$

then the subshift

$$\Lambda = \Sigma_A = \{\mathbf{s} \in \Sigma_n \mid A(s_i, s_{i+1}) = 1\ \forall\, i \in \mathbb{Z}\} \quad \text{with } \sigma_A := \sigma\restriction_{\Sigma_A}$$

is called a *topological Markov chain*.

We say that the transition from i to j is *allowed* if the entry $a_{ij} := A(i,j)$ of the *transition matrix* A is 1. By assumption, each row and each column have a nonzero entry, and such a matrix is called an *adjacency matrix*.

The metrics we defined above are not the only ones that are often used for symbolic dynamics. One simply needs a metric that induces the product topology. Other metrics that are used are of the form

$$d'_a(\mathbf{s},\mathbf{t}) = \sum_{n=-\infty}^{\infty} \frac{|s_n - t_n|}{(a)^{|n|}} \tag{1.9.2}$$

where $a > 1$. It is not hard to see that these different metrics are equivalent metrics for shift spaces with the product topology. We will typically use the metrics in the definition, but at times we will use the above definition since it is easier for certain computations.

Example 1.9.2. A subshift (not of finite type) is given by the sequences in $\{0,1\}^{\mathbb{Z}}$ that contain at most one occurrence of the symbol 1. It consists of a fixed point (the sequence of zeros) and an orbit whose α- and ω-limit sets are the fixed point. This is a discrete-time counterpart of Example 1.3.13. A discrete-time counterpart to Example 1.3.8 is given by the (1-step) subshift of finite type defined by the adjacency matrix $A = \left(\begin{smallmatrix} 1 & 1 \\ 0 & 1 \end{smallmatrix}\right)$.

Remark 1.9.3. The description of a k-step topological Markov chain may be useful for the study of coding-related features, but as dynamical systems these are topologically equivalent to topological Markov chains over the alphabet $\mathcal{A}_n^k$ (Exercise 1.30). So unless stated otherwise we will assume that a subshift of finite type is a 1-step subshift of finite type.

Remark 1.9.4. Although we will not use it, there are times when it is useful to consider the full shift on $\Sigma = \mathcal{A}^{\mathbb{Z}}$, where $\mathcal{A}$ is a compact topological (or metric) space or a probability space, using, respectively, the (compact) product topology or the product (probability) measure. It is pertinent that the definition of d_a in Definition 1.9.1 does not use the underlying topology, and it induces the product topology in that context only because the alphabet is finite and naturally comes with the discrete topology. The product topology then is that of a perfect totally disconnected space, that is, a Cantor set.

Topological Markov chains provide the base for the special flows (Definition 1.2.11) that are the subject of this section.

Definition 1.9.5. For a subshift Λ and a positive continuous function $f: \Lambda \to \mathbb{R}$, the *symbolic flow* φ_f^t is the flow over Λ under the function f. When Λ is a topological Markov chain and there exists some $a > 1$ such that the roof function is Lipschitz with respect to the metric d_a, the symbolic flow φ_f^t is called a *hyperbolic symbolic flow*.

In fact, we should extend here the notion of Lipschitz continuity (Definition 1.1.21) to a regularity notion that is particularly natural for hyperbolic flows (Definition B.1.1):

Definition 1.9.6. A map f between metric spaces is said to be Hölder continuous with exponent $\alpha \in (0,1]$ or α-Hölder if $d(f(x), f(y)) \le (d(x,y))^\alpha$ for nearby x and y.[30]

The essential feature is that hyperbolic behavior is connected with exponential growth or decay of distances between orbits, and Hölder continuity is well adapted to this because exponentially small differences in inputs result in exponentially small differences in outputs. For symbolic flows, different natural choices of distance functions are related by Hölder regularity of the identity (and its inverse), and when a roof function is Hölder continuous, the resulting flow has a natural Hölder structure. The essential phenomenon is (via composition) described by (bi-)Hölder continuity of the identity from (Σ_n, d_a) to (Σ_n, d_b): if $\alpha := \log b/\log a$, then

$$d_b(\mathrm{Id}(\mathbf{s}), \mathrm{Id}(\mathbf{t})) = d_b(\mathbf{s}, \mathbf{t}) = b^{-N(\mathbf{s},\mathbf{t})} = (a^\alpha)^{-N(\mathbf{s},\mathbf{t})} = (d_a(\mathbf{s}, \mathbf{t}))^\alpha.$$

Specifically, let

$$X = \{(\mathbf{s}, t) \mid t \in [0, f(\mathbf{s})],\ \mathbf{s} \in \Lambda\} \subset \Lambda \times \mathbb{R},$$

and identify the points $(\mathbf{s}, f(\mathbf{s}))$ and $(\sigma(\mathbf{s}), 0)$ for all $\mathbf{s} \in \Lambda$. On this identification space $\Lambda(f)$ the special flow over Λ with roof function f is described as follows (Definition 1.2.11). Let $\pi: X \to \Lambda(f)$ be the quotient map. Then $\varphi_f^t(\pi(\mathbf{s}, t_0)) = \pi(\sigma^k(\mathbf{s}, \tilde{t}))$ where $k \ge 0$ satisfies

$$\tilde{t} = t + t_0 - \sum_{j=0}^{k-1} f(\sigma^j(\mathbf{s}))$$

with $0 \le \tilde{t} < f(\sigma^k(\mathbf{s}))$.

[30]See also Definition 6.3.1. A 1-Hölder map is Lipschitz-continuous.

For these flows under functions it is of interest to connect dynamical properties of the base to those of the flow.

We are primarily interested in symbolic flows over subshifts of finite type. In this setting many of the dynamical properties of the subshift of finite type can be recovered from properties of the adjacency matrix A. For an adjacency matrix A there is an associated graph $\mathcal{G}_A$ on n vertices such that there is an edge from i to j if and only if $a_{ij} \neq 0$.[31] The reader is encouraged to draw the graphs for the matrices

$$A = \begin{pmatrix} 0&1&0&0\\ 0&0&0&1\\ 1&0&0&0\\ 0&0&1&0 \end{pmatrix}, \quad A = \begin{pmatrix} 1&1\\ 0&1 \end{pmatrix}, \quad \text{and} \quad A = \begin{pmatrix} 1&1\\ 1&0 \end{pmatrix}.$$

Lemma 1.9.7. *Let A be an adjacency matrix and $\mathcal{G}_A$ be the associated graph on n vertices. If $i, j \in \mathcal{A}_n$, then the number $\#^m_{ij}$ of distinct paths on $\mathcal{G}_A$ of length $m \in \mathbb{N}$ from i to j equals the (i, j) entry a^m_{ij} of A^m (the product of m copies of A).*

Proof. The case $m = 0$ (or $m = 1$) is clear. The induction step is

$$\#^{m+1}_{ij} = \sum_{k \in \mathcal{A}_n} \#^m_{ik} a_{kj},$$

which holds because for every $k \in \mathcal{A}_n$ every admissible path of length m connecting i and k produces exactly one admissible path of length $m + 1$ connecting i and j by adding j to it, if and only if $a_{kj} = 1$. □

Corollary 1.9.8 (Periodic-orbit growth). $\overline{\lim}_{n\to\infty} \frac{1}{n} \operatorname{card} \operatorname{Fix}(\sigma_A^n) = r(A)$, *where $r(A)$ is the spectral radius (Definition* B.3.1*) of A, that is, the largest absolute value of an eigenvalue.*

Remark 1.9.9. If we let $\mathcal{W}$ be the set of finite-length sequences that appear in Σ_A, then $w \in \mathcal{W}$ if and only if there is a corresponding allowed path on $\mathcal{G}_A$ following the prescribed vertices. We call such a finite sequence w an *allowed word* in Σ_A.

A matrix A with nonnegative integer entries is *irreducible* if for each $i, j \in \{1, \dots, n\}$ there exists an $N = N(i, j)$ such that $a^N_{ij} \neq 0$.

Proposition 1.9.10. *A symbolic flow $(\Lambda(f), \sigma_f)$ over a subshift of finite type Σ_A has dense periodic points if A is irreducible. Furthermore, $\Lambda(f)$ is transitive if and only if A is irreducible.*

Proof. By Proposition 1.6.30 it suffices to prove this for Σ_A.

To prove that the periodic points are dense in Σ_A, let $\mathbf{s} \in \Sigma_A$ and $\epsilon > 0$. Fix $N \in \mathbb{N}$ such that $a^{-N} < \epsilon$ where a is the constant in the metric. Let $w = \mathbf{s}_{-N} \cdots \mathbf{s}_N$. For the

[31] The graphs we consider are directed, allow "loops," that is, an edge from a vertex to itself, and each vertex has at least one entering and one exiting edge (because otherwise it can't occur in a bi-infinite sequence).

elements $\mathbf{s}_{-N}, \mathbf{s}_N \in \mathcal{A}_n$ there is an $n \geq 2$ such that $a^n_{\mathbf{s}_N \mathbf{s}_{-N}} \neq 0$ and hence an allowed word w' of length $n-2$ such that $ww'w$ is an allowed word. We can then define an $(N+n-2)$-periodic $\hat{\mathbf{s}} \in \Sigma_A$ by $\hat{s}_{-N} \cdots \hat{s}_{N+n-2} = ww'$. Then $d_a(\mathbf{s}, \hat{\mathbf{s}}) < \epsilon$, so periodic points are dense in Σ_A.

To show that Σ_A is transitive if A is irreducible, note that for $i, j \in \mathcal{A}_n$ there is a word $s_0 \cdots s_n \in \mathcal{W}$ that begins with i and ends with j, and let $w^{ij} := s_1 \cdots s_{n-1}$. Now order $\mathcal{W}$ by first enumerating the words of length 1 (symbols in $\mathcal{A}_n$), then all the words of length 2, then all the words of length 3, etc. To find a point with a dense forward orbit, "connect" the enumerated words as follows: if w_k and w_{k+1} are successive words in this list of allowed words, denote by $i \in \mathcal{A}_n$ the final symbol of w_k and by $j \in \mathcal{A}_n$ the first symbol in w_{k+1}. Then $w_k w^{ij} w_{k+1}$ is an allowed word. Recursively, this gives a forward-infinite sequence containing all allowed words in Σ_A. Take $\mathbf{s} \in \Sigma_A$ such that the forward sequence of terms in $\mathbf{s}$ agrees with this forward-infinite sequence. Then the forward orbit of $\mathbf{s}$ is dense in Σ_A.

Conversely, given $i, j \in \mathcal{A}_n$, transitivity gives an $s \in \sigma_A$ that goes from the cylinder set $\{s_0 = i\}$ to the cylinder set $\{s_0 = j\}$, so iwj is an allowed word for some w. Thus $a_{ij} \neq 0$. □

Example 1.9.11. For a permutation matrix A (that is, a matrix with a single 1 in each row and each column), each symbol has a unique successor, so Σ_A consists of periodic orbits (one for each cycle of the permutation) and is hence transitive if and only if there is only one such orbit, that is, the permutation is cyclic and A is irreducible, such as

$$A = \begin{pmatrix} 0&1&0&0 \\ 0&0&0&1 \\ 1&0&0&0 \\ 0&0&1&0 \end{pmatrix}.$$

In fact, permutation matrices give the only cases of subshifts of finite type with finite cardinality.

Example 1.9.12. The matrix $A = \left(\begin{smallmatrix} 1&1 \\ 0&1 \end{smallmatrix}\right)$ is reducible. For a roof function f the suspension flow Φ_f on $\Lambda(f)$ of Σ_A has a dense orbit, but no dense forward orbit. This flow consists of two periodic orbits (coming from fixed points of Σ_A) and a "heteroclinic" orbit, whose α-limit set is one of the periodic orbits and ω-limit set is the other periodic orbit. This flow is topologically conjugate to the cartesian product of the flow in Example 1.3.8 with that in Example 1.1.6.

Example 1.9.13. An irreducible matrix that appears similar to the previous example, but whose associated topological Markov chain has different dynamical properties, is given by $A = \left(\begin{smallmatrix} 1&1 \\ 1&0 \end{smallmatrix}\right)$ with $A^2 = \left(\begin{smallmatrix} 2&1 \\ 1&1 \end{smallmatrix}\right)$. So not only is there an N for each i and j with $a^N_{ij} \neq 0$, but $N = 2$ works simultaneously for all ij-pairs.

Definition 1.9.14. An integer matrix A is *positive* if each entry is positive and *eventually positive* or *aperiodic* if there is an $N \in \mathbb{N}$ such that A^N is positive.

Then the proof of Proposition 1.9.10 gives the following result:

Proposition 1.9.15. *If A is eventually positive, then Σ_A is topologically mixing.*

The next two results could have been proven earlier, and connect the results in this section to the results in the previous section.

Theorem 1.9.16. *Subshifts are expansive.*

Proof. Let Λ be a subshift, $\epsilon < 1$, and $\mathbf{s}, \hat{\mathbf{s}} \in \Lambda$. Then there is an $i \in \mathbb{Z}$ with $s_i \neq \hat{s}_i$, so $d_a(\sigma^i(\mathbf{s}), \sigma^i(\hat{\mathbf{s}})) = 1 > \epsilon$. □

The following proposition is an immediate consequence of this and Theorem 1.9.16:

Proposition 1.9.17. *Symbolic flows are expansive.*

One of the main uses of symbolic flows will be in coding invariant sets for flows; this is typically a semiconjugacy and thus produces a symbolic extension (Definition 1.3.1). As such, it does not preserve all of the topological properties of the original flow. However, the symbolic extension is usually easier to investigate and reflects the salient properties faithfully enough to be useful. We now provide a few examples of coding constructions. The existence of symbolic extensions will be established in greater generality in Section 6.6.

Example 1.9.18. In Example 1.5.24, the dynamics on Λ is topologically conjugate to the full 2-shift by labeling the two image pieces overlapping with Δ as 0 and 1 and associating points and their itineraries. The flow is thus topologically conjugate to the symbolic flow over the full 2-shift with roof function equal to 1. Variants with more crossings in Δ are topologically conjugate to a full shift on more symbols. Therefore, the set Λ has a dense set of periodic points and is topologically transitive.[32]

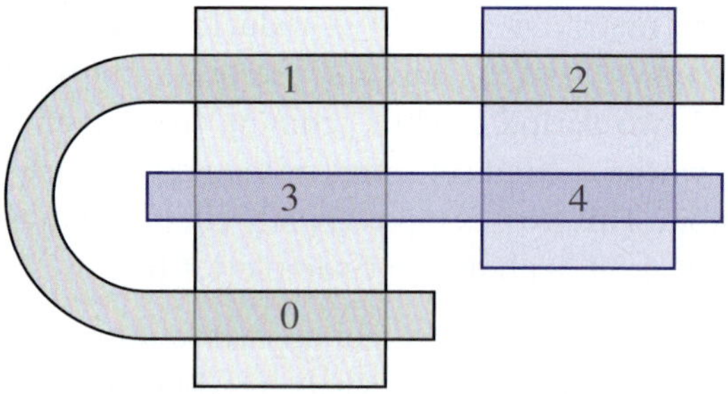

Figure 1.9.1. Coding pieces in linked horseshoes (Example 1.9.19).

[32]And has positive topological entropy (Section 4.2).

Example 1.9.19. In Example 1.5.25, the dynamics on the natural invariant (Cantor) set is topologically conjugate to a shift on five symbols by proceeding analogously using the five overlap rectangles in Figure 1.5.8. With respect to the labeling from Figure 1.9.1, the symbols 0, 1, and 3 can each be followed by the symbols 0, 1, and 2, because they lie in the left rectangle, which is mapped across the elongated horseshoe that includes the pieces numbered 0, 1, and 2. Likewise, the symbols 2 and 4 can each be followed by the symbols 3 and 4. These allowed transitions are encoded in the matrix

$$\mathbb{A} = \begin{pmatrix} 1&1&1&0&0 \\ 1&1&1&0&0 \\ 0&0&0&1&1 \\ 1&1&1&0&0 \\ 0&0&0&1&1 \end{pmatrix} \tag{1.9.3}$$

This gives a "coding" by $\mathbb{A}$, that is, a homeomorphism h between the suspensions of $\Sigma_{\mathbb{A}}$ and of the invariant Cantor set in Figure 1.5.8 that intertwines the flows.

Example 1.9.20. Suspensions of hyperbolic toral automorphisms are factors of symbolic flows. Consider the suspension of the map

$$F_A(x, y) = (2x + y,\ x + y) \qquad (\text{mod } 1) \tag{1.9.4}$$

of the 2-torus from Example 1.5.26. Draw segments of the two eigenlines at the origin

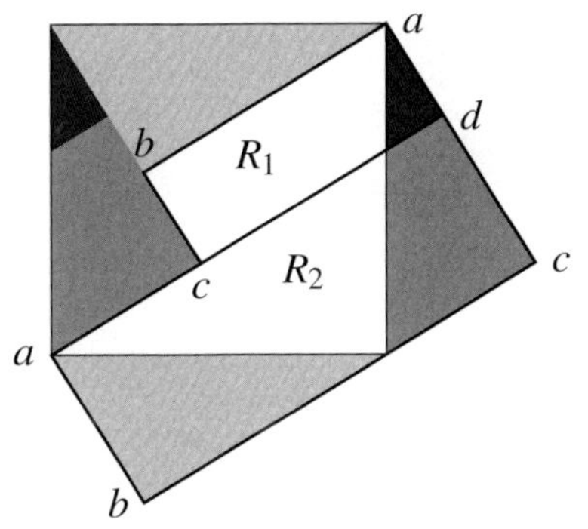

Figure 1.9.2. Partitioning the torus.

until they cross sufficiently many times and separate the torus into disjoint rectangles. Although this prescription contains an ambiguity, direct inspection shows that it can be effected by taking a segment of the contracting line in the fourth quadrant until it intersects the segment of the expanding line twice in the first quadrant and once in the third quadrant (see Figure 1.9.2). The resulting configuration is a decomposition of the torus into two rectangles R_1 and R_2. Three pairs among the seven vertices of the plane configuration are identified, so there are only four different points on the torus which serve as vertices of the rectangles. This agrees with our description: those vertices are exactly the origin and three intersection points.

Even without explicit calculation one can see that the image $F_A(R_i)$ ($i = 1, 2$) consists of several "horizontal" rectangles of "full length." The union of the boundaries $\partial R_1 \cup \partial R_2$ consists of the segments of the two eigenlines at the origin just described. The image of the contracting segment is a part of that segment. Thus, the images of R_1 and R_2 have to be "anchored" at parts of their "vertical" sides, that is, once one of the images "enters" either R_1 or R_2 it has to stretch all the way through it. Tracking where A sends integer points shows that $F_A(R_1)$ consists of three components, two in R_1 and one in R_2. The image of R_2 has two components, one in each rectangle (see Figure 1.9.3). We can use these five components $\Delta_0, \Delta_1, \Delta_2, \Delta_3, \Delta_4$ (or their

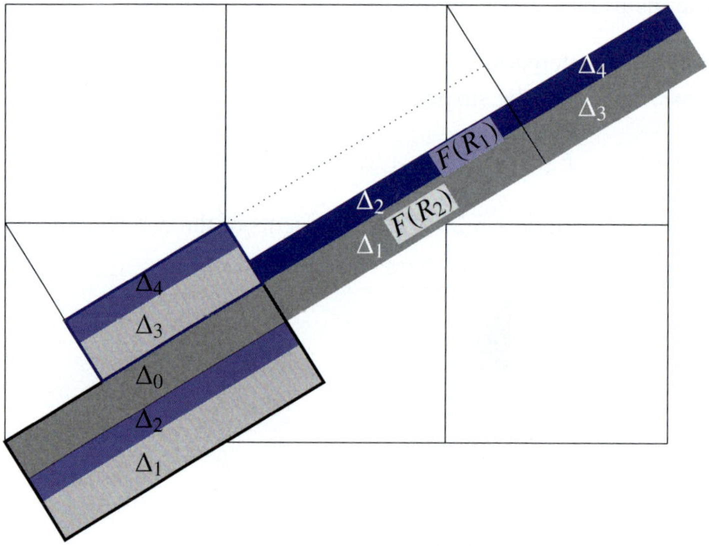

Figure 1.9.3. The image of the partition.

preimages) as the pieces in our coding construction. Due to the contraction of F_A in the "vertical" direction and contraction of F_A^{-1} in the "horizontal" direction, each intersection

$$\bigcap_{n \in \mathbb{Z}} F_A^{-n}(R_{\omega_n})$$

contains no more than one point. On the other hand, the "Markov" property, that is, the images going full length through rectangles, implies that if $\omega \in \Sigma_5$ and $F_A(\operatorname{Int} \Delta_{\omega_n}) \cap \operatorname{Int} \Delta_{\omega_{n+1}} \neq \varnothing$ for all $n \in \mathbb{Z}$, then $\bigcap_{n \in \mathbb{Z}} \overline{F_A^n(\operatorname{Int} \Delta_{\omega_n})} \neq \varnothing$. In other words, we have a "coding," that is, a continuous map $h\colon \Sigma_{\mathbb{A}} \to \mathbb{T}^2$ with $\mathbb{A}$ from (1.9.3) such that

$$F_A \circ h = h \circ \sigma.$$

Thus, F_A is a (topological) factor of $\Sigma_{\mathbb{A}}$; in this case the term "semiconjugacy" for h is apt, because we will see that it is "mostly" bijective. Every point $q \in \mathbb{T}^2$ whose positive and negative iterates avoid the boundaries ∂R_1 and ∂R_2 has a unique preimage and vice versa. The points of Σ_A whose images are on those boundaries or their iterates under F_A fall into three categories corresponding to the three segments of stable and unstable sets through 0 which define parts of the boundary. Thus, sequences are identified in the following cases: they have a constant infinite right (future) tail consisting of 0s or 4s, and agree otherwise, or else an infinite left (past) tail (of 0s and 1s, or of 4s) and agree otherwise. We summarize some of the properties of the coding.

Proposition 1.9.21. *The induced factor map between the suspensions of $\sigma\restriction_{\Sigma_{\mathbb{A}}}$ and F_A is one-to-one on all periodic points (except for those coming from fixed points). The number of preimages of any point not negatively asymptotic to the suspension of the fixed point is bounded.*

This instance of coding illustrates a general phenomenon in hyperbolic flows, and we will produce a like construction for them (Section 6.6). But we mention that this holds in the present generality.

Theorem 1.9.22 ([64, Theorem 10]). *An expansive flow without fixed points on a compact metric space is the quotient of a suspension of a subshift with the quotient map being a homeomorphism between certain invariant dense G_δ-sets.*

We conclude this chapter by emphasizing that Chapter 5 could be subsumed into the present chapter in its entirety by viewing it as the theory of expansive flows with the shadowing property (Definition 5.3.1), which is expressed in purely topological terms.[33]

Exercises

1.1. For a flow Φ on a space X and a point $x \in X$ prove that exactly one of the following holds:

(1) $t \mapsto \varphi^t(x)$ is one-to-one,

(2) there exists a smallest $t_0 > 0$ such that $\varphi^{t_0+t}(x) = \varphi^t(x)$ for all $t \in \mathbb{R}$,

(3) $x = \varphi^t(x)$ for all $t \in \mathbb{R}$.

[33]Using a metric, but no smooth structure.

1.2. Remark 1.1.22 asserts that Proposition 1.1.15 fails for (1.1.3) (unless it reduces to (1.1.1)). Describe exactly what in the proof of Proposition 1.1.15 goes wrong in that case.

1.3. If $g\colon \mathbb{R} \to \mathbb{R}$ is continuous, then writing $v = \frac{dx}{dt}$ (velocity) converts the second-order differential equation $\frac{d^2x}{dt^2} + g(x) = 0$ to the system

$$\begin{cases} \dfrac{dx}{dt} = v, \\ \dfrac{dv}{dt} = -g(x), \end{cases}$$

of first-order differential equations. Show that the *total energy* given by $H(x, v) = \frac{1}{2}v^2 + \int_0^x g(s)\,ds$ is a constant of motion.

1.4. With the notation of Remark 1.2.7 show that if ξ is a cocycle over Φ, then $\zeta(x,t) := \xi(x, \alpha(x,t))$ defines a cocycle over Ψ.

1.5. Show that if ξ from Exercise 1.4 is a Φ-coboundary, that is, $\xi(x,t) \equiv f(\varphi^t(x)) - f(x)$ for some continuous f, then ζ from Exercise 1.4 is a Ψ-coboundary.

1.6. Prove the converse of Theorem 1.4.3.

1.7. Use the reasoning in Remark 1.3.11 to show that if flows Φ and Ψ on $\mathbb{R}$ commute, and h conjugates Φ to a linear flow on $\mathbb{R}$, then h conjugates Ψ to a linear flow on $\mathbb{R}$ as well.

1.8. Show that Examples 1.3.8 and 1.3.13 are related by a semiconjugacy. Which is a factor of which?

1.9. Show that Example 1.3.9 on S^1 and Example 1.3.8 are related by a semiconjugacy. Which is a factor of which?

1.10. Are any two of the flows in Figure 1.3.3 orbit equivalent? Find all factor–extension pairs.

1.11. Carry out the "straightforward calculation" in the proof of Theorem 1.4.7.

1.12. Find all Lyapunov functions for the north–south flow (Example 1.3.9) and the south–south flow (Example 1.3.13).

1.13. Prove Theorem 1.4.23.

1.14. Prove Corollary 1.4.24.

1.15. Prove Corollary 1.4.25.

1.16. In a compact metric space, show that $\{x\}$ is attracting per Definition 1.4.17 if and only if x is attracting per Definition 1.4.1.

1.17. Show that $W^s(x)$ (Definition 1.3.26) and $W^s(\{x\})$ (Definition 1.5.5) agree. (This is a preview of Theorem 5.3.25.)

1.18. Prove Proposition 1.3.28.

1.19. Show that a topologically transitive flow is regionally recurrent.

1.20. In the context of Proposition 1.6.9 show that each of the four statements is equivalent to the existence of a *bitransitive* orbit, that is, of a point x with $\varphi^{(0,\infty)}(x)$ and $\varphi^{(-\infty,0)}(x)$ both dense.

1.21. Show that topological conjugacy (Definition 1.3.1) defines an equivalence relation among continuous flows.

1.22. Carry out the "illuminating" proof in Example 1.3.15.

1.23. Suppose $f, g\colon \mathbb{R} \to \mathbb{R}$ are expanding maps with $|f'|$ bounded and $\|f-g\|_{C^1} < \infty$. Show that there is a unique $h\colon \mathbb{R} \to \mathbb{R}$ with $h - \mathrm{Id}$ bounded such that $f \circ g = g \circ f$ and that $h_n := f^{-n} \circ g^n \xrightarrow[n\to\infty]{} h$ uniformly and $\|h_n - \mathrm{Id}\,\|_\infty \le K\|f-g\|_\infty \le K\|f-g\|_{C^1}$ for some $K > 0$.

1.24. Show that orbit equivalence (Definition 1.3.22) defines an equivalence relation among continuous flows.

1.25. As suggested in Remark 1.6.17 show that any two versions of Figure 1.4.1 (for different damping parameters) are topologically conjugate by refining the ideas in the proof of Proposition 1.4.5.

1.26. Find the stable and unstable sets (Definition 1.3.26) of a fixed point of a topological Markov chain.

1.27. Find the stable and unstable sets (Definition 1.3.26) of a point in a topological Markov chain.

1.28. Find the stable and unstable sets (Definition 1.3.26) of a periodic point in a symbolic flow.

1.29. Find the stable and unstable sets (Definition 1.3.26) of a point in a symbolic flow.

1.30. Prove the assertion in Remark 1.9.3 that k-step topological Markov chains are topologically equivalent to topological Markov chains over the alphabet $\mathcal{A}_n^k$.

1.31. Determine $\mathcal{L}$ (Definition 1.5.1), $\mathcal{B}$ (Definition 1.5.11), NW (Definition 1.5.13), $\mathcal{R}$ (Definition 1.5.33), and $\mathcal{AR}$ (Definition 1.4.18), as well as the chain decomposition (Definition 1.5.33) in Examples 1.1.5, 1.1.7, 1.1.8, 1.3.7, 1.3.8, 1.3.13, 1.3.15, 1.3.16, 1.3.17, 1.4.16, 1.5.16, 1.5.26, and 1.6.2 and Figures 1.1.4, 1.3.3, 1.4.1, 1.5.4, and 1.5.11.

1.32. Find each basin of attraction and basin of repulsion (Definition 1.5.5) of any compact invariant sets that are apparent in Figures 1.1.4, 1.3.3, 1.4.1, 1.5.4, and 1.5.11.

1.33. Determine $\mathrm{NW}(\Phi)$, $\mathrm{NW}\big(\Phi\restriction_{\mathrm{NW}(\Phi)}\big)$, $\mathrm{NW}\big(\Phi\restriction_{\mathrm{NW}(\Phi\restriction_{\mathrm{NW}(\Phi)})}\big)$ in Figures 1.1.4, 1.3.3, 1.4.1, 1.5.3, 1.5.4, and 1.5.11.

1.34. Find examples to show that each inclusion in Proposition 1.5.37 can be strict. (They can be found among examples presented in this chapter.)

1.35. In Figure 1.5.3 find the prolongational limit sets of any points not on the top line.

1.36. In the context of Remark 1.5.43, describe all possible trapping regions and attractor–repeller pairs.

1.37. In light of Proposition 1.6.7 prove or give a counterexample: if $\omega(x) \neq \varnothing$ then $\Phi\restriction_{\omega(x)}$ is topologically transitive.

1.38. Show that the complement of $\mathcal{R}(\Phi)$ is open.

1.39. In Conley's example (Figure 1.5.11) show that $\mathrm{GR}(\Phi) \neq \mathcal{R}(\Phi)$ (Remark 1.5.47).

1.40. Find $\mathrm{GR}(\Phi)$ and $\mathrm{NW}(\Phi)$ for the flow Φ on $[0,1]^2/\mathbb{Z}^2$ defined by $x' = \cos^2(2\pi y)$, $y' = \sin^2(\pi y)$.

1.41. Show that $\mathcal{R}(\Phi)/\sim$ (the space of chain-equivalence classes) is a Hausdorff topological space.

1.42. Show that the sum of two Hölder-continuous functions is itself Hölder continuous (Definition 1.9.6). Give a lower bound for the Hölder exponent.

1.43. Is the Cantor function $\mathfrak{c}$ (Remark 1.3.6) Hölder continuous? How about $x \mapsto x + \mathfrak{c}(x)$? In either case, if so, find a (sharp) Hölder exponent.

1.44. Show that the composition of two Hölder-continuous maps is itself Hölder continuous (Definition 1.9.6). Give a lower bound for the Hölder exponent of the composition.

1.45. Can the composition of two Hölder-continuous maps have larger Hölder exponent than one of them?

1.46. Find the Hölder exponent of the identity from (Σ_n, d_a) to (Σ_n, d_b) (Definition 1.9.1). (A suggestive computation was done in the text. Make it part of a cogent argument.)

1.47. Show that d_a and d'_a are equivalent metrics on Σ_n (see Definition 1.9.1 and (1.9.2)).

1.48. Is the identity $(\Sigma_n, d_a) \to (\Sigma_n, d'_a)$ (see Definition 1.9.1 and (1.9.2)) Hölder continuous? Is the inverse?

1.49. Show that a separating flow (Definition 1.8.3) is not equicontinuous (and hence not periodic).

1.50. In the context of Proposition 1.8.7(4) show that a transitive separating flow Φ either admits a sequence $t_n \to \infty$ such that $\varphi^t \to \mathrm{Id}$, or otherwise, for every $\epsilon > 0$ there is a $\gamma > 0$ such that if f is a homeomorphism that commutes with Φ, $d_{C^0}(f, \mathrm{Id}) < \gamma$, then $f = \varphi^\tau$ for some $\tau \in (-\epsilon, \epsilon)$.

1.51. Derive a counterpart of the characterization (3) in Theorem 1.7.5 of expansivity under the assumption of kinematic expansivity.

2

Hyperbolic geodesic flow*

Having developed tools for describing complicated flows we now pick up again from Section 1.1.c to describe and study geodesic flows on hyperbolic surfaces. We will see later that these are the standard examples of hyperbolic flows. This chapter may be omitted, but it provides details on the classical example that provided the impetus for studying hyperbolic flows.

Some parts of this chapter assume a basic knowledge of differential geometry. We will review some of the concepts, especially ones we will need for the dynamics of surfaces with negative curvature.

We begin with a description of the upper half-plane model of a hyperbolic metric with emphasis on the geometry and isometries of this model to have the tools we need for describing the dynamics of the geodesic flow, and we introduce the Poincaré disk as another standard model for hyperbolic geometry. We then describe the dynamics on the upper half-plane model and explain how we obtain compact factors of the Poincaré disk and hence flows on compact spaces with nontrivial recurrence. These compact factors are the classical examples of hyperbolic flows and illustrate many of the notions that we will develop in the second half of the book.

If one wants to study only the flows that have hyperbolic properties then one would study Sections 2.1.a, 2.1.b, and 2.2.a, together with Sections 2.3 and 2.4.

2.1 Isometries, geodesics, and horocycles of the hyperbolic plane and disk

The upper half-plane

$$\mathbb{H} := \{z \in \mathbb{C} \mid \operatorname{Im} z > 0\} \subset \mathbb{C}$$

is an open subset of $\mathbb{C} \sim \mathbb{R}^2$, hence a smooth manifold, and

$$\langle u + iv, u' + iv' \rangle_z := \operatorname{Re} \frac{(u + iv)(u' - iv')}{(\operatorname{Im} z)^2}$$

for $z \in \mathbb{H}$, $u + iv$, $u' + iv' \in T_z\mathbb{H}$ is symmetric, $\mathbb{R}$-bilinear, and positive definite, hence a Riemannian metric $\langle \cdot, \cdot \rangle$, called the *hyperbolic metric*. The half-plane $\mathbb{H}$ with this metric is called the *Poincaré upper half-plane* (or the Klein model or the *Lobachevsky plane*). The hyperbolic metric differs from the Euclidean metric $\mathrm{Re}(u+iv)(u'-iv')$ only by the scalar factor $(\mathrm{Im}\, z)^2$, so hyperbolic angles coincide with Euclidean angles.

Lemma 2.1.1. *The imaginary axis* $I := i\cdot(0,\infty)$ *is a geodesic with unit-speed parametrization* $t \mapsto ie^t$.

Proof. The imaginary axis I minimizes length between any two of its points: the length of a curve $t \mapsto c(t) = x(t) + iy(t)$, $x(0) = x(1) = 0$, $y(0) = y_0$, $y(1) = y_1$ connecting iy_0 to iy_1 is

$$\ell(c) = \int_0^1 \sqrt{\langle \dot c(t), \dot c(t)\rangle_{c(t)}}dt = \int_0^1 \sqrt{\frac{(\dot x(t))^2 + (\dot y(t))^2}{(y(t))^2}}dt \ge \int_0^1 \frac{\frac{dy}{dt}}{y}dt = \ell(\gamma),$$

where γ is a parametrization of the segment $i[y_0, y_1] \subset I$. □

2.1.a Isometries. The principal tool for understanding the geometry of $\mathbb{H}$ is its isometries. We begin with *linear fractional transformations*. Denote by $\mathrm{GL}_+(2,\mathbb{R})$ the collection of real 2×2 matrices with positive determinant and associate to each $\left(\begin{smallmatrix} a & b\\ c & d\end{smallmatrix}\right) \in \mathrm{GL}_+(2,\mathbb{R})$ the map

$$T := \psi\begin{pmatrix} a & b\\ c & d\end{pmatrix} : \mathbb{H} \to \mathbb{H}, \quad z \mapsto \frac{az+b}{cz+d}. \tag{2.1.1}$$

Then $T'(z) = \frac{ad-bc}{(cz+d)^2}$ and hence

$$\begin{aligned}\mathrm{Im}\, T(z) &= \frac{1}{2i}\left(\frac{az+b}{cz+d} - \frac{a\bar z+b}{c\bar z+d}\right)\\ &= \frac{(az+b)(c\bar z+d) - (a\bar z+b)(cz+d)}{2i(cz+d)(c\bar z+d)} = |T'(z)|\,\mathrm{Im}(z),\end{aligned}$$

so T maps $\mathbb{H}$ to itself. Further, $\mathcal{M} := \psi\big(\mathrm{GL}_+(2,\mathbb{R})\big)$ is a group under composition and ψ is a homomorphism with kernel $\mathbb{R}\,\mathrm{Id}$. As a matrix group, this is $\mathrm{PSL}(2,\mathbb{R})$.

Lemma 2.1.2. *The maps* $T \in \mathcal{M}$ *are isometries of the hyperbolic metric.*

Proof.

$$\underbrace{\mathrm{Re}\frac{T'(z)(u+iv)\overline{T'(z)(u'+iv')}}{(\mathrm{Im}\, T(z))^2}}_{=\langle T'(z)(u+iv), T'(z)(u'+iv')\rangle_{T(z)}} = \underbrace{\frac{T'(z)\overline{T'(z)}}{|T'(z)|^2}}_{=1}\underbrace{\mathrm{Re}\frac{(u+iv)(u'-iv')}{(\mathrm{Im}(z))^2}}_{=\langle u+iv, u'+iv'\rangle_z}.$$

□

All $T \in \mathcal{M}$ extend to $\mathbb{H} \cup \mathbb{R} \cup \{\infty\}$ by setting $T(-d/c) = \infty$ and $T(\infty) = a/c$ (or $T(\infty) = \infty$ if $c = 0$). Examples of linear fractional transformations are $z \mapsto -1/z$, $z \mapsto z + b$ ($b \in \mathbb{R}$), and $z \mapsto az$ ($a > 0$). They represent correspondingly three types of linear fractional transformation from the point of view of the intrinsic geometry of the Lobachevsky plane: *elliptic* (direct counterparts of Euclidean rotations), with a single fixed point inside the plane, *parabolic*, with no fixed points on the plane and no invariant geodesic, and *hyperbolic*, with no fixed points but a unique fixed geodesic (the axis). On $\mathbb{H}$ a parabolic map has a unique fixed point on $\mathbb{R} \cup \{\infty\}$ and a hyperbolic map has two fixed points on $\mathbb{R} \cup \{\infty\}$. Both parabolic and hyperbolic maps are counterparts of translations of the Euclidean plane.

There are also isometries other than linear fractional transformations. Clearly $z \mapsto -\bar{z}$ and $z \to 1/\bar{z}$ are examples. Geometrically the former is the reflection in the imaginary axis and the latter is the inversion with respect to the unit circle. We use linear fractional transformations now to study geodesics. Lemma 2.1.1 suggests to examine isometric images of the imaginary axis I (parametrized with unit speed by $t \mapsto ie^t$).

Lemma 2.1.3. *If C is a vertical line or a semicircle with center on the real line, then there exists a $T \in \mathcal{M}$ with $TI = C$. Furthermore, given any unit tangent vector v at a point of C one can take T such that it maps the upward vertical vector i at $i \in I$ to v.*

Proof. If C is the vertical line $\{z \mid \mathrm{Re}(z) = b\}$, take $T(z) = z + b$. If C is a semicircle with endpoints $x, x + r \in \mathbb{R}$ then note that $T_1 \colon z \mapsto z/(z + 1)$ maps I to the semicircle with endpoints 0 and 1 (since $\left|\frac{it}{1+it} - \frac{1}{2}\right| = \left|\frac{2it-(1+it)}{2(1+it)}\right| = \frac{1}{2}$) and let $T_2(z) = rz$, $T_3(z) = z + x$, and $T = T_3 \circ T_2 \circ T_1$. To map tangent vectors as desired note that there is a linear fractional transformation T_0 such that $DT_0(\mathrm{i}) = DT^{-1}(v)$, namely, either $T_0(z) = cz$ or $T_0(z) = -\frac{c}{z}$ for some $c \in \mathbb{R}_+$. Then $T \circ T_0$ is as desired. □

Corollary 2.1.4. *The group $\mathcal{M}$ acts transitively on the* unit tangent bundle $S\mathbb{H}$ *of* $\mathbb{H}$: *if $v \in T_z\mathbb{H}$, $w \in T_{z'}\mathbb{H}$, $\|v\| = 1 = \|w\|$, then there is a $T \in \mathcal{M}$ with $T(z) = z'$ and $T'(z)v = w$.*

Remark 2.1.5. Since any vertical line or semicircle with center on the real axis parametrized with unit speed is obtained via a linear fractional transformation from I parametrized by $t \mapsto ie^t$, they are all geodesics, and transitivity on $S\mathbb{H}$ implies that we have identified all geodesics. We note that the endpoints of $\psi\left(\begin{smallmatrix} a & b \\ c & d \end{smallmatrix}\right)(I)$ are $\frac{a\cdot 0+b}{c\cdot 0+d} = \frac{b}{d}$ and $\frac{a\cdot i\infty+b}{c\cdot i\infty+b+d} = \frac{a}{c}$.

2.1.b Geodesics and geodesic flow. We are now able to describe the geodesic flow on the upper half-plane.

Theorem 2.1.6. *The geodesics of the Poincaré upper half-plane are precisely the vertical half-lines and the semicircles with center on the real axis.*

Remark 2.1.7. We also have a natural identification of $\mathrm{PSL}(2,\mathbb{R})$ and $S\mathbb{H}$ given by $\gamma \sim v := \gamma\mathrm{i}$, where i is as in Lemma 2.1.3. Equivalently, set

$$\phi\colon S\mathbb{H} \to \mathrm{PSL}(2,\mathbb{R}) \quad \text{by} \quad D\left(\psi(\phi(v))\right)(\mathrm{i}) = v,$$

where Ψ is as in (2.1.1). With respect to this identification, the geodesic flow is given by $\gamma \mapsto \gamma\begin{pmatrix} e^{t/2} & 0 \\ 0 & e^{-t/2} \end{pmatrix}$.

Since they are isometries and hence send geodesics to geodesics, we also have the following proposition:

Proposition 2.1.8. *If C is a vertical line or a circle with center on the real axis and $\phi \in \mathcal{M}$ or $\phi(z) = -\bar{z}$ then $\phi(C)$ is a vertical line or a circle with center on the real axis.*

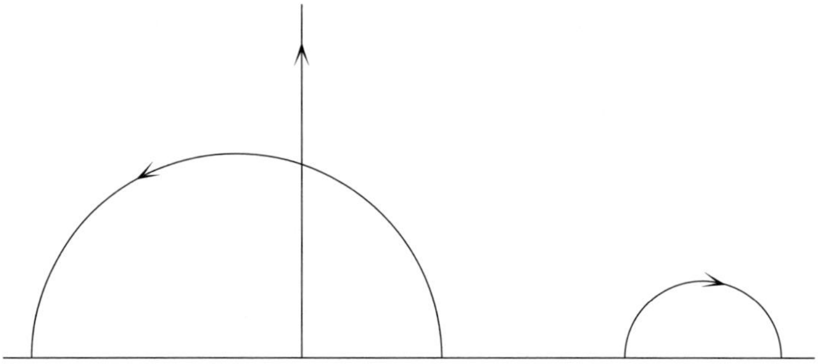

Figure 2.1.1. Geodesics on the Lobachevsky plane. [Reprinted from [213] (© Cambridge University Press, all rights reserved) with permission.]

The group Γ generated by the group $\mathcal{M}$ of linear fractional transformations and the transformation $S\colon x \mapsto -\bar{z}$ is the isometry group:

Proposition 2.1.9. *The group of isometries of $\mathbb{H}$ is generated by $\mathcal{M}$ and the symmetry $S\colon z \mapsto -\bar{z}$.*

Proof. Let ϕ be any isometry of $\mathbb{H}$. Any isometry that preserves a geodesic and a tangent vector to it is the identity on that geodesic. Since $\phi(I)$ is a geodesic, Theorem 2.1.6 and Lemma 2.1.3 give a $T \in \mathcal{M}$ such that $T^{-1}\phi_{\restriction_I} = \mathrm{Id}_{\restriction_I}$. It suffices to show that $T^{-1}\phi$ is either the identity on $\mathbb{H}$ or coincides with the symmetry $S\colon z \mapsto -\bar{z}$. Consider the geodesic C with endpoints $-r$ and r. It contains the point $ir \in I$ and hence so does $T^{-1}\phi(C)$ (since $T^{-1}\phi_{\restriction_I} = \mathrm{Id}_{\restriction_I}$). Since $T^{-1}\phi$ preserves angles, both these geodesics are orthogonal to I at ir. Hence they coincide up to orientation, that is, we either have $T^{-1}\phi(z) = z$ for $z \in C$ or $T^{-1}\phi(z) = -\bar{z}$ for $z \in C$, and hence the derivative of $T^{-1}\phi$ at ir is either the identity or the reflection in I. Since isometries are smooth, the same case occurs for all points on I; hence the same choice was made for all such geodesics, that is, $T^{-1}\phi = \mathrm{Id}$ or $T^{-1}\phi = S$ on $\mathbb{H}$. So $\phi \in \mathcal{M}$ or $\phi \circ S \in \mathcal{M}$. □

Proposition 2.1.10 (Stable manifolds). *The orbits of upward vertical unit vectors at points $x + i \in \mathbb{R} + i$ are pairwise exponentially positively asymptotic under the geodesic flow $g^t \colon S\mathbb{H} \to S\mathbb{H}$.*

Proof. We use the canonical distance on $S\mathbb{H}$: If $z, z' \in \mathbb{H}$, $v \in S_z\mathbb{H}$, $w \in S_{z'}\mathbb{H}$, then there is a geodesic $\gamma\colon [0,1] \to \mathbb{H}$ (unique if $z \neq z'$) connecting z and z', and a unique continuous vector field X along γ such that $X(0) = v$ and $\measuredangle X(t), \dot\gamma(t) = \measuredangle v, \dot\gamma(0)$ for all $t \in [0,1]$. Then

$$d(v,w) := \sqrt{(\measuredangle X(1), w)^2 + (d(z,z'))^2}.$$

Geometrically, this amounts to parallel translating v along γ to $z' \in \mathbb{H}$ and measuring angles there.

In particular, if $v \in T_{x+iy}\mathbb{H}$, $w \in T_{x+d+iy}\mathbb{H}$ are vertical unit vectors then the angle term in this distance function is $2\tan^{-1}\frac{d/2}{y} \le \frac{d}{y}$, and an upper bound for the length of the connecting geodesic is given by the length of the connecting line segment, which is d/y. Thus,

$$d(v,w) < \sqrt{2}d/y. \tag{2.1.2}$$

The orbit of the upward vertical unit vector w at $x + i \in \mathbb{H}$ projects to the geodesic $t \mapsto x + ie^t$, and the distance between the corresponding upward unit vectors i_t at ie^t and w_t at $x + ie^t$ is bounded by $\sqrt{2}xe^{-t}$. □

Remark 2.1.11. By using the transformation $z \mapsto -1/z$ one also sees then that the orbits of the outward unit normals to the circle of radius $1/2$ centered at $i/2$ are negatively asymptotic to that of i. Together, we have thus identified the stable and unstable sets for the geodesic flow. Furthermore, these sets form foliations explicitly, which we will much later produce in proper generality (Theorem 6.1.1).

Remark 2.1.12. We also note that in the proof of Proposition 2.1.10 one can let $y \to 0$ and conclude that two such vertical geodesics separate exponentially as $t \to -\infty$. In particular, geodesic arcs limiting on distinct boundary points diverge (exponentially) from each other. Contrariwise, if γ, η are geodesics such that $\{d(\gamma(t), \eta(t))\}_{t \ge 0}$ is bounded, then there is a $c \in \mathbb{R}$ such that $d(\gamma(t+c), \eta(t)) \xrightarrow[t \to +\infty]{} 0$. This also implies that if γ, η are geodesics such that $\{d(\gamma(t), \eta(t))\}_{t \in \mathbb{R}}$ is bounded, then there is a $c \in \mathbb{R}$ such that $\gamma(t + c) = \eta(t)$ for all $t \in \mathbb{R}$.

2.1.c Horocycle flow. We are now able to define the horocycle flow for the upper half-plane model. It is far from hyperbolic but nonetheless shares some topological features with the geodesic flow, with which it is tightly interwoven. The dynamics of both, in turn, is connected in important ways to the geometry.

Definition 2.1.13. Horizontal lines $\mathbb{R} + ir = \{t + ir \mid t \in \mathbb{R}\}$ are called *horocycles* centered at ∞. Circles tangent to $\mathbb{R}$ at $x \in \mathbb{R}$ are called *horocycles* centered at x. If $\gamma\colon \mathbb{R} \to \mathbb{H}$ is a geodesic then $\gamma(-\infty), \gamma(\infty) \in \mathbb{R} \cup \{\infty\}$ are the limit points of γ as $t \to -\infty$ and $t \to +\infty$, respectively. If $v \in T_z\mathbb{H}$ then let $\pi(v) := z$.

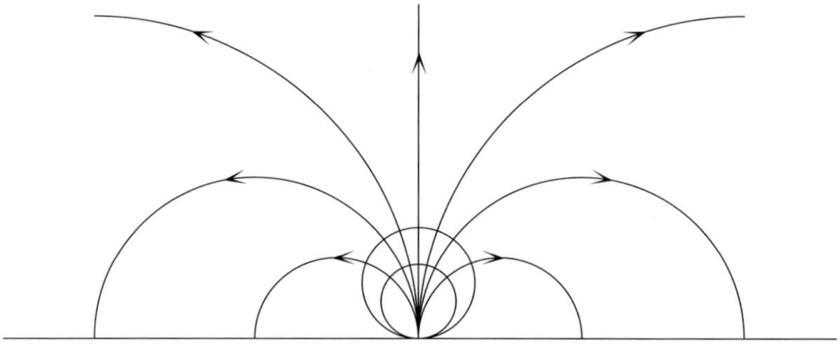

Figure 2.1.2. Geodesics and horocycles in the hyperbolic plane. [Reprinted from [213] (© Cambridge University Press, all rights reserved) with permission.]

Lemma 2.1.14. *For every horocycle H there is a $T \in \mathcal{M}$ with $T(\mathbb{R} + i) = H$.*

Proof. If $H = \mathbb{R} + ir$ take $T(z) = rz$. If H is centered at $x \in \mathbb{R}$ and of Euclidean diameter r take $T_1(z) = -1/z$, $T_2(z) = rz$, $T_3(z) = z + x$, and $T = T_3 \circ T_2 \circ T_1$. □

Remark 2.1.15. With the identification from Remark 2.1.7, these horocycles are the orbits of the *horocycle flow* $h^s : \gamma \mapsto \gamma\left(\begin{smallmatrix} 1 & s \\ 0 & 1 \end{smallmatrix}\right)$.

Example 2.1.16. The horocycle flow on a compact factor of the Poincaré disk (Section 2.3) is topologically transitive; indeed, the orbit of every g^t-periodic point is dense [191, Theorem 2.2] (see also Exercises 2.6 and 6.8).

For some purposes it is useful to have an alternative model of the Lobachevsky plane (Figure 2.1.3).

Proposition 2.1.17 (Poincaré disk). *The map $f\colon \mathbb{H} \to \mathbb{C}$, $z \mapsto \frac{z-i}{z+i}$ maps the Poincaré upper half-plane $\mathbb{H}$ onto the open unit disk $\mathbb{D}$ in $\mathbb{C}$ bounded by the unit circle $S^1 = \{z \in \mathbb{C} \mid |z| = 1\}$ since $|f(z)| = 1$ when $z \in \mathbb{R}$ and $f(i) = 0$. Pushing forward*

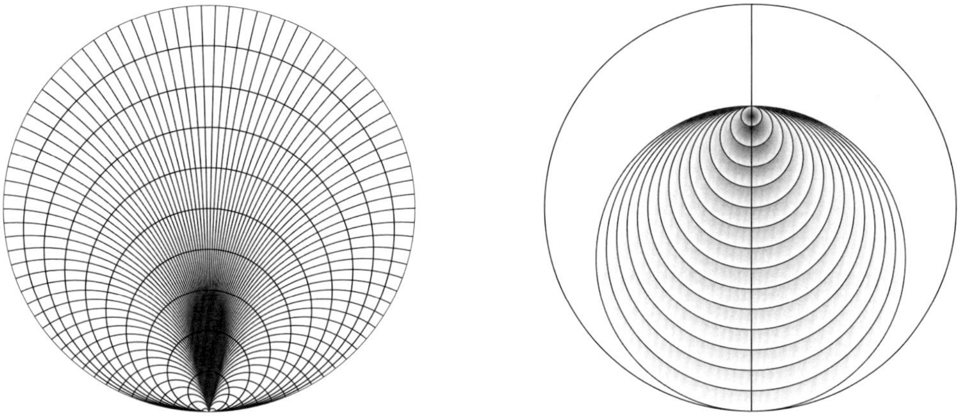

Figure 2.1.3. Geodesics and horocycles in the Poincaré disk with a common boundary point (Proposition 2.1.17), and a horocycle as a limit circle (Remark 2.1.18).

the hyperbolic Riemannian metric $\langle\cdot,\cdot\rangle$ *on* $\mathbb{H}$ *to the metric given by*

$$\langle v, w\rangle := \langle Df^{-1}v, Df^{-1}w\rangle$$

on the unit disk makes f *an isometry. The unit disk with this metric is called the* Poincaré disk. *Since* f *maps lines and circles into lines and circles and preserves angles, the geodesics in the Poincaré disk are diameters of* S^1 *and arcs of circles perpendicular to* S^1*, and the horocycles are circles tangent to* S^1 (*Figure* 2.1.3).

Remark 2.1.18 (Busemann function). It is useful to note that the word "horocycle" (sometimes "oricycle") means "limit circle," which is due to the fact that these are limits of circles as follows: For a point ξ at infinity and $x \in \mathbb{D}$ consider the geodesic $\gamma = \gamma_{\xi,x}$ with $\gamma(0) = x$ and $\gamma(t) \xrightarrow[t\to+\infty]{} \xi$. The nested union $\bigcup_{t>0} B(\gamma(t), t)$ of disks is bounded by the horocycle through x determined by ξ (Figure 2.1.3). Alternatively it can be described as the set of points $y \in \mathbb{D}$ such that $d(\gamma(t), y) - t \xrightarrow[t\to+\infty]{} 0$. Indeed, more generally, the horocycles determined by ξ are the level sets of the *Busemann function*

$$b_{\xi,x}(y) := \lim_{t\to+\infty} d(\gamma_{\xi,x}(t), y) - t$$

illustrated by Figure 2.1.3. Busemann functions are Lipschitz continuous by the triangle inequality.[1] Furthermore, this description is altogether independent of having constant curvature.

[1]Thus, this pointwise limit is uniform on compact sets by Dini's Theorem: if a monotone sequence of continuous functions on a compact space converges pointwise to a continuous function, then the convergence is uniform.

Remark 2.1.19. Horocycles are *lines* because the point on the boundary of the Poincaré disk is not included. In fact, the dynamically natural objects are their normal vector fields (in $\mathrm{PSL}(2,\mathbb{R})$ or $S\mathbb{D}$) because they define the pairwise asymptotic geodesics—positively or negatively asymptotic according to whether one considers the normal vector field pointing into or out of the horocycle. With this point of view, one can then moreover consolidate *all* unit vectors pointing to a common boundary point into a plane in $\mathrm{PSL}(2,\mathbb{R})$, and likewise with vectors pointing away from a boundary point. Each of these two sets of planes is parametrized by the boundary circle, and Figure 2.1.4 shows them in a natural presentation in $\mathrm{PSL}(2,\mathbb{R})$.

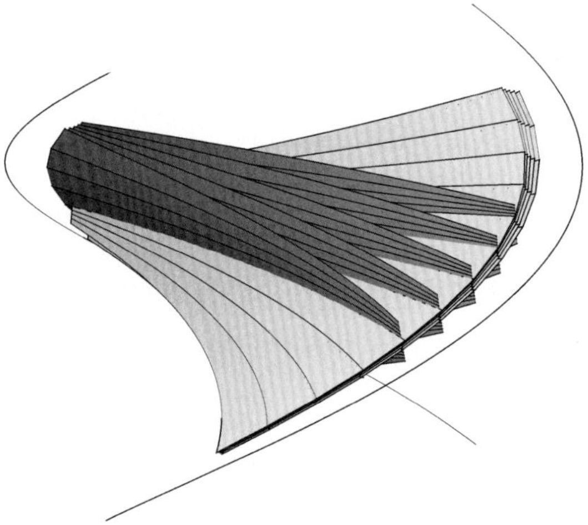

Figure 2.1.4. Horocycle foliations in $\mathrm{PSL}(2,\mathbb{R})$ (after Tsuboi). [Animations at http://www.tsuboiweb.matrix.jp/showroom/public_html/animations/gif/T3image/T3image8.html, http://www.tsuboiweb.matrix.jp/showroom/public_html/animations/gif/geodflow/geodflowconft.html, and http://www.tsuboiweb.matrix.jp/showroom/public_html/animations/gif/geodflow/geodflowconftes.html.]

2.2 Dynamics of the natural flows

We now explore some of the dynamics for the geodesic flow and horocycle flow. We begin with the geodesic flow.

2.2.a Dynamics of the geodesic flow. To further study the dynamics of the geodesic flow on $\mathbb{H}$ one can parametrize the set $S\mathbb{H}$ of unit vectors on $\mathbb{H}$ by $t, u, v \in \mathbb{R}$ as follows: Given a fixed reference vector $q \in S\mathbb{H}$ and $p \in S\mathbb{H}$ let H_p be the horocycle with p as inward normal vector, H_q the horocycle with p as outward normal vector, γ the geodesic connecting the centers of H_q and H_p (that is, the points of tangency on the real axis), v the oriented hyperbolic length of the arc of H_p between $\gamma \cap H_p$ and the footpoint $\pi(p)$ of p, t the oriented arc length of the segment of γ between H_q and H_p, and u the oriented length of the arc of H_q between $\gamma \cap H_q$ and $\pi(q)$. It is easy to see that *locally* $\phi\colon (t,u,v) \mapsto p$ is a diffeomorphism between $\mathbb{R}^3$ and $S\mathbb{H}$ (Figure 2.2.1). This does not parametrize any vector p for which the center of H_p coincides with that of H_q, but a chart based at $-q$ would cover these (see also Remark 2.2.9). In these charts the geodesic flow acts by $(\tau,u,v) \mapsto (\tau+t,u,v)$.

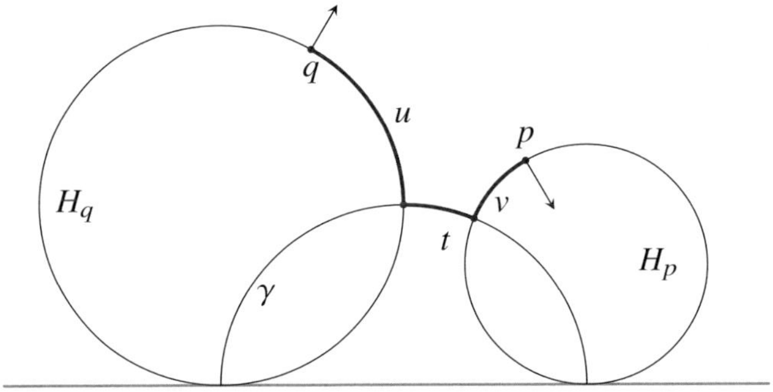

Figure 2.2.1. Local coordinates.

If $W^s(p)$ denotes the collection of inward (or upward) unit normal vectors to H_p (in this case the stable set is a manifold and so we refer to it as the *stable manifold* of p), then the orbit of any $p' \in W^s(p)$ is positively asymptotic to that of p by Proposition 2.1.9, since the orbits of upward vertical unit vectors to $\mathbb{R}+i$ have pairwise asymptotic orbits. Note that $W^s(p)$ is a level set of (t,u). Indeed $W^s(q) = \phi(\{0\} \times \{0\} \times \mathbb{R})$. The set $W^{s0}(q) := \phi(\mathbb{R} \times \{0\} \times \mathbb{R})$ the *center-stable manifold* of q. Likewise the points of $W^u(p) := -W^s(-p)$ (the *unstable manifold* of p, outward unit vectors to H_{-p}) have negatively asymptotic orbits and $W^u(q) = \phi(\{0\} \times \mathbb{R} \times \{0\})$. The set $W^{u0}(q) := \phi(\mathbb{R} \times \mathbb{R} \times \{0\})$ is called the *center-unstable manifold* of q. For vertically downward vectors we have to use the corresponding chart starting with $-q$ to make these definitions.

Proposition 2.1.10, particularly the estimate (2.1.2) of the decay of the distance between vertical tangent vectors combined with the fact that $t \mapsto x + ie^t$ is a geodesic, Definition 2.1.13, Lemma 2.1.14, and the preceding notions are summarized as follows:

Proposition 2.2.1. *The stable manifold of $v \in S\mathbb{H}$ with respect to the geodesic flow g^t is the unit normal vector field containing v to the horocycle centered at $\gamma_v(\infty)$. The unstable manifold of $v \in S\mathbb{H}$ is the unit normal vector field containing v to the* *horocycle centered at $\gamma_v(-\infty)$. In particular, all stable and unstable manifolds are 1-dimensional and the contraction and expansion rates are e^{-1} and e.*

Remark 2.2.2 (Hyperbolicity from the structure equations). One can see the hyperbolic behavior of these geodesic flows directly from their algebraic structure. The unit tangent bundle has a framing by a vertical vector field V, a horizontal vector field H, and the vector field X that generates the geodesic flow. With respect to the representation in terms of $\mathrm{PSL}(2,\mathbb{R})$ they are given by elements of the Lie algebra (that is, traceless matrices) as follows: V is the initial derivative of the rotational flow (in unit tangent circles) given by the matrices $\begin{pmatrix} \cos t/2 & -\sin t/2 \\ \sin t/2 & \cos t/2 \end{pmatrix}$,[2] so $V \sim \begin{pmatrix} 0 & -1/2 \\ 1/2 & 0 \end{pmatrix}$, while $X \sim \begin{pmatrix} 1/2 & 0 \\ 0 & -1/2 \end{pmatrix}$ is the initial derivative of $\begin{pmatrix} e^{t/2} & 0 \\ 0 & e^{-t/2} \end{pmatrix}$, so taking

$$H := [V,X] \sim \begin{pmatrix} 0 & -1/2 \\ 1/2 & 0 \end{pmatrix}\begin{pmatrix} 1/2 & 0 \\ 0 & -1/2 \end{pmatrix} - \begin{pmatrix} 1/2 & 0 \\ 0 & -1/2 \end{pmatrix}\begin{pmatrix} 0 & -1/2 \\ 1/2 & 0 \end{pmatrix} = \begin{pmatrix} 0 & 1/2 \\ 1/2 & 0 \end{pmatrix}$$

gives the *canonical framing* X, H, V and the *structure equations*

$$[V,X] = H, \quad [H,X] = V, \quad [H,V] = X. \tag{2.2.1}$$

One can check (2.2.1) by using that in the $\mathrm{PSL}(2,\mathbb{R})$-representation of $S\tilde{\Sigma}$, the vector fields of the canonical framing are given by

$$X \sim \begin{pmatrix} 1/2 & 0 \\ 0 & -1/2 \end{pmatrix}, \quad H \sim \begin{pmatrix} 0 & 1/2 \\ 1/2 & 0 \end{pmatrix}, \quad V \sim \begin{pmatrix} 0 & -1/2 \\ 1/2 & 0 \end{pmatrix}.$$

A dynamically natural variant of this framing is the one by X and

$$H_\pm := H \pm V, \quad \text{that is,} \quad H_+ \sim \begin{pmatrix} 0 & 0 \\ 1 & 0 \end{pmatrix}, \quad H_- \sim \begin{pmatrix} 0 & 1 \\ 0 & 0 \end{pmatrix}$$

with the corresponding bracket relations

$$[X,H_\pm] = [X,H] \pm [X,V] = \mp H_\pm \text{ and } [H_+,H_-] = \underbrace{[H+V,H-V]}_{=-2[H,V]} = -2X. \tag{2.2.2}$$

A vector field $fH_\pm$ invariant under the geodesic flow satisfies

$$0 = [X, fH_\pm] = (\dot{f} \mp f)H_\pm,$$

which means that $\dot{f} = \pm f$, so $f = e^{\pm t}$. Thus, the differential of the geodesic flow expands and contracts, respectively, the directions $H_\pm$; this is the defining feature of hyperbolicity (Definition 5.1.1).

[2] We encountered this in Example 1.1.29 as an extreme magnetic flow; see also Remark 2.2.10 below.

2.2.b Dynamics of the horocycle flow.

Example 2.2.3 (The horocycle flow). The vector fields X and $H_\pm$ each generate a flow we can describe explicitly (Remark 2.1.15):

$$X \rightsquigarrow \exp\left(\begin{pmatrix} 1/2 & 0 \\ 0 & -1/2 \end{pmatrix} t\right) = \begin{pmatrix} e^{t/2} & 0 \\ 0 & e^{-t/2} \end{pmatrix} \sim g^t,$$

$$H_+ \rightsquigarrow \exp\left(\begin{pmatrix} 0 & 0 \\ 1 & 0 \end{pmatrix} t\right) = \begin{pmatrix} 1 & 0 \\ t & 1 \end{pmatrix} \sim h_+^t,$$

$$H_- \rightsquigarrow \exp\left(\begin{pmatrix} 0 & 1 \\ 0 & 0 \end{pmatrix} t\right) = \begin{pmatrix} 1 & t \\ 0 & 1 \end{pmatrix} \sim h_-^t.$$

The first is (again) the geodesic flow, and, as previewed in Remark 2.1.15, the latter flows are called the horocycle flows. Note that the matrix action is on the *right* (Remark 2.1.15).

Early on (Propositions 2.1.10, 2.2.1 and Remark 2.1.15) we noted that h_-^s parametrizes the stable manifold of Id, and a matrix computation gives the commutation relation

$$\begin{pmatrix} e^{-t/2} & 0 \\ 0 & e^{t/2} \end{pmatrix} \begin{pmatrix} 1 & s \\ 0 & 1 \end{pmatrix} \begin{pmatrix} e^{t/2} & 0 \\ 0 & e^{-t/2} \end{pmatrix} = \begin{pmatrix} 1 & se^{-t} \\ 0 & 1 \end{pmatrix} \quad \text{or} \quad g^t h_-^s g^{-t} = h_-^{se^{-t}}, \tag{2.2.3}$$

which reflects the fact that geodesic flow contracts (or expands) orbits of the horocycle flow with the constant coefficient e^t. This plays important roles in the study of asymptotic behavior of both flows.[3] In addition to implying hyperbolicity of the geodesic flow, this also shows that the horocycle flow is *parabolic*, that is, characterized by polynomial behavior:

$$0 = [H_+, aX + bH_+ + cH_-] = (\dot{a} - 2c)X + (\dot{b} + a)H_+ + \dot{c}H_-$$

implies that as a function of t, c is constant ($\dot{c} = 0$), a is linear ($\dot{a} = 2c$), and b is quadratic ($\dot{b} = -a$).

We note that the bracket relation $[H_+, H_-] = -2X$ is also important because of its finitary counterpart, the *quadrilateral formula*:

$$\begin{pmatrix} 1 & 0 \\ -\epsilon s & 1 \end{pmatrix} \begin{pmatrix} 1 & (s-1)/\epsilon \\ 0 & 1 \end{pmatrix} \begin{pmatrix} 1 & 0 \\ \epsilon & 1 \end{pmatrix} \begin{pmatrix} 1 & (\frac{1}{s}-1)/\epsilon \\ 0 & 1 \end{pmatrix} = \begin{pmatrix} s & 0 \\ 0 & 1/s \end{pmatrix}$$
$$\text{or} \quad h_-^{(\frac{1}{s}-1)/\epsilon} h_+^{\epsilon} h_-^{(s-1)/\epsilon} h_+^{-\epsilon s} = g^{2\ln s}, \tag{2.2.4}$$

which is crucial below for mixing properties of the geodesic flow. Geometrically, this gives a *quadrilateral* argument: For $s = 1 + \epsilon^2$ this says that a quadrilateral with

[3]And well beyond this algebraic context (Section 8.6).

$h_\pm$-sides about ϵ causes a $2\epsilon^2$ displacement along a geodesic: we approximately have $h_-^{-\epsilon} h_+^{\epsilon} h_-^{\epsilon} h_+^{-\epsilon} \approx g^{2\epsilon^2}$. Moreover, for large s, this gives useful information by way of highly elongated quadrilaterals (Proposition 3.3.19).

In a different vein we note that h_+^s and h_-^s generate $\mathrm{PSL}(2,\mathbb{R})$ by (2.2.4).

Example 2.2.4 (The horizontal flow). The structure equations (2.2.1) are invariant under the exchange of $X \leftrightarrow H$, $V \leftrightarrow -V$, so $\xi^\pm := V \pm X = \mp[H,\xi^\pm]$, which implies hyperbolicity of the flow generated by H. It is given by

$$e^{\begin{pmatrix} 0 & 1/2 \\ 1/2 & 0 \end{pmatrix} t} = \begin{pmatrix} \cosh t/2 & \sinh t/2 \\ \sinh t/2 & \cosh t/2 \end{pmatrix},$$

which sends I to the semicircle with endpoints $\coth \frac{t}{2}$ and $\tanh \frac{t}{2}$ (these are reciprocals),[4] and the image $z(t) = \frac{i\cosh\frac{t}{2} + \sinh\frac{t}{2}}{i\sinh\frac{t}{2} + \cosh\frac{t}{2}}$ of i ranges over the upper half of the unit circle as t ranges over $\mathbb{R}$:

$$\overline{z(t)} = \overline{\frac{i\cosh\frac{t}{2} + \sinh\frac{t}{2}}{i\sinh\frac{t}{2} + \cosh\frac{t}{2}}} = \frac{i\sinh\frac{t}{2} + \cosh\frac{t}{2}}{i\cosh\frac{t}{2} + \sinh\frac{t}{2}} = \frac{1}{z(t)},$$

so $z(t)$ is indeed on the unit circle, and surjectivity is clear from

$$z(t) = \frac{i\cosh\frac{t}{2} + \sinh\frac{t}{2}}{i\sinh\frac{t}{2} + \cosh\frac{t}{2}} \frac{\cosh\frac{t}{2} - i\sinh\frac{t}{2}}{\cosh\frac{t}{2} - i\sinh\frac{t}{2}} = \frac{2\sqrt{1+\sinh^2\frac{t}{2}}\,\sinh\frac{t}{2} + i}{1 + 2\sinh^2\frac{t}{2}} \xrightarrow[t\to\pm\infty]{} \pm 1.$$

Geometrically, this flow can be described as follows: rotate a unit vector by $-\pi/2$, follow the corresponding geodesic for time t, then rotate the tangent vector back by $\pi/2$. Put differently, transport perpendicular vectors along geodesics. Presented this way, one sees that there is nothing special about "perpendicular" (Example 2.2.7).

2.2.c Reeb flow.

Let us describe a structure possessed by all geodesic flows, which is in the present case particularly easy to discern because of its algebraic nature.

Definition 2.2.5 (Contact form, Reeb flow). An (antisymmetric) n-form A on a smooth manifold M is a smooth map $A\colon TM^n \to \mathbb{R}$ that is linear in each fiber argument and antisymmetric. The exterior derivative dA of a 1-form A is the 2-form defined by

$$dA(X,Y) := \mathcal{L}_X A(Y) - \mathcal{L}_Y A(X) - A([X,Y]),$$

[4] Thus, the dynamics induced on the boundary circle $\mathbb{R} \cup \{\infty\}$ is north–south dynamics (Example 1.3.9, Figure 2.3.2).

where $\mathcal{L}$ is the Lie derivative and $[X,Y]$ is the Lie bracket. The *contraction operator* inserts a vector field in the first slot of a differential form:

$$\iota_X A := A(X,\dots) =: A \lrcorner X.$$

A 1-form A on a 3-manifold M is called a *contact form* if

$$(A \wedge dA)(X,Y,Z) := A(X)dA(Y,Z) - A(Y)dA(X,Z) + A(Y)dA(Z,X)$$

defines a volume form, that is, is nonzero at every point. (See also Section 2.6.d.) The associated plane field $\xi := \ker A$ is said to be a (cooriented) *contact structure*.

The *Reeb vector field* R_A associated to a contact form A is defined by $\iota_{R_A} A = A(R_A) = 1$ and $\iota_{R_A} dA = dA(R_A,\cdot) = 0$.[5] Its flow is called the *Reeb flow* (and it preserves the contact form because $\mathcal{L}_{R_A} A = \iota_{R_A} dA = 0$). Equivalently, R_A is the unique (up to a constant scalar factor) vector field that generates a flow which preserves the contact form. A *contact flow* is a flow that preserves a contact form.

In the case at hand, we can define a 1-form A uniquely by

$$A(X) = 1 \quad \text{and} \quad A(V) = 0 = A(H). \tag{2.2.5}$$

For $Z \in \{V,H\}$ we then have

$$dA(X,Z) = \underbrace{\mathcal{L}_X \overbrace{A(Z)}^{\equiv 0}}_{=0} - \underbrace{\mathcal{L}_Z \overbrace{A(X)}^{\equiv 1}}_{=0} - \underbrace{A(\overbrace{[X,Z]}^{\in -\{V,H\}})}_{=0} = 0,$$

so $\iota_X dA := dA(X,\cdot) \equiv 0$, and $X = R_A$, while $A \wedge dA(X,V,H) = A(X)dA(V,H) = 1$ since

$$dA(V,H) = \underbrace{\mathcal{L}_V \overbrace{A(H)}^{\equiv 0}}_{=0} - \underbrace{\mathcal{L}_H \overbrace{A(V)}^{\equiv 0}}_{=0} - \underbrace{A(\overbrace{[V,H]}^{=-X})}_{=-1} = 1.$$

Thus, $A \wedge dA$ is indeed a volume, in fact a volume particularly well adapted to this canonical framing. We have shown that the geodesic flow on $\mathbb{H}$ is a contact flow with A the canonical contact form.

The aforementioned symmetry of the structure equations implies that the horizontal flow from Example 2.2.4 is also a contact flow: Set $B(H) = 1$, $B(V) = 0 = B(X)$ and either repeat the preceding calculations or observe that by symmetry they work out to the same effect, notably $B \wedge dB(H,X,V) = 1$.[6] (Compare Exercise 2.3 below.)

[5] This is unique: the second condition determines R_A up to a scalar since dA is nondegenerate, and the first then fixes the scalar. Note that the Reeb vector field is associated to a contact form α rather than the contact structure: if $\alpha' = f\alpha$ with $f \in C^\infty(M, \mathbb{R} \setminus \{0\})$, then $d\alpha' = df \wedge \alpha + f d\alpha$, and the condition $\iota_{R_{\alpha'}} d\alpha' = 0$ implies that R_α and $R_{\alpha'}$ are not collinear unless f is constant. A Reeb field on a contact manifold (M,ξ) is the Reeb field of a (or any) contact form α with $\xi = \ker \alpha$. These are exactly the nowhere-vanishing vector fields transverse to ξ whose flow preserves ξ.

[6] In fact, $B = \iota_V dA$.

Example 2.2.6 (The vertical or fiber flow). In the PSL(2, $\mathbb{R}$)-representation of $S\tilde{\Sigma}$, the three flows corresponding to the vector fields of the canonical framing are given by

$$X \rightsquigarrow \exp\left(\begin{pmatrix} 1/2 & 0 \\ 0 & -1/2 \end{pmatrix} t\right) = \begin{pmatrix} e^{t/2} & 0 \\ 0 & e^{-t/2} \end{pmatrix},$$
$$H \rightsquigarrow \exp\left(\begin{pmatrix} 0 & 1/2 \\ 1/2 & 0 \end{pmatrix} t\right) = \begin{pmatrix} \cosh t/2 & \sinh t/2 \\ \sinh t/2 & \cosh t/2 \end{pmatrix},$$
$$V \rightsquigarrow \exp\left(\begin{pmatrix} 0 & -1/2 \\ 1/2 & 0 \end{pmatrix} t\right) = \begin{pmatrix} \cos t/2 & -\sin t/2 \\ \sin t/2 & \cos t/2 \end{pmatrix}.$$

We explore the dynamics of X and H below (Remark 2.2.11). The last of these three flows is called the *vertical* or *fiber* flow. Unlike the other two it is not hyperbolic because of a sign change in the symmetry; this is reflected in the trigonometric functions in its representation. It is a periodic flow because it consists of "spinning" around the tangent fibers.[7] The arguments $\frac{t}{2}$ give the right period, by the way: $\begin{pmatrix} \cos 2\pi/2 & -\sin 2\pi/2 \\ \sin 2\pi/2 & \cos 2\pi/2 \end{pmatrix} = \begin{pmatrix} -1 & 0 \\ 0 & -1 \end{pmatrix} \sim \begin{pmatrix} 1 & 0 \\ 0 & 1 \end{pmatrix} \bmod \pm 1$. The horizontal flow has an easy geometric interpretation:

$$\begin{pmatrix} \cos \pi/4 & -\sin \pi/4 \\ \sin \pi/4 & \cos \pi/4 \end{pmatrix}\begin{pmatrix} e^{t/2} & 0 \\ 0 & e^{-t/2} \end{pmatrix}\begin{pmatrix} \cos -\pi/4 & -\sin -\pi/4 \\ \sin -\pi/4 & \cos -\pi/4 \end{pmatrix} = \begin{pmatrix} \cosh t/2 & \sinh t/2 \\ \sinh t/2 & \cosh t/2 \end{pmatrix},$$

so "rotate $\pi/2$, follow the geodesic, rotate back $\pi/2$" or, in other words, translate a normal (rather than tangent) vector along a geodesic.

As before, one can check that $C := dA(H, \cdot)$ is a contact form invariant under the fiber flow generated by V but its Reeb field is $-V$, and $C \wedge dC(V, H, X) = -1$, so this volume has the opposite orientation from the ones defined by A and B and is hence not isotopic to either of them. The geodesic flow and the fiber flow are isotopic, however, via the magnetic flow construction from Example 1.1.29 (Remark 2.2.10), and Example 2.2.3 shows that the horocycle flow is the "midpoint" of an isotopy between the horizontal flow and the fiber flow (Remark 2.2.11) and the largest perturbation whose orbits reach the boundary (Figure 2.2.2).

Example 2.2.7 (A family of hyperbolic Reeb flows). After (2.2.2) and in Example 2.2.4 we noted that X and H generate hyperbolic flows, and that they are the Reeb flows for A and B, respectively. More generally,

$$E := E_\theta := \cos\theta A + \sin\theta B$$

[7]The counterpart to the earlier calculations is that $\xi^\pm := X \pm iH$ satisfies $[V, \xi^\pm] = \mp i\xi^\pm$, so a vector field $f\xi^\pm$ is V-invariant if and only if $0 = [V, f\xi^\pm] = \dot{f}\xi^\pm \mp if\xi^\pm$, that is, $\dot{f} = \pm if$ or $f = e^{\pm it}$.

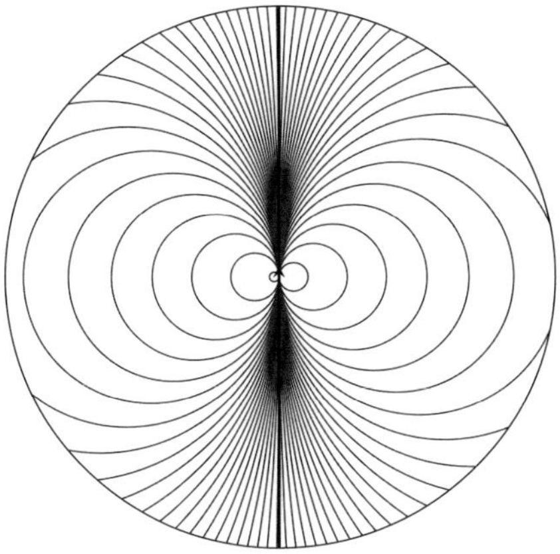

Figure 2.2.2. Magnetic perturbations of a geodesic diameter.

is a contact form with $R_E = P := \cos\theta X + \sin\theta H$, and $\zeta^\pm := \cos\theta H - \sin\theta X \pm V$ gives

$$[P, \zeta^\pm] = \mp\zeta^\pm \quad \text{so } 0 = [P, f\zeta^\pm] = (\dot f \mp f)\zeta^\pm \Rightarrow f = \text{const.}\, e^{\pm t},$$

as before. Thus R_E generates a family of hyperbolic flows parametrized by S^1. As suggested by our previous description of the horizontal flow, these consist of parallel translation along geodesics of vectors making an angle θ with the geodesic.

Remark 2.2.8. As a by-product of Example 2.2.7 we note that the horocycle flows as well are each part of an S^1-family of natural flows generated by $\zeta^\pm$.

Remark 2.2.9 (Hopf coordinates). We complement the infinitesimal version of hyperbolicity in Remark 2.2.2 by a description in *Hopf coordinates*. These are given by a homeomorphism

$$S\mathbb{D} \to \big(S^1 \times S^1 \smallsetminus \text{diagonal}\big) \times \mathbb{R}, \quad v \mapsto (v^-, v^+, \beta_{v^+}(0, \pi(v))),$$

where $v^\pm := \lim_{t\to\pm\infty} \gamma_v(t) \in \partial\mathbb{D} \sim S^1$, $\pi\colon S\mathbb{D} \to \mathbb{D}$ is the footpoint projection, and β is the *Busemann cocycle*

$$\mathbb{D} \times \mathbb{D} \times \partial\mathbb{D}, \quad x, y, \xi \mapsto \beta_\xi(x, y) := \lim_{t\to\infty} d(x, \xi_x(t)) - d(y, \xi_x(t)).$$

Here ξ_x is the geodesic with $\xi_x(0) = x$ and $\xi_x(t) \xrightarrow[t\to+\infty]{} \xi$. In these coordinates, the geodesic flow is given by $g^t(v^-, v^+, \tau) = (v^-, v^+, \tau + t)$, and it contracts the stable manifold (see Proposition 2.2.1 and Remark 2.1.7)

$$W^s(v^-, v^+, \tau) := \{(\xi, v^+, \tau) \mid \xi \in \partial\mathbb{D} \smallsetminus \{v^+\}\}.$$

Remark 2.2.10 (Magnetic flows). As in Example 1.1.29 (and as promised in Example 2.2.6) one can interpolate between the geodesic and horocycle flows as follows. The geodesic flow takes a tangent vector along a unit-speed curve with zero geodesic curvature, and the horocycle flow does the same thing along curves with geodesic curvature 1. The interpolation is to choose a different (constant) geodesic curvature to obtain other defining curves for a flow. This does, in fact, have a physical motivation in that while the geodesic flow models the motion of a force-free particle, constant nonzero geodesic curvature corresponds to the effect of a magnetic field perpendicular to the plane or disk on a charged particle, which is to produce constant acceleration perpendicular to the direction of motion and translates to constant geodesic curvature. These flows are called *magnetic flows*. (Note that depending on the orientation of the magnetic field one could drift right or left, which corresponds to making a consistent choice of horocycle, of which there are two through each tangent vector.) For a given initial tangent vector, increasing the intensity of the magnetic field (that is, geodesic curvature) produces ever smaller circles, which for curvature ± 1 just barely touch the boundary. These are the horocycles, and when one transports the normal rather than tangent unit vector, this is the horocycle flow (Example 2.2.3). A magnetic field that produces geodesic curvature greater than 1 produces motion along circles too small to reach the boundary, and therefore all orbits are periodic (as in Example 1.1.29), whereas none of the orbits are periodic for flows along curves of geodesic curvature between 1 and -1. To get periodic orbits for the geodesic flow requires passing to a compact factor. (We briefly return to magnetic flows on page 275.)

Let us briefly remark that in the spirit of Remark 1.6.17 we have here a continuous family (Definition 1.6.18) of flows and may be interested in how the dynamics changes as we make these deformations. Until the magnetic field, that is, the deviation from geodesic motion, becomes rather large, the flows look rather similar to each other. We will indeed see that for weak magnetic fields, any two of these flows are pairwise topologically orbit equivalent (Theorem 5.4.5).

Remark 2.2.11. In summary, a linear combination of X, H, V can be written as a linear combination of E_θ (from Example 2.2.7) and V and generates a flow whose orbits project to curves on $\mathbb{H}$ with constant geodesic curvature given in terms of the coefficient of V with vectors transported along them that make an angle θ with the tangent vector of the curve in the surface (the size of which is determined by the coefficient of E)—unless the linear combination is just V, in which the curve in the surface is a point (zero speed since E has coefficient 0). The special cases we noted earlier are generated by X, V, $H_\pm$, and H. We also noted that X, H, V generate contact flows but that the contact form for V is not isotopic to either of the other two.

2.3 Compact factors

Stable and unstable manifolds made their appearance earlier in Example 1.5.26 and Example 1.5.27, where they appeared as families of lines with irrational slope invariant under a hyperbolic automorphism of the 2-torus and its suspension. The existence of families of stable and unstable manifolds is a hallmark of global hyperbolic behavior; flows on compact manifolds with such behavior are called *Anosov flows* (Definition 5.1.1).

Therefore it is natural to utilize our understanding of hyperbolic behavior of the geodesics in the hyperbolic disk in order to construct first examples of such flows. All we need is to construct a compact factor of the hyperbolic disk and project the geodesic flow to that factor. We accomplish this by factoring out by a discrete group of isometries.

Draw a regular (hyperbolic) octagon Q in the Poincaré disk in $\mathbb{C}$ with vertices $v_k = de^{-k\pi i/4}$, $k = 0, \dots, 7$, joined by arcs of circles perpendicular to the unit circle (see Figure 2.3.1). We have $d \in (0, 1)$ and as $d \to 1$, the sum of the internal angles converges to 0, and it goes to 6π, the value for the Euclidean octagon, as $d \to 0$. This becomes clear by keeping d fixed and increasing the size of the Poincaré disk indefinitely so that the arcs of circles approach line segments. Thus, we can fix d

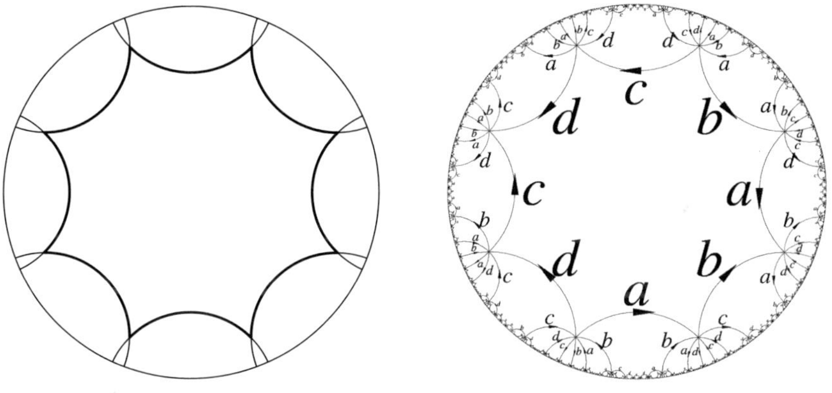

Figure 2.3.1. A hyperbolic octagon, identifications, and tiling by translates. [Reproduced from [213] (© 1995 Cambridge University Press, all rights reserved) and http://topologygeometry.blogspot.ch/2010/06/notes-from-062310.html with permission.]

such that the internal angles add up to 2π. The identification space obtained from labeling and identifying the edges as in Figure 2.3.1 is a surface Σ of genus 2. Since the internal angles of Q add up to 2π, the identification map is smooth at the vertices (which are all identified to one point), and we can therefore push the metric on Q down to Σ. We obtain a compact manifold which is locally isometric to $\mathbb{H}$. Topologically this manifold is homeomorphic to the double torus or the sphere with two handles: the half with labels a, b is a torus with a hole, and so is the other half; the hole is the common diameter along which these tori are glued together. One can also show that Σ is the space obtained by identifying orbits of the group Γ generated by the isometries mapping an edge to the one with which it is identified. In other words, the fundamental group of Σ can be identified with a discrete group Γ of hyperbolic linear fractional transformations.

Replacing 8 arcs here by $4g \geq 8$ arcs gives a metric locally isometric to that of $\mathbb{H}$ on the orientable surface of genus g (sphere with g handles).

If a linear fractional transformation γ preserves a geodesic then such a geodesic is unique and it is called the axis of γ. In fact, every $\gamma \in \Gamma$ has an axis. The projections of these geodesics to $M := \Gamma \backslash \mathbb{D}$ are precisely the closed geodesics of M. These are, of course, the projections of the closed orbits of the geodesic flow from the tangent bundle to M. The dynamics of any such γ (under iteration) restricts to a translation of the axis, and the action on the boundary circle is of north–south type much like in Example 1.3.9 as shown in Figure 2.3.2. The endpoints of the axis are fixed points, one repelling (γ^-) and one attracting (γ^+). Indeed,

$$\lim_{n \to \pm\infty} \gamma^n(x) = \gamma^{\pm} \text{ for every } x \in \mathbb{D} \cup \partial\mathbb{D} \smallsetminus \{\gamma^{\mp}\}.$$

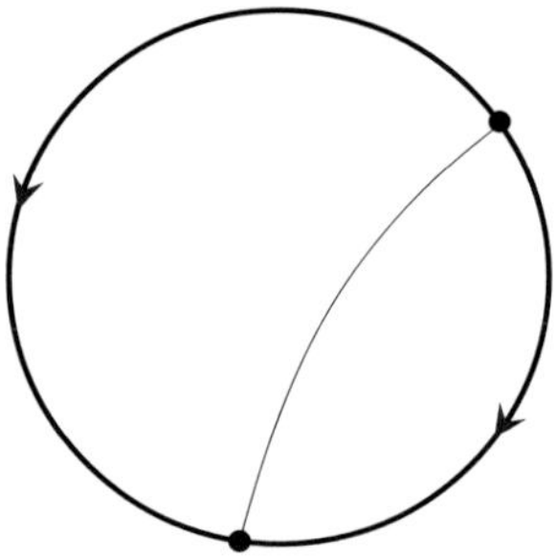

Figure 2.3.2. North–south dynamics on the boundary.

Associated to any C^2 Riemannian metric on a surface is the Gaussian curvature of the metric, an isometry-invariant real-valued function. Since the isometry group of $\mathbb{D}$ is transitive, the curvature of $\mathbb{D}$ is a constant k. Thus the induced metric on the compact factor Σ of genus 2 constructed from the octagonal fundamental domain has constant curvature k as well. The Gauss–Bonnet Theorem

$$k \cdot \operatorname{vol} M = 2\pi\chi$$

then shows that $k < 0$ because the Euler characteristic $\chi = 2 - 2g$ of Σ is negative. Conversely this then shows that any compact factor of $\mathbb{D}$ has negative Euler characteristic and hence genus at least 2. Thus the compact factors of $\mathbb{D}$ are homeomorphic to spheres with several handles attached. In fact, any compact orientable surface with a metric of constant negative curvature is isometric to a factor $\Gamma\backslash\mathbb{D}$ of $\mathbb{D}$ by a discrete group Γ of isometries of $\mathbb{D}$.

To see how the picture developed for the octagonal fundamental domain looks in the general case, consider a discrete group of orientation-preserving isometries of the Poincaré disk $\mathbb{D}$ which produces a compact factor. One can choose a fundamental domain for Γ by considering the *Dirichlet domain*

$$D := D_p := \{x \in \mathbb{D} \mid d(x,p) \leq d(x,\gamma p) \text{ for all } \gamma \in \Gamma\}$$

for any given point $p \in \mathbb{D}$. For any $\gamma \in \Gamma$ we evidently have $D_{\gamma p} = \gamma(D_p)$. The interiors of D_p and $D_{\gamma p}$ are disjoint when $\gamma \neq \mathrm{Id}$ and since Γ is discrete, there are only finitely many $\gamma \in \Gamma$ such that $D_p \cap D_{\gamma p} \neq \varnothing$. If $\gamma \in \Gamma$ is one of these elements, then $D_p \cap D_{\gamma p}$ consists of the points equidistant from p and γp, a geodesic segment. Thus D is a hyperbolic polygon, that is, bounded by finitely many geodesic arcs. The assumption that $\Gamma\backslash\mathbb{D}$ is compact means that D is compact. By construction we also observe that the sets $D_{\gamma p}$ cover $\mathbb{D}$, so we have, in fact, tessellated $\mathbb{D}$ by the images of D under Γ.

Compact factors of the hyperbolic plane cannot be embedded isometrically in $\mathbb{R}^3$ because a compact embedded surface has positive curvature at the points of contact with a circumscribed sphere. An illustration of an isometrically embedded surface of constant negative curvature is given by the pseudosphere in Figure 2.3.3.

2.4 The geodesic flow on compact hyperbolic surfaces

Unlike the geodesic flow on the round sphere and the flat torus considered in Section 1.1.c, where the dynamics turned out to be rather simple, compact factors of the hyperbolic plane have geodesic flows of a complicated dynamical nature rather similar to hyperbolic symbolic flows. The full extent of this similarity will become clear as we develop the theory of hyperbolic dynamical systems, and indeed, these very geodesic flows were and still are among the primary motivations for studying hyperbolic dynamical systems. Therefore their study here is a precursor to the central object and Part II of this book. Thus we now establish for the geodesic flow on compact factors of the hyperbolic plane some of the properties that we tend to consider typical for complicated dynamical behavior, namely, density of closed orbits, and topological transitivity.

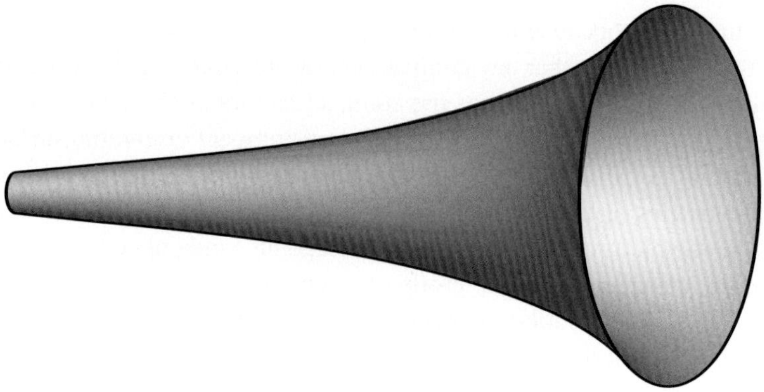

Figure 2.3.3. The pseudosphere. [Reprinted from [213] (© Cambridge University Press, all rights reserved) with permission.]

We first prove density of closed orbits:

Theorem 2.4.1 (Periodic orbits are dense). *Let Γ be a discrete group of fixed-point-free isometries of $\mathbb{D}$ such that $M := \Gamma\backslash\mathbb{D}$ is compact. Then the periodic orbits of the geodesic flow on SM are dense in SM.*

Proof. We use the model of the Poincaré disk $\mathbb{D}$. Let $v \in SM$, take a Dirichlet domain D for Γ, and let $w \in S\mathbb{D}$ be a lift of v with footpoint in D. Let c be the geodesic with $\dot{c}(0) = w$ in $\mathbb{D}$ and let x and y be the endpoints of c on the boundary of the Poincaré disk. Our strategy is to find a hyperbolic element $\gamma \in \Gamma$ such that the endpoints of its axis lie in given small δ-neighborhoods U and V, respectively, of the points $x =: c(-\infty)$ and $y =: c(\infty)$. Then among the tangent vectors to this axis one can find a vector that is close to w. The projection of the axis to M is the desired closed geodesic.

Minimality of the action of Γ on $\partial\mathbb{D}$ is the first step:

Lemma 2.4.2. *No proper closed subset of $\partial\mathbb{D}$ is invariant under the action of Γ.*

Proof. If $F \subset \partial\mathbb{D}$ is closed and Γ-invariant, then so is its convex hull E in $\mathbb{D}$, that is, the intersection of all hyperbolic half-spaces that contain F, and so also is the function $\delta(x) := d(x, E)$ on $\mathbb{D}$. Thus, δ is well defined on the quotient and hence bounded—and identically zero (otherwise it is positive on a point of a geodesic orthogonal to the boundary of E and hence unbounded). Thus $F = \partial\mathbb{D}$. □

This implies that the set of endpoints of axes in $\partial\mathbb{D}$ is dense, so we can find $\gamma, \eta \in \Gamma$ such that $\gamma^+ \in U$ and $\eta^- \in V$. If $\gamma = \eta$, we are done. Otherwise we may assume that $\gamma^{\pm}$, $\eta^{\pm}$ are four distinct points, and we will show that $\gamma^n\eta^n$ for large enough n is the desired isometry by using the north–south dynamics from Figure 2.3.2. If $W_\gamma \subset \partial\mathbb{D}$ is a neighborhood of γ^- and $W_\eta \subset \partial\mathbb{D}$ is a neighborhood of η^+ such that the closures of both of these and of U and V are pairwise disjoint, take $n \in \mathbb{N}$ such that $\eta^n(U) \subset W_\eta$ and $\gamma^n(\partial\mathbb{D} \smallsetminus W_\gamma) \subset U$, then $\gamma^n\eta^n(U) \subset U$, so $\gamma^n\eta^n$ has a (necessarily attracting) fixed point in U. Likewise $\eta^{-n}\gamma^{-n}$ (for possibly larger n) has an attracting fixed point in V. □

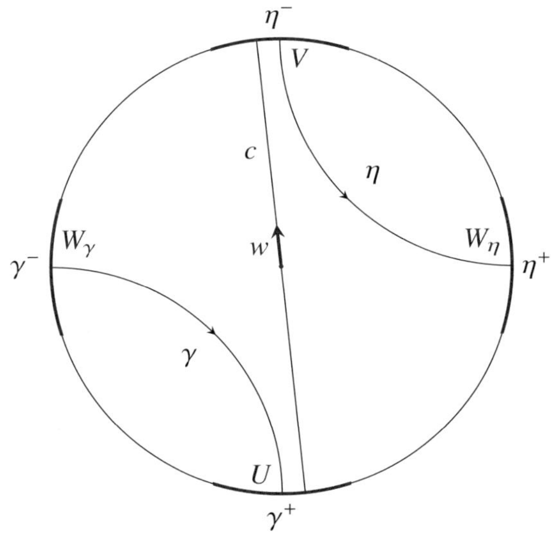

Figure 2.4.1. Ping-pong in the proof of Theorem 2.4.1.

Remark 2.4.3. The interaction of γ and η is sometimes called "playing ping-pong."

In the present context the multitude of closed orbits is "organized" rather neatly by the topology of the surface: since these orbits are based on parametrized geodesics, they can be represented by those geodesics on the manifold itself, and as a consequence of having negative curvature there is at most one closed geodesic in each free homotopy class of loops, in fact, exactly one by a curve-shortening argument in each class. This means that likewise the periodic orbits in the unit tangent bundle are in pairwise different free homotopy classes except for the duplication of a geodesic with its reverse (the homotopy being the rotation of the tangent vector by π).

We emphasize that density of closed geodesics as orbits in the *phase space* is rather stronger than density of closed geodesics on the underlying surface—indeed, the latter is generic (Definition 1.5.19) for *any* surface [201, Remark 1.6].

Theorem 2.4.4 (Transitivity). *Let Γ be a discrete group of fixed-point-free isometries of $\mathbb{D}$ such that $M := \Gamma\backslash\mathbb{D}$ is compact. Then the geodesic flow on SM is topologically transitive (see Proposition* 1.6.9).

Remark 2.4.5. Again we emphasize that this asserts the existence of an orbit that is dense in the unit tangent bundle, which is stronger than the assertion that it traces a dense geodesic in the surface.

Proof. By Theorem 2.4.1 and Proposition 1.6.9 it is sufficient to show that for any two periodic points $u, v \in SM$ (whose lifts to $\mathbb{D}$ we also denote by u and v) and neighborhoods U, V of u, v, respectively, there is $t \in \mathbb{R}$ such that $g^t(U) \cap V \neq \varnothing$. Take the geodesics c_u and c_v in $\mathbb{D}$ with $\dot{c}_u(0) = u$ and $\dot{c}_v(0) = v$. Replacing, if necessary, u by γu assume that $c_u(-\infty) \neq c_v(\infty)$, then denote by c the geodesic with endpoints $c(-\infty) = c_u(-\infty)$ and $c(\infty) = c_v(\infty)$. By Proposition 2.2.1 we can find

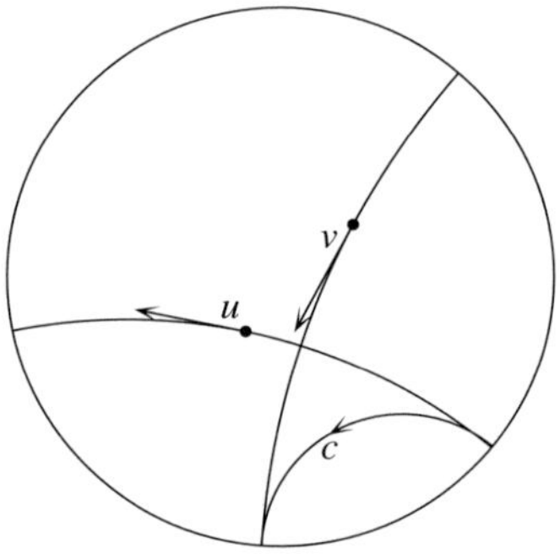

Figure 2.4.2. Transitivity of the geodesic flow. [Reprinted from [213] (© Cambridge University Press, all rights reserved) with permission.]

for each $t \in \mathbb{R}$ numbers $f(t), g(t) \in \mathbb{R}$ such that $d(\dot{c}_u(f(t)), c(t)) \xrightarrow[t\to-\infty]{\text{exponentially}} 0$ and $d(\dot{c}_v(g(t)), c(t)) \xrightarrow[t\to\infty]{\text{exponentially}} 0$. Since $\dot{c}_u$ and $\dot{c}_v$ project to closed orbits of the geodesic flow this shows that there exist t_1 and t_2 such that the projection of $\dot{c}(t_1)$ to SM is in U and the projection of $\dot{c}(t_2)$ to SM is in V. This then yields the claim. □

Remark 2.4.6 (Mixing). As noted earlier, this geodesic flow is actually topologically mixing (Exercise 2.7, Remark 7.1.13, and Corollary 8.1.6) [191, Theorem 3.1].

Furthermore, Remark 2.1.12 implies the following theorem:

Theorem 2.4.7 (Expansivity). *The geodesic flow on $\mathbb{H}$ or $\mathbb{D}$ or any factor is* expansive (*Definition* 1.7.2).

Returning attention from compact factors to the universal cover, it is instructive to go further and consider the universal cover of the unit circle bundle $S\mathbb{D}$ (rather than the circle bundle of the universal cover $\mathbb{D}$). Unrolling the circle fibers shows that topologically this is $\mathbb{D} \times \mathbb{R}$, and Figure 2.4.3 shows a way to here visualize the sets of geodesics positively or negatively asymptotic to a given boundary point. We choose to represent the set of geodesics positively asymptotic to a given boundary point as a $\mathbb{D}$-slice in the picture, shown here in red with those geodesics rendered as straight lines. In that case, the set of geodesics negatively asymptotic should be represented as in the green cross section in the figure to show that the boundary point to which each geodesic is negatively asymptotic varies over the boundary circle in a way that corresponds to an interval's worth of red slices. An interesting consequence is that the red and green sets shown here do not intersect; each green slice meets a bounded interval of red slices, and, vice versa, each red slice meets a circle minus a point worth of green sections. This is in contrast with the global product structure of a suspension (Remark 1.5.28).

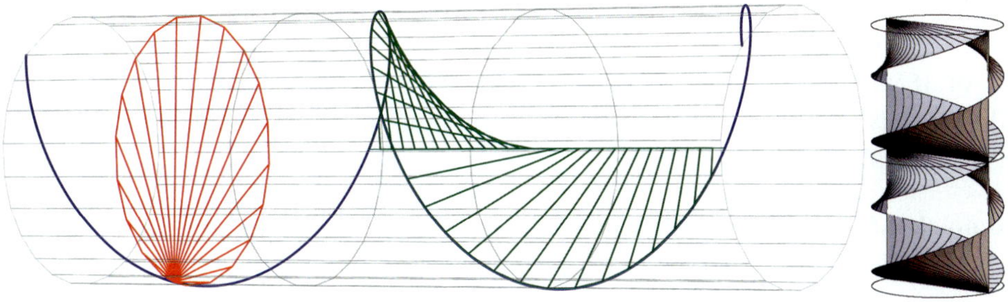

Figure 2.4.3. The universal cover of $S\mathbb{D}$. Left (after Barthelmé): the (red) "flat" and (green) "spindle" fans do not intersect. Right (after Tsuboi): still picture from animation. [Animation at http://www.tsuboiweb.matrix.jp/showroom/public_html/animations/gif/geodflow/geodflowconf.html.]

Remark 2.4.8. This geodesic flow is the original instance of an *Anosov flow* (Definition 5.1.1), which we study more carefully below. In that context, topological transitivity implies density of periodic orbits via a mechanism central to the study of their dynamics (shadowing, Section 5.3). Moreover, this geodesic flow is not only topologically transitive, but has strong ergodic properties (see for example, Theorem 7.1.12). These in turn imply that it is also topologically mixing (Definition 1.6.31).

In addition, we will later describe surgeries that produce new (contact) flows from the three flows in Remark 2.2.2 and Example 2.2.4. Those turn out to have some profoundly different features from the ones we studied here.

2.5 Symmetric spaces

An important class of manifolds of negative curvature is obtained by an algebraic construction which generalizes the algebraic description of surfaces of constant negative curvature. This involves a substantial amount of differential geometry and Lie theory and is not required for other parts of this book.

The geometric property that enabled us to describe the geodesic flow on the sphere, the torus, and the hyperbolic plane was the presence of an isometry group that is transitive on unit tangent vectors. In general, such spaces are called (globally) symmetric spaces. We begin with the traditional definition and then prove transitivity of the isometry group in the case of nonvanishing curvature.

Definition 2.5.1. A *Riemannian locally symmetric space* is a connected Riemannian manifold M such that for all $p \in M$ there is a neighborhood U on which $\exp_p \circ (-\mathrm{Id}) \circ \exp_p^{-1} \colon U \to M$ is an isometry. We call M a *globally symmetric space* if this local isometry extends to an isometry of M, that is, for every $p \in M$ there is an isometry σ_p of M with $\sigma_p(p) = p$ and $D\sigma_{p\,|p} = -\mathrm{Id}$. We call σ_p the *(global) symmetry* at p. The space is said to have *rank* 1 if there is no isometrically embedded totally geodesic Euclidean plane.

Remark 2.5.2. (1) An alternative definition is that the curvature tensor is parallel, that is, $\nabla R = 0$ and the space is simply connected.

(2) Since the endpoints of any geodesic segment are exchanged by the symmetry at the midpoint and any two points are connected by a broken geodesic, the isometry group of a globally symmetric space or compact locally symmetric space is clearly transitive on points.

(3) Having rank 1 implies that all sectional curvatures are nonzero.

(4) S^n, $\mathbb{R}^n$, $\mathbb{H} = \mathbb{RH}^2$ are globally symmetric spaces; $\mathbb{T}^n$ is locally symmetric.[8]

(5) A complete simply connected locally symmetric space is globally symmetric.

(6) Thus the universal cover of a complete locally symmetric space is a globally symmetric space.

Proposition 2.5.3. *If M is a rank-*1 *symmetric space then the isometry group is transitive on SM.*

[8] $\mathbb{T}^n$ is (globally) symmetric if one adopts the existence of a global symmetry as the definition, but not if simple-connectedness is required.

Proof. Since transitivity on points is known we only need to show that the isometry group is transitive on any particular unit sphere S_pM. To that end it suffices to show that for every 2-plane $\Pi \subset T_pM$ the isometry group is transitive on $\Pi \cap S_pM$, which in turn follows once we see that there exists an $\epsilon > 0$ such that for $v \in \Pi \cap S_pM$ there exists a family of isometries such that the images of v under their differentials cover an arc of length ϵ in $\Pi \cap S_pM$.

To that end consider a disk $D = \exp_p B(0, \delta)$ and a triangle in D with p as one vertex and interior angles α, β, γ. Consider the isometry I obtained by composing the three symmetries about the midpoints of the edges (in cyclic order). Since isometries preserve angles one easily sees by a picture that the angle between v and $DI(v)$ is $\alpha + \beta + \gamma$. Since Π has nonzero curvature the sum $\alpha + \beta + \gamma$ converges to π as the diameter of the triangle tends to 0 but it never equals π. Thus we obtain an arc of images whose size is independent of v. □

All symmetric spaces arise from an algebraic construction which generalizes the construction in the preceding subsections. To give an indication of how this comes about we begin with a direct generalization of a *geometric* construction of the hyperbolic space.

The Poincaré disk with the group of Möbius transformations can be obtained as follows. Consider the upper sheet $\mathcal{H}$ of the hyperboloid in $\mathbb{R}^3$ given by $Q(x) := x_1^2 + x_2^2 - x_3^2 = -1$, $x_3 > 0$. The group $\mathrm{SO}(2,1)$ of real 3×3 matrices preserving the indefinite quadratic form Q acts on the hyperboloid, and the index-2 subgroup preserving $x_3 > 0$ therefore acts on $\mathcal{H}$. Since the action is linear in $\mathbb{R}^3$ it sends planes through 0 (that is, planes given by $ax_1 + bx_2 - cx_3 = 0$) to planes through 0 and hence the family C of curves given by the intersection of such planes with $\mathcal{H}$ is preserved.

If we change variables to $\eta_1 = x_1/x_3$, $\eta_2 = x_2/x_3$, $\eta_3 = 1/x_3$ the hyperboloid becomes the hemisphere $\eta_1^2 + \eta_2^2 + \eta_3^2 = 1$, $\eta_3 > 0$ and a plane $ax_1 + bx_2 - cx_3 = 0$ is mapped to the plane $a\eta_1 + b\eta_2 = c$ perpendicular to the η_1–η_2 plane. Thus curves from C are mapped to circles orthogonal to the equator $\eta_3 = 0$. Finally, apply the stereographic projection centered at $(0, 0, -1)$ from the upper hemisphere to the disk $\eta_1^2 + \eta_2^2 < 1$. It is known to be conformal, so the curves from C now are (lines and) circles perpendicular to the boundary, that is, the geodesics of the Poincaré disk. One can show that the transformations that arise from $\mathrm{SO}(2,1)$ in this process are exactly the Möbius transformations. In fact, the hyperboloid is an isometric embedding of the Poincaré disk into Minkowski space $(\mathbb{R}^3, q)$ with the pseudometric q induced by the form Q.

This geometric construction generalizes to give n-dimensional real hyperbolic spaces $\mathbb{RH}^n$. Consider the upper sheet of the hyperboloid $\mathcal{H}$ in $\mathbb{R}^{n+1}$ given by $Q(x) := x_1^2 + \cdots + x_n^2 - x_{n+1}^2 = -1$, $x_{n+1} > 0$. Again let C be the family of curves that lie on planes through 0, that is, on planes given by n simultaneous equations of the form $a_1x_1 + \cdots + a_nx_n - a_{n+1}x_{n+1} = 0$. The group $\mathrm{SO}(n,1)$ of matrices preserving Q acts on $\mathcal{H}$. Change variables to $\eta_1 = x_1/x_{n+1}, \ldots, \eta_n = x_n/x_{n+1}, \eta_{n+1} = 1/x_{n+1}$ and then apply the stereographic projection centered at $(0,\ldots,0,-1)$ to map the resulting hemisphere to the open unit ball in $\mathbb{R}^n$. As before curves in C map to (lines and) circles perpendicular to the boundary of the unit ball $\mathbb{RH}^n$.

These spaces $\mathbb{RH}^n$ have (sectional) curvature -1 as well. This is clear for all tangent planes Π at $(0,\ldots,0,1)$ since in the 3-dimensional subspace of $\mathbb{R}^{n+1}$ containing Π the entire picture looks like the description of $\mathbb{RH}^2$.

For purposes of generalization it is more convenient to view $\mathbb{RH}^n$ as a subset of the n-dimensional real projective space $\mathbb{RP}^n$ of lines through 0 in $\mathbb{R}^{n+1}$ by identifying a point p on the upper hyperboloid with the line through 0 containing p. The Riemannian metric is, of course, not the induced one, but the tangent vectors to $\mathbb{RH}^n$ are tangent vectors of $\mathbb{RP}^n$. Hyperbolic distances are given as follows. Two points in this space correspond to two lines in $\mathbb{R}^{n+1}$. The plane defined by these intersects the cone $Q = 0$ in two more lines. The hyperbolic distance is given by the logarithm of the cross ratio of the four points in projective space determined by these four lines.

This latter description works over the complex field $\mathbb{C}$ as well. We obtain the n-dimensional complex hyperbolic space $\mathbb{CH}^n$ as a subset of complex projective space $\mathbb{CP}^n$, that is, the space of complex lines through the origin of $\mathbb{C}^{n+1}$, with a distance similarly defined by cross ratios. There is an important new phenomenon, however. Any tangent space can be viewed simultaneously as an n-dimensional complex linear space or a $2n$-dimensional real linear space. Thus a real vector v in a tangent space can be multiplied by $i = \sqrt{-1}$ to give a unique direction that is perpendicular to v with respect to the real structure but collinear to v with respect to the complex structure. One can check that this real 2-dimensional subspace has (sectional) curvature -4 and that multiplication by i is an isometry of the unit tangent bundle. Thus one has a natural real 1-dimensional subbundle on the unit tangent bundle $S\mathbb{CH}^n$ given by these directions. There is naturally a complementary subbundle defined by the vectors that are complex orthogonal to v and iv. Inside this subbundle all sectional curvatures are -1. This subbundle turns out to be nonintegrable.

For the geodesic flow these subbundles correspond to subbundles of vectors with expansion rates e^{2t} and e^t, respectively, and corresponding contraction rates.

For the quaternions $\mathbb{Q}$ one obtains hyperbolic spaces $\mathbb{QH}^n$ with a similar structure, but here one obtains a (real) 3-dimensional subbundle corresponding to planes of curvature -4. Even for the octonians $\mathbb{O}$ (Cayley numbers) one obtains a hyperbolic plane $\mathbb{OH}^2$, here with a corresponding 7-dimensional subbundle. The last construction, however, does not extend to higher dimension due to nonassociativity of the Cayley numbers. These examples in fact exhaust the list of Riemannian globally symmetric spaces of negative curvature. All of these spaces admit compact Riemannian factors obtained by the left action of a uniform lattice in the isometry group, so the geodesic flows on such factors provide examples of Anosov geodesic flows.

We now give, also without proof, an indication of the general *algebraic* description of globally symmetric spaces.

Proposition 2.5.4. *If M is a globally symmetric space, then the identity component G of the isometry group of M acts transitively on M and the isotropy group K of any point is compact.*

Definition 2.5.5. A globally symmetric space M is said to be of *noncompact type* if G is semisimple with no compact factors and K is a maximal compact subgroup of G.

Remark 2.5.6. Unlike in the case of $\mathbb{RH}^2$ the group G for other globally symmetric spaces of rank 1 is substantially larger than the unit tangent bundle of the manifold we are considering.

Conversely, for every connected semisimple Lie group with no compact factors and a maximal compact subgroup K (which is unique up to conjugacy by an inner automorphism of G) there is a natural globally symmetric structure on $M := G/K$, namely, every left-invariant Riemannian metric on G that is right invariant under K then makes M a Riemannian manifold and the quotient of M under the left action of a lattice Γ in G is a compact Riemannian factor of M. This is the analog of the torus and compact factors of the hyperbolic plane $\mathbb{RH}^2$.

In this model geodesics through Id are given by 1-parameter subgroups of G/K.

The general algebraic description of the geodesic flow on rank-1 Riemannian symmetric spaces of noncompact type is as follows. Let G be a simple noncompact Lie group of real rank 1. Such groups are $\mathrm{SO}(n,1)$, $\mathrm{SU}(n,1)$, $Sp(n,1)$, and F_4. Let K be a maximal compact subgroup of G. Then G/K is a globally symmetric space and its unit tangent bundle is of the form G/T, where T is a compact subgroup of K (namely, the isotropy subgroup of a tangent vector). The symmetric spaces are, correspondingly, n-dimensional real, complex, and quaternionic hyperbolic spaces and the 2-dimensional hyperbolic Cayley plane. The geodesic flow corresponds to the right action of a 1-parameter subgroup that commutes with T. (Note that in the 2-dimensional case $T = \{\mathrm{Id}\}$.)

The algebraic description of the geodesic flow on the hyperbolic plane and its factors allows another remarkable generalization to higher dimension. The idea is simply to replace $\mathrm{SL}(2,\mathbb{R})$ with $\mathrm{SL}(n,\mathbb{R})$ for larger n. For $n = 2$, as we have seen, the geodesic flow appears as the action of the positive diagonal subgroup; the natural generalization would be the following:

Definition 2.5.7. The right action of the positive diagonal subgroup

$$D_n^+ = \Bigg\{ \underbrace{\begin{pmatrix} \exp t_1 & & \\ & \ddots & \\ & & \exp t_n \end{pmatrix}}_{=:\mathrm{diag}(\exp t_1,\dots,\exp t_n)} \mid (t_1,\dots,t_n) \in \mathbb{R}^n,\ \textstyle\sum_{k=1}^n t_k = 0 \Bigg\} \cong \mathbb{R}^{n-1}$$

on $\mathrm{SL}(n,\mathbb{R})$ and its compact factors is called the *Weyl-chamber flow*.

This is our first example of an action of a higher-rank abelian group. Since it appears as a generalization of the geodesic flow on a surface $\Gamma/\mathbb{H}^2$ (an Anosov flow, Definition 5.1.1) it is natural to expect that its elements exhibit hyperbolic behavior. Note first that since all diagonal matrices commute, every element of the Weyl-chamber flow acts by isometries with respect to any left-invariant metric on $\mathrm{SL}(n,\mathbb{R})$ and hence to its projection to $\Gamma/\mathrm{SL}(n,\mathbb{R})$. Thus we should expect hyperbolicity *transverse* to the orbit direction.

Consider the 1-parameter unipotent subgroup $u_{ij}(t) = \mathrm{Id} + tn_{ij}$ where the (i,j) entry of n_{ij} is 1 and all others are 0, and let W_{ij} be the foliation into left cosets of this subgroup. An explicit calculation gives

$$\underbrace{\mathrm{diag}(e^{t_1},\dots,e^{t_n})}_{=:G_{t_1,\dots,t_n}} u_{ij}(s)\,\mathrm{diag}(e^{-t_1},\dots,e^{-t_n}) = u_{ij}(se^{t_i-t_j}),$$

that is,

$$G_{t_1,\dots,t_n} H_s^{ij} G_{-t_1,\dots,-t_n} = H^{ij}_{s\exp(t_i-t_j)}, \tag{2.5.1}$$

if we denote by H_t^{ij} the right multiplication by $u_{ij}(t)$. The dynamical interpretation of (2.5.1) is that the element $G_{t_1,\dots,t_n}$ of the Weyl-chamber flow preserves the foliation W_{ij} and expands or contracts its leaves with coefficient $e^{t_i-t_j}$ depending on whether $i > j$ or $i < j$—much as the geodesic flow expands the horocycles from one family and contracts those from the other.[9]

Thus for all elements $G_{t_1,\dots,t_n}$ with $t_1 > t_2 > \dots > t_n$ the stable and unstable foliations are the same; the set of such elements (or their indices) is called the *positive Weyl-chamber*. When the sign of $t_i - t_j$ changes, the pair of foliations W_{ij} and W_{ji} switch roles. This is an essential higher-rank effect. A *Weyl-chamber* is the subset of $D_n^+ \cong \mathbb{R}^{n-1}$ where all differences $t_i - t_j$ are nonzero and have the same sign.

The Weyl-chamber flow is a generalization of the geodesic flow on the symmetric space of the group $\mathrm{SL}(2,\mathbb{R})$, which can be described as the homogeneous space $\mathrm{SL}(2,\mathbb{R})/\mathrm{SO}(2)$ provided with a Riemannian metric that is projected from a left-invariant metric on $\mathrm{SL}(2,\mathbb{R})$ that is also $\mathrm{SO}(2)$ right invariant. Hence one may

[9] If the t_i for $i = 1,\dots n$ are pairwise different, then $G_{t_1,\dots,t_n}$ is partially hyperbolic in the sense of Definition B.5.1 with the neutral direction being that of the orbits of Weyl-chamber flow.

naturally ask about connections between the Weyl-chamber flow and the geodesic flow on the symmetric space $SL(n,\mathbb{R})/SO(n)$ provided with a Riemannian metric that is projected from a left-invariant metric on $SL(n,\mathbb{R})$ that is also $SO(n)$ right invariant. It turns out that the latter geodesic flow has $n-1$ commuting (Definition 2.6.18) first integrals (Definition 1.1.24) and the restriction of the geodesic flow to any regular value of those integrals is smoothly conjugate to a Weyl-chamber flow with properly chosen generators.

Moreover, the Weyl-chamber flow provides the main instance of an algebraic $\mathbb{R}^k$-action, whose smooth rigidity is established in Theorem 9.1.16.

2.6 Hamiltonian systems

Both in this algebraic instance and when they appear in greater generality, it is useful to have a framework for describing geodesic flows as mechanical or Hamiltonian systems rather than solely focusing on their geometric origin. This section gives a brief axiomatic introduction to the modern approach to Hamiltonian dynamics.

2.6.a Symplectic geometry. The natural geometry for describing Hamiltonian systems is an antisymmetric counterpart to a Riemannian metric. Accordingly, we begin with nondegenerate antisymmetric 2-forms on linear spaces.

Definition 2.6.1. Let E be a linear space. A 2-tensor $\alpha\colon E\times E\to\mathbb{R}$ is said to be *nondegenerate* if $\alpha^\flat\colon v\mapsto\alpha(v,\cdot)$ is an isomorphism from E to its dual space E^*. It is said to be *antisymmetric* or *skew symmetric* if $\alpha(v,w)=-\alpha(w,v)$. A nondegenerate antisymmetric 2-form is called a *symplectic* form. A linear space with a symplectic form is called a *symplectic* vector space. If (E,α), (F,β) are symplectic vector spaces then a linear map $T\colon E\to F$ is said to be *symplectic* if $T^*\beta=\alpha$.

Remark 2.6.2. If a scalar product $\langle\cdot,\cdot\rangle$ on E is fixed we can write $\alpha(\cdot,\cdot)=\langle\cdot,A\cdot\rangle$, so we identify the tensor with its matrix representation with respect to a given basis.

Proposition 2.6.3. *Let E be a linear space. If α is a symplectic form on E then* $\dim E=2n$ *for some $n\in\mathbb{N}$ and there is a basis $e_1,\dots,e_{2n}$ of E such that $\alpha(e_i,e_{n+i})=1$ if $i=1,\dots,n$ and $\alpha(e_i,e_j)=0$ if $|i-j|\neq n$. Hence, if one fixes a scalar product with respect to which $e_1,\dots,e_{2n}$ is an orthonormal basis, then $A=\begin{pmatrix}0&I\\-I&0\end{pmatrix}$ with respect to this basis, where I is the $n\times n$ identity matrix.*

Proof. Since α is nondegenerate there exist e_1, e_{n+1} such that $\alpha(e_1,e_{n+1})\neq0$, and we may take $\alpha(e_1,e_{n+1})=1$. By antisymmetry $\alpha(e_1,e_1)=\alpha(e_{n+1},e_{n+1})=0$ and

$\alpha(e_{n+1}, e_1) = -1$, so the matrix of $\alpha\restriction_{E_1}$, where $E_1 = \operatorname{span}\{e_1, e_{n+1}\}$, with respect to (e_1, e_{n+1}) is $\left(\begin{smallmatrix} 0 & 1 \\ -1 & 0 \end{smallmatrix}\right)$. The claim follows by induction on dimension: If $v \in E$, then

$$v - \alpha(v, e_{n+1})e_1 + \alpha(v, e_1)e_{n+1} \in E_2 := \{v \in E \mid \alpha(v, w) = 0 \text{ for all } w \in E_1\},$$

so $E_1 \oplus E_2 = E$ since $E_1 \cap E_2 = \{0\}$. □

Definition 2.6.4. A subspace V of a symplectic linear space (E, α) is said to be *isotropic* if $\alpha\restriction_V = 0$ and *Lagrangian* if furthermore $\dim V = \dim E/2$.

Remark 2.6.5. Thus the "adapted" basis of Proposition 2.6.3 gives a decomposition of E as a direct sum of two Lagrangian subspaces. Note that by nondegeneracy of α an isotropic subspace has dimension at most $\dim E/2$, so Lagrangian subspaces are maximal isotropic subspaces.

An interesting description of nondegeneracy is the following:

Proposition 2.6.6. *An antisymmetric 2-form α on a linear space E is nondegenerate if and only if $\dim E = 2n$ and the nth exterior power α^n of α is not zero.*

Proof. "$\Leftarrow$": If α is degenerate then $\alpha^\flat$ has nontrivial kernel, that is, there is a vector v such that $\alpha(v, w) = 0$ for all w, hence
$\alpha^n(v, v_2, \dots, v_n) = 0$ for all $v_2, \dots, v_n$.
"$\Rightarrow$": If α is nondegenerate write $\alpha = \sum_{i=1}^n dx_i \wedge dx_{i+n}$ by Proposition 2.6.3. Then

$$\alpha^n = \sum_{i_1,\dots,i_n=1}^{n} dx_{i_1} \wedge dx_{i_1+n} \wedge \dots \wedge dx_{i_n} \wedge dx_{i_n+n} = n!(-1)^{[n/2]} dx_1 \wedge \dots \wedge dx_{2n} \neq 0. \quad \square$$

The following proposition is an immediate observation from the preceding results:

Proposition 2.6.7. *If $T: (E, \alpha) \to (F, \beta)$ is a symplectic map, then T preserves volume and orientation. In particular T is invertible with Jacobian* 1.

Thus the set of symplectic maps $(E, \alpha) \to (E, \alpha)$ is a group which we call the *symplectic group* of (E, α). Assume a scalar product $\langle \cdot, \cdot \rangle$ is fixed and α is in standard form $J = \left(\begin{smallmatrix} 0 & I \\ -I & 0 \end{smallmatrix}\right)$. Here are some further simple properties of symplectic maps.

Proposition 2.6.8. *Suppose (E, α) is a symplectic vector space and $T: (E, \alpha) \to (E, \alpha)$ a symplectic map. If λ is an eigenvalue of T, then so are $\bar{\lambda}$, $1/\lambda$, $1/\bar{\lambda}$. If T has the form $\left(\begin{smallmatrix} A & B \\ C & D \end{smallmatrix}\right)$ with respect to a basis for which $\alpha(v, w) = \langle v, Jw \rangle$, then A^tC and B^tD are symmetric, and $A^tD - C^tB = I$.*

Proof. If T preserves α, and $\alpha(v,w) = \langle v, Jw\rangle$ then symplecticity means $T^t JT = J$. By calculation this implies that $A^t C$ and $B^t D$ are symmetric and $A^t D - C^t B = I$. If λ is an eigenvalue then so is $\bar{\lambda}$ since the characteristic polynomial $P(\lambda) = \det(T - \lambda I)$ has real coefficients. Furthermore $JTJ^{-1} = (T^{-1})^t$, so

$$\begin{aligned} P(\lambda) = \det(T - \lambda I) &= \det(J(T - \lambda I)J^{-1}) = \det(T^{-1})^t(I - \lambda T^t) \\ &= \det((I - \lambda T)T^{-1}) = \det(\lambda(\lambda^{-1}I - T)) = \lambda^{2n}P(\lambda^{-1}); \end{aligned}$$

hence, since 0 is not an eigenvalue, $P(\lambda) = 0$ if and only if $P(1/\lambda) = 0$. □

Exercise 2.13 gives an appropriate version of a converse to this result.

Now we discuss symplectic forms on manifolds.

Definition 2.6.9. Let M be a smooth manifold. A differential 2-form ω is a smooth map from M to the space $\bigwedge^2 T^*M$ of antisymmetric 2-tensor fields, that is, it assigns to each $x \in M$ an antisymmetric 2-tensor on T_xM. A differential 2-form ω is said to be *nondegenerate* if it is nondegenerate at every point. A nondegenerate 2-form ω with $d\omega = 0$ is called a *symplectic form.* A pair (M, ω) of a smooth manifold and a symplectic form is called a *symplectic manifold.* If (M, ω) is a symplectic manifold then a subbundle of the tangent bundle TM of M is said to be *isotropic* if at every point $p \in M$ it defines an isotropic subspace of T_pM, and *Lagrangian* if at every point $p \in M$ it defines a Lagrangian subspace of T_pM. A smooth submanifold of a symplectic manifold is said to be *isotropic* if its tangent bundle is an isotropic subbundle, and *Lagrangian* if its tangent bundle is a *Lagrangian* subbundle of TM. A diffeomorphism $f: (M, \omega) \to (N, \eta)$ between symplectic manifolds such that $f^*\eta = \omega$ is said to be a *symplectic diffeomorphism* or *symplectomorphism.* If $(M, \omega) = (N, \eta)$ it is also called a *canonical transformation.*

Symplectic C^r-diffeomorphisms of a symplectic manifold (M, ω) form a closed subset of $\mathrm{Diff}^r(M)$ with the C^r-topology. Proposition 2.6.6 immediately yields a further result:

Proposition 2.6.10. *If (M, ω) is a symplectic manifold then M is even-dimensional and ω^n is a volume form. In particular M is orientable.*

By Proposition 2.6.3 we can find coordinates around any given point x such that in T_xM the induced coordinates bring the symplectic form into standard form. This can be done by introducing any coordinate system and making an appropriate linear coordinate change in that system. Unlike in the case of a Riemannian metric, it is, however, possible to find a local chart such that the symplectic form is in standard form at *every* point of the chart. The proof uses an argument due to Moser sometimes called the "homotopy trick."

Theorem 2.6.11 (Darboux Theorem). *Let (M,ω) be a symplectic manifold and $x \in M$. There is a neighborhood U of x and coordinates $\varphi\colon U \longrightarrow \mathbb{R}^{2n}$ such that at every point $y \in U$ ω is in standard form with respect to the basis $\{\frac{\partial}{\partial x_1}, \dots, \frac{\partial}{\partial x_{2n}}\}$.*

These coordinates are referred to as *Darboux* or *symplectic coordinates*.

Proof (Moser homotopy trick). As noted, we may assume that we already have coordinates such that $M = \mathbb{R}^{2n}$ and ω is in standard form at $x = 0$ with respect to the basis $\{\frac{\partial}{\partial x_1}, \dots, \frac{\partial}{\partial x_{2n}}\}$. Thus we need to find coordinates in which ω is constant. Denote by α the form with matrix $J = \left(\begin{smallmatrix} 0 & I \\ -I & 0 \end{smallmatrix}\right)$. Let $\omega' := \alpha - \omega$ and $\omega_t = \omega + t\omega'$ for $t \in [0,1]$. Then there is a ball around 0 on which all ω_t are nondegenerate (since there is such a ball for every t and it depends continuously on t). Thus $\omega' = d\theta$ for some 1-form θ by the Poincaré Lemma, and without loss of generality $\theta(0) = 0$.

Since ω_t is nondegenerate, there is a unique (smooth) vector field X_t such that $\omega_t \lrcorner X_t := X_t \lrcorner \omega_t := \iota_{X_t}\omega_t := \omega_t(X_t, \cdot) = -\theta$. Since $X_t(0) = 0$ one can integrate X_t on a small ball around 0 to get a 1-parameter family of diffeomorphisms $\{\varphi^t\}_{t\in[0,1]}$ such that $\dot\varphi^t = X_t$ and $\varphi^0 = \mathrm{Id}$. Then

$$\frac{d}{dt}\varphi^{t*}\omega_t = \varphi^{t*}(\mathcal{L}_{X_t}\omega_t) + \varphi^{t*}\frac{d}{dt}\omega_t = \varphi^{t*}d(\omega_t \lrcorner X_t) + \varphi^{t*}\omega' = \varphi^{t*}(-d\theta + \omega') = 0,$$

so $\varphi^{1*}\omega_1 = \varphi^{0*}\omega_0 = \omega$, that is, φ^1 is the desired coordinate change. □

Remark 2.6.12. As mentioned before, this result is in contrast to the situation for Riemannian metrics, for which such charts exist only for flat metrics. An explanation is that the condition $d\omega = 0$ here may be considered an analog of flatness of a Riemannian metric.

2.6.b Cotangent bundles. We now describe an important class of spaces with a canonical symplectic structure, the cotangent bundle of a smooth manifold. Not only does a cotangent bundle have a canonical symplectic structure, but furthermore the natural coordinates induced by coordinates on the underlying manifold are symplectic coordinates.

Let M be a smooth manifold and consider local coordinates $\{q_1, \dots, q_n\}$. On the cotangent bundle these induce coordinates $\{q_1, \dots, q_n, p_1, \dots, p_n\}$. Define a 1-form θ by setting

$$\theta = -\sum_{i=1}^{n} p_i dq_i. \tag{2.6.1}$$

Then its exterior derivative is $\omega = \sum_{i=1}^{n} dq_i \wedge dp_i$, that is, a symplectic form in Darboux coordinates. The next lemma shows that this definition does not depend on the choice of coordinates on the manifold. Alternatively it shows that diffeomorphisms of the manifold induce symplectomorphisms of the cotangent bundle:

Lemma 2.6.13. *Let M be a smooth manifold and $f: M \to M$ a diffeomorphism. Then the coderivative D^*f acting on the cotangent bundle T^*M preserves θ and ω.*

Proof. If we write $(Q_1, \dots, Q_n) = f(q_1, \dots, q_n)$ then

$$D^*f(q_1, \dots, q_n, p_1, \dots, p_n) = (Q_1, \dots, Q_n, P_1, \dots, P_n),$$

where $p_j = \sum_{i=1}^n \frac{\partial Q_i}{\partial q_j} P_i$. Thus

$$\sum_{i=1}^n P_i dQ_i = \sum_{i,j=1}^n P_i \frac{\partial Q_i}{\partial q_j} dq_j = \sum_{j=1}^n p_j dq_j$$

and θ, hence ω, is preserved. □

2.6.c Hamiltonian vector fields and flows. Now we can begin to study the Hamiltonian equations.

Definition 2.6.14. Let (M, ω) be a symplectic manifold, and $H: M \to \mathbb{R}$ a smooth function. Then the vector field $X_H = dH^\#$ defined by $\omega \lrcorner X_H = dH$ is called the *Hamiltonian vector field* associated with H or the *symplectic gradient* of H. The flow Φ with $\dot{\varphi}^t = X_H$ is called the *Hamiltonian flow* of H.

A Hamiltonian vector field is C^r if and only if the Hamiltonian function is C^{r+1}. Thus one can identify the space of C^r Hamiltonian flows, which is a closed linear subspace of $\Gamma^r(TM)$, with the space $C^{r+1}(M, \mathbb{R})$ modulo additive constants.

This is indeed a formulation of the usual Hamiltonian equations

$$\dot{q}_i = \frac{\partial H}{\partial p_i}, \quad \dot{p}_i = -\frac{\partial H}{\partial q_i};$$

to see that $\dot{\varphi}^t = X_H$ gives these, we check that $X_H := \left(\frac{\partial H}{\partial p_i}, -\frac{\partial H}{\partial q_i}\right)$ satisfies $\omega \lrcorner X_H = dH$ in Darboux (symplectic) coordinates. But

$$\omega \lrcorner X_H = \sum_{i=1}^n (dq_i \wedge dp_i) \lrcorner X_H = \sum_{i=1}^n \underbrace{(dq_i \lrcorner X_H)}_{=\partial H/\partial p_i} \wedge dp_i - \sum_{i=1}^n dq_i \wedge \underbrace{(dp_i \lrcorner X_H)}_{=-\partial H/\partial q_i} = dH.$$

Remark 2.6.15. This can be restated as saying that a Hamiltonian flow is a skew-gradient flow in that X_H is orthogonal to the gradient of H (and has the same norm). This makes the next two propositions natural.

It is easy to see that Hamiltonian flows are instances of 1-parameter groups of canonical transformations (Definition 2.6.9):

Proposition 2.6.16 (Liouville Theorem). *Hamiltonian flows are symplectic and hence volume preserving.*

Proof. Let (M,ω) be a symplectic manifold, $H\colon M \to \mathbb{R}$ a smooth function, $\omega \lrcorner X_H = dH$, and $\dot{\varphi}^t = X_H$. Then

$$\frac{d}{dt}\varphi^{t*}\omega = \varphi^{t*}(\mathcal{L}_{X_H}\omega) = \varphi^{t*}(\overbrace{d(\underbrace{\omega \lrcorner X_H}_{=dH})}^{=ddH} + (\underbrace{d\omega}_{=0} \lrcorner X_H)) = \varphi^{t*}(ddH) = 0. \qquad \square$$

The converse is not true, that is, there are symplectic flows that are not Hamiltonian: A linear flow on the 2-dimensional torus with the standard volume 2-form $dx \wedge dy$ preserves area and is hence symplectic. Its velocity vector field is constant $\neq 0$. Thus if it were a Hamiltonian flow the Hamiltonian would have to have constant nonzero gradient. On the other hand the Hamiltonian attains its maximum and thus has a critical point, a contradiction. Note, incidentally, that the lift of the linear flow to $\mathbb{R}^2$ is indeed Hamiltonian. If a vector field X generates a symplectic flow, the calculation above shows that the 1-form $\omega \lrcorner X$ is closed. Thus the obstruction to being Hamiltonian is, in fact, of a topological nature (namely, vanishing of the cohomology class of the closed 1-form $\omega \lrcorner X$). See Exercise 2.14 for a discussion of a related phenomenon.

We note that the Hamiltonian is itself a constant of motion.

Proposition 2.6.17. *Let (M,ω) be a symplectic manifold, $H\colon M \to \mathbb{R}$ a smooth function, $\omega \lrcorner X_H = dH$, and $\dot{\varphi}^t = X_H$. Then $H(\varphi^t(x))$ does not depend on t.*

Proof.

$$\frac{d}{dt}H\big|_{\varphi^t(x)} = dH(\varphi^t(x))\dot{\varphi}^t(x) = \omega(X_H(\varphi^t(x)), \underbrace{\dot{\varphi}^t(x)}_{=X_H(\varphi^t(x))}) = 0. \qquad \square$$

The Poisson bracket predates the symplectic approach to Hamiltonian mechanics and was traditionally used in coordinate calculations, but also illuminates the Lie algebraic structure underlying the geometry.

Definition 2.6.18. Let (M,ω) be a symplectic manifold and $f,g\colon M \to \mathbb{R}$ smooth functions. Then the *Poisson bracket* of f and g is defined by

$$\{f,g\} := \omega(X_f, X_g) = df(X_g),$$

where $X_f = df^{\#}$ and $X_g = dg^{\#}$ (cf. Definition 2.6.14), that is, $\omega \lrcorner X_f = df$ and $\omega \lrcorner X_g = dg$. The functions f and g are said to *commute* or be *in involution* if their Poisson bracket vanishes.

Proposition 2.6.19. *In symplectic coordinates* $\{q_1,\dots,q_n,p_1,\dots,p_n\}$ *we have*

$$\{f,g\} = \sum_{i=1}^{n}\Big(\frac{\partial f}{\partial q_i}\frac{\partial g}{\partial p_i} - \frac{\partial f}{\partial p_i}\frac{\partial g}{\partial q_i}\Big). \tag{2.6.2}$$

The Poisson bracket is antisymmetric and $\{\cdot,f\} = \mathcal{L}_{X_f}$*. Also,* f *is an integral of the Hamiltonian flow of* H *if and only if* $\{f,H\} = 0$.

Proof. Equation (2.6.2) follows by definition using $X_g = (\partial g/\partial p_i, -\partial g/\partial q_i)$. Antisymmetry follows from antisymmetry of ω. Next, $\{\cdot,f\} = \mathcal{L}_{X_f}$ since

$$\mathcal{L}_{X_f} g = dg \lrcorner X_f = (\omega \lrcorner X_g) \lrcorner X_f = \omega(X_g, X_f) = \{g,f\}.$$

If φ^t is the Hamiltonian flow for H then $(d/dt) f \circ \varphi^t = {\varphi^t}^* \mathcal{L}_{X_H} f = {\varphi^t}^*\{f,H\}$ vanishes if and only if $\{f,H\}$ does. □

Remark 2.6.20. In particular we have re-proved invariance of H since $\{H,H\} = 0$.

This gives a well-known result about Hamiltonian systems with symmetries:

Theorem 2.6.21 (Noether). *Let* (M,ω) *be a symplectic manifold,* $H\colon M \to \mathbb{R}$ *smooth,* $\omega \lrcorner X_H = dH$*, and* $\dot\varphi^t = X_H$*. If* H *is invariant under the Hamiltonian flow for* f*, then* f *is a constant of motion of* φ^t.

Proof. The hypothesis is that H is an integral for the flow of f, that is, $\{f,H\} = 0$, so conversely f is an integral for the flow of H. □

Remark 2.6.22. An interesting instance may arise when the phase space of the system is a cotangent bundle and the Hamiltonian is invariant under the action on the cotangent bundle of a 1-parameter family of diffeomorphisms of the configuration space. Since such symmetries tend to be easy to detect, this result gives an easy way to find integrals of this sort.

Example 2.6.23. Consider the central-force or *Kepler problem* of two bodies moving freely, but subject to mutual gravitational attraction. In coordinates centered at the center of mass of the system the position of one body is $x \in \mathbb{R}^3 \setminus \{0\}$ and its velocity is $v \in \mathbb{R}^3$. The potential energy of the gravitational field is given by $V(x) = -1/\|x\|$, so Newton's equation $F = ma$ becomes

$$\ddot x = \nabla \frac{1}{\|x\|} = -\frac{x}{\|x\|^3}$$

or

$$\dot{x} = v,$$
$$\dot{v} = -\frac{x}{\|x\|^3}.$$

The Hamiltonian $H(x,v) = \langle v,v\rangle/2 - 1/\|x\|$ (total energy) is invariant under rotations around the origin. In particular, it is invariant under rotations in the x–y plane, which are generated by the Hamiltonian $q_1p_2 - q_2p_1$, if we choose to label the coordinates (q_1,q_2). Thus $q_1p_2 - q_2p_1$ is a first integral. It happens to be the z-component of angular momentum. The other two components are invariant by invariance under rotations in the other planes.

Definition 2.6.24. Let M be a smooth manifold. If X, Y are vector fields on M then the *Lie bracket* $[X,Y]$ is the unique vector field with $\mathcal{L}_{[X,Y]} = \mathcal{L}_Y\mathcal{L}_X - \mathcal{L}_X\mathcal{L}_Y$.

Remark 2.6.25. The Lie bracket measures to which extent the flows of two vector fields fail to commute. Indeed the Lie bracket of two vector fields vanishes identically if and only if the corresponding flows commute.

From the point of view of classical mechanics the most important (or at least the most traditional) symplectic manifolds are $\mathbb{R}^{2n}$ with the standard symplectic structure and the cotangent bundle of a differentiable manifold M (the *configuration space* of a mechanical system) with the symplectic form ω described in Section 2.6.b, notably with the invariant 1-form (2.6.1). In both cases the symplectic manifold (phase space) itself is not compact, although in the second case the configuration space M may be compact; this is true in many important classical problems such as the motion of a rigid body. Of course $\mathbb{R}^{2n}$ can also be viewed as $T^*\mathbb{R}^n$, so the first case is a particular instance of the second.

In this book we primarily consider dynamical systems with compact phase space, and to apply our concepts and methods to a Hamiltonian system with Hamiltonian H one considers the restriction of the dynamics to the hypersurfaces $H = c$, which are compact in many situations, for example, for a geodesic flow on a compact Riemannian manifold, where those hypersurfaces are sphere bundles over the configuration space. Sometimes one can make a further reduction using the first integrals other than energy. If c is not a critical value of the Hamiltonian and the hypersurface $H_c := \{x \mid H(x) = c\}$ is compact then the Hamiltonian system preserves a nondegenerate $(2n-1)$-form ω_c.[10]

[10]This can be described as follows. One can locally decompose the $2n$-dimensional measure generated by ω into $(2n-1)$-dimensional measures on $H_{c+\delta}$ for all sufficiently small $|\delta|$ and consider the conditional measures, each of which is defined up to a multiplicative constant. Thus in this case, due to Proposition 2.6.16, one can apply the Poincaré Recurrence Theorem (Theorem 3.2.1), the Birkhoff Ergodic Theorem (Theorem 3.2.15), and other facts from ergodic theory to the restriction of the Hamiltonian system to H_c.

2.6.d Contact forms. There is an important situation when the invariant $(2n-1)$-forms can be described in a particularly natural way. In the case of both $\mathbb{R}^{2n}$ and T^*M the form ω is not only closed, but also *exact*. The 1-form θ defined by $\sum_{i=1}^{n} p_i dq_i$—globally in the first case, locally in the second—obviously satisfies $d\theta = \omega$. The calculation in the proof of Lemma 2.6.13 shows that θ is defined on T^*M independently of the choice of local coordinates. Of course, in general a Hamiltonian system on T^*M does not preserve θ or any other 1-form whose exterior derivative is equal to ω. Let us see what conditions the invariance of θ imposes on the Hamiltonian:

$$\mathcal{L}_{X_H}\theta = d\theta \lrcorner X_H + d(\theta \lrcorner X_H) = dH + d(\theta \lrcorner X_H) = 0 \text{ if } \underbrace{\theta \lrcorner X_H}_{=-\sum p_i \frac{\partial H}{\partial p_i}} = -H + \text{const}.$$

Since the choice of Hamiltonian for a given vector field X_H is unique up to an additive constant, we have proved the following result:

Proposition 2.6.26. *The Hamiltonian vector field X_H on T^*M preserves the 1-form θ if and only if the Hamiltonian can be chosen as positively homogeneous in p of degree 1, that is, $H(q, \lambda p) = \lambda H(q, p)$ for $\lambda > 0$.*

The restriction of the form θ to the surface $H = c$ for a noncritical value of c of H is an example of a 1-form such that $\theta \wedge (d\theta)^{n-1}$ is nondegenerate. This motivates the following definition (see also Definition 2.2.5).

Definition 2.6.27. An alternating multilinear n-form ω on a smooth manifold is a map on n-tuples of vector fields X_i such that

$$\omega(X_{\sigma(1)}, \dots, X_{\sigma(n)}) = \operatorname{sign} \sigma \omega(X_1, \dots, X_n),$$

where $\operatorname{sign} \sigma$ is the sign of the permutation σ, and ω is $C(\mathbb{R})$-linear in each entry. The *exterior product* or *wedge product* of a j-form α and a k-form β is defined by

$$\alpha \wedge \beta(X_1, \dots, X_{j+k}) := \sum_{\substack{\sigma(1)<\cdots<\sigma(j) \\ \sigma(j+1)<\cdots<\sigma(j+k)}} \operatorname{sign} \sigma \alpha(X_{\sigma(1)}, \dots, X_{\sigma(j)}) \cdot \beta(X_{\sigma(1)}, \dots, X_{\sigma(k)}).$$

A 1-form θ on a $(2n-1)$-dimensional orientable manifold M is called a *contact form* if the $(2n-1)$-form $\theta \wedge (d\theta)^{n-1}$ (the power being with respect to the wedge product) is nondegenerate. Accordingly a pair (M, θ) of a smooth manifold with a contact form is said to be a *contact manifold*. A *contact flow* is a flow on M that preserves the contact form on M. The *Reeb flow* or *characteristic flow* of a contact form θ is the flow generated by the *Reeb vector field* R_θ defined by $\theta \lrcorner R_\theta = 1$ and $d\theta \lrcorner R_\theta = 0$. A diffeomorphism preserving the contact form is called a *contact diffeomorphism*.

Unlike a symplectic manifold, which admits a variety of Hamiltonian vector fields, a contact manifold (M, θ) comes with the canonical vector Reeb vector field R_θ, which is unique because the kernel of $d\theta$ is 1-dimensional and disjoint from that of θ by the nondegeneracy assumption. Note that $\mathcal{L}_{R_\theta}\theta \equiv 0$ since $\theta \lrcorner R_\theta = \text{const.}$, so the *Reeb flow* of the contact form preserves θ and hence all structures defined in terms of θ, in particular the volume. Thus the Reeb flow provides a canonical example of a volume-preserving flow.

Suppose now that X is a vector field generating a flow preserving the contact form θ. Then it preserves $\ker d\theta$ as well and hence commutes with the Reeb flow of θ. Thus contact flows always arise as flows commuting with the Reeb flow of a contact form.[11] Combined with hyperbolicity this usually means that X is the Reeb field up to constant scaling (page 534).

Furthermore, if the contact manifold is a level set for a homogeneous Hamiltonian then R_θ is exactly the Hamiltonian vector field. In fact, contact forms always arise in this manner from generalized homogeneous Hamiltonians (see Proposition 2.6.29). Conversely, we note the following proposition:

Proposition 2.6.28. *Geodesic flows are Hamiltonian flows with a homogeneous Hamiltonian. Hamiltonian flows for homogeneous Hamiltonians, in particular geodesic flows, are Reeb flows, hence contact flows.*

Proposition 2.6.29. *Suppose (M, θ) is a contact manifold. Then M can be embedded into a symplectic manifold (N, ω) in such a way that the restriction of the ambient symplectic form to M is $d\theta$.*

Proof. If $N = M \times \mathbb{R}$ and $\omega_{x,t} = d(e^t\theta_x)$ then $\omega^n = e^{nt}(ndt \wedge \theta \wedge (d\theta)^{n-1})$ is a volume, so (N, ω) is a symplectic manifold and ω restricted to $M \times \{0\}$ is $d\theta$. □

Locally, a contact form, similarly to a symplectic form, can be brought into a standard form. The following result is a simple consequence of the Darboux Theorem (Theorem 2.6.11) for symplectic forms.

Theorem 2.6.30 (Darboux Theorem for Contact Forms). *Let*

$$\theta_0 = x_1 dy_1 + \cdots + x_n dy_n + dz$$

be the canonical contact form on $\mathbb{R}^{2n+1}$ and (M, θ) a contact $(2n+1)$-manifold. Then for $x \in M$ there exists a neighborhood U of x with coordinates in which $\theta = \theta_0$.

Proof. For $x \in M$ pick a neighborhood V_0 of 0 in $\ker \theta_x$ and let $V = V_0 \times (-\epsilon, \epsilon)$, $U' = \exp V$, $U'_t = \exp(V_0 \times \{t\}) \subset M$. Since $d\theta$ restricted to U'_t is a symplectic

[11] The same holds for a contact diffeomorphism.

form so by the Darboux Theorem (Theorem 2.6.11), each $y \in U_t'$ has a neighborhood $U_t \subset U_t'$ on which there are Darboux coordinates $x_1, \dots, x_n, y_1, \dots, y_n, z$, that is, $d\theta = \sum dx_i \wedge dy_i$. On $U := \bigcup_{-\epsilon<t<\epsilon} U_t$ we thus have $d(\theta - \sum dx_i \wedge dy_i) = 0$ whence $\theta = \sum dx_i \wedge dy_i + dz$ and $x_1, \dots, x_n, y_1, \dots, y_n, z$ are the desired coordinates. □

Exercises

2.1. Adapt the computations after (2.2.5) to B defined by $B(H) = 1$, $B(V) = 0 = B(X)$ to check that $B \wedge dB(H, X, V) = 1$ and $B = dA(V, \cdot)$.

2.2. Check the claims in Example 2.2.7.

2.3. Check the claims in Example 2.2.6: The discussion after Example 2.2.4 and Definition 2.2.5 showed that the geodesic flow and the horizontal flow each preserve a contact form (A and $B = dA(V, \cdot)$, respectively). Check that $C := dA(H, \cdot)$ is a contact form invariant under the fiber flow V in Remark 2.2.2 but that its Reeb field is $-V$ and that $C \wedge dC(V, H, X) = -1$, so this volume has the opposite orientation from the ones defined by A and B (and E in Example 2.2.7) and is hence not isotopic to either of them.

2.4. Since the generators $H_\pm$ for the horocycle flows are linear combinations of H and V, check whether a linear combination of the 1-forms B and C in Example 2.2.6 and before it is an invariant 1-form and if so, whether it is a contact form.

2.5. Prove the existence of the defining limit in Remark 2.1.18.

2.6. Show that the horocycle flow is minimal (by combining Example 2.1.16 with the use of homogeneity as in Example 1.6.2).

2.7. As in Theorem 2.4.4, let Γ be a discrete group of fixed-point-free isometries of $\mathbb{D}$ such that $M := \Gamma\backslash\mathbb{D}$ is compact. Prove that the geodesic flow is topologically mixing by combining Example 2.1.16 and the argument for (2) $\Rightarrow$ (1) in Theorem 6.2.11 below.

A hypersurface M in $\mathbb{R}^n$ is said to be *star shaped* if there exists a point c such that every half-line from c to ∞ intersects M in exactly one point.

2.8. Prove that any star-shaped hypersurface in $\mathbb{R}^{2n}$ provided with the standard symplectic structure is of contact type.

2.9. Describe the contact form and the Reeb vector field on $S^{2n-1} \subset \mathbb{R}^{2n}$ corresponding to the vector field $\xi = \frac{1}{2} \sum_{i=1}^n \left(p_i \frac{\partial}{\partial p_i} + q_i \frac{\partial}{\partial q_i}\right)$.

2.10. Consider a hypersurface M in the cotangent bundle T^*N of a smooth manifold that intersects each fiber in a star-shaped hypersurface. Prove that M is of contact type with respect to the standard symplectic structure.

2.11. Let L be a Lagrangian subspace of a symplectic vector space E. Prove that L has a *Lagrangian complement*, that is, a Lagrangian subspace M such that $L \cap M = \{0\}$.

2.12. Prove that in a Lagrangian subspace $L \subset E$ the basis $e_1, \dots, e_{2n}$ from Proposition 2.6.3 can be chosen in such a way that $e_1, \dots, e_n \in L$.

2.13. Let $\Lambda = (\lambda_1, \dots, \lambda_{2n})$ be a collection of nonzero complex numbers with the following properties:

(1) Λ contains an even number of 1s and an even number of -1s.

(2) If $\lambda \in \Lambda$ is real, $\lambda \neq \pm 1$, then $1/\lambda \in \Lambda$ (with the same multiplicity).

(3) If $\lambda \in \Lambda$, $|\lambda| = 1$, and $\lambda \neq \pm 1$ then $\bar{\lambda} \in \Lambda$ (with the same multiplicity).

(4) If $\lambda \in \Lambda$, $|\lambda| \neq 1$, $\lambda \notin \mathbb{R}$ then $\lambda^{-1}, \bar{\lambda}, \bar{\lambda}^{-1} \in \Lambda$ (with the same multiplicities).

Prove that there exists a symplectic linear map $T\colon (\mathbb{R}^{2n}, \omega) \to (\mathbb{R}^{2n}, \omega)$, where ω is the standard symplectic form, such that Λ is the set of eigenvalues of T (with multiplicities).

2.14. Prove that the 2-cohomology class of any nondegenerate closed 2-form on a $2n$-dimensional compact manifold M is nonzero.

2.15. Prove that there is no symplectic structure on the $2n$-sphere for $n \geq 2$, that is, there is no symplectic manifold (S^{2n}, ω).

2.16. Suppose $\{\omega_t\}_{0 \leq t \leq 1}$ is a family of nondegenerate closed differential 2-forms on a compact manifold M. Prove that there exists a family of diffeomorphisms $\varphi_t\colon M \to M$ such that $\varphi_t^* \omega_t = \omega_0$ if and only if the cohomology classes of the forms ω_t are the same.

2.17. Show that the geodesic flow on any surface of revolution has a first integral independent of the total energy. This integral is called the Clairaut integral.

2.18. Prove that any discrete subgroup of $\mathbb{R}^n$ is isomorphic to $\mathbb{Z}^k$ for some $k \leq n$.

2.19. Let (M, ω) be a symplectic manifold and $\{\varphi^t\}$ a Hamiltonian flow all of whose orbits are periodic with the same minimal period T. Fix a value c of the Hamiltonian and consider the factor space N of the level surface M_c by the action of the flow. Show that the restriction of ω to M_c projects to a nondegenerate 2-form on N.

2.20. Show that the geodesic flow on the standard n-dimensional sphere satisfies the conditions of the previous exercise. Apply the procedure from that exercise to obtain a $(2n-2)$-dimensional symplectic manifold. Describe that manifold in detail for $n = 2$.

3

Ergodic theory

Two important strands of 19th-century mathematics and physics led to the evolution of the theory of dynamical systems as we know it today. Celestial mechanics was a central motivation for Poincaré in his development of topological approaches to the study of dynamical systems (including the invention of topology itself). Statistical mechanics motivated a probabilistic approach to mechanical systems, where the preservation of volume provides a natural measure. This motivated von Neumann to formalize the foundations of ergodic theory [265, 266], and this approach is broadly applicable, since there are usually many important invariant measures besides volume. Accordingly, we now study flows defined on probability spaces—in full generality. In Chapter 7 we examine how these notions apply to hyperbolic flows.

This chapter introduces basic notions in measure theory, and then studies basic properties of measures invariant under a flow. We then examine the existence of time averages of functions along orbits (ergodic theorems). Next, we introduce ergodicity, which can be viewed as analogous to transitivity in topological dynamics, and a range of mixing properties much broader than topological mixing. Mixing is quite sensitive to time changes, and this leads into a careful study of basic issues specific to continuous time: time changes and special flows. This is significantly different from the theory of mixing for discrete systems. We conclude with spectral theory, but that section is optional since we will not use it in studying hyperbolic flows.

A number of the subjects in this chapter exhibit significant differences between the discrete-time and continuous-time setting, such as invariant measures for time changes and special flows, some aspects of ergodicity, properties of mixing, and measure-theoretic entropy. The latter also requires measure theory beyond what is usually done in an introductory graduate course, and we summarize those ideas in Section A.1. Likewise, we will refer to Appendix A for other notions that are either background material or described there for maps instead of flows.

3.1 Flow-invariant measures and measure-preserving transformations

We begin with a summary of basic notions of measures and measure spaces. Let X be a set and $\mathcal{T} \subset 2^X$ a σ-algebra (that is, $\varnothing, X \in \mathcal{T}$ and $\mathcal{T}$ is closed under complements and countable unions of sets). Then $(X, \mathcal{T})$ is called a *measurable space* and the elements of $\mathcal{T}$ are referred to as measurable sets. A *measure* is a function $\mu\colon \mathcal{T} \to [0, \infty]$ such that $\mu(\bigcup_{i=1}^{\infty} A_i) = \sum_{i=1}^{\infty} \mu(A_i)$ for pairwise disjoint $A_i \in \mathcal{T}$. A set is a *null set* if $\mu(A) = 0$. The σ-algebra $\mathcal{T}$ is *complete* if any subset of a μ-null set is in $\mathcal{T}$. The completion of a σ-algebra $\mathcal{T}$ is the σ-algebra $\overline{\mathcal{T}}$ generated by $\mathcal{T}$ and the μ-null sets. A complete σ-algebra $\mathcal{T}$ is said to be separable if it is the completion of a σ-algebra generated by a countable family of sets. A set A has *full measure* or is *conull* if $\mu(A^c) = 0$.[1] For a topological space the smallest σ-algebra containing all open sets is called the *Borel σ-algebra*. An assertion is said to be *essentially* true if it holds on a set of full measure (such as an essentially constant function, an essential bound for a function, a flow with essentially no fixed points).

A measure space $(X, \mathcal{T}, \mu)$ is *σ-finite* if X is a countable union of sets of finite measure. If $\mu(X) = 1$, then we call μ a *probability measure*. A point $x \in X$ is called an *atom* if $\mu(x) > 0$, and a measure μ is *atomic* if X contains atoms for μ, *nonatomic* otherwise.

Two measures μ and ν on a measure space $(X, \mathcal{T})$ are *singular* or mutually singular if there exist disjoint measurable sets A and B such that $A \cup B = X$ and $\mu(B) = 0 = \nu(A)$. This is denoted $\mu \perp \nu$. For two measures μ and ν on a measure space $(X, \mathcal{T})$ the measure ν is *absolutely continuous* with respect to μ if $\nu(A) = 0$ for any $A \in \mathcal{T}$ such that $\mu(A) = 0$. This is denoted $\nu \ll \mu$.

We usually assume that μ is a probability measure and $\mathcal{T}$ is complete. We now define the basic notion of a measurable dynamical system.

Definition 3.1.1 (Measurability, measure-preservation, isomorphism). A map between measurable spaces is *measurable* if the preimage of any measurable set is measurable. For a map $T\colon (X, \mathcal{T}, \mu) \to (Y, \mathcal{A})$ such that the preimage of a set in $\mathcal{A}$ is a set in $\mathcal{T}$ we define the *pushforward* of μ by

$$T_*\mu(A) := \mu(T^{-1}(A)).$$

A measurable map $T\colon (X, \mathcal{T}, \mu) \to (Y, \mathcal{A}, \nu)$ is *nonsingular* if the preimage of any null set is a null set, that is, $T_*\mu \ll \nu$, and *measure preserving* if $T_*\mu = \nu$.

Two measure spaces are *isomorphic* if there exist sets $X' \subset X$ and $Y' \subset Y$ each of full measure and a measurable measure-preserving bijection $T\colon X' \to Y'$ with measurable inverse; the map T is called an *isomorphism*. An isomorphism of a set to itself is called an *automorphism*.

A flow Φ of a measure space (X, μ) is *measure preserving* if each φ^t is measurable with $\mu(\varphi^t(A)) = \mu(A)$ for all measurable A.

[1] A^c denotes the complement of A.

A *function* $f\colon (X,\mathcal{T}) \to \mathbb{R}$ is said to be *measurable* if the preimage of any open set is measurable. Two measurable functions are equivalent if they coincide on a set of full measure. If $p \in [1,\infty)$, then

$$L^p(X) := \Big\{ f\colon X \to \mathbb{R} \text{ measurable} \;\Big|\; \int |f|^p < \infty \Big\} \Big/ \stackrel{\text{ae}}{=}$$

is the linear space of equivalence classes of functions whose pth power is integrable; $L^2(X,\mu)$ is a Hilbert space with inner product $\langle f, g\rangle := \int fg\,d\mu$, and

$$L^\infty(X) := \Big\{ f\colon X \to \mathbb{R} \;\Big|\; \|f\|_\infty := \underbrace{\operatorname{essup}|f|}_{:=\inf\{M \mid \mu(|f|^{-1}((M,\infty])=0\}} < \infty \Big\} \Big/ \stackrel{\text{ae}}{=}$$

consists of the equivalence classes of essentially bounded measurable functions.

A function $f\colon X \to \mathbb{R}$ is said to be *essentially Φ-invariant* if there is a null set N off which $f \circ \varphi^t = f$ *for all* $t \in \mathbb{R}$. The important point is that N does not depend on t.[2] A set is said to be *essentially Φ-invariant* if its characteristic function is. The σ-algebra of these sets is denoted by $\mathcal{I}$.

Remark 3.1.2. We remark that for a measure-preserving flow a.e. orbit is a measurable set; this is most easily seen from Theorem 3.6.2 below because orbits of measure-preserving suspensions are measurable. This is different from the discrete setting where *every* orbit is countable, and hence measurable.

Definition 3.1.3. A complete measure μ is said to be a *Borel measure* if it is defined on the Borel σ-algebra. It is a *Radon measure* if furthermore the measure of a compact set is finite, and a *Borel probability measure* if $\mu(X) = 1$. A measure μ is *regular* if for each measurable set B and all $\epsilon > 0$ there exist an open set U_ϵ and a closed set C_ϵ such that $C_\epsilon \subset B \subset U_\epsilon$ and $\mu(U_\epsilon \smallsetminus C_\epsilon) < \epsilon$.

Example 3.1.4. We now illustrate these notions with previous examples.

- The flow in Example 1.1.5 (linear translations on $\mathbb{R}$) preserves Lebesgue measure on $\mathbb{R}$.
- Similarly, the translation flow on the circle from Example 1.1.6 preserves Lebesgue measure on the circle.
- Example 1.6.2 is a flow on the torus that preserves Lebesgue measure.
- If a flow has a fixed point x then the *Dirac measure* δ_x on it is invariant, where $\delta_x(A) := \chi_A(x)$.

---

[2]It is not required, but in the situation at hand, we can ultimately choose the exceptional set to be invariant, that is, f is measurable with respect to the completion of the σ-algebra of (properly) invariant sets.

- For the flow on the circle with a single fixed point (Example 1.3.13), the Dirac measure at the fixed point is the only flow-invariant Borel probability measure: any interval that does not contain the fixed point in its closure has countably many disjoint images and preimages of pairwise equal measure by invariance, hence is a null set.

- The suspension of Example 1.9.2 has a single invariant Borel probability measure that is the Lebesgue measure on the single periodic orbit. Definition 3.6.1 and the discussion following will clarify invariant measures for suspension flows and how these relate to invariant measures on the base space.

Theorem 3.1.5. *Every Borel probability measure μ on a metric space is regular.*

Proof. We let $\mathcal{A}$ be the collection of $A \subset X$ such that for all $\epsilon > 0$ there exist a closed set C_ϵ and an open set U_ϵ with $C_\epsilon \subset A \subset U_\epsilon$ and $\mu(U_\epsilon \smallsetminus C_\epsilon) < \epsilon$ (so $\mathcal{A}$ is the collection of regular sets). We will show that $\mathcal{A}$ is a σ-algebra and contains all open sets. It is then clear that $\mathcal{A}$ is complete, so it is the completion of the Borel σ-algebra.

First, $\varnothing, X \in \mathcal{A}$ since both are simultaneously open and closed. If $A \in \mathcal{A}$, $\epsilon > 0$, and C_ϵ and U_ϵ are as in the definition of regular, then $X \smallsetminus U_\epsilon \subset X \smallsetminus A$ is closed, and $X \smallsetminus C_\epsilon \supset X \smallsetminus A$ is open, and

$$\mu((X \smallsetminus C_\epsilon) \smallsetminus (X \smallsetminus U_\epsilon)) = \mu(U_\epsilon \smallsetminus C_\epsilon) < \epsilon.$$

Thus $X \smallsetminus A \in \mathcal{A}$.

Let $A_1, \dots \in \mathcal{A}$ and $A = \bigcup_{i=1}^\infty A_i$. Fix $\epsilon > 0$ and for each i let $U_{i,\epsilon}$ be an open set and $C_{i,\epsilon}$ be a closed set such that $C_{i,\epsilon} \subset A_i \subset U_{i,\epsilon}$ and $\mu(U_{i,\epsilon} \smallsetminus C_{i,\epsilon}) < \epsilon/3^i$. Let $\hat{C}_\epsilon := \bigcup_{i=1}^\infty C_{i,\epsilon}$ and $C_\epsilon := \bigcup_{i=1}^k C_{i,\epsilon}$, where k is such that $\mu(\hat{C}_\epsilon \smallsetminus C_\epsilon) < \epsilon/2$. Then $C_\epsilon \subset A$ is closed, $U_\epsilon := \bigcup_{i=1}^\infty U_{i,\epsilon} \supset A$ is open, and

$$\mu(U_\epsilon \smallsetminus C_\epsilon) \le \mu(U_\epsilon \smallsetminus \hat{C}_\epsilon) + \mu(\hat{C}_\epsilon \smallsetminus C_\epsilon) \le \underbrace{\sum_{i=1}^\infty \mu(U_{i,\epsilon} \smallsetminus C_{i,\epsilon})}_{<\sum_{i=1}^\infty \epsilon/3^i = \epsilon/2} + \underbrace{\mu(\hat{C}_\epsilon \smallsetminus C_\epsilon)}_{<\epsilon/2} < \epsilon.$$

So $A \in \mathcal{A}$, and $\mathcal{A}$ is closed under countable unions, hence a σ-algebra.

To finish the proof we show that $\mathcal{A}$ contains all closed $C \subset X$. Let $\epsilon > 0$. The $U_i := \{x \in X \mid d(x,C) < 1/i\} \subset C$ are open with $\varnothing = \bigcap_{i=1}^\infty U_i \smallsetminus C$, so $\mu(U_k \smallsetminus C) \to 0$, and there is a $k \in \mathbb{N}$ with $\mu(U_k \smallsetminus C) < \epsilon$. Taking $C = C_\epsilon$ and $U_\epsilon = U_k$ gives $C \in \mathcal{A}$. □

Corollary 3.1.6. *If μ is a Borel probability measure on a metric space X and B a measurable set, then*

$$\mu(B) = \sup_{C \subset B \text{ closed}} \mu(C) \quad (\mu \textit{ is } \text{inner regular})$$

and

$$\mu(B) = \inf_{U \supset B \text{ open}} \mu(U) \quad (\mu \textit{ is } \text{outer regular}).$$

Borel measures are identified by the integrals of continuous functions:

Theorem 3.1.7. *Let μ, ν be Borel probability measures on a metric space X. If $\int f \, d\mu = \int f \, d\nu$ for all continuous functions f, then $\mu = \nu$.*

Proof. This can be proved directly from the Riesz Representation Theorem for continuous functions (Theorem 3.1.10), but we provide an independent proof. Let C be a closed subset of X and fix $\epsilon > 0$. Then there exists an open set U_ϵ such that $C \subset U_\epsilon$ and $\mu(U_\epsilon \setminus C) < \epsilon$. Then $f \colon X \to [0,1]$ defined by

$$f(x) := \frac{d(x, X \setminus U_\epsilon)}{d(x, X \setminus U_\epsilon) + d(x, C)}$$

is a continuous function such that $f = 0$ on $X \setminus U_\epsilon$, $f = 1$ on C. Hence,

$$\nu(C) \le \int f \, d\nu = \int f \, d\mu \le \mu(U_\epsilon) < \mu(C) + \epsilon;$$

thus, $\nu(C) \le \mu(C)$ since ϵ is arbitrary. Switching μ and ν gives the opposite inequality, so $\mu = \nu$ for closed, hence measurable, sets. □

Proposition 3.1.8. *For a Borel measure μ on a separable metrizable space X,*

(1) *the* support $\operatorname{supp} \mu := \{x \in X \mid \mu(U) > 0 \text{ if } x \in U, U \text{ open}\}$ *of μ is closed*;

(2) $\mu(X \setminus \operatorname{supp} \mu) = 0$;

(3) *any set of full measure is dense in* $\operatorname{supp} \mu$.

Proof. (1) If $x \notin \operatorname{supp} \mu$ take $U_x \ni x$ open with $\mu(U_x) = 0$. Then $U_x \cap \operatorname{supp} \mu = \varnothing$.

(2) Since X is separable, $X \setminus \operatorname{supp} \mu$ is covered by countably many U_x as above, so $\mu(X \setminus \operatorname{supp} \mu) = 0$ by σ-additivity of μ.

(3) Contraposition: if $A \subset X$, $\varnothing \ne U := \operatorname{supp} \mu \setminus \bar{A}$ then $\mu(X \setminus A) \ge \mu(U) > 0$. □

Remark 3.1.9. If $\operatorname{supp} \mu = X$ then we say that μ has *full support* or is positive on open sets. In Example 3.1.4 we showed that the support of the sole invariant measure for Example 1.9.2 is the fixed point. We will see more interesting connections between (properties of) invariant measures and the topological dynamics on their support (Theorem 3.3.29, Exercise 3.11, Proposition 3.4.12).

Theorem 3.1.7 is related to the fact that measures define (positive) linear functionals. The converse, that positive linear functionals arise from measures, is the content of the Riesz Representation Theorem from analysis, and this will give another way to obtain invariant measures.

Theorem 3.1.10 (Riesz Representation Theorem). *Let X be a compact Hausdorff space. Then for each bounded linear functional F on $C^0(X)$ there exists a unique mutually singular pair μ, ν of finite Borel measures (Definition* 3.1.3) *such that $F(g) = \int g\,d\mu - \int g d\nu$ for all $g \in C^0(X)$.*

Remark 3.1.11. In particular, when F is positive (that is, nonnegative on positive functions) there is a unique finite Borel measure μ such that $F(\varphi) = \int \varphi\,d\mu$. This is an important class of functionals in this book. It is especially useful that the collection $\mathfrak{M}(X)$ of Borel probability measures on a compact metrizable space is a convex norm-bounded subset of the dual to $C(X)$. Further, $\mathfrak{M}$ is closed with respect to the *weak* topology* (the product topology of setwise convergence) defined by $\mu_n \to \mu :\Leftrightarrow \int_X \varphi\,d\mu_n \to \int \varphi\,d\mu\ \forall \varphi \in C(X)$ (we say that μ_n *equidistributes* to μ), and is compact and sequentially compact by the Banach–Alaoglu Theorem.[3]

We continue our study of measure-preserving flows (Definition 3.1.1) by restating what it means to preserve a measure.

Theorem 3.1.12. *Let Φ be a measurable flow of a measure space $(X, \mathcal{T}, \mu)$. Then $\int f\,d(\varphi^t_* \mu) = \int f \circ \varphi^t\,d\mu$ for all $f \in L^1(X, \mu)$ and all t.*

Proof. By definition, this holds for characteristic functions of Borel sets, hence for simple functions (linearity) and for nonnegative measurable functions (pointwise limits of increasing sequences of simple functions). Considering positive and negative parts gives the theorem. □

Corollary 3.1.13. *Let Φ be a measure-preserving flow of a measure space $(X, \mathcal{T}, \mu)$ and $f : X \to \mathbb{R}$ (or $\mathbb{C}$) integrable. Then $\int_X f(x)\,d\mu = \int_X f(\varphi^t x)\,d\mu$ for all $t \in \mathbb{R}$.*

Together with Theorem 3.1.7, this implies the following proposition:

Proposition 3.1.14. *$\mu \in \mathfrak{M}(X)$ is Φ-invariant iff $\int f \circ \varphi^t\,d\mu = \int f\,d\mu$ for all $f \in C(X)$.*

The next result can be proved in more generality, but this version will be sufficient for our needs.

---

[3]The (norm-) unit ball in the dual of a normed linear space B is weak*-compact (proved using the Tychonoff Theorem on compact products), and sequentially compact if B is separable (proved by a diagonal argument)—this implies that norm-bounded weak*-closed sets are compact/sequentially compact.

Theorem 3.1.15 (Krylov–Bogolubov Theorem). *Any continuous flow on a metrizable compact space has an invariant Borel probability measure.*

Proof. If Φ is continuous and $\mu \in \mathfrak{M}(X)$, then by Remark 3.1.11 there is a weak* accumulation point μ' of $\frac{1}{T}\int_0^T \varphi_*^t\mu \in \mathfrak{M}(X)$, and μ' is φ_*^t-invariant. □

Theorem 3.1.16. *If Φ is a continuous flow of a compact metric space then the set $\mathfrak{M}(\Phi)$ of Φ-invariant Borel probability measures is a closed, hence compact, convex subset of $\mathfrak{M}(X)$.*

Proof. If $\{\mu_n\}_{n=1}^\infty \subset \mathfrak{M}(\Phi)$ and $\mu_n \longrightarrow \mu$ in $\mathfrak{M}(X)$, then

$$\int f\, d(\varphi_*^t\mu) = \int f \circ \varphi^t\, d\mu = \lim_{n\to\infty} \int f \circ \varphi^t\, d\mu_n = \lim_{n\to\infty} \int f\, d\mu_n = \int f\, d\mu$$

for all continuous functions $f\colon X \longrightarrow \mathbb{R}$ and all $t > 0$. So $\mu \in \mathfrak{M}(\Phi)$. Convexity is clear since $\mathfrak{M}(X)$ is convex. □

Definition 3.1.17. A continuous flow on a metrizable compact space is said to be *uniquely ergodic* if it has exactly one invariant Borel probability measure, and *strictly ergodic* if it is furthermore minimal.

Remark 3.1.18. If the measure μ used in the proof of Theorem 3.1.15 is invariant, then the process becomes trivial because the accumulating family is constant, yielding $\mu' = \mu$. Indeed, a number of invariant measures often arise in an obvious way. Dirac measures on fixed points (see Example 3.1.4) are the most self-evident. If p is periodic with period ℓ, then $\delta_{\mathcal{O}(p)} := \frac{1}{\ell}\int_0^\ell \delta_{\varphi^t(p)}\, dt$ is an invariant Borel probability measure, as are convex combinations of any number of invariant Borel probability measures.[4]

For a suspension flow over a μ-preserving transformation on X, the product of μ with Lebesgue measure on $[0,1]$ defines an invariant Borel probability measure. For a flow under a function r on X, likewise for continuous $F\colon \Lambda(r) \longrightarrow \mathbb{R}$, the equation

$$\int_{X_r} F\, d\mu_r = \frac{\int_X \left(\int_0^{r(x)} F(x,t)\, dt\right) d\mu(x)}{\int_X r(x)\, d\mu(x)}$$

defines an invariant Borel probability measure μ_r. We revisit this in Definition 3.6.1 below, where it turns out that any invariant Borel probability measure for a flow can be seen as arising in this way (Theorem 3.6.2).

The next theorem connects some of the notions from topological dynamics of Chapter 1 to the set of invariant measures for a flow.

[4] A convex combination is a linear combination with nonnegative coefficients that sum to 1, that is, a weighted average.

Theorem 3.1.19 ([253]). *If Ψ is a time change of a continuous flow Φ without fixed points, then there is an affine bijection between $\mathfrak{M}(\Phi)$ and $\mathfrak{M}(\Psi)$.*[5]

Definition 3.1.20. A flow Φ of a measure space (X, μ) is *measure-theoretically isomorphic* to a flow Ψ of a measure space (Y, ν) if there is an isomorphism $h\colon X \to Y$ such that $\psi^t \circ h \overset{\text{ae}}{=} h \circ \varphi^t$ for all $t \in \mathbb{R}$. These flows are *orbit equivalent* if there is an isomorphism $h\colon X \to Y$ that sends orbits of Φ to orbits of Ψ. A flow Ψ on Y is a *factor* of Φ on X if there is a measure-preserving essentially surjective $h\colon X \to Y$ such that $\psi^t \circ h \overset{\text{ae}}{=} h \circ \varphi^t$ for all $t \in \mathbb{R}$.

Remark 3.1.21. For continuous flows the notion of orbit equivalence proved natural, but the notion of measurable orbit equivalence in the sense of "same orbits" is too weak to be interesting, as the next result illustrates.

Theorem 3.1.22 (Dye's Theorem [124, 125]). *For any two ergodic measure-preserving flows on nonatomic probability spaces, there is a measurable isomorphism between the two probability spaces that sends orbits to orbits.*

What is missing is any control of time along orbits under this isomorphism. An important equivalence relation retains just enough control by requiring the isomorphism to be monotone along orbits.

Definition 3.1.23 (Monotone (or Kakutani) equivalence [208, 209]). Two flows are *monotonically* or *Kakutani equivalent* if one of them is measurably isomorphic to the other after a time change which is smooth along orbits.[6]

Note that this does not only provide monotonicity in the orbit direction but average control of the speed change as well.

We are motivated by continuous flows on compact metric spaces X. Although we will not need the following result, we mention that it is not very restrictive to focus on this context because there is the device of *continuous representation*:

Theorem 3.1.24 (Ambrose–Kakutani [7, Theorem 5]). *A measure-preserving flow Φ on a Lebesgue space* (Definition A.1.1) *with essentially no fixed points is measure-theoretically isomorphic to a continuous special flow on a separable metric space with an invariant Borel probability measure.*

Remark 3.1.25. It is a natural and rather deeper question whether any probability-preserving flow can be realized as a *volume*-preserving flow, as conjectured by von Neumann in his foundational paper [265].

[5] Thus, a time change of a uniquely ergodic flow (Definition 3.1.17) without fixed points is uniquely ergodic; however, there are uniquely ergodic flows (with a fixed point) for which some time change is not uniquely ergodic [253].

[6] Specifically, its derivative along orbits is in $L^1(X, \mu)$. (If it is identically 1, then there is no time change, and the flows are isomorphic.)

The next example introduces the ideal representation of deterministic randomness, and we will refer to it often.

Example 3.1.26 (Bernoulli flow). Consider the full shift (Definition 1.9.1) and endow the shift space σ_n with the Borel measure μ for which $\mu(C_i^0) = p_i$ with $\sum_i p_i = 1$ (see (1.9.1)). Together with shift invariance, this uniquely defines a probability measure by Theorem 3.1.7; in fact, this is the product measure on $\mathcal{A}_n^{\mathbb{Z}}$, where $\nu(\{i\}) = p_i$ for $i \in \mathcal{A}_n$. The full shift with this measure is called a *Bernoulli shift*, and a flow is called a *Bernoulli flow* or said to have the *Bernoulli property* if every time-t map for $t \neq 0$ is measure-theoretically isomorphic to a Bernoulli shift (see Definition 3.4.3).

3.2 Ergodic theorems

The purpose of studying invariant measures is to be able to meaningfully investigate probabilities in a statistical approach to long-term evolution. This necessitates knowing that such long-term statistics exist, and theorems to this effect are called ergodic theorems. The first of these was proved by von Neumann, and it served to crystallize the notion of ergodicity. Spurred by von Neumann's article, Birkhoff established a pointwise counterpart. We begin with a precursor to these.

Note that we prove these ergodic theorems without defining ergodicity because that notion is not required; the theorems hold in greater generality. Later we will explain what additional conclusions they yield in the context of ergodicity.

Poincaré viewed recurrence as a weaker form of stability, and he had the insight that this is ubiquitous in celestial mechanics, and indeed all mechanical systems, as a simple consequence of preserving a probability measure (Corollary 3.2.2).

Theorem 3.2.1 (Poincaré Recurrence Theorem). *Let Φ be a measure-preserving flow of a probability space $(X, \mathcal{T}, \mu)$. If A is measurable and $T \geq 0$, then for almost every $x \in A$ there exists $t > T$ such that $\varphi^t(x) \in A$ (that is, there are $t_i \to \infty$ with $\varphi^{t_i}(x) \in A$).*

Proof. Since $B := \{x \in A \mid \varphi^{iT}(x) \in A^c \text{ for all } i \in \mathbb{N}\} = A \smallsetminus \left(\bigcup_{i \in \mathbb{N}} \varphi^{-iT}(A)\right)$ is measurable and the $\varphi^{-iT}(B)$ are pairwise disjoint and have the same measure as B, $\mu(B) = 0$ since $\mu(X) = 1$. □

Corollary 3.2.2. *If X is a separable metric space, Φ is a continuous flow on X, and μ is a Φ-invariant Borel probability measure, then $\mu(\mathcal{B}(\Phi)) = 1$ (Definition 1.5.11), hence $\mu(\mathcal{L}(\Phi)) = \mu(\mathrm{NW}(\Phi)) = 1$ by Proposition 1.5.37.*

Proof. For a countable base $\{U_1, U_2, \dots\}$ of open subsets of X the set of all points $x \in U_m$ with $\varphi^{t_i}(x) \in U_m$ for some $t_i \to \infty$ has full measure by the Poincaré Recurrence Theorem (Theorem 3.2.1). □

Remark 3.2.3. This corollary is not in all cases as interesting as it seems. If μ is the Dirac measure on a fixed point, then essentially all points are fixed no matter how much orbit complexity there might be elsewhere.

While the Poincaré Recurrence Theorem establishes recurrence, a qualitative phenomenon, ergodic theorems are about using statistics. We present the von Neumann (convergence in the mean) and Birkhoff (pointwise convergence) Ergodic Theorems.

Theorem 3.2.4 (Von Neumann Mean Ergodic Theorem). *If Φ is a measure-preserving flow of a measure space $(X, \mathcal{T}, \mu)$ and $f \in L^2(X, \mathcal{T}, \mu)$, then*

$$\frac{1}{T}\int_0^T f \circ \varphi^t \, dt \xrightarrow[T \to \infty]{L^2} P_\Phi(f),$$

where P_Φ is the orthogonal projection to the space $L^2(X, \mathcal{I}, \mu_{\restriction_{\mathcal{I}}})$ of Φ-invariant functions.

This theorem does not require the measure space to be a probability space. It follows from a Hilbert-space lemma, for which it is useful that one can associate with a measure-preserving map an isometric operator, and hence a 1-parameter family of such operators to a flow.

Definition 3.2.5 (Koopman operator). For $p \geq 1$ one associates to a measure-preserving map $f : (X, \mathcal{T}, \mu) \to (Y, \mathcal{V}, \nu)$ an isometric operator

$$U_f : L^p(Y, \mathcal{V}, \nu) \to L^p(X, \mathcal{T}, \mu), \quad g \mapsto g \circ f$$

on complex-valued functions, the *Koopman operator*. For a measure-preserving flow Φ we have $U^t_\Phi := U^t_{\varphi^1}$, so we sometimes write $U_\Phi := U_{\varphi^1}$ and

$$U_{\varphi^t}(f) = f \circ \varphi^t.$$

Remark 3.2.6. The case $p = 2$ is of particular interest. If $f : X \to X$ is invertible then so is U_f and in this case U_f defines a unitary operator on L^2. In particular, U_{φ^t} is a 1-parameter family of unitary operators on $L^2(X, \mathcal{T}, \mu)$ if Φ is a μ-preserving flow on X.

Remark 3.2.7. When $(X, \mathcal{T}, \mu)$ is a compact metric probability space and f is continuous, then $g \mapsto U_f g$ is continuous in the norm topology—this is clear for uniformly continuous g, and the subspace of uniformly continuous functions is dense.

Theorem 3.2.8 (Alaoglu–Birkhoff Abstract Ergodic Theorem). *Suppose H is a Hilbert space, G a group of unitary operators, and P_{H_G} the orthogonal projection to H_G, the space of its common fixed points. If $v \in H$, then $P_{H_G}(v)$ is the unique element of the closed convex hull $\overline{\text{co}}\, Gv$ of $Gv := \{gv \mid g \in G\}$ of minimal norm.*[7]

Proof. As a nonempty closed convex subset of a Hilbert space, $\overline{\text{co}}\, Gv$ contains a unique norm-minimizing element F. Since $\frac{1}{2}gF + \frac{1}{2}F \in \overline{\text{co}}\, Gv$ cannot have smaller norm, we have $gF = F$, that is, $F \in H_G$. To see that $F = P_{H_G}(v)$ we show that $v - F \perp H_G$. For $h \in H_G$ the set $\{w \in H \mid \langle w - F, h\rangle = \langle v - F, h\rangle\} \ni v$ is closed and convex and contains Gv (since each g is unitary) and hence F. Thus $\langle v - F, h\rangle = 0$ (and indeed, $\{P_{H_G}(v)\} = H_G \cap \overline{\text{co}}\, Gv$). □

Proof of the von Neumann Ergodic Theorem. Take $v \in L^2$ and $\epsilon > 0$. By the Alaoglu–Birkhoff Abstract Ergodic Theorem there is a finite convex combination $v_\epsilon = \sum_{i=1}^n U_{\varphi^{t_i}} v$ with $\|v_\epsilon - P_\Phi v\| < \epsilon$, hence $\|\frac{1}{T}\int_0^T U_\Phi^t v_\epsilon\, dt - P_\Phi v\| < \epsilon$ for any $T > 0$, and $\overline{\lim}_{T\to\infty} \|\frac{1}{T}\int_0^T U_\Phi^t v\, dt - P_\Phi v\| < 2\epsilon$. □

The Birkhoff Ergodic Theorem addresses the question of the existence of the time averages in a probability space in the sense of pointwise convergence.

Theorem 3.2.9 (Radon–Nikodym). *If $(X, \mathcal{S}, \mu)$ and $(X, \mathcal{T}, \nu)$ are σ-finite signed measure spaces and $\nu \ll \mu$, then there is a μ-a.e. unique* density *or* Radon–Nikodym derivative $\left[\frac{d\nu}{d\mu}\right] := \rho\colon X \to \mathbb{R}$ *of ν with respect to μ that is measurable with respect to the completion $\overline{\mathcal{S}}$ of $\mathcal{S}$ and such that $\nu(A) = \int_A \rho\, d\bar{\mu}$, where $\bar{\mu}$ is the completion of μ, for every A in the completion of $\mathcal{T}$.*

In particular, $\overline{\mathcal{T}} \subset \overline{\mathcal{S}}$.

Corollary 3.2.10 (Conditional expectation). *Suppose $(X, \mathcal{S}, \lambda)$ is a σ-finite measure space, $\mathcal{T} \subset \mathcal{S}$ a σ-algebra, $g \in L^1(X, \mathcal{S}, \lambda)$. Denote by $\lambda_{\restriction\mathcal{T}}$ the restriction, that is, $\lambda_{\restriction\mathcal{T}}(A) = \lambda(A)$ for all $A \in \mathcal{T} \subset \mathcal{S}$. Then the* conditional expectation

$$E(g \mid \mathcal{T}) := g_{\mathcal{T}} := \left[\frac{d(g\lambda)_{\restriction\mathcal{T}}}{d\lambda_{\restriction\mathcal{T}}}\right] \in L^1(X, \mathcal{T}, \lambda_{\restriction\mathcal{T}})$$

of g on $\mathcal{T}$ is defined λ-a.e. uniquely by $\int_A g_{\mathcal{T}}\, d\lambda = \int_A g\, d\lambda$ for all $A \in \mathcal{T}$.

Proof. Apply Theorem 3.2.9 to $\lambda_{\restriction\mathcal{T}} \gg \nu := (g\lambda)_{\restriction\mathcal{T}}$, $A \mapsto \int g\chi_A\, d\lambda$ for $A \in \mathcal{T}$. □

[7] We reproduce a proof from Terence Tao's blog.

Proposition 3.2.11. *(1) $E(\cdot \mid \mathcal{T}) =: \pi_{\mathcal{T}}\colon L^1(\mu) \to L^1(\mu_{\restriction \mathcal{T}}) \subset L^1(\mu)$ is a projection.*

(2) $\pi_{\mathcal{T}}$ is linear and positive, that is, $f \geq 0 \Rightarrow f_{\mathcal{T}} \geq 0$.

(3) If g is $\mathcal{T}$-measurable and bounded, then $E(gf \mid \mathcal{T}) = gE(f \mid \mathcal{T})$.

(4) If $\mathcal{T}_2 \subset \mathcal{T}_1$ then $E(\cdot \mid \mathcal{T}_2) \circ E(\cdot \mid \mathcal{T}_1) = E(\cdot \mid \mathcal{T}_2)$.

The proof is straightforward; we note that (1) follows from (4) but more directly from $\pi_{\mathcal{S}} = \mathrm{Id}$.

We digress briefly to a contemplation of how this plays out in L^2.

Definition 3.2.12. Suppose H is a Hilbert space and $L \subset H$ is a closed subspace. Then each $v \in H$ uniquely[8] decomposes as $v = v_0 + v_\perp$, where $v_0 \in L$ and $v_\perp \perp L$, that is, $v_\perp \perp w$ for all $w \in L$, and the *orthogonal projection to* L is defined by

$$\pi_L\colon H \to L, \quad v_0 + v_\perp \mapsto v_0.$$

Proposition 3.2.13. *If $v \in H$, $w \in L$, then $\|v - \pi(v)\| \leq \|v - w\|$ and $\langle v, w\rangle = \langle \pi_l(v), w\rangle$.*

Proof. $\|v - w\|^2 = \|v_0 + v_\perp - w\|^2 = \|v_0 - w\|^2$ is minimal iff $w = v_0 = \pi(v)$, and $\langle v, w\rangle = \langle v_0 + v_\perp, w\rangle = \langle v_0, w\rangle = \langle \pi_l(v), w\rangle$. □

Example 3.2.14. Suppose $(X, \mathcal{T}, \mu)$ is a probability space and $\mathcal{S} \subset \mathcal{T}$ is a σ-algebra in $\mathcal{T}$. Then $L := L^2(X, \mathcal{S}, \mu) \subset H := L^2(X, \mathcal{T}, \mu)$ is a closed subspace. For $f \in L^2(X, \mathcal{T}, \mu)$ and $A \in \mathcal{S}$ we then have $\chi_A \in L$ and hence by Proposition 3.2.13,

$$\int_A f\, d\mu = \langle f, \chi_A\rangle = \langle \pi_L(f), \chi_A\rangle = \int_A \pi_L(f)\, d\mu.$$

In light of uniqueness in Corollary 3.2.10 we see that $\pi_{L^2(X,\mathcal{S},\mu)} = E(\cdot \mid \mathcal{S})_{\restriction L^2(X,\mathcal{T},\mu)}$, that is, the orthogonal projection to $L^2(X, \mathcal{S}, \mu)$ is given by conditional expectation.

We next prove the Birkhoff Ergodic Theorem for discrete time. The continuous-time counterpart (Theorem 3.2.17) then follows easily. If T is a measure-preserving transformation of a measure space $(\mathcal{B}, \mu)$, denote by $\mathcal{I} := \mathcal{I}_T := \{A \in \mathcal{B} \mid T^{-1}(A) = A\}$ the invariant σ-algebra .

[8] $v_0 + v_\perp = w_0 + w_\perp \Rightarrow v_0 - w_0 = w^\perp - v^\perp \in L \cap L^\perp = \{0\}$.

Theorem 3.2.15 (Birkhoff Ergodic Theorem). *If $(X, \mathcal{T}, \mu)$ is a probability space, $T: X \to X$ is μ-preserving, and $f \in L^1(X, \mu)$, then the time average exists:*

$$f_T := \lim_{n\to\infty} \frac{1}{n} \sum_{k=0}^{n-1} f \circ T^k = f_{\mathcal{I}_T} \quad \mu\text{-a.e.}$$

In particular, f_T is measurable and T-invariant, and

$$\int f_T \, d\mu = \int f_{\mathcal{I}} \, d\mu = \int f \, d\mu. \tag{3.2.1}$$

Proof. If $g \in L^1(\mu)$, then $G_n := \max_{k \le n} \sum_{i=0}^{k-1} g \circ T^i \in L^1(\mu)$ is nondecreasing in n, and

$$\overline{\lim_{n\to\infty}} \frac{1}{n} \sum_{k=0}^{n-1} g \circ T^k \le \overline{\lim_{n\to\infty}} \frac{G_n}{n} \le 0 \quad \text{outside} \quad A := \{x \mid G_n(x) \to \infty\} \in \mathcal{I}. \tag{3.2.2}$$

Further, $G_{n+1} = g + G_n \circ T \Leftrightarrow G_n \circ T \ge 0$, so $G_{n+1} - G_n \circ T = g - \min(0, G_n \circ T) \searrow g$ on A, and

$$0 \le \int_A (G_{n+1} - G_n) \, d\mu = \int_A (G_{n+1} - G_n \circ T) \, d\mu \xrightarrow[\text{Theorem}]{\substack{\text{Monotone} \\ \text{Convergence}}} \int_A g \, d\mu = \int_A g_{\mathcal{I}} \, d\mu\restriction_{\mathcal{I}},$$

so $g_{\mathcal{I}} < 0 \Rightarrow \mu(A) = 0$. If $g := f - f_{\mathcal{I}} - \epsilon$, then $g_{\mathcal{I}} = -\epsilon < 0$, so (3.2.2) becomes

$$\overline{\lim_{n\to\infty}} \frac{1}{n} \sum_{k=0}^{n-1} (f \circ T^k) - f_{\mathcal{I}} - \epsilon \le 0 \ \mu\text{-a.e.} \quad \text{with } \epsilon > 0 \text{ arbitrary.}$$

Replacing here f by $-f$ gives $\underline{\lim}_{n\to\infty} \frac{1}{n} \sum_{k=0}^{n-1} f \circ T^k \ge f_{\mathcal{I}} - \epsilon$ μ-a.e.[9] □

Now consider a measurable map $(t, x) \mapsto \varphi^t(x)$. To obtain the corresponding Birkhoff Ergodic Theorem for a flow Φ, we apply the Birkhoff Ergodic Theorem (Theorem 3.2.15) to the measure-preserving transformation φ^1 and the function $f_1 := \int_0^1 f \circ \varphi^s \, ds$.

Proposition 3.2.16. *If $(X, \mathcal{T}, \mu)$ is a probability space, Φ a μ-preserving flow on X, and $f \in L^1(X, \mathcal{T}, \mu)$, then $\frac{1}{n} \int_0^1 f \circ \varphi^{n+s} \, ds \xrightarrow[n\to\infty]{\text{a.e.}} 0$.*

[9]This proof from [213] incorporates a shortcut by A. Fieldsteel and B. Bassler compared to the version originally communicated to us by Uwe Schmock who had first seen it in lecture notes by Erwin Bolthausen (at the time of this writing a managing editor of this book series) with attribution to Jacques Neveu. (Neveu in turn explicitly told us that he was unaware of having given any such proof.)

Proof. The Birkhoff Ergodic Theorem (Theorem 3.2.15) applied to the measure-preserving transformation φ^1 and the function $f_1 := \int_0^1 f \circ \varphi^s \, ds \in L^1(X, \mathcal{T}, \mu)$ gives

$$\underbrace{\frac{1}{n}\int_0^1 f \circ \varphi^{n+s} \, ds}_{=f_1 \circ \varphi^n} = \frac{n+1}{n}\left[\frac{1}{n+1}\sum_{k=0}^{n} f_1 \circ \varphi^k\right] - \frac{1}{n}\sum_{k=0}^{n-1} f_1 \circ \varphi^k \xrightarrow[n\to\infty]{\text{a.e.}} 1\cdot(f_1)_{\mathcal{I}} - (f_1)_{\mathcal{I}} \overset{\text{ae}}{=} 0.$$

□

Theorem 3.2.17 (Birkhoff Ergodic Theorem for flows). *If $(X, \mathcal{T}, \mu)$ is a probability space, Φ a μ-preserving flow, and $f \in L^1(X, \mathcal{T}, \mu)$, then the time average exists:*

$$f_\Phi(x) := \lim_{t\to\infty} \frac{1}{t}\int_0^t f \circ \varphi^s ds = f_{\mathcal{I}} \quad \mu\text{-a.e.}$$

Remark 3.2.18. In the proof and later, we use the Bachmann–Landau *"little O" notation*: $f(t) \in o(g(t)) :\Leftrightarrow \frac{f(t)}{g(t)} \xrightarrow[t\to a]{} 0$, where a is usually clear from the context and most often equal to 0 or ∞. The corresponding *"big O" notation* is $f(t) \in O(g(t)) :\Leftrightarrow \frac{f(t)}{g(t)}$ is bounded for t near a. We sometimes write $f(t) = o(g(t))$ and $f(t) = O(g(t))$.

Proof. We apply the Birkhoff Ergodic Theorem (Theorem 3.2.15) to establish the existence of the limit and then show that it is $f_{\mathcal{I}}$. As a minor convenience we assume $f \geq 0$; the result follows from this by considering positive and negative parts.

First note that by Tonelli's Theorem,

$$\infty > n\int f \, d\mu = \int_0^n \int_X f(\varphi^s(x)) \, d\mu \, ds = \int_X \int_0^n f(\varphi^s(x)) \, ds \, d\mu,$$

so $0 \leq f_n := \int_0^n f \circ \varphi^s \, ds$ is well defined (and finite) off a null set E_n with $\int f_1 = n \int f$. The Birkhoff Ergodic Theorem (Theorem 3.2.15) gives

$$\frac{1}{n}\int_0^n f \circ \varphi^s ds = \frac{1}{n}\sum_{k=0}^{n-1} f_1 \circ \varphi^k \xrightarrow[n\to\infty]{} E(f_1 \mid \mathcal{I}_{\varphi^1}) \text{ off a null set } F. \tag{3.2.3}$$

To pass from integer times to others, consider x outside the null set N defined as the union of the set F in (3.2.3), all the E_n above and the null set implicit in Proposition 3.2.16. Then Proposition 3.2.16 and $f \geq 0$ imply

$$0 \leq \int_0^{t-\lfloor t\rfloor} f(\varphi^s(\varphi^{\lfloor t\rfloor}(x)))ds \leq f_1(\varphi^{\lfloor t\rfloor}(x)) \in o(t),$$

so (3.2.3) gives

$$\underbrace{f_\Phi(x)}_{=\lim_{t\to\infty}\frac{1}{t}\int_0^t f(\varphi^s(x))\,ds} = \lim_{t\to\infty} \frac{\lfloor t\rfloor}{t}\frac{1}{\lfloor t\rfloor}\sum_{k=0}^{\lfloor t\rfloor-1} f_1(\varphi^k(x)) + \lim_{t\to\infty}\frac{1}{t}\int_0^{t-\lfloor t\rfloor} f(\varphi^s(\varphi^{\lfloor t\rfloor}(x)))\,ds$$
$$= (f_1)_{\mathcal{I}_{\varphi^1}} + 0.$$

Thus $\int f_\Phi = \int f$. Now apply what we proved so far to $g := f\chi_A$ for any $A \in \mathcal{I}$:

$$\int_A f_\Phi = \int f_\Phi \chi_A = \int f_\Phi(\chi_A)_\varphi = \int (f\chi_A)_\varphi = \int g_\varphi = \int g = \int f\chi_A = \int_A f,$$

and this, together with Φ-invariance, is the very definition of $f_\Phi = f_{\mathcal{I}}$.[10] □

The Birkhoff Ergodic Theorem also yields almost-everywhere convergence of negative and two-sided time averages:

Proposition 3.2.19.

$$\overline{f}_\Phi := \lim_{t\to\infty}\frac{1}{t}\int_0^t f\circ\varphi^{-s}\,ds = f_{\mathcal{I}} \overset{\text{ae}}{=} f_\Phi \quad \textit{and} \quad \frac{1}{2t}\int_{-t}^t f\circ\varphi^s\,ds \xrightarrow{\text{a.e.}} f_{\mathcal{I}}.$$

Remark 3.2.20. The Birkhoff Ergodic Theorem says that $f \mapsto f_\Phi$ is a projection to the Φ-invariant functions.

Remark 3.2.21 (Empirical measure). Another perspective on existence of time averages is given by the *empirical measure* $\epsilon_{x,T} := \frac{1}{T}\int_0^T \delta_{\varphi^s(x)}\,ds$ for a given $x \in X$.

If $f \in L^1(X,\mathcal{T},\mu)$, then $\epsilon_{x,T}(f)$ converges for μ-a.e. x by the Birkhoff Ergodic Theorem. Thus, if $L^1(X,\mathcal{T},\mu)$ is separable, then $\epsilon_{x,T}$ converges weakly for μ-a.e. x.

The exceptional set where the positive or negative time averages do not exist may, of course, depend on the function f. However, it is negligible for any invariant measure.

Definition 3.2.22. Given a continuous flow Φ of a metric space X, we say that a subset $A \subset X$ has *total measure* if A has full measure with respect to *any* Φ-invariant Borel probability measure on X.

Corollary 3.2.23. *Let X be compact metrizable, and Φ a continuous flow. Then*

$$\{x \in X \mid \lim_{t\to\infty}\tfrac{1}{t}\textstyle\int_0^t f\circ\varphi^k(x)\,ds \text{ exists for all continuous functions } f\}$$

[10] This proof follows one in ETH lecture notes by Oscar Lanford.

has total measure, as does

$$\left\{x \in X \mid \lim_{t\to\infty} \tfrac{1}{t}\int_0^t f \circ \varphi^s(x)\,ds = \lim_{t\to\infty} \tfrac{1}{t}\int_0^t f \circ \varphi^{-s}(x)\,ds \text{ for } f \in C(X)\right\}.$$

In particular, for a set of points of total measure the associated empirical measures converge weakly.

Proof. For each f_j in a countable dense set of functions the averages converge on a set E_i of total measure. Lipschitz continuity of $f \mapsto \frac{1}{t}\int_0^t f \circ \varphi^s\,ds$ implies convergence on $\bigcap_i E_i$ for all continuous f, and having total measure is stable under countable intersection. □

By the Krylov–Bogolubov Theorem (Theorem 3.1.15) a set of total measure is nonempty:

Corollary 3.2.24. *For any continuous flow Φ on a compact metric space X there is an $x \in X$ such that for every continuous function f on X the time averages $\frac{1}{t}\int_0^t f \circ \varphi^s(x)\,ds$ and $\frac{1}{t}\int_0^t f \circ \varphi^{-s}(x)\,ds$ both converge and have the same limit.*

Remark 3.2.25. We emphasize that while Corollary 3.2.23 produces an apparently large set of points whose Birkhoff averages exist, we have encountered instances of dynamical systems with a paucity of invariant measures; if these are moreover atomic, then the set promised by Corollary 3.2.23 may not look very large. For instance, in the south–south flow (Example 1.3.13) the fixed point is a set of total measure (see Exercise 3.6). Ruelle proposed calling points "historic" if they do not have a Birkhoff average, the idea being that the running average $\frac{1}{t}\int_0^t f \circ \varphi^s\,ds$ fluctuates significantly over time and thus in a vague way associates with a given average a time t or, rather, an "era" in the "history" of the orbit.[11] Among the questions one can raise when Lebesgue measure is defined on X but not invariant under a given flow, are whether Lebesgue-almost all points have Birkhoff averages or whether instead there is a set of historic points with positive Lebesgue measure. While for hyperbolic flows this does not happen (Remark 7.4.11), a variant of Figure 1.5.4 shows a situation where this is indeed the case: taking Figure 1.1.4 to represent a flow on $\mathbb{R}^2$, alter it, so orbits spiral out from the neutral fixed point to the homoclinic loop that connects adjacent saddles (Figure 3.2.1). Bowen observed that if f is a continuous function with different values at the adjacent saddles, then for any of those points whose orbits spiral toward the homoclinic loop, the Birkhoff averages do not exist—because the time spent near each of the saddles grows exponentially and therefore always moves the running averages back toward that value of f. We thus have an open set of historic points.[12]

[11] As Ruelle put it, "This absence of limit is what we want to call historical behaviour. This means that, as the time... tends to ∞, the point... keeps having new ideas about what it wants to do." [319]

[12] This example does not persist under typical perturbations because the homoclinic loop can disappear or become tangled instead; there are persistent examples, however, using homoclinic tangencies [230].

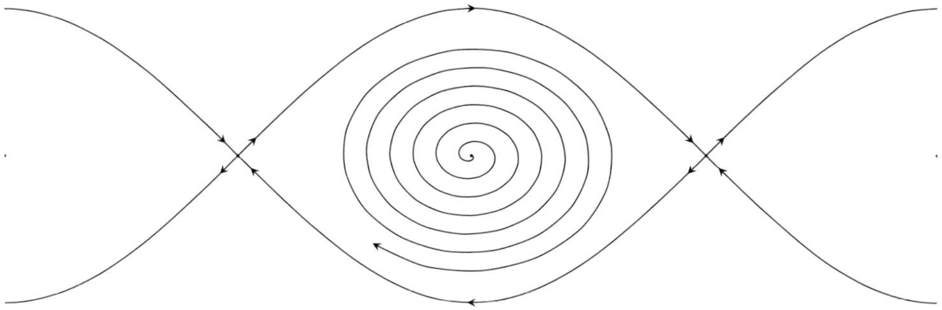

Figure 3.2.1. Spiraling toward a homoclinic loop.

3.3 Ergodicity

We now introduce a central notion of this chapter. We discussed in Section 0.3.c that ergodic theory arose from the desire for equality of time averages, on whose existence we just elaborated, with the space average of an observable. Ergodicity of an invariant Borel probability measure is the very indecomposability notion which produces this circumstance. Despite their names, the ergodic theorems in the previous section do not presuppose the measure to be ergodic, and we will show how these general theorems specialize to ergodic systems to give in particular the equality of time and space averages (Corollary 3.3.11).

Definition 3.3.1. A measure μ is said to be *ergodic* with respect to Φ, or one says that Φ is ergodic with respect to μ, if for any measurable $A \subset X$ with $\varphi^{-t}(A) = A$ for all $t \in \mathbb{R}$ either $\mu(A) = 0$ or $\mu(X \setminus A) = 0$.

Remark 3.3.2. The Φ-invariance of μ is not needed for this definition. Dirac measures are trivially ergodic, as is $\delta_{\mathcal{O}(p)}$ in Remark 3.1.18 ($\mathcal{O}(p)$ has no proper invariant subsets) and hence Lebesgue measure for the translation flow on the circle from Example 1.1.6. The first nontrivial instances are given by Propositions 3.3.6 and 3.3.7 below. It is clear from the definition (or from Proposition 3.3.12 below) that if μ is ergodic and $\nu \ll \mu \ll \nu$, then so is ν.

Ergodicity can be reformulated in functional language:

Proposition 3.3.3 (Characterization of ergodicity). *The following are equivalent*:

(1) *Φ is ergodic with respect to μ.*

(2) *Any measurable Φ-invariant $f: X \to \mathbb{C}$ is constant μ-a.e.*

(3) *Any bounded measurable Φ-invariant $f: X \to \mathbb{R}$ is constant μ-a.e.*

(4) *Any Φ-invariant $f \in L^p(X, \mu)$ is constant μ-a.e.*

(5) *Any nonnegative measurable Φ-invariant $f: X \to \mathbb{C}$ is constant μ-a.e.*

Remark 3.3.4. Ergodicity of a probability measure μ is also characterized by each of the following:

- $f \in C(X) \Rightarrow f_\Phi = \text{const.}$ μ-a.e. (Theorem 3.3.10).
- $\mu \gg \nu \in \mathfrak{M}(\Phi) \Rightarrow \mu = \nu$ (Proposition 3.3.12).
- μ is an extreme point of $\mathfrak{M}(\Phi)$ (Proposition 3.3.26).
- μ is ergodic for the time-τ map for all but countably many τ (Theorem 3.3.13, Proposition 3.3.14).

Proof of Proposition 3.3.3. These (and other) characterizations arise from the following implications: Φ is not ergodic $\Rightarrow$ there is an invariant characteristic function (namely, of an invariant set of intermediate measure) that is not constant a.e. $\Rightarrow$ there is a nonnegative bounded invariant measurable function that is not constant a.e. $\Rightarrow$ there is a nonconstant invariant $f \in L^p \Rightarrow$ there is an invariant measurable $\mathbb{C}$-valued function that is not constant a.e. $\Rightarrow$ Φ is not ergodic (because either the real or the imaginary part is a Φ-invariant measurable function $f: X \to \mathbb{R}$ and not constant almost everywhere, so there exists an $a \in \mathbb{R}$ such that $\mu(f^{-1}((a, \infty))) \notin \{0, 1\}$, and this set is invariant). □

Remark 3.3.5. With any of these characterizations and keeping in mind that invariance of the measure is not needed to define ergodicity, it is easy to see that ergodicity is preserved by time change, orbit equivalence (both for the given measure or the one induced from it by the time change or the orbit equivalence), measure-theoretic isomorphism, and passing to factors or suspensions.[13] (To which *invariant* measures these various modifications lead is an altogether different and harder question.)

As we mentioned earlier, ergodicity can be thought of as the measurable analog to transitivity. Similarly to the above, transitivity is preserved by time change, conjugacy, orbit equivalence, and passing to factors or suspensions.

Ergodicity can be (and has been) viewed as having no *measurable* constant of motion. This is different from not having constants of motion, which follows from transitivity. Ergodicity does not follow from transitivity, even if the measure is a smooth volume, and there are even minimal nonergodic systems, though such examples are not easy to construct [213, Corollary 12.6.4].

[13] A suspension is ergodic if and only if the base transformation is.

Thanks to Proposition 3.3.3, the proof of Proposition 1.6.15 yields the following results:

Proposition 3.3.6. *A linear flow $x \mapsto x + tv$ on $\mathbb{T}^n$ is ergodic with respect to Lebesgue measure if and only if the components of v are rationally independent.*

Proposition 3.3.7. *Consider $A \in \mathrm{GL}(m, \mathbb{Z})$, that is, an $m \times m$-matrix with integer entries and determinant ± 1, and assume that no eigenvalue of A is a root of unity. Then the suspension of the toral automorphism $F_A : \mathbb{T}^m \to \mathbb{T}^m$ induced by A is ergodic.*

Remark 3.3.8 (Walters). Note that the hypotheses hold for $\left(\begin{smallmatrix} 2 & 1 \\ 1 & 1 \end{smallmatrix}\right)$ and indeed any hyperbolic automorphism, but also for

$$W := \begin{pmatrix} 0 & 0 & 0 & -1 \\ 1 & 0 & 0 & 8 \\ 0 & 1 & 0 & -6 \\ 0 & 0 & 1 & 8 \end{pmatrix} \in \mathrm{GL}(4, \mathbb{Z}).$$

Its characteristic polynomial $q(\lambda) := \lambda^4 - 8\lambda^3 + 6\lambda^2 - 8\lambda + 1$ is irreducible over $\mathbb{Q}$ because so is $q(\lambda - 1) = \lambda^4 - 12\lambda^3 + 36\lambda^2 - 48\lambda + 24$ by Eisenstein's Criterion (the prime $p = 3$ divides all coefficients other than the leading one, but $p^2 = 9$ does not). The eigenvalues $2 - \sqrt{3} \pm i\sqrt{4\sqrt{3} - 6}$ lie on the unit circle, and the remaining two are real and off the unit circle. Therefore q is not a factor of $\lambda^n - 1$ for any n; since q is irreducible, the eigenvalues on the unit circle are thus not roots of unity.

Proof. A bounded measurable invariant function f does not depend on t, hence is naturally written as an F_A-invariant function on $\mathbb{T}^m$.[14] Fourier expansion gives

$$\sum_{k \in \mathbb{Z}^m} f_k \exp(2\pi i \langle k, x \rangle) = f(x) \overset{\text{ae}}{=} f(F_A(x)) = \sum_{k \in \mathbb{Z}^m} f_k \exp(2\pi i \underbrace{\langle k, Ax \rangle}_{=\langle A^t k, x \rangle}).$$

Uniqueness of the Fourier expansion implies that $f_k = f_{(A^t)^n k}$ for $n \in \mathbb{N}$. Since no root of unity is an eigenvalue of A and hence of the transpose A^t, $(A^t)^l - \mathrm{Id}$ is invertible for every $l \in \mathbb{Z} \setminus \{0\}$. So, for $k \in \mathbb{Z}^m \setminus \{0\}$ the $(A^t)^n k$ (for $n \in \mathbb{Z}$) are pairwise distinct, that is, there are infinitely many $l \in \mathbb{Z}^m$ with $f_l = f_k$. But $f \in L^1$ implies $|f_k| = |f_l| \xrightarrow[|l| \to \infty]{} 0$, so $f_k = 0$. This means that $f \overset{\text{ae}}{=} f_0$, a constant. □

Remark 3.3.9. Proposition 3.3.3 simply states in various function spaces that the subspace of Φ-invariant functions is the space of constant functions. Remark 3.2.20 lets us determine the space of Φ-invariant functions as the range of the projection $f \mapsto f_\Phi$, and doing so for a dense set of functions gives the needed information. If X is a metric space, then density of $C(X)$ in L^p gives the following theorem:

[14] Equivalently, we could invoke Remark 3.3.5.

Theorem 3.3.10. *If* $f_\Phi = \text{const.}$ *μ-a.e. for every $f \in C(X)$, then μ is ergodic.*

The converse (that the time average equals the space average—to which we alluded at the start of this section) is an important corollary of the Birkhoff Ergodic Theorem (Theorem 3.2.15).

Corollary 3.3.11 (Strong Law of Large Numbers). *If $\mu(X) = 1$, Φ is an ergodic μ-preserving flow, and $f \in L^1(X,\mu)$, then*

$$f_\Phi(x) = \lim_{T\to\infty} \frac{1}{T} \int_0^T f(\varphi^t(x))\,dt = \int_X f\,d\mu$$

for every x outside a set of measure 0.

Proof. The function f_Φ is Φ-invariant, so constant a.e. By (3.2.1) the constant is $\int f\,d\mu$. □

Thus, an invariant measure determines the asymptotic distribution of μ-almost every point if it is ergodic. A nonergodic invariant measure μ may also determine the asymptotic distribution of *some* orbits, but such orbits are always a set of μ-measure 0.

Considering densities gives the following proposition:

Proposition 3.3.12. *The measure $\mu \in \mathfrak{M}(\Phi)$ is ergodic if and only if $\mu \gg \nu \in \mathfrak{M}(\Phi) \Rightarrow \mu = \nu$.*

Proof. The measure $\mu \gg \nu \in \mathfrak{M}(\Phi) \Leftrightarrow \nu = \rho \cdot \mu$, where $\rho \in L^1(\nu)$ is the (unique hence Φ-invariant) Radon–Nikodym derivative. This is always constant ($\equiv 1$) iff ν is ergodic. □

The argument in the proof of Theorem 1.6.24 also establishes the following theorem:

Theorem 3.3.13. *If a probability measure is ergodic for a flow then for all but countably many τ it is ergodic for the time-τ map.*

Conversely, we clearly have the following proposition:

Proposition 3.3.14. *If the time-t map φ^t is ergodic for some t, then Φ is ergodic.*

Example 3.3.15. The time-t maps of the circle flow (Example 1.1.6) are ergodic (with respect to Lebesgue measure) exactly for irrational t. This can be seen via the Fourier decomposition of an invariant function f:

$$\sum_{i\in\mathbb{Z}} a_i e^{ix} = f(x) = f(x+t) = \sum_{i\in\mathbb{Z}} a_i e^{it} e^{ix} \Rightarrow \forall i \in \mathbb{Z} \quad a_i = a_i e^{it},$$

so either $a_i = 0$ for all $i \neq 0$ (so $f \equiv \text{const.}$) or $it \in \mathbb{Z}$ for some $i \in \mathbb{Z}$, hence $t \in \mathbb{Q}$.

In light of Proposition 3.3.14, ergodicity of the geodesic flow in Section 2.3 with respect to the *Liouville measure* defined by the invariant contact form in (2.2.5) follows from the next result:

Theorem 3.3.16. *For $t \neq 0$ the time-t-map of the geodesic flow on a finite-volume factor of the Poincaré disk* (*Section* 2.3) *is ergodic.*

Proof. If $f \circ g^t = f \in L^2$ (for fixed t), then $f \circ h_+^s - f$ (and likewise for h_-^s):

$$\|\underbrace{f \circ g^{nt}}_{=f} \circ h_+^s - \underbrace{f \circ g^{nt}}_{=f}\| \overset{(2.2.3)}{=\!=\!=} \|f \circ h_+^{se^{nt}} \circ g^{nt} - f \circ g^{nt}\|$$
$$= \|f \circ h_+^{se^{nt}} - f\| \xrightarrow[nt\to-\infty]{\text{Remark 3.2.7}} 0.$$

Then, g^t, h_+^s and h_-^s generate $SL(2,\mathbb{R})$, so f is $PSL(2,\mathbb{R})$-invariant, that is, for all $g \in PSL(2,\mathbb{R})$, $f \circ g \overset{\text{ae}}{=} f$, or, by the Fubini Theorem, for a.e. x we have $f(g(x)) = f(x)$ for a.e. $g \in PSL(2,\mathbb{R})$. Thus, there is an x_0 with $f(g(x_0)) = f(x_0)$ for all $g \in PSL(2,\mathbb{R})$, so $f \overset{\text{ae}}{=}$ const. □

Corollary 3.3.17. *The geodesic flow on a finite-volume factor of the Poincaré disk* (*Section* 2.3) *is ergodic with respect to the Liouville measure.*

Remark 3.3.18. Indeed, Theorem 3.3.16 implies more than ergodicity by Proposition 3.4.40 (Theorem 3.4.43). Yet stronger ergodic properties are obtained below with refined arguments (Theorem 3.4.32, and later on Theorem 7.1.12).

The horocycle flow from Theorem 3.3.16 is ergodic as well, and this is tightly connected with ergodicity of the geodesic flow.

Proposition 3.3.19. *An L^2-function invariant under a time-τ-map of the horocycle flow h_-* (*Example* 2.2.3) *on a finite-volume factor of the Poincaré disk* (*Section* 2.3) *is invariant under the time-*(2 ln 2) *map of the geodesic flow.* (*Likewise for h_+.*)

Proof. If $f \in L^2$ is invariant under h_-^τ, then (2.2.4) with $\epsilon := 1/n\tau$ and $s = 2$ gives

$$\|f \circ g^{2\ln 2} - f\|_2 = \|\underbrace{f \circ h_-^{-2n\tau}}_{=f} h_+^{1/n\tau} h_-^{n\tau} h_+^{-2/n\tau} - f\|_2$$
$$= \|f \circ h_+^{1/n\tau} h_-^{n\tau} h_+^{-2/n\tau} - \underbrace{f \circ h_-^{n\tau}}_{=f} h_+^{-2/n\tau} + f \circ h_+^{-2/n\tau} - f\|_2$$
$$\leq \|f \circ h_+^{1/n\tau} - f\|_2 + \|f \circ h_+^{-2/n\tau} - f\|_2 \xrightarrow[n\tau\to+\infty]{\text{Remark 3.2.7}} 0. \qquad \square$$

Corollary 3.3.20. *Each time-τ-map of the horocycle flow $h_\pm$* (*Example* 2.2.3) *on a finite-volume factor of the Poincaré disk is ergodic.*

Remark 3.3.21. In fact, the horocycle flow is uniquely ergodic and hence strictly ergodic (Definition 3.1.17, Exercise 6.8, Corollary 3.4.35): Birkhoff averages converge uniformly (Theorem 3.3.32). (And more—see Theorem 3.4.44 and Section 8.6.)

One can strengthen the statement that functions invariant under an ergodic flow are constant, via the following simple observation:

Proposition 3.3.22. *If Φ is a μ-preserving flow and $f\colon X \to \mathbb{R}$ satisfies $f \circ \varphi^t \le f$ ("subinvariance"), then f is Φ-invariant.*

Proof. By assumption, $A_r := \{x \in X \mid f(x) \le r\} \supset \{x \in X \mid f(\varphi^t(x)) \le r\} = \varphi^{-t}(A_r)$, while $\mu(\varphi^{-t}(A_r)) = \mu(A_r)$. Thus $\varphi^{-t}(A_r) \overset{\text{ae}}{=} A_r$ for all $r \in \mathbb{R}$. □

This and Proposition 3.3.3 yield the following corollary:

Corollary 3.3.23 (Subinvariance). *If μ is an ergodic Φ-invariant probability measure, $f\colon X \to \mathbb{R}$, and $f \circ \varphi^t \le f$, then f is constant μ-a.e.*

Proposition 3.3.24. *A probability-preserving flow Φ is ergodic iff*

$$\int_X \Big(\frac{1}{T}\int_0^T f \circ \varphi^t \, dt\Big) g \xrightarrow[T\to\infty]{} \int_X f \int_X g \tag{3.3.1}$$

for all $f, g \in L^2$, that is, if and only if $\frac{1}{T}\int_0^T f \circ \varphi^t \xrightarrow[T\to\infty]{\text{weakly}}$ const. for all $f \in L^2$.

Remark 3.3.25. For $f = \chi_A$ and $g = \chi_B$, (3.3.1) becomes

$$\frac{1}{T}\int_0^T \mu(\varphi^{-t}(A) \cap B) - \mu(A)\mu(B) \, dt \xrightarrow[T\to\infty]{} 0. \tag{3.3.2}$$

Proof. If $f = f \circ \varphi^t$, then $f = \frac{1}{T}\int_0^T f \circ \varphi^t \, dt \xrightarrow[T\to\infty]{\text{weakly}}$ const., so Φ is ergodic.

If Φ is ergodic, then Corollary 3.3.11 and the Vitali Convergence Theorem (Theorem A.3.31) give (3.3.1) for all $f, g \in L^2$. □

Corollary 3.3.11 leads to the question of whether every continuous flow has an ergodic invariant measure. This becomes clear with an alternate characterization.

Proposition 3.3.26. *Ergodic measures are the extreme points of $\mathfrak{M}(\Phi)$: $\mu \in \mathfrak{M}(\Phi)$ is not ergodic iff there exist $\mu_1 \ne \mu_2 \in \mathfrak{M}(\Phi)$ and $0 < \lambda < 1$ such that $\mu = \lambda\mu_1 + (1-\lambda)\mu_2$.*

Proof. If $\varphi^{-t}(A) = A$ and $0 < \mu(A) < 1$, then $\mu = \mu(A)\mu_1 + (1 - \mu(A))\mu_2$, where

$$\mu_1(B) := \mu_A(B) := \mu(B \mid A) := \frac{\mu(B \cap A)}{\mu(A)} \tag{3.3.3}$$

is the *density of B in A*, and $\mu_2 := \mu_{X \smallsetminus A} \perp \mu_A = \mu_1$.

Since $\mu_i \ll \mu$ for $i = 1, 2$, so the Radon–Nikodym Theorem gives Φ-invariant $\rho_i \in L^1(\mu)$ with $\int f\,d\mu_i \equiv \int \rho_i f\,d\mu$. By assumption, $\lambda\rho_1 + (1-\lambda)\rho_2 = 1 = \int \rho_1\,d\mu = \int \rho_2\,d\mu$, so $\mu_1 \neq \mu_2 \Rightarrow \rho_1 \neq \rho_2 \Rightarrow \rho_1 \not\equiv \text{const.}$, and μ is not ergodic. □

Theorem 3.3.27 (Existence of ergodic measures). *Every continuous flow on a metrizable compact space has an ergodic invariant Borel probability measure.*

Proof. By the Krein–Milman Theorem,[15] $\mathfrak{M}(\Phi) \neq \varnothing$ has extreme points. □

Corollary 3.3.28. *A uniquely ergodic flow* (*Definition* 3.1.17) *is ergodic.*

Proof. Since $\mathfrak{M}(\Phi) = \{\mu\}$, so μ is extreme, hence ergodic. □

By the Krylov–Bogolubov Theorem (Theorem 3.1.15) every minimal set is the support of an invariant measure, so we have the following theorem:

Theorem 3.3.29. *A uniquely ergodic action has only one minimal set; in particular a topologically transitive uniquely ergodic action is minimal.*

Remark 3.3.30. Exercise 3.11 provides a related inference for ("plain") ergodicity.

Example 3.3.31. The flow in Example 1.3.7 is uniquely ergodic, so unique ergodicity is compatible with trivial recurrence—but only for Dirac measures.

The circle flow (Example 1.1.6) is uniquely ergodic. To see this note that the interval $[0, \frac{1}{n})$ has measure $1/n$ because all translates by multiples of $1/n$ have the same measure, and they sum to 1. By additivity, the measure of intervals with rational endpoints is their length; this defines Lebesgue measure.

The following theorem is a more generally useful criterion:

Theorem 3.3.32. *A continuous flow is uniquely ergodic if the time averages of continuous functions converge uniformly to a constant.*

Proof. If f is a continuous function, then $\frac{1}{T}\int_0^T f \circ \varphi^t \xrightarrow[T\to\infty]{\text{uniformly}} f_0 \in \mathbb{R}$. If μ is a Φ-invariant Borel probability measure, then $\int f\,d\mu = \int f_0\,d\mu = f_0$, so μ is uniquely defined on $C(X)$ and hence unique. □

We have the converse:

Proposition 3.3.33. *If Φ is uniquely ergodic, then for every continuous function f the time averages $\frac{1}{T}\int_0^T f(\varphi^t(x))\,dt$ converge uniformly* (*to a constant*).

[15] A compact convex set in a locally convex topological vector space is the closed convex hull of its extreme points, that is, $C = \overline{\text{co}}\,\text{ex}(C)$; less than this will do, of course, when only the existence of an extreme point is needed: Define a *face* F of a compact convex set K by $x + (0,1)(x-y) \subset F \Rightarrow x + [0,1](x-y) \subset F$; K itself has this property. The Hausdorff maximal principle gives a minimal face, and the Hahn–Banach Extension Theorem shows that it must be a point, hence an extreme point.

Proof. If f is a continuous function for which this fails, then there are $a < b$, sequences of points $x_k, y_k \in X$, $k = 1, 2, \dots$, and a sequence $n_k \to \infty$ such that

$$\frac{1}{n_k}\int_0^{n_k} f(\varphi^t(x_k))\,dt < a, \qquad \frac{1}{n_k}\int_0^{n_k} f(\varphi^t(y_k)) > b.$$

A diagonal argument gives a subsequence n_{k_j} such that for every $g \in C(X)$ both

$$J_1(g) = \lim_{j\to\infty}\frac{1}{n_{k_j}}\int_0^{n_{k_j}} g(\varphi^t(x_{k_j}))\,dt \quad\text{and}\quad J_2(g) = \lim_{j\to\infty}\frac{1}{n_{k_j}}\int_0^{n_{k_j}} g(\varphi^t(y_{k_j}))\,dt$$

exist, where J_1 and J_2 are bounded linear positive Φ-invariant functionals. Thus $J_1(g) = \int g\,d\mu_1$, $J_2(g) = \int g\,d\mu_2$ for Φ-invariant probability measures μ_1 and μ_2. Since $J_1(f) \le a < b \le J_2(f)$ we have $\mu_1 \ne \mu_2$ so Φ is not uniquely ergodic. □

Theorem 3.3.34. *If a flow is uniquely ergodic then the time-τ maps for all but countably many τ are uniquely ergodic.*

Proof (Veech). By Theorem 3.3.13, the unique Φ-invariant Borel probability measure μ is φ^τ-ergodic for all but countably many τ. To show that such a φ^τ is uniquely ergodic, let ν be any φ^τ-invariant Borel probability measure. Then $\int_0^\tau \varphi^s(\nu)\,ds$ is Φ-invariant, so $\mu = \int_0^\tau \varphi^s(\nu)\,ds$ by unique ergodicity of Φ. However, μ is ergodic for φ^τ and hence an extreme point of the set of invariant measures, so the convex combination $\mu = \int_0^\tau \varphi^s(\nu)\,ds$ must be trivial, that is, $\varphi^s(\nu) = \mu$. Since μ is φ^s-invariant, this implies $\nu = \mu$, which establishes the claim. □

Remark 3.3.35. The examples of uniquely ergodic flows (as well as the majority of those one encounters in the early pertinent literature) suggest that unique ergodicity (and hence minimality) is closely tied to simple dynamics. This turns out to be wrong in the strongest possible way. Not only are there natural examples of uniquely ergodic weakly mixing flows (Definition 3.4.1, Theorem 3.4.44, Corollary 3.4.35), but by the Jewett–Krieger Theorem, *every* ergodic flow is measure-theoretically isomorphic to a uniquely ergodic one [202, 111].

Proposition 3.3.26 connects decomposability of a measure (by convex combination) and decomposability of the space. One can sharpen that connection:

Proposition 3.3.36. *Different invariant ergodic probability measures for the same flow are mutually singular.*

Proof. Call them $\nu, \mu = \mu^{\ll} + \mu^{\perp}$ with $\mu^{\ll} \ll \nu \perp \mu^{\perp}$ (invariantly by uniqueness of Lebesgue decomposition); since μ is ergodic, hence extreme, we have either $\mu = \mu^{\perp}$ or $\mu = \mu^{\ll} = \nu$ by ergodicity of ν and Proposition 3.3.12. □

Proposition 3.3.36 means that any convex combination of finitely many ergodic measures produces a corresponding nontrivial finite partition (Definition A.1.6) of the space.

Moreover, every invariant measure for a measure-preserving transformation can be decomposed into—possibly uncountably many—*ergodic components*.

Theorem 3.3.37 (Ergodic Decomposition [98, Theorem 15, p. 152]). *Every invariant Borel probability measure for a continuous flow* Φ *of a metrizable compact space* X *decomposes into an integral of ergodic invariant Borel probability measures in the following sense: there is a partition (modulo null sets) of* X *into* Φ*-invariant subsets* X_α, $\alpha \in A$, *called the* ergodic components *of* (Φ, μ), *with* A *a Lebesgue space, and each* X_α *carrying a* Φ*-invariant ergodic measure* μ_α *such that* $\int f\,d\mu = \iint f\,d\mu_\alpha\,d\alpha$ *for any function* f.

Remark 3.3.38. In metric spaces there is an explicit description of the ergodic decomposition: For each ergodic measure consider the G_δ set of typical points with respect to all continuous functions, for example, points for which the Birkhoff averages for each continuous function converge to the integral of this function with respect to the measure in question (Theorem 3.2.15). This is a null set for all other ergodic measures and these sets are evidently pairwise disjoint. They are called *ergodic sets.* This essential uniqueness of the ergodic decomposition shows that $\mathfrak{M}(\Phi)$ is essentially a simplex.

3.4 Mixing

As the circle flow (Example 1.1.6) illustrates, ergodicity is compatible with fairly uncomplicated behavior. Notions of *mixing* provide stronger stochastic properties, and the relation to ergodicity is most apparent by comparison with (3.3.2).

Unlike in the topological setting there are various notions of mixing used in the measure-theoretic setting. We first review the various definitions and list them in order of increasing strength.

Definition 3.4.1 (Mixing). A measure-preserving flow Φ of a measure space $(X, \mathcal{T}, \mu)$ is *weakly mixing* or has *continuous spectrum* (Remark 3.7.16) if for any two measurable sets A, B,

$$\frac{1}{T}\int_0^T |\mu(A \cap \varphi^{-t}(B)) - \mu(A)\mu(B)|\,dt \xrightarrow[T\to\infty]{} 0. \tag{3.4.1}$$

It is said to be *mixing* if for any two measurable sets A, B,

$$\mu(A \cap \varphi^{-t}(B)) \xrightarrow[t\to\infty]{} \mu(A)\cdot\mu(B). \tag{3.4.2}$$

It is said to be *mixing of order N* if for any $N+1$ measurable sets A_i and with $t_0 := 0$,

$$\mu\Big(\bigcap_{i=0}^{N}\varphi^{-t_i}(A_i)\Big) \xrightarrow[t_i - t_{i-1}\to\infty]{} \prod_{i=0}^{N}\mu(A_i). \tag{3.4.3}$$

It is said to be *multiply mixing* or *mixing of all orders* if it is mixing of order N for all $N \in \mathbb{N}$.

The next notion was introduced by Kolmogorov (under a different name) and is thus often referred to as the Kolmogorov property, or *K-property* for a flow.[16]

Definition 3.4.2 (K-mixing). A Φ-invariant Borel probability measure is said to be *K-mixing* or Φ is said to be a *K-flow* if for any measurable sets $A_0, \dots, A_m$ we have

$$\lim_{t\to\infty} \sup_{B\in\mathcal{A}_t(A_1,\dots,A_m)} |\mu(A_0\cap B) - \mu(A_0)\mu(B)| = 0,$$

where $\mathcal{A}_t(A_1,\dots,A_m)$ is the σ-algebra generated by the $\varphi^s(A_i)$ for $s \geq t$ and $1 \leq i \leq m$. Equivalently (Definition A.1.16),

$$\lim_{N\to\infty} \sup\big\{\big|\mu(A\cap B) - \mu(A)\mu(B)\big| \mid B \in \mathcal{A}(\textstyle\bigvee_{i=n}^{N} T^i\xi)\big\} \xrightarrow[n\to\infty]{} 0$$

for every measurable A and finite partition ξ.

Definition 3.4.3. A measure-preserving flow Φ is *Bernoulli* or said to have the *Bernoulli property* if for all $t \neq 0$ the time-t map is measure-theoretically isomorphic to a Bernoulli shift (Example 3.1.26).

Remark 3.4.4. A few comments on these notions and the relations between them:

- The circle flow (Example 1.1.6) is not weakly mixing: for $A = B = [0, 1/2)$ and $T \in \mathbb{N}$, the integral in (3.4.1) is $1/8 \neq 0$.
- Mixing is mixing of order 1.
- One can restate (3.4.2) as $\mu_B(\varphi^{-t}(A)) \xrightarrow[n\to\infty]{} \mu(A)$, that is, asymptotically $\varphi^{-t}(A)$ and B are independent sets.
- Clearly mixing implies weak mixing, so weak mixing is a weakened (average) version of the statement about asymptotic independence.
- By taking A invariant and $B := X \setminus A$ (or by comparing (3.4.1) and (3.3.1)) we find that weak mixing implies ergodicity. Thus, ergodicity is the weakest statement of this sort.

[16]Kolmogorov used "K" as an abbreviation for "quasiregular," which begins with a "K" in Russian, but it was quickly interpreted as the first letter of "Kolmogorov."

- One sharp distinction between ergodicity and these mixing notions is that ergodicity is a "transverse" property, whereas "longitudinal" issues (such as time changes) affect mixing. This step-up from ergodicity comes into sharp relief in Proposition 3.4.9: suspensions are never even weakly mixing.

- To clarify the intent of (3.4.3), we rewrite it for $N = 2$ as

$$\mu(\varphi^{-t}(A) \cap \varphi^{-s-t}(B) \cap C) \xrightarrow[s\to\infty \text{ and } t\to\infty]{} \mu(A)\mu(B)\mu(C).$$

- K-mixing means that the evolution of A_0 is eventually independent of anything involving the other A_i; this implies mixing of all orders but does not follow from it.

- The most effective criterion for the K-property is existence of a σ-algebra of measurable sets $\mathcal{A}$ such that $\mathcal{A} \subset \varphi^t \mathcal{A}$ for $t > 0$, $\bigcup_{t\ge 0} \varphi^t \mathcal{A}$ is dense in the σ-algebra $\mathcal{B}$ of all measurable sets, and $\bigcap_{t\ge 0} \varphi^{-t} \mathcal{A} = \mathcal{N}$, the trivial subalgebra of null sets and their complements. Equivalently, one can show the existence of a generator (Definition A.3.6) with trivial tail.

- The K-property is also equivalent to triviality of the *Pinsker algebra* from entropy theory. We will have an opportunity to show how this is useful (Remark 7.3.20).

- The Kolmogorov zero-one law for random variables can be used to show that the Bernoulli property implies K-mixing. There are, however, K-mixing flows that are not Bernoulli [272].

- Weak mixing does not imply mixing, and there is a significant gap between these. If one uses the weak topology on the space of measure-preserving flows on a given probability space, then the weakly mixing ones form a set of second Baire category, while mixing ones form a set of first category, that is, in this sense most flows are weakly mixing and few are (strongly) mixing.

- However, for hyperbolic flows these mixing notions are usually conflated, that is, once a hyperbolic flow is known to be weakly mixing, the various stronger mixing properties hold as well (Remark 7.3.20, Theorem 7.4.20, Remark 7.4.21)—because, for instance, there is a generator with trivial tail. Accordingly, readers focused on hyperbolic flows might choose to skip, for instance, the discussion of weak mixing on the following pages (for example, Propositions 3.4.5, 3.4.18, 3.4.19, 3.4.38, and 3.4.40), save for statements that have implications for mixing.

- We nonetheless explore these various notions of mixing with some care because there are occasions when we can explain how specifically a stronger mixing property can be proved directly (for example, in Theorem 7.1.12), and because at times weak mixing can be obtained with no additional effort over establishing ergodicity (such as in Proposition 7.3.17 or Theorem 3.4.44). This relies on some of the characterizations of weak mixing that we develop here (notably, Theorem 3.4.19, Proposition 3.4.40).

- It turns out that up to constant rescaling of time, any two Bernoulli flows are measure-theoretically isomorphic (Ornstein Isomorphism Theorem).

We mention the next result without proof as it provides a good interpretation of weak mixing as a mixing condition away from a "negligible" set of times:

Proposition 3.4.5. *A measure-preserving flow is weakly mixing if and only if for any two measurable sets A, B,*

$$\text{there is an } E \subset \mathbb{R}^+ \text{ of density } 0 \text{ such that} \\ \lim_{E \not\ni t \to \infty} \mu(\varphi^{-t}(A) \cap B) = \mu(A) \cdot \mu(B). \tag{3.4.4}$$

Here we used the following notion and fact:

Definition 3.4.6. If $\lambda(E \cap [0, s]) - ds \xrightarrow[s \to \infty]{} 0$, then we say that $E \subset \mathbb{R}^+$ has *density* d. In particular, it has density 0 if $\lambda(E \cap [0, s]) \xrightarrow[s \to \infty]{} 0$.

We note the following proposition for later use:

Proposition 3.4.7. *If* $f : \mathbb{R}^+ \to \mathbb{R}$ *is bounded, then*

$$\lim_{T \to \infty} \frac{1}{T} \int_0^T |f| = 0 \quad \text{if and only if} \quad \lim_{T \to \infty} \frac{1}{T} \int_0^T f^2 = 0.$$

Proof. This follows from Lemma 3.4.8: $\lim_{E \not\ni t \to \infty} f(t) = 0 \Leftrightarrow \lim_{E \not\ni t \to \infty} f(t)^2 = 0$. □

Lemma 3.4.8. *If* $f : \mathbb{R}^+ \to \mathbb{R}$ *is bounded, then* $\lim_{T \to \infty} \frac{1}{T} \int_0^T |f| = 0$ *iff there is an* $E \subset \mathbb{R}^+$ *of density* 0 *such that* $\lim_{E \not\ni t \to \infty} f(t) = 0$, *that is,* $0 = \lim_{t \to \infty} \begin{cases} f(t) & \text{if } t \notin E, \\ 0 & \text{if } t \in E. \end{cases}$

Proof. "If": For $M := \|f\|_\infty$ and $\epsilon > 0$ there is an $S \in \mathbb{R}^+$ such that for $T \geq S$ we have

- $\int_{[0,T] \smallsetminus E} |f(t)|\, dt < \frac{\epsilon}{M+1}$, and
- $d_T(E) := \frac{1}{T} \lambda(E \cap [0, T]) < \frac{\epsilon}{M+1}$,

so $\frac{1}{T} \int_0^T |f|\, dt = \frac{1}{T} \Big(\int_{[0,T] \cap E} |f(t)|\, dt + \int_{[0,T] \smallsetminus E} |f(t)|\, dt \Big) < M d_T(E) + \frac{\epsilon}{M+1} < \epsilon$.

"Only if": Since $E_k := \{t \in \mathbb{R}^+ \mid |f(t)| \ge 1/k\} \subset E_{k+1}$ satisfies

$$d_T(E_k) = \frac{1}{T}\int_0^T \chi_{E_k}\,dt \le \frac{k}{T}\int_0^G |f(t)|\,dt \xrightarrow[T\to\infty]{} 0,$$

recursively take $l_k \ge l_{k-1}$ such that $d_T(E_k) < 1/k$ for $T \ge l_k$. Let $E := \bigcup_{k\in\mathbb{N}} E_k \cap [l_{k-1}, l_k)$ and $\epsilon > 0$. If $k > 1/\epsilon$ and $l_{k-1} < T \notin E$, then $T \notin E_k$, and $|f(T)| < 1/k < \epsilon$. To show $d_n(E) \to 0$ take $K > 2/\epsilon$, $T \ge l_K$, and $k \ge K$ such that $l_k \le T < l_{k+1}$. Since

$$E \cap [0,T) = (E \cap [0,l_k)) \cup (E \cap [l_k,T)) \subset \underbrace{(E_k \cap [0,l_k))}_{\subset E_k\cap[0,T)} \cup \underbrace{(E_{k+1} \cap [l_k,T))}_{\subset E_{k+1}\cap[0,T)},$$

we get $d_T(E) \le \frac{1}{T}\big(T d_T(E_k) + T d_T(E_{k+1})\big) < \frac{1}{k} + \frac{1}{k+1} < \frac{2}{k} < \epsilon$. □

Unlike ergodicity, mixing properties are sensitive to timing. This is starkly illustrated by the contrast between Remark 3.3.5 and the next result.

Proposition 3.4.9. *Suspensions are not weakly mixing.*

Proof. For $A = B = X \times [0,1/2)$ and $T \in \mathbb{N}$, the integral in (3.4.1) is $1/8 \ne 0$. □

Corollary 3.4.10. *Linear flows on tori are not weakly mixing.*

Proof. They are suspensions of translations. □

Clearly, mixing and weak mixing are invariants of measure-theoretic isomorphism, and indeed one can easily show the following proposition:

Proposition 3.4.11. *If a flow is mixing (or weakly mixing), then so is any factor (Definition 3.1.20).*

This is another reason suspensions are not weakly mixing: they have circle flows (Example 1.1.6) as a factor (Proposition 1.3.3).

The next result relates the measure-theoretic and topological notions of mixing.

Proposition 3.4.12. *If μ is a mixing invariant measure for a continuous flow Φ, then $\Phi{\restriction}_{\operatorname{supp}\mu}$ is topologically mixing.*

Proof. If $A, B \subset \operatorname{supp}\mu$ are open, then $\mu(\varphi^{-t}(A) \cap B) > 0$ for all large t. □

Now we prove a criterion for mixing that allows us to use particularly convenient sets when checking mixing for specific dynamical systems.

Definition 3.4.13. A collection $C \subset \mathcal{S}$ in a measure space $(X, \mathcal{S}, \mu)$ is said to be *sufficient* if finite disjoint unions of elements of C form a dense collection with respect to the symmetric-difference metric

$$d(A,B) := d_\mu(A,B) := \mu(A \vartriangle B) \in (0,\infty]. \tag{3.4.5}$$

Remark 3.4.14. This is closely related to the Rokhlin metric from (A.2.9); see Proposition A.2.20 and Remark A.2.21.

Proposition 3.4.15. *Suppose C is a sufficient collection of sets. Then*

(1) *Φ is mixing if* (3.4.2) *holds for any $A, B \in C$,*

(2) *Φ is weakly mixing if* (3.4.1) *or* (3.4.4) *holds for any $A, B \in C$,*

(3) *Φ is ergodic if* (3.3.1) *holds for any $A, B \in C$,*

(4) *Φ is mixing of order N if* (3.4.3) *holds for any $A_i \in C$.*

Proof. We prove (1) using Proposition 3.3.24; the other parts have like proofs. Let

$$A_1, \dots, A_k,\ B_1, \dots, B_l \in C, \quad A_i \cap A_{i'} = \varnothing \text{ for } i \neq i', \quad B_j \cap B_{j'} = \varnothing \text{ for } j \neq j',$$

and $A := \bigcup_{i=1}^k A_i$, $B := \bigcup_{j=1}^l B_j$. Then $\mu(A) = \sum_{i=1}^k \mu(A_i)$, $\mu(B) = \sum_{j=1}^l \mu(B_j)$, and by assumption,

$$\mu(\varphi^{-t}(A) \cap B) = \sum_{i=1}^k \sum_{j=1}^l \mu(\varphi^{-t}(A_i) \cap B_j) \xrightarrow[t \to \infty]{} \sum_{i=1}^k \sum_{j=1}^l \mu(A_i) \cdot \mu(B_j) = \mu(A) \cdot \mu(B).$$

Thus (3.4.2) holds for any elements of the dense collection $\mathfrak{U}$ formed by finite disjoint unions of elements of C. Now let A, B be arbitrary measurable sets. Find A', $B' \in \mathfrak{U}$ such that $\mu(A \vartriangle A') < \epsilon/4$, $\mu(B \vartriangle B') < \epsilon/4$. Then by the triangle inequality,

$$\begin{aligned} |\mu(\varphi^{-t}(A) \cap B) - \mu(A)\mu(B)| &\le \mu(\varphi^{-t}(A \vartriangle A') \cap B) + \mu(\varphi^{-t}(A') \cap (B \vartriangle B')) \\ &\quad + |\mu(\varphi^{-t}(A') \cap B') - \mu(A')\mu(B')| \\ &\quad + \mu(A) \cdot \mu(B \vartriangle B') + \mu(B') \cdot \mu(A \vartriangle A') \\ &\le |\mu(\varphi^{-t}(A') \cap B') - \mu(A') \cdot \mu(B')| + \epsilon. \end{aligned}$$

Since $\epsilon > 0$ can be chosen arbitrarily small, this implies (3.4.2). □

It is not only with respect to the sets in question, but also in the conclusion that a suitable approximation is good enough.

Proposition 3.4.16. *Let Φ be a continuous flow on a compact metric space X and μ a Φ-invariant Borel probability measure for which there exist constants $c, C > 0$ such that*

$$c\mu(P)\mu(Q) \le \underline{\lim}_{t \to \infty} \mu(P \cap \varphi^{-t}(Q)) \le \overline{\lim}_{t \to \infty} \mu(P \cap \varphi^{-t}(Q)) \le C\mu(P)\mu(Q) \qquad (3.4.6)$$

for all Borel sets $P, Q \subset X$. Then μ is mixing.

Proof. The left inequality in (3.4.6) implies that $\Phi \times \Phi$ is ergodic with respect to $\mu \times \mu$: If $A, B, C, D \subset X$ are Borel sets, then

$$\varliminf_{t\to\infty} \underbrace{(\mu \times \mu)((\varphi \times \varphi)^t(A \times C) \cap (B \times D))}_{=\mu(\varphi^t(A)\cap B)\cdot\mu(\varphi^t(C)\cap D)} \geq c^2 \underbrace{\mu(A)\cdot\mu(B)}_{=(\mu\times\mu)(A\times B)} \cdot \underbrace{\mu(C)\cdot\mu(D)}_{=(\mu\times\mu)(C\times D)}.$$

The same inequality holds if we replace $A \times C$ and $B \times D$ by finite disjoint unions of product sets, and such sets approximate every measurable $P, Q \subset X \times X$. Thus,

$$\varliminf_{t\to\infty} (\mu \times \mu)((\varphi \times \varphi)^t(P) \cap Q) \geq c^2(\mu \times \mu)(P) \cdot (\mu \times \mu)(Q),$$

and $\Phi \times \Phi$ is ergodic with respect to $\mu \times \mu$. (So Φ is weakly mixing by Theorem 3.4.19 below.)

Now let ν be the diagonal measure in $X \times X$ given by $\nu(E) = \mu(\pi_1(E \cap \Delta))$, where $\Delta = \{(x,x) \mid x \in X\}$ and $\pi_1 \colon X \times X \to X$ is the projection to the first coordinate. The measure ν as well as its shift ν_t under the map $\varphi^t \times \mathrm{Id}$ are $(\Phi \times \Phi)$-invariant. Explicitly, $\nu_t(A \times B) = \mu(\varphi^t(A) \cap B)$. By the right inequality in (3.4.6) we have

$$\varlimsup_{t\to\infty} \nu_t(A \times B) = \varlimsup_{t\to\infty} \mu(\varphi^t(A) \cap B) < C\mu(A) \cdot \mu(B) = C(\mu \times \mu)(A \times B). \quad (3.4.7)$$

Let η be any weak limit point of the sequence ν_t. If $A, B \subset X$ are closed sets then $\eta(A \times B) \leq C(\mu \times \mu)(A \times B)$ by (3.4.7). Approximation by disjoint unions of products of closed sets gives $\eta(P) < C(\mu \times \mu)(P)$ for any Borel set $P \subset X \times X$, so $\eta \ll \mu \times \mu$, and $\eta = \mu \times \mu$ by Proposition 3.3.12 since η is $(\Phi \times \Phi)$-invariant and $\mu \times \mu$ is ergodic. For closed A, B with $\mu(\partial A) = \mu(\partial B) = 0$ we have

$$\mu(\varphi^t(A) \cap B) = \nu_t(A \times B) \xrightarrow[t\to\infty]{} (\mu \times \mu)(A \times B) = \mu(A) \cdot \mu(B).$$

Since the collection of all such sets is sufficient, Φ is mixing with respect to μ by Proposition 3.4.15. □

The notions of mixing and weak mixing are remarkably well behaved when passing to products:

Proposition 3.4.17. *A measure-preserving flow Φ on (X, μ) is mixing (weakly mixing) if and only if $\Phi \times \Phi$ is.*

Proof. If $\Phi \times \Phi$ is weakly mixing and $A, B \subset X$ then by Proposition 3.4.5 there is a set $E \subset \mathbb{N}$ of density 0 such that

$$\underbrace{\mu(\varphi^{-t}(A) \cap B)}_{=(\mu\times\mu)\left((\varphi\times\varphi)^{-t}(A\times X)\cap(B\times X)\right)} \xrightarrow[E\not\ni t\to\infty]{} (\mu \times \mu)(A \times X) \cdot (\mu \times \mu)(B \times X) = \mu(A)\mu(B),$$

so Φ is weakly mixing. Taking $E = \varnothing$ proves that $\Phi \times \Phi$ mixing $\Rightarrow$ Φ mixing.

Suppose now that Φ is weakly mixing. Then for measurable $A_1, A_2, B_1, B_2 \subset X$ there exist sets $E_1, E_2 \subset \mathbb{R}^+$ of density 0 such that

$$\lim_{E_i \not\ni t \to \infty} \mu(\varphi^{-t}(A_i) \cap B_i) = \mu(A_i) \cdot \mu(B_i)$$

for $i = 1, 2$. Taking $E := E_1 \cup E_2$ we find that

$$\underbrace{(\mu \times \mu)\big(\overbrace{(\varphi \times \varphi)^{-t}(A_1 \times A_2) \cap (B_1 \times B_2)}^{=(\varphi^{-t}(A_1)\cap B_1)\times(\varphi^{-t}(A_2)\cap B_2)}\big)}_{=\mu(\varphi^{-t}(A_1)\cap B_1)\mu(\varphi^{-t}(A_2)\cap B_2)} \xrightarrow[E \not\ni t \to \infty]{} \underbrace{\mu(A_1)\mu(B_1)\mu(A_2)\mu(B_2)}_{=(\mu\times\mu)(A_1\times A_2)(\mu\times\mu)(B_1\times B_2)}.$$

Since sets of the form $A \times B$ form a sufficient collection, Proposition 3.4.15(2) implies that $\Phi \times \Phi$ is weakly mixing. Taking $E_1 = E_2 = \varnothing$ gives Φ mixing $\Rightarrow$ $\Phi \times \Phi$ mixing. □

One of the implications in Proposition 3.4.17 is easy to strengthen:

Proposition 3.4.18. *If Φ is a measure-preserving flow on X and $\Phi \times \Phi$ is ergodic, then Φ is weakly mixing.*

Proof. If $\Phi \times \Phi$ is ergodic and A, B measurable, then

$$\frac{1}{T}\int_0^T \underbrace{\mu(\varphi^{-t}(A) \cap B)}_{=(\mu\times\mu)((\varphi\times\varphi)^{-t}(A\times X)\cap(B\times X))} \xrightarrow[n\to\infty]{\text{Proposition 3.3.24}} (\mu \times \mu)(A \times X)(\mu \times \mu)(B \times X) = \mu(A)\mu(B)$$

and

$$\frac{1}{T}\int_0^T \underbrace{\mu(\varphi^{-t}(A) \cap B)^2}_{=(\mu\times\mu)((\varphi\times\varphi)^{-t}(A\times A)\cap(B\times B))} \xrightarrow[n\to\infty]{\text{Proposition 3.3.24}} (\mu \times \mu)(A \times A)(\mu \times \mu)(B \times B) = \mu(A)^2\mu(B)^2,$$

so

$$\frac{1}{T}\int_0^T \underbrace{\Big(\mu(\varphi^{-t}(A) \cap B) - \mu(A)\mu(B)\Big)^2}_{=\mu(\varphi^{-t}(A)\cap B)^2 - 2\mu(\varphi^{-t}(A)\cap B)\mu(A)\mu(B) + \mu(A)^2\mu(B)^2} \xrightarrow[T\to\infty]{} 0.$$

Therefore, Φ is weakly mixing by Propositions 3.3.24 and 3.4.7. □

In fact, we have the following theorem:

Theorem 3.4.19. *The following are equivalent*:

(1) *Φ is weakly mixing,*

(2) *$\Phi \times \Phi$ is weakly mixing,*

(3) $\Phi \times \Phi$ *is ergodic,*

(4) $\Phi \times \Psi$ *is ergodic whenever* Ψ *is.*

Proof. The first three properties are equivalent by Propositions 3.4.17 and 3.4.18. (4) $\Rightarrow$ (3): If $\Phi \times \Psi$ is ergodic whenever Ψ is, then for the constant flow Ψ on a single point, this implies ergodicity of Φ, so with $\Psi = \Phi$ we find that $\Phi \times \Phi$ is ergodic. To show (1) $\Rightarrow$ (4), we use Proposition 3.4.15:

$$\left|\frac{1}{T}\int_0^T \overbrace{(\mu\times\nu)((\varphi\times\psi)^{-t}(A_1\times A_2)\cap B_1\times B_2)}^{=\overbrace{\mu(\varphi^{-t}(A_1)\cap B_1)}^{=:x_t}\overbrace{\nu(\psi^{-t}(A_2)\cap B_2)}^{=:y_t}} - \overbrace{(\mu\times\nu)(A_1\times A_2)(\mu\times\nu)(B_1\times B_2)}^{=\overbrace{\mu(A_1)\mu(B_1)}^{=:x}\overbrace{\nu(A_2)\nu(B_2)}^{=:y}}\right|$$
$$= \frac{1}{T}\left|\int_0^T x_t y_t - xy\,dt\right|$$
$$\le \underbrace{\frac{1}{T}\int_0^T |x_t - x|\cdot y_t\,dt}_{\le(\sup_t y_t)\frac{1}{T}\int_0^T |x_t-x|\,dt\to 0\ (\Phi\text{ weakly mixing})} + x\cdot\underbrace{\left|\frac{1}{T}\int_0^T y_t - y\,dt\right|}_{\to 0\ (\Psi\text{ ergodic})} \xrightarrow[T\to\infty]{} 0. \qquad \square$$

Remark 3.4.20. Item (4) motivates saying that a flow Φ is *mildly mixing* if $\Phi \times \Psi$ is ergodic whenever Ψ has a *possibly infinite* ergodic invariant measure.

Corollary 3.4.21. *If* Φ *is weakly mixing then so is* $\Phi \times \Phi \times \cdots \times \Phi$ *for any finite number of products.*

Proof. Recursively taking $\Psi = \Phi\times\Phi$, $\Psi = \Phi\times\Phi\times\Phi$ and so on in Theorem 3.4.19(4) shows that if Φ is mixing, then $\Phi \times \Phi \times \cdots \times \Phi$ is ergodic. Using (3) $\Rightarrow$ (1) with $2n$ copies of Φ then shows that the product of n copies is weakly mixing. $\square$

It may be interesting to give an evidently equivalent formulation:

Corollary 3.4.22. *If* $\Phi\times\Phi$ *is ergodic then so is* $\Phi\times\Phi\times\cdots\times\Phi$ *for any finite number of products.*

Just as ergodicity can be expressed in terms of functions rather than sets, so can the various notions of mixing. In probabilistic terms, sets are events and functions are random variables. The preceding notions of ergodicity and mixing involve various forms of eventual independence of events, and they can be recast in terms of eventual independence of random variables using the *covariance* of L^2-functions.

Definition 3.4.23. The *covariance* of $f, g \in L^2$ is defined as

$$\operatorname{cov}(f,g) := \underbrace{\langle f - \langle f,1\rangle, g - \langle g,1\rangle\rangle}_{=\int \left(f-\int f\right)\overline{\left(g-\int g\right)}} = \underbrace{\langle f,g\rangle - \langle f,1\rangle\langle 1,g\rangle}_{\int f\bar{g}-\int f\int \bar{g}}.$$

That is, we project both functions to the orthocomplement $1^\perp \subset L^2$ of the constant functions, that is, the space of functions with zero integral, by subtracting their average (to focus on their variation) and then take the inner product.

Remark 3.4.24. Like the inner product, the covariance is sesquilinear (linear in the first entry and antilinear in the second) and invariant under isometric operators (that is, $\langle U\cdot, U\cdot\rangle = \langle\cdot,\cdot\rangle \Rightarrow \operatorname{cov}(U\cdot, U\cdot) = \operatorname{cov}$). If either of the functions is constant, then the covariance is zero, so it is unaffected by the addition of constants to either function. For many statements about covariance, this allows us to assume without loss of generality that the functions in question have zero average, that is, are in $1^\perp$. Indeed, "*polarization*"[17] allows us to consider the same function in both entries:

$$\operatorname{cov}(f,g) = \tfrac{1}{4}[\operatorname{cov}(f+g, f+g) - \operatorname{cov}(f-g, f-g)].$$

The covariance also satisfies the Cauchy–Schwarz inequality: $|\operatorname{cov}(f,g)| \le \|f\|\|g\|$.

Proposition 3.4.25. *If $\Xi \subset L^2$ is a complete system, that is, $\overline{\operatorname{span}(\Xi)} = L^2$, then*

- *Φ is ergodic if and only if $\frac{1}{T}\int_0^T \operatorname{cov}(U^t_\Phi(f), g) \xrightarrow[T\to\infty]{} 0$ for all $f,g \in \Xi$,*
- *Φ is weakly mixing if and only if*

$$\frac{1}{T}\int_0^T \left|\operatorname{cov}(U^t_\Phi(f), g)\right| \xrightarrow[T\to\infty]{} 0 \tag{3.4.8}$$

  for all $f,g \in \Xi$,

- *Φ is weakly mixing if and only if for all $f,g \in \Xi$, there exists an $E \subset \mathbb{R}^+$ of density* 0 *(Definition* 3.4.6*) such that*

$$\operatorname{cov}(U^t_\Phi(f), g) \xrightarrow[E\not\ni t\to\infty]{} 0,$$

- *Φ is mixing if and only if*

$$\operatorname{cov}(U^t_\Phi(f), g) \xrightarrow[t\to\infty]{} 0 \tag{3.4.9}$$

  for all $f,g \in \Xi$,

- *Φ is* mixing of order *N if*

$$\int \prod_{i=0}^{N} f_i \circ \varphi^{t_i}\, d\mu \xrightarrow[t_i - t_{i-1}\to\infty]{} \prod_{i=0}^{N} \int f_i\, d\mu$$

  for any $\{f_0, \dots, f_N\} \subset \Xi$.

[17] $\|u+v\|^2 - \|u-v\|^2 = 4\langle u, v\rangle$.

Proof. To see how to pass from a complete system to L^2 note first that sesquilinearity of covariance means that checking any of these statements for all $f, g \in \Xi$ implies the same for all $f, g \in \operatorname{span}(\Xi)$. Now take arbitrary $f, g \in L^2$ and $f', g' \in \operatorname{span}(\Phi)$ such that $\|g - g'\| < \epsilon/2\|f\|$ and $\|f - f'\| < \epsilon/2\|g'\|$. Then

$$\begin{aligned}|\operatorname{cov}(U_\Phi^t(f), g)| &= |\operatorname{cov}(U_\Phi^t(f), g - g') + \operatorname{cov}(U_\Phi^t(f) - U_\Phi^t(f'), g') + \operatorname{cov}(U_\Phi^t(f'), g')| \\ &\le |\operatorname{cov}(U_\Phi^t(f'), g')| + \epsilon.\end{aligned}$$

Now, for each of these statements, knowing it for all $f, g \in L^2$ implies the corresponding mixing property by taking $f = \chi_A$ and $g = \chi_B$ for measurable sets A, B.

To see the converse note that characteristic functions of measurable sets (or linear combinations of those of a sufficient collection) form a complete system in L^2 for which the statement about covariance boils down to the respective mixing property. □

Remark 3.4.26. Note that we have in particular re-proved Proposition 3.4.15.

Remark 3.4.27. The characterization of mixing in (3.4.9) invites the question of how fast the covariance goes to 0 with t. This depends on the two functions involved, and the convergence can be arbitrarily slow. However, for a smooth dynamical system and functions selected from a suitable class—smooth, Lipschitz or Hölder continuous (Definition 1.9.6), for instance—hyperbolicity can produce exponential convergence to 0. This is known as exponential *decay of correlations*. Since the classes of functions needed for this are not preserved by measure-theoretic isomorphism, neither is this property, so it is a meaningful property of a smooth dynamical system rather than of its measure-theoretic isomorphism class. Note that this is sensitive to time changes; for instance, suspensions are not mixing and hence have no correlation decay at all. We elaborate on this in Section 7.7.

Parsing Definition 3.4.23, this characterization of mixing can be restated thus:

Proposition 3.4.28. *If $\Xi \subset L^2$ is a complete system, that is,* $\operatorname{span}(\Xi)$ *is dense in L^2, then Φ is mixing if and only if $U_\Phi^t(f) \xrightarrow[t\to\infty]{\text{weakly}} \int f$ for all $f \in \Xi$.*

Likewise we have a further proposition:

Proposition 3.4.29. *A Φ-invariant probability measure μ is N-mixing if and only if given any $f_i \in L^2(\mu)$, any weak accumulation point $\psi_n \xrightarrow{\text{weakly}} \psi$ of $\prod_{i=1}^N f_i \circ \varphi^{t_i}$ (with $t_i - t_{i-1} \xrightarrow[t\to\infty]{} \infty$) is constant.*

Proof. "Only if" is clear. To get "if," we recursively verify that the constant is correct.

First, take $f_i \equiv 1$ for $i \ne 1$, including taking the test function $f_0 \equiv 1$. Then the weak-accumulation statement becomes

$$\int f_1 = \int f_1 \circ \varphi^t \cdot 1 \to \text{const.} \int 1 = \text{const.},$$

so the constant is $\int f_1$ for each such subsequence, and thus $f_1 \circ \varphi^t \xrightarrow{\text{weakly}} \int f_1$. By symmetry, $f_i \circ \varphi^t \xrightarrow{\text{weakly}} \int f_i$ for all i. In particular, $f_2 \circ \varphi^{t_2-t_1} \xrightarrow[t_2-t_1\to\infty]{\text{weakly}} \int f_2$. Supposing next that $f_i \equiv 1$ for $i \notin \{1,2\}$, this implies

$$\int f_1 \circ \varphi^{t_1} f_2 \circ \varphi^{t_2} \cdot 1 = \int f_2 \circ \varphi^{t_2-t_1} f_1 \xrightarrow[t_2-t_1\to\infty]{} \int \Big(\int f_2\Big) f_1 = \int f_1 \int f_2,$$

so $f_2 \circ \varphi^{t_2} f_1 \circ \varphi^{t_1} \xrightarrow[t_2-t_1\to\infty]{\text{weakly}} \int f_1 \int f_2$ with like statements for any pair of the f_i. This can be continued. □

The next proposition is suggested by Remark 3.4.24:

Proposition 3.4.30. *In each of the statements in Proposition* 3.4.25 *one can replace* $\operatorname{cov}(U^t_\Phi(f), g)$ *by* $\operatorname{cov}(U^t_\Phi(f), f)$ *or by* $\langle U^t_\Phi(f), f\rangle$ *if* $f \perp 1$. *For instance,* Φ *is mixing if and only if*

$$\operatorname{cov}(U^t_\Phi(f), f) \xrightarrow[t\to\infty]{} 0$$

for all f *in a complete set* L^2, *which happens if and only if*

$$\langle U^t_\Phi(f), f\rangle \xrightarrow[t\to\infty]{} 0$$

for all f *in a complete set for* $1^\perp$.

Proof. While Remark 3.4.24 applies if the hypothesis is known for all $f \in L^2$, the step from a complete system to L^2 requires attention because $f \mapsto \operatorname{cov}(U^t_\Phi(f), f)$ is not linear. The following lemma covers the mixing case, and the others are analogous. The last statement follows directly from Remark 3.4.24. □

Lemma 3.4.31. *If* $\operatorname{cov}(U^t_\Phi(f), f) \xrightarrow[t\to\infty]{} 0$, *then* $\operatorname{cov}(U^t_\Phi(f), g) \xrightarrow[t\to\infty]{} 0$ *for all* $g \in L^2$.

Proof. $M_f := \{g \in L^2 \mid \operatorname{cov}(U^t_\Phi(f), g) \xrightarrow[t\to\infty]{} 0\}$ is a closed subspace of L^2 that contains 1 and f, and $U_\Phi M_f \subset M_f$: if $g \in M_f$ and $t \in \mathbb{R}^+$, then, since U_Φ is an isometry, $\langle U^t_\Phi(f), U_\Phi(g)\rangle = \langle U_\Phi(U^{t-1}_\Phi(f)), U_\Phi(g)\rangle = \langle U^{t-1}_\Phi(f), g\rangle \to 0$. Thus,

$$M_f \supset m_f := \bigcap \big\{E \subset L^2 \text{ closed} \mid 1, f \in E,\ U_\Phi(E) \subset E\big\} \supset U_\Phi(m_f).$$

If $g \in m_f^\perp$, then $\langle 1, g\rangle = 0$ and $\langle U^t_\Phi(f), g\rangle = 0$ for all t since $U^t_\Phi(f) \in U^t_\Phi(m_f) \subset m_f$, so $g \in M_f$. Thus, $L^2 = m_f \oplus m_f^\perp \subset M_f \oplus M_f = M_f$. □

We show the following theorem as an application of Proposition 3.4.28 and (2.2.3):

Theorem 3.4.32. *The geodesic flow on a finite-volume factor of the Poincaré disk* (*Section* 2.3) *is mixing.*

Lemma 3.4.33 (Mautner phenomenon).

$$f \circ g^{t_i} \xrightarrow[t_i \to \infty]{\text{weakly}} f_0 \in L^2 \Rightarrow f_0 \circ h_-^s = f_0.$$

Proof.

$$\underbrace{\| f \circ g^{t_i} h_-^s - f g^{t_i} \|}_{\xrightarrow[i\to\infty]{\text{weakly}} f_0 \circ h_-^s - f_0} \overset{(2.2.3)}{=\!=} \| f \circ h_-^{se^{-t}} g^{t_i} - f \circ g^{t_i} \|$$

$$= \| f \circ h_-^{se^{-t}} - f \| \xrightarrow[t\to\infty]{\text{Remark 3.2.7}} 0. \qquad \square$$

Proof of Theorem 3.4.32. If $f \circ g^{t_i} \xrightarrow[t_i\to\infty]{\text{weakly}} f_0 \in L^2$, then f_0 is h_--invariant, hence by Corollary 3.3.20 constant, so $f_0 = \int f$. Thus, $f \circ g^t \xrightarrow[t\to\infty]{\text{weakly}} \int f$, which gives the claim by Proposition 3.4.28. □

The mixing property has applications in this case. We obtain the following proposition first:

Proposition 3.4.34. *If $f\colon S\Sigma \to \mathbb{R}$ is continuous on the unit tangent bundle of a compact factor Σ of the Poincaré disk (Section 2.3), then for $x \in S\Sigma$,*

$$\frac{1}{2}\int_{-1}^{1} f \circ g^t(h_-^s(x))\, ds \xrightarrow[t\to+\infty]{\text{uniformly}} \int_{S\Sigma} f.$$

Proof outline. To apply mixing thicken the arc $h_-^{[-1,1]}(x)$ to $U := B^{cu} \times h_-^{[-1,1]}(x)$ of positive volume (using local product coordinates); here $B^{cu} \sim h_+^{[-\epsilon,\epsilon]}(x) \times g^{[-\epsilon,\epsilon]}(x)$. Then $\operatorname{area}(B^{cu}) \cdot \frac{1}{2}\int_{-1}^{1} f \circ g^t(h_-^s(x))\, ds \approx \int f \circ g^{-t}(y)\chi_U(y) \xrightarrow[t\to+\infty]{\text{mixing}} \operatorname{vol}(U)\int f$. It is essential here that g^{-t} does not expand in the B^{cu}-direction; this ensures uniformity of the approximation in t and equicontinuity of $x \mapsto \frac{1}{2}\int_{-1}^{1} f \circ g^t(h_-^s(x))\, ds$ for $t \geq 0$. Since $\operatorname{vol}(U) \approx \operatorname{area}(B^{cu})$, the claim follows by letting $\epsilon \to 0$. □

Corollary 3.4.35. *The horocycle flow on a compact factor of the Poincaré disk is uniquely ergodic.*

Remark 3.4.36. Compactness is not needed [323].

Proof. By Theorem 3.3.32 we check uniform convergence of Birkhoff averages to a constant using (2.2.3) and Proposition 3.4.34:

$$\frac{1}{2e^t}\int_{-e^t}^{e^t} f \circ h_-^s(g^{-t}(x))\, ds = \frac{1}{2}\int_{-1}^{1} f \circ g^t(h_-^s(x))\, ds \xrightarrow[t\to+\infty]{\text{uniformly}} \int_{S\Sigma} f. \qquad \square$$

We now take a spectral point of view.

Definition 3.4.37. A complex $f \in L^2(\mu) \setminus \{0\}$ is said to be an *eigenfunction* of a measure-preserving flow Φ on a probability space $(X, \mathcal{T}, \mu)$ if $f \circ \varphi^t = e^{2\pi i \omega t} f$ for all $t \in \mathbb{R}$; in this case ω is called the corresponding eigenfrequency and $e^{2\pi i \omega}$ the corresponding eigenvalue.

Thus, a flow is ergodic if 1 is a simple eigenvalue, and *weakly mixing* if every eigenfunction is constant almost everywhere:

Proposition 3.4.38. *Eigenfunctions of a weakly mixing measure-preserving flow are constant, that is, if $f \in L^2$ and $f \circ \varphi^t = e^{it\omega} f$ for some $\omega \in \mathbb{R}$, then $f = \text{const.}$—and hence $\omega = 0$, so Φ has only one eigenvalue.*

Proof. If $f \in L^2$ and $f \circ \varphi^t = e^{it\omega} f$, then either $e^{it\omega} \equiv 1$, so $f = \text{const.}$ by ergodicity (which follows from weak mixing), or $\omega \neq 0$, in which case

$$\langle f, 1 \rangle = \int f = \int f \circ \varphi^t = \int e^{it\omega} f = e^{it\omega} \langle f, 1 \rangle$$

implies $\langle f, 1 \rangle = 0$ and hence

$$\begin{aligned} \int |f|^2 = \frac{1}{T} \int_0^T |e^{it\omega}| \int f \bar{f} \, dt &= \frac{1}{T} \int_0^T \Big| \int e^{it\omega} f \bar{f} \Big| dt \\ &= \frac{1}{T} \int_0^T \Big| \int (f \circ \varphi^t) \bar{f} \Big| dt \xrightarrow[T \to \infty]{} 0 \end{aligned}$$

by (3.4.8) since Φ is weakly mixing. □

Remark 3.4.39. This provides yet another reason linear flows on tori are not weakly mixing: the coordinate projections are nonconstant eigenfunctions.

Proposition 3.4.40. *A flow is weakly mixing iff* every *time-t map for $t \neq 0$ is ergodic.*

Remark 3.4.41. Compare this with Theorem 3.3.13.

Proof. "$\Leftarrow$": If $f \circ \varphi^t = e^{2\pi i \omega t} f$ for all $t \in \mathbb{R}$ then either $\omega = 0$ and f is an invariant function for φ^1 (say), or $\omega \neq 0$ and f is an invariant function for $\varphi^{1/\omega}$. Either way, $f = \text{const.}$
"$\Rightarrow$": If $f \circ \varphi^\tau = f \perp 1$, then $g_k := \int_0^\tau e^{2\pi i k t/\tau} f \circ \varphi^t \, dt$ is φ^t-invariant, hence equals $\int g_k = \int \int_0^\tau e^{2\pi i k t/\tau} f \circ \varphi^t \, dt = \int_0^\tau e^{2\pi i k t/\tau} (\int f \circ \varphi^t) \, dt = 0$, so $\mathfrak{f}_x(t) := f(\varphi^t(x))$ satisfies $0 \overset{\text{ae}}{=} g_k(x) = \int_0^\tau e^{2\pi i k t/\tau} \mathfrak{f}_x(t) \, dt$ for all $k \in \mathbb{Z}$, so $\mathfrak{f}_x \overset{\text{ae}}{=} 0$, and $f \overset{\text{ae}}{=} 0$. □

Remark 3.4.42. This is another reason why suspensions are not weakly mixing.

Together with Theorem 3.3.16, this has the following corollary:

Theorem 3.4.43. *The geodesic flow on a compact factor of the Poincaré disk is weakly mixing with respect to the Liouville measure (hence not a suspension).*

Likewise, Proposition 3.4.40 and Proposition 3.3.19 give weak mixing of the horocycle flow.

Theorem 3.4.44. *The horocycle flow from Remark* 2.1.15 *is weakly mixing with respect to Lebesgue measure (hence not a suspension).*

Of course, we already saw that stronger mixing holds for the geodesic flow (Theorem 3.4.32). Theorem 7.1.12 and even more so Theorem 7.4.20 go further, and this is a good time to lay some of the ground work. We will do more with the horocycle flow later on to find that it is indeed mixing of all orders with respect to Lebesgue measure, and that this is not tied to the algebraic nature of this flow but to the "commutation" relation with the geodesic flow (Section 8.6).

The Bernoulli property differs from the other mixing notions in that verifying it appears to require finding a symbolic flow as well as a measure-theoretic isomorphism to it. It is easy to believe that this can be challenging, so it is important to have criteria for the Bernoulli property that can be verified in ways more in line with the other mixing properties. We provide an important one (without proof).

To define this new notion, we first weaken the notion of "almost everywhere" to an approximate one: we say that a property holds for ϵ-a.e. point of a set E in a measure space $(X,\mathcal{T},\mu)$ if the set B where it fails satisfies $\mu_E(B) < \epsilon$, using the conditional measure from (3.3.3). Likewise, an invertible map $f\colon (X,\mathcal{T},\mu) \to (Y,\mathcal{V},\nu)$ of measure spaces (not necessarily probability spaces) is said to be ϵ-measure-preserving if there is a $B \subset X$ with $\mu_X(B) < \epsilon$ and $|\nu(f(A))/\mu(A) - 1| < \epsilon$ for all $A \subset X \setminus B$.

Definition 3.4.45. With the notions and notation from Definition A.1.6, a measure-preserving flow Φ is said to be *very weak Bernoulli* if $f := \varphi^t$ is very weak Bernoulli for every $t \neq 0$ which in turn means that f admits arbitrarily fine partitions (or a generating partition) $\xi = \{C_1, \dots, C_k\}$ that are very weak Bernoulli as follows: Define $\alpha\colon X \to \{1,\dots,k\}$ by $\xi(x) = C_{\alpha(x)}$ and suppose that for $\epsilon > 0$ there is an $N \in \mathbb{N}$ such that for all $n \ge N$ and ϵ-a.e. atom E of $\bigvee_{j=N}^{n} f^j(\xi)$ (Definition A.1.16) there is an ϵ-measure-preserving map $\theta\colon (E \times [0,1], \mu_E \times m) \to (X,\mu)$ with

$$\overline{\lim_{k\to\infty}} \sum_{j=1}^{k} |\alpha(f^j(x)) - \alpha(f^j(\theta(x,u)))| < \epsilon$$

for ϵ-a.e. $(x,u) \in E \times [0,1]$.

This says that x and $\theta(x,u)$ have on average almost exactly the same future as described by itineraries with respect to ξ. With respect to θ, the extra flexibility from considering $E \times [0,1]$ is helpful.

Remark 3.4.46. This notion was introduced as a broader condition under which two systems with the same entropy are measurably isomorphic [271]; previously Ornstein had shown this for Bernoulli shifts. Specifically, the new development was that if a generating partition is very weak Bernoulli, then it is "finitely determined," another notion original to that paper, and that if two measure-preserving transformations acting on Lebesgue spaces have finitely determined generating partitions and the same entropy, then they are isomorphic. Four years later, Ornstein proved the following result, which reveals this property to provide an easier way of establishing the Bernoulli property.

Theorem 3.4.47 ([272, 204]). *A measure is Bernoulli if it has the* very weak Bernoulli *property (Definition* 3.4.45*).*

3.5 Invariant measures under time change

Time changes of flows were first described in Proposition 1.2.2. Definition 1.2.1 explained that this is equivalent to scaling the generating vector field, that is, passing from a generating vector field V to the generating vector field $W = \rho V$ for some $\rho\colon X \to (0,\infty)$. While this is straightforward for smooth flows, we now make this explicit for measurable flows and furthermore examine how an invariant measure for a flow corresponds to an (absolutely continuous) invariant measure for a time change of that flow.

Let us first do the straightforward calculation for smooth flows. A volume form ω is preserved by a flow generated by X if $\mathcal{L}_X\omega = 0$.

Proposition 3.5.1. *If V preserves the volume ω, then $W = \rho V$ preserves the volume ω/ρ: $\mathcal{L}_{\rho V}(\omega/\rho) = 0$.*

Proof. For scalar functions α, the *Cartan formula*

$$\mathcal{L}_X(\alpha\omega) = \iota_X \underbrace{d(\alpha\omega)}_{=0\text{ (maximum rank)}} + d(\iota_X\alpha\omega) = d(\alpha\iota_X\omega) = d\alpha \wedge \iota_X\omega$$

implies that

$$\mathcal{L}_{\rho V}(\alpha\omega) = \rho \underbrace{\mathcal{L}_V(\alpha\omega)}_{=d\alpha\wedge\iota_V\omega} + \underbrace{d\rho \wedge \iota_X(\alpha\omega)}_{=\alpha\, d\rho\wedge\iota_V\omega} = \underbrace{(\rho\, d\alpha + \alpha\, d\rho)}_{=d(\alpha\rho)} \wedge \iota_X\omega = 0$$

when $\alpha\rho = \text{const}$. □

Remark 3.5.2. The last line reflects the fact that constant rescaling of a vector field does not affect whether it preserves a given volume, and constant rescaling of a volume does not affect invariance under a given vector field.

In the measurable context, we first note that such "scaling of the generating vector field" gives a cocycle α as in Proposition 1.2.2 in the context of a μ-preserving flow Φ on X. Since Proposition 1.2.2 produces a cocycle over the time-changed flow, we here effectively study a "backward" time change, which explains the apparent mismatch between Theorem 3.5.4 and Proposition 3.5.1. It is easiest to read Theorem 3.5.4 as saying that Proposition 3.5.1 holds in the measurable context.

Proposition 3.5.3. *If $0 < \rho \in L^1(X,\mu)$, then for $t \geq 0$ and a.e. $x \in X$ the equation*

$$\int_0^\alpha \rho(\varphi^\tau(x))\,d\tau = t$$

has a unique solution $\alpha = \alpha(t,x) \geq 0$. So then does $-\int_\alpha^0 \rho(\varphi^\tau(x))\,d\tau = t$ for $t < 0$, here with $\alpha < 0$, and clearly $t \mapsto \alpha(t,x)$ is strictly increasing, $\alpha(0,x) \equiv 0$, and $\lim_{t\to\pm\infty}\alpha(t,x) = \pm\infty$.

Proof. Since $\alpha \mapsto \int_0^\alpha \rho(\varphi^\tau(x))\,d\tau = t$ is continuous and strictly increasing, the conclusion holds for all x such that $\lim_{\alpha\to\infty}\int_0^\alpha \rho(\varphi^\tau(x))\,d\tau = \infty$, and we show that this is a set of full measure by showing that $\rho_\Phi := \lim_{\alpha\to\infty}\frac{1}{\alpha}\int_0^\alpha \rho(\varphi^\tau(x))\,d\tau > 0$ a.e.

By the Birkhoff Ergodic Theorem, this latter limit exists on a Φ-invariant conull set X', and we claim that the (Φ-invariant) set $E := \{x \in X' \mid \rho_\Phi(x) = 0\}$ is a null set. We have $\int_{X'}\rho_\Phi\,d\mu = \int_{X'}\rho\,d\mu$, and by the Birkhoff Ergodic Theorem,

$$\int_{X'\setminus E}\rho_\Phi\,d\mu = \int_X (\rho\chi_{X'\setminus E})_\Phi\,d\mu = \int_X \rho\chi X'\setminus E\,d\mu = \int_{X'\setminus E}\rho\,d\mu,$$

so $\int_E \rho\,d\mu = \int_E \rho_\Phi\,d\mu = 0$. This implies $\mu(E) = 0$ since $\rho > 0$. □

Note that $x \mapsto \alpha(t,x)$ is measurable since

$$\{x \in X \mid \alpha(t,x) < \alpha\} = \{x \in X \mid \textstyle\int_0^\alpha \rho(\varphi^\tau(x))\,d\tau > t\} \text{ for } \alpha > 0.$$

Then the "backward" time change, $\varphi_\rho^t(x) := \varphi^{\alpha(t,x)}(x)$, defines a measurable flow because $\alpha(t_1,x) + \alpha(t_2,\varphi^{t_1}(x)) = \alpha(t_1+t_2,x)$ by uniqueness:[18]

$$\underbrace{\int_0^{\alpha(t_1,x)+\alpha(t_2,\varphi^{t_1}(x))}\rho(\varphi^\tau(x))\,d\tau}_{=\int_0^{\alpha(t_1,x)}\rho(\varphi^\tau(x))\,d\tau+\int_0^{\alpha(t_2,\varphi^{t_1}(x))}\rho(\varphi^\tau(\varphi^{t_1}(x)))\,d\tau} = t_1 + t_2 = \int_0^{\alpha(t_1+t_2,x)}\rho(\varphi^\tau(x))\,d\tau.$$

[18] And because $(x,t) \mapsto (x,\alpha(t,x))$ is measurable on $X \times \mathbb{R}$.

Theorem 3.5.4 (Measurable Proposition 3.5.1). *If Φ preserves μ and $0 < \rho \in L^1(\mu)$, then the time change Φ_ρ preserves the probability measure $d\mu_\rho = \rho\, d\mu / \int \rho\, d\mu$.*

Proof. We show that $\int f \circ \varphi_\rho^\tau \, d\mu_\rho = \int f \, d\mu_\rho$ for $f \in L^\infty(\mu_\rho)$:

$$f_t(x) := \frac{1}{t}\int_0^t \overbrace{f(\varphi_\rho^\tau(x))}^{f(\varphi^{\alpha(\tau,x)}(x))}\, d\tau = \frac{1}{t}\overbrace{\int_0^{\alpha(t,x)} f(\varphi^v(x))\rho(\varphi^v(x))\, dv}^{v=\alpha(\tau,x),\quad dv=\frac{d\alpha(\tau,x)}{d\tau}d\tau=\frac{d\tau}{\rho(\varphi^v(x))}}$$

$$= \overbrace{\frac{1}{t}\alpha(t,x)}^{\xrightarrow[t\to\infty]{}\lim_{\alpha\to\infty}\frac{\alpha}{\int_0^\alpha \rho(\varphi^\tau(x))\,d\tau}=\frac{1}{\rho_\Phi(x)}\ \text{(time average)}}\ \underbrace{\frac{1}{\alpha(t,x)}\int_0^{\alpha(t,x)} (\rho f)(\varphi^v(x))\, dv}_{\xrightarrow[t\to\infty]{}(\rho f)_\Phi(x)\ \text{(time average)}} \xrightarrow[t\to\infty]{} \frac{(\rho f)_\Phi}{\rho_\Phi}(x),$$

so $\rho_\Phi f_t \xrightarrow[t\to\infty]{\text{a.e.}} (\rho f)_\Phi$ and hence

$$\int_X \rho_\Phi f_t \, d\mu \xrightarrow[t\to\infty]{} \int_X (\rho f)_\Phi \, d\mu = \int_X \rho f \, d\mu = \int_X \rho \, d\mu \int_X f \, d\mu_\rho.$$

Applying this to $f \circ \varphi_\rho^\tau$ instead of f gives

$$\int_X \rho_\Phi f_t \circ \varphi_\rho^\tau \, d\mu \xrightarrow[t\to\infty]{} \int_X \rho \, d\mu \int_X f \circ \varphi_\rho^\tau \, d\mu_\rho.$$

The right-hand sides agree because the two left-hand sides have the same limit:

$$\Big|\int_X \rho_\Phi f_t \circ \varphi_\rho^\tau \, d\mu - \int_X \rho_\Phi f_t \, d\mu\Big| \le \frac{1}{t}\int_X \rho_\Phi \underbrace{\Big|\int_0^t f \circ \varphi_\rho^{\tau+s} - f \circ \varphi_\rho^s \, ds\Big|}_{\le 2\tau \|f\|_\infty} d\mu \xrightarrow[t\to\infty]{} 0.$$

□

Corollary 3.5.5. *A continuous time change of a uniquely ergodic continuous flow on a compact metric space is itself uniquely ergodic* (*Definition* 3.1.17).

3.6 Flows under a function

We now introduce special flows (Definition 1.2.11) in the measurable context. We will see that many of the results of this section do not have counterparts for the topological setting.

Definition 3.6.1. Let F be an invertible measure-preserving transformation of a finite measure space (Y, μ) and $0 < r \in L^1(Y)$ a "roof" function. On

$$X := \{(y, s) \in Y \times (0, \infty) \mid s < r(y)\}$$

with the measure ν induced by $\mu \times m$ recursively define the *flow under* r by

$$\varphi^t(y, s) := \begin{cases} (y, s + t) & \text{if } 0 \le s + t < r(y), \\ \varphi^{s+t-r(y)}(F(y), 0) & \text{if } s + t > r(y), \\ \varphi^{s+t+r(F^{-1}(y))}(F^{-1}(y), 0) & \text{if } s + t < 0. \end{cases}$$

This is a ν-preserving flow on X. Note that ν may not be a probability measure even if μ is, but ν is finite and given by the formula

$$\int_X f\, d\nu = \int_Y \left(\int_0^{r(x)} f(x, t)\, dt \right) d\mu(x) \tag{3.6.1}$$

for bounded measurable functions f.

Also, such a flow has no fixed points, or at least the set of these has measure 0. It turns out that this property of having essentially no fixed points is sufficient for being of this form. This next theorem has no counterpart for special flows in the topological setting.

Theorem 3.6.2 (Ambrose–Kakutani–Rokhlin Special-Flow Representation [6]). *Let Φ be a measure-preserving flow on a Lebesgue space* (*Definition* A.1.1) *with essentially no fixed points. Then Φ is measure-theoretically isomorphic to a special flow* (*that is, represented as a special flow*).

Remark 3.6.3. For an aperiodic flow (that is, it has essentially no closed orbits) one can choose this special representation in such a way that the roof function is arbitrarily close to a given constant in the uniform topology.

This can be viewed as a global counterpart to the local construction of flow boxes in Proposition 1.1.14, but even locally, this is a nontrivial insight into the structure of a flow. Notably, it implies that the time dependence is quite regular, which is not apparent from the definition of a measurable flow. In particular, this implies that the orbit of a.e. point is a measurable set. We note, however, that being global, this is different from topological dynamics, where being a special flow constrains the topology of the underlying manifold and important fixed-point-free flows are not of this type, for example, geodesic flows.

We mention without proof (see [97] for one) that it is possible to strengthen this result as follows.

Theorem 3.6.4 (Rudolph). *If Φ is ergodic and $p, q, s > 0$ with $p/q \notin \mathbb{Q}$ and $s < 1$, then the roof function in the representation as a special flow can be chosen such that $r(Y) = \{p, q\}$ and $\mu(r^{-1}(\{p\})) = s$.*

The proof of the Ambrose–Kakutani–Rokhlin Special-Flow Representation Theorem proceeds in two main steps. Proposition 3.6.6 produces the "geometry" of a special flow, that is, the partition of the space by the orbit segments which (a posteriori) run from the base to the roof. Proposition 3.6.7 then builds the dynamics accordingly. The needed properties of the partition are as follows.

Definition 3.6.5. A partition ξ of X is said to be an *orbit-segment* partition for Φ if

(1) each partition element is an orbit segment $C = \{\varphi^\tau(x) \mid 0 \le \tau < l\}$ in such a way that the representation of any $y \in C$ as $y = \varphi^\tau(x)$ with $0 \le \tau < l$ is unique (we call x the *bottom endpoint* and l the *length* of the orbit segment), and

(2) the function $C \ni \varphi^\tau(x) = y \mapsto (L, T)(y) := (l, \tau)$ is measurable.

Proposition 3.6.6. *A measure-preserving flow Φ on a Lebesgue space with essentially no fixed points admits an invariant set of positive measure with an orbit-segment partition for which $L \ge c$ for some $c > 0$.*

Proposition 3.6.7. *If a Lebesgue space $(X, \mathcal{T}, \mu)$ with a μ-preserving flow Φ has an orbit-segment partition, then Φ is measure-theoretically isomorphic to a special flow.*

Proof of Theorem 3.6.2. Proposition 3.6.6 gives an invariant set E of positive measure with an orbit-segment partition, on which the flow then is measure-theoretically isomorphic to a special flow by Proposition 3.6.7. We now apply this recursively.

Let $C_0 = \varnothing$, and for $i \ge 1$ there either is a set $E \subset X \setminus \bigcup_{j<i} C_j$ of measure at least $1/i$ on which Φ is measure-theoretically isomorphic to a special flow, and we let C_i be the (without loss of generality countable) union of such an E, or else we set $C_i = \varnothing$. Then the restriction of Φ to $C := \bigcup_{i \in \mathbb{N}} C_i$ is measure-theoretically isomorphic to a special flow, and C has full measure. Otherwise, Propositions 3.6.6 and 3.6.7 yield an $E \subset X \setminus C$ with $\mu(E) > 1/i$ for some $i \in \mathbb{N}$, hence $E \cap C \neq \varnothing$ by construction of C, a contradiction. □

Proof of Proposition 3.6.6. We will find two disjoint sets both of which orbits revisit indefinitely; the "exit points" from one of these on the way to the other form a good candidate for the base of a special flow. We will use averaging in the flow direction:

$$\tau \mapsto \operatorname{avg}_A^\alpha(\tau, x) := \frac{1}{\alpha} \int_0^\alpha \chi_A \circ \varphi^{t+\tau}(x)\, dt \quad \text{is} \quad \tfrac{2}{\alpha}\text{-Lipschitz for } x \in X \tag{3.6.2}$$

because
$$\frac{1}{\alpha}\Big|\underbrace{\int_0^\alpha \chi_A\circ\varphi^{t+\tau_1}\,dt-\int_0^\alpha \chi_A\circ\varphi^{t+\tau_2}\,dt}_{=\int_{\tau_1}^{\tau_1+\alpha}\chi_A\circ\varphi^t\,dt-\int_{\tau_2}^{\tau_2+\alpha}\chi_A\circ\varphi^t\,dt}\Big|\le\frac{2}{\alpha}|\tau_1-\tau_2|.$$

Since Φ has essentially no fixed points, we can find a measurable set A and a $t_0\in\mathbb{R}$ such that $\delta:=\mu((X\smallsetminus A)\cap\varphi^{t_0}(A))>0$. Let

$$E_1:=\{x\in X \mid \operatorname{avg}_A^\alpha(0,x)<1/4\},\quad E_2:=\{x\in X \mid \operatorname{avg}_A^\alpha(0,x)>3/4\},$$
$$E:=E_1\cap\varphi^{t_0}(E_2),$$

with (by Lemma 3.6.8) $\alpha>0$ small enough that

$$\mu((X\smallsetminus A)\triangle E_1)<\delta/2\quad\text{and}\quad\mu(A\triangle E_2)<\delta/2.$$

Lemma 3.6.8 (Wiener). *If $f\in L^\infty(X,\mu)$, then $\frac{1}{t}\int_0^t f\circ\varphi^s\,ds\xrightarrow[t\to 0]{L^2\text{ \& in measure}} f$.*

Proof. It suffices to prove convergence in L^2, and to that end we use the *spectral measure* σ with $\langle U_\Phi^t f,f\rangle=\int_X f\circ\varphi^t\cdot\bar f=\int_{\mathbb{R}} e^{i\lambda t}\,d\sigma(\lambda)$ (Definition 3.7.8) to get

$$\begin{aligned}
&\Big\|\frac{1}{t}\int_0^t f\circ\varphi^s\,ds-f\Big\|^2\\
&=\int_X\underbrace{\Big(\frac{1}{t}\int_0^t f\circ\varphi^s\,ds-f\Big)\overline{\Big(\frac{1}{t}\int_0^t f\circ\varphi^s\,ds-f\Big)}}_{=\frac{1}{t^2}\int_0^t f\circ\varphi^s\,ds\overline{\int_0^t f\circ\varphi^r\,dr}-\frac{1}{t}\int_0^t\bar f f\circ\varphi^s\,ds-\frac{1}{t}\int_0^t f\overline{f\circ\varphi^r}\,dr+f\bar f}\\
&\underset{\text{Fubini}}{=}\frac{1}{t^2}\int_0^t\int_0^t\overbrace{\int_{\mathbb{R}} e^{i\lambda(s-r)}\,d\sigma(\lambda)}^{=\langle U_\Phi^s f,U_\Phi^r f\rangle}\,ds\,dr\\
&\qquad-\frac{1}{t}\int_0^t\overbrace{\int_{\mathbb{R}}\overline{e^{i\lambda s}}\,d\sigma(\lambda)}^{=\overline{\langle U_\Phi^s f,f\rangle}}\,ds-\frac{1}{t}\int_0^t\int_{\mathbb{R}} e^{i\lambda r}\,d\sigma(\lambda)\,dr+\overbrace{\int_{\mathbb{R}} d\sigma(\lambda)}^{=\langle U_\Phi^0 f,f\rangle}\\
&\underset{\text{Fubini \& (3.6.3)}}{=}\int_{\mathbb{R}}\Big|\frac{e^{i\lambda t}-1}{i\lambda t}-1\Big|^2 d\sigma(\lambda)\xrightarrow[t\to 0]{\text{Lebesgue Dominated-Convergence Theorem}}0
\end{aligned}$$

because

$$\frac{1}{t^2}\int_0^t\int_0^t e^{i\lambda s}\overline{e^{i\lambda r}}\,ds\,dr=\overbrace{\frac{1}{t}\int_0^t e^{i\lambda s}\,ds}^{=\frac{1}{t}\frac{e^{i\lambda s}}{i\lambda}\Big|_0^t=\frac{e^{i\lambda t}-1}{i\lambda t}\xrightarrow[t\to 0]{}1}\overline{\frac{1}{t}\int_0^t e^{i\lambda r}\,dr}.\tag{3.6.3}$$

$\square$

As a consequence,

$$\mu(E) \geq \mu(\underbrace{(X \smallsetminus A)}_{\approx E_1} \cap \underbrace{\varphi^{t_0}(A)}_{\approx \varphi^{t_0}(E_2)}) - \mu((X \smallsetminus A)\triangle E_1) - \mu(A \triangle E_2) > 0,$$

so there is an invariant set E' with positive measure of points that visit E for arbitrarily large positive and negative times. Also,

$$E'_1 := \varphi^{[0,\alpha/8]}(E_1) \quad \text{and} \quad E'_2 := \varphi^{[0,\alpha/8]}(E_2)$$

are disjoint because if $x \in E'_1 \cap E'_2$, then $x_i := \varphi^{\tau_i}(x) \in E_i$ for suitable $\tau_i \in [0, \alpha/8]$, so

$$\frac{1}{2} < \underbrace{|\operatorname{avg}_A^\alpha(0, x_1) - \operatorname{avg}_A^\alpha(0, x_2)|}_{=\operatorname{avg}_A^\alpha(-\tau_1, x) - \operatorname{avg}_A^\alpha(-\tau_2, x)} \leq \frac{2}{\alpha}|\tau_1 - \tau_2| \leq \frac{2}{\alpha}\frac{\alpha}{8} = \frac{1}{4}$$

by (3.6.2), a contradiction.[19] The "points of exit from E'_1 on the way to E'_2" now form the base Y of a special-flow representation as follows.

For any orbit in E', the set of times when it hits E'_1 is open and without upper or lower bound, and its connected components ("E'_1-intervals") have length at least $\alpha/8$. Likewise for E'_2-intervals, and no E'_1-interval overlaps with any E'_2-interval, so for every point in E' there is a well-defined nearest such interval "below" and likewise "above." We then take $Y \subset E'$ to be the set of top endpoints of E'_1-intervals with the additional property that the nearest interval above is an E'_2-interval. This ensures that it takes more time than $\alpha/8$ to return to Y, that is, $f(x) := \min\{t > 0 \mid \varphi^t(x) \in Y\} \geq \alpha/8$ for $x \in Y$.

The desired partition elements are now given by $\varphi^{[0,f(x)]}(x)$ for $x \in Y$, and $(L,T)(\varphi^s(x)) = (f(x), s)$, so it only remains to show measurability of L and T. For T this follows from measurability of

$$B_k := \{x \in X \mid \tfrac{k}{n} \leq T(x) < \tfrac{k+1}{n}\} \quad \text{for } n \geq 8/\alpha,\ k \in \mathbb{N},$$

which reduces to that of B_0 since $B_{k+1} = \varphi^{1/n}(B_k) \smallsetminus B_0$. Note that B_0 is measurable because $x \in B_0$ means that x just exited E'_1 on the way to E'_2, that is, $x \notin E'_1 \ni \varphi^{-1/n}(x)$ and there is an $i \in \mathbb{N}$ with $\varphi^{1/n}(x), \varphi^{2/n}(x), \dots, \varphi^{i/n}(x) \notin E'_1 \cup E'_2$ but $\varphi^{i+1/n}(x) \in E'_2$. Finally, L is measurable because $L^{-1}((c,\infty)) = \varphi^{\mathbb{Q}\cap[0,c]}(T^{-1}((c,\infty))) \cup \varphi^{-c}(T^{-1}([c,\infty)))$. This completes the proof of Proposition 3.6.6 □

Remark 3.6.9. We could arrange for L to be bounded by refining the partition elements $\varphi^{[0,L(x))}(x)$ into the $\varphi^{[ic,(i+1)c)}(x)$ and $\varphi^{[kc,L(x))}(x)$, where $c := \inf L < L(x) - kc < 2c$. The new roof function L' then satisfies $c \leq L' < 2c$.

[19] The E_i are measurable: continuity of $\tau \mapsto \operatorname{avg}_A^\alpha(0, \varphi^\tau(x)) = \operatorname{avg}_A^\alpha(-\tau, x)$ implies $E'_i = \varphi^{\mathbb{Q}\cap[0,\alpha/8]}(E_i)$.

Proof of Proposition 3.6.7. The essence of the argument is that the bottom endpoints of the partition elements form the base. For checking the details of why this works it is convenient to reparametrize the flow in a piecewise linear way so all partition elements are traversed in unit time, that is, by multiplying the "speed" by $L(x)$ on the partition element containing x. Another way of putting this is to write (for $y := \varphi^{L(x)-T(x)}(x)$, and $t'(x) = \frac{L(x)-T(x)}{L(x)}$)

$$\overline{\varphi}^t(x) = \begin{cases} \varphi^{tL(x)}(x) & \text{for } 0 \le t < t'(x), \\ \varphi^{(t-t'(x))L(y)}(y) & \text{for } 1 > t \ge t'(x). \end{cases}$$

This defines a measurable flow $\overline{\Phi}$ on $(X, \mathcal{A})$ with the same orbit-segment partition, and by Theorem 3.5.4 it preserves the measure $\overline{\mu} = L\mu / \int L\, d\mu$. It suffices to show that this flow is measure-theoretically isomorphic to a special flow with $\overline{T} = T/L$ and roof function $\overline{L} \equiv 1$, that is, a suspension.

To that end note that the map $\pi\colon X \to X_1$ that sends each point to the bottom endpoint of its partition element makes $(X_1, \mathcal{A}_1 := \pi_* \mathcal{A} := \{A \subset X_1 \mid \pi^{-1}(A) \in \mathcal{A}\}, \mu_1 := \pi_*(\overline{\mu}))$ a measure space, where $\mu_1(A) = \overline{\mu}(\pi^{-1}(A))$ is preserved by the base transformation $F := \overline{\varphi}^1$. This is represented as a suspension flow $\psi^t = h \circ \overline{\varphi}^t \circ h^{-1}$ via the bijection $h\colon X \to X' := X_1 \times [0,1), \overline{\varphi}^t(x) \mapsto (x,t)$.

Lemma 3.6.10. *$\mu' := h(\mu) = \mu_1 \times$ Lebesgue $=: \overline{\nu}$ on $\mathcal{A}' := h(\mathcal{A}) = \mathcal{A}_1 \times \mathcal{B}$, the product σ-algebra in $X' = X_1 \times [0,1)$.*

Up to reversing the above time change, this proves Proposition 3.6.7 □

The proof of Lemma 3.6.10 involves careful applications of the basic notions of measure theory more than dynamical ideas, and the main effort is to show that $\mathcal{A}_1 \times \mathcal{B} = \mathcal{A}'$. The inclusion $\mathcal{A}_1 \times \mathcal{B} \subset \mathcal{A}'$ is clear: if $\mathcal{A}_1 \times \mathcal{B} \ni A = A_1 \times [t_1, t_2)$ with $A_1 \in \mathcal{A}_1$, then $h^{-1}(A) = \{y \mid \pi(y) \in A_1,\ t_1 \le \overline{T}(y) < t_2\} \in \mathcal{A}$. Therefore the main effort is the reverse inclusion, and here it is central that we are dealing with Lebesgue spaces. This is put to use via notions of "open" and "boundary" in the absence of any topology by using the flow parameter as follows: if $E \subset \mathbb{R}$ and $x \in X$, then the (flow-)closure of $C := \varphi^E(x)$ is $\overline{C} := \varphi^{\overline{E}}(x)$ and $\partial C := \varphi^{\partial E}(x)$ is the (flow-)boundary of C. More generally, then, the flow-closure and flow-boundary of $A \subset X$ are defined by

$$\varphi^{\mathbb{R}}(x) \cap \overline{A} = \overline{\varphi^{\mathbb{R}}(x) \cap A} \quad \text{and} \quad \varphi^{\mathbb{R}}(x) \cap \partial A = \partial(\varphi^{\mathbb{R}}(x) \cap A)$$

for all $x \in X$. Also, $A \subset X$ is said to be (flow-)open if $\{t \mid \varphi^t(x) \in A\}$ is open for all $x \in X$. Then we can approximate measurable sets by flow-open ones as follows.

Lemma 3.6.11. *For $A \in \mathcal{A}$ and $\epsilon > 0$ there is an $A_\epsilon \in \mathcal{A}$ such that*

(1) *A_ϵ is flow open,*

(2) *$\mu(\partial A_\epsilon) = 0$, and*

(3) *$\mu(A \triangle A_\epsilon) < \epsilon$.*

Proof. We take $A_\epsilon := A_{n,\beta} := \{x \in X \mid \mathrm{avg}_A^{1/n}(0,x) > \beta\}$ for $n \in \mathbb{N}$ and $\beta \in (0,1)$ to be determined. Note that $\{t \mid \varphi^t(x) \in A_{n,\beta}\}$ is open for all $x \in X$ by (3.6.2), and that $\partial A_{n,\beta} \subset \{x \in X \mid \mathrm{avg}_A^{1/n}(0,x) = \beta\}$, so there is a $\beta \in (0,1)$ with $\mu(\partial A_{n,\beta}) = 0$ for all $n \in \mathbb{N}$. By Lemma 3.6.8 we can choose n such that $\mu(A \triangle A_{n,\beta}) < \epsilon$. □

Lemma 3.6.12. *If $A \in \mathcal{A}$ is flow open, then $\pi(A) \in \mathcal{A}_1$, that is, $\overline{A} := \pi^{-1}(\pi(A)) \in \mathcal{A}$.*

Proof. This, and $\mu_1(\pi(A)) = \overline{\mu}(\overline{A})$, follow from

$$\overline{A} = \Big(\bigcup_{n=1}^{\infty}\bigcup_{k=1}^{2^{n+1}-2} \overline{A}_{n,k}\Big) \cup \Big(\bigcup_{n=1}^{\infty}\bigcup_{k=1}^{2^{n}-1} \overline{A}_{k/2^n}\Big),$$

where $A_{n,k} := \mathcal{A} \cap \overline{T}^{-1}((\frac{k}{2^{n+1}}, \frac{k+1}{2^{n+1}}))$,

$$\begin{aligned}\overline{A}_{n,k} := \pi^{-1}(\pi(A_{n,k})) = \Big[\overline{T}^{-1}((\tfrac{1}{2^{n+1}}, \infty)) \cap \overline{\varphi}^{\mathbb{Q}\cap[0,1-\frac{k}{2^{n+1}})}(A_{n,k})\Big] \\ \cup \Big[\overline{T}^{-1}([0, 1 - \tfrac{1}{2^{n+1}})) \cap \overline{\varphi}^{\mathbb{Q}\cap[-\frac{k+1}{2^{n+1}},0]}(A_{n,k})\Big],\end{aligned}$$

$A_{k/2^n} := \mathcal{A} \cap \overline{T}^{-1}(\{\frac{k}{2^n}\})$, and $\overline{A}_{k/2^{n+1}} := \pi^{-1}(\pi(A_{k/2^{n+1}})) = \overline{\varphi}^{[-\frac{k}{2^{n+1}},1-\frac{k}{2^{n+1}}]}(A_{k/2^{n+1}})$. □

We note also that

$$\begin{aligned} A \subset D := \Big[\bigcap_{l\in\mathbb{N}}\bigcup_{n\geq l}\bigcup_{k=1}^{2^{n+1}-2} \pi(A_{n,k}) \times \big(\frac{k}{2^{n+1}}, \frac{k+1}{2^{n+1}}\big)\Big] \\ \cup \Big[\bigcup_{n=1}^{\infty}\bigcup_{k=0}^{2^{n}-1} A_{k/2^n}\Big] \subset \overline{A} \quad \text{(flow-closure)}.\end{aligned}$$

The point is that $h(D) \in \mathcal{A}_1 \times \mathcal{B}$ since the $\pi(A_{n,k}) \in \mathcal{A}_1$ by Lemma 3.6.12 since the $A_{n,k}$ are flow open.

Proof of Lemma 3.6.10. $\overline{\mu}(A) = \overline{\nu}(h(A))$ for every $A \in \mathcal{M} := \{h^{-1}(A) \mid A \in \mathcal{A}_1 \times \mathcal{B}\} \subset 2^X$ because the σ-algebra $\mathcal{M}$ is generated by sets A for which $h(A) = A_1 \times I$, where $A_1 \in \mathcal{A}_1$ and $I \subset [0, 1)$ is an interval, and for such sets this is clear. Thus, $\mathcal{M}$ is complete with respect to μ. The preceding observation and Lemma 3.6.12 imply that for any $A \in \mathcal{A}$ and $\epsilon > 0$ there is an $A_\epsilon \in \mathcal{M}$ with $\mu(A \triangle A_\epsilon) < \epsilon$. Since X is a Lebesgue space and $\mathcal{M}$ is complete, this implies $\mathcal{M} = \mathcal{A}$—and we also have $\overline{\nu} = h(\overline{\mu})$. □

Even though the proof of Theorem 3.6.2 is not entirely constructive and hence does not give a straightforward explicit representation as a special flow, the mere existence of such a representation is useful, notably with respect to studying the interplay between entropy and time changes (Theorems 4.1.8 and 4.1.9).

As mentioned at the beginning of the section the measure ν given by (3.6.1) is not necessarily a probability measure. Therefore, an invariant probability measure for the special flow can be defined by

$$\int_X f\, d\mu_r = \frac{\int_Y \left(\int_0^{r(x)} f(x,t)\, dt \right) d\mu(x)}{\int_Y r(x)\, d\mu(x)} \tag{3.6.4}$$

for any bounded measurable function f.

In the context of special flows it is possible to produce a flow counterpart to the Kac Lemma (Lemma A.3.35), a basic result in discrete-time ergodic theory, which involves the *return time*. For measure-preserving flows this issue is far trickier than in discrete time because for a set that is far from open, closed, or convex, even defining "return time" is challenging. For returns to the base of a special flow, however, there is a simple analog of the Kac Lemma.

Proposition 3.6.13 (Flow Kac Lemma [344, Corollary 1]). *If F is a μ-preserving map on a topological probability space (X, μ), $0 < r \in L^1(\mu)$, $\mu(A) > 0$, then*

$$\int_A T_A(x)\, d\mu_A = \frac{1}{\mu(A)} \int_X r\, d\mu,$$

with μ_A the conditional measure from (3.3.3) *and $T_A(x) := \min\{t > 0 \mid \varphi_r^t(x,0) \in A\}$.*

3.7 Spectral theory*

Although we will hardly use it in our study of hyperbolic flows, we describe here some elements of the spectral approach to ergodic theory. The central idea is to connect properties of the Koopman operator (Definition 3.2.5) for a flow with dynamical properties of the flow.

Note first that for a μ-preserving flow Φ on X the operators $U^t_\Phi = U_{\varphi^t}$ associated with a flow form a 1-parameter group of unitary operators on $L^2(X, \mathcal{T}, \mu)$—and here it is useful to consider complex-valued functions.[20] Then 1 is always an eigenvalue, because constant functions are invariant. Therefore it is often natural to restrict attention to $1^\perp \subset L^2$, the space of functions with integral 0. We will usually assume that μ is a Borel probability measure, in which case $L^2(X, \mathcal{T}, \mu)$ is separable. This turns out to imply that U^t_Φ is a *continuous* group of unitary operators. One useful simple property of these operators that makes them special beyond linearity is that $U_\Phi(fg) = U_\Phi(f)U_\Phi(g)$.

An easy connection to the classification problem is that if a flow Φ on $(X, \mathcal{T}, \mu)$ and a flow Ψ on $(Y, \mathcal{V}, \nu)$ are measure-theoretically isomorphic, then their Koopman operators are conjugate (or, as one says in this context, unitarily equivalent): let $h \colon X \to Y$ be an isomorphism such that $h \circ \varphi^t = \psi^t \circ h$, then $U_{\varphi^t} \circ U_h = U_{h\circ\varphi^t} = U_{\psi^t\circ h} = U_h \circ U_{\psi^t}$. It is interesting when one can go the other way around: if one can show that the unitary operators for Φ and Ψ are conjugate, then one may hope to utilize this somehow to show that Φ and Ψ are measure-theoretically isomorphic. This is, of course, not always so.

Thus, spectral invariants of U_f, for example, eigenvalues with their multiplicities or the spectrum, are invariants of measure-theoretic isomorphism of f.

Definition 3.7.1. Two measure-preserving transformations are said to be *spectrally isomorphic* if their Koopman operators are unitarily equivalent. An invariant of spectral isomorphism is called a *spectral invariant*.

Let us illustrate how dynamical properties might be expressible in terms of the spectrum of the Koopman operator.

Proposition 3.7.2. *A μ-preserving flow Φ on X is ergodic if and only if* 1 *is a simple eigenvalue of the associated Koopman operator.*

Proof. We noted that 1 is always an eigenvalue, and simplicity of this eigenvalue is equivalent to saying that U_{φ^t}-invariant functions are constant, which is equivalent to ergodicity. □

From this, we conclude the following result:

Proposition 3.7.3. *Ergodicity is a spectral invariant.*

Definition B.3.1 formally introduces the spectrum in this context.

[20] The results we obtain in this context can be used for real linear spaces E by passing to the complexification $E_{\mathbb{C}}$ (that is, the space $E \otimes \mathbb{C}$ obtained by allowing complex scalars) and then suitably restricting attention to the real part.

Proposition 3.7.4. *If Φ is a probability-preserving flow, then*

(1) *the eigenvalues of U_Φ lie on the unit circle,*

(2) *the spectrum of U_Φ lies on the unit circle,*

(3) *the spectrum of U_Φ is compact,*

(4) *the eigenvalues of U_Φ form a subgroup of the unit circle, and*

(5) *the eigenspaces of U_Φ are pairwise orthogonal.*

Proof. (1): If A is an isometry and $Av = \lambda v$, then $\|v\| = \|Av\| = \|\lambda v\| = |\lambda|\|v\|$.
(2): If A is unitary then $r(A^{\pm 1}) \le \|A^{\pm 1}\| = 1$, so $\sigma(A^{\pm 1}) \subset \{\lambda \mid |\lambda| \le 1\}$. Then $A \in \mathrm{Aut}(V)$ implies $0 \notin \sigma(A)$ and hence $\sigma(A^{-1}) = \{\lambda^{-1} \mid \lambda \in \sigma(A)\}$ because $(1/\lambda)I - A^{-1}$ is invertible if and only if $-\lambda A[(1/\lambda)I - A^{-1}] = \lambda I - A$ is.
(3): Lemma B.3.5.
(4): If $U_\Phi(f) = \lambda f$ and $U_\Phi(g) = \mu g$, then $\mu\lambda^{-1}$ is also an eigenvalue:

$$U_\Phi(f \cdot \bar{g}) = U_\Phi(f)\overline{U_\Phi(g)} = \mu\bar{\lambda} \cdot f \cdot \bar{g} = \mu\lambda^{-1} \cdot f \cdot \bar{g}.$$

This shows closure under inverses (take $\mu = 1 = g$) and then under multiplication.
(5): If $U_\Phi(f) = \lambda f$ and $U_\Phi(g) = \mu g$, then

$$\langle f, g\rangle = \langle U_\Phi(f), U_\Phi(g)\rangle = \langle \lambda f, \mu g\rangle = \lambda\bar{\mu}\langle f, g\rangle = \lambda\mu^{-1}\langle f, g\rangle,$$

so $\lambda\mu^{-1} = 1$ or $\langle f, g\rangle = 0$. □

Remark 3.7.5. We emphasize that we are here considering eigenvalues of $U_\Phi = U_{\varphi^1}$. If $e^{i\alpha}$ is an eigenvalue of U_Φ, then there is an eigenfunction f with $U_{\varphi^t}(f) = e^{i\alpha t} f$ for all $t \in \mathbb{R}$. This itself produces a multiplicative subgroup, so for 1-parameter groups of unitary operators it is conventional to call $\alpha \in \mathbb{R}$ an eigenvalue of $(U^t)_{t\in\mathbb{R}}$ if $e^{i\alpha}$ is an eigenvalue of U^1. Then Proposition 3.7.4 tells us that the eigenvalues of a 1-parameter group of unitary operators are an additive subgroup of $\mathbb{R}$ (with pairwise orthogonal eigenspaces). See also Definition 3.4.37.

Ergodicity easily provides information about other eigenspaces.

Proposition 3.7.6. *For a measure-preserving flow Φ the following are equivalent:*

(1) *Φ is ergodic,*

(2) *all eigenfunctions have constant absolute value, and*

(3) *all eigenspaces are 1-dimensional.*

Proof. (3) ⇒ (2): $U_\Phi(|f|^2) = U_\Phi(f\overline{f}) = \lambda\overline{\lambda}f\overline{f} = |f|^2$, so $|f|$ is in the eigenspace for 1, which contains the constants and is 1-dimensional.

(2) ⇒ (1): Contraposition: if A is a proper Φ-invariant set, then χ_A is an eigenfunction with nonconstant absolute value.

(1) ⇒ (3): if f, g are nonzero eigenfunctions for λ, then $f\overline{f}$ and $g\overline{g}$ are invariant, hence constant, so f/g is a well-defined invariant function, hence constant. □

It is also easy to see the following.

Proposition 3.7.7. *Mixing is a spectral invariant (Definition* 3.7.1).

Proof. Suppose Φ on $(X,\mathcal{T},\mu)$ is mixing, Ψ ν-preserving on Y, $W \circ U_\Phi = U_\Psi \circ W$, W unitary, and $f_i = W(g_i) \in L^2(Y,\mathcal{V},\nu)$ for $i = 1,2$. Then $W1 = 1$ by Proposition 3.7.3, and

$$\begin{aligned}\langle U_\Psi^t(f_1), f_2\rangle &= \langle U_\Psi^t(W(g_1)), W(g_2)\rangle = \langle W(U_\Phi^t(g_1)), W(g_2)\rangle = \langle U_\Phi^t(g_1), g_2\rangle \\ &\xrightarrow[t\to\infty]{} \langle g_1, 1\rangle\langle g_2, 1\rangle = \langle Wg_1, W1\rangle\langle Wg_2, W1\rangle = \langle f_1, 1\rangle\langle f_2, 1\rangle.\end{aligned}$$ □

Both because this was used in the proof of Lemma 3.6.8 and because it is an important aspect of studying spectral properties, we now introduce spectral measures, which are defined by something much like a Fourier transform.

Definition 3.7.8. If Φ is a measure-preserving flow on a Lebesgue space $(X,\mathcal{T},\mu)$ and $f \in L^2(X,\mathcal{T},\mu)$, the *spectral measure* σ_f of f on $\mathbb{R}$ is defined by

$$\langle U_\Phi^t f, f\rangle = \int_{\mathbb{R}} e^{it\lambda} d\sigma_f(\lambda).$$

By taking $t = 0$ we find that $\sigma_f(\mathbb{R}) = \|f\|_2^2$.[21]

Example 3.7.10. If f is an eigenfunction of U_Φ with eigenvalue $\lambda = e^{i\alpha}$, then

$$e^{it\alpha}\|f\|_2^2 = \langle U_{\varphi^t}(f), f\rangle = \int_{\mathbb{R}} e^{it\lambda} d\sigma_f(\lambda), \tag{3.7.1}$$

which is equivalent to $\sigma_f = \|f\|_2^2\delta_\alpha$, a Dirac measure at α. Conversely, (3.7.1) implies $|\langle U_{\varphi^t}(f), f\rangle| = \|f\|_2^2 = \|U_{\varphi^t}(f)\|\|f\|$, so $U_{\varphi^t}(f)$ and f are proportional by

[21] By Theorem 3.7.9, σ_f exists since $t \mapsto \langle U_\Phi^t f, f\rangle$ is positive definite: if $(z_1,\dots,z_m) \in \mathbb{C}^m$ and $(t_1,\dots,t_m) \in \mathbb{R}^m$, then

$$0 \le \Big\|\sum_{k=1}^m z_k U_\Phi^{t_k}(f)\Big\|^2 = \Big\langle \sum_{i=1}^m z_i U_\Phi^{t_i}(f), \sum_{j=1}^m z_j U_\Phi^{t_j}(f)\Big\rangle = \sum_{i,j=1}^m \langle U_\Phi^{t_i-t_j} f, f\rangle z_i \overline{z}_j.$$

Theorem 3.7.9 (Herglotz). *A sequence $(a_i)_{i\in\mathbb{Z}}$ is the sequence of Fourier coefficients of a Borel measure on S^1 if and only if $a_{-i} = \overline{a}_i$ for all $i \in \mathbb{Z}$ and $0 \le \sum_{|i|,|j|\le N} a_{i-j}x_j\overline{x}_i$ for any $(x_i)_{i\in\mathbb{Z}}$ and $N \in \mathbb{N}$.*

the equality case of the Cauchy–Schwarz inequality, so f is an eigenfunction—with the eigenvalue given by (3.7.1). Thus, weak mixing is equivalent to the following: every $f \in L^2(X,\mu)$ whose spectral measure is a point mass is constant. Likewise, ergodicity is equivalent to the following: every $f \in L^2(X,\mu)$ whose spectral measure is δ_0 is constant.

The following notion is natural for describing a situation in which a measure-preserving transformation is "spectrally rigid."

Definition 3.7.11. We say that Φ has *pure point spectrum* or *discrete spectrum* if Φ is ergodic and there is a basis of eigenfunctions of U_Φ.

Remark 3.7.12. The terminology goes back to that in Definition B.3.1 in that the spectrum consists entirely of eigenvalues. Note also that by Proposition 3.7.6 these λ are pairwise distinct; this produces enough information for spectral isomorphism.

Proposition 3.7.13. *Ergodic measure-preserving flows with discrete spectrum and with the same eigenvalues are spectrally isomorphic.*

Proof. For each eigenvalue, map the corresponding eigenfunction for one transformation to that for the other (see Proposition 3.7.6); extend by linearity and continuity. □

Remark 3.7.14. In this case the dynamics of U_Φ consists of a product of rotations of the eigenspaces; the essential information is contained in what happens to normalized eigenfunctions. This can be exploited to show that, in fact, here the eigenvalues determine Φ up to a measure-theoretic isomorphism.

Although we omit the proof (other than Proposition 3.4.38), one can characterize weakly mixing flows analogously to the way we previously characterized ergodicity.

Proposition 3.7.15. *For a measure-preserving flow Φ the following are equivalent:*

- *Φ is weakly mixing,*
- *all eigenfunctions are constant,*
- *σ_f is nonatomic ("continuous") for every $f \perp 1$.*

Remark 3.7.16. The third of these versions is the reason one also describes this property as having continuous spectrum.

Exercises

3.1. Determine $\mathfrak{M}(\Phi)$ in Examples 1.1.5, 1.1.7, 1.3.7, 1.3.8, 1.3.9, 1.3.13, 1.3.14, 1.3.15, and 1.4.16, and in Figures 1.1.4, 1.4.1, 1.5.4, and 1.5.11.

3.2. Prove that the "derivative" of the Cantor function $\mathfrak{c}$ (Remark 1.3.6) defines a probability measure on $[0,1]$ that is supported on the ternary Cantor set (Figure 1.3.1) by setting $\mu_{\mathfrak{c}}([0,x]) = \mathfrak{c}(x)$ and extending this to a measure.

3.3. Find a nonatomic invariant Borel probability measure for the Akin flow (Example 1.3.16).

3.4. Find all ergodic invariant Borel probability measures in Examples 1.1.7, 1.3.8, 1.3.9, 1.3.13, 1.3.14, 1.3.15, 1.3.16, and 1.4.16 and Figures 1.4.1, 1.5.4, 1.5.11, and 1.1.4.[22]

3.5. For each measure obtained in Exercise 3.4 determine which of the higher mixing properties from Definitions 3.4.1, 3.4.2, and 3.4.3 hold.

3.6. Describe what invariant sets of total measure look like in Examples 1.1.7, 1.3.8, 1.3.9, 1.3.13, 1.3.14, 1.3.15, and 1.4.16 and Figures 1.4.1, 1.5.4, 1.5.11, and 1.1.4.

3.7. Describe the set of historic points in the previous examples (Remark 3.2.25).

3.8. Prove that unique ergodicity is invariant under topological conjugacy.

3.9. Prove the assertion after Corollary 3.3.11: a nonergodic invariant measure μ may also determine the asymptotic distribution of *some* orbits, but such orbits are always a set of μ-measure 0.

3.10. Construct a time change of the flow on $\mathbb{T}^2$ in Example 1.6.2 with irrational slope α such that the Dirac measure $\delta_0\colon A \mapsto \chi_{\{A\}}(0)$ is the only invariant Borel probability measure.

3.11. Prove that if μ is an ergodic invariant measure for a continuous flow Φ, then $\Phi\restriction_{\operatorname{supp}\mu}$ is topologically transitive. (Compare Proposition 3.4.12.)

3.12. Prove Proposition 3.2.11.

3.13. With the notation of Theorem 3.2.8 take $H = \mathbb{C}^3$, G generated by one rotation around the x-axis, and $v = (1,1,1)$. Determine H_G, $P_{H_G}(v)$, and $\overline{\operatorname{co}}\, Gv$ to verify the conclusion of the theorem.

3.14. Describe the class of functions $f\colon [0,1] \to \mathbb{R}$ that are measurable with respect to the σ-algebra $\mathcal{S}$ generated by $[0,1/2]$.

[22] Exercise 3.17 can help.

3.15. For the σ-algebra $\mathcal{S}$ from Exercise 3.14 describe the projection $\pi_{\mathcal{S}}$ on the space of Lebesgue-integrable functions.

3.16. Show that measurable isomorphism (Definition 3.1.20) and monotone equivalence (Definition 3.1.23) of flows are equivalence relations.

3.17. Prove that if μ is an ergodic invariant measure for a continuous flow Φ, then the orbit of μ-a.e. x is dense in $\operatorname{supp}\mu$.

3.18. Theorem 3.3.13 and Proposition 3.4.40 combine to imply that for a suspension of an ergodic probability-preserving transformation there is countable set of $\tau \in \mathbb{R}$ for which the time-τ map is not ergodic. Describe this exceptional set.

3.19. Suppose a special flow over a measure-preserving transformation F (Definition 3.6.1) is ergodic with respect to the measure μ_r from (3.6.1) (or the measure ν from Definition 3.6.1). Show that F is ergodic with respect to μ.

3.20. Prove Proposition 3.4.11.

3.21. Show that K-mixing implies mixing (Definition 3.4.1).

3.22. Show that K-mixing implies multiple mixing (Definition 3.4.1).

3.23. Show that the Bernoulli property implies mixing (Definition 3.4.1).

3.24. Show that the Bernoulli property implies multiple mixing (Definition 3.4.1).

3.25. Show that the Bernoulli property implies K-mixing (Definition 3.4.1).

3.26. Show that the spectrum of a weakly mixing flow on a nonatomic probability space is the whole unit circle.

3.27. (Uniform set recurrence). Without resorting to using ergodic theorems prove that if Φ is a probability-preserving flow and $\mu(A) > 0$, then $E := \{t \in \mathbb{R} \mid \mu(\varphi^t(A) \cap A) > 0\}$ has bounded gaps, that is, there is a $\tau \in \mathbb{R}$ such that $E \cap [t, t+\tau] \neq \varnothing$ for all $t \in \mathbb{R}$.[23]

3.28. If μ is a nonatomic ergodic probability measure for a flow Φ on X and $0 < \delta(t) \xrightarrow[t\to\infty]{} 0$, find a bounded measurable f on X with $\overline{\lim}_{T\to\infty} \frac{|\frac{1}{T}\int_0^T f\circ\varphi^t - \int f|}{\delta(t)} \stackrel{\text{ae}}{=} +\infty$. Thus, there is no guaranteed speed of convergence in the ergodic theorems.

3.29. Give direct proofs of those implications in Proposition 3.7.6 that were not established in its proof.

[23] This can be refined to the following theorem:

Theorem 3.7.17 (Khintchine [289, Theorem 3.3, p. 37]). *For any measurable A and $\epsilon > 0$ the set $\{t \in \mathbb{R} \mid \mu(\varphi^t(A) \cap A) \geq (\mu(A))^2 - \epsilon\}$ has bounded gaps.*

4

Entropy, pressure, and equilibrium states

The preceding chapters developed important notions for the study of qualitative features of dynamical systems in topological and probabilistic ways. We now introduce quantitative notions for describing the complexity of a dynamical system. The principal notion is entropy. Its probabilistic version measures complexity on an exponential scale by an approach modeled on information theory. The topological version was developed in analogy to measure-theoretic entropy and turns out to be closely connected to other measures of orbit complexity, such as the growth of periodic orbits. Inspired by the study of thermodynamics, a notion of pressure builds on these notions, and connecting these various notions in turn provides new ways of constructing measures of particular dynamical interest.

4.1 Measure-theoretic entropy

The measure-theoretic entropy of a flow is usually defined in terms of the action of its time-1 map.[1]

Definition 4.1.1 (Measure-theoretic entropy). If Φ is a μ-preserving flow, then the measure-theoretic (or Kolmogorov–Sinai) entropy of Φ is defined by $h_\mu(\Phi) := h_\mu(\varphi^1)$ (see Definition A.2.27).

While it would be desirable to have a definition for flows that avoids passing to the time-1 map, the definition of entropy in terms of the action on measurable partitions of a measure space $(X, \mathcal{T}, \mu)$ does not translate naturally to continuous time. Properties of entropy for maps can be found in Appendix A; those unfamiliar with entropy for maps are encouraged to read that chapter in connection with the current chapter. We outline the approach for entropy of maps to illustrate the reason this does not translate naturally to continuous time. The entropy of a partition (Definition A.1.6) is defined by

$$H_\mu(\xi) = -\sum_{C\in\xi} \mu(C)\log\mu(C),$$

[1] Or as $\sup_{t>0} \frac{1}{t} h_\mu(\varphi^t)$.

where $0 \log 0 := 0$. We denote by $\mathcal{P}_H$ the collection of measurable partitions (mod 0) with finite entropy, and we refer to these as *finite-entropy partitions*.

For two measurable partitions ξ and η of X the joint partition is

$$\xi \vee \eta := \{C \cap D \mid C \in \xi,\ D \in \eta\}.$$

For a measurable partition ξ and a measure-preserving (not necessarily invertible) transformation f we define the *joint partition for f of ξ* as follows. For $I \subset \mathbb{R}$ set

$$\xi_I^f := \bigvee_{i \in I \cap \mathbb{Z}} f^i(\xi), \qquad \xi_{-n}^f := \xi_{[-n,0)}^f, \qquad \xi_-^f := \xi_{(-\infty,0)}^f.$$

The measure-theoretic entropy of a measure-preserving transformation $f : X \to X$ relative to the partition ξ is $h(f,\xi) := h_\mu(f,\xi) := \lim_{n\to\infty} H(\xi_{-n}^f)/n$. The *entropy* of f with respect to μ (or the entropy of μ) is

$$h(f) := h_\mu(f) := \sup\left\{h_\mu(f,\xi) \mid \xi \in \mathcal{P}_H\right\}.$$

The difficulty with continuous-time systems is that the join of a partition over an interval in $\mathbb{R}$ does not lend itself to defining a natural notion of complexity. Accordingly, as we mentioned above we outline the definitions and properties of entropy for maps in Appendix A.

The focus of this book is on continuous flows on compact metric spaces, and for those that are fixed-point-free we can take a different approach to define measure-theoretic entropy directly rather than via time-1 maps.

Definition 4.1.2 (Measure-theoretic entropy of a flow [340]). For a continuous fixed-point-free flow Φ on a compact metric space X and $t \in \mathbb{R}$ define the (t,ϵ,Φ)-ball around $x \in M$ as

$$B(x,t,\epsilon,\Phi) := \{y \in X \mid \exists\, \alpha \in \operatorname{Rep}([0,1]),\ d(\varphi^{\alpha(s)t}x, \varphi^{st}y) < \epsilon \text{ for } 0 \le s \le 1\},$$

where

$$\operatorname{Rep}([0,a]) := \{\alpha : [0,a] \to \mathbb{R} \text{ strictly increasing continuous with } \alpha(0) = 0\}$$

is the set of all reparametrizations.

For an ergodic Φ-invariant Borel measure μ and $\delta \in (0,1)$ let $N(\delta,t,\epsilon,\Phi)$ be the minimum number of (t,ϵ,Φ)-balls whose union has measure at least $1-\delta$ and define

$$\bar{h}_\mu(\Phi) := \lim_{\epsilon\to 0} \overline{\lim_{t\to\infty}} \frac{1}{t} \log N(\delta,t,\epsilon,\Phi).$$

(This is indeed independent of δ.)

This formulation of measure-theoretic entropy for flows does not require using the time-1 map, and for a continuous flow on a compact metric space without fixed points coincides with $h_\mu(\Phi)$; see [340].

Theorem 4.1.3. *Let $\Phi\colon X \to X$ be a continuous flow on a compact metric space. If $\mu, \nu \in \mathfrak{M}(\Phi)$ and $p \in [0,1]$, then*

$$h_{p\mu+(1-p)\nu}(\Phi) = ph_\mu(\Phi) + (1-p)h_\nu(\Phi).$$

Proof. If $\xi = \{\mathbb{C}_i\}_i$ is a finite partition, then Lemma A.2.15 with $a_1 = p = 1 - a_2$, $x_1 = \mu(C_i)$, $x_2 = \nu(C_i)$ gives

$$\begin{aligned} 0 &\le H_{p\mu+(1-p)\nu}(\xi) - pH_\mu(\xi) - (1-p)H_\nu(\xi) \\ &\le -(p\log p + (1-p)\log(1-p)) \\ &\le \log 2. \end{aligned}$$

When η is a finite partition and $\xi := \bigvee_{i=0}^{n-1} \varphi^{-i}\eta$, this implies that

$$h_{p\mu+(1-p)\nu}(\Phi,\eta) = ph_\mu(\Phi,\eta) + (1-p)h_\nu(\Phi,\eta).$$

On one hand, taking the supremum over η gives

$$h_{p\mu+(1-p)\nu}(\Phi) \le ph_\mu(\Phi) + (1-p)h_\nu(\Phi).$$

For the reverse inequality, take $c_\mu < h_\mu(\Phi)$, $c_\nu < h_\nu(\Phi)$ and partitions ξ_μ,ξ_ν such that $h_\mu(\Phi,\xi_1) > c_\mu$ and $h_\nu(\Phi,\xi_1) > c_\nu$. Then $\xi := \xi_\mu \vee \xi_\nu$ satisfies

$$\begin{aligned} h_{p\mu+(1-p)\nu}(\Phi,\xi) &= ph_\mu(\Phi,\xi) + (1-p)h_\nu(\Phi,\xi) \\ &\ge ph_\mu(\Phi,\xi_m u) + (1-p)h_\nu(\Phi,\xi_n u) > pc_\mu + (1-p)c_\nu. \end{aligned}$$

Thus, $h_{p\mu+(1-p)\nu}(\Phi,\xi) \ge ph_\mu(\Phi) + (1-p)h_\nu(\Phi)$ since c_μ, c_ν were arbitrary. □

We now describe how to obtain the entropy of a flow under a function (Definition 3.6.1) from the entropy of the base map and relevant information about the function. (If the invariant measure is not normalized, then the entropy will be computed using the associated normalized measure (3.6.4).)

Theorem 4.1.4 (Abramov). *With the notation from Definition* 3.6.1, *consider a μ-preserving transformation $F\colon (Y,\mu) \to (Y,\mu)$, where μ is a probability measure, and let $\Phi = \Phi_{F,r}$ be the special flow under the roof function r. Suppose there is an $r_0 > 0$ such that $r(y) \ge r_0$ for all $y \in Y$. Then*

$$h_{\mu_r}(\Phi) = h_\mu(F)/\int r\,d\mu. \tag{4.1.1}$$

Proof. Proposition A.3.15(4) lets us scale t by any rational number, so assume without loss of generality that $0 < t < r_0$ and set $X_t := Y \times [0,t) \subset X$. The map of X_t induced by Φ is of the form $\Phi_{X_t}(y,s) = (F(y), s + r(y) - t\lfloor r(y)/t \rfloor)$. So Theorem A.3.36 and Proposition A.3.32 give

$$h_{\mu_r}(\varphi^t) = h_{(\mu_r)_{X_t}}(\Phi_{X_t})\mu_r(X_t) = h_\mu(F) \cdot \frac{t}{\int r\,d\mu}. \qquad \square$$

For the special case of suspensions (that is, $r \equiv 1$) we have the following corollary:

Corollary 4.1.5. $h_{\mu \times m}(F_\circ) = h_\mu(F)$.

Example 4.1.6. Consider F acting on two copies $A = B = (Y, \mu)$, and write $h := h_\mu(F)$. Set $r \equiv a$ on A, $r \equiv b$ on B, and $\Phi_A := \Phi_{\restriction_A}$, $\Phi_B := \Phi_{\restriction_B}$. Then, with self-explanatory notation, $h_{\mu_A}(\Phi_A) = h/a$, $h_{\mu_B}(\Phi_B) = h/b$, and Proposition A.3.15(2) gives

$$\begin{aligned} h_\mu(\Phi) &= \frac{\nu_A(X)h_\mu(\Phi_A) + \nu_B(X)h_\mu(\Phi_A)}{\nu(X)} \\ &= \frac{2}{a+b}\Big(\frac{a}{2}\frac{h}{a} + \frac{b}{2}\frac{h}{b}\Big) = \frac{2h}{a+b} = \frac{h}{\int r} = h_\mu(\Phi) \end{aligned}$$

by Abramov's formula (4.1.1).

The context of the Kac Lemma (Proposition 3.6.13) provides an application of Abramov's formula to special flows.

Proposition 4.1.7 ([344, Corollary 1]). *If F is a μ-preserving map on a topological probability space (X, μ), $0 < r \in L^1(\mu)$, $\mu(A) > 0$, then*

$$\int_A T_A(x)\,d\mu_A = \frac{h_{\mu_A}(F_A)}{h_{\mu_r}(\Phi_r)},$$

where μ_A is the conditional measure from (3.3.3), $T_A(x) := \min\{t > 0 \mid \varphi_r^t(x,0) \in A\}$, *and F_A the return map from* (A.3.6).

Abramov's formula also provides insights into the effect of time changes.

Theorem 4.1.8 (Abramov). *If $0 < \rho \in L^1(X, \nu)$, then the time change Φ_ρ (Theorem* 3.5.4) *of the special flow $\Phi = \Phi_{F,r}$ satisfies*

$$h_{\nu_\rho}(\Phi_\rho) = h_\nu(\Phi) \int \rho\,d\nu.$$

Proof. The time change Φ_ρ is measure-theoretically isomorphic to a special flow over F with a roof function r_ρ that satisfies

$$r(y) = \int_0^{r_\rho(y)} \rho,$$

which is the "distance" traveled by Φ_ρ in time $r_\rho(y)$. The Fubini Theorem gives

$$\int r\,d\mu = \int_Y \int_0^{r_\rho(y)} \rho\,d\mu = \int \rho\,d\nu,$$

and we can apply Abramov's formula (4.1.1). □

By the Ambrose–Kakutani–Rokhlin Special-Flow Representation Theorem (Theorem 3.6.2), every measurable flow with essentially no fixed points and with an invariant Borel probability measure is measure-theoretically isomorphic to a special flow, so the preceding result implies one for time changes in full generality.

Theorem 4.1.9 (Abramov). *If Φ is a measurable flow on (X,μ) with essentially no fixed points and $0 < \rho \in L^1(\mu)$, then $h_{\mu_\rho}(\Phi_\rho) = h_\mu(\Phi)\int \rho\,d\mu$.*

Corollary 4.1.10. *If Φ is a measurable flow on (X,μ), then $h_\mu(\varphi^T) = |T|h_\mu(\varphi^1)$.*

Proof. The map φ^T is the time-1 map of the flow $\Psi\colon t \mapsto \psi^t := \varphi^{tT}$. If $T > 0$, then

$$h_\mu(\varphi^T) = h_\mu(\Psi) \overset{\text{Theorem 4.1.9}}{=\!=\!=} Th_\mu(\Phi) = Th_\mu(\varphi^1). \qquad \square$$

Corollary 4.1.5 puts us in a position to study the entropy in a familiar example.

Example 4.1.11. Consider the toral automorphism from Example 1.5.26 with Lebesgue measure m as the invariant Borel probability measure. To simplify notation write F for this hyperbolic toral automorphism and denote the maximal eigenvalue by λ. Let ξ be a finite partition of $\mathbb{T}^2$ into elements of diameter less than $1/10$. We estimate $H(\bigvee_{k=-n}^n F^k(\xi)) = H(\xi_{-2n-1}^F)$ from below by estimating from above the diameter and hence the Lebesgue measure of the elements of ξ_{-2n-1}^F. Let $C \in \bigvee_{k=-n}^n F^k(\xi)$ and $x, y \in C$. Consider the line parallel to the eigenvector with eigenvalue $\lambda > 1$ passing through the point x and the line parallel to the second eigenvector passing through y. Define z as the first point of intersection of these lines. Then $d(F^k(x), F^k(y)) \le d(F^k(x), F^k(z)) + d(F^k(z), F^k(y))$. First, let $k > 0$. Then $d(F^k(z), F^k(y)) = \lambda^{-k} d(z,y) \le \lambda^{-k} d(x,y) < \lambda^{-k}/10$. Since for $k = 1, \dots, n$ the points $F^k(x)$, $F^k(y)$ belong to the same element of the partition ξ we have $d(F^k(x), F^k(y)) < 1/10$ and hence $d(F^k(x), F^k(z)) < 1/10 + \lambda^{-k}/10 < 1/5$. This implies by induction that the length of the line segment connecting $F^k(x)$ and $F^k(z)$ is also less than $1/5$. Hence $d(x,z) = \lambda^{-n} d(F^n(x), F^n(z)) < \lambda^{-n}/5$. A similar argument for negative k shows that $d(y,z) < \lambda^{-n}/5$ and hence we have $d(x,y) < 2\lambda^{-n}/5$. Thus the diameter of any element of $\bigvee_{-n}^n F^{-k}(\xi)$ is at most $2\lambda^{-n}/5$ and hence by the isoperimetric inequality its Lebesgue measure is at most $2\pi\lambda^{-2n}/25$. Thus the left inequality in Proposition A.3.1(1) gives $h(F,\xi) \ge \log\lambda$ and hence

$$h_m(F) \ge \log\lambda$$

for Lebesgue measure m. This is also the entropy of the suspension. Furthermore, comparison with Proposition 4.2.17 and Corollary 4.3.9 implies that we actually have equality.

4.2 Topological entropy

We now return to topological dynamics to introduce a counterpart of entropy in this setting, topological entropy. One way of looking at this is to ask, naively, how many orbits are observable at time-t with finite resolution." Since we have already studied periodic points as anchors for nearby dynamical behavior, counting periodic orbits is a way to seek additional information. This plays out in slightly different ways for flows than for discrete-time dynamical systems because in the latter case the lengths of periodic orbits are integers, so one can simply note how many periodic points there are for a given integer.

Furthermore, for a flow it makes no sense to try to count periodic *points* because there are either none or uncountably many of them, so we count periodic *orbits* instead.

This can be done in two different ways. It would be closest to the discrete-time case to count periodic orbits weighted by their length, which is what counting of periodic points amounts to in that case (Corollary 1.9.8). On the other hand, one can count just the number of periodic orbits without weighting by their lengths. If, however, the number of periodic orbits grows exponentially then the distinction is immaterial because most orbits of length up to T have length close to T, so the growth rate is the same.

Definition 4.2.1 (Periodic orbit growth). Let $P_T(\Phi)$ be the number of all periodic orbits of period up to T and

$$p(\Phi) := \overline{\lim_{T\to\infty}} \frac{1}{T} \log(\max(P_T(\Phi), 1))$$

the exponential growth rate of the number of periodic orbits for a flow.

Going beyond periodic points, topological entropy is the most important numerical invariant related to the orbit growth and represents the exponential growth rate of the number of orbit segments distinguishable with arbitrarily fine but finite precision. In a sense, the topological entropy describes in a crude but suggestive way the total (rather than average) exponential complexity of the orbit structure with a single number.

Let Φ be a continuous flow on a compact metric space (X, d). The family of metrics d_t^Φ defined by

$$d_t^\Phi(x, y) := \max_{0\le\tau\le t} d(\varphi^\tau(x), \varphi^\tau(y))$$

measures the distance between the orbit segments $\varphi^{[0,t]}(x)$ and $\varphi^{[0,t]}(y)$ and defines the *Bowen balls*

$$B_\Phi(x,\epsilon,t) := \{y \in X \mid d_t^\Phi(x,y) < \epsilon\}. \tag{4.2.1}$$

A set $E \subset X$ is said to be *(t,ϵ)-spanning* or *(t,ϵ)-dense* if $X \subset \bigcup_{x\in E} B_\Phi(x,\epsilon,t)$. Let $S_d(\Phi,\epsilon,t)$ be the minimal cardinality of a (t,ϵ)-spanning set, or equivalently the cardinality of a *minimal (t,ϵ)-spanning set*. This is the minimal number of initial conditions whose behavior up to time t approximates the behavior of *any* initial condition up to ϵ. Consider its exponential growth rate

$$h_d(\Phi,\epsilon) := \overline{\lim_{t\to\infty}} \frac{1}{t} \log S_d(\Phi,\epsilon,t). \tag{4.2.2}$$

Obviously $h_d(\Phi,\epsilon)$ does not decrease with ϵ.

Definition 4.2.2. The *topological entropy* of Φ is

$$h_{\text{top}}(\Phi) := h(\Phi) := h_d(\Phi) = \lim_{\epsilon\to 0} h_d(\Phi,\epsilon).$$

Remark 4.2.3. For future reference we note that for compact $K \subset X$ we can likewise define the entropy $h_{\text{top}}(\Phi,K)$ of Φ on K: replace $S_d(\Phi,\epsilon,t)$ by the minimal cardinality of an $E \subset K$ that is (t,ϵ)-dense in K, then take the exponential growth rate and let $\epsilon \to 0$.

Topological entropy is defined in terms of the metric d but does not depend on it:

Proposition 4.2.4. *If d' is another metric on X which defines the same topology as d, then $h_{d'}(\Phi) = h_d(\Phi)$.*

Proof. Consider the set D_ϵ of all pairs $(x_1,x_2) \in X \times X$ for which $d(x_1,x_2) \geq \epsilon$. This is a compact subset of $X \times X$ with the product topology. The function d' is continuous on $X \times X$ in that topology and consequently it reaches its minimum $\delta(\epsilon)$ on D_ϵ. This minimum is positive; otherwise there would be points $x_1 \neq x_2$ such that $d'(x_1,x_2) = 0$. Thus, if $d'(x_1,x_2) < \delta(\epsilon)$, then $d(x_1,x_2) < \epsilon$, that is, any $\delta(\epsilon)$-ball in the metric d' is contained in an ϵ-ball in the metric d. This argument extends immediately to the metrics d'^Φ_t and d^Φ_t. Thus for every t we have $S_{d'}(\Phi,\delta(\epsilon),t) \geq S_d(\Phi,\epsilon,t)$ so $h_{d'}(\Phi,\delta(\epsilon)) \geq h_d(\Phi,\epsilon)$ and $h_{d'}(\Phi) \geq \lim_{\epsilon\to 0} h_{d'}(\Phi,\delta(\epsilon)) \geq \lim_{\epsilon\to 0} h_d(\Phi,\epsilon) = h_d(\Phi)$. Interchanging the metrics d and d' one obtains $h_d(\Phi) \geq h_{d'}(\Phi)$. □

Corollary 4.2.5. *The topological entropy is an invariant of topological conjugacy.*

Proof. Let $\Phi\colon X \to X$, $\Psi\colon Y \to Y$ be topologically conjugate via a homeomorphism $h\colon X \to Y$. Fix a metric d on X and define d' on Y as the pullback of d, that is, $d'(y_1,y_2) = d(h^{-1}(y_1),h^{-1}(y_2))$. Then h is an isometry so $h_d(\Phi) = h_{d'}(\Psi)$. □

There are several quantities similar to $S_d(\Phi, \epsilon, t)$ that can be used to define topological entropy. For example, let $D_d(\Phi, \epsilon, t)$ be the minimal number of sets whose d_t^Φ-diameter is at most ϵ and whose union covers X. The diameter of an ϵ-ball is at most 2ϵ so every covering by ϵ-balls is a covering by sets of diameter $\le 2\epsilon$, that is,

$$D_d(\Phi, 2\epsilon, t) \le S_d(\Phi, \epsilon, t). \tag{4.2.3}$$

On the other hand, any set of diameter $\le \epsilon$ is contained in the ϵ-ball around each of its points so

$$S_d(\Phi, \epsilon, t) \le D_d(\Phi, \epsilon, t). \tag{4.2.4}$$

Lemma 4.2.6. *For any $\epsilon > 0$ the limit $\lim_{t\to\infty}(1/t)\log D_d(\Phi, \epsilon, t) < \infty$ exists.*

Proof. We show that the sequence $a_n := \log D_d(\Phi, \epsilon, n)$ is subadditive: $a_{m+n} \le a_n + a_m$. Then $\lim_{n\to\infty} a_n/n$ exists by Lemma 4.2.7 below. This implies the claim by monotonicity of $t \mapsto D_d(\Phi, \epsilon, t)$.

To prove that $D_d(\Phi, \epsilon, s+t) \le D_d(\Phi, \epsilon, t) \cdot D_d(\Phi, \epsilon, s)$ for all s, t, note that if A has d_t^Φ-diameter less than ϵ and B has d_s^Φ-diameter less than ϵ, then $A \cap \varphi^{-t}(B)$ has $d_{s+t}^{\varphi^t}$-diameter less than ϵ. Thus if $\mathfrak{A}$ is a cover of X by $D_d(\Phi, \epsilon, t)$ sets of d_t^Φ-diameter less than ϵ and $\mathfrak{B}$ is a cover of X by $D_d(\Phi, \epsilon, s)$ sets of d_s^Φ-diameter less than ϵ, then the cover by all sets $A \cap \varphi^{-t}(B)$, where $A \in \mathfrak{A}$, $B \in \mathfrak{B}$, which contains at most $D_d(\Phi, \epsilon, t) \cdot D_d(\Phi, \epsilon, s)$ sets, is a cover by sets of $d_{s+t}^{\varphi^t}$-diameter less than ϵ. □

Lemma 4.2.7 (Bowen–Fekete Lemma, subadditivity). *If there are k, L such that $a_{m+n} \le a_m + a_{n+k} + L$ for all $m, n \in \mathbb{N}$ then $\frac{a_n}{n} \xrightarrow[n\to\infty]{} \inf_{n\in\mathbb{N}} \frac{a_{n+k}+L}{n} \in [-\infty, \infty)$.*

Proof. Setting $l = r + in$ with $0 \le r < n$ gives $\frac{a_l}{l} \le \frac{a_r + i(a_{n+k}+L)}{r+in}$. If $l \to \infty$ (with n fixed, so $i \to \infty$), then $\overline{\lim}_{l\to\infty} \frac{a_l}{l} \le \inf_n \frac{a_{n+k}+L}{n} \le \underline{\lim}_{n\to\infty} \frac{a_{n+k}+L}{n} = \underline{\lim}_{l\to\infty} \frac{a_l}{l}$.[2] □

From (4.2.3) and (4.2.4) we see that

$$\tilde{h}_d(\Phi, \epsilon) := \lim_{n\to\infty} (1/t)\log D_d(\Phi, \epsilon, t) \ge h_d(\Phi, \epsilon) \ge \tilde{h}_d(\Phi, 2\epsilon),$$

and similarly for $\underline{h}_d(\Phi, \epsilon) := \underline{\lim}_{t\to\infty}(1/t)\log S_d(\Phi, \epsilon, t)$ instead of $h_d(\Phi, \epsilon)$. Thus,

$$\lim_{\epsilon\to 0} \tilde{h}_d(\Phi, \epsilon) = \lim_{\epsilon\to 0} \underline{h}_d(\Phi, \epsilon) = h(\Phi),$$

and $\lim_{\epsilon\to 0}\big(h_d(\Phi, \epsilon) - \underline{h}_d(\Phi, \epsilon)\big) = 0$. So the topological entropy can be defined in terms of the number of open sets whose d_t^Φ-diameter is at most ϵ and whose union covers X.

[2] This extends to $k, L \ge 0$ depending on n so long as both are $o(n)$.

One can define topological entropy via the maximal number $N_d(\Phi,\epsilon,t)$ of points in X with pairwise d_n^{Φ}-distances at least ϵ. We call such a set of points (t,ϵ)-*separated*. Such points generate the maximal number of orbit segments of length t that are

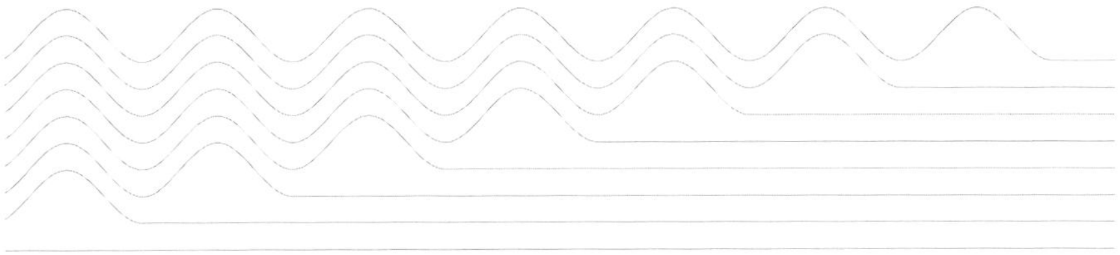

Figure 4.2.1. A separated set. [Reprinted from [177] (© Cambridge University Press, all rights reserved) with permission.]

distinguishable with precision ϵ. A maximal (t,ϵ)-separated set is a (t,ϵ)-spanning set, that is, for any such set of points the ϵ-balls around them cover X, because otherwise it would be possible to increase the set by adding any point not covered. Thus

$$N_d(\Phi,\epsilon,t) \geq S_d(\Phi,\epsilon,t). \tag{4.2.5}$$

On the other hand, no ϵ-ball can contain two points 2ϵ apart. Thus

$$S_d(\Phi,\epsilon,t) \geq N_d(\Phi,2\epsilon,t). \tag{4.2.6}$$

Using (4.2.5) and (4.2.6) we obtain

$$\underline{h}_d(\Phi,\epsilon) \leq \varliminf_{t\to\infty} \frac{1}{t}\log N_d(\Phi,2\epsilon,t) \leq \varlimsup_{t\to\infty} \frac{1}{t}\log N_d(\Phi,2\epsilon,t) \leq h_d(\Phi,\epsilon)$$

and hence

$$h_{\text{top}}(\Phi) = \lim_{\epsilon\to 0}\varlimsup_{t\to\infty} \frac{1}{t}\log N_d(\Phi,\epsilon,t) = \lim_{\epsilon\to 0}\varliminf_{t\to\infty} \frac{1}{t}\log N_d(\Phi,\epsilon,t),$$

justifying the description as "the exponential growth rate of the number of orbit segments distinguishable with arbitrarily fine but finite precision."

For a map $f\colon X\to X$ the corresponding family of metrics is given by

$$d_n^f(x,y) := \max_{0\leq i\leq n} d(f^i(x), f^i(y)), \tag{4.2.7}$$

and the topological entropy is similarly defined as

$$h_{\text{top}}(f) := h_d(f) = \lim_{\epsilon\to 0} h_d(f,\epsilon) = \lim_{\epsilon\to 0}\varlimsup_{n\to\infty}\frac{1}{n}\log S_d(\Phi,\epsilon,n).$$

Equicontinuity of $\{\varphi^s \mid |s|\leq 1\}$ implies the following result:

Proposition 4.2.8. $h_{\text{top}}(\Phi) = h_{\text{top}}(\varphi^1)$.

Remark 4.2.9. See also Proposition 4.3.7.

Proof. Let $\epsilon > 0$ and $\delta > 0$ such that $d(\varphi^t x, \varphi^t y) < \epsilon$ for $0 < t \le 1$ when $d(x, y) < \delta$. An (n, δ)-spanning set E for φ^1 is (t, ϵ)-spanning for Φ, where $t \le n$. Then $S_d(\Phi, \epsilon, t) \le S_d(\varphi^1, \delta, n)$ for $t \le n$, so

$$\overline{\lim_{t\to\infty}} \frac{1}{t} \log S_d(\Phi, \epsilon, t) \le \overline{\lim_{n\to\infty}} \frac{1}{n} \log S_d(\varphi^1, \delta, n).$$

As $\epsilon \to 0$ this becomes $h_{\text{top}}(\Phi) \le h_{\text{top}}(\varphi^1)$.

The other direction follows directly from the definitions. Indeed,

$$\overline{\lim_{t\to\infty}} \frac{1}{t} \log S_d(\Phi, \epsilon, t) \ge \overline{\lim_{n\to\infty}} \frac{1}{n} \log S_d(\Phi, \epsilon, n) \ge \overline{\lim_{n\to\infty}} \frac{1}{n} \log S_d(\varphi^1, \epsilon, n),$$

so $h_d(\Phi, \epsilon) \ge h_d(\varphi^1, \epsilon)$ for each $\epsilon > 0$, and $h_{\text{top}}(\Phi) \ge h_{\text{top}}(\varphi^1)$. □

Corollary 4.2.10. *For $t \in \mathbb{R}$ we have $h_{\text{top}}(\varphi^t) = |t| h_{\text{top}}(\varphi^1) = |t| h_{\text{top}}(\Phi)$.*

Proof. For $\epsilon > 0$ there exists $\delta(\epsilon) > 0$ such that $d(x, y) < \delta(\epsilon) \Rightarrow d(\varphi^r(x), \varphi^r(y)) < \epsilon$ for $0 \le r \le s$. If E is (n, δ)-spanning for φ^s, then E is (m, ϵ)-spanning for φ^t so long as $mt \le ns$. So $S_d(\varphi^t, m, \epsilon) \le S_d(\varphi^1, \delta, \lfloor \frac{mt}{s} \rfloor + 1)$, hence

$$\begin{aligned}\overline{\lim_{m\to\infty}} \frac{1}{m} \log S_d(\varphi^t, \epsilon, m) &\le \overline{\lim_{m\to\infty}} \frac{1}{m} \log S_d\Big(\varphi^s, \delta, \lfloor \frac{mt}{s} \rfloor + 1\Big) \\ &= \overline{\lim_{m\to\infty}} \Big(\frac{1}{m}(\lfloor \frac{mt}{s} \rfloor + 1)\Big) h_d(\varphi^s, \delta) = \frac{t}{s} h_d(\varphi^s, \delta).\end{aligned}$$

So $s h_{\text{top}}(\varphi^t) \le t h_{\text{top}}(\varphi^s)$. By symmetry we have equality, and setting $s = 1$ gives the claim for $t \ge 0$. Finally, the image of a (t, ϵ)-separated set for Φ is (t, ϵ)-separated for the inverse flow and vice versa. So $h_{\text{top}}(\varphi^{-1}) = h_{\text{top}}(\varphi^1)$. The result now follows. □

Proposition 4.2.11. *If Ψ is a factor (Definition 1.3.1) of Φ, then $h_{\text{top}}(\Psi) \le h_{\text{top}}(\Phi)$.*

Proof. Let $\Phi\colon X \to X$, $\Psi\colon Y \to Y$, $\pi\colon X \to Y$, $\pi \circ \Phi = \Psi \circ \pi$, $\pi(X) = Y$, and d_X, d_Y be the distance functions in X and Y, correspondingly.

We have that π is uniformly continuous, so for any $\epsilon > 0$ there is $\delta(\epsilon) > 0$ such that if $d_X(x_1, x_2) < \delta(\epsilon)$, then $d_Y(\pi(x_1), \pi(x_2)) < \epsilon$. Thus the image of any $(d_X)_t^\Phi$ ball of radius $\delta(\epsilon)$ lies inside a $(d_Y)_t^\Psi$ ball of radius ϵ, that is,

$$S_{d_X}(\Phi, \delta(\epsilon), t) \ge S_{d_Y}(\Psi, \epsilon, t).$$

Taking logarithms and limits, we obtain the result. □

We amplify this with Theorem 4.2.13 below.

We now present a few elementary properties of topological entropy. The proofs demonstrate the usefulness of switching back and forth from one of the three definitions to another.

Proposition 4.2.12. (1) *$h_{\mathrm{top}}(\Phi\restriction_\Lambda) \le h_{\mathrm{top}}(\Phi)$ if Λ is closed and Φ-invariant.*

(2) *If $X = \bigcup_{i=1}^m \Lambda_i$, where Λ_i $(i = 1,\dots,m)$ are closed Φ-invariant sets, then $h_{\mathrm{top}}(\Phi) = \max_{1\le i\le m} h_{\mathrm{top}}(\Phi\restriction_{\Lambda_i})$.*

(3) *$h_{\mathrm{top}}(\varphi^{mt}) = |m| h_{\mathrm{top}}(\varphi^t)$.*

(4) *$h_{\mathrm{top}}(\Phi\times\Psi) = h_{\mathrm{top}}(\Phi) + h_{\mathrm{top}}(\Psi)$.*
Here, if $\Phi\colon X\to X$, $\Psi\colon Y\to Y$, then $\Phi\times\Psi\colon X\times Y\to X\times Y$ is defined by $(\varphi^t\times\psi^t)(x,y) = (\varphi^t(x),\psi^t(y))$.

We note that (3) is the best we can do: there is no Abramov-like theorem for topological entropy as in the measurable case.

Proof. (1): Every cover of X by sets of d_t^Φ-diameter less than ϵ covers Λ.
(2): The union of covers of $\Lambda_1,\dots,\Lambda_m$ by sets of diameter less than ϵ covers X, so

$$D_d(\Phi,\epsilon,t) \le \sum_{i=1}^m D_d(\Phi\restriction_{\Lambda_i},\epsilon,t),$$

that is, for at least one i,

$$D_d(\Phi\restriction_{\Lambda_i},\epsilon,t) \ge \frac{1}{m} D_d(\Phi,\epsilon,t).$$

Since there are only finitely many i, at least one i works for infinitely many t, so

$$\overline{\lim_{t\to\infty}} \frac{\log D_d(\Phi\restriction_{\Lambda_i},\epsilon,t)}{t} \ge \overline{\lim_{t\to\infty}} \frac{\log D_d(\Phi,\epsilon,t) - \log m}{t} = \tilde h_d(\Phi,\epsilon).$$

This proves (2).
(3): If $m > 0$, then $d_t^{\varphi^{mt}} = d_{mt}^{\varphi^t}$, hence (3). If $m = -1$, then $B_{\varphi^t}(x,\epsilon,t) = B_{\varphi^{-t}}(\varphi^t(x),\epsilon,t)$ and $S_d(\varphi^t,\epsilon,t) = S_d(\varphi^{-t},\epsilon,t)$, so $h_{\mathrm{top}}(\varphi^t) = h_{\mathrm{top}}(\varphi^{-t})$. Together, these imply (3) for all m.
(4): Balls in the product metric

$$d((x_1,y_1),(x_2,y_2)) := \max(d_X(x_1,x_2), d_Y(y_1,y_2))$$

on $X\times Y$ are products of balls on X and Y. The same is true for balls in $d_t^{\varphi^t\times\psi^t}$. Thus

$$S_d(\Phi\times\Psi,\epsilon,t) \le S_{d_X}(\Phi,\epsilon,t) S_{d_Y}(\Psi,\epsilon,t)$$

and $h_{\mathrm{top}}(\Phi \times \Psi) \leq h_{\mathrm{top}}(\Phi) + h_{\mathrm{top}}(\Psi)$. On the other hand, the product of any (t, ϵ)-separated set in X for Φ and any (t, ϵ)-separated set in Y for Ψ is a (t, ϵ)-separated set for $\Phi \times \Psi$. Thus

$$N_d(\Phi \times \Psi, \epsilon, t) \geq N_{d_X}(\Phi, \epsilon, t) \times N_{d_Y}(\Psi, \epsilon, t)$$

and hence $h_{\mathrm{top}}(\Phi \times \Psi) \geq h_{\mathrm{top}}(\Phi) + h_{\mathrm{top}}(\Psi)$. □

We will see later that in the case of hyperbolic flows one of the standard methods to compute entropy is to find an extension that is uniformly finite-to-one and whose entropy is easier to compute. This works because the entropies of the two systems are equal:

Theorem 4.2.13 (Entropy of finite factors). *If $\Phi\colon X \to X$ and $\Psi\colon Y \to Y$ are continuous flows on compact metric spaces and $\pi\colon X \to Y$ is a semiconjugacy from Φ to Ψ that is uniformly finite-to-one, that is, there is a bound on the number of preimages of any point, then $h_{\mathrm{top}}(\Phi) = h_{\mathrm{top}}(\Psi)$.*

Proof. Proposition 4.2.11 gives $h(\Phi) \geq h(\Psi)$. Proposition 4.2.8 reduces showing that $h(\Phi) \leq h(\Psi)$ to proving that $h(\varphi^1) \geq h(\psi^1)$, and using time-1 maps lets us set up a combinatorial argument in discrete time for effective control of the number of orbits; X, Y come with metrics d, d', respectively.

For $\epsilon > 0$, $C \geq \max_{y \in Y} \#\pi^{-1}(y)$, $m \in \mathbb{N}$, $y \in Y$ let

$$U_y = U_{y,n,\epsilon} = \{x \in X \mid d_n^{\varphi^1}(x, z) < \epsilon \text{ for some } z \in \pi^{-1}(y)\} \supset \pi^{-1}(y).$$

Since φ is continuous there is an open neighborhood W_y of y such that $\pi^{-1}(W_y) \subset U_y$.

Since Y is compact there is a finite cover $\{W_{y_1}, \dots, W_{y_p}\}$. Let $\beta > 0$ be a Lebesgue number of this cover (that is, if $y \in Y$ there exists W_{y_j} such that $\overline{B_\beta(y)} \subset W_{y_j}$). For sufficiently small $\epsilon > 0$ we will show that

$$\frac{1}{n}\log(N_d(\varphi^1, 2\epsilon, n)) \leq \frac{1}{n}\log(S_{d'}(\psi^1, \beta, n)) + \frac{1}{m}\log C + \frac{1}{n}\log C. \tag{4.2.8}$$

This completes the proof because then

$$h_d(\varphi^1, 2\epsilon) \leq h_{d'}(\psi^1, \beta) + \frac{1}{m}\log C, \quad \text{hence } h_d(\varphi^1, 2\epsilon) \leq h_{d'}(\psi^1, \beta)$$

since m is arbitrary. If $\epsilon \to 0$, then $\beta \to 0$, so indeed $h(\varphi^1) \leq h(\psi^1)$.

So, for $n \in \mathbb{N}$ let $\ell \in \mathbb{N}$ such that $(\ell - 1)m < n \leq \ell m$. Let $A \subset X$ be a maximal $(n, 2\epsilon)$-separated set for φ^1 and $B \subset Y$ be a minimal (n, β)-spanning set for ψ^1.

For $y \in B$ let $q(j, y) \in \{y_1, \dots, y_p\}$ such that $\overline{B_\beta(\varphi^j(y))} \subset W_{q(j,y)}$. Now define

$$\pi_\ell \colon A \to B \times X^\ell \text{ by } \pi_\ell(x) = (y, x_0, \dots, x_{\ell-1})$$

where $d'_n(y, \pi(x)) \le \beta$, $y \in B$, and $x_s \in \pi^{-1}(q(sm, y))$ such that $d_m(\varphi^{sm}(x), x_s) < \epsilon$ for all $0 \le s < \ell$; this is possible since

$$\pi \circ \varphi^{sm}(x) = \psi^{sm} \circ \pi(x) \in \overline{B_\beta(\varphi^{sm}(y))} \subset W_{q(sm,y)}$$

implies

$$\varphi^{sm}(x) \in \pi^{-1}(W_{q(sm,y)}) \subset U_{q(sm,y),m,\epsilon}.$$

Claim 4.2.14. *π_ℓ is one-to-one.*

Proof. If $\pi_\ell(x) = \pi_\ell(x')$, $0 \le t < m$, and $0 \le s \le \ell$, then

$$d(\varphi^{sm+t}(x), \varphi^{sm+t}(x')) \le d_m(\varphi^{sm}(x), x_s) + d_m(x_s, \varphi^{sm}(x')) \le \epsilon + \epsilon = 2\epsilon.$$

Since $m\ell \ge n$ we get $d_n(x, x') \le 2\epsilon$, hence $x = x'$ since A is 2ϵ-separating. □

This gives (4.2.8): if $y \in B$, then

$$\underbrace{\#\Big(\pi_\ell(A) \cap \big(\{y\} \times X^\ell\big)\Big)}_{\le \#(\pi_\ell(A)) = N_d(\varphi^1, 2\epsilon, n)} \le \prod_{s=0}^{\ell-1} \#\Big(\pi^{-1}(q(sm, y))\Big) \le C^\ell.$$

There are $\#(B) = S_{d'}(\psi^1, \beta, n)$ choices of y, so $N_d(\varphi^1, 2\epsilon, n) \le S_{d'}(\psi^1, \beta, n) C^\ell$, and

$$\frac{1}{n} \log(N_d(\varphi^1, 2\epsilon, n)) \le \frac{1}{n} \log(S_{d'}(\psi^1, \beta, n)) + \underbrace{\frac{1}{n} \log C^\ell}_{= \frac{\ell m}{nm} \log C \le \frac{n+m}{nm} \log C}.$$ □

From Theorem 4.2.13 and Proposition 1.9.21 we see that the suspension of the symbolic flow constructed for the toral automorphism F_A in (1.9.4) and the suspension of F_A itself have the same entropy. We will see later that this is a more general result for codings of hyperbolic flows. In the more general setting it can be difficult to compute the entropy for the hyperbolic flow, but easier to compute the entropy for the symbolic coding.

Furthermore, Proposition A.3.15(4) connects the measure-theoretic entropy of a map and that of the special flow with a roof function. There is no corresponding connection between the *topological* entropy of the special flow and the topological entropy of the base—except for (constant-time) suspensions.

Proposition 4.2.15. *The topological entropy of a suspension flow equals that of the base.*

Proof. By Proposition 4.2.8 we want to show that the entropy of the time-1 map is that of the base. The time-1 map is the cartesian product of the base and the identity; the latter has zero entropy because there is no dependence on n in (4.2.7), so the discrete-time counterpart of Proposition 4.2.12(4) yields the claim. □

This helps compute the topological entropy for several suspensions. Notably, we can apply the following to the corresponding suspensions (compare Corollary 1.9.8):

Proposition 4.2.16. *If Σ_A is a topological Markov chain, then* $h_{\text{top}}(\sigma\restriction_{\Sigma_A}) = \log|\lambda_A^{\max}|$.

Proof. We endow the space Σ_N with the metric $d = d_{10N}$ given by

$$d'_{10N}(\omega,\omega') = \sum_{n=-\infty}^{\infty} \frac{|\omega_n - \omega'_n|}{(10N)^{|n|}}.$$

Then for $\alpha = (\alpha_{-m},\dots,\alpha_m)$ the symmetric cylinder $C_\alpha^m = \{\omega \in \Sigma_N \mid \omega_i = \alpha_i \text{ for } |i| \le m\}$ is at the same time the ball of radius $\epsilon_m = (10N)^{-m}/2$ around each of its points. Similarly, if we fix numbers $\alpha_{-m},\dots,\alpha_{m+n}$, the cylinder

$$C^{-m,\dots,n+m}_{\alpha_{-m},\dots,\alpha_{n+m}} = \{\omega \in \Sigma_N \mid \omega_i = \alpha_i \text{ for } -m \le i \le m+n\} \tag{4.2.9}$$

is at the same time the ball of radius ϵ_m around each of its points with respect to the metric d_n associated with the shift σ. Thus, any two d_n balls of radius ϵ_m are either identical or disjoint and there are exactly N^{n+2m+1} different ones of the form (4.2.9). The covering of Σ_N by those balls is thus minimal, so $S_{d'_{10N}}(\sigma,\epsilon_m,n) = N^{n+2m+1}$ and

$$h_{\text{top}}(\sigma\restriction_{\Sigma_N}) = \lim_{m\to\infty}\lim_{n\to\infty}\frac{1}{n}\log N^{n+2m+1} = \log N.$$

Similarly, for the topological Markov chain Σ_A, we have $S_{d'_{10N}}(\sigma,\epsilon_m,n)$ equals the number of those cylinders (4.2.9) that have nonempty intersection with the set Σ_A. Assume each row of the matrix A contains at least one 1. Since the number of admissible paths of length n that begin with the symbol i and end with the symbol j is equal to the entry a_{ij}^n of the matrix A^n, the number of nonempty cylinders of rank $n+1$ in Σ_A is equal to $\sum_{i,j=0}^{N-1} a_{ij}^n < C \cdot \|A^n\|$ for some constant C. On the other hand, since all numbers a_{ij}^n are nonnegative, $\sum_{i,j=0}^{N-1} a_{ij}^n > c\|A^n\|$ for another constant $c > 0$. Thus, we have

$$S_{d'_{10N}}(\sigma,\epsilon_m,n) = \sum_{i,j=0}^{N-1} a_{ij}^{n+2m}$$

and

$$\log|\lambda_A^{\max}| = \log r(A) = \lim_{n\to\infty}\frac{1}{n}\log\|A^n\| = \lim_{n\to\infty}\frac{1}{n}\log\|A^{n+2m}\|$$
$$= \lim_{n\to\infty}\frac{1}{n}\log S_{d'_{10N}}(\sigma,\epsilon_m,n) = h_{\text{top}}(\sigma\restriction_{\Sigma_A}),$$

where $r(A)$ is the spectral radius of the matrix A (Definition B.3.1). □

As we noted before, Theorem 4.2.13 and Proposition 1.9.21 imply that the suspension of the symbolic flow constructed for the toral automorphism F_A in (1.9.4) and the suspension of F_A itself have the same entropy, so Proposition 4.2.16 now enables us to compute the entropy of the latter. We can at the same time determine the growth of the number of periodic orbits (Definition 4.2.1).

Proposition 4.2.17. *If* $F = F_A\colon \mathbb{T}^2 \to \mathbb{T}^2$ *is given by* $F(x,y) = (2x+y, x+y)$ (mod 1), *then its suspension* $F_\circ$ *satisfies*

$$h_{\text{top}}(F_\circ) = p(F_\circ) = \log\frac{3+\sqrt{5}}{2},$$

the larger eigenvalue of $A = \left(\begin{smallmatrix}2&1\\1&1\end{smallmatrix}\right)$ *from Example* 1.5.26.

Proof. To show that $h_{\text{top}}(F_\circ) = \frac{3+\sqrt{5}}{2}$ we show that $\left(\begin{smallmatrix}2&1\\1&1\end{smallmatrix}\right)$ and

$$A = \begin{pmatrix}1&1&0&1&0\\1&1&0&1&0\\1&1&0&1&0\\0&0&1&0&1\\0&0&1&0&1\end{pmatrix}$$

from (1.9.3) have the same maximal eigenvalue (in fact, the same set of nonzero eigenvalues): subtract column 4 of $A - \lambda\,\mathrm{Id}$ from the first two columns and column 5 from the third, then add rows 1 and 2 to row 4 and row 3 to row 5:

$$\begin{pmatrix}1-\lambda&1&0&1&0\\1&1-\lambda&0&1&0\\1&1&-\lambda&1&0\\0&0&1&-\lambda&1\\0&0&1&0&1-\lambda\end{pmatrix}\to\begin{pmatrix}-\lambda&0&0&1&0\\0&-\lambda&0&1&0\\0&0&-\lambda&1&0\\\lambda&\lambda&0&-\lambda&1\\0&0&\lambda&0&1-\lambda\end{pmatrix}\to\begin{pmatrix}-\lambda&0&0&1&0\\0&-\lambda&0&1&0\\0&0&-\lambda&1&0\\0&0&0&2-\lambda&1\\0&0&0&1&1-\lambda\end{pmatrix}.$$

To determine the growth of periodic orbits, let $\lambda = \frac{3+\sqrt{5}}{2}$ and $G = F^n - \mathrm{Id}$. Then $\mathrm{Fix}(F_A^n) = G^{-1}(0,0)$ is parametrized by $\mathbb{Z}^2 \cap (A^n - \mathrm{Id})([0,1)\times[0,1))$. The cardinality is

$$\text{area}\left((A^n - \text{Id})([0,1]\times[0,1])\right) = |\det(A^n - \text{Id})|$$
$$= |(\lambda^n - 1)(\lambda^{-n} - 1)| = \lambda^n + \lambda^{-n} - 2,$$

which has exponential growth rate $\log \lambda$. □

The coincidence of topological entropy with the growth rate of periodic orbits (Definition 4.2.1) is not accidental but due to expansivity (Remark 4.2.25).

Proposition 4.2.18. *Let Φ be a continuous expansive flow on a compact metric space X, and 2η an expansivity constant. Then for $\epsilon \in (0,\eta)$, $\delta > 0$ there is a $C_{\delta,\epsilon}$ such that $N_d(\Phi,\delta,t) \le C_{\delta,\epsilon} N_d(\Phi,\epsilon,t)$ for all $t > 0$.*

Proof. Proposition 1.8.4 and equicontinuity of $\Phi_{\restriction [-T,T]\times X}$ give $T, \alpha > 0$ with

$$d_{2T}^{\Phi}(\varphi^{-T}(x), \varphi^{-T}(y)) \le 2\epsilon \Rightarrow d(x,y) < \delta$$

and

$$d(x,y) < \alpha \Rightarrow d_{2T}^{\Phi}(\varphi^{-T}(x), \varphi^{-T}(y)) \le \delta.$$

Let E be a maximal (t,δ)-separated set and F a maximal (t,ϵ)-separated set. For $x \in E$ there is a $z = S(x) \in F$ with $d_t^{\Phi}(x,z) < \epsilon$, so $\operatorname{card} E \le \sum_{z\in F} \operatorname{card} S^{-1}(\{z\})$, and we bound $\operatorname{card} S^{-1}(\{z\})$ as follows.

If $x \ne y \in S^{-1}(\{z\})$, then $d_t^{\Phi}(x,y) \le 2\epsilon$ by definition of S, so $d(\varphi^s(x), \varphi^s(y)) \le \delta$ for $s \in [T, t-T]$ by choice of T, and the choice of α implies that either $d(x,y) > \alpha$ or $d(\varphi^t(x), \varphi^t(y)) > \alpha$. Thus,

$$\operatorname{card} S^{-1}(\{z\}) = \operatorname{card}\left\{(x, \varphi^t(x)) \mid S(x) = z\right\}$$
$$\le \max\left\{\operatorname{card} A \mid A \subset X \times X \text{ and } d(a,b) > \alpha \text{ for } a, b \in A\right\} =: C_{\delta,\epsilon},$$

since the $(x, \varphi^t(x))$ form just such a separated set. □

With (4.2.2), (4.2.5), (4.2.6), and Definition 4.2.2 this implies the following theorem:

Theorem 4.2.19 (Entropy of expansive flows). *If Φ is a continuous expansive flow on a compact metric space and 4δ is an expansivity constant, then $h_{\text{top}}(\Phi) = h_d(\Phi,\delta)$.*

Remark 4.2.20 (Entropy-expansiveness). Although we do not prove it, expansivity can be replaced in these applications to entropy (and pressure) by a broader notion called entropy-expansiveness (or h-expansiveness) defined as follows: Φ is *entropy expansive* if

$$h_{\text{top}}^*(\Phi,\epsilon) := \sup_{x\in X} h_{\text{top}}\Big(\Phi, \bigcap_{t\in\mathbb{R}} \varphi^{-t}(B_\epsilon(\varphi^t(x)))\Big) = 0$$

(Remark 4.2.3) for some $\epsilon > 0$, which is then called an h-expansivity constant. (Expansivity is a special case in which the intersection is a short orbit segment of x.) In particular, Theorem 4.2.19 has the following counterpart:

Theorem 4.2.21 ([57, Corollary 2.5]). *$h_{\mathrm{top}}(\Phi) \le h_d(\Phi,\epsilon) + h^*_{\mathrm{top}}(\Phi,\epsilon)$, so if ϵ is an h-expansivity constant, then $h_{\mathrm{top}}(\Phi) = h_d(\Phi,\epsilon)$.*

It is useful to augment our notation beyond Definition 4.2.1:

Definition 4.2.22. Denote by $\mathbb{O}_t(T)$ the set of periodic orbits γ of Φ for which a period $\pi(\gamma)$ is in $[T-t,T+t]$ (this is finite by expansivity, and π is well defined on $\mathbb{O}_t(T)$), and let $\mathbb{P}_t(T) := \bigcup_{\gamma\in\mathbb{O}_t(T)}\gamma$ be the set of points with these periods. For a periodic orbit γ denote by $\pi'(\gamma)$ its shortest or prime period and $\mathbb{O}'_t(T) := (\pi')^{-1}([T-t,T+t])$.

Proposition 4.2.23 (Periodic points are separated). *With α as in Theorem 1.7.5(3), taking one point from each $\gamma \in \mathbb{P}_{\alpha/2}(t)$ gives a (t,α)-separated set.*

Proof. If $x, y \in \mathbb{P}_{\alpha/2}(t)$ with periods a, b, respectively, are not (t,α)-separated, set $t_{pm+q} = pa + q\alpha$ and $u_{pm+q} = pb + q\alpha$ with $0 \le q < m := 1 + \lfloor \frac{t-\alpha/2}{\alpha} \rfloor$ to get

$$d(\varphi^{t_{pm+q}}(x), \varphi^{u_{pm+q}}(y)) = d(\varphi^{q\alpha}(x), \varphi^{q\alpha}(y)) \le \alpha,$$

so x, y are on the same orbit by Theorem 1.7.5(3). □

Theorem 4.2.24 (Entropy and orbit growth). *If Φ is an expansive flow, then $p(\Phi) \le h_{\mathrm{top}}(\Phi)$.*[3]

Proof. If α is as in Theorem 1.7.5(3), then $\operatorname{card}\mathbb{O}_{\alpha/2}(t) \le N_d(\Phi,\alpha/2,t)$ by Proposition 4.2.23, so with the notation of Definition 4.2.1,

$$P_t(\Phi) \le \sum_{n=1}^{\lfloor t/\alpha \rfloor} \operatorname{card}\mathbb{O}_{\alpha/2}(n\alpha) \le \frac{t}{\alpha} N_d(\Phi,\alpha/2,t),$$

since $t \mapsto N_d(\Phi,\alpha/2,t)$ is nondecreasing. As $t \to \infty$ invoke Theorem 4.2.19. □

Remark 4.2.25. Remark 7.3.14 gives a sufficient condition for equality in Theorem 4.2.24, the specification property. This means that for hyperbolic flows, topological entropy is the exponential growth rate of periodic orbits. Theorem 7.6.9 refines this substantially.

Theorem 4.2.26. *Let Φ be a continuous flow on a compact metric space,* NW(Φ) *its nonwandering set. Then $h_{\mathrm{top}}(\Phi) = h_{\mathrm{top}}(\Phi\restriction_{\mathrm{NW}(\Phi)})$.*

[3] In particular, Φ has only countably many closed orbits.

Remark 4.2.27. Looking ahead, this is a corollary of the Variational Principle (Theorem 4.3.8) because Φ-invariant probability measures are supported on $\mathrm{NW}(\Phi)$.

Proof. Since $\mathrm{NW}(\Phi) \subset X$, we have $h_{\mathrm{top}}(\Phi\restriction_{\mathrm{NW}(\Phi)}) \le h_{\mathrm{top}}(\Phi)$. To show the other inequality we use a combinatorial argument, so we switch to the time-1 map as in the proof of Theorem 4.2.13.

Fix $n \ge 1$ and $\epsilon > 0$. Let A be an (n,ϵ)-spanning set of minimum cardinality for $\varphi^1\restriction_{\mathrm{NW}(\Phi)}$. Let

$$U = \{x \in X \mid d_n(x,y) < \epsilon \text{ for some } y \in A\}.$$

So U is an open neighborhood of $\mathrm{NW}(\Phi)$. Since $U^c = X \smallsetminus U$ is compact and all its points are wandering, there exists a uniform $\beta > 0$ such that $0 < \beta \le \epsilon$ and for all $y \in U^c$ we have

$$\varphi^m(B_\beta(y)) \cap B_\beta(y) = \varnothing \quad \text{for all } m \ge 1.$$

Now take a minimal (n,β)-spanning set B for U^c. Then $C := A \cup B$ is an (n,ϵ)-spanning set for X. Let $l \in \mathbb{N}$ and define $\pi_l \colon X \to C^l$ by $\pi_l(x) = (y_0,\dots,y_{l-1})$ with $d_n(\varphi^{in}(x), y_i) < \begin{cases} \epsilon \text{ and } y_i \in A & \text{if } \varphi^{in}(x) \in U, \\ \beta \text{ and } y_i \in B & \text{if } \varphi^{in}(x) \in U^c. \end{cases}$

Claim 4.2.28. *If $\pi_l(x) = (y_0,\dots,y_{l-1})$, then $y_i \in B$ does not repeat in the l-tuple.*

Proof. Since $B_\beta(y_i)$ is wandering the result follows. □

Claim 4.2.29. *If $ln \ge m$, then π_l is one-to-one for $(m,2\epsilon)$-separated points.*

Proof. If $\pi_l(x) = \pi_l(x')$, then for $0 \le j \le n$ and $0 \le i < l$ we have

$$d(\varphi^{in+j}(x), \varphi^{in+j}(x')) \le d_n(\varphi^{in}(x), y_i) + d_n(y_i, \varphi^{in}(x')) < \epsilon + \epsilon = 2\epsilon,$$

so $d_m(x,x') \le d_{ln}(x,x') < 2\epsilon$. □

Claim 4.2.30. *Let q be the minimum cardinality of an (n,β)-spanning set of U^c for φ^1 and p be the minimum cardinality of an (n,ϵ)-spanning set of* $\mathrm{NW}(\Phi)$ *for φ^1. Then $\#(\pi_l(E)) \le (q+1)!\,l^q p^l$ for an $(n,2\epsilon)$-separated set E.*

Proof. Let I_j be the subset of l-tuples in $\pi_l(E)$ with exactly j of the $y_i \in B$. Since y_i cannot be repeated in $\pi_l(x)$ we have $j \le q$, and there are $\binom{q}{j}$ ways of picking the j points $y_i \in B$, $l!/(l-j)!$ ways of arranging the choices of positions, and at most $p^{l-j} \le p^l$ ways of picking the remaining terms. So

$$\#(I_j) \le \binom{q}{j} \frac{l!}{(l-j)!} p^l$$

and

$$\#(\pi_l(E)) = \sum_{j=0}^{q} I_j \le \sum_{j=0}^{q} \underbrace{\binom{q}{j}}_{\le q!} \underbrace{\frac{l!}{(l-j)!}}_{\le l_j \le l^q} p^l \le (q+1)!\,l^q p^l. \qquad \square$$

We now return to the proof of the theorem. Since $(q+1)!\,l^q$ grows at most polynomially in l, wandering points do not contribute to the entropy:

$$\begin{aligned} h_{2\epsilon}(\varphi^1) &= \overline{\lim_{n\to\infty}} \frac{1}{n} \log(N_d(\varphi^1, 2\epsilon, n)) \\ &\le \overline{\lim_{l\to\infty}} \frac{\log((q+1)!) + q\log(l) + l\log(p)}{(l-1)n} \\ &= \frac{\log(p)}{n} = \frac{\log(S_d(\varphi^1 \restriction_{\mathrm{NW}(\Phi)}, \epsilon, n))}{n} \xrightarrow[n\to\infty]{} h_\epsilon(\varphi^1 \restriction_{\mathrm{NW}(\Phi)}), \end{aligned}$$

so $h_{\mathrm{top}}(\Phi) = h_{\mathrm{top}}(\varphi^1) \le h_{\mathrm{top}}(\varphi^1 \restriction_{\mathrm{NW}(\Phi)}) = h_{\mathrm{top}}(\Phi\restriction_{\mathrm{NW}(\Phi)})$ by letting $\epsilon \to 0$. $\square$

Example 4.2.31. Example 1.1.8 is a flow of isometries, so $d_t = d$ for all $t \ge 0$, and $S_d(\Phi, \epsilon, t) = S_d(\Phi, \epsilon, 0)$ for all $t \ge 0$. Hence, the topological entropy is zero.

Examples 1.3.8 and 1.3.13 are flows for which the nonwandering set is finite. By Theorem 4.2.26, the entropy is zero.

We now show that if the flow is sufficiently regular and the dimension of the space is finite, then the topological entropy is finite and bounded by a product of the dimension of the space and the Lipschitz constant of the map.

Definition 4.2.32. Let (X, d) be a metric space. A flow $\Phi\colon X \to X$ is *Lipschitz continuous* if

$$L(\Phi) := \exp\Big\{ \sup_{0<t\le 1} \sup_{x\ne y} \frac{1}{t} \log\Big(\frac{d(\varphi^t(x), \varphi^t(y))}{d(x,y)}\Big)\Big\} < \infty.$$

The constant $L(\Phi)$ is the *Lipschitz constant* of Φ.

This definition implies $d(\varphi^t x, \varphi^t y) \le L(\Phi)^t d(x,y)$ for $x, y \in X$ and $t \in [0,1]$.

Remark 4.2.33. If V is a Lipschitz continuous vector field on a compact Riemannian manifold, then it generates a flow Φ via (1.1.1) for which

$$\frac{d(\varphi'(x), \varphi'(y))}{d(x,y)} \le L,$$

where L is a Lipschitz constant of V. Then a straightforward computation shows that the flow Φ is Lipschitz continuous. Here, orbits are solutions of (1.1.1) and hence C^1 curves, which makes the Lipschitz assumption on V a more stringent condition than Definition 4.2.32.

Definition 4.2.34. Let (X,d) be a compact metric space and $\#B_d(\epsilon)$ be the minimum cardinality of a covering of X by ϵ-balls. The *box dimension*[4] of X is

$$\mathrm{BD}(X) := \overline{\lim_{\epsilon\to 0}} \frac{\log \#B_d(\epsilon)}{|\log\epsilon|} \in [0,\infty].$$

Remark 4.2.35. It is easy to see that this is invariant under bi-Lipschitz maps and that $\mathrm{BD}(\bigcup_{i=1}^n X_i) = \max_i \mathrm{BD}(X_i)$. Thus, $\mathrm{BD}([0,1]^n) = n$, so the box dimension of a Riemannian manifold is the topological dimension.

Theorem 4.2.36 (Entropy is finite). *Let (X,d) be a compact metric space with finite box dimension* $\mathrm{BD}(X)$ *and $\Phi\colon X \to X$ a Lipschitz continuous flow on X. Then*

$$h_{\mathrm{top}}(\Phi) \le \mathrm{BD}(X)\log\max\{1, L(\Phi)\}.$$

Proof. Let $L = \max\{1, L(\Phi)\}$. Then $B_d(x, L^{-t}\epsilon) \subset B_{d_t^\Phi}(x,\epsilon)$ for all $x \in X$, $t \ge 0$, $\epsilon > 0$. This implies $S_d(\Phi,\epsilon,t) \le \#B_d(L^{-t}\epsilon)$. Now $|\log(L^{-t}\epsilon)| = t\log L + \log\epsilon$, so

$$\begin{aligned} t = \frac{|\log L^{-t}\epsilon| - \log\epsilon}{\log L} &= \frac{|\log L^{-t}\epsilon|}{\log L}\Big(1 - \frac{\log\epsilon}{|\log L^{-t}\epsilon|}\Big) \\ &= \frac{|\log L^{-t}\epsilon|}{\log L}\Big(1 + O\Big(\frac{1}{t}\Big)\Big), \end{aligned}$$

and

$$\begin{aligned} \overline{\lim_{t\to\infty}} \frac{\log S_d(\Phi,\epsilon,t)}{t} \le \overline{\lim_{t\to\infty}} \log\frac{\#B_d(L^{-t}\epsilon)}{t} &= (\log L)\overline{\lim_{t\to\infty}} \frac{\log \#B_d(L^{-t}\epsilon)}{|\log L^{-t}\epsilon|} \\ &= \mathrm{BD}(X)\log L. \end{aligned}$$ □

Corollary 4.2.37. *If Φ is a Lipschitz continuous flow on a compact Riemannian manifold M, then $h_{\mathrm{top}}(\Phi) < \infty$.*

The topological entropy for a flow is obviously invariant under flow equivalence (time-preserving conjugacy). It changes under time change and hence under orbit equivalence in a rather complicated way. However, arguing similarly to the proof of Proposition 7.6.15 one can show that if a continuous flow without fixed points has zero (or finite) topological entropy, then so does any time change (Theorem 4.3.16). Let us comment on the much harder question of how topological entropy changes

[4] Or box-counting dimension, Minkowski dimension, upper box dimension, entropy dimension, Kolmogorov dimension, Kolmogorov capacity, limit capacity, upper Minkowski dimension.

under perturbation of a flow. This dependence need not even be continuous[5] and even in discrete time the picture is quite subtle [214, p. 584]. For hyperbolic flows, the subject of this book, this plays out exceptionally well, however (Theorem 5.4.27).

As a preview of theory to be presented further on, we present its historic precursor here. In 1964 William Parry proved the following result:

Theorem 4.2.38 (Parry [213, Proposition 4.4.2]). *The topological entropy of a topologically mixing* (*Proposition* 1.9.15) *shift* Σ_A (*Definition* 1.9.1) *is the maximal eigenvalue* λ *of* A (*Proposition* 4.2.16). *If* v *is a corresponding positive right eigenvector,* P *is defined by* $\lambda_i v_i P_{ij} = A_{ij} v_j$ *and* p *is the probability vector with* $pP = p$*, then the* σ_A*-invariant Markov measure* m_P *defined on cylinders* (1.9.1) *by*

$$m_P\big(C^{0,\dots,k}_{i_1,\dots,i_k}\big) = p_{i_0} P_{i_0 i_1} \cdots P_{i_{k-1} i_k}$$

has entropy λ*, and all other* σ_A*-invariant Borel probability measures have smaller entropy. It is called the* Parry measure.

Thus, in this case the entropy of each measure is at most the topological entropy, this upper bound is attained, and by a unique invariant Borel probability measure.

Remark 4.2.39 (Topological entropy on noncompact spaces). Compactness of the phase space underlies the very definition of entropy we used here, and we do not go beyond this context. There are reasons for being interested in topological entropy of dynamical systems whose phase space is not compact, and there are definitions of topological entropy that make sense beyond the compact context. Restricted to the context of compact phase space, these all agree with the entropy obtained from the definitions we gave here, as they should. However, outside that context these definitions might give different values. An essential next level of case distinction is whether the dynamical system in question is uniformly continuous or not—in the former case these various definitions tend to coincide, but in the latter they can diverge wildly [179]. Accordingly, for such dynamical systems the question arises, case by case, of what about exponential orbit complexity one wants to study, quantify, or investigate. Billiard maps provide a natural instance (Remark 5.2.33) that illustrates how these difficulties arise naturally: the billiard map is defined on a compact metric space, and it is natural to inquire about exponential orbit complexity. But it has discontinuities, which undermines the approach to topological entropy in a fundamental way. Restricting to the continuity set, which is substantial, is natural, but while this set is dense, it is not compact, and the billiard map is not uniformly continuous on it (otherwise it would extend continuously to the closure).

[5]The easiest example is in discrete time and noninvertible: $z \mapsto \lambda z^2$ on the unit disk in $\mathbb{C}$ has entropy $\log 2$ for $\lambda = 1$ and 0 for $\lambda < 1$.

4.3 Topological pressure and equilibrium states

To extend the notion of entropy recall that it is calculated by counting the elements of a maximal (t,ϵ)-separated set, that is, by summing 1 over the elements of the set. It is natural instead to allow weighted sums over separating or spanning sets. This leads to the notion of pressure, a term motivated by statistical mechanics.

Remark 4.3.1 (Thermodynamical motivation). One of the fundamental laws of thermodynamics is that the entropy of an isolated system can never decrease in time. An isolated system then approaches a state where the entropy cannot increase and so therefore remains constant. This relates to the notion of a measure of maximal entropy (Definition 4.3.11) and the states in the support of the measure are the points in the state space where the energy has been maximized.

If a system is not isolated, for instance, if the system is placed inside a heat bath, then it will tend toward an equilibrium, which is called a thermodynamic equilibrium. The free energy of a system is the amount of work that a thermodynamic system can perform, so the free energy is the internal energy of a system minus the amount of energy that cannot perform work. In thermodynamics the entropy is just the unusable energy multiplied by the temperature. So, in an isolated system, maximizing the entropy is equivalent to minimizing the free energy. The system will then evolve to a state where the free energy cannot decrease and so remains constant. The measures associated with this equilibrium are a generalization of the measures of maximal entropy and are called equilibrium states (Definition 4.3.11).

In our context, this notion is defined using a continuous $f\colon X \to \mathbb{R}$, called a *potential* or *observable*. The term "observable" reflects the fact that an observation of a system usually yields a real number (the measurement) that depends on the state of the system, that is, a point in phase space. We use these functions as weights in sums over spanning sets.

Definition 4.3.2. Let X be a compact metric space and Φ be a continuous flow on X. For $f \in C^0(X)$, and $t \geq 0$ set $S_t f := \int_0^t f \circ \varphi^\tau \, d\tau$ and

$$N_d(\Phi, f, \epsilon, t) := \sup\left\{ \textstyle\sum_{x\in E} e^{S_t f(x)} \;\middle|\; E \subset X \text{ is } (t,\epsilon)\text{-separated}\right\},$$

$$S_d(\Phi, f, \epsilon, t) := \inf\left\{ \textstyle\sum_{x\in E} e^{S_t f(x)} \;\middle|\; X = \bigcup_{x\in E} B_\Phi(x,\epsilon,t)\right\},$$

$$\begin{aligned} D_d(\Phi, f, \epsilon, t) := \inf\Big\{ & \textstyle\sum_{C\in E} \inf_{x\in C} e^{S_t f(x)} \;\Big|\; X \subset \bigcup_{C\in E} C \\ & \text{and } \operatorname{diam}_{d_t^\Phi}(C) \leq \epsilon \text{ for } C \in E \subset 2^X \Big\}. \end{aligned}$$

The expressions $\sum_{x\in E} e^{S_t f(x)}$ are sometimes called *statistical sums*.[6] Then

$$P(f) := P(\Phi, f) := \lim_{\epsilon\to 0} \overline{\lim_{t\to\infty}} \frac{1}{t} \log S_d(\Phi, f, \epsilon, t)$$

is called the *topological pressure* of Φ with respect to f.

Remark 4.3.3. The definition implies that $P(f + c) = P(f) + c$ for any $c \in \mathbb{R}$, and if f and g are cohomologous (Definition 1.3.18), then $P(f) = P(g)$.

Analogously to (4.2.5), (4.2.6), (4.2.3), and (4.2.4) we have

$$\begin{aligned} N_d(\Phi, f, 2\epsilon, t) \le S_d(\Phi, f, \epsilon, t) \le N_d(\Phi, f, \epsilon, t), \\ D_d(\Phi, f, 2\epsilon, t) \le S_d(\Phi, f, \epsilon, t) \le D_d(\Phi, f, \epsilon, t), \end{aligned} \tag{4.3.1}$$

which shows that

$$P(\Phi, f) = \lim_{\epsilon\to 0} \overline{\lim_{t\to\infty}} \frac{1}{t} \log N_d(\Phi, f, \epsilon, t) = \lim_{\epsilon\to 0} \underline{\lim_{t\to\infty}} \frac{1}{t} \log N_d(\Phi, f, \epsilon, t), \tag{4.3.2}$$

and

$$P(\Phi, f) = \lim_{\epsilon\to 0} \lim_{t\to\infty} \frac{1}{t} \log D_d(\Phi, f, \epsilon, t) \tag{4.3.3}$$

by an argument similar to that following Lemma 4.2.6, since $D_d(\Phi, f, \epsilon, t)$ is submultiplicative similarly to Lemma 4.2.6.

Remark 4.3.4. When $f = 0$, Definition 4.3.2 gives topological entropy: $P(\Phi, 0) = h_{\text{top}}(\Phi)$. If $c \in \mathbb{R}$, then $S_t(f + c) = tc + S_t f$, so $S_d(\Phi, f + c, \epsilon, t) = e^{tc} S_d(\Phi, 0, \epsilon, t)$ and $P(f + c) = P(f) + c$. We also have $S_d(\Phi, f, \epsilon, t) \le \|e^{S_t f}\|_{C^0} \cdot S_d(\Phi, \epsilon, t)$ and thus

$$\overline{\lim_{t\to\infty}} \frac{1}{t} \log S_d(\Phi, f, \epsilon, t) \le \|f\|_{C^0} + \overline{\lim_{t\to\infty}} \frac{1}{t} \log S_d(\Phi^t, \epsilon, t).$$

This is finite if Φ is a smooth flow on a compact manifold and f is continuous. Finally, $t \mapsto N_d(\Phi, f, \epsilon, t)$ and $t \mapsto S_d(\Phi, f, \epsilon, t)$ are nondecreasing, so $S_d(\Phi, f, \epsilon, \lfloor t \rfloor) \le S_d(\Phi, f, \epsilon, t) \le S_d(\Phi, f, \epsilon, \lceil t \rceil)$ and hence

$$P(\Phi, f) = \lim_{\epsilon\to 0} \overline{\lim_{\mathbb{N}\ni n\to\infty}} \frac{1}{n} \log S_d(\Phi, f, \epsilon, n).$$

Remark 4.3.5. The proof of Proposition 4.2.4 extends to show that pressure is independent of the metric (inducing a given topology) used to define it, thus justifying some of our notation. This implies that pressure is invariant under topological conjugacy: if $\varphi^t = \pi^{-1} \circ \psi^t \circ \pi$ and $g = f \circ \pi$ then $P(\varphi^t, f) = P(\psi^t, g)$.

[6] Or *partition sums*.

For what follows it is convenient to work with the time-1 map for a while.

Definition 4.3.6. $P(\varphi^1, f) := \lim_{\epsilon\to 0}\overline{\lim}_{n\to\infty}\frac{1}{n}\log S_d(\varphi^1, f, \epsilon, n)$, where[7]

$$S_d(\varphi^1, f, \epsilon, n) := \inf\Big\{\sum_{x\in E} e^{S_n f(x)} \;\Big|\; X = \bigcup_{x\in E} \underbrace{B_{\varphi^1}(x,\epsilon,n)}_{:=\{y\in X \mid d_t^{\varphi^1}(x,y):=\max_{0\le k\le n} d(\varphi^k(x),\varphi^k(y))<\epsilon\}}\Big\}.$$

Proposition 4.3.7. $P(\Phi, f) = P(\Phi, S_1 f) = P(\varphi^1, f)$.

Proof. Equicontinuity of $\{\varphi^t \mid |t| \le s\}$ implies $P(\Phi, f) = P(\Phi, f\circ\varphi^s)$ for any $s\in\mathbb{R}$, and a computation shows that $\|S_t S_1 f - S_{t-2} f\circ\varphi^1\|_\infty \le \|f\|_\infty$ for all $t\ge 2$. □

Theorem 4.3.8 (Variational Principle). *If Φ is a continuous flow on a compact metric space X and $f\colon X\to\mathbb{R}$ is continuous, then*

$$P(\Phi, f) = \sup_{\mu\in\mathfrak{M}(\Phi)}\big[h_\mu(\Phi) + \textstyle\int f\,d\mu\big]. \tag{4.3.4}$$

When $f\equiv 0$ this gives the Variational Principle for entropy.

Corollary 4.3.9. *If Φ is a continuous flow on a compact metric space X, then*

$$h_{\mathrm{top}}(\Phi) = \sup_{\mu\in\mathfrak{M}(\Phi)} h_\mu(\Phi).$$

Remark 4.3.10. The quantity $h_\mu(\Phi)+\int f\,d\mu$ is called the *free energy*. The topological pressure then tries to maximize the free energy. Up to a change in sign this is the same as the free energy we described previously. Due to the change in sign, thermodynamics tries to minimize the free energy.

Before we prove Theorem 4.3.8, we explore some related results.

Definition 4.3.11. A measure $\mu\in\mathfrak{M}(\Phi)$ such that $P(\Phi, f) = h_\mu(\Phi) + \int f\,d\mu$ is an *equilibrium measure* or *equilibrium state* for Φ associated with f. A measure $\mu\in\mathfrak{M}(\Phi)$ such that $h_{\mathrm{top}}(\Phi) = h_\mu(\Phi)$ is a *measure of maximal entropy* for Φ.

Example 4.3.12. Example 4.1.11 and Proposition 4.2.17 show that Lebesgue measure is a measure of maximal entropy for the suspension of the toral automorphism from Example 1.5.26; in particular, the supremum in Corollary 4.3.9 is attained in this case.

We briefly describe the collection of equilibrium states.

Theorem 4.3.13 (Set of equilibrium states). *Let $\Phi\colon X\to X$ be a continuous flow of a compact metric space and $f\colon X\to\mathbb{R}$ be continuous. Then*

[7] Although we did not adapt the notation accordingly, this is a definition of pressure for a homeomorphism.

(1) *the set* $\mathfrak{M}_f(\Phi)$ *of equilibrium states for* f *is convex,*

(2) *if* $f - g$ *is cohomologous to* $c \in \mathbb{R}$ *(Definition* 1.3.18*), then* $\mathfrak{M}_f(\Phi) = \mathfrak{M}_g(\Phi)$,[8]

(3) *if* $h_{\text{top}}(\Phi) < \infty$ *and* $\mathfrak{M}_f(\Phi) \neq \varnothing$*, then the extreme points of* $\mathfrak{M}_f(\Phi)$ *are exactly the ergodic equilibrium states, and*

(4) *if the entropy map* $\mu \mapsto h_\mu(\Phi)$ *is upper semicontinuous,*[9] *then* $\mathfrak{M}_f(\Phi)$ *is nonempty and compact.*

Proof. (1) follows from the affine property of entropy (Theorem 4.1.3): if $\mu, \nu \in \mathfrak{M}_f(\Phi)$ and $p \in [0,1]$, then

$$\underbrace{h_{p\mu+(1-p)\nu}(\Phi)}_{=ph_\mu(\Phi)+(1-p)h_\nu(\Phi)} + \underbrace{\int f\, d(p\mu + (1-p)\nu)}_{=p\int f\, d\mu+(1-p)\int f\, d\nu} = pP(\Phi, f) + (1-p)P(\Phi, f) = P(\Phi, f).$$

(2): The assumption is that f is cohomologous to $g + c$, so on both sides of (4.3.4) replacing f by $g + c$ amounts to adding c.

(3): If $\mu \in \mathfrak{M}_f(\Phi)$ is ergodic, then it is an extreme point of $\mathfrak{M}(\Phi)$ and hence of $\mathfrak{M}_f(\Phi)$. Conversely, if $\mu = p\mu_1 + (1-p)\mu_2 \in \mathfrak{M}_f(\Phi)$ is an extreme point of $\mathfrak{M}_f(\Phi)$ with $\mu_1, \mu_2 \in \mathfrak{M}(\Phi)$, then

$$\begin{aligned} P(\Phi, f) &= h_\mu(\Phi) + \int f\, d\mu \\ &= ph_{\mu_1}(\Phi) + (1-p)h_{\mu_2}(\Phi) + p\int f\, d\mu_1 + (1-p)\int f\, d\mu_2 \le P(\Phi, f), \end{aligned}$$

so $\mu_1, \mu_2 \in \mathfrak{M}_f(\Phi)$. Thus, $\mu = \mu_1 = \mu_2$, and μ is an extreme point of $\mathfrak{M}(\Phi)$, hence ergodic by Theorem 3.1.16.

(4): If $\mu \mapsto h_\mu(\Phi)$ is upper semicontinuous, then so is $\mu \mapsto h_\mu(\Phi) + \int f\, d\mu$. An upper semicontinuous function on a compact space has a maximum, so $\mathfrak{M}_f(\Phi) \neq \varnothing$, and upper semicontinuity further implies that $\mathfrak{M}_f(\Phi)$ is compact. □

Proposition 4.3.14 (High-pressure measure). *Let* (X, d) *be a compact metric space,* Φ *a continuous flow on* X*,* $f \in C^0(X)$*,* $E_n \subset X$ *an* (n, ϵ)*-separated set,*

$$\nu_n := \Big(\sum_{x\in E_n} e^{S_n f(x)}\Big)^{-1} \sum_{x\in E_n} e^{S_n f(x)} \delta_x, \quad \textit{and} \quad \mu_n := \frac{1}{n}\int_0^n \varphi_*^s \nu_n\, ds.$$

[8] The converse is Theorem 7.3.24.

[9] This is the case if Φ is expansive (Corollary A.3.14).

Then there exists a weak-accumulation point $\mu \in \mathfrak{M}(\Phi)$ of $\{\mu_n\}_{n\in\mathbb{N}}$ that satisfies*

$$\varlimsup_{n\to\infty} \frac{1}{n} \log \sum_{x\in E_n} e^{S_n f(x)} \le h_\mu(\Phi) + \int f\,d\mu.$$

Before proving this, we note that it implies half of the Variational Principle.

Corollary 4.3.15. $P(\Phi, f) \le \sup_{\mu\in\mathfrak{M}(\Phi)} h_\mu(\Phi) + \int f\,d\mu$.

Proof. Fix $\delta > 0$ and let $\{E_n\}_{n\in\mathbb{N}}$ be (n,ϵ)-separated sets in X such that

$$\sum_{x\in E_n} e^{S_n f(x)} \ge N_d(\Phi, f, \epsilon, n) - \delta.$$

For an accumulation point μ of the μ_n in Proposition 4.3.14 we then have

$$\varlimsup_{n\to\infty} \frac{1}{n} \log N_d(\Phi, f, \epsilon, n) \le h_\mu(\Phi) + \int f\,d\mu.$$

Taking here the supremum over μ and letting $\epsilon \to 0$ gives the claim. □

Proof of Proposition 4.3.14. Let n_k be a subsequence such that

$$\lim_{k\to\infty} \log \sum_{x\in E_{n_k}} e^{S_{n_k} f(x)} = \varlimsup_{n\to\infty} \log \sum_{x\in E_n} e^{S_n f(x)}$$

and μ a weak*-accumulation point of μ_{n_k}. That $\mu_n = \frac{1}{n}\int_0^n \varphi_*^s \nu_n\,ds$ implies that μ is Φ-invariant as in the proof of Theorem 3.1.15.

Let ξ be a partition whose elements have diameter less than ϵ and $\mu(\partial\xi) = 0$. Let $E_n = \{x_1, \dots, x_m\}$ be an (n,ϵ)-separated set. Then (see (4.1.1) and Definition A.2.1)

$$\begin{aligned} H_{\nu_n}(\xi_{-n}^{\varphi^1}) + \underbrace{n\int f\,d\mu_n}_{=\int S_n f\,d\nu_n} &= \sum_{x\in E_n} [-\nu_n(\{x\})\log(\nu_n(\{x\})) + \nu_n(\{x\}) S_n f(x)] \\ &= \log \sum_{x\in E_n} e^{S_n f(x)}. \end{aligned}$$

Here the last equality is a simple computation or an application of (the easy "=" part of) Lemma A.2.14 below. If $a(k) = \lfloor (n-k)/q \rfloor$ for $0 \le k < q < n$, then this gives

$$\frac{q}{n}\log \sum_{x\in E_n} e^{S_n f(x)} = \underbrace{\frac{q}{n} H_{\nu_n}(\xi_{-n}^{\varphi^1})}_{=\frac{1}{n}\sum_{k=0}^{q-1} H_{\nu_n}(\xi_{-n}^{\varphi^1})} + q\int f\,d\mu_n$$

$$\text{(Proposition A.2.6(4))} \quad \le \sum_{k=0}^{q-1}\Bigl(\sum_{r=0}^{a(k)-1} \frac{1}{n} H_{\varphi_*^{rq+k}\nu_n}(\xi_{-q}^{\varphi^1}) + \frac{2q}{n}\log\#(\xi)\Bigr) + q\int f\,d\mu_n$$

$$\text{(Proposition A.2.6(6))} \quad \le H_{\mu_n}(\xi_{-q}^{\varphi^1}) + \frac{2q^2}{n}\log\#(\xi) + q\int f\,d\mu_n.$$

Hence, $\lim_{k\to\infty}\frac{1}{n_k}\log\sum_{x\in E_{n_k}} e^{S_{n_k}f(x)} \le \frac{1}{q}\lim_{k\to\infty} H_{\mu_{n_k}}(\xi_{-q}^{\varphi^1}) + \int f\,d\mu_{n_k} = \frac{1}{q}H_\mu(\xi_{-q}^{\varphi^1}) + \int f\,d\mu$ and $\overline{\lim}_{n\to\infty}\frac{1}{n}\log\sum_{x\in E_n} e^{S_n f(x)} \le h_\mu(\varphi^1,\xi) + \int f\,d\mu \le h_\mu(\varphi^1) + \int f\,d\mu = P_\mu(f,\Phi)$. □

Proof of Theorem 4.3.8. In light of Corollary 4.3.15, it remains to show that

$$h_\mu(\Phi) + \int f\,d\mu \le P(\Phi,f) \quad \text{for every } \mu \in \mathfrak{M}(\Phi). \tag{4.3.5}$$

Let $\xi = \{C_1,\dots,C_k\}$ be a measurable partition of X. Then $\mu(C_i) = \sup\bigl\{\mu(B) \mid B \subset C_i \text{ is closed}\bigr\}$, so there are compact $B_i \subset C_i$ (think of these as the "islands") such that $H_\mu(\xi|\mathcal{B}) \le 1$ for $\mathcal{B} = \{B_0,\dots,B_k\}$ with $B_0 = X \setminus (\bigcup_{j=1}^k B_i)$ (think of this as "the sea"), and

$$h_\mu(\Phi,\xi) \le h_\mu(\Phi,\mathcal{B}) + H_\mu(\xi|\mathcal{B}) \le h_\mu(\Phi,\mathcal{B}) + 1.$$

Now let $d = \min\{d(B_i,B_j) \mid i,j \in \{1,\dots,k\},\ i \ne j\} > 0$ and $\delta \in (0,d/2)$ such that $|f(x)-f(y)| < 1$ whenever $d(x,y) < \delta$. Let $E \subset X$ be an (n,δ)-spanning set. For $C \in \mathcal{B}_{-n}^{\varphi^1}$ there is an $x_C \in \overline{C}$ such that $(S_n f)(x_C) = \sup\bigl\{S_n f(x) \mid x \in C\bigr\}$ and a $y_C \in E$ such that $d_n^\Phi(x_C,y_C) \le \delta$, so $S_n f(x_C) \le S_n f(y_C) + n$. Then

$$H_\mu(\mathcal{B}_{-n}^{\varphi^1}) + \int S_n f\,d\mu$$
$$\le \underbrace{\sum_{C\in\mathcal{B}_{-n}^{\varphi^1}} \mu(C)(-\log\mu(C) + \underbrace{S_n f(x_C)}_{\le S_n f(y_C)+n})}_{\le \log\sum_{C\in\mathcal{B}_{-n}^{\varphi^1}} e^{S_n f(y_C)+n} \text{ by Lemma A.2.14}} \le n + \log\Bigl(\underbrace{2^n \sum_{x\in E} e^{S_n f(x)}}_{\delta<\frac{d}{2},\ y\in E \Rightarrow \#\{C\in\mathcal{B}_{-n}^{\varphi^1} \mid y_C = y\} \le 2^n}\Bigr),$$

and

$$\frac{1}{n}H_\mu(\mathcal{B}_{-n}^{\varphi^1}) + \underbrace{\int f\,d\mu}_{=\frac{1}{n}\int S_n f\,d\mu} \le 1 + \log 2 + \frac{1}{n}\log\sum_{x\in E} e^{S_n f(x)}.$$

Therefore, $h_\mu(\Phi,\xi) + \int f\,d\mu \le h_\mu(\Phi,\mathcal{B}) + 1 + \int f\,d\mu \le 2 + \log 2 + P(\Phi, f)$, and hence $h_\mu(\varphi^1) + \int f\,d\mu \le 2 + \log 2 + P(\Phi, f)$. Applying this to φ^n and $S_n f$ gives

$$h_\mu(\Phi) + \int f\,d\mu \le P(\Phi, f) + (2 + \log 2)/n \xrightarrow[n\to\infty]{} P(\Phi, f). \qquad \square$$

If Φ_r is a special flow, then the Variational Principle (Theorem 4.3.8) and the Abramov Theorem (Theorem 4.1.4) imply

$$h_{\text{top}}(\Phi_r) = \sup_{\mu_r \in \mathfrak{M}(\Phi_r)} h_{\mu_r}(\Phi_r) = \sup_{\mu\in\mathfrak{M}(\sigma)} \frac{h_\mu(\sigma)}{\int r\,d\mu}. \tag{4.3.6}$$

This is useful with respect to an earlier question.

Theorem 4.3.16 (Entropy and time change). *If a continuous flow without fixed points has zero (or finite) topological entropy, then so does any time change.*

Proof outline. Because there are no fixed points, assume without loss of generality that the flows Φ, Ψ are special flows (Theorem 3.6.2) over the same base transformation σ under roof functions r_Φ, r_Ψ with (by compactness) bounded logarithms. Then the right-hand side of (4.3.6) is the same for both Φ and Ψ. If $h_{\text{top}}(\Phi) < \infty$, then this supremum is finite, and hence $h_{\text{top}}(\Psi) < \infty$; if $h_{\text{top}}(\Phi) = 0$ then likewise, $h_{\text{top}}(\Psi) = 0$. $\square$

We next relate the pressure of a special flow to the pressure of the base dynamics, provided there is a unique equilibrium state for the base dynamics (for which the discrete-time counterpart of Theorem 7.3.6 gives sufficient conditions).

Proposition 4.3.17. *Let X be a compact metric space, F a homeomorphism on X with $h_{\text{top}}(F) < \infty$, $r\colon X \to (0,\infty)$ continuous, Φ_r the special flow on X_r, $G \in C(X_r)$, and $g(x) := \int_0^{r(x)} G(x,t)\,dt \in C(X)$. Then the following hold:*

- *There is a unique $c \in \mathbb{R}$ with $P(F, g - cr) = 0$.*
- *If F has a unique equilibrium state m for $g - cr$, then m_r (from (3.6.4)) is the unique equilibrium state of G for Φ_r, and $c = P(\Phi_r, G)$.*
- *If F has a unique equilibrium state m for $-h_{\text{top}}(\Phi_r)r$, then m_r is the unique measure of maximal entropy for Φ_r on X_r.*

Proof. We first show that the continuous map $c \mapsto P(F, g - cr)$ is strictly decreasing. If μ is an F-invariant measure and $c_1 < c_2$, then continuity of r together with $r > 0$ and $h_{\text{top}}(F) < \infty$ imply

$$\begin{aligned} h_\mu(F) + \int (g - c_1 r)\, d\mu &= h_\mu(F) + \int g\, d\mu - c_1 \int r\, d\mu \\ &> h_\mu(F) + \int g\, d\mu - c_2 \int r\, d\mu = h_\mu(F) + \int (g - c_2 r)\, d\mu. \end{aligned}$$

Let μ_n be a sequence of F-invariant probability measures with $\lim_{n\to\infty} h_{\mu_n}(F) + \int g - c_2 r\, d\mu_n = P(F, g - c_2 r)$. By taking a subsequence assume that $\mu_n \xrightarrow[n\to\infty]{\text{weak*}} \mu$. Then

$$\begin{aligned} P(F, g - c_1 r) &\geq \lim_{n\to\infty} h_{\mu_n}(F) + \int (g - c_1 r)\, d\mu_n \\ &= \lim_{n\to\infty} h_{\mu_n}(F) + \int (g - c_2 r)\, d\mu_n + (c_2 - c_1) \int r\, d\mu_n \\ &= P(F, g - c_2 r) + (c_2 - c_1) \int r\, d\mu > P(F, g - c_2 r), \end{aligned}$$

so $c \mapsto P(F, g - cr)$ is strictly decreasing. Furthermore, $\lim_{c\to\pm\infty} P(F, g - cr) = \mp\infty$ since $h_{\text{top}}(F) < \infty$, so there is a unique $c \in \mathbb{R}$ with $P(F, g - cr) = 0$.

If m is an equilibrium state of F for $g - cr$, then $h_m(F) + \int (g - cr)\, dm = P(F, g - cr) = 0$, hence

$$c = \frac{h_m(F)}{\int r dm} + \frac{\int g dm}{\int r dm} \overset{\text{Theorem 4.1.4}}{=\!=\!=} h_{m_r}(\Phi_r) + \int G\, dm_r.$$

If m is the unique equilibrium state of F for $g - cr$, then m_r is the unique equilibrium state of Φ_r for G because for any F-invariant probability measure $\mu \neq m$ we have $0 > h_\mu(F) + \int (g - cr)\, d\mu$, so

$$h_{m_r}(\Phi_r) + \int G\, dm_r = c > \frac{h_\mu(F) + \int g\, d\mu}{\int r\, d\mu} \overset{\text{Theorem 4.1.4}}{=\!=\!=} h_{\mu_r}(\Phi_r) + \int G\, d\mu_r.$$

This also implies that $c = h_{m_r}(\Phi_r) + \int G\, dm_r = P(\Phi_r, G)$. Furthermore, by Theorem 3.6.2, each measure μ_r arises from a measure μ concluding the proof of the second part of the theorem.

If $G = 0$, then $g = 0$, so $P(F, 0 - cr) = 0$ when $c = P(\Phi_r, 0) = h_{\text{top}}(\Phi_r)$. So, if there is a unique equilibrium state m for $-h_{\text{top}}(F) r$, then m_r is a unique measure of maximal entropy. □

Remark 4.3.18. Under the assumptions of Theorem 4.3.13 and Theorem 4.3.8, $h_\mu(\Phi)$ and hence $h_\mu(\Phi) + \int f\,d\mu$ on the right-hand side of (4.3.4) is upper semicontinuous if Φ is expansive (Corollary A.3.14)—thus Theorem 4.3.13(4) gives existence of an equilibrium measure. This is notable, but the importance of equilibrium states rests in great part on our ability to study them carefully, and a nonconstructive existence result is of limited use in this respect. Therefore, we now give more restrictive sufficient conditions for the existence of equilibrium states because they allow us to construct them explicitly (Theorem 4.3.23). These involve controlling the "dynamical distortion" of the potential (Definition 4.3.19). We will later see that in the principal context of this book, equilibrium states are unique (Theorem 7.3.6).

Definition 4.3.19. Let X be a metric space, Φ a flow. With the notation from Definition 4.3.2, the set $V(\Phi)$ of *Bowen-bounded* functions [60, p. 193] for Φ is

$$\{f \in C^0(X) \mid \exists\, K, \epsilon > 0\ \forall\, t > 0\colon d_t^\Phi(x,y) < \epsilon \Rightarrow |S_t f(x) - S_t f(y)| < K\}, \quad (4.3.7)$$

and the set $V_0(\Phi)$ of *Walters-continuous* functions [346, p. 125] for Φ is

$$\{f \in C^0(X) \mid \forall\, \epsilon > 0\ \exists\, \delta > 0\ \forall\, t > 0\colon d_t^\Phi(x,y) < \delta \Rightarrow |S_t f(x) - S_t f(y)| < \epsilon\}. \quad (4.3.8)$$

These regularity conditions may look technical but arise naturally in hyperbolic flows: Hölder-continuous functions (Definition 1.9.6) are Walters continuous (and hence Bowen bounded) for a hyperbolic flow (Proposition 7.3.1) due to a quantitative (exponential) version of Proposition 1.8.4 (Theorem 6.2.4). Periodic data determine a Walters-continuous function, or rather its cohomology (Theorem 5.3.23). The utility of Bowen-boundedness lies in the following, which makes Proposition 4.3.14 the main step in the construction of equilibrium states.

Lemma 4.3.20. *Let Φ be an expansive flow on a compact metric space X with expansivity constant δ_0 (cf. Definition 1.7.1). Then for $f \in V(\Phi)$, $\epsilon \in (0, \delta_0/2)$, and $\delta > 0$ there exists $C_{\delta,\epsilon}$ such that (for all $t > 0$)*

$$N_d(\Phi, f, \delta, t) \le C_{\delta,\epsilon} N_d(\Phi, f, \epsilon, t).$$

Remark 4.3.21. If $\delta > \epsilon$ we can take $C_{\delta,\epsilon} = 1$.

Proof. For $0 < \epsilon < \delta_0/2$ expansivity gives a $T > 0$ such that

$$d_{2T}^\Phi(\varphi^{-T}(x), \varphi^{-T}(y)) \le 2\epsilon \Rightarrow d(x,y) < \delta, \quad (4.3.9)$$

and equicontinuity gives $\alpha > 0$ such that $d(x,y) \le \alpha \Rightarrow d_{2T}^\Phi(\varphi^{-T}(x), \varphi^{-T}(y)) \le \delta$. If E is a maximal (t,δ)-separated set and F a maximal (t,ϵ)-separated set, then for $x \in E$ there is a $z(x) \in F$ such that $d_t^\Phi(x, z(x)) < \epsilon$. The cardinality of

$E_z := \{x \in E \mid z(x) = z\}$ is bounded uniformly in t: If $x, y \in E_z$ then $d_t^{\Phi}(x,y) \le 2\epsilon$ by definition of E_z, hence $d(\varphi^s(x), \varphi^s(y)) \le \delta$ for $s \in [T, t-T)$ by choice of T, and thus, by choice of α and since $\{x, y\}$ is (t,δ)-separated, $d(x,y) > \alpha$ or $d(\varphi^t(x), \varphi^t(y)) > \alpha$. Therefore

$$\begin{aligned}\operatorname{card}(E_z) &= \operatorname{card}\{(x,\varphi^t(x)) \mid x \in E_z\} \\ &\le \max\{\operatorname{card} A \mid A \subset X \times X \text{ and } (a,b) \in A,\ a \ne b \Rightarrow d(a,b) > \alpha\} =: M\end{aligned}$$

since the $(x, \varphi^t(x))$ form just such an α-separated set. Now take ϵ, K as in (4.3.7) so that $|S_t f(x) - S_t f(z)| \le K$ for $x \in E_z$ and

$$\sum_{x\in E} e^{S_t f(x)} \le \sum_{z\in F} \underbrace{\operatorname{card} E_z}_{\le M} e^K e^{S_t f(z)} \le \underbrace{M e^K}_{=:C_{\delta,\epsilon}} N_d(\Phi, f, \epsilon, t). \qquad \square$$

Together with (4.3.1) and (4.3.2) this gives the following proposition:

Proposition 4.3.22. *If Φ is expansive, 3ϵ an expansivity constant, $f \in V(\Phi)$, then*

$$P(\Phi, f) = \lim_{t\to\infty} \frac{1}{t} \log S_d(\Phi, f, \epsilon, t).$$

With Proposition 4.3.14, this in turn gives existence of equilibrium states:

Theorem 4.3.23 (Equilibrium states exist). *If Φ be an expansive flow on a compact metric space, $f \in V(\Phi)$, and $P(\Phi, f) < \infty$ (Definition* 4.3.2*), then every weak*-accumulation point of the μ_n in Proposition* 4.3.14 *is an equilibrium state for Φ associated with f.*

Remark 4.3.24. One can weaken the assumption of expansivity in the spirit of Section 1.8 (Exercise 4.25).

While we will be able to establish uniqueness of equilibrium states for hyperbolic flows later, this may not hold for systems beyond this context, even though much progress has recently been made for nonuniformly hyperbolic dynamical systems.[10] We will revisit Theorem 4.3.23 in a context where equilibrium states are unique (Theorem 7.3.6). For that work and elsewhere, another way of singling out an invariant Borel probability measure is important, and we now define this property and connect it to equilibrium states.

Definition 4.3.25 (Gibbs measure). For a continuous flow Φ on a compact metric space X and a potential function $f : X \to \mathbb{R}$, a measure $\mu \in \mathfrak{M}(\Phi)$ is a *Gibbs measure* for f with constant P if for $\delta > 0$ there is a constant $C > 0$ such that for $x \in X$ and $t > 0$ we have

$$\frac{1}{C} \le \frac{\mu(B_{\Phi}(x,t,\delta))}{\exp(S_t f(x) - tP)} \le C.$$

[10] Systems with more than one equilibrium state are said to be in a *phase transition* [145].

For hyperbolic Φ, Proposition 7.3.15 says that for each $f \in V(\Phi)$ there is a Gibbs measure with constant $P = P(\Phi, f)$ (Definition 4.3.2). Our present object is to show that this is an alternate way of producing equilibrium states.

Theorem 4.3.26 (Gibbs implies equilibrium). *If Φ is a continuous flow on a compact metric space X and $f \in V(\Phi)$, then a Gibbs measure with constant $P = P(\Phi, f)$ is an equilibrium state for f.*

Proof. Fix $t > 0$ and $\epsilon > 0$. Let $\beta > 0$ such that $d(x, y) \le \beta$ implies that $d_t(x, y) \le \epsilon$. Let $\xi = \{B_1, \dots, B_m\}$ be a measurable partition of X such that $\operatorname{diam}(B_i) \le \beta$ for all $1 \le i \le m$ and hence $\operatorname{diam} \varphi^s(A) \le \epsilon$ for all $A \in \xi_{-n}^{\varphi^t}$, $n \in \mathbb{N}$ and $s \in [0, nt]$. By the Gibbs property we have $\mu(A) \le C \exp(S_{nt} f(x) - ntP)$ for any $x \in A$, and hence $H(\xi_{-n}^{\varphi^t}) \ge -\log C + Pnt - \int S_{nt} f \, d\mu$ by Proposition A.2.6(1). Thus,

$$
\begin{aligned}
tP(\Phi, f) = Pt &\le \underbrace{\lim_{n\to\infty} \frac{1}{n} H(\xi_{-n}^{\varphi^t})}_{=h_\mu(\varphi^t, \xi)} + \underbrace{\frac{1}{n} \int S_{nt} f(x) \, d\mu}_{=\int S_t f(x)\, d\mu} \\
&\le \underbrace{h_\mu(\varphi^t)}_{=t h_\mu(\Phi)} + \underbrace{\int S_t f(x) \, d\mu}_{=t \int f \, d\mu} \\
&= tP(\Phi, f),
\end{aligned}
$$

so $h_\mu(\Phi) + \int f \, d\mu = P(\Phi, f)$. □

Additional assumptions on a flow imply uniqueness of equilibrium states (Theorem 7.3.6) for Bowen-bounded potentials. Instead of presenting this strengthening here, we defer it to the context of hyperbolic flows, where Bowen-boundedness is particularly natural—and invariant under topological equivalence, unlike in the present context.

We close by remarking that while we gave a motivation for the study of equilibrium states in terms of thermodynamical concepts that can be transferred to dynamical systems, the principal motivation of dynamicists in studying them is that they provide a collection of measures (rather than just the measure of maximal entropy) that are deeply connected to the dynamics of a flow and have strong stochastic properties (Remark 7.3.20). Even in the case of the measure of maximal entropy, the possibility of establishing these deep connections with the dynamics are closely connected with the arguments that establish uniqueness; the existence result we have in hand now conveys little information about properties of such a measure beyond its entropy. This is starkly apparent outside the context of hyperbolic dynamical systems when one considers those with zero topological entropy, for which then all invariant measures maximize entropy. Among hyperbolic flows, billiard systems provide an illuminating instance. Billiard flows (Figure 0.2.2) are continuous and hence have a measure of maximal entropy, but little is known about such measures. The associated billiard map is not even continuous in the case of dispersing billiards (Section 5.2.d), so the very definition of topological entropy becomes problematic, and even though the technology for discrete-time systems is better developed, the first steps in this direction are being taken only at the time of this writing.

Also notable among equilibrium states is the one for a special potential, the *geometric potential*, which is of exceptional interest in its own right: This Sinai–Ruelle–Bowen measure is central to the description of hyperbolic attractors (Theorem 7.4.10) and stands in for volume when this is not invariant; as a corollary, volume is an equilibrium state if invariant, and enjoys the stochastic properties we derive in full generality—and sometimes even more (Theorem 7.4.20). Lastly, this Sinai–Ruelle–Bowen measure can be a tool for establishing results about smooth dynamics whose statements make no reference to probabilistic aspects (Theorem 9.2.7), and the question of when it coincides with the measure of maximal entropy leads to interesting rigidity results (Section 9.3). This theory has recently been developed also for geodesic flows on noncompact negatively curved manifolds [283].

4.4 Equilibrium states for time-t maps*

We digress to connect equilibrium states for a flow and for its time-t maps. The problem is that the set of invariant measures for the time-t map of a flow may be larger than the set of invariant measures for the flow. We begin with measures of maximal entropy.

In Proposition 4.2.8 we showed that the topological entropy of a flow is equal to the topological entropy of the time-1 map of the flow, and Corollary 4.2.10 gives $|t|h_{\text{top}}(\Phi) = h_{\text{top}}(\varphi^t)$ for any t. Therefore, any measure of maximal entropy for Φ is a measure of maximal entropy φ^t. However, there may be measures of maximal entropy for the time-t map that are not measures of maximal entropy for the flow. For instance, if we start with a map $f: X \to X$ with a measure of maximal entropy and a constant-time suspension with roof function 1, then the time-1 map will have an invariant measure supported on each $X \times \{c\}$ for $0 \le c < 1$, but these are not flow invariant and hence not measures of maximal entropy for the flow. Weak mixing avoids this problem:

Theorem 4.4.1. *If Φ has a unique measure μ of maximal entropy and μ is weakly mixing, and if $t > 0$, then μ is the unique measure of maximal entropy for the time-t map of Φ.*

Proof (communicated by Federico Rodriguez Hertz). Let ν be a measure of maximal entropy for the time-t map φ^t. Then the measure $\int_0^t \varphi_*^s \nu \, ds$ is Φ-invariant and a measure of maximal entropy since $\varphi_*^s \nu$ is invariant under the time-t map and has the same entropy as ν and μ. So

$$\int_0^t \varphi_*^s \nu \, ds = \mu,$$

and μ is a linear combination (by the integral) of φ^t-invariant measures. But by Proposition 3.4.40, φ^t is μ-ergodic, so $\varphi_*^s \nu = \mu$ for every s, in particular $\nu = \mu$. □

For an equilibrium state associated with a potential function $f \colon X \to \mathbb{R}$ the Variational Principle implies that for the time-1 map we have

$$\begin{aligned} P(\Phi, f) &= \sup_{\mu \in \mathfrak{M}(\Phi)} \overbrace{h_\mu(\Phi)}^{=h_\mu(\varphi^1)} + \int f \, d\mu \\ &\le \sup_{\mu \in \mathfrak{M}(\varphi^1)} h_\mu(\varphi^1) + \int f \, d\mu = P(\varphi^1, f). \end{aligned}$$

More generally, for the time-t map one usually replaces the potential function f by $f_t = \int_0^t f(\varphi^s x) \, ds$ and considers $P(\varphi^t, f_t) = \sup_{\mu \in \mathfrak{M}(\varphi^t)} h_\mu(\varphi^t) + \int f_t \, d\mu$. Neither $P(\varphi^t, f_t)$ nor $P(\varphi^1, f_1)$ relates straightforwardly to $P(\Phi, f)$ unless f is Φ-invariant.

Exercises

4.1. Compute the topological entropy of the flows in Examples 1.1.8, 1.3.8, 1.3.13, 1.3.15, 1.3.16, 1.4.16, and 1.5.26, and in Figures 1.3.3 and 1.5.11.

4.2. Find all ergodic invariant Borel probability measures of maximal entropy for the flows in Examples 1.1.8, 1.3.8, 1.3.13, 1.3.15, 1.3.16, and 1.4.16, and in Figures 1.3.3 and 1.5.11.

4.3. Compute the entropy of each invariant Borel probability measure found in Exercise 3.1.

4.4. Compute the topological entropy of the suspension flow obtained analogously to Example 1.5.26 from $\left(\begin{smallmatrix} 5 & 3 \\ 3 & 2 \end{smallmatrix}\right)$.

4.5. Compute the topological entropy of the suspension flow obtained analogously to Example 1.5.26 from $\left(\begin{smallmatrix} 1 & 1 \\ 3 & 4 \end{smallmatrix}\right)$.

4.6. Compute the topological entropy of the suspension flow obtained analogously to Example 1.5.26 from $\begin{pmatrix} 0 & 0 & 1 \\ 1 & 0 & 4 \\ 0 & 1 & 3 \end{pmatrix}$.

4.7. Compute the entropy of Lebesgue measure for the suspension of the toral automorphism from Example 1.5.26.

4.8. Compute the entropy of Lebesgue measure for the special flow over the toral automorphism from Example 1.5.26 under the roof function $r = 1 + \sin 2\pi x$.

4.9. Use entropy considerations to explain why instead of the partition in Figure 1.9.2 we used the partition in Figure 1.9.3 for coding the map and suspension flow given by $F(x, y) = (2x + y, x + y) \pmod 1$ in (1.9.4).

4.10. Prove that no partition by two pieces can be used for coding the map given by $F(x, y) = (2x + y, x + y) \pmod 1$ in (1.9.4).

4.11. Is there a partition by four pieces that can be used for coding the map given by $F(x, y) = (2x + y, x + y) \pmod 1$ in (1.9.4)?

4.12. Is there a 3-piece coding for the toral automorphism defined by $\begin{pmatrix} 1 & 1 \\ 1 & 0 \end{pmatrix}$?

4.13. Compute the entropy of the "$(p, 1 - p)$-Bernoulli" measure for the suspension of the full 2-shift Σ_2. That measure is defined as the product measure of the measure on $\{0, 1\}$ that assigns measure p to $\{0\}$ and $1 - p$ to $\{1\}$, that is, a cylinder defined by a word with k occurrences of 0 and $n - k$ occurrences of 1 has measure $p^k(1 - p)^{n-k}$.

4.14. Compute the entropy of the measure induced by the $(p, 1-p)$-Bernoulli measure for the special flow over the full 2-shift Σ_2 under the roof function $r \colon \omega \mapsto 1 + \omega_0$.

4.15. Generalizing Exercise 4.13, compute the entropy of the $\vec{p}$-Bernoulli measure for the suspension of the full n-shift, that is, the product measure of the measures that assign probability p_i to $\{i\}$, with $\sum_i p_i = 1$.

4.16. Compute the pressure of the potential $f \colon (\omega, t) \mapsto \omega_0$ for the "$(p, 1-p)$-Bernoulli" measure and the suspension of the full 2-shift Σ_2 (see Exercise 4.13).

4.17. Compute the pressure of the potential $f \colon (\omega, t) \mapsto \omega_0$ for the $\vec{p}$-Bernoulli measure and the suspension of the full n-shift.

4.18. Compute the topological pressure of the potential $f \colon (\omega, t) \mapsto \omega_0$ for the suspension of the full 2-shift Σ_2.

4.19. Prove the assertions in Remark 4.3.3.

4.20. Prove (4.3.3).

4.21. Prove the assertions in Remark 4.3.5.

4.22. Carry out the computation in the proof of Proposition 4.3.7.

4.23. Find the equilibrium state of the suspension of the full 2-shift for the function $f\colon (\omega,t) \mapsto \omega_0$.

4.24. Find the equilibrium state of the suspension of the full 2-shift for the function $f\colon (\omega,t) \mapsto \omega_0 \cdot \omega_1$.

4.25. Show that Lemma 4.3.20 and hence Proposition 4.3.22 and Theorem 4.3.23 hold for kinematic-expansive flows (Definition 1.8.3).

Part II

Hyperbolic flows

Introduction to Part II

We now come to the principal subject matter of this book, hyperbolic dynamics. The next three chapters contain the general theory of hyperbolic flows, and these chapters develop topological, measurable, and differentiable properties for these flows. The last two chapters of this section investigate properties of Anosov flows, and these two chapters contain numerous recent results.

Chapter 5 defines hyperbolicity and develops its essential features, as well as a range of new examples—several of which have not previously appeared outside the research literature. Many of the results in this chapter are actually consequences of two properties of hyperbolicity—expansivity and the shadowing property. Following a strategy of Anosov, Katok, and Bowen, we make a point in this chapter to highlight the number of results that can be obtained from just these two aspects of hyperbolic flows. Chapter 6 introduces the concept of stable and unstable manifolds. The resulting foliation structure refines our understanding of hyperbolic dynamics. Chapter 7 studies measurable and statistical aspects of hyperbolic flows; hyperbolic dynamics is deterministic but of such complexity that a probabilistic approach is natural. Related regularity issues are refined in the chapter on rigidity (Chapter 9).

The concluding chapters are (mainly) dedicated to Anosov flows, and we pursue two topics further. A topological study (Chapter 8) explores dynamical and structural features of Anosov flows as well as new examples of them. Chapter 9 is more focused on smooth dynamics and explores among other things a range of situations in which the generally rare circumstance of smooth conjugacy (or orbit equivalence) arises in natural contexts from the coincidence of some dynamical features with those of an algebraic counterpart. Most of the results in these last two chapters come with a proof, but unlike the previous parts of this book we do not strive to prove all the results we use. At the least we provide references where the proofs can be found.

5

Hyperbolicity

This chapter introduces the main subject of the book by defining a hyperbolic set, and then investigating properties of these sets. The Alekseev Cone Criterion shows that hyperbolicity is a robust property (persists under perturbation). This criterion is also effective in checking hyperbolicity in a collection of "physical" examples such as geodesic flows, billiards, gases, and linkages.

We then implement the Anosov–Katok–Bowen program to establish many of the qualitative dynamical features of a hyperbolic flow from the *shadowing property* [207, §2], also known as the *pseudo-orbit tracing property* (Section 5.3). A core result is that in the hyperbolic context the chain decomposition, which in Section 1.5 seemed somewhat theoretical, can be explicitly used to describe the overall orbit structure of a hyperbolic set. Specifically, there are finitely many chain components, and each contains a dense orbit as well as a dense collection of periodic orbits. This also leads us to a natural global definition of hyperbolicity of a flow, which is implicit in the literature (Definition 5.3.50) but does not seem to have been taken up as a definition.

Shadowing also allows us to investigate the persistence of the orbit structure of a hyperbolic set: for a perturbation of a flow, not only does the presence of hyperbolic behavior persist, but the entire orbit structure arising from hyperbolicity is indestructible even under C^0-perturbations. Furthermore, C^1-perturbations of a hyperbolic set do not create additional hyperbolicity, and the full dynamics of a hyperbolic set are present in the perturbation (structural stability). A remarkable feature unique to hyperbolic flows is that the recurrent part of the orbit structure is in its entirety rigid under C^1-perturbations—the perturbation has exactly the same recurrent dynamics, no more, and no less (Ω-stability).

We emphasize that this chapter does not use the existence of stable and unstable manifolds. These are central to the hyperbolic theory, but we chose to emphasize how much of the core dynamics can be obtained from shadowing alone. The invariant foliations will be introduced and immediately put to use in Chapter 6. Remarks 5.4.30 and 5.4.31 will put this purely topological approach in perspective.

5.1 Hyperbolic sets and basic properties

The geodesic flow on compact factors of the hyperbolic plane (Remark 2.2.2) and its horizontal twin (Example 2.2.4) are iconic examples of the kind of flow in which we are interested (Remark 5.1.3). This helps give context for the definition and provides intuition for the defining properties of these sets.

Definition 5.1.1 (Hyperbolic set). Let M be a smooth manifold[1] and Φ a smooth flow on M. A compact Φ-invariant set Λ is a *hyperbolic set* for Φ if there exist a finite number of hyperbolic fixed points $\{p_1, \dots, p_k\}$, a closed set Λ' with a splitting $T_{\Lambda'}M = E^s \oplus E^c \oplus E^u$, and constants $C \geq 1$, $\lambda \in (0,1)$, $\mu > 1$ such that $\Lambda = \Lambda' \cup \{p_1, \dots, p_k\}$ and

- $E^c(x) := \mathbb{R}V(x) \neq \{0\}$ for all $x \in \Lambda'$, where $V := \dot{\varphi}$ as in (1.1.2),
- $\|D\varphi^t \restriction_{E^s_x}\| \leq C\lambda^t$ for all $t > 0$ and all $x \in \Lambda'$, and
- $\|D\varphi^{-t} \restriction_{E^u_x}\| \leq C\mu^{-t}$ for all $t > 0$ and all $x \in \Lambda'$.

A smooth flow Φ on a closed[2] connected manifold M is said to be an *Anosov flow* (or *hyperbolic flow*) if M is hyperbolic for Φ. If $\dim M = 3$, then such a flow is called an Anosov 3-flow.

Remark 5.1.2. Compactness is an essential part of this definition because it ensures nontrivial recurrence, which in combination with hyperbolicity produces the dynamical phenomena this book studies. We note that one can (rather artificially, of course) define a Riemannian metric on $\mathbb{R}^3$ with respect to which the flow $(x, y, z) \mapsto (x+t, y, z)$ is uniformly hyperbolic [350]. This leads to (largely recent) work on these maps, which has some echoes with our study of Anosov flows on the universal cover in Section 8.5, say [160].

This definition of a hyperbolic set allows the existence of (isolated) fixed points, in contrast to what is often done elsewhere. Allowing fixed points gives greater generality, and we will find that the main results are no different. The inclusion of fixed points is also a natural adaptation for the study of stability properties later.

Remark 5.1.3 (Examples). Numerous prior examples are of this kind:

- The suspensions in Examples 1.5.26 and 1.5.27 are Anosov flows.
- By Theorem 5.1.11 below, so are time changes of these, hence all special flows over these automorphisms.

[1] Implicitly assumed connected throughout this book.

[2] That is, compact and without boundary.

- So are geodesic flows on compact factors of the hyperbolic plane—the discussion in Remark 2.2.2 establishes the requirements of Definition 5.1.1.

- Example 2.2.4 does so for the horizontal flow generated by H, which therefore gives yet another example of an Anosov flow.

More hyperbolic flows appear in Remark 5.1.10 and Sections 5.2, 6.5, 6.7, 8.2, and 8.3.

Anosov flows were conceived as a codification of the salient features of geodesic flows of compact manifolds with negative curvature. These in turn were studied as the first examples of ergodic, indeed chaotic, mechanical systems and hence remain the primary continuous-time example in this theory. Theorem 5.2.4 below establishes that these are indeed Anosov flows, whether or not the curvature is constant as in Chapter 2. Its proof addresses the fact that we usually cannot identify the contracting and expanding subspaces as readily as in the algebraic case (for example, as in Remark 2.2.2). This idea likewise underlies the other "physical" examples in Section 5.2.

Proposition 5.1.4. *Let* Λ *be a hyperbolic set for a flow* Φ, $\tau \in \{u, s, c, cs, cu\}$. *Then*

- $x \mapsto E_x^\tau$ *is* Φ*-invariant and continuous,*

- $\dim E_x^\tau$ *is locally constant, and*

- *the subspaces* E_x^τ *are pairwise uniformly transverse for* $\tau = u, s, c$: *there is* $\alpha_0 > 0$ *such that for any* $x \in \Lambda$, *the angle between* $\xi \in E_x^\tau$ *and* $\eta \in E_x^{\tau'}$ *is at least* α_0 *when* $\tau \neq \tau'$.

Proof. This holds trivially at any fixed point. Elsewhere, the inequalities $\|D\varphi^t \xi\| \leq C\lambda^t \|\xi\|$ invariantly characterize E_x^s, and similarly for $\tau \in \{u, cs, cu\}$. By continuity of $D\varphi^t$ the set of (x, ξ) on which they hold is closed, so $\lim_{x \to x_0} E_x^\tau \subset E_{x_0}^\tau$. Then $\dim E_{x_0}^u + \dim E_{x_0}^s = \dim M - 1 = E_x^u + E_x^s$ implies that neither inclusion is strict, so $E_{x_0}^\tau = \lim_{x \to x_0} E_x^\tau$.

Since the angle between $\xi \in E_x^\tau$ and $\eta \in E_x^{\tau'}$ is continuous and positive ($E_x^\tau \cap E_x^{\tau'} = \{0\}$) it has a positive minimum. □

We note that one can do better than continuity: Theorem 5.1.16 establishes Hölder continuity (Definition 1.9.6). The next lemma produces a metric such that we can take $C = 1$ in Definition 5.1.1. Such a metric is called an *adapted metric* or *Lyapunov metric*.

Proposition 5.1.5 (Adapted metric). *Let Λ be a hyperbolic set for a flow Φ with λ, μ, C as in Definition 5.1.1 and $\underline{\lambda} \in (\lambda, 1)$, $\underline{\mu} \in (1, \mu)$. Then there is a continuous Riemannian metric such that for the induced norm $\|\cdot\|^*$, for $t \geq 0$ and for $x \in \Lambda$ we have*

$$\|D\varphi^t\restriction_{E_x^s}\|^* \leq \underline{\lambda}^t \quad \textit{and} \quad \|D\varphi^{-t}\restriction_{E_x^u}\|^* \leq \underline{\mu}^{-t}.$$

Proof. We adapt the norm on each of the spaces E^s and E^u. For $v \in E_x^s$ define

$$(\|v\|_x^s)^2 := \underbrace{\int_0^\infty \underline{\lambda}^{-2s} \left(\|D\varphi^s v\|_{\varphi^s(x)}\right)^2 ds}_{\leq \int_0^\infty \underline{\lambda}^{-2s} C\lambda^{2s} \|v\|_x ds < \infty}.$$

As an integral of quadratic forms, this is a quadratic form and hence the norm arises from an inner product. This is the desired norm on E_x^s because if $v \in E_x^s$ and $t > 0$, then

$$\left(\|D\varphi^t v\|_{\varphi^t x}^s\right)^2 = \underbrace{\int_0^\infty \underline{\lambda}^{-2s} \left(\|D\varphi^{t+s} v\|_{\varphi^{t+s} x}\right)^2 ds}_{= \underline{\lambda}^{2t} \int_0^\infty \underline{\lambda}^{-2(t+s)} (\|D\varphi^{t+s} v\|_{\varphi^{t+s} x})^2 ds}$$

$$= \underline{\lambda}^{2t} \underbrace{\int_t^\infty \underline{\lambda}^{-2s} \left(\|D\varphi^s v\|_{\varphi^s(x)}\right)^2 ds}_{\leq \int_0^\infty \underline{\lambda}^{-2s} (\|D\varphi^s v\|_{\varphi^s(x)})^2 ds} \leq \underline{\lambda}^{2t} \left(\|v\|_x^s\right)^2.$$

Similarly, the desired metric on E^u is

$$(\|v\|_x^u)^2 = \int_0^\infty \underline{\mu}^{-2s} \left(\|D\varphi^{-s} v\|_{\varphi^{-s}(x)}\right)^2 ds.$$

For $v = v_s + v_u \in E_x^s \oplus E_x^u$, where $v_\tau \in E_x^\tau$, let $(\|v\|_x^*)^2 := (\|v_s\|_x^s)^2 + (\|v_u\|_x^u)^2$; this is a metric on $E_x^s \oplus E_x^u$ with $E^s \perp E^u$. For the nonfixed points in Λ one can extend this to a metric in the center direction. For $v = v_c + v_s + v_u \in T_x M$ where $v_s \in E_x^s$, $v_u \in E_x^u$, and $v_c \in E_x^c$, let $(\|v\|_x^*)^2 := (\|v_s\|_x^s)^2 + (\|v_u\|_x^u)^2 + (\|v_c\|_x)^2$. This induces a metric on $T_x M$, which is continuous since the components are continuous.

Furthermore, this metric can be extended to a continuous metric on all of M and changed into a smooth Riemannian metric on all of M by a perturbation so small as to preserve the defining inequalities. □

Checking that a given set is hyperbolic for a flow involves the challenge of finding the invariant subbundles E^u and E^s. Outside of algebraic situations, it is not clear how to go about that. Fortunately, it turns out that approximate knowledge of these suffices, and in practice, one can establish that a set is hyperbolic by using cone fields.

Definition 5.1.6. For a set $X \subset M$ with a splitting $T_xM = E_x \oplus F_x$ for each $x \in X$, and for $\beta \in (0,1)$ the *β-cone field* consists of the *β-cone*

$$C_\beta(E,F) = \{v + w \mid v \in E,\ w \in F,\ \|w\| < \beta\|v\|\}$$

of E_x and F_x at each $x \in X$.

Proposition 5.1.7 (Alekseev Cone Field Criterion). *A compact Φ-invariant set Λ is hyperbolic if and only if there exist constants $\lambda, \beta \in (0,1)$, $C \geq 1$, and a decomposition $T_x\Lambda = S_x \oplus E^c \oplus U_x$ for each $x \in \Lambda$ such that for all $x \in \Lambda$ and all $t > 0$ we have*

- *$E^c_x = \mathbb{R}X(x) \neq \{0\}$ for all nonfixed points $x \in \Lambda$ and $E^c_x = \{0\}$ for the fixed points, where X is the generating vector field,*
- $D\varphi^t\left(\overline{C_\beta(U_x, E^c_x \oplus S_x)}\right) \subset C_\beta(U_{\varphi^t(x)}, E^c_{\varphi^t(x)} \oplus S_{\varphi^t(x)})$,
- $D\varphi^{-t}\left(\overline{C_\beta(S_x, E^c_x \oplus U_x)}\right) \subset C_\beta(S_{\varphi^t(x)}, E^c_{\varphi^t(x)} \oplus U_{\varphi^t(x)})$,
- $\|D\varphi^t\xi\| \leq C\lambda^t\|\xi\|$ *for* $\xi \in C_\beta(S_x, E^c_x \oplus U_x)$, *and*
- $\|D\varphi^{-t}\xi\| \leq C\lambda^t\|\xi\|$ *for* $\xi \in C_\beta(U_x, E^c_x \oplus S_x)$.

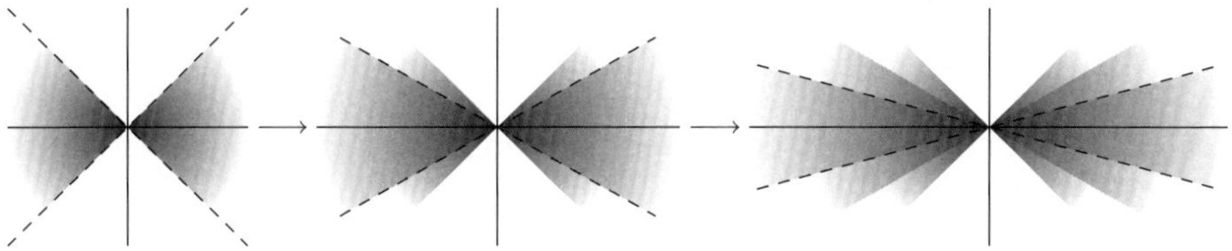

Figure 5.1.1. Invariant cones.

Proof. "Only if" is an easy consequence of the definitions.

"If": We show that

$$E^u_x := \bigcap_{t>0} D\varphi^t\left(\overline{C_\beta(U_{\varphi^{-t}(x)}, E^c_{\varphi^{-t}(x)} \oplus S_{\varphi^{-t}(x)})}\right)$$

and

$$E^s_x := \bigcap_{t>0} D\varphi^{-t}\left(\overline{C_\beta(S_{\varphi^t(x)}, E^c_{\varphi^t(x)} \oplus U_{\varphi^t(x)})}\right)$$

are as in Definition 5.1.1. By construction, they are expanded and contracted, respectively, and equivariant, so we need only show that these are linear subspaces of the right dimension. To that end, let S_x^∞ be an accumulation point of $(D\varphi^t(S_{\varphi^{-t}(x)}))_{t>0}$ in the following sense: by compactness of the unit sphere, orthonormal bases in $D\varphi^t(S_{\varphi^{-t}(x)})$ accumulate to a frame, and we denote its linear hull by $S_x^\infty \subset E_x^u$. Then $\dim S_x^\infty = \dim S_x$. Defining $T_x^\infty \subset E_x^s$ in like manner, we now show that with the definitions above, $E_x^u = S_x^\infty$, and a like argument then gives $E_x^s = T_x^\infty$.

If $v \in E_x^u$, then $v = v^u + v^{cs}$, where $v^u \in S_x^\infty$ and $v^{cs} \in E_x^c \oplus E^s$, and there is a $K \in \mathbb{R}$ such that

$$\|v^{cs}\| = \|D\varphi^t(D\varphi^{-t}(v - v^u))\| \le K \underbrace{\|D\varphi^{-t}(v - v^u)\|}_{=\|D\varphi^{-t}(v) - D\varphi^{-t}(v^u)\| \le C\lambda^t(\|v\| + \|v^u\|)} \xrightarrow[t\to\infty]{} 0. \qquad \square$$

The subbundles S and T in Proposition 5.1.7 need not be invariant. They simply need to be close to an invariant subbundle by a factor of β. This flexibility makes them easily extendable with the same defining properties, so while it follows directly from the definitions that every closed invariant subset of a hyperbolic set for Φ is also a hyperbolic set, the Cone Field Criterion allows us to conclude more interestingly, that one can sometimes envelop a given hyperbolic set by a larger one.

Proposition 5.1.8 (Persistence of hyperbolicity). *A compact hyperbolic set $\Lambda \subset X$ for a flow Φ has a neighborhood $U \subset X$ such that $\Lambda_\Phi^U := \bigcap_{t\in\mathbb{R}} \varphi^t(\overline{U})$ is a hyperbolic set*[3] *and, moreover, so is $\Lambda_\Psi^U := \bigcap_{t\in\mathbb{R}} \psi_t(\overline{U})$ when Ψ is sufficiently C^1-close to Φ.*

Proof. Let Λ be a hyperbolic set for a flow Φ. First assume we have an adapted metric on Λ with hyperbolic constants $\lambda \in (0,1)$ and $\mu > 1$, and fix $\underline{\lambda} \in (\lambda, 1)$ and $\underline{\mu} \in (1, \mu)$. Extend the splitting on Λ to a continuous splitting (not necessarily invariant) in a sufficiently small neighborhood V of Λ, and fix $\beta > 0$ sufficiently small and cones $C_\beta(E_x^s, E_x^c \oplus E_x^u)$ and $C_\beta(E_x^u, E_x^c \oplus E_x^s)$. If $x \in \Lambda$ and $t > 0$, then

$$D\varphi^{-t} C_\beta(E_x^s, E_x^c \oplus E_x^u) \subset C_{\lambda^t\beta}(E_x^s, E_x^c \oplus E_x^u)$$

and

$$D\varphi^t C_\beta(E_x^u, E_x^c \oplus E_x^s) \subset C_{\mu^{-t}\beta}(E_x^u, E_x^c \oplus E_x^s).$$

Also, we can choose V and β such that

$$\|D\varphi^{-t}\xi\| \le \underline{\mu}^{-t}\|\xi\| \text{ for } \xi \in C_\beta(E_x^u, E_x^c \oplus E_x^s)$$

and

$$\|D\varphi^t\eta\| \le \underline{\lambda}^t\|\eta\| \text{ for } \eta \in C_\beta(E_x^s, E_x^c \oplus E_x^u).$$

For a possibly smaller neighborhood U of Λ and $x \in \overline{U}$ the conditions in Proposition 5.1.7 hold not only for Φ, but also any flow Ψ that is C^1-close to Φ. $\square$

[3] With the notation of Definition 1.4.18, $\Lambda_\Phi^U = A_U \cap R_U$.

Although Proposition 5.1.8 does not assert that $\Lambda_\Psi^U \neq \varnothing$, the next result is a direct consequence.

Corollary 5.1.9. *Any sufficiently small C^1-perturbation of an Anosov flow is an Anosov flow.*

Remark 5.1.10. Thus, the magnetic flows from Remark 2.2.10 are Anosov flows when the magnetic field is weak enough.

Unfortunately, in this observation and in Proposition 5.1.8 itself, there is no control over how large a perturbation one can allow. Nor is there in Theorem 5.2.19 below, and this is generally a difficulty in hyperbolic dynamics. (But Theorem 5.2.20 is a rare instance where this is possible.)

The last "generic" application of the Alekseev Cone Criterion is rather basic:

Theorem 5.1.11 (Hyperbolicity of time changes). *Let Λ be a hyperbolic set for a flow Φ. If Ψ is a smooth time change of Φ, then Λ is a hyperbolic set for Ψ.*

Proof. Write $\psi^t(x) = \varphi^{\alpha(t,x)}(x)$ as in Proposition 1.2.2 with $\alpha(0,\cdot) = 0$. Choose for each $x \in \Lambda$ local coordinates $x = (x^c, x^u, x^s)$ centered at x and adapted to the splitting $T_xM = E_x^c \oplus E_x^u \oplus E_x^s$ so that with respect to these coordinates

$$D\varphi^t(0) = \begin{pmatrix} 1 & 0 & 0 \\ 0 & A_t & 0 \\ 0 & 0 & B_t \end{pmatrix}$$

with $\|B_t\| \le \lambda^t < 1$ and $\|A_t^{-1}\| \le \mu^{-t} < 1$. In these coordinates

$$D\psi^t(0) = \begin{pmatrix} 1 & \alpha_{x^u}(t,x) & \alpha_{x^s}(t,x) \\ 0 & A_{\alpha(t,x)} & 0 \\ 0 & 0 & B_{\alpha(t,x)} \end{pmatrix},$$

where $\alpha_{x^u}(t,x)$ and $\alpha_{x^s}(t,x)$ are the partial derivatives of α with respect to x^u and x^s, respectively. By compactness of Λ there is a linear upper bound $Kt > 0$ for their size when $t > 0$. To prove hyperbolicity of ψ^t we use the Cone Criterion. We write vectors in $T_x\Lambda = E_x^c \oplus E_x^u \oplus E_x^s$ as (u, v, w) with $u \in E_x^c$, $v \in E_x^u$, $w \in E_x^s$ and let

$$\|u, v, w\|^2 := \epsilon^2\|u\|^2 + \|v\|^2 + \|w\|^2,$$

where a sufficiently small $\epsilon > 0$ will be specified later. For $\gamma < \sqrt{\mu^2 - 1}$ we now check whether the γ-cone given by

$$\epsilon^2\|u\|^2 + \|w\|^2 \le \gamma^2\|v\|^2$$

is $D\psi^t$-invariant for $t \in [0,1]$. Take ϵ such that

$$K^2t^2\epsilon^2 + \lambda^{2\alpha(t,x)} \le 1 \text{ for } t \in [0,1].$$

If $(u', v', w') = D\psi^t(u, v, w)$ then

$$\begin{aligned}
\epsilon^2\|\underbrace{u'}_{\substack{=u+\alpha_x u\, v \\ +\alpha_x s\, w}}\|^2 + \|\underbrace{w'}_{=B_{\alpha(t,x)}w}\|^2 &\le \epsilon^2\,(\|u\| + Kt\|v\| + Kt\|w\|)^2 + \lambda^{2\alpha(t,x)}\|w\|^2 \\
&= \overbrace{\epsilon^2\|u\|^2 + (\underbrace{K^2t^2\epsilon^2 + \lambda^{2\alpha(t,x)}}_{\le 1})\|w\|^2}^{\le\gamma^2\|v\|^2} \\
&\quad + \epsilon^2 Kt\Big(Kt\|v\|^2 + 2\underbrace{\|u\|}_{\le\gamma\|v\|/\epsilon}\|v\| + 2\underbrace{\|u\|}_{\le\gamma\|v\|/\epsilon}\underbrace{\|w\|}_{\le\gamma\|v\|} + 2Kt\|v\|\underbrace{\|w\|}_{\le\gamma\|v\|}\Big) \\
&\le \gamma^2\Big(1 + \frac{\epsilon Kt}{\gamma^2}(\epsilon Kt(1+2\gamma) + 2\gamma(1+\gamma))\Big)\|v\|^2 \\
&< \gamma^2\mu^{2\alpha(t,x)}\|v\|^2 \le \gamma^2\|v'\|
\end{aligned}$$

for sufficiently small $\epsilon > 0$ and $t \in (0,1]$. Thus γ-cones are ψ^t-invariant. To check that vectors in γ-cones expand, note that $\epsilon^2\|u'\|^2 + \|w'\|^2 \ge \delta^{\alpha(t,x)}(\epsilon^2\|u\|^2 + \|w\|^2)$ for some $\delta > 0$ and take $\gamma > 0$ small enough so that

$$\frac{\mu^{2\beta} + \delta^\beta\gamma^2}{1+\gamma^2} \ge \eta^\beta, \quad \text{that is, } (\delta^\beta - \eta^\beta)\gamma^2 + \mu^{2\beta} \ge \eta^\beta \tag{5.1.1}$$

for some $\eta > 1$ and all $\beta > 0$. Then if $\epsilon^2\|u\|^2 + \|w\|^2 \le \gamma^2\|v\|^2$ we have

$$\begin{aligned}
\underbrace{\epsilon^2\|u'\|^2 + \|w'\|^2}_{\ge\delta^{\alpha(t,x)}(\epsilon^2\|u\|^2+\|w\|^2)} + \underbrace{\|v'\|^2}_{\ge\mu^{2\alpha(t,x)}\|v\|^2} &\ge \overbrace{\delta^{\alpha(t,x)}}^{=\eta^{\alpha(t,x)}+(\delta^{\alpha(t,x)}-\eta^{\alpha(t,x)})}\underbrace{(\epsilon^2\|u\|^2 + \|w\|^2)}_{\le\gamma^2\|v\|^2} + +\mu^{2\alpha(t,x)}\|v\|^2 \\
&\ge \eta^{\alpha(t,x)}(\epsilon^2\|u\|^2 + \|w\|^2) \\
&\quad + \underbrace{[(\delta^{\alpha(t,x)} - \eta^{\alpha(t,x)})\gamma^2 + \mu^{2\alpha(t,x)}]}_{\ge\eta^{\alpha(t,x)} \text{ by (5.1.1)}}\|v\|^2 \\
&\ge \eta^{\alpha(t,x)}(\epsilon^2\|u\|^2 + \|v\|^2 + \|w\|^2).
\end{aligned}$$

Since ψ^{-t} is a time change of φ^{-t} there is a corresponding cone family for ψ^{-t}. □

Remark 5.1.12. We give another proof later (page 333) that uses stable and unstable manifolds.

We conclude with variations on the Cone Criterion. The first is a mere change in formalism, which expresses the Alekseev Cone Criterion in terms of Lorentz metrics that behave analogously to Lyapunov functions or metrics. The second is a simplification for area-preserving flows.

Definition 5.1.13. A *Lorentz metric* is a nondegenerate bilinear form g with signature $(n-1,1)$, so $n-1$ eigenvalues of the quadratic form $Q(x) := g(x,x)$ are positive and one is negative.[4]

Proposition 5.1.14. *A smooth flow $\varphi^t : M \to M$ of a 3-manifold M is an Anosov flow if and only if there are two continuous Lorentz metrics Q^+ and Q^- on M and constants $a, b, c, T > 0$ such that*

(1) *for all $v \in T_xM$, $t > T$, if $Q^\pm(v) > 0$ then $Q^\pm(D_x\varphi^{\pm t}(v)) > ae^{bt}Q^\pm(v)$,*

(2) *$C^+ \cap C^- = \varnothing$, where $C^\pm$ is the $Q^\pm$-positive cone,*

(3) *$Q^\pm(X) = -c$ where X is the generating vector field,*

(4) *$D_x\varphi^{\pm T}(\overline{C^\pm(x)}) \smallsetminus \{0\} \subset C^\pm(\varphi^{\pm T}(x))$.*

Proof. If φ^t is an Anosov flow we can choose disjoint cones around the strong-stable and strong-unstable directions, neither of which contains X. These define (up to a factor) the Lorentz metrics, and choosing $c = 1$ fixes the metrics; we omit the details.

Assume now the above conditions for two continuous Lorentz metrics $Q^\pm$ and constants $a, b, c, T > 0$. The cone fields $C^\pm$ induce fields $\mathcal{E}^\pm$ of ellipses in the projectivization PTM of TM, and φ^t acts on fields of ellipses by $(\varphi^t_*\mathcal{E})(x) := PD_{\varphi^{-t}(x)}\varphi^t(\mathcal{E}(\varphi^{-t}(x)))$. Then

- condition (2) implies that $\mathcal{E}_t^+(x) \cap \mathcal{E}_t^-(x) = \varnothing$,
- condition (4) implies that $\overline{\mathcal{E}_T^\pm(x)} \subset \operatorname{int}\mathcal{E}^\pm(x)$.

If we endow each $\mathcal{E}^\pm(x)$ with the Hilbert metric then this last property (strict nesting) implies that $D\varphi^{\pm T}$ induces contractions $\mathcal{E}^\pm(x) \to \mathcal{E}^\pm(\varphi^{\pm T}(x))$ of the Hilbert metrics with a factor that can be chosen uniformly by compactness of M. Thus, the diameter of $\mathcal{E}_t^\pm(x) \subset \mathcal{E}^\pm(x)$ as measured by the Hilbert metric on $\mathcal{E}^\pm(x)$ goes to 0 exponentially, so $\Delta^\pm(x) := \bigcap_{t>T}\mathcal{E}_t^\pm(x)$ are points, and $\Delta^+(x) \neq \Delta^-(x)$ for all $x \in M$ since $\mathcal{E}_t^+(x) \cap \mathcal{E}_t^-(x) = \varnothing$.

Clearly $\Delta^\pm$ define φ^t-invariant line fields $E^\pm$, and since $X \notin C^\pm$ by condition (3), $\Delta^+(x) \neq X(x) \neq \Delta^-(x)$.

[4]By Sylvester's law of inertia, this is independent of the choice of such a basis. A Riemannian metric is a like form of signature $(n, 0)$, that is, positive definite.

Now choose a continuous Riemannian metric on M whose unit spheres intersect $E^\pm$ in points for which $Q^\pm = 1$. Then condition (1) implies that $E^\pm$ are exponentially expanding and contracting, respectively, as required. □

Lastly, we consider a variant of the Cone Criterion, which in turn is useful in other applications. The Cone Criterion requires us to check two things: invariance of cones and expansion of vectors in them. In the case of planar cones, it seems plausible that (strict) invariance of cones under *area-preserving* dynamics would cause vectors in cones to expand under the dynamics as area is "squeezed out." This is indeed the case and will be useful later. We establish this in dimension 2 using convenient local coordinates in which the invariant cones are given as the first and third quadrant.

Theorem 5.1.15 (Wojtkowski Cones). *Suppose $A_k = \begin{pmatrix} a_k & b_k \\ c_k & d_k \end{pmatrix}$ are matrices such that for some $\epsilon > 0$, all $k \in \mathbb{Z}$ and all $v = \binom{x}{y}$ with $xy > 0$ we have $|\det A_k| \geq 1$ and*

$$A_k v \in C_\epsilon := \left\{\binom{x}{y} \in \mathbb{R}^2 \mid \epsilon y \leq x \leq y/\epsilon\right\}.$$

Then there are $c > 0$ and $\lambda > 1$ such that

$$\|A_{k-1} \dots A_{k-i} v\| \geq c\lambda^i \|v\|$$

for all $k \in \mathbb{Z}$, $i \in \mathbb{N}$ and $v \in C_\epsilon$.

Proof (Wojtkowski, Kourganoff). Since $\frac{\epsilon}{2}(x^2 + y^2) \leq P(\binom{x}{y}) := xy \leq \frac{2}{\epsilon}(x^2 + y^2)$ for $\binom{x}{y} \in C_\epsilon$, we check the conclusion for $\sqrt{P}$ instead of $\|\cdot\|$. Specifically, we show $P(A_k v) \geq \frac{1}{1-\epsilon^2} P(v)$ for $v \in C_\epsilon$ and $k \in \mathbb{Z}$.

Without loss of generality $\det A_k > 0$ (otherwise left-multiply A_k by $\left(\begin{smallmatrix} 0 & 1 \\ 1 & 0 \end{smallmatrix}\right)$) and all entries of A_k are positive (otherwise multiply by $-\mathrm{Id}$), so

$$1 \leq a_k d_k - b_k c_k \leq \frac{1}{\epsilon} b_k \frac{1}{\epsilon} c_k - b_k c_k = \left(\frac{1}{\epsilon^2} - 1\right) b_k c_k = \frac{1 - \epsilon^2}{\epsilon^2} b_k c_k$$

since $\binom{a_k}{c_k} = A_k \binom{1}{0} \in C_\epsilon$ and $\binom{b_k}{d_k} = A_k \binom{0}{1} \in C_\epsilon$ by continuity. This implies that $b_k c_k \geq \frac{\epsilon^2}{1-\epsilon^2} = \frac{1}{1-\epsilon^2} - 1 > \frac{1/2}{1-\epsilon^2} - \frac{1}{2}$. For $v = \binom{x}{y} \in C_\epsilon$ we thus have

$$\begin{aligned} P(A_k v) &= (a_k x + b_k y)(c_k x + d_k y) \geq (a_k d_k + b_k c_k) xy \\ &= \underbrace{(a_k d_k - b_k c_k)}_{\geq 1} xy + 2 b_k c_k xy \geq (1 + 2 b_k c_k) P(v) \geq \frac{1}{1 - \epsilon^2} P(v). \end{aligned}$$ □

Proposition 5.1.4 noted that the defining subbundles are automatically continuous, and we now establish that they are indeed Hölder continuous. This result is formulated in discrete time in such a way as to be applicable to the time-1 map of a flow and to give information about both the weak- and strong-(un)stable subbundles. In this form it is due to Brin and Stuck [70].

Theorem 5.1.16. *Suppose $f\colon M \to M$ is a $C^{1+\alpha}$-diffeomorphism of a compact manifold and admits a (λ, μ)-splitting*:

$$\|Df^n(x)v^s\| \le C\lambda^n\|v^s\| \quad \textit{and} \quad \|Df^n(x)v^u\| \ge C^{-1}\mu^n\|v^u\|$$

for $v^s \in E^s(x)$, $v^u \in E^u(x)$, and $n \in \mathbb{N}$. If $b := b_1^2+H$, where $b_1 := \max_{x\in M}\|Df(x)\| \ge 1$ and H is the Hölder coefficient of Df, then E^s is $\alpha\frac{\log\mu-\log\lambda}{\log b-\log\lambda}$-Hölder continuous.

Remark 5.1.17. We note that it is not necessary to assume $\lambda < 1$ or $\mu > 1$.

First we show that the "contracting" subspaces of close maps must be close.

Lemma 5.1.18. *If $L_n^i\colon \mathbb{R}^N \to \mathbb{R}^N$ are linear maps for $i = 1, 2$, $n \in \mathbb{N}$ and there are $b > 0$, $\delta \in (0, 1)$, $C > 1$, $\mu > \lambda < b$ and subspaces $E^1, E^2 \subset \mathbb{R}^N$ such that $\|L_n^1 - L_n^2\| \le \delta b^n$,*

$$\|L_n^i(v)\| \begin{cases} \le C\lambda^n\|v\| & \textit{for } v \in E^i, \\ \ge C^{-1}\mu^n\|v\| & \textit{for } v \perp E^i, \end{cases}$$

then $d(E^1, E^2) \le 3C^2\frac{\mu}{\lambda}\delta^{\frac{\log\mu-\log\lambda}{\log b-\log\lambda}}$.

Proof. Since $\gamma := \lambda/b < 1$ there is a unique $n \in \mathbb{N}$ for which $\gamma^{n+1} < \delta \le \gamma^n$. For $v \in E^2$ we then have

$$\begin{aligned}\|L_n^1(v)\| \le \|L_n^2(v)\| + \|L_n^1 - L_n^2\|\|v\| &\le C\lambda^n\|v\| + \delta b^n\|v\| \\ &\le (C\lambda^n + (b\gamma)^n)\|v\| \le 2C\lambda^n\|v\|.\end{aligned}$$

Writing $v = v^1 + v^\perp \in E^1 \oplus E^{1^\perp}$ gives

$$\|L_n^1(v)\| = \|L_n^1(v^1 + v^\perp)\| \ge \|L_n^1(v^\perp)\| - \|L_n^1(v^1)\| \ge C^{-1}\mu^n\|v^\perp\| - C\lambda^n\|v^1\|,$$

hence

$$\|v - v^1\| = \|v^\perp\| \le C\mu^{-n}(\|L_n^1(v)\| + C\lambda^n\|v^1\|) \le 3C^2\Big(\frac{\lambda}{\mu}\Big)^n\|v\|,$$

so $d(v, E^1) \le 3C^2(\lambda/\mu)^n\|v\|$, which implies $d(E^1, E^2) \le 3C^2\frac{\mu}{\lambda}\big(\frac{\lambda}{\mu}\big)^{n+1}$ by symmetry. Now $x := \frac{\log\mu-\log\lambda}{\log b-\log\lambda} \Rightarrow \frac{\lambda}{\mu} = \gamma^x \Rightarrow \big(\frac{\lambda}{\mu}\big)^{n+1} = (\gamma^{n+1})^x < \delta^x$, hence the claim. □

In preparation for our application of this lemma we note how the bound "δb^n" arises from a $C^{1+\alpha}$-diffeomorphism.

Lemma 5.1.19. *If f is a $C^{1+\alpha}$-diffeomorphism of a manifold $M \subset \mathbb{R}^N$, $\|x-y\|<1$, and $n \in \mathbb{N}$, then*

$$\|Df^n(x) - Df^n(y)\| \le b^n \|x-y\|^\alpha$$

with $b := b_1^2 + H$, where $b_1 := \|Df(\cdot)\|_\infty \ge 1$ and H is the Hölder coefficient of Df.

Proof. First a remark about Hölder continuity: if $\|g(x) - g(y)\| \le H\|x-y\|^\alpha$ whenever $\|x-y\| \le 2$, then for $\|x-y\| > 1$ we subdivide a path from x to y into $\lfloor \|x-y\| \rfloor + 1$ equal pieces to get

$$\|g(x) - g(y)\| \le \sum_{i=0}^{\lfloor \|x-y\| \rfloor} \|g(x_i) - g(x_{i+1})\| \le \sum_{i=0}^{\lfloor \|x-y\| \rfloor} H \le H\|x-y\|,$$

so Hölder continuity is characterized by

$$\|g(x) - g(y)\| \le H \max(\|x-y\|^\alpha, \|x-y\|) \text{ for all } x, y.$$

We now prove the claim by induction, noting that the case $n=1$ is clear and that $\|f^n(x) - f^n(y)\| \le b_1^n \|x-y\|$. Then

$$\begin{aligned}
&\|Df^{n+1}(x) - Df^{n+1}(y)\| \\
&\quad \le \|Df(f^n(x))\|\|Df^n(x) - Df^n(y)\| + \|Df(f^n(x)) - Df(f^n(y))\|\|Df^n(y))\| \\
&\quad \le b_1 b^n \|x-y\|^\alpha + H\max((b_1^n\|x-y\|)^\alpha, b_1^n\|x-y\|)b_1^n \\
&\quad \le \underbrace{[b_1 b^n + H b_1^{2n}]}_{=b^n[b_1 + H(b_1^2/b)^n] < b^n[b_1+H] \le b^n[b_1^2+H] = b^{n+1}} \|x-y\|^\alpha.
\end{aligned}$$

□

Having intimated how the first inequality in Lemma 5.1.18 arises, hyperbolicity now uses the other two for the main result.

Proof of Theorem 5.1.16. As implied by Lemma 5.1.19, we assume $M \subset \mathbb{R}^N$ (without loss of generality by the Whitney Embedding Theorem) and prove smoothness of $E^s \oplus TM^\perp$ by applying our lemmas to $L(x) := Df(x) \oplus 0\colon T_xM \oplus T_xM^\perp \to \mathbb{R}^N$ as follows. Set $L_n(x) := L(f^{n-1}(x)) \circ \cdots \circ L(f(x)) \circ L(x)$ and note that on one hand $L_n(x)\restriction_{T_xM} = Df^n(x)$ and on the other hand by compactness of M we can adjust C such that $v \perp E^s(x) \Rightarrow \|Df^n(x)(v)\| \ge C^{-1}\mu^n\|v\|$. Thus, by Lemma 5.1.19 we can apply Lemma 5.1.18 for $\delta = \|x_1 - x_2\|^\alpha < 1$ to $L_n^i := L_n(x_i)$ and $E^i := E^s(x_i)$. □

In the spirit of Remark 5.1.17 we point out another circumstance where a much weaker assumption than hyperbolicity produces information about invariant subbundles. Specifically, pointwise hyperbolicity implies uniform hyperbolicity as follows.

Theorem 5.1.20 ([181]). *Let f be a C^1-diffeomorphism of a compact smooth manifold M with a compact invariant set K on which there is a proper Baire cone family*[5] *$C_x \subset T_xM$, and*

$$\varliminf_{n\to\infty} \frac{1}{n} \min_{v\in C_x^1} \log(\|D_x f^n(v)\|) > 0$$

for all $x \in K$. Then there exist $c, \chi > 0$ and a continuous cone family $\{K_x\}_{\in K}$ such that

$$\|D_x f^n(v)\| \geq c \cdot e^{\chi n} \text{ for every } x \in K,\ v \in K_x, \text{ and } n \in \mathbb{N}.$$

Indeed, for every $x \in K$ there is a subspace $E_x \subset T_xM$ such that $Df(E_x) = E_{f(x)}$ and

$$\|D_x f^n(v)\| \geq c \cdot e^{\chi n} \text{ for every } v \in C_x \text{ and } n \in \mathbb{N}.$$

(The obvious corollary that together with a like assumption on a stable counterpart this yields uniform hyperbolicity—of a flow by using the time-1 map—is much older [322].) This means that any reasonable notion of nonuniform hyperbolicity must allow nonhyperbolic points.

5.2 Physical flows: Geodesic flows, magnetic flows, billiards, gases, and linkages

While in the examples of Remark 5.1.3 the definition of hyperbolicity was easily checkable directly, the Cone Criterion in particular provides a convenient way to establish hyperbolicity of various classes of "mechanical" flows (beyond Remark 2.2.10), and these are explored in this section. Specifically, we show that geodesic flows of negatively curved manifolds are Anosov flows (Theorem 5.2.4) and substantially weaken the needed hypotheses in the case of surfaces (Theorem 5.2.8); the same approach then establishes hyperbolicity of dispersing billiards (Theorem 5.2.26), and these are in turn connected to the gas models that motivated Maxwell and Boltzmann (Theorem 5.2.41). Finally, we describe Anosov flows which are mechanical in a way that could be made into an actual desktop model (Theorem 5.2.47).

5.2.a Geodesic flows. Hyperbolic geodesic flows were studied extensively in Chapter 2. Those flows were of an algebraic nature, which makes the defining conditions easy to check directly (Remark 2.2.2). We now go beyond the geodesic flow on the hyperbolic plane and its compact factors in Chapter 2. This requires a little differential geometry (which is less important for our purposes than the results). Geodesic

[5] A cone family that is the pointwise limit of continuous cone families.

flows of negatively curved manifolds are an important example both historically and mathematically. Indeed, as mentioned in the preface, the concept of an Anosov flow arose as Anosov axiomatized the arguments used in working with geodesic flow on manifolds of negative sectional curvature. Indeed, the motivation was that these are mechanical (in particular, physical) systems because this represents the motion of a free particle on the manifold. From that point of view it is natural to think of geodesic flows as Hamiltonian flows for the Hamiltonian $H(x,v) = \frac{1}{2}g(v,v)$, which is the kinetic energy (only). Here g is the Riemannian metric.

To formally introduce the geodesic flow in full generality, that is, on any compact Riemannian manifold M, we take a more direct geometric approach as follows. The *geodesic equation* is a suitable way to write $\ddot{\gamma} = 0$, that is, zero acceleration, which corresponds to free-particle motion, where $\dot{\gamma}$ is the tangent vector to a curve $t \mapsto \gamma(t)$, and the second derivative can be expressed using the *Levi-Civita connection* ∇ or *Riemannian covariant derivative*:

$$\nabla_{\dot{\gamma}}\dot{\gamma} = 0. \tag{5.2.1}$$

This, then, defines a flow on the unit tangent bundle of M as before.

To introduce curvature, which has an essential effect on the dynamics, let R be the curvature tensor defined by

$$R(u,v)w := \nabla_v\nabla_u w - \nabla_u\nabla_v w + \nabla_{[u,v]}w.$$

Then $\langle R(u,v)w,x\rangle = \langle R(w,x)u,v\rangle$ and $R(u,u) = 0$ for $u,v,w,x \in T_pM$. If $u,v \in T_pM$ are linearly independent, then the *sectional curvature*

$$K(S) := \frac{\langle R(u,v)u,v\rangle}{\langle u,u\rangle\langle v,v\rangle - \langle u,v\rangle^2}$$

depends only on the 2-plane $S \subset T_pM$ spanned by the vectors[6] u and v and is the Gaussian curvature at p of the 2-manifold $\exp_p S$ with respect to the Riemannian metric induced from M, where exp is the Riemannian exponential map.[7] We usually assume that this is always negative and hence, by compactness, bounded from above by $-k^2 < 0$.

Jacobi fields help discern the effect of curvature on the dynamics of the geodesic flow. For a geodesic $\gamma : \mathbb{R} \to M$, a Jacobi field $Y : t \mapsto Y(t) = \frac{\partial V}{\partial s}$ is an "*infinitesimal variation*" arising from a *geodesic variation* $V : \mathbb{R} \times (a,b) \to M$, where each $\gamma_s := V(\cdot,s)$ is a geodesic with $\gamma_0 = \gamma$.

[6]Because changing to a base (u',v') of S can be accomplished by repeated application of the steps $(u,v) \mapsto (v,u)$, $(u,v) \mapsto (au,v)$ and $(u,v) \mapsto (u+av,v)$, none of which change K.

[7]Defined by $\exp_p(tv) = \gamma_v(t)$ for unit vectors v and t near 0, where γ_v is the geodesic with $\gamma_v(0) = p$, $\dot{\gamma}_v(0) = v$.

Proposition 5.2.1. *An infinitesimal variation is a solution of the Jacobi equation*

$$\ddot{Y}(t) + \underbrace{K(t)}_{:=R(\dot\gamma(t),\cdot)\dot\gamma(t)} Y(t) = 0, \tag{5.2.2}$$

where dots denote differentiation with respect to t.

Proof. Since $[\frac{\partial V}{\partial t}, \frac{\partial V}{\partial s}] = 0$ we have $\nabla_{\frac{\partial V}{\partial t}} \frac{\partial V}{\partial s} = \nabla_{\frac{\partial V}{\partial s}} \frac{\partial V}{\partial t}$. Thus, for $s = 0$ we have

$$\begin{aligned}
\ddot{Y} = \nabla_{\frac{\partial V}{\partial t}} \nabla_{\frac{\partial V}{\partial t}} \frac{\partial V}{\partial s} &= \nabla_{\frac{\partial V}{\partial t}} \nabla_{\frac{\partial V}{\partial s}} \frac{\partial V}{\partial t} \\
&= -\Big(\underbrace{\nabla_{\frac{\partial V}{\partial s}} \nabla_{\frac{\partial V}{\partial t}} \frac{\partial V}{\partial t}}_{=0 \text{ (geodesic equation)}} - \nabla_{\frac{\partial V}{\partial t}} \nabla_{\frac{\partial V}{\partial s}} \frac{\partial V}{\partial t} - \nabla_{[\frac{\partial V}{\partial s}, \frac{\partial V}{\partial t}]} \frac{\partial V}{\partial t}\Big) \\
&= -R\Big(\frac{\partial V}{\partial s}, \frac{\partial V}{\partial t}\Big)\frac{\partial V}{\partial t} = -R(\dot\gamma, Y)\dot\gamma.
\end{aligned}$$

□

Conversely, we have the following proposition:

Proposition 5.2.2. *Solutions of the Jacobi equation are infinitesimal variations.*

Proof. If Y is a solution of the Jacobi equation along γ, let $h_i(s)$ for $|s| < \epsilon$ be curves with $(h_i(0), h_i'(0)) = (\gamma(t_i), Y(t_i))$ for $i = 1, 2$. If ϵ and $t_1 - t_2$ are small enough then for all s there is a unique shortest geodesic $V(\cdot, s)$ from $h_1(s)$ to $h_2(s)$. Also, Y and the vector field $X = \frac{\partial V}{\partial s}$ along γ are solutions of the Jacobi equation that agree at t_1 and t_2, hence everywhere because they solve the same second-order differential equation. □

Remark 5.2.3 (Orthogonal Jacobi fields). A tangential Jacobi field is of the form $Y(t) = f(t)\dot\gamma(t)$ with $\ddot f(t) = 0$ (since $\ddot\gamma(t) = 0 = K(t)\dot\gamma(t)$) and hence linear in time. On the other hand, the projection Y_T onto $\mathbb{R}\dot\gamma$ of any Jacobi field Y is of the same form with $f(t) = \langle Y(t), \dot\gamma(t)\rangle$. But $\ddot f = \langle \ddot Y, \dot\gamma\rangle = -\langle K\dot\gamma, Y\rangle = 0$ and thus the tangential projection Y^T of Y is a Jacobi field. By linearity of the Jacobi equation the same holds for $Y^\perp := Y - Y^T \perp \dot\gamma$. Another way to represent *orthogonal Jacobi fields* is to note that if $Y(t)$ is a Jacobi field along a geodesic γ and both $Y(t_0)$ and $\dot Y(t_0)$ are orthogonal to $\dot\gamma(t_0)$ for some t_0, then $Y(t)$ and $\dot Y(t)$ are orthogonal to $\dot\gamma(t)$ for all t. We denote the set of orthogonal Jacobi fields by $\mathcal{J}(\gamma)$.

If $\dim(M) = n$ and γ is a geodesic in M, then the dimension of the space of Jacobi fields along γ is $2n$. The space of orthogonal Jacobi fields is then $(2n-2)$-dimensional since the space of tangential Jacobi fields is 2-dimensional.[8]

[8] If we restrict to the unit tangent bundle, that is, to unit-speed geodesics, then the dimension of the space of Jacobi fields is $2n - 1$ and the space of tangential Jacobi fields is 1-dimensional, so the space of orthogonal Jacobi fields is $(2n - 2)$-dimensional in either case.

We now make more precise how the behavior of Jacobi fields reflects the dynamics of the geodesic flow g^t. For $p \in M$, $v \in T_pM$ denote by γ_v the geodesic with $\gamma_v(0) = p$, $\dot\gamma_v(0) = v$. Then there are isomorphisms

$$\psi_v\colon T_vTM \to T_pM \oplus T_pM, \quad \xi \mapsto (x, x') \quad \text{with } \psi_{g^t v}(Dg^t\xi) = \big(Y(t), \dot Y(t)\big),$$

where Y is the Jacobi field along γ_v with $Y(0) = x$ and $\dot Y(0) = x'$.

Theorem 5.2.4. *The geodesic flow of a compact Riemannian manifold with negative sectional curvature is an Anosov flow.*

Proof. We establish the cone conditions for Proposition 5.1.7 by connecting curvature and the Jacobi equation (5.2.2), with Lemma 5.2.5 as the key step.

Let M be a compact Riemannian manifold with tangent bundle TM, unit tangent bundle $SM := \{v \in TM \mid \|v\| = 1\}$, and geodesic flow $g^t : SM \to SM$. Its dynamics can be described in terms of the evolution of Jacobi fields, that is, we can describe an action of g^t (or Dg^t, rather) on Jacobi fields. Two linearly independent tangential Jacobi fields with linear growth correspond to affine reparametrizations of the geodesic, that is, shifts of the initial point and uniform changes of speed. The first variation corresponds to the flow direction for the geodesic flow on the unit tangent bundle SM; the second is transverse to SM. Thus, in order to establish that the geodesic flow on SM is an Anosov flow it is sufficient to show that the space of orthogonal Jacobi fields admits a splitting into exponentially contracting and exponentially expanding invariant subspaces.

To study orthogonal Jacobi fields it suffices to know that they are solutions of the Jacobi equation (5.2.2) and that the operator K in that equation is negative definite and symmetric: the curvature assumption together with compactness implies the existence of $k, \kappa > 0$ such that $-k^2$ is an upper bound for the sectional curvature and

$$\langle KY, Y\rangle \le -k^2\langle Y, Y\rangle \text{ when } Y \perp \dot\gamma, \quad \text{and} \quad \langle KY, KY\rangle < \frac{1}{\kappa^2} \text{ for } Y \in SM.$$

To show hyperbolicity of the geodesic flow define a new norm on $T_pM \oplus T_pM$ by $\|u, v\| := \sqrt{\langle u, u\rangle + \epsilon\langle v, v\rangle}$ for $u, v \in T_pM$ and for some fixed $\epsilon < 1/\kappa$, and note that $0 \le \langle u - v, u - v\rangle = \langle u, u\rangle - 2\langle u, v\rangle + \langle v, v\rangle$ and hence

$$2\epsilon\langle u, v\rangle \le \epsilon\langle u, u\rangle + \epsilon\langle v, v\rangle \le \|u, v\|^2.$$

Then $\langle Y, \dot Y\rangle / \|Y, \dot Y\|^2 \ge \delta$ defines a cone in the sense of Definition B.5.6, and from the discussion above we see that the cone families $C^+ = \{Y \in \mathcal{J}(\gamma) \mid \langle Y, \dot Y\rangle \ge 0\}$ and $C^- = \{Y \in \mathcal{J}(\gamma) \mid \langle Y, \dot Y\rangle \le 0\}$ can be equivalently defined by

$$C^\pm_{(p,v)} = \{(x, x') \in T_pM \oplus T_pM \mid \langle x, \dot\gamma_v(0)\rangle = 0,\ \langle x', \dot\gamma_v(0)\rangle = 0,\ \pm\langle x, x'\rangle \ge 0\}.$$

Lemma 5.2.5. *The family of cones C^+_δ given by $\langle Y,\dot Y\rangle \ge 0$ is strictly invariant, and vectors in it grow exponentially in time.*

Proof. Since

$$\frac{d}{dt}\langle Y,\dot Y\rangle = \langle \dot Y,\dot Y\rangle + \underbrace{\langle Y,\ddot Y\rangle}_{=-\langle R(Y,\dot\gamma)\dot\gamma,Y\rangle=-\langle K(t)Y,Y\rangle\ge k^2\langle Y,Y\rangle} \ge \overbrace{\langle \dot Y,\dot Y\rangle + k^2\langle Y,Y\rangle}^{>0 \text{ unless } Y=0=\dot Y} \underbrace{\ge}_{0\le\langle \dot Y-kY,\dot Y-kY\rangle=\langle\dot Y,\dot Y\rangle-2k\langle Y,\dot Y\rangle+k^2\langle Y,Y\rangle} 2k\langle Y,\dot Y\rangle \ge 0,$$

$(d\varphi^t)(C^+(x)) \subset \operatorname{int}\left(C^+_{\varphi^t x}\right)$ and[9]

$$\|Y(t),\dot Y(t)\|^2 \ge \frac{1}{2\epsilon}\langle Y(t),\dot Y(t)\rangle \ge \frac{1}{2\epsilon}e^{2kt}\langle Y(0),\dot Y(0)\rangle \ge \frac{\delta}{2\epsilon}e^{2kt}\|Y(0),\dot Y(0)\|. \qquad \square$$

One could likewise show that C^- is strictly invariant and expanding in negative time, but this now follows from reversibility of the geodesic flow (Remark 1.1.30): by definition,

$$g^{-t}(v) = -g^t(-v). \tag{5.2.3}$$

We thus obtain a splitting $T_vSM = S_v \oplus E^c_v \oplus U_v$ and cones satisfying the conditions of Proposition 5.1.7 to obtain Theorem 5.2.4. □

Remark 5.2.6. Jacobi fields not only determine cone fields as above but also the stable and unstable subbundles. To that end consider the orthogonal Jacobi vector field determined (uniquely) by the boundary-value problem $Y_s(0) = v$, $Y_s(s) = 0$ for any $v \perp \dot\gamma(0)$. Then $Y := \lim_{s\to+\infty} Y_s$ (pointwise) is a *stable Jacobi field*, that is, with $Y(t) \xrightarrow[t\to+\infty]{} 0$. Stable Jacobi fields define infinitesimal variations of pairwise forward-asymptotic geodesics, that is, stable vectors. A like construction gives unstable Jacobi fields.

Later on (Section 6.2) we likewise obtain stable and unstable manifolds (sets of positively or negatively asymptotic geodesics), whereas Propositions 2.1.10 and 2.2.1 did so by using the algebraic structure in an essential way.

It is plausible that having negative curvature everywhere is not strictly needed for Theorem 5.2.4, and in his seminal papers on ergodicity of geodesic flows of negatively curved surfaces, Hopf recognized the essential features of hyperbolicity and commented on the possibility of even allowing some positive curvature.[10] We instead explore how much flatness can be allowed for surfaces by developing more carefully the mechanism that gives hyperbolicity (Theorem 5.2.8).

[9] Note that k appears below, when k^2 arose as a curvature bound; the dynamical growth and contraction rates are indeed related to curvature data via square roots.

[10] He specifically illustrated this by giving explicit finitary geometric criteria to control the effects of positive curvature [198, p. 593f].

The technical ingredient is to "projectivize" the action on Jacobi fields. In the 2-dimensional case orthogonal Jacobi fields are represented by the scalars $y = \langle Y, n\rangle$, where n is a unit normal vector field to the geodesic, and the Jacobi equation becomes $\ddot{y} + Ky = 0$. Where $y \neq 0$ we can projectivize this to $u := \dot{y}/y$, which then satisfies the *Riccati equation*

$$\dot{u} = \frac{d}{dt}\frac{\dot{y}}{y} = \frac{\ddot{y}y - (\dot{y})^2}{y^2} = -K - u^2,$$

with $K(t)$ the Gauss curvature at $\gamma(t)$, as before.[11]

Proposition 5.2.7. *The geodesic flow g^t of a closed surface M is an Anosov flow if there is an $m > 0$ such that for any solution u of the Riccati equation along any geodesic $\gamma\colon [0,1] \to M$ with $u(0) = 0$, we have $u(1) \geq m$ (and u is defined on $[0,1]$).*

Proof. For $v \in S_xM$ let $\gamma = \gamma_v$ be the geodesic with $\gamma(0) = x$ and $\dot{\gamma}(0) = v$ and choose a smooth orthogonal basis $(\dot{\gamma}, e_1, e_2)$ at each $\gamma(t)$. It suffices to check that

$$A_k = D_{(\gamma(k),\dot{\gamma}(k))}g^1 \text{ on } \dot{\gamma}(k)^\perp$$

with respect to the basis (e_1, e_2) is as in Theorem 5.1.15, and since $|\det A_k| = 1$ (g^t is volume preserving), it suffices to show that with $\epsilon := \min(1/4, K_{\max}, m) > 0$ all solutions u of the Riccati equation along a geodesic $\gamma\colon [0,1] \to M$ with $u(0) > 0$ are defined on $[0,1]$ and satisfy $\epsilon \leq u(1) \leq 1/\epsilon$. Here $-K_{\max}$ is the minimum of the Gauss curvature. (Theorem 5.1.15 gives expanding cones, and (5.2.3) then gives the contracting ones.)

The easy direction is that $u(1) \geq \epsilon$: if u_0 is the solution with $u_0(0) = 0$, then $u(1) \geq u_0(1) \geq m \geq \epsilon$ by assumption.

The other inequality follows by contradiction: Suppose $u(1) > 1/\epsilon$. Then $u(t) > 1/\epsilon$ for $t \in [0,1]$ because $u(t) > 1/\epsilon \Rightarrow \dot{u}(t) \leq (1/\epsilon) - u(t)^2 < 0$. Thus, $\dot{u} \leq (1/\epsilon) - u^2$ and hence

$$-\frac{d}{dt}\frac{1}{u} = \frac{\dot{u}}{u^2} \leq \frac{1}{\epsilon u^2} - 1 \leq -\frac{1}{2}, \quad \text{so} \quad \frac{1}{u(1)} > \frac{1}{u(1)} - \frac{1}{u(0)} \geq \frac{1}{2} > \epsilon,$$

contrary to our assumption. □

We now give a curvature condition that implies the hypotheses of Proposition 5.2.7 and hence hyperbolicity. If the Gauss curvature is zero at each point of a geodesic, then the Jacobi equation shows that this geodesic is not hyperbolic. Remarkably, the existence of such a geodesic is the only obstruction to hyperbolicity of the geodesic flow:

[11] This generalizes to higher dimension by considering a symmetric operator U defined by $\dot{Y} = UY$ and obtaining a Riccati equation for it.

Theorem 5.2.8 ([126, Corollary 3.6], [228]). *The geodesic flow of a closed nonpositively curved Riemannian surface is Anosov if (and only if) every geodesic contains a point where the curvature is negative.*

Lemma 5.2.9. *With this assumption there are $M, T > 0$ such that every unit-speed geodesic γ satisfies $\int_0^t K(\gamma(s))\,ds \le -M$ for $t \ge T$.*

Proof. Otherwise there are geodesics γ_n on $[-n,n]$ with $\int_{-n}^{n} K(\gamma(t))\,dt \ge -1/n$. By the Arzelà–Ascoli Theorem a subsequence converges uniformly on each $[-n,n]$ to a geodesic γ on $\mathbb{R}$ with $\int_{\mathbb{R}} K(\gamma(t))\,dt = 0$ (Dominated-Convergence Theorem). □

Proof of Theorem 5.2.8 (Kourganoff). Take $M, T < 1$ as in Lemma 5.2.9 ($T < 1$ by possibly scaling the metric). To check the hypotheses of Proposition 5.2.7 let u be the solution of the Riccati equation along a geodesic γ for which $u(0) = 0$. Showing that u is defined on (at least) $[0,1]$ is the main effort and yields a uniform lower bound for $u(1)$ as a by-product.

If u is defined on $[0,1]$, then let $t_1 = 1$; otherwise there is a $t_1 \in (0,1]$ such that $[0,t_1)$ is the maximal interval on which u is defined.

Let $t_2 := \sup\{t \in [0,t_1] \mid u(t) \ge M\} \in [0,t_1]$ and $t \in [t_2,t_1)$. Then

$$u(t) = u(t_2) + \int_{t_2}^{t} \dot u(s)\,ds = u(t_2) - \int_{t_2}^{t} K(s)\,ds - \int_{t_2}^{t} u^2(s)\,ds.$$

If $t_2 = 0$, this gives

$$u(t) = 0 - \int_{t_2}^{t} \underbrace{K(s)}_{\le 0}\,ds - \int_{t_2}^{t} u^2(s)\,ds \ge \begin{cases} -M^2, & \\ M - M^2 & \text{if } t = 1 \text{ (Lemma 5.2.9).}\end{cases}$$

Otherwise,

$$u(t) = \underbrace{u(t_2)}_{\ge M} - \int_{t_2}^{t} \underbrace{K(s)}_{\le 0}\,ds - \int_{t_2}^{t} \underbrace{u^2(s)}_{\le M^2}\,ds \ge M - M^2;$$

that is, $u(t) \ge -M^2$ in either case, which means that u is defined on an open interval around t of uniform size, so $t_1 = 1$, and u is defined on $[0,1]$.

With this in hand, the preceding shows that $u(1) \ge M - M^2 > 0$, as needed. □

One reason one might in the case of surfaces be interested in the extent to which *positive* curvature is allowed is that a compact surface isometrically embedded in $\mathbb{R}^3$ has points of positive curvature (for instance, any point touching a smallest sphere that contains the embedded surface). Michael Herman asked whether there are compact surfaces in $\mathbb{R}^3$ with Anosov geodesic flow. The answer turns out to be affirmative [122].

If one takes a sufficiently large and sufficiently thin spherical shell with sufficiently many hyperboloid-like holes drilled through from outside to inside in a dense-enough pattern, the hyperbolicity produced from the negative curvature in the holes outweighs the small positive curvature between encounters with such holes. This raises new questions. Can this be done with surfaces of low genus? Can it be done with genus 2?[12]

The following does not address the question as posed but provides a visually appealing counterpart *in* S^3. The spherical billiard shown on the left of Figure 5.2.1 is uniformly hyperbolic on its regular set by Theorem 5.2.26 below, and Theorem 5.2.49 below then implies that a sufficiently "thin" surface as shown on the right of Figure 5.2.1 (embedded isometrically in S^3 and presented here in stereographic projection to $\mathbb{R}^3$) has Anosov geodesic flow [227]. Regrettably, one cannot make this construction work in $\mathbb{R}^3$; there are necessarily conjugate points.[13]

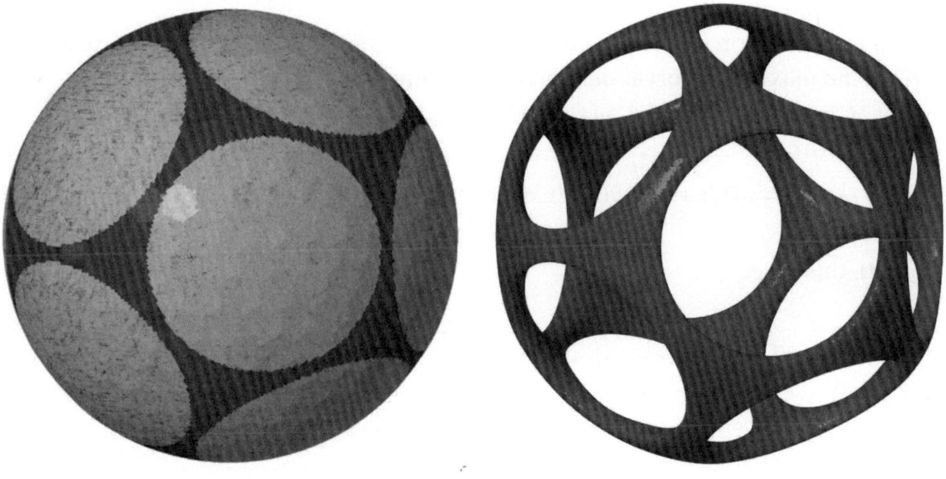

Figure 5.2.1. A genus-11 Anosov surface projected stereographically from S^3 (from http://mickael.kourganoff.fr/images/poster.pdf; see also [227]).

Remark 5.2.10. One may ask which smooth manifolds admit metrics whose geodesic flow is Anosov. Klingenberg ([222], see also [281, Section 2.6]) proved that for such a manifold M there are no conjugate points, every closed geodesic has index 0, the universal cover is a disk, $\pi_1(M)$ has exponential growth, every nontrivial abelian subgroup of $\pi_1(M)$ is infinite cyclic, the geodesic flow is ergodic, and periodic orbits

[12] Progress toward bounding the needed genus has been made only just now [121].

[13] Two distinct zeros of a Jacobi field are referred to as *conjugate points*.

are dense. These are the main properties known for geodesic flows of negatively curved manifolds, so this result suggests that Anosov manifolds may admit a negatively curved metric. (It should be noted that Anosov metrics need not be negatively curved [197].)

Eberlein [126] has done much work to characterize Anosov metrics. For example, for compact (M, g) without conjugate points he describes stable and unstable spaces of perpendicular Jacobi fields and shows equivalence of complementarity of these, the metric being Anosov, and vanishing of all bounded perpendicular Jacobi fields. If g has no focal points, that is, the length of a perpendicular Jacobi field with initial value zero is increasing, then uniform exponential growth of all such perpendicular Jacobi fields implies that g is Anosov. Other connections between hyperbolic dynamics and Riemannian geometry are surveyed in [224].

5.2.b Benoist–Hilbert geodesic flows. We now introduce geodesic flows that differ significantly from those we previously encountered because they do not arise from a Riemannian metric, and they manifest the distinction in remarkable ways. However, Chapter 2 provides good preparation, and the approach here is similarly explicit in the way it uses "pedestrian" computations. We will introduce these flows in the proper context and establish that they are indeed Anosov flows. Without entering their study more deeply, we point to some of their particularly interesting features.

Definition 5.2.11 (Projective convexity, divisibility). Let $\mathrm{PGL}(\mathbb{R}^m)$ be the group of projective transformations of the projective space $\mathbb{P}(\mathbb{R}^m)$,[14] that is, $\mathrm{GL}(\mathbb{R}^m)$ modulo homotheties. An open set $\Omega \subset \mathbb{P}(\mathbb{R}^m)$ is said to be *convex* if it intersects each projective line in a connected set, *projectively (or properly) convex*, if there is moreover a projective hyperplane that does not intersect the closure of Ω, and *strictly convex* if every projective line intersects the boundary $\partial\Omega$ in at most two points. A projectively convex Ω is said to be *divisible* if there is a discrete torsion-free[15] subgroup $\Gamma < \mathrm{PGL}(\mathbb{R}^m)$ that preserves Ω and with compact quotient $M = \Gamma\backslash\Omega$ (following Furstenberg and Benoist we say that Γ *divides* Ω.) One can prove and we will use that $\partial\Omega$ is C^1.

Example 5.2.12. The *ellipsoid* $\Omega_0 := \{[v] \in \mathbb{P}(\mathbb{R}^m) \mid q(v) > 0\}$, where q is a quadratic form on $\mathbb{R}^m$ with signature $(1, m-1)$ is strictly convex and divided by any cocompact lattice in its isometry group $\mathrm{SO}(q)$.

[14] This can be viewed as the space of lines through $0 \in \mathbb{R}^m$, the points on the unit sphere in $\mathbb{R}^m$, or the equivalence classes of $\mathbb{R}^m \smallsetminus \{0\}$ modulo collinearity.

[15] That is, only the identity has finite order.

Definition 5.2.13 (Hilbert distance and geodesic flow). The *Hilbert distance* d_Ω on a projectively convex $\Omega \subset \mathbb{P}(\mathbb{R}^m)$ is defined by $d_\Omega(x,y) := |\log((a,b;x,y))|$, where

$$(a,b;x,y) := \frac{x-a}{x-b}\Big/\frac{y-a}{y-b} = \frac{(x-a)(y-b)}{(x-b)(y-a)}$$

is the *cross ratio* with $a,b \in \partial\Omega$ such that a, b, x, y lie on the line $\langle x,y\rangle$ through $x \neq y$. The differences are interpreted as real numbers (Euclidean distances), and this

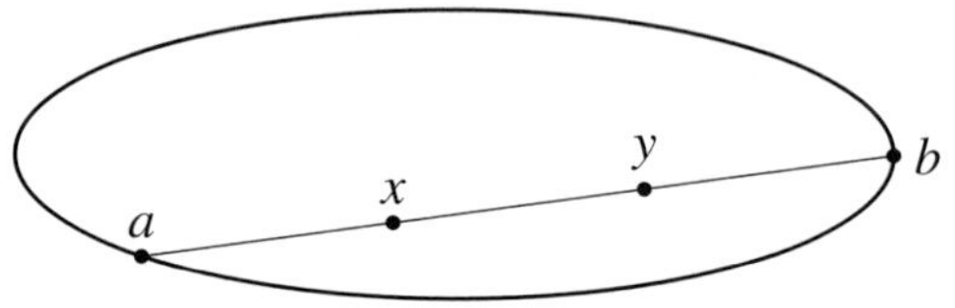

Figure 5.2.2. The cross ratio.

distance is invariant under all Ω-preserving projective transformations. This implies that the shortest curve between any two points of Ω is a line segment. The *geodesic flow* on Ω is defined by $\tilde{g}^t(x,\xi)$ being the unit tangent vector in the direction of ξ at the point x_t at distance t from x on the line through x defined by ξ. Its projection g^t is called the geodesic flow on $M = \Gamma\backslash\Omega$.

As promised, we will prove that these geodesic flows are in scope for us:

Theorem 5.2.14. *The geodesic flow of a compact factor of a divisible strictly convex subset of* $\mathbb{P}(\mathbb{R}^m)$ *is an Anosov flow.*

We will prove this result later in this section. We now explore the dynamical features of these flows, analogously to Chapter 2. Theorem 5.2.14 will then be a rather easy consequence.

We begin by making the notions from Definition 5.2.13 more explicit and amenable to computation. First we take a global affine chart, that is, we may assume that $\overline{\Omega} \subset \mathbb{R}^{m-1} \subset \mathbb{P}(\mathbb{R}^m)$, a suitable affine hyperplane, so Ω is a bounded convex subset of $\mathbb{R}^{m-1}$, and the tangent bundle is $\Omega \times \mathbb{R}^{m-1}$. Define C^1-maps

$$\left.\begin{aligned} p\colon T\Omega &\to \Omega,\\ p^\pm\colon T\Omega &\to \partial\Omega,\\ \sigma^\pm\colon T\Omega &\to (0,\infty),\end{aligned}\right\} \text{ by } \sigma^+(w)(p^+(w)-x) = \xi = \sigma^-(w)(x-p^-(w)) \text{ and } p(\underbrace{w}_{=(x,\xi)\in T\Omega\smallsetminus\{0\}}) = x.$$

This allows us to define the *Hilbert norm* $\|w\|_\Omega := \sigma^+(w) + \sigma^-(w)$ of vectors $w = (x,\xi) \in T\Omega \smallsetminus \{0\}$. The maps $p, p^\pm, \|\cdot\|_\Omega$ are independent of the affine chart, and the *unit tangent bundle* is $SM = \{w \in T\Omega \mid \|w\|_\Omega = 1\}$. One can check that the geodesic flow (unit-speed motion along lines) on $S\Omega$ is thus given by

$$\tilde{g}^t(w) =: w_t = (x_t, \xi_t) = \left(x + \frac{e^t - 1}{\sigma^+(w)e^t + \sigma^-(w)}\xi, \frac{e^t}{(\sigma^+(w)e^t + \sigma^-(w))^2}\xi\right).$$

Remark 5.2.15. This shows in particular, that this is a C^1-flow—and no more regular than that unless the boundary is more smooth. One can improve this by way of reparametrization as follows. For a smooth Γ-invariant Riemannian metric g on Ω follow Hilbert geodesics with constant g-speed.

The following will turn out to be a foliation by center-stable sets for the geodesic flow:

$$\widetilde{W}^{cs}(w) := (p^+ \restriction_{S\Omega})^{-1}(p^+(w)) \quad \text{for } w \in S\Omega,$$

the collection of unit vectors pointing to the same boundary point. Isolating strong-stable leaves geometrically requires a little extra work.

Claim 5.2.16.

$$\begin{aligned}\widetilde{W}^s(w) &:= \{v_1 = (x_1, \xi_1) \in \widetilde{W}^{cs}(w) \mid w_1 = w \text{ or } \langle x, x_1\rangle \cap \langle p^-(w), p^-(w_2)\rangle \subset \underbrace{\mathcal{T}_{p^+(w)}}_{:=\text{tangent space to the boundary}}\} \\ &= \{v_1 = (x_1, \xi_1) \in S\Omega \mid d_\Omega(\underbrace{p(\tilde{g}^t(w))}_{=:x_t}, \underbrace{p(\tilde{g}^t(w_1))}_{=:x_{1,t}}) \xrightarrow[t\to\infty]{} 0\}.\end{aligned}$$

Proof. This is a nice application of the fact that a cross ratio is naturally defined for a set of lines in the following sense: if points a, b, x, y are collinear and four lines are drawn from a distinct point $\{q\}$ to these, then for any other line not through q with corresponding intersection points A, B, X, Y, the cross ratios agree, that is, $(a, b; x, y) = (A, B; X, Y)$, and conversely, their agreement implies that the four lines through a and A etc. are concurrent.

Note first that $p^+(w) \neq p^+(w_1) \Rightarrow d_\Omega(x_t, x_{1,t}) \xrightarrow[t\to\infty]{} \infty$, so we may assume $p^+(w) = p^+(w_1) =: p_1^+$. Then

$$(p_1^+, p^-(w); x, x_t) = e^t = (p_1^+, p^-(w_1); x_1, x_{1,t})$$

implies (see Figure 5.2.3) that

$$\langle x_t, x_{1,t}\rangle \ni q := \langle p^-(w), p^-(w_1)\rangle \cap \langle x, x_1\rangle.$$

Since the line $\langle x_t, x_{1,t}\rangle$ converges to $\langle p_1^+, q\rangle$, this implies the claim:

$$d_\Omega(x_t, x_{1,t}) \xrightarrow[t\to\infty]{} 0 \Leftrightarrow \langle p_1^+, q\rangle \text{ is tangent to } \partial\Omega \Leftrightarrow q \in \mathcal{T}_{p_1^+}. \qquad \square$$

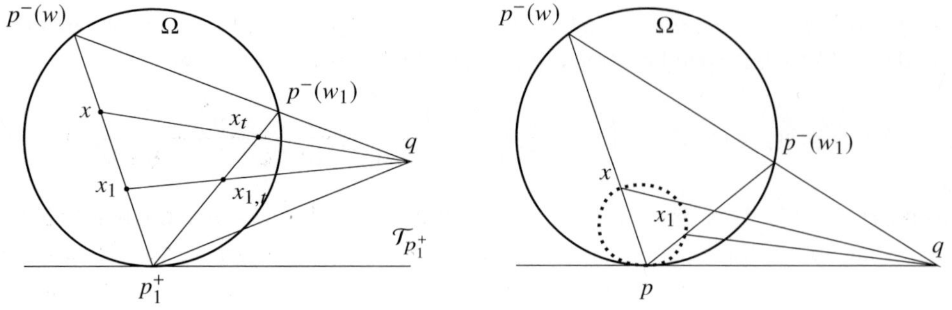

Figure 5.2.3. Strong-stable leaves and Busemann functions.

The (footpoint) projection in Ω of a strong-stable set (leaf) is a horocycle, and we now give alternate descriptions of this. One is as the limit of d_Ω-spheres through $x \in \Omega$ as their centers tend to $p \in \partial\Omega$. Another is as a 0-level set of a *Busemann function*:

$$\mathcal{H}_{(x,\xi)} = \{x_1 \in \Omega \mid b_x(x_1, p^+(x,\xi)) = 0\},$$

where the Busemann function b is defined on $\Omega \times \Omega \times \partial\Omega$ by

$$b_{x_1}(x_2, p) := \lim_{x \to p} \big(d_\Omega(x_1, x) - d_\Omega(x_2, x)\big)$$

or equivalently (see Figure 5.2.3) as the logarithm of the cross ratio of the four lines through $q := \mathcal{T}_p \cap \langle p_1^-, p_2^- \rangle$ and p, p_1^-, x_2, x_1, respectively, where p_i^- is the other boundary point on $\langle x_i, p \rangle$.

While the *definition* of the stable subbundle E^s as the tangent bundle of the stable foliation is universal, the explicit formulas in this context give an equally explicit representation of this subbundle:

$$E^s_w = \{(y, -\sigma^+(w)y) \in T_w S\Omega \mid y \in T_x \mathcal{H}_w\},$$
$$E^u_w = \{(y, \sigma^-(w)y) \in T_w S\Omega \mid y \in T_x \mathcal{H}_w\}.$$

By construction, these are Γ- and g^t-invariant, and $TS\Omega = E^s \oplus \mathbb{R}X \oplus E^u$.

Proof of Theorem 5.2.14. The flip map $v \mapsto -v$ conjugates the geodesic to its reverse, so it suffices to check that vectors in E^s contract exponentially. To that end we reduce to considerations in $T\Omega$ by observing that the existence of a compact factor implies that for any Riemannian norm $\|\cdot\|$ on $S\Omega$ there is a $C \geq 1$ such that

$$\frac{1}{C}\|(y, -\sigma^+(w)y)\| \leq \|y\|_\Omega \leq C\|(y, -\sigma^+(w)y)\|.$$

Thus, it suffices to show that for $\lambda \in (0,1)$ there is a $T > 0$ as follows: if $(y, -\sigma^+(w)y) \in E^s_w$, hence

$$D\tilde{g}^t((y, -\sigma^+(w)y)) = (y_t, -\sigma^+(\tilde{g}^t(w))y_t),$$

then $\|y_T\|_\Omega \le \lambda \|y\|_\Omega$.

To that end an explicit description of y_t suggested by Figure 5.2.3 helps: writing $w = (x,\xi)$ and $w_t := \tilde{g}^t(w) = (x_t, \xi_t)$, we find that y_t is the unique vector tangent to the horosphere $\mathcal{H}_{w_t}$ such that $p^+(w)$, $x + y$, and $x_t + y_t$ lie on a line.

Since $\partial\Omega$ is strictly convex, the map $t \mapsto \|y_t\|_\Omega = \sigma^+(y_t) + \sigma^-(y_t)$ is strictly decreasing, and indeed to 0 as $t \to \infty$ since $\partial\Omega$ is C^1. Thus, writing $E^s_1 := \{v \in E^s \mid \|v\| = 1\}$, the function $F\colon E^s_1 \times \mathbb{R} \to \mathbb{R}$, $(v,t) \mapsto \|y_t\|_\Omega / \|y\|_\Omega$ is continuous, decreasing in t with $F(\cdot, 0) \equiv 1$ and $F(v,t) \xrightarrow[t\to\infty]{} 0$, so there is a unique (and continuous and Γ-invariant) $\tau\colon E^s_1 \to (0,\infty)$ such that $F(v, \tau(v)) = \lambda$. Take $T := \max \tau$. □

Remark 5.2.17. It is nontrivial to show that there are examples of this type beyond Example 5.2.12; this is a substantial part of the work of Benoist [45, 39, 41, 40, 42, 43, 44]. Another is the investigation of what can occur if one does not require strict convexity. Benoist showed that there are nontrivial instances of this and studied their features. From the dynamical point of view, including their smooth ergodic theory, the definitive study at this time is due to Bray [65].

5.2.c Magnetic and Finsler geodesic flows.

We can now develop Remark 2.2.10 and Remark 5.1.10 by studying magnetic flows in greater generality.

Definition 5.2.18 (Magnetic flow). On a Riemannian manifold M with an antisymmetric tensor $\mathfrak{m}\colon TM \to TM$ (that is, $\langle \mathfrak{m}v_1, v_2\rangle + \langle v_1, \mathfrak{m}v_2\rangle = 0$ for all $v_1, v_2 \in T_xM$ and all $x \in M$)[16] consider the flow defined on SM by the following counterpart to the geodesic equation (5.2.1):

$$\nabla_{\dot\gamma}\dot\gamma = \mathfrak{m}\dot\gamma.$$

We note that unlike geodesic flows, which are reversible (see (5.2.3)), these flows are not.

The size of $\mathfrak{m}$ is a natural measure of the size of the magnetic perturbation to the geodesic flow, and Corollary 5.1.9 implies that for small $\mathfrak{m}$ a magnetic flow built from an Anosov geodesic flow is itself an Anosov flow:

Theorem 5.2.19. *A magnetic flow on a Riemannian manifold whose geodesic flow is Anosov is itself Anosov for sufficiently small* $\mathfrak{m}$ *(Definition* 5.2.18*).*

Thus, we have broadened the class of "physical" Anosov flows beyond geodesic flows.

[16] In Remark 2.2.10 this was a 90° rotation combined with a constant scaling.

It is natural to ask how large this magnetic perturbation $\mathfrak{m}$ can be without losing hyperbolicity. While this kind of question is usually difficult and indeed nigh impossible to answer (this is the central difficulty in bounding the genus of the examples at the end of Section 5.2.a), it turns out that here we have the rare case of explicit control of the hyperbolicity domain (in terms of curvature bounds).

Theorem 5.2.20 ([159, Théorème 4.1]). *If the sectional curvatures K of a closed Riemannian manifold M satisfy $-k_2^2 \le K \le -k_1^2 < 0$, then magnetic flows with $\frac{5}{4}\|\mathfrak{m}\|_\infty^2 + \|\nabla\mathfrak{m}\|_\infty < k_1^2$ are Anosov flows.*

Riemannian geodesic flows can be generalized in other directions than magnetic flows. We next note that the notion of a Riemannian metric is more restrictive than Riemann himself initially intended; he suggested that instead of using an inner product in each tangent space, a norm in each tangent space should suffice.[17] This idea was pushed further by Paul Finsler.

Definition 5.2.21. A C^r *Finsler manifold* is a smooth manifold with a (not necessarily symmetric) norm on each tangent space, which depends C^r on the footpoint. More precisely, a C^r Finsler metric is a continuous $F\colon TM \to (0,\infty)$ that is

- C^r on nonzero vectors,
- positive definite ($F(x,v) \ge 0$ with "="$\Leftrightarrow$ $v = 0$),
- positively homogeneous ($F(x,\lambda v) = \lambda F(x,v)$ when $\lambda > 0$),
- strongly convex ($\left(\frac{\partial^2 F^2}{\partial v_i \partial v_j}\right)_{i,j}$ is positive definite).

Further, F is *reversible* if $F(x,-v) = F(x,v)$ for all $(x,v) \in TM$.

Remark 5.2.22. Positive homogeneity ensures that the Finsler length of a curve does not depend on the parametrization. Of course, Riemannian metrics are Finsler metrics.

One way to think of perturbing a Riemannian metric to a Finsler metric is to represent a Finsler metric by its unit balls, a smooth family of strongly convex sets containing 0 in each tangent space; perturbations of Riemannian balls then give examples. An important category of such deformations consists of Randers metrics

[17] But he decided not to pursue this more general framework: "*Die Untersuchung dieser allgemeinern Gattung würde zwar keine wesentlich andere Principien erfordern, aber ziemlich zeitraubend sein und verhältnissmässig auf die Lehre vom Raume wenig neues Licht werfen, zumal da sich die Resultate nicht geometrisch ausdrücken lassen; ich beschränke mich daher auf die Mannigfaltigkeiten, wo das Linienelement durch die Quadratwurzel aus einem Differentialausdruck zweiten Grades ausgedrückt wird.*"—"The study of this more general class would not require any essentially new principles, but be rather time consuming and shed little new light on the theory of space, particularly since the results cannot be expressed geometrically; therefore I confine myself to the manifolds where the line element is expressed by the square root of a differential expression of degree two."

$F = \sqrt{g} + \theta$, where g is a Riemannian metric and θ a 1-form with norm at most 1. One can think of θ as a "wind" that deflects otherwise freely moving particles. Accordingly, these metrics are only reversible when $\theta \equiv 0$.

The Hilbert metrics in Section 5.2.b arise from Finsler metrics as follows. For $(x, v) \in TM$ let $v^{\pm}$ be the points where the projective line determined by v intersects the boundary and set

$$F(x, v) := \frac{1}{\|v\|}\Big(\frac{1}{|x - v^+|} + \frac{1}{|x - v^-|}\Big).$$

Definition 5.2.23. The geodesic flow of a Finsler metric is defined by constant-speed locally distance-minimizing curves or equivalently as the Hamiltonian flow for the Hamiltonian $\frac{1}{2}F^2$ with respect to the usual symplectic form $\omega = -d\theta$, where θ the usual Liouville contact 1-form.

Interestingly, magnetic flows (for sufficiently weak magnetic fields) are time changes of Finsler geodesic flows, where the Finsler metric is a suitable perturbation of the underlying Riemannian metric [96, Corollary 2].

In terms of the second description in Definition 5.2.23, one can add a magnetic field to a Finsler geodesic flow by considering the Hamiltonian flow for the Hamiltonian $\frac{1}{2}F^2$ with respect to the symplectic form $\omega + \pi^*\Omega$ for a smooth closed 2-form Ω. We illustrate that this is a legitimate subject of investigation with the following cocycle-rigidity result.

Theorem 5.2.24 ([107, Theorem B]). *Suppose the magnetic flow Φ of (F, Ω) on a closed connected Finsler manifold M is Anosov and generated by the vector field X. If $h\colon M \to \mathbb{R}$ is smooth and θ is a smooth 1-form on M such that $h + \theta(v) \equiv Xu$ for some smooth $u\colon SM \to \mathbb{R}$, then $h \equiv 0$ and θ is exact.*

This result implies other rigidity results [107] and can be extended beyond Anosov flows[18] [106].

5.2.d Billiards. Billiard flows provided our first example of flows that are naturally represented as flows under a function (Example 1.2.13) because in the cases that then came to mind, the boundary of the billiard table is a global section (Figure 0.2.2). To discuss billiards with hyperbolic behavior we begin with a formal definition of a billiard.

Definition 5.2.25. A smooth *billiard* table B in $R = \mathbb{T}^2$ (which is a good model for motion in a periodic crystal) or $R = \mathbb{R}^2$ is the closure of an open set B° of R whose boundary is a finite disjoint union of smoothly embedded circles called the walls of B.

[18] To "no conjugate points."

A billiard is said to be *dispersing* if every wall γ has negative curvature (that is, if T is the tangent vector of γ and N the normal vector pointing into the table, then $\langle \frac{\partial T}{\partial s}, N\rangle < 0$, where s is the arc-length parameter; Figure 0.2.2 instead has a boundary with positive curvature). The phase space of the billiard is the unit tangent bundle SB° with the billiard flow φ^t defined as in Example 1.2.13 (straight-line motion with optical reflection) except for

- t such that φ^t is at the boundary (while one could adjust the definition in such a way as to make the flow continuous at such points, it cannot be differentiable; moreover, the billiard *map* is discontinuous at such grazing collisions),
- $t \geq T$ if φ^T is tangent to the boundary (this is a removable discontinuity but necessarily a failure of differentiability).

We define the *regular set* Ω to be those points in SB° for which the second possibility (grazing collisions) occurs for no positive or negative time; this is a residual conull flow-invariant set, and φ^t is smooth on it. The billiard table B has *finite horizon* if the boundary is a global section, that is, every orbit meets the boundary.

The *Sinai billiard* is $\mathbb{T}^2$ minus a disk; it is dispersing with infinite horizon. Removing instead disks of diameter $\frac{3}{10}$ and $\frac{19}{20}$ around $(0,0)$ and $(\frac{1}{2}, \frac{1}{2})$, respectively, gives finite horizon. Note the finite-horizon dispersing spherical billiard in Figure 5.2.1 and Exercise 5.17.

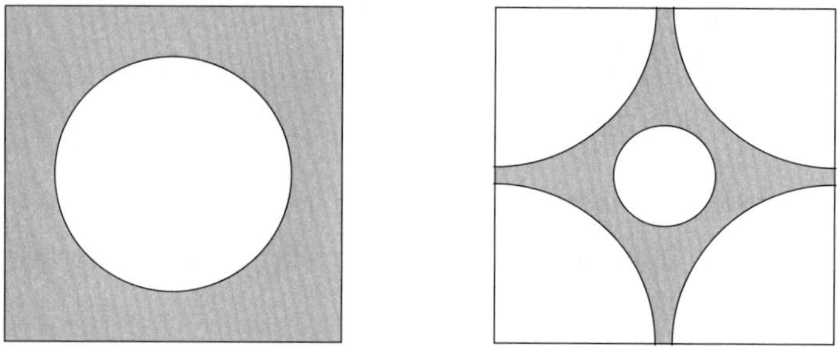

Figure 5.2.4. Dispersing billiards on $\mathbb{T}^2$ (the particle moves in the shaded region): the Sinai billiard and a finite-horizon billiard.

Similarly to Theorem 5.2.8, one can show that finite-horizon dispersing billiards are uniformly hyperbolic away from the collision singularities:

Theorem 5.2.26 ([228]). *The regular set of a finite-horizon dispersing billiard is uniformly hyperbolic, that is, it has all the required properties from Definition* 5.1.1 *except for compactness.*[19]

[19] And there are no fixed points.

Remark 5.2.27. We noted earlier that without compactness, the definition of uniform hyperbolicity may lose much of its interest (Remark 5.1.2). But billiard flows have a compact phase space, so they exhibit the combination of recurrence and hyperbolicity (on a dense hence precompact set) to which Remark 5.1.2 pointed.

Theorem 5.2.26 is obtained by introducing Jacobi fields and a Riccati equation, but in this case hyperbolicity comes from the collisions.

Let $V\colon (a,b)\times(c,d) \longrightarrow M$ be a variation of a billiard orbit $\gamma = V(\cdot,0)$, that is, $V(\cdot,s)$ is a unit-speed billiard orbit for each s with collision times $t_i(s)$. Then $Y := \frac{\partial V}{\partial s}$ is called a Jacobi field where defined. Away from collisions $\ddot{Y} = 0$, and we now investigate the jump discontinuities at collisions. The reflection of a billiard orbit rotates the tangent vector by 2θ, where θ is the angle of incidence; we write $R_{2\theta}$ for rotation by 2θ and now show the counterpart for Jacobi fields.

Lemma 5.2.28. *If Y^- and Y^+ are the values of the Jacobi field before and after collision with incidence angle θ, then $Y^+ = -R_{2\theta}Y^-$.*

Corollary 5.2.29. *Orthogonal Jacobi fields remain orthogonal after a collision.*

Proof. Denote by $\tau(s)$ the time of collision of $s \mapsto V(\cdot,s)$ with a point $\Gamma(r(s))$ of a boundary piece Γ parametrized by arc length. Denote by $\omega^\pm(s)$ the angle between the horizontal axis and $\frac{\partial}{\partial t}\big|_{t=t^\pm} V(t,s)$ (before and after collision). Then $\theta = \frac{1}{2}(\omega^+ - \omega^-)$, and $\psi := \frac{1}{2}(\omega^+ + \omega^-)$ is the angle between the horizontal axis and the tangent to Γ at the collision point. For small s and $t^\pm$ near τ, we then have

$$V(t^\pm,s) = \Gamma(r(s)) + R_{\omega^\pm}(t^\pm - \tau(s)).$$

Differentiating with respect to s at $s = 0$ then gives

$$\begin{aligned} Y(t^\pm) &= \frac{\partial r}{\partial s} R_{\psi(s)}\begin{pmatrix}1\\0\end{pmatrix} - \frac{\partial \tau}{\partial s} R_{\omega^\pm(s)}\begin{pmatrix}1\\0\end{pmatrix} + R_{\pi/2}R_{\omega^\pm}(t^\pm - \tau(s))\frac{\partial \omega^\pm(s)}{\partial s} \\ &\xrightarrow[t^\pm\to\tau]{} \frac{\partial r}{\partial s} R_{\psi}\begin{pmatrix}1\\0\end{pmatrix} - \frac{\partial \tau}{\partial s} R_{\omega^\pm}\begin{pmatrix}1\\0\end{pmatrix}, \end{aligned}$$

so

$$Y^+ + R_{2\theta}Y^- = \frac{\partial r}{\partial s} \underbrace{R_\psi(\mathrm{Id} + R_{2\theta})}_{=2\cos\theta R_{\omega^+}}\begin{pmatrix}1\\0\end{pmatrix} - 2\underbrace{\frac{\partial \tau}{\partial s}}_{=\frac{\partial\tau}{\partial r}\frac{\partial r}{\partial s}} R_{\omega^+}\begin{pmatrix}1\\0\end{pmatrix} = 0. \qquad \square$$

As with geodesic flows, orthogonal Jacobi fields are described by a scalar y using a unit vector field orthogonal to the orbit, and writing $u = \dot{y}/y$, we obtain the Riccati equation $\dot{u} = -u^2$ between collisions, and we have the following lemma:

Lemma 5.2.30. *At a collision, $y^+ = -y^-$, $\dot{y}^+ = -\dot{y}^- + \frac{2\kappa y^-}{\sin\theta}$, and $u^+ = u^- - \frac{2\kappa}{\sin\theta}$, where κ is the curvature of Γ at the collision point (negative for dispersing billiards).*

Proof. That $y^+ = -y^-$ follows from the previous lemma. Next,

$$\dot{y}^+ + \dot{y}^- = \frac{\partial(\omega^+ + \omega^-)}{\partial s} = 2\frac{\partial\psi}{\partial s} = 2\frac{\partial\psi}{\partial r}\frac{\partial r}{\partial s} = 2\kappa\frac{y^-}{\sin\theta},$$

and $u^+ - u^- = \frac{\dot{y}^+}{y^+} - \frac{\dot{y}^-}{y^-} = -\frac{\dot{y}^+ + \dot{y}^-}{y^-} = -\frac{2\kappa}{\sin\theta}$. □

Corollary 5.2.31. *In a dispersing billiard, $u(0) \geq 0 \Rightarrow u(t) \geq 0$ for all $t \geq 0$.*

Proof. If there is no collision between time 0 and t_1, then $\dot{u} = -u^2$, so either $u(0) = 0$ and hence $u \equiv 0$ on $[0, t_1]$ or $u(0) > 0$ and hence

$$\frac{d}{dt}\frac{1}{u} = -\frac{\dot{u}}{u^2} = 1,$$

so $\frac{1}{u}$ is increasing, hence positive on $[0, t_1]$. And the previous lemma shows that collisions increase u. □

Analogously to Lemma 5.2.9, collision times are bounded.

Lemma 5.2.32. *For a finite-horizon billiard there is a $T > 0$ such that every unit-speed billiard orbit has a collision in $[0, T]$.*

Proof. Otherwise, there are billiard orbits γ_n without collision on $[-n, n]$, and by compactness (and suitable choice of parametrizations) a subsequence of $(x_n, v_n) := (\gamma_n(0), \dot{\gamma}_n(0))$ converges to $(x, v) \in SB^\circ$, which then defines a limit geodesic γ, necessarily periodic, and with period τ, say (because it is contained in the billiard table B, hence not dense in $\mathbb{T}^2$). If γ has no collision, we are done. Otherwise, there is a ball $B_0 \subset \mathbb{T}^2 \smallsetminus B$ tangent to γ and also a ball $B_1 \subset \mathbb{T}^2 \smallsetminus B$ tangent to γ *on the other side*, because if not then a geodesic with initial vector (x', v) close to $(x, v) := (\gamma(0), \dot{\gamma}(0))$ for x' close enough to x on that other side is collision-free. If $v_n = v$ for any $n \geq \tau$, then γ_n, being τ-periodic, is collision-free, so $v_n \neq v$ for all $n \in \mathbb{N}$. This, however, implies that for large enough n, γ_n intersects B_0 or B_1 on $[-2\tau, 2\tau]$, contrary to their construction. □

Proof of Theorem 5.2.26. For $(x, v) \in SB^\circ$ write $W_{(x,v)} := v^\perp \subset T_{(x,v)}B^\circ$. For an orbit γ and $(w, w') \in W_{(\gamma(0),\dot{\gamma}(0))}$ there is an orthogonal Jacobi field Y along γ with $(Y, \dot{Y}) = (w, w')$. Denoting by t_k the collision times and $\tilde{t}_k := \frac{1}{2}(t_k + t_{k+1})$, the linear maps

$$A_k := D_{(\gamma(\tilde{t}_k),\dot{\gamma}(\tilde{t}_k))}\varphi^{\tilde{t}_{k+1}-\tilde{t}_k} : W_{(\gamma(\tilde{t}_k),\dot{\gamma}(\tilde{t}_k))} \to W_{(\gamma(\tilde{t}_{k+1}),\dot{\gamma}(\tilde{t}_{k+1}))}$$

have determinant ± 1 since the billiard flow φ^t preserves volume, so uniform hyperbolicity follows from Theorem 5.1.15 once we show that $u := \dot{y}/y$ satisfies

$$T_1 \le \frac{1}{u(\tilde{t}_{k+1})} \le T_2 - \frac{1}{\kappa_{\max}},$$

where $0 < T_1 \le \frac{1}{2}(t_{k+1} - t_k) \le T_2 < \infty$ for all k, and $\kappa_{\max} < 0$ is the minimum curvature of the boundary. To see this note that $\dot{u} = -u^2$ on $(t_k, \tilde{t}_k)$, so $\frac{d}{dt}\frac{1}{u} \equiv 1$, and

$$\frac{1}{u(\tilde{t}_{k+1})} = \overbrace{\frac{1}{u(t^+_{k+1})}}^{\le -1/\kappa_{\max}} + \overbrace{\tilde{t}_{k+1} - t_{k+1}}^{\in [T_1, T_2]}. \qquad \square$$

Remark 5.2.33. Not surprisingly, the singular set of a dispersing billiard causes considerable difficulties in studying the ergodic theory of the Liouville measure. These have been overcome to a remarkable extent. Other invariant measures (of which by hyperbolicity there are many) have been far less well understood. In this context the differences between the billiard flow and the associated billiard map (Example 1.2.13) are particularly stark. The billiard flow is continuous, though not smooth at tangential collisions. This implies that the Variational Principle (Theorem 4.3.8) holds, though the theory of equilibrium states has not been developed in this context. For the billiard *map*, the situation is far worse because tangential collisions are discontinuities, so our definition of topological entropy is not even applicable—though it turned out recently that there is a natural way of defining it in that particular context, and that this yields a variational principle and an interesting measure of maximal entropy [21].

5.2.e Gases of particles. So far billiards have appeared solely as "toy models," and while this is sufficient motivation for studying them, their role in dynamical systems is much larger because, as we now show following [100], the natural microscopic model of a gas of hard particles is itself a billiard problem. We first illustrate this in the simplest nontrivial case.

Example 5.2.34 (2-disks billiard). Consider two disks of unit mass and with radius r moving freely on $\mathbb{T}^2$ and colliding elastically with each other. With respect to a frame with origin at their joint center of mass, their positions are opposites, so one of them describes a configuration completely. The possible configurations are those in which the disks do not overlap, that is, the centers are at least $2r$ apart. In our choice of coordinates, this system is modeled by free motion of a point mass in $\mathbb{T}^2$ with a disk of radius $2r$ removed. This is the configuration space of a dispersing billiard, though it remains to check that the direction changes at collisions correspond to reflection in this model.

To address the latter point in this example, let us formally define billiards in arbitrary dimension.

Definition 5.2.35. A billiard table is a compact Riemannian manifold B with boundary. A billiard orbit is a unit-speed geodesic with reflection in $T_p\partial B \subset T_pB$ at points $p \in \partial B$. Here in an inner-product space E we define reflection in a codimension-1 subspace V by $x \mapsto x - 2\langle x, u\rangle u$ for a unit vector $u \perp V$.

Our objective in what follows is to verify that particles moving freely and with elastic collisions are indeed billiard systems as in Definition 5.2.35. To that end it is helpful to clearly describe what we mean by mechanical systems with collisions.

First, the configuration space for a mechanical system with collisions consists of a subset $B = \bigcap_i D_i^{-1}([0,\infty))$ of an n-manifold M, where the $D_i\colon M \to \mathbb{R}$ are piecewise C^1-functions with nonzero differential.

Collisions occur on $\partial B \subset \bigcap_i D_i^{-1}(0)$, which has a well-defined tangent space away from the intersections of two such level sets and away from those points where a D_i is not C^1, the singular points. A collision at such a point is said to be singular, and orbits are not defined beyond such times. Other collisions are regular, and the regular set consists of those orbits that never have a singular collision.

In the case of a multiparticle system the D_i are the signed pairwise distances between particles. Singularities correspond to multiple (simultaneous) particle collisions or to collisions between two particles that involve more than one point of contact.

Definition 5.2.36. Free motion in a mechanical system is geodesic motion with respect to the Riemannian metric whose norm is the total kinetic energy of the system, called the *kinetic-energy metric*.

Thus, conservation of energy corresponds to constant speed, which is essential for optical reflection.

Collisions at regular points $p \in \partial B$ are described in terms of the D_i with $D_i(p) = 0$ by a map $R_p\colon V_- \to V_+$, where $V_\pm := \{v \in T_pM \mid \pm dD(v) \le 0\}$; then $R_p(v) = V \Rightarrow v \in V_- \cap V_+ = \ker dD$, and D_i is decreasing before the collision and increasing thereafter. We extend R_p to T_pM by imposing $R_p(-v) = -R(v)$ and now describe further properties of R that determine it explicitly.

Definition 5.2.37 (Elastic collision). A linear map $R\colon E \to E$ of an n-dimensional inner product space E is an *elastic collision* if

(1) R preserves the norm,

(2) there is a vector N such that $R(V_\pm) = V_\mp := \{v \in E \mid \mp\langle v, n\rangle \ge 0\}$,

(3) there is a full-rank linear map $L\colon E \to \mathbb{R}^{n-1}$ with $LR = L$ (called a *sufficient set of linear invariants*).

As we indicated previously, the first property reflects conservation of (kinetic) energy, and the second one says that $N^\perp$ is not crossed. We will obtain the linear invariants from conservation of momentum. For now we note that elastic collisions are reflections in $N^\perp$.

Proposition 5.2.38. *An elastic collision is a reflection in $N^\perp$ (Definition 5.2.37).*

Proof. Fix a unit vector $u \in \ker L$. For $v \in E$ we have $LRv = Lv$, hence $Rv - v = \tau(v)u \in \ker L$ defines $\tau \in E^*$, so

$$\langle v, v\rangle = \langle Rv, Rv\rangle = \langle v + \tau(v)u, v + \tau(v)u\rangle = \langle v, v\rangle + 2\tau(v)\langle u, v\rangle + (\tau(v))^2, \quad (5.2.4)$$

that is, $\tau(v) = 0$ or $\tau(v) = -2\langle u, v\rangle$. In light of Definition 5.2.35 we show that the latter possibility holds for all $v \in E$ and that $\ker L = \mathbb{R}N$. The reason is that

$$\tau(v) = 0 \Rightarrow Rv = v \Rightarrow v \in V_+ \cap V_- = N^\perp \quad (5.2.5)$$

by Definition 5.2.37(2). Thus $v \perp \ker L \Rightarrow \langle u, v\rangle = 0 \xRightarrow{(5.2.4)} \tau(v) = 0 \Rightarrow v \perp N$, so $(\ker L)^\perp \subset N^\perp$ hence $(\ker L)^\perp = N^\perp$ since both are 1-dimensional. Thus, $\ker L = \mathbb{R}N$. Equation (5.2.5) also implies $\tau(v) = 0 \Rightarrow v \in N^\perp = (\ker L)^\perp \Rightarrow \langle u, v\rangle = 0 = -\frac{1}{2}\tau(v) \Rightarrow \tau(v) = -2\langle u, v\rangle$. □

Corollary 5.2.39. *A mechanical system with collisions whose regular collisions are elastic can be modeled as a billiard system.*

Proof. By definition of "mechanical" the motion is along geodesics, and by Proposition 5.2.38 the boundary collisions are reflections in $(\ker dD_i(p))^\perp = (T_p(\partial B))^\perp$, where p is the collision point and D_i is such that $D_i(p) = 0$. □

This has further implications: R preserves velocity components tangent to ∂B, that is, if π is the projection to $\ker dD_p$, then $\pi \circ R_p = \pi$. And $F\colon T_pM \to \mathbb{R}$ is collision invariant if and only if $FRv \equiv Rv$ if and only if $\ker F \ni Rv - v = -2\langle u, v\rangle u$ for all v if and only if $(\ker dD_p)^\perp \subset \ker F$.

Example 5.2.40 (Two point masses on the interval). We illustrate this formalism in the particularly simple example of two point masses m_1, m_2 at $x_1, x_2 \in [0, 1]$, respectively, that collide elastically with each other and with the endpoints. We describe the configuration space as $M = \mathbb{R}^2$ with $D_1 = 1 - x_1$ (that is, $x_1 \le 1$), $D_2 = x_2$ (that is, $x_2 \ge 0$), and $D_3 = x_1 - x_2$ (that is, x_1 is to the right of x_2). Thus, $B = \bigcap_{i=1}^3 D_i^{-1}([0, \infty))$ with inner product $\langle\langle (v_1, v_2), (w_1, w_2)\rangle\rangle = m_1 v_1 w_1 + m_2 v_2 w_2$.

A collision between the two masses is described by $D_3 = 0$, and the linear-invariants map is the linear momentum $L(v) = m_1 v_1 + m_2 v_2$, so $\ker L = \mathbb{R}(m_2, -m_1)$, and with $u := \frac{1}{\sqrt{m_1^2+m_2^2}}(m_2, -m_1)$ we get

$$R(v) = v - 2\langle u, v\rangle u = (v_1, v_2) - \underbrace{2\frac{m_2 v_1 - m_1 v_2}{m_1^2 + m_2^2}}_{=:C(v)}(m_2, -m_1)$$
$$= (v_1 - C(v)m_2, v_2 + C(v)m_1).$$

We note that the more common approach to this example is to use the standard inner product rather than the one giving kinetic energy and to accordingly change the configuration space to a triangle with sides $\sqrt{m_1}$ and $\sqrt{m_2}$.

The main result from these endeavors is a theorem:

Theorem 5.2.41 (Cowan [100]). *The gas of hard particles is a point billiard.*

Remark 5.2.42. The particles are not assumed to be spherical, so angular momentum and its transfer between particles is in scope.

Proof. The gas of hard particles is modeled by N piecewise smooth rigid bodies B_i moving freely in $\mathbb{R}^3$ and having nonsingular inertial tensor. The "position" of B_i is a point in $M_i := F_i \times G_i$, where $F_i \sim \mathbb{R}^3$ parametrizes possible locations of the center of mass and $G_i \sim \mathrm{SO}(3)$ describes the orientation of B_i. Thus, the configuration space is $B := \{p \in M := M_1 \times \cdots \times M_N \mid D_{ij}(p) \geq 0 \text{ for } 1 \leq i, j \leq N\}$ using the signed distances $D_{ij} \colon M \to \mathbb{R}$ between ∂B_i and ∂B_j chosen to be positive away from overlaps. If, furthermore, each B_i is constrained to remain in $A \subset \mathbb{R}^3$ with piecewise smooth boundary $\partial A = \bigcup_{k=1}^m A_k$ we instead have

$$B = \{p \in M \mid D_{ij}(p) \geq 0,\ D'_{ik} \geq 0 \text{ for } 1 \leq i, j \leq N,\ 1 \leq k \leq m\},$$

where the D'_{ik} are the signed distances between ∂B_i and A_k. Regular boundary points are those where only one inequality among these fails to be strict; other (singular) boundary points represent double collisions or configurations where more than one point of a particle is in contact with another particle or with ∂A.

The total kinetic energy combines translational and rotational energy, and for the latter, the intertial tensors I_i of the B_i are the counterparts of mass:

$$\langle (v, \omega), (v', \omega')\rangle = \sum_{i=1}^{N} m_i v_i v'_i + I_i \omega_i \omega'_i.$$

We note that this defines free motion in a non-Euclidean space, so the motion is not along lines.

We now show that the collisions between these particles are elastic. For two-particle collisions consider B_1 and B_2 to fix ideas. We need to find the linear-invariants map $L\colon T_pM \to \mathbb{R}^{6N-1}$. A total of $6(N-2)$ obvious linear invariants are given by the velocity components of the B_i for $i > 2$. The needed 11 additional linear invariants are

- 3 components L_1, L_2, L_3 of total linear momentum,
- 3 components each $(L_4, \dots, L_9)$ of angular momentum of B_1 and B_2 with respect to the collision point because relative to that point the torque from the collision is zero,
- 2 velocity components L_{10}, L_{11} of B_1 projected to the collision plane.

To check that the L built from these has maximal rank, we adduce the velocity L_{12} of B_1 normal to the collision plane, and show that the resulting extension L' has trivial kernel: the trivial invariants tell us that the B_i for $i > 2$ have zero velocities. If $L_{10} = L_{11} = L_{12} = 0$, then the translational velocity of B_1 is zero, and $L_1 = L_2 = L_3 = 0$ implies the same for B_2, so the angular momenta come from angular velocities, which are therefore zero.

To conclude we note that collisions with ∂A play out the same way because the 5 nontrivial invariants are L_4, L_5, L_6, L_{10}, L_{11} from the previous arguments. Thus, Corollary 5.2.39 gives Theorem 5.2.41. □

We conclude with brief remarks on what are called *no-slip billiards*, which can be imagined as involving "sticky" disks, which also brings in exchanges of angular momentum, even between spherical particles [73]. While its governing rules are not quite as principled as the one in the previous subsection, they have a sound physical justification and they have been studied significantly more, with some surprising results.[20] The central definition is a little less abstract than Definition 5.2.37:

Definition 5.2.43. Let M be the configuration manifold of two rigid bodies with smooth boundaries in $\mathbb{R}^n$ endowed with the kinetic-energy metric. The collision map $C\colon T_qM \to T_qM$ is said to be *strict* if

(1) kinetic energy is preserved,

(2) linear and angular momentum are conserved,

(3) C is an involution (time reversibility),

(4) collision forces act only at the point of impact.

[20] For instance, the motion of a "sticky" disk bouncing between parallel lines is bounded, and the no-slip "stadium" billiard is not ergodic, in contrast to the conventional stadium billiard: .

In $\mathbb{R}^2$, these requirements imply that collisions are either elastic or of the no-slip type [102, Theorem 1.1];[21] this is a central motivation for the definition of no-slip billiards, as is the fact that this system preserves the usual Liouville measure [102, Theorem 1.2]. These systems have been studied to great effect by Feres and collaborators, and we recommend the richly illustrated introduction [102].

5.2.f Linkages. We illustrate our next class of systems with a particularly salient example. Linkages consist of rods connected by joints at each of which there may or may not be a mass. Instead of formalizing that definition, we present the instance of interest.

Definition 5.2.44 (Kourganoff linkage). The *Kourganoff linkage* consists of points $(a,0)$, $(b,0)$, $(0,c)$, (d,e), $(-2,f)$, $(2,g) \in \mathbb{R}^2$ and connected by massless rods of lengths 1, l, and r (Figure 5.2.5) subject to

$$(l-2)^2 + r^2 < 1 \text{ and } 3 - l < r < 1/2 \tag{5.2.6}$$

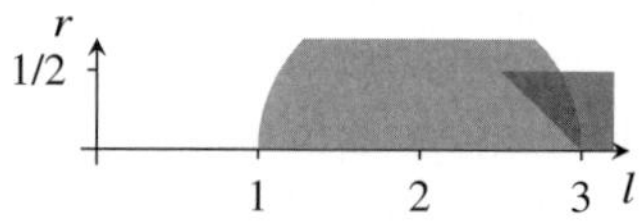

(for instance, $l = 11/4$, $r = 1/3$) with mass ϵ^2 at $(0,c)$, no mass at (d,e), and unit masses at the other joints. Note that the massive joints are constrained to motion along lines. Its configuration space C is the set of $(a,b,c,d,e,f,g) \in \mathbb{R}^7$ with

$$\begin{aligned}(a+2)^2 + f^2 &= 1 \ = (b-2)^2 + g^2,\\ (a-d)^2 + e^2 &= l^2 = (b-d)^2 + e^2,\\ (c-e)^2 + d^2 &= r^2,\end{aligned}$$

and endowed with the kinetic-energy metric (Definition 5.2.36), whose geodesic flow represents free motion of the linkage.

Remark 5.2.45. Here is a little physical intuition about why this system has "collisions" when $\epsilon = 0$ and not otherwise: when $\epsilon = 0$ it is possible to reach a configuration with $a = l = b$ with positive speed of the masses at $(a,0)$ and $(b,0)$, and that speed will be instantly reversed, a collision. For positive ϵ this is impossible because at that instant c necessarily has infinite derivative, which necessitates infinite kinetic energy.

Remark 5.2.46. A look at Figure 5.2.5 shows that the configuration space is described by the two circles that parametrize the orientation of the unit-length rods plus a real

[21] In $\mathbb{R}^n$, one obtains the disjoint union of orthogonal Grassmannian manifolds $\mathrm{Gr}(k, n-1)$, $k = 0, \dots, n-1$ of all k-dimensional planes in $\mathbb{R}^{n-1}$.

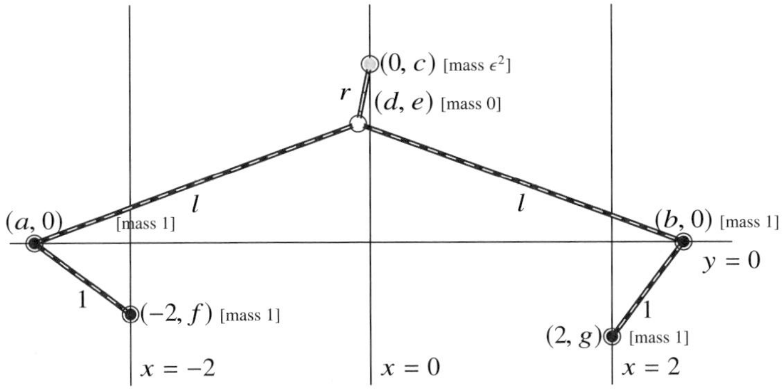

Figure 5.2.5. The Kourganoff linkage.

parameter for the shortest rod. That is to say, this linkage has a 2-dimensional configuration space: a 2-torus parametrizes the orientation of the two lower pairs of vertices (the ends of the unit-length rods), and once these are fixed, (d, e) is fixed modulo the sign of e (completely fixed if $e = 0$), for either of which there are two possibilities for c. This naturally immerses the configuration space in $\mathbb{T}^2 \times \mathbb{R}$ and defines a natural projection to $\mathbb{T}^2$ with at most four preimages per point (Figure 5.2.7). More formally, since $(a + 2, f)$ and $(b - 2, g)$ lie on the unit circle (Figure 5.2.5), the configuration space C in Definition 5.2.44 is contained in $\mathbb{T}^2 \times \mathbb{R}^3$ parametrized by

$$(\theta, \phi, c, d, e) \mapsto (a, b, c, d, e, f, g) = (-\cos\theta - 2, \cos\phi + 2, c, d, e, \sin\theta, \sin\phi).$$

One can imagine the physical construction of such a linkage using rotational joints and with five vertices attached to *prismatic joints*, frictionless sleeves that slide along the respective constraint lines (Figure 5.2.8). (Those sleeve joints are a mere convenience, and linkage purists can replace each by a massless *Peaucellier* or *Hart linkage* (Figure 5.2.6), which produces straight-line motion using only rotational joints, or they can instead be approximated by arcs of vast circles traced by the ends of additional long rods.)

Theorem 5.2.47. *For sufficiently small ϵ the free motion of the Kourganoff linkage is an Anosov flow.*

Remark 5.2.48. This, finally, is a realistic physical system in the sense of a buildable mechanical device whose dynamics is Anosov. Specifically, since the Anosov property is persistent, a Kourganoff linkage with rods of sufficiently small (rather than zero) mass is Anosov, and if constructed with sufficiently small friction will exhibit corresponding dynamics. It should be noted that ϵ itself arises from a like use of stability and is

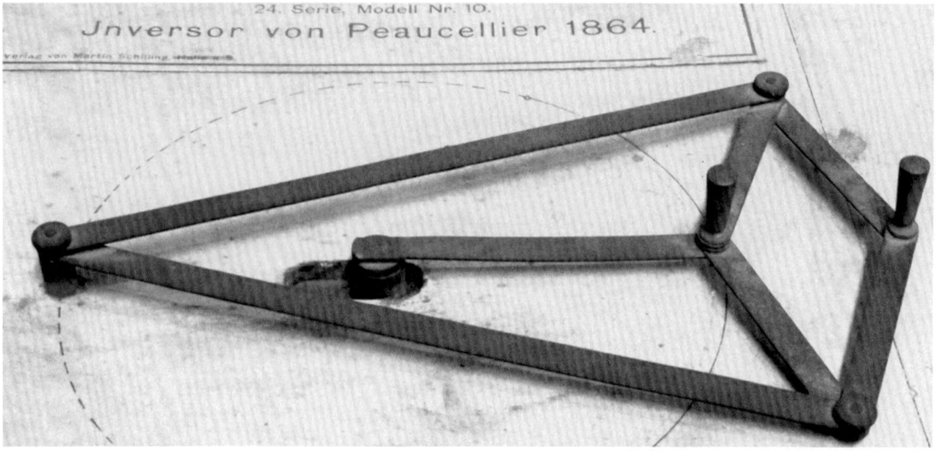

Figure 5.2.6. The Peaucellier linkage at the University of Tokyo [http://www.ms.u-tokyo.ac.jp/models/models/invertors.pdf].

therefore not explicit. Nonetheless, unlike any previously known Anosov linkages, the geometry of the Kourganoff linkage is completely explicit [226, 199].

We note as well that the point of this is not merely the existence of an Anosov linkage—a universality theorem asserts that any compact Riemannian manifold is the configuration space of a linkage, and this includes negatively curved ones. However, the linkages obtained from the application of that theorem are astronomically more complicated than the one here. This is instead a "realistic" linkage.

The proof strategy is to establish that for small enough ϵ the configuration space with the kinetic-energy metric is so close to a hyperbolic billiard that the geodesic flow is necessarily Anosov. This involves a result to the effect that "compressing" a surface in $\mathbb{T}^2 \times \mathbb{R}$ along the z-direction asymptotically gives a billiard in $\mathbb{T}^2$ in the sense that the geodesic flow uniformly converges to the billiard dynamics under this procedure. If the limiting billiard is hyperbolic (such as by Theorem 5.2.26), then stability of hyperbolicity implies that the geodesic flow of the surface is Anosov once it is sufficiently compressed toward $\mathbb{T}^2$.

This "flattening idea" goes some time back. In the 1920s Birkhoff noted that if one of the principal axes of an ellipsoid tends to 0, then the geodesic flow of this ellipsoid appears to tend to the billiard flow of the limiting ellipse.[22] Arnold suggested

[22]"In order to see how the theorem of Poincaré and its generalization can be applied, we will consider first a special but highly typical system of this sort, namely that afforded by the motion of a billiard ball upon a convex billiard table. This system is very illuminating for the following reason: any Lagrangian system with two degrees of freedom is isomorphic with the motion of a particle on a smooth surface rotating uniformly about a fixed axis

a reverse idea in the 1960s in hopes that it would establish hyperbolicity of dispersing billiards: that a dispersing billiard in $\mathbb{T}^2$ can be approximated by the geodesic flow of a surface of negative curvature made by gluing together two copies of the billiard (ergodicity of that billiard was later proved by Sinai using a different approach).[23] This makes these ideas explicit:

Theorem 5.2.49 ([226, Theorem 5]). *Let $B \subset \mathbb{T}^2$ be a finite-horizon dispersing billiard with smooth scatterers and $\Sigma \subset E := \mathbb{T}^2 \times \mathbb{R}$ an immersed surface such that $B = \pi(\Sigma)$, where π is the natural projection to $\mathbb{T}^2$. Then the Euclidean metric on E induces a Riemannian metric h_ϵ on $\Sigma_\epsilon := f_\epsilon(\Sigma)$, where*

$$f_\epsilon\colon E \to E, \quad (x, y, z) \mapsto (x, y, \epsilon z),$$

which in turn induces the Riemannian metric $g_\epsilon := f_\epsilon^*(h_\epsilon)$ on Σ. Suppose

(1) *the surface $\pi^{-1}(\operatorname{Int} B) \cap \Sigma$ is transverse to the fibers of π (no vertical tangent planes) and*

(2) *the curvature of $\Sigma \cap V$ is nonzero at $q \in \pi^{-1}(\partial B) \cap \Sigma$ (nondegenerate boundary projection),*[24]

where V is a neighborhood of q in the vertical affine plane through q that is perpendicular to $T_q\Sigma$. Then for small enough ϵ, the geodesic flow of (Σ, g_ϵ) is Anosov.

Proof of Theorem 5.2.47. Figure 5.2.7 makes it plausible that the hypotheses of Theorem 5.2.49 hold—the third picture shows a billiard that is clearly dispersing with finite horizon. The proof consists of verifying this explicitly.

We first check that the configuration space C is a smooth submanifold of $\mathbb{T}^2 \times \mathbb{R}$ as follows.

and carrying a conservative field of force with it. In particular if the surface is not rotating and if the field of force is lacking, the paths of the particle will be geodesics. If the surface is now flattened to the form of a plane convex curve C, the 'billiard ball problem' results. But in this problem the formal side, usually so formidable in dynamics, almost completely disappears, and only the interesting qualitative questions need to be considered. If C is an ellipse an integrable problem results, namely the limiting case of an ellipsoid treated by Jacobi." [50, p. 169f]

[23]"In precisely the same way a torus billiard table can be regarded as a two-sided torus with a hole on which the point moves along a geodesic. But if the two-sided ellipse is an oblate ellipsoid, the two-sided torus with a hole will be an oblate 'Kringel' [this is the northern German term for 'pretzel'] (of genus 2). Thus, motion on our torus billiard table is a limiting case of motion along a geodesic on the knot-shaped surface... Thus, a two-sided torus billiard table can be regarded as an oblate surface with negative curvature everywhere: on flattening, all the curvature is accumulated along the circumference." [15, Chapter VI, §4].

[24]This ensures that any geodesic in that preimage is unstable, that is, has sensitive dependence on initial conditions.

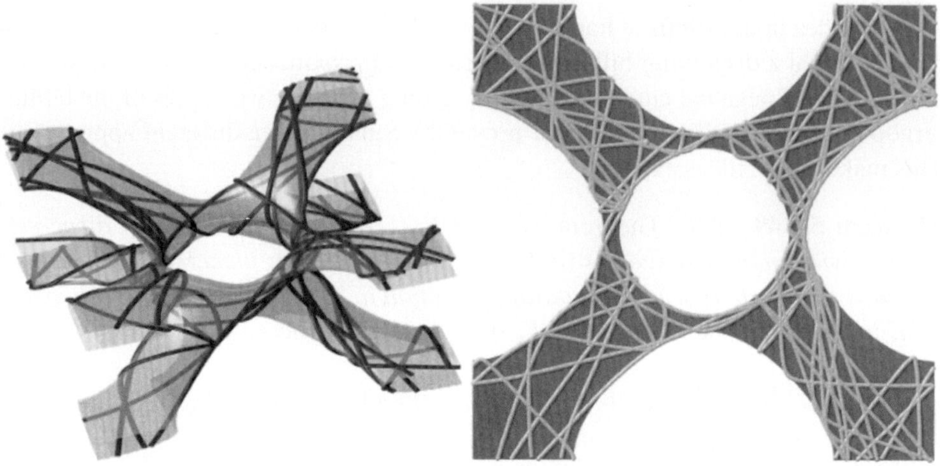

Figure 5.2.7. The configuration space with an orbit, and its projection to the torus (see http://mickael.kourganoff.fr/images/poster.pdf).

- Near $p \in C$ with $c \neq e \neq 0$, C is the graph over $\mathbb{T}^2$ (that is, over θ, ϕ) of

$$\begin{aligned} d &= \frac{-\cos\theta + \cos\phi}{2}, \\ e &= \pm\sqrt{l^2 - \Big(\frac{\cos\theta + \cos\phi}{2} + 2\Big)^2}, \\ c &= e \pm \sqrt{r^2 - \Big(\frac{\cos\theta - \cos\phi}{2}\Big)^2}, \end{aligned} \tag{5.2.7}$$

  with "±" depending on p.

- Near $p \in C$ where $\phi \neq 0 \bmod \pi$ and $(-\cos\theta - 2, 0)$, (d, e), and $(0, c)$ are not aligned, C is a graph over θ and c: d and e are simple roots of a second-order polynomial, hence depend smoothly on θ and c, and $\phi = \pm\cos^{-1}(2d + \cos\theta)$.

- Likewise, near $p \in C$ where $\theta \neq 0 \bmod \pi$ and $(\cos\theta + 2, 0)$, (d, e), and $(0, c)$ are not aligned, C is a graph over ϕ and c.

For each $p \in C$ at least one of these scenarios applies: if the latter two scenarios do not apply, suppose $\theta = 0 \bmod \pi$ and $\phi = 0 \bmod \pi$, so $\theta = \phi \bmod 2\pi$ since $r < 1/2$, hence $\theta = \phi = \pi \bmod 2\pi$ since $l < 3$, so $c \neq e \neq 0$ contrary to our assumption. Thus (by symmetry without loss of generality) instead $(-\cos\theta - 2, 0)$, (d, e), and $(0, c)$ are aligned, and failure of the first scenario implies $e \in \{0, c\}$, so $(-\cos\theta - 2, 0)$, (d, e), and $(0, c)$ are all on the x-axis, contrary to $l + r > 3$.

The kinetic-energy metric on C is given by

$$g_\epsilon = da^2 + df^2 + db^2 + dg^2 + \epsilon^2 dc^2 = d\theta^2 + d\phi^2 + \epsilon^2 dc^2,$$

and it is nondegenerate on C because of the local embedding as a graph. As mentioned before, its geodesic flow is the free motion of the Kourganoff linkage.

The nature of the embedding further implies that the projection

$$p\colon \mathbb{T}^2 \times \mathbb{R}^3 \to \mathbb{T}^2 \times \mathbb{R}, \quad (\theta, \phi, c, d, e) \mapsto (\theta, \phi, c)$$

restricts to an isometric immersion of C to a surface Σ in $\mathbb{T}^2 \times \mathbb{R}$ with the metric $g_\epsilon = d\theta^2 + d\phi^2 + \epsilon^2 dc^2$. The projection $\pi\colon \mathbb{T}^2 \times \mathbb{R} \to \mathbb{T}^2$ maps Σ to

$$B = \pi(C) = \big\{(\theta, \phi) \in \mathbb{T}^2 \mid |\cos\theta - \cos\phi| \le 2r,\ \cos\theta + \cos\phi \le 2l - 4\big\} \quad (5.2.8)$$

with boundary

$$\big\{\cos\theta - \cos\phi = 2r\big\} \cup \big\{\cos\phi - \cos\theta = 2r\big\} \cup \big\{\cos\theta + \cos\phi = 2l - 4\big\}.$$

We later show that this is a finite-horizon dispersing billiard but first establish Theorem 5.2.49(1) (this is clear) and Theorem 5.2.49(2). Consider one of the boundary components and suppose $\pi(q) \in \big\{(\theta, \phi) \in \mathbb{T}^2 \mid \cos\theta + \cos\phi = 2l - 4\big\}$.

Denoting by N the normal vector at q and by subscripts θ and ϕ the corresponding projections, parametrize the (θ, ϕ)-projection of a normal line by $\theta(t) := q_\theta + tN_\theta$, $\phi(t) := q_\phi + tN_\phi$, $F(\theta, \phi) := \frac{\cos\theta + \cos\phi}{2} + 2$, and

$$c(t) = \pm\sqrt{l^2 - (F(\theta(t)\phi(t)))^2} \pm \sqrt{r^2 - \Big(\frac{\cos\theta(t) - \cos\phi(t)}{2}\Big)^2},$$

so that $(\theta(t), \phi(t), c(t)) \in \Sigma$ according to (5.2.7) and $(\theta(0), \phi(0), c(0)) = q$ by choosing "$\pm$" appropriately. For t near 0 we then have

$$c(t) = \pm\sqrt{t\frac{d}{dt}\Big|_{t=0} F(\theta(t), \phi(t)) + O(t^2)} \pm \sqrt{r^2 - \Big(\frac{\cos\theta(0) - \cos\phi(0)}{2}\Big)^2} + O(t).$$

It suffices to show that $t \mapsto c(t)$ is invertible near $c(0)$ and that the inverse has nonzero second derivative. To that end note that

$$(c(t) - c(0))^2 = \pm t \underbrace{\frac{d}{dt}\Big|_{t=0} F(\theta(t), \phi(t))}_{=\begin{pmatrix}\theta'(0)\\ \phi'(0)\end{pmatrix} \nabla F(\theta(0), \phi(0)) \neq 0 \text{ since } 2<l<3} + o(t)$$

and hence

$$t = \pm \frac{1}{\frac{d}{dt}\big|_{t=0} F(\theta(t), \phi(t))} (c(t) - c(0))^2 + o(c(t) - c(0))^2,$$

as required.

Finally, we show that B is a finite-horizon dispersing billiard—which is easy to believe from Figure 5.2.7. As to dispersion, consider the boundary component $F(\theta,\phi) := \cos\theta + \cos\phi = 2l - 4 \in (0,2)$ (since $2 < l < 3$). Its curvature is

$$\nabla \cdot \frac{\nabla F}{\|\nabla F\|} = -\nabla \cdot \frac{1}{\sqrt{\sin^2\theta + \sin^2\phi}} \begin{pmatrix} \sin\theta \\ \sin\phi \end{pmatrix} = -\frac{\cos\theta\sin^2\phi + \sin^2\theta\cos\phi}{\sqrt{\sin^2\theta + \sin^2\phi}^{\,3}}.$$

As required, the numerator is positive because it equals

$$\begin{aligned}\cos\theta(1-\cos^2\phi) + \cos\phi(1-\cos^2\theta) &= (\cos\theta + \cos\phi)\cos^2\theta \\ &\quad - (\cos^2\theta + 2\cos\theta\cos\phi + \cos^2\phi)\cos\theta \\ &\quad + (\cos\theta + \cos\phi) \\ &= \underbrace{(2l-4)}_{>0}\big[\cos^2\theta - \underbrace{(2l-4)}_{<2}\cos\theta + 1\big] > 0.\end{aligned}$$

For the boundary component $\{\cos\phi - \cos\theta = 2r\}$ the corresponding numerator is

$$\sin^2\phi\cos\theta - \sin^2\theta\cos\phi = 2r\big[\cos^2\theta + \underbrace{2r}_{<2}\cos\theta + 1\big] > 0,$$

and similarly for $\{\cos\theta - \cos\phi = 2r\}$.

We conclude the proof by showing that if $(l-2)^2 + r^2 < 1$ and $r < 1/2$, then B has finite horizon.

Otherwise there is a bi-infinite geodesic $(\theta(t), \phi(t))$, and we first show that it has slope ± 1. Up to exchanging θ and ϕ we may assume the slope is in $[-1,1]$, so there is a t_0 with $\theta(t_0) = 0 \bmod 2\pi$, so $G := \{\phi(t) - \phi(t_0) \bmod 2\pi \mid t \in \mathbb{R},\ \theta(t) = 0 \bmod 2\pi\}$ is a subgroup of $\mathbb{R}/2\pi\mathbb{Z}$, and for $t \in G$ we have $|\cos\theta(t) - \cos\phi(t)| \le 2r$ by (5.2.8), so $\cos\phi(t) \ge 2 - 2r > 0$ and hence $G \subset (-\frac{\pi}{2}, \frac{\pi}{2}) \bmod 2\pi$, which means that G is a point and the slope is in $\{0, \pm 1\}$. The slope cannot be 0 because in that case taking t such that $\cos(t) = -1$ and (5.2.8) give $-1 + 2r \ge \cos\phi(t) \ge 1 - 2r > 0$, contrary to $r < 1/2$.

Thus, the slope is 1 (up to replacing θ by $-\theta$). Therefore, there are t_1, t_2 with

$$\begin{aligned}\phi(t_1) + \theta(t_1) &= \pi \bmod 2\pi, \\ \phi(t_2) + \theta(t_2) &= 0 \bmod 2\pi.\end{aligned}$$

Averaging these two equations and using $\theta(t_2) - \theta(t_1) = \phi(t_2) - \phi(t_1) \bmod 2\pi$ (slope 1) gives $\phi(t_2) - \phi(t_1) = \frac{\pi}{2} \bmod \pi$, and hence $\cos\phi(t_2)\cos\phi(t_1) = -\sin\phi(t_2)\sin\phi(t_1)$. Squaring both sides here gives

$$\begin{aligned}\cos^2\phi(t_2)\cos^2\phi(t_1) &= (1 - \cos^2\phi(t_2))(1 - \cos^2\phi(t_1)) \\ &= 1 - \cos^2\phi(t_2) - \cos^2\phi(t_1) + \cos^2\phi(t_2)\cos^2\phi(t_1),\end{aligned}$$

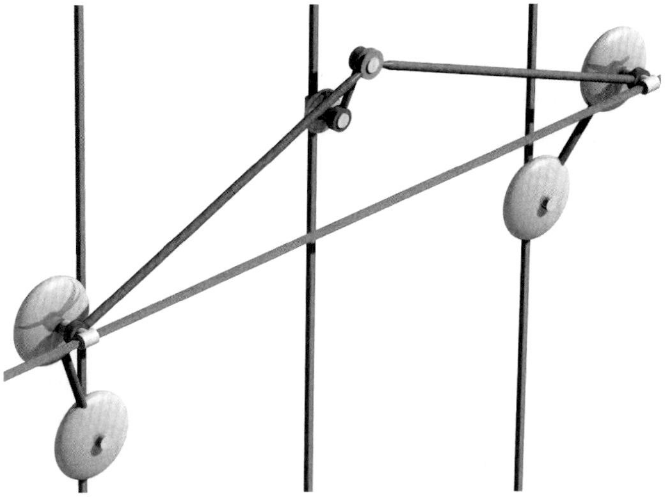

Figure 5.2.8. The Kourganoff linkage, a mechanical Anosov system animated by Jos Leys at http://mickael.kourganoff.fr/videos/anosov-linkage.mov; see also https://icerm.brown.edu/video_archive/?play=1138 (picture adapted from http://mickael.kourganoff.fr/images/poster.pdf).

so $\cos^2\phi(t_2) + \cos^2\phi(t_1) = 1$. The choice of t_1 implies that

$$\underbrace{\cos\phi(t_1)}_{=-\cos\theta(t_1)} = \tfrac{1}{2}(\cos\phi(t_1) - \cos\theta(t_1)) \le \tfrac{1}{2}2r = r$$

by (5.2.8), and the choice of t_2 and (5.2.8) imply

$$\underbrace{\cos\phi(t_2)}_{=\cos\theta(t_2)} = \tfrac{1}{2}(\cos\phi(t_2) + \cos\theta(t_2)) \le \tfrac{1}{2}(2l - 4) = l - 2.$$

Thus, $1 = \cos^2\phi(t_2) + \cos^2\phi(t_1) \le r^2 + (l-2)^2 < 1$ (by (5.2.6)), a contradiction. □

5.3 Shadowing, expansivity, closing, specification, and Axiom A

The orbit structure of hyperbolic dynamical systems has a distinctive and iconic richness and complexity, and these features can be derived from what thereby appears as a core feature of hyperbolic dynamics: the shadowing of orbits. From shadowing we will see that in a hyperbolic system anything one can imagine approximately happening is, to good approximation, actually happening in the system. This section shows that the Shadowing Lemma (Theorem 5.3.3) produces the essential richness of the orbit structure of a hyperbolic dynamical system: expansivity (Corollary 5.3.5), the Anosov Closing Lemma (Theorem 5.3.11), specification (Theorem 5.3.61), spectral decomposition (Theorem 5.3.37), and a natural definition of hyperbolicity (Theorem 5.3.47), as well as topological stability (Theorems 5.3.7, 5.3.8).

The stronger Anosov Shadowing Theorem (Theorem 5.4.1) further implies structural stability (Theorem 5.4.5).[25] We reserve this for the next section and emphasize that Section 5.4 (through to Theorem 5.4.11) is independent of this one, that is, a reader can learn about structural stability directly without working through the present section first.

Definition 5.3.1 (Shadowing). Let Φ be a flow on a metric space M and g be an ϵ-pseudo-orbit for Φ (Definition 1.5.29). Then g is said to be *δ-shadowed* if there exists a point $p \in M$ and a homeomorphism $\alpha : \mathbb{R} \to \mathbb{R}$ such that $\alpha(t) - t$ has Lipschitz constant δ and $d(g(t), \varphi^{\alpha(t)}(p)) \le \delta$ for all $t \in \mathbb{R}$. A set $Y \subset M$ has the *shadowing property* if for any $\delta > 0$ there is an $\epsilon > 0$ such that any ϵ-pseudo-orbit in Y is δ-shadowed by a point $p \in M$. We say that Φ has the shadowing property if this holds for $Y = M$. A set $Y \subset M$ has *L-Lipschitz shadowing for $\epsilon_0 > 0$* if any ϵ-pseudo-orbit in Y with $\epsilon \le \epsilon_0$ is $L\epsilon$-shadowed by a point $p \in M$.

Remark 5.3.2. For clarity we emphasize that the pseudo-orbit is shadowed by a "slightly misparametrized" orbit: the speed may have to be adjusted slightly to keep near the pseudo-orbit.

Theorem 5.4.1 below implies that hyperbolic sets have the shadowing property:

Theorem 5.3.3 (Shadowing Lemma). *A hyperbolic set for a flow has* a neighborhood *with L-Lipschitz shadowing for some $\epsilon_0 > 0$ and for some $L > 0$. The shadowing* point *need not be unique because neither is the choice of the parametrization. But the shadowing* orbit *is unique and the shadowing* point *is determined up to a small shift within that orbit.*

Remark 5.3.4. Implicitly, Theorem 5.3.3, or, rather, Definition 5.3.1, controls the timing of the shadowing orbit to within a percentage error, where the percentage is small for small ϵ.

The uniqueness assertion of Theorem 5.3.3 implies that no two orbits can shadow each other:

Corollary 5.3.5. *The restriction of a flow to a* (*sufficiently small neighborhood of a*) *hyperbolic set is expansive* (*Definition* 1.7.2).

---

[25] ... and symbolic descriptions, which we do not include here [213, Theorem 18.2.5].

Proof. Let Λ be a hyperbolic set for a flow Φ and let ϵ_0 and L be given by Theorem 5.3.3. If x and y are as in the definition of expansivity note that the refinement in Remark 1.7.7 of the characterization Theorem 1.7.5(1) implies that $\mathcal{O}(x)$ and $\mathcal{O}(y)$ α-shadow each other for suitably chosen α. Then both α-shadow the pseudo-orbit $\mathcal{O}(x)$ and hence are on the same orbit by uniqueness of shadowing (for small enough α). □

Remark 5.3.6. We continue to derive consequences of the Shadowing Lemma, but the reader is encouraged to verify that these consequences can equivalently be obtained by combining the shadowing property (without the uniqueness assertion) with expansivity (Remark 1.7.12).

As a preview of coming attractions, we note that the Shadowing Lemma implies stability:

Theorem 5.3.7 (Topological stability). *Anosov flows are topologically stable, that is, any sufficiently C^0-close flow is an extension* (*Definition* 1.3.1).

The proof *idea* is straightforward: the orbits of the perturbation are pseudo-orbits for the given flow and hence shadowed by genuine orbits of the flow; this correspondence between orbits of the perturbation and those of the given flow gives the factor map—but one needs to check that it is continuous [347]. While this is possible, we will instead step up from the Shadowing Lemma to the Shadowing Theorem (Theorem 5.4.1) where such continuous dependence is built into the conclusion. In passing, we note that topological stability implies a nontrivial variant of structural stability (Theorem 5.4.5) for C^0-perturbations:

Theorem 5.3.8 (Topological structural stability). *Any two sufficiently C^0-close Anosov flows are orbit equivalent.*

Proof. The factor map in Theorem 5.3.7 is injective because the orbit of the perturbation that shadows a given one is unique by expansivity (from the Anosov property) of the perturbation. □

Remark 5.3.9. The argument actually shows, of course, that Anosov flows are C^0-structurally stable (Definition 5.4.4) among *expansive* flows.

When the Anosov flows are geodesic flows, this last observation has a remarkable refinement.[26]

Theorem 5.3.10 ([147, Théorème B]). *All Anosov geodesic flows of a given closed manifold that supports a Riemannian metric with constant negative curvature are pairwise topologically orbit equivalent.*

[26]For which the "averaging" idea underlying Proposition 1.3.30 was developed.

We now apply the Shadowing Lemma to study the structure of hyperbolic sets. The uniqueness assertion of Theorem 5.3.3 implies not only expansivity but also that the shadowing orbit is periodic when one starts with a periodic pseudo-orbit.

Theorem 5.3.11 (Anosov Closing Lemma). *If Λ is a hyperbolic set for a flow Φ then there are a neighborhood U of Λ and numbers $\epsilon_0, L > 0$ such that for $\epsilon \le \epsilon_0$ any compact ϵ-pseudo-orbit in U is $L\epsilon$-shadowed by a unique compact orbit for Φ.*

Remark 5.3.12. The definition of hyperbolicity allows isolated hyperbolic fixed points (Definition 5.1.1), and these are pseudo-orbits shadowed only by themselves.

Proof. Except for "periodic," this is just Lipschitz shadowing. Uniqueness forces the shadowing orbit to close up: If the pseudo-orbit has period T and $\mathcal{O}(x)$ is a shadowing orbit, then so is $\mathcal{O}(\varphi^{nT}(x))$ for $n \in \mathbb{Z}$. If $nT > L\epsilon/\min \|X\|$, then uniqueness gives $\varphi^{T'}(x) = x$ for T' near nT. □

This can also be proved directly rather than as a corollary of Theorem 5.3.3.

Remark 5.3.13. Except for particularly short pseudo-orbits one can take $n = 1$ in the proof of Theorem 5.3.11, so the shadowing orbit has a length comparable with that of the pseudo-orbit; the accuracy of shadowing controls a percentage difference in orbit length beyond the "misparametrization" of the pseudo-orbit itself, which is reflected by α in Definition 5.3.1. The latter effect is apparent for a pseudo-orbit $t \mapsto \varphi^{1.1t}(x)$ of φ^t for which the shadowing orbit is $t \mapsto \varphi^t(x)$, which has a 10% difference in speed. In important applications, the pseudo-orbit is an almost periodic orbit segment, for which it may be desirable to control the timing more finely, and Theorem 6.2.4 below (a quantitative version of Proposition 1.8.4) does so by bounding an *absolute* error instead (Remark 6.2.5), which is critical for several of those applications.

The following proposition is what we have actually proved:

Proposition 5.3.14. *If Φ is expansive with shadowing, then $\overline{\mathrm{Per}(\Phi)} = \mathcal{R}(\Phi)$.*

Corollary 5.3.15. *Let Φ be a smooth flow on a compact manifold M. Then,*

(1) *if $\mathcal{R}(\Phi)$ is hyperbolic, then $\overline{\mathrm{Per}(\Phi)} = \mathcal{B}(\Phi) = \mathcal{L}(\Phi) = \mathrm{NW}(\Phi) = \mathcal{R}(\Phi)$,*

(2) *if $\mathrm{NW}(\Phi)$ is hyperbolic, then $\overline{\mathrm{Per}(\Phi)} = \mathrm{NW}(\Phi|_{\mathrm{NW}(\Phi)})$,*

(3) *if the limit set $\mathcal{L}(\Phi)$ is hyperbolic, then $\overline{\mathrm{Per}(\Phi)} = \mathcal{L}(\Phi)$,*

(4) *if Λ is a hyperbolic set for Φ and V a neighborhood of Λ such that Λ_Φ^V (Proposition 5.1.8) is hyperbolic, then $\overline{\mathrm{Per}(\Phi_{\restriction \Lambda_\Phi^V})} = \mathrm{NW}(\Phi_{\restriction \Lambda_\Phi^V})$.*

Proof. (1): For all δ each $x \in \mathcal{R}(\Phi)$ is in a periodic δ-chain in $\mathcal{R}(\Phi)$ (Theorem 1.5.39), which is $L\delta$-shadowed by a periodic p (Theorem 5.3.11), so $x \in \overline{\mathrm{Per}(\Phi)}$, and $\mathcal{R}(\Phi) \subset \overline{\mathrm{Per}(\Phi)} \subset \mathcal{B}(\Phi) \subset \mathcal{L}(\Phi) \subset \mathrm{NW}(\Phi) \subset \mathcal{R}(\Phi)$ (Proposition 1.5.37).

(2): "$\subset$" is clear. "$\supset$": $x \in \mathrm{NW}(\Phi|_{\mathrm{NW}(\Phi)})$ implies that x is arbitrarily near periodic pseudo-orbits in $\mathrm{NW}(\Phi)$, hence in $\overline{\mathrm{Per}(\Phi)}$ by Theorem 5.3.11 applied to the hyperbolic set $\mathrm{NW}(\Phi)$.

(3): "$\subset$" is Remark 1.5.12. "$\supset$": It suffices to show that $x \in \omega(y) \Rightarrow x \in \overline{\mathrm{Per}(\Phi)}$ (Definition 1.5.1). Also, $d(\varphi^t(y), \omega(y)) \xrightarrow[t\to\infty]{} 0$ by Proposition 1.5.7(3). Given $\delta > 0$ there exist $t_0, t_1 > 0$ with $d(\varphi^{t_0}(y), \varphi^{t_0+t_1}(y)) < \delta$, $d(\varphi^{t_0}(y), x) < \delta$, and $d(\varphi^t(y), \omega(y)) < \delta$ for $t_0 \le t \le t_0+t_1$. The periodic δ-chain $\varphi^{[t_0,t_0+t_1]}(y)$ is within δ of $\omega(y)$ and shadowed by a periodic orbit $\mathcal{O}$ with $d(x, \mathcal{O}) < \delta + L\delta$. Thus, $x \in \overline{\mathrm{Per}(\Phi)}$.

(4): For sufficiently small $\epsilon > 0$ denote by U_ϵ the $\epsilon/(2L+1)$-neighborhood of $x \in \mathrm{NW}(\varphi\restriction_{\Lambda_\Phi^V})$ in Λ_Φ^V, where L is as in the Closing Lemma. For some $T > 1$ there exists a $y \in \varphi^T(U_\epsilon) \cap U_\epsilon$, and then $d(\varphi^T(y), y) < 2\epsilon/(2L+1)$, so the Closing Lemma gives a periodic $z \in \Lambda_\Phi^V$ with $d(\varphi^t(z), \varphi^t(y)) < 2L\epsilon/(2L+1)$ for $0 \le t < T$. Then $d(x,z) \le d(x,y) + d(y,z) \le \frac{(2L+1)\epsilon}{2L+1} = \epsilon$. □

It happens that Λ and Λ_Φ^V coincide in most of our examples, and this is useful.

Definition 5.3.16 (Local maximality, basic set). A hyperbolic set Λ for Φ is said to be *locally maximal* or *isolated* if there is a neighborhood V of Λ (an *isolating neighborhood*) such that $\Lambda = \Lambda_\Phi^V$ (Proposition 5.1.8). If furthermore $\varphi^t\restriction_\Lambda$ has a positive semiorbit that is dense in Λ, then Λ is said to be a *basic set*.[27]

Remark 5.3.17 (Basic sets are regionally recurrent). If Λ is a basic set (Corollary 5.3.15(4))then $\mathrm{NW}(\varphi^t\restriction_\Lambda) = \Lambda$.

Example 5.3.18. A natural example of a closed invariant hyperbolic set that is not locally maximal is given by a hyperbolic periodic orbit together with the orbit of a transverse homoclinic point (see Figure 6.5.1; dynamically this is similar to Example 1.3.13 with a periodic orbit rather than a fixed point at the center of attention).

This situation appears in the horseshoe (Figure 1.5.6), for example, coded by the set Λ_0 of sequences of 0s and 1s that have no more than one 1. This set is not locally maximal since for every $N \in \mathbb{N}$ it is contained in the closed set Λ_0^N consisting of all sequences such that any two 1s are separated by at least N 0s and for any open neighborhood V of Λ_0 we have $\Lambda_0^N \subset V$ for sufficiently large N.

It is not hard to see that Λ_0^N is indeed locally maximal, and for any neighborhood V of Λ_0 there is an invariant locally maximal hyperbolic set $\widetilde{\Lambda}$ such that $\Lambda_0 \subset \widetilde{\Lambda} \subset V$.

[27] This notion appears to go back to Anosov [11].

Indeed, although any closed invariant subset of the horseshoe is hyperbolic and may have an extremely complicated structure, it can always be enveloped by a locally maximal one (such as Λ_V^f for an appropriate open neighborhood V as in Proposition 5.1.8).

Remark 5.3.19 (Reader beware!). In general, however, if Λ is a hyperbolic set and V an open neighborhood of Λ, there may not exist a locally maximal hyperbolic invariant set $\widetilde{\Lambda}$ such that $\Lambda \subset \widetilde{\Lambda} \subset V$ (Theorem 6.7.1). Since this is a rather recent discovery, and it is easy to believe otherwise, it is often assumed implicitly in the literature that a hyperbolic set is either locally maximal or included in a locally maximal set, so readers may need to check whether this is used as an unstated assumption.

We note that Theorem 5.3.37, one of the central results of this chapter, implies local maximality.

Proposition 5.3.20. *If Λ is a locally maximal hyperbolic set, then for sufficiently small $\delta > 0$ there exist $\gamma > 0$ and $\epsilon \in (0, \gamma)$ such that any ϵ-pseudo orbit that stays within γ of Λ is δ-shadowed by a point* in Λ.

Proof. Let U be an isolating neighborhood of Λ, and $\eta > 0$ such that $\bigcup_{x\in\Lambda} B_\eta(x) \subset U$. Let $\delta = \eta/2$ and fix $\epsilon_1 > 0$ such that any ϵ_1-pseudo orbit in Λ is $\delta/2$-shadowed. By uniform continuity of Φ there exists $\gamma \in (0, \delta/4)$ and $\epsilon \in (0, \gamma)$ such that any ϵ-pseudo orbit $g : I \to X$ that stays within γ of Λ has an ϵ_1-pseudo orbit $g' : I \to X$ such that $d(g(t), g'(t)) < \delta/2$ for all $t \in I$. Then g' is $\delta/2$-shadowed by a point in Λ, and this implies that the pseudo orbit g is δ-shadowed by a point in Λ. □

We now have the following immediate consequence:

Corollary 5.3.21. *The restriction of a flow to a locally maximal hyperbolic set has the shadowing property.*

This shows that if V is sufficiently small and Λ is a basic set that does not consist of a single fixed point, then the shadowing orbits in all prior results are in Λ and Λ has many periodic orbits.

Corollary 5.3.22. *If Λ is a locally maximal hyperbolic set for Φ, then periodic points are dense in* $\mathrm{NW}(\Phi_{\restriction\Lambda})$. *In particular, periodic points are dense in basic sets.*

To give another expression of the abundance of closed orbits, we show a precursor of a result that periodic data determine whether a function[28] is a coboundary (Theorem 6.3.2). Suppose f is null cohomologous (Definition 1.3.18). Then $\varphi^T(x) = x \Rightarrow \int_0^T f(\varphi^t(x)\,dt = F(\varphi^T(x)) - F(x) = 0$. For Walters-continuous functions, this necessary condition for being null cohomologous is sufficient.

[28] Or, rather, a cocycle.

Theorem 5.3.23 (Topological Livshitz Theorem). *Let Λ be a basic set for a flow Φ generated by a vector field X, f Walters-continuous for Φ (Definition 4.3.19). If $\varphi^T(x) = x$ implies $\int_0^T f(\varphi^t(x))\,dt = 0$, then f is null cohomologous (Definition 1.3.18), that is, there is a continuous $F\colon \Lambda \to \mathbb{R}$ with $f = XF$, the derivative in the flow direction and F is unique up to an additive constant.*

Proof. Uniqueness: If $\Lambda = \overline{\mathcal{O}(x_0)}$ and $XF = XF'$, then $X(F - F') \equiv 0$, so $F - F'$ is constant on the dense orbit, hence constant. Existence: Set $F(\varphi^t(x_0)) := \int_0^t f(\varphi^s(x_0))\,ds$. We show that F is uniformly continuous on $\mathcal{O}(x_0)$. This implies that F has a unique continuous extension to $\Lambda = \overline{\mathcal{O}(x_0)}$, and since f and XF are continuous and agree on a dense set, they coincide, concluding the proof.

Given $\epsilon > 0$ take $\delta < \epsilon/2\|f\|_\infty$ as in Bowen-boundedness (4.3.8) for $\epsilon/2$ and $\eta = \delta/L$ with L as in the Anosov Closing Lemma (Theorem 5.3.11). If $t_1 < t_2$ and $d(\varphi^{t_1}(x_0), \varphi^{t_2}(x_0)) < \eta$, then $\varphi^{[t_1,t_2]}(x_0)$ is δ-shadowed by a T-periodic point y with $|T - t_2 + t_1| < \delta$, so $d^{\Phi}_{t_2-t_1}(x_0, y) < \delta$. Then

$$|\underbrace{S_{t_2-t_1} f(x_0)}_{=F(\varphi^{t_2}(x_0))-F(\varphi^{t_1}(x_0))} - \underbrace{S_{t_2-t_1} f(y)}_{=S_{T-t_2+t_1} f(y)}| < \epsilon/2,$$

and $|F(\varphi^{t_2}(x_0)) - F(\varphi^{t_1}(x_0))| < \epsilon/2 + |S_{T-t_2+t_1} f(y)| < \epsilon/2 + \delta\|f\|_\infty < \epsilon$. □

Remark 5.3.24. While interesting, this result is limited in applications because we do not yet have a ready supply of Walters-continuous functions. We soon will; see Theorem 6.3.2 and Proposition 7.3.1.

The next consequence of shadowing is that being asymptotic to a compact locally maximal hyperbolic set implies being asymptotic to a specific point in that set. To formalize this, the *local* counterparts of the stable and unstable sets of a point (Definition 1.3.26) are defined by

$$W^s_\epsilon(x) := \{y \in W^s(x) \mid d(\varphi^t(x), \varphi^t(y)) \le \epsilon \text{ for } t \ge 0\},$$

and

$$W^u_\epsilon(x) := \{y \in W^u(x) \mid d(\varphi^t(x), \varphi^t(y)) \le \epsilon \text{ for } t \le 0\}$$

and are called the *local stable and unstable sets* respectively, or more precisely the *ϵ-local stable and unstable sets.*

Theorem 5.3.25 (In-Phase Theorem). *If Λ is a compact locally maximal hyperbolic set for Φ on M, then with the terminology of Definition* 1.5.5 *and* (1.3.1),

$$W^s(\Lambda) = \bigcup_{x\in\Lambda} W^s(x) \quad \textit{and} \quad W^u(\Lambda) = \bigcup_{x\in\Lambda} W^u(x),$$

and for each $\epsilon > 0$, Λ *has a neighborhood* U *with* $\bigcap_{t\geq 0}\varphi^{-t}(U) \subset W^s_\epsilon(\Lambda) := \bigcup_{x\in\Lambda} W^s_\epsilon(x)$ *(and analogously for* W^u*).*

Remark 5.3.26. Here "$\supset$" follows from the definition, and "$\subset$" says that a point asymptotic to Λ approaches Λ in a way that is "in phase" with an orbit of Λ.

Proof. If $y \in W^s(\Lambda)$ and $\eta > 0$, then there is a $T > 0$ such that for all $t \geq T$ we have an $x_t \in \Lambda$ with $d(\varphi^t(y), x_t) < \eta$ (Proposition 1.5.7(4)). If $\epsilon > 0$ and δ is as in the Shadowing Lemma (Theorem 5.3.3), then by uniform continuity of φ^1 we can choose η such that

$$d(\varphi^1(x_t), x_{t+1}) \leq d(\varphi^1(x_t), \varphi^1(\varphi^t(y))) + d(\varphi^{t+1}(y), x_{t+1}) < \delta,$$

so $(x_t)_{t\geq T}$ is ϵ-shadowed by some $x \in \Lambda$. Then $y \in W^s(x)$ because

$$t \geq T \Rightarrow d(\varphi^t(y), \varphi^t(x)) \leq d(\varphi^t(y), x_t) + d(x_t, \varphi^t(x)) \leq \delta + \epsilon. \qquad \square$$

We note from this that unstable sets for attractors are contained in the attractor.

Theorem 5.3.27 (Attractor, unstable set). *If* Λ *is a hyperbolic attractor for a flow* Φ, *then* $W^u(\Lambda) \subset \Lambda$.

Proof. There are a trapping region U for Λ and $\epsilon > 0$ such that $W^u_\epsilon(\varphi^t(x)) \subset U$ for each $x \in \Lambda$ and $t \in \mathbb{R}$. Then $W^u(x) \subset \bigcap_{t\geq 0}\varphi^t(U) = \Lambda$ since

$$t \geq 0 \Rightarrow W^u(x) = \varphi^t(\underbrace{W^u(\varphi^{-t}(x))}_{=\bigcup_{s\geq 0}\varphi^s(W^u_\epsilon(\varphi^{-s-t}(x)))\subset U}) \subset \varphi^t(U). \qquad \square$$

Arguing as in the proof of Theorem 5.3.25 we obtain the following proposition:

Proposition 5.3.28. *A hyperbolic set is locally maximal if and only if the restriction to it has the shadowing property.*

In his seminal paper, Smale introduced the following property to focus on dynamical systems for which hyperbolicity is the dominant feature:

Definition 5.3.29 (Axiom A). A flow Φ satisfies *Axiom A* if $\mathrm{NW}(\Phi)$ is hyperbolic and $\overline{\mathrm{Per}(\Phi)} = \mathrm{NW}(\Phi)$.

Remark 5.3.30. Analogously to Remark 5.1.2, our definition of Axiom A allows for hyperbolic fixed points, whereas Smale's original definition of Axiom A excluded singularities (he used "Axiom A′" as the name for our Axiom A). Our choice follows Bowen's terminology.

The second feature in this axiom is slightly stronger than what the Anosov Closing Lemma would imply from the first one; Smale thought it possible that it is a consequence of the hyperbolicity of NW(Φ), and he was "generically right": although any manifold of dimension at least 4 supports a flow whose nonwandering set is hyperbolic, but which is not Axiom A [108], for C^1-generic flows the nonwandering set is the closure of the periodic points (Theorem 1.5.22), so if the nonwandering set is hyperbolic, then it generically satisfies Axiom A. Corollary 5.3.15(2) implies the following proposition:

Proposition 5.3.31. *If a flow* Φ *satisfies* Axiom A, *then* $\mathrm{NW}(\Phi\restriction_{\mathrm{NW}(\Phi)}) = \mathrm{NW}(\Phi)$.

Corollary 5.3.15(1) implies the following result (see Definitions 1.5.33, 1.5.1, and 1.5.11):

Proposition 5.3.32. *If* $\mathcal{R}(\Phi)$ *is hyperbolic, then* Φ *satisfies Axiom A and* $\overline{\mathrm{Per}(\Phi)} = \mathcal{B}(\Phi) = \mathcal{L}(\Phi) = \mathrm{NW}(\Phi|_{\mathrm{NW}(\Phi)}) = \mathrm{NW}(\Phi) = \mathcal{R}(\Phi)$ *(and more; see Theorem* 5.3.44*).*

Transitive Anosov flows satisfy Axiom A by the Anosov Closing Lemma. The suspension of an Axiom-A diffeomorphism (defined analogously) is an Axiom-A flow.

The chain decomposition (Proposition 1.5.35) is particularly effective here because of the following observation.

Proposition 5.3.33. *If* $\Lambda = \overline{\mathrm{Per}(\Phi)}$ *is hyperbolic,*[29] *then there is an* $\epsilon > 0$ *such that any periodic points* p, q *with* $d(p,q) < \epsilon$ *are chain equivalent. In particular, the chain components* Λ_i *of* $\Phi\restriction_\Lambda$ *are open and hence finite in number. Moreover, each* Λ_i *is topologically transitive.*

Proof. If ϵ is small enough for Theorem 5.3.3, then the concatenation of $\varphi^{(-\infty,0)}(p)$ and $\varphi^{[0,\infty)}(q)$ is $L\epsilon$-shadowed by a ("heteroclinic") $\mathcal{O}(z)$. Uniqueness and Proposition 1.8.4 give $\alpha(z)\cap\mathcal{O}(p) \neq \varnothing \neq \omega(z)\cap\mathcal{O}(q)$. For any desired $\rho > 0$, concatenation of $\varphi^{(-\infty,T_1)}(p)$, $\varphi^{[T_1,T_2)}(z)$, and $\varphi^{[T_2,\infty)}(q)$ for suitable T_1, T_2 then includes a ρ-chain from p to q.

Note that Λ_i is topologically transitive because if $U, V \subset \Lambda_i$ are open and $\epsilon > 0$, there is a periodic ϵ-chain in Λ that meets both, and for small enough ϵ, so does the shadowing periodic orbit $\mathcal{O} \subset \mathrm{Per}(\Phi) \subset \overline{\mathrm{Per}(\Phi)} = \Lambda$ from Theorem 5.3.11. □

Remark 5.3.34. A pertinent variant of chain equivalence (Definition 1.5.33) would be $x \sim_\epsilon y :\Leftrightarrow x \in \mathcal{R}_\epsilon(y)\ \&\ y \in \mathcal{R}_\epsilon(x)$, and in this case the equivalence classes are obviously open. Proposition 5.3.33 shows that this stabilizes in the present context, that is, "$\sim$"="$\sim_\epsilon$" for small ϵ.

[29] Or Φ is expansive with shadowing.

Remark 5.3.35. The arguments for Proposition 5.3.33 prove that if Λ is hyperbolic and periodic points are dense in Λ, then Λ has a finite chain decomposition into clopen transitive sets.

Proposition 5.3.14, Theorem 1.5.39, and Proposition 5.3.33 imply the following corollary:

Corollary 5.3.36. *If either Φ or $\Phi_{\restriction \mathcal{R}(\Phi)}$ is expansive with shadowing, then the chain components of Φ are open in $\mathcal{R}(\Phi)$, so by compactness they are finite in number and admit a filtration* (Theorem 1.5.51).

We now show that the chain components are basic sets.

Theorem 5.3.37 (Spectral Decomposition, Smale [339]). *In each of the following situations Λ is a finite disjoint union of basic sets Λ_i (hence locally maximal).*

(1) $\Lambda = \mathrm{NW}(\Phi_{\restriction K})$ *for some compact locally maximal hyperbolic set K.*

(2) $\Lambda = \mathrm{NW}(\Phi)$ *and Φ satisfies Axiom A.*

(3) $\Lambda = \mathcal{R}(\Phi)$ *is hyperbolic.*

(4) $\Lambda = \mathcal{L}(\Phi)$ *is hyperbolic.*

(5) $\Lambda = \overline{\mathrm{Per}(\Phi)}$ *is hyperbolic.*

Proof. We first show that Λ is a finite union of transitive hyperbolic Λ_i.

In case (1), the Λ_i are the intersections of Λ with the chain components of $\Phi_{\restriction K}$ (which is expansive with shadowing, so Corollary 5.3.36 applies). To see that they are transitive suppose $U, V \subset \Lambda_i$ are open and $\epsilon > 0$. There is a periodic ϵ-chain in K that meets both, and for small enough ϵ, so does the shadowing periodic orbit $\mathcal{O}$ from Theorem 5.3.11, which lies in an isolating neighborhood, hence in K by local maximality, then in $\Lambda = \mathrm{NW}(\Phi_{\restriction K})$ by periodicity. Thus, Proposition 1.6.9(4) holds.

In the remaining cases $\Lambda = \overline{\mathrm{Per}(\Phi)}$ (Corollary 5.3.15), so the chain components Λ_i of $\Phi_{\restriction \Lambda}$ are open, finite in number, and transitive (Proposition 5.3.33).

We will show local maximality of Λ_i in a similar fashion in all cases. We find a sufficiently small neighborhood U of Λ_i such that Λ_Φ^U is hyperbolic and show that any $x \in \Lambda_\Phi^U$ accumulates on an $x^+ \in \Lambda_i$ in positive time and on an $x^- \in \Lambda_i$ in negative time; the periodic chain made from this and an orbit segment in Λ_i from near x^+ to near x^- then is shadowed by a closed orbit that helps show $x \in \Lambda_i$.

In case (1), take U disjoint from the other Λ_j. If $x \in \Lambda_\Phi^U$, then $x \in K$ by local maximality of K, so $\varnothing \neq \omega(x) \subset K$, that is, $\mathcal{O}^+(x)$ accumulates on an $x^+ \in \Lambda = \mathrm{NW}(\Phi_{\restriction K})$ and likewise with an x^- in the α-limit set, so with a segment of a dense orbit in Λ_i from near x^+ to near x^- we get a closed chain, and by Theorem 5.3.11, $x \in \overline{\operatorname{Per}(\Phi_{\upharpoonright_K})} \subset \operatorname{NW}(\Phi_{\upharpoonright_K}) = \Lambda$, hence $x \in \Lambda_i$.

In cases (2), (3), and (4), we find that $x \in \Lambda_\Phi^U \Rightarrow \omega(x) \cup \alpha(x) \subset \mathcal{L}(\Phi) \subset \Lambda$, so $\omega(x) \cup \alpha(x) \subset \Lambda_i$, and as before, $O^+(x)$ accumulates on an $x^+ \in \Lambda_i$ and $O^-(x)$ accumulates on an $x^- \in \Lambda_i$, so with a segment of a dense orbit in Λ_i from near x^+ to near x^- we get a closed chain, and by Theorem 5.3.11, $x \in \overline{\operatorname{Per}(\Phi)} = \Lambda$, hence $x \in \Lambda_i$.

While case (5) implies cases (2), (3), and (4), we argued those separately because in this case the argument is slightly more involved. Let V be a neighborhood of Λ that is disjoint from the other Λ_j and such that Λ_Φ^V is hyperbolic. If U is a neighborhood of Λ such that $\overline{U} \subset V$ and $x \in \Lambda_\Phi^u$, then $\omega(x) \cup \alpha(x) \subset \Lambda_i$ as follows. If $x^+ \in \omega(x)$ then either x^+ is fixed and hence in $\overline{\operatorname{Per}(\Phi)} = \Lambda$, so in Λ_i. Otherwise, taking $t_n \longrightarrow +\infty$ such that $\varphi^{t_n}(x) \longrightarrow x^+$ gives closed ϵ_n-chains through x^+ with $\epsilon_n \longrightarrow 0$ using $\varphi^{(t_n, t_{n+1})}(x)$, which are $L\epsilon_n$-shadowed by closed orbits, so $x^+ \in \overline{\operatorname{Per}(\Phi)} = \Lambda$, and $x^+ \in \Lambda_i$. Likewise with $\alpha(x)$. Now the previous argument applies: $O^+(x)$ accumulates on an $x^+ \in \Lambda_i$ and $O^-(x)$ accumulates on an $x^- \in \Lambda_i$, so with a segment of a dense orbit in Λ_i from near x^+ to near x^- we get a closed chain, and by Theorem 5.3.11, $x \in \overline{\operatorname{Per}(\Phi)} = \Lambda$, hence $x \in \Lambda_i$. □

Remark 5.3.38. The Λ_i can also be described by an equivalence relation defined in terms other than chain equivalence (Section 6.2). A remarkable variant of the spectral decomposition appears in Theorem 8.3.5 and the constructions described thereafter.

Remark 5.3.39. This is a good moment to emphasize a distinction from the corresponding decomposition for discrete-time dynamics. In that context, the basic sets are topologically transitive, but can be further decomposed into topologically mixing components. For flows this is in general *not* possible, as illustrated by suspensions.

Proposition 5.3.40. *Let Φ be a flow such that either $\mathcal{L}(\Phi)$ or $\mathcal{R}(\Phi)$ is hyperbolic or Φ is Axiom A (Definition* 5.3.29*). Then $M = \bigcup_{i=1}^m W^s(\Lambda_i) = \bigcup_{i=1}^m W^u(\Lambda_i)$ with each union disjoint, where $\{\Lambda_i\}_{i=1}^k$ is the Spectral Decomposition Theorem (Theorem* 5.3.37*).*

Proof. There are pairwise disjoint open $U_i \supset \Lambda_i$ for $i \in \{1, \dots, k\}$. If $x \in M$, then (Proposition 1.5.17) $\omega(x) \subset \Lambda = \bigcup_{i=1}^k \Lambda_i \subset \bigcup_{i=1}^k U_i$ is connected (Proposition 1.5.7(4)), so there is a unique i with $\omega(x) \subset U_i$ (and hence $x \in W^s(\Lambda_i)$). Reversing the flow shows the same for W^u. □

Remark 5.3.41. Proposition 5.3.40 and Theorem 5.3.25 imply that if we have a spectral decomposition of Λ (which is the case if $\mathcal{L}(\Phi)$ or $\mathcal{R}(\Phi)$ is hyperbolic or Φ is Axiom A) and $x \in M$, then $x \in W^s(y)$ and $x \in W^u(z)$ for some $y, z \in \Lambda$. So there are nontrivial stable and unstable sets for any point of M even if these points may not be contained in Λ.

The last few results played out quite similarly when $\mathcal{L}(\Phi)$ or $\mathcal{R}(\Phi)$ is hyperbolic or Φ is Axiom A, even though these are not equivalent. It turns out that there is a common underlying notion—which then makes a natural definition of hyperbolicity—obtained by adding an extra condition under which these three scenarios become equivalent. Specifically, we show that hyperbolicity of the chain recurrent set is equivalent to the flow satisfying Axiom A (Definition 5.3.29) and having no cycles among the basic sets as defined below.

Definition 5.3.42 (Cycles). Suppose Φ is a flow on a compact manifold M satisfying Axiom A or such that either $\mathcal{L}(\Phi)$ or $\mathcal{R}(\Phi)$ is hyperbolic. Define a partial ordering $\gg$ on the basic sets $\Lambda_1, \dots, \Lambda_n$ from the Spectral Decomposition Theorem (Theorem 5.3.37) by

$$\Lambda_i \gg \Lambda_j \text{ if } \big(W^u(\Lambda_i) \smallsetminus \Lambda_i\big) \cap \big(W^s(\Lambda_j) \smallsetminus \Lambda_j\big) \neq \varnothing.$$

A *k-cycle* consists of a sequence $\Lambda_{i_1} \gg \Lambda_{i_2} \gg \cdots \gg \Lambda_{i_k} \gg \Lambda_{i_1}$ of basic sets. The flow Φ *has no cycles* if this happens for no k.

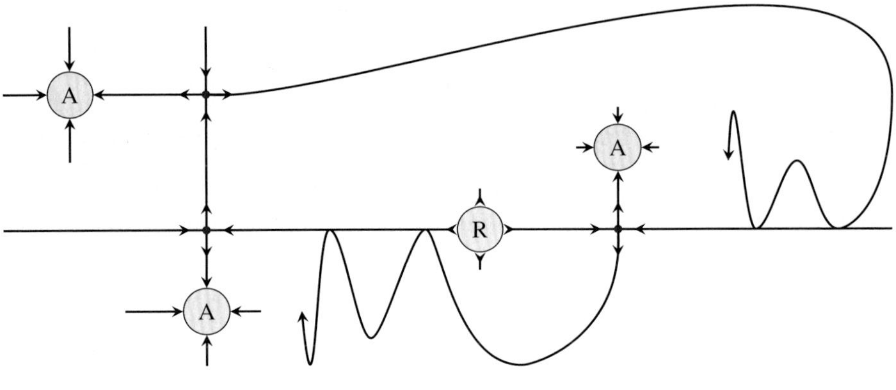

Figure 5.3.1. Axiom A with a cycle.

Remark 5.3.43. To see how the presence of a cycle is compatible with Axiom A, note that for a flow with a section as shown in Figure 5.3.1 (suggested to us by Hayashi) the nonwandering set is finite and includes a 3-cycle of saddles; the intersections of stable and unstable manifolds of successive saddles are either an interval or a tangency; the attractors and repeller (including a repeller at ∞) are strategically placed to keep the nonwandering set finite.

Cycles are precluded by having a filtration, so Corollary 5.3.36 gives the following result:

Theorem 5.3.44. *If Φ is a flow on a compact manifold M and $\mathcal{R}(\Phi)$ is hyperbolic, then Φ has no cycles.*

Theorem 5.3.45. *If $\mathcal{L}(\Phi)$ is hyperbolic and Φ has no cycles, then $\mathcal{L}(\Phi) = \mathcal{R}(\Phi)$.*

Lemma 5.3.46. *If $p \in \mathcal{R}(\Phi) \setminus \mathcal{L}(\Phi)$, then $p \in W^s(\Lambda_i)$ for some $i \in \{1,\dots,k\}$ (Remark 5.3.41), and there is a $q \in W^u(\Lambda_i) \cap \big(\mathcal{R}(\Phi) \setminus \mathcal{L}(\Phi)\big)$.*

Before proving the above lemma we show how it implies Theorem 5.3.45.

Proof of Theorem 5.3.45. We proceed by contraposition: if there is an $x_1 \in \mathcal{R}(\Phi) \setminus \mathcal{L}(\Phi)$, then $x_1 \in W^s(\Lambda_{i_1})$ for some $i_1 \in \{1,\dots,k\}$. By Lemma 5.3.46 there is an $x_2 \in W^u(\Lambda_{i_1}) \cap \big(\mathcal{R}(\Phi) \setminus \mathcal{L}(\Phi)\big)$ and hence an i_2 such that $x_2 \in W^s(\Lambda_{i_2})$. By Lemma 5.3.46, there is an $x_3 \in W^u(\Lambda_{i_2}) \cap \big(\mathcal{R}(\Phi) \setminus \mathcal{L}(\Phi)\big)$—and so on. Since there are only finitely many Λ_i, this sequence contains a cycle. □

Proof of Lemma 5.3.46. We first pick a compact neighborhood U of Λ_i such that $p \notin U$ and $\varphi^t(U) \cap \Lambda_j = \varnothing$ for $j \neq i$ and $0 \le t \le 1$.

For $n \in \mathbb{N}$ we fix a $1/n$-chain $g_n : I_n \to M$ that starts and ends at p, and a point p_n in g_n that is closest to Λ_i. Since $p \in W^s(\Lambda_i)$, we know by possibly taking a subsequence that $d(p_n, \Lambda_i) \to 0$ as $n \to \infty$. For each n let $t_n \in I_n$ such that $g_n(t_n) = p_n$. Then for n large there exists some $s_n \ge 1$ such that $g_n(t_n + s) \in \operatorname{int}(U)$ for $0 \le s < s_n$, but $q_n := g_n(t_n + s_n) \notin \operatorname{int}(U)$.

Since $p_n \xrightarrow[n\to\infty]{} \Lambda_i$ we have $s_n \xrightarrow[n\to\infty]{} \infty$. Let q be a limit point of q_n. Then $q \in \mathcal{R}(\Phi) \setminus \mathcal{L}(\Phi)$. By construction, $\varphi^t(q) \in U$ for $t < 0$. So $\alpha(q) \subset \Lambda_i$, and $q \in W^u(\Lambda_i)$. □

Theorem 5.3.47 (Axiom A, no cycles). *For a C^1-flow Φ, the following are equivalent:*

(1) *Φ satisfies Axiom A and has no cycles.*

(2) *$\mathcal{L}(\Phi)$ is hyperbolic and has no cycles.*

(3) *$\mathcal{R}(\Phi)$ is hyperbolic.*

Remark 5.3.48. By Proposition 5.3.32, the pertinent hyperbolic set is the same in these three equivalent cases: $\overline{\operatorname{Per}(\Phi)} = \mathcal{B}(\Phi) = \mathcal{L}(\Phi) = \operatorname{NW}(\Phi|_{\operatorname{NW}(\Phi)}) = \operatorname{NW}(\Phi) = \mathcal{R}(\Phi)$.

Proof. (1) ⇒ (2) from the definition (Definition 5.3.29 and Proposition 1.5.37).

(2) implies that $\mathcal{L}(\Phi)$ is the closure of the periodic points, has a spectral decomposition, and no cycles, so Theorem 5.3.45 gives $\mathcal{L}(\Phi) = \mathcal{R}(\Phi)$, hence (3).

(3) ⇒ (1) because $\mathcal{R}(\Phi) = \overline{\operatorname{Per}(\Phi)}$ (Corollary 5.3.15) and has a spectral decomposition (Theorem 5.3.37) without cycles (Theorem 5.3.44). □

Remark 5.3.49. In the literature one variously finds the assumption of Axiom A with no cycles, or of hyperbolic chain recurrent set, or of hyperbolic limit set with no cycles. By Theorem 5.3.47 these are equivalent. The variety of such usage also underscores the importance of this concept, so we make it our definition of hyperbolicity.

Definition 5.3.50 (Hyperbolic flow). A flow Φ is said to be *hyperbolic* if one of the following equivalent conditions holds:

- Φ satisfies Axiom A and has no cycles.
- $\mathcal{L}(\Phi)$ is hyperbolic and has no cycles.
- $\mathcal{R}(\Phi)$ is hyperbolic.

Following Bowen, we write

$$\mathcal{A} := \{\Phi \mid \Phi \text{ is hyperbolic}\}. \tag{5.3.1}$$

Remark 5.3.51. This notion is of a global nature compared to Definition 5.1.1. Therefore, it is less apparent that this is an open condition. Our other results about persistence of hyperbolicity are either potentially vacuous (Proposition 5.1.8), highly specialized (Corollary 5.1.9), or only imply that the *presence* of hyperbolicity is an open condition (Theorems 5.3.7, 5.4.5). However, a global counterpart (Theorem 5.4.15) to Theorem 5.4.5 does control the entire chain-recurrent set and thus finally establishes that C^1-perturbations of hyperbolic flows are themselves hyperbolic (Corollary 5.4.16).

We note an obvious consequence of spectral decomposition for Anosov flows.

Theorem 5.3.52 (Anosov flow, transitive). *For an Anosov flow Φ on a manifold M the following are equivalent*:

(1) *The spectral decomposition of Φ is $\{M\}$.*

(2) Φ *is* regionally recurrent (*Definition* 1.5.13).

(3) Φ *is topologically transitive.*

(4) *Periodic points are dense in M.*

We develop this further in Theorem 6.2.10 but mention a related observation.

Theorem 5.3.53. *The interior of the nonwandering set of an Anosov flow is either empty or the whole manifold.*

Proof [292, Lemma 4.2]. If Λ_0 is a basic set with nonempty interior, then it contains a periodic point p and a neighborhood U of p. The weak-stable and weak-unstable leaves of p are dense in Λ_0, and $W^{cu}(p) \subset \bigcup_{t\geq 0} \varphi^t(U)$, $W^{cs}(p) \subset \bigcup_{t\leq 0} \varphi^t(U)$, so $W^{cu}(p) \cup W^{cs}(p) \subset \Lambda_0$, and $\overline{W^{cu}(p)} \cup \overline{W^{cs}(p)} \subset \Lambda_0$, so Λ_0 is W^{cu}- and W^{cs}-saturated. For any $x \in \Lambda_0$ we thus have $W^{cu}(x) \cup W^{cs}(x) \subset \Lambda_0$, so density of $W^{cu}(x)$ in Λ_0, hence in $W^{cs}(x)$ implies that Λ_0 contains a product neighborhood of x, so Λ_0 is open and closed in M, hence $\Lambda_0 = M$. □

The above result can be strengthened to show that if Φ is a flow containing a transitive hyperbolic set with nonempty interior, then Φ is Anosov. We now formalize an observation from Corollary 5.3.36:

Theorem 5.3.54 (Filtration). *Let $\Phi \in \mathcal{A}$ and $\Lambda_1, \dots, \Lambda_k$ be the spectral decomposition. Then there is a filtration* $\mathbf{M}$ *of M composed of $M_0 \subset M_1 \subset \cdots \subset M_k$ such that $\Lambda_i = K_i^{\Phi}(\mathbf{M})$ for each $i \in \{1, \dots, k\}$.*

In particular, we have the following theorem:

Theorem 5.3.55 (Attracting basic set). *If $\Lambda_1, \dots, \Lambda_k$ is the spectral decomposition of a hyperbolic flow Φ, then Λ_i is an attractor if there is no j with $\Lambda_i \gg \Lambda_j$.*

Remark 5.3.56. The constructions in Theorem 8.3.13 virtually reverse-engineer this by gluing together filtrating neighborhoods of hyperbolic sets in order to produce examples of hyperbolic flows.

Volume-preserving hyperbolic flows have neither an attractor nor any cycles:

Corollary 5.3.57. *The spectral decomposition of a volume-preserving hyperbolic flow has only one piece.*

We have come a long way, and we repeat that the preceding are all consequences of the Shadowing Lemma.[30] Of course, we also have yet to prove the Shadowing Lemma. We will do so presently. First we combine shadowing with transitivity.

Bowen introduced *specification* as a notion that formalizes how shadowing and transitivity make it possible to prescribe the evolution of an orbit to the extent of specifying a finite collection of arbitrarily long orbit segments and any fixed precision: as long as one allows for enough time between the specified segments one can find a single (periodic) orbit approximating this entire itinerary and the time between the segments depends only on the quality of the approximation and not on the length of the specified segments.

Definition 5.3.58 (Specification). Let X be a compact metric space and Φ be a flow on X. Then Φ satisfies *specification* if for any $\epsilon > 0$ there exists some T_ϵ such that

[30] Or, alternatively, shadowing and expansivity.

given any finite collection of points $x_0, \dots, x_n \in X$ and times $t_0, \dots, t_n \in [0, \infty)$ there exists a point $y \in X$, and $s_0, \dots, s_n \in [0, T_\epsilon]$, and for each $i \in \{0, \dots, n\}$ we have

$$d(\varphi^t y, \varphi^{t-\sum_{j=0}^{i-1} t_j + s_j} x_i) < \epsilon \quad \text{for } t \in [0, t_i] + \sum_{j=0}^{i-1} t_j + s_j, \quad \text{and} \quad \varphi^{\sum_{i=0}^{n} t_i + s_i}(y) = y.$$

Remark 5.3.59. The "transition" times s_j here are controlled only to the extent that they need not be very long—depending on the desired accuracy of the approximation. With both more tools and stronger assumptions, we will later be able to prescribe the transition times exactly (Theorem 7.3.4).

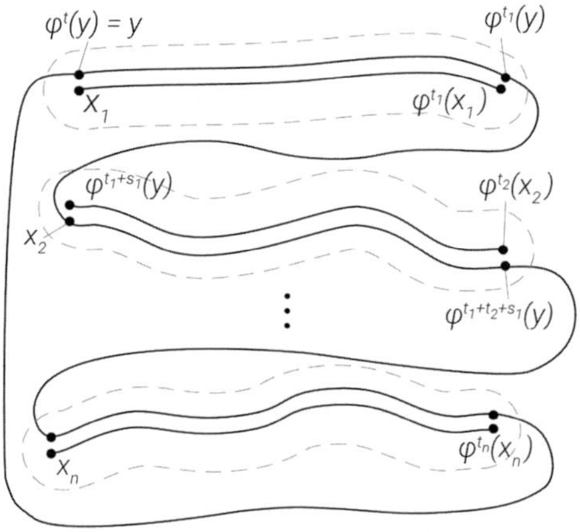

Figure 5.3.2. Specification of orbit segments.

The idea is to associate the orbit segment $\varphi^{[0,t_i]}(x_i)$ with the pair $(x_i, t_i) \in X \times [0, \infty)$. Such a collection of orbit segments has specification if given $\epsilon > 0$ there is a closed orbit that stays within ϵ of each orbit segment in turn provided we allow a transition time between the orbit segments, which can be chosen to be no more than T_ϵ.

There are variants of this definition in the literature. The orbit y might not be required to be a closed orbit, or it is required that the transition time is equal between each of the orbit segments. For the hyperbolic case, any of the various versions hold (possibly subject to assuming topological mixing), but in other situations one may need to choose one specific variant. A stronger counterpart requires that the transition time (rather than bounds on it) is prescribed (Definition 7.3.2).

Combined with expansivity, specification forces exponential orbit complexity:

Proposition 5.3.60. *An expansive continuous flow Φ with specification on a compact space X with more than one orbit has exponential growth of periodic orbits* (*Definition* 4.2.1) *and hence positive topological entropy* (*Theorem* 4.2.24).

Proof. Since $X \neq \varnothing$ there is an $x_0 \in X$, and by the specification property there is a closed orbit p that starts near x_0. Denote by T_0 its least period. Suppose $\epsilon > 0$ is an expansivity constant. With the notation of Definition 4.2.22 it suffices to show that $\#\mathbb{O}'_{T_0+T_\epsilon}(T + 2T_0 + 2T_\epsilon) \geq 2\#\mathbb{O}'_{T_0+T_\epsilon}(T)$ for all $T > T_0 + T_\epsilon$. To see this, consider $q \in \mathbb{O}'_{T_0+T_\epsilon}(T)$ and apply the specification property with the specifications q, p and q, p, p to get two (by expansivity distinct) elements of $\mathbb{O}'_{T_0+T_\epsilon}(T + 2T_0 + 2T_\epsilon)$. □

Theorem 5.3.61 (Specification Theorem). *If Λ is a basic set for a flow Φ, then $\Phi_{\restriction\Lambda}$ has the specification property.*

Proof. By Remark 1.6.11 the orbit segments of the specification can be interpolated to a closed $\epsilon/2L$-orbit by orbit segments whose length is bounded in terms of ϵ as follows. The first interpolating segment begins within $\epsilon/2L$ of $\varphi^{t_1}(x_1)$ and ends after time t_1' within $\epsilon/2L$ of x_2. The next one begins within $\epsilon/2L$ of $\varphi^{t_1+t_1'+t_2}(x_2)$ and ends within $\epsilon/2L$ of x_3, and so on; the last one ends within $\epsilon/2L$ of x_1. By Theorem 5.3.3, this pseudo-orbit is $\epsilon/2$-shadowed by an orbit, which is then as desired. □

Remark 5.3.62. The proof reveals that shadowing and transitivity together imply specification (and that shadowing and specification both hold for basic sets), though more is needed for the stronger specification property in Theorem 7.3.4: on one hand, mixing rather than just transitivity is needed, and on the other hand, finer control using the invariant foliations is essential. Conversely, however, specification implies transitivity but not shadowing because of the required transition times (strong specification implies topological mixing, but not shadowing).

Bowen's Specification Theorem, suitably strengthened, is a useful tool for the study of statistical properties of orbits within hyperbolic sets (Theorem 7.3.6). Proposition 5.3.60 gives a much simpler application:

Proposition 5.3.63. *A basic set that does not consist of a single orbit has exponential growth of periodic orbits and hence positive topological entropy.*[31]

5.4 The Anosov Shadowing Theorem, structural and Ω-stability

We finally present the shadowing result, which makes the proof of Theorem 5.3.7 easier, implies the Shadowing Lemma (Theorem 5.3.3), and leads to structural stability (Corollary 5.4.7).

[31] See also Remark 4.2.25.

Theorem 5.4.1 (Anosov Shadowing Theorem). *If M is a Riemannian manifold, Φ a C^1-flow, then any compact hyperbolic set $\Lambda \subset M$ for Φ has a neighborhood V and $\epsilon_0, \delta_0, C > 0$ such that if*

- $\psi^t : V \to M$ *is generated by a vector field* X,
- $d_{C^1}(\varphi^t, \psi^t) < \epsilon_0$ *for* $|t| \le 1$,
- N *is a topological space,*
- $\sigma^t : N \to N$ *a continuous flow,*
- $\alpha \in C^0(N, V)$ *such that* $Y := \frac{d}{dt}|_0 \alpha \circ \sigma^t$ *exists, and*
- $d_{C^0}(Y, X \circ \alpha) < \epsilon < \epsilon_0$,

then there are $\beta \in C^0(N, \Lambda_V^\psi)$ *and* $\tau\colon N \times \mathbb{R} \to \mathbb{R}$ *with*

- $\beta \circ \sigma^t = \psi^{\tau(\cdot,t)} \circ \beta$,
- $d_{C^0}(\alpha, \beta) < C\epsilon$, *and*
- $\mathrm{Lip}(\tau(x,\cdot) - \mathrm{Id}) < C\epsilon$.

Moreover, β is locally transversely unique: $\overline{\beta} \circ \sigma^t = \psi^{\tilde{\tau}(\cdot,t)} \circ \overline{\beta}$ and $d_{C^0}(\alpha, \overline{\beta}) < \delta_0 \Rightarrow \overline{\beta}(x) = \psi^{\theta(x)}(\beta(x))$ for some continuous $\theta\colon N \to [-C\delta_0, C\delta_0]$.

Remark 5.4.2. The Shadowing Lemma (Theorem 5.3.3) follows by taking $\Psi = \Phi$, $N = \mathbb{R}$, $\sigma^t(x) = x + t$, $\alpha(t) = g(t)$ (differentiable with derivative near $X \circ \alpha$; see Remark 1.5.31).

Proof. By the Whitney Embedding Theorem $M \subset \mathbb{R}^n$ for suitable n, so we take $M = \mathbb{R}^n$ without loss of generality: if the result is known for $\mathbb{R}^n$, embed M and augment M to a tubular neighborhood $U' \subset \mathbb{R}^n$ while extending Φ and a C^1-close Ψ to U' by the same contraction normal to M and apply the result. It gives a β consisting of full orbits of the extension of Ψ, so $\beta(N) \subset M$ because Ψ contracts normally to M and indeed, $\beta(N) \subset V$, hence $\beta(N) \subset \Lambda_V^\Psi$ because it consists of orbits.

Hyperbolicity is the central ingredient in the proof, and it will be used in a standard way to set up a contraction, whose fixed point is the desired object. However, hyperbolicity plays out transversely to the flow direction, so we isolate this transverse behavior with the following device. Let $X^\perp$ be the normal bundle, that is, $X_p^\perp + p = X^\perp(p)$ is the hyperplane through $p \in V$ orthogonal to X, and $X_\epsilon^\perp$ denotes the ϵ-ball around p in $X(p)$. For small enough ϵ_0 this gives well-defined projections $\pi_p\colon \psi^{[-1,1]}(X_{\epsilon_0}^\perp) \to X^\perp, \psi^t(x) \mapsto x$ for $|t| \le 1$, $x \in X_{\epsilon_0}^\perp$. We also denote by $C_\alpha^\perp(N, V)$ the space of continuous $\beta\colon N \to V$ such that $\beta(x) \in X^\perp(\alpha(x))$ for $x \in N$.

We then seek a fixed point of

$$F\colon C^{\perp}_{\alpha}(N,V) \to C^{\perp}_{\alpha}(N,\mathbb{R}^n),$$
$$\beta \mapsto \pi_\alpha \circ \psi^1 \circ \beta \circ \sigma^{-1}\colon x \mapsto \pi_{\alpha(x)}(\psi^1(\beta(\sigma^{-1}(x)))) \in X^{\perp}(\alpha(x)).$$

Represent $\beta \in C^0(Y,\mathbb{R}^n)$ by the vector field $v_\beta := \beta - \alpha \in X^{\perp}_{\alpha(\cdot)}$ (a section of the bundle $\{(y, X^{\perp}_{\alpha(y)}) \mid y \in N\}$ over N). In these terms F is represented by

$$F_\alpha\colon v \mapsto \pi_\alpha \circ \psi^1(\alpha \circ \sigma^{-1} + v \circ \sigma^{-1}) - \alpha =: (\underbrace{DF_\alpha|_0}_{\text{linear part}} + \underbrace{H}_{\text{higher-order terms}})(v).$$

Then $v = F_\alpha(v) = (DF_\alpha|_0 + H)(v) \Leftrightarrow v = -((DF_\alpha)_0 - \mathrm{Id})^{-1}H(v) =: T(v)$.

Lemma 5.4.3. *There are a neighborhood $V \supset \Lambda$, $\epsilon_0, \epsilon > 0$, and $R > 0$ independent of N, Ψ, α with $\|((DF_\alpha)_0 - \mathrm{Id})^{-1}\| < R$ when $d_{C^1}(\Phi,\Psi) < \epsilon_0$, $d_{C^0}(Y, X \circ \alpha) < \epsilon$.*

Proof. For $\delta > 0$ there are $\epsilon_0 > 0$, $\mu < 1$, and a neighborhood $V \supset \Lambda$ to which the splitting $T_\Lambda M = E^u \oplus E^s \oplus X$ extends (maybe not invariantly) for Ψ such that $d_{C^1}(\Phi,\Psi) < \epsilon_0$. Then with respect to $E^u \oplus E^s \oplus X$, we have

$$D\psi^1 = \begin{pmatrix} a_{uu} & a_{uu} & * \\ a_{us} & a_{ss} & * \\ * & * & b \end{pmatrix},$$

where $\|a_{uu}\|^{-1} < \mu$, $\|a_{ss}\| < \mu$, $\|a_{uu}\| < \delta^2\mu$, $\|a_{us}\| < \delta^2\mu$, and $\|*\| < \delta^2\mu$. With respect to the decomposition into unstable and stable vector fields in $X^{\perp}$,

$$((DF_\alpha)_0\xi)(y) = d\pi_\alpha D\psi^1|_{\alpha(\sigma^{-1}(y))}\xi(\sigma^{-1}(y)) \quad \text{splits into} \quad (DF_\alpha)_0 = \begin{pmatrix} A_{uu} & A_{uu} \\ A_{us} & A_{ss} \end{pmatrix},$$

where $d_{C^0}(\alpha, \psi^1\alpha\sigma^{-1}) < \epsilon$ and $d_{C^1}(\Phi,\Psi) < \epsilon_0$ imply

$$\|A_{uu}\|^{-1} < \frac{1+\mu}{2}, \quad \|A_{uu}\| < \delta\mu, \quad \|A_{us}\| < \delta\mu, \quad \|A_{ss}\| < \frac{1+\mu}{2}. \qquad \square$$

To show that T contracts, we control H. If $k_i(t) := H_i(v + th)$ (components with respect to the canonical basis in $\mathbb{R}^n$) then $k_i(1) - k_i(0) = \int_0^1 k_i'(t)\,dt$ gives

$$H(v+h) - H(v) = h\int_0^1 DH(v+th)\,dt = h\int_0^1 DF_\alpha|_{v+th} - DF_\alpha|_0\,dt.$$

Since ψ^1, hence F_α, is C^1, so there is a δ_0 with $\|DF_{\alpha}|_{v+th} - DF_{\alpha}|_0\| \le \frac{1}{2R}$ if $\|v\|$, $\|v + h\| < \delta_0, t \in [0, 1]$.[32] Thus

$$\|v_1\|, \|v_2\| < \delta_0 \Rightarrow \|T(v_1) - T(v_2)\| < \tfrac{1}{2}\|v_1 - v_2\|.$$

With $\theta = \min(\delta, \delta_0)$ and $\epsilon < \theta/(2R)$ as in Lemma 5.4.3, $d_{C^0}(\alpha\sigma^1, \psi^1\alpha) < \epsilon$, which follows from $d_{C^0}(Y, X \circ \alpha) < \epsilon$, gives

$$\begin{aligned}\|T(0)\| < R\|H(0)\| &= R\|\psi^1 \circ \alpha \circ \sigma^{-1} - \alpha\| \\ &= Rd_{C^0}(\alpha, \psi^1\alpha\sigma^{-1}) = Rd_{C^0}(\alpha\sigma^1, \psi^1\alpha) \le \frac{\theta}{2},\end{aligned}$$

so T is a 1/2-contraction on the closed ball of $X^\perp$-vector fields of norm up to δ_0. By the Contraction Mapping Principle (Proposition B.1.3), T has a unique fixed point, which yields the desired β. Uniqueness of the fixed point implies transverse uniqueness; θ in the statement of the theorem is continuous and small because β, $\overline{\beta}$, and Ψ are continuous, Ψ has positive speed, and β, $\overline{\beta}$ are close. □

As promised, we first use the Anosov Shadowing Theorem to prove topological stability.

Proof of Theorem 5.3.7. Note $h := \beta$ from Theorem 5.4.1 with $\Lambda = M = N$, $\Psi = \Phi$, $\sigma = \Phi'$ sufficiently C^0-close to Φ, and $\alpha = \mathrm{Id}$ is the desired factor map. To see that the map is surjective we note that it is a continuous perturbation of Id. □

If we could apply the same reasoning to Φ' to get a factor map the other way around, we would expect it to be the inverse of the h in Theorem 5.3.7, which would then be a homeomorphism. This can indeed be done if Φ' is hyperbolic but that requires C^1-closeness, rather than C^0-closeness. Accordingly, while the previous application focused on approximate orbits by taking $\Phi = \Psi$ in Theorem 5.4.1, we now obtain a profound strengthening of Proposition 5.1.8 by C^1-perturbing Φ, which is interesting because it implies that all features of the orbit structure of a hyperbolic flow are robust, that is, they persist under small perturbations of the system and hence are not sensitive to the parameters of a model from which the flow arises. In this respect it is especially interesting if the entire orbit structure is topologically unchanged under small perturbations.

Definition 5.4.4 (Structural stability). A flow Φ is said to be *structurally stable* if there is a C^1-neighborhood (Definition 1.6.18) U of Φ in the class of C^1-flows such that any flow in U is orbit equivalent to Φ (see Definition 1.3.22).

[32] δ_0 is determined by R alone, which is a measure of hyperbolicity. "More hyperbolicity" means smaller R and hence less constraint on δ_0.

We note that we have already encountered flows with this feature. This is explicit in restricted form in Proposition 1.4.5 and suggested as an exercise (about the damped pendulum) in Remark 1.6.17. When expressed globally rather than locally, the Hartman–Grobman Theorem (Corollary B.4.11 for continuous time or Theorem 5.5.3 for perturbations of linear maps obtained from localization (Theorem B.4.12)) is a similar instance. In the present context, the dynamics is incomparably more complicated, however.

Interest in structural stability first arose from Smale's agenda of classifying dynamical systems up to topological equivalence because this gives open (hence manageable) equivalence classes. Theorem 5.4.1 implies that hyperbolic systems are structurally stable, though not in the exact same sense, except in the case of Anosov flows:

Theorem 5.4.5 (Strong structural stability of hyperbolic sets). *Suppose Λ is a compact hyperbolic set for a C^1-flow Φ on M. Then there are*

- *a C^1-neighborhood U of Φ,*
- *a C^0-neighborhood V of the inclusion ι of Λ in M (which can be viewed as the identity) and*
- *a continuous map $h\colon U \to C(\Lambda, M)$, $\Psi \mapsto h_\Psi$*

such that $h_\Phi = \iota$ and for each $\Psi \in U$,

(1) *h_Ψ is a continuous embedding,*

(2) *h_Ψ is the transversely unique map in V for which $\psi^{\tau(t)} \circ h_\Psi = h_\Psi \circ \varphi^t \restriction_\Lambda$, where τ is as in Theorem 5.4.1,*

(3) *$\Lambda_\Psi := h_\Psi(\Lambda)$ is a hyperbolic set for Ψ.*

Definition 5.4.6. The map $\Psi \mapsto \Lambda_\Psi$ is called the *continuation* of Λ.

Proof. We use symmetry and uniqueness by applying the Shadowing Theorem (Theorem 5.4.1) three times.

With $0 < \epsilon < \delta_0/2$ as in the Shadowing Theorem, $Y = \Lambda$, $\alpha = \mathrm{Id}\restriction_\Lambda$, $\sigma = \Phi$, we get a transversely unique $\beta\colon \Lambda \to V$ and a monotone τ with $\beta \circ \varphi^t = \psi^{\tau(t)} \circ \beta$. By Proposition 5.1.8, $\Lambda' := \beta(\Lambda)$ is hyperbolic.

To show that β is injective, apply the Shadowing Theorem the other way around: with ϵ as before, $Y = \Lambda'$, $\alpha' = \mathrm{Id}\restriction_{\Lambda'}$, interchange Φ and Ψ (which we can do if ϵ is small enough) to obtain β' with $\beta' \circ \psi^t \restriction_\Lambda = \varphi^{\tau'(t)} \circ \beta'$.

To see that β is a homeomorphism note that $h := \beta' \circ \beta$ satisfies

$$\beta' \circ \beta \circ \varphi^t = \beta' \circ \psi^{\tau(t)} \circ \beta = \varphi^{\tau'(\tau(t))} \circ \beta' \circ \beta,$$

where $\tau'(\tau(t))$ is increasing and close to t. Since Id does the same ($\mathrm{Id} \circ \varphi^t = \psi^t \circ \mathrm{Id}$), transverse uniqueness in Theorem 5.4.1 implies that $\beta'(\beta(x)) = \psi^{\theta(x)}\,\mathrm{Id}(x)$, where θ

can be taken increasing by the last argument in the proof of Theorem 1.7.5, and then $\beta' \circ \beta$ is surjective by continuity of θ. □

Corollary 5.4.7. *Anosov flows are structurally stable. The orbit equivalence is unique up to small time shifts when chosen near the identity.*

Remark 5.4.8. This proof of structural stability ultimately relies on the Contraction Mapping Principle because this was the main device used in the proof of the Shadowing Theorem. The fixed point of a contraction depends smoothly on the contraction when this is meaningful in a given application, and accordingly, the conjugacy given by structural stability of a hyperbolic C^{k+1}-embedding depends C^k on the perturbation (in the C^0-topology for conjugacies) [248].[33]

Remark 5.4.9. We emphasize that the orbit equivalence in Corollary 5.4.7 is rarely a conjugacy (Remark 9.2.2).

Remark 5.4.10. It is natural to ask whether for small enough perturbations the conjugacy is more regular. A suggestive result by Palis–Viana [275] and de la Llave (unpublished) says that in dimension 2 (and discrete time) the Hölder exponent of the conjugacy is close to 1, that is, for each $\alpha \in (0, 1)$ there is a neighborhood of the dynamical system such that the conjugacy between the respective hyperbolic sets has Hölder exponent α for each perturbation in this neighborhood. De la Llave also showed, however, that this fails in higher dimension.

Another question about structural stability is how large a perturbation can be without ceasing to be orbit equivalent to the given flow. Numerous subjects in dynamics would benefit from general results along these lines, but there are few. Among them is a remarkable criterion for magnetic flows (Definition 5.2.18).

Theorem 5.4.11 ([159, Théorème 3.2]). *If M is a Riemannian manifold whose sectional curvatures K satisfy $-k_2^2 \le K \le -k_1^2 < 0$, then all magnetic flows with $\|\mathfrak{m}\|_\infty < k_1$ are pairwise orbit equivalent.*

We later address the question of how regular (beyond continuity) the orbit equivalence is (that is, the homeomorphism in its definition). Note that this is not a well-defined question because it is not "the" homeomorphism—we only have transverse uniqueness. Therefore the question is how regular this h can be chosen. Proposition 1.3.30 describes the extent to which altering a given choice of orbit equivalence can improve the regularity of the dependence on time. Transverse regularity is fixed by transverse uniqueness, and we will study that later (Theorem 6.4.3).

[33] Here the loss of one derivative is related to the fact that an operator such as $\beta \mapsto g \circ \beta \circ \sigma^{-1}$ in the proof of the Shadowing Theorem that involves a composition is C^k if the maps in question are C^{k+1}.

One of the crowning achievements in dynamical systems is well beyond the scope of this text but we state it here to complete the picture: hyperbolicity is indeed *equivalent* to structural stability as follows:

Theorem 5.4.12 (Hayashi [349, 187, 188]). *A C^1-flow Φ is C^1-structurally stable iff Φ satisfies Axiom A and* strong transversality:[34] $W^s(\gamma_1) \cap W^u(\gamma_2) = W^s(\gamma_1) \pitchfork W^u(\gamma_2)$ *for any orbits γ_1 and γ_2 in* $\mathrm{NW}(\Phi)$.

Remark 5.4.13. A counterpart for C^r-stability with $r > 1$ seems quite out of reach, even for surface diffeomorphisms. Hyperbolicity of the nonwandering set is obtained from stability using periodic orbits, so a C^2 closing lemma or C^2 general density theorem would be needed—which is among the most difficult problems in dynamics. Furthermore, uniform (weak) hyperbolicity over all periodic orbits is derived from the so-called Franks Lemma, which is also possible only in the C^1-topology.

We next produce a corresponding "global" stability result (Theorem 5.4.15), albeit only the "if" part. (The major achievement of Hayashi is the much harder "only if" direction.) This is a proper "global" counterpart to the Structural Stability Theorem (Theorem 5.4.5). Not only does a compact locally maximal hyperbolic set persist under perturbation, and with topologically identical dynamics, but for an Axiom-A flow with the no-cycles property the nonwandering set of a perturbation is topologically and dynamically the same as that for the original flow. This is called Ω-stability because the original notation for the nonwandering set was Ω. Our preceding work easily produces a nominally stronger version of this classical result (stability of $\mathcal{R}(\Phi)$ rather than just of $\mathrm{NW}(\Phi)$), but we refer to it by the original name.

Definition 5.4.14. For $r \geq 1$ a C^r-flow Φ is said to be (C^r-) *Ω-stable*, and $\mathcal{R}(\Phi)$ is said to be (C^r-) *stable* if for any flow Ψ that is sufficiently C^r-close to Φ (see Definition 1.6.18) there is an orbit equivalence (in a given neighborhood of the identity) between $\Phi\restriction_{\mathcal{R}(\Phi)}$ and $\Psi\restriction_{\mathcal{R}(\Psi)}$.

Theorem 5.4.15 (Ω-Stability Theorem). *C^1 hyperbolic flows*[35] *are C^1-Ω-stable.*

Proof. By the Spectral Decomposition Theorem (Theorem 5.3.37) there are disjoint basic sets $\Lambda_1, \dots, \Lambda_m$ for Φ such that $\mathcal{R}(\Phi) = \bigcup_{i=1}^m \Lambda_i$, and without cycles. Furthermore, $\Lambda_i = W^u(\Lambda_i) \cap W^s(\Lambda_i)$ for each i, and we choose isolating open sets U_i for Λ_i with pairwise disjoint closures.

If Ψ is a flow sufficiently C^1-close to Φ, then by the Structural Stability Theorem (Theorem 5.4.5) there are disjoint hyperbolic basic sets $\tilde{\Lambda}_i \subset U_i$ for Ψ that are

[34] This notion uses stable and unstable *manifolds*; see Theorem 6.1.1.

[35] As in Definition 5.3.50.

orbit equivalent to Λ_i for each $i \in \{1,\dots,m\}$, and for which the U_i are isolating neighborhoods. Therefore, it suffices to show that $\mathcal{R}(\Psi) \subset \bigcup_{i=1}^m U_i$.

We show $\mathrm{NW}(\Psi) \subset \bigcup_{i=1}^m U_i$ (with a proof different from [302]), so Ψ satisfies Axiom A with no cycles, and $\mathcal{R}(\Psi) = \mathrm{NW}(\Psi) \subset \bigcup_{i=1}^m U_i$ (Remark 5.3.48).

Suppose to the contrary that there is a sequence of flows $\Phi_n \xrightarrow[n\to\infty]{C^r} \Phi$ such that for each n there is a nonwandering point of Φ_n in $M \smallsetminus \bigcup_{i=1}^m U_i$. Since $M \smallsetminus \bigcup_{i=1}^m U_i$ is compact, there exists a point $p \in M \smallsetminus \bigcup_{i=1}^m U_i$ and a sequence of points $p_n, q_n \in M$ converging to p (by possibly replacing Φ_n with a subsequence) and a sequence of times $t_n \xrightarrow[n\to\infty]{} \infty$ with $\varphi_n^{t_n}(p_n) = q_n$. Then there exist some $i_0 \neq i_1 \in \{1,\dots,m\}$ with $p \in W^u(\Lambda_{i_0}) \cap W^s(\Lambda_{i_1})$ by Proposition 5.3.40.

By continuity of the flow and passing to a subsequence there is a sequence of times $t_n^1 \in (0,t_n)$ such that $\varphi_n^{t_n^1}(p_n) \xrightarrow[n\to\infty]{} \Lambda_{i_1}$. Since $\varphi_n^{t_n}(p_n) = q_n \to p$, there is a sequence $T_n^1 \in (t_n^1, t_n)$ such that $\varphi_n^{T_n^1}(p_n) \in \partial U_{i_1}$ and $\varphi_n^{[t_n^1,T_n^1)}(p_n) \subset U_i$. Since Λ_{i_1} is Φ-invariant and $\varphi_n^{T_n^1}(p_n) \notin U_{i_1}$, we have $T_n^1 - t_n^1 \xrightarrow[n\to\infty]{} \infty$.

By compactness of ∂U_{i_1} there is a subsequence with $\varphi_n^{T_n^1}(p_n) \to x_1 \in \partial U_{i_1}$. Since $T_n^1 - t_n^1 \to \infty$ we know that $\mathcal{O}^-(x_1) \subset U_{i_1}$ and so $x_1 \in W^u(\Lambda_{i_1})$. Since $x_1 \notin \mathcal{R}(\Phi)$ there is an i_2 with $x_1 \in W^s(\Lambda_{i_2})$ and $i_2 \notin \{i_0, i_1\}$ by construction: $i_2 \neq i_1$ because $x_1 \in W^u(\Lambda_{i_1}) \smallsetminus \Lambda_{i_1} \Rightarrow x_1 \notin W^s(\Lambda_{i_1})$, and $i_2 \neq i_0$ because otherwise we have a cycle and are done.

Arguing likewise with x_1 gives a sequence of times $t_n^2 \xrightarrow[n\to\infty]{} \infty$ such that $t_n^2 \in (T_n^1, t_n^1)$ and $\varphi_n^{t_n^2}(p_n) \xrightarrow[n\to\infty]{} \Lambda_{i_2}$, and a sequence of times $T_n^2 \in (t_n^2, t_n)$ such that $\varphi_n^{T_n^2}(p_n) \in \partial U_{i_2}$ and $\varphi_n^t(p_n) \in U_{i_2}$ for $t \in [t_n^2, T_n^2)$. Furthermore, as before we have $T_n^2 - t_n^2 \to \infty$ as $n \to \infty$ by possibly taking a subsequence.

As before, by taking a subsequence as necessary we obtain $\varphi_n^{T_n^2}(p_n) \xrightarrow[n\to\infty]{} x_2 \in \partial U_{i_2}$, and $x_2 \in W^u(\Lambda_{i_2}) \cap W^s(\Lambda_{i_3})$ for some $i_3 \notin \{i_0, i_1, i_2\}$: $i_3 \neq i_2$ because $W^u(\Lambda_{i_2}) \cap W^s(\Lambda_{i_2}) = \Lambda_{i_2}$ and $i_3 \notin \{i_0, i_1, i_2\}$ because otherwise we have a cycle and are done.

Continuing in this manner we obtain a sequence of points $\{x_n\}$, which for $n \geq m$ contradicts the no-cycles assumption. Hence, $\mathcal{R}(\Psi) = \bigcup_{i=1}^m \tilde{\Lambda}_i$ for Ψ sufficiently C^r-close to Φ. □

Corollary 5.4.16. *Hyperbolicity* (*Definition* 5.3.50) *is a C^1-open condition.*[36]

We previously mentioned the Structural Stability Theorem (Theorem 5.4.12) as a high point in dynamics, and it comes with a counterpart to Theorem 5.4.15—C^1-Ω-stability characterizes hyperbolicity:

---

[36] See also Theorem 6.1.6.

Theorem 5.4.17 (Hayashi [349, 187, 188]). *C^1-Ω-stable flows are hyperbolic.*[37]

These results (together with the discrete-time counterparts) were long known as the (Palis–)Smale Stability Conjectures. While a proof is far outside our scope, we briefly discuss some of the ingredients.

Proof ingredients. The principal contribution was Mañé's proof of the discrete-time counterpart to Theorem 5.4.12, on which Palis quickly built his proof of the discrete-time counterpart of Theorem 5.4.17. Hayashi overcame the formidable additional difficulties for flows.

A promising approach to Theorem 5.4.17 is to use contraposition: use any failure of hyperbolicity to make a change in the orbit structure that disproves Ω-stability. This is still a formidable task, and Axiom A (for instance) also requires density of periodic points $\mathrm{NW}(\Phi)$.

To see this, a simple general principle is helpful: any C^1-generic property of $\mathrm{NW}(\Phi)$ that is invariant under orbit equivalence holds for Ω-stable flows. This principle holds because invariance means that it either holds for all flows in a C^1-neighborhood (and hence for the Ω-stable flow itself) or it fails for every flow in a neighborhood—contrary to its genericity. Applying this to Pugh's General Density Theorem (Theorem 1.5.22) tells us that one part of Axiom A automatically holds for Ω-stable flows: $\overline{\mathrm{Per}(\Phi)} = \mathrm{NW}(\Phi)$.

With Theorem 6.1.6, this principle further tells us that fixed and periodic points of Ω-stable flows are hyperbolic. With this, it looks like we are close. Yet lots of hard work awaits—eased by having hyperbolicity at our disposal.

The fundamental hurdle in the case of flows, which has no discrete-time counterpart, is the need to rule out fixed points in the closure of the periodic ones (Definition 5.1.1). Thus, in order to otherwise follow the strategy implemented by his predecessors in the discrete-time context, Hayashi needed to prove the following theorem:

Theorem 5.4.18 ([188]). *The closure of the periodic orbits (not fixed points) contains no fixed points.*

It is here that his most prominent contribution enters.

Theorem 5.4.19 (Hayashi Connecting Lemma [188]). *Consider a flow Φ with an isolated hyperbolic set Λ which has an almost homoclinic point, that is to say, $\big(\overline{W^s(\Lambda)} \cap W^u(\Lambda)\big) \cup \big(\overline{W^u(\Lambda)} \cap W^s(\Lambda)\big) \smallsetminus \Lambda \neq \varnothing$.*

Then for every C^1-neighborhood U of Φ there is a $\Psi \in U$ that agrees with Φ on a neighborhood of Λ and has a homoclinic point, that is, $\big(W^u(\Lambda) \cap W^s(\Lambda)\big) \smallsetminus \Lambda \neq \varnothing$.

[37] As in Definition 5.3.50.

The proof uses delicate perturbations along the lines of the Pugh Closing Lemma (Theorem 1.5.21). This helps prove Theorem 5.4.18 as follows. If periodic points accumulate on a fixed point, then that (hyperbolic) fixed point comes with an almost-homoclinic point (Figure 5.4.1); the Connecting Lemma perturbs this to a homoclinic

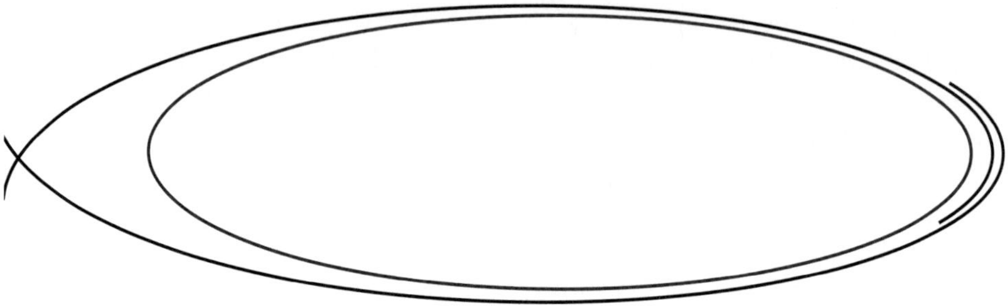

Figure 5.4.1. A periodic orbit close to a hyperbolic fixed point.

point, which is nonwandering but not transverse, an unstable phenomenon that thereby rules out Ω-stability. □

The combination of shadowing and expansivity is central to the preceding arguments, and we now remark (without proof) on the implications of expansivity and related properties per se rather than with the additional strength in which we saw them, in some cases with the additional strong requirement that these properties hold robustly. That is, we examine the implications of expansivity, sometimes together with shadowing without Lipschitz shadowing or uniqueness, and sometimes in restriction to the interior of the set of systems possessing these features.

We first illustrate the potential power of expansivity alone:

Theorem 5.4.20 (Mañé [250]). *For a C^1-residual set of flows on a given compact manifold (that is, C^1-generically),*

- *expansivity implies hyperbolicity* [328] (*in fact, a like conclusion holds assuming only "measure expansivity"* [239], *that is, one allows a null set of exceptions to expansivity in a suitable way*),
- *if the homoclinic class of a hyperbolic periodic orbit is expansive and isolated, then it is hyperbolic* [240],
- *a measure-expansive locally maximal homoclinic class is hyperbolic* [239].

Since we have not discussed genericity results much, we should note that genericity in the C^1-topology is appealing in that it can produce interesting phenomena as well as results like this one, but that genericity in finer topologies is much harder, so there is often a desire to go from C^1-genericity results either to analogous results in a finer topology or a more explicit description of the exceptions to the generic circumstance.

Results about all systems (rather than generic ones) can be obtained, for instance, if instead of expansivity one assumes stable (or robust) expansivity. Put differently, these would be results not about the collection of expansive systems, but about the interior of this collection, usually in the C^1-topology. The first such result is in the volume-preserving category:

Theorem 5.4.21 (Robust expansivity; [3, 4]). *C^1-robustly expansive volume-preserving flows are Anosov.*

Absent volume preservation, there are a few characterizations of robust expansivity with or without shadowing:

Theorem 5.4.22 ([261]). *Consider a flow Φ on a compact manifold. The following are equivalent:*

- *Φ is C^1-robustly expansive and has the shadowing property;*
- *Φ is C^1-robustly expansive and structurally stable;*
- *Φ is Anosov.*

The following are equivalent:

- *Φ is C^1-robustly expansive;*
- *Φ is quasi-Anosov;*
- *Φ has no fixed points and is Axiom A and quasi-transverse.*

Here, we used the following definition:

Definition 5.4.23. An invariant set Λ of a flow Φ is said to be *quasi-hyperbolic* if $\|D\varphi^t(v)\|_{t\in\mathbb{R}}$ is unbounded for $T_\Lambda M \ni v \perp \frac{d\varphi^t}{dt}|_{t=0}$ and *quasi-transverse* if $T_xW^u(x) \cap T_xW^s(x) = \{0\}$ for $x \in \Lambda$. If $\Lambda = M$, then Φ is quasi-Anosov and quasi-transverse, respectively.

It should be noted that robustness of any dynamical property is a severe strengthening over just assuming that property by itself, and this, rather than merely the strength of expansivity, is manifest in the preceding results. We illustrate this with a 3-dimensional counterpart that involves topological transitivity:

Theorem 5.4.24 ([115]). *Robustly transitive 3-flows are Anosov.*

We now remark on our definition of hyperbolicity of a flow as hyperbolicity of the chain-recurrent set (Definition 5.3.50). It turns out that there is a sufficient (and obviously necessary) criterion for this that may be easier to verify.

Theorem 5.4.25 ([92, 322]). *A compact Φ-invariant set Λ is hyperbolic if it is quasi-hyperbolic and $\Phi_{\restriction_\Lambda}$ is chain recurrent.*

Indeed, this gives an alternative proof that the geodesic flow of a negatively curved Riemannian manifold is Anosov [92, Theorem 4.1].

At the end of Section 4.2 we raised the question of how topological entropy depends on a flow, promising that for hyperbolic flows this plays out better than it does in general. Structural stability is the key ingredient.

Theorem 5.4.26 (Continuity of entropy). *The topological entropy of a hyperbolic flow changes continuously under C^1-perturbations.*

We should say that this is an aspect of dynamical systems that plays out rather differently in discrete time: structural stability gives a conjugacy in that case, so topological entropy is locally constant. We are now investigating a subject that is quite specific to flows.

For Anosov flows, one can go well beyond continuous dependence. Structural stability is a central ingredient, though some pertinent results can be obtained without it. In this context, however, our approach to structural stability shows a weakness: our proof gives limited information on how the orbit equivalence depends on the perturbation, other than continuously. Another proof does; it is due to Moser and obtains the orbit equivalence by applying the Contraction Principle/Implicit Function Theorem directly to the problem rather than taking a detour through shadowing (see page 328). Corollary B.4.11 is another exemplar of this approach, and the detailed information on how the fixed point of a contraction varies with the contraction (Proposition B.1.3) indicates that with this approach one expects the orbit equivalence to depend smoothly on an Anosov flow in a suitable sense—and when chosen properly, of course; as we noted, there is flexibility in the orbit direction. The implications for topological entropy are astonishing:

Theorem 5.4.27 (Smoothness of entropy [216, 215, 214]). *Suppose $s \mapsto \Phi_s$ is a 1-parameter family of Anosov flows for $s \in (-\epsilon, \epsilon)$.*

- *If $s \mapsto \Phi_s$ and Φ_0 are C^{k+1} and $1 \le k \le \infty$, then $s \mapsto h_{\text{top}}(\Phi_s)$ is C^k.*
- *If $s \mapsto \Phi_s$ and Φ_0 are C^1, then so is $s \mapsto h_{\text{top}}(\Phi_s)$.*
- *If $s \mapsto \Phi_s$ and Φ_0 are analytic, then so is $s \mapsto h_{\text{top}}(\Phi_s)$.*

That geodesic flows of negatively curved manifolds are Anosov flows gives immediate applications in that context. However, for these, one can strengthen the results.

Theorem 5.4.28 ([216, Theorem 1]). *Suppose (M,g) is a C^2-closed Riemannian manifold without conjugate points, and $s \mapsto g_s$ is a C^1 (!) perturbation by metrics without conjugate points. Then $s \mapsto h_{\text{top}}(g_s)$ is Lipschitz continuous.*[38]

Theorem 5.4.29 ([216, Theorem 3 & Remark c)], [223]). *If (M,g_0) is a C^2-closed Riemannian manifold of nonpositive sectional curvature and $s \mapsto g_s$ is a C^2-perturbation, then $s \mapsto h_{\text{top}}(g_s)$ is C^1.*

Once a function is differentiable, one can aim to compute and use the derivative. Indeed, [216] obtains derivative formulas, and criteria for vanishing of the derivative give interesting precursors to rigidity results in Section 9.3.

Remark 5.4.30 (Shadowing and expansivity). We emphasized the shadowing property as an essential mechanism for a number of core features of hyperbolic dynamics, and while we clarified that what was mainly being used was the combination of shadowing and expansivity, we often used them in the full strength provided by hyperbolicity in the first place (in contrast to Section 1.8).[39]

The exposition in this chapter was guided by wanting to show the power of the Anosov–Katok approach in which shadowing (with uniqueness or expansivity) is used to develop the topological dynamics and stability of hyperbolic sets [207]. Another approach to doing so without first introducing the invariant foliations directly uses the Contraction Principle to produce the desired structures and traces back to Moser and Mather; it is well presented in the concise introduction by Yoccoz [353] and illustrated in Sections 5.5 and 5.6. Finally, we have on occasion brushed up against the limitations of the shadowing approach, and these are averted by building on the invariant foliations, as we do in Chapter 6.

We emphasize that an alternative purely topological approach to hyperbolicity can be built on a topological counterpart of the invariant foliations as follows.

Remark 5.4.31 (Smale flows). One can define a topological equivalent to a locally maximal hyperbolic set (Definition 5.3.16) using stable and unstable *sets*: a *Smale flow* or topological Anosov flow is an expansive flow on a compact metric space whose local stable and unstable sets have a unique point of intersection (giving a well-defined Bowen bracket as in Proposition 6.2.2). Much of the theory of hyperbolic sets of flows holds for Smale flows *with a Hölder structure* [298].

[38] Here we used shorthand: $h_{\text{top}}(g_s)$ is the topological entropy of the geodesic flow of g_s.

[39] Lipschitz shadowing implies structural stability [290], but the proof is on a completely different level from anything we have done here, in that it uses Theorem 5.4.12.

5.5 Local linearization: The Hartman–Grobman Theorem

Looking now at the local context, we explore how well the dynamics near a hyperbolic fixed point (Definition 1.1.25) is described by the dynamics of the linearization. This can be viewed as a local counterpart to structural stability, but there are interesting contrasts to point out. One is that there is an extension that is not at all perturbative, and the other is that, being local to a fixed point, this result can produce a conjugacy rather than an orbit equivalence. Additionally, while structural stability holds for the hyperbolic set itself, for a hyperbolic fixed point we obtain a conjugacy on a neighborhood of the point (Theorem 5.5.3).

Specifically, we now show how the Hartman–Grobman Theorem (Theorem B.4.14) for discrete-time systems translates to a corresponding result for flows. One such translation is too straightforward to state as a separate result: the return map to a transversal through a hyperbolic periodic point of a flow is a map with a hyperbolic fixed point, so the Hartman–Grobman Theorem (Theorem B.4.14) applies directly to the return map. We here develop the other application, to hyperbolic fixed points of flows. The purpose is twofold: it provides insight into the dynamics of a flow near a hyperbolic fixed point, but it also illustrates a mechanism by which conjugacies between time-1 maps are conjugacies between flows in some generality.

Theorem 5.5.1 (Hartman–Grobman). *Let M be a smooth manifold, Φ a continuously differentiable flow on M, and $p \in M$ a hyperbolic fixed point of Φ. Then for each $T > 0$ there exist neighborhoods U of p and V of $0 \in T_pM$ as well as a homeomorphism $h\colon U \to V$ such that $\varphi^t = h^{-1} \circ D\varphi^t_p \circ h$ on U for all $t \in [-T,T]$.*

Proof. Without loss of generality (Theorem B.4.12) assume that $M = \mathbb{R}^n$, $p = 0$, $L^t := D_0\varphi^t$ is a hyperbolic linear map for $t \neq 0$, with $\Delta := \varphi^1 - L^1$ bounded, ϵ as in (B.4.1), and $\ell := L(\Delta) < \epsilon$. Corollary B.4.11 gives a unique homeomorphism $h\colon E \to E$ with $h - \mathrm{Id}$ bounded and $h \circ L^1 = \varphi^1 \circ h$. It suffices to check that

$$\varphi^t \circ h \circ L^{-t} = h \quad \text{for all } t \in \mathbb{R} \tag{5.5.1}$$

because this establishes that h is a conjugacy between any of the time-t maps and its linearization; localization then translates this into the conclusion of Theorem 5.5.1. To check (5.5.1) for a given $t \in \mathbb{R}$ note that $\varphi^t \circ h \circ L^{-t}$ conjugates φ^1 and L^1:

$$\varphi^1 \circ [\varphi^t \circ h \circ L^{-t}] \circ L^{-1} = \varphi^t \circ \underbrace{[\varphi^1 \circ h \circ L^{-1}]}_{=h} \circ L^{-t} = \varphi^t \circ h \circ L^{-t}.$$

Uniqueness of such a conjugacy implies $\varphi^t \circ h \circ L^{-t} = h$ by boundedness of

$$\varphi^t \circ h \circ L^{-t} - \mathrm{Id} = \underbrace{(\varphi^t - L^t)}_{=0 \text{ outside a compact set}} \circ h \circ L^{-t} + L^t \circ \underbrace{(h - \mathrm{Id})}_{\text{bounded}} \circ L^{-t}.$$

□

Remark 5.5.2. A notable consequence of such a local conjugacy to the linear part is that the stable and unstable sets of a hyperbolic fixed point are (topological) manifolds. This in itself is important, and the central additional tool we will develop in Chapter 6 is a twofold strengthening of this observation: the stable and unstable sets are indeed *smooth* manifolds, and this fact holds in a "nonstationary" way—all points of a hyperbolic set have (smooth) stable and unstable manifolds. This is the basis for substantial developments beyond those in the present chapter, but it is also a description of geometric features that arise from the dynamics—and of which Chapter 2 gave a preview.

The Hartman–Grobman Theorem (Theorem 5.5.1) can be "globalized" as a purely topological statement by going beyond the need to match linear parts. More precisely, the index of a hyperbolic fixed point classifies local topological conjugacy classes nearby.

Theorem 5.5.3. *Let M, N be smooth manifolds, Φ, Ψ continuously differentiable flows on M and N, respectively, and $p \in M$, $q \in N$ hyperbolic fixed points of Φ, Ψ, respectively, with the same indices. Then for each $T > 0$ there exist neighborhoods U of p and V of q, and a homeomorphism $h\colon U \to V$ such that $h \circ \varphi^t = \psi^t \circ h$ on U for all $t \in [-T, T]$.*

Proof. By the Hartman–Grobman Theorem (Theorem 5.5.1), both flows are locally conjugate to their linear parts, and by Theorem 1.4.7, these are conjugate. □

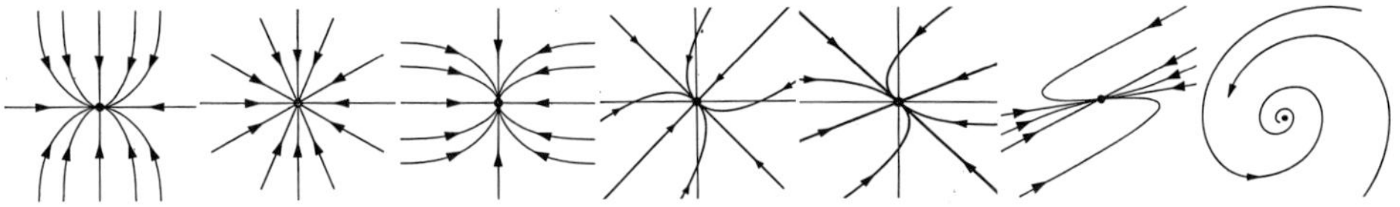

Figure 5.5.1. Pairwise locally topologically conjugate attracting points.

Remark 5.5.4 (Refinements). While it is appealing to have topologically classified hyperbolic fixed points, Theorem 5.5.3 shows that the price is that this equivalence may carry little *geometric* information—literally only the dimensions of the expanding and contracting subspaces. This is the case even for an attracting fixed point (Figure 5.5.1) and raises the question of whether the homeomorphism in Theorem 5.5.1 has enough regularity to provide a geometric picture.

Actually, the Hartman–Grobman conjugacy is Hölder continuous (Definition 1.9.6) [312, 33], a fact we will later use. Indeed, the conjugacy can be taken to have Hölder exponent arbitrarily close to 1 [357] when the flow is C^∞.

Furthermore, Hartman proved that the conjugacy is C^1 if the manifold is a surface (and it is clear in that case that the derivative at the fixed point is the identity). This is of interest because it tells us more than the topological dynamics near the fixed point: Figure 5.5.1 illustrates that the phase portrait of an attracting fixed point can look quite different depending on whether there are two distinct eigenvalues of the linear part, and if not, on whether there are two linearly independent eigenvectors or not; the eigenvectors themselves affect the phase portrait. Topologically, none of these distinctions can be detected (Theorem 5.5.3), but having a differentiable conjugacy whose linear part is the identity tells us that the phase portrait for the flow looks much like that of the linearization in every respect, save for gentle bending of the orbits as we move away from the fixed point. This means that often a fairly good phase portrait of a flow on a surface can be obtained by starting with thumbnails of linearized phase portraits at each hyperbolic fixed point. Example 6.9.4 shows how useful this is.

Regrettably, this convenient fact fails in higher dimension; Hartman gave examples in dimension 3 whose linearizing homeomorphism is not C^1. In Theorems 6.8.10 and 6.8.20 we see that it is possible for a stable or unstable hyperbolic fixed point to have a smooth linearization if the eigenvalues satisfy certain conditions. Only much more recently was it proved that even without such additional assumptions, the conjugacy can be taken differentiable *at the fixed point* (and with derivative equal to the identity, Theorem 6.9.1), which has the desired consequence for drawing phase portraits: a thumbnail of the phase portrait of the linearization pasted in at the fixed point gives a geometrically accurate representation of the actual phase portrait in a neighborhood as in Example 6.9.4.

5.6 The Mather–Moser method*

Sections 5.3 and 5.4 developed the topological orbit structure of hyperbolic sets, including the notion of a hyperbolic flow itself, plus structural (and Ω-) stability, and an underlying agenda was to do so using shadowing and expansivity as the source of all these phenomena. We now show an alternate route to structural stability. The most focused presentation of this approach due in large part to Moser and Mather develops the core theory in a self-contained way and establishes the Hartman–Grobman Theorem,[40] expansivity, structural stability, the Shadowing Lemma, stable/unstable manifolds, local product structure, spectral decomposition in that order (and in discrete time) [353] (see also [329]). In this section we limit ourselves to structural stability for illustrative purposes, but we use the same approach to establish Corollary B.4.11, which we also use here to establish expansivity.

[40] See our proof of Theorem B.4.10.

The beginning of this section introduces notation, terminology, and basic facts that will be useful elsewhere as well; the core starts with Remark 5.6.9.

It is not essential, but helpful, to define partial hyperbolicity here (for the discrete-time context). It will be convenient to use the following notation.

Definition 5.6.1 (Conorm). We define the *conorm* $\llbracket A \rrbracket$ of a linear map A by

$$\llbracket A \rrbracket := \inf \{ \|Av\|/\|v\| \mid \|v\| = 1 \}.$$

This is complementary to the usual norm $\|A\| := \sup \{ \|Av\|/\|v\| \mid \|v\| = 1 \}$.

Definition 5.6.2. An embedding f is said to be *partially hyperbolic* on Λ (in the narrow sense) if there exists a Riemannian metric called a *Lyapunov metric* in an open neighborhood U of Λ for which there are numbers[41]

$$0 < \lambda < \zeta \le \xi < \mu \quad \text{with } \lambda < 1 < \mu \tag{5.6.1}$$

and a pairwise orthogonal invariant splitting into stable, center, and unstable directions

$$T_xM = E^s(x) \oplus E^c(x) \oplus E^u(x), \quad d_x f E^\tau(x) = E^\tau(f(x)),\ \tau = s, c, u \tag{5.6.2}$$

such that

$$\begin{aligned} \|d_x f \restriction E^s(x)\| \le \lambda < \zeta \le \llbracket d_x f \restriction E^c(x) \rrbracket \\ \le \|d_x f \restriction E^c(x)\| \le \xi < \mu \le \llbracket d_x f \restriction E^u(x) \rrbracket. \end{aligned}$$

In this case we set $E^{cs} := E^c \oplus E^s$ and $E^{cu} := E^c \oplus E^u$.

Remark 5.6.3. This is equivalent to requiring that for any Riemannian metric there is a constant C for which there are numbers λ_i, μ_i, $i = 1, 2, 3$ as in (5.6.1) and an invariant splitting (5.6.2) such that

$$\begin{aligned} \|d_x f^n \restriction E^s(x)\| &\le C\lambda^n, \\ C^{-1}\zeta^n \le \llbracket d_x f \restriction E^c(x) \rrbracket &\le \|d_x f \restriction E^c(x)\| \le C\xi^n, \\ C^{-1}\mu^n &\le \llbracket d_x f \restriction E^u(x) \rrbracket. \end{aligned}$$

Example 5.6.4. The time-1 map of an Anosov flow is partially hyperbolic.

It is useful to have a characterization of (partial) hyperbolicity in terms of the action of the differential on vector fields.

[41] We chose the letters ζ and ξ for the middle numbers because ξ looks "just a little bigger" than ζ.

Theorem 5.6.5 (Mather). *Let M be a smooth manifold, $U \subset M$ an open subset, $f: U \to M$ a C^1-embedding, and $\Lambda \subset U$ a compact f-invariant set. Denote by Γ_b the set of bounded vector fields on Λ and by $\Gamma_c \subset \Gamma_b$ the set of continuous vector fields on Λ (these are sections of the bundle $T_\Lambda M := TM_{\restriction \Lambda}$), and for a vector field X on Λ define $\mathcal{F}(X)$ by*

$$\mathcal{F}(X)(f(x)) := Df_x(X(x)).$$

Then for $\ell^- < \ell^+$ the following are equivalent:

(1) *There exist $\lambda < \ell^-$ and $\mu > \ell^+$ such that Λ is (partially)* hyperbolic *with λ, μ as in Definition* 5.6.2.

(2) $\mathrm{sp}(\mathcal{F}_{\restriction \Gamma_b}) \cap \{z \in \mathbb{C} \mid \ell^- \le |z| \le \ell^+\} = \varnothing$.

(3) $\mathrm{sp}(\mathcal{F}_{\restriction \Gamma_c}) \cap \{z \in \mathbb{C} \mid \ell^- \le |z| \le \ell^+\} = \varnothing$.

Proof. (1) $\Rightarrow$ (2): Check that the splitting $\Gamma_b(T_\Lambda M) = \Gamma_b(E^\lambda) \oplus \Gamma_b(E^\mu)$ has the desired properties.

(2) $\Rightarrow$ (3): Since $\Gamma_c \subset \Gamma_b$ is an invariant Banach subspace, $\mathrm{sp}(\mathcal{F}_{\restriction \Gamma_c}) \subset \mathrm{sp}(\mathcal{F}_{\restriction \Gamma_b})$.

(3) $\Rightarrow$ (1): This involves two simple steps.

Lemma 5.6.6. *The projections $\pi^\pm$ that define the splitting $\Gamma_c = \mathcal{E}^\lambda \oplus \mathcal{E}^\mu$ are $C^0(\Lambda)$-linear.*

We delay the proof of the above lemma and proceed with the proof of the theorem.

A map $L: \Gamma_c \to \Gamma_c$ is said to be $C^0(\Lambda)$-linear if $L(\varphi X) = \varphi \cdot L(X)$ for all $\varphi \in C^0(\Lambda)$. This lets us apply a general fact about continuous maps of bundles.

Lemma 5.6.7. *A $C^0(\Lambda)$-linear map $L: \Gamma_c \to \Gamma_c$ is pointwise defined, that is, there is a continuous family $(L_x: T_x M \to T_x M)_{x \in \Lambda}$ of linear maps such that $L(X)(x) = L_x(X(x))$ for all $x \in \Lambda$.*

Now, Lemma 5.6.6 provides the hypotheses for Lemma 5.6.7 applied to $\pi^\pm$, so we obtain fiberwise linear maps $\pi_x^\pm$, and these are complementary projections since $\pi^\pm$ are (check that $(\pi^\pm)^2 = \pi^\pm$ and $\pi^- + \pi^+ = \mathrm{Id}$ imply the same for $\pi_x^\pm$). This gives continuous subbundles $E_x^\lambda := \pi_x^+(T_x M)$ and $E_x^\mu := \pi_x^-(T_x M)$ with the desired properties. □

Proof of Lemma 5.6.6. The main point is that the subspaces $\mathcal{E}^\lambda$ and $\mathcal{E}^\mu$ are $C^0(\Lambda)$-closed: If $X \in \mathcal{E}^\lambda$ and $\varphi: \Lambda \to \mathbb{R}$ is continuous (hence bounded), then $\varphi X \in \mathcal{E}^\lambda$ because $\mathcal{F}^n(\varphi X) = \varphi \circ f^{-n} \cdot \mathcal{F}^n(X)$. Thus $\Gamma_c = \mathcal{E}^\lambda \oplus \mathcal{E}^\mu$ as $C^0(\Lambda)$-modules; since $\pi^\pm$ is $C^0(\Lambda)$-linear on $\mathcal{E}^\lambda$ and $\mathcal{E}^\mu$ (it is 0 or Id), the claim follows. □

Proof of Lemma 5.6.7. If $X \equiv 0$ on an open set U then $\pi^{\pm}(X) = 0$ on U: For $x \in U$ take $\varphi \in C^0(\Lambda)$ such that $\varphi(x) = 1$ and $\varphi X \equiv 0$ to get

$$\pi^{\pm}(X)(x) = 1 \cdot \pi^{\pm}(X)(x) = \varphi(x) \cdot \pi^{\pm}(X)(x) = \pi^{\pm}(\varphi X)(x) = \pi^{\pm}(0)(x) = 0.$$

If $X \in \Gamma_c$ and $X(x) = 0$ take $X_n \to X$ with $X_n = 0$ on $B(x, \frac{1}{n})$ and hence $\pi^{\pm}(X)(x) = \lim \pi^{\pm}(X_n)(x) = 0$.

If $(x, v) \in T_\Lambda M$, $X \in \Gamma_c$ and $X(x) = v$, then $\pi_x^{\pm}(v) := \pi^{\pm}(X)(v)$ is thus independent of such X. □

Definition 5.6.8 (Fibered linear automorphisms). Suppose K is a compact metric space, $\pi\colon E \to K$ a finite-dimensional vector bundle, and $f\colon K \to K$ a homeomorphism. Then $F\colon E \to E$ is called a linear automorphism of E fibered over f if for every $x \in K$ the restriction F_x of F to $E_x := \pi^{-1}(x)$ is a linear isomorphism onto $E_{f(x)}$ depending continuously on x. We denote by $\Gamma_b(E)$ and $\Gamma_c(E)$ the (Banach) space of bounded, respectively continuous, sections of π. The action $\mathcal{F}$ of F on sections X of π is given by

$$F_x(X(x)) = \mathcal{F}(X)(f(x)).$$

(The action $\mathcal{F}$ is linear and preserves Γ_b and Γ_c.)

Remark 5.6.9 (Transverse bundle). The setting of Theorem 5.6.5 is an instance of this situation, with E being the tangent bundle and $F = Df$. For invariant sets $\Lambda \subset M$ of flows, a related useful bundle is the *transverse bundle* $T_\Lambda^\Phi M$ defined by $T_x^\Phi M = T_x M/\dot{\varphi}$, the linear space $T_x M$ modulo the flow direction, which inherits a norm or inner product from $T_x M$. Theorem 5.6.5 tells us that time-t maps of hyperbolic flows induce a *hyperbolic* action F on the transverse bundle E, that is, there is an invariant splitting

$$E(x) = E^s(x) \oplus E^u(x), \quad FE^\tau(x) = E^\tau(f(x)),\ \tau = s, u$$

such that

$$\|F\restriction_{E^s(x)}\| \le \lambda < 1 < \mu \le [\![F\restriction_{E^u(x)}]\!].$$

Theorem B.4.8 applied to $\mathcal{G}$ defined by $G_x(X(x)) = \mathcal{G}(X)(f(x))$ gives the proof of the next result.

Theorem 5.6.10 (Invariant section). *If $F\colon E \to E$ is a hyperbolic linear automorphism fibered over a homeomorphism f of K and $G\colon E \to E$ is fibered over f such that $\ell := L(G - F) < \epsilon := \min(1 - \lambda, 1 - \mu^{-1})$ (see Definition* B.1.1), *then there is a unique bounded section X of E such that $\mathcal{G}(X) = X$, and X is continuous with $\|X\| \le (\epsilon - \ell)^{-1} \sup_{x \in K} \|G_x(0)\|$.*

Localization (Theorem B.4.12) provides applications of results like this to a compact hyperbolic set of diffeomorphisms, with G being the localization of Df—variously on the tangent bundle or the transverse bundle. For instance, the Hartman–Grobman Theorem implies expansivity without first establishing shadowing with uniqueness: with $K := \Lambda_\Phi^U$ (as in Proposition 5.1.8) hyperbolic, the localization G on the transverse bundle of $D\varphi^t$ from Theorem B.4.12 is Lipschitz-close to $D\varphi^t$ on the transverse bundle, so we can apply the Hartman–Grobman Theorem (Corollary B.4.11) to conclude that for $x \in K$, $v \in E_x \setminus \{0\}$ the $\mathcal{G}$-orbit of the section X with $X(x) = v$, $X(y) = 0$ for $y \neq x$ is unbounded, so the orbit of $\exp_x v$ does not stay in localization neighborhoods around the orbit of x.

Structural stability is an application of Theorem 5.6.10.

Mather–Moser proof of Theorem 5.4.5. By assumption, we can localize Ψ to a fibered action on the transverse bundle which is Lipschitz-close to that of $D\Phi$. The unique bounded (and then continuous) section X from Theorem 5.6.10 is continuous in Ψ and gives the orbit equivalence h by

$$h(x) = \exp_x(X(x). \tag{5.6.3}$$

As in Theorem 5.3.8, h is injective by expansivity. □

Exercises

5.1. Show that in the proof of Proposition 5.1.5 one directly obtains a smooth adapted metric by taking $\left(\|v\|_x^s\right)^2 := \int_0^S \underline{\lambda}^{-2s}\left(\|D\varphi^s v\|_{\varphi^s(x)}\right)^2 ds$ for large enough S.

5.2. In Definition 5.2.13 show that d_Ω is a distance function.

5.3. In Definition 5.2.13 show that if x, y, z are collinear with z between x and y, then $d_\Omega(x,y) = d_\Omega(x,z) + d_\Omega(z,y)$.

5.4. Show that the Smale Horseshoe presented in Example 1.5.24 is locally maximal (Definition 5.3.16).

5.5. Show that hyperbolic attractors are locally maximal.

5.6. Show that the stable and unstable sets of a hyperbolic fixed point (Definition 1.3.26) are topological manifolds.

5.7. Show that h in (5.6.3) is as claimed in Theorem 5.4.5.

5.8. Show directly that a hyperbolic symbolic flow (Definition 1.9.5) has the shadowing property (Definition 5.3.1).

5.9. Show directly that the Anosov Closing Lemma (Theorem 5.3.11) holds for a hyperbolic symbolic flow (Definition 1.9.5).

5.10. Show directly that a transitive hyperbolic symbolic flow (Definition 1.9.5) has the specification property (Definition 5.3.58).

5.11. Show directly that $\overline{\mathrm{Per}(\Phi)} = \mathrm{NW}(\Phi)$ for a hyperbolic symbolic flow Φ (Definition 1.9.5).

5.12. Show directly that a hyperbolic symbolic flow (Definition 1.9.5) that does not consist of a single orbit has exponential growth of periodic orbits and hence positive topological entropy.

5.13. Let M be an m-dimensional Riemannian manifold with sectional curvatures in $[-K^2, -k^2]$. Prove that

$$k(m-1) \leq v(M) := \lim_{r\to\infty} \frac{1}{r} \log \mathrm{vol}(B(x,r)) \leq K(m-1)$$

(volume growth).

5.14. Prove that the fundamental group $\pi_1(M)$ of a compact manifold that admits a metric of negative sectional curvature has exponential growth, that is, for any given system Γ of generators of $\pi_1(M)$ the number of elements $\gamma \in \pi_1(M)$ that can be represented by words of length at most n grows exponentially with n.

5.15. Prove that the universal cover of a manifold of negative sectional curvature is diffeomorphic to Euclidean space.

5.16. Prove that all geodesics c on a manifold of negative sectional curvature are minimal, that is, for any two points on the lift of c the segment of the lift between these points is the shortest curve between its ends.

5.17. Figure 5.2.4 shows dispersing billiards, one with a single scatterer and one with finite horizon. Design a dispersing finite-horizon billiard on $\mathbb{T}^2$ with only one scatterer.

6

Invariant foliations

A key objective of Chapter 5 was to show how much of the orbit structure in hyperbolic dynamics can be discerned solely from shadowing and expansivity. In so doing, we did not need the flow to be differentiable; most of the results hold for a continuous flow of a compact metric space with shadowing and expansivity. The discrete-time counterpart—a homeomorphism of a compact metric space with shadowing and expansivity—was termed a *Smale space* by Ruelle. Aoki and Hiraide [13] proved that for Smale spaces one cannot only obtain most of the results from the previous chapter, but also some of the results we present in this chapter, such as Theorem 6.2.7 on the existence of a local product structure (Definition 6.2.6) and Theorem 6.6.5 on the existence of a Markov partition (Definition 6.6.4). It is reasonable to expect that the results of Aoki and Hiraide can be extended from maps to flows, but our purposes are better served by introducing and using more subtle geometric structures, notably the invariant foliations.

The Hartman–Grobman Theorem (Theorem B.4.14) implies that the stable and unstable sets of a hyperbolic fixed point are topological manifolds. As we previewed at the time, the central geometric property we introduce in this chapter is that the stable and unstable sets in fact are smoothly immersed manifolds and these manifolds form (flow-invariant) foliations tangent to the stable and unstable subbundles. Proofs using these foliations are both illuminating and sometimes easier than the counterparts based on shadowing and expansivity.

Moreover, as we saw in the case of geodesic flows (Chapter 2), when the dynamics are intertwined with the geometry, these invariant foliations arise naturally and are of interest in their own right. Then the full force of geometric information can be brought to bear on dynamical investigations by using these structures; results in [223, 283] are but some outstanding examples of this.

These foliations are not only of interest in their own right and essential to many of the results in this chapter but also indispensable in major parts of the ergodic theory of hyperbolic flows (Chapter 7). For instance, there are characterizations of topological transitivity and mixing in terms of the invariant foliations (Theorem 6.2.11), and whether a flow is of algebraic type or not (Section 9.7).

A number of results in this chapter are easier to prove in the discrete-time setting and so we at times refer to Appendix B. Readers unfamiliar with properties of stable and unstable manifolds for diffeomorphisms are encouraged to read that chapter in connection with this one.

6.1 Stable and unstable foliations

We begin this chapter by showing that the stable and unstable sets of points in a hyperbolic flow are smooth manifolds.

Theorem 6.1.1 (Stable- and Unstable-Manifold Theorem). *Let Λ be a hyperbolic set for a C^r-flow Φ on M, $r \in \mathbb{N}$, C, λ, μ as in Definition* 5.1.1, *and $t_0 > 0$. Then for each $x \in \Lambda$ there is a pair of embedded C^r-disks $W^s_{\mathrm{loc}}(x)$, $W^u_{\mathrm{loc}}(x)$, depending continuously on x in the C^1-topology and called the* local strong-stable manifold *and the* local strong-unstable manifold *of x, respectively, such that*

(1) $T_x W^s_{\mathrm{loc}}(x) = E^s_x$, $T_x W^u_{\mathrm{loc}}(x) = E^u_x$,

(2) $\varphi^t(W^s_{\mathrm{loc}}(x)) \subset W^s_{\mathrm{loc}}(\varphi^t(x))$ *and* $\varphi^{-t}(W^u_{\mathrm{loc}}(x)) \subset W^u_{\mathrm{loc}}(\varphi^{-t}(x))$ *for* $t \geq t_0$,

(3) *for every $\delta > 0$ there exists $C(\delta)$ such that*

$$\begin{aligned} d(\varphi^t(x), \varphi^t(y)) &< C(\delta)(\lambda + \delta)^t d(x, y) && \text{for } y \in W^s_{\mathrm{loc}}(x),\ t > 0, \\ d(\varphi^{-t}(x), \varphi^{-t}(y)) &< C(\delta)(\mu - \delta)^{-t} d(x, y) && \text{for } y \in W^u_{\mathrm{loc}}(x),\ t > 0, \end{aligned}$$

(4) *there exists a continuous family U_x of neighborhoods of $x \in \Lambda$ such that*

$$\begin{aligned} W^s_{\mathrm{loc}}(x) &= \{y \mid \varphi^t(y) \in U_{\varphi^t(x)} \text{ for } t > 0,\ d(\varphi^t(x), \varphi^t(y)) \xrightarrow[t \to +\infty]{} 0\}, \\ W^u_{\mathrm{loc}}(x) &= \{y \mid \varphi^{-t}(y) \in U_{\varphi^{-t}(x)} \text{ for } t > 0,\ d(\varphi^{-t}(x), \varphi^{-t}(y)) \xrightarrow[t \to +\infty]{} 0\}. \end{aligned}$$

Proof. The Hadamard–Perron Theorem (Theorem B.5.2) applied to the time-t_0 map φ^{t_0} with $T_x M = E^s \oplus (E^c_x \oplus E^u_x)$ yields the existence of $W^s_{\mathrm{loc}}(x) \in C^r$ satisfying (1)–(4) for $t \in \mathbb{N}t_0$. The same with $T_x M = (E^s_x \oplus E^c_x) \oplus E^u_x$ yields $W^u_{\mathrm{loc}}(x) \in C^r$ satisfying (1)–(4) with $-t \in \mathbb{N}t_0$.

Observe now that (4) holds for positive multiples of t_0 if and only if it holds for real t. Once (3) holds for $t \in \mathbb{N}t_0$ it trivially holds for $t > 0$ by adjusting the constant $C(\delta)$ since $\{\varphi^t\}_{t \in [0, t_0]}$ is equicontinuous and M is compact. □

Remark 6.1.2. With a little care one can replace the condition $t \geq t_0$ in (2) by $t > 0$.

The sets

$$\begin{aligned} W^s(x) &:= \bigcup_{t>0} \varphi^{-t}(W^s_{\mathrm{loc}}(\varphi^t(x))) = \big\{y \in M \mid d(\varphi^t(x), \varphi^t(y)) \xrightarrow[t \to \infty]{} 0\big\}, \\ W^u(x) &:= \bigcup_{t>0} \varphi^t(W^u_{\mathrm{loc}}(\varphi^{-t}(x))) = \big\{y \in M \mid d(\varphi^{-t}(x), \varphi^{-t}(y)) \xrightarrow[t \to \infty]{} 0\big\} \end{aligned}$$

are defined independently of a particular choice of local stable and unstable manifolds, and are smooth injectively immersed manifolds called the global *strong-stable* and *strong-unstable* manifolds. The manifolds

$$W^{cs}(x) := \bigcup_{t\in\mathbb{R}} \varphi^t(W^s(x)) \text{ and } W^{cu}(x) := \bigcup_{t\in\mathbb{R}} \varphi^t(W^u(x))$$

are called the *weak-stable* and *weak-unstable* manifolds (or *center-stable* and *center-unstable* manifolds) of x. Note that $T_xW^{cs} = E^s_x \oplus E^c_x$, $T_xW^{cu} = E^c_x \oplus E^u_x$.

Locally, then we have a picture like Figure 6.1.1 for each nonfixed point $p \in \Lambda$.

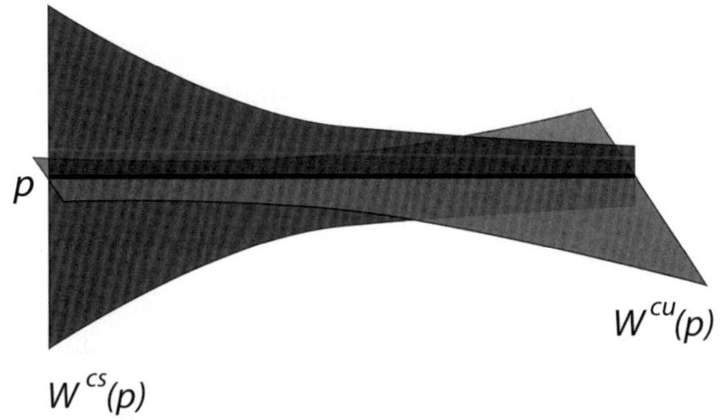

Figure 6.1.1. Local center-stable and center-unstable leaves.

Proposition 5.1.4 and compactness imply the next result.

Proposition 6.1.3. *Let Λ be a hyperbolic set for a flow Φ. Then there are a neighborhood U of Λ and $\alpha > 0$ such that if $x, y \in \Lambda$ and $z \in W^s(x) \cap W^u(y) \cap U$, then for any $\xi \in T_zW^s(x)$ and $\eta \in T_zW^u(y)$ the angle between ξ and η is at least α.*

Remark 6.1.4. We proved hyperbolicity of time changes (Theorem 5.1.11) with just the Alekseev Cone Criterion much earlier, but this is an opportunity to show an alternative argument that tracks what happens to strong-stable leaves under a time change.

Second proof of Theorem 5.1.11. Since Ψ and Φ have the same orbits, Λ is Ψ-invariant, as are the center-stable and center-unstable manifolds for Φ. We use this to determine stable (and unstable) manifolds for Ψ. For $x \in \Lambda$ and $y \in W^s_\Phi(x)$ (the strong-stable manifold of x for Φ) we have

$$d(\varphi^t(x), \varphi^t(y)) \xrightarrow[t\to\infty]{\text{exponentially}} 0 \quad \text{hence } d(\underbrace{\varphi^{\alpha(t,x)}(x)}_{=\psi^t(x)}, \varphi^{\alpha(t,x)}(y)) \xrightarrow[t\to\infty]{\text{exponentially}} 0,$$

where α is as in (1.2.1). Here, $\varphi^{\alpha(t,x)}(y)$ is a parametrization of the orbit of y which asymptotically lags (or leads) the corresponding Ψ-orbit by a constant:

$$\alpha(t,x) - \alpha(t,y) = \int_0^t \underbrace{\rho(\varphi^s(x)) - \rho(\varphi^s(y))}_{\text{exponentially small}}\, ds \xrightarrow[t\to\infty]{\text{exponentially}} a(y) \in \mathbb{R},$$

so the triangle inequality gives $d(\psi^t(x), \psi^{t-a(y)}(y)) \xrightarrow[t\to\infty]{\text{exponentially}} 0$, and by the theorem about differentiation inside an integral,[1] $a(y)$ depends smoothly on $y \in W^s_\Phi(x)$. Thus,

$$W^s_\Psi(x) = \{\psi^{-a(y)}(y) \mid y \in W^s_\Phi(x)\}$$

is the Ψ-stable manifold of x for Ψ. Due to the exponential contraction, its tangent space at x is the Ψ-stable subspace at x. A similar argument for the unstable direction establishes hyperbolicity of Λ for Ψ. □

Since we now have the needed terminology we digress to point out that hyperbolic flows are generic in the following sense.

Definition 6.1.5. A fixed point p of a local flow is said to be *transverse* if the differential at p of any time-t map for $t \neq 0$ does not have 1 as an eigenvalue. Equivalently, the linear part of the vector field at p does not have 0 as an eigenvalue.

A periodic point p of period $t > 0$ for a flow is said to be *transverse* if 1 is a simple eigenvalue of the differential at p of the time-t map of the flow. Equivalently, p is a transverse fixed point for the Poincaré map on a transversal to the flow near p.

A smooth flow is said to be a *Kupka–Smale flow to order* t if all fixed points and all periodic orbits of period less than t are hyperbolic and the t-balls in their stable and unstable manifolds are pairwise transverse. It is called a *Kupka–Smale flow* if it is a Kupka–Smale flow to order t for all $t > 0$.

Theorem 6.1.6 (Kupka–Smale Theorem). *Let $0 < r \le k \le \infty$ and M a compact C^k-manifold. Then for any $t > 0$, Kupka–Smale flows of order t are a C^r-dense C^1-open set and hence Kupka–Smale flows are a C^r-dense C^1-G_δ set in the space of C^r-flows.*

The *Inclination Lemma* illuminates the local geometry of the stable and unstable manifolds near a hyperbolic periodic point and can be useful in proving a number of results on the structure of hyperbolic sets. We note, however, that in those arguments it often suffices instead to invoke only smoothness and continuous dependence of the invariant manifolds. Thus, the remainder of this section is optional reading.

[1] $\frac{d}{dx}\int_a^b f(x,y)\,dy = \int_a^b \frac{d}{dx} f(x,y)\,dy$.

We here obtain consequences for flows of the Inclination Lemma (or λ-Lemma) for diffeomorphisms (Theorem B.6.1) or of its proof. The first application is the most direct one: applying Theorem B.6.1 to the diffeomorphism $f = \varphi^1$ gives the following proposition:

Proposition 6.1.7 (Inclination Lemma for fixed points). *Suppose p is a hyperbolic fixed point of a smooth flow Φ and $\mathcal{D}$ is a disk that transversely intersects $W^s(p)$. Then the sets $\varphi^t(\mathcal{D})$ accumulate on $W^u(p)$ in the C^1-topology as $t \to +\infty$. Specifically, for any disk Δ in $W^u(p)$ and any $\epsilon > 0$ there is a $t > 0$ and a $\mathcal{D}' \subset \mathcal{D}$ such that $d_{C^1}(\varphi^t(\mathcal{D}'), \Delta) < \epsilon$.*

Since our interest in flows centers on those for which fixed points are isolated, it is more interesting to have analogous statements for hyperbolic periodic points. The first restricts attention to a section.

Proposition 6.1.8 (Inclination Lemma in a section). *Consider a hyperbolic periodic point p for a C^r-flow Φ on a manifold M, and a C^r Poincaré section S transverse to the flow containing p. Then the Poincaré return map is a local diffeomorphism f, and if $\mathcal{D} \subset S$ is a disk that transversely intersects $W^s_f(p) \subset S$, then the $f^n(\mathcal{D})$ accumulate on $W^u_f(p) \subset S$ in the C^1-topology as $n \to +\infty$, that is, for any disk Δ in $W^u_f(p)$ and any $\epsilon > 0$ there is an $n \in \mathbb{N}$ and a $\mathcal{D}' \subset \mathcal{D}$ such that $d_{C^1}(f^n(\mathcal{D}'), \Delta) < \epsilon$.*

Proof. This is a direct application of Theorem B.6.1, except that here f is a local diffeomorphism. So one can either check that the proof of Theorem B.6.1 works in this context or extend f to a diffeomorphism of Euclidean space by Theorem B.4.12. □

Remark 6.1.9. It is not really needed that the point of transverse intersection lies in the local section because one can always achieve that by first applying φ^t for sufficiently large t because Φ preserves transversality.

Finally, a version of the Inclination Lemma that treats periodic orbits of flows directly rather than through sections uses not the Inclination Lemma for diffeomorphisms but its proof.

Proposition 6.1.10 (Inclination Lemma for flows). *Let p be a hyperbolic periodic orbit of least period $T > 0$ for a flow Φ of a manifold M with a splitting $T_pM = E^s \oplus E^c \oplus E^u$. Let D be an embedded disk intersecting $W^s(p)$ transversely at some point $q \in W^s(p)$ such that $\dim(D) = \dim(E^u) + 1$. Then for any $\epsilon > 0$ there exists an $N \in \mathbb{N}$ such that for each $n \geq N$ there is an embedded disk $D_n \subset D$ containing q such that $\varphi^{Tn}(D_n)$ is C^1 ϵ close to $W^{cu}(p)$.*

Proof. Although the diffeomorphism φ^T does not quite satisfy the hypotheses of the Inclination Lemma (Theorem B.6.1), the proof produces this result nonetheless (Remark B.6.3) because no expansion is needed in order to establish that iterates of $\mathcal{D}$ are C^1-close to the center-unstable manifold of p. It is used solely to assert that large disks in $W^{cu}(p)$ are approximated by images of $\mathcal{D}$, and Proposition 6.1.10 makes no such claim. □

Remark 6.1.11. This argument illuminates what one should expect: that any disk in $W^u(p)$ has a neighborhood Δ in $W^{cu}(p)$ such that for any $\epsilon > 0$ there is an $n \in \mathbb{N}$ and a $\mathcal{D}_n \subset \mathcal{D}$ with $d_{C^1}(\varphi^{Tn}(\mathcal{D}_n), \Delta) < \epsilon$.

Remark 6.1.12. We repeat that many invocations of the Inclination Lemma can be averted by instead using that center-stable and center-unstable manifolds are C^r and depend continuously on the basepoint.

6.2 Global foliations, local maximality, Bowen bracket

The presence of the invariant foliations for a hyperbolic flow allows us to complement the previous dynamical insights for hyperbolic sets by a geometric understanding, which in turn augments our description of the dynamics on these sets. This section and those that follow give a panorama of ways in which this can be done. For instance, Remark 5.3.41 indicates that the invariant foliations are meaningful well beyond a neighborhood of a hyperbolic set.

In the case of geodesic flows, the global foliations can be described geometrically. As we do this it may be instructive to revisit the discussion of surfaces of constant negative curvature at the end of Section 2.1 and Theorem 5.2.4 as well as Remark 5.2.6. The universal cover $\widetilde{M}$ of a negatively curved Riemannian manifold M is diffeomorphic to $\mathbb{R}^n$ (Exercise 5.15). We begin with unstable manifolds. Fix $v \in S\widetilde{M}$ and let

$$B_T := \{\gamma(0) \mid \gamma \text{ geodesic}, \gamma(-T) = \gamma_v(-T)\},$$
$$W_T := \{\dot\gamma(0) \mid \gamma \text{ geodesic}, \gamma(-T) = \gamma_v(-T)\},$$

the outside unit normal vectors to B_T. Here, W_T is a smooth submanifold of $S\widetilde{M}$ of dimension $n-1$, where $n = \dim M$.

Consider any curve in W_T. Associated with the corresponding geodesic variation is a Jacobi field Y with $Y(-T) = 0$ (see the proof of Theorem 5.2.4). Unless $Y = 0$ we have $\langle Y(t), \dot Y(t)\rangle > 0$ for $t > -T$ since $\dot Y(-T) \neq 0$ and $Y(t-T) = t\dot Y(-T) + o(t)$ whence $\langle Y(t-T), \dot Y(t-T)\rangle > 0$ for small positive values of t. But we showed that this must then hold for all $t > 0$.

Thus, every tangent vector to W_T is contained in a cone from the invariant family. As in the Hadamard–Perron Theorem this implies that $W_T \xrightarrow[T\to\infty]{} W^u(v)$, a smooth $(n-1)$-dimensional submanifold of $S\widetilde{M}$. Since the projection $\pi\colon S\widetilde{M} \to \widetilde{M}$ is smooth, the spheres B_T converge to a smooth submanifold B_∞ called a *horosphere* (which means limit sphere).

In fact, $W^u(v)$ consists of the outward unit normals to B_∞, which itself can be described as $\{\gamma(0) \mid \gamma \text{ geodesic}, d(\gamma(t), \gamma_v(t)) \xrightarrow[t\to-\infty]{} 0\}$.

Remark 6.2.1. If M is oriented then one can orient unstable leaves consistently by taking for each $v \in SM$ a positive orthonormal frame whose first vector is v. This orients a horosphere, and this orientation lifts to the corresponding unstable leaf.

Another example in which we can explicitly see the properties of the stable and unstable manifolds is the suspension of a hyperbolic toral automorphism (Example 1.5.26): for any point the stable and unstable manifolds are lines obtained as the projections of the contracting and expanding eigendirections translated to the basepoint.

We now characterize local maximality through local stable and stable manifolds (local product structure), and then revisit the spectral decomposition in this light.

Proposition 6.2.2 (Bowen bracket). *For a hyperbolic set Λ for a flow Φ and $\epsilon > 0$ sufficiently small there exists a $\delta > 0$ such that if $x, y \in \Lambda$ such that $d(x,y) < \epsilon$, then there exists some $t = t(x,y) \in (-\epsilon, \epsilon)$ such that*

$$W^s_\epsilon(\varphi^t(x)) \cap W^u_\epsilon(y) = \{[x,y]\}$$

consists of a single point. This intersection point $[x,y]$ of $W^{cs}_\epsilon(x)$ and $W^u_\epsilon(y)$ is called the Bowen bracket[2] *of x and y, and there exists $C_0 = C_0(\delta) > 0$ such that if $x, y \in \Lambda$ and $d(x,y) < \delta$, then $d_s(\varphi^{t(x,y)}(x), [x,y]) < C_0 d(x,y)$ and $d_u(y, [x,y]) < C_0 d(x,y)$, where d_s and d_u denote the distances along the stable and unstable manifolds.*

Remark 6.2.3. Thus, the Bowen bracket is defined on the ϵ-neighborhood of the diagonal in $\Lambda\times\Lambda$ and maps to an ϵ-neighborhood of Λ in the manifold. Equation (2.2.4) is a particularly concrete context for this concept.

This definition produces one of two possible Bowen brackets; a complementary choice would be $\{[x,y]\} = W^s_\epsilon(x) \cap W^{cu}_\epsilon(y)$. Unless otherwise indicated, we will use the notation $[x,y]$ for the construct in Proposition 6.2.2 rather than this one.

Proof. Proposition 6.1.3 implies uniform transversality of W^u and W^{cs} and that there is exactly one point z of intersection of $W^{cs}_\epsilon(x)$ and $W^u_\epsilon(y)$, which, as we now show, depends continuously on x and y. By continuous dependence of $W^{cs}_\epsilon(x)$ and

[2]Or Smale bracket [298].

$W^u_\epsilon(y)$, we can choose a chart to $R^l \oplus \mathbb{R}^k$ near z in which leaves of $W^{cs}_\epsilon(x')$ are graphs of Lipschitz maps $F_{x'}(\cdot)$ over $\mathbb{R}^l$ and leaves of $W^u_\epsilon(y')$ are graphs of Lipschitz maps $G_{y'}(\cdot)$ over $\mathbb{R}^k$, in both cases with Lipschitz constants less than 1. Then $W^{cs}_\epsilon(x') \cap W^u_\epsilon(y') = \{(x_0, y_0)\} = \{(G_{x'}(F_{y'}(x_0)), G_{x'}(F_{y'}(y_0)))\}$, so both coordinates are fixed points of contractions depending continuously on (x', y'). Since Λ is compact we obtain uniform ϵ and δ. □

This lets us improve on shadowing (and on Proposition 1.8.4) as suggested by Figure 6.1.1. Then $d(\varphi^{t+\tau}(x), \varphi^t(y)) \le d(\varphi^{t+\tau}(x), \varphi^t([x,y])) + d(\varphi^t([x,y]), \varphi^t(y))$ implies that for λ, μ as in Definition 5.1.1, $\eta \ge \max(\lambda, \mu^{-1})$, there are $\delta, C > 0$ such that if $x \in \Lambda$, $d(\varphi^t(x), \varphi^t(y)) < \delta$ for $|t| \le T$, then

$$d(\varphi^{t+\tau}(x), \varphi^t(y)) \le C\eta^{T-|t|} \cdot \left(d(\varphi^{-T}(x), \varphi^{-T}(y)) + d(\varphi^T(x), \varphi^T(y))\right),$$

where $\tau = t(x,y) \in (-\delta, \delta)$ and $[\cdot,\cdot]$ are as in Proposition 6.2.2. This can be amplified to provide information about similar closeness of shadowing orbits, including their timing.

Theorem 6.2.4 (Exponential shadowing). *For a hyperbolic set Λ of a flow Φ on a closed manifold $\exists\, c, \eta > 0\ \forall\, \epsilon > 0\ \exists\, \delta > 0$: if $x, y \in \Lambda$, $s\colon \mathbb{R} \to \mathbb{R}$ continuous, $s(0) = 0$ and $d(\varphi^t(x), \varphi^{s(t)}(y)) < \delta$ for all $|t| \le T$, then*

(1) $|t - s(t)| \le 3\epsilon$ *for all* $|t| \le T$,

(2) *if $t(x,y)$ is as in Proposition* 6.2.2 *and* $|t| \le T$, *then*

$$d(\varphi^t(y), \varphi^t(\varphi^{t(x,y)}(x))) < ce^{-\eta(T-|t|)}.$$

Proof. Let U be a neighborhood of Λ such that Λ_U is hyperbolic, d an adapted metric for a neighborhood of Λ_U, and $\eta := -\log\min(\lambda, 1/\mu)$ with λ, μ as in Definition 5.1.1 for Λ_U, $\epsilon_0 > 0$ such that $\bigcup_{x\in\Lambda} B_{\epsilon_0}(x) \subset U$, $\delta_0 > 0$ as in Proposition 6.2.2 (for ϵ_0), $\epsilon < \epsilon_0/8$ such that

$$\sup\left\{d(\varphi^t(x), x) \mid x \in \Lambda,\ |t| \le 4\epsilon\right\} \le \epsilon_0/8,$$

$\delta < \delta_0$ from Proposition 6.2.2, correspondingly, $x, y \in \Lambda$, $s\colon \mathbb{R} \to \mathbb{R}$ continuous with $s(0) = 0$ and $d(\varphi^{s(t)}(y), \varphi^t(x)) < \delta$ for $|t| \le T$, and

$$A := \left\{t \in [0,T] \mid |t - s(t)| \ge 3\epsilon \text{ or } d(\varphi^t(y), \varphi^t(\underbrace{[x,y]}_{=:w})) \ge \epsilon_0/2\right\},$$

$$B := \left\{t \in [0,T] \mid |t - s(-t)| \ge 3\epsilon \text{ or } d(\varphi^{-t+t(x,y)}(x), \varphi^{-t}(w)) \ge \epsilon_0/2\right\}.$$

To see that $A = \varnothing$ suppose $A \neq \varnothing$ and let $t_1 := \inf A$. We will show $|s(t_1) - t_1| < \epsilon_0/2$ and $d(\varphi^{t_1}(w), \varphi^{t_1}(y)) < \epsilon_0/2$, which is impossible.

First, $\varphi^{s(t_1)}(w) \in W^u_{\epsilon_0}(\varphi^{s(t_1)}(y))$ since

$$\underbrace{d(\varphi^{s(t_1)-t}(y), \varphi^{s(t_1)-t}(w))}_{\leq d(\varphi^{s(t_1)-t}(y),\varphi^{t_1-t}(y)) + d(\varphi^{t_1-t}(y),\varphi^{t_1-t}(w)) + d(\varphi^{t_1-t}(w),\varphi^{s(t_1)-t}(w))} < \frac{3\epsilon_0}{4}$$

$$\underbrace{d(\varphi^{s(t_1)-t}(y),\varphi^{t_1-t}(y))}_{<\epsilon_0/8 \text{ since } |t_1-s(t_1)|<4\epsilon} + \underbrace{d(\varphi^{t_1-t}(y),\varphi^{t_1-t}(w))}_{\leq \epsilon_0/2\text{—choice of } \epsilon_0 \text{ \& } d \text{ adapted}} + \underbrace{d(\varphi^{t_1-t}(w),\varphi^{s(t_1)-t}(w))}_{<\epsilon_0/8 \text{ since } |t_1-s(t_1)|<4\epsilon}$$

for $t \geq 0$. Likewise, $\varphi^{s(t_1)}(w) \in W^s_{\epsilon_0}(\varphi^{s(t_1)+t(x,y)}(x))$, hence

$$\varphi^{s(t_1)}(w) \in W^s_{\epsilon_0}(\varphi^{s(t_1)+t(x,y)}(x)) \cap W^u_{\epsilon_0}(\varphi^{s(t_1)}(y)),$$

because if $t \geq 0$, then

$$\underbrace{d(\varphi^{s(t_1)+t}(w), \varphi^{s(t_1)+t(x,y)+t}(x))}_{\leq d(\varphi^{s(t_1)+t}(w),\varphi^{t_1+t(x,y)+t}(w)) + d(\varphi^{t_1+t}(w),\varphi^{t_1+t(x,y)+t}(x)) + d(\varphi^{t_1+t(x,y)+t}(x),\varphi^{s(t_1)+t(x,y)+t}(x))} \leq \frac{3\epsilon_0}{8}.$$

$$\underbrace{d(\varphi^{s(t_1)+t}(w),\varphi^{t_1+t(x,y)+t}(w))}_{<\epsilon_0/8 \text{ since } |t_1-s(t_1)|<4\epsilon} + \underbrace{d(\varphi^{t_1+t}(w),\varphi^{t_1+t(x,y)+t}(x))}_{<\epsilon<\epsilon_0/8 \text{ since } t_1+t\geq 0} + \underbrace{d(\varphi^{t_1+t(x,y)+t}(x),\varphi^{s(t_1)+t(x,y)+t}(x))}_{<\epsilon_0/8 \text{ since } |t_1-s(t_1)|<4\epsilon}$$

With $|s(t_1)-t_1+t(x,y)| \leq |t_1-s(t_1)|+|t(x,y)| \leq 4\epsilon < \epsilon_0$ and $d(\varphi^{s(t_1)}(y), \varphi^{t_1}(x)) < \delta$, this implies

$$t(\varphi^{t_1}(x), \varphi^{s(t_1)}(y)) = s(t_1) - t_1 + t(x,y), \text{ and } \varphi^{s(t_1)}(w) = [\varphi^{t_1}(x), \varphi^{s(t_1)}(y)],$$

and thus, actually, $|s(t_1) - t_1 + t(x,y)| \leq \epsilon$, so

$$\underbrace{d(\varphi^{t_1}(w), \varphi^{t_1}(y))}_{\leq d(\varphi^{t_1}(w),\varphi^{s(t_1)}(w)) + d(\varphi^{s(t_1)}(w),\varphi^{s(t_1)}(y)) + d(\varphi^{s(t_1)}(y),\varphi^{t_1}(y))} < \epsilon + 2\Big(\frac{\epsilon_0}{8}\Big) \leq \frac{3\epsilon_0}{8} < \frac{\epsilon_0}{2},$$

$$\underbrace{d(\varphi^{t_1}(w),\varphi^{s(t_1)}(w))}_{<\epsilon_0/8 \text{ since } |t_1-s(t_1)|<4\epsilon} + \underbrace{d(\varphi^{s(t_1)}(w),\varphi^{s(t_1)}(y))}_{<\epsilon} + \underbrace{d(\varphi^{s(t_1)}(y),\varphi^{t_1}(y))}_{<\epsilon_0/8 \text{ since } |t_1-s(t_1)|<4\epsilon}$$

as well as $|s(t_1) - t_1| \leq |s(t_1) - t_1 + t(x,y)| + |t(x,y)| \leq 2\epsilon < 3\epsilon$, as claimed.

Similarly, one shows that $B = \varnothing$, so $|s(t) - t| \leq 3\epsilon < \epsilon_0/2$ for $t \in [-T,T]$. Additionally, $\varphi^T(w) \in W^u_{\epsilon_0/2}(\varphi^T(y))$ and $\varphi^{-T}(w) \in W^s_{\epsilon_0/2}(\varphi^{-T+t(x,y)}(x))$ because d is adapted (Remark 6.1.2). Therefore, if $|t| \leq T$ and $c := \epsilon_0$, then

$$d(\varphi^t(w), \varphi^s(y)) \leq \frac{c}{2} e^{-\eta(T-|t|)} \text{ and } d(\varphi^t(w), \varphi^{t+t(x,y)}(x)) \leq \frac{c}{2} e^{-\eta(T-|t|)}. \qquad \square$$

Remark 6.2.5. This improves the Anosov Closing Lemma (Theorem 5.3.11) in situations where the closed pseudo-orbit consists of a small number of orbit segments, such as a single orbit segment that almost closes up: the percentage error in periods from the Closing Lemma is improved to a small *absolute* difference. Claim 6.3.3 is such an application. Proposition 6.2.18 below is an instance where this is critical (Remark 6.2.19).

Definition 6.2.6. A hyperbolic set Λ has a *local product structure* if $[x, y] \in \Lambda$ for x, y and (sufficiently small) ϵ as in Proposition 6.2.2.

Theorem 6.2.7. *A hyperbolic set for a flow is locally maximal if and only if it has a local product structure.*

Proof. If Λ is locally maximal, fix an adapted metric in a neighborhood V of Λ, an isolating neighborhood $U \subset V$ of Λ, and $\epsilon > 0$ such that $\bigcup_{x\in\Lambda} B_\epsilon(x) \subset U$ and such that there is a δ such that if $x, y \in \Lambda$ and $d(x, y) < \delta$, then $z := [x, y]$ as in Proposition 6.2.2 is well defined—and by definition both forward and backward asymptotic to Λ. Since we use an adapted metric, $\mathcal{O}(z) \subset U$. But U is an isolating neighborhood, so $z \in \Lambda$, and Λ has a local product structure.

If Λ has a local product structure, fix $\epsilon > 0$, $\delta > 0$, and $C_0 > 0$ as in Proposition 6.2.2. To show that Λ is locally maximal, we establish that if an orbit is sufficiently close to Λ then it must be in Λ.

If V is a sufficiently small neighborhood of Λ, then Λ_V is hyperbolic and with the same constants ϵ, δ, and C_0 as above, that is, if $x, y \in \Lambda_V$ and $d(x, y) < \delta$, then $d_s(\varphi^{t(x,y)}(x), [x, y]) < C_0 d(x, y)$ and $d_u(y, [x, y]) < C_0 d(x, y)$.

Fix $\alpha < \frac{\delta}{3}\min(1, 2C_0)$ such that $d(\varphi^t(x), \varphi^t(y)) < \frac{\delta}{3}$ for $0 \le t \le 1$ whenever $x \in \Lambda$, $y \in W^u_\alpha(x)$. If furthermore $d(\mathcal{O}^+(y), \Lambda) < \alpha/C_0$, then for each $n \in \mathbb{N}$ there is a $y_n \in \Lambda$ with $d(y_n, \varphi^n(y)) < \alpha/C_0$. Since $\varphi^1(x), y_1 \in \Lambda$ and $d(\varphi^1(x), y_1) \le d(\varphi^1(x), \varphi^1(y)) + d(\varphi^1(y), y_1) < \delta/3 + \alpha/C_0$ we see that $x_1 = [y_1, \varphi^1(x)] \in \Lambda$ and $\varphi^1 y \in W^u_\alpha(x_1)$. Continuing, we construct a sequence of points $x_n = [y_n, \varphi^1(x_{n-1})] \in \Lambda$ with $\varphi^n(y) \in W^u_\alpha(x_n)$. Let $z_n = \varphi^{-n}(x_n)$. Then $z_n \xrightarrow[n\to\infty]{} y$ by the uniform contraction in the local unstable manifolds. Thus $y \in \Lambda$ since Λ is closed. Similarly, if $y \in W^s_\alpha(x)$ and α is analogously defined for $-1 \le t \le 0$, and $\mathcal{O}^-(y)$ stays within α/C_0 of Λ, then $y \in \Lambda$.

If $\alpha_1 \in (0, \alpha)$ is sufficiently small and $d(\mathcal{O}(y), \Lambda) < \alpha_1$ then $\Lambda \cup \mathcal{O}(y)$ is a hyperbolic set by Proposition 5.1.8. Furthermore, for α_1 possibly smaller, if we let $x \in \Lambda$ such that $d(x, y) < \alpha_1$, then we can define $z_1 = [x, y]$ and $z_2 = [y, x]$, and the above argument shows that $z_1, z_2 \in \Lambda$. Hence, $y \in \Lambda$ by the product structure of Λ. □

We now connect the spectral decomposition to the invariant foliations; this will have relevance later when we investigate Markov partitions. This also explains the name of the components of the decomposition.

Second proof of Theorem 5.3.37. We first define a relation on the periodic points contained in a locally maximal hyperbolic set Λ. Let $p, q \in \mathrm{Per}(\Phi_{\restriction\Lambda})$. Then $p \sim q :\Leftrightarrow$ "$W^{cu}(p) \cap W^{cs}(q) \cap \Lambda \ne \varnothing \ne W^{cs}(p) \cap W^{cu}(q) \cap \Lambda$ with both intersections transverse in at least one point." This relation is clearly reflexive and symmetric, and we presently show that it is also transitive. We will then see that this equivalence relation defines the set Λ_i as the closure of an equivalence class.

Transitivity: If $x, y, z \in \mathrm{Per}(\Phi_{\upharpoonright_\Lambda})$ and $p \in W^{cu}(x) \cap W^{cs}(y) \cap \Lambda$, $q \in W^{cu}(y) \cap W^{cs}(z) \cap \Lambda$ are transverse intersection points, then by continuity of unstable leaves the images of a ball around p in $W^{cu}(p) = W^{cu}(x)$ accumulate on $W^{cu}(y)$ so $W^{cu}(x)$ and $W^{cs}(z)$ have a transverse intersection in Λ.

By Theorem 6.2.7 the equivalence classes are open, so by compactness there are finitely many of them, and we denote by $\Lambda_1, \dots, \Lambda_m$ their (pairwise disjoint) closures. By Corollary 5.3.15(4), $\mathrm{NW}(f_{\upharpoonright_\Lambda}) \subset \overline{\mathrm{Per}(\Phi_{\upharpoonright_\Lambda})}$ since Λ is locally maximal, so $\bigcup_{i=1}^m \Lambda_i = \mathrm{NW}(\Phi_{\upharpoonright_\Lambda})$.

To show that $\varphi^t{}_{\upharpoonright_{\Lambda_i}}$ is topologically transitive note first that if $p \in \Lambda_i$ is periodic and $p \sim q$ with q periodic, then there is by definition a $z \in W^{cu}(p) \cap W^{cs}(q) \cap \Lambda$ that is a point of transverse intersection, so $W^{cu}(p)$ accumulates on the orbit of q. Thus, $W^{cu}(p) \cap \Lambda$ is dense in $\Lambda_i \cap \mathrm{Per}(\varphi^t{}_{\upharpoonright_\Lambda})$, hence in its closure Λ_i for every periodic $p \in \Lambda_i$.

We conclude by showing that a hyperbolic set $\Lambda = \mathrm{NW}(\varphi^t{}_{\upharpoonright_\Lambda})$ is topologically transitive if $W^{cu}(p) \cap \Lambda$ is dense in Λ for every periodic $p \in \Lambda$. We need to check that for any two open sets V and W in Λ there exists a $T \in \mathbb{R}^+$ such that $\varphi^t(V) \cap W \neq \varnothing$ for all $t \geq T$ (Definition 1.6.31). For open $V, W \subset \Lambda$ density of periodic points implies the existence of a periodic point $p \in V \cap \Lambda$. Since V is open there exists $\delta > 0$ such that $W^u_\delta(p) \subset V$. Since $W^{cu}(p) \cap \Lambda$ is dense there exists $T_0 \in \mathbb{R}$ such that $W \cap \varphi^{T_0}(W^u_\delta(p)) \neq \varnothing$. Let T_1 be the period of p. Then for $n \in \mathbb{N}$ and $t = T_0 + nT_1$ we have $W \cap \varphi^t(V) \neq \varnothing$. □

This proof yields a new corollary of Theorem 5.3.37:

Proposition 6.2.8. *Let Λ be a basic set for Φ and $p \in \Lambda$ be a periodic point. Then*

- *the center-stable manifold of p is dense in $W^s(\Lambda)$, and*
- *the center-unstable manifold of p is dense in $W^u(\Lambda)$.*

Proof. Let $p, q \in \Lambda$ be periodic points. Then $W^{cu}(p) \pitchfork W^{cs}(q) \neq \varnothing$ and $W^{cs}(p) \pitchfork W^{cu}(q) \neq \varnothing$. Furthermore, the local stable manifold of the orbit of p accumulates on the local stable manifold for q. By iteration this implies that the stable manifold of the orbit of p accumulates on the stable manifold of q. Since the periodic points q are dense, Theorem 5.3.25 says that the stable manifold of p is dense in $W^s(\Lambda)$. Reversing the flow proves the same for unstable manifolds. □

Corollary 6.2.9. *A compact locally maximal hyperbolic set Λ is topologically transitive if and only if periodic points are dense in Λ and the center-unstable manifold of every periodic point is dense in Λ.*

Proof. The spectral decomposition is trivial. □

We can refine this (and Theorem 5.3.52) in the case of Anosov flows.

Theorem 6.2.10. *For Anosov flows the following are equivalent*:

(1) *The spectral decomposition has one piece* (*the whole manifold*).

(2) *The flow is* regionally recurrent (*Definition* 1.5.13).

(3) *The flow is topologically transitive.*

(4) *Periodic points are dense.*

(5) *All center-unstable leaves are dense.*

(6) *All center-stable leaves are dense.*

Proof. Equivalence of (1), (2), (3), (4) is Theorem 5.3.52. We show (4) $\Rightarrow$ (6) $\Rightarrow$ (2), and this for the reversed flow establishes (4) $\Rightarrow$ (5) $\Rightarrow$ (2).

(4) $\Rightarrow$ (6): If $x \in M$, then $M = X := \overline{W^{cu}(x)}$ because X is open as follows. By (4), it suffices to check that $z \in X$, $\varphi^T(p) = p \in R_\epsilon(z) := \{[a,b] \mid a \in W^u_\epsilon(z),\ b \in W^{cs}_\epsilon(z)\} \Rightarrow p \in X$, and this is clear because $X \ni \varphi^{-iT}([p,z]) \xrightarrow[i\to\infty]{} p$, so $p \in \overline{X} = X$.

(6) $\Rightarrow$ (2) follows if $W^{cu}(p)$ is dense for some periodic point p because this implies (2) since $W^{cs}(p)$ is dense by (6), so homoclinic points are dense, and they are nonwandering (see for example, Corollary 6.5.3 below).

Suppose to the contrary that $M \neq X := \overline{W^{cu}(p)}$, with X a union of unstable leaves and Φ-invariant. Take $\epsilon > 0$ such that

$$U := \bigcup_{x\in X} W^s_\epsilon(x) \neq M.$$

Then $\varphi^t(U) \subset \bigcup_{x\in X} W^u_{\lambda^t\epsilon}(x)$ for $t \geq 0$ and a suitable $\lambda \in (0,1)$, and $X = \bigcap_{t\geq 0} \varphi^t(U)$ (nested intersection) and $Y := \bigcap_{t\geq 0} \varphi^{-t}(M \setminus U)$ are disjoint, and Y is closed and Φ-invariant, so there is a $\delta > 0$ with $d(x,y) \geq \delta$ for all $x \in X$ and $y \in Y$. Thus Y is a union of stable leaves because if $y \in Y$ and $z \in W^s(y)$ then $z \in Y$ because otherwise $Y \ni \varphi^t(z) \xrightarrow[t\to\infty]{} X$ contrary to (6) since $\varphi^t(x) \in Y$ and $d(\varphi^t(x), \varphi^t(y)) \xrightarrow[t\to\infty]{} 0$. □

The following is a counterpart for the mixing case (and note the pertinent dichotomy in Theorem 8.1.3).

Theorem 6.2.11. *For a locally maximal hyperbolic set* Λ *the following are equivalent*:

(1) $\Phi_{\restriction\Lambda}$ *is topologically mixing.*

(2) *The periodic points of* Λ *are dense in* Λ *and for each periodic point* $p \in \Lambda$ *we have* $\overline{W^s(p)} \cap \Lambda = \Lambda$ *and* $\overline{W^u(p)} \cap \Lambda = \Lambda$.

(3) $\Phi_{\restriction\Lambda}$ *is transitive and each open set contains periodic points with incommensurate periods.*[3]

[3]By Proposition 6.2.18 we can replace "each open set contains" with "has."

Definition 6.2.12. Here we say that the periods of a set of periodic points are *commensurate* if they are all in $p\mathbb{Z}$ for some $p > 0$; incommensurate otherwise (that is, they generate a dense subgroup of $\mathbb{R}$).

Proof. We prove (3) $\Leftrightarrow$(1) $\Leftrightarrow$(2).

- (1) $\Rightarrow$ (2): This strengthening of the contrapositive is of independent interest:

Proposition 6.2.13. *If $p \in \Lambda$ is a periodic point of Φ and $W^u(p)$ is not dense, then $\Phi_{\restriction \Lambda}$ is the suspension of a homeomorphism f of $K := \Lambda \cap \overline{W^u(p)}$.*

Proof. Let T be the period of p. Analogously to Proposition 1.6.27 there is a minimal nonempty $L \subset K$ such that

$$L \text{ is closed, } W^u\text{-saturated and } \varphi^T\text{-invariant.} \tag{6.2.1}$$

The compact, hence closed, set $\varphi^{[0,T]}(L)$ is W^{cu}-saturated, hence equal to Λ by density of weak-unstable manifolds.

Claim 6.2.14. *If $\varphi^t(L) \cap L \neq \varnothing$ then $\varphi^t(L) = L$.*

Proof. $L \cap \varphi^t(L)$ satisfies (6.2.1), so $L \cap \varphi^t(L) = L$ by minimality, that is, $L \subset \varphi^t(L)$, so $\varphi^{-t}(L) \subset L$. However, $\varphi^{-t}(L)$ also satisfies (6.2.1), so $\varphi^{-t}(L) = L$ by minimality. Apply φ^t to get the claim. □

Claim 6.2.15. *There is a smallest $s > 0$ such that $\varphi^s(L) \cap L \neq \varnothing$ (and hence $\varphi^s(L) = L$).*

Proof. Since $\mathcal{S} := \{s \geq 0 \mid \varphi^s(L) \cap L \neq \varnothing\}$ is closed, we just need to find a positive lower bound. We show that if there is no positive lower bound, then $L = \Lambda$.

Note that recursively, if $t \in \mathcal{S}$ and hence $\varphi^t(L) = L$, then $\varphi^{nt}(L) = L$ for all $n \in \mathbb{Z}$. If there are arbitrarily small such t, then $\mathcal{S}$ is therefore dense in $\mathbb{R}$ and hence equal to $\mathbb{R}$, which implies that L is W^{cu}-saturated, hence dense, and hence equal to Λ. □

This choice of s gives $\Lambda = \varphi^{[0,s)}(L)$, and this is a disjoint union. This allows us to recognize the suspension as follows: Λ is a bundle over S^1 using the projection

$$\pi : \Lambda \to S^1, \quad \varphi^t(x) \mapsto t \bmod s,$$

where $x \in L$, and $f = \varphi^s{}_{\restriction L}$ is the base map.

Since L is W^u-saturated, $W^u(p)$ must lie in some $\varphi^t(L)$, so $L = K$. □

- (2) $\Rightarrow$ (1): The main step is to establish uniformity of density.

Lemma 6.2.16. *For $\epsilon > 0$ and each periodic point p there is an $R > 0$ such that the R-disk $W^u_R(p)$ in $W^u(p)$ is ϵ-dense in Λ.*

Proof. Otherwise there are ϵ and p such that for each $N \in \mathbb{N}$ there is an $x_n \in \Lambda$ such that $B(x_n, \epsilon)$ and the N-ball in $W^u(p)$ are disjoint. Passing to a subsequence, let $x = \lim_{n\to\infty} x_n$ and note that $x \notin \overline{W^u(p)}$. □

The same conclusion holds uniformly for all $q \in \mathcal{O}(p)$: Let T be a period of p, $\epsilon > 0$, and $\epsilon' > 0$ such that if $t \in [0, T]$, then the preimage under φ^t of an ϵ-ball contains an ϵ'-ball. Let $R' > 0$ be as in Lemma 6.2.16 for ϵ', and $R > 0$ such that $W^u_{R'}(p) \subset \varphi^{-t}(W^u_R(\varphi^t(p)))$ when $0 \le t \le T$. Now, if $q = \varphi^{t_q}(p) \in \mathcal{O}(P)$ with $0 \le t_q \le T$, and B is an ϵ-ball, then $B \cap W^u_R(q) \ne \varnothing$.

Now suppose $U, V \subset \Lambda$ are (relatively) open and choose $\epsilon > 0$ such that V contains an ϵ-ball and U contains an ϵ-neighborhood of a periodic point p. Take $T > 0$ such that if $t \ge T$, then $\varphi^t(W^u_\epsilon(p))$ contains an R-disk in $W^u(\varphi^t(p))$. Then $\varphi^t(U) \cap V \ne \varnothing$ for $t \ge T$.

- (1) $\Rightarrow$ (3): Contraposition. Let $O \subset \Lambda$ be an open set such that the period of each periodic point in O is in $\pi\mathbb{Z}$. Consider a flow box (see Proposition 1.1.14) $U = \varphi^{(-\pi/4, \pi/4)}(D_x) \subset O$ of diameter less than $\delta < \pi v/2L$, where D_x is a local transversal through $x \in O$, $v := \min |X| > 0$ is the minimum speed, and L is as in Theorem 5.3.11. Suppose $\varphi^T(U) \cap U \ne \varnothing$, that is, there is a $p \in U$ such that $\varphi^T(p) \in U$. Then $\varphi^{[0,T]}(p)$ defines a periodic δ-pseudo-orbit and is hence $L\delta$ shadowed by a periodic orbit, hence $|T - n\pi| \le L\delta/v < \pi/2$ for some $n \in \mathbb{N}$, so Φ is not mixing.

- (3) $\Rightarrow$ (1): Suppose $U, V \subset \Lambda$ are (relatively) open and $p, q \in U$ are periodic with incommensurate periods π_p and π_q. By Proposition 6.2.8 there is, analogously to Lemma 6.2.16, an R such that $W^{cu}_R(p) \cap V \ne \varnothing$, that is, there is an open neighborhood W of some $p' \in \mathcal{O}(p)$ such that $W^u_R(x) \cap V \ne \varnothing$ for all $x \in W$. Also, there is a $q' \in \mathcal{O}(q)$ such that $W^s_{\text{loc}}(p') \cap W^u_{\text{loc}}(q') \ne \varnothing$. Thus, there are $\epsilon > 0$ and $n \in \mathbb{N}$ such that $\varphi^{[n\pi_q - \epsilon, n\pi_q + \epsilon]}(q') \subset W$ and $\varphi^{[n\pi_p - \epsilon, n\pi_p + \epsilon]}(p') \subset W$ for all $n \ge N$. Since π_p and π_q are incommensurate, this shows that for all sufficiently large t there are $x \in W \subset U$ with $\varphi^t(x) \in V$. □

Proposition 6.2.17. *If Λ is a compact locally maximal hyperbolic set, periodic points are dense in Λ, and $W^u(p)$ or $W^s(p)$ is dense for some periodic p, then $W^u(z)$ and $W^u(z)$ are dense for all $z \in \Lambda$.*

Proof. The contrapositive of Proposition 6.2.13 and its counterpart for W^s show that $W^u(p)$ and $W^s(p)$ are dense for all periodic p. To show that $W^u(z)$ (and hence likewise $W^s(z)$) is dense for all $z \in \Lambda$, introduce convenient local neighborhoods $O_r(x) := \bigcup_{y \in W^u_r(x)} W^{cs}_r(y)$ for $r > 0$ and $x \in \Lambda$. We can choose a $\delta(r)$ independently of x such that $B(x, \delta(r)) \subset O_r(x)$ and moreover, every W^u-leaf that meets $B(x, \delta(r))$

"goes across $O_r(x)$": $W^u(z) \cap B(x, \delta(r)) \neq \varnothing \Rightarrow W^u(z) \cap W_r^{cs}(y) \neq \varnothing$ for all $y \in W_r^u(x)$.[4]

For $x \in \Lambda$ and $\epsilon > 0$ we will find a point of $W := W^u(z)$ within ϵ of x. Take periodic points $p_1, \ldots, p_k$ such that the balls $B(p_i, \delta(\epsilon/2))$ cover Λ and note that if $T > 0$ is large enough, then $\varphi^T(W^u_{\epsilon/2}(p_i)) \cap B(x, \epsilon/2) \neq \varnothing$ for $1 \leq i \leq k$, and that there is an i such that $\varphi^{-T}(W) \cap B(p_i, \delta(\epsilon/2)) \neq \varnothing$. By choice of T there is a $q \in W^u_{\epsilon/2}(p_i) \cap \varphi^{-T}(B(x, \epsilon/2))$. Since these points are in $O_{\epsilon/2}(p_i)$, there is also a $y \in \varphi^{-T}(W) \cap W^{cs}_{\epsilon/2}(q)$.

Then $d(q, y) < \epsilon/2$, $d(\varphi^T(q), \varphi^T(y)) < \epsilon/2$, and

$$d(x, \varphi^T(y)) \leq d(x, \varphi^T(q)) + d(\varphi^T(q), \varphi^T(y)) < \frac{\epsilon}{2} + \frac{\epsilon}{2},$$

so $\varphi^T(y) \in W$ is the desired point. □

The following answers a natural question arising from Theorem 6.2.11: its third characterization is equivalent to the set of all periods being incommensurate.

Proposition 6.2.18. *If the periods of a basic set are incommensurate, then so are those of periodic points that intersect a given open set.*

Proof. We need to show that the subgroup P_O of $\mathbb{R}$ generated by periods of periodic points in O is dense in $\mathbb{R}$ given that the subgroup P generated by all periods is. To that end we show that if O is open, $\epsilon > 0$, and p is ρ-periodic, then there is a $\tau \in \mathbb{R}$ such that for all $n \in \mathbb{N}$ there are periodic points $p_n \in O$ whose period is within $\epsilon/2$ of $\tau + n\rho$. This implies the claim because it shows that P_O contains elements ϵ-close to any element of P, so density of P implies density of P_O.

By transitivity, there is an orbit segment from within $\epsilon/2L$ of p (where L is as in the Anosov Closing Lemma, Theorem 5.3.11) to the center of an $\epsilon/2$-ball in O and an orbit segment from within $\epsilon/2L$ of that point to within $\epsilon/2L$ of p. Denote by τ the sum of their lengths. For each $n \in \mathbb{N}$ the $\epsilon/2L$-pseudo-orbit consisting of these orbit segments and n periods of p is $\epsilon/2$-shadowed by a periodic orbit that contains a point $p_n \in O$ and has the desired period. □

Example 1.6.35 used the above result to establish topological mixing for a special flow over the hyperbolic toral automorphism (Example 1.5.26).

Remark 6.2.19. Theorem 5.3.11 is less explicit about the timing of the shadowing orbit than needed for this proof, and Theorem 6.2.4 implies that the periods we obtain are as asserted.

[4] We only use the existence of such a y.

Remark 6.2.20. If all periods are a multiple of $k \in \mathbb{R}$, then the ζ-function from (7.6.9) is a product of rational functions of $z = e^{-ks}$.

From the preceding, one can extract the following (compare Theorem 6.2.11):

Theorem 6.2.21 ([61, p. 77],[59]). *For a basic set Λ there are three mutually exclusive possibilities*:

- *Λ is a point,*
- *the restriction of the flow to Λ is a (constant-time) suspension,*

  $\Leftrightarrow$ *the restriction of the flow to Λ is not topologically mixing,*

  $\Leftrightarrow$ *the periods of the restriction of the flow to Λ are commensurate,*
- *every strong stable or unstable manifold is dense in Λ.*

The equivalence of the formulations of the second possibility and the Ω-Stability Theorem imply genericity of mixing:

Corollary 6.2.22. *The subset of $\mathcal{A}$ (see (5.3.1)) for which an infinite basic set is not topologically mixing is of the first category in the C^r-topology for $r \geq 1$.*

Proof. By Theorem 6.2.21 it suffices to show that having commensurate periods is a first-category phenomenon in $\mathcal{A}$. Specifically, denote by $V_n \subset \mathcal{A}$ the flows such that for some (infinite) element X of the spectral decomposition there is a $\tau \geq 1/n$ for which $\varphi^t(x) = x \in X \Rightarrow t \in \mathbb{Z}\tau$. Then it suffices to show that V_n is closed and nowhere dense (that is, closed with dense complement) in $\mathcal{A}$, since $\bigcup_n V_n$ is the set of flows in $\mathcal{A}$ with commensurate periods.

To see that V_n is closed, suppose $V_n \ni \Phi_k \xrightarrow[C^1]{} \Psi \in \mathcal{A}$. The Ω-Stability Theorem (Theorem 5.4.15) implies that for large enough k there is an orbit equivalence between $\Phi_k\restriction_{\mathrm{NW}(\Phi_k)}$ and $\Psi\restriction_{\mathrm{NW}(\Psi)}$, and this orbit equivalence identifies the spectral decompositions $\mathrm{NW}(\Phi_k) = \bigcup_i \Lambda_i^k$ and $\mathrm{NW}(\Psi) = \bigcup_i \Lambda_i^\Psi$, so we can pass to a subsequence and find a fixed i such that there is a $\tau_k \geq 1/n$ for which $\varphi^t(x) = x \in X_k := \Lambda_i^k \Rightarrow t \in \mathbb{Z}\tau_k$ (and each X_k is identified with $X := \Lambda_i^\Psi$).

Ω-stability further implies that each Ψ-periodic orbit γ in X corresponds to periodic orbits γ_k in X_k whose (least) periods converge to that of γ as $k \to \infty$. Since these are each upper bounds for τ_k, we can take $\tau_k \in [1/n, M]$ for some $M \in \mathbb{R}$ and hence $\tau_k \to \tau \in [1/n, M]$ after passing to a subsequence. That done, we now find for arbitrary closed orbits γ in X^Ψ that, by the assumption on X_k, the corresponding periodic orbits γ^k in X_k satisfy $n_k\tau_k = \mathrm{period}(\gamma_k) \xrightarrow[k\to\infty]{} \mathrm{period}(\gamma)$, so $\gamma \in \mathbb{Z}\tau$, hence $\Psi \in V_n$. Thus, V_n is closed.

That the complement of V_n is dense is the easy part; indeed the complement of $\bigcup_n V_n$ is dense: take two distinct periodic orbits γ_1, γ_2 in a basic set and disjoint neighborhoods of them. Perturb the flow in one of them (only) in an arbitrarily C^r-small way so that $\frac{\text{period}(\gamma_1)}{\text{period}(\gamma_2)} \notin \mathbb{Q}$. □

These arguments can be varied to show that often no two periods are commensurate.

Proposition 6.2.23. *For a C^r-generic ($r \in \mathbb{N}$) element of $\mathcal{A}$ (see (5.3.1)) the periods of closed orbits are pairwise incommensurate.*

Proof. We show that the set of such flows is a C^r-dense intersection of C^r-open subsets of $\mathcal{A}$.

To show density enumerate the periodic orbits of a $\Phi_0 \in \mathcal{A}$ as $\{p_n \mid n \in \mathbb{N}\}$; this enumeration carries to any Φ sufficiently near Φ_0. For $\epsilon > 0$ such that $\mathcal{A}$ contains the C^r-ϵ-ball around Φ_0 and $i \in \mathbb{N}$ recursively define a time change Φ_i of Φ_{i-1} localized in the complement of $\bigcup_{j\leq i} p_j$ and C^r-$\epsilon 2^{-i-1}$-close to Φ_{i-1} for which the periods of the p_j for $j \leq i+1$ are pairwise incommensurate. The limit is within ϵ of Φ_0 in the C^r-topology and as desired.

Consider the set V_n of $\Phi \in \mathcal{A}$ for which a pair of distinct closed orbits has periods at most n and a period ratio p/q with $p, q \in \mathbb{N}$ and $q \leq n$. This is closed, and their union over n is the set of Φ with a commensurate pair of closed orbits. Thus, the complement is C^r-generic. □

The next result (by Mañé) provides a criterion for a flow to be Anosov. It can be used in Section 8.3 to provide examples of nontransitive Anosov flows.

Theorem 6.2.24 (Mañé Criterion). *A smooth flow on a compact manifold is an Anosov flow if and only if the chain-recurrent set is hyperbolic, the (weak) stable and unstable manifolds intersect transversely at one (hence every) point of each orbit, and their dimension is constant.*

Proof. That Anosov flows satisfy the criterion is clear. To show the other direction we use that if the chain recurrent set is hyperbolic then the flow is Axiom A with no cycles. If the (weak) stable and unstable manifolds intersect transversely at every point of each orbit, then we have strong transversality. Now assume that the dimension of the stable and unstable splitting is constant on the manifold. By strong transversality, each point $p \in M$ is contained in a strong-stable manifold and strong-unstable manifold, and the sum of the dimensions of these is $n-1$, where n is the dimension of M. Therefore, we have a splitting $T_pM = E^s \oplus E^c \oplus E^u$ that is continuous and flow invariant. For the basic sets $\Lambda_1, \dots, \Lambda_n$ there are adapted metrics and constants of hyperbolicity. Furthermore, the constants can be extended to neighborhoods of the basic sets $O_1, \dots, O_n$ as we have done previously. Let $O = \bigcup_{i=1}^n O_i$. Since

$X := M \setminus O$ is compact, there is a uniform bound $T > 0$ such that no orbit will spend more than the amount of time T in X. Therefore there exists a constant $C > 0$ that compensates for the time spent in X using the constants of hyperbolicity for the basic sets. Hence, M is a hyperbolic set, and the flow is Anosov. □

6.3 Livshitz theory

In this section we present a geometric take on the abundance of periodic orbits, namely that by a combination of exponential closing and Hölder regularity, periodic data determine a global quantity (Theorem 6.3.2). We will see that the leading regularity notion in hyperbolic dynamics is Hölder continuity, and the essential reason is that both it and hyperbolicity are described by exponential behavior.

Definition 6.3.1. A map f between metric spaces is said to be Hölder continuous with exponent $\alpha \in (0, 1]$ if $d(f(x), f(y)) \le (d(x, y))^\alpha$ for nearby x and y.

A map $f : \mathbb{R} \to \mathbb{R}$ is said to be β-*Hölder* continuous, written $f \in C^\beta$, if $f \in C^{[\beta]}$ and $f^{([\beta])}$ is Hölder continuous with exponent $\beta - [\beta]$. (In the literature this is usually denoted by $C^{[\beta],\beta-[\beta]}$ or $C^{[\beta]+(\beta-[\beta])}$; if $\beta < 1$ this coincides with the usual definition given above.)

We denote $C^{\beta-} := \bigcap_{\epsilon>0} C^{\beta-\epsilon}$ as the space of functions which are Hölder of all orders less than β and $C^{\beta+} := \bigcup_{\epsilon>0} C^{\beta+\epsilon}$ denotes the functions which are $C^{\beta+\epsilon}$ for some ϵ. Also, $C^{1,\mathrm{Lip}} \subset C^{2-\epsilon}$ is the space of functions with *Lipschitz* (that is, 1-Hölder) derivative.[5] A subbundle is said to be C^β (resp. $C^{\beta-}$, $C^{\beta+}$) if it is spanned by vector fields with C^β (resp. $C^{\beta-}$, $C^{\beta+}$) components in a chart.

The connection with exponential behavior is that exponentially small differences in the inputs result in exponentially small differences in the outputs of a Hölder-continuous map. On one hand, this causes natural structures associated with hyperbolic dynamics to be Hölder continuous, and on the other hand, Hölder-continuous functions play well with hyperbolic dynamics (see, for example, Proposition 7.3.1). This section showcases this interplay nicely by combining exponential closing and Hölder regularity to great effect.

We have approached the abundance of periodic points in hyperbolic flows in a variety of ways. They are dense, and the specification property (Definition 7.3.2) says in essence that all conceivable orbit behaviors are exhibited by periodic orbits. We will also see their abundance in the context of ergodic theory—the measures discussed in Chapter 7 are all obtained as weak limits of measures on periodic orbits. Here, we present a much more geometric take on the abundance of periodic orbits,

[5] See also Definitions 4.2.32 and B.1.1.

namely that by a combination of exponential closing and Hölder regularity, periodic data determine a global quantity. We previously saw this in Theorem 5.3.23, but that result was ineffective due to a lack of Walters-continuous functions. While that lack is easy to remedy (Proposition 7.3.1), we here work directly with Hölder-continuous functions, because this class is particularly well adapted to the arguments below and yields more information about the solutions of a cohomological equation:

Theorem 6.3.2 (Livshitz Theorem). *Let Λ be a compact locally maximal hyperbolic set for a flow Φ generated by a vector field X on a manifold M and $a\colon \Lambda \to \mathbb{R}$ Hölder continuous such that $\varphi^T(x) = x \Rightarrow \int_0^T a(\varphi^t(x))\,dt = 0$. If there is a dense orbit in Λ, then there is a continuous $A\colon \Lambda \to \mathbb{R}$ such that $a = XA$, the derivative in the flow direction. Moreover, A is unique up to an additive constant and Hölder continuous with the same Hölder exponent as a. Furthermore, if $a \in C^1$, then $A \in C^1$.*

Proof. If $\Lambda = \overline{\mathcal{O}(x_0)}$, set $A(\varphi^t(x_0)) := \int_0^t a(\varphi^s(x_0))\,ds$. The next result implies that A is uniformly continuous and hence has a unique continuous extension to $\Lambda = \overline{\mathcal{O}(x_0)}$, which then clearly has the same Hölder exponent as well.

Claim 6.3.3. *A is Hölder continuous on $\mathcal{O}(x_0)$ with the same Hölder exponent as a.*

Proof. Suppose $|a(x) - a(y)| \le Hd(x,y)^\alpha$ for small $d(x,y)$. If $t_1 < t_2$ are such that $\epsilon := d(\varphi^{t_1}(x_0), \varphi^{t_2}(x_0))$ is as in the Anosov Closing Lemma (Theorem 5.3.11) and as in Theorem 6.2.4, then a T-periodic orbit with $|T - t_2 + t_1| < \epsilon$ $L\epsilon$-shadows $\varphi^{[t_1,t_2]}(x_0)$ and contains a point y such that

$$d(\varphi^{t_1+s}(x_0), \varphi^s(y)) \le C\epsilon\eta^{\min(s,T-s)} \quad \text{for } s \in [0,T].$$

Thus,

$$\begin{aligned}
\big|A(\varphi^{t_1}(x_0)) - A(\varphi^{t_1+T}(x_0))\big| &= \Big|\int_0^T a(\varphi^{T+s}(x_0))\,ds\Big| \\
&= \Big|\int_0^T a(\varphi^{T+s}(x_0)) - a(\varphi^s(y))\,ds + \underbrace{\int_0^T a(\varphi^s(y))\,ds}_{=0}\Big| \\
&\le \int_0^T \underbrace{\big|a(\varphi^{T+s}(x_0)) - a(\varphi^s(y))\big|}_{\le HC^\alpha\epsilon^\alpha\eta^{\alpha\min(s,T-s)}}\,ds \\
&\le 2HC^\alpha\epsilon^\alpha \int_0^T \eta^{\alpha s}\,ds.
\end{aligned}$$

On the other hand, $\big|A(\varphi^{t_2}(x_0)) - A(\varphi^{t_1+T}(x_0))\big| = \big|\int_{t_2}^{t_1+T} a(\varphi^s(x_0))\,ds\big| \le \epsilon\|a\|_\infty$. □

We now return to the proof of the theorem and prove the uniqueness of A. If $XA = XA'$, then $X(A - A') \equiv 0$, so $A - A'$ is a constant of motion (Definition 1.1.24), hence constant on the dense orbit, hence constant. Finally, a and XA are continuous and agree on a dense set and therefore coincide.

Now suppose $a \in C^1$. By the preceding, A is Lipschitz continuous (hence differentiable a.e., but we need more). We show that the derivatives of A along stable (hence by symmetry unstable) leaves exist and are continuous. Since A is also continuously differentiable along orbits ($XA = a$), it is then C^1 by Lemma 6.3.4 below. If $y \in W^s(x)$, then

$$\begin{aligned} A(y) - A(x) &= \lim_{T\to+\infty} \Big(-\int_0^T a(\varphi^s(y)) - a(\varphi^s(x))\,ds \Big) + A\big(\varphi^T(x)\big) - A\big(\varphi^T(y)\big) \\ &= -\int_0^\infty a(\varphi^s(y)) - a(\varphi^s(x))\,ds. \end{aligned}$$

Differentiating with respect to $y = x + tv$ at $t = 0$ (in local coordinates) gives $D_v A(x) = -\int_0^\infty D_{v_s} a(\varphi^s(x)) D_v(\varphi^s)(x)$ by the chain rule, where $v_s := D\varphi^s v$. Both factors of the integrand are exponentially small in s (since the v_s are exponentially small and a is C^1, and since v is a stable vector), so the improper integral converges uniformly, hence to a well-defined continuous function, which thus agrees with the derivative of the left-hand side. □

Here we used the following elementary precursor of Theorem 9.2.9:

Lemma 6.3.4. *If $f\colon \mathbb{R}^n \to \mathbb{R}$ is C^1 along the leaves of k continuous transverse foliations, then f is C^1.*

Proof. For y near x we need to find a linear map L_x such that $f(y) - f(x) = L_x(x - y)$ up to higher-order terms in $|y - x|$. We do this for two foliations A and B; for larger k the proof differs mainly by notation. For $z \in A(x) \cap B(y)$ we have

$$f(y) - f(x) = f(y) - f(z) + f(z) - f(x) = L_z^B(y - z) + L_x^A(z - x)$$

for linear maps L^A and L^B that depend continuously on the basepoint, so $L_z^B \to L_x^B$ as $y \to x$, hence $z \to x$ and thus $L_z^B(y - z) = L_x^B(y - z)$ up to higher order, and we can take L_x to be the linear map that restricts to L^A on $TA(x)$ and to L^B on $TB(x)$. □

Remark 6.3.5. We saw that any two solutions of the cohomological equation differ by a constant of motion (Proposition 1.6.5). If we drop the assumption that there is a dense orbit, then we can apply the Livshitz Theorem on each transitive component of the spectral decomposition, which gives a solution of the cohomological equation that is unique up to the addition of a function that is constant on each component.

Part of the interest in Theorem 6.3.2 and its proof is the regularity of the solution A of the cohomological equation. Since the existence of a bounded solution implies the vanishing of periodic data, we get a result that improves the regularity (and without using transitivity):

Corollary 6.3.6. *Let Λ be a compact hyperbolic set for a flow Φ generated by a vector field X on a manifold M and $A_0\colon M \to \mathbb{R}$ bounded. If $a := XA_0{\restriction_\Lambda}$ is Hölder continuous, then there is an $A\colon \Lambda \to \mathbb{R}$ (unique up to the addition of a constant of motion) such that $XA_0 = XA$, and A is Hölder continuous with the same Hölder exponent as a, and C^1 if a is.*

The nature of the uniqueness assertion is such that for *continuous* A_0, the conclusion is about A_0 itself.

Corollary 6.3.7. *Let Λ be a compact hyperbolic set for a flow Φ generated by a vector field X on a manifold M and $A\colon M \to \mathbb{R}$ continuous. If $XA{\restriction_\Lambda}$ is Hölder continuous or C^1, then so is A.*

To see an application of this theory, recall that we noted in Section 1.3 that conjugacy (flow equivalence) of flows preserves the periods of closed orbits. For time changes of hyperbolic flows, there is an easy converse: If the periods of closed orbits are unaffected by a time change, then the time-changed flow is conjugate to the original one.

Proposition 6.3.8. *Let Φ be a flow on a manifold M with a compact topologically transitive hyperbolic set Λ and Ψ a time change of Φ with a Hölder-continuous α (as in Proposition 1.2.2). If the periods of all periodic orbits of Φ and Ψ agree then Φ and Ψ are conjugate on Λ via a homeomorphism which is Hölder continuous and C^1 if the time change is C^1.*

Proof. Recall from page 45 that a time change $\psi^t(x) = \varphi^{\alpha(t,x)}(x)$ arises from a conjugacy, that is, is trivial, if there is a function $\beta\colon \Lambda \to \mathbb{R}$, differentiable along orbits, such that $\alpha(t,x) - t = \beta(x) - \beta(\varphi^t(x))$. But by assumption the values of the cocycle on the left-hand side over periodic orbits are zero, so by the Livshitz Theorem (Theorem 6.3.2) there is a Hölder solution β which is C^1 if α is. □

Theorem 6.3.9. *Suppose Φ and Ψ are orbit equivalent on hyperbolic sets Λ, Λ', respectively, and that the periods of corresponding periodic orbits in Λ and Λ' agree. Then Φ and Ψ are conjugate (Definition 1.3.1).*

Proof. By Theorem 6.4.3 below, the orbit equivalence h can be taken to be Hölder continuous. Thus $h \circ \varphi^t \circ h^{-1}$ is a Hölder-continuous time change of ψ^t with the same periods as ψ^t; hence by Proposition 6.3.8 it is Hölder conjugate to ψ^t and hence so is φ. □

Let us state without proof a remarkable strengthening of the regularity statement in Theorem 6.3.2 or Corollary 6.3.6:

Theorem 6.3.10 (Livshitz regularity). *Let Λ be a compact hyperbolic set for a C^∞-flow Φ generated by a vector field X on a manifold M and $A_0: M \to \mathbb{R}$ bounded. If $a := XA_0\restriction_\Lambda$ is C^∞, then there is a C^∞-function $A: \Lambda \to \mathbb{R}$ such that $XA_0 = XA$.*

Thus, Corollary 6.3.7 has a C^∞ counterpart, so we get high regularity "for free." (There are suitable C^k counterparts to this.) Here is an easy application.

Theorem 6.3.11. *If a C^∞ Anosov flow Φ preserves a continuous volume $\mu = \rho\, d\,\mathrm{vol}$, then $\rho \in C^\infty$.*

Proof. $X \log \rho = \eta$, where $\mathcal{L}_{\dot\varphi}(d\,\mathrm{vol}) = \eta\, d\,\mathrm{vol}$. □

Of course, the question of the existence of an invariant volume in the first place can be answered using Theorem 6.3.2:

Theorem 6.3.12 (Livshitz–Sinai). *A transitive Anosov flow Φ has an invariant volume if and only if $\varphi^\tau(x) = x \Rightarrow \det d\varphi^\tau(x) = 1$.*

We also note that the Livshitz Theorem plays a central role in the classification of equilibrium states (Theorem 7.3.24).

Finally, there is a strengthening of the Livshitz Theorem complementary to that which obtains higher regularity of the solution: the existence of a merely measurable solution is enough.

Theorem 6.3.13 (Measurable Livshitz Theorem). *If X generates a volume-preserving C^2 Anosov flow and $A_0: \Lambda \to \mathbb{R}$ is measurable and such that $a := XA_0$ is α-Hölder continuous, then there is an α-Hölder $A: \Lambda \to \mathbb{R}$ such that $A = A_0$ a.e.*

6.4 Hölder continuity of orbit equivalence

The Livshitz Theorem illustrates the importance of Hölder continuity (Definition 6.3.1) in hyperbolic dynamics, and Theorem 5.1.16 illustrates that invariant structures in hyperbolic dynamics are Hölder continuous. This section shows that the class of Hölder-continuous functions on a hyperbolic set is preserved by orbit equivalence; we do this by showing that orbit equivalences can be chosen Hölder continuous. The class of Hölder-continuous functions is deeply important for the ergodic theory of hyperbolic flows through the definition of pressure, and the study of pressure in turn gives us insights into the behavior of smooth invariant measures—showing their ergodicity, for example. On the other hand, the class of Hölder-continuous functions is a natural class of functions to study, since the principal structures associated with hyperbolicity are Hölder continuous with respect to the smooth structure—although they usually do not possess any higher regularity (such as Lipschitz continuity or C^1).

Here and elsewhere we obtain Hölder continuity by establishing Hölder continuity along the flow direction and stable and unstable leaves separately. This is sufficient because the stable and unstable manifolds are uniformly transverse continuously varying Lipschitz submanifolds, and we give this result as an abstract lemma about metric spaces, where we think of stable and unstable leaves as ("vertical" and "horizontal") equivalence classes. (Recursively, this also works for three or more foliations.)

Proposition 6.4.1. *Let Λ be a metric space with two equivalence relations $\sim_h$ and $\sim_v$ for which there exist $\epsilon > 0$, $K_1 \in \mathbb{R}$ such that if $x \sim_h y \sim_v z$ and $d(x,z) < \epsilon$, then*

$$d(x,y)^2 + d(y,z)^2 \le K_1 d(x,z)^2,$$

and such that for sufficiently nearby $x, y \in \Lambda$ there exists a w such that $x \sim_h w \sim_v y$. Let $\varphi\colon \Lambda \to X$ be a map to a metric space X with some $K_2, \alpha > 0$ for which $d(x,y) < \epsilon$ and $x \sim_h y$ or $x \sim_v y$ imply

$$d(\varphi(x), \varphi(y)) \le K_2 d(x,y)^\alpha.$$

Then $d(\varphi(x), \varphi(y)) \le 2K_1K_2 d(x,y)^\alpha$ for all sufficiently close $x, y \in \Lambda$.[6]

Proof. First note that for $(x,y) \in \mathbb{R}^2$ one has

$$(|x|^\alpha + |y|^\alpha)^{1/\alpha} \le 2^{1/\alpha}(x^2 + y^2)^{1/2}.$$

To see this, note (for example, by drawing $\{(x,y) \in \mathbb{R}^2 \mid |x|^\alpha + |y|^\alpha = 1\}$) that the discrepancy is greatest for $x = y$, in which case

$$\frac{(|x|^\alpha + |y|^\alpha)^{1/\alpha}}{(x^2 + y^2)^{1/2}} = \frac{2^{1/\alpha}|x|}{2^{1/2}|x|} < 2^{1/\alpha}.$$

For $x, y \in \Lambda$ take w such that $x \sim_h w \sim_v y$ and note that

$$\begin{aligned} d(\varphi(x), \varphi(y)) &\le d(\varphi(x), \varphi(w)) + d(\varphi(w), \varphi(y)) \le K_2(d(x,w)^\alpha + d(w,y)^\alpha) \\ &\le 2K_2(d(x,w)^2 + d(w,y)^2)^{\alpha/2} \le 2K_1K_2 d(x,y)^\alpha. \end{aligned}$$

□

Theorem 6.4.2. *Let Λ and Λ' be compact hyperbolic sets for diffeomorphisms f and f', respectively, and $h = f'hf^{-1}\colon \Lambda \to \Lambda'$ a topological conjugacy. Then both h and h^{-1} are Hölder continuous.*

[6] Theorem 9.2.9 is a remarkable related result.

Proof. Since f and f' appear symmetrically in the statement it suffices to check that h itself is Hölder continuous. Furthermore we just showed that it is indeed enough to show Hölder continuity of h along stable and unstable manifolds. Since h also conjugates f^{-1} and f'^{-1} (for which stable and unstable manifolds reverse roles) it is, in fact, enough to prove that $h_{\upharpoonright W^u(x)\cap\Lambda}$ is Hölder continuous for every $x \in \Lambda$ (with uniform constant and exponent).

To this end take $c < 1 < C$ such that C is a Lipschitz constant for f and c is a Lipschitz constant for $f'^{-1}{}_{\upharpoonright W^u}$ and let $\alpha > 0$ be such that $cC^\alpha < 1$. Fix $\epsilon_0 > 0$. Since Λ is compact and h is continuous, hence uniformly continuous, there exists $\delta_0 > 0$ such that $d(x, y) < \delta_0$ implies $d(h(x), h(y)) < \epsilon_0$.

Now if $x, y \in \Lambda$, $y \in W^u(x)$, and $\delta := d(x, y)$ is sufficiently small, then there exists $n \in \mathbb{N}$ such that

$$d(f^n(x), f^n(y)) \le C^n\delta < \delta_0 \le C^{n+1}\delta.$$

Hence $d(h(f^n(x)), h(f^n(y))) < \epsilon_0$, by choice of δ_0, so using $cC^\alpha < 1$ we have

$$\begin{aligned} d(h(x), h(y)) &= d(f'^{-n}hf^n(x), f'^{-n}hf^n(y)) < c^n\epsilon_0 \\ &= c^n\delta_0^\alpha \cdot \epsilon_0/\delta_0^\alpha \le (cC^\alpha)^n C^\alpha(\epsilon_0/\delta_0^\alpha)\delta^\alpha < C^\alpha(\epsilon_0/\delta_0^\alpha)(d(x, y))^\alpha. \quad \square \end{aligned}$$

An analog of the preceding result applies to flows as well by similar reasoning. There is, however, a new aspect to be taken into account here, namely, the lack of uniqueness in the flow direction. We thus obtain the following result:

Theorem 6.4.3. *Let $\Lambda \subset M$, $\Lambda' \subset M'$ be compact hyperbolic sets for flows Φ and Ψ, respectively, and suppose that Φ and Ψ are orbit equivalent via $h\colon \Lambda \to \Lambda'$. Then arbitrarily C^0-close to h there is an orbit equivalence that is Hölder continuous together with its inverse.*

Proof. We begin with a local construction of a Hölder orbit equivalence. Take small smooth transversals $\mathcal{T}$ at $p \in \Lambda$ and $\mathcal{T}'$ at $q = h(p) \in \Lambda'$. Then locally $h(\mathcal{T})$ projects canonically to $\mathcal{T}'$ along the orbits of Ψ and the composition of h with this projection is Hölder continuous by the same arguments as in the proof of Theorem 6.4.2 using the intersections of $\mathcal{T}$ with weak-unstable and weak-stable foliations as the equivalence classes in Proposition 6.4.1.

Now fix some $\delta > 0$ and cover Λ by flow boxes whose floors are small smooth transversals and fix corresponding smooth transversals in Λ'. From the Hölder-continuous map between these transversals construct local conjugacies on the flow boxes by taking them to be time preserving. This gives local homeomorphisms from these flow boxes to Λ'. To assemble these into one global map take a smooth partition of unity on Λ subordinate to the covering by these flow boxes. Now all images of a point $x \in \Lambda$ lie on an orbit segment and thus one can take the average of the corresponding time parameters weighted by the values of the members of the partition of unity at x. This gives a well-defined Hölder-continuous map $\tilde{h}$ which is C^0-close to h and takes orbits of Φ to orbits of Ψ. Further, $\tilde{h}$ is also differentiable along the orbits of Φ. The remaining problem is that $\tilde{h}$ may not be monotone along orbits.

To find a homeomorphism with the desired properties we use the fact that $\tilde{h}$ is as C^0-close to the homeomorphism h as we please, so long as δ is sufficiently small. This implies that there is an $\eta > 0$ such that for any $x \in \Lambda$ and $t > \eta$ we have $\tilde{h}(\varphi^t(x)) = \psi^s(\tilde{h}(x))$ with $s > 0$. This implies that defining $h'(x) := (1/\eta)\int_0^\eta \tilde{h}(\varphi^t(x))dt$ (the integral interpreted as one involving the real parameter along the orbit of x) gives a homeomorphism h' with all desired properties. □

6.5 Horseshoes and attractors

As discussed in the preface, in his study of the 3-body problem, Poincaré came upon complicated orbit structures: he noticed that if p is a hyperbolic periodic orbit whose center-stable and center-unstable manifolds intersect transversely at a point not on the orbit of p, then there are complicated tangles as shown (for a Poincaré section) in Figure 6.5.1. Birkhoff noticed that the transverse intersection point off of the orbit of p was in the closure of periodic orbits. Smale introduced the notion of the horseshoe as a dynamical object that encapsulates the complexity seen near the transverse intersection.

Nonlinear versions of Example 1.5.24 have the same qualitative features (both in the base and the roof function)—this is the content of the Structural Stability Theorem (Theorem 5.4.5). Thus, we now define a horseshoe with this in mind as a model and show how horseshoes naturally arise for transverse homoclinic points.

Definition 6.5.1. Let Φ be a flow on a manifold M. A transitive hyperbolic set for Φ with at least two distinct periodic orbits is a *horseshoe* if it is orbit equivalent to a transitive hyperbolic symbolic flow with at least two distinct periodic orbits (Definition 1.9.5).

In a Poincaré section for a periodic orbit, the periodic orbit gives rise to a fixed point for the return map to the section. If the stable and unstable manifolds of this fixed point intersect transversely at a second point, then this is a transverse homoclinic point, and this generates a horseshoe as described below. In particular, this occurs for suspensions of maps with transverse homoclinic points as in Figure 6.5.1. When the stable and unstable curves of a hyperbolic fixed point intersect transversely, they produce tangles (Figure 6.5.1), and these in turn produce horseshoes for an iterate of the map; this is the Birkhoff–Smale Theorem, illustrated in Figure 6.5.2.

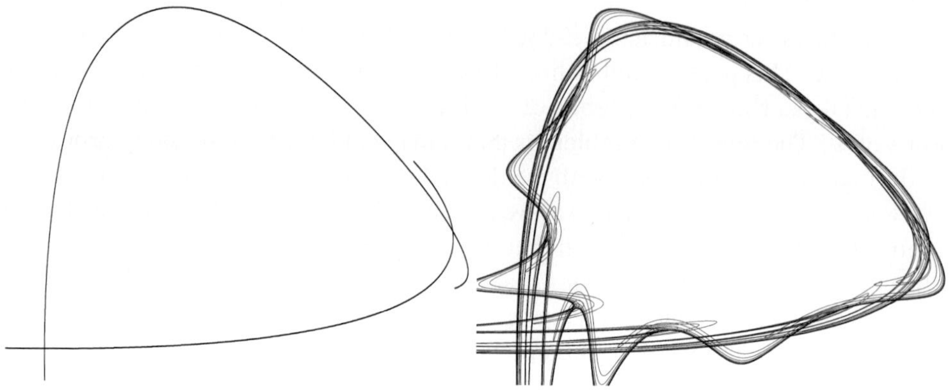

Figure 6.5.1. Transverse homoclinic point and homoclinic tangles. [Reprinted from [213] (© Cambridge University Press, all rights reserved) with permission.]

Theorem 6.5.2 (Birkhoff–Smale Theorem). *Let M be a compact Riemannian manifold and Φ be a smooth flow on M. If p is a hyperbolic periodic orbit for Φ, q is a transverse homoclinic point for p in a smooth section for the flow near p with return map f, and U is a neighborhood of $\{p, q\}$, then there is an $n \in \mathbb{N}$ such that f^n contains a hyperbolic invariant set $\Lambda \subset U$ topologically conjugate to the full 2-shift. Furthermore, the suspension of Λ is orbit equivalent to a hyperbolic flow over the 2-shift.*

Corollary 6.5.3. *Every transverse homoclinic point for a hyperbolic periodic orbit of a smooth flow is in the closure of the periodic points and is hence a nonwandering point for the flow. A flow with a transverse homoclinic point for a hyperbolic periodic orbit has positive topological entropy and exponential growth of periodic orbits.*

Theorem 5.3.55 indicates that hyperbolic attractors arise somewhat naturally, but it is useful to exhibit nontrivial examples explicitly. To that end, we now produce the derived-from-Anosov or *DA flow*, which is in its own right a prominent example, but also plays a pivotal role in a later construction of an anomalous Anosov flow (Theorem 8.3.3), and we derive from it the *Plykin attractor*, on which in turn rests the construction of hyperbolic sets that are not enveloped by a locally maximal one (Definition 6.5.8, Section 6.7).

Let F be the Anosov diffeomorphism of $\mathbb{T}^2$ induced by $\left(\begin{smallmatrix} 2 & 1 \\ 1 & 1 \end{smallmatrix}\right)$. We are going to modify it by deforming it near the origin to make that a repelling fixed point. Then the complement of a small closed disk around the origin is a trapping region, and we will describe the associated attractor (for the suspension).

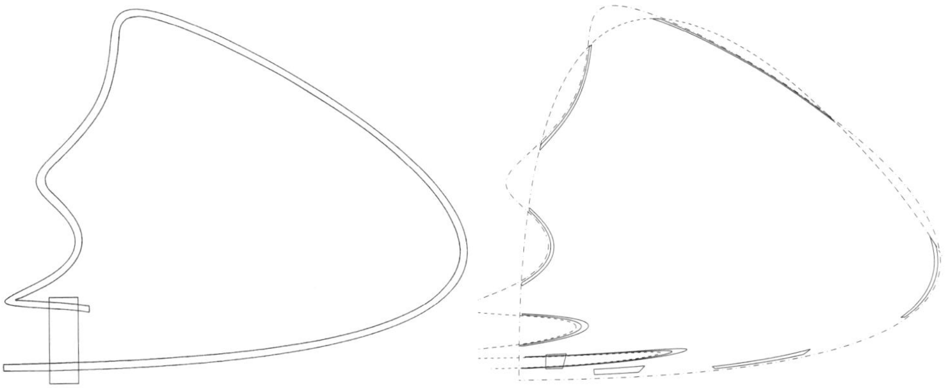

Figure 6.5.2. Horseshoes from tangles. [Reprinted from [213, 177] (© Cambridge University Press, all rights reserved) with permission.]

Denote by v^u and v^s the normalized eigenvectors corresponding to the eigenvalues $\lambda_1 = (3+\sqrt{5})/2$ and $\lambda_2 = \lambda_1^{-1} = (3-\sqrt{5})/2$, respectively, and let e^u and e^s be the stable and unstable vector fields obtained from v^u and v^s by parallel translation. Then $E^u(p) = \mathbb{R}e^u(p)$ and $E^s(p) = \mathbb{R}e^s(p)$ and $DF_p e^u(p) = \lambda_1 e^u(F(p))$ and $DF_p e^s(p) = \lambda_2 e^s(F(p))$ for all $p \in \mathbb{T}^2$.

On a disk U centered at 0 introduce coordinates (x_1, x_2) diagonalizing A, that is, such that $F(x_1, x_2) = (\lambda_1 x_1, \lambda_2 x_2)$ on U.

Definition 6.5.4. The derived-from-Anosov flow or DA flow is the suspension of the diffeomorphism $f\colon \mathbb{T}^2 \to \mathbb{T}^2$ defined by (6.5.1) below.

Remark 6.5.5. We next establish that f has a hyperbolic attractor Λ whose complement W is dense in $\mathbb{T}^2$ and is the basin of the attracting fixed point 0 for f^{-1}. Thus, the DA flow has a closed hyperbolic attractor with empty interior.

To construct f let $\phi\colon \mathbb{R} \to [0,1]$ be an even C^∞-function such that

$$\phi(t) = \begin{cases} 1 & \text{if } |t| \le 1/8, \\ 0 & \text{if } |t| \ge 1/4, \end{cases}$$

$$\phi'(t) < 0 \text{ if } 1/8 < t < 1/4,$$

and note that

$$g(x) := \frac{x\phi(x)}{1+bx^2} \quad \text{with } b > \frac{\min \phi'}{\frac{1}{4} - \lambda_2} \text{ sufficiently large}$$

is odd with

$$g'(x) = \underbrace{\frac{x\phi'(x)}{1+bx^2}}_{\frac{|x|}{1+bx^2}\le\frac{1}{2b}} + \phi(x)\underbrace{\frac{1-bx^2}{(1+bx^2)^2}}_{\frac{1-z}{(1+z)^2}\ge -\frac{1}{8}} \ge \frac{\min\phi'}{2b} - \frac{1}{8} > -\frac{\lambda_2}{2}.$$

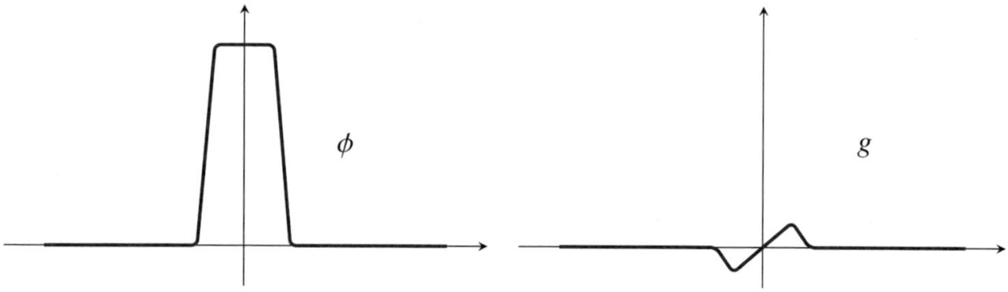

Figure 6.5.3. Bump functions.

Let

$$f\colon \mathbb{T}^2 \to \mathbb{T}^2, \quad x \mapsto \begin{cases} F(x) & \text{if } x \notin U, \\ F(x_1,x_2) + (0, \phi(x_1)g(x_2)) & \text{if } x = (x_1,x_2) \in U. \end{cases} \tag{6.5.1}$$

The fixed points of f are in U, where $(x_1,x_2) = f(x_1,x_2)$ is equivalent to

$$\begin{aligned} x_1 &= \lambda_1 x_1, \\ x_2 &= \lambda_2 x_2 + \phi(x_1)g(x_2). \end{aligned}$$

The first equation implies $x_1 = 0$ so the second equation reduces to

$$\left(\lambda_2 - 1 + \frac{\phi(x_2)}{1+bx_2^2}\right) x_2 = 0$$

with solutions $x_2 = 0, \pm\bar{x}$, where $\phi(\bar{x}) = (1-\lambda_2)(1+b\bar{x}^2)$. In order to determine the nature of these fixed points we note that

$$Df_{(x_1,x_2)} = \begin{pmatrix} \lambda_1 & 0 \\ \phi'(x_1)g(x_2) & h(x_1,x_2) \end{pmatrix} = \begin{cases} \begin{pmatrix} \lambda_1 & 0 \\ 0 & \lambda_2+1 \end{pmatrix} & \text{when } (x_1,x_2) = (0,0), \\ \begin{pmatrix} \lambda_1 & 0 \\ 0 & \lambda_2 + g'(\pm\bar{x}) \end{pmatrix} & \text{when } (x_1,x_2) = (0,\pm\bar{x}), \end{cases}$$

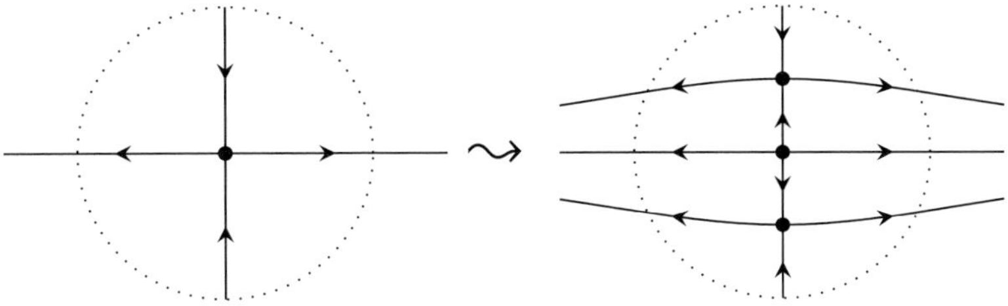

Figure 6.5.4. The DA-bifurcation of the fixed point.

with $h(x_1, x_2) := \lambda_2 + \phi(x_1)g'(x_2)$ and $g'(\pm\bar{x}) = \frac{\pm\bar{x}\phi'(\pm\bar{x})}{1+b\bar{x}^2} + (1-\lambda_2)\frac{1-b\bar{x}^2}{1+b\bar{x}^2} < 0 + 1 - \lambda_2$. Thus, $(0,0)$ is a repelling fixed point and $(0, \pm\bar{x})$ are hyperbolic fixed points.

The stable manifold of 0 is f-invariant, and Df preserves the stable subbundle E^s of F although it may not contract vectors in E^s everywhere, and in fact permutes the stable manifolds for F in the same way as F does. The unstable manifold

$$W = W^u(0) = \{p \in \mathbb{T}^2 \mid \alpha(p) = \{0\}\} = \bigcup_{n\in\mathbb{N}} f^n(U_0)$$

of 0, where U_0 is a sufficiently small open neighborhood of 0, is open and we show later that it is dense.

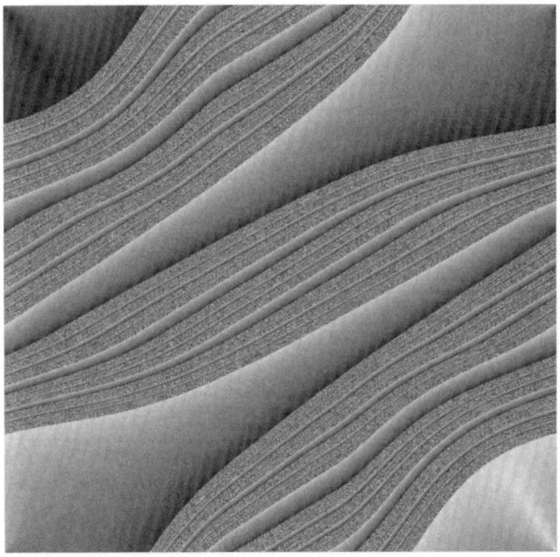

Figure 6.5.5. The unstable manifold of 0. [Picture after Yves Coudène, from https://www.lpsm.paris/pageperso/coudene/dyn1.html.]

Proposition 6.5.6. *If b is sufficiently large, then $\Lambda := \mathbb{T}^2 \setminus W$ is a hyperbolic attractor.*

Lemma 6.5.7. *There exists $\lambda' < 1$ such that $h(x_1, x_2) < \lambda'$ on Λ.*

Proof. By compactness it suffices to show that $h < 1$ on Λ. Let

$$V := \{(x_1, x_2) \in U \mid h(x_1, x_2) \geq 1\} = \bigcup_{(x_1,x_2)\in U} (\{x_1\} \times V_{x_1}),$$

where $V_{x_1} = \{x_2 \mid h(x_1, x_2) \geq 1\}$. Note that $h(x_1, t) = \lambda_2 + \phi(x_1) g'(t) \geq 1$ if and only if $g'(t) \geq \frac{1-\lambda_2}{\phi(x_1)}$. Since g' is an even function it follows that V_{x_1} is a symmetric interval for all x_1 and, furthermore, if $x > y \geq 0$ then $\phi(x) \leq \phi(y)$ and $V_x \subset V_y$. On the other hand if we write $f(x_1, x_2) = (f_1(x_1, x_2), f_2(x_1, x_2)) = (x_1', x_2')$ then $h(x_1, x_2) = \frac{\partial f_2}{\partial x_2}(x_1, x_2)$ and $f^{-1}(\{x_1'\} \times V_{x_1'}) = \{x_1\} \times V_{x_1}'$ (since f permutes unstable leaves of F) with V_{x_1}' symmetric and of length no more than that of V_{x_1} (since $h(x_1', x_2') \geq 1$). Thus $V_{x_1}' \subset V_{x_1}$ and since $f(x_1, 0) = F(x_1, 0)$ we conclude that $f^{-1}(V) \subset V$. Moreover, since $\{0\} \times V_0 \subset \mathbb{T}^2 \setminus \Lambda$ we find $\{x_1\} \times V_{x_1} \subset \mathbb{T}^2 \setminus \Lambda$ for all x_1 sufficiently close to 0. Consequently $f^{-n}(v) \subset \mathbb{T}^2 \setminus \Lambda$ for some $n \in \mathbb{N}$ and hence $V \subset \mathbb{T}^2 \setminus \Lambda$. □

Proof of Proposition 6.5.6. Note that $\Lambda := \mathbb{T}^2 \setminus W$ is an attractor by definition (and $\mathbb{T}^2 \setminus \overline{U_0}$ is a trapping region for it). The lemma shows that on Λ the diagonal elements of Df are λ_1 and a function bounded from above by $\lambda' < 1$. For large b the off-diagonal element is close to zero since $\phi'(x_1)$ is bounded and $g(x_2) \leq \frac{x}{1+bx^2} \leq \frac{1}{2b}$. One obtains hyperbolicity of Λ as follows. If we write $Df_x = \begin{pmatrix} \lambda_1 & 0 \\ G(x) & H(x) \end{pmatrix}$ and take $\epsilon \in (0, \sqrt{\lambda_1^2 - 2})$ such that $\frac{1}{\lambda_1 - \lambda'} < \sqrt{1 + 1/\epsilon^2} - 1$, then $|G(x)| < \epsilon$ for all $x \in \Lambda$ if b is sufficiently large. Consider now horizontal cones of the form $|v| < \gamma |u|$ with $\frac{\epsilon}{\lambda_1 - \lambda'} < \gamma < \sqrt{\epsilon^2 + 1} - \epsilon < 1$. Note that these are invariant under Df since if we let $(u', v') := Df_x(u, v)$ then

$$|v'| = |G(x)u + H(x)v| < \epsilon|u| + \lambda'|v| < (\lambda'\gamma + \epsilon)|u| < \gamma\lambda_1|u| = \gamma|u'|$$

since $\epsilon < (\lambda_1 - \lambda')\gamma$. To see that vectors in γ-cones expand, note that

$$\begin{aligned}
|(u', v')|^2 &= u'^2 + v'^2 = \lambda_1^2 u^2 + (G(x)u + H(x)v)^2 \\
&\geq \lambda_1^2 u^2 + G^2(x)u^2 - 2\underbrace{|G(x)|\, H(x)\, |u||v|}_{\leq \gamma |G(x)|\, H(x)\, u^2} + \underbrace{H^2(x)v^2}_{= -[1-H^2(x)]v^2 + v^2 \geq -\gamma^2[1-H^2(x)]u^2 + v^2} \\
&\geq [\lambda_1^2 + \underbrace{G^2(x)}_{> -\epsilon^2} - 2\gamma \underbrace{|G(x)|}_{<\epsilon}\, \underbrace{H(x)}_{<1} - \gamma^2 \underbrace{(1 - H^2(x))}_{<1}]u^2 + v^2 \\
&> (\lambda_1^2 - \underbrace{(\epsilon^2 + 2\gamma\epsilon + \gamma^2)}_{=(\epsilon+\gamma)^2 < \epsilon^2 + 1 < \lambda_1^2 - 2 + 1 = \lambda_1^2 - 1})u^2 + v^2 > u^2 + v^2.
\end{aligned}$$

Since the stable manifolds are given, hyperbolicity follows from Proposition 5.1.7 without studying vertical cones. □

To show that $W = \mathbb{T}^2 \setminus \Lambda$ is dense in $\mathbb{T}^2$, consider $p \in \Lambda$ and any open neighborhood U_p of p. Then there is a point $q \in U_p$ that is periodic for F, with period n, say. The stable manifold L of q (under F) is thus f^n-invariant and dense. Density of W follows if we can find $N \in \mathbb{N}$ such that $f^{-Nn}(L^1) \cap W \neq \varnothing$, where $L^1 = L \cap U_p$. But this is necessarily the case, since otherwise $L_f := \bigcup_{n\in\mathbb{N}} f^{-Nn}(L^1) \subset \Lambda$ and by hyperbolicity f^{-n} expands L_f so $L_f = L$. But L is dense in $\mathbb{T}^2$, so we would have $\Lambda = \mathbb{T}^2$, a contradiction. Thus Λ is the complement of an open dense set.

We have thus produced a hyperbolic attractor on $\mathbb{T}^2$.

We conclude this section with the presentation of a closely related attractor on a sphere or a disk.

Definition 6.5.8. The *Plykin attractor* is the suspension of the attractor on S^2 or a disk obtained from the DA attractor by identification of x and $-x$ as shown in Figure 6.5.7 and described below.

The DA map f is invariant under $J\colon \mathbb{T}^2 \to \mathbb{T}^2$, $J(x) = -x \pmod 1$, that is, $f \circ J = J \circ f$, and $(1/2, 1/2)$ is a periodic point of f since $f(1/2, 1/2) = F(1/2, 1/2) = (1/2, 0)$, $f(1/2, 0) = (0, 1/2)$, and $f(0, 1/2) = (1/2, 1/2)$. Thus, undertaking a like construction on those three additional points gives a map $f_P\colon \mathbb{T}^2 \to \mathbb{T}^2$, which commutes with J, has a single repelling fixed point, a repelling period-3 orbit, and a hyperbolic attractor Λ (Figure 6.5.6).

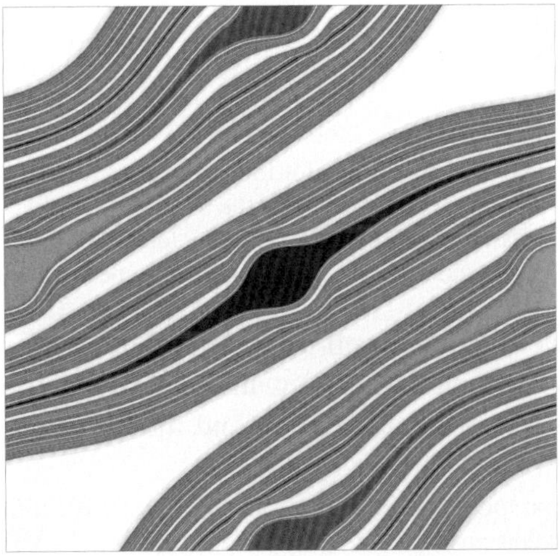

Figure 6.5.6. The DA construction repeated on a period-3 orbit. [Picture after Yves Coudène, from https://www.lpsm.paris/pageperso/coudene/dyn1.html.]

Note that J fixes these four repelling points:

$$(0,0),\quad -\left(\frac{1}{2},\frac{1}{2}\right)=\left(\frac{1}{2},\frac{1}{2}\right),\quad -\left(\frac{1}{2},0\right)=\left(\frac{1}{2},0\right),\quad -\left(0,\frac{1}{2}\right)=\left(0,\frac{1}{2}\right).$$

Thus, if V_i, $i=1,\dots,4$, are disks around $(0,0),(1/2,1/2),(1/2,0),(0,1/2)$, respectively, of the same radius and contained in $\mathbb{T}^2\setminus\Lambda$, then $M:=(\mathbb{T}^2\setminus\bigcup_{i=1}^4 V_i)/(x\sim -x)$ is a smooth manifold, indeed a 2-sphere with four holes. Since $f_P(-x)=-f_P(x)$, it induces a map $f'\colon M\to M$ which is smooth and injective. Filling $S^2\setminus M$ with four repellers (one fixed and one period-3 cycle) gives a diffeomorphism $\tilde{f}\colon S^2\to S^2$ with a hyperbolic attractor (obtained by projecting Λ onto M). This is the (discrete-time) *Plykin attractor* on S^2, shown in Figure 6.5.7.[7]

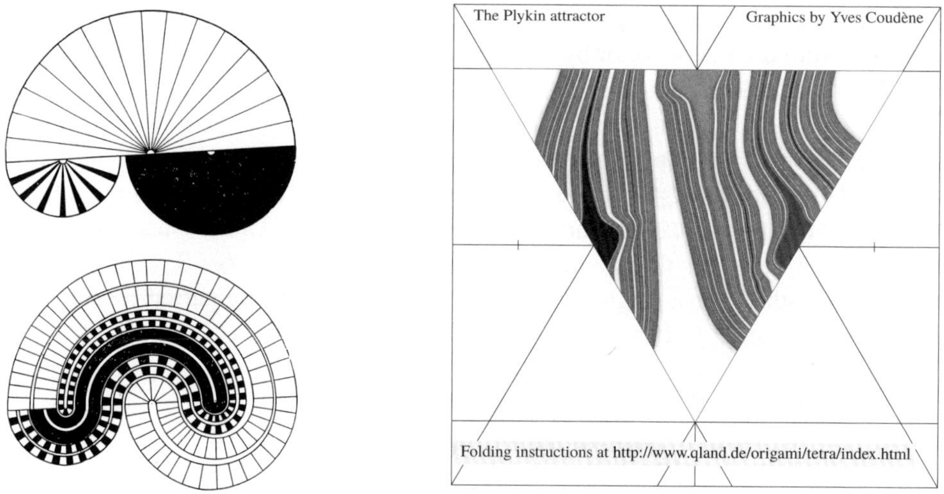

Figure 6.5.7. The Plykin map and a template for folding a realistic rendition onto a tetrahedron $\simeq$ sphere. (See Exercise 6.10 or, more simply, cut out the triangle and fold along the lines connecting the three periodic points to gather the white vertices at the top.) [Picture on left from [294], picture on right after Yves Coudène, from https://www.lpsm.paris/pageperso/coudene/dyn1.html.]

Furthermore, if we remove a well-chosen neighborhood of the fixed point we obtain a closed 2-disk D such that $f(D)\subset \operatorname{int}(D)$, and the nonwandering set of the restriction to D consists of the attractor Λ and the repelling period-3 orbit. The suspension of this flow is a flow on a solid torus.

In both cases, the suspension is the attractor in Definition 6.5.8.

[7] Topologically, this is also interesting because the three pieces of the basin of the repelling periodic orbit are "Wada basins," an example of *lakes of Wada*: they are three disjoint connected open sets in the plane or sphere that share the same boundary—the Plykin attractor in this case. This notion was introduced by Kunizô Yoneyama who credits Takeo Wada with the idea (on page 60 of "Theory of continuous set of points (not finished)," Tohoku Mathematical Journal **12**, 43–158; see https://www.jstage.jst.go.jp/article/tmj1911/12/0/12_0_43/_pdf/-char/en).

6.6 Markov partitions

We now construct Markov partitions for hyperbolic sets. The construction for flows and maps are similar, but there are nontrivial difficulties for flows due to the flow direction. The essential idea is to construct a fine mesh of many local sections and to consider the return map (Figure 0.2.1) to their union, which is a global section. The flow is a special flow over this return map, and we will construct an analog of a Markov partition for the return map. (This requires, of course, that there are no fixed points.) This establishes a connection with a symbolic system that is defined as a special flow (Definition 1.2.11) over a topological Markov chain with a roof function that corresponds to the "travel times" between the local sections for the smooth system. Due to the nature of this construction, these are also often called *Markov sections*.

Definition 6.6.1. Let $D \subset M$ be a smooth codimension-1 disk transverse to the flow with $\Lambda \cap D \neq \varnothing$ whose diameter is so small that D is included in a ξ-flow box (Definition 1.1.13) $U_\xi(D) := \varphi^{[-\xi,\xi]}(D)$ for some small $\xi > 0$. Within this flow box we define the projection to D by $\pi_D(\varphi^t(x)) := x \in D$ for $t \in [-\xi, \xi]$. This is smooth and induces a Bowen bracket on $\Lambda \cap D$: if $B \subset \Lambda \cap D$ is closed (with respect to the subspace topology) with $d(B, \partial D) > 0$ and with diameter sufficiently small (relative to $d(B, \partial D)$), then define

$$[\cdot,\cdot]_D \colon B \times B \to \Lambda \cap D, \quad [x,y]_D := \pi_D([x,y]).$$

If $[B,B]_D \subset B$ then B is called a *rectangle*, in which case we can write $[\cdot,\cdot]_B := [\cdot,\cdot]_D\restriction_{B\times B}$, and

$$W^s(x,B) := \{[x,y]_B \mid y \in B\} = B \cap \pi_D(U_\xi(D) \cap W^s_\xi(x)),$$
$$W^u(x,B) := \{[z,x]_B \mid z \in B\} = B \cap \pi_D(U_\xi(D) \cap W^u_\xi(x))$$

for $x \in B$, provided $\operatorname{diam}(B)$ is small enough. Further, B is said to be *proper* if $B = \overline{\operatorname{Int}_{\Lambda\cap D} B}$ where the interior is with respect to the subspace topology.

This gives local product parametrizations defined by the homeomorphisms

$$P_x \colon W^u(x,B) \times W^s(x,B) \to B, \quad (u,v) \to [u,v]_B.$$

We write

$$\partial^s B := P_x(\partial W^u(x,B) \times W^s(x,B)),$$
$$\partial^u B := P_x(W^u(x,B) \times \partial W^s(x,B)),$$

and $\partial W^i(x,B)$ is relative to $W^i(x, \Lambda \cap D)$ for $i = u, s$.

It is useful to observe the following:

Lemma 6.6.2. *If B is a rectangle then $\partial B = \partial^s B \cup \partial^u B$.*

Proof. We have $x \in \operatorname{Int}_\Lambda B \Rightarrow x \in \operatorname{Int}_{\Lambda\cap W^u_\eta(x)}(B \cap W^u_\eta(x) \cap \Lambda) = \operatorname{Int}_{\Lambda\cap W^u_\eta(x)} W^u_B(x)$ since B is a neighborhood of x in Λ. Thus $\partial^s B \subset \partial B$. Likewise $\partial^u B \subset \partial B$. Conversely, if

$$x \in (\operatorname{Int}_{\Lambda\cap W^s_\eta(x)} W^s_B(x)) \cap (\operatorname{Int}_{\Lambda\cap W^u_\eta(x)} W^u_B(x))$$

then by continuity of $[\cdot,\cdot]_D$ there is a neighborhood U of x in Λ such that for all $y \in U$ we have $[x, y]_D, [y, x]_D \in B$ hence

$$y' := [[y, x]_D, [x, y]_D] \in B \cap W^s_\eta(x) \cap W^u_\eta(y) \subset W^s_\eta(x) \cap W^u_\eta(y) \subset \{y\},$$

so $x \in \operatorname{Int}_\Lambda B$. □

From the above we see that $\partial^s B$ and $\partial^u B$ are independent of x.

Definition 6.6.3. A *proper family* of size η is a finite collection $\{T_0, \dots, T_{n-1}\}$ of closed subsets of Λ such that $\Lambda = \varphi^{[-\eta,0]}(\mathcal{T})$, where $\mathcal{T} := \bigcup_{i=0}^{n-1} T_i$ and there are closed differentiable transversals $D_0, \dots, D_{n-1}$ that are pairwise disjoint such that

(1) $\dim D_i = \dim M - 1$,

(2) $\operatorname{diam} D_i < \eta$,

(3) $T_i = \overline{\operatorname{Int}_{\Lambda\cap D_i} T_i} \subset D_i$,

(4) if $i \neq j$, then at least one of the sets $D_i \cap \varphi^{[0,\eta]}(D_j) = \varnothing$ or $D_j \cap \varphi^{[0,\eta]}(D_i) = \varnothing$.

Now for η sufficiently small (at least smaller than the expansive constant for the flow) and a proper family $\mathcal{T}$ of size η, each $x \in \mathcal{T}$ admits a first positive $f(x) \le \eta$ such that $\varphi^{f(x)}(x) \in \mathcal{T}$. Since the D_i are pairwise disjoint, closed, transverse to the flow, and there are only finitely many of them, there exists some $\beta > 0$ such that $f(x) \ge \beta$ for all $x \in \mathcal{T}$. Furthermore, f is Hölder continuous on each domain of continuity.

For a proper family, we define a return map

$$F_{\mathcal{T}}\colon \mathcal{T} \to \mathcal{T}, \quad x \mapsto \varphi^{f(x)}(x). \tag{6.6.1}$$

This map is a bijection, but neither $f(x)$ or $F_{\mathcal{T}}(x)$ need be continuous on $\mathcal{T}$. However, the restriction of $F_{\mathcal{T}}$ to

$$\mathcal{T}^* := \bigcap_{i\in\mathbb{Z}} F_{\mathcal{T}}^{-i}\Big(\bigcup_{i=0}^{n-1} \operatorname{Int}_{\Lambda\cap D_i} T_i\Big).$$

is continuous. By the Baire Theorem, $\mathcal{T}^*$ is dense in $\mathcal{T}$.

Definition 6.6.4. A proper family $\{R_0, \dots, R_{n-1}\}$ of sufficiently small size η is said to be a *Markov partition* if each R_i is a rectangle, $\mathcal{R} = \bigcup_{i=0}^{n-1} R_i$, the set

$$\mathcal{R}^* := \bigcap_{i \in \mathbb{Z}} F_{\mathcal{R}}^{-i}\Big(\bigcup_{i=0}^{n-1} \operatorname{Int}_{\Lambda \cap D_i} R_i\Big)$$

has a continuous return map and is dense as described above, and

(1) if $x \in U_{ij} := \overline{\mathcal{R}^* \cap R_i \cap F_{\mathcal{R}}^{-1}(R_j)}$ then $W^s(x, R_i) \subset U_{ij}$,

(2) if $y \in V_{ki} := \overline{\mathcal{R}^* \cap F_{\mathcal{R}}(R_k) \cap R_i}$ then $W^u(x, R_i) \subset V_{ki}$.

For $x \in \mathcal{R}$ let $i(x)$ be the unique $i \in \{1, \dots, n\}$ such that $x \in R_i$. By construction, the map $i \colon \mathcal{R} \to \{1, \dots, n\}$ is continuous, and so is the *itinerary map*

$$I \colon \mathcal{R}^* \to \Sigma_n, \quad x \mapsto (i(F_{\mathcal{T}}^j(x)))_{j \in \mathbb{Z}}.$$

Since η was chosen less than the expansive constant, the map I is injective, but may of course not be bijective. Then

$$\Lambda_{\mathcal{R}} = \overline{\{I(x) \mid x \in \mathcal{R}^*\}} \subset \Sigma_n$$

is a subshift of Σ_n because $\sigma(I(x)) = I(F_{\mathcal{R}}(x))$ for $x \in \mathcal{R}^*$ implies that $\sigma(\Lambda_{\mathcal{R}}) = \Lambda_{\mathcal{R}}$.

Now we are ready to prove the anticipated result.

Theorem 6.6.5 (Markov partitions). *Let Λ be a hyperbolic set without fixed points and U be a neighborhood of Λ. Then there exists a hyperbolic set $\tilde{\Lambda}$ with $\Lambda \subset \tilde{\Lambda} \subset U$ that has a Markov partition. That is to say, there is a semiconjugacy from a hyperbolic symbolic flow to $\Phi\restriction_{\tilde{\Lambda}}$ that is finite-to-one and one-to-one on a residual set of points, where the roof function for the subshift of finite type corresponds to the travel times between the local sections for the smooth system.*

Proof. Replacing U by $U \setminus \{\text{fixed points}\}$ if necessary, we may assume that Λ_U is a hyperbolic set with no fixed points. We fix an adapted metric on Λ_U.

By compactness of Λ_U and continuity of the exponential map and Φ, there is an $r > 0$ such that $\bigcup_{x \in \Lambda} B_{2r}(x) \subset U$ for all $x \in \Lambda_U$, $y \in D_x := \exp(E_r^s(x) \oplus E_r^u(x))$ the angle between $T_y D$ and $\dot{\varphi}^t(y)$ is in $(\pi/2 - \eta, \pi/2 + \eta)$ for suitable η, and $\varphi^{(-r,r)}(D_x)$ is a flow box.

Fix $\delta \in (0, r)$ less than an expansive constant for Λ_U, $\epsilon \in (0, \delta/2)$ as in the Shadowing Lemma (Theorem 5.3.3) for Λ_U, and $\gamma < \epsilon/2$ such that if $z_1, z_2 \in \Lambda$ satisfy

- $0 < d(z_1, z_2) < \gamma$,

- there exists some $0 < t < r$ such that $\varphi^t(z_1) \in D_{z_2}$,
- $x, y \in \Lambda_U \cap D_{z_1}$, and
- $d(x, y) < \gamma$,

then $d(x', y') < \epsilon/2$ for the corresponding points $x', y' \in D_{z_2}$ in the forward orbit of x and y that are the first intersection with D_{z_2}. Choose a set $P := \{p_0, \dots, p_{N-1}\}$ that is γ-dense in Λ such that $D_{p_i} \cap D_{p_j} = \varnothing$ for $i \neq j$, and let $\mathcal{T} = \bigcup_{i=0}^{N-1} D_{p_i}$. Define the return map $F_{\mathcal{T}}$ as in (6.6.1) on the set $\mathcal{T}$. Then

$$\Sigma_P := \left\{ \omega \in \Sigma_N \mid d(F_{\mathcal{T}}(p_{\omega_i}), p_{\omega_{i+1}}) < \epsilon \right\}$$

is a topological Markov chain. The choices of r, δ, ϵ, γ, and P above imply

$$\Lambda \cap \mathcal{T} \subset X := \left\{ y \in \mathcal{T} \mid F^i_{\mathcal{T}}(y) \text{ exists for all } i \in \mathbb{Z} \right\} \subset \Lambda_U \cap \mathcal{T}.$$

Let Int_X be the interior of X with the subspace topology given by the disks. By the Baire Theorem the set

$$X^* := \left\{ x \in X \mid F^i_{\mathcal{T}}(x) \in \mathrm{Int}_X \ \forall\, i \in \mathbb{Z} \right\}$$

is a dense G_δ in X. For each $x \in X^*$ there exists a unique itinerary $\omega \in \Sigma_P$ such that $F^i_{\mathcal{T}}(x) \in D_{p_{\omega_i}}$ for all $i \in \mathbb{Z}$ and the set of such ω is a dense G_δ set Σ^*_P in Σ_P. The inverse $\beta^* \colon \Sigma^*_P \to X^*$ of the itinerary map is uniformly continuous because the transverse disks and the flow are smooth, and hence extends to a map β onto a set $\hat{\Lambda} \subset X$ such that $\beta \colon \Sigma_P \to \hat{\Lambda}$ is a semiconjugacy.

Also, for each $\omega \in \Sigma^*_P$ there exists a unique minimal $f(\beta(\omega)) > 0$ such that

$$F_{\mathcal{T}}(\beta(\omega)) = \beta(\sigma(\omega)) = \varphi^{f(\beta(\omega))}(\beta(\omega)).$$

By continuity we can extend the function $f(\beta(\omega))$ to Σ_P. Since the D_i are differentiable, f is Lipschitz (since the D_i are Lipschitz and the flow is smooth). Now,

$$\tilde{\Lambda} = \bigcup_{x \in \hat{\Lambda}} \Big(\bigcup_{0 \le s < f(x)} \varphi^s(x) \Big) \subset \Lambda_U$$

is a hyperbolic set containing Λ. We will show that $\tilde{\Lambda}$ is the desired hyperbolic set with a Markov partition. To do this we first establish the connection between the shadowing orbits and the points in Σ_P. The special flow φ^t_f over Σ_P with roof function f then has an induced semiconjugacy $\hat{\beta} \colon \Sigma_{P,f} \to \Lambda'$ where $\Sigma_{P,f}$ is the identification space of the special flow over Σ_P with roof f.

The construction and the choice of constants imply the following result:

Claim 6.6.6. *The semiconjugacy $\hat{\beta}$ is given by the unique point shadowing the pseudo-orbit $g\colon \mathbb{R} \to M, t \mapsto \varphi^s(p_{\omega_i})$, where $s \in [0, f(\beta(\sigma^i\omega)))$ and $\sum_{j=0}^{i} f(\beta(\sigma^j\omega)) \le t < \sum_{j=0}^{i+1} f(\beta(\sigma^j\omega))$.*

For $\omega, \omega' \in \Sigma_P$ with $\omega_0 = \omega_0'$ we let

$$[\omega,\omega']_i := \begin{cases} \omega_i & \text{for } i \ge 0, \\ \omega_i' & \text{for } i \le 0. \end{cases}$$

Then $[\cdot,\cdot]$ commutes with β, that is,

$$\beta([\omega,\omega']) = [\beta(\omega), \beta(\omega')]_{D_{p_{\omega_0}}}.$$

If $R_i' := \{\beta(\omega) \mid \omega_0 = i\}$ then $\mathcal{R}' := \{R_i' \mid 0 \le i < N\}$ satisfies a condition similar to Definition 6.6.4(1): if $x = \beta(\omega)$ with $(\omega_0, \omega_1) = (i,j)$ and $y = \beta(\omega') \in W^s_{R_i'}(x)$ with $\omega_0' = i$, then $F_{\mathcal{R}}(y) \in W^s_\eta(F_{\mathcal{R}}(x))$ (see (6.6.1)) but also $y = [x,y] = [\beta(\omega), \beta(\omega')] = \beta([\omega,\omega'])$ and hence $F_{\mathcal{R}}(y) \in \beta(\sigma([\omega,\omega'])) \in R_j'$, so $F_{\mathcal{R}}(y) \in W^s_{R_j'}(F_{\mathcal{T}}(x))$. This proves half of the analog of Definition 6.6.4(2) and the other half follows similarly, that is,

$$F_{\mathcal{R}}(W^s_{R_i'}(x)) \subset W^s_{R_j'}(F_{\mathcal{R}}(x)) \text{ and } W^u_{R_j'}(F_{\mathcal{R}}(x)) \subset F_{\mathcal{R}}(W^u_{R_i'}(x)).$$

Note also that by continuity of β the R_i' are compact, hence closed. This yields the existence of a Markov partition for $\tilde{\Lambda}$.

The following lemma is also useful to note:

Lemma 6.6.7. *With the above notation $F_{\mathcal{R}}(\partial^s\mathcal{R}) \subset \partial^s\mathcal{R}$ and $\partial^u\mathcal{R} \subset F_{\mathcal{R}}(\partial^u\mathcal{R})$.*

Proof. For $x \in R_i$ there exist $j, x_n \in \operatorname{Int} R_i \cap f^{-1}(R_j)$ such that $x_n \to x \in R_i \cap F_{\mathcal{R}}^{-1}(R_j)$, hence $W^u_{R_j}(F_{\mathcal{R}}(x)) \subset F_{\mathcal{R}}(W^u_{R_i}(x))$. Thus, if $x \notin \partial^s\mathcal{R}$, hence $W^u_{R_j}(F_{\mathcal{R}}(x))$ is a neighborhood of $F_{\mathcal{R}}(x)$ in $W^s_\eta(F_{\mathcal{R}}(x)) \cap \Lambda$, then $W^u_{R_i}(X)$ is a neighborhood of x in $W^s_\eta(x) \cap \Lambda$, so $x \notin \partial^s\mathcal{R}$. The other inclusion follows by considering $F_{\mathcal{R}}^{-1}$. □

Theorem 6.2.4 and differentiability of the D_i imply that there exist constants $a > 1$ and $K > 0$ such that if $s, t \in \Sigma_P$ satisfy $s_i = t_i$ for all $|i| \le N$ then $d(s,t) \le Ka^{-N}$. Thus, β is Lipschitz with respect to the metric d_a from Definition 1.9.1. For $s \in \Sigma_A$, define the roof function as $f(\beta(s))$. Since the D_i are differentiable, f depends differentiably, hence Lipschitz continuously on the basepoint, so $f \circ \beta$ is Lipschitz on Σ_P since β is. Therefore, the symbolic flow is a hyperbolic symbolic flow. □

Another intuitive and useful consequence is that topological transitivity of the hyperbolic set and of its Markov model are equivalent. If the symbolic model is transitive then so is the hyperbolic set $\tilde{\Lambda}$. The next result establishes the other direction.

Proposition 6.6.8. *If Λ is a fixed-point-free hyperbolic set for the flow then the topological Markov chain Σ_P obtained from the coding above is topologically transitive if $F_{\mathcal{T}}\restriction_{\tilde{\Lambda}}$ is topologically transitive.*

Proof. If $\varnothing \neq U, V \in \Sigma_P$ are open there exist $\omega, \omega' \in \Sigma_P$ and $m \in \mathbb{N}$ such that $U' := C^{-m,\dots,m}_{\omega_{-m},\dots,\omega_m} \subset U$ and $V' := C^{-m,\dots,m}_{\omega'_{-m},\dots,\omega'_m} \subset V$, hence

$$U_{\Lambda'} := \operatorname{Int} \bigcap_{-m\le i\le m} F_{\mathcal{T}}^{-i}(\operatorname{Int} R_{\omega_i}) \neq \varnothing \neq V_{\tilde{\Lambda}} := \operatorname{Int} \bigcap_{-m\le i\le m} F_{\mathcal{T}}^{-i}(\operatorname{Int} R_{\omega'_i})$$

and, of course, $(\beta)^{-1}(U_{\tilde{\Lambda}}) \subset U' \subset U$ and $(\beta)^{-1}(V_{\Lambda'}) \subset V' \subset V$. Thus topological transitivity of $F_{\mathcal{R}}$ implies the corresponding property for σ restricted to Σ_P. □

That, under the factor β, points of X^* have only one preimage is one way of saying that the factor map is close to injective. We push this a little further now by noting that points have at most finitely many preimages.

Before proceeding we let Σ_A be a subshift of finite type with adjacency matrix A and alphabet $\{0,\dots,m-1\}$ such that the hyperbolic symbolic flow $\Sigma_{A,f}$ generates a Markov partition for a hyperbolic set Λ with semiconjugacy given by $\hat{\beta}$. Then define

$$R_i^+ = \hat{\beta}\big(\{(s,t) \mid s \in \Sigma_A,\ s_0 = i,\ t \in [0, f(s))\}\big).$$

For $i, j \in \{0,\dots,m-1\}$ we say that $i \sim j :\Leftrightarrow R_i^+ \cap R_j^+ \neq \varnothing$. This induces an equivalence relation on Σ_A given by

$$\omega \sim_A \omega' :\Leftrightarrow \omega_i \sim \omega'_i \text{ for all } i.$$

Then $\omega \sim_A \omega'$ if and only if there is an $s \in [0, f(\omega))$ such that $\hat{\beta}(\omega, s) = \hat{\beta}(\omega', 0)$ or an $s \in [0, f(\omega'))$ such that $\hat{\beta}(\omega, 0) = \hat{\beta}(\omega', s)$.

Claim 6.6.9. *If $N \in \mathbb{N}_0$ and $\omega, \omega' \in \Sigma_A$ satisfy*

(1) $\omega_i \sim \omega'_i$ *when* $0 \le i \le N$,

(2) $\omega_0 = \omega'_0$,

(3) $\omega_N = \omega'_N$,

then $\omega_i = \omega'_i$ for $0 \le i \le N$.

Proof of Claim 6.6.9. We have $x = \beta(\omega) \in \bigcap_{i=0}^{N} F^{-i}(\overline{\mathrm{Int}(R_{\omega_i})}) = \overline{\bigcap_{i=0}^{N} F^{-i}(\mathrm{Int}(R_{\omega_i}))}$, so after changing x slightly, if necessary, we may assume $x \in \bigcap_{i=0}^{N} f^{-i}(\mathrm{Int}(R_{\omega_i}))$. Note that by construction this leaves $(\omega_0, \dots, \omega_N)$ (the actual subject of the claim) unchanged. Likewise, we take $y = \beta(\omega')$ such that $y \in \bigcap_{i=0}^{N} F^{-i}(\mathrm{Int}(R_{\omega'_i}))$.

To prove the claim we now take $z := [x, y]$ and show that $F^i_{\mathcal{R}}(z) \in \mathrm{Int}\, R_{\omega_i} \cap \mathrm{Int}\, R_{\omega'_i}$ (hence $\omega_i = \omega'_i$) for $0 \le i \le N$.

Since $x \in \mathrm{Int}\, R_{\omega_0}$ and $y \in \mathrm{Int}\, R_{\omega'_0} = \mathrm{Int}\, R_{\omega_0}$, we have

$$z = [x, y]_D \in W^s(x, R_{\omega_0}),$$

and actually, $z \in \mathrm{Int}\, W^s(x, R_{\omega_0})$. Furthermore, there exists $\hat{\omega} \in \Sigma_A$ such that $\hat{\omega} = [\omega, \omega']$ and $\beta\hat{\omega} = z$. Then $\omega_i = \hat{\omega}_i$ for all $i \ge 0$. Applying the Markov property recursively, we therefore conclude that indeed

$$F^i_{\mathcal{R}}(z) \in \mathrm{Int}\, W^s(F^i_{\mathcal{R}}(x), R_{\omega_i}) \subset \mathrm{Int}\, R_{\omega_i} \text{ when } 0 \le i \le N.$$

For $0 \le i \le N$ we have $\omega_i \sim \omega'_i$, hence $R^+_{\omega_i} \cap R^+_{\omega'_i} \ne \varnothing$, so $d(F^i_{\mathcal{R}}(x), F^i_{\mathcal{R}}(y))$ is small enough to conclude that

$$F^i_{\mathcal{R}}(z) = F^i_{\mathcal{R}}([x, y]) = [F^i_{\mathcal{R}}(x), F^i_{\mathcal{R}}(y)] = \beta(\sigma^i(\hat{\omega})).$$

We also have

$$F^N_{\mathcal{R}}(z) = F^N_{\mathcal{R}}([x, y]) = F^N_{\mathcal{R}}([\sigma^N \omega, \sigma^N \omega']) = [F^i_{\mathcal{R}}(x), F^i_{\mathcal{R}}(y)] = \beta(\sigma^N(\hat{\omega})).$$

Then $\hat{\omega}_i = \omega'_i$ for all $i \le N$. □

Proposition 6.6.10. *If Λ is a fixed-point-free hyperbolic set and $\mathcal{R} = \{R_0, \dots, R_{m-1}\}$ is a Markov partition of diameter less than half an expansivity constant, then under the factor map, β no point has more than m^2 preimages.*

Proof. If $x \in \Lambda$ and $(\omega^1, t^1), \dots, (\omega^K, t^K) \in (\hat{\beta})^{-1}(\{x\})$ are pairwise distinct, then there is an $l_{ij} \in \mathbb{Z}$ for which $\omega^i_{l_{ij}} \ne \omega^j_{l_{ij}}$. For any $n \ge |l_{ij}|$ we then have $(\omega^i_{-n}, \dots, \omega^i_n) \ne (\omega^j_{-n}, \dots, \omega^j_n)$. Since $\omega^i \sim \omega^j$ the claim then implies that $(\omega^i_{-n}, \omega^i_n) \ne (\omega^j_{-n}, \omega^j_n)$.

Taking $n_0 := \max_{ij} |l_{ij}|$ we find that the K pairs $(\omega^i_{-n_0}, \omega^i_{n_0}) \in \{0, \dots, m-1\}^2$ are all distinct, which implies that $K \le m^2$. □

Let us note a related consequence.

Proposition 6.6.11. *If x is* bitransitive, *that is, x has dense positive and negative semiorbits, then x has only one preimage under the coding.*

Proof. Given $(\omega,t),(\omega',t') \in (\hat{\beta})^{-1}(x)$, there exist sequences $s_k \to -\infty$ and $t_k \to \infty$ such that $\varphi^{s_k}(x) \in \operatorname{Int}(R_j^+)$ for some j and $\varphi^{t_k}(x) \in \operatorname{Int}(R_i^+)$ for some i. Then there exist sequences $n_k, m_k \to \infty$ such that $\omega_{-n_k} = \omega'_{-n_k}$ and $\omega_{m_k} = \omega'_{m_k}$. Then $\omega = \omega'$ and hence $t = t'$. □

Another easy consequence of Proposition 6.6.10 is the following corollary:

Corollary 6.6.12. *If x is periodic then so is every code, that is, every point of $(\hat{\beta})^{-1}(x)$.*

Shifting by common periods of preimages, this in turn implies a further result:

Corollary 6.6.13. *If x is periodic and $(\omega,t),(\omega',t') \in (\hat{\beta})^{-1}(x)$ satisfy $\omega_i = \omega_i$ for some i, then $\omega = \omega'$ and $t = t'$.*

Remark 6.6.14. By Theorem 4.2.13 the symbolic flow associated with a Markov partition for a hyperbolic set Λ of a flow Φ has the same topological entropy as $\Phi_{\upharpoonright\Lambda}$.

With Propositions 6.6.10 and 6.6.11 in mind, one may ask when the semiconjugacy obtained is a conjugacy. Clearly it is necessary that $\tilde{\Lambda}$ be totally disconnected, because Σ_A is. By now we have enough machinery to show that this condition is indeed sufficient.

Proposition 6.6.15. *Let Λ be a totally disconnected hyperbolic set with no fixed points and with a Markov partition. Then $F_{\mathcal{R}\upharpoonright\Lambda}$ is topologically conjugate to a topological Markov chain.*

Proof. We outline the proof, only skipping details that are easy to fill in.

Construct a cover by rectangles that are both open and closed. This is accomplished by taking closed open sets on stable and unstable manifolds of a point and using $[\cdot,\cdot]$ to produce such a rectangle. Thus we have a partition of Λ into disjoint closed rectangles and hence there is a $\gamma > 0$ such that if two points are closer than γ then they are in the same rectangle and their bracket is in the same rectangle. Now code according to this partition. The coding map is an injective continuous map. The image is a closed invariant subset of the full shift which has a local product structure by the previous remark. Thus the image is a locally maximal subset of the full shift (Exercise 6.9). Finally, the image of Λ under the coding is an N-step topological Markov chain, and any N-step topological Markov chain is topologically conjugate to a topological Markov chain. □

6.7 Failure of local maximality*

A long-standing question in smooth dynamical systems was whether every hyperbolic set Λ for a flow is included in a locally maximal hyperbolic set $\tilde{\Lambda}$. A stronger version of the question is whether, given any hyperbolic set Λ and any neighborhood U of Λ, there exists a locally maximal hyperbolic set $\tilde{\Lambda}$ such that $\Lambda \subset \tilde{\Lambda} \subset U$ [213, p. 272]. Anosov called such sets *premaximal* (in analogy to precompactness).

By Theorem 6.6.5, given a hyperbolic set Λ and a neighborhood U there exists a hyperbolic set $\tilde{\Lambda}$ with $\Lambda \subset \tilde{\Lambda} \subset U$ that has a Markov partition. Although a locally maximal hyperbolic set has a Markov partition, this does not imply that any hyperbolic set with a Markov partition is locally maximal.

Anosov proved that 1-dimensional hyperbolic sets for a flow are premaximal: if Λ is a topologically 1-dimensional hyperbolic set and U is a neighborhood of Λ, then there exists a locally maximal hyperbolic set $\tilde{\Lambda}$ with $\Lambda \subset \tilde{\Lambda} \subset U$.

We now answer the original question in the negative: in general, not only are hyperbolic sets not premaximal, but they may not even be included in *any* locally maximal hyperbolic set—and robustly so. The first example of this was due to Crovisier [104] and gave an open set of diffeomorphisms of the 3-torus each of which contains a hyperbolic set not contained in a locally maximal hyperbolic set. For flows we have the following.

Theorem 6.7.1. *Let M be a closed C^r-manifold* $(1 \leq r \leq \infty)$ *with* $\dim M \geq 3$. *Then there is a C^k-open set of C^k-flows* $(1 \leq k \leq r)$ *on M, each of which contains a hyperbolic set that is not contained in a locally maximal one.*

Proof. We will modify the suspension of the Plykin attractor from Definition 6.5.8 as in [134] to get a hyperbolic set not included in any locally maximal set. Finally, we show how this example can be extended to higher-dimensional settings.

Recall that the Plykin attractor from Definition 6.5.8 on the solid 2-torus is the suspension of a map of a closed disk D_0 (containing the discrete-time Plykin attractor Λ) strictly into itself, so the suspension is a smooth flow of a solid 2-torus T_0 with a hyperbolic attractor Λ_a (the suspension of the Plykin attractor) such that $\varphi^t(T_0) \subset \operatorname{int}(T_0)$ for all $t > 0$. Inside T_0 the nonwandering set consists of Λ_a and a repelling periodic orbit, of period 3, which is the suspension of a repelling period-3 orbit.

We now extend the flow to a larger torus T_1 so that $\phi^t(x) \longrightarrow \Lambda_a$ as $t \longrightarrow \infty$ for all $x \in \operatorname{int}(T_1)$, and so that every $x \in \partial T_1$ is 1-periodic. To do this it is sufficient to extend the map $f\colon D_0 \longrightarrow D_0$ to a map f of a closed 2-disk D_1 containing D_0 in its interior. The extension is carried out by using a bump function so that the map remains the same in D_0 and such that the map is the identity on the boundary of D_1

while $f^n(x) \to \Lambda$ as $n \to \infty$ for each $x \in \operatorname{int}(D_1)$. The flow Φ on T_1 is simply the suspension of the map $f\colon D_1 \to D_1$ (with constant roof function 1).

We now thicken T_1 to a larger solid 2-torus T and extend the flow to T so that the flow is the identity in a neighborhood of the boundary of T. To do this we again first extend the flow so that in $T - T_1$ each point has period 1 and then use a bump function so that we obtain the desired flow on T. Then let D be a 3-dimensional disk such that T is contained in the interior of D and extend the flow to be the identity on $D \smallsetminus T$.

Fix some $p \in \partial T_1$ and an open neighborhood U of $\mathcal{O}(p)$ small enough to be disjoint from ∂T and Λ, and alter Φ in U so that p is a hyperbolic saddle periodic point with $W^s(p) \subset U \cap \partial T_1$. Also, $W^{cu}(p) \setminus \mathcal{O}(p)$ contains two components, one of which is contained in $T - T_1$ and the other, denoted $W^*(p)$ is contained strictly in the interior of T_1.

Let $q \in \Lambda$ be a periodic point. Then $\overline{W^{cs}(q)} = \overline{W^s(\Lambda_a)} = \operatorname{int}(T_1)$, so for any $z \in W^*(p)$ there is a point in $W^{cs}(q)$ arbitrarily close to z. Perturb the flow in a neighborhood of z so that $z \in W^{cs}(q) \pitchfork W^*(p)$. This can be done since z is a wandering point for the flow. Indeed, for an arbitrarily small neighborhood U of z that contains a point of $W^{cs}(q)$ there is a $t > 0$ such that $\varphi^{-t}(U) \cap U = \varnothing$ since z is wandering, so if we perturb φ in $\varphi^{-t}(U)$ we can adjust $W^*(p)$ without changing $W^{cs}(q)$ in U.

Now we need two definitions. A hyperbolic set Λ for a C^1-flow has a *heteroclinic tangency* if there exist $x, y \in \Lambda$ such that $W^s(x) \cap W^u(y)$ contains a point of tangency. A point of *quadratic tangency* for a C^2-flow is defined as a point of heteroclinic tangency where the curvature of the stable and unstable manifolds differs at the point of tangency.

Take $w \in W^*(p)$ not in the orbit of z and $q' \in W^u(q)$ not in the orbit of q. Another perturbation as above can ensure that $w \in W^*(p) \cap W^{cs}(q')$ and that, furthermore, $W^u(p)$ and $W^s(q')$ have a quadratic tangency at w. Let I be the segment of $W^u(p)$ from z to w, and let J be the segment of $W^u(q)$ from q to q'. Figure 6.7.1 shows a cross section of the constructed flow.

We will show that the resulting flow Φ on the disk D contains a hyperbolic set that cannot be contained in a locally maximal set. Let $\Lambda = \Lambda_a \cup \mathcal{O}(p) \cup \mathcal{O}(z)$. The sets $\mathcal{O}(p)$, $\mathcal{O}(z)$, and Λ_a are invariant under Φ, by definition. By construction, $\varphi^t(z)$ converges to $\mathcal{O}(q)$ as $t \to \infty$ and converges to $\mathcal{O}(p)$ as $t \to -\infty$. Since these are both in Λ and Λ_a is closed, so is Λ.

For z we let $E^s_z = T_z W^s(q)$ and $E^u_z = T_z W^u(p)$. For $x = \varphi^t(z)$ for some $t \in \mathbb{R}$ we let $E^s_x = T_x W^s(\varphi^t(q))$ and $E^u_x = T_x W^u(\varphi^t(p))$. This gives an invariant splitting on the closed invariant set Λ and by the Inclination Lemma, this splitting varies continuously.

Let $\lambda_p, \lambda_a \in (0, 1)$ be constants that guarantee hyperbolic behavior in $\mathcal{O}(p)$ and Λ_a, respectively. Let y be an arbitrary element of $\mathcal{O}(p)$. We assume an adapted metric, so

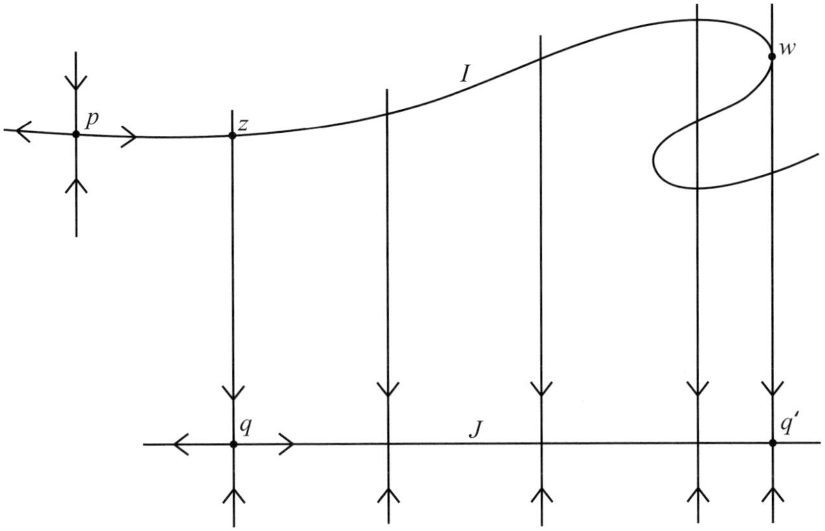

Figure 6.7.1. Intervals I and J.

for $t > 0$ and $v \in E^s(y)$ we have $\|D\varphi^t(y)v\| \le \lambda_p^t\|v\|$, and for $t > 0$ and $v \in E^u(y)$ we have $\|D\varphi^{-t}(y)v\| \le \lambda_p^t\|v\|$. Similarly, for Λ_a if $y \in \Lambda_a$ we have $\|D\varphi^t(y)v\| \le \lambda_a^t\|v\|$ for $v \in E^s(y)$, and for $t > 0$ and $v \in E^u(y)$ we have $\|D\varphi^{-t}(y)v\| \le \lambda_p^t\|v\|$. Let $\lambda_{\max} = \max\{\lambda_p, \lambda_a\}$ and $\lambda \in (\lambda_{\max}, 1)$. Furthermore, there exists an $\epsilon > 0$ such that for all $x \in \Lambda$ where $d(x, \mathcal{O}(p)) < \epsilon$ we have contraction in E_x^s by at least λ and expansion in E_x^u by at least λ^{-1}. The same holds for points in Λ sufficiently close to $\mathcal{O}(q)$.

Now, $\lim_{t\to\infty} \varphi^t(z) = \mathcal{O}(q)$ and $\lim_{t\to\infty} \varphi^{-t}(z) = \mathcal{O}(p)$, so given $\epsilon > 0$ there exists a $T > 0$ such that $\varphi^t(z)$ is ϵ-close to $\mathcal{O}(q)$ and $\varphi^{-t}(z)$ is ϵ-close to $\mathcal{O}(p)$ for all $t \ge T$. By the previous paragraph, we can guarantee hyperbolicity in ϵ-neighborhoods of $\mathcal{O}(p)$ and $\mathcal{O}(q)$ for ϵ sufficiently small. Now we only need guarantee hyperbolicity outside those neighborhoods. For $t \in [-T, T]$ and $v \in \{a \in \mathbb{R}^3 : \|a\| = 1\}$, the function $\|D\varphi^t(z)v\|$ is bounded because it is a continuous function on a compact domain. So, there exists a $C \ge 1$ such that, for any $x \in \Lambda$ and $t > 0$, we have $\|D\varphi^t(x)v\| \le C\lambda^t\|v\|$ for all $v \in E^s(x)$, and $\|D\varphi^{-t}(x)v\| \le C\lambda^t\|v\|$ for all $v \in E^u(x)$. Hence, Λ is hyperbolic for Φ.

Now suppose $\Lambda \subset \Lambda'$, where Λ' is a locally maximal hyperbolic set. We will see that this implies that $w \in \Lambda'$, but since $W^s(p)$ and $W^u(q')$ are not transverse at w this implies that the hyperbolic splitting for Λ' is not transverse, a contradiction. So to conclude the proof we show that w is necessarily in Λ'.

Fix $\epsilon > 0$ and $\delta > 0$, constants from the local product structure on Λ'. By construction, every point in I is in the stable manifold of exactly one point in J. For each $x \in I$ we then fix x' the unique point in J associated with x. Fix $T > 0$ such that $t \geq T$ implies $d(\varphi^t(x), \varphi^t(x')) < \delta/2$ for all $x \in I$. Such a T exists since I and J are compact and the stable manifolds connecting I to J depend continuously on points in I. Let $I' = I \cap \Lambda'$. Then $z \in I'$ and for $t \geq T$ we see that a small neighborhood of z in I is contained in I', to see this just take the bracket of $\varphi^t(z)$ and points in $\varphi^t(J)$ near $\varphi^t(q)$. Continuing to use the product structure we see that the entire interval I must be contained in Λ', a contradiction.

We now show the construction is robust under perturbation. Since transversality is trivially open, and hyperbolicity is open, it is sufficient to show that there remains a point $\tilde{w} \in W^u(\tilde{p}) \cap W^s(\tilde{u})$ for some $\tilde{u} \in W^{cu}(\tilde{q})$ where $\tilde{p}$ is the continuation of p and $\tilde{q}$ is the continuation of q for the perturbed flow. By construction, the stable manifolds for all the $x \in W^{cu}(\tilde{q})$ locally foliate the region, so there must exist a point $\tilde{u} \in W^{cu}_{\text{loc}}(\tilde{q})$ and a point $\tilde{w} \in W^{cs}(\tilde{u}) \cap W^u(\tilde{p})$ such that the 1-dimensional path $W^u(\tilde{p})$ remains tangent to the 2-dimensional plane $W^{cs}(\tilde{u})$ at $\tilde{w}$. Specifically, we have $T_{\tilde{w}} W^u(\tilde{p}) \subsetneq T_{\tilde{w}} W^{cs}(\tilde{u})$.

For any 3-dimensional manifold we simply embed the above example into a chart for the manifold and let the flow on the rest of the manifold be the identity. We then obtain an open set of flows on the manifold that contain a hyperbolic set that is never contained in a locally maximal hyperbolic set.

For higher-dimensional examples we let $n = \dim(M)$ where $n > 3$. Then let D_n be a closed n-dimensional disk. For 3 dimensions we let the flow be defined as above. In the other directions we let the flow contract in the interior of T_1 stronger than any construction in the closed 3-disk and such that the closed 3-disk is invariant under the flow. Using a bump function we allow the flow to be the identity in a neighborhood of the boundary of D_n. One can check that for the resulting flow the set Λ is hyperbolic and has a stable splitting of dimension $n - 2$ and a 1-dimensional unstable splitting. The previous construction shows that Λ can still not be included in a locally maximal hyperbolic set. Then as before we can show that Λ is never included in a locally maximal hyperbolic set. To include this in a manifold we again simply embed the flow in a local coordinate chart for the manifold and extend the flow to be the identity on the rest of the manifold. □

6.8 Smooth linearization and normal forms*

We have already seen in Chapter 5 that *topologically* a hyperbolic fixed point of a (map or) flow is linear (Theorem 5.5.1). We are now equipped to go a little more deeply than that, and we do so in different ways. The most natural improvement of the Hartman–Grobman Theorem (Theorem 5.5.1) would seem to be that hyperbolic fixed points can be linearized smoothly rather than only topologically. Although the obvious obstructions to smooth conjugacy (inequivalent linear parts) do not enter here, there are obstructions of a different nature.

We describe two ways of dealing with these obstructions. One of them is to produce a smooth conjugacy (near the hyperbolic fixed point) not to a linear system but to one with few nonzero terms in its Taylor expansion, a normal form (Theorem 6.8.11). This can be seen as analogous to the failure of a matrix to be diagonalizable and the Jordan normal form as the next best thing. The other possibility is to make a "nonresonance" assumption on the eigenvalues of the linear part at the hyperbolic fixed point which implies that the normal form is linear after all, which then implies smooth linearization (Theorem 6.8.10). Although we do not describe it here, we mention a "hybrid" approach in which a less restrictive assumption (which excludes resonances only up to some finite order) results in a linearization with a corresponding degree of finite differentiability.

We proceed as follows. We begin with an introduction that provides the basic concepts (notably jets) and outlines the strategy. This strategy is then carried out: the must illuminating part is the formal analysis, which clarifies what normal form to expect and how it is connected to resonances. To make the formal analysis effective we then either work in the analytic category and assume that the power series for the desired conjugacy converges, or we introduce the so-called Borel Lemma in the smooth category, which provides a smooth function with a prescribed power series. This gives the two aforementioned main results. In light of future applications we then develop these results a little further for flows: if a flow or map is brought into a normal form or linearized, then so is, simultaneously, any dynamical system that commutes with it. This plays an important part when we study commuting flows (Section 9.1). The next section uses results based on normal forms to refine the assertion of the Hartman–Grobman Theorem.

6.8.a Jets, formal power series, and smooth equivalence. In hyperbolic flows the linearized system locally serves as a model for the behavior of a nonlinear flow, thus implying that nonlinear terms constitute a perturbation which should be kept under control. A natural next step for local analysis is to consider the terms of order higher than linear in a more systematic way and determine more precisely the extent to which their influence has to be taken into account or whether it can be disregarded altogether. A hyperbolic periodic orbit is the best setting for such an analysis. The key phenomena here are "resonances" between the eigenvalues of the linearized map. Their presence or absence determines what higher-order terms must be taken into account.

In Example 1.3.10 we saw that eigenvalues at a fixed point are preserved under smooth conjugacy (since the differential of the conjugacy conjugates the linear parts of the maps). This motivates thinking about whether the conjugacy in the Hartman–Grobman Theorem (Theorem B.4.14) can be taken smooth. We introduced the refined Hartman–Grobman Theorem above, and we expand on these ideas now in order to both give a larger picture and present some of the ideas we use in the proof of differentiability of the Hartman–Grobman linearization.

First we should find out whether there are other *infinitesimal* invariants of local smooth conjugacy besides those coming from the linear part of the map at the fixed point. For that purpose, fix local coordinates near a fixed point p of a map f and consider the coefficients of the kth Taylor polynomial of f for $k = 2, 3, \dots$. This set of coefficients is called the kth *jet* or *k-jet* $J_p^k(f)$ of f at p. Thus two C^k-maps f and g have the same k-jet at a (not necessarily fixed) point p if $\|f(x) - g(x)\| = o(\|x - p\|^k)$. Obviously the first jet of a map is determined by the value of the map and its linear part. A real-analytic map is the limit of its kth Taylor polynomials as $k \to \infty$. For C^∞-maps one can write down a Taylor series, but it may not converge at more than one point. Thus we are led to consider *formal* power series, that is, a formal expression consisting of an infinite sum of monomials. The algebraic operations and substitutions are performed by applying the rules familiar from convergent power series.

Obviously the (formal) Taylor series of a C^∞-map at a point determines all jets at that point. The k-jets of a map at a point can be identified with polynomial maps in a local coordinate system and their composition is defined by taking the usual composition and discarding higher-order terms. Furthermore, near a fixed point a local C^k-conjugacy between maps produces a conjugacy between the k-jets at the fixed point. Thus the conjugacy classes of these jets are the infinitesimal invariants we set out to find. For C^∞-maps, C^∞ local conjugacy implies that the formal Taylor series at the reference point are conjugate, where the composition of the formal power series is obtained by substitution.

Now we can outline our strategy for solving the smooth conjugacy problem as follows. First we look for invariants of conjugacy of k-jets for any $k \in \mathbb{N}$. It turns out that all those invariants are completely determined by the linear part of the map, that is, its first jet. Let us say that two C^∞-maps f and g are *C^∞-tangent at p* if $J_p^k(f) = J_p^k(g)$ for all $k \in \mathbb{N}$, or $\|f(x) - g(x)\| = o(\|x - p\|^k)$ for all $k \in \mathbb{N}$, or equivalently if the formal Taylor series for f and g at p coincide. The second step is to show that if all jets of f and g are conjugate then there exists a C^∞-map h such that f and $f' := h^{-1} \circ g \circ h$ are C^∞-tangent. Finally, if the linear part of f at p is hyperbolic and f' is C^∞-tangent to f at p then f and f' are locally C^∞-conjugate via a local diffeomorphism C^∞-tangent to the identity. This last step can be carried out in various ways. If the linear part of the map f is contracting one can consider the conjugacy equation as the fixed-point equation $h = (f')^{-1} \circ h \circ f$ and use the Contraction Mapping Principle to show that the solution h is C^∞. In the properly hyperbolic case one can use a refined version of the fundamental-domain method called the Sternberg wedge method. We use a version of the homotopy trick, which was used to prove Theorem 2.6.11, to accomplish this final step (Theorem 6.8.9). This reduces the problem to analyzing the solution of a twisted cohomological equation.

6.8.b Formal analysis of smooth conjugacy. Since we want to conveniently manipulate power series in n variables, we use multi-indices systematically. Thus, for $k = (k_1, \dots, k_n) \in \mathbb{N}_0^n$ we define the *size* of k to be $|k| := \sum_{i=1}^n k_i$ and if $x = (x_1, \dots, x_n) \in \mathbb{R}^n$ we let $x^k := \prod_{i=1}^n x_i^{k_i}$.

6.8.b.1 Formal linearization in the nonresonance case.

Definition 6.8.1. Let $\lambda = (\lambda_1, \dots, \lambda_n) \in \mathbb{R}^n$. A relation of the form $\lambda_i = \lambda^k$ is called a *resonance* and correspondingly the condition "$\lambda_i \neq \lambda^k$ for all i and all $k \in \mathbb{N}_0^n$" is called a *nonresonance assumption*.

Proposition 6.8.2. *Consider a formal power series f given by*

$$f_i(x) = \sum_{k \in \mathbb{N}^n} f_{i,k} x^k$$

with vanishing constant term and whose linear part g is $\operatorname{diag} \lambda$ *(that is, the diagonal matrix with eigenvalues $\lambda_1, \dots, \lambda_n$) where λ satisfies the nonresonance assumption. Then there exists a formal power series h solving the conjugacy equation $h \circ f = g \circ h$.*

Proof. We write $f_i(x) = \sum_{k \in \mathbb{N}^n} f_{i,k} x^k = \lambda_i x_i + \sum_{|k|>1} f_{i,k} x^k$ as the linear part plus terms of higher order. The nonresonance assumption for $|k| = 1$ implies that the λ_i are pairwise distinct. Since the linear part of h commutes with that of f (because the latter coincides with that of g), it is diagonal, so

$$h_i(x) = \sum_{k \in \mathbb{N}^n} h_{i,k} x^k = \alpha_i x_i + \sum_{|k|>1} h_{i,k} x^k.$$

The ith coordinate of the conjugacy equation $h \circ f(x) = g \circ h(x)$ now becomes

$$\sum_{k \in \mathbb{N}^n} h_{i,k} (f(x))^k = \lambda_i h_i(x),$$

or, by splitting off the linear parts,

$$\alpha_i f_i(x) + \sum_{|k|>1} h_{i,k} \Big(\lambda_1 x_1 + \sum_{|j_1|>1} f_{1,j_1} x^{j_1} \Big)^{k_1} \cdots \Big(\lambda_n x_n + \sum_{|j_n|>1} f_{n,j_n} x^{j_n} \Big)^{k_n} = \lambda_i h_i(x).$$

We want to solve this equation inductively in $m := |k|$ for the coefficients $h_{i,k}$ of h in terms of the coefficients $f_{i,k}$ and the λ_i. In other words, we want to construct a conjugacy for m-jets given a conjugacy for $(m-1)$-jets. For $m = 1$ the choices are arbitrary, for example, one can take $Dh = \mathrm{Id}$. Suppose $m \in \mathbb{N}$ and we have determined all $h_{i,k}$ as desired for all $|k| < m$. Then for any k such that $|k| = m$ consider the coefficients of the terms involving x^k. Comparing them yields

$$\alpha_i f_{i,k} + \lambda^k h_{i,k} = \lambda_i h_{i,k} + C_{i,k}, \tag{6.8.1}$$

where $C_{i,k}$ involves only coefficients with indices j of size $|j| < m$, which are thus entirely determined by the previous steps. Thus we solve for $h_{i,k}$ by taking

$$h_{i,k} = \frac{\alpha_i f_{i,k} - C_{i,k}}{\lambda_i - \lambda^k}, \tag{6.8.2}$$

which is possible by the nonresonance assumption. □

6.8.b.2 Normal forms in the resonance case. Now suppose we are in the situation of the previous proposition but there are $k \in \mathbb{N}^n$ such that $\lambda_i = \lambda^k$ for some i. Then terms with $h_{i,k}$ in (6.8.1) disappear, that is, one cannot remove the term involving x^k from the ith coordinate function. This observation leads to a study of *normal forms*, which are the natural generalizations of the linear part of a map.

Definition 6.8.3. If $\lambda_i = \lambda^k$, a nonzero term $c \cdot x^k$ in the ith coordinate function of a formal power series is called a *resonance term*.

A *formal normal form* of a power series f is a formal power series g with the same linear part whose power series contains only linear and resonance terms such that the conjugacy equation

$$h \circ f = g \circ h \tag{6.8.3}$$

holds for a formal power series h.

The previous process can be generalized to the situation where g has nonlinear terms, that is, $g_i(x) = \sum_{k \in \mathbb{N}^n} g_{i,k} x^k = \lambda_i x_i + \sum_{|k|>1} g_{i,k} x^k$. Then (6.8.1) becomes

$$\alpha_i f_{i,k} + \lambda^k h_{i,k} = \lambda_i h_{i,k} + \alpha^k g_{i,k} + C_{i,k}, \tag{6.8.4}$$

where the $C_{i,k}$ are determined by the previous steps and may involve lower-order terms of g. Suppose now that g has *only* resonance terms and the linear part of h is fixed, say, to be the identity. Then (6.8.2) still holds for the nonresonance terms, and for resonance terms, (6.8.4) implies $g_{i,k} = f_{i,k} - C_{i,k}$, that is, the resonance terms of g are uniquely defined if there is a formal conjugacy. Thus, within a formal conjugacy class (with diagonal linear part) the normal form is indeed uniquely defined up to choice of linear part. The following proposition summarizes the situation:

Proposition 6.8.4. *For any formal power series with diagonal linear part there exists a uniquely defined formal normal form.*

Even if f is an analytic map its formal normal form may not converge. There are, however, cases when it does, even for formal power series. The first one is the nonresonance case. Another is contracting maps. In this case there are only finitely many resonances (namely, no more than $-\log r(Df^{-1})/\log r(Df)$, where r denotes spectral radius) and hence the normal form is a polynomial. Thus it makes sense to look for normal forms in the smooth or analytic categories. The analytic case is more straightforward.

Definition 6.8.5. A *normal form* for an analytic map f is its formal normal form g whose power series converges and the solution h of the formal conjugacy equation (6.8.3) is given by a convergent power series.

In the smooth case we need to take into account that a smooth function is not determined by its Taylor series.

Definition 6.8.6. A *normal form* for a smooth map f is its formal normal form g whose power series converges and the conjugacy equation $h \circ f = g \circ h$ has a smooth solution h.

6.8.b.3 Examples of resonance cases. Interest in these nonlinear normal forms is motivated by the fact that in several natural settings one has "built-in" resonances, that is, in certain natural classes of maps there is a natural collection of resonances exhibited by every one of these maps. The main examples are area-preserving and symplectic maps.

If p is a fixed point of a map preserving a positive absolutely continuous measure then the determinant of the differential Df_p is ± 1. Thus, if $\lambda_1, \dots, \lambda_n$ are the eigenvalues of Df_p counted with multiplicities then

$$\lambda_1 \cdots \lambda_n = \pm 1$$

and there are resonance relations

$$\lambda_i = \lambda_1 \cdots \lambda_{i-1} \lambda_i^2 \lambda_{i+1} \cdots \lambda_n \quad \text{or} \quad \lambda_i = \lambda_1^2 \cdots \lambda_{i-1}^2 \lambda_i^3 \lambda_{i+1}^2 \cdots \lambda_n^2.$$

Similarly, if p is a fixed point of a symplectic diffeomorphism then according to Proposition 2.6.8 the eigenvalues can be split into pairs of mutually reciprocal numbers, that is, the vector of eigenvalues can be arranged to look like

$$(\lambda_1, \lambda_1^{-1}, \lambda_2, \lambda_2^{-1}, \dots, \lambda_n, \lambda_n^{-1})$$

and hence there are n resonances

$$\lambda_i = \lambda_i^2 \lambda_i^{-1}, \quad i = 1, \ldots, n.$$

Now we describe the possible resonances in the 2-dimensional hyperbolic case. There are two eigenvalues which we denote by λ_- and λ_+ such that $|\lambda_-| < 1 < |\lambda_+|$. A resonance has the form $\lambda_- = \lambda_-^k \lambda_+^l$ or $\lambda_+ = \lambda_+^k \lambda_-^l$, that is, $\lambda_-^{k-1} = \lambda_+^{-l}$ or $\lambda_+^{k-1} = \lambda_-^{-l}$, where $k, l \in \mathbb{N}_0$ are nonnegative integers. Thus if $\varphi_\pm = |\log|\lambda_\pm||$ then the resonance implies $\varphi_-/\varphi_+ \in \mathbb{Q}$. Conversely, it is easy to see that this implies that there is a resonance. If $\varphi_-/\varphi_+ = p/q$ with $p, q \in \mathbb{N}$ relatively prime then we say that there is a (p,q)-resonance. To describe the normal forms in this case let us assume for simplicity that both eigenvalues are positive. Then $\lambda_-^q \lambda_+^p = 1$ and if $\lambda_- = \lambda_-^k \lambda_+^l$ then $k = mq + 1$, $l = mp$ for some $m \in \mathbb{N}$. Similarly, $\lambda_+ = \lambda_+^k \lambda_-^l$ implies $k = mp + 1$, $l = mq$. Hence if f is a normal form then

$$f(x,y) = \Big(\lambda_- x\Big(1 + \sum_{m=1}^{\infty} a_m (x^q y^p)^m\Big), \lambda_+ y\Big(1 + \sum_{m=1}^{\infty} b_m (x^q y^p)^m\Big)\Big).$$

Note that in the area-preserving case there is a $(1,1)$-resonance.

6.8.c The hyperbolic smooth case. One way to justify the formal manipulations with power series is by showing that all power series involved have positive radius of convergence. Although the absence of resonances may not be sufficient for analytic linearization even in the hyperbolic case it is so in the C^∞ category. The basic idea is that for any formal power series there is a C^∞-function whose Taylor series coincides with the given power series.

Proposition 6.8.7 (Borel). *For any sequence $(a_k)_{k \in \mathbb{N}_0^n} \in \mathbb{R}^{\mathbb{N}_0^n}$ there exists a C^∞-function $f : \mathbb{R}^n \to \mathbb{R}^n$ with Taylor coefficients $(a_k)_{k \in \mathbb{N}_0^n}$.*

Proof. We use the broadly useful notion of a *bump function*, a smooth nonzero function with compact support, such as

$$b_1 : \mathbb{R} \to \mathbb{R}, \quad x \mapsto \begin{cases} e^{2-(x+1)^{-2}-(x-1)^{-2}} & \text{when } |x| \le 1, \\ 0 & \text{when } |x| > 1, \end{cases}$$

which vanishes to all orders at ± 1. We will use the bump function $b_2(x) := \int_{|x|}^{\infty} b_1(t-2)\,dt \Big/ \int_{-1}^{1} b_1(t)\,dt$, which is 1 on $[-1,1]$: The series

$$f(x) := \sum_{k \in \mathbb{N}_0^n} a_k x^k b_2(|k|! C_{|k|} \|x\|^2),$$

with $C_N := \sum_{|i|\le N} |a_i|$, converges since for each $x \neq 0$ there are only finitely many nonzero terms. It converges uniformly since

$$\begin{aligned}\left\|a_k x^k b_2\left(|k|!\,C_{|k|}\|x\|^2\right)\right\| &\le |a_k|\|x\|^{|k|} \underbrace{b_2(|k|!\,C_{|k|}\|x\|^2)}_{=0 \text{ or } |k|!C_{|k|}\|x\|^2\le 2} \\ &\le |a_k|\Big(\frac{2}{C_{|k|}|k|!}\Big)^{\frac{|k|}{2}} \le \Big(\frac{2}{|k|!}\Big)^{\frac{|k|}{2}}.\end{aligned} \tag{6.8.5}$$

To evaluate the derivatives of order N consider points x such that $|k|!C_{|k|}\|x\|^2 < 1$ for $|k| \le N$. For these points we have

$$f(x) = \sum_{|k|\le N} a_k x^k + \sum_{|k|>N} a_k x^k b_2(|k|!C_{|k|}\|x\|^2).$$

By (6.8.5) the second sum is a bounded multiple of the sum of its lowest-order terms (by factoring them out), so the remainder is of order higher than N and the derivatives up to order N yield the required coefficients. □

Corollary 6.8.8. *Suppose f is a C^∞-map with a fixed point p such that the linear part is* diag λ *(that is, the diagonal matrix with eigenvalues $\lambda_1, \dots, \lambda_n$) satisfying the nonresonance condition $\lambda_i \neq \lambda^k$ for all i and all $k \in \mathbb{N}_0^n$. Then there exists a local C^∞-map h such that $h \circ f \circ h^{-1}$ is C^∞-tangent to the linear part of f.*

Proof. By Proposition 6.8.7 there is a C^∞-map h with the Taylor series from Proposition 6.8.2. □

Therefore local smooth linearization for hyperbolic fixed points with no resonances follows from the next result:

Theorem 6.8.9. *Let f be a C^∞-map with a hyperbolic fixed point p and g any C^∞-map C^∞-tangent to f. Then there is a neighborhood U of p and a C^∞-diffeomorphism h which is C^∞-tangent to the identity such that $h \circ f = g \circ h$.*

Proof. First, using Theorem B.5.3 introduce adapted local coordinates with p at the origin such that the stable and unstable manifolds of p are the coordinate spaces $\mathbb{R}^k$ and $\mathbb{R}^{n-k}$, respectively. Since the stable and unstable manifolds for g are C^∞-tangent to those for f one can conjugate g by a diffeomorphism C^∞-tangent to the identity such that the resulting stable and unstable manifolds coincide with those for f. Next, by the Extension Theorem (Theorem B.4.12) we can construct C^∞-diffeomorphisms of $\mathbb{R}^n$ that fix the origin and coincide with the coordinate representations of f and g, respectively, in a smaller neighborhood of the origin and with the linear part of f and g outside a larger neighborhood, preserve $\mathbb{R}^k$ and $\mathbb{R}^{n-k}$, and are C^1-close to the linear part. We still denote these maps by f and g. Then $\alpha := f - g$ has zero jets of all orders at the origin and vanishes outside some neighborhood of 0. Next we show that α can be decomposed as

$$\alpha = \alpha^+ + \alpha^-,$$

where α^+ and all its jets vanish on $\mathbb{R}^k$ and α^- and its jets vanish on $\mathbb{R}^{n-k}$. We construct conjugacies C^∞-tangent to the identity between f and $w := f + \alpha^-$ and between w and g.

To obtain α^- take a C^∞-function ρ on the unit sphere S such that $\rho \equiv 1$ on the intersection of S with the horizontal cone $H_{1/2}$ and $\rho = 0$ on the intersection $S \cap V_{1/2}$ and set

$$\alpha^-(x) = \alpha(x)\rho\Big(\frac{x}{\|x\|}\Big) \quad \text{for } x \neq 0$$

and $\alpha^-(0) = 0$. Then set $\alpha^+ = \alpha - \alpha^-$. Clearly these are as desired, except that we need to verify that both are C^∞ at the origin. Note that for $k \in \mathbb{N}_0^n$ and $m \in \mathbb{N}$, we have $\|D^k\alpha(x)\| = o(\|c\|^m)$ and that the derivatives of ρ are bounded. Using the chain rule one sees that the expression for $D^k\alpha^-$ outside the origin is a polynomial in the derivatives of α, ρ, and $\|x\|^{-1}$, and that each monomial contains α or some of its derivatives. This implies that $\|D^k\alpha^-(x)\| = o(\|x\|^m)$ and hence α^- is a C^∞-function.

Now let $f_t := f + t\alpha^-$ for $t \in [0,1]$. We look for a family of C^∞-diffeomorphisms h_t such that

$$f_0 = h_t^{-1} \circ f_t \circ h_t.$$

This family is generated by the family of vector fields

$$v_t = \frac{d(h_s h_t^{-1})}{ds}\Big|_{s=t}.$$

Differentiating the relation $h_s \circ h_t^{-1} \circ f_t = f_s \circ h_s \circ h_t^{-1}$ with respect to s we obtain

$$v_t \circ f_t - Df_t(v_t) = \alpha^- \quad \text{or} \quad v_t - (f_t)_* v_t = \alpha^- \circ f_t^{-1},$$

where $f_* v = Df(v \circ F^{-1})$. Inverting the operator $\mathrm{Id} - (f_t)_*$ formally using the geometric series we obtain

$$v_t = \sum_{m=0}^{\infty} (f_t)_*^m \alpha^- \circ f_t^{-1} = \sum_{m=0}^{\infty} Df_t^m \alpha^- \circ f_t^{-m-1}.$$

To show that v_t is a C^∞ vector field in a neighborhood of the origin we need to show that this sum converges in the C^∞-topology, that is, that the sum of kth derivatives converges for every $k \in \mathbb{N}_0^n$. Such observations were first made in step (5) in the proof of the Hadamard–Perron Theorem (Theorem B.5.2). Note first that by the chain rule and product rule the kth derivative of an m-fold composition grows at a rate of at most $C^m m^{|k|}$, where C is an upper bound for the derivatives up to order $|k|$ of the individual terms. Thus the kth derivative of f^{-m-1} grows at most exponentially with m. Next consider the kth derivative of $\alpha^- \circ f_t^{-m-1}$. By the chain rule this is a polynomial in derivatives of α^- and f_t^{-n-1} and each term contains a derivative of α^-, or α^- itself, evaluated at f^{-m-1}, that is, exponentially close to $\mathbb{R}^k$. Thus these factors are superexponentially small by construction of α^- and hence the kth derivative of $\alpha^- \circ f_t^{-m-1}$ converges to zero superexponentially as $m \to \infty$. Again, the kth derivative of the entire summand is a polynomial whose terms each contain a derivative of $\alpha^- \circ f_t^{-m-1}$. Thus each term is superexponentially small (because it consists of a superexponentially small factor and m bounded factors), so in fact the kth derivatives of the summands go to zero superexponentially.

Thus we obtain the desired family h_t and hence the conjugacy between f and $f + \alpha^-$. The second conjugacy between $f + \alpha^-$ and g is constructed similarly using positive iterates of f and α^+ instead of α^-. □

Thus we have proved the following theorem:

Theorem 6.8.10 (Sternberg Linearization). *Suppose f is a C^∞-diffeomorphism with a hyperbolic fixed point p such that the linear part of f at p has no resonances. Then near p, f is C^∞-conjugate to its linear part.*

In fact, the previous arguments give results about C^∞-conjugacy even in the presence of resonances:

Theorem 6.8.11 (Sternberg Normal Form). *Suppose that f is a C^∞-diffeomorphism with a hyperbolic fixed point p such that the linear part of f at p is diagonal and the normal form of f near p is a convergent power series. Then f is locally C^∞-conjugate to its normal form.*

Proof. Proposition 6.8.4 gives a formal conjugacy between f and its normal form, so by Proposition 6.8.7 we obtain a conjugacy to a map C^∞-tangent to the normal form, which by Theorem 6.8.9 gives the result. □

As mentioned before, like results hold under slightly less restrictive assumptions. For instance, the following special case of Theorem 6.8.10 holds with a relaxed nonresonance assumption: for a linear map $A\colon \mathbb{R}^n \to \mathbb{R}^n$ we say that the matrix is nonresonant if $\operatorname{Re}\lambda_i \neq \operatorname{Re}\lambda_1^{m_1} \cdots \operatorname{Re}\lambda_n^{m_n}$ whenever the m_j are nonnegative integers with $\sum m_j \geq 2$.

Theorem 6.8.12. *If $f\colon \mathbb{R}^n \to \mathbb{R}^n$ is a C^∞-diffeomorphism and the origin is a hyperbolic sink for f with $Df(0)$ nonresonant, then there is a C^∞-smooth linearization of f.*

Even for analytic maps the normal form may not be convergent. As we pointed out before, for contracting maps the formal normal form is always a polynomial and hence converges. In particular, we can rule out all resonances altogether by a *bunching assumption*:

Corollary 6.8.13. *Suppose that f is a C^∞-diffeomorphism with a contracting fixed point p such that $Df_{|p}$ is diagonal. Then f is C^∞-conjugate to a polynomial normal form.*

If in addition $-\log r(Df^{-1})/\log r(Df) < 2$ then f is C^∞-linearizable at p.

This condition is sharp (the map $(x, y) \mapsto (\lambda x, \lambda^2 y + ax^2)$ cannot be C^2-linearized for any $\lambda \in (0, 1)$ and $a \neq 0$ [171, Part 4]). In particular equality cannot be allowed in the bunching condition.

The standing assumption of diagonalizability of the linear part was only used in the formal part of our arguments and these can, in fact, be modified to work in the case of nontrivial Jordan normal forms in the linear part. In particular, Corollary 6.8.13 holds without the diagonalizability assumption.

A simple special case illustrates the possibility of obtaining a linearization of intermediate regularity by ruling out only "low-order" resonances.

Theorem 6.8.14 ([171, Part 3]). *A hyperbolic saddle fixed point in dimension 2 can be $C^{1+\epsilon}$-linearized.*

Example 6.9.4 shows why this is interesting.

The arguments for Theorem 6.8.11 can be used to provide useful information even if the normal form does not converge. To this end we use Proposition 6.8.7 to get a local normal form.

Definition 6.8.15. A *local normal form* for a smooth map f is a smooth map g whose Taylor series at the fixed point coincides with the formal normal form of f.

Theorem 6.8.16. *Suppose that f is a C^∞-diffeomorphism with a hyperbolic fixed point p such that the linear part of f at p is diagonal. Then f is locally C^∞-conjugate to its local normal form.*

Proof. First, Proposition 6.8.4 gives a formal conjugacy between f and its normal form, so by Proposition 6.8.7 applied to the formal conjugacy and the normal form we obtain a conjugacy to a map C^∞-tangent to the local normal form, which then by Theorem 6.8.9 gives the result. □

We remark that the preceding arguments can be carried out for maps of finite differentiability as well, but with a loss of several degrees of differentiability. This in itself is not a surprising observation, but finding optimal results where this loss is minimal requires much more careful estimates. Before taking a closer look, we outline an application of normal-form theory that sharpens Corollary 6.8.13 in a peculiar way: a C^1-conjugacy to a quadratic map.

Theorem 6.8.17 (Bronstein–Kopanskiĭ). *If $f\colon \mathbb{R}^d \to \mathbb{R}^d$ is a C^∞-diffeomorphism such that $f(0) = 0$ and $L := Df(0)$ is hyperbolic, then on a neighborhood of 0 there is a C^1-conjugacy $H^+ = \mathrm{Id} + \Delta h$ with $\|\Delta h(x)\| \in o(\|x\|)$[8] between f and a quadratic polynomial $\mathfrak{f}(x) := Lx + Q(x)$ with only weakly resonant terms (Definition 6.8.18).*

Proof. This follows from [72, Theorem 11.9] once one verifies that monomials of order greater than 2 satisfy their condition $\mathcal{A}(1)$. Per [72, Remark 7.6], $\mathcal{A}(1)$ in turn follows from their condition $S(1)$ (see (6.8.6)), and this is relatively straightforward to check [72, bottom p. 191]. Indeed, denote by $-\lambda_l < \cdots < -\lambda_1 < 0 < \mu_1 < \cdots < \mu_m$ the distinct values of $\log|\nu|$, where ν is an eigenvalue of L. For a multi-index $\tau = (\alpha^l, \dots, \alpha^1, \beta^1, \dots, \beta^m)$ the condition $S(1)$ [72, p. 111] is that

$$\text{either } \sum_{i=1}^{r} \alpha^i \lambda_i > \lambda_r \text{ for some } r \le l \text{ or } \sum_{j=1}^{s} \beta^j \mu_j > \mu_s \text{ for some } s \le m. \qquad (6.8.6)$$

For terms of order at least 3 the multi-index exponent satisfies $|\tau| := \sum \alpha^i + \sum \beta^j \ge 3$ and hence $S(1)$ because either $\sum \alpha^i \ge 2$, in which case $r = \max\{i \mid \alpha^i > 0\}$ works, or else $s = \max\{j \mid \beta^j > 0\}$ does. □

Definition 6.8.18. Note that Q contains only *weakly resonant* terms if the following holds: if in coordinates adapted to the decomposition into generalized eigenspaces (Definition B.2.1) or *root spaces* the ith component of Q contains a term $a \cdot xy$ with $a \ne 0$ then the eigenvalues λ and μ associated with the root spaces corresponding to x and y, respectively, are related to the eigenvalue η associated with the root space containing the ith unit vector by $|\eta| = |\lambda||\mu|$.

We now turn our attention to flows, and for our purposes it suffices to consider the contracting part. First, the appropriate notion of resonance for a matrix which *generates* a flow is a little different from that in discrete time—think of the exponential as going from this continuous-time version to the discrete-time counterpart:

Definition 6.8.19. Denote the spectrum of an $n \times n$ matrix A by $\Sigma(A) = \{\lambda_1, \dots, \lambda_n\}$, the eigenvalues of A repeated with multiplicity. Then A is said to be *stable hyperbolic* if $\operatorname{Re}\lambda < 0$ for all $\lambda \in \Sigma(A)$. We say that a stable hyperbolic matrix is *nonresonant* if $\operatorname{Re}\gamma(\lambda, m) \ne 0$ for any m where $|m| \ge 2$ and any $\lambda \in \Sigma(A)$.

The continuous-time counterpart of the Sternberg Linearization Theorem (Theorem 6.8.10; restricted to contracting flows), then is the following:

[8] See Remark 3.2.18.

Theorem 6.8.20. *If $f \in C^\infty$ and $x' = f(x) = Ax + R(x)$ where A is a nonresonant stable hyperbolic matrix and $D^k R(0) = 0$ for $k = 0, 1$, then there exists a C^∞-smooth linearization.*

There is a Lie-group structure behind normal forms, and one manifestation is that commuting maps have the same normal form (likewise for flows); the direct way to see this is to bring one of them into normal form and see that commutation implies being in the same normal form. For our applications it suffices to establish this when the normal form of a contracting flow is linear (Theorems 6.8.21, 6.8.22), which implies that the smooth linearization from Theorem 6.8.12 for the stable manifold of a sink simultaneously smoothly linearizes any smooth map in the centralizer.

Theorem 6.8.21 (Simultaneous linearization). *Let $A\colon \mathbb{R}^n \to \mathbb{R}^n$ be a nonresonant stable hyperbolic matrix and Φ_A be the linear flow generated by A. If g is a C^∞ homeomorphism such that $g\varphi_A^t = \varphi_A^t g$ for all $t \in \mathbb{R}$, then g is linear.*

Proof. Let $\overline{\lambda} = \max\{\operatorname{Re}\lambda \mid \lambda \in \Sigma(A)\}$ and $\underline{\lambda} = \min\{\operatorname{Re}\lambda \mid \lambda \in \Sigma(A)\}$. So $\underline{\lambda} < \overline{\lambda} < 0$. Fix $N \in \mathbb{N}$ to be the smallest natural number such that $N\overline{\lambda} < \underline{\lambda}$. Let $g = P + R$ where P is a polynomial of degree less than N and R is the remainder term such that $\|R(x)\|/\|x\|^N$ is bounded in a neighborhood of the origin. Linearity of φ^t implies that

$$g = \varphi_A^{-t} g \varphi_A^t = \varphi_A^{-t} P \varphi_A^t + \varphi_A^{-t} R \varphi_A^t$$

for any $t \in \mathbb{R}$. Also, since Φ_A is linear, $P = \varphi_A^{-t} P \varphi_A^t$ and $R = \varphi_A^{-t} R \varphi_A^t$.

Since $\|R(x)\| = \|x\|^N h(x)$, where h is bounded as x approaches infinity,

$$\begin{aligned}\|\varphi_A^{-t} R(\varphi_A^t x)\| &\le \|\varphi_A^{-t}\| \cdot \|\varphi_A^t x\|^N h(\varphi_A^t x)\\ &\le \|\varphi_A^{-t}\| \cdot \|\varphi_A^t\|^N \|x\|^N h(\varphi_A^t x).\end{aligned}$$

Since $N\overline{\lambda} - \underline{\lambda} < 0$ we have

$$\lim_{t\to\infty} \|\Phi_A^{-t}\| \, \|\Phi_A^t\|^N = \lim_{t\to\infty} e^{(N\overline{\lambda}-\underline{\lambda})t} = 0.$$

Since Φ_A is a stable linear flow and $h \to 0$ as $x \to 0$, $h(\Phi_A^t x) \xrightarrow[t\to\infty]{} 0$. Hence, $R(x) = 0$ and g is a polynomial of degree less than N.

Now we use the nonresonant conditions on the eigenvalues of A to show that the P is in fact linear. First of all, the nonresonant condition implies that there are n distinct eigenvalues for A and so the real Jordan form of A is block diagonal. Also, the nonresonant condition implies that if $\alpha_i = \operatorname{Re}\lambda_i$ for each i, then

$$e^{\alpha_i} \neq e^{\alpha_1 m_1} \cdots e^{\alpha_n m_n}$$

where each m_j is a nonnegative integer and $\sum_{m_j} \geq 2$. Then A can be written in the form

$$A(x_1, x_1', \dots, x_k, x_k', x_{k+1}, \dots, x_\ell) = (\Lambda_1(x_1, x_1'), \dots, \Lambda_k(x_k, x_k'), \lambda_{k+1}x_{k+1}, \lambda_\ell, x_\ell)$$

where $\Lambda_j = \left(\begin{smallmatrix} \alpha_j & -\mu_j \\ \mu_j & \alpha_j \end{smallmatrix}\right)$ for each eigenvalue $\lambda_j = \alpha_j + \mu_j$ of A where $\mu_j > 0$. Then the linear flow Φ_A is represented by

$$\begin{aligned} &e^{At}(x_1, x_1', \dots, x_k, x_k', x_{k+1}, \dots, x_\ell) \\ &\quad = (e^{\Lambda_1 t}(x_1, x_1'), \dots, e^{\Lambda_k t}(x_k, x_k'), e^{\lambda_{k+1}t}x_{k+1}, \dots, e^{\lambda_\ell t}x_\ell) \end{aligned}$$

where

$$e^{\Lambda_j t} = \begin{pmatrix} e^{\alpha_j t}\cos\mu_j t & -e^{\alpha_j t}\sin\mu_j t \\ e^{\alpha_j t}\sin\mu_j t & e^{\alpha_j t}\cos\mu_j t \end{pmatrix}.$$

Now a straightforward, but somewhat long computation, shows that if $e^{-At}Pe^{At} = P$ for all $t \in \mathbb{R}$ and the eigenvalues are nonresonant, then P has to be linear. □

The discrete-time counterpart has a similar proof:

Theorem 6.8.22. *Let $A\colon \mathbb{R}^n \to \mathbb{R}^n$ be a linear map such that the origin is a stable hyperbolic fixed point for the map and assume that the eigenvalues of A are nonresonant. If g is a C^∞ homeomorphism such that $gA = Ag$, then g is linear.*

It will be useful to spell this out in greater detail:

Proposition 6.8.23. *Let $A\colon \mathbb{R}^n \to \mathbb{R}^n$ be a nonresonant stable hyperbolic matrix and Φ_A be the linear flow generated by A.*

(1) *If*

$$A(x_1, x_1', \dots, x_k, x_k', x_{k+1}, \dots, x_\ell) = (\Lambda_1(x_1, x_1'), \dots, \Lambda_k(x_k, x_k'), \lambda_{k+1}x_{k+1}, \lambda_\ell x_\ell)$$

and $\lambda_\ell > \alpha_i$ for all $1 \leq i < \ell - 1$ and $\alpha_i = \operatorname{Re}\lambda_i$, and g is a C^∞ homeomorphism of $\mathbb{R}^n$ such that $ge^{At} = e^{At}g$ for all $t \in \mathbb{R}$ and g preserves the orbits of the flow e^{At}, then $c \neq 0$ and Σ_c the hyperplane $x_\ell = c$ there is an induced map $\bar{g}\colon \Sigma_c \to \Sigma_c$ defined by $\bar{g}(x) = g(\mathcal{O}(x) \cap \Sigma_c)$ and $\bar{g}$ is of the form

$$\begin{aligned} &\bar{g}(x_1, x_1', \dots, x_k, x_k', x_{k+1}, \dots, x_{\ell-1}) \\ &\quad = (B_1(x_1, x_1'), \dots, B_k(x_k, x_k'), a_{k+1}x_{k+1}, \dots, a_{\ell-1}, x_{\ell-1}), \end{aligned}$$

where

$$B_j = \begin{pmatrix} a_j & -b_j \\ b_j & a_j \end{pmatrix} \quad \textit{for all } 1 \leq j \leq k.$$

(2) *If*

$$A(x_1,\dots,x_{k-1},x_k,x_k',\dots,x_\ell,x_\ell') \\ = (\lambda_1 x_1,\dots,\lambda_{k-1}x_{k-1},\Lambda_k(x_k,x_k'),\dots,\Lambda_\ell(x_\ell,x_\ell'))$$

and $\alpha_\ell > \alpha_i$ for all $1 \le i < \ell-1$ and $\alpha_i = \operatorname{Re}\lambda_i$, and g is a C^∞ homeomorphism of $\mathbb{R}^n$ such that $ge^{At} = e^{At}g$ for all $t \in \mathbb{R}$ and g preserves the orbits of the flow e^{At}, then for $r > 0$ and $C(r)$ the cylinder $x_\ell = r\cos\theta, x_\ell' = r\sin\theta$ there is an induced map $\bar{g}\colon C(r) \to C(r)$ defined by $\bar{g}(x,\theta) = g(O(x,\theta) \cap C(r))$ and $\bar{g}$ is of the form

$$\bar{g}(x_1,x_{k-1},x_k,x_k',\dots,x_{\ell-1},x_{\ell-1}',\theta) \\ = (a_1 k_1,\dots,a_{k-1}x_{k-1},B_k(x_k,x_k'),\dots,B_{\ell-1}(x_{\ell-1},x_{\ell-1}'),\theta+\theta_0),$$

where

$$B_j = \begin{pmatrix} a_j & -b_j \\ b_j & a_j \end{pmatrix} \quad \text{for all } k \le j \le \ell-1.$$

Proof. Since g commutes with A, Theorem 6.8.21 implies that g is linear. Furthermore, since the matrix A is hyperbolic we can modify the proof of Proposition 1.8.7 to show that g takes orbits of the linear flow e^{At} to orbits.

Let $\mathbb{R}^n = \bigoplus_{j=1}^{\ell} E_j$ where E_j is 1- or 2-dimensional depending on whether λ_j has zero imaginary part or nonzero imaginary part, respectively. Then $g_j(x) = a_j x$ where $a_j \in \mathbb{R}$ if $\lambda_j \in \mathbb{R}$, or $g_j(x) = B_j x$ where

$$B_j = \begin{pmatrix} a_j & -b_j \\ b_j & a_j \end{pmatrix} \quad \text{if } \lambda_j \notin \mathbb{R}.$$

Now assume that $\lambda_\ell \in \mathbb{R}$. Fix $c \in \mathbb{R}$ where $c \neq 0$ and let $\bar{g}$ be the induced map on Σ_c. Then the trajectory of e^{At} through the point

$$(\bar{x},c) = (x_1,x_1',\dots,x_k,x_k',x_{k+1},\dots,x_{\ell-1},c)$$

takes the form $x_\ell(t,\bar{x}) = ce^{\lambda_\ell t}$ and if $1 \le j \le \ell-1$, then $x_j(t,\bar{x}) = e^{\lambda_j t}x_j$ if $\lambda_j \in \mathbb{R}$ and

$$x_j(t,\bar{x}) = e^{\alpha_j t}B(\beta_j t)(x_j,x_j')$$

when $\lambda_j = \alpha_j + i\beta_j$ where $\beta_j \neq 0$, and

$$B(t) = \begin{pmatrix} \cos t & -\sin t \\ \sin t & \cos t \end{pmatrix}.$$

Now if $x_\ell = ce^{\lambda_\ell t}$, then $t = \frac{\ln x_\ell - \ln c}{\lambda_\ell}$.

If $\lambda_j \in \mathbb{R}$ let $\bar{\lambda}_j = \lambda_j/\lambda_\ell$. If $\lambda_j = \alpha_j + i\beta_j$ is not real, then let $\bar{\alpha}_j = \alpha_j/\lambda_\ell$ and $\bar{\beta}_j = \beta_j/\lambda_\ell$. Then we can define $\bar{x}_j(x_\ell, \bar{x})$ by

$$\bar{x}_j(x_\ell, \bar{x}) = \begin{cases} \left(\frac{x_\ell}{c}\right)^{\bar{\lambda}_j} x_j & \text{when } \lambda_j \in \mathbb{R}, \\ \left(\frac{x_\ell}{c}\right)^{\bar{\alpha}_j} B\left(\bar{\beta}_j \ln(x_\ell/c)\right)(x_j, x_j') & \text{when } \lambda_j \text{ is not real.} \end{cases}$$

Let

$$\tau(x_\ell, \bar{x}) = (x_1(x_\ell, \bar{x}), \dots, x_{\ell-1}(x_\ell, \bar{x}), x_\ell).$$

Then we have $g(\tau(x_\ell, \bar{x})) = \tau(a_\ell x_\ell, \bar{g}(\bar{x}))$. This implies that

$$g_j(\tau(x_\ell, \bar{x})) = a_j(x_j(x_\ell, \bar{x})) = a_\ell \left(\frac{x_\ell}{c}\right)^{\bar{\lambda}_j} x_j = x_j(a_\ell x_\ell, \bar{g}(\bar{x})) = \left(\frac{a_\ell x_\ell}{c}\right)^{\bar{\lambda}_j} \bar{g}_j(\bar{x})$$

for $\lambda_j \in \mathbb{R}$. Then $\bar{g}_j(\bar{x}) = a_\ell^{1-\bar{\lambda}_j} x_j$. Similarly, we have

$$\begin{aligned} g_j(\tau(x_\ell, \bar{x})) &= B_j(x_j(x_\ell, \bar{x})) \\ &= B_j\left(\left(\frac{x_\ell}{c}\right)^{\bar{\alpha}_j} B\left(\bar{\beta}_j \ln(x_\ell/c)\right)(x_j, x_j')\right) \\ &= x_j(a_\ell x_\ell, \bar{g}(\bar{x})) \\ &= \left(\frac{a_\ell x_\ell}{c}\right)^{\bar{\alpha}_j} B(\bar{\beta}_j \ln(a_\ell x_\ell/c))\bar{g}_j(\bar{x}) \end{aligned}$$

for $\lambda_j \notin \mathbb{R}$. Solving for $\bar{g}_j(\bar{x})$ we see that $\bar{g}_j$ is linear and of the desired form.

Now we assume that $\lambda_\ell \notin \mathbb{R}$. Then the orbit of e^{At} for the point

$$(\bar{x}, \theta) = (x_1, x_{k-1}, x_k, x_k', \dots, x_{\ell-1}, x_{\ell-1}', \theta)$$

takes the form $x_\ell(t, (\bar{x}, \theta)) = e^{\alpha_\ell t} B(\beta_j t)(r\cos\theta, r\sin\theta)$ and if $1 \le j \le \ell - 1$ we have $x_j(t, (\bar{x}, \theta)) = e^{\lambda_j t} x_j$ if $\lambda_j \in \mathbb{R}$ and

$$x_j(t, (\bar{x}, \theta)) = e^{\alpha_j t} B(\beta_j t)(x_j, x_j')$$

when $\lambda_j = \alpha_j + i\beta_j$ where $\beta_j \neq 0$, and

$$B(t) = \begin{pmatrix} \cos t & -\sin t \\ \sin t & \cos t \end{pmatrix}.$$

Now as above we see that $t = \frac{\ln \|(x_\ell, x_\ell')\|}{\alpha_\ell}$. Let $s = \|(x_\ell, x_\ell')\|$. Then we have $x_\ell(s, (\bar{x}, \theta)) = sB(\bar{\beta}_j \ln s)(r\cos\theta, r\sin\theta)$ where $\bar{\beta}_j = \beta_\ell/\alpha_\ell$. Furthermore, we have $x_j(s, (\bar{x}, \theta)) = s^{\bar{\alpha}_j} x_j$ if $\lambda_j \in \mathbb{R}$ and $\bar{\alpha}_j = \lambda_j/\alpha_\ell$, and $x_j(s, (\bar{x}, \theta)) = s^{\bar{\alpha}_j} B(\bar{\beta}_j \ln s)(x_j, x_j')$ if $\lambda_j \notin \mathbb{R}$ where $\bar{\alpha}_j = \alpha_j/\alpha_\ell$ and $\bar{\beta}_j = \beta_j/\alpha_\ell$. Arguing similarly to the previous case we can show that $\bar{g}$ is linear and of the desired form. □

6.9 Differentiability in the Hartman–Grobman Theorem*

We now apply the theory of normal forms to refine the Hartman–Grobman Theorem (Theorem 5.5.1). This is of interest in its own right, and this section can be read by readers who skipped Section 6.8, because we invoke only one specific result from it. Remark 5.5.4 noted that the Hartman–Grobman Theorem provides a purely topological conclusion, but (without the nonresonance assumptions needed for smooth linearization) it can be refined to yield a local conjugacy that is differentiable at the fixed point. Although we are interested in flows, the proof of Theorem 5.5.1 shows that it suffices to prove this in discrete time because the conjugacy we obtain that way is by uniqueness the conjugacy between the flow and its linearization.

Theorem 6.9.1 (Differentiable Hartman–Grobman Theorem [356]). *If the flow in the Hartman–Grobman Theorem* (Theorem 5.5.1) *is C^2 then the linearizing homeomorphism is differentiable at the fixed point, and its derivative at the fixed point is the identity.*[9]

Remark 6.9.2. Even the C^2 assumption is a little more than needed. It suffices for the flow to be C^1 near the equilibrium and for the differential to have sufficiently high Hölder exponent at the equilibrium. However, assuming C^1 only is definitely not enough. The proof we give here assumes that the flow is C^∞ in order to use a reduction to a dynamical system given by a (particular kind of) second-order polynomial.[10]

Example 6.9.4 shows an application of this result. We mention a different improvement on the Hartman–Grobman Theorem with like applications:

Theorem 6.9.3 ([357]). *If the flow in the Hartman–Grobman Theorem* (Theorem 5.5.1) *is C^∞ then for each $\alpha \in (0,1)$ there is a neighborhood of the fixed point on which the linearizing homeomorphism and its inverse are α-Hölder continuous.*

By Theorem 6.8.20 a *nonresonant* asymptotically stable fixed point for a flow has a smooth linearization nearby. However, neither this nor Theorem 6.8.10 implies Theorem 6.9.1 because we allow resonances.

Proof outline. We outline the proof (from [165]) here; the rest of the section is taken up by the full proof. The reader may decide that this outline is sufficient and take a look at Example 6.9.4 to see how this result is used.

---

[9] It was at the ETH (in conversations with Jürgen Pöschel) where the second author's interest in proving this result was sparked.

[10] Invoking Theorem 6.8.17 from [72] is the only reason we make the C^∞ assumption, and, even in this context it can be replaced by a C^k assumption, where k depends on the spectrum of the linear part. As this dependence is complicated, we assume C^∞, but a keen reader may decide to study the precise assumptions of [72, Theorem 11.9]—or the proof in [356], which assumes far less.

First of all, as we showed in proving Theorem 5.5.1 from its discrete-time counterpart, we can consider diffeomorphisms rather than flows. The conclusion of Theorem 6.9.1 is that the conjugacy is within $o(x)$ of the identity. We use a result of Bronstein and Kopanskiĭ that reduces the time-1 map to its second-order expansion via a C^1-conjugacy (Theorem 6.8.17), that is, it tells us that without loss of generality the diffeomorphism is a quadratic polynomial—of a particular kind.

The iterates of this map are polynomials, and we estimate their coefficients first recursively, and then definitively in Lemma 6.9.8. Unsurprisingly, we define the conjugacy via a natural correspondence between orbits under the linear and nonlinear dynamics; it has the product form $h := (h_+, h_-)$ with the components defined in (6.9.2). To show that this produces a map within $o(x)$ of the identity (Lemma 6.9.14) requires us to control the difference between orbits for the linear and nonlinear maps by combining these coefficient estimates with an estimate as to how close to the origin a point must be in order to stay in a given neighborhood for a specified time (Lemmas 6.9.10, 6.9.11, and 6.9.12). □

Example 6.9.4. To illustrate the application of differentiability in the Hartman–Grobman Theorem in a simple context, we draw a phase portrait of the system of differential equations $\begin{cases} dx/dt=2x-2x^2-xy, \\ dy/dt=4y-2xy-y^2, \end{cases}$ as follows.[11] Setting the right-hand side to 0 to find the equilibria gives

$$0 = 2x - 2x^2 - xy = x(2 - 2x - y),$$
$$0 = 4y - 2xy - y^2 = y(4 - 2x - y),$$

so the equilibria are $(0,0)$, $(0,4)$, $(1,0)$. To linearize at each of these, note that the matrix of partial derivatives of the right-hand side is $\begin{pmatrix} 2-4x-y & -x \\ -2y & 4-2x-2y \end{pmatrix}$, so the linearizations are

- $(0,0)$: $\begin{pmatrix} 2 & 0 \\ 0 & 4 \end{pmatrix}$ with eigenvalue–eigenvector pairs $\left(2, \begin{pmatrix} 1 \\ 0 \end{pmatrix}\right)$, $\left(4, \begin{pmatrix} 0 \\ 1 \end{pmatrix}\right)$, so for the linearization the origin is a repeller;

- $(0,4)$: $\begin{pmatrix} -2 & 0 \\ -8 & -4 \end{pmatrix}$ with eigenvalue–eigenvector pairs $\left(-2, \begin{pmatrix} 1 \\ -4 \end{pmatrix}\right)$, $\left(-4, \begin{pmatrix} 0 \\ 1 \end{pmatrix}\right)$, so the origin is an attractor;

- $(1,0)$: $\begin{pmatrix} -2 & -1 \\ 0 & 2 \end{pmatrix}$ with eigenvalue–eigenvector pairs $\left(-2, \begin{pmatrix} 1 \\ 0 \end{pmatrix}\right)$, $\left(2, \begin{pmatrix} 1 \\ -4 \end{pmatrix}\right)$, so this is a hyperbolic fixed point of saddle type.

[11] Being 2-dimensional, this is, of course, an application of the much older result Theorem 6.8.14.

Moreover, the phase portraits of the linearizations are as follows:

(0,0) (0,4) (1,0)

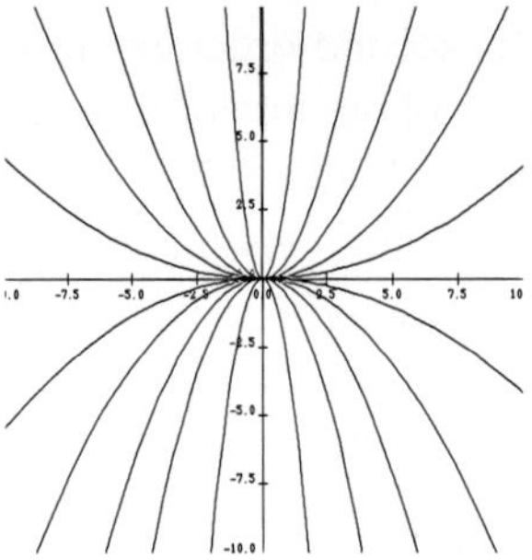

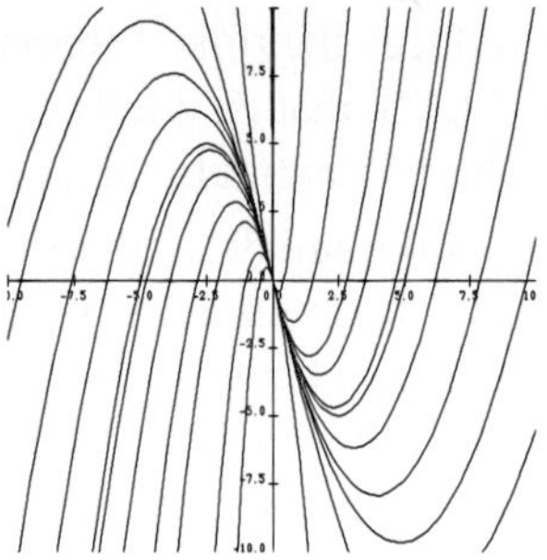

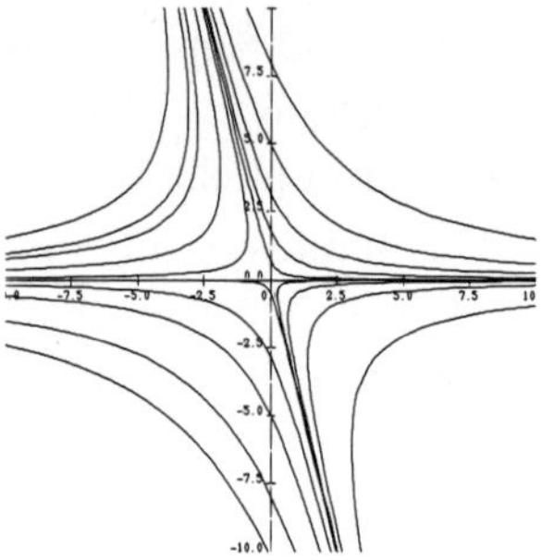

Because the linearizations show that the Hartman–Grobman Theorem applies to each equilibrium, we can now place each of these pictures on the respective fixed point as a thumbnail of the actual phase portrait and plausibly connect these. The resulting phase portrait is shown in Figure 6.9.1. Note that the local appearance near each equilibrium

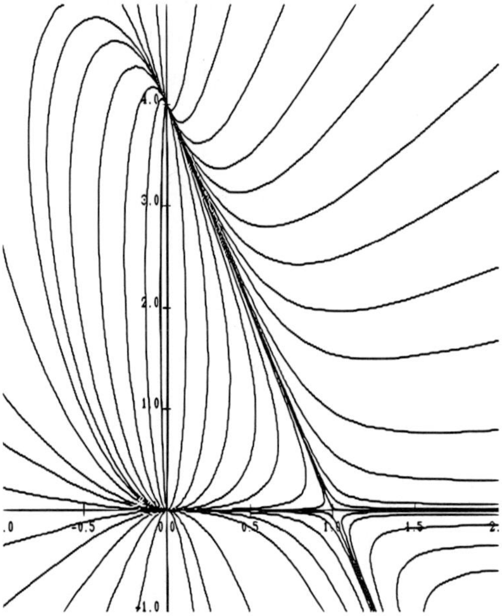

Figure 6.9.1. Phase portrait in the plane pieced together from Hartman–Grobman patches.

is *exactly* rather than just topologically that of the corresponding linearization—in contrast to Figure 5.5.1.

We now present the full proof of Theorem 6.9.1 using Theorem 6.8.17.

Proof of Theorem 6.9.1. That Q in Theorem 6.8.17 contains only weakly resonant terms (Definition 6.8.18) implies that if we consider the root space decomposition of $L := Df(0)$ and define a linear map D to be a scaling by a on each root space for an eigenvalue of absolute value a, then we have the following result.

Lemma 6.9.5. $Q \circ D = DQ$.

Proof. This restates that Q contains only weakly resonant terms (Theorem 6.8.17). By definition, this means that if (in coordinates adapted to the root space decomposition) the ith component of Q contains a term $a \cdot xy$ with $a \neq 0$ then the eigenvalues λ and μ associated with the root spaces corresponding to x and y, respectively, are related to the eigenvalue η associated with the root space containing the ith unit vector by $|\eta| = |\lambda|\,|\mu|$. Note that this is exactly the claim. □

Define J by $L = DJ$ and note (for example via the Jordan normal form of L) that $DJ = JD$, all eigenvalues of J have absolute value 1, and that the entries of J^n are polynomials in n. We assume our coordinate system is adapted to the root space decomposition, so J is block diagonal. We occasionally write $Q(x)$ for $Q(x,x)$.

Remark 6.9.6. The iterates $\mathfrak{f}^n$ of $\mathfrak{f}$ are polynomials, as are the $h_n := L^{-n}\mathfrak{f}^n =: \mathrm{Id} + \mathrm{NL}_n$.[12] The conjugacy will be constructed from the h_n, and accordingly, we wish to estimate the coefficients of the nonlinear terms. To that end denote by $b_{n,m}$ the sum of the absolute values of all mth-order coefficients in the coordinate representation of h_n (this is the *height* of the mth-order term of h_n). The first step is a recursive estimate.

Lemma 6.9.7. $\exists\, N, k \in \mathbb{N}\ \forall\, n, m \in \mathbb{N}\colon\ b_{n+1,m} \leq b_{n,m} + Nn^{k-1} \sum_{p=1}^{m-1} b_{n,p} b_{n,m-p}$.

Proof. $\mathrm{Id} + \mathrm{NL}_{n+1} = h_{n+1} = L^{-n-1}\mathfrak{f} \circ \mathfrak{f}^n = L^{-n-1}(L+Q)L^n(\mathrm{Id} + \mathrm{NL}_n)$, so

$$\begin{aligned} \mathrm{NL}_{n+1} &= \mathrm{NL}_n + L^{-n-1}Q(L^n(\mathrm{Id} + \mathrm{NL}_n)) \\ &= \mathrm{NL}_n + J^{-n-1}D^{-n-1}Q(D^n J^n(\mathrm{Id} + \mathrm{NL}_n)) \qquad (6.9.1)\\ \text{Lemma 6.9.5 and } DJ = JD \Rightarrow \quad &= \mathrm{NL}_n + D^{-1}J^{-n-1}Q(J^n(\mathrm{Id} + \mathrm{NL}_n)). \end{aligned}$$

To bound $b_{n+1,m}$ note first that terms of a given order m in a coordinate representation of $J^n(\mathrm{Id} + \mathrm{NL}_n)$ come in linear combinations with polynomial coefficients of mth-order terms in $\mathrm{Id} + \mathrm{NL}_n$. These polynomial coefficients arise from entries of J^n, and the form of the linear combinations is otherwise independent of n. Thus, the sum $\beta_{n,p}$ of the absolute values of all pth-order coefficients in the coordinate representation of

[12] Here, "NL" stands for "nonlinear terms."

$J^n(\mathrm{Id} + \mathbb{N}\mathrm{L}_n)$ is at most $P_1(n)b_{n,p}$ for some polynomial P_1 that is independent of p because it encodes only the action of J^n.

Likewise, in a coordinate representation of $D^{-1}J^{-n-1}Q(J^n(\mathrm{Id} + \mathbb{N}\mathrm{L}_n))$ terms of a given order m come in linear combinations with polynomial coefficients of mth-order terms in $Q(J^n(\mathrm{Id} + \mathbb{N}\mathrm{L}_n))$. These coefficients arise from entries of J^{-n-1}, and the form of the linear combinations is otherwise independent of n. Thus, the sum of mth-order coefficients in $D^{-1}J^{-n-1}Q(J^n(\mathrm{Id} + \mathbb{N}\mathrm{L}_n))$ is bounded in terms of that in $Q(J^n(\mathrm{Id} + \mathbb{N}\mathrm{L}_n))$ by including a polynomial multiplier $P_2(n)$.

Sorting by the order m of terms, the ith component of $Q(J^n(\mathrm{Id} + \mathbb{N}\mathrm{L}_n))$ is

$$\begin{aligned}\sum_{j,l} a_{ijl}\big[J^n(\mathrm{Id} + \mathbb{N}\mathrm{L}_n)\big]_j\big[J^n(\mathrm{Id} + \mathbb{N}\mathrm{L}_n)\big]_l \\ = \sum_m \sum_{p=1}^{m-1} \sum_{j,l} a_{ijl} \sum_{|\rho|=p} \alpha_{j,n,\rho}x^\rho \sum_{|\tau|=m-p} \alpha_{l,n,\tau}x^\tau.\end{aligned}$$

Thus, the absolute values of all mth-order coefficients in $Q(J^n(\mathrm{Id} + \mathbb{N}\mathrm{L}_n))$ sum to

$$\sum_i \sum_{p=1}^{m-1} \sum_{j,l} |a_{ijl}| \Big[\sum_{|\rho|=p} |\alpha_{j,n,\rho}|\Big]\Big[\sum_{|\tau|=m-p} |\alpha_{l,n,\tau}|\Big] \le c_0 \sum_p \beta_{n,p}\beta_{n,m-p},$$

using $\sum_{|\rho|=p,|\tau|=m-p} |\alpha_{j,n,\rho}\alpha_{l,n,\tau}| = \sum_{|\rho|=p} |\alpha_{j,n,\rho}| \sum_{|\tau|=m-p} |\alpha_{l,n,\tau}|$. Take $N, k \in \mathbb{N}$ such that $P_2(n)c_0P_1^2(n) \le Nn^{k-1}$ for all $n \in \mathbb{N}$ to get the claim by (6.9.1). □

This recursion relation allows us to bound the coefficients inductively.

Lemma 6.9.8. *For any $\alpha > 0$ there is a $C > 0$ such that $b_{n,m} \le e^{\alpha(n+C)m}$.*

Proof. If $\alpha C \ge \sup_n 2\log N + 2k\log n - \alpha n$ then $(N^2n^{2k})^m \le e^{\alpha(n+C)m}$, so it suffices to show inductively that if N, k are as in Lemma 6.9.7 then $b_{n,m} \le (Nn^k)^{2m-1}$. This is clear for $n = 1$, and if true for $b_{i,m}$ with $i \le n$ and $m \in \mathbb{N}$ then by Lemma 6.9.7,

$$\begin{aligned} b_{n+1,m} &\le (Nn^k)^{2m-1} + Nn^{k-1}\sum_{p=1}^{m-1} \underbrace{(Nn^k)^{2p-1}(Nn^k)^{2(m-p)-1}}_{=N^{2m-2}n^{2(m-2)k}} \\ &\le N^{2m-1}\big(n^{(2m-1)k} + (2m-1)kn^{(2m-1)k-1}\big) \le N^{2m-1}(n+1)^{(2m-1)k} \end{aligned}$$

by recognizing the leading terms in the binomial expansion of $(n+1)^{(2m-1)k}$. □

Remark 6.9.9. These arguments establish these same results also for the case where the coefficients are bounded functions; one merely replaces the coefficients by their bounds.

Denote now by $-\lambda_l < \cdots < -\lambda_1 < 0 < \mu_1 < \cdots < \mu_m$ the distinct values of $\log|\nu|$, where ν is an eigenvalue of L, and define

$$R_+ := \{(\alpha^1,\dots,\alpha^l,\beta^1,\dots,\beta^m) \mid \sum_{i=1}^m \beta^i\mu_i - \sum_{i=1}^l \alpha^i\lambda_i = \mu_s \text{ for some } s \le m\}.$$

Lemma 6.9.10. *There exists $p \in (0,1/2)$ such that $\sum_{i=1}^m \beta_i \ge p|\tau|$ for any $\tau \in R_+$.*

Proof. $\sum_{i=1}^m \beta^i\mu_i - \sum_{i=1}^l \alpha^i\lambda_i = \mu_s > 0$ gives $\lambda_1 \sum_{i=1}^l \alpha^i \le \sum_{i=1}^l \alpha^i\lambda_i < \sum_{i=1}^m \beta^i\mu_i \le \mu_m \sum_{i=1}^m \beta^i$ and $|\tau| = \sum_{i=1}^l \alpha^i + \sum_{i=1}^m \beta^i < (1+\frac{\mu_m}{\lambda_1})\sum_{i=1}^m \beta^i$. Thus, any $p \le \frac{1}{1+(\mu_m/\lambda_1)}$ is as required. □

These estimates now control how close orbits are to the origin.

Lemma 6.9.11. *There are $H, \gamma_1 > 0$ such that if δ is sufficiently small and $\|\mathfrak{f}^j(x)\| < \delta$ for $j = 0,\dots,n$ then $|x_i| < e^{H^2\log\delta - \gamma_1 n}$ for each expanding coordinate x_i.*

Proof. Denote by $(\cdot)_\pm$ the coordinate projections to the expanding and contracting directions, respectively. We will use that the conjugacy h to $\mathfrak{f}$ is H-Hölder continuous together with its inverse (Remark 5.5.4) and preserves the contracting subspace. Let $(x_+, x_-) = h((x'_+, x'_-))$. If $\|\mathfrak{f}^j(x)\| < \delta$ for $j = 0,\dots,n$ then $\|L^j(x')\| < \delta^H$ for $j = 0,\dots,n$. For the linear part of $\mathfrak{f}$ we have $\|x'_+\| < \delta^H e^{-\mu_1 n}$, so there is a $\gamma_1 > 0$ such that $\|x_+\| \le \|h((x'_+, x'_-)) - h((0, x'_-))\| < \delta^{H^2} e^{-\gamma_1 n}$. □

Lemma 6.9.12. *There exists $\gamma > 0$ such that if $\|\mathfrak{f}^j(x)\| < \delta$ for $j = 0,\dots,n$ then*

$$c_{n,m}(x) := \max\{|x^\tau| \mid \tau \in R_+,\ |\tau| = m\} \le e^{-\gamma(n+C_\delta)m}\|x\| \quad \textit{for } m \ge 2,$$

where $C_\delta \to \infty$ as $\delta \to 0$.

Proof. Note that x^τ has over pm expanding components (counted with multiplicity) by Lemma 6.9.10. Applying Lemma 6.9.11 to these and $|x_i| \le \|x\|$ to the remaining $m - (\lfloor pm\rfloor + 1) > m - (\frac{m}{2}+1) \ge \frac{m}{2} - 1 \ge 0$ gives $|x^\tau| < e^{-pm(\gamma_1 n - H^2\log\delta)}\|x\|$. □

Remark 6.9.13. By Lemma 6.9.8 there exists $C > 0$ such that $b_{n,m} \le e^{\gamma(n+C)m/2}$, and we henceforth take $\delta > 0$ such that $C_\delta > C$.

We now construct the desired conjugacy, starting with the Bronstein–Kopanskiĭ conjugacy H^+ from Theorem 6.8.17 of the given map to $\mathfrak{f}$.

Consider a nonincreasing C^∞ "bump" function $\varphi\colon [0,1] \to [0,1]$ with $\varphi(1/4) = 1$ and $\varphi(3/4) = 0$, and multiply the quadratic terms of $\mathfrak{f}$ by $\varphi(\|x\|/\delta)$. Near 0 this new map $\tilde{\mathfrak{f}}$ is conjugate to $\mathfrak{f}$ by the identity. If $\|x\| \ge \delta$ then $\tilde{\mathfrak{f}}(x) = Lx$.

Take $\delta' < \delta$ sufficiently small, and for $\|x\| < \delta'$ define

$$n_+(x) := 1 + \max\{n \in \mathbb{N} \mid \|\tilde{\mathfrak{f}}^i(x)\| < \delta \text{ for } 0 \le i \le n\}.$$

Then $\lim_{\|x\|\to 0} n_+(x) = \infty$ and $n_+(\tilde{\mathfrak{f}}(x)) = n_+(x) - 1$. Since the linear part of f is hyperbolic, n_+ is finite off the contracting direction. For $\|x\| < \delta'$ and with $h_n := L^{-n}\tilde{\mathfrak{f}}^n$ as in Remark 6.9.6 we set $h_+ = 0$ on the contracting direction and

$$h_+(x) := (h_{n_+(H^+(x))}(H^+(x)))_+. \tag{6.9.2}$$

Continuity on the contracting direction follows from the last sentence in the proof of Lemma 6.9.14. Discontinuities of n_+ do not produce discontinuities of h_+ because $h_{n_+(H^+(x))}(H^+(x)) = h_{n_+(H^+(x))+1}(H^+(x))$.

Theorem 6.8.17 applied to f^{-1} yields a conjugacy to a quadratic polynomial. To this our intermediate results also apply, and hence we can define n_- and h_- analogously to n_+ and h_+ and set $h := (h_+, h_-)$. The next two results for h_+ combined with the analogous ones for h_- (which we omit) show that h is differentiable at 0 and is a conjugacy between f and its linear part L.

Lemma 6.9.14. $Dh_+(0) = \mathrm{Id}$, *that is,* $\|(h_+(x) - x)_+\| \in o(\|x\|)$.[13]

Proof. Since $\|H^+(x)-x\| = \|\Delta h(x)\| \in o(\|x\|)$ by Theorem 6.8.17, and $o(\|H^+(x)\|) \subset o(\|x\|)$ we show $\|(h_{n_+(x)}(x) - x)_+\| \in o(\|x\|)$. With $c_{n,m}$ from Lemma 6.9.12,

$$\begin{aligned}\|(h_{n_+(x)}(x) - x)_+\| &\le \sum_{\substack{\text{expanding}\\ \text{coordinates } i}} \sum_{\substack{\tau\in R_+\\ |\tau|\ge 2}} |\alpha_{i,n_+,\tau}|\cdot|x^\tau| \le \sum_{m\ge 2}\sum_{\substack{|\tau|=m\\ \tau\in R_+}} |\alpha_{i,n_+,\tau}|\cdot c_{n_+,m}(x)\\ &\le \sum_{m\ge 2} c_{n_+,m}(x) b_{n_+,m} \le \sum_{m\ge 2} e^{-\gamma(n_++C_\delta)m} e^{\gamma(n_++C)m/2}\|x\|\end{aligned}$$

by Remark 6.9.9 and Lemma 6.9.8 with $\alpha = \gamma/2$. Remark 6.9.13 and $q := e^{-\gamma(n_++C)/2}$ give $\|(h_{n_+(x)}(x)-x)_+\| \le \sum_{m\ge 2} q^m\|x\| = \|x\|q^2/(1-q)$. If $\|x_+\| \to 0$ then $n_+(x) \to \infty$ and $q \to 0$, so $\|(h_{n_+(x)}(x) - x)_+\| \in o(\|x\|)$, hence $\|(h(x) - x)_+\| \in o(\|x\|)$. □

Analogously to our other notation we write $L = L_+ \oplus L_-$. Then

$$\begin{aligned}(h(f(x)))_+ &= (L^{-n_+(H^+(f(x)))}\,\tilde{\mathfrak{f}}^{n_+(H^+(f(x)))}(H^+(f(x))))_+\\ &= (L^{-n_+(\tilde{\mathfrak{f}}(H^+(x)))}\,\tilde{\mathfrak{f}}^{n_+(\tilde{\mathfrak{f}}(H^+(x)))}\,(\tilde{\mathfrak{f}}(H^+(x))))_+\\ &= (L^{-n_+(H^+(x))+1}\,\tilde{\mathfrak{f}}^{n_+(H^+(x))}\quad (H^+(x)))_+\\ &= L_+^{-n_+(H^+(x))+1}\,(\tilde{\mathfrak{f}}^{n_+(H^+(x))}\quad (H^+(x)))_+\\ &= L_+(L^{-n_+(H^+(x))}\tilde{\mathfrak{f}}^{n_+(H^+(x))}\quad (H^+(x)))_+ = L_+h_+(x) = (Lh(x))_+.\end{aligned}$$

This observation and its counterpart for h_- give $h \circ f = L \circ h$ for small x, that is, h conjugates f and L in a neighborhood of 0. Since Lemma 6.9.14 and its counterpart for h_- imply invertibility near 0, this completes the proof of Theorem 6.9.1. □

[13] See Remark 3.2.18.

Exercises

6.1. Prove that for suspensions $t(x, y) \equiv 0$ in Proposition 6.2.2.

6.2. Describe the Bowen bracket (Proposition 6.2.2) explicitly for symbolic flows (see Exercise 1.29).

6.3. Show that the Smale Horseshoe in Example 1.5.24 has local product structure.

6.4. Show that a hyperbolic attractor has local product structure.

6.5. In light of Theorem 6.2.7, Theorem 5.3.37 produces a local product structure in five scenarios. Obtain this local product structure directly by using Bowen brackets instead of shadowing in the corresponding parts of the proof of Theorem 5.3.37.

6.6. Show that a compact factor of $\mathbb{H}$ (Section 2.4) has closed geodesics of incommensurate lengths.

6.7. For a compact factor Σ of $\mathbb{H}$ (Section 2.4) suppose v is a unit vector generating a closed geodesic γ_v. Show that there are unit vectors $v_i \to v$ such that each v_i generates a closed geodesic whose length is incommensurate with that of γ_v.

6.8. Show that the horocycle flow (Example 2.2.3) on a compact factor of $\mathbb{H}$ is minimal (Definition 1.6.21). (This replaces the use of Example 2.1.16 elsewhere.)

6.9. Show that Theorem 6.2.7 holds for shifts (using the notation from the proof of Proposition 6.6.15).

6.10. Fold the Plykin attractor onto a tetrahedron using the template in Figure 6.5.7 following the instructions from http://www.qland.de/origami/tetra/index.html.

7

Ergodic theory of hyperbolic sets

This chapter is devoted to invariant probability measures for hyperbolic sets. In Section 7.1, we show by use of the Hopf argument that any volume-preserving Anosov flow on a compact manifold is ergodic and indeed multiply mixing with respect to the volume form. Expanding on the Hopf argument we discuss the important concept of stable ergodicity (Section 7.2), which holds for volume-preserving Anosov flows.

The ergodic notions (entropy, pressure, and equilibrium states) from Chapter 4 have been well studied in the context of hyperbolic sets (Section 7.3). We prove that a transitive locally maximal hyperbolic set carries a unique measure of maximal entropy, which is therefore ergodic. This follows from a construction due to Bowen that focuses on periodic points as the carriers of information on entropy (and along the way gives rather precise asymptotics for the growth rate of periodic orbits). Applying the Bowen construction more generally we study the existence and uniqueness of equilibrium states for Hölder-continuous potentials. This naturally leads to the investigation of the Sinai–Ruelle–Bowen measure (Section 7.4), which is the equilibrium state for a natural dynamical potential, the geometric potential (Definition 7.4.1), and which also provides a "physical measure" in that it determines the asymptotic distribution of Lebesgue-a.e. orbit.

We also discuss rates of mixing for hyperbolic sets (Section 7.7) but can only prove illustrative results for discrete-time dynamics. The picture is less clear for hyperbolic flows than for maps, and the technical challenges are far greater, so we give only a descriptive outline of the current state of the art.

Hamenstädt and Margulis provided another construction of the measure of maximal entropy for transitive Anosov flows. In Section 7.5 we present this construction, which centers on a scaling property of the conditionals of this measure and has interesting connections to dimension theory. It also gives an even more precise (multiplicative) asymptotic of the growth rate of the number of periodic orbits than the Bowen construction (Theorem 7.6.9).

We must mention here that the theory of Perron–Frobenius operators or transfer operators appears here only minimally (Section 7.7); this is a powerful approach for the production and study of invariant measures, notably equilibrium states. It underlies much of the work on decay of correlations but predates it and is much more broadly useful (such as in [190, 62, 256]).

7.1 The Hopf argument, absolute continuity, mixing

In Chapters 3 and 4 we developed the ergodic theory of flows, and we are now prepared to examine ergodic properties of hyperbolic dynamical systems beyond the homogeneous examples we have been able to treat so far. As indicated in the preface, ergodicity with respect to a volume form was a major motivation for studying hyperbolic flows, and we present the two main ways of establishing ergodicity as well as stronger mixing properties. The first of these is the Hopf argument, the original method for establishing ergodicity of volume in hyperbolic dynamical systems that are not of an algebraic nature. Indeed, Hopf grasped the fundamental way in which hyperbolicity provides the very mechanism that produces ergodicity (Theorem 7.1.26), and we will furthermore see that it produces mixing properties as well (Corollary 7.1.10, Theorem 7.1.12, Theorem 7.1.28). The other method is the construction of equilibrium states, whose intimate connection with periodic orbits produces ergodicity (Theorem 7.3.6) and indeed strong mixing (Remark 7.3.20). Both methods have their own importance. The Hopf method has proved useful as research developed beyond uniform hyperbolicity more quickly than the theory of equilibrium states, while the latter retains its central role in producing a multitude of interesting, tractable, and mixing measures for hyperbolic dynamical systems.

The essential idea of the Hopf argument is to use Theorem 3.3.10 by showing that Birkhoff averages of continuous functions are constant on stable leaves and on unstable leaves, and then to show that this implies that any invariant function is constant almost everywhere.

Definition 7.1.1. A function $f\colon X \to \mathbb{R}$ is *subordinate* to W^s or W^s*-saturated* if there is a set $G \subset X$ with $\mu(G) = 1$ such that $x, y \in G$ and $y \in W^s(x)$ imply $f(x) = f(y)$, and likewise for W^u.

In this case $f^s(x) := \begin{cases} 0 & \text{if } W^s(x)\cap G=\varnothing, \\ f(y) & \text{if } y\in G\cap W^s(x), \end{cases} \overset{\text{ae}}{=} f$ is (everywhere!) constant on stable sets.

Theorem 7.1.2 (Hopf argument I). *If (X, μ) is a metric Borel probability space and the flow Φ is μ-preserving, then any Φ-invariant $f \in L^1(\mu)$ is W^s-saturated and W^u-saturated.*

Proof. Suppose f is uniformly continuous. Then f_Φ is W^s-saturated: If $f_\Phi(x)$ exists and $y \in W^s(x)$, then

$$\frac{1}{T}\int_0^T f(\varphi^t(x))\,dt - \frac{1}{T}\int_0^T f(\varphi^t(y))\,dt = \frac{1}{T}\int_0^T f(\varphi^t(x)) - f(\varphi^t(y))\,dt \xrightarrow[T\to\infty]{} 0$$

since $d(\varphi^t(x), \varphi^t(y)) \xrightarrow[t\to+\infty]{} 0$ and f is uniformly continuous. Thus $f_\Phi(y) = f_\Phi(x)$, as claimed. (In particular, f_Φ is *defined* and constant on $W^s(x)$.)

If $f \in L^1(\mu)$ there are uniformly continuous g^n such that $\|f - g^n\|_p < 1/n$ and hence $\|f_\Phi - g^n_\Phi\|_p < \epsilon$,[1] that is, $g^n_\Phi \xrightarrow[t\to+\infty]{L^1} f_\Phi$, so a subsequence of the g^n_Φ converges a.e. to f_Φ. Since the g^n_Φ are W^s-saturated, so is f_Φ. Then W^{us}-saturation follows by reversing time. □

Corollary 7.1.3. *If (X, μ) is a metric Borel probability space and Φ is μ-preserving such that any Φ-invariant W^s- and W^u-saturated $f \in L^1(\mu)$ is constant, then Φ is ergodic.*

We first put this to use with a "traditional" application—to the suspension of a hyperbolic toral automorphism.

Proposition 7.1.4. *If $A \in \mathrm{GL}(m, \mathbb{Z})$ is hyperbolic, then the suspension of the induced automorphism F_A of $\mathbb{T}^m$ is ergodic with respect to Lebesgue measure.*

Proof. For $q \in \mathbb{T}^m$ the *stable subspace* of (q,t) is $W^s(q,t) = \pi(E^- + q) \times \{t\}$, where E^- is the contracting subspace of A and $\pi\colon \mathbb{R}^m \to \mathbb{T}^m$ is the projection. Likewise, $W^u(q,t) = \pi(E^+ + q) \times \{t\}$.

To apply Corollary 7.1.3 consider a Φ-invariant $f \in L^1$ for which there is a set $G \subset \mathbb{T}^n$ of measure 1 with $x, y \in G \times \{t\}$, $y \in W^s(x) \Rightarrow f(x) = f(y)$ and $x, y \in G \times \{t\}$, $y \in W^u(x) \Rightarrow f(x) = f(y)$. If we can conclude that $f \overset{\text{ae}}{=}$ const., then Corollary 7.1.3 implies ergodicity.

Let $D^\pm \subset E^\pm$ be small disks and $q \in \mathbb{T}^m$. Then q has a neighborhood that is up to rotation and translation of the form $D^- \times D^+$, and

$$C := G \cap (D^- \times D^+)$$

has full Lebesgue measure in $D^- \times D^+$, that is, if $\mu^\pm$ denotes the normalized Lebesgue measure on $D^\pm$ and $\mu = \mu^- \times \mu^+$, then $\int_{D^-\times D^+} \chi_C \, d\mu = 1$. By the Fubini Theorem,

$$1 = \int_{D^-\times D^+} \chi_C \, d\mu = \int_{D^-} \int_{D^+} \chi_C \, d\mu^+ \, d\mu^-, \quad \text{so} \int_{D^+} \chi_C(u, \cdot) \, d\mu^+ \overset{\mu^-\text{-a.e.}}{=\!=\!=} 1.$$

Fix such a $u_0 \in D^-$, and note that by construction $C^- := D^- \times \big(C \cap (\{u_0\} \times D^+)\big)$ has full Lebesgue measure. If $(u, v), (u', v') \in C^- \cap C$, a set of full measure, then

$$f(u, v, t) = f(u_0, v, t) = f(u_0, v', t) = f(u', v', t).$$

This applies to any such neighborhood of an arbitrary $q \in \mathbb{T}^n$, so $f \overset{\text{ae}}{=}$ const. □

[1] Both f_Φ and g^n_Φ exist a.e. by the Birkhoff Ergodic Theorem (Theorem 3.2.15).

Remark 7.1.5. This application of the Hopf argument yields a weaker conclusion than Proposition 3.3.7, and with more effort. However, the simplicity of the proof of Proposition 3.3.7 relied entirely on the linearity of this system, whereas the Hopf argument does not and supports other applications. The argument above uses the Fubini Theorem, however, and to that end relies on smooth local charts in which Lebesgue measure is a product measure. We will develop ways to sidestep this use of linearity, leading to Theorem 7.1.26.

We now amplify the Hopf argument in the direction of mixing. The underlying technical ingredient is a Hilbert-space lemma that "upgrades" weak convergence to pointwise convergence.

Theorem 7.1.6 (Banach–Saks). *If* $x_n \xrightarrow{\text{weakly}} x$ *in a Hilbert space, then there is a subsequence* x_{n_k} *such that* $z_j := \frac{1}{j}\sum_{k=1}^{j} x_{n_k} \xrightarrow{\|\cdot\|} x$. *In* L^2, z_{j^2} *then converges a.e. (Borel–Cantelli).*

Proof. Passing to $x_n - x$ assume $x_n \xrightarrow{\text{weakly}} 0$ and hence recursively choose $x_{n_1} = x_1$ and x_{n_k} such that $|\langle x_{n_k}, x_{n_i}\rangle| < 1/k$ for $1 \le i < k$. Then $\|x_{n_k}\| \le C$ for some $C \in \mathbb{R}$ (weakly convergent $\Rightarrow$ norm-bounded), and

$$\Big\|\frac{1}{j}\sum_{k=1}^{j} x_{n_k}\Big\|^2 = \frac{1}{j^2}\sum_{i,k=1}^{j}\langle x_{n_k}, x_{n_i}\rangle$$

$$= \frac{1}{j^2}\Bigg[\underbrace{\sum_{k=1}^{j}\|x_{n_k}\|^2}_{\le jC^2} + 2\underbrace{\sum_{k=1}^{j}\overbrace{\sum_{1\le i<k}\underbrace{\langle x_{n_k}, x_{n_i}\rangle}_{<1/k}}^{<1}}_{<j}\Bigg] \in O\Big(\frac{1}{j}\Big). \qquad \square$$

The Banach–Saks Theorem gives the following amplified Hopf argument.

Theorem 7.1.7. *If* X *is a metric space,* Φ *is a continuous flow on* X, μ *is a* Φ-*invariant Borel probability measure, and* $f_i \in L^2(\mu)$, *then any weak accumulation point* $F_n = \prod_{i=1}^{N} f_i \circ \varphi^{t_{i,n}} \xrightarrow[n\to\infty]{\text{weakly}} F$ *of* $\prod_{i=1}^{N} f_i \circ \varphi^{t_i}$ *is* W^s-*subordinate.*

Proposition 3.4.29 gives a strong immediate consequence of Theorem 7.1.7 to the effect that a far stronger assumption than joint ergodicity (Remark 7.1.11) gives multiple mixing:

Corollary 7.1.8. Φ *is multiply mixing if every* W^s-*subordinate function* $f \in L^2$ *is constant a.e.*

(Remark 8.6.12 gives an application of this.)

Proof of Theorem 7.1.7 (Coudène). By the Banach–Saks Theorem (Theorem 7.1.6) an accumulating subsequence $F_n \xrightarrow[n\to\infty]{L^2\text{-weakly}} F$ admits sequences m_l, n_{i_k} with

$$\mathcal{F}_l := \frac{1}{m_l}\sum_{k=0}^{m_l-1} F_{n_{i_k}} \xrightarrow{\text{a.e.}} F.$$

We passed to pointwise convergence because this is W^s-subordinate for bounded uniformly continuous f_i: if $p_{ij}^l := f_i(\varphi^{(t_i)_l}(x_j))$ for $j = 1,2$ with $x_2 \in W^s(x_1)$, then

$$\prod_{i=1}^N p_{i2}^l - \prod_{i=1}^N p_{i1}^l = \sum_{j=1}^N \underbrace{\prod_{i<j} p_{i2}^l}_{\text{bounded}} \underbrace{(p_{j2}^l - p_{j1}^l)}_{\xrightarrow[l\to\infty]{}0} \underbrace{\prod_{i>j} p_{i1}^l}_{\text{bounded}} \xrightarrow[l\to\infty]{} 0.$$

Now approximate $f_i^0 \in L^2 \cap L^\infty$ by bounded uniformly continuous functions f_i^k such that $\|f_i^0 - f_i^k\|_2 < 1/k$ and this time let $p_{ij}^l := f_i^j \circ \varphi^{(t_i)_l}$ to find that weak limits (of subsequences if necessary) satisfy

$$\begin{aligned}\|F - F^k\| &\le \varliminf_{l\to\infty} \Big\|\prod_{i=1}^N p_{ik}^l - \prod_{i=1}^N p_{i0}^l\Big\| \\ &\le \sum_{j=1}^N \underbrace{\prod_{i<j} \|p_{ik}^l\|_\infty}_{\text{bounded}} \underbrace{\|p_{jk}^l - p_{j0}^l\|_2}_{\xrightarrow[k\to\infty]{}0} \underbrace{\prod_{i>j} \|p_{i0}^l\|_\infty}_{\text{bounded}} \xrightarrow[k\to\infty]{} 0,\end{aligned}$$

so, passing to a subsequence, $F^k \xrightarrow{\text{a.e.}} F$ and this is W^s-subordinate. □

Theorem 7.1.9 (Babillot–Hopf argument). *If (X,μ) is a metric Borel probability space, Φ is μ-preserving, and $f \in L^2(\mu)$, then weak-accumulation points of $\{f \circ \varphi^t \mid t \ge 0\}$ are W^s-saturated and W^u-saturated.*

Proof (Babillot–Coudène). If $f \perp I \subset L^2(\mu)$, the (closed) subspace of functions subordinate to W^u, and $f \circ \varphi^{t_i} \xrightarrow[i\to\infty]{\text{weakly}} g$, then Theorem 7.1.7 applied to φ^{-t} gives a subsequence $t_{i_k} \to \infty$ with $g \circ \varphi^{-t_{i_k}} \xrightarrow[k\to\infty]{\text{weakly}} g' \in I$, so $\langle g, g\rangle = \lim_{k\to\infty}\langle f \circ \varphi^{t_{i_k}}, g\rangle = \lim_{k\to\infty}\langle f, g \circ \varphi^{-t_{i_k}}\rangle = \langle f, g'\rangle = 0$, that is, $g = 0$, so $f \circ \varphi^t \xrightarrow[t\to\infty]{\text{weakly}} 0$.

For an arbitrary $f = f_I + f^\perp \in I \oplus I^\perp = L^2$ we then have $f^\perp \circ \varphi^t \xrightarrow[t\to\infty]{\text{weakly}} 0$, so the accumulation points of $f \circ \varphi^t$ are accumulation points of $f_I \circ \varphi^t \in I$, hence W^s-saturated and W^u-saturated. □

Since mixing is characterized in terms of weak convergence (Proposition 3.4.28), we obtain the next result.

Corollary 7.1.10 (Hopf argument II). *If (X, μ) is a metric Borel probability space and Φ is μ-preserving such that any W^s- and W^u-saturated $f \in L^2(\mu)$ is constant, then Φ is mixing.*

Remark 7.1.11. The condition that any W^s- and W^u-saturated $f \in L^2(\mu)$ is constant is referred to as *joint ergodicity* of W^s and W^u, and Corollary 7.1.10 says that this implies mixing. We say that W^s is ergodic if any W^s-saturated $f \in L^2(\mu)$ is constant, and Corollary 7.1.8 shows that this implies multiple mixing. We now apply this to geodesic flows, with Corollary 3.3.20 as the main ingredient—by Theorem 3.4.44, W^s is ergodic, and Corollary 7.1.8 implies the following result:

Theorem 7.1.12. *The geodesic flow on a compact surface of constant negative curvature is multiply mixing.*

Remark 7.1.13. This strengthens Corollary 3.3.17 and Theorems 3.4.43 and 3.4.32. Stronger mixing follows from Theorem 7.4.20 in greater generality than here. This is analogous to Theorems 6.2.11 and 8.1.3 in the context of topological dynamics. Since the Liouville measure is positive on open sets, this in particular strengthens Theorem 2.4.4: this geodesic flow is topologically mixing (see also Exercise 2.7 and Corollary 8.1.6). We note more generally (than Theorem 3.4.32) that any volume-preserving Anosov flow (Definition 5.1.1) that is not a suspension is mixing of all orders (Theorem 7.3.22).

The use of the Fubini Theorem in the proof of Proposition 7.1.4 was pivotal, and this requires looking for ways to check the assumed implication in Corollary 7.1.3, that is, to more generally see that "constant on stable and unstable $\Rightarrow$ constant" from hyperbolicity without having the Fubini Theorem available. The reason is that the Fubini Theorem often does fail in that the foliations are not sufficiently smooth to apply the theorem. We will see this in Example 7.1.23. In particular, as we briefly mentioned in the introduction, Hopf ran into problems proving ergodicity for certain geodesic flows since he could not establish the foliations were C^1. It took over 30 years for Anosov to find a mechanism to establish ergodicity. In doing so he observed that even if the foliations are not C^1, there is a property they satisfy that is sufficient to establish ergodicity. The property that replaces the use of the Fubini Theorem is absolute continuity of the invariant foliations with respect to Riemannian volume: that a set of full measure intersects almost every stable leaf in a set of full leaf-measure. In order to formally define absolute continuity we first introduce some definitions from differential geometry.

Definition 7.1.14. A C^r d-dimensional *foliation* of an n-dimensional manifold M consists of a covering by open sets U_i and maps $h_i \colon U_i \longrightarrow \mathbb{R}^n$ such that the transition functions $h_{ij}\colon \mathbb{R}^n \to \mathbb{R}^n$ defined by $h_{ij} = h_j h_i^{-1}$ are of the form

$$h_{ij}(x, y) = (h^1_{ij}(x), h^2_{ij}(x, y)),$$

where x is the first $n - d$ coordinates and y denotes the last d coordinates. So

$$h^1_{ij}\colon \mathbb{R}^{n-d} \to \mathbb{R}^{n-d},$$
$$h^2_{ij}\colon \mathbb{R}^{n} \to \mathbb{R}^{d},$$

where the h^1_{ij} are C^r and the h^2_{ij} are continuous. In the chart U_i the strips $x =$ const. align with the corresponding strips in U_j. The strips connect together from chart to chart to form a leaf of the foliation which is a maximal connected immersed submanifold.

Definition 7.1.15. A *foliation box* of a foliation $\mathcal{F}$ with d-dimensional leaves is the domain of a foliation chart, that is, it is the image O of a homeomorphism $\mathbb{R}^{n-d} \times \mathbb{R}^d \to M$ that sends each $\{x\} \times \mathbb{R}^d$ into a leaf of $\mathcal{F}$; the image that contains $z \in O$ is denoted by $\mathcal{F}_{\mathrm{loc}}(z)$. For $\mathcal{F} = W^s$ or $\mathcal{F} = W^u$ we will always choose foliation boxes small enough that these images are *local leaves* of $\mathcal{F}$.

Remark 7.1.16. This definition of a foliation differs from the one in geometry in that the h^2_{ij} are only taken to be continuous and not smooth. Additionally, the holonomy maps are defined differently from what is usually done in geometry, where parallel translation along loops is involved [161]. Therefore, one must take care when trying to apply results about foliations and holonomy maps in geometry directly to flows.

The first notion of absolute continuity is of a transverse nature.

Definition 7.1.17. The holonomies coming from a foliation $\mathcal{F}$ are locally defined by maps h_{τ_1,τ_2} between two nearby smooth transversals τ_1 and τ_2 that send $x \in \tau_1$ to the unique intersection point of τ_2 with the local leaf of $\mathcal{F}$ through x.

A foliation $\mathcal{F}$ with smooth leaves is *transversely absolutely continuous with bounded Jacobians* if for any $\alpha \in (0, \pi/2]$ there are $C, R > 0$ such that for any foliation box O of diameter at most R and any smooth transversals τ_1 and τ_2 to $\mathcal{F}$ that lie in O and make an angle at least α with $\mathcal{F}$ we have

$$m_{\tau_1}(A)/C \le m_{\tau_2}(h_{\tau_1,\tau_2}(A)) \le C m_{\tau_1}(A)$$

for every m_{τ_1}-measurable set $A \subset \tau_1$. Here m_N denotes the volume on a smooth submanifold $N \subset M$ induced by the Riemannian metric. Equivalently, there is a bounded positive measurable function J_h such that for every m_{τ_1}-measurable $A \subset \tau_1$ we have

$$m_{\tau_2}(h_{\tau_1,\tau_2}(A)) = \int_A J_h(x)\, dm_{\tau_1}(x).$$

We now define absolute continuity (without "transversely").

Definition 7.1.18. We say that a foliation $\mathcal{F}$ with smooth leaves is *absolutely continuous with bounded Jacobians* if for any $\alpha \in (0, \pi/2]$ there are $C, R > 0$ such that for any foliation box O of diameter at most R and any smooth transversal τ to $\mathcal{F}$ that lies in O and makes an angle at least α with $\mathcal{F}$ we have

$$\frac{m(A)}{C} \le \int_\tau m_{\mathcal{F}_{\mathrm{loc}}(z)}(A \cap \mathcal{F}_{\mathrm{loc}}(z))\, dm_\tau(z) \le Cm(A)$$

for every measurable $A \subset O$. More precisely, there is a measurable family of functions (called *conditional densities*) $\delta_z : \mathcal{F}_{\mathrm{loc}}(z) \to [1/C, C]$ such that

$$m(A) = \int_\tau \int_{\mathcal{F}_{\mathrm{loc}}(z)} \chi_A(z, y)\delta_z(y)\, dm_{\mathcal{F}_{\mathrm{loc}}(z)}(y)\, dm_\tau(z)$$

for every measurable $A \subset O$.

Remark 7.1.19. Theorem 7.1.25 below establishes absolute continuity for $C^{1+\alpha}$ hyperbolic flows, and by Corollary 7.1.3 this implies that Anosov flows are ergodic (Theorem 7.1.26). (This is further strengthened in Theorems 7.1.12, 7.1.28, Remark 7.1.29, and Theorem 7.4.20.)

Equilibrium states of hyperbolic flows are "product-like" in a way that directly implies this property (Theorem 7.3.28); this includes the Bowen–Margulis measure (page 450), Lebesgue measure, and Sinai–Ruelle–Bowen measures (Remark 7.1.21).

Proposition 7.1.20. *If $\mathcal{F}$ is transversely absolutely continuous with bounded Jacobians then $\mathcal{F}$ is absolutely continuous with bounded Jacobians.*

Proof. Take τ and O as in Definition 7.1.18 and include τ in a C^1-foliation $\mathcal{G}$ of O, that is, fixing $z \in \tau$ we have $\mathcal{G}_{\mathrm{loc}}(z) = \tau$ and $O = \bigcup_{y \in \mathcal{F}_{\mathrm{loc}}(z)} \mathcal{G}_{\mathrm{loc}}(y)$. Since $\mathcal{G}$ is a C^1-foliation it is absolutely continuous and transversely absolutely continuous (both with bounded Jacobians), and indeed there are continuous conditional densities $\delta_y(\cdot)$ for which

$$m(A) = \int_{\mathcal{F}_{\mathrm{loc}}(z)} \int_{\mathcal{G}_{\mathrm{loc}}(y)} \chi_A(y, x)\delta_y(x)\, dm_{\mathcal{G}_{\mathrm{loc}}(y)}(x)\, dm_{\mathcal{F}_{\mathrm{loc}}(z)}(y). \tag{7.1.1}$$

Since we are using y as the parameter for leaves of $\mathcal{G}$, we now denote by h_y the holonomy from τ to $\mathcal{G}_{\mathrm{loc}}(y)$ and by J_y its Jacobian, that is to say, the inner integral above can be written as

$$\int_{\mathcal{G}_{\mathrm{loc}}(y)} \chi_A(y, x)\delta_y(x)\, dm_{\mathcal{G}_{\mathrm{loc}}(y)}(x) = \int_\tau \chi_A(h_y(s))J_y(s)\delta_y(h_y(s))\, dm_\tau(s).$$

This is where we use the transverse absolute continuity hypothesis.

Substituting the right-hand side into (7.1.1) gives

$$m(A) = \int_{\mathcal{F}_{\mathrm{loc}}(z)} \int_\tau \chi_A(h_y(s)) J_y(s) \delta_y(h_y(s))\, dm_\tau(s)\, dm_{\mathcal{F}_{\mathrm{loc}}(z)}(y).$$

This is an integral with respect to a product measure, so we can exchange the order of integration to get

$$m(A) = \int_\tau \int_{\mathcal{F}_{\mathrm{loc}}(z)} \chi_A(h_y(s)) J_y(s) \delta_y(h_y(s))\, dm_{\mathcal{F}_{\mathrm{loc}}(z)}(y)\, dm_\tau(s). \tag{7.1.2}$$

Using transverse absolute continuity of $\mathcal{G}$ we now rewrite the new inner integral. To that end denote by $\bar{h}_s$ the holonomy along leaves of $\mathcal{G}$ from $\mathcal{F}_{\mathrm{loc}}(z)$ to $\mathcal{F}_{\mathrm{loc}}(s)$ for $s \in \tau$, and let $\bar{J}_s$ be its Jacobian. Note that $h_y(s) = \bar{h}_s(y)$, so if we write $r := h_y(s)$ then $y = \bar{h}_s^{-1}(r)$ and we can make the corresponding change of variables:

$$\int_{\mathcal{F}_{\mathrm{loc}}(z)} \chi_A(h_y(s)) J_y(s) \delta_y(h_y(s))\, dm_{\mathcal{F}_{\mathrm{loc}}(z)}(y) = \int_{\mathcal{F}_{\mathrm{loc}}(s)} \chi_A(r) J_y(s) \delta_y(r) \bar{J}_s^{-1}(r)\, dm_{\mathcal{F}_{\mathrm{loc}}(s)}(r).$$

Inserting this into (7.1.2) gives

$$m(A) = \int_\tau \int_{\mathcal{F}_{\mathrm{loc}}(s)} \chi_A(r) J_y(s) \delta_y(r) \bar{J}_s^{-1}(r)\, dm_{\mathcal{F}_{\mathrm{loc}}(s)}(r)\, dm_\tau(s).$$

This proves absolute continuity with density functions $J_y(s)\delta_y(\cdot)\bar{J}_s^{-1}(\cdot)$. □

Remark 7.1.21. By fixing $r = r_0$ in this last integral, we see that locally, m (and likewise, the SRB-measure in Definition 7.4.1 below) is equivalent, in the sense of mutual absolute continuity (with bounded Jacobians), to a product measure.

Example 7.1.22. A simple example may serve to illuminate some distinctions here. Consider the square $[0,1]^2$ with the foliation $\mathcal{F}$ defined as follows: each leaf is a line segment with endpoints $(x,0)$ and $(h(x),1)$, where $h\colon [0,1] \to [0,1]$ is the homeomorphism that sends the ternary Cantor set C to a "fat" Cantor set with positive Lebesgue measure and endpoints 0 and 1, and that is linear on the gaps. This foliation is not transversely absolutely continuous: take $\tau_1 = [0,1] \times \{0\}$, $\tau_2 = [0,1] \times \{1\}$ and $A = C$ in Definition 7.1.17. Note, though, that it only "barely" fails to be so because if τ_1 and τ_2 are both disjoint from $[0,1] \times \{0\}$, then the conclusion of Definition 7.1.17 holds.

Taking a closer look at Definition 7.1.18 we note that when A is taken to be the $\mathcal{F}$-saturation of C, it has positive Lebesgue measure while the integral in the first version of this definition is 0 if we take the "pathological" transversal $\tau = [0,1] \times \{0\}$. However, the second (and "official") version applies here; indeed $dm_{\mathcal{F}_{\mathrm{loc}}}$ could be represented by the natural measure on a generic smooth transversal. Accordingly, this is an example of an absolutely continuous foliation that is not transversely absolutely continuous—albeit only just.

Absolute continuity is not a consequence of the definition of a foliation:

Example 7.1.23 (Katok's foliation foiling Fubini [259]). If $p \in (0,1)$, then

$$f_p \colon [0,1] \to [0,1], \quad x \mapsto \begin{cases} \dfrac{x}{p} & \text{for } x \in [0,p), \\ \dfrac{x-p}{1-p} & \text{for } x \in [p,1), \end{cases}$$

(Figure 7.1.1) preserves Lebesgue measure on $[0,1]$ and is ergodic, so $(\chi_{[0,p)})_{f_p}(x) = p$

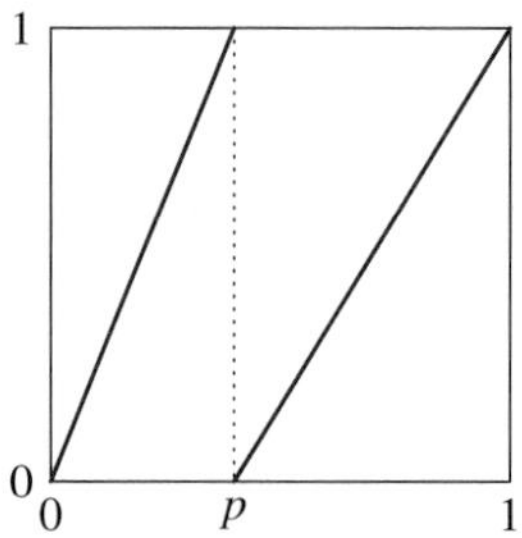

Figure 7.1.1. f_p.

for a.e. x (time average) by Corollary 3.3.11, that is, almost every x lands in $[0,p)$ about p of the time. Therefore the set

$$A := \{(p,x) \mid (\chi_{[0,p)})_{f_p}(x) = p\} \subset S := \{(p,x) \mid p \in (0,1),\ x \in [0,1]\}$$

has Lebesgue measure 1. On the other hand, there is a homeomorphism $h \colon S \to S$ such that $h_p := h(p,\cdot)$ is increasing, and $f_p \circ h_p = h_p \circ f_{1/2}$. (In particular, $h_p(1/2) = p$.) Then

$$f_p^n(h_p(x)) < p \Leftrightarrow h_p(f_{1/2}^n(x)) < p \Leftrightarrow f_{1/2}^n(x) < 1/2,$$

so $(\chi_{[0,p)})_{f_p}(h_p(x)) = (\chi_{[0,1/2)})_{f_{1/2}}(x)$ is independent of p. Thus, $A_z := \operatorname{graph}(h(\cdot,z))$ intersects A in at most one point, while the A_z are leaves of a continuous foliation of S by continuous curves.

Remark 7.1.24. Katok's original example was created in 1992, possibly on the spot, and had a 2-dimensional Anosov base. It was distilled down to a 1-dimensional expanding base by Burns and Flaminio. From Burns the idea traveled via Wilkinson,

Shub, and Pugh to Milnor, who published it with this explicit choice of expanding base and noted that Yorke independently had an example as well. *In Section* B.7.a *we* produce the original Katok example as passed on to us by Keith Burns [180, p. 31]. It should be noted that the leaves of the foliation in Example 7.1.23 are in fact analytic [259].

Theorems B.7.5 and B.7.6 apply to the time-1 map of a flow restricted to a hyperbolic set to give the following theorem:

Theorem 7.1.25 (Absolute continuity). *The strong and weak, stable and unstable foliations of a $C^{1+\alpha}$ hyperbolic flow are (transversely) absolutely continuous with bounded Jacobians. (And the orbit foliation is so because it is differentiable.)*

Corollary 7.1.3 then implies the following theorem:

Theorem 7.1.26. *Volume-preserving $C^{1+\alpha}$ Anosov flows are ergodic.*

A pertinent dichotomy will later prove useful:

Theorem 7.1.27. *For a continuous n-form θ invariant under a C^2 Anosov flow Φ on a compact n-manifold M there are two possibilities:*

- $\theta \equiv 0$ *or*
- *the measure determined by θ on M is equivalent to the Riemannian volume and ergodic for Φ.*

Proof [292, Corollary 4.6]. Unless $\theta \equiv 0$, there is an $x \in M$ at which $\theta_x \neq 0$, and then $x \in \mathrm{NW}(\Phi)$ by the Poincaré Recurrence Theorem (Theorem 3.2.1) so $\mathrm{NW}(\Phi)$ has nonempty interior, and hence $\mathrm{NW}(\Phi) = M$ (Theorem 5.3.53). Thus Φ is topologically transitive (Theorem 5.3.52), and $\theta_x \neq 0$ on an open dense set, so it defines a volume equivalent to the Riemannian volume; being invariant and finite, it is ergodic (Theorem 7.1.26). □

To get stronger mixing properties from, say, Corollary 7.1.10 one needs to exclude suspensions in a suitable way. Theorem 7.1.12 is an instance of this, and without proof, we give here a general result based on this insight.

Theorem 7.1.28. *Weakly mixing volume-preserving $C^{1+\alpha}$ Anosov flows are mixing.*

Remark 7.1.29. Multiple mixing actually holds in this case: the *Anosov Alternative* [10], a measurable counterpart to the Plante Alternative (Theorem 8.1.3) for volume-preserving Anosov flows, is that the strong (un)stable foliation either is of suspension type, that is, the Anosov flow is not weakly mixing, or is ergodic (Remark 7.1.11), in which case the flow is multiply mixing (Corollary 7.1.8). Note that contact Anosov flows are always of this latter kind.

It is reasonable to hope that for hyperbolic billiards (Theorem 5.2.26) one obtains ergodic properties in the same fashion as for geodesic flows, and this is not far off. However, that the regular set is neither compact nor a manifold makes for extra technical work. The preceding ergodicity arguments and those to come rely on the expansion of unstable manifolds under the flow, and while in the case of billiards, that may be differentiably true, the manifolds themselves are truncated by encounters with singularities. Balancing this against the local expansion is delicate indeed, and it takes a book to build a suitable machinery [220]. The core insight is easy to state: that the measure of a neighborhood of the singularity set must be bounded by a power of its thickness. Theorem 7.4.20 below does work in that context to give strong mixing properties of finite-horizon dispersing billiards [335, 270, 22]. The questions for the billiard flow versus the associated billiard map are, of course, closely related. Some results for billiard maps presented in the spirit of this section can be found in [174].

Remark 7.1.30. It is far harder to see that a gas of hard particles (Section 0.3.c) is ergodic. While this is a billiard system (Theorem 5.2.41), considering nonspherical particles is a nonstarter because the resulting billiards are not hyperbolic [101]. For spherical particles this ergodicity question is known as the Boltzmann–Sinai hypothesis, and its solution was announced by Sinai in 1963 and eventually proved by generations of others 50 years later [332]. The daunting challenges involved are far beyond the scope of this book. The difficulty that is easiest to spot is that while two particles collide, all others can move freely to some extent, so the scatterers in the configuration space are cylindrical and hence only semidispersing.

7.2 Stable ergodicity*

Theorem 7.1.26 invites a digression to a rather modern subject in dynamical systems that combines topological and ergodic concerns while also greatly pushing the Hopf argument. Questions of stability and persistence are usually raised in topological and smooth dynamics, such as the persistence of the Anosov property (Corollary 5.1.9). Ergodic theory tends to study a system at hand, or relations between given systems. Stable ergodicity concerns the robustness of an ergodic property under perturbations of a smooth system. It is nicely illustrated by the following result.

Proposition 7.2.1. *Volume-preserving $C^{1+\alpha}$ Anosov flows are stably ergodic: a volume-preserving $C^{1+\alpha}$ Anosov flow has a C^1-neighborhood among volume-preserving $C^{1+\alpha}$ Anosov flows that consists of flows for which volume is ergodic.*

Proof. Volume-preserving $C^{1+\alpha}$ Anosov flows are ergodic (Theorem 7.1.26), and so are $C^{1+\alpha}$ volume-preserving C^1-perturbations because they are themselves volume-preserving $C^{1+\alpha}$ Anosov flows. □

This simple argument hides a few subtle aspects of this issue. First of all, structural stability is more powerful than persistence of hyperbolicity, but it does not produce stable ergodicity by itself because on one hand it gives an orbit equivalence rather than a conjugacy, and on the other hand the orbit equivalence is a homeomorphism and may not be absolutely continuous (Remark 9.2.8). Both of these mean that the conjugacy cannot be expected to send volume to volume, so it produces ergodicity of an unspecified invariant measure for the perturbation. The underlying issue is the abundance of invariant measures and our focus on a specific one.

The question about stable ergodicity we speak to is whether hyperbolicity can produce stable ergodicity when it is less complete than in the Anosov situation. Weakening hyperbolicity from Anosov to our definition of a hyperbolic flow does not broaden the perspective: volume preservation means that the chain-recurrent set is the whole manifold, so our definition of hyperbolicity of a flow (Definition 5.3.50) implies that the flow is a transitive Anosov flow.

Instead, dynamicists turned to weakening hyperbolicity to partial hyperbolicity (Definition 5.6.2), where the center direction in the definition of a hyperbolic set (Definition 5.1.1) is allowed to have dimension higher than 1, that is, it is a subbundle of vectors that experience less expansion and contraction than stable and unstable vectors. Partial hyperbolicity and stable ergodicity are active research areas, but beyond our intended scope. This section presents highlights of this research area, and we recommend, for instance, [288, 352, 83] for further reading.

Remark 7.2.2. In light of the regularity hypothesis required even in Proposition 7.2.1, we make the standing assumption in the remainder of this section that all flows under consideration are at least C^2.

We formalized partial hyperbolicity for discrete time in Definition 5.6.2.

Remark 7.2.3. Each of the subbundles E^u, E^s, E^c, E^{cu}, E^{cs} in Definition 5.6.2 is Hölder continuous (Theorem 5.1.16). The center subbundle turns out to be more fragile than the others, however. The Hirsch–Pugh–Shub theory of normal hyperbolicity [194] helps control both the (moderate) regularity of its leaves and provide some robustness under perturbation—once a center foliation is known to exist! Existence is a rather delicate matter and is known only under several rather stringent assumptions, while nonexistence is an open property. The Katok example in Section B.7.a shows that if there is a center foliation at all it may fail to be absolutely continuous. It turns out that this is not at all exceptional. For instance, in a volume-preserving perturbation of the time-1 map of the geodesic flow of a compact surface with negative curvature, the Liouville measure either has atomic disintegration along the center foliation, or the perturbation is itself the time-1 map of a smooth volume-preserving flow (and hence the disintegration is Lebesgue) [321, 195, 20]. It is conjectured that the center foliation of a "typical" partially hyperbolic diffeomorphism with negative central exponents is not absolutely continuous[2] and that, moreover, arbitrarily close to a partially hyperbolic dynamical system whose central Lyapunov exponents are 0, there are partially hyperbolic dynamical systems with negative central exponents.

Example 7.2.4. The time-1 map of an Anosov flow is partially hyperbolic. Note that this is obviously not ergodic when the flow is a suspension.

This weakening thus poses challenges for the Hopf argument because it makes it much easier for the now lower-dimensional stable and unstable leaves to fail to intersect. This problem does not go away by restricting to flows, because the product of a volume-preserving Anosov flow with the identity is a partially hyperbolic flow for which volume is a nonergodic invariant measure. The presence of such intersections (or some irreducibility condition that implies it) thus becomes a necessary additional assumption. But while partial hyperbolicity persists under C^1-perturbation (by a cone criterion or by showing that the Mather spectrum—the spectrum of $\mathcal{F}$ in Theorem 5.6.5—various continuously), it is not obvious that the intersections of the stable and unstable leaves will do so.

Let us explore the complementary situation first.

Definition 7.2.5. The foliations W^u and W^s (or the subbundles E^u and E^s) are said to be *jointly integrable* if every point of M is contained in a local product neighborhood for which

$$\mathcal{H}_{y,z}(W^s(u) \cap W^{cs}_{\delta'}(y)) \subset W^s(\mathcal{H}_{y,z}(u)) \cap W^{cs}_{\delta}(z),$$

where $\mathcal{H}_{y,z}\colon W^{cs}_{\delta}(y) \to W^{cs}(z)$ is the *holonomy* along strong-unstable leaves defined by

$$\mathcal{H}_{y,z}(a) \in W^u_{\text{loc}}(a) \cap W^{cs}(z)$$

for $z \in W^u_{\text{loc}}(y)$ and $a \in W^{cs}_{\delta}(y)$.

A characterization of joint integrability versus not joint integrability can be expressed in terms of su-quadrilaterals. Fix a point $p \in M$. First move to a point $q_0 \in W^s(p)$, next a point $q_1 \in W^u(q_0)$, then a point $q_2 \in W^s(q_1)$ such that

[2] Mañé proved that if the center foliation is 1-dimensional and has compact leaves then this foliation is not absolutely continuous provided the Lyapunov exponent in the center direction is nonzero on a set of positive measure. (The only record of this appears to be a fax with statement and proof he sent to Michael Shub on 13 September 1993, of which we received a scan from Amie Wilkinson; it is notable that 1993 is in the very early era of this subject.)

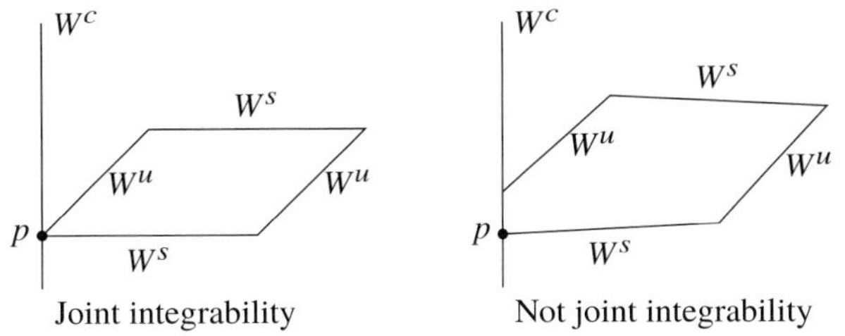

Figure 7.2.1. Joint integrability versus not joint integrability.

$O(p) \cap W^u(q_2) \neq \varnothing$. If we finish the quadrilateral and move back to the original point p then the stable and unstable foliations may be jointly integrable. If we finish the quadrilateral and move back to another point on the orbit of p (in the center manifold $W^c(p)$), then the stable and unstable foliations are not jointly integrable.

Example 7.2.6 (Frame flows). Example 2.2.7 pointed out that parallel-translating a vector along a geodesic can be construed to define a flow, and combined with just tangent and normal vectors this defines a partially hyperbolic flow as follows. Consider the bundle of oriented orthonormal 2-frames on a surface Σ. This produces a fiber bundle $\pi\colon N \to M := S\Sigma$, where the natural projection π takes a frame into its first vector. Over n-dimensional manifolds we get a fiber bundle $\pi\colon N \to M$ where the associated structure group $\mathrm{SO}(n-1)$ acts on fibers by rotating the frames, keeping the first vector fixed. Therefore, we can identify each fiber N_x with $\mathrm{SO}(n-1)$. The *frame flow* Φ acts on frames by moving their first vectors by the geodesic flow and moving the other vectors by parallel translation along the geodesic defined by the first vector. Thus, $\pi \circ \varphi^t = g^t \circ \pi$ for each t. The frame flow preserves the measure that is locally the product of the Liouville measure with normalized Haar measure on $\mathrm{SO}(n-1)$. The frame flow is a partially hyperbolic flow. The center bundle has dimension $1 + \dim \mathrm{SO}(n-1)$ and is spanned by the flow direction and the fiber direction.

With this, the question of stable ergodicity of volume-preserving partially hyperbolic flows becomes one of finding conditions under which the stable and unstable leaves intersect. The notion introduced by Brin and Pesin is called accessibility and stable accessibility. One needs to home in on a notion of accessibility that is strong enough to make the Hopf argument work and robust enough to (often) be stable under C^1-perturbations—while not being altogether so restrictive as to be unsatisfying. Making the Hopf argument work required significant breakthroughs in geometric measure theory, while stability of accessibility involves subtle hyperbolic dynamics. Before pursuing this further, we note another need when considering partially hyperbolic sets rather than the "partial" counterpart of an Anosov flow: for this line of reasoning we need partial hyperbolicity to persist under perturbation. There are two other general situations in which this is the case. One is that of partially hyperbolic attractors because attractors are stable under perturbation. The other is that of a partially hyperbolic set that is a normally hyperbolic manifold [288, Remark 2.20].[3]

Let us briefly examine the issues that have to be addressed in this approach. First, to use the Hopf argument, one needs absolute continuity of the stable and unstable foliations in order to use the Fubini argument analogously to the way we have already done. Fortunately, our Theorem B.7.5 is for the partially hyperbolic case.[4] Next, accessibility needs to be defined in order to support the Hopf argument, which in this case needs to look like our arguments for weak mixing, because invariance of a function does not make it constant on the now larger center foliation (additional reasons to bypass the center are that there may be no center foliation, without expansion or contraction Birkhoff averages need not be constant on it, and it is often not absolutely continuous). Accordingly, while in hyperbolic flows any two points are connected by a heteroclinic one (the strong-unstable manifold of one of them intersects the weak-stable of the other or vice versa), it is sufficient if any two points are connected by an "su-path," that is, a path concatenated from segments, each of which lies in a stable or an unstable leaf. Put differently, call these two points equivalent when this happens, and assume that there is only one equivalence class. As in the hyperbolic case, the formal definition does not refer to actual paths.

Definition 7.2.7. Let Φ be a partially hyperbolic flow on a compact Riemannian manifold M.

Two points $p, q \in M$ are said to be *accessible*, if there are points $z_i \in M$ with $z_0 = p$, $z_\ell = q$, such that $z_i \in W^\sigma_{\text{loc}}(z_{i-1})$ for $i = 1, \dots, \ell$ and $\sigma = s$ or u. The collection of points $z_0, z_1, \dots, z_\ell$ is called the *us-path* connecting p and q and is denoted variously by $[p,q]_\Phi = [p,q] = [z_0, z_1, \dots, z_\ell]$.[5]

Accessibility is an equivalence relation and the collection of points accessible from a given point p is called the *accessibility class* of p.

A flow Φ is said to have the *accessibility property* if any two points are accessible (so the sole accessibility class is the whole manifold).

We say that Φ has the *essential accessibility property* if the partition of M by the accessibility classes is trivial in the measure-theoretical sense, that is, any measurable set that consists of accessibility classes has measure 0 or 1.

[3] We rely here on a prior allusion to cone arguments, but the actual argument is semicontinuity of the Mather spectrum [288, Theorem 3.4].

[4] It is stated for a diffeomorphism and, applied to the time-1 map, gives the same conclusion.

[5] There is an actual path from p to q that consists of pieces of smooth curves on local stable or unstable manifolds with the z_i as endpoints.

If Φ has the accessibility property then the subbundle $E^s \oplus E^u$ is not integrable (so the stable and unstable foliations are not jointly integrable). Otherwise, the accessibility class of any $p \in M$ would be the leaf of the corresponding foliation passing through p.

We digress to mention that essential accessibility is also interesting with respect to topological dynamics:

Theorem 7.2.8 ([77]). *An essentially accessible volume-preserving partially hyperbolic diffeomorphism is topologically transitive.*

Is (essential) accessibility stable? Does it hold often?

Definition 7.2.9. A partially hyperbolic flow Φ is said to be *stably accessible* if there is a C^1-neighborhood $\mathcal{U}$ of Φ (possibly in the space of flows preserving a Φ-invariant Borel probability measure ν) such that any flow $\Psi \in \mathcal{U}$ has the accessibility property.

Corollary 6.2.22 (with Theorem 6.2.21) is encouraging; it even shows genericity of having dense strong leaves in our context. Our line of reasoning does not apply since it uses incommensurability of periodic orbits, which is explicitly limited to helping in the orbit direction, but some of what we did elsewhere is suggestive. The *quadrilateral formula* (2.2.4) we used in an algebraic context suggests geometric arguments outside of the algebraic context: it says that a quadrilateral with $h_\pm$-sides about ϵ causes a $2\epsilon^2$ displacement along a geodesic, and for an Anosov flow one would wish to show that small su-quadrilaterals produce some displacement in the flow direction.

In partially hyperbolic dynamics, this *Brin quadrilateral argument* establishes robustness of accessibility as follows: if a small such quadrilateral from x ends on $W^c(x) \setminus \{x\}$, then homotoping it (by shrinking each edge) to length 0 gives a continuous curve in $W^c(x)$ of endpoints that ends at x, and this circumstance should be robust.

Let us repeat this more carefully, assuming for simplicity that the central subbundle E^c is integrable. At $p \in M$ consider a 4-legged us-path $[p = z_0, z_1, z_2, z_3, z_4]$ and connect z_{i-1} with z_i by a geodesic γ_i in the corresponding stable or unstable manifold to obtain the curve $\Gamma_p = \bigcup_{1 \le i \le 4} \gamma_i$. We parametrize it by $t \in [0, 1]$ with $\Gamma_p(0) = p$.

If the subbundle $E^s \oplus E^u$ were integrable (and hence, f not accessible), the endpoint $z_4 = \Gamma_p(1)$ would lie on the leaf through p of the corresponding foliation. Therefore, one can hope to achieve accessibility if one can arrange a 4-legged us-path in such a way that $\Gamma_p(1) \in W^c(p)$ and $\Gamma_p(1) \neq p$. In this case the path Γ_p can be homotoped through 4-legged us-paths originating at p to the trivial path so that the endpoints stay in $W^c(p)$ during the homotopy and form a continuous curve. Such a situation is usually persistent under small perturbations of Φ and hence leads to stable accessibility.

What Corollary 6.2.22 (with Theorem 6.2.21) suggests, is actually true: genericity of accessibility.

Theorem 7.2.10 ([120][6]). *Let $q \geq 1$ and Φ be a partially hyperbolic C^q-flow (possibly preserving a smooth measure ν on M). Then in every C^1-neighborhood $\mathcal{U}$ (possibly of μ-preserving flows) of Φ there is a* stably *accessible C^q-flow Ψ.*

Before getting to the main ergodicity theorem, let us remark on time-t maps. For an Anosov flow Φ on a compact smooth Riemannian manifold M stable accessibility of the time-1 map depends on whether the subbundle $E^s \oplus E^u$ is integrable, that is, whether the stable and unstable foliations of the time-1 map are jointly integrable.

To make this notion precise, fix $\varepsilon > 0$ and for a point $x \in M$ consider a local smooth submanifold

$$\Pi(x) = \bigcup_{y \in B^u(x,\varepsilon)} \bigcup_{-\varepsilon \leq \tau \leq \varepsilon} \varphi^\tau(y)$$

through x. For $x, x' \in M$ let $\pi_{x,x'} \colon \Pi(x) \to \Pi(x')$ be the holonomy map generated by the family of local stable manifolds. The foliations W^s and W^u are *jointly integrable* if for every $y \in \Pi(x)$ the image of the local unstable leaf $W^{uu}_{\text{loc}}(y)$ under $\pi_{x,x'}$ is the local unstable leaf $W^{uu}_{\text{loc}}(\pi_{x,x'}(y))$.

Theorem 7.2.11 ([80]). *If the stable and unstable foliations of an Anosov flow are not jointly integrable, then the time-1 map is stably accessible.*

This suggests that being a "suspension" is the sole obstruction to stable ergodicity in this context.

Corollary 7.2.12. *The time-1 map is stably accessible for*

(1) *geodesic flows of negatively curved manifolds (indeed, contact Anosov flows, Theorem* 8.1.4) [217];

(2) *C^2 volume-preserving Anosov flows on compact 3-manifolds that are not suspensions* [80].

By now we see that the expected ingredients for stable ergodicity are in place for a substantial class of dynamical systems. It is time to state the ergodicity theorem (as yet without "stable").

Theorem 7.2.13 (Grayson–Pugh–Shub–Burns–Wilkinson Ergodicity Theorem). *A C^2 volume-preserving partially hyperbolic essentially accessible center-bunched diffeomorphism is ergodic.*

Remark 7.2.14. As we noted, one can replace "center-bunched" by "dim $E^c = 1$."

[6] An outline of the argument for 1-dimensional center direction can be found in [288, Theorem 8.5].

Grayson, Pugh, and Shub [157] proved ergodicity for small perturbations of the time-1 map of the geodesic flow on a surface of constant negative curvature. Wilkinson's thesis extended this to small perturbations of the time-1 map of the geodesic flow on a surface of possibly variable negative curvature. Pugh and Shub [304, 305, 306] then proved ergodicity of dynamically coherent diffeomorphisms under a stronger center-bunching condition. Burns and Wilkinson weakened the needed center-bunching condition (to Definition 7.2.15) [82] and then removed the need for dynamical coherence [83].

Definition 7.2.15 ([82]). We say that a partially hyperbolic diffeomorphism f is *center bunched* if we can take $\max\{\mu_1, \lambda_3^{-1}\} < \lambda_2/\mu_2$ in (see Definition 5.6.1)

$$\begin{aligned}
\lambda_1 &\le \llbracket d_x f \upharpoonright E^s(x) \rrbracket \le \|d_x f \upharpoonright E^s(x)\| \le \mu_1,\\
\lambda_2 &\le \llbracket d_x f \upharpoonright E^c(x) \rrbracket \le \|d_x f \upharpoonright E^c(x)\| \le \mu_2,\\
\lambda_3 &\le \llbracket d_x f \upharpoonright E^u(x) \rrbracket \le \|d_x f \upharpoonright E^u(x)\| \le \mu_3.
\end{aligned}$$

More precisely, we require merely "pointwise center bunching":

$$\max\{\mu_1(p), \lambda_3^{-1}(p)\} < \lambda_2(p)/\mu_2(p) \quad \text{for every } p,$$

where $\lambda_i(p)$ and $\mu_i(p)$ are *pointwise* bounds on rates of expansion and contraction. (This is automatic if $\dim E^c = 1$.)

This Burns–Wilkinson center bunching imposes a much weaker constraint than earlier versions.

Although it is no longer needed for this result, we provide the definition of dynamical coherence in its most economical form; it guarantees not only integrability of the central subbundle, but its unique integrability:

Definition 7.2.16 (Burns–Wilkinson coherence). A partially hyperbolic embedding is said to be *dynamically coherent* if E^{cs} and E^{cu} are integrable to foliations W^{cs} and W^{cu}, respectively.

Consider the hypotheses of Theorem 7.2.13 in light of our desire for stable ergodicity. Most of them are robust: partial hyperbolicity is a C^1-open condition, and so is center bunching by semicontinuity (though this is not trivial). We are restricting to volume-preserving perturbations, so stability of accessibility is the only additional assumption we need.

Theorem 7.2.17 (Grayson–Pugh–Shub–Burns–Wilkinson Stable Ergodicity Theorem). *A C^2-diffeomorphism is stably ergodic if it is volume preserving, partially hyperbolic, stably (essentially) accessible, and center bunched.*

The need to assume *stable* accessibility motivates two conjectures about accessibility.

Conjecture 7.2.18. *A partially hyperbolic dynamical system with the accessibility property is stably accessible.*

An example by Brin shows that this conjecture fails if one replaces accessibility by essential accessibility [79].

Conjecture 7.2.19. *The space of stably accessible partially hyperbolic dynamical systems is open and dense in the C^r-topology for any $r \geq 1$. (This is known for $r = 1$ [120].)*

Fortunately, we established stable accessibility in a few salient situations. For instance, Theorem 7.2.11 now becomes the following:

Theorem 7.2.20 ([80]). *If the stable and unstable foliations of a volume-preserving Anosov flow are not jointly integrable, then the time-1 map is stably ergodic.*

In the special case of flows on 3-manifolds one can strengthen this result:

Theorem 7.2.21. *The time-1 map of a volume-preserving Anosov 3-flow is stably ergodic if and only if the flow is not the suspension of an Anosov diffeomorphism.*

In particular, we have the following proposition:

Proposition 7.2.22. *The time-1 map of a volume-preserving topologically mixing C^2 Anosov 3-flow is stably ergodic.*

This theory applies to the aforementioned frame flows (Example 7.2.6) as well.

Theorem 7.2.23. *Let Φ be the frame flow on an n-dimensional compact smooth Riemannian manifold with sectional curvatures between $-\Lambda^2$ and $-\lambda^2$. Then in each of the following cases the flow is ergodic (indeed, Bernoulli by Theorem 7.4.20), and the time-1 map of the frame flow is stably ergodic (and stably K-mixing by Remark 7.3.20):*

(1) *if the curvature is constant* [66];

(2) *for a set of metrics of negative curvature which is open and dense in the C^3-topology* [66];

(3) *if n is odd and $n \neq 7$* [67];

(4) *if n is even, $n \neq 8$, and $\lambda/\Lambda > 0.93$* [68];

(5) *if $n = 7$ and $\lambda/\Lambda > 0.99023\dots$* [78];

(6) *if $n = 8$ and $\lambda/\Lambda > 0.99023\dots$* [78].

Ergodicity of the frame flow was proved by the respective cited authors; [78] pointed out K-mixing and the Bernoulli property and used [81, Corollary 1.2] (which relies on [69]) to deduce those of the time-1 maps across all cases.

We now give an indication of what is involved in proving Theorem 7.2.13. We begin with recasting the Hopf argument for the present situation. Let f be a partially hyperbolic C^2-diffeomorphism of a smooth compact Riemannian manifold M that preserves a smooth measure ν. To recast the Hopf argument say that $x, y \in M$ are stably equivalent if

$$\rho(f^n(x), f^n(y)) \to 0 \text{ as } n \to +\infty,$$

and unstably equivalent if

$$\rho(f^n(x), f^n(y)) \to 0 \text{ as } n \to -\infty.$$

Stable and unstable equivalence classes induce two partitions of M, and we denote by $\mathcal{S}$ and $\mathcal{U}$, the Borel σ-algebras they generate. Recall that for an algebra $\mathcal{A} \subset \mathcal{B} :=$ the Borel σ-algebra, its saturated algebra is the set

$$\mathrm{Sat}(\mathcal{A}) = \{B \in \mathcal{B} \colon \text{there exists } A \in \mathcal{A} \text{ with } \nu(A \triangle B) = 0\}.$$

The Hopf argument says that f is ergodic if

$$\mathrm{Sat}(\mathcal{S}) \cap \mathrm{Sat}(\mathcal{U}) = \mathcal{T} := \text{the trivial algebra.} \tag{7.2.1}$$

If f is an Anosov diffeomorphism, then the stable equivalence class containing a point x is its stable leaf $W^s(x)$. Similarly, the unstable equivalence class containing x is its unstable leaf $W^u(x)$. The σ-algebra $\mathcal{S}$ consists of those Borel sets S for which $W^s(x) \subset S$ whenever $x \in S$, and the σ-algebra $\mathcal{U}$ consists of those Borel sets U for which $W^u(x) \subset U$ whenever $x \in U$. The relation (7.2.1) holds by absolute continuity of stable and unstable foliations, which proves ergodicity of Anosov diffeomorphisms.

If f is partially hyperbolic, then the stable and unstable foliations W^s and W^u of M also generate Borel σ-algebras $\mathcal{M}^s$ and $\mathcal{M}^u$, respectively, so $\mathcal{S} \subset \mathcal{M}^s$ and $\mathcal{U} \subset \mathcal{M}^u$ (note that stable and unstable sets containing a point x may be larger than $W^s(x)$ and $W^u(x)$ due to possible contractions and expansions along the central directions). It follows that

$$\mathrm{Sat}(\mathcal{S}) \cap \mathrm{Sat}(\mathcal{U}) \subset \mathrm{Sat}(\mathcal{M}^s) \cap \mathrm{Sat}(\mathcal{M}^u).$$

If f is accessible then $\mathrm{Sat}(\mathcal{M}^s \cap \mathcal{M}^u) = \mathcal{T}$ (in fact, it suffices to assume essential accessibility (Definition 7.2.7)), and ergodicity then follows from

$$\mathrm{Sat}(\mathcal{M}^s) \cap \mathrm{Sat}(\mathcal{M}^u) \subset \mathrm{Sat}(\mathcal{M}^s \cap \mathcal{M}^u)$$

(the opposite inclusion is obvious)—and the assumptions of Theorem 7.2.13 turn out to guarantee this.

7.3 Specification, uniqueness of equilibrium states

We now return to the study of equilibrium states, which began in Section 4.3. First we show that for hyperbolic flows the class of Bowen-bounded functions, and indeed that of Walters-continuous functions (Definition 4.3.19), is substantial: it contains all Hölder-continuous functions.

Proposition 7.3.1. *Let M be a Riemannian manifold, Φ a smooth flow, and $\Lambda \subset M$ a locally maximal hyperbolic set for Φ. Then every Hölder-continuous function on Λ is Walters continuous, hence Bowen bounded* (*Definition* 4.3.19).

Proof. Suppose f is α-Hölder: $|f(x) - f(y)| \le cd(x,y)^\alpha$ and $\epsilon > 0$. With C', η as in Theorem 6.2.4 choose $\delta > 0$ such that $2\delta\|f\|_\infty + 2c(C'\delta)^\alpha \int_0^\infty \eta^{\alpha s}\,ds < \epsilon$. If $d_t^\Phi(x,y) < \delta$, then Proposition 6.2.2 gives

$$|S_t f(x) - S_t f(y)| \le \underbrace{|S_t(f - f\circ\varphi^\tau)(x)|}_{=|S_\tau f(x) - S_\tau f(\varphi^t(x))| \le 2\delta\|f\|_\infty} + \underbrace{|S_t f\circ\varphi^\tau(x) - S_t f(y)|}_{<c\int_0^t (C'(\eta^s+\eta^{t-s})\delta)^\alpha\,ds \le 2c(C'\delta)^\alpha \int_0^\infty \eta^{\alpha s}\,ds} < \epsilon.$$

□

By Theorem 6.4.3 the class of Hölder-continuous functions is invariant under orbit equivalence of hyperbolic sets and hence a natural class of functions to consider.

By expansivity of hyperbolic sets, Theorem 4.3.23 implies that if we restrict a continuous potential to a hyperbolic set, then there exists an equilibrium state. We now establish conditions to ensure uniqueness of equilibrium states for hyperbolic flows, with specification as a central tool, albeit in a stronger form than Definition 5.3.58 which we here obtain assuming topological mixing rather than mere transitivity as in Theorem 5.3.61. The essential reason for this improvement is Proposition 6.2.17, though we do not actually invoke density of strong leaves in the proof.

Definition 7.3.2 (Strong specification). A flow Φ on a compact metric space X satisfies *specification* if for any $\epsilon > 0$ there is a T_ϵ such that for $x_0, \dots, x_n \in X$ and $t_0, \dots, t_{n+1} \in [0,\infty)$[7] with $t_{i+1} \ge t_i + T_\epsilon$ there is a y with

$$d(\varphi^t(y), \varphi^{t-t_i}(x_i)) < \epsilon \quad \text{for } t \in [t_i, t_{i+1} - T_\epsilon], \tag{7.3.1}$$

and we can take $y \in \mathbb{P}_\epsilon(t_{n+1} - t_0)$ (Definition 4.2.22).

Remark 7.3.3. • A compact locally maximal hyperbolic set with this property is necessarily topologically mixing because it produces incommensurate periods (Theorem 6.2.11 and Proposition 6.2.18).

[7] Note that the t_i play a different role here than in Definition 5.3.58; they are the prescribed start times for each segment.

- Of course, a bare-handed argument verifies in full generality that the strong specification property implies topological mixing: for any two ϵ-balls pick x_0 in one, x_1 in the other, $t_0 = 0$, and $t_1 \geq T_\epsilon$ arbitrary to obtain the statement of Definition 1.6.31.
- By choosing $t_i \in \tau\mathbb{Z}$, this property directly implies the corresponding property for the time-τ map of Φ except that one might not be able to choose y periodic.

Theorem 7.3.4 (Specification Theorem). *If Λ is a topologically mixing compact locally maximal hyperbolic set for a flow Φ, then $\Phi_{\restriction_\Lambda}$ has the (strong) specification property (Definition 7.3.2).*

Remark 7.3.5. This result is interesting even if the shadowing orbit is not required to be closed. The reader will note that in this case the proof even works when the first or last specified segment is infinite. Also, the arguments here implicitly re-prove the conclusion of Proposition 6.2.17; while redundant, this line of reasoning provides more intuitive and direct control of the construction of the shadowing orbit.

Proof. For $\epsilon > 0$ and $L \geq 1$ as in Theorem 5.3.11 take $\delta := \epsilon/8L$ and $T_\epsilon > 0$ such that if $x, y \in \Lambda$ then for an adapted metric (Proposition 5.1.5) $\varphi^t(B(x,\delta)) \cap B(y,\delta) \neq \varnothing$ for all $t \geq T_\epsilon$ (Remark 1.6.32) and furthermore, if $y \in W^u_\epsilon(x)$ and $t \geq T_\epsilon$, then $d(\varphi^{-t}(x), \varphi^{-t}(y)) < \frac{1}{2}d(x,y)$. Let $y_0 := \varphi^{-t_0}(x_0)$.

The choice of T_ϵ gives a $z \in \varphi^{T_\epsilon}\big(B(\varphi^{t_1-T_\epsilon}(y_0),\delta)\big) \cap B(x_1,\delta)$, hence y_1 with

$$\varphi^{t_1-T_\epsilon}(y_1) \in W^u_\delta(\varphi^{t_1-T_\epsilon}(y_0)) \cap W^{cs}_\delta(\varphi^{-T_\epsilon}(z)).$$

The choice of T_ϵ gives a $z \in \varphi^{T_\epsilon}\big(B(\varphi^{t_2-T_\epsilon}(y_1),\delta)\big) \cap B(x_2,\delta)$, hence y_2 with

$$\varphi^{t_2-T_\epsilon}(y_2) \in W^u_\delta(\varphi^{t_2-T_\epsilon}(y_1)) \cap W^{cs}_\delta(\varphi^{-T_\epsilon}(z)).$$

Eventually this gives a $z \in \varphi^{T_\epsilon}\big(B(\varphi^{t_{n+1}-T_\epsilon}(y_n),\delta)\big) \cap B(x_0,\delta)$, hence y_{n+1} with

$$\varphi^{t_{n+1}-T_\epsilon}(y_{n+1}) \in W^u_\delta(\varphi^{t_{n+1}-T_\epsilon}(y_n)) \cap W^{cs}_\delta(\varphi^{-T_\epsilon}(z)).$$

Then $\varphi^{[t_1,t_{n+1}]}(y_{n+1})$ closes within $\epsilon/2L$ and shadows the specification within $\epsilon/4L$. The closed orbit from Theorem 5.3.11 is within $\epsilon/2$ of $\varphi^{[t_1,t_{n+1}]}(y_{n+1})$ and hence shadows the specification within ϵ, as required. □

Specification was introduced originally by Bowen to establish uniqueness of equilibrium states:

Theorem 7.3.6 (Bowen). *If Φ is an expansive flow with (strong) specification and with finite topological entropy on a compact metric space, $f \in V(\Phi)$, then the equilibrium state for Φ associated with f is unique and weakly mixing.*[8]

[8] In particular, Φ is not a suspension (Proposition 3.4.9).

Remark 7.3.7. We will study equilibrium states to a significant extent, and in light of Remark 7.3.3, this means that at those times we (sometimes implicitly) assume that the flow in question is topologically mixing.

Corollary 7.3.8. *If Φ is a smooth flow on a compact manifold, Λ a topologically mixing basic set, and $f \in V(\Phi_{\restriction_\Lambda})$,*[9] *then there is a unique equilibrium state for $\Phi_{\restriction_\Lambda}$ associated with f, and it is weakly mixing—indeed, K-mixing* (*Remark* 7.3.20).

Our proof strategy is as follows. We will show (Proposition 7.3.15) that for each Bowen-bounded potential function f there is a measure μ_f (7.3.3) with the Gibbs property (Definition 4.3.25), where $P = P(\Phi, f)$ (Definition 4.3.2), which is hence an equilibrium state (Theorem 4.3.26) and also ergodic (Proposition 7.3.17). We finally (on page 430) establish that μ_f is the unique equilibrium state for Φ with respect to f by showing that $\nu \perp \mu \Rightarrow P_\nu(\Phi, f) < P(\Phi, f)$. Specification is central to this because the measures whose weak limit is the equilibrium state, in (7.3.3), are defined on periodic orbits. Since the Gibbs property (of the measure in (7.3.3)) is defined by precise exponential behavior (Definition 4.3.25), we begin by producing precise exponential bounds for various of the salient quantities (Propositions 7.3.10, 7.3.13, 7.3.13). Throughout, we assume that Φ is an expansive flow (with expansivity constant δ_0) with specification and that $f \in V(\Phi)$ with ϵ, K as in the definition of $V(\Phi)$ (Definition 4.3.19). Specifically, the first step is careful control of the growth of statistical sums (Proposition 7.3.10). Next, the same control for statistical sums over periodic points, defined in (7.3.2), is obtained in Proposition 7.3.13. Because these two quantities grow in lockstep, the candidate measure from (7.3.3) is a Gibbs measure (Proposition 7.3.15), hence an equilibrium state (Theorem 4.3.26). Moreover, this allows us to show ergodicity, and that in turn reduces the proof of uniqueness to showing that any invariant Borel probability measure that is singular with respect to the measure from (7.3.3) has lower pressure.

Lemma 7.3.9. *For $\epsilon \in (0, \delta_0/3)$ there are $k_\epsilon K_\epsilon > 0$ such that if $t_1, t_2 > 0$, then*

$$\begin{aligned} k_\epsilon N_d(\Phi, f, \epsilon, t_1) N_d(\Phi, f, \epsilon, t_2) &\le N_d\big(\Phi, f, \epsilon, t_1 + t_2\big) \\ &\le K_\epsilon N_d(\Phi, f, \epsilon, t_1) N_d(\Phi, f, \epsilon, t_2). \end{aligned}$$

Proof. (i) If E is $(t_1 + t_2, \epsilon)$-separated and F_j a maximal $(t_j, \epsilon/2)$-separated set, then for $x \in E$ there is a unique $z(x) := (z_1(x), z_2(x)) \in F_1 \times F_2$ such that $d^\Phi_{t_1}(x, z_1(x)) \le \epsilon/2$ and $d^\Phi_{t_2}(\varphi^{t_1}(x), z_2(x)) \le \epsilon/2$, and $z(\cdot)$ is injective. Since furthermore

$$\begin{aligned} |S_{t_1+t_2} f(x) - S_{t_1} f(z_1(x)) - S_{t_2} f(z_2(x))| &\le |S_{t_1} f(x) - S_{t_1} f(z_1(x))| \\ &\quad + |S_{t_2} f(\varphi^{t_1}(x)) - S_{t_2} f(z_2(x))| \le 2K, \end{aligned}$$

[9] For example, f is Hölder continuous; see Proposition 7.3.1.

we have

$$\sum_{x\in E} \exp(S_{t_1+t_2}f(x)) \le e^{2K} N_d\Big(\Phi, f, \frac{\epsilon}{2}, t_1\Big) N_d\Big(\Phi, f, \frac{\epsilon}{2}, t_2\Big).$$

Now invoke Proposition 4.2.18.

(ii) If E_j is $(t_j, 3\epsilon)$-separated, $a_1 = 0$, $a_2 = t_1 + T_\epsilon$, with T_ϵ as in the specification property, and $I_j = [a_j, a_j + t_j]$, then specification implies that for $x := (x_1, x_2) \in E_1 \times E_2$ there is a $z = z(x)$ such that $d^{\Phi}_{t_j}(\varphi^{a_j}(z), x_j) < \epsilon$. Note that by construction $E := \big\{z(x) \mid x \in E_1 \times E_2\big\}$ is $(a_2 + t_2, \epsilon)$-separated. Since furthermore

$$S_{a_2+t_2} f(z(x)) \ge -T_\epsilon \|f\|_{C^0} - 2K + S_{t_1} f(x_1) + S_{t_2} f(x_2),$$

we have

$$N_d(\Phi, f, \epsilon, a_2 + t_2) \ge e^{-T_\epsilon \|f\|_{C^0} - 2K} N_d(\Phi, f, 3\epsilon, t_1) N_d(\Phi, f, 3\epsilon, t_2).$$

Meanwhile, $a_2 + t_2 = t_1 + t_2 + T_\epsilon$ and part (i) yield

$$N_d(\Phi, f, \epsilon, a_2 + t_2) \le K_\epsilon N_d(\Phi, f, \epsilon, t_1 + t_2) N_d(\Phi, f, \epsilon, T_\epsilon),$$

so by Lemma 4.3.20,

$$N_d(\Phi, f, \epsilon, t_1 + t_2) \ge N_d(\Phi, f, 3\epsilon, t_1) N_d(\Phi, f, 3\epsilon, t_2) / K_\epsilon N_d(\Phi, f, \epsilon, T_\epsilon) e^{T_\epsilon \|f\|_{C^0} + 2K}.$$

□

Proposition 7.3.10. *Let X be a compact metric space and Φ a flow with expansivity constant δ_0 and with the specification property. If $0 < \epsilon < \delta_0/3$, $t \in \mathbb{N}$, and k_ϵ and K_ϵ are as in Lemma 7.3.9, then*

$$\frac{1}{K_\epsilon} e^{tP(f)} \le N_d(\Phi, f, \epsilon, t) \le \frac{1}{k_\epsilon} e^{tP(f)}.$$

Proof. Proposition 4.3.22 and (4.3.1) give $P(f) = \lim_{t\to\infty} \frac{1}{t} \log N_d(\Phi, f, \epsilon, t)$ for $\epsilon < \frac{\delta_0}{2}$. By Lemma 7.3.9 we can apply the Bowen–Fekete Lemma (Lemma 4.2.7) to $\log K_\epsilon N_d(\Phi, f, \epsilon, t)$ and $-\log k_\epsilon N_d(\Phi, f, \epsilon, t)$. □

Unlike in Proposition 4.3.14, we define the desired equilibrium state using statistical sums over *periodic* points.

Definition 7.3.11. With the notation of Definition 4.2.22 define

$$\mu_{\epsilon,t} := \frac{1}{P(\Phi, f, t, \epsilon)} \sum_{\gamma\in\mathbb{O}_\epsilon(t)} e^{S_{\pi(\gamma)} f(x)} \delta_\gamma \in \mathfrak{M}(\Phi),$$

where δ_γ is the Lebesgue measure on γ (that is, induced from Lebesgue measure on $[0, \pi'(\gamma)]$ by $t \mapsto \varphi^t(x)$ for $x \in \gamma$) and

$$P(\Phi, f, t, \epsilon) := \sum_{\gamma \in \mathbb{O}_\epsilon(t)} \pi'(\gamma) e^{S_{\pi(\gamma)} f(x)} \tag{7.3.2}$$

is the normalizing factor. Weak*-compactness of $\mathfrak{M}(\Phi)$ gives an accumulation point

$$\mu_f = \lim_{t_i \to \infty} \mu_{\epsilon, t_i} \in \mathfrak{M}(\Phi). \tag{7.3.3}$$

We will show that this is an equilibrium state and that it is the only one (so, a posteriori, this is a proper limit).

We first use the characterization Theorem 1.7.5(3) of expansivity to connect periodicity and separation.

Proposition 7.3.12. *If ϵ and α are as in Theorem 1.7.5(3), $q \geq \epsilon$, $\rho \leq \alpha/2$, $\theta \leq \alpha$, then $x, y \in \mathbb{P}_\rho(t)$ and $x \notin \varphi^{[-q,q]}(y) \Rightarrow x$ and y are (t, θ)-separated.*

Proof. Otherwise, $x, y \in \mathbb{P}_{\alpha/2}(t)$ are not (t, α)-separated, so there exist $a, b \in [t - \frac{\alpha}{2}, t + \frac{\alpha}{2}]$ with $\varphi^a(x) = x$ and $\varphi^b(y) = y$, and

$$d(\varphi^{t_{pm+q}}(x), \varphi^{u_{pm+q}}(y)) = d(\varphi^{q\alpha}(x), \varphi^{q\alpha}(y)) \leq \alpha$$

for $0 \leq q < m := 1 + \lfloor \frac{1}{\alpha}(t - \frac{\alpha}{2}) \rfloor$, where $t_{pm+q} := pa + q\alpha$ and $u_{pm+q} := pb + q\alpha$, so $y = \varphi^t(x)$ with $|t| < \epsilon$ by Theorem 1.7.5(3). □

Proposition 7.3.13. *Let X be a compact metric space, Φ an expansive flow with the specification property, $f \in V(\Phi)$, and $\epsilon > 0$ as in Definition 4.3.19 sufficiently small. Then there exist $c_1, c_2 > 0$ such that for sufficiently large t,*

$$c_1 e^{tP(f)} \leq P(\Phi, f, t, \epsilon) \leq c_2 e^{tP(f)}.$$

Remark 7.3.14. With $f \equiv 0$ (and the notation of Definition 7.3.11) this becomes

$$c_1 e^{t h_{\text{top}}(\Phi)} \leq \sum_{\gamma \in \mathbb{O}_\epsilon(t)} \pi'(\gamma) \leq c_2 e^{t h_{\text{top}}(\Phi)},$$

so $p(\Phi) = h_{\text{top}}(\Phi)$ (Definition 4.2.1).

Proof. If $q > 0$ is such that $x, y \in \mathbb{P}_\epsilon(t)$ are (t, ϵ)-separated unless $x \in \varphi^{[-q,q]}(y)$ (see Proposition 7.3.12), then we can find a (t, ϵ)-separated subset E_γ of $\gamma \in \mathbb{O}_\epsilon(t) =: \{\gamma_1, \dots, \gamma_r\}$ with $\pi'(\gamma) \leq 2q \operatorname{card} E_\gamma$, so

$$P(\Phi, f, t, \epsilon) \leq \sum_{i=1}^{r} 2q \sum_{z \in E_{\gamma_i}} e^{S_{\pi(\gamma_i)} f(z)} \leq 2q e^{\epsilon \|f\|} \sum_{z \in E} e^{S_t f(z)} \underset{\substack{\square \\ \text{Proposition 7.3.10}}}{\leq} \frac{2q e^{\epsilon \|f\|}}{k_\epsilon} e^{tP(f)}$$

since $E := \bigcup_{i=1}^{r} E_{\gamma_i}$ is (t, ϵ)-separated. This gives the upper bound.

If E is a $(t - T_\epsilon, 3\epsilon)$-separated set with T_ϵ as in specification, and $x \in E$ there exists $z = z(x) \in \mathbb{P}_\epsilon(t)$ with $d^\Phi_{t-T_\epsilon}(x,z) \le \epsilon$, and $z(\cdot)$ is "close enough" to injective: If $\beta > 0$ is such that $|s| < 3\beta \Rightarrow d(x, \varphi^s(x)) < \epsilon$ for all $x \in X$, then

$$x \ne x' \Rightarrow \varphi^{[-\beta,\beta]}(z(x)) \cap \varphi^{[-\beta,\beta]}(z(x')) = \varnothing,$$

because $x \ne x' \Rightarrow d(\varphi^s(z(x)), \varphi^s(z(x'))) > \epsilon$ for some $s \in [0, t - T_\epsilon]$ by the triangle inequality, so $z(x') \notin \varphi^{[-3\beta,3\beta]}(z(x))$. This implies that

$$P(\Phi, f, t, \epsilon) \ge \sum_{x\in E} 2\beta \exp\big(\underbrace{S_{\pi(y)} f(z(x))}_{\ge S_{t-T_\epsilon} f(x) - K - (T_\epsilon+\epsilon)\|f\|}\big) \ge 2\beta e^{-K-(T_\epsilon+\epsilon)\|f\|} \underbrace{N_d(\Phi, f, t - T_\epsilon, 3\epsilon)}_{\ge N_d(\Phi, f, t-T_\epsilon, \epsilon)/C_{\epsilon,3\epsilon}}$$

by Lemma 4.3.20 with K as in (4.3.7). Proposition 7.3.10 gives the lower bound. □

We can now show that μ_f is a Gibbs measure (Definition 4.3.25) for $P(f)$.

Proposition 7.3.15 (Gibbs property). *Let (X, d) be a compact metric space, Φ an expansive flow with specification, $f \in V(\Phi)$, μ_f as in* (7.3.3), *and $\epsilon > 0$ as in Definition* 4.3.19. *Then there are $A_\epsilon, B_\epsilon > 0$ such that for $x \in X$ and $t > 0$ (with the notation from* (4.2.1)*) we have*

$$A_\epsilon e^{S_t f(x) - tP(f)} \le \mu_f(B_\Phi(x, \epsilon, t)) \le B_\epsilon e^{S_t f(x) - tP(f)}.$$

In particular, for $f \equiv 0$, we have $P(f) = h$ (topological entropy), so the measure μ_0 of maximal entropy satisfies

$$A_\epsilon \le e^{ht} \mu_0(B_\Phi(x, \epsilon, t)) \le B_\epsilon. \tag{7.3.4}$$

We recall that by Theorem 4.3.26, μ_f is then an equilibrium state for Φ with respect to f.

Proof. If s is (very) large and Q_s is (s, ϵ)-separated with

$$\sum_{y\in Q_s} e^{S_s f(y)} = N_d(\Phi, f, s, \epsilon) \ge e^{sP(f)}/K_\epsilon$$

(by Proposition 7.3.10), and $\alpha < \epsilon/6$ is such that

$$d(\varphi^u(z), z) < \frac{\epsilon}{3} \quad \text{whenever } |u| \le 2\alpha \text{ and } z \in X, \tag{7.3.5}$$

then specification gives

$$w(y) \in \mathbb{O}_{2\alpha}(t + s + 2M_\alpha)$$
$$\text{with } \begin{cases} d(\varphi^r(w(y)), \varphi^r(x)) < \alpha \text{ for } r \in [0, t], \\ d(\varphi^{p_v+\sigma}(w(y)), \varphi^v(y)) < \alpha \text{ for } v \in [0, s],\ p_v := v + t + M_\alpha, \end{cases}$$

with $|\sigma| < \alpha$. Then (7.3.5) implies that $y \neq y' \in Q_s \Rightarrow \varphi^{[0,\alpha]}(w(y)) \cap \varphi^{[0,\alpha]}(w(y')) = \varnothing$: there is a $v \in [0,s]$ such that $d(\varphi^v(y), \varphi^v(y')) > \epsilon$ and hence

$$d(\varphi^{p_c+\sigma}(w(y)), \varphi^{p_v+\sigma}(w(y'))) > \epsilon - 2\alpha > 2\frac{\epsilon}{3}$$

by the triangle inequality, so (7.3.5) implies $d(\varphi^{p_v+u}(w(y)), \varphi^{p_v}(w(y'))) > 0$ for $|u| \leq \alpha$ and hence $\varphi^{p_v+[0,\alpha]}(w(y)) \cap \varphi^{p_v+[0,\alpha]}(w(y')) = \varnothing$, which implies the claim.

Furthermore, $d(\varphi^r(\varphi^u(w(y))), \varphi^r(x)) < \frac{\epsilon}{3} + \alpha < \frac{\epsilon}{2}$ for $u \in [0,\alpha]$, $r \in [0,t]$, so $\bigcup_{y \in Q_s} \varphi^{[0,\alpha]}(w(y)) \subset B_\Phi(x,\epsilon,t)$, and

$$\begin{aligned}\mu_{\alpha,t+s+2M_\alpha} &\geq \underbrace{\frac{1}{P(\Phi, f, t+s+2M_\alpha, \epsilon)}}_{\geq e^{-P(f)(t+s+2M_\alpha)}/c_2} \sum_{y \in Q_s} \alpha \underbrace{\exp(S_{\pi(w(y))} f(w(y)))}_{\geq S_t f(x) - (2M_\alpha+\alpha)\|f\| - 2K + S_s f(y)} \\ &\geq A_\epsilon e^{S_t f(x) - tP(f)}\end{aligned}$$

by Proposition 7.3.13 with K as in (4.3.7). Now we let $s \to \infty$:

$$\begin{aligned}\mu_f(B_\Phi(x,\epsilon,t)) &\geq \lim_{i \to \infty} \mu_{\alpha,t_i}(B_\Phi(x,\epsilon,t)) \\ &\geq \lim_{s \to \infty} \mu_{\alpha,t+s+2M_\alpha}(B_\Phi(x,\epsilon,t)) \geq A_\epsilon e^{S_t f(x) - tP(f)}.\end{aligned}$$

The following, with $V = B_\Phi(x,\epsilon,t)$, $\theta = 2\epsilon$, $\rho = \epsilon$, gives the reverse inequality:

Lemma 7.3.16. *$\forall \theta, \rho, \eta > 0\ \exists S > 0\ \forall t' > 0, V \subset X\ \exists A \subset V \cap \mathbb{P}_\rho(t')$ and $R\colon A \to [-\eta, \eta]$ such that $\varphi^{R_x}(x) \subset V$, $\varphi^s(A)$ is (t',θ)-separated for any s (so, if $V = B_\Phi(x,\epsilon,t) \supset A$, then $\varphi^t(A)$ is moreover $(t-t', 2\epsilon)$-separated), and*

$$\begin{aligned}\frac{1}{2(S+1)}\mu_{\rho,t'}(V) &\leq \mu_{\rho,t'}\Big(\bigcup_{x \in A} \varphi^{R_x}(x)\Big) \\ &= \overbrace{\frac{1}{P(\Phi, f, t', \rho)}}^{\leq e^{-t'P(f)}/c_1} \underbrace{\sum_{x \in A} \lambda(R_x) \exp(S_{\pi(x)} f(x))}_{\substack{\leq 2\eta \sup\{\sum_{z \in B} e^{S_{\pi(z)} f(z)} \mid B\ (t'-t, 2\epsilon)\text{-separated}\} \\ \leq 2\eta e^{S_t f(z) + \epsilon\|f\|} N_d(\Phi, f, t'-t, 2\epsilon)}}.\end{aligned}$$

Thus, Proposition 7.3.10 gives $\mu_{\epsilon,t'}(B_\Phi(x,\epsilon,t)) \leq B_\epsilon e^{S_t f(z) - tP(f)}$, and

$$\mu_f(B_\Phi(x,\epsilon,t)) = \lim_{i \to \infty} \mu_{\epsilon,t_i}(B_\Phi(x,\epsilon,t)) \geq \lim_{t' \to \infty} \mu_{\epsilon,t'}(B_\Phi(x,\epsilon,t)) \leq B_\epsilon e^{S_t f(x) - tP(f)}.$$

□

Proof of Lemma 7.3.16. By Proposition 7.3.12 there is a $q > 0$ such that $x, y \in \mathbb{P}_\rho(t'), x \notin \varphi^{[-q,q]}(y) \Rightarrow x, y$ are (t', θ)-separated. Let $\eta' := \min(\eta, q)$, $S := \lceil 2q/\eta' \rceil$, and $A := \bigcup_{\gamma \in \mathbb{P}_\epsilon(t')} A_\gamma$, where the choice of A_γ will now be described.

Partition $\gamma \in \mathbb{P}_\rho(t')$ into closed segments $I_1, \dots, I_m$ of length $l \in (\eta'/2, \eta)$. Then $I_i \cap \varphi^{[-q,q]}(I_j) = \varnothing$ whenever $|i - j| > S$ in the sense of cyclic ordering. To get separation, and at the same time control the percentage of time spent in V, group these i into $E_1, \dots, E_{2(S+1)}$ such that $i, j \in E_k, i \neq j \Rightarrow |i-j| > S$ (by taking $E_1 = \{1, \dots, S\}$, $E_2 = \{S+1, \dots, 2S\}$, and so on but putting the last at most $S+1$ indices into singletons E_k). Then

$$\mu_{\rho,t'}(V \cap \gamma) = \sum_{k=1}^{2(S+1)} \mu_{\rho,t'}\Big(V \cap \gamma \cap \bigcup_{i \in E_k} I_i\Big),$$

so

$$\mu_{\rho,t'}\Big(V \cap \gamma \cap \bigcup_{i \in E_{k^*}} I_i\Big) \geq \frac{1}{2(S+1)} \mu_{\rho,t'}(V \cap \gamma)$$

for some k^*, and we define A_γ to consist of one $x_i \in V \cap I_i$ for each $i \in E_{k^*}$, and

$$R_{x_i} := \big\{t \mid \varphi^t(x_i) \in V \cap I_i\big\} \subset [-\eta, \eta].$$

If $x \in A := \bigcup_{\gamma \in \mathbb{P}_\epsilon(t')} A_\gamma$, then $\varphi^{R_x}(x) \subset V$, and $\varphi^s(A)$ is (t', θ)-separated for all s. And

$$\begin{aligned}
\mu_{\rho,t'}\Big(\bigcup_{x \in A} \varphi^{R_x}(x)\Big) &= \frac{1}{P(\Phi, f, t', \rho)} \sum_{x \in A} \lambda(R_x) e^{S_{\pi(x)} f(x)} \\
&= \sum_{\gamma \in \mathbb{P}_\rho(t')} \underbrace{\frac{1}{P(\Phi, f, t', \rho)} \sum_{x \in A_\gamma} \lambda(R_x) e^{S_{\pi(x)} f(x)}}_{= \mu_{\rho,t'}(V \cap \gamma \cap \bigcup_{i \in E_{k^*}} I_i) \geq \frac{1}{2(S+1)} \mu_{\rho,t'}(V \cap \gamma)} \\
&\geq \frac{1}{2(S+1)} \mu_{\rho,t'}(V).
\end{aligned}$$

□

Proposition 7.3.17. *The measure μ_f in (7.3.3) is ergodic (indeed, weakly mixing).*

Proof. As in the proof of Proposition 3.4.16, we derive this from something like the lower bound in that result. Specifically, we show that there is a $c > 0$ with

$$c\mu_f(P)\mu_f(Q) \leq \varliminf_{r \to \infty} \mu_f(P \cap \varphi^{[-r-\beta, -r+\beta]}(Q)) \tag{7.3.6}$$

for all measurable sets $P, Q \subset X$. This implies ergodicity because if $Q = X \setminus P$ is a Φ-invariant set of intermediate measure, then the intersection on the right-hand side is empty, so (7.3.6) fails. We note that as in the proof of Proposition 3.4.16, this argument gives ergodicity of $\Phi \times \Phi$ and hence weak mixing by Theorem 3.4.19.

It suffices to show $\underline{\lim}_{r\to\infty} \mu(\overline{U}_1 \cap \varphi^{[-r-\beta,-r+\beta]}(\overline{U}_2)) \geq c\,\overline{\lim}_{t\to\infty} \mu_{\epsilon,t}(V_1)\mu_{\epsilon,t}(V_2)$ for $\delta > 0$, $V_1, V_2 \subset X$ compact, and δ-neighborhoods U_1, U_2 of V_1, V_2, respectively (Borel regularity).

To do so, let $\alpha, \beta > 0$ and $\eta \in (0, \beta/4)$ such that

$$\sup\left\{d(x,\varphi^s(x)) \mid x \in X,\ |s| \leq 4\eta\right\} < \alpha,$$

$\alpha^* < \alpha$ such that $2\alpha^*$ is an expansivity constant for η (Definition 1.7.1), $\epsilon \in (0, \alpha^*/2)$ such that

$$\sup\left\{d(x,\varphi^s(x)) \mid x \in X,\ |s| \leq 6\epsilon\right\} < \alpha^*,$$

$c^* > 0$ such that $\sup\left\{d(\varphi^s(x),\varphi^s(y)) \mid d(x,y) \leq c^*,\ |s| \leq \eta\right\} < \delta$, and (Proposition 1.8.4) $T > 0$ such that

$$d(\varphi^s(x),\varphi^s(y)) \leq 2\alpha^* \text{ for } |s| < T \Rightarrow d(\varphi^s(x), y) \leq c^* \text{ for some } s \in [-\eta,\eta].$$

From now suppose $t \geq 2T$ and $\tau_1, \tau_2 > 0$ (at the end of the proof these will tend to ∞), and let

$$t_0 := -T, \quad t_1 := t_0 + t + T_\epsilon, \quad t_2 := t_1 + \tau_1 + T_\epsilon, \quad t_3 := t_2 + t + T_\epsilon, \quad t_4 := t_3 + \tau_2 + T_\epsilon,$$

where T_ϵ comes from specification (Definition 7.3.2, Theorem 7.3.4). Let E_i be $(\tau_i, 3\alpha)$-separated sets for $i = 1, 2$, and A_i the sets (and R_x the function) obtained from Lemma 7.3.16 applied to V_i with $\theta = 3\alpha$, $\rho = 3\epsilon$.

For $\mathbf{z} = (z_0, \ldots, z_3) \in E := \varphi^{-T}(A_1) \times E_1 \times \varphi^{-T}(A_2) \times E_2$ there is a $w(\mathbf{z}) \in \mathbb{P}_{3\epsilon}(t_4 - t_0)$ with $d(\varphi^{t_i+u+s(t_i+u)}(w), \varphi^u(z_i)) < \epsilon$ for $u \in [0, t_{i+1} - t_i - T_\epsilon]$ (notation as in (7.3.1)).

Claim 7.3.18. *$\varphi^{-T}(\{w(\mathbf{z})\colon \mathbf{z} \in E\})$ is $(t_4 - T_\epsilon + 3\epsilon, \alpha)$-separated.*

Proof. If $\mathbf{z} \neq \mathbf{z}'$, then $z_k \neq z'_k$ for some k and hence a $u \in [0, t_{k+1} - t_k - T_\epsilon]$ such that (with $t = t_k + u \in [-T, t_4 - T_\epsilon]$)

$$\begin{aligned} 3\alpha &< d(\varphi^u(z_k), \varphi^u(z'_k)) \\ &\leq \underbrace{d(\varphi^u(z_k), \varphi^{t+s(t)}(w(\mathbf{z})))}_{\leq\epsilon} + d(\varphi^{t+s(t)}(w(\mathbf{z})), \varphi^{t+s(t)}(w(\mathbf{z}'))) \\ &\quad + \underbrace{d(\varphi^{t+s(t)}(w(\mathbf{z}')), \varphi^{t+s'(t)}(w(\mathbf{z}')))}_{\leq\alpha^*} + \underbrace{d(\varphi^{t+s'(t)}(w(\mathbf{z}')), \varphi^u(z'_k))}_{\leq\epsilon}, \end{aligned}$$

that is, $d(\varphi^{t+s(t)}(w(\mathbf{z})), \varphi^{t+s(t)}(w(\mathbf{z}'))) \geq 3\alpha - \alpha^* - 2\epsilon > \alpha$. □

If $w = w(\mathbf{z})$, $x = \varphi^T(z_0) \in A_1$, then $d(\varphi^p(w), \varphi^p(x)) < \epsilon < 2\alpha^*$ whenever $|p| \le T$ by choice of T, so $d(\varphi^{u(w)}(w), x) \le c^*$ with $|u(w)| \le \eta$.

Claim 7.3.19. *The $\varphi^{u(w)+R_x}(w)$ are pairwise disjoint (for different $w = w(\mathbf{z})$ with $\mathbf{z} \in E$) subsets of $U_1 \cap \varphi^{[-r-\beta,-r+\beta]}(U_2)$, where $r := t + \tau_1 + 2T_\epsilon$.*

Proof. $\varphi^{u(w)+R_x}(w)$ is contained in a δ-neighborhood of $\varphi^{R_x}(x)$ and hence in U_1 since $R_x \subset [-\eta, \eta]$ and by choice of c^*.

If $x_2 := \varphi^T(z_2) \in A_2$, and $|p| \le T$, then $d(\varphi^{p+r+s(p+r)}(w), \varphi^p(x_2)) < \epsilon$, so

$$d(\varphi^{p+r}(w), \varphi^p(x_2)) < \epsilon + \alpha^* < 2\alpha^*$$

by choice of ϵ since $|s(p+r)| \le 3\epsilon$. Thus,

$$f_r(w) \in f_{[-\eta,\eta]} \underbrace{B_{c^*}(V_2)}_{c^*\text{-neighborhood of } V_2 \subset U_2} \subset \varphi^{[-\eta,\eta]}(U_2).$$

Since $u(w) \in [-\eta, \eta]$ and $R_x \subset [-\eta, \eta]$, this implies $\varphi^r(\varphi^{u(w)+R_x}(w)) \subset \varphi^{[-3\eta,3\eta]}(U_2)$, hence, since $\beta \ge 3\eta$,

$$\varphi^{u(w)+R_x}(w) \subset \varphi^{[-r-\beta,-r+\beta]}(U_2).$$

To see disjointness, suppose $\varphi^{u(w)+R_x}(w) \cap \varphi^{u(w')+R'_x}(w') \ne \varnothing$. Then $w' = \varphi^v(w)$ with $|v| \le 4\eta$, hence $d(\varphi^s(w), \varphi^s(w')) = d(\varphi^s(w), \varphi^s(\varphi^v(w))) < \alpha$ for all s by definition of η, so $w = w'$ since the w are (T, α)-separated. □

This gives the desired lower bound: if $t' := t_4 - t_0 = \tau_1 + \tau_2 + 2T + 4T_\epsilon$, then

$$\begin{aligned}
&\mu_{3\epsilon,t'}(U_1 \cap \varphi^{[-r-\beta,-r+\beta]}(U_2)) \\
&\quad \ge \frac{1}{P(\Phi, f, t', 3\epsilon)} \sum_w \lambda(R_x) \exp(S_{\pi(w)} f(w)) \\
&\quad \ge \frac{e^{-4K-4T_\epsilon \|f\| - 3\epsilon}}{P(\Phi, f, t', 3\epsilon)} \underbrace{N_d(\Phi, f, \tau_1, 3\alpha) \sum_{z_2} e^{S_{\pi()} f(z_2)}}_{\ge \frac{1}{2\eta} P(\Phi, f, t, 3\alpha) M \mu_{3\alpha,t}(V_2)} \underbrace{N_d(\Phi, f, \tau_2, 3\alpha) \sum_{z_0} e^{S_{\pi()} f(z_0)} \lambda(R_x)}_{\ge P(\Phi, f, t, 3\alpha) M \mu_{3\alpha,t}(V_1)} \\
&\quad \ge c \mu_{3\alpha,t}(V_1) \mu_{3\alpha,t}(V_2),
\end{aligned}$$

where $M := \frac{1}{2(S+1)}$ is as in Lemma 7.3.16. Fixing $t \ge 2T$ and τ_1, let $\tau_2 \to \infty$ (with $t' = t_i$) to get

$$\mu_f \overline{U}_1 \cap \varphi^{[-r-\beta,-r+\beta]}(\overline{U}_2) \ge \overline{\lim_{i \to \infty}} \mu_{3\epsilon,t_i} U_1 \cap \varphi^{[-r-\beta,-r+\beta]}(U_2) \ge c \mu_{3\alpha,t}(V_1) \mu_{3\alpha,t}(V_2).$$

Now letting $\tau_1 \to \infty$ (with $r = s_k$) gives

$$\underline{\lim}_{k\to\infty} \mu_f \overline{U}_1 \cap \varphi^{[-s_k-\beta,-s_k+\beta]}(\overline{U}_2) \geq c\mu_{3\alpha,t}(V_1)\mu_{3\alpha,t}(V_2),$$

which, as $t \to \infty$, becomes

$$\underline{\lim}_{r\to\infty} \mu_f \overline{U}_1 \cap \varphi^{[-r-\beta,-r+\beta]}(\overline{U}_2) \geq c\,\overline{\lim}_{t\to\infty} \mu_{3\alpha,t}(V_1)\mu_{3\alpha,t}(V_2). \qquad \square$$

Proof of Theorem 7.3.6. We show $P_\nu(\Phi, f) = P(\Phi, f) \Rightarrow \nu = \mu$; this implies $P_\mu(\Phi, f) = P(\Phi, f)$ and that there is only one accumulation point. By Proposition 7.3.17, μ in (7.3.3) is ergodic, so it suffices to show $\nu \perp \mu \Rightarrow P_\nu(\Phi, f) < P(\Phi, f)$.

For $t \in \mathbb{R}^+$ and a maximal $(t, 2\epsilon)$-separated set E_t take Borel sets β_x such that $B_\Phi(x,\epsilon,t) \subset \beta_x \subset B_\Phi(x, 2\epsilon, t)$, $\mathfrak{B}_t := \{\beta_x \mid x \in E_t\}$ is a partition, and $(\mu+\nu)(\partial\mathfrak{B}_t) = 0$. Since Φ is expansive, $\operatorname{diam}\varphi^{-t/2}(\mathfrak{B}_t) \xrightarrow[t\to\infty]{} 0$ so if $\varphi^t(B) = B \subset X$ such that $\mu(B) = 0$ and $\nu(B) = 1$ then there exist finite unions C_t of elements of $\mathfrak{B}_t$ such that

$$(\mu + \nu)(C_t \vartriangle B) = (\mu + \nu)(\Phi^{-t/2}(C_t) \vartriangle B) \xrightarrow[t\to\infty]{} 0.$$

Furthermore, if $\epsilon < \delta_0/2$ then $\mathfrak{B}_t$ is generating for Φ^t, that is, $th_\nu(\Phi) = h_\nu(\varphi^t) = h_\nu(\varphi^t, \mathfrak{B}_t) \leq H_\nu(\mathfrak{B}_t)$. By possibly shrinking ϵ we can ensure that $S_t\varphi \leq K + S_t\varphi(x)$ on $B_f(x, 2\epsilon, t)$ and hence on β_x (see Definition 4.3.19). This yields

$$\begin{aligned}
tP_\nu(\Phi, f) &\leq -\sum_{x\in E_t}\Big(\nu(\beta_x)\log(\nu(\beta_x)) + \int_{\beta_x} S_t\varphi\, d\nu\Big)\\
&\leq K + \sum_{x\in E_t;\beta_x\subset C_t} \nu(\beta_x)(S_t\varphi(x) - \log\nu(\beta_x))\\
&\quad + \sum_{x\in E_t;\beta_x\cap C_t=\varnothing} \nu(\beta_x)(S_t\varphi(x) - \log\nu(\beta_x))\\
&\leq K + \nu(C_t)\log \sum_{x\in E_t;\beta_x\subset C_t} e^{S_t\varphi(x)}\\
&\quad + \nu(X \smallsetminus C_t)\log \sum_{x\in E_t;\beta_x\cap C_t=\varnothing} e^{S_t\varphi(x)} + \frac{2}{e},
\end{aligned}$$

where the last estimate used (A.2.12). We now apply Proposition 7.3.15 to get

$$\begin{aligned}
&t\big(P_\nu(\Phi, f) - P(\Phi, f)\big) - \frac{2}{e} - K\\
&\quad\leq \nu(C_t)\log\Big(\sum_{x\in E_t;\beta_x\subset C_t} e^{S_t\varphi(x)-tP(\Phi,f)}\Big)\\
&\quad\quad + \nu(X \smallsetminus C_t)\log\Big(\sum_{x\in E_t;\beta_x\cap C_t=\varnothing} e^{S_t\varphi(x)-tP(\Phi,f)}\Big)
\end{aligned}$$

$$\leq \nu(C_t)\log(A_\epsilon^{-1}\mu(C_t)) + \nu(X \setminus C_t)\log(A_\epsilon^{-1}\mu(X \setminus C_t)) \xrightarrow[t\to\infty]{} -\infty,$$

since $\nu(C_t) \to 1$ and $\mu(C_t) \to 0$. Thus $P_\nu(\Phi, f) < P(\Phi, f)$, hence uniqueness. □

Remark 7.3.20. Theorem 7.3.6 asserts that the equilibrium state is weakly mixing, and stronger mixing properties are suggested by Remark 3.4.4, which implies K-mixing as follows ([233, Proposition 1.4]; see also [270], Theorem 7.4.20, and Remark 8.6.12). If $\Phi' := \Phi$ on $X' := X$, then $\Phi \times \Phi'$ has a unique equilibrium state for $F(x, x') := f(x) + f(x') \in C^{\Phi\times\Phi'}(X \times X')$.[10] Define a probability measure on $X \times X'$ by extending

$$\bar{\mu}(A \times B) := \int_A \underbrace{E(\chi_B \mid \pi(\Phi))}_{\substack{\text{conditional expectation from Corollary 3.2.10} \\ \text{with respect to the Pinsker partition of } \Phi}} d\mu,$$

and check that

$$h_{\bar{\mu}}(\Phi \times \Phi') = 2h_\mu(\Phi) = h_{\mu\times\mu}(\Phi \times \Phi') \text{ and } \int F\, d\bar{\mu} = 2\int f\, d\mu = \int F\, d\mu \times \mu',$$

so $\bar{\mu} = \mu \times \mu$ by uniqueness, hence $\pi(\Phi)$ is trivial, which characterizes K-mixing. (For Anosov flows, one even obtains the Bernoulli property [310].)

Remark 7.3.21. We noted earlier that kinematic expansivity (Definition 1.8.3) is sufficient for the results so far. Exercise 4.25 shows that it suffices for existence. The reader is encouraged to verify that kinematic expansivity suffices for uniqueness as follows. Lemma 7.3.9 holds when one assumes kinematic expansivity instead of expansivity (Exercise 7.10), as do Propositions 7.3.10 (Exercise 7.11) and 7.3.12 (Exercise 7.12) by following the line of reasoning of the proof of Theorem 1.7.5(3) from expansivity, using that here there are only minor timing differences. Since Proposition 7.3.12 holds assuming only kinematic expansivity, so do Propositions 7.3.13, 7.3.15, and 7.3.17 (Exercise 7.13)—and hence Theorem 7.3.6.

We summarize Theorem 6.2.21, Remark 7.3.20, and Corollary 7.3.8:

Theorem 7.3.22 (Bowen–Anosov Alternative). *For a basic set Λ of a flow Φ there are three mutually exclusive possibilities*:

- *Λ is a point*;

[10] In discrete time this is true because the product system is expansive with specification, but product flows are not expansive, so here a rather more careful argument is required for existence and uniqueness of equilibrium states.

- $\Phi_{\restriction\Lambda}$ *is topologically a (constant-time) suspension (smoothly so, and of an Anosov diffeomorphism, if* Φ *is Anosov); if there is more than one orbit, then this is generically not the case (Corollary* 6.2.22)

  $\Leftrightarrow$ $\Phi_{\restriction\Lambda}$ *is not topologically mixing*

  $\Leftrightarrow$ *the periods of* $\Phi_{\restriction\Lambda}$ *are commensurate*;

- – *every strong stable or unstable manifold is dense in* Λ,
  – $\Phi_{\restriction\Lambda}$ *is topologically mixing,*
  – *every Bowen-bounded* $f\colon \Lambda \to \mathbb{R}$ *has a unique equilibrium state* μ_f *for* $\Phi_{\restriction\Lambda}$,
  – μ_f *is K-mixing (and Bernoulli if* Φ *is Anosov; Theorem* 7.4.20).

Proposition 4.3.17 about equilibrium states of special flows can be formulated more nicely in the context of (the discrete-time counterpart of) Theorem 7.3.6:

Proposition 7.3.23. *On a compact metric space X, let F be an expansive homeomorphism with specification,* $h_{\text{top}}(F) < \infty$, $0 < r \in V(F)$, Φ_r *the special flow on* X_r. *Then*

- $h_{\text{top}}(\Phi_r)$ *is the unique solution of* $P(F, -cr) = 0$;
- *if m is the unique equilibrium state of F for* $-h_{\text{top}}(\Phi_r)r$, *then* m_r *is the unique measure of maximal entropy for* Φ_r *on* X_r.

If $G \in V(\Phi_r)$, *and* $g(x) := \int_0^{r(x)} G(x,t)\,dt \in V(F)$, *then*

- $P(\Phi_r, G)$ *is the unique solution of* $P(F, g - cr) = 0$;
- *if m is the unique equilibrium state for* $g - cr$, *then* m_r *(from* (3.6.4)) *is the unique equilibrium state of G for* Φ_r.

Uniqueness in Theorem 7.3.6 gives a map $f \mapsto \mu_f$ that associates to a Hölder potential the corresponding equilibrium state. Is it injective? There are different potentials that give rise to the same equilibrium state: adding a constant to a function f changes $P(f)$ and $P_\mu(f)$ by the same additive constant and thus produces the same equilibrium state, that is, $\mu_{f+c} = \mu_f$ for any constant c. Furthermore, one sees directly from Definition 7.3.11 that two potentials f, g give rise to the same equilibrium state if their statistical sums $S_{\pi(\gamma)}f(x)$ and $S_{\pi(\gamma)}g(x)$ coincide for all periodic orbits, which is the case when f and g are cohomologous (see Proposition 1.3.19, Theorem 4.3.13). However, by the Livshitz Theorem, these are the only ways in which two potentials give rise to the same equilibrium state.

Theorem 7.3.24 (Classification of equilibrium states). *Let Λ be a topologically mixing compact locally maximal hyperbolic set for a flow Φ generated by X and $f, g\colon \Lambda \to \mathbb{R}$ Hölder continuous. Then $\mu_f = \mu_g$ if and only if $g(x) = f(x) + c + Xk$ for some Hölder-continuous $k\colon \Lambda \to \mathbb{R}$, that is, if and only if f and g are cohomologous up to a constant.*

Proof. Adjust g by an additive constant such that $P(f) = P(g)$ (Remark 4.3.4). By the Livshitz Theorem (Theorem 6.3.2) and by symmetry, it suffices to show that $S_t f(x) \le S_t g(x)$ whenever $\varphi^t(x) = x$. The Gibbs property (Proposition 7.3.15) implies that $A^f_\epsilon e^{S_t f(x)} \le B^g_\epsilon e^{S_t g(x)}$ hence $S_t f(x) + \log A^f_\epsilon \le S_t g(x) + \log B^g_\epsilon$ whenever $\varphi^t(x) = x$, so $S_t f(x) = \lim_{n\to\infty} S_{nt} f(x)/n \le \lim_{n\to\infty} S_{nt} g(x)/n = S_t g(x)$ since $\varphi^{nt}(x) = x$. □

Let us add another observation about equilibrium states for Anosov flows—equilibrium states vary continuously with the Anosov flow.

Theorem 7.3.25 (Weak continuity of equilibrium states). *On a manifold M consider C^1-functions f_n and Anosov flows Φ_n such that $\Phi_n \xrightarrow[n\to\infty]{C^1} \Phi$ and $f_n \xrightarrow[n\to\infty]{C^1} f$. Then $\mu_n \xrightarrow[n\to\infty]{\text{weakly}} \mu$, where the μ_n are the equilibrium states for f_n of Φ_n and μ is the equilibrium state for f of Φ.*

Proof. If $\mu^* = \lim_{k\to\infty} \mu_{n_k}$ is any weak accumulation point of the μ_n, then

$$
\begin{aligned}
P(\Phi, f) &\overset{(4.3.5)}{\ge} h_{\mu^*}(\Phi) + \int f\, d\mu^* \\
&\overset{\substack{\text{Remark}\\ 4.3.18}}{\ge} \underbrace{\overline{\lim_{k\to\infty}}\, h_{\mu_{n_k}}(\Phi_{n_k}) + \int f_{n_k}\, d\mu_{n_k}}_{=P(\Phi_{n_k}, f_{n_k})} \overset{\substack{\text{structural}\\ \text{stability}}}{=\!=\!=} P(\Phi, f),
\end{aligned}
$$

so μ^* is an equilibrium state and by uniqueness the weak limit of the μ_n. □

Remark 7.3.26. Continuous dependence suggests a way of explaining terminology we touched upon in passing earlier. Suppose we have, say, a 1-parameter family of flows which are hyperbolic except for one ("bifurcation") value of the parameter. The continuously varying equilibrium states for flows on either side of this exceptional parameter might converge to two different equilibrium states for the nonhyperbolic flow with the bifurcation parameter, and this marks the transitional nature of this flow on the boundary of hyperbolicity. Inspired by this, flows with more than one equilibrium state are said to be in a *phase transition* [145].

Remark 7.3.27. We emphasize that among the reasons equilibrium states are particularly interesting in smooth dynamics are various special properties that are both distinctive and helpful in further investigations. The Gibbs property (Proposition 7.3.15) has already proved useful. An important additional one is a local product structure which, for instance, makes equilibrium states amenable to the Hopf argument.

Theorem 7.3.28 (Local product structure of equilibrium states [190, Theorem 2]). *If Λ is a basic set for a flow Φ, then the equilibrium state μ_f for a Hölder-continuous potential $f\colon \Lambda \to \mathbb{R}$ is (up to a normalizing factor) locally given by the product $\mu^s \times \mu^u \times dt$, where μ^u and μ^s are measures on (strong) unstable and stable leaves,*[11] *respectively, such that*

- $(\varphi^t)^*\mu^u = e^{\int_0^t f\circ\varphi^s - P(f)\,ds}\mu^u$,
- $(\varphi^t)^*\mu^s = e^{\int_t^0 f\circ\varphi^s - P(f)\,ds}\mu^s$,
- $d\mu^u(y) = e^{\int_0^\infty f\circ\varphi^s\circ\rho_{x,y} - f\circ\varphi^s\,ds}\,d\rho_{x,y}^*\mu^u(x)$,
- $d\mu^s(y) = e^{\int_{-\infty}^0 f\circ\varphi^s\circ\rho'_{x,y} - f\circ\varphi^s\,ds}\,d{\rho'}_{x,y}^*\mu^s(x)$,

where $\rho_{x,y}\colon W^u_\epsilon(x) \to W^u_\epsilon(y), z \mapsto [z,y]$ (Proposition 6.2.2) is the holonomy along center-stable leaves and analogously ${\rho'}_{x,y}\colon W^s_\epsilon(x) \to W^s_\epsilon(y)$ (Remark 6.2.3) moves points along center-unstable manifolds.

While a proof of this is beyond our scope, Section 7.5 develops the case $f \equiv 0$, the Bowen–Margulis measure.

The proof of Theorem 7.3.6 may seem like a long slog through estimates, but the core approach of using expansivity and specification is quite elegant—while also covering the smooth and symbolic situations at the same time. Expansivity, and even more so specification, are closely connected to the uniformity of our hyperbolicity assumption, and this approach fell somewhat to the wayside in research beyond uniform hyperbolicity. A 1980 paper by Katok [210] is a striking exception which combines the Bowen approach with Pesin theory to great effect. The results were indeed remarkable (such as that positive entropy in any 2-dimensional diffeomorphism is entirely attributable to horseshoes) and this paper is still cited frequently.[12] Yet, it remained singular, and it focused on shadowing rather than specification. Only much more recently, a renaissance of Bowen's approach began with a collaboration of Climenhaga and Thompson, whose feat of imagination and technical mastery revealed that Bowen's approach, suitably adapted and combined with other advances since, turns out to provide a powerful machinery ready for application in nonuniformly hyperbolic dynamics [94]. The underlying idea is to decompose every orbit segment into "good" and "bad" parts, where the "good" parts satisfy Bowen's conditions, and the "bad" parts carry smaller topological pressure than the whole flow.

[11] The conditionals from Definition A.1.13.

[12] Indeed, Katok was prouder of this paper than any other.

To illustrate, consider the geodesic flow on a compact (rank-1) nonpositively curved Riemannian manifold. The rank-1 condition on the Riemannian metric is that there is a geodesic for which the tangent vector field is the only parallel Jacobi field, and the (open dense invariant) regular set is the set of tangent vectors to such geodesics; the singular set is its complement (and empty only in the well-understood case of negative curvature). This kind of geodesic flow is the original exemplar that motivated the development of the theory of nonuniform hyperbolicity because here one obtains (in a nonuniform way) complete hyperbolicity. Twenty years prior, Gerhard Knieper had been able to establish in this context that there is a unique measure of maximal entropy [223]; this was a huge breakthrough and only possible at the time because Knieper was able to fully utilize all the structure provided by the geometry of this flow. Now, a mere by-product of the applications of the Climenhaga–Thompson technique is a direct and purely dynamical proof of this fact. To single out a specific issue, it is a corollary of Knieper's uniqueness proof that the singular set carries less topological entropy than the regular set—but this does not produce a direct constructive proof of the entropy gap. The Climenhaga–Thompson machinery produces such an argument: approximate singular orbit segments by regular orbit segments having the specification property, and use these to build a collection of orbits with greater topological entropy than the singular set; this re-proves the result by Knieper using dynamical methods [76]. More generally, they also obtain uniqueness of equilibrium states in this context. Before stating the result we mention that an ergodic invariant probability measure is *hyperbolic* if all of the Lyapunov exponents are nonzero.

Theorem 7.3.29 (Burns–Climenhaga–Fisher–Thompson [76]). *The geodesic flow of a compact rank-*1 *nonpositively curved Riemannian manifold has a unique equilibrium state for any Hölder-continuous potential with a* pressure gap, *that is, if the singular set does not carry full pressure, meaning that the restriction of the geodesic flow to it has smaller pressure than on the complement or, equivalently, in total. (This is automatic for potentials that are locally constant on a neighborhood of the singular set.) The unique equilibrium state is hyperbolic, fully supported, and is the weak* limit of weighted regular closed geodesics.*

The pressure gap is essential: without it, there is at least one equilibrium state supported on the singular set. On the other hand, only equilibrium states supported on the regular set are hyperbolic. There have been further extensions, by the way, such as to closed surfaces without conjugate points and genus at least 2 [93].

7.4 Sinai–Ruelle–Bowen measures

We have studied invariant measures in some generality, with particular attention to equilibrium states. These were motivated by the measure of maximal entropy on one hand, which we study further in the next section. On the other hand, in the case of smooth flows it is natural to be interested in smooth invariant measures. These two measures are special when invariant measures abound. However, not all smooth dynamical systems preserve a smooth measure, and we now introduce an equilibrium state that is always present and is smooth when there is a smooth invariant measure and of unique interest for attracting hyperbolic sets (Remark 7.4.11). It is thus the natural generalization of a smooth invariant measure. This realizes one aim in developing the theory of equilibrium states to the extent that we have done so.

Definition 7.4.1. If Λ is a topologically mixing compact hyperbolic set for a flow Φ, then the *Sinai–Ruelle–Bowen measure* or *SRB-measure* for Φ (on Λ) is the equilibrium state $\mu_{\text{SRB}} := \mu_J$ for the *geometric potential*

$$J := J^u := -\frac{d \log j_t}{dt}\Big|_{t=0} = -\frac{dj_t}{dt}\Big|_{t=0},$$

where $j_t(x)$ is the Jacobian of $d\varphi^t : E^u(x) \longrightarrow E^u(\varphi^t(x))$ with respect to the volume defined by the Riemannian metric.

Remark 7.4.2. The geometric potential J is Hölder continuous since E^u is. Since $j_{s+t}(x) = j_t(\varphi^s(x)) j_s(x)$, we have $-\log j_t(\varphi^s(x)) = -\log j_{s+t}(x) + j_s(x)$, hence $J(\varphi^s(x)) = -\frac{d \log j_\tau(x)}{d\tau}\big|_{\tau=s}$, and $\int_0^t J(\varphi^s(x))\, ds = -\log j_t(x)$.

We assume that the Riemannian metric is adapted to Φ (Proposition 5.1.5).

This particular measure is of great interest with respect to the study of attractors,[13] and the main theorem (Theorem 7.4.10) is that a *Lebesgue*-a.e. point near an attractor will be a typical point for its SRB measure. This does, in effect, say that the toner density produced by printing a computed orbit starting near an attractor will reflect the density of the SRB measure. For an Anosov flow, the entire manifold is a global attractor, so any smooth invariant measure is, in fact, the SRB measure (Theorem 7.4.14) and hence unique and (provided the flow is topologically mixing) mixing by Theorem 7.3.6 (actually, K-mixing by Remark 7.3.20, and we explain at the end that one indeed obtains the Bernoulli property (Theorem 7.4.20)). Its entropy then is the average infinitesimal unstable volume-expansion rate (the Pesin Entropy Formula, Remark 7.4.15). If no smooth measure is invariant, then the flow itself dissipates a volume to a forward and a backward SRB measure (Theorem 7.4.13). In the midst of this we will also demonstrate the utility of this measure with respect to questions that appear to have no connection to statistical methods at all (Theorem 7.4.19).

[13] To which Theorems 5.3.27 and 5.3.25 are pertinent.

Proposition 7.4.3 (Volume Lemma). *If Λ is a compact locally maximal hyperbolic set for a C^2-flow Φ and $\epsilon > 0$ is sufficiently small then there exist $C_\epsilon, D_\epsilon > 0$ such that*

$$D_\epsilon \le j_t(x)\lambda(B_\Phi(x,\epsilon,t)) \le C_\epsilon \tag{7.4.1}$$

for all $t \ge 0$, where λ denotes the Riemannian volume.[14]

Proof. Let $m = \dim M$. We first replace the balls $B_\Phi(x,\epsilon,t)$ by sets that are easier to handle. On a neighborhood V_x of each $x \in M$ introduce adapted coordinates $\psi_x\colon \underbrace{B_\epsilon(0)}_{\subset T_x M} \to M$ in which

$$\psi(\underbrace{E^u(x)}_{\sim\mathbb{R}^k\times\{0\}}) \subset W^u(x), \qquad \psi(\underbrace{E^{cs}(x)}_{\sim\{0\}\times\mathbb{R}^{m-k}}) \subset W^{cs}(x),$$

and $\varphi_x := \psi^{-1}_{\varphi^1(x)} \circ \varphi^1 \circ \psi_x$ is tangent to $D_x\varphi^1$ at 0. If $y \in V_x$ is parametrized by $(y_1, y_2) \in \mathbb{R}^k \times \mathbb{R}^{m-k}$ then there are constants $c_1, c_2 > 0$ independent of x and y with

$$c_1 \max(\|y_1\|, \|y_2\|) \le d(x,y) \le c_2 \max(\|y_1\|, \|y_2\|).$$

Let

$$\begin{aligned} B_\epsilon(x) := \big\{ y \in V_x \mid V_{\varphi^\tau(x)} \ni \varphi^\tau(y) = (y_1^{(\tau)}, y_2^{(\tau)}),& \\ \max(\|y_1^{(\tau)}\|, \|y_2^{(\tau)}\|) \le \epsilon \text{ for } 0 \le \tau \le t \big\}&. \end{aligned}$$

Then $B_{\epsilon/c_2}(x) \subset B_\Phi(x,\epsilon,n) \subset B_{\epsilon/c_1}(x)$ and in order to prove (7.4.1) it suffices to prove a like estimate for $\lambda(B_\epsilon(x))$. Furthermore, the measure λ in V_x is given by a density with bounded logarithm. Thus instead of estimating $\lambda(B_\epsilon(x))$ we estimate the volume $\mathrm{vol}^m(B_\epsilon(x))$ of $B_\epsilon(x)$ in adapted coordinates. Similarly the Jacobian, which is taken with respect to a fixed Riemannian metric, can be replaced by the Jacobian of φ^{-t} from adapted local coordinates near $\varphi^t(x)$ to those near x.

Then, $B_\epsilon(x)$ contains a piece S of the stable manifold $W^s_\epsilon(x)$ of size ϵ in adapted coordinates, whose $(m-k)$-dimensional volume then is of order ϵ^{m-k}. In local coordinates fix $y = (0, y_2) \in S$, and let $U_{y_2} := \{z \in B_\epsilon(x) \mid z = (y_1, y_2)\}$, that is, the horizontal "slice" of $B_\epsilon(x)$ through y. Then

$$\mathrm{vol}^m(B_\epsilon(x)) = \int_S \mathrm{vol}^k(U_{y_2})\, dy_2. \tag{7.4.2}$$

To estimate the k-dimensional volume of the sets U_{y_2} note that for small enough ϵ the tangent space $T_z U_{y_2}$ is close to the unstable subspace E^u_z and in particular inside

[14] Compare (7.4.1) to the Gibbs property (Definition 4.3.25).

an invariant cone family around E^u. This implies that if in local coordinates around $\varphi^\tau(x)$ we write $\varphi^\tau(z) = (z_1^{(\tau)}, z_2^{(\tau)})$, then $\|z_1^{(\tau)}\| \geq \mu^r \|z_1^{(\tau-r)}\|$ for some $\mu > 1$ whenever $0 \leq r \leq \tau$. Hence $\|z_1^{(\tau)}\|$ is maximal for $\tau = t$, and the image $\varphi^t(U_{y_2})$ is the graph of a Lipschitz function in the adapted coordinates around $\varphi^t(x)$ defined over the whole ϵ-ball around the origin. Thus $\mathrm{vol}^k(\varphi^t(U_{y_2}))$ is of order ϵ^k. Now if ω is the k-dimensional volume element on $\varphi^t(U_{y_2})$ then

$$\mathrm{vol}^k(U_{y_2}) = \int_{\varphi^t(U_{y_2})} \tilde{j}_{-t}\, d\omega,$$

where $\tilde{j}_t$ is the Jacobian of $d\varphi^t : TU_{y_2} \to TU_{z_2^{(\tau)}}$.

Together with (7.4.2) this means that Proposition 7.4.3 follows from Proposition 7.3.1: at $w := \varphi^t(z) \in \varphi^t(U_{y_2})$ the ratio of the Jacobian of φ^{-t} on $T\varphi^t(U_{y_2})$ and the Jacobian of φ^{-t} on $E^u_{\varphi^t(x)}$ (which is $j_{-t}(\varphi^t(x))$) is bounded from above and below by positive constants. □

Proposition 7.4.4. *If Λ is a compact locally maximal hyperbolic set for a C^2-flow Φ, $\epsilon > 0$ sufficiently small and $B_\Phi(\epsilon,t) := \bigcup_{x\in\Lambda} B_\Phi(x,\epsilon,t)$, then*

$$P(J) = \lim_{t\to\infty} \frac{1}{t} \log \lambda(B_\Phi(\epsilon,t)).$$

Proof. Fix $\delta \leq \epsilon$ and a maximal (t,δ)-separated set $E \subset \Lambda$ (see Section 4.2). If $x \in \Lambda$ then $x \in B_\Phi(y,\delta,n)$ for some $y \in E$ by maximality, and hence

$$B_\Phi(y,\delta/2,t) \subset B_\Phi(x,\epsilon,t) \subset B_\Phi(y,\delta+\epsilon,t).$$

Since the $B_\Phi(y,\delta/2,t)$ are pairwise disjoint, Proposition 7.4.3 yields

$$D_{\delta/2} \sum_{y\in E} j_{-t}(\varphi^t(x)) \leq \lambda(B(\epsilon,n)) \leq C_{\delta+\epsilon} \sum_{y\in E} j_{-t}(\varphi^t(x)).$$

Thus, $P(\Phi, J) = \lim_{n\to\infty} \frac{1}{n} \log \lambda(B(\epsilon,n))$ by (4.3.2). □

Corollary 7.4.5. *Under the hypotheses of Proposition* 7.4.4 *we have $P(J) \leq 0$, and if $\lambda(W^s_\epsilon(\Lambda)) > 0$ then $P(J) = 0$, so $h_{\mu_{\mathrm{SRB}}}(\Phi) = -\int J\, d\mu_{\mathrm{SRB}}$.*

Proof. $\lambda(B(\epsilon,n)) \leq \lambda(M) < \infty \Rightarrow P(J) = \lim_{t\to\infty} \frac{1}{t} \log \lambda(B(\epsilon,t)) \leq 0$, while

$$\begin{aligned} &0 < \lambda(W^s_\epsilon(\Lambda)) \leq \lambda(B(\epsilon,t)) \\ &\Rightarrow 0 = \lim_{t\to\infty} \frac{1}{t} \log \lambda(B(\epsilon,t)) = P(J) = h_{\mu_{\mathrm{SRB}}}(\Phi) + \int J\, d\mu_{\mathrm{SRB}}. \end{aligned}$$ □

For context we note that the second implication is reversible, and more:

Theorem 7.4.6 ([63, Proposition 5.4, Theorem 5.6]). *The SRB measure of a basic set Λ depends continuously (in the weak topology) on the C^2-flow (in the C^1-topology),*[15] *as do its entropy and the pressure of the unstable Jacobian, and the following are equivalent:*

- *Λ is an attractor,*
- *$\lambda(W^s_\Lambda) > 0$,*
- *$P(\Phi_{\restriction\Lambda}, J^u) = 0$.*

Corollary 7.4.7. *If Φ is a C^2 hyperbolic flow on a compact manifold M, then the closures of the basins of the attractors cover M, and λ-almost all points approach some attractor. If Λ is a basic set with $\lambda(\Lambda) > 0$, then Λ is a connected component of M, and $\Phi_{\restriction\Lambda}$ is an Anosov flow.*

Proof. Hyperbolicity implies that $M = \bigcup \{W^s(\Lambda) \mid \Lambda \text{ is a basic set}\}$, so the complement of the basins of the attractors is $\bigcup \{W^s(\Lambda) \mid \Lambda \text{ is not an attractor}\}$, which is a null set and hence also has empty interior.

We have $\lambda(\Lambda) > 0$ implies $\lambda(W^s(\Lambda)) > 0$, so Λ is an attractor, hence $W^u(\Lambda) = \Lambda$ (Theorem 5.3.27). Likewise, $\lambda(W^u(\Lambda)) > 0$, so Λ is a repeller, hence $W^u(\Lambda) = \Lambda$ is open; being also closed and connected (since $W^{cs}(p)$ is dense for periodic $p \in \Lambda$), it is a connected component. □

When Corollary 7.4.5 gives $P(J) = 0$, one can use Proposition 7.3.15 to show the following result:

Theorem 7.4.8. *If Λ is a compact hyperbolic attractor of a C^2-flow Φ then the conditionals of the Sinai–Ruelle–Bowen measure on unstable leaves (Definition* A.1.13*) are equivalent to the Riemannian volume on the leaves.*

Remark 7.4.9 (Pesin Entropy Formula). The attractor case of Corollary 7.4.5 is worth emphasizing because it says something of interest about volume entropy: the measure-theoretic entropy of a volume-preserving hyperbolic flow is the average volume-expansion rate in the unstable subbundle. Indeed, since equilibrium states (hence, in this case, volume, for which we could also invoke the Hopf argument) are ergodic, the word "average" can be omitted or replaced by "almost-everywhere value of." This fact turns out to hold in rather greater generality and is known as the *Pesin Entropy Formula* (Remark 7.4.15), which in turn has been further extended by Ledrappier and Young [236, 237, 238].

[15] The dependence is C^1 for C^3 Axiom-A flows when viewed as a functional on C^2-functions [320].

We now come to the main theorem of this section.

Theorem 7.4.10 (Sinai–Ruelle–Bowen measures are physical measures). *Suppose Λ is a compact hyperbolic attractor of a C^2-flow Φ on M and $f\colon M \to \mathbb{R}$ is continuous. Then*

$$\lim_{T\to\infty}\int_0^T f(\varphi^t(x)) = \int f\, d\mu_{\mathrm{SRB}}$$

for λ-almost all $x \in W^s(\Lambda)$.

Remark 7.4.11. The conclusion of this result looks much like that of the Birkhoff Ergodic Theorem (Theorem 3.2.15). Indeed, for μ_{SRB}-almost every point this is a restatement of the Birkhoff Ergodic Theorem. The crucial difference is that, while μ_{SRB} is supported on the attractor, the conclusion here holds for almost every point with respect to the *Riemannian measure* on a neighborhood of the attractor (in fact by Corollary 7.4.7 for almost every point altogether, except that the asymptotics vary depending on which basin a point is in). This means that a point picked "at random" (with respect to the *Riemannian volume*) in a neighborhood of the attractor will be generic for the *Sinai–Ruelle–Bowen measure*, that is, its distribution will represent the Sinai–Ruelle–Bowen measure exactly. In short, a Lebesgue-a.e. point is SRB-generic, or, for a Lebesgue-a.e. point x the empirical measure (Remark 3.2.21) converges weakly to the SRB-measure. This is why physicists call a measure with this property an *observable* or *physical measure*. Note that by the Birkhoff Ergodic Theorem (Theorem 3.2.15), volume is a physical measure if invariant.

We should note as well that the mere existence of Birkhoff averages is a nontrivial part of this result, as noted already in Remark 3.2.25. To emphasize, we have shown that for a hyperbolic flow a Lebesgue-a.e. point has a Birkhoff average (or, in the terminology of Remark 3.2.25, almost no point is historic).

Proof. We restate the result as follows. For some $\delta > 0$, we let

$$\begin{aligned}
f(T,x) &:= \frac{1}{T}\int_0^T f(\varphi^t(x))\,dt \quad\text{and}\quad \bar f := \int f\, d\mu_{\mathrm{SRB}},\\
C_T(f,\delta) &:= \{x \in M \mid |f(T,x) - \bar f| > \delta\},\\
C(f,\delta) &:= \bigcap_{N=1}^{\infty}\bigcup_{n=N}^{\infty} C_n(f,\delta)\\
&= \big\{x \in M \mid |f(n,x) - \bar f| > \delta \text{ for infinitely many } n \in \mathbb{N}\big\},
\end{aligned}$$

and aim to show $\lambda(C(f,\delta)\cap W^s(\Lambda)) \to 0$ for all continuous f.

Take $\epsilon > 0$ such that $|f(x) - f(y)| < \delta$ whenever $d(x,y) < \epsilon$, and fix $N \in \mathbb{N}$. We construct $(n,2\epsilon)$-separated sets recursively in a way that keeps "old" points when passing to larger n: for $n \ge N$ let $R_n \subset \Lambda \cap C_n(f,2\delta)$ be a maximal set for which

(1) $B(x,\epsilon,n) \cap B(y,\epsilon,n) = \varnothing$ if $x \in R_n$, $y \in R_k$, and $N \le k < n$,

(2) $B(x,\epsilon,n) \cap B(y,\epsilon,n) = \varnothing$ if $x, y \in R_n$ and $x \neq y$.

Set $V_{N,\epsilon} := \bigcup_{k=N}^{\infty} \bigcup_{x \in R_k} B(x,\epsilon,k)$.

Claim 7.4.12. $\bigcup_{n=N}^{\infty} C_n(f,3\delta) \cap W^s_\epsilon(\Lambda) \subset V_{N,2\epsilon}$.

Proof. If $x \in C_n(f,3\delta) \cap W^s_\epsilon(\Lambda)$ for some $n \ge N$ then there is a $z \in \Lambda$ such that $x \in W^s_\epsilon(z)$ and hence $z \in C_n(f,2\delta)$ by choice of ϵ. Since R_n is maximal, there is a $y \in R_k$ for some k between N and n such that $B(z,\epsilon,n) \cap B(y,\epsilon,k) \neq \varnothing$ and hence $x \in B(x,\epsilon,n) \subset B(z,2\epsilon,k)$. □

This claim and Proposition 7.4.3 imply that

$$\begin{aligned} \lambda\Big(\bigcup_{n=N}^{\infty} C_n(f,3\delta) \cap W^s_\epsilon(\Lambda)\Big) &\le C_{2\epsilon} \sum_{k=N}^{\infty} \sum_{x \in R_k} j_{-t}(\varphi^t(x)) \\ &\le C_{2\epsilon}\mu_{\mathrm{SRB}}(V_{N,\epsilon})/A_\epsilon \end{aligned} \tag{7.4.3}$$

using Corollary 7.4.5, Proposition 7.3.15, and that $V_{N,\epsilon}$ is defined by a disjoint union (by maximality of R_n).

If $x \in R_k \subset C_k(f,2\delta)$ then $B(x,\epsilon,k) \subset C_k(f,\delta)$ and hence $V_{N,\epsilon} \subset \bigcup_{k=N}^{\infty} C_k(f,\delta)$. Ergodicity of μ_{SRB} (Theorem 7.3.6) and the Birkhoff Ergodic Theorem (Theorem 3.2.15) then imply

$$0 = \mu_{\mathrm{SRB}}(C(f,\delta)) = \lim_{N\to\infty} \mu_{\mathrm{SRB}}\Big(\bigcup_{n=N}^{\infty} C_n(f,\delta)\Big) \ge \lim_{N\to\infty} \mu_{\mathrm{SRB}}(V_{N,\epsilon}).$$

Inequality (7.4.3) then gives $\lambda(\bigcup_{n=N}^{\infty} C_n(f,3\delta) \cap W^s_\epsilon(\Lambda)) \xrightarrow[N\to\infty]{} 0$, so $\lambda(C(f,3\delta) \cap W^s_\epsilon(\Lambda)) = 0$.

To replace $\lambda(W^s_\epsilon(\Lambda))$ by $\lambda(W^s(\Lambda))$ note that if $\delta' \le 3\delta$ then

$$C(f,\delta') \cap \varphi^{-n}(W^s_\epsilon(\Lambda)) \subset \varphi^{-n}(C(f,3\delta) \cap W^s_\epsilon(\Lambda)),$$

which is a set of Riemannian measure 0 since φ^n is a diffeomorphism and hence maps λ-null sets to λ-null sets. Definition 1.5.5 then implies that

$$\lambda(C(f,\delta') \cap W^s_\epsilon(\Lambda)) \le \sum_{n=0}^{\infty} \lambda(C(f,\delta') \cap \varphi^{-n}(W^s_\epsilon(\Lambda))) = 0.$$

This is the desired result. Indeed, since δ' is arbitrary, this implies that

$$\lim_{n\to\infty} f(n,x) = \bar{f} \text{ for all } x \in W^s(\Lambda) \text{ outside a } \lambda\text{-null set } N(f).$$

If $f \in C(M) = \overline{\{f_n\}_{n\in\mathbb{N}}}$ then $\lim_{n\to\infty} f(n,\cdot) = \bar{f}$ on $W^s(\Lambda) \setminus \bigcup_{n\in\mathbb{N}} N(f_n)$. □

Acting on measures gives a parallel version of Theorem 7.4.10: volume dissipates toward the SRB measure.

Theorem 7.4.13. *Suppose Λ is a mixing attractor for a C^2-flow Φ and ν is a probability measure with support in $W^s(\Lambda)$. Then $\varphi^{t*}(\nu) \xrightarrow[t\to\infty]{\text{weak*}} \mu_{\text{SRB}}$, that is, $\int f \circ \varphi^t \, d\nu \xrightarrow[t\to\infty]{} \int f \, d\mu_{\text{SRB}}$ for continuous $f\colon M \to \mathbb{R}$.*

Proof. Approximation arguments show that we can assume without loss of generality that $\nu = \rho\lambda$ for a bounded density $\rho \geq 0$ and that the support of μ is in an open set whose closure is in $W^s(\Lambda)$. Then for a given $\epsilon > 0$ there is a τ such that if $t \geq \tau$, then $\operatorname{supp} \varphi^{t*}(\nu)$ is in an ϵ-neighborhood of Λ, in particular, given a maximal (T,ϵ)-separated set $E \subset \Lambda$ for $\Phi_{\restriction \Lambda}$, $\operatorname{supp} \varphi^{t*}(\nu) \subset \bigcup_{x\in E} B(x, 2\epsilon, T)$. Now the probability measure

$$\nu_{\epsilon,T} := \sum_{x\in E} \left(\frac{\int \psi_x \, d(\varphi^{t(\epsilon)*}(\nu))}{\int \chi_{B(x,\epsilon,T)} \, d\mu} \right) \chi_{B(x,\epsilon,T)} \mu_{\text{SRB}} \ll \mu_{\text{SRB}},$$

where $\{\psi_x\}_{x\in E}$ is a measurable partition of unity on $\operatorname{supp} \varphi^{t*}(\nu)$ subordinate to $\{B(x, 2\epsilon, T)\}_{x\in E}$, can be written as $\nu_{\epsilon,T} = \rho_{\epsilon,T} \mu_{\text{SRB}}$ with $\|\rho_{\epsilon,T}\|_\infty$ bounded independently of T because

$$\frac{\int \psi_x \, d(\varphi^{t(\epsilon)*}(\nu))}{\int \chi_{B(x,\epsilon,T)} \, d\mu} \leq \|r_{t(\epsilon)}\|_\infty \underbrace{\frac{\lambda(B(x,2\epsilon,T))}{\mu_{\text{SRB}}(B(x,\epsilon,T))}}_{\text{bounded by Propositions 7.3.15, 7.4.4, 7.4.3}}$$

if $\varphi^{t(\epsilon)*}(\nu) = r_{t(\epsilon)}\lambda$. This approximates the effect of the flow on ν in the following sense: $\nu_{\epsilon,T}$ redistributes the weight of $\varphi^{t(\epsilon)*}(\nu)$ in such a way that everything that goes into $B(x,\epsilon,T)$ comes from $B(x,2\epsilon,T)$. Likewise, $\varphi^{t*}(\nu_{\epsilon,T})$ redistributes the weight of $\varphi^{t+t(\epsilon)*}(\nu)$ such that what goes into $\varphi^t(B(x,\epsilon,T))$ comes from $\varphi^t(B(x,2\epsilon,T))$, and for $t \in [0,T]$ the diameter of $\varphi^t(B(x,2\epsilon,T))$ is at most 4ϵ. Thus, for a given closed weak neighborhood N of 0 we can choose ϵ such that

$$\varphi^{t+t(\epsilon)*}(\nu) - \varphi^{t*}(\nu_{\epsilon,T}) \in N \quad \text{whenever } t \in [0,T].$$

Since $\|\rho_{\epsilon,T}\|_\infty$ is bounded independently of T, we can find $T_n \to \infty$ such that $\rho_{\epsilon,T} \to \rho_\epsilon$ weakly in L^∞ (as the dual of L^1), so

$$\varphi^{t+t(\epsilon)*}(\nu) - \varphi^{t*}(\rho_\epsilon \mu_{\text{SRB}}) \in N \quad \text{whenever } t \geq 0.$$

On the other hand, μ_{SRB} is mixing, so $\rho_\epsilon \circ \varphi^t \xrightarrow[t\to\infty]{\text{weakly}} \int \rho_e \, d\mu_{\text{SRB}} = 1$ by Proposition 3.4.28, that is, there is a $t_N \in \mathbb{R}$ such that $t \geq t_N \Rightarrow \varphi^{t*}(\rho_\epsilon \mu_{\text{SRB}}) - \mu_{\text{SRB}} \in N$, hence

$$\varphi^{t*}(\nu) - \mu_{\text{SRB}} \in 2N \quad \text{for } t \geq t(\epsilon) + t_N. \qquad \square$$

We now describe the ergodic theory of smooth invariant measures for Anosov flows by using the results about the Sinai–Ruelle–Bowen measure.

Theorem 7.4.14. *Let M be a compact connected smooth Riemannian manifold and Φ a mixing Anosov flow. Then Φ has at most one smooth invariant measure: if ν is a smooth invariant measure then ν is the Sinai–Ruelle–Bowen measure for Φ (hence is mixing by Theorem* 7.3.6*), and $h_\nu(\Phi) = -\int \log J\, d\nu$ (the* Pesin Entropy Formula*).*

Remark 7.4.15. The minus sign in front of the integral that gives entropy combines with that in the definition of J, so the entropy of the SRB measure, in particular of an invariant volume, is given by the average infinitesimal unstable volume-expansion rate. Example 4.1.11 illustrates this.

Proof. If Φ has a smooth invariant measure ν then the Poincaré Recurrence Theorem (Theorem 3.2.1) implies that a ν-a.e. point is nonwandering. Since $\mathrm{NW}(\Phi)$ is closed and ν is positive on open sets, this means that $\mathrm{NW}(\Phi) = M$.

If $f\colon M \to \mathbb{R}$ is continuous, the Birkhoff Ergodic Theorem (Theorem 3.2.15) gives

$$\lim_{T\to\infty} \frac{1}{T}\int_0^T f \circ \varphi^t \overset{\nu\text{-a.e.}}{=\!=\!=} f_{I_\Phi},$$

while Theorem 7.4.10 implies that

$$\lim_{T\to\infty} \frac{1}{T}\int_0^T f \circ \varphi^t \overset{\lambda\text{-a.e.}}{=\!=\!=} \int f\, d\mu_{\mathrm{SRB}}.$$

Since $\nu \ll \lambda$, this latter equality holds ν-a.e., so $f_{I_\Phi} \overset{\nu\text{-a.e.}}{=\!=\!=} \int f\, d\mu_{\mathrm{SRB}}$, and hence

$$\int f\, d\nu = \int f_{I_\Phi}\, d\nu = \int f\, d\mu_{\mathrm{SRB}}.$$

That this holds for all continuous φ means that $\nu = \mu_{\mathrm{SRB}}$.

The entropy is given by Corollary 7.4.5. □

Here is a simple application:

Proposition 7.4.16. *The topological and Liouville entropies of an m-manifold of curvature -1 (Chapter 2) are both $m-1$.*

Proof. The assertion about the Liouville entropy is the Pesin Entropy Formula. This coincides with the topological entropy because the Liouville measure coincides with the measure of maximal entropy: the Jacobian is constant, so the equilibrium state is the same as that for potential 0. □

Remark 7.4.17. This generalizes to locally symmetric spaces of nonconstant curvature because here as well the Jacobian is constant, and so the Liouville measure has maximal entropy. When curvature is normalized to have maximum -1, the entropy of (the geodesic flow of) complex hyperbolic m-space is $2m$ because there is a 1-dimensional direction with expansion rate 2, and this adds 1 to the entropy of the Liouville measure by the Pesin Entropy Formula. For quaternionic hyperbolic m-spaces it is $4m + 2$ because 3 of the $4m - 1$ expanding directions have a rate of 2, and for the Cayley plane it is 22, accounting for 7 of 15 directions having expansion rate 2.

Theorem 7.4.18 ([246, Corollary 4.4]). *For a C^k Anosov 3-flow the densities (from Theorem 7.4.8) of the SRB-measure in each unstable (or stable) leaf are C^k.*

Proof. The densities are defined up to a multiplicative constant and if $y \in W^u(x)$ then one can show that

$$\frac{\rho(y)}{\rho(x)} = \lim_{n\to\infty} \frac{J^u f^{-n}(x)}{J^u f^{-n}(y)},$$

and the latter expression is in fact C^k along the leaves, and all derivatives are continuous. A similar argument holds for the stable leaves. □

We now show an application to Anosov 3-flows that preserve a contact structure (see Section 6.2).

Theorem 7.4.19. *Orbit-equivalent contact Anosov flows on a 3-dimensional manifold are C^1-conjugate if the periods of corresponding periodic orbits coincide.*

Proof. The flows are Hölder conjugate by Theorem 6.3.9. The conjugacy preserves the strong-stable and strong-unstable foliations and the orbits. The weak-stable and weak-unstable foliations are C^1 by Corollary 9.4.11. The strong-stable and strong-unstable subbundles are the intersections of the weak-stable and weak-unstable subbundles with the kernel of the contact form by Lemma 8.1.5. Since the contact form is C^1 this implies that the strong foliations are C^1.

By Theorem 7.4.14 the equilibrium state for the unstable Jacobian is the invariant volume induced by the contact form. This implies that the conjugacy preserves volume, and since it preserves the three 1-dimensional foliations (orbits and the strong foliations), which are C^1, it preserves the conditional measures on those foliations and hence it is C^1. □

Finally, we explain how one obtains the Bernoulli property from K-mixing and hyperbolicity [270, 91].

Theorem 7.4.20. *The Sinai–Ruelle–Bowen measure is Bernoulli. In particular, weakly mixing volume-preserving Anosov flows*[16] *are Bernoulli.*

[16] And, more generally, equilibrium states for Anosov flows [310].

Proof outline [270]. By Theorem 3.4.47 we need to check the very weak Bernoulli property from Definition 3.4.45. Let ξ be a finite generating (or very fine) partition. We will show that it has the desired property by an analysis of suitable small local-product neighborhoods. Consider a local-product neighborhood of the type unstable $\times$ center stable on which the SRB measure μ_{SRB} is equivalent to a product measure $\bar{\mu}$ (Remark 7.1.21) with bounded Radon–Nikodym derivative r. Being measurable, r is close to a step function over measurable rectangles, so for $\epsilon > 0$ there is a smaller local-product neighborhood $P \sim P^u \times P^{cs}$ and an $r_0 > 0$ such that

$$B := \{x \in P \mid |r_0 - r(x)| \geq \epsilon^2\}$$

satisfies $(\mu_{\text{SRB}})_P(B) \leq \epsilon^2$ and $\bar{\mu}_P(B) \leq \epsilon^2$; in a possibly smaller B we can further assume that most center-stable fibers have almost the same conditional measure. This means that there are ϵ-measure-preserving maps $\theta_z \colon P^u \times \{z\} \times [0,1] \to P$, $(p,z,x) \mapsto (p, \vartheta(x))$ along weak-stable leaves, where $\vartheta \colon [0,1] \to P^{cs}$ (with normalized measure) is a measurable isomorphism.

We will use these θ_z to construct the map θ required by Definition 3.4.45. In anticipation, we note that if θ_z were to slide along *strong* stable leaves, we would be assured that future itineraries match as required by Definition 3.4.45, but the construction of θ_z ensures only that the forward orbits of a point and its image stay close rather than being positively asymptotic. This is sufficient for our purposes: take P to be thin enough in the flow direction (depending on the chosen partition ξ) such that the set of times where future itineraries mismatch has density less than ϵ (by ergodicity).

We now map P by iterates of $f := \varphi^t$ to get local product neighborhoods $P_n = f^n(P)$ with induced ϵ-measure-preserving maps θ_n from almost every unstable leaf in P_n onto P_n. By ergodicity, there is an $N \in \mathbb{N}$ such that each point is contained in about equally many P_n with $1 \leq n \leq N$, and by K-mixing there is an $M \in \mathbb{N}$ such that if $m \geq M$, then ϵ-a.e. atom E of $\bigvee_{j=-m}^{-M} f^{-j}(\xi)$ intersects each P_n in a set of $(\mu_{\text{SRB}})_E$-measure close to $\mu_{\text{SRB}}(P)$ and further, $\sum \chi_{P_n}$ is ϵ-close to $N\mu(P)$ ϵ-a.e. on E. Then $X \times [0,1]$ is ϵ-approximated by $\bigcup_{n=0}^{N} P_n \times J_n$, where the $J_n \subset [0,1]$ are disjoint intervals of length $1/N\mu(P)$, and the restriction of the various θ_n to $E \times [0,1]$ gives, modulo these approximations, the required map in Definition 3.4.45. □

Remark 7.4.21. This indicates that the Bernoulli property is central to the smooth ergodic theory of hyperbolic flows, and this conclusion is quite broadly true, that is, even when the hyperbolicity is nonuniform or the system has singularities [270, §1]. More remarkably yet, hyperbolic techniques can be brought to bear beyond situations in which hyperbolicity per se is assumed:

Theorem 7.4.22 ([235]). *Every equilibrium measure of a contact* 3*-flow for a Hölder continuous potential has at most countably many ergodic components with positive entropy, and each of them is Bernoulli.*

This implies strong stochastic properties of geodesic flows on surfaces in a generality quite beyond Theorem 5.2.8:

Corollary 7.4.23 ([235]). *The geodesic flow of a nonflat compact smooth orientable closed surface of nonpositive curvature is Bernoulli with respect to its* (*unique*) *measure of maximal entropy.*

In Theorem 7.3.25 we noted that equilibrium states depend continuously on the dynamical system, and this section focuses on a preferred equilibrium state. Even though in some ways the Sinai–Ruelle–Bowen measure is harder to obtain than some equilibrium states for other more regular potentials, it depends more nicely on the dynamical system than stated in Theorem 7.3.25, in part because the potential itself is tied to the dynamics:

Theorem 7.4.24 ([320, 86]). *The Sinai–Ruelle–Bowen varies C^1 with an Axiom-A flow.*

At the end of our discussion at the end of Section 7.3 about extensions of the Bowen techniques beyond uniform hyperbolicity, we stated a result about Hölder potentials on rank-1 manifolds (Theorem 7.3.29). The same work also produced insights related to Sinai–Ruelle–Bowen measures:

Theorem 7.4.25 (Burns–Climenhaga–Fisher–Thompson). *For $q < 1$ and the geometric potential qJ^u* (*Definition* 7.4.1), *the geodesic flow of a closed rank-*1 *surface has a unique equilibrium state μ_q, which is hyperbolic, Bernoulli* (*Theorem* 7.4.22), *fully supported, and the weak* limit of weighted regular closed geodesics. The function $q \mapsto P(qJ^u)$ is C^1.*

Remark 7.4.26. It is conspicuous that SRB-measure is excluded in Theorem 7.4.25 because q cannot be 1. This is because in that case the pressure is $h(\mu) - \int \varphi^u = 0$ because (for a surface) the integrand is 0 a.e., and so is the entropy on the singular set. This means that the restriction to the singular set is h-expansive and hence has an equilibrium state (Remark 4.2.20)—which, being supported on the singular set, differs from the equilibrium state on the regular set.

7.5 Hamenstädt–Margulis measure*

This section presents an alternative construction due initially to Margulis of the unique measure of maximal entropy in the case of topologically mixing Anosov flows

(Theorem 7.5.2). Unlike the Bowen construction (Theorem 7.3.6), which produces this measure as a limit distribution of periodic orbits, the Hamenstädt–Margulis construction builds on geometrically defined measures on unstable leaves. This alternative description allows us to obtain a much more precise asymptotic of the growth rate for the number of periodic orbits (Theorem 7.6.9). Throughout this section we assume that Φ is a topologically mixing Anosov flow on M (Theorem 7.3.22). We first establish some notation.

Definition 7.5.1. For $i = s, u, cs, cu$ we write $A \subset W^i$ when $A \subset W^i(p)$ for some $p \in M$ and use the notions of openness, compactness, continuity, and measurability for sets and functions defined on an unstable leaf. The distance function on a leaf $W^i(p)$ induced by the Riemannian structure of this leaf is denoted by d^i, and $B^i(p,r) := \{q \in W^i(p) \mid d^i(p,q) < r\}$ is the r-ball around p in $W^i(p)$.

If $x, y \in \Lambda$ are sufficiently close, then there is a well-defined *holonomy map* $\mathcal{H} = \mathcal{H}^s \colon B^{cu}(x,\epsilon) \to W^{cu}(y), z \mapsto [z,y] \in B^{cu}(y,\delta)$ as in Remark 6.2.3, where ϵ and δ depend on x and y. We say that $A, B \subset W^{cu}$ are *ϵ-equivalent* if there is a local-product rectangle $\mathcal{P} \colon A \times D \to M$ with $a, b \in D$ such that $\mathcal{P}(A \times \{a\}) = A$, $\mathcal{P}(A \times \{b\}) = B$, and $\operatorname{diam}^s(\mathcal{P}(\{x\} \times D)) < \epsilon$ for all $x \in A$; we write $\mathcal{H}(\mathcal{P}((x,a))) = \mathcal{P}(x,b)$, and likewise for $A, B \subset W^u$ and the holonomy $\mathcal{H}^{cs}$ along center-stable leaves.

Theorem 7.5.2 (Bowen–Margulis measure). *For a C^2 Anosov flow with dense strong-stable and strong-unstable leaves the Bowen measure coincides with the* Hamenstädt–Margulis measure *(Definition* 7.5.19*), an invariant Borel probability measure μ with full support such that the conditionals μ^u_z on unstable leaves $W^u(z)$ are fully supported holonomy-invariant Radon measures, and $\mu^u_{\varphi^t(z)} \circ \varphi^t = e^{h_{\mathrm{top}}(\Phi)t} \cdot \mu^u_z$.*

Throughout, we take a as large as possible such that $\mu = e^a$ is as in Definition 5.1.1, except that we assume that the Riemannian metric is adapted to the dynamics as in Proposition 5.1.5, that is, that for all $t > 0$ and all $x \in M$ we have

$$\|D\varphi^t \restriction_{E^s_x}\|^* \le e^{-at} \quad \text{and} \quad \|D\varphi^{-t} \restriction_{E^u_x}\|^* \le e^{-at}.$$

Definition 7.5.3 (Hamenstädt metric). Fix $R \in \mathbb{R}$, and for $x, y \in W^u(z)$ define

$$\eta(x,y) := \eta_{z,R}(x,y) := e^{-\sup\{t \in \mathbb{R} \mid d_{\varphi^t(z)}(\varphi^t(x),\varphi^t(y)) \le R\}},$$

and call η^a the *Hamenstädt metric*. For $x \in W^u(z)$ denote by $B_\eta(x,\epsilon)$ the ϵ-ball for η around x.

Remark 7.5.4. It is clear that this metric is entwined with the dynamics of the flow, and the connection is related to topological entropy in particular since

$$B_{\eta_{x,R}}(x,e^{-t}) = B^u_\Phi(x,R,t) := B_\Phi(x,R,t) \cap W^u_R(x) \quad \text{(Bowen ball).} \tag{7.5.1}$$

Lemma 7.5.5. *η^a is a distance on $W^u(z)$, $\eta_z = \eta_{z'}$ for $z' \in W^u(z)$, $\eta \circ \varphi^t = e^t\eta$, and*

$$\varphi^{-t}(B_\eta(x,\epsilon)) = B_\eta(\varphi^{-t}(x), e^{-t}\epsilon). \tag{7.5.2}$$

Proof. The triangle inequality $\eta(x_1,x_2)^a \le \eta(x_1,y)^a + \eta(y,x_2)^a$ is clear if any $r_i := d_{\varphi^t(z)}(\varphi^t(x_i),\varphi^t(y)) > R$ for $i = 1,2$, where $t := -\log\eta(x_1,x_2)$, because then $\eta^a(x_i,y) > \eta^a(x_1,x_2)$. If both $r_i < R$, then

$$\eta^a(x_1,x_2) = e^{-at} \underbrace{\le}_{r_1+r_2 \ge d_{\varphi^t(z)}(\varphi^t(x_1),\varphi^t(x_2))=R} \frac{r_1+r_2}{R} e^{-at} \underbrace{\le}_{\text{Claim 7.5.6 with } \Delta=R,\ \delta=r_i,\ x=x_i} \eta^a(x_1,y) + \eta^a(x_2,y).$$

Further, $\eta \circ \varphi^t = e^t\eta$ implies

$$\varphi^{-t}(B_\eta(x,\epsilon)) = \{y \mid \underbrace{\eta(\varphi^t(y),x)}_{=e^t\eta(y,\varphi^{-t}(x))} < \epsilon\} = B_\eta(\varphi^{-t}(x), e^{-t}\epsilon).$$ □

Claim 7.5.6. *$\eta^a_\Delta(x,y) \ge (\delta/\Delta)e^{-at}$, where $\delta := d_{\varphi^t(z)}(\varphi^t(x),\varphi^t(y))$.*

Proof. Since $d_{\varphi^{t+\tau}(z)}(\varphi^{t+\tau}(x),\varphi^{t+\tau}(y)) \ge \delta e^{a\tau} = \Delta$ if $e^{a\tau} = \Delta/\delta$, so $\eta_\Delta(x,y) \ge e^{-(t+\tau)} = (\delta/\Delta)^{1/a}e^{-t}$. □

Claim 7.5.6 with $\delta = r \le \Delta = R$, $t = -\log\eta_r(x,y)$ also gives the following corollary:

Corollary 7.5.7. *$\eta_R \le \eta_r \le (R/r)^{1/a}\eta_R$.*

Remark 7.5.8. For geodesic flows of Riemannian metrics whose curvature is bounded above by -1, we can take $a = 1$, so η is a proper distance. We suppose henceforth that having rescaled time by a if necessary, we can take $a = 1$, making η a distance.[17]

Definition 7.5.9 (Spherical measure). For $h := h_{\text{top}}(\Phi)$ and $S \subset W^u(z)$ let

$$\sigma_\epsilon(S) := \inf\left\{ \textstyle\sum_{i\in\mathbb{N}} \epsilon_i^h \mid S \subset \bigcup_{i\in\mathbb{N}} B_\eta(x_i,\epsilon_i) \text{ with } x_i \in W^u(z),\ \epsilon_i \le \epsilon\right\},$$
$$\sigma(S) := \sigma_z(S) := \sup_{\epsilon>0} \sigma_\epsilon(S).$$

Remark 7.5.10. Here σ is the h-dimensional spherical measure on W^u associated with η. If $a \ne 1$, this might not be a distance, and one could view σ as the h/a-dimensional spherical measure associated with the Hamenstädt distance η^a.

[17] This understates the importance of this issue, which is a significant part of, for instance, the Lyapunov-rigidity theorem of Butler (Theorem 9.3.14), who points out that control of time is less important when the stable and unstable foliations have dimension at least 2; this is one of the principles on which Mostow rigidity [48] works, in which homotheties (modeling the effect of time changes) do not change the property of being quasiconformal, which is a trivial property in dimension 1.

Our first few arguments do not depend on a particular choice of h, but readers familiar with the notion of Hausdorff dimension will guess that there is at most one choice of h for which σ defines a nonzero Radon measure; Theorem 7.5.12 shows that $h = h_{\text{top}}(\Phi)$ is that parameter, and it also establishes a property of independent interest (see Theorem 7.5.23).

Definition 7.5.11. A metric measure space X, μ, d is said to be *Ahlfors-regular* with exponent N if $r^{-N}\mu(B_d(x,r))$ has positive upper and lower bounds.

Theorem 7.5.12. *$(W^u(z), \mu_0^u, \eta_{z,R})$ is Ahlfors-regular (uniformly in z) with exponent $h = h_{\text{top}}(\Phi)$, where μ_0^u is the unstable conditional of μ_0 (Bowen measure of maximal entropy).*

Remark 7.5.13. Since the unstable foliation does not define a measurable partition (globally), these conditional measures are uniquely defined on local product rectangles (Definition A.1.13) modulo the normalization of the rectangle to a probability space; by using a finite cover by product rectangles, the conditionals are well defined modulo a finite set of constant scale factors; this does not affect the conclusion of Theorem 7.5.12. A posteriori, Theorem 7.5.2 establishes, of course, that these conditional measures are, in fact, entirely canonical.

Proof. The definition of conditional measure, (7.5.1), and the Gibbs property (7.3.4) imply $A_R \le e^{ht}\mu_0^u(B_{\eta_{x,R}}(x, e^{-t})) \le B_R$. □

This allows us to replace ϵ_i^h in the definition of σ by the conditional Bowen measure of a ball, which gives the following result:

Corollary 7.5.14. *$A_R\sigma \le \mu_0^u \le B_R\sigma$, so σ is a fully supported Radon measure.*

Theorem 7.5.15 (Scaling). *$\sigma(\varphi^t(S)) = e^{ht}\sigma(S)$.*

Proof. Writing $\epsilon_i' = e^{-t}\epsilon_i$, and $x_i' = \varphi^{-t}(x_i)$ gives

$$\begin{aligned}
&\sigma_\epsilon(\varphi^t(S)) \\
&\quad = \inf\Big\{ \textstyle\sum_{i\in\mathbb{N}} \epsilon_i^h \;\Big|\; \varphi^t(S) \subset \bigcup_{i\in\mathbb{N}} B_\eta(x_i,\epsilon_i) \text{ with } x_i \in W^u(z),\ \epsilon_i \le \epsilon \Big\} \\
&\quad = \inf\Big\{ \textstyle\sum_{i\in\mathbb{N}} \underbrace{\epsilon_i^h}_{=e^{ht}\epsilon_i'} \;\Big|\; S \subset \bigcup_{i\in\mathbb{N}} \underbrace{\varphi^{-t}(B_\eta(x_i,\epsilon_i))}_{\overset{(7.5.2)}{=\!=}B_\eta(\varphi^{-t}(x_i),e^{-t}\epsilon_i)=B_\eta(x_i',\epsilon_i')} \text{ with } x_i' \in W^u(\varphi^{-t}(z)),\ \epsilon_i' \le e^{-t}\epsilon \Big\} \\
&\quad = e^{ht}\sigma_{e^{-t}\epsilon}(S).
\end{aligned}$$

□

Lemma 7.5.16. *For $\delta > 0$ there is an $\epsilon > 0$ such that if $S, S' \subset W^u$ are ϵ-equivalent (Definition 7.5.1), then $(1-\delta)\sigma(S) \le \sigma(S') \le (1+\delta)\sigma(S)$.*

Proof. By symmetry it suffices to prove $\sigma(S') \le \big(\frac{R+\theta(C\epsilon)}{R}\big)^{h/a}\sigma(S)$ with $\theta(\epsilon) \xrightarrow[\epsilon\to 0]{} 0$. To that end suppose $S \subset \bigcup_{i\in\mathbb{N}} B_{\eta_z}(x_i,\delta_i)$ with $\sum_i \delta_i^h \le \sigma(S)+\delta$ and $\delta_i < \delta < 1$. Now, if $\{x\}$, $\{x'\}$ are ϵ-equivalent, then $x'' := W^{cu}(x') \cap W^s(x) = \varphi^\tau(x')$ for some $\tau \in \mathbb{R}$, so there is a (uniform) $C \in \mathbb{R}$ for which $\varphi^t(x)$, $\varphi^t(x')$ are $C\epsilon$-equivalent for $t > 0$. Thus, if $y \in S \cap B_{\eta_z}(x,\delta)$, then $d_{\varphi^{-\log\delta}(z)}(\varphi^{-\log\delta}(x), \varphi^{-\log\delta}(y)) < R$ and

$$d_{\mathcal{H}(\varphi^{-\log\delta}(z))}(\mathcal{H}(\varphi^{-\log\delta}(x)), \mathcal{H}(\varphi^{-\log\delta}(y))) < R + \theta(C\epsilon)$$

with $\theta(\epsilon) \xrightarrow[\epsilon\to 0]{} 0$ by uniform continuity of E^u, so Corollary 7.5.7 implies

$$\eta_{\mathcal{H}(z)}(\mathcal{H}(x), \mathcal{H}(y)) < \Big(\frac{R+\theta(C\epsilon)}{R}\Big)^{1/a}\delta.$$

Then $\mathcal{H}(S \cap B_{\eta_z}(x_i,\delta_i)) \subset S' \cap B_{\eta_{\mathcal{H}(z)}}\big(\mathcal{H}(x_i), \big(\frac{R+\theta(C\epsilon)}{R}\big)^{1/a}\delta_i\big)$, hence the claim. □

Setting $\sigma =: \mu^u$ induces a family of nonzero Radon measures μ^{cu} on leaves of W^{cu} by setting $\mu^{cu}(\varphi^{[0,\epsilon)}(S)) = \epsilon\mu^u(S)$ for $S \subset W^u$ and extending this to a measure—which then also has the scaling property from Theorem 7.5.15:

$$\mu^{cu}(\varphi^t(S)) = e^{ht}\mu^{cu}(S). \tag{7.5.3}$$

Theorem 7.5.17. *If $U_1, U_2 \in W^{cu}$ are ϵ-equivalent for some $\epsilon > 0$, then $\mu^{cu}(U_1) = \mu^{cu}(U_2)$.*

Proof. It suffices to consider $U_1 = \varphi^{[0,\tau]}(U)$ with $U \subset W^u$. Then $\mu^{cu}(U_1) = \tau\mu^u(U)$, and

$$U_2 = \mathcal{H}^s(U_1) = \mathcal{H}^s(\varphi^{[0,\tau]}(U)) = \varphi^{[0,\tau]}(\mathcal{H}^s(U)) = \bigcup_{x\in U}\varphi^{[f(x),f(x)+\tau]}(\mathcal{H}^{cs}(U)),$$

hence $\mu^{cu}(U_2) = \tau\mu^u(\mathcal{H}^{cs}(U))$. For any $\delta > 0$ take $\epsilon > 0$ as in Lemma 7.5.16 and $T \in \mathbb{R}$ such that $\varphi^T(U)$ and $\varphi^T(\mathcal{H}^{cs}(U))$ are ϵ-equivalent. Then

$$\frac{\mu^{cu}(U_1)}{\mu^{cu}(U_2)} = \frac{\mu^u(U)}{\mu^u(\mathcal{H}^{cs}(U))} \overset{\text{Theorem 7.5.15}}{=\!=\!=\!=} \frac{\mu^u(\varphi^T(U))}{\mu^u(\varphi^T(\mathcal{H}^{cs}(U)))} \in [1-\delta, 1+\delta]$$

by Lemma 7.5.16. Since δ was arbitrary, this proves Theorem 7.5.17. □

Proof of Theorem 7.5.2. The construction in Definition 7.5.9 also produces a spherical measure μ^s on W^s, and with the property

$$\mu^s(\varphi^t(S)) = e^{-ht}\mu^s(S) \tag{7.5.4}$$

from Theorem 7.5.15. From these measures on leaves we now construct a finite Φ-invariant measure on M by locally defining a *weighted product measure* as follows.

Every $p \in M$ has a neighborhood $U(p)$ which is a *rectangle*, that is, using the local product structure we can write $U(p)$ as $U^{cu}(p) \times U^s(p) \subset W^{cu}(p) \times W^s(p)$. If $O \subset U(p)$ let

$$f_O(q) := \mu^s((\{q\} \times U^s(p)) \cap O) \quad (q \in U^{cu}(p)).$$

Lemma 7.5.18. *f_O is upper semicontinuous (hence locally integrable).*

Proof. For $x \in U^{cu}$ and $\epsilon > 0$ there exists $\delta > 0$ such that $d^{cu}(x, y) < \delta \Rightarrow$

$$\frac{\mu^s\left((\pi^s\left(O \cap (\{x\} \times U^s)\right) \times U^{cu}) \cap (\{y\} \times U^s)\right)}{\mu^s((\{y\} \times U^s) \cap O)} > 1 - \epsilon,$$

where π^s is the projection to W^s. Now apply Lemma 7.5.16. □

For $q \in U^s(p)$, $A \subset U^{cu}(p)$ let $\mu_q(A) := \mu^{cu}(A \times \{q\})$ wherever defined. By Theorem 7.5.17 this is independent of $q \in U^s(p)$. Together with Lemma 7.5.18 this shows that

$$\mu(O) := \int f_O(x)\, d\mu_q(x) \tag{7.5.5}$$

is well defined. Extending this to the completion of the Borel σ-algebra gives a measure which is finite since M is compact and (local product) rectangles have finite measure and are Φ-invariant: the scaling properties (7.5.3) and (7.5.4) combine with (7.5.5) to give

$$\mu(\varphi^t(A)) = \mu(A)$$

for all measurable $A \in M$. Lemma 7.5.16 and Theorem 7.5.17 establish holonomy invariance of the conditionals. After normalizing to a probability measure, Corollary 7.5.14 then implies Theorem 7.5.2. □

Definition 7.5.19. The Φ-invariant Borel probability measure obtained from (7.5.5) by extension and normalization is called the *Hamenstädt–Margulis measure* or *Margulis measure* for Φ.

Remark 7.5.20. Holonomy invariance of the measure of maximal entropy holds not only for mixing Anosov flows, but for mixing Axiom-A flows (Theorem 7.3.28 [62]). Thus, the measure of maximal entropy is widely referred to also as the *Bowen–Margulis measure*.

Remark 7.5.21 (Butler). A fully supported measure μ on a 1-manifold defines a distance by setting $d(x, y)$ to be the measure of the arc from x to y. This means that for an Anosov 3-flow (such as the geodesic flow of a surface), the Bowen–Margulis conditionals define distances on stable and unstable leaves—which are each bi-Lipschitz to the corresponding η from Definition 7.5.3. Thus, in this case, the Hamenstädt metric provides no additional information beyond the Bowen–Margulis measure.

Example 7.5.22 (Butler). To think through the Hamenstädt construction in concrete terms and in a different situation, consider the suspension φ^t of a linear toral automorphism f. Suppose E^u is 2-dimensional and f has eigenvalues $b_1 = e < e^2 = b_2$ on E^u and hence positive Lyapunov exponents 1 and 2. The unstable foliation W^u is smoothly subfoliated by foliations W_1 and W_2 into lines that correspond to these two exponents. With the parameter $a = 1$ in the Hamenstadt metric η, check through scaling arguments that $\eta\restriction_{W^u}$ is bi-Lipschitz equivalent to the metric on the leaf induced by the norm (in orthonormal coordinates (x, y) on the leaf with the foliations W_1 and W_2 being the x- and y-axes respectively)

$$\|(x, y)\| = \sqrt{|x|^2 + |y|}.$$

The point is that this norm has the same e^t-scaling as the Hamenstädt metric since $\varphi^t(x, y) = (e^t x, e^{2t} y)$ in these coordinates.

Show from this description that η has Hausdorff dimension $1 + 2 = 3$. On the other hand, we can also compute that this flow has topological entropy $(\log b_1 + \log b_2)/\log b_1 = 1 + 2$.

This can be generalized to any linear toral automorphism with simple spectrum on its unstable bundle, although the formulas become more complicated, and it is closely related to a standard means of analyzing the Carnot–Carathéodory metric on the Heisenberg group using the Koranyi norm.[18]

Remark 7.5.10 alluded to the notion of Hausdorff dimension, and this is indeed closely connected to the subject at hand:

Theorem 7.5.23. *An Ahlfors N-regular space has Hausdorff dimension N.*

This means that we have shown the Hamenstädt distance to have Hausdorff dimension $h = h_{\text{top}}(\Phi)$.

There are much deeper connections between dynamics, entropy, and fractal dimensions (Definition 4.2.34) [287],[19] and they are of a rather different nature because of the rather special nature of the Hamenstädt metrics. An illuminating instance arises in connection with the Pesin Entropy Formula (Remark 7.4.9) according to which the entropy of volume (if invariant) is given by the sum of the positive Lyapunov exponents (Remark 7.4.15). For other invariant measures, this can be

[18] See Section 4 of http://www.math.montana.edu/wildrick/courses/597-fall-15/lectures/Lecture_12.pdf.
[19] We first saw a connection in Theorem 4.2.36.

extended, provided each Lyapunov exponent is weighted with the fractal dimension of the measure conditioned on the corresponding unstable subleaf [236, 237, 238]. For Lebesgue measure, this is just the multiplicity, but for other measures this puts the fractal dimension at the center of connecting expansion and complexity.

7.6 Asymptotic orbit growth*

Following Charles H. Toll, we establish a *multiplicative asymptotic* of the growth of periodic orbits for flows, that is, finer bounds than a determination of the exponential rate (Theorem 7.6.9).[20] The main ingredient is a description of "local product flow boxes" and their *full components* of intersection. These provide the context in which the equality of Bowen measure and Margulis measure as well as the earlier estimates give the multiplicative asymptotic of the growth of periodic points. Throughout this section we again assume that Φ is a topologically mixing Anosov flow on a compact Riemannian manifold M.

Local product flow boxes are simple to describe and involve two size parameters ϵ and δ. We first obtain several technically useful properties that depend on δ being sufficiently small with respect to ϵ.

Take $0 < \epsilon < 1/2 \min\{1, \delta_0\}$, where δ_0 is an expansivity constant of Φ (see Definition 1.7.2), that is, if $d(\varphi^t(x), \varphi^t(y)) < \delta_0$ for all $t \in \mathbb{R}$ then $y \in \mathcal{O}(x)$. Assume furthermore that ϵ is less than the least period of orbits of Φ. Given $p \in M$ let $C := B^{cu}(p) := \bigcup_{0 \le t \le \epsilon} \varphi^t(\overline{B^u(p,\delta)})$. If ϵ is sufficiently small then $B := \bigcup_{z \in C} B^s(z)$ is a local product rectangle, where $B^s(z) \subset \overline{B^s(z,\delta)}$. For $x \in B$ denote by $B^u(x)$ the connected component of $W^u(x) \cap B$ containing x and similarly define $B^{cu}(x)$ and $B^s(x)$. Define π_C by $C \cap B^s(x) = \{\pi_C(x)\}$. If $z \in C$ then clearly B contains an orbit segment of (parameter) length ϵ. This is true for all $x \in B$:

Lemma 7.6.1. *If $x \in B$ then there exists $t_0 \in [0, \epsilon]$ such that $\{\varphi^t(x) \mid t \in [t_0 - \epsilon, t_0]\} \subset B$.*

Proof. Let $z = \pi_C(x)$. If $\varphi^t(z) \in C$ then $\varphi^t(B^s(x)) = B^s(\varphi^t(x))$, hence $\{\varphi^t(x)\} = B^s(\varphi^t(z)) \cap B^{cu}(x) \subset B$. □

From Section 6.2 we recall the following theorem:

Theorem 7.6.2. *There are η, $\gamma > 0$ such that if $d(x,y) < \eta$ then there is a unique $\theta = \theta(x,y) \in (-\gamma, \gamma)$ such that $\varnothing \neq B^s(x,\gamma) \cap B^u(\varphi^\theta(y)) =: \{[x,y]\}$. Further, θ and $[\cdot,\cdot]$ are continuous on $\{(x,y) \in M \times M \mid d(x,y) < \eta\}$.*

[20] A finer asymptotic in greater generality was contemporaneous with Toll's work [278].

Define $\tau \colon B \to [0,\epsilon]$ by $z \in \varphi^{\tau(z)}(B^u(p))$ when $z \in C$ and $\tau(x) := \tau(\pi_C(x))$ for $x \in B$. For sufficiently small δ uniform continuity of the unstable foliation yields

$$\tau(B^u(x)) \subset [\tau(x) - \epsilon^2, \tau(x) + \epsilon^2] \tag{7.6.1}$$

for $x \in B$. We furthermore assume that δ is also small enough that if $y \in B^u(x)$ then $B^s(x)$ and $B^s(y)$ are ζ-equivalent (Definition 7.5.1) with ζ small enough that (Lemma 7.5.16)

$$\left| \frac{\mu^s(B^s(x))}{\mu^s(B^s(y))} - 1 \right| < \epsilon.$$

Lemma 7.6.3. *There exists $K > 0$ (independent of ϵ and δ) such that if $x, y \in B$ then*

$$\left| \frac{\mu^s(B^s(x))}{\mu^s(B^s(y))} - 1 \right| < K\epsilon.$$

Proof. Suppose $\tau(x) = t$, $\tau(y) = t'$, $\tau(w) = 0$. Then

$$\frac{\mu^s(B^s(x))}{\mu^s(B^s(y))} = \frac{\mu^s(B^s(x))}{\mu^s(B^s(\varphi^t(w)))} \underbrace{\frac{\mu^s(B^s(\varphi^t(w)))}{\mu^s(B^s(\varphi^{t'}(w)))}}_{=e^{h(t-t')}} \frac{\mu^s(B^s(\varphi^{t'}(w)))}{\mu^s(B^s(y))}.$$

The remaining fractions differ from 1 by at most ϵ, so we obtain the claim. □

We write $x \sim y$ if $x, y \in B$ lie on a common orbit segment contained in B and let $[x] := \{y \in B \mid y \sim x\}$. Relation (7.6.1) shows that if $y \in B^{cu}(x)$ and $\tau(x), \tau(y) \in (\epsilon^2, \epsilon - \epsilon^2)$ then $\bigcup_{z \in B^u(x)}[z] = \bigcup_{z \in B^u(y)}[z]$. For $x \in B$ we set $\Delta(x) := \sup\{r > 0 \mid B^u(x,r) \subset B\}$. For $T > 0$ there exists $r(T)$ (exponentially decreasing as $T \to \infty$) such that if $x, \varphi^T(x) \in B$, $\Delta(x) > r(T)$ then $B^u(\varphi^T(x)) \subset \varphi^T(B)$.

Definition 7.6.4. Let $B^\circ(T) := \{x \in B \mid \epsilon^2 < \tau(x) < \epsilon - \epsilon^2,\ \Delta(x) > r(T)\}$. If Δ_0 is a connected component of $B^\circ(T) \cap \varphi^T(B^\circ(T))$ then $\Delta := \bigcup_{x \in \Delta_0}[x] \cap \varphi^T(B)$ is called a *full component of intersection*.

An important observation is that full components of intersection essentially correspond bijectively to periodic orbits. This is the content of the next two lemmas.

Lemma 7.6.5. *If Δ is a full component of intersection of $B \cap \varphi^T(B)$ then Δ intersects a unique orbit of period in $[T - \epsilon, T + \epsilon]$.*

Proof. Consider the action of the iterates of φ^T on the projection of Δ to $B/\!\sim$ to obtain a unique fixed point which corresponds to the desired orbit. □

Conversely it is easy to check the following lemma:

Lemma 7.6.6. *Each orbit segment in $B^\circ(T)$ of length $\epsilon - 2\epsilon^2$ that belongs to a periodic orbit of period in $[T-(\epsilon-2\epsilon^2), T+(\epsilon-2\epsilon^2)]$ intersects a unique full component of intersection of $B \cap \varphi^T(B)$.*

Next we estimate the number of full components of intersection:

Proposition 7.6.7. *Let $\Delta(T)$ be the number of full components of intersection in $B \cap \varphi^T(B)$. Then $\Delta(T) = 2e^{hT}\mu(B)(1+O(\epsilon))(1+o(T^0))$, with $O(\epsilon)$ independent of B.*

Proof. We will calculate $\mu(\varphi^T(B))$ in (7.6.3) and $\mu(B \cap \varphi^T(B))$ in (7.6.6). Since μ is mixing (Theorem 7.3.6), this yields the claim. If $x, y \in C$ then $B^s(x)$ and $B^s(y)$ are ζ-equivalent, hence $\mu^s(B^s(x)) = (1+O(\epsilon))\mu^s(B^s(y))$, and by the scaling property $\mu^s(\varphi^T(B^s(x))) = (1+O(\epsilon))\mu^s(\varphi^T(B^s(y)))$ as well and hence

$$\mu^s(\varphi^T(B^s(x))) = C(T)(1+O(\epsilon)), \tag{7.6.2}$$

with $C(T)$ independent of $x \in B$. Since $\mu^{cu}(\varphi^T(C)) = e^{hT}\mu^{cu}(C)$ we get

$$\mu(\varphi^T(B)) = e^{hT}\mu^{cu}(C)C(T)(1+O(\epsilon)), \tag{7.6.3}$$

with $O(\epsilon)$ independent of B.

To calculate $\mu(B \cap \varphi^T(B))$ we first show that it suffices to consider full components of intersection. Any point of $B \cap \varphi^T(B)$ not contained in a full component of intersection is in

$$A_T := \Big(\varphi^T\Big(\bigcup_{0 \le t \le \epsilon^2} B_t\Big) \cap \bigcup_{\epsilon-\epsilon^2 \le t \le \epsilon} B_t\Big) \cup \Big(\varphi^T\Big(\bigcup_{\epsilon-\epsilon^2 \le t \le \epsilon} B_t\Big) \cap \bigcup_{0 \le t \le \epsilon^2} B_t\Big)$$
$$\cup \Big(\big\{x \in B \mid \Delta(x) \le r(x)\big\} \cap \varphi^T(B_t)\Big),$$

because it is too close to the boundary of B either in the time direction or in an unstable leaf. By mixing, each of the first two sets has measure $\epsilon^2\mu(B)^2(1+o(T^0))$. The measure of the third set decreases exponentially with T, so by absorbing it into the error we have

$$\mu(A_T) \le 2\epsilon^2\mu(B)^2(1+o(T^0)). \tag{7.6.4}$$

To prove the proposition it is clearly useful to calculate the measure of full components of intersection via $\Delta(T)$ and their average measure. To calculate the average measure note that a full component of intersection Δ is of the form $\Delta = B \cap \varphi^T(\Delta')$, where $\Delta' = \bigcup\big\{[x] \mid x \in B,\ \varphi^T(x) \in \Delta\big\}$ and either $\varphi^T(\Delta' \cap B_\epsilon) \subset \Delta$ (a "front intersection component") or $\varphi^T(\Delta' \cap B_0) \subset \Delta$ (a "back intersection component"). In the first case we define the thickness of Δ by

$$\theta(\Delta) := \inf\big\{\tau(x) \mid x \in \varphi^T(\Delta' \cap B_\epsilon)\big\},$$

and in the second by

$$\theta(\Delta) := \epsilon - \sup\{\tau(x) \mid x \in \varphi^T(\Delta' \cap B_0)\}.$$

By (7.6.1) every orbit segment in Δ has (parameter) length $\epsilon(1 + O(\epsilon))$, so (7.6.2) gives

$$\mu(\Delta) = \frac{1}{\epsilon}\theta(\Delta)\mu^{cu}(C)C(T)(1 + O(\epsilon)). \tag{7.6.5}$$

Naturally we have the following result:

Lemma 7.6.8. *The average thickness of full components of intersection of $B \cap \varphi^T(B)$ is $(\epsilon/2)(1 + O(\epsilon))(1 + o(T^0))$.*

Proof. Partition B into $n := [1/\epsilon]$ sets $S_i := \bigcup\{B_t \mid \epsilon j \le tn \le \epsilon(j+1)\}$ $(0 \le j < n)$. For sufficiently large T the number of components of $\varphi^T(S_{n-1}) \cap S_i$ is independent of j by mixing, so the average thickness is $(\epsilon/2)(1 + O(1/n))(1 + o(T^0))$ as claimed. □

Since the error in (7.6.4) can be absorbed into $O(\epsilon)(1 + o(T^0))$ we obtain from (7.6.5) and Lemma 7.6.8,

$$\mu(B \cap \varphi^T(B)) = \tfrac{1}{2}\Delta(T)C(T)\mu^{cu}(C)(1 + O(\epsilon))(1 + o(T^0)). \tag{7.6.6}$$

Since μ is mixing, (7.6.3) yields

$$\mu(B \cap \varphi^T(B)) = \mu(B)\mu^{cu}(C)e^{hT}C(T)(1 + O(\epsilon))(1 + o(T^0)),$$

which, with (7.6.6), yields the claim. □

We now give the promised multiplicative asymptotic for the growth of the number $P_t(\Phi)$ of periodic orbits of Φ with period at most t.

Theorem 7.6.9 (Multiplicative orbit-growth asymptotic). *If Φ is a topologically mixing Anosov flow on a compact Riemannian manifold M, then*

$$\lim_{t\to\infty} t h_{\text{top}}(\Phi) P_t(\Phi) e^{-t h_{\text{top}}(\Phi)} = 1, \quad \textit{that is,}\quad P_t(\Phi) \sim \frac{e^{t h_{\text{top}}(\Phi)}}{t h_{\text{top}}(\Phi)}.$$

Remark 7.6.10. That $e^{t h_{\text{top}}(\Phi)}/t$ is the growth rate of $P_t(\Phi)$ was previously known [58] but Margulis was the first to show that $\lim_{t\to\infty} t h_{\text{top}}(\Phi) P_t(\Phi) e^{-t h_{\text{top}}(\Phi)}$ exists [256]; Toll was the first to determine this limit.

Proof. We write $P_{T,\epsilon} := \operatorname{card} \mathbb{O}_t(T)$ (Definition 4.2.22) and let μ_B denote the Bowen measure. Then $\mu_B(B) = \mu_B(B^\circ(T))(1 + O(\epsilon))$ for sufficiently large T, and Lemmas 7.6.5 and 7.6.6 imply

$$\begin{aligned}
\frac{\epsilon\Delta(T)}{P_{T,\epsilon}} &= \frac{1}{P_{T,\epsilon}} \sum_{O\in\mathbb{O}_\epsilon(T)} \delta_O(B) \\
&\le \mu_B(B)(1 + o(T^0)) = \mu_B(B^\circ(T))(1 + O(\epsilon))(1 + o(T^0)) \\
&= \frac{1}{P_{T,\epsilon-2\epsilon^2}} \sum_{O\in\mathbb{O}_{\epsilon-2\epsilon^2}(T)} \delta_O(B^\circ(T))(1 + O(\epsilon)) \le \frac{\epsilon\Delta(T)}{P_{T,\epsilon-2\epsilon^2}}(1 + O(\epsilon)).
\end{aligned}$$

By Proposition 7.6.7 this yields

$$P_{t,\epsilon-2\epsilon^2} \le \frac{2\epsilon e^{hT}\mu(B)}{\mu_B(B)}(1 + O(\epsilon))(1 + o(T^0)) \le P_{t,\epsilon}.$$

Replacing ϵ by ϵ' with $\epsilon' - 2\epsilon'^2 = \epsilon$ introduces another factor of $1 + O(\epsilon)$ and by Theorem 7.5.2 we get

$$P_{T,\epsilon} = 2\epsilon e^{hT}(1 + O(\epsilon))(1 + o(T^0)). \tag{7.6.7}$$

Since $P_{T,\epsilon}$ is the number of periodic orbits with *a* period in $(T-\epsilon, T+\epsilon]$, we have

$$P_{T,\epsilon} = T_1(P_{T+\epsilon}(\Phi) - P_{T-\epsilon}(\Phi)) + T_2(P_{(T+\epsilon)/2}(\Phi) - P_{(T-\epsilon)/2}(\Phi)) + \cdots,$$

where $iT_i \in [T-\epsilon, T+\epsilon]$. By (7.6.7) this simplifies to $P_{T,\epsilon} = T_1(P_{T+\epsilon}(\Phi) - P_{T-\epsilon}(\Phi))(1 + o(T^0))$. Using $T_1 = T(1 + o(T^0))$ and (7.6.7) we find

$$P_{T+\epsilon}(\Phi) - P_{T-\epsilon}(\Phi) = \frac{2\epsilon}{T} e^{hT}(1 + O(\epsilon))(1 + o(T^0)). \tag{7.6.8}$$

Now fix $T_0 > 1/h$ such that $|o(T^0)| < \epsilon$ for all $T \ge T_0$ on the right-hand side. Writing

$$\begin{aligned}
P_{T+\epsilon}(\Phi) = {}& (P_{T+\epsilon}(\Phi) - P_{T-\epsilon}(\Phi)) + (P_{T-\epsilon}(\Phi) - P_{T-3\epsilon}(\Phi)) \\
&+ \cdots + (P_{T-2j\epsilon+\epsilon}(\Phi) - P_{T-2j\epsilon-\epsilon}(\Phi)) + P_{T-2j\epsilon-\epsilon}(\Phi)
\end{aligned}$$

for $T - (2j+1)\epsilon \le T_0 < t - 2j\epsilon$ and estimating the differences by (7.6.8) gives

$$P_{T+\epsilon}(\Phi) = \frac{e^{hT}}{T} S(T)(1 + O(\epsilon))(1 + o(T^0)),$$

where $S(T) := 2\epsilon\big(1 + \frac{T}{T-2\epsilon}e^{-2\epsilon h} + \cdots + \frac{T}{T-2j\epsilon}e^{-2j\epsilon h}\big)$ and we absorbed $P_{T-2j\epsilon-\epsilon}(\Phi)$ into $(1 + o(T^0))$. Observe now that $S(T)$ is a Riemann sum for

$$\underbrace{\int_0^{T-T_0} \frac{T}{T-x} e^{-hx}\,dx}_{=Te^{-hT}\int_{T_0}^T \frac{e^{hu}}{u}du} = Te^{-hT}\left(\frac{e^{hT}}{hT} - \frac{e^{hT_0}}{hT_0} - T\int_{T_0}^T \frac{e^{hu}}{hu^2}du\right) = \frac{1}{h}(1 + o(T^0)).$$

The integrand decreases on $[0, T - T_0]$, so $S(T) = \frac{1}{h}(1 + O(\epsilon))(1 + o(T^0))$ and hence

$$\begin{aligned} P_T(\varphi) &= P_{T+\epsilon}(\Phi)(1 + O(\epsilon)) \\ &= \frac{e^{hT}}{T} S(T)(1 + O(\epsilon))(1 + o(T^0)) = \frac{e^{hT}}{hT}(1 + O(\epsilon))(1 + o(T^0)), \end{aligned}$$

that is, $\lim_{T\to\infty} \frac{P_T(\varphi)}{e^{ht}/hT} = (1 + O(\epsilon))$, proving the claim. □

Theorem 7.6.11 (Margulis). *Let M be a compact Riemannian manifold of negative sectional curvature, $G(t)$ the number of different closed geodesics of length at most t, and h the topological entropy of the geodesic flow. Then $G(t)2the^{-th} \xrightarrow[t\to\infty]{} 1$.*

Proof. Each closed geodesic of length t defines exactly two period-t orbits of the geodesic flow. The geodesic flow is a contact Anosov flow, hence topologically mixing by Theorem 8.1.4. Thus Theorem 7.6.9 applies. □

While this is a remarkably fine asymptotic, the understanding of the growth of periodic orbits has become even more refined. For instance, for geodesic flows Φ on smooth negatively curved compact surfaces and sufficiently strongly pinched negatively curved manifolds,

$$P_t(\Phi) = \int_2^{e^{h_{\text{top}}(\Phi)t}} (\log u)^{-1} du + O(e^{(h_{\text{top}}(\Phi)-\epsilon)t})$$

as $t \to +\infty$ [119, 300]. The published version of the Margulis asymptotic includes a broad survey of related investigations [256, 8, 151]. The dynamical zeta function encodes all periods of a flow in one function and is an important ingredient of these refinements of orbit counting as well as an important tool in several areas of hyperbolic dynamics [279, 256, 8, 151].

Definition 7.6.12 (ζ-function). The *ζ-function* is defined by

$$\zeta_\Phi(s) = \prod_\gamma \left(1 - e^{-s\,l(\gamma)}\right)^{-1}, \tag{7.6.9}$$

where the product is taken over all (nonfixed) closed orbits γ of the flow and $l(\gamma)$ is the smallest positive period ("length") of γ.

If $P_T(\Phi) < \infty$ (Definition 4.2.1) for all T, then a standard convergence criterion for infinite products (or studying its logarithm) shows that this product converges for $\operatorname{Re} s > p(\Phi)$ and has singularities on the critical line $\operatorname{Re} s = p(\Phi)$.

Remark 7.6.13 (Discrete-time ζ-function). We digress to connect this form of the ζ-function to the one more commonly used for discrete-time systems. To that end, let $P_n(f) := \operatorname{card}\operatorname{Fix}(f^n)$ be the number of periodic points of a map f, and define the ζ-function for f by

$$\zeta_f(z) = \exp \sum_{n\in\mathbb{N}} \frac{P_n(f)}{n} z^n,$$

where $z \in \mathbb{C}$. By the ratio test this converges for $|z| < \exp(-p(f))$ and has singularities on the circle $|z| = \exp(-p(f))$. The coefficient $\frac{P_n(f)}{n}$ is the number of n-periodic *orbits*, so

$$\zeta_f(z) = \exp \sum_{n\in\mathbb{N}} \sum_{\substack{n\text{-periodic}\\ \text{orbits}}} z^n.$$

Here, for instance, a 3-periodic orbit appears again as a 6-periodic orbit, a 9-periodic orbit, and so on, and this invites reexpressing the sums in terms of sums over *prime* (or shortest) orbits:

$$\sum_{n\in\mathbb{N}} \sum_{\substack{n\text{-periodic}\\ \text{orbits}}} z^n = \sum_{\gamma\text{ prime}} \sum_{k\in\mathbb{N}} \frac{z^{kl(\gamma)}}{k},$$

where the denominator accounts for the overcounting in this rearranged sum, so

$$\begin{aligned} \zeta_f(z) = \exp \underbrace{\sum_{\gamma\text{ prime}} \sum_{k\in\mathbb{N}} \frac{z^{kl(\gamma)}}{k}}_{=-\log(1-z^{l(\gamma)})} &= \prod_{\gamma\text{ prime}} \exp(-\log(1-z^{l(\gamma)})) \\ &= \prod_{\gamma\text{ prime}} (1-z^{l(\gamma)})^{-1}. \end{aligned} \tag{7.6.10}$$

Taking $z = e^{-s}$ here gives (7.6.9). And this form does not require the periods to be integers. Moreover, for flows it is natural to work with prime orbits, because these are the canonical representatives of the orbits as geometric objects.

Remark 7.6.14. Equation (7.6.9) shows the analogy with the Riemann ζ-function

$$\zeta(s) = \sum_{n=1}^{\infty} \frac{1}{n^s} \overset{\text{Euler}}{\underset{\substack{\text{product}\\ \text{formula}}}{=\!=\!=}} \prod_{p\text{ prime}} (1-p^{-s})^{-1},$$

where the Euler product formula is seen by an argument due to Euler:

$$\frac{1}{2^s}\zeta(s) = \sum_{n\text{ even}} \frac{1}{n^s}, \text{ so } \left(1-\frac{1}{2^s}\right)\zeta(s) = \sum_{2\nmid n} \frac{1}{n^s}$$

and likewise

$$\Bigl(1-\frac{1}{2^s}\Bigr)\Bigl(1-\frac{1}{3^s}\Bigr)\zeta(s)=\sum_{2\nmid n,\ 3\nmid n}\frac{1}{n^s},$$

and so on through all primes, giving $\prod_p(1-p^{-s})\zeta(s)=1$.

The numbers $P_T(\Phi)$ and the ζ-function are obviously invariants of conjugacy. In general, they are *not* invariant under orbit equivalence. Although an orbit equivalence takes periodic orbits into periodic orbits, it may change their period. However, a cruder property survives a time change.

Proposition 7.6.15. *Let X, Y be compact metric spaces and $\varphi^t\colon X\to X$ and $\psi^t\colon Y\to Y$ continuous flows without fixed points. Suppose that the flows are orbit equivalent and $p(\Psi)=0$. Then $p(\Phi)=0$.*

Proof. Let $h\colon X\to Y$ be a homeomorphism that maps orbits of the flow φ^t onto orbits of ψ^t. Then $h(\varphi^1(x))=\psi^{\alpha(x)}h(x)$ with α continuous and positive; hence it is bounded from above, say $\alpha(x)\le M$. Thus the image of any orbit segment for Φ of length 1 is an orbit segment of Ψ of length at most M. Hence the image of a segment of length at most T has length less than $(T+1)M$ which for $T\ge 1$ is less than $2MT$. That means that the image of any periodic orbit of period $\le T$ has period $\le 2MT$, that is,

$$P_{2MT}(\Psi)\ge P_T(\Phi),$$

and $p(\Psi)\ge p(\Phi)/2M$. □

Example 7.6.16 ([318]). For the flow Φ under the function r over the full 2-shift defined by $r=r_i$ on the cylinder $\omega_0=i$ for $i=1,2$ we have (see (7.6.10))

$$\begin{aligned}\zeta(s)&=\exp\Bigl[-\sum_{n\in\mathbb{N}}\frac{1}{n}\sum_{\omega=\sigma^n(\omega)}\exp\Bigl(\sum_{k=0}^{n-1}r(\sigma^k(\omega))\Bigr)\Bigr]\\&=\exp\Bigl[-\sum_{n\in\mathbb{N}}\frac{1}{n}\sum_{p=0}^{n}\binom{n}{p}\exp\bigl(-s(pr_0+(n-p)r_1)\bigr)\Bigr]=1-e^{-r_0s}-e^{-r_1s}.\end{aligned}$$

We remark briefly how the ζ-function has been used to count periodic orbits. Initially, this was done for flows over subshifts of finite type with a roof function such that the periods are incommensurate. The pertinent "weight" is $N(\gamma)=e^{h\ell(\gamma)}$, where ℓ is the (least) period of the periodic orbit γ and $h=h_{\text{top}}(\Phi)$. Then

$$\pi(T):=\operatorname{card}\bigl\{\gamma\mid \gamma\text{ periodic with }N(\gamma)\le T\bigr\}\sim\frac{T}{\log T},$$

that is, $\pi(T)\log T/T \xrightarrow[T\to\infty]{} 1$, if r is locally constant [280] or Hölder continuous [277], which via coding then applies to hyperbolic flows (with incommensurate periods), including mixing Anosov flows, thus generalizing Theorem 7.6.9.

The proof of the Prime Number Theorem (that the number of primes of size up to N is asymptotically $N/\log N$) uses analytic properties of the Riemann ζ-function, and in the present context this is mirrored by establishing analogous properties of the dynamical ζ-function. For locally constant roof functions $r(x) = r(x_0, x_1)$ over a subshift Σ_A this is easy to describe: for $s \in \mathbb{C}$ let $P^s_{ij} := A_{ij}e^{-sr(i,j)}$ to get $\zeta(s) = \det(\mathrm{Id} - P^s)^{-1}$. For Hölder-continuous roof functions, one instead uses Ruelle transfer operators—which also play a central role in studying correlation decay (Section 7.7), a subject which is thereby closely connected to orbit counting.

7.7 Rates of mixing*

We pick up from Remark 3.4.27 to explore quantitative measures of mixing. Proposition 3.4.25 expresses the various mixing properties in terms of functions (or observables or random variables) by way of covariance (Definition 3.4.23). Specifically, mixing means that $\operatorname{cov}(U^t_\Phi(f), g) \xrightarrow[t\to\infty]{} 0$ for any $f, g \in L^2$, and we now consider the question of how rapid this convergence might be. Since covariance is closely related to correlation, this is known as the rate of *decay of correlations*.

We first explore this issue in the context of a toy problem in discrete time to show what conclusions one might expect and want to reach. When we consider the matter for flows, we will concentrate on pointing out the issues that make this situation different from this simple example. These are twofold. On one hand, the toy problem is purely expanding, which makes it much simpler but noninvertible. The correct discrete-time counterpart of a flow is an invertible map. This necessitates dealing with contracting directions in addition to expanding ones, which alone adds significant difficulty. Next, we have already noted that the notion of mixing for flows is quite sensitive to timing issues; for instance, being a suspension precludes mixing altogether. Accordingly, subtleties in the orbit direction caused this matter to make substantial progress (in the case of flows) only recently.

We now attend to our introductory discrete-time toy problem.

Proposition 7.7.1. *Consider the expanding endomorphism $E_m : x \mapsto mx \bmod 1$ on S^1 for $|m| \geq 2$ with Lebesgue measure and suppose $\varphi, \psi : [0,1] \to \mathbb{R}$ are α-Hölder-continuous functions with coefficient L, that is, $|\psi(x) - \psi(y)| \leq L|x-y|^\alpha$. Then correlations decay with the exponential rate $m^{-\alpha}$:*

$$|\operatorname{cov}(U^n_{E_m}(\varphi), \psi)| \leq L m^{-\alpha n}\|\varphi\|.$$

Remark 7.7.2. We note that the parameters that affect the decay rate of correlations are the expansion rate of the transformation as well as the Hölder exponent of the functions under consideration. In particular, for Lipschitz-continuous functions the decay rate is the reciprocal of the expansion rate of the transformation.

Proof. Assume without loss of generality that $\varphi \perp 1$. Then

$$\begin{aligned}
|\operatorname{cov}(U^n_{E_m}(\varphi),\psi)| &= \Big|\sum_{k=0}^{m^n-1}\int_{\frac{k}{m^n}}^{\frac{k+1}{m^n}} \varphi\circ E^n_m\cdot\bar{\psi}\Big| \\
&\le \Big|\sum_{k=0}^{m^n-1}\int_{\frac{k}{m^n}}^{\frac{k+1}{m^n}} \varphi\circ E^n_m\cdot\overline{\Big(\psi-\int_{\frac{k}{m^n}}^{\frac{k+1}{m^n}}\psi\Big)}\Big| \\
&\quad+\Big|\sum_{k=0}^{m^n-1}\int_{\frac{k}{m^n}}^{\frac{k+1}{m^n}} \varphi\circ E^n_m\cdot\overline{\int_{\frac{k}{m^n}}^{\frac{k+1}{m^n}}\psi}\Big| \\
&\le \frac{L}{m^{\alpha n}}\sum_{k=0}^{m^n-1}\int_{\frac{k}{m^n}}^{\frac{k+1}{m^n}} |\varphi\circ E^n_m| + \Big|\overline{\int_{\frac{k}{m^n}}^{\frac{k+1}{m^n}}\psi}\sum_{k=0}^{m^n-1}\underbrace{\int_0^1\frac{\varphi}{m^n}}_{=0}\Big| \\
&= Lm^{-\alpha n}\|\varphi\|.
\end{aligned}$$

□

For smoother functions we get even more rapid decay of correlations:

Proposition 7.7.3. *Consider the expanding endomorphism E_m for $|m|\ge 2$ with Lebesgue measure and suppose $\varphi,\psi\colon[0,1]\to\mathbb{R}$ are C^r-functions. Then correlations decay with the exponential rate m^{-r}: $\operatorname{cov}(U^n_{E_m}(\varphi),\psi)=O(m^{-rn})$. (See Remark 3.2.18.) In particular, analytic functions have superexponential decay of correlations.*

Proof. If $\varphi(x)=\sum_{k\in\mathbb{Z}}\varphi_k\exp(2\pi ikx)\perp 1$ and $\psi(x)=\sum_{l\in\mathbb{Z}}\psi_l\exp(2\pi ilx)\perp 1$, then

$$\Big|\int U^n_{E_m}\varphi\cdot\psi\Big| = \Big|\sum_{k,l\in\mathbb{Z}}\varphi_k\psi_l\int\exp(2\pi i(m^nk+l)x)\Big| = \sum_k\varphi_k\psi_{-m^nk},$$

and $|\psi_l|\le C|l|^{-r}$, hence $|\psi_{-m^nk}|\le C(m^n|k|)^{-r}=Cm^{-rn}|k|^{-r}$, so

$$\Big|\sum_k\varphi_k\psi_{-m^nk}\Big|\le\sqrt{\sum|\varphi_k|^2}\sqrt{\sum|\psi_{-m^nk}|^2}\le\|\psi\|_2\cdot C'm^{-rn}.$$

□

We note from these that rates of mixing depend on regularity of the functions involved; the converse to speeding up the rate of mixing by looking at smoother functions is that in, say, L^2 there is no hope to get a specific rate. This dependence also shows that rates of mixing are associated with a smooth flow rather than a class modulo measurable isomorphism. (This is also a reason to express these rates in terms of functions rather than sets.) On the other hand, we note that the rate of correlation decay in both results is exponential, regardless of the class of functions under consideration, so long as the class is contained in that of Hölder-continuous functions. In particular, it is fair to say that this discrete-time dynamical system has exponential decay of correlations for Hölder functions, and this means that having exponential decay of correlations (or failing to do so) is an interesting notion for hyperbolic flows since the class of Hölder-continuous functions is invariant under orbit equivalence (Theorem 6.4.3).

In both of these results one can see expansion as an essential mechanism manifested in either the stretching out of subintervals to $[0, 1]$ or in the "stretching" of Fourier coefficients. Trying to apply the same steps to a hyperbolic map immediately fails, and not only do the expressions fail to improve that need to be controlled, but they indeed get much worse from the contraction. The reader may try to see how this works out in Example 1.5.26.

As this subject came to the fore, the principal approach was to use a Markov partition in order to represent a hyperbolic diffeomorphism as a shift and then to use a result that one can replace a function on the shift (modulo cohomology) by one that depends only on the past, that is, a function of the expanding direction only. For such functions, arguments like the preceding ones again get traction.

There is a cost to the symbolic approach in that one potentially loses geometric information when passing to the shift; this invites the question of whether a direct approach might yield sharper results. These issues also arise for flows.

There is a much more serious difficulty for flows, illustrated by a problem with the earlier statement that "having exponential decay of correlations is an interesting notion for hyperbolic flows since the class of Hölder-continuous functions is invariant under orbit equivalence (Theorem 6.4.3)": a mixing flow can be orbit equivalent to a suspension, which is never mixing. While this statement can be fixed by invoking a conjugacy rather than an orbit equivalence, the underlying sensitivity to timing issues is an essential and profound problem: while hyperbolicity helps mix up sets in the phase space, this is only fully effective if the structure of the flow provides a mechanism by which this can force mixing in the flow direction as well. Simple explicit examples illustrate the need for this:

Proposition 7.7.4 ([318]). *The flow from Example* 7.6.16 *with* $r_0/r_1 \notin \mathbb{Q}$ *is mixing, but, for functions independent of the base variable, not exponentially so.*

The flow counterpart to our toy example, or rather to its invertible cousin in Example 1.5.26, is the geodesic flow of a surface of constant negative curvature (Chapter 2), and exponential mixing was indeed established for these in the 1980s

[260, 95, 311]. The pertinent counterpart to Fourier analysis is representation theory. The 3-dimensional case was then treated in the 1990s by Pollicott [299] who established the existence of a spectral gap of the associated transfer operators as a sufficient condition for exponential mixing. These developments notwithstanding, one can say that for three decades there was no fundamental progress on the question of decay of correlations. Meanwhile, however, Chernov introduced uniform nonintegrability of the strong foliations (which is illustrated by Proposition 9.2.3 below) as the mechanism that forces mixing in the flow direction even in the absence of a complete algebraic description, which led to (subexponential) rates of mixing for geodesic flows of surfaces of *variable* negative curvature [89, 90], and he conjectured that the correlation decay is exponential for these. This restarted this research area, and it was then transformed by Dolgopyat [116].[21] Among other things, this allowed him to prove the following result:

Theorem 7.7.5 (Dolgopyat). *Mixing Anosov flows with C^1-foliations have exponential decay of correlations* [116], *C^∞ Axiom-A flows are rapidly mixing if there are two periodic orbits whose periods have a diophantine ratio* [117], *and suspensions over shifts are generically exponentially mixing* [118].

To outline the ideas involved, we follow [245] and first describe the state of the art from the 1980s [297]. Consider the *transfer operator* on probability measures (viewed as functionals) associated to a flow Φ by

$$\mu \mapsto \mathcal{L}^t(\mu) \text{ with } \mathcal{L}^t(\mu)(f) = \mu(f \circ \varphi^t).$$

Clearly, invariant measures are fixed points. Moreover, if $f \in C^r(M)$ with $m(f) = 1$, where m is Lebesgue measure, then $\mathcal{L}^t(m_f) \xrightarrow[t\to\infty]{} \mu_{\text{SRB}}$, and we aim to show that this happens at an exponential rate.

Note $t \mapsto \mathcal{L}^t$ embeds $[0, \infty)$ as a strongly continuous semigroup of bounded linear operators on $\mathfrak{M}(\Phi)$, which hence has a generator Z (a bounded linear operator) with resolvent (Definition B.3.1)

$$R(z) := (z\,\mathrm{Id} - Z)^{-1} = \int_0^\infty e^{-zt}\mathcal{L}^t\,dt \tag{7.7.1}$$

in terms of which one can write (with $z = a + ib$ and large enough a)

$$\mathcal{L}^t = \lim_{L\to\infty} \int_{-L}^{L} e^{zt} R(z)\,db. \tag{7.7.2}$$

Studying R instead of the transfer operator itself directly addresses the problem of the absence of hyperbolicity in the flow direction; the time integral in (7.7.1) smooths

[21] Among the important other work that followed is the development of methods to study correlation decay directly for the flow in hand rather than a symbolic model [244].

functions out along the flow direction, and R can be handled in ways similar to the transfer operators in discrete time where the challenge in the flow direction is absent. Equation (7.7.2) then produces the required exponential convergence rate for the transfer operator itself.

For $z \in (0,\infty) + \mathbb{R}i$ there is a $\sigma_z > 0$ such that if $f, g \in C^r(M)$ with $m(f) = 1$, then

$$R(z)m_f(g) = \mu_{\mathrm{SRB}}(g)/z + \hat{R}(z)m_f(g),$$

with

$$|\hat{R}^n(z)^n m_f(g)| \leq C_{z,f,g}(\Re(z) + \sigma_z)^{-n},$$

so $z \mapsto \hat{R}(z)m_f(g)$ is analytic near 0 and on $(0,\infty) + i\mathbb{R}$, that is, for $L > M \geq 0$ there is an $\omega_M > 0$ such that the path

$$\gamma_{L,M}(s) := \begin{cases} a + i(s + a + \omega_m), & -L - a - \omega_M \leq s \leq -M - a - \omega_M, \\ -M - \omega_M - s - iM, & -M - a - \omega_M \leq s \leq -M, \\ -\omega_M + is, & -M \leq s \leq M, \\ -M - \omega_M + s + iM, & M \leq s \leq M + a + \omega_M, \\ a + i(s - a - \omega_M), & M + a + \omega_M \leq s \leq L + a + \omega_M, \end{cases}$$

is in the domain of analyticity of $\hat{R}(\cdot)$ and hence (writing again $z = a + ib$ and using

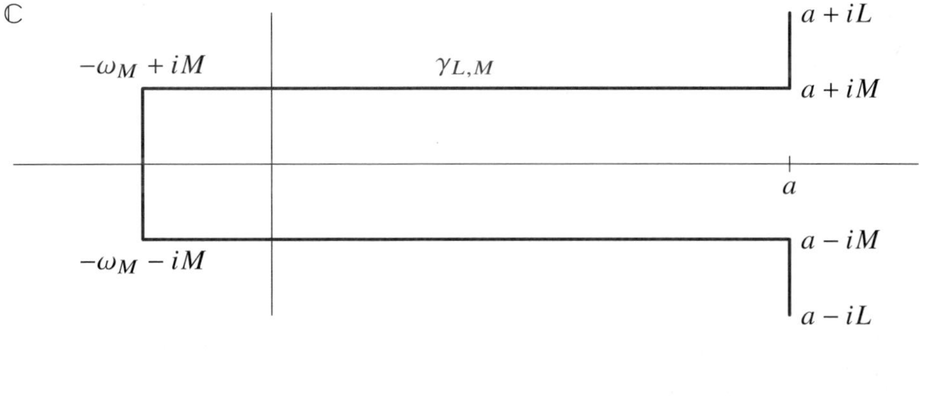

Figure 7.7.1. The path $\gamma_{L,M}$.

$R(z) = \sum_{k=0}^{r} z^{-k-1} Z^k + R(z)Z^r)$ we have

$$\begin{aligned}
\mathcal{L}^t(m_f(g)) &= \mu_{\text{SRB}}(g) + \lim_{L\to\infty} \int_{-L}^{L} e^{zt}\hat{R}(z)m_f(g)\,db\\
&= \mu_{\text{SRB}}(g) + \lim_{L\to\infty} \int_{\gamma_{L,M}} e^{zt}\hat{R}(z)m_f(g)\,dz\\
&= \mu_{\text{SRB}}(g) + \lim_{L\to\infty} \int_{\gamma_{L,M}} \frac{e^{zt}}{z^r}\hat{R}(z)Z^r m_f(g)\,dz\\
&= \mu_{\text{SRB}}(g) + e^{-\omega_M t} \lim_{L\to\infty} \int_{-M}^{M} \frac{e^{ibt}}{-\omega_M + ib}\hat{R}(-\omega_M + ib)Z^r m_f(g)\,db\\
&\quad + \lim_{L\to\infty} \int_{M\le|b|\le L} \frac{e^{at+ibt}}{(a+ib)^r}\hat{R}(a+ib)Z^r m_f(g)\,db\\
&\quad - \int_{-\omega_M}^{a} \frac{e^{iMt}}{(x+iM)^r}\hat{R}(x+iM)Z^r m_f(g)\,dx\\
&\quad + \int_{-\omega_M}^{a} \frac{e^{-iMt}}{(x-iM)^r}\hat{R}(x-iM)Z^r m_f(g)\,dx.
\end{aligned}$$

Therefore, there is a $C_{M,f,g,r}$ such that

$$|\mathcal{L}^t m_f(g) - \mu_{\text{SRB}}(g)| \le C_{M,f,g,r}(M^{-r+1} + e^{-\omega_M t}).$$

Although the right-hand side suggests exponential decay, the fact that we do not know how $C_{M,f,g,r}$ and ω_M depend on M means that a substantial new idea is needed. A method that yields a positive lower bound for ω_M combined with a lower bound M^α for $C_{M,f,g,r}$ would suffice. Overcoming this challenge opened up this subject at the end of the 20th century.

Theorem 7.7.6 (Dolgopyat inequality [116]). *If the stable and unstable foliations are C^1 and uniformly not jointly integrable, then there are $a, \alpha, \beta > 0$ such that*

$$R(a+ib)^{\beta \log|b|} m_f(g) = O(|b|^{-\alpha}|f|_u|g|_s)$$

for large $|b|$, where $|f|_u = |f|_\infty + |\partial_u f|_\infty$, $|g|_s = |g|_\infty + |\partial_s g|_\infty$ with ∂_u, ∂_s the derivatives in the strong-unstable and strong-stable directions.

With moderate effort this implies the next corollary, which can be used in the previous estimates to imply exponential decay of correlations.

Corollary 7.7.7. $|R(-\omega + ib)m_f(g)| \le C_{f,g}|b|^{\beta \log(\alpha+\omega)}$.

We continue with a description of the new ideas that produce Theorem 7.7.6. The core insight is that the claim is akin to a quantitative version of the Riemann–Lebesgue Lemma, because the assertion is that the contributions for large b, which are the "high-frequency contributions," are small [343].

The starting point is

$$R(z)^n \nu_\psi(\varphi) = \frac{1}{(n-1)!} \int_0^\infty t^{n-1} e^{-zt} \nu_\psi(\varphi \circ \phi_t)\, dt.$$

The Stirling formula (A.3.5) shows that the contribution of the integral from 0 to cn is $\lesssim \frac{(cn)^n}{n!} \lesssim \left(\frac{c}{e}\right)^n$, hence negligible for small c. For large n we can then assume that φ is essentially constant along stable leaves and ν_ψ is essentially constant along strong-unstable leaves, and one can furthermore disintegrate ν_ψ along unstable leaves to reduce the problem to estimates of

$$\frac{1}{(n-1)!} \int_{cn}^\infty t^{n-1} e^{-zt} \int_W \varphi \circ \phi_t,$$

where W is a small local strong-unstable leaf. Finally, partitioning the integral over time into time intervals of fixed length gives integrals

$$\int_{W_c} e^{-zt} \varphi \circ \phi_l$$

with W_c a local center-unstable leaf (of a fixed size) and $l \geq cn$. By changing variable, the above integral in turn becomes an integral over $\phi_l W_c$, which is a large manifold in the strong-unstable direction. Partitioning it into manifolds W_i of fixed size gives

$$\int_{W_c} e^{-zt} \varphi \circ \phi_l = \sum_i \int_{W_i} e^{-zt} \varphi J_i,$$

where the J_i reflect the Jacobian of the change of variables and a partition of unity is used to smoothly subdivide the integral. The essential improvement over prior developments is that the integral on the right-hand side can be shown to be $O(|z|^{-1})$ when the stable and unstable foliations are uniformly not jointly integrable.

For a finite cover, we can group the W_i according to which element of the covering each meets, and we aim to bound the sum over each such subcollection. We consider a covering by balls, and for large n each ball U will be "packed" by many tightly stacked W_i, each of which is related to one leaf W_U of this collection near the center of U. The integral over each of these W_i can then be reexpressed as an integral over W_U as follows. The unstable and flow coordinates v, s on W_U are sent to the corresponding coordinates $(u,t) = \Psi_i(v,s)$ on W_i by the strong-stable holonomy Ψ_i in such a way that (because the holonomy is assumed C^1) $t \sim s + a_i v$ with a_i proportional to the distance between W_U and W_i by a factor that is nonzero by the assumption of uniform nonintegrability (a quadrilateral from $p \in W_U$ to $\Psi_i(p) \in W_i$, then along the strong-unstable direction, then back to W_U by Ψ_i^{-1}, and back to the orbit of p along the strong-unstable direction produces a uniform displacement in the flow direction). Thus, with a suitable Jacobian $\tilde{J}_i$,

$$\sum_i \int_{W_i} e^{-zt}\varphi J_i = \sum_i \int_{W_U} e^{-z(s+a_i w)}\varphi \tilde{J}_i + O(|\partial_s \varphi|_\infty)$$

$$\begin{bmatrix}\text{Cauchy–Schwarz}\\ \text{inequality}\end{bmatrix} \leq |\varphi|_\infty \sqrt{\sum_{i,j\in W_U} \int_{W_U} e^{-z(a_i-a_j)w}\tilde{J}_j\tilde{J}_i + O(|\partial_s\varphi|_\infty)} = O(|z|^{-1/2})$$

since

$$\int_{W_U} e^{-z(a_i-a_j)w}\tilde{J}_j\tilde{J}_i \sim |\tilde{J}_j\tilde{J}_i|_{C^1}|z|^{-1}|a_i-a_j|^{-1}$$

and subject to controlling the $a_i - a_j$ by bounding how closely the W_i can be packed. This gives the desired control in Theorem 7.7.6 for large imaginary parts of z.

The preceding discussion eluded a major technical point we mentioned beforehand: the work it describes was carried out at the symbolic level via Markov partitions, and it is natural to expect that some information may be lost in the coding. Accordingly, Liverani led the way in developing methods to obtain these results without coding, that is, by direct "geometric" methods. Proposition 7.7.3 indicates a grave technical difficulty with this (even in discrete time): that argument is centered on the effects of expansion to smooth out the observables and decorrelate them. In the contracting direction the very opposite happens instead. Liverani's approach is to construct Banach spaces of "functions" in which these arguments work. The elements of these spaces are smooth functions in the expanding subspaces and distributions (in the sense of generalized functions) in the contracting subspaces. The latter are dual to smooth functions, and therefore the decorrelation is aided rather than confounded by the contractive effects [244].

Altogether, this describes the core advances on this subject at the beginning of the 21st century.

Exercises

7.1. A symbolic flow (Definition 1.9.5) is said to be *sofic* if it is a factor of a symbolic flow over a shift of finite type. Prove that every topologically transitive sofic flow that is a factor of a hyperbolic symbolic flow has a unique measure of maximal entropy.

7.2. Prove directly that the strong specification property implies topological mixing.

7.3. Calculate the pressure for the full N-shift σ_N for the function $\varphi(\omega) = \omega_0$.

7.4. Show that if h is surjective, $h \circ \varphi^t = \psi^t \circ h$, and $f = g \circ h$, then $P(\varphi, f) \geq P(\psi, g)$.

7.5. Let $p = (p_0, \dots, p_{m-1})$, $p_i > 0$ for $i = 1, \dots, m$, $\sum_{i=0}^{m-1} p_i = 1$. Prove that the product (Bernoulli) measure μ_p is an equilibrium state for the full shift σ_m and some Hölder-continuous function f.

7.6. Let Φ be an expansive flow on a compact metric space X with the specification property, and f be Bowen bounded. Prove that $\operatorname{supp} \mu_f = X$.

7.7. In the context of Exercise 7.6 prove that $h_{\mu_f} > 0$ unless X is a point.

7.8. Show that if a flow preserves volume, then volume is the Sinai–Ruelle–Bowen measure.

7.9. Determine the Hausdorff dimension of the ternary Cantor set (Remark 1.3.6) (with the standard metric on the real line).

7.10. Prove that Lemma 7.3.9 holds when one assumes kinematic expansivity (Definition 1.8.3) instead of expansivity.

7.11. Prove that Proposition 7.3.10 holds when one assumes kinematic expansivity (Definition 1.8.3) instead of expansivity.

7.12. Prove that Proposition 7.3.12 holds when one assumes kinematic expansivity (Definition 1.8.3) instead of expansivity (Exercise 1.51).

7.13. Using that Proposition 7.3.12 holds assuming only kinematic expansivity (Exercise 7.12), show that so do Propositions 7.3.13, 7.3.15, and 7.3.17.

8

Anosov flows

One of the purposes of this book is to examine those features of hyperbolic flows that are different from the discrete-time counterparts, and Anosov systems are one of the primary classes where differences appear. While there are many parallels in fundamental aspects of Anosov flows and Anosov diffeomorphisms—provided one accounts for longitudinal effects in flows (Section 8.1)—there are also fundamental differences that appear between these two classes of systems.

For instance, is an Anosov system topologically transitive? In discrete time this question remains unanswered, but for all known Anosov diffeomorphisms this was answered affirmatively over 40 years ago. However, a complete answer is not expected any time soon. This is related to the problem of classifying all Anosov systems, and in discrete time there has been little progress for decades. So far, it appears that Anosov diffeomorphisms are rather rigid objects in terms of topological conjugacy; all known examples are topologically equivalent to an algebraic diffeomorphism. Specifically, all known Anosov diffeomorphisms are conjugate to a hyperbolic toral automorphism or a counterpart on infranilmanifolds. However, it is not known whether there are other Anosov diffeomorphisms.

For Anosov flows the situation is quite different. While the question about their transitivity took some efforts to resolve (Section 8.3), the answer is no, and this is but one manifestation of the proliferation of examples of Anosov *flows* that have arisen by surgery constructions. One of the fundamental distinctions between Anosov diffeomorphisms and Anosov flows is the following: surgery constructions are naturally suited for the creation of new flows from old, and have no equally natural discrete-time counterpart. Accordingly, some sections of this chapter are devoted to surgeries that produce new kinds of Anosov flows. First, even among geodesic flows of negatively curved Riemannian manifolds (Theorem 5.2.4) there are instances where the fundamental group is different from that of any locally symmetric space. In this case, the phase space of the geodesic flow is topologically distinct from the unit tangent bundle of any locally symmetric space, so these geodesic flows are not orbit equivalent to an algebraic system. Furthermore, already in low-dimensional settings, dimension 3, there are compact manifolds that admit Anosov flows without being homeomorphic to a mapping torus (suspension manifold) or the unit tangent bundle of any surface (Section 8.2). Among those examples there are even Anosov flows with nowhere-dense nonwandering sets which are therefore dynamically quite different from any volume-preserving flow (Section 8.3). In this context it is the wealth of examples that poses a challenge for any classification, but we present a range of insights pertinent to structural analysis of Anosov flows. In contrast to the discrete-time context, one sees progress toward a comprehensive understanding of Anosov flows, even though that goal is presently out of reach.

As noted, Anosov flows can be topologically transitive or not, which is established by our prior examples combined with those from Section 8.3. There is another dichotomy among Anosov flows: they may be $\mathbb{R}$-*covered* (Definition 8.4.4) or not, and in this dichotomy we present this new kind of Anosov flow in Section 8.3. These dichotomies overlap partially—being $\mathbb{R}$-covered implies topological transitivity (Proposition 8.4.6)—but not fully. Sections 8.4 and 8.5 introduce not only this latter dichotomy and the pertinent terminology but also substantial tools for structural analysis.

Section 8.6 explores ergodic properties of geodesic and horocycle flows. Earlier, such as in Proposition 3.3.19, we used their "commutation" relations (2.2.3) and (2.2.4) as the central tool for understanding ergodic properties of both flows, and here we show how much further one can extend this relation and those arguments. What happens here is that hyperbolicity provides tools for studying a parabolic flow (Section 1.1.d) by a deep entanglement of the flows. This demonstrates that the sections in this chapter that are centered on geodesic flows can be used as a starting point for a broader and deeper understanding of other flows. On the other hand, this study of parabolic flows provides a counterpoint to the hyperbolic effects that dominate this book.

The style of this chapter is a little different from previous chapters in that some results are presented more in the style of a survey. We define the necessary notions, and give full proofs of many of the results, but we are more liberal in stating results without proof.

8.1 Anosov diffeomorphisms, suspensions, and mixing

This section has a dual purpose. We briefly present the structural theory of Anosov diffeomorphisms as a counterpoint to our later developments for Anosov flows. This discrete-time theory is the base for understanding Anosov flows that are suspensions or special flows, and the bulk of this section shows that suspensions are characterized by not being mixing

The main classification theorem for Anosov diffeomorphisms is the following one:

Theorem 8.1.1 ([141, 252]). *Every Anosov diffeomorphism of an infranilmanifold is topologically conjugate to a hyperbolic automorphism.*

This classifies all Anosov diffeomorphisms hitherto known. Franks [141] proved this result on tori and assuming regional recurrence (which is definitely necessary). Manning [252] generalized the arguments to the infranilmanifold setting and showed that transitivity is automatic.

Another classified category is that of codimension-1 Anosov diffeomorphisms.

Theorem 8.1.2 ([141, 267]). *Codimension-*1 *Anosov diffeomorphisms are conjugate to linear toral automorphisms.*

Franks [141] showed that regionally recurrent codimension-1 Anosov diffeomorphisms (that is, those with 1-dimensional stable or unstable manifolds) are topologically conjugate to a linear toral automorphism; Newhouse [267] showed that regional recurrence follows from the codimension-1 assumption.

Let us emphasize the unknowns here: it is not known whether there are Anosov diffeomorphisms on manifolds other than infranilmanifolds, and it is not known whether Anosov diffeomorphisms are always topologically transitive. We will see that the picture for Anosov flows is quite different in both respects. The classification problem for Anosov flows suffers from a surfeit rather than a dearth of examples. We will mention related conjectures as we go (Conjecture 8.4.18).

As the next step in the program described in the introduction, we study the relationship between being topologically transitive versus topologically mixing. In discrete time these coincide in the Anosov case: transitive Anosov diffeomorphisms are mixing, and this situation corresponds to density of unstable (or stable) manifolds of periodic points. There is a corresponding fact for transitive Anosov flows—weak-unstable manifolds of periodic points are dense (Corollary 6.2.9)—but for the suspension of a transitive Anosov diffeomorphism only *weak* unstable manifolds are dense (by Theorem 6.2.11 and Example 1.6.34). Indeed, density of weak-unstable manifolds versus density of strong-unstable manifolds is the very difference that corresponds to the difference between the flow being transitive versus mixing. This is the subject of the present section: transitive Anosov flows are mixing unless they are suspensions—and (unlike in Theorem 7.3.22) not only topologically but smoothly so.

Theorem 8.1.3 (Plante Alternative). *For a transitive Anosov flow either*

(1) *each strong-stable and each strong-unstable manifold is dense* (*so the flow is topologically mixing by Theorem* 6.2.11), *or*

(2) *the flow is a suspension of an Anosov diffeomorphism* (*and hence not topologically mixing by Example* 1.6.34).

This raises the question of how one rules out a suspension to obtain mixing for a flow. We give an important example that includes geodesic flows.

Theorem 8.1.4. *Contact Anosov flows are topologically mixing.*

Proof. The contact form θ provides a smooth Φ-invariant measure via the volume element $\theta \wedge (d\theta)^{m-1}$, so $\mathrm{NW}(\Phi) = M$ by the Poincaré Recurrence Theorem (Theorem 3.2.1). The suspension scenario in Theorem 8.1.3 is ruled out by the next result.

Lemma 8.1.5. $\ker\theta = E^u \oplus E^s$.

Proof. The contact form θ vanishes on (strong) stable vectors: the unit tangent bundle of M is compact, so by continuity θ is bounded on it, that is, if v is any vector then $|\theta(v)| \leq \text{const.}\, \|v\|$. If $v \in E^s(x)$, then Φ-invariance of θ yields $|\theta(v)| = |\varphi^t_*\theta(v)| = |\theta(\varphi^t(v))| \leq \text{const.}\, \|\varphi^t(v)\| \xrightarrow[t\to\infty]{} 0$, so $\theta(v) = 0$. Likewise, $\theta = 0$ on unstable vectors. □

Now, if Φ is a suspension, then $\ker\theta = E^u \oplus E^s$ is tangent to the foliation into level sets of t of the suspension manifold, hence $\theta = f\,dt$ with Φ-invariant, hence constant smooth f (Proposition 1.6.4), and $d\theta = f \cdot ddt = 0$. But then $\theta \wedge d\theta \equiv 0$, contrary to the contact property. □

Corollary 8.1.6. *Geodesic flows of compact negatively curved Riemannian manifolds are topologically mixing.*

Proof. They are contact Anosov flows (Proposition 2.6.28). □

We note a complementary theorem by Plante.

Theorem 8.1.7. *If Φ is a C^2 Anosov 3-flow whose invariant splitting is C^1, then Φ is topologically transitive and either a suspension (of an Anosov diffeomorphism of $\mathbb{T}^2$ because $E^s \oplus E^u$ is integrable) or a contact flow in the sense that the canonical invariant 1-form A defined by $A(\dot\varphi) \equiv 1$, $A\restriction_{E^u \oplus E^s} = 0$ defines an invariant ergodic measure equivalent to the Riemannian volume.*

Proof [292, Theorem 4.7]. By assumption, the canonical form is C^1, so $\theta := A \wedge dA$ is a well-defined Φ-invariant 3-form. Unless $\theta \equiv 0$, Theorem 7.1.27 implies the second conclusion (and transitivity). If $\theta \equiv 0$, then by the Frobenius Theorem $\ker A = E^s \oplus E^u$ is integrable, that is, tangent to a C^1-foliation: if X, Y are sections of $\ker A$ then so is $[X,Y]$ because with $V := \dot\varphi$ we have

$$0 \equiv \theta(V,X,Y) = A(V)dA(X,Y) = 1 \cdot (\mathcal{L}_X A(Y) - \mathcal{L}_Y A(X) - A([X,Y])) = -A([X,Y]).$$

The argument after Lemma 8.1.12 shows that Φ is the suspension of an Anosov diffeomorphism of $\mathbb{T}^2$, hence topologically transitive. □

This and Theorem 7.3.22 include an important ergodic counterpart to Theorem 8.1.3:

Theorem 8.1.8 (Anosov Alternative [10, Theorem 14]). *For a volume-preserving Anosov flow Φ either*

(1) *the strong-stable and strong-unstable foliations are ergodic and Φ is mixing of all orders* (*Corollary* 7.1.8) *and indeed Bernoulli* (*Theorem* 7.4.20), *or*

(2) *the flow is a suspension of an Anosov diffeomorphism* (*and hence not weakly mixing by Proposition* 3.4.9).

Turning to the proof of Theorem 8.1.3, we note that in contrast to its ergodic counterpart (Definition 7.1.1) the present context does not allow exceptional null sets with respect to saturation.

Definition 8.1.9. A set is said to be *saturated* by a foliation if it is a union of leaves.

Proof of Theorem 8.1.3. Proposition 6.2.13 shows that if there is a periodic point whose strong-unstable manifold is not dense, then the flow is a suspension. By Proposition 6.2.17 the only other possibility is that all strong-unstable manifolds are dense (not just those of periodic points). It remains to show that the suspension obtained in the first case is not just topological, that is, to show that the set over which we suspend is indeed a C^1 submanifold.

To that end recall that for a suspension flow each strong-unstable leaf densely fills a codimension-1 submanifold that is (up to a shift by the flow) the manifold over which we suspend. Since the same goes for strong-stable leaves, joint integrability (Definition 7.2.5) is clearly necessary for being a suspension.

Lemma 8.1.10. *W^u and W^s are jointly integrable* (*Definition* 7.2.5) *unless all strong-unstable and all strong-stable leaves are dense.*

Proof. If any one strong-unstable or strong-stable leaf fails to be dense, then by Proposition 6.2.17 this is the case for a leaf of a periodic point, and then the flow is a suspension of a map $f = \varphi^s \restriction_K$ of the W^u-saturated set K in Proposition 6.2.13.

Claim 8.1.11. *K is W^s-saturated.*

Proof. If $x \in K$, $y \in W^s(x)$ take t such that $y \in \varphi^t(K)$. To see that $t = 0$ note that K and $\varphi^t(K)$ are φ^s-invariant and compact while $d((\varphi^s)^n(x), (\varphi^s)^n(y)) \xrightarrow[n\to\infty]{} 0$. □

The claim implies that $W^s(y) \cap W^{cs}_{\delta'}(y)$ lies entirely in some $\varphi^t(K)$, and moreover, that $\mathcal{H}_{y,z}(W^s(y) \cap W^{cs}_{\delta'}(y))$ also lies in $\varphi^t(K)$.

If the desired inclusion fails, that is,

$$\mathcal{H}_{y,z}(W^s(u) \cap W^{cs}_{\delta'}(y)) \not\subset W^s(\mathcal{H}_{y,z}(u)) \cap W^{cs}_{\delta}(z),$$

the mismatch can only be in the time direction, which means that there is an open interval of $s \in \mathbb{R}$ for which $\varphi^s(K) \cap K \neq \varnothing$, contrary to Proposition 6.2.13. □

It is now easy to see that this defines a smooth structure:

Lemma 8.1.12. *If W^u and W^s are jointly integrable, then $E^u \oplus E^s$ is tangent to a C^1-foliation.*[1]

Remark 8.1.13. It is important that the foliation in question does not just have smooth leaves but C^1-foliation charts.

Proof. One can see that through each point there is a C^1 submanifold to which $E^u \oplus E^s$ is tangent. To see that the resulting foliation has C^1-foliation charts we use that for any ball B in a leaf we have a local parametrization $(-\delta, \delta) \times B \to M$, $(s, x) \mapsto \varphi^s(x)$. Once one identifies B with a Euclidean ball, this collection of parametrizations defines the required C^1-atlas of M. □

To conclude the proof of Theorem 8.1.3 we need to show that the leaves of the C^1-foliation in the preceding lemma are the $\varphi^t(K)$. To that end take a leaf L and note first that since L contains a strong-unstable manifold it is dense in some $\varphi^t(K)$. This reduces the problem to showing that L is a compact submanifold of M.

If this is not so, then there are points in L that are far apart in the intrinsic distance, yet close in the metric on M. Since L is a codimension-1 submanifold for which the projection π from Proposition 6.2.13 is locally injective, this means that these points are separated by a small time shift. However, this gives small values of s for which $\varphi^{t+s}(K) \cap \varphi^t(P) \neq \varnothing$, contrary to Proposition 6.2.13. □

8.2 Foulon–Handel–Thurston surgery

In this section, we modify the geodesic flow of a compact surface of constant negative curvature (Chapter 2) by a surgery construction to obtain flows of an entirely new topological type (Theorem 8.2.3). These are transitive (indeed, mixing) and hence do not add directly to the investigation of the transitivity question, but they introduce us to a much broader class of $\mathbb{R}$-covered flows, as we will call them, than geodesic flows and suspensions alone.

[1]This does not imply that E^u or E^s are C^1, only that $E^u \oplus E^s$ is.

It may help the intuition not only for this section, but also for the structural analysis later in the chapter to revisit some features of those geodesic flows with particular dynamical aspects in mind. This discussion is not needed in a technical sense, so the reader can skip ahead to Definition 8.2.2 and Theorem 8.2.3.

Early on we introduced sections for a flow (Figure 0.2.1), and when studying geodesic flows on the sphere, Birkhoff introduced a local section that can also be helpful in our context. He considered those tangent vectors at points on the equator of the sphere that point into the northern hemisphere. In the unit tangent bundle this is an annulus (S^1 in the base times an open interval of angles), and at times it is convenient to consider the closure, which additionally includes two periodic orbits as the boundary components (the equator traveled east and west, respectively).

Definition 8.2.1. A *Birkhoff annulus* is an embedded annulus whose interior is transverse to the flow and whose boundary consists of two periodic orbits.

Birkhoff annuli abound for geodesic flows of a negatively curved surface because this construction works for any geodesic, the annulus being those unit vectors along it whose inner product with a chosen normal vector field is nonnegative. Although for the geodesic flow this may seem overcomplicated, let us give a purely dynamical description of this construction. To that end, consider the upper half-plane model $\mathbb{H}$ and suppose the geodesic in question is the imaginary axis I (with upward vectors). Taking the rightward horizontal unit vector field to determine the Birkhoff annulus means that we select those points whose orbits limit in positive time on ∞ or a nonnegative real point. The following alternative description eliminates the need for the normal vector field as a starting point: For every such orbit, its unstable leaf (outward unit vectors on a circle tangent to $\mathbb{R}$ at the nonnegative positive-time limit) contains a point of the stable manifold of *I to the right of I*. Likewise, each stable leaf of such a point contains a point in the right half of the unstable leaf of I. In an Anosov 3-flow the weak-stable or weak-unstable leaf of an orbit is bisected by the orbit; choosing one half for each gives a description of the Birkhoff annulus construction without appeal to an underlying geometric structure. (Not every pairing gives a nontrivial set; consider the right unstable and left stable halves of I, for example.)

Returning to the choice of the right half of the stable manifold of I and the right half of the unstable manifold of I, the set of orbits it encompassed can be viewed as a "rectangle" with these halves as its sides because every orbit is characterized by its pair of (weak) stable and unstable leaves. This rectangle includes I and its reverse as vertices, but the two other vertices are missing because there is no orbit that is positively asymptotic to I and negatively asymptotic to its reverse because there is no geodesic with both (asymptotic) endpoints at 0 (Figure 8.5.5).

This description was framed with $S\mathbb{H}$ in mind, but as noted in the discussion represented in Figure 2.4.3, the proper context is the universal cover of $S\mathbb{H}$ (or of $S\mathbb{D}$).

Here one can pass to the orbit space and obtain literal rectangles from this construction. In either context, one can now translate the joining of a Birkhoff annulus defined by a closed geodesic to the one defined by its reverse, and in this rectangle picture, this corresponds to producing a like rectangle that shares exactly the vertex that corresponds to the reverse closed geodesic. In $S\mathbb{H}$ the vertices corresponding to I are identified, but notably, in the universal cover of $S\mathbb{H}$ they are not, so this unfolds to a bi-infinite string of such rectangles joined by vertices (Figure 8.5.6).

Beyond the context of geodesic flows one might hope to obtain a Birkhoff annulus from a pair of isotopic periodic orbits, but one needs to find an isotopy through circles transverse to the flow. For a geodesic flow, moreover, Birkhoff annuli come in pairs corresponding to the two choices of normal vector field, and taken together, such a pair gives a *Birkhoff torus*, which consists of all the unit tangent vectors to a closed geodesic and is transverse to the geodesic flow except on the two periodic orbits.

Whether there are embedded *transverse* tori is an entirely different matter. We will see that geodesic flows have none, while the Franks–Williams flow (Section 8.3) does—as the central piece of the construction.

With this in mind, one can describe the essential purpose of the present section as performing surgery on a Birkhoff annulus that produces new (contact) Anosov flows. In contrast, Section 8.3 modifies an Anosov flow in such a way that there is a transverse torus (which breaks the Anosov property, so surgery then restores it to give an Anosov flow with a transverse torus).

Definition 8.2.2. An Anosov flow on a 3-manifold is said to be of *algebraic type* if it is finitely covered by the geodesic flow of a compact surface (Section 2.3) or the suspension of a diffeomorphism of the 2-torus (Example 1.5.26).

The Reeb flow of a contact form (Definition 2.2.5) is *nondegenerate* if all its periodic orbits are transverse (1 is not an eigenvalue of the differential of the first-return map) and a *contact Anosov flow* if it is an Anosov flow.

Geodesic flows of negatively curved surfaces are the primary example of contact Anosov flows (on 3-manifolds), and here we describe a construction to obtain topologically new flows of this type.

Theorem 8.2.3 ([137]). *Consider a negatively curved oriented surface Σ and a closed geodesic $\mathfrak{c}\colon S^1 \to \Sigma$, $s \mapsto \mathfrak{c}(s)$. For $q \in \mathbb{Z}$ there is a $(1,q)$-Dehn surgery (described below) that produces a 3-manifold that (unless $q = 0$) is not homeomorphic to the unit tangent bundle of* any *surface or to a suspension manifold, and a volume-preserving flow on it that is therefore not orbit equivalent to any algebraic flow. This flow is Anosov if and only if $q \geq 0$. If $\mathfrak{c}$ is* filling *and* indivisible,[2] *then for all but finitely*

[2] A closed curve c in a surface *fills* the surface if $\alpha \cap c \neq \varnothing$ whenever α is a closed curve that is not null-homotopic. It (and, more generally, a closed orbit) is said to be *indivisible* if it is not the same geodesic traversed more than once.

many $q > 0$ *the resulting* 3*-manifold is hyperbolic, and each free homotopy class of closed orbits of the new flow is* (*countably*) *infinite* (*Theorem* 8.5.1).[3]

Remark 8.2.4. Given our focus on hyperbolic flows, this surgery is most interesting when $q > 0$, but we note that the flows obtained for $q < 0$ are of interest as well, even though we will not study them here.

Proof. Let $s \mapsto \gamma(s)$ be the unit vector field perpendicular to the closed geodesic $\mathfrak{c}: S^1 \to \Sigma$, $s \mapsto \mathfrak{c}(s)$ given by $\theta = \pi/2$, that is, rotated in the positive direction. If $\mathfrak{c}$ is simple and separating as in Figure 8.2.1, we denote the unit tangent bundles of the two components of the surface by M_1 and M_2, and the common boundary of M_1 and

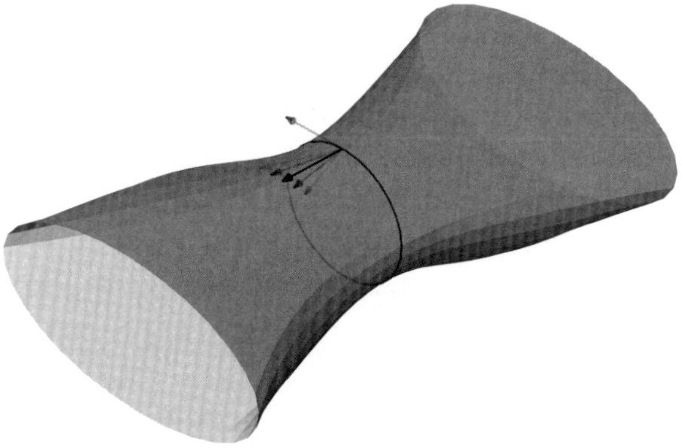

Figure 8.2.1. Near-normal vectors to a geodesic. [Reproduced from Foulon and Hasselblatt [137].]

M_2 is a torus $S^1 \times S^1$ parametrized by the parameter s of the geodesic $\mathfrak{c}$ and the angle θ with the tangent vector of the geodesic. Either way we change the geodesic flow g^t on the unit tangent bundle $S\Sigma$ in a neighborhood Λ of γ. To parametrize Λ, linearize the angle θ with the tangent vector field to $\mathfrak{c}$ by taking $w := \frac{\ell}{2\pi} \cos\theta$ for θ near $-\pi/2$, where ℓ is the length of $\mathfrak{c}$. This gives parameters

$$(t, s, w) \in \Omega := (-\eta, \eta) \times S^1 \times (-\epsilon, +\epsilon),$$

where $t \in (-\eta, \eta)$ parametrizes the flow direction and $s \in S^1$ is the parameter along $\mathfrak{c}$; γ is parametrized by $\{0\} \times S^1 \times \{0\}$. The parameters t, s, w arise from the flows generated by the canonical framing (2.2.1) on $S\mathbb{H}$, so they produce Darboux coordinates:

[3]For algebraic flows, free homotopy classes of closed orbits have finite cardinality, for geodesic flows no two (parametrized) orbits are homotopic, though by rotating the tangent vector though π, each is isotopic to its flip, which has the same *image* as another orbit (the same geodesic run backward), and only in suspensions are all free homotopy classes of *images* of orbits singletons [36].

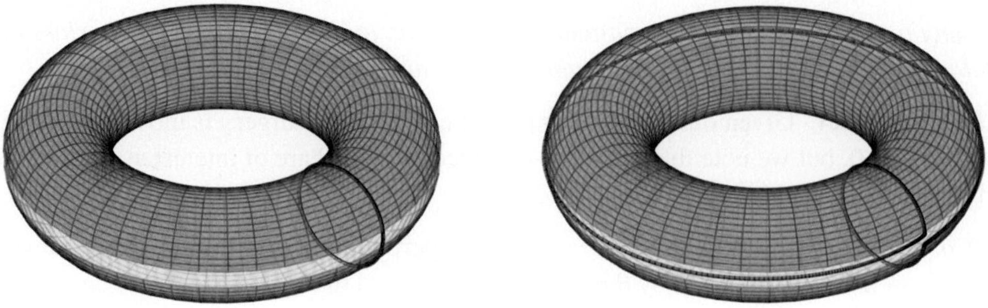

Figure 8.2.2. Surgery annulus before and after surgery ($q = 1$). [Reproduced from Foulon and Hasselblatt [137].]

Lemma 8.2.5. *The standard contact form A in this chart satisfies*

$$A = dt + w\,ds, \quad dA = dw \wedge ds, \quad \textit{and} \quad A \wedge dA = dt \wedge dw \wedge ds.$$

Proof. If g_s denotes the Riemannian metric at $(0, s) \in \Sigma$ and we write $(0, s, \theta) = (x, u) \in S\Sigma$, then $d\pi(\partial/\partial\theta) = 0$ implies that for a vector $Z = a\frac{\partial}{\partial t} + b\frac{\partial}{\partial s} + c\frac{\partial}{\partial \theta}$ we have

$$A_{(0,s,\theta)}(Z) = g_s(u, d\pi(Z)) = g_s\Big(u, a\,d\pi\Big(\frac{\partial}{\partial t}\Big) + b\,d\pi\Big(\frac{\partial}{\partial s}\Big)\Big) = a + bg_s\Big(u, \frac{\partial}{\partial s}\Big).$$

Taking S^1 to have length 2π we necessarily obtain $\|\partial/\partial s\| = \ell/2\pi$. Since a priori $A_{(t,s,\theta)} = dt + g\,ds + h\,d\theta$ with functions g and h, this implies $A_{(t,s,\theta)} = dt + \frac{\ell}{2\pi}\cos\theta\,ds$, that is, $A = dt + w\,ds$. The other claims immediately follow. □

These conventions are natural with respect to the canonical framing from page 122 by X, the vertical vector field $V = \partial/\partial\theta$ and the horizontal vector field $H = [V, X]$ in that $dw(Y) > 0$, $dw(V)\,ds(H) = dw \wedge ds(V, H) = dA(V, H) = 1$ and E^+ is spanned by a vector $V + H$ in the first quadrant (because $[X, V + H] = [X, V] + [X, H] = H + V$).

The $(1, q)$-Dehn surgery can be described as splitting apart an annulus in the manifold and gluing both copies of this annulus back together with a shear, that is, by identifying the two annuli via the shear map. Although this shear map is a homeomorphism of the annulus, this defines a discontinuous operation since the resulting space is no longer homeomorphic to the original one. It is important that the surgery yield a smooth manifold and that the resulting vector field is well defined on this manifold.

Lemma 8.2.5 gives an annulus that *contains* γ and is uniformly transverse to the flow. We split the flow-box chart from Lemma 8.2.5 into two one-sided flow-box neighborhoods of the surgery annulus, and while the initial transition map between these on $\{0\} \times S^1 \times (-\epsilon, +\epsilon)$ is the identity, the surgered manifold is defined by imposing the desired shear as the transition map on this annulus:

$$F\colon S^1 \times (-\epsilon, \epsilon) \to S^1 \times (-\epsilon, \epsilon), \quad (s, w) \mapsto (s + f(w), w) \tag{8.2.1}$$

with $f\colon [-\epsilon, \epsilon] \to S^1$, $w \mapsto \exp(iqg(w/\epsilon))$, $q \in \mathbb{N}$, $g\colon \mathbb{R} \to [0, 2\pi]$ nondecreasing smooth, $0 \le g' \le 4$ even, and $g((-\infty, -1]) = \{0\}$, $g([1, \infty)) = \{2\pi\}$. The use of flow-box charts ensures that the original vector field defining the contact Anosov flow defines a smooth vector field on the surgered manifold, that is, that the orbits are reglued to smooth curves.

Proposition 8.2.6. *The new flow preserves the Liouville volume defined by $A \wedge dA$.*

Proof. Since $d(s + f(w)) = ds + f'(w)\,dw$ we have

$$F_*A = dt + w\,d(s + f(w)) = A + w\,f'(w)\,dw,$$
$$F_*dA = dw \wedge d(s + f(w)) = dw \wedge ds = dA,$$

and

$$F_*A \wedge dA = A \wedge dA,$$

so $A \wedge dA$ is a well-defined volume on M_F. The vector field X on M induces a vector field X_F on M_F. Its flow clearly preserves $A \wedge dA$. □

It remains to show that for $q > 0$ the new flow is an Anosov flow, and we use the formulation of hyperbolicity in terms of suitable Lyapunov–Lorentz metrics as described in Proposition 5.1.14; this is a reformulation of the usual cone criterion for hyperbolicity.

By Proposition 5.1.14 there is a pair of Lyapunov–Lorentz metrics for φ^t. We deform these to work as needed for φ_h^t. First, as in the proof of Theorem 5.2.4, we arrange for the Lyapunov–Lorentz metrics for φ^t to have the form

$$Q^{\pm} = \pm dw\,ds - c\,dt^2$$

in Λ, where c is chosen sufficiently small to ensure that the positive $Q^{\pm}$-cone contains $E^{\pm}$. Choose $\beta\colon \mathbb{R} \to \mathbb{R}^+$ smooth with $\beta((-\infty, 0]) = \{1\}$, $\beta([\eta, \infty)) = \{0\}$, and $\beta' < 0$ on $(-\eta, \eta)$:

Claim 8.2.7. *Taking $Q_0^{\pm}$ and $Q_1^{\pm}$ to be the old Lyapunov–Lorentz metrics outside Λ and*

$$Q_i^{\pm} := \pm(dw\,ds - i\beta(t) f'(w)\,dw^2) - c\,dt^2$$

inside defines Lyapunov–Lorentz metrics for φ_h^t. Here, $i = 0$ on one side of the surgery and $i = 1$ on the other.

Proof. Our choice of f and β ensures that these are smooth metrics.

These choices fit together, that is, F sends the choice on one side to that on the other, because for $t = 0$,

$$\begin{aligned} F_*Q_0^\pm = F_*\big(\pm\, dw\, ds - c\, dt^2\big) &= \pm dw\, d(s - f(w)) - c\, dt^2 \\ &= \pm(dw\, ds - f'(w)\, dw\, dw) - c\, dt^2 \\ &= \pm(dw\, ds - \beta(0) f'(w)\, dw^2) - c\, dt^2 = Q_1^\pm. \end{aligned}$$ □

Of the required properties in Proposition 5.1.14, (2) and (3) are clear. We wish to check that (1) and (4) in Proposition 5.1.14 are inherited from the same properties for φ^t. Outside Λ this is given since $\varphi = \varphi_h$.

On Λ we have been using a flow-box chart for φ_t, so the flow is represented by a shift in time. This makes it a $Q_0^\pm$-isometry. We need to see how the surgery and the choice of β affect invariance, that is, Proposition 5.1.14(4). To that end it is helpful to restrict attention to the trace of these cones in the s–w plane. Here, the $Q_0^\pm$-cones show as quadrants since

$$0 = Q_0^\pm\Big(a\frac{\partial}{\partial s} + b\frac{\partial}{\partial w}\Big) = ab$$

implies $a = 0$ or $b = 0$. On the immediate other side of the surgery ($t = 0$) they are given by

$$0 = Q_1^\pm\Big(a\frac{\partial}{\partial s} + b\frac{\partial}{\partial w}\Big) = ab - f'(w)b^2 = (a - f'(w))b,$$

which implies $a = f'(w)b$ or $b = 0$; since $f' \le 0$, this describes a subcone of the first and third quadrants that shares the horizontal axis. Since Λ is a flow-box, φ_h leaves these cones exactly invariant, which means that strict monotonicity of β produces a strictly invariant cone field that connects smoothly at $t = \eta$. This gives Proposition 5.1.14.(4).

To obtain Proposition 5.1.14(1) note that, Λ being a flow-box chart,

$$\pm Q_1^\pm\Big(D\varphi_h^t\Big(a\frac{\partial}{\partial s} + b\frac{\partial}{\partial w}\Big)\Big) = \pm Q_1^\pm\Big(a\frac{\partial}{\partial s} + b\frac{\partial}{\partial w}\Big) = ab - \beta(t) f'(w) b^2$$

is increasing in t. Combined with the exponential growth outside Λ, this yields Proposition 5.1.14(1).

With Proposition 8.2.9 below, this completes the proof of Theorem 8.2.3. □

We now show that the flows thus obtained are not topologically orbit equivalent to an algebraic flow.

Definition 8.2.8. A 3-manifold is said to be Seifert fibered if it admits a decomposition into a disjoint union of circles (the fibers) such that each fiber has a tubular neighborhood diffeomorphic (in a fiber-preserving way) to the torus $D^2 \times S^1$ obtained from $D^2 \times [0,1]$ by identifying $D^2 \times \{0\}$ and $D^2 \times \{1\}$ via a rational rotation.

Proposition 8.2.9. *When the surgery is carried out using a separating curve, the resulting flow is not topologically orbit equivalent to an algebraic Anosov flow. More strongly* [341, *p.* 419], *no finite cover of the surgered manifold is a Seifert-fibered manifold (much less a sphere bundle) or a torus bundle over a circle.*

Proof. We study finite covers of the surgered manifold M by examining their fundamental group. The two pertinent facts are

(1) the fundamental group of a torus bundle over a circle is solvable; thus we wish to show that $\pi_1(M)$ is not virtually solvable, that is, has no solvable finite-index subgroup;

(2) the fundamental group of a Seifert-fibered manifold contains an infinite normal cyclic subgroup generated by a regular fiber [326, p. 432]; thus we want to show that no finite-index subgroup of $\pi_1(M)$ contains an infinite cyclic normal subgroup.

One observation that we will use for both of these items is the following.

Remark 8.2.10. If a group contains a finite-index subgroup H and a free subgroup F, then H contains a subgroup of F that is isomorphic to F.

By the van Kampen Theorem [186, Theorem 1.20], we have

$$\pi_1(M) = \pi_1(M_1) \underset{\pi_1(\partial M_1)}{*} \pi_1(M_2) = \pi_1(M_1) \underset{\pi_1(\partial M_2)}{*} \pi_1(M_2),$$

using the isomorphism $F_*\colon \pi_1(\partial M_1) \longrightarrow \pi_1(\partial M_2)$ induced by F from (8.2.1).

Puncturing a surface of genus g and retracting the remainder to its skeleton (a string of $2g$ circles) shows that the fundamental group is a free group with $2g$ generators. Thus, we see that $\pi_1(M_i) = F_i \oplus \mathbb{Z}$ for $i = 1,2$, where F_1 and F_2 are free groups.

If $H < \pi_1(M)$ has finite index, then, as remarked above, it contains a free group inherited from F_1 or F_2, and since this holds recursively, H is not solvable, hence item (1).

For item (2) suppose $\langle g \rangle < H$ is an infinite cyclic normal subgroup. This means that for every $h \in H$ there is a $p_h \in \mathbb{Z}$ such that $hgh^{-1} = g^{p_h}$. Clearly, $p_{h_1 h_2} = p_{h_1} p_{h_2}$

for any $h_1, h_2 \in H$, so $p_{\mathrm{Id}} = 1$ implies that for each $h \in H$ we have $p_h \in \{\pm 1\}$ and $p_h = p_{h^{-1}}$. Thus, after possibly passing to the index-2 subgroup $\{h \in H \mid p_h = 1\}$, we may assume without loss of generality that H is in the centralizer of g, that is, $gh = hg$ for all $h \in H$.

As remarked above, there is a free group $F_H \subset F_1 \cap H$ isomorphic to F_1 (we only need that it is large enough). We can write $g = wr^k$ with w a word in generators of F_1 and F_2 only and r the generator of the S^1-factor of $\pi_1(M_1)$: writing one of the generators of $\pi_1(\partial M_2)$ as $\alpha_a r = a = \omega_a r'^k$ with $\alpha_a \in F_1$, $\omega_a \in F_2$, and r' the generator of the S^1-factor of $\pi_1(M_2)$, we find that any occurrence of $r\omega$ with $\omega \in F_2$ can be rewritten as $\alpha_a^{-1}\omega_a\omega\omega_a^{-1}\alpha_a r$; one applies this recursively to get the claim. We thus find that for any $h \in F_H \subset F_1$ we get

$$whr^k = wr^k h = gh = hg = hwr^k,$$

that is, $wh = hw$. But for $w \neq \mathrm{Id}$, this only holds when h is a power of w. Since F_H is not a subgroup of a cyclic group, we must have $g = r^k$ for some $k \in \mathbb{Z}$.

The same reasoning using F_2 shows that $g = r'^{\ell}$ for some $\ell \in \mathbb{Z}$, where r' is the generator of the S^1-factor of $\pi_1(M_2)$. This is incompatible with the earlier observation that $g = r^k \sim (sr')^k$, where s represents the word in F_2 corresponding to the slope of the surgery—unless $s = \mathrm{Id}$, so the surgery is trivial. Hence (2). □

Remark 8.2.11. The assertion in Theorem 8.2.3 that if $\mathfrak{c}$ is *filling* and *indivisible*, then for all but finitely many $q > 0$ the resulting 3-manifold is hyperbolic in the sense of Thurston, results from a combination of two results: a folklore theorem says that removing a tubular neighborhood of $\mathfrak{c}$ in this case gives a hyperbolic manifold with boundary, and a theorem of Thurston says that then gluing a torus back in with a Dehn twist results in a closed hyperbolic for all but finitely many values of the twist parameter.

What we have presented so far is the Handel–Thurston surgery in modern form. A subsequent refinement by Foulon modifies it in such a way that the resulting flow is a contact flow.

Theorem 8.2.12 (Foulon surgery [137]). *A suitable time change of the flow obtained in Theorem* 8.2.3 *is a contact flow.*

Proof. The problem we have to address appears in the proof of Proposition 8.2.6: $F_*A = dt + w\,d(s + f(w)) = A + w\,f'(w)\,dw$ implies that there is no well-defined contact form after surgery. The additive nature of the discrepancy here also suggests a deformation which produces a contact form after surgery: take $A_h^{\mp} = A \mp dh$ for

$\pm t \geq 0$, where

$$h(t,w) := \frac{1}{2}\underbrace{\lambda(t)}_{\lambda\colon \mathbb{R}\to[0,1] \text{ is a smooth bump function}} \int_{-\epsilon}^{w} x f'(x)\,dx \text{ on } (-\eta,\eta)\times(-\epsilon,\epsilon) \text{ and } h = 0 \text{ outside,}$$

satisfies $dh = \frac{1}{2} w f'(w) dw$ on the surgery annulus and $h \equiv 0$ for t close to $\pm\eta$. Hence $F^*(A_h^+) = A_h^-$ and $A_h^\pm$ induces a contact form A_A because

$$F_*(A_h) = F_*(A - dh) = F_*A - F_*\,dh = (A + 2\,dh) - dh = A + dh = A_h.$$

Its Reeb field is a time change

$$X_h := \frac{X}{1 \pm dh(X)}$$

of X because clearly $A_h(X_h) \equiv 1$. This is well defined by Lemma 8.2.13. □

Lemma 8.2.13. *If* $0 < \epsilon < \frac{\eta}{2\pi q}$, *then* $|dh(X)| < 1$.

Proof. The relations $|f'(w)| = q\left|\frac{d}{dw} g\left(\frac{w}{\epsilon}\right)\right| = \frac{q}{\epsilon}\left|g'\left(\frac{w}{\epsilon}\right)\right|$, $|\lambda'| \leq \pi/\eta$, $0 \leq g' \leq 4$ give

$$\begin{aligned}\left|\frac{\partial h}{\partial t}\right| = \left|\frac{1}{2}\lambda'(t)\int_0^w x\, f'(x)\,dx\right| &\leq \left|\frac{q\pi}{2\eta}\right| \int_0^{\epsilon} x \frac{|g'|}{\epsilon}\,dx \\ &\leq \left|\frac{q\pi}{2\eta}\epsilon\frac{4}{\epsilon}\int_0^{\epsilon} dx\right| = \frac{2q\pi}{\eta}\epsilon < 1.\end{aligned}$$ □

We note that by contraposition, Theorem 9.3.11 below gives the following proposition:

Proposition 8.2.14. *The Bowen–Margulis measure of the contact flow from Theorem* 8.2.12 *is singular with respect to the Liouville measure, that is, the topological entropy is strictly larger than the Liouville entropy.*

Remark 8.2.15. Whether or not one establishes hyperbolicity (after Proposition 8.2.6), we have shown that this construction produces a new contact flow from the geodesic flow of a surface of constant negative curvature (and independently of whether the twist is in the direction that preserves hyperbolicity). This gives contact flows on new manifolds, and one can show that even those flows among these which are not hyperbolic necessarily exhibit exponential orbit complexity. Moreover, with suitably reconfigured annuli one can carry out a like construction starting with each of the other two flows induced from those in Example 2.2.6. While doing this with the horizontal flow does not yield anything fundamentally different from what we here obtain from the geodesic flow, the vertical (fiber) flow is strictly periodic, so *any* complexity of its surgered counterpart is entirely due to the surgery. Thus, hyperbolicity is not needed to make Theorem 8.2.3 interesting.

Remark 8.2.16 (Goodman surgery). That the surgery presented here produces hyperbolic manifolds was not known at the time of the Handel–Thurston work because the "folklore" result invoked in Remark 8.2.11 was only established later. Handel and Thurston remarked that it would be interesting to know whether one can obtain hyperbolic manifolds in this way, and this prompted Goodman to produce a Dehn-surgery construction that could at the time be seen to do this [153]. It is of independent interest. The surgery annulus is constructed near a closed orbit by a (large) deformation, and then the same kind of Dehn surgery is carried out. Specifically, taking the closed orbit to be that of the origin in the suspension of $\left(\begin{smallmatrix} 2 & 1 \\ 1 & 1 \end{smallmatrix}\right)$ (which is a figure-eight knot), whose complement is hyperbolic, Thurston's result on hyperbolicity of Dehn-surgered manifolds can be applied.

8.3 Anomalous Anosov flows

Our examples of Anosov flows so far are topologically transitive, and we will see that this is tied deeply to their structure. For Anosov geodesic flows the next theorem gives a classification of the flows.

Theorem 8.3.1 (Ghys–Plante). *If M is a closed manifold that admits a Riemannian metric of constant negative curvature and Φ, Ψ are Anosov geodesic flows for Riemannian metrics on M, then they are topologically orbit equivalent* [147].

Indeed, every Anosov flow on a circle bundle over a surface is topologically equivalent, up to finite covers, to the geodesic flow of a hyperbolic Riemannian surface, and this holds more generally on 3-manifolds whose fundamental group has nontrivial center [26]; *see also* [27, *Theorem B*].

Every Anosov flow on a torus bundle over the circle is topologically equivalent to the suspension of a toral automorphism; *indeed that manifold supports only one other Anosov flow, the suspension of the inverse* [293].

Remark 8.3.2. The manifolds in this theorem support essentially only one Anosov flow; this is quite different for the manifolds in Theorem 8.3.13.

Anosov flows that are not geodesic flows exhibit phenomena unique to flows (with no counterpart for Anosov diffeomorphisms). In this section we explore constructions of Anosov flows that are of types entirely different from what we have seen previously. The flows are distinguished by having closed transversals; this is both a device in the construction as well as an essential means to making sure they are not topologically transitive, while the $\mathbb{R}$-covered Anosov flows we study in Section 8.5, and turn out to have built in Section 8.2, are transitive and are either suspensions or akin to geodesic flows ("skewed"). The constructions in this chapter originated with Franks and Williams, with whose example we begin this section. A later construction by Bonatti and Langevin [56] turns out to be of the same type, and both are generalized by the Béguin–Bonatti–Yu construction presented at the end of this section. $\mathbb{R}$-covered Anosov flows arise from surgeries first introduced by Handel and Thurston [170] and soon generalized by Goodman [153]; this is the construction we refined in Section 8.2 to furthermore preserve the contact property. We mention here a different kind of surgery due to Fried [144] (which shows that every smooth transitive Anosov 3-flow is obtained by Dehn surgery of the suspension of a pseudo-Anosov homeomorphism) and the fact that Bonatti, Barbot, and Fenley have not only produced altogether the most complete studies of Anosov 3-flows, but also a range of additional constructions.

The examples of Anosov flows that we have seen so far invite the conjecture that Anosov flows are topologically transitive. While the counterpart for Anosov diffeomorphisms remains a question, this section presents Anosov flows that are not topologically transitive, and we show how this construction produces an infinite variety of new Anosov flows.

Theorem 8.3.3 (Franks–Williams). *There is an Anosov flow on a 3-manifold that is not topologically transitive, that is, whose chain-recurrent set is not the whole manifold.*

The construction in fact produces a flow with an attractor–repeller pair, and consists of gluing together a flow with an attractor and a time-reversed copy of the same flow. That this flow is an Anosov flow will be a consequence of either Theorem 6.2.24 or the Alekseev Cone Criterion.

An Anosov flow of a 3-manifold is topologically transitive if the splitting $TM = E^u \oplus E^s \oplus \mathbb{R}X$ is C^1 (Theorem 8.1.7), so this splitting is not C^1 for the Franks–Williams flow.

In preparation for surgery we remove from the DA flow (Definition 6.5.4) a cylindrical tube around the repelling orbit of the origin. The flow on the remainder is then well defined for positive time because it goes "inward" from the void left by the tube. Specifically, represent a neighborhood of the repelling orbit as $D \times [0, 1]$, where D is a disk with polar coordinates (r, θ) such that $r = 0$ represents points of the periodic orbit and chosen such that in these coordinates the identification map in the suspension construction has the form

$$(r, \theta, t) \mapsto (er, \theta, t + 1),$$

possibly after rescaling t (to make the radial scale factor exactly e). This means that the set B' parametrized by $(\epsilon e^t, \theta, t)$ is invariant under this identification, hence a well-defined torus in the suspension manifold. We remove the interior of this torus, including the orbit of 0.

A preliminary step ensures that gluing the remaining manifold together with a time-reversed copy produces a smooth flow on the resulting manifold. To deform the flow so the generating vector field is normal to the boundary and has unit speed there, glue a thickened torus $\mathbb{T}^2 \times [0,1]$ to the boundary B' by identifying $\mathbb{T}^2 \times \{1\}$ with B'. Isotopically extend φ^t through this neighborhood in such a way that it is transverse to each toral layer $\mathbb{T}^2 \times s$ and for $s \in [0, 1/2]$ is furthermore normal to these layers and has unit speed. In particular, this is so on the boundary $B := \mathbb{T}^2 \times \{0\}$. Finally, use the flow itself to extend the local coordinates (r, θ, t) through this collar. This gives the "attracting half" of our system.

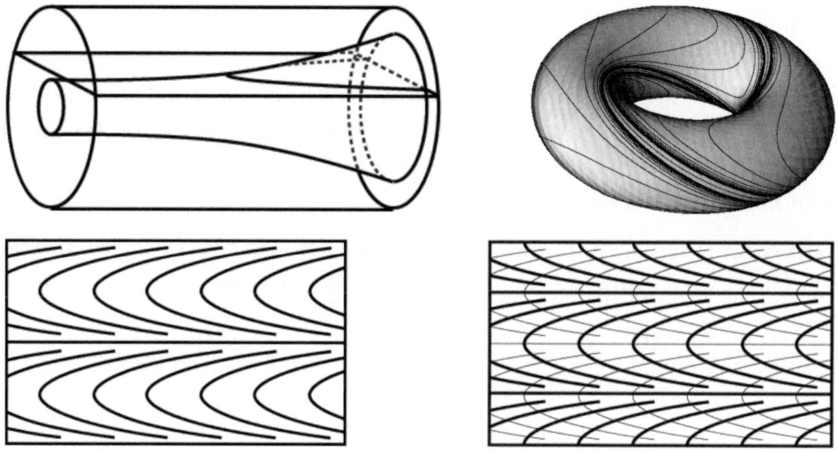

Figure 8.3.1. The excised tube, the foliation on it (shown two ways, including one by Tsuboi), and the complementary foliations. [Animation at http://www.tsuboiweb.matrix.jp/showroom/public_html/animations/gif/stereograph3e/stereograph3e.html.]

The "repelling half" is an identical copy of this attracting half, but with reversed flow direction.

Lemma 8.3.4. *Gluing the attracting and repelling halves together along the boundary tori B and $\bar{B}$ by the identification $t = \bar{t}$ and $\theta = \bar{\theta} - \pi/2$ ensures that the stable and unstable leaves of the resulting flow are transverse at points of the common boundary $B \sim \bar{B}$.*

This establishes the hypotheses of the Mañé Criterion: the chain-recurrent set of the resulting flow is the union of the attractor and the repeller, hence hyperbolic. The dimension of the stable and unstable manifolds is constant, and they are transverse at a point of each orbit: for orbits in the attractor and in the repeller, this is hyperbolicity, and all other orbits go through B, where the lemma establishes transversality. Another method of proof is to show that the construction satisfies the Cone Criterion in Proposition 5.1.7.

Proof of the lemma. The stable leaves of boundary points are those of positive semiorbits in the attracting half. These agree with those of the Anosov diffeomorphism from which the DA map was constructed, and they are therefore planes parallel to the one defined by $\theta \in \{0, \pi\}$ (up to possibly changing the choice of θ by an additive constant). The distance from the line $r = 0$ of such a plane is given by $d = r|\cos\theta|$, so its intersection with the boundary $B' = \{r = \epsilon e^t\}$ is given by $\epsilon e^t = d/|\cos\theta|$ or $\log\epsilon + t - \log d + \log|\cos\theta| = 0$. Thus, the unstable subbundle on B' is the kernel of the 1-form $dt + \tan\theta\, d\theta$, and since the coordinates and foliations extend from here to B via the flow, the same holds on B. In like manner, the unstable subbundle on $\bar{B}$, which is defined by the negative semiorbits in the repelling part, is the kernel of the 1-form $d\bar{t} + \tan\bar{\theta}\, d\bar{\theta} = dt + \cot\theta\, d\theta$. This is orthogonal to $dt + \tan\theta\, d\theta$, which makes stable and unstable leaves orthogonal on $B \sim \bar{B}$ and hence transverse. □

By construction, the torus along which we glued the two pieces of the construction separates the manifold into pieces each of which contains a basic set. It turns out that this is indicative of the structure of any nontransitive Anosov diffeomorphism, in which case the spectral decomposition comes with a corresponding topological decomposition.

Theorem 8.3.5 ([74]). *If Φ is a nontransitive Anosov flow of a 3-manifold M, then there is a finite set of disjoint (incompressible) tori in the complement of* $\mathrm{NW}(\Phi)$ *that decompose M into components each of which contains exactly one of the basic sets in the spectral decomposition of* $\mathrm{NW}(\Phi)$.

The flow direction is important in the construction. Indeed, if we try a similar construction for a map of a 3-torus we would have one of the stable or unstable splittings 2-dimensional and another 1-dimensional. The gluing procedure can then not be done while keeping the foliations transverse (there will be a point of tangency). In low dimensions and codimension-1 (so one of the subbundles has dimension 1) it is known that any Anosov diffeomorphism is transitive. As mentioned previously, for higher dimensions it is not known whether there exist nontransitive Anosov diffeomorphisms.

Moreover, the Franks–Williams construction has come to be seen as but an instance of a toolkit for the construction of Anosov flows on 3-manifolds. The two pieces of the construction are instances of the following (recall Remark 5.3.41 here).

Definition 8.3.6. A *plug* is a pair (M, V) of a 3-manifold M with boundary and a vector field V on M that is transverse to the boundary ∂M. It is *attracting* if the *exit boundary* $\partial^{\text{out}} M \subset \partial M$ (the part of ∂M where V points outward) is empty, *repelling* if the *entrance boundary* $\partial^{\text{in}} M$ (where V points inward) is empty, and *hyperbolic* if $\Lambda := \bigcap_{t \in \mathbb{R}} \varphi^t(M)$ is hyperbolic with 1-dimensional stable and unstable subbundles and no fixed points; here φ^t is the flow generated by V. In the latter case, $W^s(\Lambda)$ intersects $\partial^{\text{in}} M$ transversely (and is disjoint from $\partial^{\text{out}} M \subset \partial M$), so $\mathcal{L}_V^s := W^s(\Lambda) \cap \partial^{\text{in}} M$ is a 1-dimensional lamination, called the *entrance lamination*. Likewise, one obtains the *exit lamination* $\mathcal{L}_V^u := W^u(\Lambda) \cap \partial^{\text{out}} M$.

If (M, V) is a hyperbolic attracting plug, then $\mathcal{L}_V^s$ is a foliation of $\partial^{\text{in}} M$—by lines due to transversality—so every component of $\mathcal{L}_V^s$ is a 2-torus.[4]

Proposition 8.3.7 (Béguin–Bonatti–Yu). *The entrance and exit laminations are* Morse–Smale laminations, *that is, they have finitely many compact leaves, every half noncompact leaf is asymptotic to a compact leaf, and each compact leaf can be oriented so as to be "attracting" (the contracting orientation).*

With this terminology, the arguments in the Franks–Williams construction imply the following result:

Theorem 8.3.8 (Franks–Williams Plugs). *If (M, V) and (N, W) are hyperbolic plugs, B^{out} is a union of connected components of $\partial^{\text{out}} M$, B^{in} is a union of connected components of $\partial^{\text{in}} N$, and $f \colon B^{\text{out}} \to B^{\text{in}}$ is a diffeomorphism such that $f_*(\mathcal{L}_V^u)$ is transverse to $\mathcal{L}_W^s$, then (P, X) is a hyperbolic plug, where X is the vector field induced by V and W on $P := M \cup N / f$ (in particular, there is a differentiable structure on P that is compatible with M and N and such that X is differentiable).*

As in the Franks–Williams construction hyperbolicity is established in the above situation by Proposition 5.1.7 or Theorem 6.2.24. The Franks–Williams construction uses an attracting plug and a repelling plug with one boundary component each, so the resulting plug has no boundary.

An additional property of the entrance and exit laminations is reminiscent of the DA construction and central to the creation of Anosov flows via these ideas.

Definition 8.3.9. A Morse–Smale lamination is said to be *filling* if every connected component of its complement is a "strip" as in Figure 6.5.5, that is, a topological disk bounded by two distinct noncompact leaves that are asymptotic to each other at both ends.[5]

[4] Or possibly a Klein bottle if M is not orientable.

[5] "Bounded by" is a little more subtle than the topological notion of boundary, which in this case gives the whole complement; we refer to the *accessible boundary*, which is the collection of points in the topological boundary that are endpoints of open segments in the set.

This name reflects the fact that such laminations can in an obvious way be "filled in" to foliations; as a consequence, a surface with such a lamination is a torus. Less obviously, if one of the entrance or exit laminations of a hyperbolic plug is filling, then so is the other. In that case we speak of a hyperbolic plug with filling Morse–Smale laminations.

There is one more needed strengthening of the notions so far:

Definition 8.3.10. Two laminations $\mathcal{L}_1$, $\mathcal{L}_2$ of a surface S are said to be *strongly transverse* if they are transverse and if each connected component of $S \smallsetminus (\mathcal{L}_1 \cup \mathcal{L}_2)$ is a "rectangle" whose boundary consists of two segments each from $\mathcal{L}_1$ and $\mathcal{L}_2$. A diffeomorphism $f\colon B^{\text{out}} \to B^{\text{in}}$ as in Theorem 8.3.8 is said to be strongly transverse if $f_*(\mathcal{L}^u_V)$ is strongly transverse to $\mathcal{L}^s_W$.

Proposition 8.3.11. *If* $\mathcal{L}^u_V$ *and* $\mathcal{L}^s_W$ *are filling and* f *is strongly transverse in Theorem* 8.3.8, *then* P *has filling Morse–Smale laminations.*

While Theorem 8.3.8 suggests an additive approach, where examples are built in a step-by-step assembly of plugs, Anosov flows can be obtained even if the two plugs in Theorem 8.3.8 are the same. Here, the step from "hyperbolic" to "Anosov" is nontrivial and necessitates an additional hypothesis to make sure that the gluing diffeomorphism can be chosen well.

Also, while based on the Franks–Williams idea, this is a rather different assembly because of the self-gluing and the absence of attractors and repellers. Nonetheless, this construction can also produce nontransitive Anosov flows, and there is a remarkable criterion for transitivity in the spirit of spectral decomposition and the no-cycles condition.

Theorem 8.3.12 (Béguin–Bonatti–Yu). *If* (M, V) *is a hyperbolic plug with filling Morse–Smale laminations whose maximal invariant set contains no attractors or repellers, then a strongly transverse gluing diffeomorphism* $f\colon \partial^{\text{out}} M \to \partial^{\text{in}} M$ *can be isotoped in such a way that the vector field* X *induced on* M/f *by* V *is Anosov.*

Further, X *is transitive if and only if any two vertices of the following directed graph can be joined by a directed path: the vertices are the basic sets in the spectral decomposition of* V*, and an edge goes from* Λ_i *to* Λ_j *if either* $W^u_V(\Lambda_i) \cap W^s_V(\Lambda_j) \neq \varnothing$ *or* $f_*(W^u_V(\Lambda_i) \cap \partial^{\text{out}} M) \cap (W^s_V(\Lambda_j) \cap \partial^{\text{in}} M) \neq \varnothing$.

Theorem 8.3.12 (or a several-plugs counterpart of it based on Theorem 8.3.8) produces a veritable machine or "Lego set" for recursively building new Anosov flows from known ones if one prepends the first part of the Franks–Williams construction—making a plug from an Anosov flow.

One implementation of this "machine" or "construction game" is the Béguin–Bonatti–Yu blow-up–excise–glue construction starting from a transitive Anosov flow:

- "Blow up" two closed orbits, one each by a repelling DA construction and an attracting DA construction to get a flow with these periodic orbits and a saddle hyperbolic set Λ as basic pieces.

- Excise tubular neighborhoods of the two closed orbits chosen such that a hyperbolic plug remains.

- Glue the two boundary tori as in Theorem 8.3.12 to get a new transitive Anosov flow; it turns out that its restriction to Λ has the initial flow as a factor.

The last item implies that this construction can be applied repeatedly to produce "ever more complicated" transitive Anosov flows. Here are some remarkable examples of what variants of this construction can produce.

Theorem 8.3.13 (Béguin–Bonatti–Yu). *There is a closed 3-manifold that carries both a transitive Anosov flow and a nontransitive Anosov flow.*

For each $n \in \mathbb{N}$ there is a 3-manifold that carries n Anosov flows, no two of which are orbit equivalent.

There is a transitive Anosov flow on a 3-manifold with infinitely many pairwise nonisotopic transverse tori.[6]

The construction of a closed 3-manifold that carries both a transitive Anosov flow and a nontransitive Anosov flow goes as follows. Starting with an Anosov flow (such as the suspension of $\left(\begin{smallmatrix} 2 & 1 \\ 1 & 1 \end{smallmatrix}\right)$) select two closed orbits and perform blow-up–excise–glue surgery in two ways. First, perform DA modifications on both periodic orbits so as to make both attracting, and likewise on an identical copy, but making both expanding. Then delete tubular neighborhoods of each and glue the resulting manifold together along these tori analogously to the Franks–Williams construction; this also gives a nontransitive Anosov flow. Second, instead perform DA surgeries to make one each of these two orbits attracting and repelling, and the same in reverse on an identical copy, then excise and glue these two manifolds together to get a transitive Anosov flow.

Remark 8.3.14. Theorem 8.3.13 suggests the question of whether there is a manifold that carries infinitely many distinct Anosov flows. This has not been answered to date but is being investigated.

In light of the complementary universe of Anosov flows we study and produce in Sections 8.5 and 8.2, one might further ask about the extent to which Anosov 3-flows of these different kinds can coexist on a given manifold. This indicates a rich field of study at the boundary of 3-manifold topology and dynamical systems.

[6] This is interesting (with respect to decomposition ideas as in Theorem 8.3.5): given the construction of these examples from repeated addition of hyperbolic plugs, one might hope for a complete decomposition into basic standard pieces along finitely many embedded tori which are unique modulo isotopy—this is too much to hope for.

We close with an observation on the title of the present section. “Anomalous” was chosen to echo the title of the seminal work in this realm [143], but it also reflects the fact that the flows constructed here are of a more deeply different nature from suspensions and geodesic flows than the $\mathbb{R}$-covered flows in Section 8.2 discussed more generally in Section 8.5 below. And while “nontransitive” is the feature for which the Franks–Williams construction was first undertaken, we repeat that it does not characterize the Béguin–Bonatti–Yu flows.

8.4 Codimension-1 Anosov flows

There is something jarring about the possibility that Anosov flows need not be topologically transitive (Theorem 8.3.3). This is a reason Franks and Williams referred to their examples as anomalous Anosov flows. There are, of course, natural sufficient conditions for topological transitivity of Anosov flows. For instance, the flows constructed in Section 8.2 are also exotic (topologically different from algebraic ones), but since they are volume preserving and hence ergodic (see Section 7.1), they are topologically transitive. While volume preservation is a restrictive assumption, it actually suffices to assume that the flow in question is orbit equivalent to a volume-preserving one because orbit equivalence preserves transitivity—albeit also the fact that some smooth measure is preserved (Proposition 3.5.1). At any rate, it is natural as well to seek topological conditions for transitivity, and this section presents one of them. If the (strong) stable or unstable foliation of an Anosov flow is 1-dimensional, then the flow is topologically transitive—provided the underlying manifold has dimension greater than 3, because the Franks–Williams flow is a counterexample.[7] While this is the major result of this section we take some time discussing codimension-1 Anosov flows. Among the reasons is that these are the Anosov flows one can most hope to understand to any general extent from a topological point of view.

Definition 8.4.1. An Anosov flow is said to be a codimension-1 Anosov flow if either the strong-stable or the strong-unstable leaves are 1-dimensional.

To fix ideas we assume throughout this section that strong-unstable leaves are 1-dimensional. We note that the codimension-1 foliation is continuously differentiable by Corollary 9.4.11. Let us take a first look at a “transverse” approach to studying these kinds of Anosov flows, that is, an approach that focuses on the space of orbits.

Definition 8.4.2. For a manifold M, denote by $\pi\colon \tilde{M} \to M$ the universal cover. The

[7] By contrast, whether all Anosov *diffeomorphisms* are topologically transitive is an open question that is deemed exceedingly hard.

lifts $\widetilde{\mathcal{W}}^i$ to $\tilde{M}$ of the foliations $\mathcal{W}^i$ with $i = cs, cu, s, u$ for an Anosov flow Φ on M are called the *lifted foliations*, and the lift $\widetilde{\Phi}$ of Φ to $\tilde{M}$ is called the *lifted flow*.

The *leaf space* of a foliation is the identification space of $\tilde{M}$ whose elements are leaves. Denote by $\mathcal{O}^\Phi$, $\mathcal{L}^s$, and $\mathcal{L}^u$ the leaf spaces of the lifted orbit foliation and of $\widetilde{\mathcal{W}}^{cs}$ and $\widetilde{\mathcal{W}}^{cu}$, respectively, with canonical projections $\pi^\Phi\colon \tilde{M} \to \mathcal{O}^\Phi$, $\pi^s\colon \tilde{M} \to \mathcal{L}^s$, $\pi^u\colon \tilde{M} \to \mathcal{L}^u$.

Since the natural action on $\tilde{M}$ of the fundamental group Γ of M permutes the lifted leaves, it induces action on the three leaf spaces. This action interacts with the flow: a point $x \in \widetilde{\mathcal{W}}^{cs}$ is a leaf $\widetilde{W}^{cs}$, which projects to a leaf W^{cs} of $\mathcal{W}^s$ in M, and the Γ-stabilizer of x is the fundamental group of W^{cs} (which is trivial except for periodic x, when it is $\mathbb{Z}$). Along similar lines one proves the following:

Proposition 8.4.3. *Let Φ be a codimension-1 Anosov flow. Then*

- *the Γ-stabilizer of a point in $\mathcal{O}^\Phi$, $\mathcal{L}^u$, or $\mathcal{L}^s$ is either trivial or cyclic,*
- *if $\gamma \in \Gamma$ fixes $O \in \mathcal{O}^\Phi$, then O is a hyperbolic fixed point of γ,*
- *if $\gamma \in \Gamma$ fixes a point x of $\mathcal{L}^u$ or $\mathcal{L}^s$, then x is an attracting or repelling fixed point of γ,*
- *(Franks) the set of points in $\mathcal{L}^u$ (or $\mathcal{L}^s$) with nontrivial Γ-stabilizer is dense,*[8]
- *(Verjovsky) each leaf of $\mathcal{W}^u$ intersects each leaf of $\mathcal{W}^{cs}$ in at most one point and likewise for $\mathcal{W}^s$ and $\mathcal{W}^{cu}$, hence*
- *$\mathcal{L}^s$ (or $\mathcal{L}^u$) is a connected and simply connected 1-manifold, that is, each point disconnects it; it is a separable topological space in which each point has a neighborhood homeomorphic to an interval,* but not necessarily Hausdorff.

The possible failure of the Hausdorff separation axiom in the last item suggests calling $x \in \mathcal{L}^u$ *nonseparated* if there is a $y \in \mathcal{L}^u$ such that every neighborhood of x overlaps with every neighborhood of y.

Definition 8.4.4. A codimension-1 Anosov flow is said to be $\mathbb{R}$*-covered* (or *covered by an* $\mathbb{R}$*-foliation*) if the leaf space $\mathcal{L}^s$ is Hausdorff (hence homeomorphic to $\mathbb{R}$).

Proposition 8.4.5. *The Franks–Williams flow (Theorem* 8.3.3*) is not* $\mathbb{R}$*-covered.*

Proof. Figure 8.3.1 shows Reeb components of the stable/unstable foliation (the middle picture there shows a pair of those), and these produce nonseparated leaves.[9] □

[8] That is to say, the set of (un)stable leaves of periodic points is dense—even when periodic points themselves are not, as in Theorem 8.3.3.

[9] While Figure 8.3.1 shows leaves rather than the leaf space, this being a picture of a compact transversal that meets every orbit at most once implies that the same configuration appears in the leaf space. Note that a similar picture can occur for the orbits of a flow (Figure 1.5.3) and thus produce a non-Hausdorff orbit space—so having a "nice" orbit space in the present context is nontrivial (Proposition 8.4.19).

Proposition 8.4.5 is also an immediate corollary of the following:

Proposition 8.4.6. *An $\mathbb{R}$-covered Anosov flow is topologically transitive.*

Proof. Otherwise, there is a nondense Γ-orbit in $\mathcal{L}^s$; denote by I a connected component of the complement of its closure and let a be an endpoint. Then I has a dense set of points with nontrivial Γ-stabilizer, and each $\gamma \in \Gamma$ with a fixed point in I fixes I and hence without loss of generality a. The generator of the stabilizer of a then fixes all points of I with nontrivial stabilizer, contrary to the fixed-point set of any $\gamma \in \Gamma$ being hyperbolic, hence discrete. □

By contrast, Figure 2.4.3 leads to the following proposition:

Proposition 8.4.7. *The geodesic flow of a compact negatively curved surface is $\mathbb{R}$-covered.*

Proof. Using horocycles, stable leaves are (homeomorphically) represented by boundary points of the Poincaré disk $\mathbb{D}$ or equivalently, by unit tangent vectors at $0 \in \mathbb{D}$. Passing to the universal cover of the *phase space* "unrolls" these tangent circles to the fibers of a line bundle, and hence represents $\mathcal{L}^s$ homeomorphically as $\mathbb{R}$. □

More generally, we have the following proposition:

Proposition 8.4.8 ([28, Theorem A]). *Contact Anosov 3-flows are $\mathbb{R}$-covered.*

Remark 8.4.9. Though our present focus is not ergodic theory it is worth noting that the Margulis measure (Theorem 7.5.2) induces a Γ-invariant measure on $\mathcal{L}^s$ or $\mathcal{L}^u$, whichever is 1-dimensional, that scales homothetically under Φ.

Although this is not needed, we mention a foundational theorem by Verjovsky:

Theorem 8.4.10 ([345]). *For a codimension-1 Anosov flow on an n-manifold M,*

(1) *the closed orbits are homotopically nontrivial,*

(2) *the lifts of the codimension-1 leaves to the universal cover are diffeomorphic to $\mathbb{R}^{n-1}$,*

(3) *the universal cover of M is diffeomorphic to $\mathbb{R}^n$* [276, *Corollary* 5].[10]

Proof outline. Item (1) is a consequence of a theorem of Haefliger about codimension-1 foliations and implies (2). □

[10]The underlying fact is that $\mathbb{R}^n$ is the only simply connected n-manifold that has a foliation with leaves diffeomorphic to $\mathbb{R}^{n-1}$ [276, Corollary 3].

Oddly, it is not known whether periodic orbits of Anosov flows are always homotopically nontrivial (or whether there could be contractible ones).

We now come to the famous theorem of Verjovsky, which obtains topological transitivity without assuming that the flow is $\mathbb{R}$-covered:

Theorem 8.4.11 (Verjovsky). *A codimension-*1 *Anosov flow on a manifold with dimension at least* 4 *is topologically transitive.*

We prove this by showing that every leaf of W^{cs} is dense (see Theorem 6.2.10), so we first introduce a few notions pertinent to foliations. We assume that the weak-stable foliation W^{cs} is transversely oriented, that is, TM/TW^{cs} is oriented. If need be, this can be achieved by passing to a suitable double cover, and topological transitivity on the double cover implies topological transitivity on the manifold itself, so this is no loss of generality. The transverse orientation of W^{cs} yields a flow Θ along leaves of W^u, which is hence transverse to W^{cs}. Open saturated sets can then be studied with the following:

Definition 8.4.12 (Dippolito completion). Let E be an open saturated set for a foliation of a Riemannian manifold M, and let d_E be the distance function on E induced by restricting the Riemannian metric on M to E. Then the completion $\hat{E}$ of E with respect to the metric d_E is called the *Dippolito completion* of E, and we denote by p the canonical projection $\hat{E} \to \overline{E}$ to the topological closure, by $\hat{\mathcal{F}}$ the foliation of $\hat{E}$ induced by $\mathcal{F}$, by $\hat{\Theta}$ the flow induced by the transverse flow Θ, and by $\delta E = \delta^+ E \cup \delta^- E$ the boundary and the boundary components where Θ flows outward, respectively inward (these are unions of leaves).

To clarify this, note that d_E is not the same as the distance induced by the distance on M. For instance, if E is the complement of a leaf, then Dippolito-completing E adds two copies of that leaf. Compactness of M implies:

Proposition 8.4.13 (Dippolito). *If V^- is a leaf in $\delta^- E$, then there is a compact $A^- \subset V^-$ and a $T > 0$ such that for all $x \in V^- \setminus A^-$ there is a $t \in (0,T)$ with $\hat{\theta}^t(x) \in \delta^+ E$.*

Proof. The manifold M is covered by the interiors of finitely many sets $B_i \times I_i$ such that $B_i \times \{t\}$ is a closed ball in a leaf of $\mathcal{F}$ and $\{x\} \times I_i$ is a closed orbit segment of Θ. We identify I_i and $\{c_i\} \times I_i$, where c_i is the center of B_i. With finitely many exceptions, the connected components of $E \cap I_i$ are intervals in the interior of I_i. Set $A_i := \varnothing$ if either $I_i \subset E$ or the last point q_i of the θ-orbit segment I_i is not in E. If q_i but not I_i is contained in E, let $A_i := \{a_i\}$, where a_i is the point of $I_i \cap \partial E$ nearest q_i in I_i. Then $A^- := V^- \cap p^{-1}(\bigcup_i B_i \times A_i)$ is as desired. □

Proof of Theorem 8.4.11 (Matsumoto [258]). A *minimal* set for a foliation $\mathcal{F}$ of a manifold is a minimal element (with respect to inclusion) of the collection of nonempty, closed, $\mathcal{F}$-saturated sets (see Definition 8.1.9); it is *exceptional* if it is neither a single leaf nor the whole manifold. The existence of minimal sets is proved analogously to Proposition 1.6.27 or using Zorn's Lemma. In fact, Proposition 1.6.27 is a special case, where the foliation is the orbit foliation. In the case at hand, a minimal set of the weak-stable foliation cannot be a single leaf since minimal sets are compact. If it is the whole manifold, then all weak-stable leaves are dense, and the flow is transitive by Theorem 6.2.10. Therefore we assume that there is an exceptional minimal set and derive a contradiction.

Each leaf of W^{cs} is homeomorphic to either $\mathbb{R}^{n-1}$ or to $S^1 \times \mathbb{R}^{n-2}$, where $n = \dim M$. The latter occurs if and only if the leaf contains a periodic orbit, and we call such leaves periodic leaves.

Any connected component E of the *complement* of the exceptional minimal set is open and saturated by weak-stable leaves and hence has a Dippolito completion $\hat{E}$. The (positive) normal vector field to the weak-stable subbundle defines a transverse flow $\Theta = \{\theta^t\}_{t\in\mathbb{R}}$.

We will have to argue two cases separately, and in either one, we produce an inconsistency with the dynamics of Φ.

The first and significantly simpler case is the one in which there is a leaf V^- in $\delta^- E$ (or in $\delta^+ E$—reversing θ interchanges these) which is nonperiodic (and hence homeomorphic to $\mathbb{R}^{n-1}$). In that case take A^- as in Proposition 8.4.13 and assume (by enlarging it if necessary) that it is homeomorphic to a ball D^{n-1}. Then

$$V^- \smallsetminus A^- \subset U^- := \{x \in V^- \mid \hat{\theta}^t(x) \in \delta^+ E \text{ for some } t =: \tau(x) > 0\}$$

by Proposition 8.4.13. The continuous $\tau\colon U^- \to \mathbb{R}$ introduced hereby defines a map

$$h\colon U^- \to \delta^+ E, \quad x \mapsto \hat{\theta}^{\tau(x)}(x)$$

that nicely intertwines with the Anosov flow: V^- and $\delta^+ E$ are $\hat{\Phi}$-invariant, and $\hat{\Phi}$ sends orbits of $\hat{\Theta}$ to orbits of $\hat{\Theta}$, so U^- is $\hat{\Phi}$-invariant and

$$h(\hat{\varphi}^t(x)) = \hat{\varphi}^t(h(x)).$$

This implies $U^- = V^-$ because U^- is $\hat{\Phi}$-invariant and no orbit is contained in A^-. But then τ is bounded by Proposition 8.4.13, contrary to the fact that $\hat{\Phi}$ expands $\hat{\theta}$-orbits.

We now come to the rather harder case in which every leaf V^- in $\delta^- E$ (and in $\delta^+ E$) is periodic, that is, V^- contains a unique periodic orbit $\mathcal{O}(p)$, which we view as $S^1 \times \{0\}$ oriented by the slow direction and enlarge to a tubular neighborhood $S^1 \times D^{n-2}$ that contains the set A^- from Proposition 8.4.13 and such that

(1) $\hat{\Phi}$ is transverse to the boundary $S^1 \times S^{n-3}$ and flows inward,

(2) each fiber $\{t\} \times D^{n-2}$ lies in a strong-stable leaf (and is hence transverse to $\hat{\Phi}$).

Because $O(p)$ is the only $\hat{\Phi}$-periodic orbit in this neighborhood, the argument that $U^- = V^-$ in the first case shows that when we define U^-, τ, and h as before, we now have $V^- \smallsetminus O(p) \subset U^-$. If $U^- = V^-$, then we are done by the argument at the end of the first case so we henceforth assume $U^- = V^- \smallsetminus O(p)$. Thus U^- is connected because $n - 2 \geq 2$,[11] and therefore $h(U^-)$ is in a single leaf V^+ of $\delta^+ E$ which is homeomorphic to $S^1 \times \mathbb{R}^{n-2}$ (because otherwise we are in the previous case). Applying the argument that $U^- = V^- \smallsetminus O(p)$ to $h(U^-)$ with the flow $\hat{\theta}$ reversed implies that $h(U^-) = V^+ \smallsetminus O(q)$ for a unique periodic orbit $O(q)$ and that h is a homeomorphism.

To cover these periodic leaves by nonperiodic ones, the following auxiliary notion becomes nontrivial and helpful:

Definition 8.4.14 (Leaf holonomy). For a loop γ in a leaf of W^{cs}, the transverse flow lets us define a holonomy as follows. The map

$$\Gamma\colon [0,1] \times [-\eta,\eta] \to M, \quad (s,t) \mapsto \theta^t(\gamma(s))$$

induces a foliation $\Gamma^*\mathcal{F}$ on the rectangle $[0,1] \times [-\eta,\eta]$ whose leaves are components of the preimages of leaves of $\mathcal{F}$, and they are transverse to the "vertical" foliation $s = \text{const}$. Then, $\gamma \sim [0,1] \times \{0\}$ is one of them, so for some δ, each leaf through $(0,t)$ with $|t| < \delta$ contains a point $(1, g(t)) \in \{1\} \times [-\eta,\eta]$, and this defines an orientation-preserving homeomorphism $\mathrm{Hol}_\gamma\colon (-\delta,\delta) \to [-\eta,\eta]$ with 0 as a fixed point. It (or its germ H_γ at 0[12]) is called the *leaf holonomy along* γ.

Fix $x^-(a,b) \in S^1 \times S^{n-3}$, and let I be the $\hat{\theta}$-orbit segment from x^- to $x^+ := h(x^-)$. Then the leaf holonomy Hol_γ of the loop $\gamma = S^1 \times \{b\}$ is defined on I and surjective with x^- as an expanding, hence isolated, fixed point. We assume that there are no fixed points of Hol_γ other than x^- and x^+ in I; otherwise replace x^+ by the fixed point nearest X^-. Then for $y \in \operatorname{int} I$ we have $\mathrm{Hol}^n_\gamma(y) \xrightarrow[n\to\pm\infty]{} x^\pm$, and for any

$$z \in U := \{z \in V := \hat{W}^s(y) \mid z = \hat{\theta}(x) \text{ for some } t > 0,\ x \in U^-\}$$

there is a unique $x =: \pi(z) \in U^-$ with $z = \hat{\theta}^t(x)$, and $\pi\colon U \to U^-$ is a covering map that intertwines with $\hat{\Phi}$.

The complement of $\mathbb{R} \times S^{n-3} = \pi^{-1}(S^1 \times S^{n-3})$ in V has two connected components, the *exterior* $\pi^{-1}(V^- \smallsetminus S^1 \times D^{n-2})$ and the *interior* $\mathcal{I}$. By π-invariance, $\hat{\Phi}$ flows into $\mathcal{I}$.

[11] This is where we use that $n \geq 4$.

[12] That is, equivalence class of maps modulo agreeing on a neighborhood of 0.

Now extend π (albeit possibly not as a covering map) to

$$W := \{z \in V \mid z = \hat{\theta}(x) \text{ for some } t > 0,\ x \in V^-\} \subset V$$

and denote by J the positive $\hat{\theta}$-semiorbit of the Φ-periodic point p from the start of this case.

There are points $p' \in V \cap J$ arbitrarily close to p because $V \cap I$ contains points $\mathrm{Hol}_\gamma^{-n}(y)$ arbitrarily close to $x^- \in V^-$. Then $p' \subset B^- \subset V$ such that $\pi_{\upharpoonright B^-}$ is a homeomorphism onto $\{a\} \times D^{n-2} \ni p$. Then B^- separates $\mathcal{I}$ into two components (corresponding to $y > a$ and $t < a$), one of which accumulates on V^-, and the closure of the other of which we denote by C_-. Thus, $\hat{\Phi}$ flows into C_-: It is transverse to B_- because it is transverse to $\{a\} \times D^{n-2}$, and the forward orbit of p', being in the weak-unstable leaf of p, cannot accumulate on V^-. Since $\hat{\Phi}$ flows into $\mathcal{I}$, the orbit of p' must instead enter C_-.

We now argue likewise with $q \in V^+$ to obtain a disk B_+ bounding a component C_+ that does not accumulate on V^+ and such that $\hat{\Phi}$ flows (transversely) into C_+. But this is impossible: $C := C_- \cap C_+ \subset V$ is a ball[13] into which $\hat{\Phi}$ flows, contrary to the dynamics on a weak-stable leaf. This completes the proof of Theorem 8.4.11. □

We remark on types of Anosov flows to which transitivity of codimension-1 Anosov flows is pertinent.

If a codimension-1 Anosov flow has a global section and is hence (topologically equivalent to) a special flow, that is, topologically equivalent to a suspension, then the base transformation is topologically conjugate to an algebraic Anosov diffeomorphism (Theorem 8.1.2). Verjovsky wondered whether in this situation there always is a global section, and he proved it when the phase space is a manifold with solvable fundamental group.

Theorem 8.4.15 (Verjovsky). *A codimension-1 Anosov flow on a manifold with solvable fundamental group has a global section and is hence a suspension over an algebraic Anosov diffeomorphism.*

Barbot instead gave a dynamical criterion for being a suspension:

Theorem 8.4.16 ([25, Théorème 2.7]). *A codimension-1 Anosov flow in which every (1-dimensional) strong-unstable leaf meets every weak-stable leaf is topologically orbit equivalent to a suspension.*

[13] The boundary is an $(n-2)$-sphere (a cylinder plus two disks) and the complement is noncompact; this implies that it is a ball by the generalization of a theorem of Schönflies: if $E\colon S^{n-1} \to S^n$ is a topological embedding and A is the closure of a component of $S^n \smallsetminus E(S^{n-1})$, then $A \approx D^n$ if A is a manifold.

Verjovsky conjectured that neither this hypothesis nor the topological hypothesis of Theorem 8.4.15 is needed:

Conjecture 8.4.17 (Verjovsky–Ghys Conjecture). *A codimension-*1 *Anosov flow* (*in dimension greater than* 3) *has a global section and is hence a suspension over an algebraic Anosov diffeomorphism* (*necessarily of a torus*).

This has been proved by assuming for instance that the invariant foliations are C^1 (or Lipschitz continuous [333]), but remains open as stated. (In light of Theorem 8.4.11 one can without loss of generality assume volume preservation: a topologically transitive codimension-1 Anosov flow on a closed manifold is topologically equivalent to a C^∞ Anosov flow with a C^∞-invariant volume [17].) Recent work explored possible ways of constructing Anosov flows that are not topologically conjugate to algebraic flows (that is, geodesic flows or suspensions of algebraic Anosov diffeomorphisms), and this led to a stronger conjecture:

Conjecture 8.4.18 (Barthelmé–Bonatti–Gogolev–Hertz [35]). *Anosov flows whose stable and unstable subbundles do not have the same dimension are orbit equivalent to the suspension of an Anosov diffeomorphism.*[14]

As we mentioned at the start of this section, codimension-1 Anosov flows are more amenable to topological analysis than Anosov flows are in full generality. We now outline a way in which one can determine whether two of them are orbit equivalent by checking conjugacy of the fundamental-group actions on the orbit space. This is interesting in part because the homeomorphism involved in orbit equivalence is determined up to a shift along orbits, whereas one can hope to find a conjugacy between group actions by the Contraction Principle.

To that end we first show that the orbit space is topologically nice.

Proposition 8.4.19. *If* Φ *is a codimension-*1 *Anosov flow on an n-manifold M, then its orbit space* O^Φ *is diffeomorphic to* $\mathbb{R}^{n-1}$, *and the canonical projection* $\pi\colon \tilde{M} \to O^\Phi$ *is a locally trivial* $\mathbb{R}$*-principal fibration.*

Proof outline. First, O^Φ is a (possibly not Hausdorff) topological $(m-1)$-manifold: each orbit $\tilde{O}$ of $\widetilde{\Phi}$ admits a local transversal that meets any other orbit of $\widetilde{\Phi}$ in at most one point, and this gives local transversals whose intersections with leaves of the lifted weak foliations are connected; this yields the claim.

Next, we show by contraposition that O^Φ is Hausdorff—if O_1 and O_2 are orbits of $\widetilde{\Phi}$ that are not separated by $\widetilde{\Phi}$-invariant open sets, then they coincide. For $i = 1, 2$, let U_i be the $\tilde{\mathcal{W}}^s$ saturation of $F_i := \tilde{W}^{cu}(\tilde{O}_i)$. If these are not disjoint, then there is a $\tilde{\mathcal{W}}^s$-leaf S that meets F_i in a unique point x_i for $i = 1, 2$. If $x_1 \neq x_2$, then there are disjoint

[14] The claims in [143] to flows incompatible with this conjecture turn out to be unsupported—which prompted the work leading to this conjecture.

S-neighborhoods of the x_i, and their $\tilde{\mathcal{W}}^{cu}$-saturations separate O_1 and O_2. If $x_1 = x_2$, then O_1 and O_2 are on the same unstable leaf, so if the preceding arguments with s and u interchanged do not produce separating neighborhoods, then they lead to $O_1 = O_2$.

Thus, O^Φ is a Hausdorff topological $(n-1)$-manifold, and $\pi(\tilde{\mathcal{W}}^{cu})$ is a foliation by codimension-1 planes, so O^Φ is diffeomorphic to $\mathbb{R}^{n-1}$.[15] That π is locally trivial follows because local transversals produce local sections of π. □

Theorem 8.4.20. *If for $i = 1, 2$, Φ_i is a codimension-1 Anosov flow on a manifold M_i with fundamental group Γ_i, then Φ_1 is orbit equivalent to Φ_2 or its reverse if and only if there are an isomorphism $I\colon \Gamma_1 \to \Gamma_2$ and a homeomorphism $h\colon O^{\Phi_1} \to O^{\Phi_2}$ such that $h(\gamma O) = I(\gamma)h(O)$ for all $\gamma \in \Gamma_1$ and $O \in O^{\Phi_1}$.*

In this case, moreover, h can be chosen so as to lift to a homeomorphism $H\colon \tilde{M}_1 \to \tilde{M}_2$ with $\pi_{\Phi_2} \circ H = h \circ \pi_{\Phi_1}$, where $\pi_{\Phi_i}\colon \tilde{M}_i \to O^{\Phi_i}$ are the projections.

Proof outline. That orbit equivalence implies equivalence of the group actions is clear. For the converse we first produce the lift H in the last assertion. Choose a global (not necessarily connected) section T of Φ_1 in such a way that every connected component of its lift $\tilde{T}$ to $\tilde{M}_1$ meets every $\tilde{\Phi}_1$-orbit in at most one point and each $\tilde{\Phi}_1$-orbit meets a finite (nonzero) number of connected components of $\tilde{T}$.

Since π_{Φ_1} is a locally trivial fibration, the restriction of $h \circ \pi_{\Phi_1}$ to any connected component $\tilde{C}$ of $\tilde{T}$ lifts to a map $h_{\tilde{C}}\colon \tilde{C} \to \tilde{M}_2$ in an I-equivariant way. Now, for every $\tilde{X} \in \tilde{M}_1$ there is a connected component $\tilde{C}$, a $\tilde{y} \in \tilde{C}$, and a $t \in \mathbb{R}$ such that $\tilde{x} = \tilde{\varphi}_1^t(\tilde{y})$, and we "define" $H(\tilde{x}) := \varphi_2^t(h_{\tilde{C}}(\tilde{y}))$.

This is not quite well defined because possibly more than one connected component could be chosen here, so this method produces finitely many points—but they all lie on the same orbit; replacing this collection by a weighted average (along the orbit), where the weights come from a partition of unity on M_1, now produces a well-defined continuous equivariant map H such that $\pi_{\Phi_2} \circ H = h \circ \pi_{\Phi_1}$.

The orbit equivalence is now produced by using equivariance of H to get an induced map $\mathfrak{h}\colon M_1 \to M_2$, which sends orbits to orbits. If injective, this is indeed the orbit equivalence, but similarly to the application of Claim 1.7.6 we need to ensure monotonicity in the flow direction.

More precisely, we have $\mathfrak{h}(\varphi_1^t(x)) = \varphi_2^{\alpha(t,x)}(\mathfrak{h}(x))$ for a cocycle α, and injectivity is equivalent to strict monotonicity of $t \mapsto \alpha(t,x)$ for fixed x. A suitable "diffusion" due to Ghys and Gromov along orbits as in the proof of Proposition 1.3.30 (or a transversality argument by Matsumoto and Tsuboi) modifies $\mathfrak{h}$ in such a way as to make α strictly monotone in t. Whether α is increasing or decreasing determines whether $\mathfrak{h}$ connects Φ_1 to Φ_2 or to the reverse of Φ_2. □

[15] The only simply connected manifolds that admit plane foliations are, up to diffeomorphism, the Euclidean spaces $\mathbb{R}^n$ [276, Corollary 3].

Remark 8.4.21. Unsurprisingly, a criterion for whether one has an orbit equivalence between the flows versus between one and the reverse of the other is that $\mathfrak{h}$ sends stable leaves of Φ_1 to stable leaves of Φ_2 and likewise for unstable ones, or whether it sends stable leaves to unstable ones and vice versa. This can also be checked via the analogous dichotomy for the action of h on the projections of the foliations to the orbit space.

8.5 $\mathbb{R}$-covered Anosov 3-flows

We continue the study of codimension-1 Anosov flows by returning our attention to flows on 3-manifolds. The central ingredients are the notions from the beginning of Section 8.4. This is quite complementary to topological results using assumptions on the fundamental group (Theorem 8.4.15 is an example, though not the best).[16] We repeat that the Franks–Williams flows in Section 8.3, not being topologically transitive, are fundamentally different from the $\mathbb{R}$-covered Anosov 3-flows we henceforth attend to.

For what follows, the discussion of Birkhoff annuli at the beginning of Section 8.2 may provide useful intuition; we will see manifestations of those ideas beyond geodesic flows.

The foundational result of the global topological theory of $\mathbb{R}$-covered Anosov 3-flows is the following, and its main case distinction recalls the difference between the situations of Remark 1.5.28 and Figure 2.4.3:

Theorem 8.5.1 (Barbot, Fenley). *If an orientable Anosov 3-flow Φ is $\mathbb{R}$-covered (Definition 8.4.4), then the leaf space $\mathcal{L}^s$ is also Hausdorff. The embedding $\mathcal{O}^\Phi \to \mathcal{L}^u \times \mathcal{L}^s$ sending each orbit to its (center-) stable and (center-) unstable leaf is equivariant with respect to the diagonal action of Γ, and its image $\mathcal{O}$ is open and either*

- *$\mathcal{L}^u \times \mathcal{L}^s$, in which case we say that the flow is a product flow (and it is orbit equivalent to the suspension (Theorem 8.4.16) of a linear toral automorphism (Example 1.5.27), which is transitive but not mixing), or*
- *the open set bounded by the graphs of homeomorphisms $\alpha, \beta\colon \mathcal{L}^u \to \mathcal{L}^s$ (with $\alpha^{-1} \circ \beta$ and $\beta \circ \alpha^{-1}$ increasing), in which case we say that Φ is a* skewed *$\mathbb{R}$-covered Anosov flow.*

The action of a deck transformation g on the (un)stable leaf space either has

- *no fixed point,*

[16] This is also complementary to rigidity theory, for example, the exploitation of smoothness assumptions on the foliation such as in Theorem 9.7.16.

- *exactly one fixed point (and Φ is of product type), or*
- *infinitely many fixed points (and Φ is skewed).*

In the latter cases, g is associated with a periodic orbit in the sense that the loop to which it corresponds is freely homotopic to a closed orbit.

If Φ is skewed and M is a hyperbolic manifold, then every closed orbit is freely homotopic to infinitely many others.

Theorem 8.5.1 gives a clear description of the orbit space of a skewed ℝ-covered Anosov 3-flow. We can picture it as a strip between the graphs of two increasing (say) functions that is foliated by horizontal and vertical line segments (Figure 8.5.1).

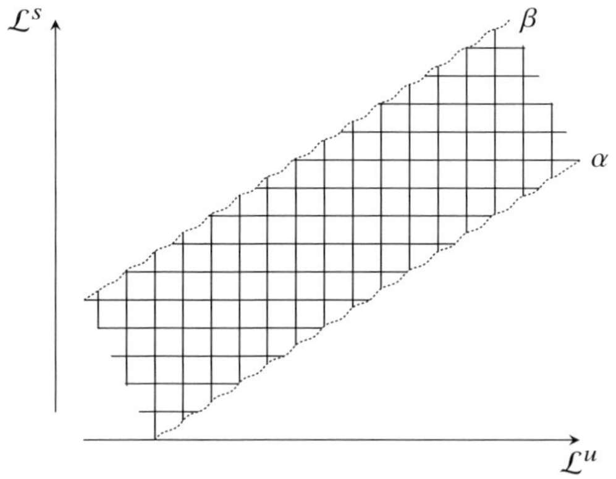

Figure 8.5.1. The orbit space of a skewed ℝ-covered Anosov 3-flow.

Remark 8.5.2. Of these two cases, product flows can be deemed completely understood since toral automorphisms are.

Proposition 8.5.3. *Geodesic flows of negatively curved surfaces and more generally, contact Anosov 3-flows are skewed ℝ-covered Anosov flows.*

Proof. They are ℝ-covered by Proposition 8.4.8 (and Proposition 8.4.7), and skewed since they are not orbit equivalent to suspensions (suspensions have a global section S, 3-flows preserving a contact form A cannot: $0 \neq \int_S dA = \int_{\partial S} A = 0$ when $\partial S = \varnothing$). □

Conjecturally this is an equivalence:

Conjecture 8.5.4 (Barthelmé). *ℝ-covered Anosov 3-flows are topologically orbit equivalent to a contact Anosov flow.*

We make some observations about the dynamics of skewed $\mathbb{R}$-covered Anosov flows:

Proposition 8.5.5. *For a skewed $\mathbb{R}$-covered Anosov flow Φ on a 3-manifold M,*

(1) *M is orientable,*

(2) *Φ is orbit equivalent to its reverse,*

(3) *no compact (boundaryless) surface is transversely immersed in M,*

(4) *Φ is topologically mixing.*

The impossibility of a transversal implies Proposition 8.4.5 and also the following result:

Corollary 8.5.6. *The Béguin–Bonatti–Yu flows (for example, Theorem* 8.3.13*) are not $\mathbb{R}$-covered.*

Remark 8.5.7. Corollary 8.5.6 shows that a flow can be topologically transitive without being $\mathbb{R}$-covered. This motivates "anomalous" (rather than "nontransitive") in the title of Section 8.3. A complementary fact is that there are manifolds that support both an $\mathbb{R}$-covered Anosov flow and an Anosov flow that is not $\mathbb{R}$-covered [27].

Proof of Proposition 8.5.5. If p^u and p^s are the coordinate projections in $\mathcal{L}^u \times \mathcal{L}^s$, then

$$\tau := (\tau^u, \tau^s) := (\alpha^{-1} \circ p^s, \beta \circ p^u) \tag{8.5.1}$$

is a Γ-equivariant homeomorphism of $\mathcal{L}^u \times \mathcal{L}^s$ that preserves $\mathcal{O}$ and exchanges the stable and unstable foliations. (Here α and β are as in Theorem 8.5.1.) Theorem 8.4.20 and Remark 8.4.21 then yield the orbit equivalence to the reverse flow. A transversely immersed surface defines a cohomology class that has nonnegative values on orbits, and this is incompatible with orbit equivalence to the reverse flow. Finally, since the orbit equivalence to the reverse interchanges the stable and unstable foliations, they are both or neither transversely orientable. Since orbits are oriented, so is M. Finally, mixing follows from Theorem 8.1.3 since Φ is not a suspension by (3). □

A much harder recent result is that the second item is a characterization:

Theorem 8.5.8 (Barthelmé–Gogolev [37]). *If Φ is an Anosov 3-flow for which one of the weak foliations is transversely orientable, then Φ is orbit equivalent to its reverse by a homeomorphism isotopic to the identity if and only if Φ is skewed $\mathbb{R}$-covered.*

The inherent symmetry of the argument for Proposition 8.5.5 implies that "half" of the data (the Γ-action on $\mathcal{L}^u$ together with τ^u from (8.5.1)) determine the flow already: $\mathcal{O}$, hence $\mathcal{O}^\Phi$, is equivariantly identified with the subset of $\mathcal{L}^u \times \mathcal{L}^u$ bounded by the graphs of τ^u and the identity.

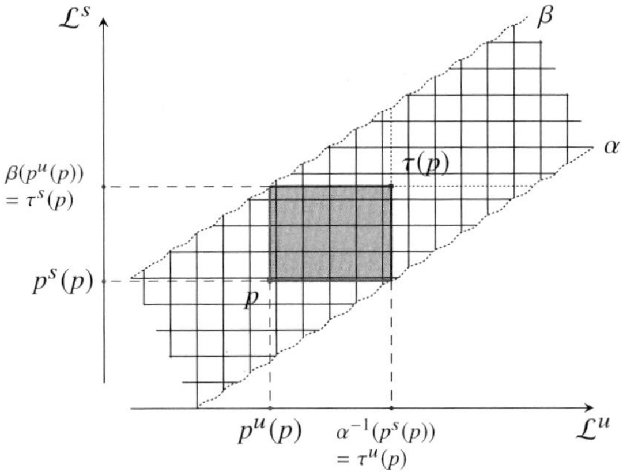

Figure 8.5.2. The proof of Proposition 8.5.5: illustration of (8.5.1).

Corollary 8.5.9. *Two skewed* ℝ*-covered Anosov* 3*-flows* Φ, Ψ *are orbit equivalent if and only if there is a homeomorphism* $f\colon \mathcal{L}^u(\Phi) \to \mathcal{L}^u(\Psi)$ *that is equivariant by the fundamental-group actions and such that* $f \circ \tau^u_\Phi = \tau^u_\Psi \circ f$.

Proof of Theorem 8.5.1 [132]. If Φ is of product type then the conclusion of Theorem 8.5.1 follows. Therefore, we assume that Φ is orientable (after possibly passing to a double cover) and not of product type. We will show that if $\mathcal{L}^u$ is Hausdorff then so is $\mathcal{L}^s$ and that Φ is skewed in that case—by constructing a "lozenge" as in Figure 8.5.2, albeit in reverse, because we are building rather than using the skewed structure. To fix ideas, identify $\mathcal{O}^\Phi$ with $H := (-1,1) \times \mathbb{R} \subset \mathbb{R}^2$ so the product and

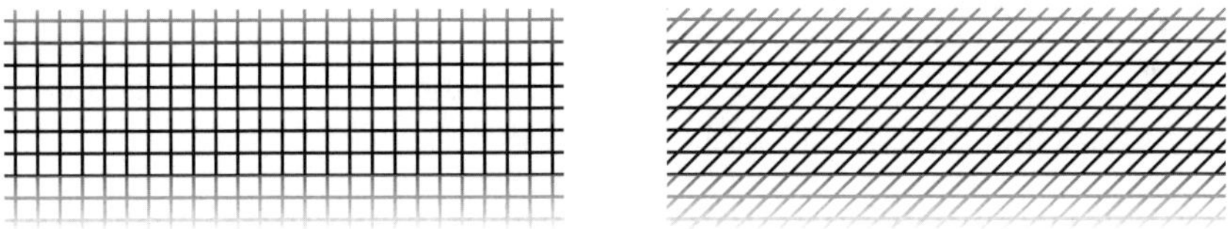

Figure 8.5.3. Product and skewed structures on $H := (-1,1) \times \mathbb{R}$.

skewed situations look as in Figure 8.5.3 with $\widetilde{\mathcal{W}}^{cu} = \{W_t \mid t \in \mathbb{R}\}$ consisting of horizontal segments and t the vertical coordinate. (Note that the right-hand picture is isotopic to Figure 8.5.1, which matches the description in the statement and will later be more convenient.)

Since Φ is not of product type, there is a $V \in \widetilde{\mathcal{W}}^s$ that does not intersect all W_t; say it misses W_u and meets W_t for $t \in (u, \bar{u})$. Its corresponding boundary point must lie on the boundary of H and (since u is finite) either on the line $x = 1$ ("positive boundary") or on the line $x = -1$ ("negative boundary"). To fix ideas and in line with the right half of Figure 8.5.3 suppose the former (Figure 8.5.4).

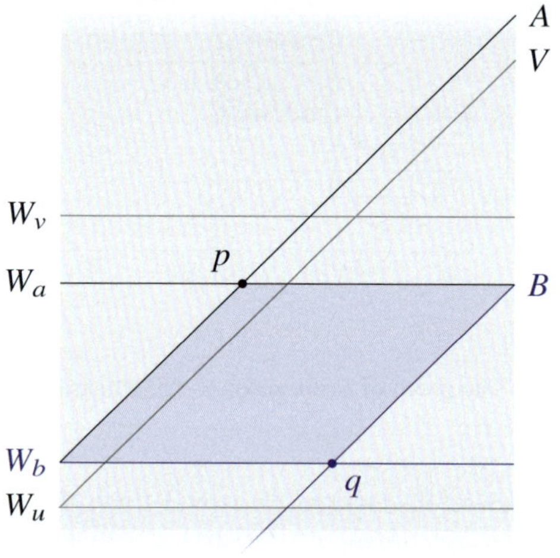

Figure 8.5.4. Building a lozenge.

Let $v \in (u, u_0]$ and choose a periodic orbit p near $V \cap W_v$ that lies right of V and below W_v. Then $p = A \cap W_a$ with $A \in \widetilde{\mathcal{W}}^u$ and $u < a < v$, so $A \cap W_u = \varnothing$, and the "negative half" of A determined by p, that is, its part below p, intersects W_t for $b < t < a$ for some $b \geq u$. Since p is periodic, there is an indivisible deck transformation α that fixes A and hence $\{W_t \mid b < t < a\}$; since α fixes W_a (because it contains p—and because this leaf space is known to be $\mathbb{R}$), α fixes W_b. Here, as in Theorem 8.2.3, a closed orbit is said to be *indivisible* if it is not a repeated closed orbit, and in that case the associated deck transformation is also said to be indivisible.

This implies that there is a closed orbit $q \in W_b$.

Claim 8.5.10. *The closed orbit q is freely homotopic to the reverse of p (Lemma* 8.5.13).

Since α expands or contracts the stable leaf B through q and leaves W_a invariant, it follows that $B \cap W_a = \varnothing$ and that α fixes no other $\widetilde{\mathcal{W}}^u$-leaf between W_a and W_b. (We note that this argument shows that in a product flow a deck transformation has at most one fixed point on the stable or unstable leaf space.) Thus, the positive ("upper") half of B meets exactly the leaves W_t with $t \in (b, a)$ and ends on the negative boundary. The quadrilateral thus determined (with sides being the respective halves of A, B, W_a, and W_b and vertices being p, q and one point on each boundary component) and like ones for other periodic points serve as the skeleton to derive the picture on the right of Figure 8.5.3—which shows that $\mathcal{L}^u$ is also $\mathbb{R}$-covered.

Remark 8.5.11. Such a quadrilateral is called a *lozenge*[17] with corners p and q, which for this terminology need not be periodic; the arguments here show that if one corner is periodic, then so is the other: a deck transformation that fixes one corner must also fix the other. Lozenges are a useful notion for pushing these arguments further. For instance, one can show this way that any Anosov 3-flow (not necessarily ℝ-covered!) that is not a suspension has periodic orbits p, q such that q is freely homotopic to p^{-1}. We emphasize as well that a lozenge is a "rectangle" but with only two vertices because the orbit space is open, so the vertices on its edge in Figure 8.5.2 are "at infinity." Thus, lozenges are more often drawn to look like in Figure 8.5.5.

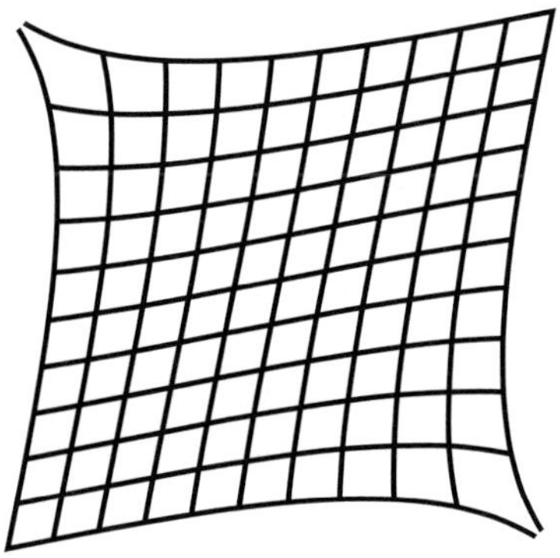

Figure 8.5.5. A lozenge by itself.

But first we digress to note that one could instead consider the negative ("lower") part of B: it similarly ends on another fixed leaf W_c, which must contain a periodic point, and so on (and likewise, when one moves upward from A.) This proves that in the skewed case a deck transformation has countably many fixed points on either leaf space. With our previous parenthetical, this proves the triple alternative in the second half of Theorem 8.5.1. Claim 8.5.10 shows that in the skewed case, these countably many orbits γ_i form two infinite free homotopy classes (even and odd indices, respectively), and the last assertion of Theorem 8.5.1 follows from arguments in 3-manifold topology to the effect that in a hyperbolic manifold no closed orbit is nontrivially freely homotopic to itself (basically, because that would produce a nontrivial 2-torus, while hyperbolicity of the manifold implies atoroidality [132, Lemma 4.3]), so the orbits we obtained are distinct. This is illustrated in Figure 8.5.6.

[17]From Wikipedia: "A lozenge (◊), often referred to as a diamond, is a form of rhombus. The definition of lozenge is not strictly fixed, and it is sometimes used simply as a synonym (from the French losange) for rhombus."

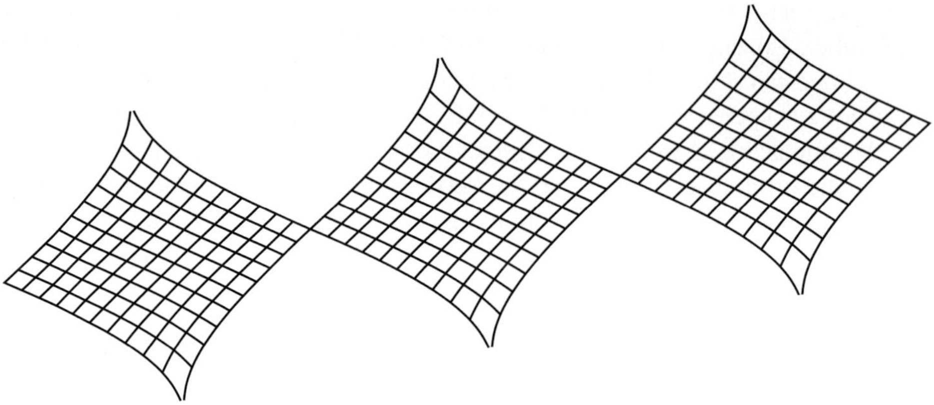

Figure 8.5.6. A string of lozenges.

To put lozenges to use, consider any $\tilde{\mathcal{W}}^s$-leaf C and $r \in C$. Although it is not needed for the theorem, we now invoke topological transitivity as a shortcut: it implies that the projection of B is dense, so there are deck transformations α_i such that $\alpha_i(B)$ lies near γ and on either side of it, so $\{t \in \mathbb{R} \mid C \cap W_t \neq \varnothing\}$ is bounded above and below, and C necessarily goes to the positive boundary at the bottom end and to the negative boundary at the upper end, which produces the same arrangement for all $\tilde{\mathcal{W}}^s$-leaves and thus shows that $\mathcal{L}^s$ is also Hausdorff.

We next show that Φ is skewed, that is, we produce the full picture on the right-hand side of Figure 8.5.3. We show that no two distinct stable leaves limit on the same unstable leaf on either boundary; together with the same result for unstable leaves limiting on distinct stable ones, this shows that every W_t is the upper/lower end of a complementary leaf, which defines the requisite pair of homeomorphisms.

For purposes of contradiction suppose W_s is the upper limit of both A and B with A to the right of B. Then A, B intersect W_t for $w \le t < s$ (but not W_s) and we can choose periodic orbits $p \in W_a$, $q \in W_b$ between A and B such that $w < a < b < s$ and with stable leaves A', B', respectively, which then necessarily also limit on W_s. Hence, A intersects W_b, and we will show that this is impossible.

One can argue as before that the indivisible deck transformation α associated with p fixes W_s, which thus contains a closed orbit r freely homotopic to the reverse of p. The very same goes for q, so p and q are freely homotopic and hence associated to the same deck transformations, which implies that α fixes W_b. Since it expands or contracts A, we must have $A \cap W_b = \varnothing$, a contradiction. □

Remark 8.5.12. Because of its central role let us describe once more the construction of a periodic lozenge in Figure 8.5.4, which is at the same time a description of (8.5.1) and Figure 8.5.2: draw complementary half-leaves from a periodic point p to the boundary of $\mathcal{O}^{\Phi}$, then draw the complementary half-leaves back inward from the endpoints; they intersect in a periodic point q.

Claim 8.5.10 above follows from the next lemma:

Lemma 8.5.13. *If p, q are indivisible closed orbits and q^n is freely homotopic to p^m (we write $q^n \simeq p^m$), then $p \simeq q$ or $q \simeq p^{-1}$. (And the deck transformation associated with an indivisible closed orbit is indivisible in the fundamental group.)*

Proof. The orbits lift to $\tilde{q} \in W_t$ and $\tilde{p} \in W_s$, and the deck transformation γ associated with q clearly preserves W_t, and γ^n preserves W_s. Since Φ is orientable, γ induces an orientation-preserving homeomorphism on $\mathcal{L}^u$, so γ itself preserves W_s, and therefore q is freely homotopic to a closed curve in the projection $\pi(W_s)$, whose fundamental group is generated by $[p]$. Thus $q \simeq p^i$ for some $i \in \mathbb{Z}$ and likewise, $p \simeq q^j$ for some $j \in \mathbb{Z}$, so $q \simeq q^{ij}$, and $ij = 1$. □

We remark that these arguments are but sampling ideas that are being pushed much further at the time of this writing; the lozenges in the proof have supported a much deeper understanding of these Anosov flows [36].[18]

To pick up some of the earlier discussion and contrast it with the present situation, let us mention the possibility of lozenges being adjacent in the sense of sharing a side rather than solely a vertex. This happens in an Anosov flow that has a transverse torus. Unless the flow is a suspension, the weak-stable and weak-unstable foliations have some closed leaves on this transverse torus. Each of these corresponds to periodic orbits that are freely homotopic (up to orientation), and a coherent lift of these orbits (that is, including a lift of the free homotopy) has lozenges sharing sides. The interior of their union is the image of the transverse torus. The Franks–Williams example has four distinct orbits in the free homotopy class, and the Béguin–Bonatti–Yu construction gives more complicated examples.

8.6 Horocycle and unstable flows*

We have seen various close interactions between the geodesic flow and the horocycle flow of a compact negatively curved surface, and there have been hints that this is part

[18]The only comprehensive exposition known to us is a set of lecture notes for the *School on Contemporary Dynamical Systems*, held in July 2017 in Montréal, available at https://sites.google.com/site/thomasbarthelme/Anosov_flows_in_3_manifolds.pdf.

of a larger picture. We are now in a position to show (following Marcus [254, 255]) how the structural theory of hyperbolic systems we developed in previous chapters can make these connections in a broader context to support the analysis of flows that generalize the horocycle flow from Chapter 2. We saw in Proposition 3.3.19 (say) how the basic renormalization commutation relation (2.2.3) between the geodesic flow g^t and the horocycle flow h_+^t on a surface of constant negative curvature,

$$g^t h_+^s g^{-t} = h_+^{se^t},$$

can be used for the study of horocycle flows, and in some more general situations. Our methods there were algebraic and explicit in nature, as is appropriate for systems of an algebraic nature.

We now leave the domain of algebraic dynamics and consider flows associated with the expanding foliations of hyperbolic systems. Even in the situations where those objects are defined unambiguously, such as horocycle flows on surfaces of variable curvature, they possess only moderate regularity; in the latter case they are C^1 but never C^2 (Theorem 9.7.12). In more general situations there is also a question of choosing a suitable parametrization that does not influence the description of invariant measures but may affect mixing and other properties. Two natural candidates are smooth parametrizations and the singular but dynamically well-defined Margulis parameter introduced in Section 7.6. As it turns out, a variety of dynamical properties can be effectively studied for both types of parametrizations.

8.6.a Definition of horocycle and unstable flows. To define horocycle flows we generalize the considerations from Section 2.1.c to variable negative curvature, although we will concentrate on 2-dimensional manifolds.

Geodesic flows can be defined on any Riemannian manifold, and if the sectional curvature of the manifold is negative, then the geodesic flow is an Anosov flow (Theorem 5.2.4). The stable and unstable foliations project to horospheric foliations on the manifold itself (Section 6.2), which means that in the case of surfaces one can represent the situation on the universal cover pictorially in much the same way as in the constant-curvature situation of Figure 2.1.3. A horocycle flow is a flow whose orbits are the horocycles.

Definition 8.6.1. Consider an oriented negatively curved surface N. A flow on $M = SN$ whose orbits are the unstable manifolds (horospheres) of the geodesic flow is called a *horocycle flow*. More generally, given an Anosov flow on a manifold M with 1-dimensional orientable unstable foliation, a flow whose orbits are the unstable leaves is called an *unstable flow*.

Remark 8.6.2. Note that this definition does not specify a parametrization, so even for any one geodesic flow this describes a class of flows (that are related by time changes or reversals). The existence of such flows is clear: orientability of the unstable foliation gives a unit vector field along it which defines a parametrization of a flow of the desired kind. In other words, we actually study a 1-dimensional oriented foliation, but our notions and tools are best adapted to viewing it as a flow.

In the case of geodesic flows Remark 6.2.1 provides this orientation. We remark that in this case the flow is a little more regular than required. Corollary 9.4.11 shows that the weak-unstable subbundle is C^1, and the strong-unstable subbundle is given by intersecting the weak-unstable subbundle with the kernel of the contact form, which is smooth. So the weak-unstable subbundle of the geodesic flow on a surface is C^1. (In fact, Theorem 9.4.10 shows that the unstable subbundle is $C^{1+\alpha}$ for all $\alpha < 1$, but no more by Theorem 9.7.12.)

The definition implies that such flows have no fixed points.

Some properties of these flows depend on a parametrization, others do not. In the former case we need to specify a parametrization.

Remark 8.6.3. Our further investigations rely on *renormalization* relations between the Anosov flow and the corresponding unstable flow. If φ^s is an unstable flow for an Anosov flow g^t then there is a function

$$\mathfrak{s}\colon \mathbb{R}\times\mathbb{R}\times M \to \mathbb{R} \quad \text{such that } g^t(\varphi^s(x)) = \varphi^{\mathfrak{s}(t,s,x)}(g^t(x)). \tag{8.6.1}$$

This can be taken to measure the rate at which g^t expands unstable manifolds as measured by the time parameter of φ^s.

We define three distinguished choices of parametrization.

Definition 8.6.4. The *Margulis parametrization* or the *conformal parametrization* of a horocycle flow is the parametrization ψ^s for which $\mathfrak{s}(t,s,x) = h^t s$ in (8.6.1).

The horocycle flow for which

$$\frac{d}{ds}\Big|_{s=0}\mathfrak{s}(t,s,x) = Y_x(t),$$

the unstable Jacobi field (Remark 5.2.6) along the geodesic defined by $x \in SM$, is called the *standard horocycle flow*.

The horocycle parametrized with unit speed is called the *unit-speed horocycle flow*.

For the Margulis parametrization the interplay between the horocycle flow and the geodesic (or Anosov) flow is particularly direct. We will show below that such a parametrization exists; the reason for the name is that this uses the Margulis measure (Definition 7.5.19).

Remark 8.6.5. Since g^t contracts unstable leaves when $t < 0$ it is easy to see that for $t < 0$ and $x \in M$ we have

$$\lim_{s\to\infty} \mathfrak{s}(t, s, x) - s = -\infty.$$

However, this also follows from the much stronger result that asymptotically we always have $\mathfrak{s}(t, s, x) = h^t s$ (Lemma 8.6.16 below).

We now explicitly describe the Margulis parametrization. The central ingredient is the Margulis measure from Definition 7.5.19, or, more to the point, its conditionals μ^u on unstable leaves. The pertinent property for us now is that they scale conformally under the geodesic flow. This determines a parametrization for a flow ψ^s by taking the direction determined by the orientation on W^u and the speed just such that the μ^u-measure of the arc between $\psi^t(x)$ and $\psi^s(x)$ is $|t - s|$. This gives an injective parametrization of leaves of W^u since μ^u is positive on open sets and unbounded on a leaf. Moreover, $s \mapsto \psi^s(s)$ is continuous since μ^u is a nonatomic Borel measure. We will obtain continuity of ψ^s as a by-product of holonomy compatibility.

We can now see that this gives the desired parametrization: by Theorem 7.5.15 we have

$$g^t \circ \psi^s = \psi^{h^t s} \circ g^t, \tag{8.6.2}$$

which is exactly (8.6.1) with the condition of Definition 8.6.4.

Next, we check the promised holonomy compatibility. Note that for $x, y \in M$ close enough and $s \in \mathbb{R}$ small enough to be in a local product neighborhood we can write

$$\psi^{k_{x,y}(s)}(x) = [\psi^s(y), x],$$

(Bowen bracket, Proposition 6.2.2). The function $k_{x,y}$ thus defined measures how holonomy affects the parametrization. We claim that

(1) $k_{x,y}(\cdot)$ is strictly increasing,

(2) $k_{x,y}(\cdot)$ is Lipschitz continuous,

(3) $\lim_{y\to x} k_{x,y}(s) = s$ uniformly in s,

(4) $\lim_{y\to x} \frac{d}{ds} k_{x,y}(s) = 1$ uniformly in s.

For (4) note that the derivative is defined almost everywhere by (2); set it to 1 where not otherwise defined.

Item (1) follows from the fact that μ^u is positive on open sets.

Items (2) and (4) follow because Lemma 7.5.16 implies

$$\left| \frac{k_{x,y}(t) - k_{x,y}(s)}{t - s} - 1 \right| < \epsilon$$

for sufficiently small $s \neq t$. Indeed, this also implies (3) because

$$\lim_{y\to x} \psi^{k_{x,y}(0)}(x) = \lim_{y\to x}[y,x] = [x,x] = x$$

tells us that $\lim_{y\to x} k_{x,y}(0) = 0$.

To check that $(x,s) \mapsto \psi^s(x)$ is continuous we write $\psi^s(x) = [\psi_{k_{x,y}(s)}(x), y]$ and note that $[\cdot,\cdot]$ and $t \mapsto \psi^t(x)$ are continuous as is $(y,s) \mapsto k_{x,y}(s)$ by (3) because

$$|k_{x,y}(s) - k_{x,x}(t)| = |k_{x,y}(s) - t| \le |k_{x,y}(s) - s| + |s - t|.$$

This concludes the description of this parametrization.

8.6.b Minimality, entropy, unique ergodicity, and mixing. Minimality of horocycle flows is easy:

Proposition 8.6.6. *Horocycle flows are minimal.*

Proof. Geodesic flows are contact (Proposition 2.6.28) and Anosov (Theorem 5.2.4), hence topologically mixing (Theorem 8.1.4), so all unstable manifolds are dense (Theorem 8.1.3), which is the claim. □

The commutation relation (8.6.1) makes it easy to obtain the entropy of unstable flows.

Theorem 8.6.7. *Horocycle flows have zero entropy.*

Proof. The standard horocycle flow has finite entropy (Corollary 4.2.37), hence so does the Margulis parametrization Ψ (Theorem 4.3.16), and

$$h_{\text{top}}(\psi) = h_{\text{top}}(\psi^1) \xlongequal{\text{Proposition 4.2.12(3)}} \frac{1}{n} h_{\text{top}}(\psi^n) \xlongequal{(8.6.4)} \frac{1}{n} \underbrace{h_{\text{top}}(\psi^1)}_{<\infty} \xrightarrow[n\to\infty]{} 0.$$

The claim follows from Theorem 4.3.16. □

Theorem 8.6.8. *A horocycle flow is uniquely ergodic.*

Remark 8.6.9. Indeed, we prove this for any unstable flow for a mixing C^2 Anosov flow with oriented 1-dimensional unstable subbundle. (Note that mixing follows from not being a suspension; see Theorem 8.1.3.)

Proof. Because unique ergodicity is independent of the parametrization (Corollary 3.5.5), we work with the Margulis parametrization.

To prove that this horocycle flow is uniquely ergodic we use (8.6.2) and Theorem 3.3.32: we will show that Birkhoff averages uniformly converge to a constant.

The first major step in this direction is the following:

Lemma 8.6.10. *If f is continuous on M then*

$$E_n f(x) := \frac{1}{2^n}\int_0^{2^n} f(\psi^s(g^{t_n}(x)))\,ds = \int_0^1 f(g^{t_n}(\psi^s(x)))\,ds$$

converges uniformly to a constant as $n \to \infty$. Here t_n is defined by $h^{t_n} = 2^n$ with h as in (8.6.2).

To see that this implies unique ergodicity, note that it implies

$$\forall f \in C(M)\, \exists c \in \mathbb{R}\, \forall \epsilon > 0\, \exists n \in \mathbb{N}\, \forall y \in M \quad |E_n f(y) - c| < \epsilon.$$

To apply this we want to move basepoints a little: For $j \in \mathbb{N}$ and $x \in M$ the quantity

$$\frac{1}{2^n j}\int_0^{2^n j} f(\psi^s(x))\,ds$$

is the average of $\{E_n f(g^{-t_n}(\psi^{2^n i}(x)))\}_{i=0}^{j-1}$ and hence lies in $(c-\epsilon, c+\epsilon)$. But since $2^n j \xrightarrow[j\to\infty]{} \infty$ in an arithmetic progression, this implies that for large enough (but uniform) t we have

$$\left|\frac{1}{t}\int_0^t f(\psi^s(x))\,ds - c\right| < 2\epsilon,$$

that is, $\frac{1}{t}\int_0^t f(\psi^s(x))\,ds \to c$ uniformly, as claimed. □

Proof of Lemma 8.6.10. The definition of E_n implies that

$$E_{n+m} f = \frac{1}{2^m}\sum_{j=0}^{2^m-1} E_n f \circ \psi^j \circ g^{t_m}, \tag{8.6.3}$$

so $c_n := \min_{x\in M} E_n f(x)$ is nondecreasing, and we can take $c := \lim_{n\to\infty} c_n$.

Claim 8.6.11. *$\{E_n f\}_{n\in\mathbb{N}}$ is equicontinuous and uniformly bounded.*

By the Arzelà–Ascoli Theorem there is a subsequence n_k such that $E_{n_k} f \to h \in C(M)$ uniformly. Then $\min h = c$, and (8.6.3) shows that given any $m \in \mathbb{N}$ we have

$$E_{n_k+m} f \to \bar{h} := \frac{1}{2^m}\sum_{j=0}^{2^m-1} h \circ \psi^j \circ g^{t_m},$$

and $\min \bar{h} = c$ as well.

The point is that if $\bar{h}(x_0) = c$, then $h(\psi^j(g^{t_m}(x_0))) = c$ for $0 \le j < 2^m$. Since m above was arbitrary, $h = c$ on a δ-dense set (Proposition 1.6.25), so $h \equiv c$ by continuity, and $E_{n_k} f \to c$ uniformly. But then (8.6.3) implies that $E_n f \to c$ uniformly, as claimed. □

Proof of Claim 8.6.11. We use that $\{g^t\restriction_{W_\eta^{0s}(x)}\}_{t\ge 0}$ is equicontinuous since it contracts. Specifically, given $x\in M$ and $\epsilon>0$, if $y\in M$ is close enough to x then for $|s|<1$ and $t\ge 0$ we have

$$|f(g^t(\psi^s(y)))-f(g^t([\psi^s(y),x]))|<\epsilon$$

as well as

$$|k_{x,y}(s)-s|<\epsilon \text{ and } \Big|\frac{d}{ds}k_{x,y}(s)-1\Big|<\epsilon.$$

To prove the claim we show that for all $n\in\mathbb{N}$ we have

$$|E_nf(x)-E_nf(y)|<\epsilon(1+3\|f\|).$$

To see this we step from $E_nf(y)$ to $E_nf(x)$ via

$$\int_0^1 f(g^{t_n}([\psi^s(y),x]))\,ds=\int_0^1 f(g^{t_n}(\psi^{k_{x,y}(s)}(x)))\,ds$$

and

$$\int_0^1\Big(\frac{d}{ds}k_{x,y}(s)\Big)\cdot(f(g^{t_n}(\psi^{k_{x,y}(s)}(x))))\,ds=\int_{k_{x,y}(0)}^{k_{x,y}(1)} f(g^{t_n}(\psi^s(x)))\,ds$$

(change of variables using that $k_{x,y}$ is Lipschitz, hence absolutely continuous).

The three resulting errors are

$$\begin{aligned}\Big|E_nf(y)-\int_0^1 f(g^{t_n}([\psi^s(y),x]))\,ds\Big|&\\ =\Big|\int_0^1 f(g^{t_n}(\psi^s(y)))-f(g^{t_n}([\psi^s(y),x]))\,ds\Big|&<\epsilon,\end{aligned}$$

$$\Big|\int_0^1 f(g^{t_n}(\psi^{k_{x,y}(s)}(x)))\,ds-\int_0^1\Big(\frac{d}{ds}k_{x,y}(s)\Big)\cdot(f(g^{t_n}(\psi^{k_{x,y}(s)}(x))))\,ds\Big|<\epsilon\|f\|,$$

and

$$\Big|\int_{k_{x,y}(0)}^{k_{x,y}(1)} f(g^{t_n}(\psi^s(x)))\,ds-E_nf(x)\Big|<\|f\|\cdot|k_{x,y}(1)-1+k_{x,y}(0)|<2\epsilon\|f\|.$$

The sum of these is bounded by $\epsilon(1+3\|f\|)$, as claimed. □

Remark 8.6.12. The Anosov flow is K-mixing with respect to the *Margulis measure* by Remark 7.3.20, but on the other hand Theorem 8.6.8 implies that it is mixing of all orders by Corollary 7.1.8.

While Theorem 8.6.8 holds independently of the parametrization chosen, we wish to point out that for the Margulis parametrization we used in the proof, the Margulis measure is not just ergodic—this extends Theorem 3.4.44, and by using essentially the same argument.

Proposition 8.6.13. *The Margulis measure is weakly mixing* (*Definition* 3.4.37) *for the horocycle flow with the Margulis parametrization.*

Proof. The Margulis measure is the sole invariant measure of the horocycle flow with the Margulis parametrization, so it is ergodic (Corollary 3.3.28). This implies that some time-t map of this horocycle flow is ergodic (Theorem 3.3.13). Since the Margulis measure is invariant under g^t, we can rewrite (8.6.2) as

$$g^{\log_h(t/s)} \circ \psi^s = \psi^t \circ g^{\log_h(t/s)} \tag{8.6.4}$$

establishes measure-theoretic isomorphisms between any two time-t maps for $t > 0$. This implies that all time-t maps are ergodic, which implies weak mixing (Proposition 3.4.40). □

This result may be surprising because our first observation about horocycle (and unstable) flows was that they are minimal. Minimality and unique ergodicity appeared first for rotations and toral translations, and none of those are mixing (Corollary 3.4.10). Our primary examples of mixing dynamical systems were of a rather different nature; mixing was due to hyperbolicity. Proposition 8.6.13 suggests that minimality and mixing are not mutually exclusive, and indeed, they are not:

Theorem 8.6.14. *Unstable flows* (*Definition* 8.6.1) *are topologically mixing.*

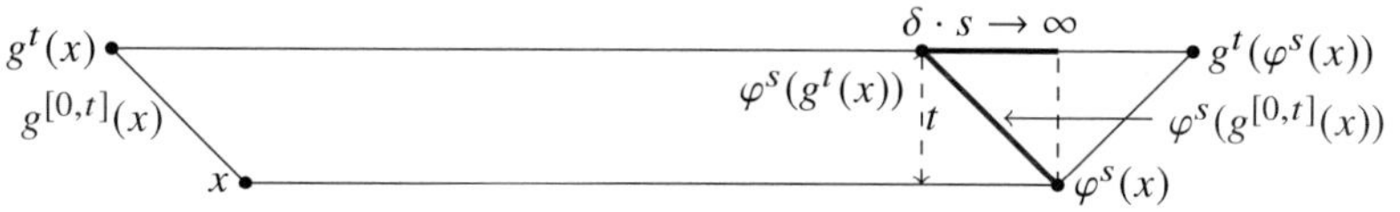

Figure 8.6.1. Stretching of a geodesic segment to a horocyclic one.

The idea is that a geodesic segment $g^{[0,t]}(x)$ is (slowly) stretched by Φ to be ever more dense because it tracks the Φ-orbit of $g^t(x)$ within about t for length $\delta s \to \infty$ (where δ depends on t), hence within about ϵt for length $\epsilon\, \delta s \to \infty$, and this Φ-orbit is dense by minimality (Figure 8.6.1).

Proof. If $U, V \subset M$ open choose a disk whose closure D is in V, and (by minimality) an $s_1 > 0$ such that

$$\varphi^{[0,s_1]}(x) \cap D \neq \varnothing \quad \text{for } x \in M. \tag{8.6.5}$$

(Here, $\varphi^A(B) := \bigcup_{s\in A}\varphi^s(B)$.) Also choose $x_0 \in U$ and $a > 0$ such that

$$g^{[0,a]}(x_0) \subset U \text{ and } \sup\{d(g^t(x),x) \mid x \in M,\ 0 \le t \le a\} < d(D, M \setminus V). \quad (8.6.6)$$

Then $\varphi^s(U) \cap V \neq \varnothing$ for all sufficiently large s because

$$\varphi^s(g^{[0,a]}(x_0)) \cap V \neq \varnothing, \quad (8.6.7)$$

by Claim 8.6.15 below. □

Claim 8.6.15. *For large enough s there is a $t \in [0,a]$ for which $\varphi^s(g^t(x_0)) \in V$.*

Proof. $\mathfrak{s}(-a,s,g^a(x_0)) < s - s_1$ for large s (Remark 8.6.5). Since $\mathfrak{s}(0,s,g^0(x_0)) < s$, this implies that $[s, s-s_1] \subset \mathcal{S} := \bigcup_{t\in[0,a]}\mathfrak{s}(-t,s,g^t(x_0))$. Equation (8.6.5) gives a $t \in [0,a]$ with

$$\varphi^{\mathfrak{s}(-t,s,g^t(x_0))}(x_0) \in D.$$

By (8.6.6) this implies $\varphi^s(g^t(x_0)) = g^t(\varphi^{\mathfrak{s}(-t,s,g^t(x_0))}(x_0)) \in V$ as claimed. □

Our next object is to establish that horocycle flows are, in fact, measure-theoretically mixing (with respect to the sole invariant measure). While the above proof of mixing demonstrates the essential ideas needed for the measure-theoretic result, there is a need for substantial refinements, and therefore we now digress to two technical results regarding the parameter change in (8.6.1). We already mentioned the first of these:

Lemma 8.6.16 (Expanding lemma). *If φ^s is an unstable flow* (*Definition* 8.6.1), *$\mathfrak{s}$ is as in* (8.6.1), *h is as in* (8.6.2), *and $a < b \in \mathbb{R}$, then*

$$\mathfrak{s}(t,s,x)/s \xrightarrow[s\to\infty]{} h^t$$

uniformly in $(x,t) \in M \times [a,b]$.

Proof. Fix $[a,b] \subset \mathbb{R}$. For the Margulis parametrization ψ^s we have equality by definition, and it is useful to compare the given parametrization φ^s with the Margulis parametrization. As in Definition 1.2.1, denote the time change by α: $\varphi^s(x) = \psi^{\alpha(s,x)}(x)$. This and (8.6.2) imply

$$\alpha(\mathfrak{s}(t,s,x), g^t(x)) = h^t\alpha(s,x), \quad (8.6.8)$$

and hence

$$\frac{\mathfrak{s}(t,s,x)}{s} = h^t\frac{\alpha(s,x)/s}{\alpha(\mathfrak{s}(t,s,x), g^t(x))/\mathfrak{s}(t,s,x)}. \quad (8.6.9)$$

Claim 8.6.17. $\alpha(s,x)/s \xrightarrow[s\to\infty]{} \alpha_0 \neq 0$ *uniformly in $x \in M$.*

The claim implies first that the numerator on the right-hand side of (8.6.9) tends to α_0. Together with (8.6.8) it also implies that $\mathfrak{s}(t,s,x) \xrightarrow[s\to\infty]{} \infty$ uniformly in $x \in M$, $t \in [a,b]$. But then the denominator on the right-hand side of (8.6.9) is of the form $\alpha(s^*,x)/s^*$ with $s^* \to \infty$, hence goes to α_0 uniformly in $x \in M, t \in [a,b]$ as well. □

Proof of Claim 8.6.17. We use unique ergodicity of Φ. It suffices to prove this for an arithmetic progression of s, so we take t such that φ^t is uniquely ergodic (Theorem 3.3.34) and write

$$\frac{\alpha(nt,x)}{n} = \frac{\sum_{i=0}^{n-1} a(\varphi^{nt}(x))}{n},$$

where $a(x) := \alpha(t,x)$. This is a Birkhoff average of a continuous function and hence converges uniformly to a constant (Proposition 3.3.33). □

The next result is a rather finer version of this statement.

Lemma 8.6.18. *If φ^s is an unstable flow such that $\frac{\partial^2 \mathfrak{s}}{\partial t\, \partial s}$ is continuous in (t,s,x), and h is as in* (8.6.2)*, then $u_{s,x}(t) := \mathfrak{s}(-t,s,g^t(x))$ satisfies*

$$\frac{1}{s}\frac{d}{dt}\Big|_{t_0} u_{s,x}(t) \xrightarrow[s\to\infty,\ t_0\to 0]{\text{uniformly in } x\in M} -\log h.$$

Proof. It is useful to note that $u_{s,x}(t)$ is the inverse of $\mathfrak{s}$ in that

$$s = \mathfrak{s}(t, u_{s,x}(t), x). \tag{8.6.10}$$

To see this, write $z = g^t(x)$ to get

$$\begin{aligned}
\varphi^s(z) &= \varphi^s(g^t(x)) = g^t[g^{-t}(\varphi^s(g^t(x)))] \\
&= g^t[\varphi^{u_{s,x}(t)}(x)] = \varphi^{\mathfrak{s}(t,u_{s,x}(t),x)}(g^t(x)) \qquad \text{by (8.6.1)} \\
&= \varphi^{\mathfrak{s}(t,u_{s,x}(t),x)}(z).
\end{aligned}$$

The point of (8.6.10) is that differentiating with respect to t gives

$$0 = \frac{d}{dt}\Big|_{t_0} \mathfrak{s}(t, u_{s,x}(t_0), x) + \underbrace{\Big[\frac{d}{ds}\Big|_{u_{s,x}(t_0)} \mathfrak{s}(t_0, s, x)\Big]}\Big[\frac{d}{dt}\Big|_{t_0} u_{s,x}(t)\Big].$$

$$= \tfrac{d}{ds}\big|_0 \mathfrak{s}(t_0, s, \varphi^{u_{s,x}(t_0)}x) \xrightarrow[t_0\to 0]{\text{uniformly in }(s,x)} \tfrac{d}{ds}\big|_0 \mathfrak{s}(0,s,\varphi^{u_{s,x}(0)}x) = \tfrac{d}{ds}\big|_0 \mathfrak{s}(0,s,x) = 1$$

Thus, Lemma 8.6.18 follows from the next claim. □

Claim 8.6.19. $\frac{1}{s}\frac{d}{dt}|_{t_0}\mathfrak{s}(t, u_{s,x}(t_0), x) \xrightarrow[s\to\infty]{} \log h$ *uniformly in* $(x,t) \in M \times [-\epsilon, \epsilon]$.

Proof. Note that $h_{s,x}(t) := \frac{\mathfrak{s}(t,u_{s,x}(t_0),x)}{u_{s,x}(t_0)} \xrightarrow[s\to\infty]{} h^t$ uniformly in (x,t) (Lemma 8.6.16), and

$$h'_{s,x}(t) = \frac{\frac{d}{dt}|_{t_0}\mathfrak{s}(t, u_{s,x}(t_0), x)}{u_{s,x}(t_0)} = \frac{\int_0^{u_{s,x}(t_0)} \frac{\partial^2}{\partial s\,\partial t}|_{t=t_0,s=0}\mathfrak{s}(t, s, \varphi^s x)\, ds}{u_{s,x}(t_0)}$$
$$\xrightarrow[s\to\infty]{\text{uniformly in }(x,t)} \int \frac{\partial^2}{\partial s\,\partial t}|_{t=t_0,s=0}\mathfrak{s}(t, s, x)\, d\mu(x) =: h(t)$$

by continuity of $\frac{\partial^2}{\partial s\,\partial t}|_{t=t_0,s=0}\mathfrak{s}(t,s,x)$ and unique ergodicity of the invariant measure μ. Since $h_{s,x}$ and $h'_{s,x}$ converge uniformly we can exchange limit and differentiation to get $h(t) = \frac{d}{dt}h^t = h^t \log h$ and (Lemma 8.6.16)

$$\frac{1}{s}\frac{d}{dt}|_{t_0}\mathfrak{s}(t, u_{s,x}(t_0), x) = h'_{s,x}(t)\cdot\frac{s}{u_{s,x}(t_0)} \xrightarrow[s\to\infty]{\text{uniformly in }(x,t)\in M\times[-\epsilon,\epsilon]} h^t \log h \cdot \frac{1}{h^t}$$
$$= \log h. \qquad \square$$

We now exhibit the essential ingredient of the proof of mixing with respect to the invariant measure. In the proof that these flows are topologically mixing we showed that for a given $a > 0$ the image $\varphi^s(g^{[0,a]}(x_0))$ under φ^s of an unstable arc is ϵ-dense for large enough s (see, for example, (8.6.7)). The idea is the same for establishing that these flows are measure-theoretically mixing, but the requirement now is that long arcs of this type be almost uniformly distributed in order to adequately reflect suitable measures of intersections of sets.

Lemma 8.6.20. *Under the hypotheses of Lemma* 8.6.18, *suppose* $f: M \to \mathbb{R}$ *is continuous and* $b > 0$. *Then*

$$\frac{1}{b}\int_0^b f(\varphi^s(g^t(x)))\, dt \xrightarrow[s\to\infty]{} \int f\, d\mu$$

uniformly in $x \in M$.

Proof. Fix $\epsilon > 0$; we will show that

$$\max_x \left|\frac{1}{b}\int_0^b f(\varphi^s(g^t(x)))\, dt - \int f\, d\mu\right| < \epsilon$$

for all sufficiently large s. First note that it suffices to do so for $b < b_0(\epsilon)$: writing

$$\frac{1}{b}\int_0^b f(\varphi^s(g^t(x)))\, dt = \frac{1}{n}\sum_{i=0}^{n-1}\frac{1}{b/n}\int_0^{b/n} f(\varphi^s(g^t(g^{ib/n}(x))))\, dt$$

shows that it suffices to bring each summand in the average to within ϵ of $\int f\, d\mu$.

Choose $b_0(\epsilon)$ such that $\sup_{t\in[0,b_0(\epsilon)]} \|f - f \circ g^t\| < \epsilon/2$, and restrict attention to $b \in [0, b_0(\epsilon)]$, for which

$$\frac{1}{b}\int_0^b f(\varphi^s(g^t(x)))\,dt = \frac{1}{b}\int_0^b f(g^t(\varphi^{u_{s,x}(t)}(x)))\,dt$$

is within $\epsilon/2$ of

$$\frac{1}{b}\int_0^b f(\varphi^{u_{s,x}(t)}(x))\,dt \underbrace{=}_{\text{[change of variables } u=u_{s,x}(t)]} \frac{1}{b}\int_s^{u_{s,x}(b)} f(\varphi^u(x))\Big[\frac{1}{\frac{d}{dt}\big|_{u_{s,x}^{-1}(u)} u_{s,x}(t)}\Big]\,du$$

$$\underbrace{=}_{\text{[Mean-Value Theorem; } t_0\in[0,b]]} \frac{1}{b}\frac{1}{\frac{d}{dt}\big|_{t_0} u_{s,x}(t)}\int_s^{u_{s,x}(b)} f(\varphi^u(x))\,du$$

$$= \Big[\frac{\frac{u_{s,x}(b)-s}{b}}{\frac{d}{dt}\big|_{t_0} u_{s,x}(t)}\Big]\cdot\Big[\frac{1}{u_{s,x}(b)-s}\int_s^{u_{s,x}(b)} f(\varphi^u(x))\,du\Big].$$

It remains to bring this within $\epsilon/2$ of $\int f\,d\mu$, and the point of these manipulations was that the last term is an ergodic average, hence by Remark 8.6.5 and unique ergodicity converges uniformly to $\int f\,d\mu$.

Thus, we only have to show that the first factor can be brought close to 1. To that end note that by the Mean-Value Theorem,

$$\frac{u_{s,x}(b)-s}{b} = \frac{d}{dt}\Big|_{t_1} u_{s,x}(t)$$

with $t_1 \in [0, b]$. Thus, we need to control

$$\frac{\frac{d}{dt}\big|_{t_1} u_{s,x}(t)}{\frac{d}{dt}\big|_{t_0} u_{s,x}(t)} \quad \text{for } t_0, t_1 \in [0, b] \tag{8.6.11}$$

uniformly in x by taking b small and then s large. This is a delicate problem and the reason for Lemma 8.6.18, which now tells us that

$$\frac{\frac{d}{dt}\big|_{t_1} u_{s,x}(t)}{\frac{d}{dt}\big|_{t_0} u_{s,x}(t)} = \frac{\frac{1}{s}\frac{d}{dt}\big|_{t_1} u_{s,x}(t)}{\frac{1}{s}\frac{d}{dt}\big|_{t_0} u_{s,x}(t)} \xrightarrow[t_0,t_1\to 0]{s\to\infty} \frac{-\log h}{-\log h} = 1,$$

so small b and large s make (8.6.11) as close to 1 as needed—uniformly in x. □

The following is straightforward to verify.

Theorem 8.6.21. *Consider a flow Φ on a measure space (X, μ) for which there is a $\lambda > 0$ such that $\mu(\varphi^t(A)) = \lambda^t \mu(A)$ for each measurable set A and every $t \in \mathbb{R}$. Suppose $S \subset X$ is measurable and such that $\Phi\restriction_{[0,a]\times S}$, $(t, x) \mapsto \varphi^t(x)$ is injective for some $a > 0$. Then $S_a := \Phi([0, a] \times S)$ is called a* flow box *over S, and*

$$\int f \, d\mu = \int_0^a \int_S \lambda^t f(\varphi^t(x)) \, d\mu_S(x) \, dt,$$

for measurable $f : S_a \to \mathbb{R}$, where $\mu_S(A) := \frac{\mu(\varphi^{[t_1,t_2]}(A))}{\int_{t_1}^{t_2} \lambda^t \, dt}$ with $0 \le t_1 < t_2 \le a$.

Note that the case $\lambda = 1$ corresponds to invariant measures.

Theorem 8.6.22. *If φ^s is an unstable flow such that $\frac{\partial^2 s}{\partial t \, \partial s}$ is continuous in (t, s, x), then the invariant measure for φ^s is mixing.*

Proof. We will show that $\int_N f \circ \varphi^s \, d\mu \xrightarrow[s\to\infty]{} \mu(N) \int f \, d\mu$ whenever $f : M \to \mathbb{R}$ is continuous and N is a local product neighborhood. Since continuous functions and linear combinations of characteristic functions of product neighborhoods are dense in $L^1(\mu)$, this implies that μ is mixing.

For a product neighborhood $N := \varphi^{[0,a]}(g^{[0,b]}(W^s_\epsilon(z)))$ Theorem 8.6.21 applied twice gives a measure μ^s on $W^s_\epsilon(z)$ such that the natural parametrization $(s, t, x) \mapsto \varphi^s(g^t(x))$ gives

$$\int_N f \, d\mu = \int_{W^s_\epsilon(z)} \int_0^b \int_0^a h^{-t} f(\varphi^\sigma(g^t(x))) \, d\sigma \, dt \, d\mu^s(x)$$

for any continuous $f : N \to \mathbb{R}$. In fact, the measure μ^s, being transverse to the unstable flow, is independent of the parametrization chosen and hence coincides with the measure μ^s from Section 7.5 (except that the roles of the stable and unstable manifolds are here interchanged).

This, plus the Fubini Theorem, allows us to rewrite

$$\int_N f \circ \varphi^s \, d\mu = \int_{W^s_\epsilon(z)} \int_0^a \int_0^b h^{-t} f(\varphi^{s+\sigma}(g^t(x))) \, dt \, d\sigma \, d\mu^s(x).$$

Claim 8.6.23. $\int_0^b h^{-t} f(\varphi^{s+\sigma}(g^t(x))) \, dt \xrightarrow[s\to\infty]{} \left[\int_0^b h^{-t} \, dt\right]\left[\int f \, d\mu\right]$.

Proof. This is essentially Lemma 8.6.20, except for the factor h^{-t}. Lemma 8.6.20 gives $\int_0^b f(\varphi^{s+\sigma}(g^t(x))) \, dt \xrightarrow[s\to\infty]{} \left[\int_0^b 1 \, dt\right]\left[\int f \, d\mu\right]$, and the trick is that it does so for arbitrarily small b. Therefore we subdivide $[0, b]$ into small subintervals on which h^{-t} is sufficiently close to constant. Piecing together the corresponding applications of Lemma 8.6.20 yields the claim. □

The claim finally yields

$$\int_N f\circ\varphi^s\,d\mu=\int_{W_\epsilon^s(z)}\int_0^a\int_0^b h^{-t}f(\varphi^{s+\sigma}(g^t(x)))\,dt\,d\sigma\,d\mu^s(x)$$
$$\xrightarrow[s\to\infty]{}\int_{W_\epsilon^s(z)}\int_0^a\Big[\int_0^b h^{-t}\,dt\Big]\Big[\int f\,d\mu\Big]\,d\sigma\,d\mu^s(x)=\mu(N)\int f\,d\mu.\ \square$$

We now check the smoothness assumption in Theorem 8.6.22 for the three parametrizations in Definition 8.6.4.

Corollary 8.6.24. *An unstable flow with the Margulis parametrization is mixing.*

Proof. $\frac{\partial^2\mathfrak{s}}{\partial t\,\partial s}=\frac{\partial^2}{\partial t\,\partial s}h^t s$ is continuous in (t,s,x). □

Corollary 8.6.25. *Standard horocycle flows are mixing.*

Proof. Using the definition and the Jacobi equation (5.2.2) we have

$$\frac{\partial^2\mathfrak{s}}{\partial t\,\partial s}=\dot Y_{\varphi^s(x)}(t)=-\int_{-\infty}^t K(g^\tau(\varphi^s(x)))Y_{\varphi^s(x)}(\tau)\,d\tau$$

is continuous in (t,s,x) because K is continuous and $Y_\cdot(t)\xrightarrow[t\to-\infty]{}0$ exponentially and uniformly. □

Corollary 8.6.26. *Unstable flows with unit-speed parametrization are mixing.*

Proof. Differentiate $g^t(\varphi^s(x))=\varphi^{\mathfrak{s}(t,s,x)}(g^t(x))$ with respect to s to get

$$\Big\|Dg^t\Big(\frac{\partial}{\partial s}\varphi^s(x)\Big)\Big\|=\frac{\partial}{\partial s}\mathfrak{s}(t,s,x)\Big\|\frac{\partial}{\partial s}\varphi^s(g^t(x))\Big\|=\frac{\partial}{\partial s}\mathfrak{s}(t,s,x),$$

since the parametrization is with unit speed. To see that the t-derivative is continuous note that Dg^t is C^1 in t, and $x\mapsto\frac{\partial}{\partial s}\varphi^s(x)$ is continuous. □

8.6.c Multiple mixing. Unstable flows with the Margulis parametrization are, in fact, mixing of all orders (Definition 3.4.37). This is an ergodic result that requires little structural information except for the renormalization relation (8.6.2).

Theorem 8.6.27. *Let M be a manifold and μ a Borel probability measure on M. If G and Ψ are continuous fixed-point-free μ-preserving flows on M such that*

(1) *Ψ is ergodic with respect to μ,*

(2) (8.6.2) *holds for some $h>1$,*

then Ψ is mixing of all orders.

We actually prove the following: for a μ-preserving flow ϕ^s on X define

$$P(N) :\Leftrightarrow \forall \{f_1, \dots, f_n\} \subset C_c(X):$$
$$\frac{1}{n-m}\int_m^n \prod_{i=1}^N f_i(\phi^{K_i u}(x))\,du \xrightarrow[\substack{n-m\to\infty \\ (K_i-K_{i-1})n\to\infty \\ 1=K_1<K_2<\cdots<K_N}]{L^2} \prod_{i=1}^N \int f_i\,d\mu.$$

Prior to proving this (for all N), we check that $P(N)$ implies mixing of order N.

To see this, we will verify mixing with respect to small flow boxes. Thus we first localize $P(N)$ as follows.

Lemma 8.6.28. *If $\mathcal{S}$ is a collection of local sections S (Theorem 8.6.21) for the flow g^t in Theorem 8.6.27 and*

- $\mu_S(S) \le 1$ *for all $S \in \mathcal{S}$,*
- $\alpha \in [0,1)$,
- ψ^s *satisfies $P(N)$, and*
- $\{f_1, \dots, f_n\} \subset C_c(X)$,

then

$$\frac{1}{n-\alpha n}\int_{\alpha n}^n \int_S \prod_{i=1}^N f_i(\phi^{K_i u}(y))\,d\mu_S(y)\,du \xrightarrow[\substack{n\to\infty \\ (K_i-K_{i-1})n\to\infty \\ 1=K_1<K_2<\cdots<K_N}]{\text{uniformly in } S\in\mathcal{S}} \mu_S(S)\prod_{i=1}^N \int f_i\,d\mu.$$

Proof. We first note that by recognizing an L^2-product,

$$\frac{1}{b}\int_0^b \underbrace{\frac{1}{n-\alpha n}\int_{\alpha n}^n \int_S \prod_{i=1}^N f_i(\psi^{K_i u}(g^t(y)))\,d\mu_S(y)\,du}_{=:h_{n,K_1,\dots,K_N,S}(t)}\,dt$$
$$= \Big\langle \frac{1}{b}\chi_{S_b}, \frac{1}{n-\alpha n}\int_{\alpha n}^n \prod_{i=1}^N f_i\circ\psi^{K_i u}\,du\Big\rangle \xrightarrow[\substack{n\to\infty \\ (K_i-K_{i-1})n\to\infty \\ 1=K_1<K_2<\cdots<K_N}]{} \mu_S(S)\cdot\prod_{i=1}^N\int f_i\,d\mu$$

by hypothesis, and convergence is uniform in S (Cauchy–Schwarz inequality using $\|\chi_{S_b}/b\|_2 \le b^{-1/2}$).

To obtain the lemma from this note that (8.6.2) and a change of variable allow us to rewrite the integrand as

$$h_{n,K_1,\dots,K_N,S}(t) = \frac{h^t}{n-\alpha n}\int_{\alpha n/h^t}^{n/h^t}\int_S \prod_{i=1}^N f_i(g^t(\psi^{K_i w}(y)))\,d\mu(y)\,dw,$$

so $h_{n,K_1,\dots,K_N,S}(t)$ is equicontinuous at 0. As b (and t) go to 0 we obtain from the above that

$$h_{n,K_1,\dots,K_N,S}(0) \xrightarrow[\substack{n\to\infty\\ (K_i-K_{i-1})n\to\infty\\ 1=K_1<K_2<\cdots<K_N}]{\text{uniformly in } S\in\mathcal{S}} \mu_S(S)\prod_{i=1}^{N}\int f_i\,d\mu,$$

which is the conclusion of the lemma. □

We now verify mixing of order N by checking the conditions of Definition 3.4.37 for $f_0 = \chi_{S_a}$ and $f_1,\dots,f_N \in C_c(X)$ such that $\max_{1\le i\le N}\|f_i\|_\infty \le 1$, $\mu_S(S)\le 1$, and $s_0 = 0$. (Linear combination and approximation then yield the general statement.)

It is enough to check "thin" flow boxes, that is, to consider S_a for small a:

$$\begin{aligned}\frac{1}{a}\int\prod_{i=0}^{N} f_i(\psi^{s_i}(x))\,d\mu &= \frac{1}{a}\int f_0(x)\prod_{i=1}^{N} f_i(\psi^{s_i}(x))\,d\mu\\ &= \frac{1}{n}\sum_{i=0}^{n-1}\frac{1}{a/n}\int \chi_{g^{ia/n}(S)_{a/n}}\prod_{i=1}^{N} f_i(\psi^{s_i}(x))\,d\mu.\end{aligned}$$

Therefore, we check mixing of order N by proving the following result:

Claim 8.6.29. *Uniformly in $T\in\{g^t(S)\mid S\in\mathcal{S},\ t\in[0,a]\}$ we have*

$$\lim_{b\to 0}\ \overline{\lim_{s_i-s_{i-1}\to\infty}}\left|\frac{1}{b}\int\chi_{T_b}(x)\prod_{i=1}^{N} f_i(\psi^{s_i}(x))\,d\mu - \mu_T(T)\prod_{i=1}^{N}\int f_i\,d\mu\right| = 0.$$

Proof. For $\epsilon>0$ we can choose b small enough such that the leftmost item,

$$\frac{1}{b}\int\chi_{T_b}(x)\prod_{i=1}^{N} f_i(\psi^{s_i}(x))\,d\mu = \frac{1}{b}\int_0^b\int_T\prod_{i=1}^{N} f_i(\underbrace{\psi^{s_i}(g^t(y))}_{=g^t(\psi^{s_i/h^t}(y))})\,d\mu(y)\,dt$$

(we used Theorem 8.6.21 and (8.6.2)) is within ϵ of

$$\mathcal{X} := \frac{1}{b}\int_0^b\int_T\prod_{i=1}^{N} f_i(\psi^{s_i/h^t}(y))\,d\mu(y)\,dt.$$

We change to the variable $u = s_1/h^t$ and use the Mean-Value Theorem to find a

$u_0 \in [s_1/h^b, s_1]$ for which

$$\mathcal{X} = \frac{1}{b}\frac{1}{u_0 \log h}\int_{s_1/h^b}^{s_1}\int_T \prod_{i=1}^N f_i(\psi^{s_i u/s_1}(y))\,d\mu(y)\,du$$
$$= \Big[\underbrace{\frac{h^b-1}{b\log h}}_{\xrightarrow[b\to 0]{} \frac{dh^b/db}{\log h}=1}\Big]\Big[\underbrace{\frac{s_1/h^b}{u_0}}_{\in[h^{-b},1]}\Big]\Big[\frac{1}{s_1-h^{-b}s_1}\int_{s_1/h^b}^{s_1}\int_T \prod_{i=1}^N f_i(\psi^{s_i u/s_1}(y))\,d\mu(y)\,du\Big].$$

This is close to $\mu_T(T)\prod_{i=1}^N \int f_i\,d\mu$ by Lemma 8.6.28 with $\alpha = h^{-b}$. □

Proof of Theorem 8.6.27. We now check inductively that ψ^s satisfies $P(N)$ for all $N \in \mathbb{N}$. For $N = 1$ this is the case by ergodicity and the Mean Ergodic Theorem (Theorem 3.2.4). Thus suppose N is such that ψ^s satisfies $P(n)$ for all $n < N$.

To reduce the clutter in limits, we choose sequences of parameters as follows: for $l \in \mathbb{N}$ take $n_l, m_l, K_{1,l}, \dots, K_{N,l}$ such that

- $n_l - m_l \xrightarrow[l\to\infty]{} \infty$,
- $(K_{i,l} - K_{i-1,l})n_l \xrightarrow[l\to\infty]{} \infty$,
- $1 = K_{1,l} < K_{2,l} < \cdots < K_{N,l}$,
- $\lim_{l\to\infty} K_{i,l} \in [0,\infty]$ exists for each i.

The last item can be arranged by passing to a subsequence.

Claim 8.6.30.

$$\frac{1}{n_l - m_l}\int_{m_l}^{n_l}\prod_{i=1}^N f_i(\psi^{K_{i,l}u}(x))\,du \xrightarrow[l\to\infty]{L^2(\mu)} \prod_{i=1}^N \int f_i\,d\mu.$$

In proving this claim we consider two cases separately: in the first case one of the $K_{i,l} \xrightarrow[l\to\infty]{} \infty$ and we let j be the smallest such index. We will see that we can split the indices into those below j and those above, and we can then use the inductive hypothesis on each piece (after rescaling the "top" part).

Write $\mathcal{F}_u^n(x) := \prod_{i=1}^n f_i(\psi^{K_{i,l}u}(x))$. Then for sufficiently large $p \in \mathbb{N}$ we have

$$|\mathcal{F}_{u_1}^{j-1}(x) - \mathcal{F}_{u_2}^{j-1}(x)| < \epsilon \quad \text{when } |u_1 - u_2| < 1/p$$

and

$$\Big|\frac{1}{p}\sum_{=1}^{p}\mathcal{F}_{q/p}^{j-1}(x_i) - \int_0^1 \mathcal{F}_u^{j-1}(x_i)\,du\Big| < \epsilon,$$

hence

$$\left|\frac{1/p}{n_l - m_l}\sum_{q=m_l p+1}^{n_l p} \mathcal{F}_{p/q}^{j-1}(x) - \frac{1}{n_l - m_l}\int_{m_l}^{n_l} \mathcal{F}_u^{j-1}(x)\,du\right| < \epsilon. \tag{8.6.12}$$

The split of indices works out as follows. In the expression $\frac{1}{n_l - m_l}\int_{m_l}^{n_l} \mathcal{F}_u^N(x)\,du$ of the claim we replace the terms below j by using (8.6.12). The product of the top terms, $\prod_{i=j}^N f_i(g^{K_{i,l}q/p}(x))$, is approximated by $b_{q,l}(x) := p\int_{(q-1)/p}^{q/p}\prod_{i=j}^N f_i(g^{K_{i,l}u}(x))\,du$, so

$$\left|\frac{1}{n_l - m_l}\int_{m_l}^{n_l} \mathcal{F}_u^N(x)\,du - \frac{1/p}{n_l - m_l}\sum_{q=m_l p+1}^{n_l p} \mathcal{F}_{p/q}^{j-1}(x)\cdot b_{q,l}(x)\right| < \epsilon. \tag{8.6.13}$$

We first deal with the "top end." A change of variable shows that

$$b_{q,l}(x) = \frac{1}{K_{j,l}/p}\int_{K_{j,l}\frac{q-1}{p}}^{K_{j,l}\frac{q}{p}} \prod_{i=j}^N f_i(\psi^{K_{i,l}w/K_{j,l}}(x))\,dw,$$

and because $K_{j,l}/p \xrightarrow[l\to\infty]{} \infty$ and $\big(\frac{K_{i,l}}{K_{j,l}} - \frac{K_{i-1,l}}{K_{j,l}}\big)K_{j,l}n_l \xrightarrow[l\to\infty]{} \infty$, we can apply the induction hypothesis to this expression and therefore replace $b_{q,l}(x)$ in (8.6.13) by $\prod_{i=j}^N \int f_i\,d\mu$ up to a small error. To the resulting expression

$$\frac{1/p}{n_l - m_l}\sum_{q=m_l p+1}^{n_l p} \mathcal{F}_{p/q}^{j-1}(x)\cdot\prod_{i=j}^N \int f_i\,d\mu$$

we can again apply (8.6.12) (now backward) to instead get

$$\begin{aligned}
&\frac{1}{n_l - m_l}\int_{m_l}^{n_l} \mathcal{F}_u^{j-1}(x)\,du\cdot\prod_{i=j}^N \int f_i\,d\mu\\
&\quad = \frac{1}{n_l - m_l}\int_{m_l}^{n_l} \prod_{i=1}^{j-1} f_i(\psi^{K_{i,l}u}(x))\,du\cdot\prod_{i=j}^N \int f_i\,d\mu
\end{aligned}$$

and a slightly larger error.

The induction hypothesis for the terms with $i < j$ makes the last expression (and hence $\frac{1}{n_l - m_l}\int_{m_l}^{n_l}\prod_{i=1}^N f_i(\psi^{K_{i,l}u}(x))\,du$) L^2-close to $\prod_{i=1}^N \int f_i\,d\mu$.

This gives the claim (and hence $P(N)$) when some $K_{j,l} \to \infty$.

In the complementary case in which all the $K_{i,l}$ are bounded, a compactness argument (in the parameter space) allows us to produce limits that are uniform in the possible choices of the $K_{i,l}$. We then conclude the inductive proof of $P(N)$ using the following general lemma.

For a manifold M denote by $\mathcal{F}$ the set of continuous flows with the topology of uniform convergence on compact sets. For a Borel probability measure μ we say that $\mathcal{E} \subset \mathcal{F}$ is μ-ergodic if μ is an ergodic invariant measure for every $\varphi \in \mathcal{E}$.

Lemma 8.6.31. *Suppose $\mathcal{E} \subset \mathcal{F}$ and there are*

- $c\colon \mathcal{E} \to \mathbb{R}^+$,
- *Borel probability measures μ and ν,*
- *a closed $\mathcal{E}$-invariant algebra $\mathcal{A} \subset C_c(M)$,*

such that

- *$\mathcal{E}$ has compact closure $\bar{\mathcal{E}}$,*
- *$\bar{\mathcal{E}}$ is μ-ergodic,*
- *$\mathcal{E}$ is $(\nu, c, \mathcal{A})$-*generic*, that is, if $f \in \mathcal{A}$, then*

$$\frac{1}{n-m}\int_m^n \int f(\varphi^u(x))\, d\nu(x)\, du \xrightarrow[c(\varphi)\cdot n\to\infty]{n-m\to\infty} \int f\, d\mu.$$

Then $\frac{1}{n-m}\int_m^n f(\varphi^u(x))\, du \xrightarrow[c(\varphi)\cdot n\to\infty]{n-m\to\infty} \int f\, d\mu$ for all $f \in \mathcal{A}$.

To apply this in the case of bounded $K_{i,l}$ take an interval $E^* \subset [1,\infty)$ such that $K_{i,l} \in E^*$ for all i, l. Let $M := X^N$, ν the diagonal measure, and $\mathcal{E}$ the set of flows on M of the form

$$\varphi^u(x_1,\dots,x_n) = (\psi^{K_1 u}(x_1),\dots,\psi^{K_N u}(x_N))$$

with $1 = K_1 < \dots < K_N \in E^*$ (which has compact closure), and $c(\varphi) := \min_{1<i\le N} K_i - K_{i-1}$. For $\mathcal{A}$ take the closure of the algebra generated by products $\prod_{i=1}^N f_i$ with $f_i \in C_c(X)$. Note that $\bar{\mathcal{E}}$ is also ergodic for the product measure μ^N by Theorem 3.4.19.

The $(\nu, c, \mathcal{A})$-genericity of $\mathcal{E}$ follows from our induction hypothesis because we showed that it implies mixing of order $N-1$, that is, the even stronger statement

$$\int \prod_{i=1}^N f_i(\varphi^{K_i n}(x))\, d\mu \xrightarrow[(K_i-K_{i-1})n\to\infty]{} \prod_{i=1}^N \int f_i\, d\mu.$$

Now note that the conclusion of the lemma with the present terminology is exactly $P(N)$. This proves Theorem 8.6.27. □

Proof of Lemma 8.6.31. Assume $f \perp 1$; the lemma follows by adding a constant.

Ergodicity plus the Mean Ergodic Theorem (Theorem 3.2.4) show that for every $\varphi \in \bar{\mathcal{E}}$ there is an m_φ such that $\int \left(\frac{1}{m_\varphi}\int_0^{m_\varphi} f(\varphi^u(x))\,du\right)^2 d\mu(x) < \epsilon$. By compactness there is a uniform choice: there are finitely many numbers m_j such that for all $\varphi \in \bar{\mathcal{E}}$ there is a j for which $h_\varphi(x) := \frac{1}{m_j}\int_0^{m_j} f(\varphi^u(x))\,du$ satisfies $\int (h_\varphi(x))^2 d\mu(x) < \epsilon$.

Since the $g_\varphi := h_\varphi^2 \in \mathcal{A}$ have common compact support and compact closure, the genericity assumption implies that there is an $M_0 \in \mathbb{N}$ such that if $n - m \ge M_0$ and $\varphi \in \mathcal{E}$ is such that $c(\varphi)n \ge M_0$ then

$$\left|\frac{1}{n-m}\int_m^n \int g_\varphi(\varphi^u(x))\,dv\,du\right| < \epsilon.$$

At the same time, for sufficiently large $n - m$ we have

$$\left|\frac{1}{n-m}\int_m^n f(\varphi^u(x)) - h_\varphi(\varphi^u(x))\,du\right| < \epsilon$$

for all $\varphi \in \mathcal{E}$ and $x \in M$. Combining these last two bounds gives

$$\left\|\frac{1}{n-m}\int_m^n f(\varphi^u(x))\,du\right\|_{L^2} \le \underbrace{\left\|\frac{1}{n-m}\int_m^n f(\varphi^u(x)) - h_\varphi(\varphi^u(x))\,du\right\|_{L^2}}_{<\epsilon}$$

$$+ \underbrace{\sqrt{\frac{1}{n-m}\int_m^n \int g_\varphi(\varphi^u(x))\,dv\,du}}_{<\sqrt{\epsilon}}\,. \qquad \square$$

Since this has been a long proof at the end of a long development through this Section 8.6, let us recapitulate: We just established Theorem 8.6.27 (page 522), which says that if continuous fixed-point-free μ-preserving flows G and Ψ satisfy the renormalization relation (8.6.2) for some $h > 1$ and Ψ is ergodic with respect to μ, then Ψ is mixing of all orders, the desired corollary being that the Margulis parametrization of an unstable flow is mixing of all orders. We worked our way up to this through intermediate results: the Margulis measure is weakly mixing (Proposition 8.6.13) and mixing (Corollary 8.6.24) for the horocycle flow.

These strong stochastic properties are particularly remarkable because at the same time unstable flows have zero entropy (Theorem 8.6.7) and are uniquely ergodic (Theorem 8.6.8).

Topologically as well, there is a contrast with other examples because properties coexist here that otherwise seem to belong to rather distinct categories: these flows are topologically mixing (Theorem 8.6.14) while at the same time being minimal (Proposition 8.6.6).

Figure 8.6.1 serves to show that, of course, while mixing happens in these flows, it happens much more slowly than we have come to expect in hyperbolic flows: instead of being exponential, it is linear. The renormalization relation (8.6.2) is the key for transferring the strong stochastic properties of the Anosov flow to the horocycle/unstable flow, but the trade-off is that exponential rates are translated to linear ones.

The purposes of this section included applying hyperbolic dynamics to study a closely intertwined family of parabolic flows and illustrating by the resulting examples that topological and ergodic properties (minimality and unique ergodicity) that we otherwise saw only in uncomplicated algebraic examples can coexist with strong topological and ergodic properties (topological mixing and multiple mixing, albeit in subexponential incarnations) that we otherwise know only from hyperbolic flows.

9

Rigidity

"Rigidity" usually refers to resistance to or the impossibility of deformation. This notion inspired definitions of rigidity in various parts of mathematics, including dynamics. A familiar instance arises in connection with functions of a complex variable: once such a function is differentiable, then it is infinitely differentiable. This is an instance of a property called smooth rigidity. Moreover, for functions of a complex variable the values on any set with an accumulation point determine the function completely, and this property is called geometric rigidity. In dynamical systems, analogs of these types of rigidity imply strong properties for the system, as we will see in this chapter. The Livshitz Theorem (Theorem 6.3.2) produced a precursor of smooth rigidity in dynamical systems; under the right circumstances a measurable solution can be replaced by a continuous one, and this then turns out to be C^∞.

We have already seen another form of rigidity when studying the structural stability of Anosov flows (Theorem 5.3.8, Corollary 5.4.7). Structural stability is also called *local topological rigidity*, because the topological type of an Anosov flow cannot be changed by a C^1-perturbation. In the discrete-time setting, a classical instance of *global topological rigidity* is that every Anosov diffeomorphism of a torus is topologically conjugate to a linear one [140, 252], [213, Theorem 18.6.1]. Theorem 8.3.1 gave a continuous-time result of a similar flavor, as it says that certain algebraic Anosov flows are globally topologically rigid.

We begin this chapter by examining the nonexistence of faithful $\mathbb{R}^2$-actions for flows: this is referred to as *higher-rank rigidity*. We examine commuting flows, and show there are essentially no faithful $\mathbb{R}^2$-actions for hyperbolic flows (Corollary 9.1.4, Theorem 9.1.14). If one suitably relaxes hyperbolicity (to the right kind of partial hyperbolicity), then there are faithful $\mathbb{R}^2$-actions, but the algebraic ones are locally smoothly rigid (Theorem 9.1.16).

Next, we encounter *smooth rigidity* of Anosov 3-flows related to conjugacies. Homeomorphisms that establish the orbit equivalence in structural stability are usually not differentiable. Sufficient conditions for its smoothness are naturally interesting, since differentiability can imply higher smoothness or infinite differentiability. Topological *conjugacy* (rather than orbit equivalence) provides an instance of smooth rigidity. A topological *conjugacy* between geodesic flows of Riemannian surfaces is a C^1-diffeomorphism (Theorem 9.2.1) and, in turn, a C^1-conjugacy between C^k Anosov

3-flows (for $k > 1$) is $C^{k-\epsilon}$ (Theorem 9.2.7). So as *smooth* dynamical systems, and not just topologically, these two flows are indistinguishable.

Next, in building up to results on geometric rigidity we examine properties of entropy. Specifically, we assume the Bowen–Margulis measure is absolutely continuous with respect to volume for the geodesic flow of a negatively curved surface, and show that this implies the manifold has constant curvature (Theorem 9.3.1). We present much of the state of the art, including Theorem 9.3.11 for contact Anosov 3-flows, as well as the astonishing Lyapunov-spectrum rigidity theorem of Butler (Theorem 9.3.15).

We dedicate several sections to smoothness of the invariant foliations in order to build up to the rigidity results in Section 9.7. Section 9.4 begins by establishing a sharp regularity result for the invariant (weak) subbundles, that is, the strongest possible improvement of Theorem 5.1.16 in this context (for hyperbolic sets rather than specifically for Anosov flows). Section 9.5 examines regularity results for the joint strong subbundles in volume-preserving Anosov 3-flows. These are not only of interest for rigidity results, but because they provide additional information that can be useful in other settings. For instance, Corollaries 9.4.11 and 9.4.12 give C^1-foliations in somewhat broad generality. At the end of Section 9.5 we see a hint at rigidity, but before pursuing this, Section 9.6 establishes that the regularity results in Section 9.4 really are sharp using a mechanism called threading. We return to smooth rigidity with Theorem 9.7.1. Theorem 9.7.12 gives results on both smooth rigidity and geometric rigidity for Anosov 3-flows. Higher-dimensional results (Theorems 9.7.16 and 9.7.17) are also given.

We conclude with two simple geometric rigidity results. For geodesic flows, topological conjugacy prompts the question of whether the underlying manifolds must be isometric. This would mean that the flows are *geometrically* indistinguishable; and this would be an instance of *geometric rigidity* in dynamics. Theorem 9.8.2 produces an isometry for surfaces. The proof of this theorem, together with the proof of Theorem 9.8.1, introduces tools for establishing such results. These are *Godbillon–Vey invariants* (a novel tool we develop fully except that we restrict attention to 3-manifolds). We also describe the Bott–Kanai connection, a central tool for the original proofs of the important results in this area, which we describe here but do not develop.

As in the previous chapter we present some material in a manner similar to a survey of the topic. Specifically, we present some results without proof to convey an impression of the pertinent landscape. While we include some higher-dimensional results, the higher-dimensional setting is far less clear. Typically, it is a matter of conjecture how rigidity questions play out in a higher dimension.

9.1 Multidimensional time: Commuting flows

While the focus of this book is flows, we now touch upon a major subject in modern rigidity theory: the action of groups other than $\mathbb{R}$ (and $\mathbb{Z}$) with some hyperbolicity. We study the question of whether there can be (faithful) $\mathbb{R}^k$-actions (with $k > 1$) by hyperbolic flows, that is, whether a hyperbolic flow can commute with another flow; this can be viewed as having symmetries. We will see that typically this does not happen. Anosov flows never commute with another flow (Corollary 9.1.4), and hyperbolic flows rarely do (Theorem 9.1.14) (the difference being that off the nonwandering set a hyperbolic flow might not be expansive; see Example 1.8.8).

While these results are definitive, they may be taken as indicating that "complete" hyperbolicity is too strict a notion and should be relaxed to hyperbolicity normal to the orbit foliation (by immersed copies of $\mathbb{R}^k$). We give an indication of the state of the art on such actions—they exist but are rigid in ways analogous to what we see in other contexts later: where $\mathbb{R}$-actions are (locally) topologically rigid, these actions are *smoothly* so (Theorem 9.1.16).

We start with the notion of a centralizer.

Definition 9.1.1. Consider a C^r-flow Φ on a closed manifold M[1] with $0 \le r \le \infty$. A C^r-diffeomorphism $f\colon M \to M$ *commutes with* Φ if $f\varphi^t = \varphi^t f$ for all $t \in \mathbb{R}$. A C^r-flow Ψ commutes with Φ if all ψ^s do, that is, $\psi^s\varphi^t = \varphi^t\psi^s$ for all $s, t \in \mathbb{R}$. The set of such flows Ψ is called the C^r*-centralizer* $Z^r(\Phi)$ of Φ. We say that the C^r-centralizer of Φ is trivial if $Z^r(\Phi)$ consists of all constant-time reparametrizations of Φ, that is, $\Psi \in Z^r(\Phi) \Rightarrow \psi^t = \varphi^{ct}$ for some $c \in \mathbb{R}$ and all $t \in \mathbb{R}$.

Remark 9.1.2. When $r_1 \ge r_2$, $Z^{r_1}(\Phi) \subset Z^{r_2}(\Phi)$.

Flows are 1-parameter subgroups of the diffeomorphism (or homeomorphism) group of a manifold, and this notion of centralizer is related to the group-theoretic notion (Remark 1.8.9)—except that we restrict here to commuting *flows* rather than commuting elements of the diffeomorphism group, which is the discrete-time counterpart.

The existence of a nontrivial C^r-centralizer for a C^r-flow (or of commuting diffeomorphisms) implies the existence of symmetries in the dynamical system, and for hyperbolic flows this is rare. Indeed, it has been conjectured in a number of ways that typical flows have trivial centralizer, where typical can mean a generic set, open and dense set, or Lebesgue almost every point of a family of flows given by a finite number of parameters—typically one does not expect nontrivial symmetries to be present in the dynamical system. Let us note, however, some obvious instances: the geodesic flow on the usual genus-2 surface has a finite symmetry group that includes

[1] On a compact topological space X when $r = 0$.

reflection and rotation isometries of the double torus (but there are 1-parameter families of symmetries as in Theorem 2.6.21). The suspension of $\left(\begin{smallmatrix}2&1\\1&1\end{smallmatrix}\right)$ has the symmetry coming from $x \mapsto -x$.

We first recall that topological features of a flow can constrain the centralizer of a flow—expansivity ensures that centralizers are trivial. With Definition 9.1.1 (and Definition 1.8.3), considering time-δ maps of Ψ in Proposition 1.8.7(4) and (8) implies the following theorem:

Theorem 9.1.3. *A transitive separating flow has trivial C^0-centralizer*[2] *as does, more generally, a kinematic-expansive flow on a connected space with at most countably many chain components, all of which are topologically transitive.*[3]

Corollary 9.1.4. *Anosov flows have trivial centralizer.*

Theorem 5.4.22 implies the next result.

Corollary 9.1.5. *A quasitransverse* (*Definition* 5.4.23) *hyperbolic flow without fixed points on a connected space has trivial centralizer.*

From Proposition 1.8.7 we deduce the following result in particular:

Proposition 9.1.6. *Let Φ be a C^r Axiom-A flow on a closed manifold M and let $\epsilon > 0$ be an expansive constant for $\Phi\restriction_{\mathrm{NW}(\Phi)}$. If $f \in \mathrm{Diff}(M)$ commutes with Φ and $d_0(f, \mathrm{Id}) < \epsilon$, where* Id *is the identity map on M, then $f(x) \in \mathcal{O}(x)$ for all $x \in \mathrm{NW}(\Phi)$.*

Proposition 9.1.6 suggests that the restriction to the nonwandering set of the centralizer of an Axiom-A flow is trivial, and the proof of Theorem 9.1.14 below formalizes the argument. The main problem is to extend Proposition 9.1.6 to wandering points—where expansivity may fail. This is the content of Theorem 9.1.13. The issue is that on the wandering set the dynamics (in regard to elements of the centralizer) is harder to control. The reason is that, similarly to Section 5.4, on a sufficiently small neighborhood of a wandering point one can change the dynamics of the unstable manifold without changing the dynamics in the stable manifold and vice versa. Hence, one can change the dynamics in the past or future without affecting the other. Here and in Section 5.4 this causes difficulties, while in Proposition 9.6.6 it is being put to use—albeit for a negative result.

As suggested by Corollary 9.1.4, the majority of results establishing triviality of the C^r-centralizer are for hyperbolic flows (though we also call attention to [269]). Our main result is Theorem 9.1.14: Axiom-A flows with the *strong transversality condition* (as in Theorem 5.4.12) generically have trivial centralizer.

[2] For Anosov flows this was asserted previously to be well known and elementary [148, p. 262].

[3] This and Proposition 1.8.7 are so elementary that the proofs occur much earlier in this book. There are few results like these, they are not widely known, and our results are new.

We begin with a number of preliminary results as we work our way from implications for periodic points (Lemma 9.1.7—the derivative at a periodic point is linearly conjugate to that of the image under any element of the centralizer; compare Example 1.3.10) to their invariant manifolds (Lemma 9.1.8), to the basins of attractors and repellers (Theorem 9.1.9), and ultimately to the whole manifold (Theorem 9.1.13, Theorem 9.1.14).

Lemma 9.1.7. *Let Φ be a C^r-flow and suppose $f \in \mathrm{Diff}^r(M)$ commutes with Φ. If $p \in M$ is a fixed point or periodic point for Φ with period T, then so is $f(p)$, and the derivatives of φ^T at p and $f(p)$ are (linearly) conjugate.*

Proof. If $\varphi^t(p) = p$, then $f(p) = f(\varphi^t(p)) = \varphi^t(f(p))$. If $p \in M$ is a fixed point, then this holds for all $t \in \mathbb{R}$ and so $f(p)$ is a fixed point for Φ. If $p \in M$ is T-periodic, then this holds for $t = T$, so $f(p)$ is T-periodic. Differentiation gives

$$D\varphi^T(f(p))Df(p) = Df(\varphi^T(p))D\varphi^T(p) = Df(p)D\varphi^T(p).$$

Hence, $D\varphi^T(f(p)) = Df(p)D\varphi^T(p)[Df(p)]^{-1}$. □

Specifically, this says that the spectrum of $D\varphi^T(p)$ and $D\varphi^T(f(p))$ are the same; later we deform $D\varphi^t$ to constrain where a commuting diffeomorphism can send a periodic orbit.

Proposition 1.3.29 implies that any element in the centralizer of an Axiom-A flow maps stable (unstable) manifolds to stable (unstable) manifolds.

Lemma 9.1.8. *Let Φ be a C^r Axiom-A flow on a closed manifold M. If $f \in \mathrm{Diff}(M)$ commutes with Φ and $x \in M$, then*

$$f(W^s(x, \Phi)) = W^s(f(x), \Phi) \quad \textit{and} \quad f(W^u(x, \Phi)) = W^u(f(x), \Phi).$$

The next two facts extend local information about the centralizer to more global information. The first is that commuting diffeomorphisms which agree on an open subset of the basin of a hyperbolic attractor agree on the basin. The second provides conditions under which commuting maps that agree on an open set coincide on the whole space. These use smooth linearization (Theorem 6.8.20) of a flow and that this linearizes the entire centralizer (Proposition 6.8.23), both of which require a nonresonance assumption (Definition 6.8.19).

Theorem 9.1.9 (Centralizer and attractors). *Let Φ be a C^∞-flow on a manifold M and $\Lambda \subset M$ be a transitive hyperbolic attractor containing a fixed or periodic point p that is nonresonant. If $f_1, f_2 \in \mathrm{Diff}^\infty(M)$ such that f_1 and f_2 commute with Φ, and there exists an open set $V \subset W^s(\Lambda)$ such that $f_1 \restriction_V = f_2 \restriction_V$, then $f_1 \restriction_{W^s(\Lambda)} = f_2 \restriction_{W^s(\Lambda)}$.*

Proof. We first suppose that p is a nonresonant stable hyperbolic fixed point for Φ. Then there exists a neighborhood U of p and $\Phi_{\restriction U}$ is linear with the appropriate smooth coordinate system (Theorem 6.8.20). Let $f = f_1 \circ f_2^{-1}$. Then f is a C^∞ homeomorphism of M and $f\varphi^t = \varphi^t f$ for all $t \in \mathbb{R}$. Hence, $f_{\restriction U}$ is linear using the same smooth coordinate system (Theorem 6.8.21).

There exists some $t > 0$ such that $\varphi^t(V) \cap U$ is an open set. By hypothesis, $f_{\restriction \varphi^t(V)}$ is the identity. So using the local coordinate system we have f a linear diffeomorphism on U that is the identity on a nonempty open subset of U. Hence, f is the identity on U. Now for any $y \in W^s(p)$ there exists some t such that $\varphi^t(y) \in U$. Then $f(y) = (\varphi^{-t} f \varphi^t)(y) = y$ and $f_{\restriction W^s(p)}$ is the identity. Hence, $f_1 = f_2$ on $W^s(p)$.

More generally, we let p be a nonresonant hyperbolic periodic point contained in Λ. Since $W^s(\mathcal{O}(p))$ is dense in $W^s(\Lambda)$ by the In-Phase Theorem (Theorem 5.3.25) and the second proof of the Spectral Decomposition Theorem, there exists some $T_0 > 0$ such that $W^s(\varphi^{T_0}(p)) \cap V$ contains an open set in $W^s(\varphi^{T_0}(p)$. Let T be the period of p. As above we define $f = f_1 \circ f_2^{-1}$. Then f is a C^∞ homeomorphism of M and $f\varphi^t = \varphi^t f$ for all $t \in \mathbb{R}$.

By Theorem 6.8.12 there exists a neighborhood U of $\varphi^{T_0}(p)$ in $W^s(\varphi^{T_0}(p))$ and a smooth coordinate system such that φ^T is linear on U. Then there is some $n \in \mathbb{N}$ such that $\varphi^{nT}(V)$ contains an open set in U. Then, as above, f is the identity in U. Hence, f is the identity on $W^s(\varphi^{T_0})(p)$.

Now for $y \in W^s(\mathcal{O}(p))$ there exists some t such that $\varphi^t(y) \in W^s(\varphi^{T_0}(p))$. Then $f(y) = (\varphi^{-t} f \varphi^t)(y) = y$. Since $W^s(\mathcal{O}(p))$ is dense in $W^s(\Lambda)$ and f is the identity on $W^s(\mathcal{O}(p))$ we see that f is the identity on $W^s(\Lambda)$ so $f_1 = f_2$ on $W^s(\Lambda)$. □

For a closed manifold M let $\mathcal{A}_S^r(M)$ be the set of C^r Axiom-A flows on M with the strong transversality condition (Theorem 5.4.12).

Remark 9.1.10. It is not hard to see that strong transversality implies the no-cycles condition for an Axiom-A flow.

We need the next lemma concerning Axiom-A flows with the strong transversality condition (as in Theorem 5.4.12) to link basins of attractors and repellers in a manner that allows us to obtain global results from results in the basins of the attractors and repellers.

Lemma 9.1.11. *Let Λ, Λ' be two transitive hyperbolic attractors for $\Phi \in \mathcal{A}_S^1(M)$ with $\overline{W^s(\Lambda)} \cap \overline{W^s(\Lambda')} \neq \varnothing$. Then there is a repeller Λ'' such that $W^s(\Lambda) \cap W^u(\Lambda'') \neq \varnothing \neq W^s(\Lambda') \cap W^u(\Lambda'')$.*

Proof. Let $x \in \overline{W^s(\Lambda)} \cap \overline{W^s(\Lambda')}$. Then $x \in W^s(\Lambda_1)$ for some basic set Λ_1 for Φ, and there is a sequence $W^s(\Lambda) \ni x_n \xrightarrow[n\to\infty]{} x$. Then $\varphi^t(x_n) \xrightarrow[n\to\infty]{} \varphi^t(x)$ for each t. If $t > 0$ is sufficiently large, then $\varphi^t(x)$ is closer to Λ_1 than the constants from the local product structure, so $W^s_\epsilon(\varphi^t(x_n))$ intersects the center-unstable manifold of a point in Λ_1 for sufficiently large n, where ϵ is the constant from the local product structure. Since $W^s(\Lambda)$ is invariant this gives $\overline{W^s(\Lambda)} \cap W^u(\Lambda_1) \neq \varnothing$.

By Axiom A there is a basic set Λ_2 such that $W^s(\Lambda_2) \cap W^u(\Lambda_1) \neq \varnothing$, and modifying the above argument gives $W^s(\Lambda_2) \cap \overline{W^s(\Lambda)} \neq \varnothing$. We can continue this process so long as the basic set we have obtained is not an attractor. However, there is a Λ_j at which the process stops since Φ has the no-cycles property.

Thus, Λ_j is an attractor with $W^s(\Lambda_j) \cap \overline{W^s(\Lambda)} \neq \varnothing$, but this implies that $W^s(\Lambda_j) \cap W^s(\Lambda) \neq \varnothing$ since the basin of an attractor is open. This implies that $\Lambda = \Lambda_j$. Since $W^s(\Lambda_i) \cap W^u(\Lambda_{i-1}) \neq \varnothing$ for $1 \leq i \leq j$ the Inclination Lemma implies that $W^s(\Lambda) \cap W^u(\Lambda_1) \neq \varnothing$ and $W^s(\Lambda_1) \subset \overline{W^s(\Lambda)}$. Similarly, $W^s(\Lambda_1) \subset \overline{W^s(\Lambda')}$. By the strong transversality condition, $W^s(\Lambda_1)$ intersects $W^u(\Lambda'')$ for some repeller Λ''. □

Theorem 9.1.12 (Centralizer rigidity). *There is an open and dense set O of $\mathcal{A}^\infty(M)$ such that if $f_1, f_2 \in \mathrm{Diff}^\infty(M)$ commute with $\Phi \in O$ and $f_1\restriction_U = f_2\restriction_U$ on some open set $U \subset M$, then $f_1 = f_2$.*

Proof. For any $T > 0$ and any Axiom-A C^∞-flow there are only a finite number of periodic orbits with period less than T, so there is an open dense subset $\mathcal{A}^\infty_T(M)$ of $\Phi \in \mathcal{A}^\infty(M)$ such that if $t < T$ and $\varphi^t(p) = p$, then $D\varphi^t(p)\restriction_{E^i(p)}$ is nonresonant for $i = u, s$ and if $\varphi^t(q) = q$, then $D\varphi^t(p)$ and $D\varphi^t(q)$ are conjugate if and only if $q \in \mathcal{O}(p)$. Thus, if

(1) $\Phi \in \mathcal{A}^\infty_T(M)$,

(2) $g \in \mathrm{Diff}^\infty(M)$ commutes with Φ, and

(3) $p \in M$ is periodic for Φ with period less than T,

then $g(p) \in \mathcal{O}(p)$.

For $\Phi \in \mathcal{A}^\infty(M)$ there is a $T_0 > 0$ such that each attractor or repeller contains a periodic point of period less than T_0. By structural stability of Axiom-A flows there is an open set $\mathcal{U}$ of $\Psi \in \mathcal{A}^\infty(M)$ such that each attractor or repeller of Ψ contains a periodic point of period less than T_0. Now let $\mathcal{V}$ be a connected component of $\mathcal{U} \cap \mathcal{A}^\infty_{T_0}(M)$. We will establish the conclusion of the theorem on $\mathcal{V}$.

Suppose $\Phi \in \mathcal{V}$, that $f_1, f_2 \in \mathrm{Diff}^\infty(M)$ commute with Φ, and that there is an open set $U \subset M$ where $f_1\restriction_U = f_2\restriction_U$. Then since there is an open and dense set of points in M that are contained in the basin of an attractor for Φ, there exists an attractor Λ_1 such that $U \cap W^s(\Lambda)$ is open. By Theorem 9.1.9, $f_1\restriction_{W^s(\Lambda_1)} = f_2\restriction_{W^s(\Lambda_1)}$.

Now let Λ_2 be a repeller such that $W^u(\Lambda_2) \cap W^s(\Lambda_1)$ contains an open set. Then $f_1 \restriction_{W^u(\Lambda_2)} = f_2 \restriction_{W^u(\Lambda_2)}$. Since there are only a finite number of attractors and repellers we can continue this process and Lemma 9.1.11 implies that f_1 and f_2 agree on an open and dense set in M, so $f_1 = f_2$ by continuity. □

We can now prove the first major result of this section, an Axiom-A counterpart to Proposition 1.8.7 and the needed extension of Proposition 9.1.6 beyond the nonwandering set. Triviality of centralizers is then a relatively easy consequence.

Theorem 9.1.13 (Discrete centralizer). *There is an open and dense subset O of $\mathcal{A}_S^\infty(M)$ such that for each $\Phi \in O$ there exists a C^0-neighborhood $\mathcal{V}$ of* Id *in* $\mathrm{Diff}^\infty(M)$ *such that if $g \in \mathcal{V}$ commutes with Φ, then $g(x) \in \mathcal{O}(x)$ for each $x \in M$, where $\mathcal{O}(x)$ is the Φ-orbit of x.*

Proof. By Proposition 9.1.6, given $\Phi \in \mathcal{A}_S(M)$ there is an $\epsilon > 0$ such that if $g \in \mathrm{Diff}^\infty(M)$ commutes with Φ and $d_0(g, \mathrm{Id}) < \epsilon$, then $g(x) \in \mathcal{O}(x)$ for all $x \in \mathrm{NW}(\Phi)$. We need to use the Sternberg Linearization Theorem (Theorem 6.8.20) together with properties of attractors and repellers for Axiom-A flows to extend this result to M.

We begin as in the proof of Theorem 9.1.12. For any $T > 0$ and any C^∞ Axiom-A flow there are only a finite number of periodic orbits with period less than T. Thus there is an open dense set $(\mathcal{A}_T)_S^\infty(M)$ of $\mathcal{A}_S^\infty(M)$ such that if $\Phi \in (\mathcal{A}_T)_S^\infty(M)$, $t < T$, and $\varphi^t(p) = p$, then $D\varphi^t(p)\restriction_{E^i(p)}$ is nonresonant for $i = u, s$, and if $\varphi^t(q) = q$, then $D\varphi^t(p)$ and $D\varphi^t(q)$ are conjugate if and only if $q \in \mathcal{O}(p)$. Thus, if

(1) $\Phi \in (\mathcal{A}_T)_S^\infty(M)$,

(2) $g \in \mathrm{Diff}^\infty(M)$ commutes with Φ, and

(3) $p \in M$ is periodic for Φ with period less than T,

then $g(p) \in \mathcal{O}(p)$.

For $\Phi \in \mathcal{A}_S^\infty(M)$ there is a $T_0 > 0$ such that each attractor or repeller contains a periodic point of period less than T_0. By structural stability of Axiom-A flows there is an open set $\mathcal{U}$ of $\mathcal{A}_S^\infty(M)$ such that each attractor or repeller in $\mathcal{U}$ contains a periodic point of period less than T_0. Now let $\mathcal{V}$ be a connected component of $\mathcal{U} \cap (\mathcal{A}_{T_0})_S^\infty(M)$. We will establish the conclusion of the theorem on $\mathcal{V}$.

We divide the proof into two cases. In the first, a transitive attracting or repelling set contains a periodic orbit. In the second, the attracting or repelling set is a fixed point.

Let $\Phi \in \mathcal{V}$ and Λ be a transitive attractor for Λ such that Λ contains a periodic point. We can choose the neighborhood of the identity map such that if p is a periodic point of Λ with period T and g is C^0-close to the identity with g commuting with Φ, then $g(p) = \varphi^t(p)$ for $0 \leq t < T/2$. Then $\varphi^{-t}g$ maps $W^s(p)$ to $W^s(p)$ and commutes with Φ. To conclude the proof in this case it is sufficient to show that $\varphi^{-t}g$ is the identity in a neighborhood of p in $W^s(p)$. To see this note that this implies that g is the identity on the orbits of Φ in a neighborhood of p in $W^s(p)$. Since g commutes with Φ we see that this implies that $\varphi^{-t}g$ is the identity on the space of orbits for all points in $W^s(\mathcal{O}(p))$. Since $W^s(\mathcal{O}(p))$ is dense in $W^s(\Lambda)$ this implies that $\varphi^{-t}g$ is the identity on the orbits of Φ for $W^s(\Lambda)$.

We now show that there is a neighborhood of p in $W^s(p)$ such that $\varphi^{-t}g$ is the identity. Since p is nonresonant, $\varphi^T \restriction_{W^s(p)}$ satisfies the Sternberg conditions (Definition 6.8.19) and so there is a neighborhood U of p in $W^s(p)$ and a C^∞-linearizing coordinate system ξ. Then $\varphi^{-t}g$ commutes with φ^T and so is simultaneously linearized by ξ. Furthermore, since p is nonresonant we may assume that the linear system given by the matrix A has a block diagonal Jordan form with distinct eigenvalues and since $\varphi^{-t}g$ and φ^T commute, the matrix B representing the linear action for $\varphi^{-t}g$ is simultaneously block diagonalizable with the same form.

If A has the form

$$A(x_1,\dots,x_{k-1},x_k,x_k',\dots,x_\ell,x_\ell') = (\lambda_1 x_1,\dots,\lambda_{k-1}x_{k-1},\Lambda_k(x_k,x_k'),\dots,\Lambda_\ell(x_\ell,x_\ell'))$$

then B has the form

$$B(x_1,\dots,x_{k-1},x_k,x_k',\dots,x_\ell,x_\ell') = (\mu_1 x_1,\dots,\mu_{k-1}x_{k-1},M_k(x_k,x_k'),\dots,M_\ell(x_\ell,x_\ell'))$$

where each Λ_j and M_j has the form

$$\begin{pmatrix} a_j & -b_j \\ b_j & a_j \end{pmatrix} \quad \text{for all } 1 \leq j \leq k.$$

Since g is chosen close to the identity we may assume that each of the diagonal entries for B is much smaller in modulus than the entries for A. Now we take each coordinate at a time and make an arbitrarily small perturbation to the vector field generating Φ so that in each coordinate B must be the identity. By Proposition 6.8.23 there is an $(\ell - 1)$-dimensional Lie group of matrices that commute with A. Fix a set of generators $\{B_1,\dots,B_{\ell-1}\}$ of the Lie group. We choose $x_1,\dots,x_{\ell-1}$ points in U and sufficiently small neighborhoods $U_1,\dots,U_{\ell-1}$ such that $x_i \in U_i$ for each $i \in \{1,\dots,\ell-1\}$. Now we modify the vector field in each U_i so that none of the matrices in the 1-dimensional subspace spanned by B_i can be the linearization of a flow that commutes with Φ.

Let h be the smooth linearization in U. It is important that h only be given by the flow Φ. Furthermore, if we make a small modification to the vector field in U_i this does not modify the matrix A, but does make a small change to the conjugacy h. Now we work in the linearized flow. Let $s \neq 0$ such that $e^{B_i s} h(x_i)$ is in $\partial h(U_i)$ or just outside the region. Fix $t \neq 0$ such that both $e^{At} h(x_i) \notin h(U_i)$ and $e^{At} e^{B_i s} h(x_i) \notin h(U_i)$; see Figure 9.1.1. Now modify the generating vector field V in a small region about x_i such that

$$h^{-1}(e^{-B_i s} e^{-At} e^{B_i s} e^{At} h(x_i)) \neq x_i.$$

As stated above, the important component of this is that h is not defined by B_i, but only by Φ and A.

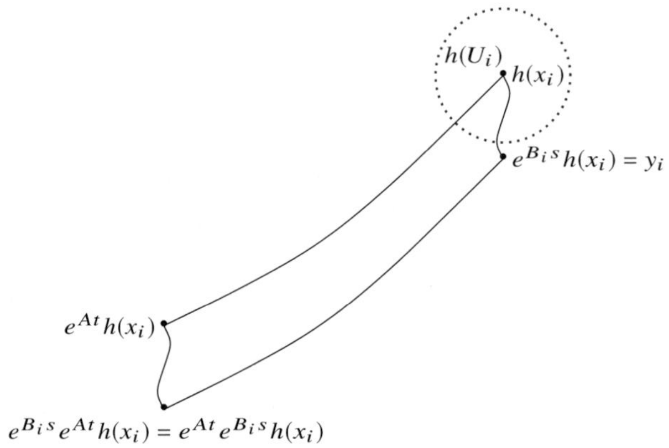

Figure 9.1.1. Commuting linear flows.

Furthermore, the perturbations above give an open condition. Namely, that if Ψ is a flow that commutes with Φ then restricted to a neighborhood of p in the strong-stable manifold the linearization of Ψ is given by a multiple of A. Hence, orbits of Φ are preserved by Ψ.

We now let p be a nonresonant stable hyperbolic fixed point for Φ and use Proposition 6.8.23 and the induced linear action on either the hyperplanes or cylinders. Then we can perturb the vector field in a small neighborhood similar to the method described above such that the induced action must be the identity.

Hence, there is an open and dense set of flows in $\mathcal{V}$ such that any C^∞-diffeomorphism in this set that is C^0-close to the identity and commutes with the flow must map points into their own orbits. □

Theorem 9.1.14 (Axiom A and trivial centralizer). *There is an open and dense subset V of the set $\mathcal{A}_S^\infty(M)$ such that each $\Phi \in V$ has trivial centralizer.*

Proof. Let Φ be in the open and dense set of $\mathcal{A}_S^\infty(M)$ from Theorem 9.1.13 and $\Psi \in Z^\infty(\Phi)$. Then Φ-orbits are Ψ-invariant for sufficiently small t by Theorem 9.1.13, that is, $\psi^t(\varphi^s(x)) = \varphi^{\alpha(s,t,x)}(x)$ for all $s \in \mathbb{R}$. Let Y be the vector field generating Ψ

and X be the vector field generating Φ. Then there is a function $\rho\colon M \setminus \mathrm{Fix}(\Phi) \to \mathbb{R}$ such that $Y = \rho X$.

If $x \in M \setminus \mathrm{Fix}(\Phi)$ and $y, y' \in W^s(x)$, then $\psi^t(W^s(x)) = W^s(\psi^t(x))$, so ρ is constant on $W^s(x)$ (and likewise on $W^u(x)$). At the same time, if $x \in M \setminus \mathrm{Fix}(\Phi)$, then $X\rho = 0$ since $XY - YX = 0$, so ρ is constant on $W^s(\mathcal{O}(x))$. For a transitive attractor Λ that is not a fixed point this implies that ρ is constant on $W^s(\Lambda)$ since $W^s(\mathcal{O}(x))$ is dense in $W^s(\Lambda)$ for any periodic point x.

Any stable fixed point or unstable fixed point satisfies Sternberg's condition (Definition 6.8.19), so φ^t and ψ^t are simultaneously linearized nearby for all sufficiently small t (Theorem 6.8.21), which implies that ρ is constant on the stable or unstable set of the fixed point.

Therefore, ρ is constant on the basin of any attractor or repeller. Their union is an open and dense set and Lemma 9.1.11 shows that the basins intersect, so ρ is constant on $M \setminus \mathrm{Fix}(\Phi)$, that is, $Y = cX$ for some constant $c \in \mathbb{R}$. □

We now explain rigidity results when the hyperbolicity assumption is adapted to allow for nontrivial $\mathbb{R}^k$-actions. The basic examples of Anosov flows are algebraic $\mathbb{R}$-actions: geodesic flows of locally symmetric spaces (such as surfaces of constant negative curvature) and suspensions of algebraic Anosov diffeomorphisms. Structural stability means that these actions are locally topologically rigid and in a strong sense: the orbit equivalence can be chosen near the identity and is transversely unique among such orbit equivalences. For $\mathbb{R}^k$-actions, we instead get local *smooth* rigidity which furthermore is time preserving up to an algebraic time change. We begin with the definition of such actions.

Definition 9.1.15 (Anosov action, algebraic action). Suppose $\mathbb{R}^k \subset H$ is a subgroup of a connected Lie group H and Λ a lattice in H. If $\mathbb{R}^k$ acts on a compact quotient H/Λ by left translations and C is a compact subgroup of H which commutes with $\mathbb{R}^k$, then the $\mathbb{R}^k$-action on H/Λ descends to an action on $C\backslash H/\Lambda$. An *algebraic* $\mathbb{R}^k$*-action* ρ is a finite factor of such an action. Let $\mathfrak{c}$ be the Lie algebra of C. The *linear part* of ρ is the representation of $\mathbb{R}^k$ on $\mathfrak{c}\backslash\mathfrak{h}$ induced by the adjoint representation of $\mathbb{R}^k$ on the Lie algebra $\mathfrak{h}$ of H.

For our purposes, the *standard* $\mathbb{R}^k$*-actions* are actions by infranilmanifold-automorphisms, Weyl-chamber flows (Definition 2.5.7), twisted Weyl-chamber flows, and some further extensions (as explicitly described in [218, Section 2.2]).

An action of a Lie group G on a compact manifold is *Anosov* if some element $g \in G$ is partially hyperbolic with the G-orbit direction as the center bundle (Definition 5.6.2).

Theorem 9.1.16 (Katok–Spatzier [219, Corollary 5]). *The standard algebraic Anosov actions of* $\mathbb{R}^k$ *for* $k > 1$ *with semisimple linear part are locally* C^∞*-rigid. Moreover,*

- *the C^∞-conjugacy φ between the action composed with an automorphism ρ and a perturbation can be chosen C^1-close to the identity;*
- *the automorphism ρ is unique and also close to the identity;*
- *φ is unique among conjugacies close to the identity up to translations in the acting group.*

Remark 9.1.17 ([29]). We note that there has been work on Anosov actions in different directions. If the Anosov element is in the nilradical of (the not necessarily abelian or nilpotent acting Lie group) G, then its stable/unstable foliations are G-invariant, and if G is nilpotent, then there is a Spectral Decomposition Theorem analogous to Theorem 5.3.37. If the Anosov element has 1-dimensional unstable subbundle and G acts on a manifold of dimension at least $\dim(G) + 3$, then the action is topologically transitive, analogously to Theorem 8.4.11.

In another digression beyond the abelian context, we note a classification.

Definition 9.1.18 (Nilpotent algebraic action). Suppose G is a Lie group, $\Gamma < G$ a torsion-free uniform lattice, $K < G$ compact, $\mathfrak{h}$ a nilpotent subalgebra of the Lie algebra of G which normalizes and trivially intersects the Lie algebra of K. Then the action of the simply connected Lie group H with Lie algebra $\mathfrak{h}$ on the compact manifold $\Gamma\backslash G/K$ by right translations is called a nilpotent algebraic action.

Theorem 9.1.19 (Barbot–Maquera–Tomter [30, 342]). *Up to finite covers and central extensions, nilpotent algebraic Anosov actions are nil-suspensions of either*

- *a suspension of a $\mathbb{Z}^k$ Anosov action on a nilmanifold ($G = N \rtimes \mathbb{R}^k$, $\Gamma = \Lambda \rtimes \mathbb{Z}^k$, $K = \{e\}$, $H = \mathbb{R}^k$), or*
- *a modified Weyl chamber action: G is noncompact semisimple with finite center, K a compact subgroup which centralizes a maximal $\mathbb{R}$-diagonalizable subalgebra $\mathfrak{a}$ of the Lie algebra of G and contains the semisimple part of $Z_G(\mathfrak{a})$, $\mathfrak{h} \supset \mathfrak{a}$ is an abelian subalgebra with* $\mathrm{Lie}(Z_G(\mathfrak{a})) = \mathfrak{h} \oplus \mathrm{Lie}(K)$.

Remark 9.1.20. As a complement we mention symmetric Anosov flows of mixed type [341, 342, 75]. "Symmetric" here includes suspensions of toral automorphisms ("solvable type") and geodesic flows of locally symmetric spaces ("semisimple type"). "Mixed" ones combine both. For instance, Tomter constructs an Anosov flow on a compact 7-manifold with both a "semisimple" and a "solvable" part, which can be viewed as a combination of the geodesic flow on a compact Riemannian surface of constant negative curvature and the suspension of a 4-dimensional toral Anosov diffeomorphism. It is obtained by letting $\mathrm{SL}(2,\mathbb{R})$ act suitably on $\mathbb{R}^4$ and taking a semidirect product. Specifically, realize $SL(2, \mathbb{R})$ as the group of unit quaternions in the even-dimensional Clifford algebra corresponding to an anisotropic quadratic form on $\mathbb{R}^3$, that is, a quadratic diophantine equation without integer solutions. Finite-volume quotients are easy to find. It takes more effort to find a uniform, discrete subgroup (find a uniform discrete subgroup of $SL(2, \mathbb{R})$ that preserves an integer lattice of $\mathbb{R}^4$ using the theory of arithmetic subgroups) [341].

9.2 Conjugacies

We have already seen that timing constraints may result in smooth rigidity; Theorem 7.4.19 is a prime example. Now we turn to smooth rigidity results arising from the presence of a conjugacy rather than merely an orbit equivalence, that is, from the preservation of time, Theorems 9.2.1 and 9.2.7. By contrast, we note an astonishing theorem of Ghys: an Anosov 3-flow on a circle bundle is topologically orbit equivalent to the geodesic flow of a Riemannian metric on the base with constant negative curvature [147, Théorème A].

Theorem 9.2.1 ([131, Theorem 5.2]). *A topological* conjugacy *between geodesic flows of Riemannian surfaces with negative curvature is a C^1-diffeomorphism.*

Remark 9.2.2. The step to differentiability is significant. This implies, for instance, that all geometric data associated with periodic points, such as eigenvalues of the return map, are preserved by the conjugacy, because the linear parts of the return maps are linearly conjugate. By contraposition, this means that changing a geodesic flow in such a way as to affect such data at even one periodic point precludes the existence of a *topological* conjugacy to the original geodesic flow.

Proof [131]. We begin by building counterparts to the various flows we developed in the case of constant negative curvature in Section 2.2. Let X be the geodesic vector field, $H_+ \neq 0$ an expanding vector field, and $H_- \neq 0$ a contracting vector field. By Corollary 9.4.11, we can take $H_\pm$ to be C^1. Analogously to the case of constant curvature, $[X, H_\pm]$ is a scalar multiple of $H_\pm$, so the orbits of g and of the horospheric flow h_- generated by H_- span the weak-stable foliation, and those of g and h_+ the unstable foliation. The presence of an invariant contact form means that $a \neq 0$ in

$$[H_-, H_+] = aX + bH_+ + cH_-.$$

We assume that X, H_+, H_- is *properly oriented*, that is, $a > 0$, and introduce functions σ, τ, ρ of $x \in M$ and small $s, t \in \mathbb{R}$ by

$$g^\rho(h_-^\tau(h_+^s(x))) = h_+^\sigma(h_-^t(x)); \tag{9.2.1}$$

this relation reflects the fact that $W^s(h_+^s(x)) \cap W^{cu}(h_-^t(x))$ and $W^u(h_-^t(x)) \cap W^{cs}(h_+^s(x))$ lie on the same g-orbit, and it implies that $\operatorname{sign}\sigma = \operatorname{sign}s$ and $\operatorname{sign}\tau = \operatorname{sign}t$. The next two lemmas give the following fact, which is needed for (9.2.3).

Proposition 9.2.3.

$$\frac{\rho}{st} \xrightarrow[s,t\to 0]{\text{uniformly on } M} a.$$

Lemma 9.2.4.

$$\sqrt{(\sigma - s)^2 + \rho^2 + (\tau - t)^2} \in O(|st|).$$

Proof. We will use that if vector fields X, Y generate flows φ^t, ψ^t, and $f\colon M \to \mathbb{R}$ is C^2, then uniformly in x the second directional derivatives are limits as follows:

$$X(Yf)(x) - Y(Xf)(x) = \lim_{s,t\to 0} \frac{1}{st}\big(f(\psi^t(\varphi^s(x))) - f(\varphi^s(\psi^t(x)))\big). \tag{9.2.2}$$

For $\Delta > 0$ there is an $A > 0$ such that $|s|, |t| < \Delta \Rightarrow d(h_+^s(h_-^t(x)), h_-^t(h_+^s(x))) \le A|st|$ for all $x \in M$, while transversality of H_+, H_-, X and compactness yield a $\Delta' \le \Delta$ and $A' \ge A$ such that

$$|s|, |t|, |r| < \Delta' \Rightarrow \forall\, x \in M\ d(h_-^t(g^r(h_+^s(y))), y) > \sqrt{T^2 + r^2 + s^2}/A'.$$

If s, t are small enough that $\max(|s|, |t|, |\sigma - s|, |\tau - t|, |\rho|) < \Delta'$, then

$$\begin{aligned}\sqrt{(\sigma - s)^2 + \rho^2 + (\tau - t)^2} &< A' d\Big(\overbrace{h_-^{t-\tau}\big(g^{-\rho}\big(\underbrace{h_+^{\sigma-s}\big(h_+^s(h_-^t}_{=h_+^\sigma h_-^t = g^\rho h_-^\tau h_+^s}(x))\big)\big)\big)}^{=h_-^t(h_+^s(x))}, h_+^s(h_-^t(x))\Big)\\ &\le AA'|st|.\end{aligned}$$

□

Lemma 9.2.5.

$$\frac{\rho}{st}X + \frac{s-\sigma}{st}H_+ + \frac{\tau - t}{st}H_- \xrightarrow[s,t\to 0]{\text{uniformly in } x} [H_-, H_+] = aX + bH_+ + cH_-.$$

Proof. For $f \in C^2(M)$ the Mean-Value Theorem and (9.2.2) give[4]

$$\begin{aligned}&\boxed{\rho X f} + \underbrace{\rho}_{\in O(st)}\big[\underbrace{Xf(g^{\rho_1}(h_-^\tau(h_+^s(\cdot)))) - Xf}_{\in o(st)}\big]\\ &= \rho X f(g^{\rho_1}(h_-^\tau(h_+^s(\cdot))))\\ &= f(\underbrace{h_+^\sigma(h_-^t}_{=g^\rho h_-^\tau h_+^s}(\cdot))) - f(h_-^\tau(h_+^s(\cdot)))\end{aligned}$$

[4]See Remark 3.2.18.

$$= \Big(\underbrace{\overbrace{f(h_+^s(h_-^t(\cdot)))}^{\text{cancellation 1}} - \overbrace{f(h_-^t(h_+^s(\cdot)))}^{\text{cancellation 2}}}_{= \boxed{[H_-, H_+] f st} + o(st)} \Big)$$

$$+ \Big(\underbrace{f(h_+^\sigma(h_-^t(\cdot))) - \overbrace{f(h_+^s(h_-^t(\cdot)))}^{\text{cancellation 1}}}_{\substack{= \int_s^\sigma H_+ f(h_+^{s'}(h_-^t(\cdot)))\, ds' = (\sigma - s) H_+ f(h_+^{s_1}(h_-^t(\cdot))) \\ = \underbrace{(\sigma - s)}_{\in O(st)} \underbrace{\big(H_+ f(h_+^{s_1}(h_-^t(\cdot))) - H_+ f \big)}_{\in o(st)} + \boxed{(\sigma - s) H_+ f}}} \Big)$$

$$- \Big(\underbrace{f(h_-^\tau(h_+^s(\cdot))) - \overbrace{f(h_-^t(h_+^s(\cdot)))}^{\text{cancellation 2}}}_{\substack{= \int_t^\tau H_- f(h_-^{t'}(h_+^s(\cdot)))\, dt' = (\tau - t) H_- f(h_-^{t_1}(h_+^s(\cdot))) \\ = \underbrace{(\tau - t)}_{\in O(st)} \underbrace{\big(H_- f(h_-^{t_1}(h_+^s(\cdot))) - H_- f \big)}_{\in o(st)} + \boxed{(\tau - t) H_- f}}} \Big).$$

Rearrange to get $[H_-, H_+] = \frac{\rho}{st} X + \frac{s-\sigma}{st} H_+ + \frac{\tau - t}{st} H_- + o(1)$ uniformly on M. □

We now turn to the conjugacy problem. Denote by h the conjugacy between the geodesic flows g and g'. (We assume that g and g' are both normalized—by scaling the metric—to have topological entropy 1.) The counterparts for g' of the various vector fields, functions, and flows just described are denoted in the same way but with primes added.

Since h preserves the strong invariant foliations, we can use local dynamical coordinates to write

$$h(h_-^t(h_+^s(g^r(x)))) = h'^{t'}_-(h'^{s'}_+(g'^{r'}(x')))$$

with $x' = h(x)$, $r = r'$, $s' = S(g^r(x), s)$, $t' = T(h_+^s(g^r(x)), t)$, where S, T are defined by $h'^{S(x,s)}_+(h(x)) = h(h_+^s(x))$ and $h'^{T(x,t)}_-(h(x)) = h(h_-^t(x))$. Since (r, s, t) and (r', s', t') are local C^1 coordinates, it will suffice to check continuous differentiability of

$$(r, s, t) \mapsto (r', s', t') = (r, S(g^r(x), s), T(h_+^s(g^r(x)), t)).$$

Lemma 9.2.6. *$s \mapsto S(\cdot, s)$ and $t \mapsto T(\cdot, t)$ are C^1.*

Proof. Uniformly on M we have $S(\cdot, s) \xrightarrow[s \to 0]{} 0$, $T(\cdot, t) \xrightarrow[s,t \to 0]{} 0$, $\frac{\rho(\cdot, s, t)}{sta} \xrightarrow[t \to 0]{\text{Proposition 9.2.3}} 1$, and $\frac{\rho'(\cdot, s', t')}{s't'a'} \xrightarrow[s', t' \to 0]{} 1$, and we apply this to both sides of

$$\rho(x, s, t) = \rho'(h(x), S(x, s), T(x, t)) :$$

For $\epsilon > 0$ choose $\delta > 0$ such that $|s|, |t| < \delta \Rightarrow \left| \frac{S(\cdot, s) T(\cdot, t) a' \circ h}{sta} - 1 \right| < \epsilon$ or

$$\left| \log \frac{S(\cdot, s)}{s} + \log \frac{T(\cdot, t)}{t} - \log \frac{a}{a' \circ h} \right| < \epsilon. \tag{9.2.3}$$

With fixed $|t| \in (0,\delta)$ this implies that $s \mapsto \log S(\cdot,s)/s$ is within ϵ of a continuous function. Since ϵ is arbitrary, $\log S(\cdot,s)/s$ uniformly converges to a continuous function as $s \to 0$, hence $s \mapsto S(\cdot,s)/s$ uniformly converges to a nonzero continuous function, and $s \mapsto S(\cdot,s)$ is differentiable at $s = 0$.

To show continuity of the derivative note that the cocycle condition $S(x, s_0 + s) = S(x, s_0) + S(h_+^{s_0}(x), s)$ implies that $s \mapsto S(\cdot, s)$ is differentiable for all s and with

$$\frac{dS(\cdot,s)}{ds}\Big|_{s=s_0} = \frac{dS(h_+^{s_0}(x),s)}{ds}\Big|_{s=0}.$$

Arguing as above from (9.2.3) at $h_+^{s_0}(\cdot)$ rather than $(\cdot)$ shows that $S(h_+^{s_0}(\cdot), s)/s$ uniformly converges as $s \to 0$ to a function that is continuous in s_0.

That $t \mapsto T(\cdot,t)$ is C^1 follows in like manner from (9.2.3). □

From this, we finally show that the conjugacy h is C^1; by symmetry, this implies that h is a C^1-diffeomorphism. To check that

$$(r,s,t) \mapsto (r',s',t') = (r, S(g^r(x),s), T(h_+^s(g^r(x)),t))$$

is C^1, we begin with $(r,s) \mapsto s' = S(g^r(x),s)$, the problem being the r-dependence in the first entry. We circumvent this by using the commutation relations to rewrite this as a composition of C^1-maps.

The reparametrization $\alpha(\cdot,\cdot,r)$ of h_+ defined by $g^r(h_+^s(x)) = h_+^{\alpha(x,s,r)}(g^r(x))$ for fixed r is C^1, as is the reverse reparametrization $\tilde\alpha$ with $\tilde\alpha(x,\alpha(x,s,r),r) = s$, and likewise for α', $\tilde\alpha'$ on the side of g'. The rendering of an (s,r)-coordinate patch in

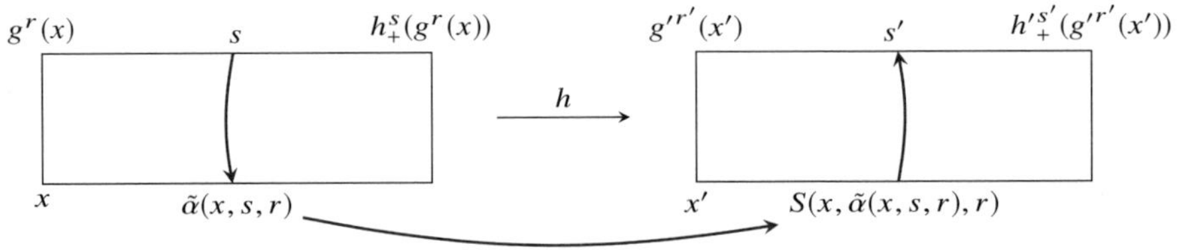

Figure 9.2.1. Smoothness of the conjugacy along unstable leaves.

Figure 9.2.1 suggests how to build the desired composition: each map is C^1 in

$$(s,r) \mapsto \tilde\alpha(x,s,r) \mapsto S(x,\tilde\alpha(x,s,r),r) \mapsto \alpha'(x', S(x,\tilde\alpha(x,s,r),r),r) = s'.$$

The same argument with T shows that $(r,s) \mapsto T(g^r(x),t)$ is C^1. This is useful, but it remains to show that $(r,s,t) \mapsto T(h_+^s(g^r(x)),t)$ is C^1, which uses a similar tactic. The function τ from (9.2.1) is C^1, as is its counterpart τ' as well as the inverse

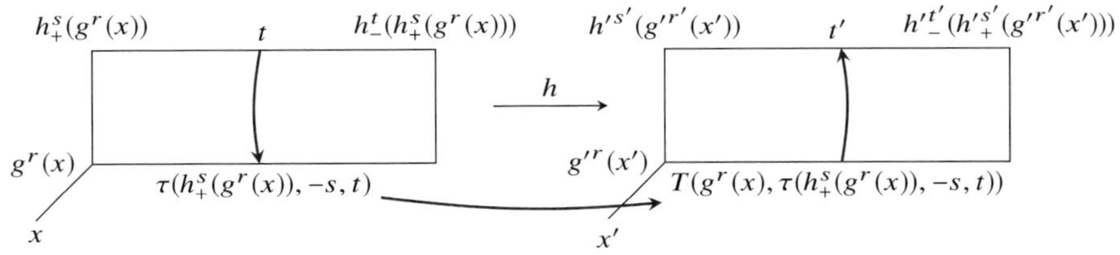

Figure 9.2.2. Smoothness of the conjugacy along stable leaves.

reparametrization $\tilde{\tau}$ of $\tau(x,s,\cdot)$ given by $\tilde{\tau}(x,s,\tau(x,s,t)) = t$ and its counterpart $\tilde{\tau}'$. Taking now our cue from the (r,s,t)-patch in Figure 9.2.2, we find that

$$(r,s,t) \mapsto t' = \tilde{\tau}'(h'^{s'}_+(g'^r(x')), -s, T(g^r(x), \tau(h_+^s(g^r(x)), -s, t)), t)$$

is C^1 by the chain rule because we noted already that $(r,s) \mapsto T(g^r(x), t)$ is C^1. □

Theorem 9.2.7. *A C^1-conjugacy between C^k Anosov 3-flows is $C^{k-\epsilon}$.*

Remark 9.2.8. A stronger result actually holds, with the same proof: a conjugacy between C^k Anosov 3-flows is $C^{k-\epsilon}$ if it is continuous and absolutely continuous.

Proof outline [246, Theorem 1.1]. A C^1-conjugacy preserves the unstable Jacobian and hence sends SRB-measure to SRB-measure, so it respects the densities of SRB-measure on unstable leaves and on stable leaves. Locally, along a leaf we thus have

$$\int_0^s \rho_0(s')\,ds' = \int_{h(0)}^{h(s)} \rho_{h(0)}(s')\,ds',$$

which upon differentiation with respect to s becomes a differential equation whose coefficient functions are C^k by Theorem 7.4.18, and so h is (uniformly) C^k when restricted to any leaf. To conclude smoothness in the (say) unstable direction without restriction, note that when pieces of unstable manifolds converge to a limiting piece, the restrictions of h to these are a C^k-bounded sequence, hence have a convergent subsequence in the $C^{k-\epsilon}$-topology; since h is continuous, this is a proper limit rather than just that of a subsequence. Thus, h is $C^{k-\epsilon}$ in the stable, unstable, and flow directions. This implies the claim by Theorem 9.2.9. □

Theorem 9.2.9 ([203]). *If f is uniformly $C^{k+\alpha}$ along the leaves of two continuous transverse foliations with smooth leaves, then f is $C^{k+\alpha}$.*

Remark 9.2.10. Remark 9.2.2 noted that Theorem 9.2.1 fixes geometric data rigidly. Theorem 9.2.7 (together with Theorem 9.2.1) goes further in that a topological conjugacy is necessarily essentially as smooth as the flows, so they are completely indistinguishable as smooth dynamical systems.

We close with an instance of *geometric rigidity* as opposed to smooth rigidity.

Theorem 9.2.11 ([159, Théorème 7.3]). *Suppose the Gauss curvature K of a closed Riemannian surface S satisfies $-k_2^2 \leq K \leq -k_1^2 < 0$ and a magnetic flow (Definition* 5.2.18) *with $\|\mathfrak{m}\|_\infty^2 + \|\nabla\mathfrak{m}\|_\infty < k_1$ is topologically conjugate to the geodesic flow of a negatively curved metric on S with the same total area. If the conjugacy is absolutely continuous, then $\mathfrak{m} \equiv 0$ and the metrics are isometric.*

9.3 Entropy and Lyapunov exponents

Next we turn to entropy rigidity. In the Variational Principle for entropy it is natural to expect that the inequality is "generically" strict. This is trivially the case in hyperbolic flows if one considers varying the invariant measure, because the measure of maximal entropy is unique, so the inequality is always strict for other measures. A more interesting question arises for hyperbolic flows preserving a smooth measure. In that case one may ask whether that smooth invariant measure has maximal entropy, or equivalently, one can ask whether the Bowen–Margulis measure is a smooth measure, that is, absolutely continuous with respect to an ambient volume. Even without having exhibited the Bowen–Margulis measure in many cases, and without having computed the entropy of many dynamical systems, one would expect this coincidence to be rare because it seems to impose rather special symmetries. If the Bowen measure is absolutely continuous, then periodic orbits are equidistributed with respect to volume, that is, geometrically. At the same time, absolute continuity of the Margulis measure implies strong homothetic behavior along the invariant foliations. We present a result of this very type that led to an important conjecture along these lines.

Theorem 9.3.1 (Katok Entropy Rigidity [211]). *If the Bowen–Margulis measure for the geodesic flow of a negatively curved metric on a compact surface is absolutely continuous, then the curvature is constant.*

We outline a proof of Theorem 9.3.2 below, which implies this. (See also Theorem 9.3.11.)

For a negatively curved C^∞ Riemannian metric g on a surface M, there is a unique C^∞-function $\rho\colon M \to (0,\infty)$ such that $g_0 := \rho g$ has constant curvature and the same volume as g, that is, $\int \rho \, d\mu_g = 1$, where μ_g is the Riemannian measure on (M,g), which is the projection of the Liouville measure λ_g on the g-unit tangent bundle S_gM of M. The Cauchy–Schwarz inequality implies that the *conformal coefficient*

$$\rho_g := \int \sqrt{\rho}\, d\mu_g \le 1$$

with equality if and only if ρ, and hence the curvature of g, is constant.

Theorem 9.3.2 ([211]). *If g is a negatively curved C^∞ Riemannian metric on a surface, then*

$$h^g_{\mathrm{vol}}/\rho_g \le h^{g_0}_{\mathrm{top}} \le \rho_g h^g_{\mathrm{top}},$$

where h^g_{vol} is the Liouville entropy of the geodesic flow of g, and h_{top} is the topological entropy of the respective geodesic flow. (Either inequality is strict unless g has constant curvature.)

This implies Theorem 9.3.1: if $h^g_{\mathrm{vol}} = h^g_{\mathrm{top}}$, then $\rho_g^2 \ge 1$, hence, as noted above, $\rho \equiv$ const., and g is constantly curved.

We note that in Theorem 9.3.2 the existence of ρ and the left inequality do not use the negative-curvature assumption and hold for any smooth metric g. The right inequality holds when g has no focal points.

Proof outline for Theorem 9.3.2 [212]. We first use that the Liouville measure for g_0 is the Bowen measure, that is, closed geodesics are equidistributed. Specifically, given a continuous function $f\colon M \to \mathbb{R}$ and $\epsilon > 0$ there is a $T > 0$ such that for $(1-\epsilon)$% of the geodesics of length at most T, the average along each of f is within ϵ of the space average $\int_{S_{g_0}M} f\, d\lambda_{g_0}$. We restate and apply this with $f = \rho^{-1/2}$: for $\epsilon > 0$ there is a $T > 0$ and a set E of cardinality at least $(1-\epsilon)P_T(g_0)$ of closed g_0-geodesics of length at most T such that the average of $\rho^{-1/2}$ along each is within ϵ of the λ_{g_0}-average

$$\int_{S_{g_0}M} \rho^{-1/2}\, d\lambda_{g_0} = \int_M \rho^{-1/2}\, d\mu_{g_0} = \int_M \sqrt{\rho}/\rho\, d\mu_{g_0} = \int_M \sqrt{\rho}\, d\mu_g = \rho_g$$

of ρ. Here, $P_T(g_0)$ is the number of closed g_0-geodesics of length at most T (compare with Definition 4.2.1).

This helps because the g-length of any $\gamma_0 \in E$ is its g_0-length times the average of $\rho^{-1/2}$ along γ_0 and hence less than $T(\rho_g + \epsilon)$. Also, each such γ_0 represents a unique free homotopy class, which in turn contains a unique closed g-geodesic which, being the g-shortest curve in that class has g-length less than the g-length of γ_0 and hence less than $T(\rho_g + \epsilon)$. This shows that

$$P_{T(\rho_g+\epsilon)}(g) \ge (1-\epsilon)P_T(g_0),$$

with $P_\bullet(g)$ defined analogously to $P_\bullet(g_0)$ above.

Since these are pairwise separated (due to expansivity, but here simply because they are in pairwise distinct homotopy classes), this determines entropy:

$$h_{\text{top}}^{g} \geq \varlimsup_{T\to\infty} \frac{\log P_{T(\rho_g+\epsilon)}(g)}{T(\rho_g+\epsilon)} \geq \varlimsup_{T\to\infty} \frac{\log(1-\epsilon)P_T(g_0)}{T(\rho_g+\epsilon)} = \frac{h_{\text{top}}^{g_0}}{\rho_g+\epsilon}.$$

Since ϵ was arbitrary, we obtain $h_{\text{top}}^{g_0} \leq \rho_g h_{\text{top}}^{g}$.

To prove the other inequality we imitate this argument with g and g_0 interchanged, and in order to make the Liouville entropy appear, we use its orbit-counting definition (Definition 4.1.2). □

The 1982 paper containing Theorem 9.3.2 [211] includes computations of the Liouville and topological entropies for the geodesics of all compact negatively curved locally symmetric spaces, which shows that in all these cases, both entropies coincide. Katok then remarked, "It looks like a reasonable conjecture that those are the only cases of manifolds of negative curvature for which the Liouville measure has maximal entropy."

Conjecture 9.3.3 (Katok Entropy Conjecture). *If the topological and Liouville entropies of the geodesic flow of a negatively curved manifold coincide, then the manifold is a locally symmetric space.*

While we do not seem at this time to be much closer to settling it than decades ago, this conjecture engendered a great amount of mathematical progress (including Theorem 5.4.27), and some of the major results established in the interim are well worth stating. Theorem 9.3.2 suggests a slightly different line of inquiry. As we noted parenthetically, both inequalities in Theorem 9.3.2 are strict unless the curvature is constant, and this gives rise to questions in arbitrary dimension: With normalized total volume, is the Liouville entropy uniquely maximized by locally symmetric metrics? It is not—there are same-volume perturbations of compact locally symmetric spaces of negative curvature with larger Liouville entropy [135]. However, the inequality and rigidity for topological entropy does indeed generalize, and this is a special case of the following major result.

Theorem 9.3.4 (Besson–Courtois–Gallot [47, 48]). *Minimizers (among homotopy-equivalent negatively curved Riemannian manifolds) of topological entropy with prescribed volume are locally symmetric spaces.*

Remark 9.3.5. Besson–Courtois–Gallot conjectured (correctly) that one can broaden this to Finsler manifolds [53], and it holds for Hilbert geometries [2].

Another consequence of their main result is worth stating. (Yet another is Mostow rigidity, which is outside the scope of this book.)

Theorem 9.3.6. *A Riemannian manifold of dimension at least* 3 *whose geodesic flow is* C^1*-conjugate to that of a compact locally symmetric manifold* M *is isometric to* M.

This is not just remarkable in itself but a booster of other rigidity results whose conclusion is the smooth conjugacy, and which are hereby amplified to yield isometry. Here is an instance (with the amplification built in):

Theorem 9.3.7 (Foulon–Labourie [138]). *A compact negatively curved Riemannian manifold whose horospheres have constant mean curvature is locally symmetric.*

In fact, together with a result of Ledrappier, this leads to a rigidity statement that has a rather similar flavor to the Katok Entropy Rigidity Conjecture in that the coincidence of the Bowen–Margulis measure with another preferred measure implies rigidity:

Theorem 9.3.8 (Ledrappier–Foulon–Labourie–Besson–Courtois–Gallot [234]). *A compact orientable Riemannian manifold whose Bowen–Margulis measure and harmonic measure coincide is locally symmetric.*

We do not define harmonic measure here but point out that it is connected with the asymptotics of Brownian motion on the manifold. Ledrappier showed that the coincidence of these measures implies asymptotic harmonicity, that is, that horospheres have constant mean curvature, from which the Foulon–Labourie result deduces smooth conjugacy to the geodesic flow of a locally symmetric space, and Theorem 9.3.6 then produces an isometry.

Another natural measure class at infinity is the Lebesgue or *visibility measure* defined by projecting the Lebesgue measure on any unit tangent sphere S_xM to the sphere at infinity along geodesics from x. A counterpart to Ledrappier's result gives the following in exactly the same manner:

Theorem 9.3.9 (Yue–Foulon–Labourie–Besson–Courtois–Gallot [354]). *A compact orientable Riemannian manifold whose visibility measure and harmonic measure coincide is locally symmetric.*

In light of this, the Katok Entropy Conjecture could be restated as follows:

Conjecture 9.3.10. *A compact orientable Riemannian manifold is locally symmetric if and only if the visibility, harmonic and Bowen–Margulis measure are not pairwise distinct, that is, if and only if two of the three coincide.*

There is an altogether different direction into which one could widen the purview of Theorem 9.3.1: instead of looking for analogs in higher dimension, one can ask whether Anosov flows for which the Bowen–Margulis measure is absolutely continuous have to be smoothly conjugate to the geodesic flow of a surface of constant curvature. Since the conclusion implies that the flow is a contact flow, the following is a natural answer to the question in dimension 3.

Theorem 9.3.11 (Foulon [136]). *A C^∞ contact Anosov flow on a closed 3-manifold whose measure-theoretic entropy is equal to its topological entropy, is, up to finite covers, C^∞-conjugate to the geodesic flow of a closed surface of constant negative curvature.*

It should be noted here that even topological conjugacy is a highly nontrivial assertion: the entropy assumption here determines even the *topological* nature of the flow, whereas in the preceding results, this was explicitly given in advance.

The entropy of a hyperbolic flow is intimately related to the contraction and expansion rates in the flow (Remark 7.4.15). Accordingly, we now indicate ways in which information about these gives rise to rigidity phenomena. A first instance is that Theorem 9.2.7 (C^1-conjugacies are smooth) is a corollary of a stronger result: instead of assuming that the conjugacy is C^1, it suffices to assume the coincidence of contraction and expansion rates: if two transitive Anosov flows on a compact 3-manifold are topologically conjugate and the contraction and expansion rates of corresponding periodic orbits agree, then the conjugating homeomorphism is analytic [247, Theorem 1.5.], [246, Theorem 1.1]. One can go much further and obtain rigidity from periodic data without first assuming a topological (time-preserving) conjugacy. To present a related result about orbit equivalence, define the *Lyapunov spectrum* of a periodic orbit as follows.

Definition 9.3.12 (Lyapunov spectrum). If Φ is a flow on an n-dimensional manifold and $\varphi^T(p) = p$, then let $\vec{\lambda}(p)$ be the n-vector whose entries are $\frac{1}{T}\log|\lambda_i|$ in increasing order, where the λ_i are the eigenvalues of $D\varphi^T{}_{\restriction p}$ repeated according to their respective multiplicities.[5]

For the geodesic flow on a locally symmetric space (X, g_0), this is independent of p and hence denoted by $\vec{\lambda}(X, g_0)$.

Example 9.3.13. A volume-preserving Anosov 3-flow has $\vec{\lambda}(p) = \xi(p)(-1, 0, 1)$ for some $\xi\colon \operatorname{Per}(\Phi) \to (0, \infty)$ with $\xi \equiv 1$ for the geodesic flow of any hyperbolic surface. This shows in particular that Theorem 9.3.15 fails for geodesic flows on surfaces—and for the same reason Theorem 9.3.6 and results like it assume dimension at least 3.

[5]These are the Lyapunov exponents of p.

Theorem 9.3.14 (Dynamical Lyapunov spectrum rigidity [84, 85]). *Let (X, g_0) be a closed negatively curved locally symmetric space with* $\dim X \geq 3$ *and* Φ *a* $C^{r+\epsilon}$ *Anosov flow on a Riemannian manifold* M *with* $r \geq 4$, $\epsilon > 0$ *such that for each* $p \in \mathrm{Per}(\Phi)$ *there is a* $\xi(p) > 0$ *with* $\vec{\lambda}(p) = \xi(p)\vec{\lambda}(X, g_0)$. *Then any orbit equivalence between* Φ *and the geodesic flow of* (X, g_0) *is flow related (Definition* 1.3.22) *to a* $C^{r-3+\epsilon}$ *orbit equivalence.*[6] *Moreover, there is a smooth conjugacy if and only if* ξ *is constant.*[7]

The result has a geometric implications. Geodesic flows of locally symmetric spaces (Section 2.5) provide the most astonishing instance.

Theorem 9.3.15 (Geometric Lyapunov spectrum rigidity [84, 85]). *If* Y *is a closed negatively curved Riemannian manifold homotopic to a closed negatively curved locally symmetric space* (X, g_0) *with* $\dim X \geq 3$, *and if for each periodic point* p *of the geodesic flow for* Y *there is a* $\xi(p) > 0$ *such that* $\vec{\lambda}(p) = \xi(p)\vec{\lambda}(X, g_0)$, *then* Y *is isometric to* (X, cg_0) *for some constant* $c > 0$.

Remark 9.3.16. A posteriori $\xi(\cdot)$ is constant, but this is not assumed. If Y is itself a locally symmetric space, then this is the Mostow Rigidity Theorem [48].

9.4 Optimal regularity of the invariant subbundles

Theorem 5.1.16 establishes Hölder continuity of the invariant subbundles in great generality, and in Proposition 9.4.3 we will give the optimal Hölder exponent for the hyperbolic case. That is, we now more closely study the regularity of the stable and unstable subbundles of hyperbolic flows in terms of local "spectral" control, namely bounds on ratios involving contraction and expansion rates of the flow (Definition 9.4.1). Theorem 9.4.10 provides such a regularity statement for the invariant subbundles (see also Theorem 9.5.3), and Theorem 9.6.2 shows that it is sharp. We conclude this section with easy corollaries for codimension-1 flows (Corollaries 9.4.11 and 9.4.12) as well as geodesic flows (Corollary 9.4.14). We also give a relatively easy instance of smooth rigidity in Theorem 9.7.1 and comment on related subjects in Remark 9.7.9.

We begin with a notion that previously appeared as a way to preclude resonances when we studied normal forms.

Definition 9.4.1 (Bunching). A compact hyperbolic set Λ for a flow Φ on a Riemannian manifold M with invariant splitting admits $C, \epsilon > 0$ such that for all $p \in \Lambda$ there exist

[6] The loss of three derivatives is not optimal.

[7] This last assertion is not stated in [84] but can be obtained from the preceding and Livshitz theory.

$\mu_f < \mu_s < 1 - \epsilon < 1 + \epsilon < \nu_s < \nu_f$ as follows: if $v \in E^s(p)$, $u \in E^u(p)$ and $t > 0$, then

$$\frac{1}{C}\mu_f^t\|v\| \le \|D\varphi^t(v)\| \le C\mu_s^t\|v\| \quad \text{and} \quad \frac{1}{C}\nu_f^{-t}\|u\| \le \|D\varphi^{-t}(u)\| \le C\nu_s^{-t}\|u\|.$$

We say that $p \in \Lambda$ is *α-bunched* if $\mu_s(p)\nu_s(p)^{-1} \le \big(\min(\mu_f(p), \nu_f(p)^{-1})\big)^\alpha$, and φ^t is *α-bunched* if $\sup_{p\in\Lambda} \mu_s\nu_s^{-1}\big(\min(\mu_f, \nu_f^{-1})\big)^{-\alpha} \le 1$. The *unstable bunching constant* is $B^u(\Phi) := \inf_{p\in\Lambda}(\log\mu_s - \log\nu_s)/\log\mu_f$.

Remark 9.4.2. Note that we chose the $\mu_f, \mu_s, \nu_s, \nu_f$ to depend on p, and the "min" above are taken over two numbers, they do *not* range over Λ. If μ_f and μ_s are chosen optimally they essentially give the maximal and minimal multipliers for the strong-stable subbundle and likewise ν_f and ν_s give the maximal and minimal expansion rates for the unstable subbundle. Remark 9.4.20 below suggests a further refinement.

For *symplectic* Anosov flows, that is, flows on odd-dimensional manifolds with a flow-invariant 2-form that is nondegenerate on transversals to the flow, the *invariant symplectic form* for the flow) the notion of bunching is clearer since $\mu_i = \nu_i^{-1}$ $(i = f, s)$ and hence α-bunching implies $\mu_s^{2/\alpha} \le \mu_f \le \mu_s$, which bunches the "spectrum" of contracting multipliers increasingly as $\alpha \to 2$. The main examples of symplectic Anosov flows are geodesic flows of negatively curved manifolds. The Riemannian structure gives a canonical contact form which renders these flows symplectic. (Alternatively symplecticity is a consequence of the Hamiltonian structure of free particle motion on a manifold.) Then a-pinching of the sectional curvature of a Riemannian manifold (Definition 9.4.13) implies $2\sqrt{a}$-bunching of the geodesic flow. This is related to the fact that the rate obtained in the proof of Lemma 5.2.5 is the square root of the curvature bound above (5.2.2).

Proposition 9.4.3. *$E^u \in C^{B^u(\Phi)}$ if $B^u(\Phi) \in (0, 1)$, and if $B^u(\Phi) \ge 1$, then E^u has modulus of continuity $O(x|\log x|)$ (see Remark 3.2.18).*

Remark 9.4.4. It is the same for E^s and $B^s(\Phi) := \inf_{p\in\Lambda} \frac{\log\nu_s - \log\mu_s}{\log\nu_f}$ (the *stable bunching constant*), and for both simultaneously with $B(\Phi) := \inf_{p\in\Lambda} \frac{\log\nu_s - \log\mu_s}{\max(\log\nu_f, -\log\mu_f)}$ (the *bunching constant*). Moreover, we later obtain differentiability with Hölder-continuous derivatives if $B^u(\Phi) > 1$ in Theorem 9.4.10. (See also Theorem 9.5.3.)

The proof is a careful application of the Hadamard graph-transform method.

Proof. First we introduce adapted coordinates.[8] For $p \in \Lambda$ take a hypersurface $\mathcal{T}_p$ transverse to $\dot\varphi$ of uniform size depending C^∞ on p. For each p let $W^u := W^u(p) \cap \mathcal{T}_p$, $W^s := W^s(p) \cap \mathcal{T}_p$, $E^u := TW^u$, and $E^s := TW^s$. Take coordinates $\Xi\colon \Lambda \times [-\epsilon, \epsilon]^{k+l} \to$

[8]The construction here is quite simple; Lemma 9.5.5 further adapts to an invariant area, and finer adaptations are possible.

M such that $\Xi_p\colon [-\epsilon,\epsilon]^{k+l} \xrightarrow{C^\infty} \mathcal{T}_p$ is continuous in p, $[-\epsilon,\epsilon]^k \times \{0\} \to W^u$, $\{0\} \times [-\epsilon,\epsilon]^l \to W^s$. Write the coordinates as (x,y) with $\Xi_p(x,0) \in W^u$ and $\Xi_p(0,y) \in W^s$.

We now projectivize the action on graphs. Denote the induced (return or Poincaré) map by $\phi^t\colon \mathcal{T}_p \to \mathcal{T}_{\varphi^t p}$ and represent $E^u(0,y)$ as the image of a linear map $\left(\begin{smallmatrix} I \\ D \end{smallmatrix}\right)\colon \mathbb{R}^k \to \mathbb{R}^{k+l}$. To obtain the desired regularity we give estimates along orbits in the stable manifold of a reference orbit. These estimates are uniform (in both reference orbit and the orbit at hand) and hence establish the desired regularity uniformly along stable manifolds. Smoothness in the unstable direction then shows that this is, in fact, the actual degree of regularity.

Lemma 9.4.5 (Action in coordinates). *Given $p \in \Lambda$, $q \sim (0,y) \in W^s$, $(0,y_t) := \phi^t(0,y)$, there exist $C > 0$ and $C_t > 0$ such that $D\phi^t = \left(\begin{smallmatrix} A_t & 0 \\ B_t & C_t \end{smallmatrix}\right)$ with $\|A_t^{-1}\| < C\nu_s(q)^{-t}$, $\|C_t\| < C\mu_s(q)^t$, $\|C_t^{-1}\| < C\mu_f(q)^{-t}$, $\|B_t\| < C_t\|y\|$, $C\|y_t\| \geq \mu_f(q)^t\|y\|$.*

Proof. $\|A_t^{-1}\| < \nu_s(q,t)^{-1}$ in coordinates centered at q. Up to a distortion factor, uniformly bounded independently of t, the linear part of the coordinate change is of the form $\left(\begin{smallmatrix} I & O \\ D & I \end{smallmatrix}\right)$, so that up to a bounded factor the representations A_t^{-1} agree in both systems, as do the ones for C_t and C_t^{-1}. We have $\|B_t\| < C_t\|y\|$ since φ^t is a diffeomorphism with B_t differentiable and vanishing at the origin of the coordinate system. For the remaining claim it is slightly easier and by boundedness of coordinate changes clearly sufficient to show $\|y\| \leq C\mu_f(p,t)^{-1}\|\phi^t(y)\|$. To this end let $\gamma_t\colon [0,1] \to \mathcal{T}_{\varphi^t p}$ be a geodesic with $\gamma_t(0) = \phi^t(p)$, $\gamma_t(1) = \phi^t(q)$, where $q \sim (0,y)$. By the Inclination Lemma (Theorem B.6.1) $\phi^{-t}\gamma_t$ converges to a smooth curve $c(\cdot) \subset \mathcal{T}_p$. If $\lim_{n\to\infty} \|\phi^t(y)\|/\mu_f(p,t)\|y\| = 0$ then by the Intermediate-Value Theorem this holds for all $c(s)$, $s \in (0,1]$. Compactness of Λ controls higher derivatives and yields uniformity in s, so $\lim_{n\to\infty} \|D\phi^t(v)\|\|v\|/\mu_f(p,t) = 0$ for $v = \dot{c}(0)$, contrary to the choice of μ_f. □

The projectivized action is that on D as follows. Let $z_t = (0,y_t)$. Then

$$\begin{pmatrix} I \\ D(z_n) \end{pmatrix} \leftrightarrow \begin{pmatrix} I \\ D(z_{n-1}) \end{pmatrix} \begin{pmatrix} A_1(z_{n-1}) & 0 \\ B_1(z_{n-1}) & C_1(z_{n-1}) \end{pmatrix} = \begin{pmatrix} A_1(z_{n-1}) \\ B_1(z_{n-1}) + C_1(z_{n-1})D(z_n) \end{pmatrix},$$

where "$\leftrightarrow$" indicates that the maps on either side have the same image. We renormalize the right-hand side in such a way that the upper block is the identity:

$$\begin{aligned}\begin{pmatrix} A_1(z_{n-1}) \\ B_1(z_{n-1}) + C_1(z_{n-1})D(z_n) \end{pmatrix} A_1^{-1}(z_{n-1}) &= \begin{pmatrix} I \\ (B_1(z_{n-1}) + C_1(z_{n-1})D(z_n))A_1^{-1}(z_{n-1}) \end{pmatrix} \\ &=: \begin{pmatrix} I \\ D(z_n) \end{pmatrix}.\end{aligned}$$

Inductively, this gives

$$\begin{aligned} D(z_n) &= [B_1(z_{n-1}) + C_1(z_{n-1})D(z_{n-1})]A_1^{-1}(z_{n-1}) \\ &= \sum_{i=0}^{n-1} C_i(z_{n-i})B_1(z_{n-i-1})A_{i+1}^{-1}(z_{n-i-1}) + C_n(z_0)D(z_0)A_n^{-1}(z_0). \end{aligned}$$

So, if Φ is α-bunched, then $\mu_s^i \nu_s^{-i} \le \mu_f^{\alpha i} = O(\|z_n\|/\|z_{n-i}\|)^\alpha$, and

$$\frac{\|D(z_n)\|}{\|z_n\|^\alpha} = O\Big(\sum_{i=0}^{n-1} \mu_s^i \nu_s^{-i} \frac{\|z_{n-i-1}\|}{\|z_n\|^\alpha}\Big) + O\Big(\frac{\mu_s^n \nu_s^{-n}}{\|z_n\|^\alpha}\Big) = O\Big(1 + \sum_{i=0}^{n-1} \|z_{n-i-1}\|^{1-\alpha}\Big).$$

If $\alpha < 1$ then the right-hand side is uniformly bounded, so the unstable subbundle is α-Hölder along W^s. If $\alpha = 1$ then the right-hand side is uniformly $O(n)$, so the unstable subbundle has modulus of continuity $O(x|\log x|)$ along W^s. Since the unstable subbundle is (uniformly) smooth along W^u it is as regular as along the stable foliation (Proposition 6.4.1 below). □

Remark 9.4.6. If Φ is α-bunched for some $\alpha > 1$, then the preceding estimates imply that the unstable subbundle is Lipschitz continuous.

As we just suggested, we can obtain stronger conclusions with stronger bunching. First among these is differentiability:

Proposition 9.4.7. *If* $\sup_{p\in M} \mu_s \nu_s^{-1} \mu_f^{-1} < 1$ *in Definition* 9.4.1 *then* $E^u\restriction_{W^s}$ *is differentiable.*

This follows from Lemma 9.4.9 below once we show that differentiability can be proved via similar estimates as before:

Claim 9.4.8. $f\colon \mathbb{R}^n \to \mathbb{R}^n$ *is differentiable if for* $v_i \in \mathbb{R}^n$ *with* $v_1 + v_2 + v_3 = 0$,

$$\frac{\big|h_2h_3f(x+v_1h_1)+h_1h_3f(x+v_2h_2)+h_1h_2f(x+v_3h_3) - (h_2h_3+h_1h_3+h_1h_2)f(x)\big|}{h_1h_2h_3} \to 0$$

uniformly as $(h_1, h_2, h_3) \to 0$ *(we write "*$o(h_1h_2h_3)$*"[9]). For a function on a manifold replace the arguments* $(x + v_ih_i)$ *by* $c_i(h_i)$*, where* c_i *is a curve with* $\dot c_i(0) = v_i$.

Proof. Setting $v_3 = 0$ shows existence of directional derivatives; then this condition shows that these depend linearly on the direction. Uniformity guarantees that this is sufficient. □

[9] See Remark 3.2.18.

In adapted coordinates consider triples of vector fields on M as follows:

$$\mathcal{B} := \big\{(v_1, v_2, v_3) \text{ with } v_i(p) \in E^s(p) \cap T\mathcal{T}_p,\ v_1 + v_2 + v_3 = 0, \\ \text{and } \|v_1\| + \|v_2\| + \|v_3\| = 1\big\}.$$

If

$$\tilde{v}_i(p) = \frac{D\phi^t v_i(\varphi^{-t} p)}{\xi_t(v_1(\varphi^{-t}p), v_2(\varphi^{-t}p), v_3(\varphi^{-t}p))}$$

with $\xi_t \in \mathbb{R}$ such that $(\tilde{v}_1, \tilde{v}_2, \tilde{v}_3) \in \mathcal{B}$ then φ^t acts on $\mathcal{B}$ by $\mathcal{P}_t(v_1, v_2, v_3) := (\tilde{v}_1, \tilde{v}_2, \tilde{v}_3)$. The sufficient condition in Claim 9.4.8 and hence Proposition 9.4.7 follows from the next lemma:

Lemma 9.4.9. *For $K > 0$ there are $T, \epsilon > 0$ such that if $\gamma_{v_i(p)}$ denote geodesics in $\mathcal{T}_p$ with $\dot{\gamma}_{v_i(p)}(0) = v_i(p)$ and for $p \in M$, $(v_1, v_2, v_3) \in \mathcal{B}$, $0 < h_1, h_2, h_3 < \epsilon$,*

$$\|h_2h_3D(\gamma_{v_1(p)}(h_1)) + h_1h_3D(\gamma_{v_2(p)}(h_2)) + h_1h_2D(\gamma_{v_3(p)}(h_3))\| < Kh_1h_2h_3,$$

then $\forall\, p \in M$, $t \in [T, 2T]$, $(\tilde{v}_1, \tilde{v}_2, \tilde{v}_3) \in \mathcal{P}_t\mathcal{B}$, $0 < h_1, h_2, h_3 < \epsilon$ and with $\tilde{h}_i := \xi_t h_i$,

$$\|\tilde{h}_2\tilde{h}_3D(\gamma_{\tilde{v}_1(p)}(\tilde{h}_1)) + \tilde{h}_1\tilde{h}_3D(\gamma_{\tilde{v}_2(p)}(\tilde{h}_2)) + \tilde{h}_1\tilde{h}_2D(\gamma_{\tilde{v}_3(p)}(\tilde{h}_3))\| < \tfrac{3}{4}K\tilde{h}_1\tilde{h}_2\tilde{h}_3.$$

Proof. For T such that $(\sup_{p\in M} \mu_s\nu_s\mu_f^{-1})^T < L^{-3}/3$ consider $\phi^t(\gamma_{v_i(\varphi^{-t}p)}(h_i))$ instead of $\gamma_{\tilde{v}_i(p)}(\xi_t h_i)$.[10] Let $\Gamma_i := \gamma_{v_i(\varphi^{-t}p)}(h_i)$, $A_i := A_t(\Gamma_i)^{-1}$, $B_i := B_t(\Gamma_i)$, $C_i := C_t(\Gamma_i)$, $h_iD_i := h_1h_2h_3D(\Gamma_i)$, $\tilde{h}_i\tilde{D}_i := \tilde{h}_1\tilde{h}_2\tilde{h}_3D(\phi^t(\Gamma_i)) = \tilde{h}_i\xi_t^2C_iD_iA_i + \tilde{h}_1\tilde{h}_2\tilde{h}_3B_iA_i$. Then

$$\begin{aligned}
\|\tilde{D}_1 + \tilde{D}_2 + \tilde{D}_3\| &\le \xi_t^2\|C_1D_1A_1 + C_2D_2A_2 + C_3D_3A_3\| \\
&\quad + \|\tilde{h}_2\tilde{h}_3B_1A_1 + \tilde{h}_1\tilde{h}_3B_2A_2 + \tilde{h}_1\tilde{h}_2B_3A_3\| \\
&\le \overbrace{\xi_t^2 \underbrace{\|C_1\|}_{\le L\mu_s^t} \cdot \underbrace{\|D_1 + D_2 + D_3\|}_{\le Kh_1h_2h_3 = K\xi_t^{-3}\tilde{h}_1\tilde{h}_2\tilde{h}_3} \cdot \underbrace{\|A_1\|}_{\le L\nu_s^{-t}}}^{L^2\mu_s^t\nu_s^{-t}\xi_t^{-1}K\tilde{h}_1\tilde{h}_2\tilde{h}_3} \\
&\quad + \xi_t^2 \underbrace{\|C_2 - C_1\|}_{\in o(h_1h_2h_3)}\|D_2\|\|A_1\| + \xi_t^2\|C_2\|\|D_2\|\underbrace{\|A_2 - A_1\|}_{\in o(h_1h_2h_3)} \\
&\quad + \xi_t^2 \underbrace{\|C_3 - C_1\|}_{\in o(h_1h_2h_3)}\|D_3\|\|A_1\| + \xi_t^2\|C_3\|\|D_3\|\underbrace{\|A_3 - A_1\|}_{\in o(h_1h_2h_3)} \\
&\quad + \xi_t^2 \underbrace{\|h_2h_3B_1A_1 + h_1h_3B_2A_2 + h_1h_2B_3A_3\|}_{\in o(h_1h_2h_3)}.
\end{aligned}$$

We have $\xi_t^{-1} \le L\mu_f^{-t}$ after possibly adjusting L, so for $t > T$ the first term is at most $\frac{1}{3}K\tilde{h}_1\tilde{h}_2\tilde{h}_3$. Since $\|D_i\| \in O(h_1h_2h_3)$ by Remark 9.4.6, the other terms are

[10] These are tangent as $h_i \to 0$, so the error is $o(\tilde{h}_1\tilde{h}_2\tilde{h}_3)$ in the notation from Remark 3.2.18.

$o(h_1h_2h_3)$=$o(\tilde h_1\tilde h_2\tilde h_3)$ uniformly for $t \in [T,2T]$. Take $\epsilon > 0$ such that their sum is at most $\frac{1}{3}K\tilde h_1\tilde h_2\tilde h_3$ for $t \in [T,2T]$. □

As mentioned previously, once the bunching constant exceeds 1, the derivatives are Hölder continuous with the Hölder exponent that one would naturally expect.

Theorem 9.4.10. *$E^u \in C^{B^u(\Phi)}$ if $B^u(\Phi) \notin \mathbb{N}$, and $E^u \in C^{B^u(\Phi)-1+x|\log x|}$ if $B^u(\Phi) \in \mathbb{N}$.*

Proof. We limit ourselves to $B^u(\Phi) \le 2$ since this is as high as one can get for stable and unstable bunching simultaneously, and because beyond this point the reader could adapt the arguments given here if needed. Thus, Propositions 9.4.3 and 9.4.7 leave us to produce the right Hölder exponent for the derivative when $B^u(\Phi) > 1$.

Then suppose that Φ is α-bunched with $\alpha > 1$. Then D is C^1 by Proposition 9.4.7, and we write $D(y) = {}^LD(y) + {}^ND(y)$ with ${}^LD(y)$ linear in y, ${}^ND(y)$ differentiable in y, $D_y\,{}^ND(0) = 0$, and $A_t^{-1}(y) = A_t^{-1}(0) + {}^LA_t^{-1}(y) + {}^NA_t^{-1}(y)$, etc.,

$$N(z) := B_1(z)A_1^{-1}(z) - {}^LB_1(z)A_1^{-1}(0) + C_1(z)\,{}^LD(z)A_1^{-1}(z) - C_1(0)\,{}^LD(z)A_1^{-1}(0) \\ \in O(\|z\|).$$

Note that $\|D_zN(z)\| \in O(\|z\|)$. As in the proof of Proposition 9.4.3 we inductively get

$$\begin{aligned} {}^ND(z_n) &= N(z_{n-1}) + C_1(z_{n-1})\,{}^ND(z_{n-1})A_1^{-1}(z_{n-1}) \\ &= \sum_{i=0}^{n-1} C_i(z_{n-i})N(z_{n-i-1})A_i^{-1}(z_{n-i-1}) + C_n(z_0)\,{}^ND(z_0)A_n^{-1}(z_0) \end{aligned}$$

and

$$\begin{aligned} D_{z_n}\,{}^ND(z_n) = \sum_{i=0}^{n-1} \big[&(D_{z_n}C_i(z_{n-i}))N(z_{n-i-1})A_i^{-1}(z_{n-i-1}) \\ &+ C_i(z_{n-i})(D_{z_n}N(z_{n-i-1}))A_i^{-1}(z_{n-i-1}) \\ &+ C_i(z_{n-i})N(z_{n-i-1})(D_{z_n}A_i^{-1}(z_{n-i-1}))\big] \\ &+ D_{z_n}\big[C_n(z_0)\,{}^ND(z_0)A_n^{-1}(z_0)\big]. \end{aligned}$$

By the product and chain rules,

$$\begin{aligned} D_{z_n}C_i(z_{n-i}) &= D_{z_n}\sum_{j=1}^{i} C_1(z_{n-j}) \\ &= \sum_{j=1}^{i}\Big(\prod_{k=1}^{j-1} C_1(z_{n-k})\Big)D_{z_n}C_1(z_{n-j})\Big(\prod_{k=j+1}^{i} C_1(z_{n-k})\Big) \\ &= \sum_{j=1}^{i} C_{j-1}(z_{n-j+1})[D_{z_{n-j}}C_1(z_{n-j})C_j^{-1}(z_{n-j})]C_{i-j}(z_{n-i}), \end{aligned}$$

so

$$\|D_{z_n} C_i(z_{n-i})\| = O\Big(\sum_{j=1}^{i} \mu_s^{i-1}\mu_f^{-j}\Big) = O(\mu_s^{i-1}\mu_f^{-i}).$$

Likewise $\|D_{z_n} A_i^{-1}(z_{n-i})\| \in O(\nu_s^{1-i}\mu_f^{-i})$ and similarly

$$\|D_{z_n} N(z_{n-i-1})\| \in O(\|z_{n-i-1}\|)\|C_{i+1}^{-1}(z_{n-i-1})\| = O(\|z_{n-i-1}\|\mu_f^{-i}).$$

Since we assume $\mu_s^i\nu_s^{-i} \le \mu_f^{\alpha i} = O(\|z_n\|/\|z_{n-i}\|)^\alpha$,

$$\begin{aligned}
&\|D_{z_n}[C_n(z_0)\,{}^N D(z_0)A_n^{-1}(z_0)]\| \\
&\quad = \big\|[D_{z_n}C_n(z_0)]\,{}^N D(z_0)A_n^{-1}(z_0) + C_n(z_0)[D_{z_n}\,{}^N D(z_0)]A_n^{-1}(z_0) \\
&\qquad\qquad + C_n(z_0)\,{}^N D(z_0)D_{z_n}A_n^{-1}(z_0)\big\| \\
&\quad = \underbrace{O(\mu_s^{n-1}\mu_f^{-n}\nu_s^{-n}) + O(\mu_s^n\mu_f^{-n}\nu_s^{-n}) + O(\mu_s^n\nu_s^{1-n}\mu_f^{-n})}_{\in O(\mu_f^{(\alpha-1)n}) = O(\|z_n\|^{\alpha-1})}.
\end{aligned}$$

Therefore, again using $\mu_s^i\nu_s^{-i} \le \mu_f^{\alpha i} = O(\|z_n\|/\|z_{n-i}\|)^\alpha$,

$$\begin{aligned}
&\|D_{z_n}\,{}^N D(z_n)\| \\
&= O\Big(\sum_{i=0}^{n-1}\big[\mu_s^{i-1}\mu_f^{-i}\|z_{n-i}\|\nu_s^{-i} + \mu_s^i\|z_{n-i}\|\mu_f^{-i}\nu_s^{-i} + \mu_s^i\|z_{n-i}\|\nu_s^{1-i}\mu_f^{-i}\big]\Big) + O(\|z_n\|^{\alpha-1}) \\
&= O\Big(\sum_{i=0}^{n-1}\|z_{n-i}\|\Big(\frac{\|z_n\|}{\|z_{n-i}\|}\Big)^{\alpha-1}\Big) + O(\|z_n\|^{\alpha-1}) = O\Big(\Big[1 + \sum_{i=0}^{n-1}\|z_{n-i}\|^{2-\alpha}\Big]\|z_n\|^{\alpha-1}\Big).
\end{aligned}$$

If $\alpha < 2$ then the right-hand side is $O(\|z_n\|^{\alpha-1})$, hence the unstable subbundle has $(\alpha-1)$-Hölder derivative along W^s. If $\alpha = 2$ then the right-hand side is $O(n\|z_n\|)$, so the derivative of the unstable subbundle has modulus of continuity $O(x\,|\log x|)$ along W^s. It is easy to see that derivatives in the unstable direction have the same (and even higher) modulus of continuity along W^s. Being uniformly smooth along W^u the unstable subbundle has the claimed regularity (Proposition 6.4.1). □

Corollary 9.4.11. *If* codim $E^u = 1$, *then* $E^u \in C^{1+\inf_{p\in M}\frac{\log \nu_s}{-\log \mu_f}-\epsilon} \subset C^1$.

Proof. By assumption $\mu_f = \mu_s$. □

Corollary 9.4.12. *If φ preserves volume and* codim $E^u = 1$ *then the Anosov splitting is* $C^{1+\inf_{p\in M}\frac{\log \nu_s}{-\log \mu_f}-\epsilon}$. *In particular it is* $C^{1+\epsilon}$.

Proof. Since $\mu_s\nu_f \le 1$ we have $\frac{\log\nu_s - \log\mu_s}{\log\nu_f} \ge \frac{\log\nu_s}{\log\nu_f} + 1 \ge 1 + \frac{\log\nu_s}{-\log\mu_s}$. Use Theorem 9.4.10 and Corollary 9.4.11. □

We now briefly explore bunching hypotheses (Definition 9.4.1) in a geometric context.

Definition 9.4.13. The sectional curvature of a compact negatively curved Riemannian manifold N is *relatively a-pinched* if $C \le$ sectional curvature $< aC$ for some $C\colon N \to -\mathbb{R}_+$. If C is constant, the curvature is said to be (absolutely) a-pinched.

The a-pinching of the sectional curvature of a Riemannian manifold implies $2\sqrt{a}$-bunching of the geodesic flow; this is closely related to the way in which the curvature bound k^2 used in Lemma 5.2.5 yields an expansion rate $2k$ in that argument. *Horospheric foliations* of a geodesic flow are defined to be the foliations of the unit tangent bundle of the manifold by the stable and unstable leaves of the geodesic flow. Thus we have the following corollary:

Corollary 9.4.14. *For $a \in [0,1)$, a-pinched negatively curved manifolds have $C^{2\sqrt{a}}$ horospheric foliations.*

Relative pinching implies bunching and hence regularity as well, but less than absolute pinching.

Theorem 9.4.15 ([146, Theorem 1.6]). *For $0 \le b \le a \le 1$, a compact b-pinched relatively a-pinched Riemannian manifold has $C^{a-b+\sqrt{(a+b)^2+4(1-a)b}}$ horospheric foliations.*

Remark 9.4.16. With

$$a - b + \underbrace{\sqrt{(a+b)^2 + 4(1-a)b}}_{=\sqrt{(a-b)^2+4b}} \begin{cases} \ge 2a & (\text{“}>\text{” when } 0 < b \le a < 1), \\ = 2\sqrt{a} & \text{if } a = b, \end{cases}$$

so the degree of regularity interpolates between powers of a. Specifically,

$$a - b + \sqrt{(a-b)^2 + 4b} \ge 2a^{1/(b+1)}$$

as follows: $\frac{d}{dx}\frac{x\log x}{x-1} = 1 - \frac{\log x}{(x-1)^2} > 1$ on $(0,1)$, and $\frac{x\log x}{x-1} \xrightarrow[x\to1]{} 1$, so $\frac{x\log x}{x-1} < 1$ on $[0,1)$ and hence $\frac{d}{db}\frac{x^2}{1-x}(1-x^b) = x^{b+1}\frac{x\log x}{x-1} < 1 = \frac{d}{db}b$ on $[0,1)$. Thus, $b \ge \frac{x^2}{1-x}(1-x^b)$ on $[0,1)$ since we have equality for $b = 0$. Therefore, $b(1-x) \ge x^2(1-x^b)$, and $b \ge x(b - x^{b+1}) + x^2$. With $x = a^{1/(b+1)}$, this gives

$$\begin{aligned}(a-b)^2 + 4b \ge (a-b)^2 + 4x(b-a) + 4x^2 &= [2x - (a-b)]^2 \\ &= [2a^{1/(b+1)} - (a-b)]^2,\end{aligned}$$

hence $\sqrt{(a+b)^2+4(1-a)b} = \sqrt{(a-b)^2+4b} \geq 2a^{1/(b+1)} - (a-b)$, as claimed. Therefore we have the following corollary:

Corollary 9.4.17. *For $0 \leq b \leq a \leq 1$, a compact b-pinched relatively a-pinched Riemannian manifold has $C^{2a^{1/(b+1)}}$ horospheric foliations.*

This (or Theorem 9.4.15) also gives the following "relative" counterpart of Corollary 9.4.14.

Proposition 9.4.18 ([172]). *For $a \in [0,1)$, relatively a-pinched negatively curved manifolds have C^{2a} horospheric foliations.*

Our focus has been exclusively on the regularity of the subbundles, but a closely related important issue is the regularity of holonomies or, equivalently, of the Bowen bracket (Proposition 6.2.2). While Lipschitz and higher regularity of the subbundles implies like regularity for the holonomies by standard results from the theory of ordinary differential equations, this is not so with regularity below Lipschitz continuity, even though the provable regularity has turned out to be the same for either category. (This arises in connection with the second part of Theorem 9.6.12 below.) We reproduce a statement about the holonomies.

Theorem 9.4.19 ([325][11]). *If Λ is a basic set for Φ and λ, μ as in Definition 5.1.1, then*

$$\underbrace{\|D\varphi^T \restriction_{E^{uu}(p)}\|^{-1}}_{=:\mu_T(p)=\text{strong contraction of }\varphi^{-T}} \leq \underbrace{\|D\varphi^{-T} \restriction_{E^{uu}(p)}\|}_{=:\lambda_T(p)=\text{weak contraction of }\varphi^{-T}}$$
$$\leq \mu^{-T} < 1 < \lambda^{-T} \leq \underbrace{\|D\varphi^T \restriction_{E^{ss}(p)}\|^{-1}}_{=:\eta_T(p)=\text{weak expansion of }\varphi^{-T}},$$

and for any

$$\alpha < \min\left(1, \overline{\lim_{T\to\infty}} \lim_{n\to\infty} \inf_{p\in\Lambda} \frac{\sum_{i=0}^n \log\lambda_T(\varphi^{-iT}(p)) - \log\eta_T(\varphi^{-iT}(p))}{\sum_{i=0}^n \log\mu_T(\varphi^{-iT}(p))}\right), \tag{9.4.1}$$

there is a $C > 0$ such that if $x, y \in \Lambda$ and $d(x,y) \leq \epsilon$ (as in Proposition 6.2.2), then

$$d(y,[y,x]) \leq C(d(x,[x,y]))^\alpha,$$

that is, moving between strong-stable leaves along weak-unstable ones is α-Hölder.

Remark 9.4.20. The expression in (9.4.1) suggests, of course, that the notion of bunching (Definition 9.4.1) could be slightly refined along similar lines. Like the

[11] Proved there for discrete time.

results about regularity of subbundles (Theorem 9.6.1), this one is provably optimal [324], but unlike the failure of regularity on large sets one sees in Anosov systems (Theorem 9.6.12), for hyperbolic Cantor sets this seems to be confined to sets of smaller Hausdorff dimension than the basic set itself or even stable or unstable slices [182, 183]. This is a surprising difference between Anosov systems and hyperbolic Cantor sets that has yet to be fully explored.

9.5 Longitudinal regularity

We next develop two new lines of inquiry: one is the regularity of the *strong* subbundles, which one expects to be similar but different (see, for example, Proposition 9.5.6), and the other is how far one can push up the regularity by bunching conditions. It is illuminating and relatively easy to do both at the same time for volume-preserving Anosov flows of 3-manifolds. Here, the combination of volume preservation and 1-dimensional strong subbundles produces the maximum bunching constraint (there is only one contraction and expansion rate each and they are reciprocals), and in a way that produces higher regularity than the counterpart to Theorem 9.4.10. However, one rate drops out in graph-transform arguments for strong subbundles because the complementary direction includes the flow direction. Accordingly, we expect the dimension assumption to yield a slight improvement over modulus of continuity $x|\log x|$.

Definition 9.5.1. A continuous function $f\colon U \to \mathbb{R}$ on an open set $U \subset \mathbb{R}$ is said to be *Zygmund regular* if there is $Z > 0$ such that $|f(x+h) + f(x-h) - 2f(x)| \le Z|h|$ for all $x \in U$ and sufficiently small h. To specify a value of Z we may refer to a function as being Z-Zygmund. The function is said to be "little Zygmund" if $|f(x+h) + f(x-h) - 2f(x)| \in o(|h|)$.[12]

Remark 9.5.2. Zygmund regularity implies modulus of continuity $O(|x\log|x||)$ and hence H-Hölder continuity for all $H < 1$. It follows from Lipschitz continuity and hence from differentiability. Being little Zygmund implies having modulus of continuity $o(|x\log|x||)$.

From Proposition 9.5.7 below we obtain the following theorem:

Theorem 9.5.3 (Longitudinal Zygmund regularity). *Let $k \ge 2$, and Φ be a C^k volume-preserving Anosov flow on a 3-manifold. Then $E^u \oplus E^s$ is Zygmund regular.*

Remark 9.5.4. The transverse counterpart (Theorem 9.7.12) involves regularity that is one degree higher.

[12] See Remark 3.2.18.

Our first ingredient is adapted coordinates that also incorporate the invariant volume and thereby require a finer argument than at the start of the proof of Proposition 9.4.3. (Lemma 9.6.4 goes further yet.)

Lemma 9.5.5. *There exist local coordinates adapted to the invariant foliations, that is, coordinate systems* $\mathfrak{C}\colon M \times (-\epsilon,\epsilon)^3 \to M$ *such that* $\mathfrak{C}_p := \mathfrak{C}(p,\cdot)$ *satisfies*

(1) $\mathfrak{C}_p$ *is* C^k *for every* $p \in M$,

(2) $\mathfrak{C}_p$ *depends (Hölder-) continuously on* p,

(3) $\mathfrak{C}_p$ *preserves volume for each* $p \in M$,

(4) $\mathfrak{C}_p(0) = p$,

(5) $\mathfrak{C}_p((-\epsilon,\epsilon)\times\{0\}\times\{0\}) = W^u_{\mathrm{loc}}(p) \cap \mathfrak{C}_p((-\epsilon,\epsilon)^3)$,

(6) $\mathfrak{C}_p^{-1}(\varphi^\delta(\mathfrak{C}_p(u,t,s))) = (u, t+\delta, s)$ *for* $|\delta| < \epsilon$,

(7) $\mathfrak{C}_p(\{0\}\times\{0\}\times(-\epsilon,\epsilon)) = W^s_{\mathrm{loc}}(p) \cap \mathfrak{C}_p((-\epsilon,\epsilon)^3)$.

Proof. Since W^u and W^s are continuous, there is a continuous family $p \mapsto \mathcal{T}_p$ of local C^k transversals, each containing $W^u_{\mathrm{loc}}(p) \cup W^s_{\mathrm{loc}}(p)$ (the family can be taken Hölder continuous once it is known that $p \mapsto W^i_{\mathrm{loc}}(p)$ are Hölder continuous). We produce local coordinates in each $U := \mathcal{T}_p$ that are adapted to area and to $W^u_{\mathrm{loc}}(p) \cup W^s_{\mathrm{loc}}(p)$; building flow boxes on these then gives the desired local coordinates.

Take $p\colon U \to \mathbb{R}$ such that $p \equiv 0$ on W^s and P defined by $dp = \alpha(P,\cdot)$ is transverse to W^u; here α is the area form on U. If P^t is the Hamiltonian flow generated by P, then $\forall\, z \in U\ \exists!\, y \in W^u,\ q(z) \in \mathbb{R}\colon z = P_1^{q(z)}(y)$, so $\{q_1, p_1\} = 1$ (Poisson bracket), $q\restriction_{W^u} = 0$, and p, q are the desired coordinates. □

Since the subbundles are invariant under the flow, the coordinate representation of the flow preserves the axes of the local coordinate system as well as volume. Note that this flow produces maps between different local coordinate patches.

The differential of φ^T at points of the stable leaf (the third coordinate axis, or s-axis) therefore takes the following form:

$$D\varphi^T(0,0,s) = \begin{pmatrix} \alpha^{-1} & 0 & 0 \\ b_1 & 1 & 0 \\ b_2 & 0 & \alpha \end{pmatrix},$$

where $b_i(s) \in O(s)$ and $|\alpha(s)| < 1$. As a warm-up and because we need it later, we put this to use for proving that the strong subbundles are Hölder continuous.

Proposition 9.5.6. *In the context of Theorem* 9.5.3, *for each* $H \in (0,1)$ *there is a* $Z > 0$ *such that the graph transform preserves the space of subbundles that are* H*-Hölder along* E^s *with constant* Z *in local coordinates.*

Thus the graph transform preserves Hölder continuity. More specifically, applying the graph transform to a subbundle that is H-Hölder with sufficiently large constant Z in local coordinates gives a subbundle with the same property (for the same Z and H). This holds for any $H < 1$. This implies immediately that E^u is H-Hölder for any $H < 1$, because it shows that the unique fixed point of the graph transform lies in the space of H-Hölder subbundles.

Proof of Proposition 9.5.6. In adapted coordinates a subbundle transverse to $E^c \oplus E^s$ is represented by graphs of linear maps from E^u to $E^c \oplus E^s$. Using the canonical representation of the tangent bundle of $\mathbb{R}^3$ we can write any subspace transverse to the t–s plane as the *image* of a linear map given by a column matrix $\begin{pmatrix} 1 \\ e \\ \overline{e} \end{pmatrix}$, with $\begin{pmatrix} 1 \\ e(s) \\ \overline{e}(s) \end{pmatrix}$, accordingly, the local matrices giving the restriction of such a subbundle to stable leaves. The advantage of this representation is that applying the derivative amounts to simple composition. The image (in the coordinates at $\varphi^T(p)$) of the subspace is the represented as the image of the linear map with matrix

$$\begin{pmatrix} \alpha^{-1} & 0 & 0 \\ b_1 & 1 & 0 \\ b_2 & 0 & \alpha \end{pmatrix} \begin{pmatrix} 1 \\ e \\ \overline{e} \end{pmatrix} = \begin{pmatrix} \alpha^{-1} \\ b_1 + e \\ b_2 + \overline{e}\alpha \end{pmatrix},$$

which is also the image of the linear map with matrix

$$\begin{pmatrix} 1 \\ \alpha(s)b_1(s) + \alpha(s)e(s) \\ \alpha(s)b_2(s) + \overline{e}(s)\alpha^2(s) \end{pmatrix} =: \begin{pmatrix} 1 \\ e(s_\varphi) \\ \overline{e}(s_\varphi) \end{pmatrix}, \tag{9.5.1}$$

where $\varphi^T(0,0,s) =: (0,0,s_\varphi)$ in local coordinates and $|\alpha(s)s| = |s_\varphi + O(s^2)| = |s_\varphi|(1 + O(s))$ by the C^2 assumption.

To prove Proposition 9.5.6 assume $|e(s)| \le Z|s|^H$ with uniform Z and H. Then

$$\begin{aligned} |e(s_\varphi)| &= |\alpha(s)||b_1(s) + e(s)| \\ &\le |\alpha(s)|\big[O(s) + Z|s|^H\big] \\ &= |\alpha(s)s|^H\big[Z|\alpha|^{1-H} + |\alpha(s)|^{1-H}O(s^{1-H})\big] \\ &= |s_\varphi|^H(1 + O(s))\big[Z|\alpha|^{1-H} + |\alpha(s)|^{1-H}O(s^{1-H})\big] \le Z|s_\varphi|^H \end{aligned} \tag{9.5.2}$$

for sufficiently large Z. Likewise, $|\overline{e}(s)| \leq Z|s|^H$ with uniform Z and $H \in (0, 1]$ gives

$$\begin{aligned}|\overline{e}(s_\varphi)| &= |\alpha(s)b_2(s) + \overline{e}(s)\alpha^2(s)| \\ &= |\alpha(s)s|^H \left[|\alpha(s)|^{1-H} O(|s|^{1-H}) + Z|\alpha(s)|^{2-H}\right] \\ &= |s_\varphi|^H (1 + O(s))\left[|\alpha(s)|^{1-H} O(|s|^{1-H}) + Z|\alpha(s)|^{2-H}\right] \leq Z|s_\varphi|^H\end{aligned}$$

for sufficiently large Z and sufficiently small s. □

We return to this argument later.

To prove Zygmund regularity of the strong-unstable subbundle we vary the strategy slightly. Instead of showing that the space of Z-Zygmund subbundles is preserved for sufficiently large Z, we show that for each Z there is a $Z' > 0$ such that repeated application of the graph transform to a Z-Zygmund subbundle always gives Z'-Zygmund subbundles. Since the space of Z'-Zygmund subbundles is closed, this proves that the unique fixed point E^u is Z'-Zygmund.

Proposition 9.5.7. *For each $Z > 0$ there is a $Z' > 0$ such that all forward images under the graph transform of the space of Z-Zygmund subbundles transverse to $E^s \oplus E^c$ lie in the space of Z'-Zygmund subbundles.*

Proof. Consider a subbundle that is represented in local coordinates as $\begin{pmatrix} 1 \\ e \\ \overline{e} \end{pmatrix}$ with $|e(s)+e(-s)| \leq Z|s|$ and $|\overline{e}(s)+\overline{e}(-s)| \leq Z|s|$ for all s in any local coordinate system.

Then $|e(s_\varphi) + e(-(s_\varphi))| \leq |e(s_\varphi) + e((-s)_\varphi)| + |e((-s)_\varphi) + e(-(s_\varphi))|$, where the last term is $O(s^{2H})$ for any $H < 1$ by Proposition 9.5.6 because $|s_\varphi + (-s)_\varphi| = \|\varphi(0,0,s) + \varphi(0,0,-s)\| \in O(s^2)$ since φ is C^2.

The other term is estimated as follows:

$$\begin{aligned}|e(s_\varphi) + e((-s)_\varphi)| &= |\alpha(s)b_1(s) + \alpha(-s)b_1(-s) + \alpha(s)e(s) + \alpha(-s)e(-s)| \\ &\leq |\alpha(s)||b_1(s) + b_1(-s)| + |b_1(-s)||\alpha(s) - \alpha(-s)| \\ &\quad + |\alpha(s)||e(s) + e(-s)| + |e(-s)||\alpha(s) - \alpha(-s)| \\ &\leq |\alpha(s)|O(s^2) + O(s)O(s) + |\alpha(s)|Z|s| + O(s^H)O(s) \\ &\leq Z(1 + O(s^H))|\alpha(s)s| \\ &= Z(1 + O(s^H))(1 + o(s))|s_\varphi| \\ &= Z(1 + \kappa(s))|s_\varphi|,\end{aligned}$$

with $\kappa(s) \in O(s^H)$ decreasing in Z.

Note that $Z' := Z\prod_{i=0}^{\infty}(1 + \kappa(s_{\varphi^i})) < \infty$ and all images of $\begin{pmatrix} 1 \\ e \\ \overline{e} \end{pmatrix}$ under the graph transform are Z'-Zygmund in local coordinates. □

The regularity in Theorem 9.5.3 is sharp: it is easy to see that differentiability of the strong stable or unstable foliation cannot be expected.

Proposition 9.5.8. *There is an obstruction to differentiability of the strong-unstable subbundle.*

Remark 9.5.9. This C^1-obstruction is actually an obstruction to slightly lower regularity (Proposition 9.7.5).

Proof. If p is a T-periodic point, then differentiating (9.5.1) at 0 gives $e'(0)\alpha(0) = \alpha(0)b_1'(0) + \alpha(0)e'(0)$, and hence $K(p,T) := b_1'(0) = 0$. □

This simple observation is the germ of a nice instance of *smooth rigidity*, which we demonstrate in Theorem 9.7.1: if this obstruction vanishes, then we get much higher smoothness.

9.6 Sharpness for transversely symplectic flows, threading

The preceding explorations drew structural conclusions from exceeding the maximum provable regularity of the invariant foliations. To complement this picture, we now return to the context of Proposition 9.4.3, where one can show that the Hölder exponent of the invariant subbundles resulting from bunching data is optimal, and in a way that produces (in symplectic contexts) a positive-codimension constraint for exceeding this regularity. Even though that constraint has not been exploited to yield structural information, it provides a strong way of establishing sharpness of Proposition 9.4.3:

Theorem 9.6.1 (Sharpness of Hölder exponent). *For an open dense set of symplectic Anosov flows the regularity of the invariant subbundles is exactly that in Theorem* 9.4.10.

Proof. We carry out the proof for the context of Proposition 9.4.3. The necessary closed condition (9.6.1) for excessive regularity (Theorem 9.6.2) means that the invariant foliations have to "thread a needle" when the regularity is higher than asserted in Proposition 9.4.3 and fails for an open set of symplectic Anosov flows. Density of this set follows from Proposition 9.6.6 which finds a point where the perturbation in Proposition 9.6.10 to break the threading condition (9.6.1) can be implemented (the "dethreading"). □

The first step to establishing sharpness is to exhibit a rare circumstance that is necessary for excessive regularity.

Theorem 9.6.2 (Threading). *If $\alpha \in (0,1)$ and a transversely symplectic flow Φ has C^α unstable subbundle E^u but is not α-bunched, then E^u satisfies the positive-codimension threading condition* (9.6.1) (*on page* 568).

Remark 9.6.3. This means that if the functions μ_f, μ_s, ν_f, ν_s are chosen optimally, then the predicted regularity is as good as it can be. At periodic points the optimal choice is to take $\mu_f(p)$, $\mu_s(p)$, $\nu_f(p)$, $\nu_s(p)$ to be moduli of eigenvalues of the differential $D\Phi_p$ of the Poincaré map of a local section.

The threading condition (9.6.1) consists of an identity which implies that part of the unstable subbundle (which is by construction determined by the "past") is determined by the "future" of the orbit. We will later use that these can be modified independently to "break" this relation.

We will use that periodic α-bunching implies α-bunching [173][13] by assuming that Φ has a periodic point p with $\mu_s(p)\nu_s(p)^{-1} > (\mu_f(p))^\alpha$ (this is the reverse of the bunching inequality in Definition 9.4.1), where $\mu_f(p)$, $\mu_s(p)$, $\nu_s(p)$ are as in Definition 9.4.1 and also moduli of eigenvalues of the differential $D\Phi_p$ of the Poincaré map of a local section.

Transverse symplecticity implies $\mu_s(p) = \nu_s(p)^{-1}$, so $\mu_s(p)\nu_s(p)^{-1} > (\mu_f(p))^\alpha$ becomes $\nu_s(p)^{-2} > (\mu_f(p))^\alpha$.

Before embarking on the proof, we produce a higher-dimensional symplectic counterpart to the adapted coordinates in Lemma 9.5.5.

Lemma 9.6.4. *Let $F\colon U \longrightarrow (\mathbb{R}^{2n}, \omega)$ symplectic, $F(0) = 0 \in U \subset \mathbb{R}^{2n}$ hyperbolic, $0 \in W_1^j \subset W_2^j \subset \cdots \subset W_n^j = W^j(0)$ submanifolds, $\dim W_i^j = i$, $j = u, s$. Then there exist coordinates $(p_1, \dots, p_n, q_1, \dots, q_n)$ on U such that*

(1) $\omega = \sum dp_i \wedge dq_i$ ($\sim \begin{pmatrix} 0 & I \\ -I & 0 \end{pmatrix}$, *where I is the (n,n)-identity*),

(2) $W_i^s = \{(0, \dots, 0, 0, \dots, 0, q_{n-i+1}, \dots, q_n) \in U\}$,

(3) $W_i^u = \{(0, \dots, 0, p_{n-i+1}, \dots, p_n, 0, \dots, 0) \in U\}$.

Proof. Let $M_i \supset W_i^s \cup W_i^u$ be a $2i$-dimensional submanifold of U, $p_1\colon U \longrightarrow \mathbb{R}$ such that $p_1\restriction_{W^s \cup M_{n-1}} = 0$, and P_1 defined by $dp_1 = \omega(P_1, \cdot)$ is transverse to a hypersurface $N_n \supset W_n^u \cup M_{n-1}$. If P_1^t is the Hamiltonian flow generated by P_1 then $\forall z \in U\ \exists!\, y \in N_n$, $q_1(z) \in \mathbb{R}$: $z = P_1^{q_1(z)}(y)$. Thus $\{q_1, p_1\} = 1$, $q_1\restriction_{N_n} = 0$, and $M_{n-1} = \{x \in U \mid p_1(x) = q_1(x) = 0\}$ since dp_1 and dq_1 are independent. If $n = 1$ this ends the proof, otherwise the $(2n-2)$-manifold $\left(M_{n-1}, \omega\restriction_{M_{n-1}}\right)$ is symplectic because its tangent spaces are the skew complement of the span of P_1 and Q_1 (defined

[13] [173] invokes the proof of a lemma from elsewhere that does not work as intended, but the implication can also be obtained from [205, Theorem 1.4] or [193, Proposition 3.5].

by $dq_1 = \omega(Q_1, \cdot))$,[14] hence has adapted coordinates $\{p_i, q_i\}_{i=2}^n$. Note $\{p_i, q_i\}_{i=1}^n$ are adapted: $\forall\, z \in U\ \exists!\ y \in M_{n-1},\ s,t \in \mathbb{R}$: $z = P_1^s(Q_1^t(y))$, where Q_1^t is the Hamiltonian flow for q_1; (2), (3) hold by construction, (1) follows from standard arguments [16, §43E]. □

Proof of Theorem 9.6.2. In adapted coordinates on $\mathcal{T}_p$, E^u is the graph of $D \in \mathrm{gl}(n, \mathbb{R})$ or the image of $\left(\begin{smallmatrix} I \\ D \end{smallmatrix}\right)\colon \mathbb{R}^n \simeq \mathbb{R}^n \times \{(0,0)\} \longrightarrow \mathbb{R}^{2n}$. At $q = (0, \dots, 0, z) =: z_0$ on the fast stable leaf through 0 the differential of Φ is $D\Phi_{|z_0} = \left(\begin{smallmatrix} A & 0 \\ B & C \end{smallmatrix}\right)$ with $C = A^{t^{-1}}$ lower block triangular. To bring $D\Phi\left(\begin{smallmatrix} I \\ D \end{smallmatrix}\right) = \left(\begin{smallmatrix} A & 0 \\ B & C \end{smallmatrix}\right)\left(\begin{smallmatrix} I \\ D \end{smallmatrix}\right) = \left(\begin{smallmatrix} A \\ B+CD \end{smallmatrix}\right)$ into the form $\left(\begin{smallmatrix} I \\ \widetilde{D} \end{smallmatrix}\right)$, right multiply with A^{-1} and use symplecticity ($A^{-1} = C^t$):

$$\begin{pmatrix} I \\ \widetilde{D} \end{pmatrix} = \Phi^* \begin{pmatrix} I \\ D \end{pmatrix} = \begin{pmatrix} A \\ B + CD \end{pmatrix} A^{-1} = \begin{pmatrix} I \\ (B + CD)C^t \end{pmatrix}.$$

Since E^u is invariant, $D(\Phi(z_0)) = (B + CD(z_0))C^t$. Denote the upper-left k-blocks of A, B, C, and D by a, b, c, and d, respectively, where k is such that $\|(c(0))^{-n}\| \le C\nu_s^n$. Since C is lower block triangular, $\widetilde{d} = c \cdot d \cdot c^t + b \cdot c^t$ or

$$d(\Phi(z_0)) = c(z_0) \cdot d(z_0) \cdot c^t(z_0) + b(z_0) \cdot c^t(z_0).$$

With d we have isolated the component of D on which the flow acts with the slowest expansion and contraction. Symplecticity allows us to "decouple" this block from the others and obtain this recursion relation.

If $z_n := \Phi^n(z_0)$, $\xi_{z_0}^n := \prod_{i=0}^{n-1} c(z_{n-i-1})$, $\Delta_{z_0}^n := -\sum_{i=0}^{n-1} (\xi_{z_0}^{i+1})^{-1} b(z_i) (\xi_{z_0}^i)^{t^{-1}}$, then

$$d(z_n) = \xi_{z_0}^n \cdot (d(z_0) - \Delta_{z_0}^n) \cdot \xi_{z_0}^{n\,t},$$

so $d(\cdot) \in C^\alpha \Rightarrow \|d(z_0) - \Delta_{z_0}^n\| \le \xi_{z_0}^{n\,-1} \cdot \|d(z_n)\| \cdot \xi_{z_0}^{n\,t^{-1}} \le C_2 \nu_s^{2n} \nu_f^{-\alpha n} \longrightarrow 0$ since $\nu_s^2 < \nu_f^\alpha$ and $\|(c(0))^{-n}\| \le C\nu_s^n$. Thus

$$d(z) = \Delta_z^\infty. \tag{9.6.1}$$

□

Remark 9.6.5. Both this obstruction and that in Proposition 9.5.8 have positive codimension, but they are of a different nature. The latter is expressible in terms of derivatives of the flow, whereas (9.6.1) arises from the geometry of the invariant subbundles. Nonetheless, being of positive codimension, both imply that "typically" Anosov flows do not have more regular subbundles than the regularity theorems assert.

[14] For $v \in T_x M_{n-1}$ we have $0 = dp_1(v) = \omega(P_1, v)$ and $0 = dq_1(v) = \omega(Q_1, v)$.

What we have obtained with the threading condition (9.6.1) is the following. If $\binom{I}{D}$ represents E^u then $d(z)$, like E^u, is determined by the past of the orbit of p. Equation (9.6.1) shows that if excessively regular, then d is simultaneously determined by the future of the orbit of p, namely by Δ_z^∞. We shall see that the past and future can be modified independently, so this is a special situation.

Accordingly, we seek a point where the past and future can be disentangled—which will allow a perturbation to cause the threading condition (9.6.1) to fail.

Proposition 9.6.6. *Let M be a Riemannian manifold, Φ an Anosov flow, $p \in M$ periodic. Then the fast stable leaf of p contains a negatively nonrecurrent point.*

Remark 9.6.7. Density of nonrecurrent points on the stable leaf is clear since any point heteroclinic to a different periodic point has this property and periodic points are dense. But the existence of heteroclinic points in the *fast* stable leaf is not as clear. The proof we give actually yields *density* of nonrecurrent points on the fast stable leaf.

Proof. Let C, $\chi > 0$ such that $\|D\varphi^t\restriction_{E^s}\| \le C \cdot e^{-\chi t}$ for $t > 0$, $\mathcal{T}$ a hypersurface through p transverse to the flow direction, $W^{fs}(p)$ the fast stable leaf of p for the return map Φ, $q \in W^{fs}(p) \setminus \{p\}$, and $B_1 = B_{\sqrt{\epsilon}}(q)$ where $\epsilon > 0$ is such that

$$T := \inf\left\{t > t_0 \mid \varphi^{-t} B_{\sqrt{\epsilon}}(q) \cap B_{\sqrt{\epsilon}}(q) \ne \varnothing\right\} > \chi^{-1} \log 4C,$$

and $t_0 := \inf\{t > 0 \mid \varphi^{-t} B_{\sqrt{\epsilon}}(q) \cap B_{\sqrt{\epsilon}}(q) = \varnothing\}$, where $B_\epsilon(q)$ is the ϵ-ball around q in M.

For $x \in \mathcal{T}$, $A \subset \mathcal{T}$ let $\operatorname{rad}_x(A) := \sup\{d(x, y)\colon\ y \in A\}$, where d is the distance on $\mathcal{T}$. Let $W^{fs}(p) := \mathbb{C}(B_1 \cap (\text{fast stable leaf of } p), p)$ (as in Definition 1.6.13) and $W^s(x) := \mathbb{C}(B_1 \cap (\text{stable leaf of } x), x)$ for $x \in B_1$. For ϵ small $B_2 := B_1 \cap \mathcal{T}$ has local product structure, that is, local stable and unstable leaves intersect in a point, so we introduce coordinates on $W^u(p)$ and $W^s(p)$ and represent $x \in B_2$ by (*coordinate of* $W^s(x) \cap W^u(p)$, *coordinate of* $W^u(x) \cap W^s(p)$).

Let D be an ϵ-ball in $\mathcal{T} \cap$ (weak-unstable leaf of q), $U_1 := \bigcup_{x \in D} B^s_\epsilon(x)$, $Q_1 := \bigcup_{x \in D} B^s_{2\epsilon}(x)$, $V_1 := \bigcup_{x \in D} \overline{B^s_{4\epsilon}(x)}$, and $W_1 := \bigcup_{x \in D} B^s_{5\epsilon}(x)$, where "$\overline{}$" denotes closure and $B^s_\epsilon(x)$ is the ϵ-ball in $W^s(x)$, $x \in B_1$. For $x \in B_1$ let $U_x := U_1 \cap W^s(x)$, $V_x := V_1 \cap W^s(x)$, $Q_x := Q_1 \cap W^s(x)$, $W_x := W_1 \cap W^s(x)$, and $S_x := V_x \setminus Q_x$. Thus we may view D as a little essentially horizontal segment and U_1, Q_1, V_1, W_1 then are essentially "rectangles" $D \times B_\epsilon$, $D \times B_{2\epsilon}$, $D \times B_{4\epsilon}$, $D \times B_{5\epsilon}$. The subscript x denotes a vertical slice through these. The S_x are spherical shells in $W^s(x)$.

It suffices to find a point in $U := U_1 \cap W^{fs}(p)$ that does not return to U_1 in negative time. Define $t\colon B_2 \to \mathbb{R}^+ \cup \{\infty\}$, $x \mapsto t(x) := \inf\{t > 0 \mid \varphi^{-t}x \in \overline{U}_1\}$ and take $x_0 \in U_2 := \overline{U}$ such that $t(x_0) = \min\{t(x)\colon x_0 \in U_2\} > T$. There is a smooth function $\tau\colon W_2 := W_{\varphi^{-t(x_0)}x_0} \to \mathbb{R}^+$ such that $\psi_1(x) := \varphi^{\tau(x)}x \in \mathcal{T}$ for $x \in W_2$ and $\tau(\varphi^{-t(x_0)}x_0) = t(x_0)$. Thus $\psi_1\colon W_2 \to \mathcal{T}$ is a diffeomorphism onto its image and $\psi_1(W_2) \subset W^s(p) = W^s(x_0)$. The intersection $U \cap S_1$ of the spherical shell $S_1 = \psi_1(S_{\varphi^{-t(x_0)}x_0}) \in W^s(p)$ with U consists of points not returning to U_1 for time t with $-T_1 := -\max\{t(x)\colon x \in V_2\} \le t < 0$, where $V_2 := V_{\varphi^{-t(x_0)}x_0}$.

Claim 9.6.8. *The intersection $U \cap S_1$ has a connected component U_3' such that $(\psi_1(W_2) \setminus S_1) \cup U_3'$ is connected.*

Proof. U is connected. Further, $\operatorname{rad}_{x_0}(U) \ge \epsilon$, and $\operatorname{rad}_{x_0}(S_1) < \epsilon$ by choice of T. So U contains points outside S_1. Since $x_0 \in \overline{U}$ is inside the shell S_1, so are some points of U. □

Take $U_3 \subset U_3'$ closed and connected such that $(\psi_1(W_2) \setminus S_1) \cup U_3$ is connected. Then $\psi_1^{-1}(U_3)$ connects the complement of $S_{\varphi^{-t(x_0)}x_0}$ in W_2 and is connected. Therefore, $\operatorname{rad}_x(\psi_1^{-1}(U_3)) \ge \epsilon$ for all $x \in \psi_1^{-1}(U_3)$. Let $t\colon W_2 \to \mathbb{R}^+ \cup \{\infty\}$, $x \mapsto t(x) := \inf\{t > 0 \mid \varphi^{-t}x \in \overline{U}_1\}$ and take $x_1 \in \psi_1^{-1}(U_3)$ such that $t(x_1) = \min\{t(x)\colon x \in \psi_1^{-1}(U_3)\} > T$. There is a smooth function $\tau\colon W_3 := W_{\varphi^{-t(x_1)}x_1} \to \mathbb{R}^+$ such that $\psi_2(x) := \varphi^{\tau(x)}x \in \mathcal{T}$ for $x \in W_3$ and $\tau(\varphi^{-t(x_1)}x_1) = t(x_1)$. Thus $\psi_2\colon W_3 \to \mathcal{T}$ is a diffeomorphism onto its image and $\psi(W_3) \subset W^s(x_1)$. The spherical shell $S_2 := \psi_2(S_{\varphi^{-t(x_1)}x_1}) \subset W^s(x_1)$ consists of points not returning to U_1 for time t with $-T_2 := -\max\{t(x)\colon x \in V_3\} \le t < 0$, where $V_3 := V_{\varphi^{-t(x_1)}x_1}$.

Claim 9.6.9. $\psi_1^{-1}(U_3) \cap S_2$ *has a connected component U_4' with $(\psi_2(W_3) \setminus S_2) \cup U_4'$ connected.*

Proof. As before, $\psi_1^{-1}(U_3)$ is connected and $\operatorname{rad}_{x_1}(\psi_1^{-1}(U_3)) \ge \epsilon$. □

No points of $U_4 := \psi_1(U_4') \subset U_3$ return to U_1 for time $t \in [-T_1 - T_2, 0) \supset [-2T, 0)$. Iterating this argument gives compact $U_1 \supset \cdots \supset U_{n+2} \supset \cdots$ with return times $\ge nT$. The nonempty intersection consists of negatively nonrecurrent points. □

Proposition 9.6.10 (Dethreading). *Failure of the threading condition* (9.6.1) (*on page* 568) *is dense among symplectic Anosov flows.*

Proof. Pick a negatively nonrecurrent (respectively nonreturning for geodesic flows) point q in the fast stable leaf of the periodic point p and inside the adapted coordinate patch. Translate the coordinates so that q is the origin. Construct a perturbation as follows: Let $\overline{\gamma}(s,t,u,v,w,z) = \frac{1}{2}\langle s,s\rangle = \frac{1}{2}\|s\|^2$ and $\gamma = \rho\overline{\gamma}$ where $\langle\cdot,\cdot\rangle$ is the standard inner product in $\mathbb{R}^k$ and $\rho \in C^\infty(B_\epsilon, \mathbb{R})$ is C^k-small and such that $\rho = \epsilon^{k+1}$ on B_{ϵ^2} and $\rho = 0$ on $B_\epsilon \setminus B_{\epsilon(1-\epsilon)}$. The vector field X with $\omega(X, \cdot) = d\gamma$ generates a complete symplectic flow G_τ.

The vector field $\overline{X}(s,t,u,v,w,z) = (0,0,0,s,0,0)^t$ satisfies $\omega(\overline{X}, \cdot) = d\overline{\gamma}$ and generates the flow $\overline{G}_\tau(s,t,u,v,w,z) = (s,t,u,v+\tau s,w,z)$ with

$$D\overline{G}_\tau = \begin{pmatrix} I & 0 \\ \begin{smallmatrix} \tau I & 0 & 0 \\ 0 & 0 & 0 \\ 0 & 0 & 0 \end{smallmatrix} & I \end{pmatrix} =: \begin{pmatrix} I & 0 \\ \tau E & I \end{pmatrix}.$$

Let $\eta \in C^\infty(\mathbb{R}, \mathbb{R}^+)$ be C^k-small such that $\eta(x) = 0$ $(x < 0)$, $\eta(x) = \epsilon^{k+1}$ $(x > \epsilon)$. Redefine $\varphi^\tau \simeq (s,x) \mapsto (s+\tau, x)$ on $B = [0,\epsilon] \times B_\epsilon$ so that $\overline{\varphi}^\tau(0,x) = (\tau, G_{\eta(\tau)}(x))$. Then, $\overline{\varphi}^\tau$ is a symplectic C^k-small perturbation of φ^τ. It causes (9.6.1) to fail:

$$D\overline{G}_\tau \begin{pmatrix} I \\ D \end{pmatrix} = \begin{pmatrix} I & 0 \\ \tau E & I \end{pmatrix} \begin{pmatrix} I \\ D \end{pmatrix} = \begin{pmatrix} I \\ \tau E + D \end{pmatrix},$$

and similarly for $DG_\tau = \left(\begin{smallmatrix} I & 0 \\ E & I \end{smallmatrix}\right)$. This ends the proof by Claim 9.6.11. □

Claim 9.6.11. *The sum $E + D$ is the unstable direction at q for the flow $\overline{\varphi}^\tau$.*

Proof. Take a subbundle $\mathcal{E}$ on $\{\varphi^t(q)\}_{t<0}$ close to E^u. Since q is nonrecurrent (respectively nonreturning for geodesic flows) $E^u(\varphi^{-\epsilon}q) = \lim_{t\to\infty} D\varphi^t(\mathcal{E}(\varphi^{-t-\epsilon}(q))) = \overline{E}^u(\varphi^{-\epsilon}q)$, and $\overline{\varphi}^t$-invariance of $\overline{E}^u$ gives the claim. □

The failure of regularity tends to be pervasive through the phase space. One way to make this explicit is the following.

Theorem 9.6.12 ([185, 184]). *For $\alpha \in (0,1)$ there is a nonempty C^1-open set of transversely symplectic C^∞ Anosov flows for which*

- *E^{cu} and E^{cs} are C^α only on a set whose complement is residual and has full measure for any fully supported ergodic invariant Borel probability measure, and*
- *the Bowen bracket* (*Proposition* 6.2.2) *is almost nowhere C^α.*

This sharpness result and the earlier ones suggest that systems with excessive regularity of the invariant subbundles are "special" in some sense, although here that merely means that they are rare in the sense of Baire category. However, there are no methods for "upgrading" this to rigidity results. This is different when cocycle obstructions vanish, as we will see soon.

9.7 Smooth invariant foliations

We begin with smooth rigidity by building on Proposition 9.5.8.

Theorem 9.7.1 (Smooth longitudinal rigidity). *Let M be a 3-manifold, $k \geq 2$, Φ a C^k volume-preserving Anosov flow. Then $E^u \oplus E^s$ is Zygmund regular, and there is an obstruction to higher regularity that can be described geometrically as the curvature of the image of a transversal under a return map. This obstruction defines the cohomology class of a cocycle (the longitudinal Katok–Anosov–Moser cocycle), and the following are equivalent:*

(1) *$E^u \oplus E^s$ is little Zygmund (see Definition* 9.5.1*);*

(2) *The longitudinal Katok–Anosov–Moser cocycle is a coboundary;*

(3) *$E^u \oplus E^s$ is Lipschitz;*

(4) *$E^u \oplus E^s \in C^{k-1}$.*

Remark 9.7.2. If $E^u \oplus E^s \in C^{k-1}$, then Φ is a suspension or contact flow (Theorem 8.1.7), so there is structural information in the conclusion of this theorem, and it can thus be viewed as a weak geometric rigidity result. We note that the last item clearly implies all the previous ones (the second one thanks to Proposition 9.5.8). Accordingly, the following results show that each of these items implies the next.

Theorem 9.2.7 is another instance of smooth rigidity.

The first step is to show that the obstruction from Proposition 9.5.8 arises from the cocycle promised in Theorem 9.7.1. To that end note that $K(p,T)$ is naturally defined as a second-order partial derivative of φ^T in adapted coordinates at any point $p \in M$. There is a natural geometric interpretation. Consider the transversals $\Delta := \psi_p((-\epsilon,\epsilon) \times \{0\} \times (-\epsilon,\epsilon))$ and $\Delta' := \psi_{\varphi^T(p)}((-\epsilon,\epsilon) \times \{0\} \times (-\epsilon,\epsilon))$. Then $\Delta' \cap \varphi^T(\Delta)$ contains local strong-stable and strong-unstable manifolds of $\varphi^T(p)$, but the two transversals are not usually identical. As one sees from the coordinate representation of the flow, the obstruction gives the off-diagonal term in the Hessian of the map $(-\delta,\delta)^2 \to \mathbb{R}$ that gives the lengths of the orbit segments between Δ' and $\varphi^T(\Delta')$. This can be viewed as the "relative curvature" of image transversal versus original transversal. Lemma 9.7.6 shows that if the obstruction vanishes we can choose transversals such that the image transversals agree to third order.

Lemma 9.7.3. *K is an additive cocycle (Definition* 1.2.4*).*

We call K the *longitudinal Katok–Anosov–Moser cocycle.*[15]

[15] Katok because this cocycle is modeled on the one that plays a central role in Theorem 9.7.12, Anosov first noted a counterpart of Proposition 9.5.8 for the weak subbundles, and Moser's normal form contains that item as the "resonance" term.

Proof. To show that $K(p, T+S) = K(\varphi^T(p), S) + K(p, T)$ for all $p \in M$, $T, S \in \mathbb{R}$, write

$$D\varphi^T(0,0,s) = \begin{pmatrix} \alpha^{-1} & 0 & 0 \\ b_1 & 1 & 0 \\ b_2 & 0 & \alpha \end{pmatrix} \text{ at } p \quad \text{and} \quad D\varphi^S(0,0,s) = \begin{pmatrix} \bar\alpha^{-1} & 0 & 0 \\ \bar b_1 & 1 & 0 \\ \bar b_2 & 0 & \bar\alpha \end{pmatrix} \text{ at } \varphi^T(p).$$

Then

$$D\varphi^{T+S}(0,0,s) = \begin{pmatrix} * & 0 & 0 \\ \bar b_1 \alpha^{-1} + b_1 & 1 & 0 \\ * & 0 & * \end{pmatrix}.$$

Using $\bar b_1(0) = 0$ this gives

$$\begin{aligned} K(p, T+S) = \frac{d}{ds}(\bar b_1 \alpha^{-1} + b_1)_{|s=0} &= \Big(\frac{d}{ds}\bar b_1(0)\Big)\alpha^{-1}(0) + \frac{d}{ds} b_1(0) \\ &= \bar b_1'(0) + b_1'(0) = K(\varphi^T(p), S) + K(T, p). \qquad \square \end{aligned}$$

Lemma 9.7.4. *The cohomology class of the longitudinal Katok–Anosov–Moser cocycle is unaffected by coordinate changes.*

Proof. Consider coordinate changes to coordinates with our desired properties. To see how the longitudinal Katok–Anosov–Moser cocycle changes we examine the change in the differential of φ^T entailed by the coordinate change. We need only study points on the stable leaf. For the coordinate calculations we agree that the coordinate change transforms variables $(\tilde u, \tilde t, \tilde s)$ to (u, t, s). Variables in coordinates at $\varphi^T(p)$ are marked by a subscript T. At a point $(0, 0, \tilde s)$ an allowed coordinate change has differential

$$\begin{pmatrix} a & 0 & 0 \\ b & 1 & 0 \\ * & 0 & a^{-1} \end{pmatrix} \text{ with inverse } \begin{pmatrix} a_T^{-1} & 0 & 0 \\ -a_T^{-1} b_T & 1 & 0 \\ * & 0 & a_T \end{pmatrix},$$

in coordinates at $\varphi^T(p)$, where the entries are evaluated at $\tilde s_T$. Note that $a = \frac{d\tilde s}{ds}$. In these new coordinates the differential of φ^T at $(0, 0, \tilde s)$ becomes

$$\begin{pmatrix} a_T^{-1} & 0 & 0 \\ -a_T^{-1} b_T & 1 & 0 \\ * & 0 & a_T \end{pmatrix} \begin{pmatrix} \alpha^{-1} & 0 & 0 \\ b_1 & 1 & 0 \\ b_2 & 0 & \alpha \end{pmatrix} \begin{pmatrix} a & 0 & 0 \\ b & 1 & 0 \\ * & 0 & a^{-1} \end{pmatrix} = \begin{pmatrix} * & 0 & 0 \\ -a\alpha^{-1} a_T^{-1} b_T + a b_1 + b & 1 & 0 \\ * & * & a^{-1}\alpha a_T \end{pmatrix}.$$

Note that therefore $\frac{d\tilde{s}_T}{d\tilde{s}} = a^{-1}\alpha a_T$. This gives

$$\begin{aligned}\tilde{K}(p,T) &:= \frac{d}{d\tilde{s}}\tilde{b}_1|_{\tilde{s}=0} \\ &= -a(0)\alpha^{-1}(0)a_T^{-1}(0)\frac{d}{d\tilde{s}}b_T(0) + a(0)\frac{d}{d\tilde{s}}b_1(0) + \frac{d}{d\tilde{s}}b(0) \qquad (9.7.1) \\ &= \frac{d}{ds}b_1(0) + \frac{d}{d\tilde{s}}b(0) - \frac{d\tilde{s}}{d\tilde{s}_T}\frac{d}{d\tilde{s}}b_T(0) = K(p,T) + b'(0) - b'_T(0),\end{aligned}$$

which is cohomologous to K. □

Actually, K is an obstruction to $E^u \oplus E^s$, being more regular than Zygmund.

Proposition 9.7.5. *If $E^u \oplus E^s$ is little Zygmund (Definition 9.5.1) then the longitudinal Katok–Anosov–Moser cocycle is null cohomologous.*

Proof. Let p be any T-periodic point. Then in our usual coordinates (9.5.1) gives

$$e(s_\varphi) = \alpha(s)(b_1(s) + e(s)) = \alpha(0)(1 + o(s))(b_1'(0)s + o(s) + e(s)).$$

If $\alpha := \alpha(0)$ then $\alpha s = s_\varphi + O(s^2)$. Since e is little Zygmund, it is H-Hölder for all $H \in (1/2, 1)$, so

$$e(\alpha s) = e(s_\varphi) + \underbrace{o(s^{2H})}_{=o(s)} = \alpha(1+o(s))(b_1'(0)s+o(s)+e(s)) = \alpha(K(p,T)s+e(s)+o(s)).$$

Recursively, this gives $e(\alpha^n s) = \alpha^n(nK(p,T)s + e(s) + \sum_{i=0}^{n-1}\alpha^{-i}o(\alpha^i s))$. Since the terms of the sum converge to 0 we get $\sum_{i=0}^{n-1}\alpha^{-i}o(\alpha^i s)/n \to 0$ as $n \to \infty$. Therefore,

$$\begin{aligned}0 &= \lim_{n\to\infty}\frac{e(\alpha^n s)}{\alpha^n s\log(\alpha^n s)} \\ &= \lim_{n\to\infty}\left[\frac{nK(p,T)}{n\log\alpha + \log s} + \frac{e(s)}{s\log(\alpha^n s)} + \frac{\sum_{i=0}^{n-1}\alpha^{-i}o(\alpha^i s)}{s\log(\alpha^n s)}\right] = \frac{K(p,T)}{\log\alpha}.\end{aligned}$$ □

We now study what happens when the longitudinal Katok–Anosov–Moser cocycle is trivial. As a first step we show that this allows more perfectly adapted coordinate systems.

Lemma 9.7.6. *If the longitudinal Katok–Anosov–Moser cocycle is a coboundary then there are nonstationary local coordinates in which it vanishes identically.*

Proof. If the longitudinal Katok–Anosov–Moser cocycle is null cohomologous then there is a smooth $k\colon M \to \mathbb{R}$ such that $K(p,T) = k(\varphi^T(p)) - k(p)$ for all $p \in M$,

$T \in \mathbb{R}$. With the notation from the proof of Lemma 9.7.4 define a coordinate change at p by

$$\begin{pmatrix} u \\ t \\ s \end{pmatrix} = \begin{pmatrix} \tilde{u} \\ \tilde{t} + k(p)\tilde{u}\tilde{s} \\ \tilde{s} \end{pmatrix}.$$

In these new coordinates $\tilde{K}(p,T) = 0$ for all $p \in M$, $t \in \mathbb{R}$ by (9.7.1). □

The first direct consequence of the existence of such coordinates is that in this case $E^u \oplus E^s$ is Lipschitz continuous.

Proposition 9.7.7. *If the longitudinal Katok–Anosov–Moser cocycle is a coboundary then $E^u \oplus E^s$ is Lipschitz continuous.*

Proof. We combine the arguments of Propositions 9.5.6 and 9.5.7, using that in our new coordinates the cocycle is trivial, that is, $b_1'(0) = 0$ in (9.5.1), so $b_1(s) \in O(s^2)$. Assume that $|e(s)| \le Z|s|$ with uniform Z. Then the first calculation, (9.5.2), in Proposition 9.5.6 becomes

$$\begin{aligned} |e(s_\varphi)| &= |\alpha(s)||b_1(s) + e(s)| \\ &\le |\alpha(s)|(O(s^2) + Z|s|) \\ &= (Z + O(s))|\alpha(s)s| = Z(1 + \kappa(s))|s_\varphi|, \end{aligned}$$

with $\kappa(s) \in O(s)$ decreasing in Z. With the argument from Proposition 9.5.6 for $\bar{e}$, which works for $H = 1$, this implies Lipschitz continuity as in Theorem 9.5.3. □

We finally show that Lipschitz continuity implies smoothness.

Proposition 9.7.8. *If the volume-preserving Anosov flow is C^k and the subbundle $E^u \oplus E^s$ is Lipschitz continuous, then $E^u \oplus E^s$ is C^{k-1}.*

Proof. By the Livshitz Theorem there is a C^k-invariant volume form Ω (Theorem 6.3.11). The canonical invariant 1-form A associated to the flow by $A(\dot{\varphi}) = 1$, $A_{\restriction E^u \oplus E^s} = 0$ is Lipschitz continuous, and we aim to prove that it is C^{k-1}.

In local charts as in Lemma 9.5.5 we have

$$A = dt + \lambda(u,s)ds + \beta(u,s)du,$$

where λ and β are Lipschitz-continuous functions independent of the variable t. Then the Anosov vector field $X = \dot{\varphi}$ $(= \partial_t)$ is in the kernel of the integrable 2-form dA and the L^1 3-form $A \wedge dA$ is also flow invariant. Ergodicity gives an $\eta \in \mathbb{R}$ such that

$$A \wedge dA \overset{\text{ae}}{=} \eta\Omega,$$

with respect to the invariant measure associated to Ω, that is, $dA \overset{\text{ae}}{=} w = \eta\Omega(X,\cdot) = \eta i_X \Omega$, a C^k-invariant 2-form. Invariance implies $0 = \mathcal{L}_X(w)/\eta = i_X\, dw + d i_X\, w$, so $dw = 0$ (it is a 3-form that vanishes on X). Moreover, w is exact by the Stokes Theorem for Lipschitz continuous forms: Let c_θ, $\theta \in [-\epsilon, \epsilon]$ be a family of regular disjoint 2-cycles such that $[c_\theta] = c \in H^2(M)$, and set $\Delta = \bigcup_{t\in[-\epsilon,\epsilon]} c_\theta$. By the Fubini Theorem w is exact:

$$2\epsilon \cdot \omega(c) = \int_{-\epsilon}^{\epsilon} \Big(\int_{c_\theta} w \Big) d\theta = \int_\Delta d\theta \wedge w = \int_\Delta d\theta \wedge dA = \int_{-\epsilon}^{\epsilon} \Big(\int_{c_\theta} dA \Big) d\theta = 0.$$

Thus, there is a C^k 1-form $\tilde{A}$ such that $w = d\tilde{A}$. Then the Lipschitz-continuous 1-form $A - \tilde{A}$ is almost everywhere closed. Choose a reference Riemannian metric g. Using the same type of Fubini–Stokes argument establishes the de Rham–Hodge Theorem (there is a harmonic form in each cohomology class) in this context: there exist a g-harmonic 1-form H in the same cohomology class and a 1-form μ such that

$$A - \tilde{A} = H + \mu, \quad d\mu \overset{\text{ae}}{=} 0, \quad \int_\gamma \mu = 0$$

for any 1-cycle γ. We aim to show that μ is exact, but instead of using the same trick as before we will be more explicit to get higher regularity.

In adapted local coordinates at $p \in M$ denote by $[0, x]$, $x = (u, t, s)$ the image of the segment by the corresponding local chart. Introduce for any $c \in \mathbb{R}$ a local Lipschitz continuous function

$$f_p^c(x) := c + \int_{[0,x]} \mu.$$

For every x, v, and small ϵ the Stokes Theorem gives

$$f_p^c(x + \epsilon v) - f_p^c(x) = \int_{[x,x+\epsilon v]} \mu - \int_{T_\epsilon} d\mu,$$

where T_ϵ is the oriented triangle $(0, x, x + \epsilon v)$. Then $df_p^c(x)(v) = \mu(x)(v)$ for almost every x and v. Since the integral of μ along any closed curve vanishes there is a global function f that locally coincides with one of the functions f_p^c, that is,

$$A \overset{\text{ae}}{=} \tilde{A} + H + df. \tag{9.7.2}$$

We conclude the proof by showing that df is C^{k-1}.

Using the Anosov vector field and the definition of the canonical 1-form A we write

$$1 - \tilde{A}(X) - H(X) \overset{\text{ae}}{=} df(X).$$

The terms on the left are C^k, so by Theorem 6.3.13, there exists a C^k-boundary $b\colon M \to \mathbb{R}$ such that

$$1 - \tilde{A}(X) - H(X) = db(X).$$

This in particular means that the Lipschitz-continuous function $\rho := b - f$ satisfies

$$d\rho(X) \stackrel{\text{ae}}{=} 0. \tag{9.7.3}$$

We show that this implies flow invariance of ρ. Choose $p \in M$ and an adapted chart in which the Anosov vector field X is expressed as ∂_t. Assume there are $x_0 = (u_0, t_0, s_0)$ and $t \in \mathbb{R}$ such that $\rho(\varphi_t(x_0)) - \rho(x_0) \neq 0$. By continuity there is an open set U in the transversal $\{(u, t_0, s)\}$ such that

$$0 < \int_U (\rho(\varphi_t(u, t_0, s)) - \rho(u, t_0, s))\, du\, ds = \int_U \Big(\int_{t_0}^{t_0+t} d\rho(X(w))dw \Big)\, du\, ds,$$

contrary to (9.7.3).

Invariance and ergodicity imply that ρ is constant (everywhere by continuity), so $db \equiv df$ and hence $A = \tilde{A} + H + db$ because both sides of (9.7.2) are continuous. Thus A is C^{k-1}, and hence so is its kernel $E^u \oplus E^s$. □

Remark 9.7.9. Theorem 9.7.1 provides a lovely instance of smooth rigidity,[16] and Remark 9.7.2 explains how its conclusion implies structural information. However, the conclusion of Theorem 9.7.1 holds for all suspensions of Anosov diffeomorphisms and for all contact Anosov flows, so this does not provide a basis for finding an instance of *geometric rigidity* in the sense of directly obtaining, say, constant curvature from this conclusion in the case of geodesic flows analogously to the corresponding "transverse" situation discussed beginning with Remark 9.7.11. Nonetheless, this is a source of geometric rigidity results, albeit of a slightly different flavor.

Theorem 9.7.10 ([105, Theorem A]). *Consider a magnetic Anosov flow (Definition 5.2.18) on a Riemannian surface with the magnetic tensor $\mathfrak{m}$ viewed as a 2-form.*

- *If $\mathfrak{m}$ is exact, then the longitudinal Katok–Anosov–Moser cocycle vanishes if and only if $\mathfrak{m} \equiv 0$, that is, the flow is geodesic after all.*
- *If $\mathfrak{m}$ is not exact, then the longitudinal Katok–Anosov–Moser cocycle vanishes if and only if $\mathfrak{m}$ is a (constant) multiple of the area form, and the curvature is constant.*

[16] Theorem 9.2.7 is another.

Remark 9.7.11. The situation is quite different when we return to considerations of the *weak* stable or unstable foliation. A volume-preserving Anosov 3-flow satisfies the strongest bunching condition possible, so Theorem 9.4.10 implies that the weak foliations are $C^{1+x|\log x|}$. Analogously to Theorem 9.5.3 it turns out that in this case the *derivatives* are Zygmund regular and that being little Zygmund implies smooth rigidity, that is, that the weak foliations are smooth. So far, this is quite analogous to Theorem 9.7.1. However, in this case one also obtains geometric rigidity: the flow must then be smoothly conjugate to either the suspension of a toral automorphism or to the geodesic flow of a surface of constant curvature. So the system is algebraic up to smooth conjugacy.

Theorem 9.7.12 (Ghys–Hurder–Katok [200]). *The weak-stable and weak-unstable subbundles of a volume-preserving C^{k+3} Anosov 3-flow have Zygmund first derivatives, and if either of them has "little Zygmund" first derivatives* (*Definition* 9.5.1), *then both are C^k. If, furthermore, the flow is a geodesic flow, then the metric has constant negative curvature.*

Remark 9.7.13. The rigidity conclusion also holds under the assumption that the derivative of one or the other of the weak subbundles has modulus of continuity $o(x|\log x|)$ [200] or is of bounded variation [164].

If the flow is a geodesic flow, a theorem by Ghys gives a smooth conjugacy to the geodesic flow of a surface of constant curvature, and Theorem 9.3.1 then implies that the metric has constant curvature (Theorem 9.8.1). (And Remark 9.7.20 is pertinent here as well.) We remark that while this structural information was obtained in the original preprint by Hurder and Katok, they assumed closeness to the algebraic model; the published version invokes his result. It is interesting here to spell out the theorem by Ghys because it does not assume the topology of a geodesic flow.

Theorem 9.7.14 ([148, 149]). *A volume-preserving C^∞ Anosov 3-flow with C^2 weak foliations is either a suspension or C^∞-conjugate* (*modulo Remark* 9.7.20) *to the geodesic flow of a constantly curved surface.*

Remark 9.7.15 ([150, Théorème 4.6]). The C^2-hypothesis can be replaced by assuming Lipschitz-continuous derivative, and for C^r-flows ($r \geq 2$) one gets a C^r-conjugacy to the geodesic flow. One can even drop volume preservation in the assumptions, but then the "algebraic" models include exotic "hybrids" (Example 9.7.25).

Theorem 9.7.12 predates smooth longitudinal rigidity (Theorem 9.7.1), and we presented the proof of the latter to illustrate the approach in the proof of Theorem 9.7.12. Indeed, the counterpart here to Proposition 9.5.8 is a much earlier observation by Anosov that there is an obstruction to the weak subbundles being C^2. In this case it involves third derivatives of the flow (while the obstruction in Proposition 9.5.8 involves second derivatives). Of course, Theorem 9.7.12 also carries structural information while Theorem 9.7.1 does not.

Following Guysinsky [164] one can explain the Anosov cocycle using local normal forms. For a smooth area-preserving Anosov diffeomorphism on $\mathbb{T}^2$ DeLatte [110, 109] showed that one can find local smooth coordinate systems around each point that depend continuously (actually C^1) on the point and bring the diffeomorphism f into the *Moser normal form* [262]

$$f(x, y) = \begin{pmatrix} \lambda_p^{-1} x/\varphi_p(xy) \\ \lambda_p y \varphi_p(xy) \end{pmatrix},$$

where (x, y) are in local coordinates around a point p and the expression on the right is in coordinates around $f(p)$. The terms involving φ_p that depend on the product xy correspond to the natural resonance $\lambda_p \lambda_p^{-1} = 1$ that arises from area preservation (actually from the family of resonances $\lambda_p = \lambda_p^{n+1} \lambda_p^{-n}$). The function φ_p is as smooth as f, and $\varphi_p(0) = 1$. Now we suppress the (continuous) dependence of λ and φ on p and apply this to the return map f of a local cross section at a periodic point of a flow. For a point $(0, y)$ we then have

$$Df = \begin{pmatrix} \lambda^{-1} xy(1/\varphi)'(xy) + \lambda^{-1}/\varphi(xy) & \lambda^{-1} x^2 (1/\varphi)'(xy) \\ \lambda y^2 \varphi'(xy) & \lambda xy \varphi'(xy) + \lambda \varphi(xy) \end{pmatrix} = \begin{pmatrix} \lambda^{-1} & 0 \\ \lambda y^2 \varphi'(0) & \lambda \end{pmatrix}.$$

In these local coordinates the (center-) unstable direction at a point $(0, y)$ on the (center-) stable leaf of p in the section is spanned by a vector $(1, a(y))$. Since this subbundle is invariant under Df and since $f(0, y) = (0, \lambda y)$, the coordinate representation of Df from above gives $a(\lambda y) = \lambda^2 y^2 \varphi'(0) + \lambda^2 a(y)$. If the unstable subbundle is C^2 then differentiating this relation twice with respect to x at 0 gives $\lambda^2 a''(0) = 2\lambda^2 \varphi'(0) + \lambda^2 a''(0)$, that is, $\varphi'(0) = 0$. This means that the Anosov obstruction is $\varphi'(0)$, where φ arises from the nonstationary Moser–DeLatte normal form.

Hurder and Katok established that this obstruction arises from a cocycle which (cohomologically) vanishes if and only if (one and hence both of) the stable and unstable subbundles have "little Zygmund" first derivatives (Definition 9.5.1). In that latter case, the cocycle lends itself to showing that the regularity is indeed C^3, from where a bootstrap kicks in to give C^∞ subbundles. Thus far, this is a smooth-rigidity result, and work of Ghys then produces a geometric-rigidity result: that the flow is smoothly conjugate to one or the other of the algebraic ones.

The geometric rigidity of Theorem 9.7.12 (indeed, of Remark 9.7.13) has been extended to higher dimension, albeit with a "regularity gap." Specifically, Theorem 9.6.1 shows that in higher dimension the invariant subbundles are more than a little less regular than C^2. Conversely, at present, regularity somewhat above C^2 is needed for rigidity phenomena.

Kanai introduced the essential tool for these rigidity results, the Bott–Kanai connection (Proposition 9.8.25), and proved a rigidity result for geodesic flows with a curvature-pinching assumption. Feres and Katok built on this by increasingly removing pinching assumptions, and then the following result altogether removed the need to start with a geodesic flow in the first place.

Theorem 9.7.16 (Benoist–Foulon–Labourie [46]). *A contact Anosov flow with sufficiently smooth*[17] *foliations is smoothly conjugate to the geodesic flow of a locally symmetric Riemannian manifold* (*modulo Remark* 9.7.20 *below*).

Indeed, Theorem 9.3.6 then implies the following result:

Theorem 9.7.17. *A Riemannian metric whose geodesic flow has smooth invariant foliations is* isometric *to a locally symmetric metric.*

The regularity required in Theorem 9.7.16 is quite a bit higher than C^2 (or a hair less), which suffices in the lowest-dimensional context. Whether C^2 turns out to suffice after all is not wholly clear, but there are tantalizing hints [355]:

Theorem 9.7.18 ([169]). *For a compact manifold of negative curvature with C^2 horospherical foliations, the Liouville entropy coincides with the topological entropy* (*Conjecture* 9.3.3 *would then imply rigidity*).

Theorem 9.7.19 ([224, Corollary 2.3]). *If (M, g_0) is a compact locally symmetric space of negative curvature and g a negatively curved metric that is conformally equivalent to g_0, then the Anosov splitting for g is C^2 if and only if g and g_0 coincide up to a constant.*

Remark 9.7.20. The statement of Theorem 9.7.16 is not quite correct. There are two modifications one can make to a contact Anosov flow, including geodesic flows, that affect neither the contact Anosov property nor smoothness of the invariant foliations, so the statement is to be understood modulo these additional modifications. One of these is to pass to finite covers (or quotients) of the phase space. The other is a trivial or *canonical time change* (Proposition 1.3.21) which, unlike most time changes, preserves the contact property.

Definition 9.7.21. A *canonical time change* is defined using a closed 1-form α by replacing the generator X of the flow by the vector field $X/(1 + \alpha(X))$, provided α is such that the denominator is positive.

[17] This depends on either the dimension or bunching information.

Proposition 9.7.22 (Cohomology class). *If α and β are cohomologous closed 1-forms with $1+\alpha(X)>0$ and $1+\beta(X)>0$ then the associated canonical time changes of X are smoothly conjugate.*

Proof. Writing $\beta = \alpha + df$ with smooth f lets us use Proposition 1.3.21:

$$\frac{X}{1+\beta(X)} = \frac{X}{1+\alpha(X)+df(X)} = \frac{\frac{X}{1+\alpha(X)}}{1+df(\frac{X}{1+\alpha(X)})} = \Big(\frac{X}{1+\alpha(X)}\Big)_f. \qquad \square$$

In the context of Definition 5.1.1, the *canonical* 1*-form* is the invariant 1-form A associated to Φ by

$$A(X)=1, \quad A(E^u \oplus E^s)=0.$$

Being a contact Anosov flow is equivalent to the canonical 1-form being smooth and nondegenerate ($A \wedge dA$ is a volume). Picking up on Definition 9.7.21 and Proposition 9.7.22 we have the following proposition:

Proposition 9.7.23 (Regularity). *Suppose X_0 generates an Anosov flow, and α is a closed 1-form such that $1+\alpha(X_0)>0$. If A_0 denotes the canonical form for X_0 then $A := A_0+\alpha$ is the canonical form for $X := \frac{X_0}{1+\alpha(X_0)}$.*

Remark 9.7.24. In particular, this shows that canonical time changes with smooth closed forms do not affect the regularity of the canonical form.

Proof. Two invariant 1-forms for an Anosov flow are proportional because both are constant on X, and a continuous invariant 1-form that vanishes on X is trivial (because an invariant 1-form A is trivial on E^s—and likewise on E^u—by invariance: $A(v) = \underbrace{A(\varphi^t(v))}_{\xrightarrow[t\to\infty]{}0}$ for $v \in E^s$).

Since α is closed we have $dA = dA_0$. Also $A(X) = \frac{A_0(X_0)+\alpha(X_0)}{1+\alpha(X_0)} = 1$, which implies that $\mathcal{L}_X A = 0$, that is, A is X-invariant and hence proportional to the canonical 1-form of X—and indeed equal to it since $A(X) \equiv 1$. $\square$

Looking back at Theorem 9.7.12, we note a rather striking feature: the regularity assumption is imposed on the *weak* foliations, which are unaffected by time changes, while the conclusion gives a conjugacy, which rigidly fixes the longitudinal behavior. Furthermore, nothing else in the assumption gives any indication that there is longitudinal control at all. This underscores the subtlety of these kinds of results.[18]

[18] In this regard it is noteworthy that a rigidity theorem along the lines of that by Benoist–Foulon–Labourie is possible for transversely symplectic flows, in which timing is less "locked down" than for contact flows. Without defining the notions involved, we state this recent result: a topologically mixing transversely symplectic Anosov flow whose weak-stable and weak-unstable subbundles are C^∞ and whose Hamenstädt metrics (Definition 7.5.9)

It turns out that volume preservation is essential not only for the arguments, but for the conclusion. This is demonstrated by a construction of Ghys, which introduces a few interesting ideas.

Example 9.7.25 (Ghys quasi-Fuchsian flows [149]). That time changes preserve the Anosov property (Theorem 5.1.11) and the weak foliation motivates thinking about an Anosov flow as a 1-dimensional foliation equipped with suitable complementary foliations. On a 3-manifold, this would be a pair of foliations with 2-dimensional leaves, and in the present context, smooth.

To "hybridize" geodesic flows of a given surface Σ, replace the unit tangent bundle (which depends on a choice of Riemannian metric) by the *homogeneous bundle* $H\Sigma$ of half-lines in $T_x\Sigma$ for each $x \in \Sigma$. For any one Riemannian metric this is naturally identified with the unit tangent bundle, but it allows us to consider different geodesic flows on the same manifold $H\Sigma$. Specifically, for any Riemannian metric g on Σ denote by W_g^s and W_g^u the weak-stable and weak-unstable foliations of the geodesic flow on $H\Sigma$. If now two smooth such Riemannian metrics g_1, g_2 of curvature -1 are sufficiently C^3-close, then the hybrid pair $W_{g_1}^s$ and $W_{g_2}^u$ is transverse, and the intersection is close to the orbits of either geodesic flow, with any smooth choice of parametrization defining an Anosov flow on this 3-manifold with smooth weak foliations (which is not conjugate to any such geodesic flow).[19] Volume preservation is the missing ingredient here.

We also note something of interest with respect to Conjecture 9.3.3:

Theorem 9.7.26 ([54]). *Applying a canonical time change to the geodesic flow of a negatively curved locally symmetric space does not change the Liouville entropy, but for small enough ϵ, the time change $X \mapsto X/(1 + \epsilon\alpha(X))$ (combined with a volume renormalization) increases the topological entropy.*

9.8 Godbillon–Vey invariants*

The purpose of this section is to introduce a tool from foliation theory that extends an invariant which was itself of interest in the context of the result from Remark 9.7.13:

are sub-Riemannian is up to finite covers and a constant change of timescale C^∞-conjugate (not just orbit equivalent!) to the geodesic flow of a locally symmetric space [128]. Here, the control of time enters largely through the sub-Riemannian hypothesis and the rescaling in the definition of the Hamenstädt metrics. Indeed, in the case of 3-flows, sub-Riemannian $\Leftrightarrow$ Riemannian, and assuming this plus $a = 1$ in Remark 7.5.10 turns out to immediately imply that the Lyapunov exponents at all periodic points are ± 1.

[19] Ghys calls these flows quasi-Fuchsian because they are parametrized by a *pair* of metrics rather than a single one, in which case "Fuchsian" would recall the underlying Fuchsian (fundamental) group.

Theorem 9.8.1 ([200]). *Negatively curved surfaces with C^2 horospheric foliations are constantly curved.*

Our proof of this will introduce the Bott–Kanai connection and also use the Godbillon–Vey invariants we present in this section.

To indicate further how these can be used to produce rigidity results in straightforward ways, we also prove another geometric-rigidity result, which follows from Theorem 9.3.1:

Theorem 9.8.2. *Suppose the geodesic flows φ^t and ψ^t of Riemannian surfaces M and S, respectively, are topologically conjugate. If S has constant curvature -1, then so does M.*

While the assumptions imply that the conjugacy is smooth (Theorem 9.2.7), the point is that smooth conjugacy controls the geometry, and that this is easy with the Godbillon–Vey invariants introduced here.

We now introduce Godbillon–Vey invariants for suitable foliations in contact 3-manifolds, study their interaction with the canonical flow associated to the contact form, explore consequences of this flow being Anosov, compute the top invariant among these for geodesic flows, and provide the main properties that underlie the rigidity results.

Specifically, if (M, A) is a contact 3-manifold and $\mathcal{F}$ a C^2 maximal isotropic foliation, we define *Godbillon–Vey invariants* GV_i for $i = 0, 1, 2$ in Definition 9.8.11. We lead up to that definition with work that introduces notions needed for the definition itself as well as for the proof that these are well defined (Proposition 9.8.12).

GV_0 is (by definition) the volume of the manifold, and for a contact Anosov flow and the associated weak-stable foliation, GV_1 is the Liouville entropy (Proposition 9.8.15).

For geodesic flows of surfaces, we compute GV_2 in Proposition 9.8.18 (the Mitsumatsu formula), and the first rigidity result is the computation of the Godbillon–Vey invariants for contact Anosov flows with absolutely continuous Margulis measure (Theorem 9.8.19), and its application to geodesic flows (Theorem 9.8.21) then follows from (merely!) the Cauchy–Schwarz inequality. This then implies both rigidity results above.

Remark 9.8.3. What we construct here as Godbillon–Vey invariants extend the classical Godbillon–Vey class and invariant, but we do not build on the classical construction. Here is a condensed outline of how that goes. If ω is a completely integrable nonsingular 1-form, then there is a 1-form η such that $d\omega = \omega \wedge \eta$ (Frobenius Theorem), and $0 = dd\omega = d(\omega \wedge \eta) = \omega \wedge d\eta$, so there is a 1-form ξ with $d\eta = \omega \wedge \xi$, hence $\eta \wedge d\eta$ is closed, and its de Rham cohomology class is independent of such a choice of η—another choice must be of the form $\eta' = \eta + u\omega$ for a function u, and then $\eta' \wedge d\eta' = \eta \wedge d\eta + d(u\, d\omega)$. Indeed, this depends only on the codimension-1 foliation

F defined by complete integrability of ω, for, any ω' defining the same foliation is a scalar multiple of ω. The cohomology class of $\eta \wedge d\eta$ is called the Godbillon–Vey class of F, and if $\dim M = 3$, then $\int \eta \wedge d\eta$ is called the Godbillon–Vey invariant of F; it is a characteristic class, depends only on the foliated cobordism class of (M, F), is nontrivial, and varies continuously and nontrivially with F. By contrast, we show that the combination of a contact structure and an orientable maximal isotropic foliation gives rise to a *sequence* of what we call Godbillon–Vey invariants. The point is that as a *sequence* they are of interest for rigidity results.

Definition 9.8.4 (Isotropic, normal bundle). Let M be a contact 3-manifold. A subspace $V \subset T_x M$ is said to be *isotropic* if $dA_x\restriction_V = 0$ and *maximal isotropic* if it furthermore has dimension 2. A subbundle is said to be (maximal) isotropic if it is so at each point, and a foliation is (maximal) isotropic if its tangent bundle is so. If M is a smooth manifold and F is a subbundle of TM, then we define the *normal bundle*

$$\mathcal{N}(F) := \{\omega \in T^*M \mid \iota_\xi \omega = 0 \text{ whenever } \xi \in F\}.$$

Lemma 9.8.5. *If F is integrable and $\omega \in \mathcal{N}(F)$, then $d\omega\restriction_F = 0$.*

Proof. If Z_1, Z_2 are tangent to F, then $0 \equiv \omega(Z_1) \equiv \omega(Z_2) \equiv \omega([Z_1, Z_2])$, so

$$d\omega(Z_1, Z_2) = \mathcal{L}_{Z_1}\omega(Z_2) - \mathcal{L}_{Z_2}\omega(Z_1) + \omega([Z_1, Z_2]) = 0 + 0 + 0. \qquad \square$$

We define the Godbillon–Vey invariants in terms of a 1-form transverse to the maximal isotropic foliation $\mathcal{F}$ (with tangent bundle F) in question. Specifically, since we assume F to be orientable, we henceforth fix an everywhere nonzero $\alpha \in \mathcal{N}(\mathcal{F})$. We will assume $\alpha \in C^2$ (specifically in the proof of Lemma 9.8.9), and hence that $\mathcal{F} \in C^2$. From here through Lemma 9.8.10 we study a 1-form β that is the key ingredient to defining our Godbillon–Vey invariants.

Proposition 9.8.6. *If α is C^1, then $d\alpha = \beta \wedge \alpha$ for a 1-form β.*

Proof. Since $\mathcal{N}(\mathcal{F})$ is 1-dimensional and contains both α and $\iota_Z\, d\alpha$ for any $Z \in F$ (Lemma 9.8.5), so the fact that α vanishes nowhere yields a $\beta(Z)$ for which $\iota_Z\, d\alpha = \beta(Z)\alpha$, and β is a 1-form on F. Now consider an extension of β to any 1-form. Then $\beta \wedge \alpha$ and $d\alpha$ can be evaluated on any pair of vectors by decomposing each vector with respect to a basis that contains a basis of F. For both $d\alpha$ and $\beta \wedge \alpha$ the only nonzero expressions that thus arise are those that include precisely one vector in F, and we just showed that $d\alpha = \beta \wedge \alpha$ for such pairs. $\square$

Remark 9.8.7. That β is uniquely defined on $\mathcal{F}$ means that it is well-defined modulo $\mathcal{N}(\mathcal{F})$, that is, we uniquely defined $[\beta] := \{\beta + \omega \mid \omega \in \mathcal{N}(\mathcal{F})\}$.

Proposition 9.8.8. *The cohomology class* $[\beta]$ *is well defined independently of the choice of* α: $\alpha' = e^f \alpha$ *with* $f\colon M \to \mathbb{R}$ *produces* $\beta' = \beta + df$.

Proof. $\beta' \wedge \alpha' = d\alpha' = d(e^f \alpha) = de^f \wedge \alpha + e^f d\alpha = e^f df \wedge \alpha + e^f \beta \wedge \alpha = (df + \beta) \wedge e^f \alpha = (df + \beta) \wedge \alpha'$. □

Accordingly, we write $[[\beta]] := \{\beta + df + \omega \mid f\colon M \to \mathbb{R},\ \omega \in \mathcal{N}(\mathcal{F})\}$.

Lemma 9.8.9. $d\beta \wedge \alpha = 0$ *and* $d\beta\restriction_{\mathcal{F}} = 0$.

Proof. $0 = dd\alpha = d\beta \wedge \alpha + \beta \wedge d\alpha = d\beta \wedge \alpha + \beta \wedge \beta \wedge \alpha = d\beta \wedge \alpha$. If $Z_1, Z_2 \in \mathcal{F}$, then $0 = \iota_{Z_1,Z_2} 0 = \iota_{Z_1,Z_2} d\beta \wedge \alpha = d\beta(Z_1, Z_2)\alpha$ because $\alpha \in \mathcal{N}(\mathcal{F})$. Then $d\beta(Z_1, Z_2) = 0$ because α is nowhere zero. □

Lemmas 9.8.5 and 9.8.9 serve to give Proposition 9.8.12 via the following result:

Lemma 9.8.10. $\omega \wedge (d\beta)^i \wedge d\omega^{p-i} \wedge dA^{1-p} = 0$ *for* $0 \le i \le p \le 1$ *and* $\omega \in \mathcal{N}(\mathcal{F})$.

Proof. Evaluating this form on three linearly independent vectors and decomposing these with respect to a basis that contains a basis for $\mathcal{F}$ gives a linear combination of expressions, each of which contains at least two elements of $\mathcal{F}$. Inserting a vector from $\mathcal{F}$ into $\omega \in \mathcal{N}(\mathcal{F})$ gives 0, and if two elements of $\mathcal{F}$ are inserted into $(d\beta)^i \wedge d\omega^{p-i} \wedge dA^{1-p}$, we get 0 because one gets 0 whenever more than one such vector is inserted into $d\beta$ (Lemma 9.8.9), $d\omega$ (Lemma 9.8.5), or dA. □

Since $[[\beta]]$ is intrinsically defined, we can now define the Godbillon–Vey invariants.

Definition 9.8.11 (Godbillon–Vey invariants). If (M, A) is a contact 3-manifold and $\mathcal{F}$ a C^2 maximal isotropic foliation, define the *Godbillon–Vey invariants* by

$$\mathrm{GV}_0 = \int_M A \wedge dA =: \mathrm{vol}_A(M), \qquad \text{(the contact volume)}$$

$$\mathrm{GV}_1 = \int_M \beta \wedge dA,$$

$$\mathrm{GV}_2 = \int_M \beta \wedge d\beta.$$

Proposition 9.8.12. *The Godbillon–Vey invariants are well defined.*

Proof. We need to show that $\mathrm{GV}_{p+1} = \int_M \beta \wedge d\beta^p \wedge dA^{1-p}$ is constant on $[[\beta]]$, that is, that replacing β by $\beta + df + \omega$ and therefore $d\beta$ by $d\beta + d\omega$ has no effect. Replacing β by $\beta + df$ makes no difference:

$$\begin{aligned}
&\int_M (\beta + df) \wedge d(\beta + df)^p \wedge dA^{1-p} - \int_M \beta \wedge d\beta^p \wedge dA^{1-p} \\
&\qquad = \int_M df \wedge d\beta^p \wedge dA^{1-p} = \int_M d\big(f \cdot d\beta^p \wedge dA^{1-p}\big) = 0.
\end{aligned}$$

To see the effect of adding ω, expand $d(\beta + \omega)^p$:

$$\begin{aligned}
&(\beta + \omega) \wedge d(\beta + \omega)^p \wedge dA^{1-p} \\
&\qquad = \beta \wedge d(\beta + \omega)^p \wedge dA^{1-p} + \underbrace{\omega \wedge d(\beta + \omega)^p \wedge dA^{1-p}}_{=0 \text{ by Lemma 9.8.10}} \\
&\qquad = \beta \wedge d\beta^p \wedge dA^{1-p} + \sum_{i=1}^{p} c_i \beta \wedge d\beta^{p-i} \wedge d\omega^i \wedge dA^{1-p},
\end{aligned}$$

so

$$\begin{aligned}
&\int_M (\beta + \omega) \wedge d(\beta + \omega)^p \wedge dA^{1-p} - \int_M \beta \wedge d\beta^p \wedge dA^{1-p} \\
&\qquad = \sum_{i=1}^{p} c_i \underbrace{\int_M \beta \wedge d\beta^{p-i} \wedge d\omega^i \wedge dA^{1-p}}_{= \underbrace{\int_M d(\omega\wedge\beta\wedge d\beta^{p-i}\wedge d\omega^{i-1}\wedge dA^{1-p})}_{=0} - \underbrace{\int_M \omega\wedge d\beta^{p-i+1}\wedge d\omega^{i-1}\wedge dA^{1-p}}_{=0 \text{ by Lemma 9.8.10}}} \\
&\qquad = 0.
\end{aligned}$$

□

9.8.a The Reeb field. While GV_0 is volume, we now identify GV_1.

Proposition 9.8.13. $\mathrm{GV}_1 = \int \beta(X) A \wedge dA$, *where* X *is the* Reeb field *of* A *defined uniquely by* $\iota_X A = 1$ *and* $\iota_X\, dA = 0$.

Proof. That $\iota_X\, dA = 0$ implies

- X is tangent to $\mathcal{F}$, so $\mathcal{F}$ is invariant under the flow generated by X,
- by duality there are a vector field η and a function λ with $\beta = \lambda A + \iota_\eta\, dA$,
- inserting X gives $\lambda = \beta(X)$, and
- the 3-form $\iota_\eta\, dA^2$ vanishes whenever X is in any slot.

Thus $\beta \wedge dA = \beta(X) A \wedge dA + \iota_\eta\, dA \wedge dA = \beta(X) A \wedge dA$. □

Now we return to dynamics, specifically the Reeb flow of A, which is the flow Φ generated by the Reeb vector field X of A, and from now we assume that this is an Anosov flow. Then $\mathcal{F} := \mathbb{R}X \oplus E^-$ is integrable to a continuous foliation $\mathcal{F}$ with smooth leaves, the weak-stable foliation, which is maximal isotropic:

Lemma 9.8.14. *$dA_{\restriction\mathcal{F}} = 0$, that is, $dA(Z_1, Z_2) = 0$ if $Z_1, Z_2 \in \mathbb{R}X \oplus E^-$.*

Proof. Using $\iota_X\, dA = 0$ reduces this to the case $Z_1, Z_2 \in E^-$, where

$$dA(Z_1, Z_2) = dA(d\varphi^t(Z_1), d\varphi^t(Z_2)) \xrightarrow[t\to+\infty]{} 0$$

since A, hence dA, is φ^t-invariant and $\|d\varphi^t(Z_i)\| \xrightarrow[t\to+\infty]{} 0$. □

Proposition 9.8.15. *If $\mathcal{F}$ is the weak-stable foliation of a contact Anosov flow, then $\mathrm{GV}_1 = h_{\mathrm{vol}}\,\mathrm{vol}_A(M)$, where h_{vol} is Liouville entropy.*

Proof. Choose $\beta = 0$ on E^+. Then $\mathcal{L}_X\alpha = \iota_X\, d\alpha = \beta(X)\alpha$, that is, $\beta(X)$ is the infinitesimal relative change of the unstable volume under the flow. Rescale A so $\mathrm{vol}_A(M) = 1$. Then the time average of $\beta(X)$, hence by ergodicity (Theorem 7.1.26) its space average GV_1, is then the average unstable infinitesimal volume expansion, and by the Pesin Entropy Formula (Remark 7.4.15), this is h_{vol}. □

We next compute GV_2 for geodesic flows of surfaces. Denote the standard vertical vector field by V. Then $H := [V, X]$ and X are horizontal, and

$$1 = A \wedge dA(X, V, H) = A(X)dA(V, H) = dA(V, H). \tag{9.8.1}$$

If K is the curvature, then the structural equations are

$$[X, V] = -H, \quad [H, V] = X, \quad [X, H] = KV.$$

If the invariant line bundle $F \cap \ker A$ is spanned by the vector field

$$\xi = uV + H,$$

then comparing coefficients in $(\dot u + K)V - uH = [X, \xi] = f\xi = fuV + fH$ implies $f = -u$ and $-u^2 = fu = \dot u + K$, which gives the *Riccati equation* $\dot u + u^2 + K = 0$.

Lemma 9.8.16. *If we choose $\alpha = \iota_\xi\, dA$, then $\alpha(H) = u$, $\alpha(V) = -1$, $\alpha(X) = 0 = \alpha(\xi)$.*

Proof. We have

$$\underbrace{\alpha(V)}_{=dA(\xi,V)=dA(uV+H,V)} = dA(H, V) \underbrace{=}_{(9.8.1)} -1, \quad \alpha(X) = 0 = \alpha(\xi) = \alpha(uV + H) = -u + \alpha(H).$$

□

Lemma 9.8.17. *If we choose $\beta(V)=0$, then $\beta(X)=-u$, $\beta(H)=\mathcal{L}_V u$.*

Proof. We have

$$\beta(X)\alpha(H) = d\alpha(X,H) = \underbrace{\mathcal{L}_X\alpha(H)}_{=\mathcal{L}_X u=\dot u} - \mathcal{L}_H \underbrace{\alpha(X)}_{\equiv 0} + \underbrace{\alpha([H,X])}_{=-K\alpha(V)=K}$$

$$= -u^2 = -u\alpha(H) \qquad \text{(Riccati equation)}$$

and

$$\underbrace{\beta(\xi)\alpha(H)}_{=d\alpha(\xi,H)} = \underbrace{\mathcal{L}_\xi\alpha(H)}_{=\mathcal{L}_\xi u = u\mathcal{L}_V u+\mathcal{L}_H u} - \mathcal{L}_H \underbrace{\alpha(\xi)}_{\equiv 0} + \underbrace{\alpha([H,\xi])}_{=u[H,V]+(\mathcal{L}_H u)V} = \underbrace{\alpha(H)}_{=u}(\mathcal{L}_V u).$$

□

Proposition 9.8.18 (Mitsumatsu formula). $\mathrm{GV}_2 = \int_M u^2 + 3(\mathcal{L}_V u)^2\, A\wedge dA$ *for maximal isotropic foliations invariant under geodesic flows of surfaces. The* Mitsumatsu defect $\int_M 3(\mathcal{L}_V u)^2\, A\wedge dA$ *is the deviation of* GV_2 *from its value for constant curvature.*

Proof. We show $\int u^2+3(\mathcal{L}_V u)^2 = \int \lambda$ for $\lambda\colon M\to\mathbb{R}$ with $\beta\wedge d\beta = \lambda A\wedge dA$, hence

$$\lambda \underbrace{A\wedge dA(X,V,H)}_{=1} = \beta\wedge d\beta(X,V,H)$$

$$= \beta(X)d\beta(V,H) + \beta(H)d\beta(X,V) + \underbrace{\beta(V)}_{=0} d\beta(H,X).$$

Here

$$d\beta(V,H) = \mathcal{L}_V \underbrace{\beta(H)}_{=\mathcal{L}_V u} - \mathcal{L}_H \underbrace{\beta(V)}_{\equiv 0} - \underbrace{\beta([V,H])}_{=\beta(-X)=u} = \mathcal{L}_V^2 u - u,$$

$$d\beta(X,V) = \mathcal{L}_X \underbrace{\beta(V)}_{\equiv 0} - \mathcal{L}_V \underbrace{\beta(X)}_{=-u} - \underbrace{\beta([X,V])}_{=\beta(-H)=-\mathcal{L}_V u} = 2\mathcal{L}_V u,$$

so

$$\lambda = \underbrace{\beta(X)}_{=-u}\underbrace{d\beta(V,H)}_{=\mathcal{L}_V^2 u-u} + \underbrace{\beta(H)}_{=\mathcal{L}_V u}\underbrace{d\beta(X,V)}_{=2\mathcal{L}_V u} = u^2 + 2(\mathcal{L}_V u)^2 - u\mathcal{L}_V^2 u$$

by (9.8.1). It remains to show that $\int (\mathcal{L}_V u)^2 + u\mathcal{L}_V^2 u = 0$ (integration by parts):

$$\underbrace{(A\wedge d\iota_V\, dA)}_{=-d(A\wedge\iota_V\, dA)=d\iota_V(A\wedge dA)=\mathcal{L}_V A\wedge dA}(X,V,H) = d(\iota_V\, dA)(V,H) = \underbrace{-\iota_V\, dA([V,H])}_{=dA(-V,[V,H])}$$

$$= dA(V,X) = 0$$

implies

$$0 = \int_M \mathcal{L}_V\big(u\mathcal{L}_V u\, A\wedge dA\big) = \int_M \mathcal{L}_V u\, \mathcal{L}_V u\, A\wedge dA + \int_M u\mathcal{L}_V\mathcal{L}_V u\, A\wedge dA. \quad □$$

Lastly, we compute specific values for the Godbillon–Vey invariants in the special case that the geometric-rigidity results focus on.

Theorem 9.8.19 (Godbillon–Vey Entropy Rigidity). $\mathrm{GV}_i = h^i \operatorname{vol}_A(M)$ *for contact Anosov flows with absolutely continuous Margulis measure, where h is topological entropy.*

Remark 9.8.20. This applies to geodesic flows of negatively curved locally symmetric spaces, in particular, of surfaces with constant negative curvature.

Proof. The (un)stable conditionals of the Margulis measure are volumes and scale with h (Theorem 7.5.15). Therefore $h\alpha = \mathcal{L}_X\alpha = \iota_X\, d\alpha + d\iota_X\alpha = \beta(X)\alpha$ (since $\iota_X\alpha \equiv 0$), so $\beta(X) \equiv h$, hence $\beta = hA + \iota_\eta\, dA$ (from the proof of Proposition 9.8.13), and

$$h\beta\wedge\alpha = h\,d\alpha = dh\alpha = d\mathcal{L}_X\alpha = \mathcal{L}_X d\alpha = \mathcal{L}_X\beta\wedge\alpha+\beta\wedge\mathcal{L}_X\alpha = \mathcal{L}_X\beta\wedge\alpha+\beta\wedge h\alpha.$$

Thus $\mathcal{L}_X\beta \wedge \alpha = 0$, hence $\mathcal{L}_X\beta(v) = 0$ for any $v \in \mathbb{R}X \oplus E^-$. Choose $\beta = 0$ on E^+, so $\beta = f(x)A$, hence $\beta = hA$. □

We now apply these invariants to geometric rigidity.

Theorem 9.8.21 (Godbillon–Vey Rigidity). *If* $\mathrm{GV}_0 = c$, $\mathrm{GV}_1 = hc$, *and* $\mathrm{GV}_2 = h^2c$ *for a negatively curved Riemannian metric on a surface, then the curvature is constant, c is the volume, and h the topological entropy.*

This is an immediate consequence of the following proposition:

Proposition 9.8.22. *For the geodesic flow of a negatively curved Riemannian metric on a surface,* $\frac{\mathrm{GV}_0\,\mathrm{GV}_2}{(\mathrm{GV}_1)^2} \geq 1$ *with equality if and only if the curvature is constant.*

Proof. Since Lemma 9.8.17 and Proposition 9.8.18 give

$$\mathrm{GV}_0 = \int_M A\wedge dA,\quad \mathrm{GV}_1 = \int_M -uA\wedge dA,\quad \mathrm{GV}_2 = \int_M u^2 + 3(\mathcal{L}_V u)^2 A\wedge dA,$$

the Cauchy–Schwarz inequality

$$\mathrm{GV}_1 = \int_M -uA\wedge dA \leq \Big(\int_M u^2 A\wedge dA\Big)^{\frac12}\Big(\int_M A\wedge dA\Big)^{\frac12} \leq (\mathrm{GV}_2)^{\frac12}(\mathrm{GV}_0)^{\frac12}$$

allows equality only if $u \equiv$ const.,[20] hence $K = -(\dot u + u^2) = -u^2 \equiv$ const. □

[20] And, redundantly, $\mathcal{L}_V u \equiv 0$.

Remark 9.8.23. By invoking the Godbillon–Vey invariants, Proposition 9.8.22 implicitly assumes that the invariant foliations are C^2, which by itself is known to imply constant curvature. This necessitates extending the definition to the case of $C^{1+\frac{1}{2}+\epsilon}$-invariant foliations for other applications. However, in both applications below, the invariant foliations are indeed C^2.

Proof of Theorem 9.8.2. The conjugacy F is $C^{k-\epsilon}$ when $\varphi^t \in C^k$ (Theorem 9.2.7), so the invariant foliations are $C^{k-\epsilon}$. A contact Anosov flow is the Reeb flow of a unique contact form. Thus F sends the contact form A for φ^t to that for ψ^t, and likewise for dA and the weak-unstable foliation—which is hence C^2. Thus, the Godbillon–Vey invariants match up, that is, $\mathrm{GV}_i^M = \mathrm{GV}_i^S$ for $i = 0, 1, 2$, so $\frac{\mathrm{GV}_0^M \mathrm{GV}_2^M}{(\mathrm{GV}_1^M)^2} = \frac{\mathrm{GV}_0^S \mathrm{GV}_2^S}{(\mathrm{GV}_1^S)^2} = 1$ by Proposition 9.8.22, which implies by Proposition 9.8.22 that M has constant curvature. □

Remark 9.8.24. This theorem is not contingent on defining Godbillon–Vey invariants for lower regularity because the conjugacy sends the smooth maximally isotropic foliation to a C^2 maximally isotropic foliation. The same goes for the next result, which recovers a special case of a rigidity result of Hurder and Katok via a remarkably simple proof.

Proof of Theorem 9.8.1. The C^2 splitting yields a Bott–Kanai connection.

Proposition 9.8.25. *There is a unique φ^t-invariant connection ∇ that parallelizes the geometric structure ($\nabla A = 0$, $\nabla dA = 0$, $\nabla E^\pm \subset E^\pm$) and with $\nabla_{Z^\mp} Z^\pm = p^\pm[Z^\mp, Z^\pm]$ and $\nabla_X Z^\pm = [X, Z^\pm] \pm \gamma Z^\pm$ for any sections $Z^\pm$ of $E^\pm$, where $p^\pm$ is the projection to $E^\pm$ given by the decomposition. If F is a ∇-parallel subbundle of TM then the (rank-1) bundle of volume forms on F has a natural flat connection induced by ∇.*[21]

With $F = E^+$, Proposition 9.8.25 gives a *parallel* unstable volume,[22] which is then holonomy invariant and hence gives the conditionals of the Bowen–Margulis measure. This establishes the hypothesis of Theorem 9.8.19 (with C^2 splitting), so Theorem 9.8.21 applies. □

[21] See [46, Proposition 2.3 & Section 3.2, Lemma 4.1].
[22] See [46, Section 4.2].

A

Measure-theoretic entropy of maps

This appendix provides results needed for Chapters 3 and 4. They are presented here both for completeness and to emphasize that there are certain aspects of the theory where one needs to use the time-t map to establish a result for flows. This is especially true when we need a countable collection of objects. For those familiar with the discrete setting, this appendix can be used as a reference for facts invoked in the main text.

A.1 Lebesgue spaces

To develop subtle notions in ergodic theory such as entropy, it is useful to have a suitable decomposition theory of a measure space, and this is the case for probability spaces adapted to an underlying topological structure. This is a surprisingly mild restriction, and we now develop this notion and some of the resulting properties, following [98], which is the definitive exposition of these topics, and with specific references to the locations for the corresponding statements in that book because that is where complete proofs are found. Since these notions are not immediately needed, readers can skip this section and refer back to it as needed.

Definition A.1.1 (Lebesgue space). A *Lebesgue space* $(X, \mathcal{A}, \mu)$ is a set with a probability measure μ on a complete σ-algebra $\mathcal{A}$ that is isomorphic to $([0,1], \overline{\mathcal{B}}, \lambda)$, that is, Lebesgue measure on the completion $\mathcal{B}$ of the Borel σ-algebra on the unit interval.

Remark A.1.2 ([98, Lemma 15.2]). If $(X, \mathcal{T}, \mu)$ is a Lebesgue space, then every complete σ-algebra $\mathcal{A} \subset \mathcal{T}$ is separable. In other words, it has a *basis* $\mathcal{B}$: a countable collection $\mathcal{B} = \{B_i\}_{i \in \mathbb{N}} \subset \mathcal{S}$ whose union is X and for which there is a null set N such that for $x, y \in X \smallsetminus N$ there exists $B \in \mathcal{B}$ such that $x \in B$, $y \notin B$.

Being a Lebesgue space is a far less restrictive condition than it seems.

Definition A.1.3. A *Polish space* is a topological space whose topology is given by a complete separable metric. A *standard Borel space* is a Borel subset of a Polish space (with the completion of the Borel σ-algebra).

Theorem A.1.4 (Isomorphism Theorem [98, Theorem 13.1]). *If X is a standard Borel space and μ a nonatomic Borel probability measure on X, then $(X, \overline{\mathcal{B}}, \mu)$ is a Lebesgue space.*

Proof. The topology of X has a countable base $\{O_i\}_{i \in \mathbb{N}}$, and

$$\varphi_1 \colon X \to \{0,1\}^{\mathbb{N}}, \quad x \mapsto (\chi_{O_i}(x))_{i \in \mathbb{N}}, \qquad \varphi_2 \{0,1\}^{\mathbb{N}} \to [0,1], \quad (a_i)_{i \in \mathbb{N}} \mapsto \sum_i \frac{a_i}{3^i}$$

are injective and Borel, so the Borel injection $\varphi_2 \circ \varphi_1$ is an isomorphism between $(X, \overline{\mathcal{B}}, \mu)$ and $([0,1], \overline{\mathcal{B}}, (\varphi_2 \circ \varphi_1)_*(\mu))$. The latter is isomorphic to $([0,1], \overline{\mathcal{B}}, \lambda)$ via $\varphi(x) := (\varphi_2 \circ \varphi_1)_*(\mu)([0,x))$. □

One important property of Lebesgue spaces is the following.

Theorem A.1.5 (Measurability lemma [98, Proposition 13.1]). *If f is a measure-preserving measurable map between Lebesgue spaces and A a measurable set with $f(A) \cap f(A^c) = \varnothing$, then $f(A)$ is measurable. In particular, if f is injective, then it is an isomorphism.*

A crucial notion for entropy theory is particularly well behaved in the context of Lebesgue spaces.

Definition A.1.6 (Measurable partition). If $(X, \mathcal{T}, \mu)$ is a measure space, then a partition ξ is a piecewise disjoint cover of X; its elements are called atoms. For $x \in X$ we define $\xi(x)$ by $x \in \xi(x) \in \xi$, and we say that two partitions essentially agree if there is a null set A such that $\xi(x) \smallsetminus A = \eta(x) \smallsetminus A$ for almost all $x \in X$.[1] A partition ξ is said to be *measurable* if it is (essentially[2]) *countably defined*: There is a *basis* for ξ, that is, a countable family of $B_n \in \mathcal{T}$ that separates the elements of ξ, that is, for $C_1 \neq C_2 \in \xi$ there is an n such that $C_1 \subset B_n$ and $C_2 \cap B_n = \varnothing$ or vice versa.

Example A.1.7. The extreme examples of measurable partitions are the trivial partition $\mathcal{N} := \{X\}$ corresponding to the trivial algebra $\mathcal{A}(\mathcal{N}) = \mathcal{N} := \{\varnothing, X\}$ and the point partition $\mathcal{E} := \{\{x\}\}_{x \in X}$ corresponding to the full algebra $\mathcal{A}(\mathcal{E}) = \mathcal{B}$.

Example A.1.8. The orbit partition of an irrational rotation is not a measurable partition. Consider $X = S^1 = \mathbb{R}/\mathbb{Z}$, $\alpha \notin \mathbb{Q}$, and the rotation $R_\alpha \colon x \mapsto x + \alpha \pmod 1$. Let $\xi := \{\{R_\alpha^i(x) \mid i \in \mathbb{Z}\}_{x \in S^1}\}$ be the partition of S^1 into the orbits of R_α. Each partition element is countable, hence measurable, but the partition is not. This can be seen by noting that the conditional measures on partition elements are R_α-invariant by uniqueness; since each partition element is a copy of $\mathbb{Z}$ with R_α acting by translation,

[1] Equivalently, $\xi = \eta \pmod 0$ if for any element $C \in \xi$ of positive measure one can find an element $D \in \eta$ such that $\mu(C \vartriangle D) = 0$. Here $\vartriangle$ means symmetric difference: $A \vartriangle B := (A \cup B) \smallsetminus (A \cap B) = (A \smallsetminus B) \cup (B \smallsetminus A)$.

[2] That is, essentially agrees with a partition with the following property.

the conditional measures are translation-invariant probability measures on $\mathbb{Z}$, which is impossible because all integers must have the same measure.[3]

Remark A.1.9. The reasoning in the preceding example in fact illustrates that the orbit partition of an ergodic transformation always has a trivial factor space.

The utility of the notion of a Lebesgue space is the correspondence between measurable partitions and various other natural constructs for ergodic theory:

Theorem A.1.10 (Rokhlin Correspondence [98, Theorem 15.1]). *If $(X, \mathbb{T}, \mu)$ is a Lebesgue space then there are bijections between*

- *measurable partitions of X,*
- *complete σ-algebras in $\mathbb{T}$,*
- *closed subalgebras of $L^0(X, \mathbb{T}, \mu)$,*
- *factors of $(X, \mathbb{T}, \mu)$ up to isomorphism.*

We now outline the nature of these various bijections.

In a Lebesgue space there is a duality between measurable partitions and complete σ-algebras. This bijection is defined as follows. For a measurable partition ξ the associated complete σ-algebra

$$\mathcal{A}(\xi) := \overline{\{A \in \mathcal{T} \mid A = \bigcup_{x \in A} \xi(x)\}}$$

is generated by the sets B_n in the definition of ξ being a measurable partition [98, Lemma 15.1]. Conversely, given a complete σ-algebra $\mathcal{A} \subset \mathcal{T}$, which is separable and hence generated by countably many sets B_n and null sets, we define the partition $\mathcal{P}(\mathcal{A})$ by

$$\mathcal{P}(\mathcal{A})(x) = \bigcap_{x \in B_n} B_n \setminus \bigcup_{x \notin B_n} B_n;$$

the B_n serve as the sets in the definition of measurability of this partition, and this is well defined independently of the choice of the B_n. The partition elements are also called the *atoms* of $\mathcal{A}$.

Proposition A.1.11 ([98, Proposition 15.1]). *Let $(X, \mathcal{T}, \mu)$ be a Lebesgue space. Then ξ is a measurable partition iff $\mathcal{P}(\mathcal{A}(\xi)) = \xi$, and $\mathcal{A} \subset \mathcal{T}$ is a complete σ-algebra iff $\mathcal{A}(\mathcal{P}(\mathcal{A})) = \mathcal{A}$.*

[3] Alternatively, note that by unique ergodicity of R_α the algebra generated by ξ consists of R_α-invariant sets and is hence the trivial algebra $\mathcal{N}$, which engenders the trivial partition, rather than the orbit partition.

Proof. If ξ is a measurable partition, then $\mathcal{A}(\xi)$ is generated by the B_n in the definition of "measurable partition," and these B_n are then used to define $\mathcal{P}(\mathcal{A}(\xi))$, which is therefore ξ itself. If $\mathcal{A} \subset \mathcal{T}$ is a complete σ-algebra, then it is separable and hence generated by countably many B_n together with null sets; these B_n then define $\mathcal{P}(\mathcal{A})$ and then also $\mathcal{A}(\mathcal{P}(\mathcal{A})) = \mathcal{A}$. □

Measurable partitions are also in an obvious bijective correspondence with factors:

Definition A.1.12 (Factor [98, Definition 15.2]). A *factor* of a Lebesgue space $(X, \mathbb{T}, \mu)$ is a Lebesgue space $(Y, \mathcal{S}, \nu)$ with a measurable map $\pi\colon X \to Y$ such that $\nu = \pi_*\mu$, the *projection*.

In this case $\{\pi^{-1}(\{y\}) \mid y \in Y\}$ is a measurable partition, and conversely, given a measurable partition ξ of $(X, \mathbb{T}, \mu)$ one can take its members to define equivalence classes and thus associate with it the factor $(X/\xi, \pi_*\mathbb{T}, \pi_*\mu)$ defined by $\pi(x) := \xi(x)$ and $\pi_*\mathbb{T} := \{A \subset X/\xi \mid \pi^{-1}(A) \in \mathbb{T}\}$. This is a Lebesgue space [98, Lemma 15.4], and these two associations are inverses of each other [98, Section 15.5].

Finally, the correspondence between σ-algebras in $\mathbb{T}$ and subalgebras of L^0 is $\mathbb{T} \ni \mathcal{A} \mapsto L^0(X, \mathcal{A}, \mu)$ and $L^0(X, \mathbb{T}, \mu) \ni \mathbf{A} \mapsto \{A \in \mathbb{T} \mid \chi_A \in \mathbf{A}\}$ [98, Proposition 15.2].

A particularly interesting special case of this correspondence involves the σ-algebra of invariant sets or the algebra of invariant functions, which results in the ergodic decomposition as the corresponding partition, with the corresponding factor describing the weights of its pieces.

Definition A.1.13 (Conditionals). A measurable partition ξ of a Lebesgue space has *conditional measures* $\{\mu_C\}_{C\in\xi}$ such that (C, μ_C) is a Lebesgue space for μ_ξ-almost every $C \in \xi$ and if $A \subset X$ is measurable then $A \cap C$ is μ_C-measurable for almost all $C \in \xi$, the function $C \mapsto \mu_C(A \cap C) =: \mu(A \mid C)$ is measurable on X/ξ and $\mu(A) = \int_\xi \mu_C(A \cap C)\, d\mu_\xi(C)$. This system of conditional measures is unique a.e., that is, if μ_C and μ'_C are conditionals for ξ then $\mu'_C = \mu_C$ for μ_ξ-a.e. $C \in \xi$.

(Whenever $\mu(C) > 0$ this is, of course, the same as in (3.3.3).)

Definition A.1.14 (Refinement). There is an obvious partial-ordering relation between partitions: $\xi \leq \eta$ if and only if for all $D \in \eta$ there exists a $C \in \xi$ such that $D \subset C$. If $\xi \leq \eta$ we say that η is a *refinement* of ξ and that ξ is *subordinate* to η.

Remark A.1.15. This ordering behaves well when passing from partitions to σ-algebras (ordered by inclusion): $\mathcal{A}(\xi) \subset \mathcal{A}(\xi') \Leftrightarrow \xi \leq \xi'$.

Definition A.1.16 (Join). If $\{\xi_i\}_{i\in I}$ is a collection of measurable partitions we define their *join* $\bigvee_{i\in I}\xi_i = \sup_{i\in I}\xi_i$ to be the smallest measurable partition that is a refinement of ξ_i for all $i \in I$; for finite partitions,

$$\xi \vee \eta = \{C \cap D \mid C \in \xi,\ D \in \eta,\ \mu(C \cap D) > 0\}.$$

Also, $\bigwedge_{i\in I}\xi_i = \inf_{i\in I}\xi_i$ is the largest partition subordinate to ξ_i for all $i \in I$, and

$$\xi_n \nearrow \xi :\Leftrightarrow \bigvee_{n\in\mathbb{N}} \xi_n = \xi \quad \text{and} \quad \xi_n \le \xi_{n+1} \text{ for all } n \in \mathbb{N},$$

$$\xi_n \searrow \xi :\Leftrightarrow \bigwedge_{n\in\mathbb{N}} \xi_n = \xi \quad \text{and} \quad \xi_{n+1} \le \xi_n \text{ for all } n \in \mathbb{N}.$$

Remark A.1.17. If for σ-algebras $\{\mathcal{A}_i\}_{i\in I}$ we define $\bigvee_{i\in I}\mathcal{A}_i$ to be the smallest σ-algebra that contains all $\mathcal{A}_i$ then

$$\mathcal{A}\Big(\bigvee_{i\in I}\xi_i\Big) = \bigvee_{i\in I}\mathcal{A}(\xi_i) \quad \text{and} \quad \mathcal{A}\Big(\bigwedge_{i\in I}\xi_i\Big) = \bigcap_{i\in I}\mathcal{A}(\xi_i).$$

One can define "$\nearrow$" and "$\searrow$" for σ-algebras analogously to the case of partitions.

An alternative description of this for L^2-functions is given in Example 3.2.14.

A.2 Entropy and conditional entropy

We now show how the notions in the previous section can be used to define the entropy of a partition for a measure-preserving transformation.

A.2.a Entropy of a partition. One way of introducing entropy is to consider information about points obtainable from a partition. Given X and a partition ξ, suppose we wish to locate a point $x \in X$. Knowing which element of ξ contains x provides some information; presumably a great deal if this element $\xi(x)$ has small measure—probabilistically speaking, this represents an unlikely event. We therefore wish to define an information function by $I[\xi](x) = \varphi(\mu(\xi(x)))$ with continuous nonnegative φ. A natural choice of φ is determined by the following consideration.

We say that finite partitions ξ and η are *independent* if

$$\mu(C \cap D) = \mu(C) \cdot \mu(D)$$

for all $C \in \xi$, $D \in \eta$. It is natural to wish the information obtained from knowledge about both partitions to be additive in this case, that is, we would like to have

$$I[\xi \vee \eta] = I[\xi] + I[\eta]$$

for independent partitions, where the *joint partition* $\xi \vee \eta$ is the smallest common refinement of ξ and η (Definition A.1.16). This implies that for two sets C, D with $\mu(C \cap D) = \mu(C) \cdot \mu(D)$ we require

$$\varphi(\mu(C) \cdot \mu(D)) = \varphi(\mu(C \cap D)) = \varphi(\mu(C)) + \varphi(\mu(D)).$$

Up to choice of a factor, this implies that $\varphi = -\log$.

Thus, the entropy of a partition is now defined as the (space) average of the (measurable) *information function*

$$x \mapsto I[\xi](x) = -\log \mu(\xi(x)). \tag{A.2.1}$$

Definition A.2.1. The *entropy* of a measurable partition ξ is

$$\begin{aligned} H(\xi) &:= H_\mu(\xi) \\ &:= \int_X I[\xi]\, d\mu = \begin{cases} -\displaystyle\sum_{\substack{C \in \xi \\ \mu(C)>0}} \mu(C) \log \mu(C) & \text{if } \mu\Big(\displaystyle\bigcup_{\substack{C \in \xi \\ \mu(C)>0}} C\Big) = 1, \\ \infty & \text{otherwise.} \end{cases} \end{aligned} \tag{A.2.2}$$

We denote by $\mathcal{P}_H$ the collection of measurable partitions (mod 0) with finite entropy, and we refer to these as *finite-entropy partitions.*

Finite-entropy partitions are (essentially) finite or countable, and for countable partitions the entropy may be infinite.

In most cases we suppress the dependence of entropy on the measure, but where more than one measure is involved in a discussion we use a subscript. If $f : X \to X$ is a measure-preserving transformation, ξ a measurable partition of X, and $f^{-1}(\xi) := \{f^{-1}(C) \mid C \in \xi\}$ then obviously

$$H(f^{-1}(\xi)) = H(\xi). \tag{A.2.3}$$

Definition (A.2.2) illuminates and makes natural the following notion of conditional entropy of a partition with respect to another partition which plays a central role in the entropy theory for measure-preserving transformations.

Definition A.2.2. Using conditional measures (see (3.3.3) or Definition A.1.13) define the (measurable) *conditional information function* by

$$I[\xi \mid \eta](x) := -\log \mu(\xi(x) \mid \eta(x)). \tag{A.2.4}$$

We define conditional entropy similarly to (A.2.2): Let ξ, η be two measurable partitions of (X, μ). The *conditional entropy* of ξ with respect to η is

$$H(\xi \mid \eta) := \int_X I[\xi \mid \eta]\, d\mu. \tag{A.2.5}$$

Remark A.2.3. It may at times be useful in connection with conditional information and entropy to think of ξ as the "numerator" and η as the "denominator" because both expressions are increasing in ξ and decreasing in η.

If $\mathcal{N} := \{X\}$ is the trivial partition then $H(\xi) = H(\xi \mid \mathcal{N})$.

Example A.2.4. Let $X = [0,1] \times [0,1]$ be the unit square with Lebesgue measure, η the partition into vertical intervals $\{x\} \times [0,1]$, and ξ the partition into vertical intervals $\{x\} \times [0, f(x)]$ and $\{x\} \times (f(x), 1]$, where $f\colon [0,1] \to [0,1]$ is a measurable function. Then

$$H(\xi \mid \eta) = -\int_0^1 [f(x)\log f(x) + (1 - f(x))\log(1 - f(x))]\, dx.$$

Remark A.2.5. Alternatively, $H(\xi \mid \eta) = -\sum_{D\in\eta}\mu(D)\sum_{C\in\xi}\mu(C \mid D)\log\mu(C \mid D)$.

For finite or countable measurable partitions note that if we denote by ξ_D the partition of D into the intersections $D \cap C$, $C \in \eta$, such that $\mu(D \cap C) > 0$ then

$$H(\xi \mid \eta) = \sum_{D\in\eta}\mu(D)H_{\mu_D}(\xi_D). \tag{A.2.6}$$

The following proposition summarizes basic properties of entropy and conditional entropy which we use systematically; this includes the behavior relative to the *joint partition* $\xi \vee \eta$ from Definition A.1.16.

Proposition A.2.6. *Let $(X, \mathcal{B}, \mu)$ be a probability space and let ξ, η, ζ be finite or countable measurable partitions of X and $\mathcal{N} = \{X\}$. Then*

(1) $0 \underbrace{\le}_{\text{“=”}\Rightarrow\xi=\mathcal{N}} -\log(\sup_{C\in\xi}\mu(C)) \le H(\xi) \underbrace{\le}_{\text{“=” if and only if all elements of } \xi \text{ have equal measure}} \log\operatorname{card}\xi$;

(2) • $0 \underbrace{\le}_{\text{“=”}\Leftrightarrow\xi\le\eta} H(\xi \mid \eta) \underbrace{\le}_{\text{“=”}\Leftrightarrow\xi \text{ and } \eta \text{ are independent}} H(\xi)$;

• *if $\zeta \ge \eta$ then $H(\xi \mid \zeta) \le H(\xi \mid \eta)$;*

(3) $I[\xi \vee \eta \mid \zeta] = I[\xi \mid \zeta] + I[\eta \mid \xi \vee \zeta]$, *thus*

• $H(\xi \vee \eta \mid \zeta) = H(\xi \mid \zeta) + H(\eta \mid \xi \vee \zeta)$ *and, in particular, $\zeta = \mathcal{N}$ gives*

$$H(\xi \vee \eta) = H(\xi) + H(\eta \mid \xi); \tag{A.2.7}$$

• $H(\xi \vee \zeta \mid \zeta) = H(\xi \mid \zeta)$;

• *if $\xi \le \eta$, then*

$$H(\eta \mid \zeta) = H(\xi \mid \zeta) + H(\eta \mid \xi \vee \zeta); \tag{A.2.8}$$

(4) $H(\xi \vee \eta \mid \zeta) \le H(\xi \mid \zeta) + H(\eta \mid \zeta)$ *and, in particular,* $H(\xi \vee \eta) \le H(\xi) + H(\eta)$;

(5) $H(\xi \mid \eta) + H(\eta \mid \zeta) \geq H(\xi \mid \zeta)$;

(6) *if* λ_i *are probability measures on* X, $a_i \geq 0$, $\sum_{i\in I} a_i = 1$, *then for every partition* ξ *measurable for all* λ_i,

$$\sum_{i\in I} a_i H_{\lambda_i}(\xi) \leq H_{\sum_{i\in I} a_i\lambda_i}(\xi) \leq \sum_{i\in I} a_i H_{\lambda_i}(\xi) + \log \operatorname{card} I,$$

and, indeed, the left inequality generalizes to $\int H_{\lambda_\alpha}(\xi)\,d\alpha \leq H_{\int \lambda_\alpha\,d\alpha}(\xi)$.

Corollary A.2.7. $H(\alpha \vee \beta) = H(\alpha) + H(\beta)$ *for independent partitions* α *and* β.

Proof. Use Proposition A.2.6(2) and (3). □

Corollary A.2.8. *For* $\xi, \eta \in \mathcal{P}_H$ *(Definition* A.2.1*) let*

$$d_R(\xi,\eta) := H(\xi \mid \eta) + H(\eta \mid \xi). \tag{A.2.9}$$

Then d_R *is a metric on* $\mathcal{P}_H$. *It is called the* Rokhlin metric.

Proof. $d_R(\xi,\eta) \geq 0$ by (2). If $d_R(\xi,\eta) = 0$ then $H(\xi \mid \eta) = H(\eta \mid \xi) = 0$. By (2) $\xi \geq \eta$ and $\eta \geq \xi$. But this immediately implies that $\xi = \eta \pmod 0$. The symmetry of d_R is immediate from (A.2.9). Finally, from (5),

$$\begin{aligned} d_R(\xi,\zeta) &= H(\xi \mid \zeta) + H(\zeta \mid \xi) \\ &\leq H(\xi \mid \eta) + H(\eta \mid \zeta) + H(\zeta \mid \eta) + H(\eta \mid \xi) = d_R(\xi,\eta) + d_R(\eta,\zeta). \end{aligned}$$ □

Several of the results in Proposition A.2.6 are consequences of convexity of the function $x \log x$, so we begin with pertinent convexity lemmas.

Definition A.2.9. $\phi\colon (a,b) \to \mathbb{R}$ is said to be *convex* if $x, y \in (a,b)$, $\lambda \in [0,1] \Rightarrow$

$$\phi(\lambda x + (1-\lambda y)) \leq \lambda\phi(x) + (1-\lambda)\phi(y)$$

and ϕ is said to be *strictly convex* if equality implies $x = y$ or $\lambda \in \{0,1\}$. Equivalently, the set of points in $\mathbb{R}^2$ above the graph of ϕ is convex, that is (recursively),

$$\phi\Big(\sum \alpha_i x_i\Big) \leq \sum \alpha_i \phi(x_i) \tag{A.2.10}$$

whenever $x_i \in (a,b)$, $\alpha_i \geq 0$, and $\sum \alpha_i = 1$. If ϕ is strictly convex then equality in (A.2.10) implies that the convex combination is trivial, that is, all x_i such that $\alpha_i \neq 0$ are equal.

Indeed, we have a continuous analog of (A.2.10):

Proposition A.2.10 (Jensen inequality). *If X is a probability space, $g \in L^1(X)$, and ϕ is convex on $\mathbb{R}$, then $\int_X \phi(g(x))\,dx \ge \phi\big(\int_X g(x)\,dx\big)$.*[4]

Proof. $t \mapsto \frac{\phi(t)-\phi(t_0)}{t-t_0}$ is nondecreasing, so ϕ has one-sided derivatives at each point, and the left derivative never exceeds the right derivative. For m between the left and right derivatives of ϕ at $t_0 = \int_X g$, we then have $\phi(t) \ge m(t-t_0) + \phi(t_0)$ for all t. Set $t = g(x)$ and integrate over X. □

Proposition A.2.11. *If $\phi'' > 0$ on (a,b), then ϕ is strictly convex.*

Proof. Fix $y > x$ and $\alpha, \beta \in (0,1)$ such that $\alpha + \beta = 1$. By the Mean-Value Theorem $\phi(\alpha x+\beta y) - \phi(x) = \phi'(\bar z)\beta(y-x)$ for some $\bar z \in (x, \alpha x + \beta y)$ and $\phi(y) - \phi(\alpha x + \beta y) = \phi'(z)\alpha(y-x)$ for some $z \in (\alpha x + \beta y, y)$ with $\phi'(\bar z) < \phi'(z)$ since $\phi'' > 0$. Then

$$\begin{aligned}\beta\big(\phi(y) - \phi(\alpha x + \beta y)\big) &= \phi'(z)\alpha\beta(y-x)\\ &> \phi'(\bar z)\alpha\beta(y-x) = \alpha\big(\phi(\alpha x+\beta y) - \phi(x)\big),\end{aligned}$$

hence $\phi(\alpha x + \beta y) < \alpha\phi(x) + \beta\phi(y)$. □

Proposition A.2.12. *The function $\phi\colon [0,\infty) \to \mathbb{R}$ defined by*

$$\phi(x) := \begin{cases} x\log x & \text{if } x \ge 0, \\ 0 & \text{if } x = 0, \end{cases} \tag{A.2.11}$$

is strictly convex.

Proof. $\phi'(x) = 1 + \log x$ and $\phi''(x) = 1/x > 0$ for $x \in (0,\infty)$. □

Lemma A.2.13. *If $b_i \in \mathbb{R}$ and $x_i \ge 0$ for $i = 1, \dots, m$ then*

$$\begin{aligned}\sum_{i=1}^m x_i(b_i - \log x_i) &\le \sum_{i=1}^m x_i \log\Big(\sum_{j=1}^m e^{b_j}\Big) - \phi\Big(\sum_{i=1}^m x_i\Big)\\ &\le \sum_{i=1}^m x_i \log\Big(\sum_{j=1}^m e^{b_j}\Big) + \frac{1}{e}.\end{aligned} \tag{A.2.12}$$

The first inequality is strict unless $x_i = ce^{b_i}$ with c independent of i. In particular, if $b_i = 0$ for all i, this gives

$$-\sum_{i=1}^m x_i \log x_i \le \sum_{i=1}^m x_i \log m + \frac{1}{e}.$$

[4]For strictly convex ϕ, equality implies that g is constant.

Proof. Let $a_i := e^{b_i}$ for all i, $A := \sum_{i=1}^m a_i$. Strict convexity of $\phi(x) = x \log x$ implies

$$\begin{aligned}\frac{1}{A}\sum_{i=1}^m x_i \log\Big(\frac{x_i}{a_i}\Big) &= \sum_{i=1}^m \frac{a_i}{A}\phi\Big(\frac{x_i}{a_i}\Big) \ge \phi\Big(\sum_{i=1}^m \frac{a_i}{A}\frac{x_i}{a_i}\Big) = \phi\Big(\frac{\sum_{i=1}^m x_i}{A}\Big)\\ &= \frac{\sum_{i=1}^m x_i}{A}\log\Big(\frac{\sum_{i=1}^m x_i}{A}\Big) = \frac{1}{A}\Big[\phi\Big(\sum_{i=1}^m x_i\Big) - \sum_{i=1}^m x_i \log A\Big]\end{aligned}$$

with equality iff $x_i/e^{b_i} = \text{const}$. Since $\phi(x) \ge -1/e$, this yields the claim. □

Since $\phi(1) = 0$, Lemma A.2.13 implies the following result:

Lemma A.2.14. *If $\sum_{i=1}^m x_i = 1$ in Lemma* A.2.13, *then*

$$\sum_{i=1}^m x_i(b_i - \log x_i) \le \log \sum_{i=1}^m e^{b_i} \quad \textit{with equality if and only if } x_i \sum_{i=1}^m e^{b_i} = e^{b_i}$$

(because $1 = \sum_i x_i = c \sum_i e^{b_i}$). If $b_i = 0$ for all i, this reduces to

$$-\sum_{i=1}^m x_i \log x_i \le \log m \quad \textit{with equality if and only if all } x_i = 1/m.$$

Lemma A.2.15. *If $x_i, a_i \ge 0$, $\sum_i a_i = 1$, then*

$$\sum_i x_i\phi(a_i) \le \phi\Big(\sum_i a_i x_i\Big) - \sum_i a_i\phi(x_i) \le 0.$$

Proof. The second inequality is (A.2.10),[5] and the first, monotonicity of logarithms:

$$\begin{aligned}&\phi\Big(\sum_i a_i x_i\Big) - \sum_i a_i\phi(x_i) - \sum_i x_i\phi(a_i)\\ &= \sum_i \underbrace{a_i x_i}_{\ge 0}\Big[\overbrace{\log\Big(\sum_i a_i x_i\Big) - \log x_i - \log a_i}^{=\log(\sum_i a_i x_i) - \log(x_i a_i) \ge 0 \text{ because log is increasing}}\Big] \ge 0.\end{aligned}$$

□

Proof of Proposition A.2.6. (1) μ is a probability measure, so (A.2.2) implies

$$0 \le -\log\Big(\sup_{C\in\xi}\mu(C)\Big) = \inf I[\xi] \le \int_X I[\xi]\,d\mu = H_\mu(\xi).$$

The statement $H(\xi) \le \log \operatorname{card}\xi$ is vacuous unless $\xi = (C_1, \dots, C_k)$ is finite, in which case Lemma A.2.14 yields $H(\xi) \le \log k$ with equality if and only if $\mu(C_i) = 1/k$ for all i.

[5] And hence generalizes as in Proposition A.2.10.

(2) The inequality follows from convexity of ϕ:

$$0 \le H(\xi \mid \eta) = -\sum_{D\in\eta} \mu(D) \sum_{C\in\xi} \phi(\mu(C \mid D))$$
$$= -\sum_{C\in\xi} \underbrace{\sum_{D\in\eta} \mu(D)\phi(\mu(C \mid D))}_{\ge\phi(\sum_{D\in\eta}\mu(D)\mu(C|D))=\phi(\mu(C))} \le H(\xi). \tag{A.2.13}$$

Now $\phi(x) < 0$ for $0 < x < 1$, so if $H(\xi \mid \eta) = 0$ then for every β with $\mu(D) > 0$ we have $\phi(\mu(C \mid D)) = 0$ for all $C \in \xi$ and consequently $\xi \le \eta$. If $H(\xi \mid \eta) = H(\xi)$ then we must have equality in (A.2.13) for each term of the summation over α, that is,

$$\phi(\mu(C)) = \phi\Big(\sum_{\substack{D\in\eta\\ \mu(D)>0}} \mu(D)\mu(C \mid D)\Big) = \sum_{\substack{D\in\eta\\ \mu(D)>0}} \mu(D)\phi\big(\mu(C \mid D)\big).$$

By strict convexity of the function ϕ this implies that if $\mu(D) > 0$ and $\mu(C) > 0$ then $\mu(C \mid D) = \mu(C)$, that is, $\mu(C \cap D) = \mu(C) \cdot \mu(D)$.

Applying the inequality $H_{\mu_D}(\xi \mid \zeta) \le H_{\mu_D}(\xi)$ to the conditional measures μ_D on each element D of the partition η and integrating over that partition we obtain $H(\xi \mid \zeta) = H(\xi \mid \zeta \vee \eta) \le H(\xi \mid \eta)$.

(3)
$$I[\xi \vee \eta \mid \zeta](x) = -\log \frac{\mu(\xi(x) \cap \eta(x) \cap \zeta(x))}{\mu(\zeta(x))}$$
$$= -\log \frac{\mu(\xi(x) \cap \zeta(x))}{\mu(\zeta(x))} - \log \frac{\mu(\xi(x) \cap \eta(x) \cap \zeta(x))}{\mu(\xi(x) \cap \zeta(x))}$$
$$= I[\xi \mid \zeta](x) + I[\eta \mid \xi \vee \zeta](x).$$

Now integrate with respect to x.

(4) This follows from (3) and the inequality $H(\eta \mid \xi \vee \zeta) \le H(\eta \mid \zeta)$ which in turn follows from (2) since $\xi \vee \zeta \ge \zeta$.

(5) By (3) and (4) we have $H(\zeta \mid \xi \vee \eta) = H(\xi \vee \zeta \mid \eta) - H(\xi \mid \eta) \le H(\zeta \mid \eta)$. Then using (3) several times we obtain

$$\begin{aligned} H(\zeta \mid \eta) + H(\eta \mid \zeta) &= H(\xi \vee \eta) + H(\eta \vee \zeta) - H(\eta) - H(\zeta)\\ &= H(\xi \vee \eta) + H(\zeta \mid \eta) - H(\zeta)\\ &= H(\xi \vee \eta \vee \zeta) - H(\zeta \mid \xi \vee \eta) + H(\zeta \mid \eta) - H(\zeta)\\ &\ge H(\xi \vee \eta \vee \zeta) - H(\zeta) \ge H(\xi \vee \zeta) - H(\zeta) = H(\xi \mid \zeta). \end{aligned}$$

(6) If $C \in \xi$ then (with $\phi(x) = x \log x$ as in (A.2.11)) Lemma A.2.15 gives

$$0 \le -\phi\Big(\sum_i a_i\lambda_i(C)\Big) + \sum_i a_i\phi(\lambda_i(C)) \le -\sum_i \lambda_i(C)\phi(a_i).$$

Summing over $C \in \xi$ this yields

$$0 \le H_{\sum_i a_i\lambda_i}(\xi) - \sum_i a_i H_{\lambda_i}(\xi) \le -\sum \phi(a_i) \le \log \operatorname{card} I$$

by (1). The continuous generalization of the left inequality is Proposition A.2.10 implemented in Lemma A.2.15. □

Remark A.2.16. The conditional expectation in Corollary 3.2.10 is also defined λ-a.e. uniquely by

$$\varphi_{\mathcal{T}}\restriction_C = \int_C \varphi \, d\lambda_C \quad \text{for all } C \in \pi(\mathcal{T}),$$

where $\pi(\mathcal{T})$ is as in Definition A.1.12.[6]

Remark A.2.17. At times we apply this result to a σ-algebra $\mathcal{T}$ that arises from a partition ξ; in that case we may write $E(\varphi \mid \xi)$ instead of $E(\varphi \mid \mathcal{A}(\xi))$.

With the notation of Definition A.1.16 we have the following theorem:

Theorem A.2.18. *Let ξ be a finite partition of a probability space $(X, \mathcal{B}, \mu)$.*

- *If $\eta_n \nearrow \eta$, then $H(\xi \mid \eta_n) \searrow H(\xi \mid \eta)$ as $n \to \infty$.*
- *If $\eta_n \searrow \eta$, then $H(\xi \mid \eta_n) \nearrow H(\xi \mid \eta)$ as $n \to \infty$.*

Proof. The monotonicity assertions are clear from Proposition A.2.6(2), and by (A.2.5) the limits are obtained by showing that

$$\int I[\xi \mid \eta_n] \to \int I[\xi \mid \eta].$$

Recall that $I[\xi \mid \eta](x) = -\log \mu(\xi(x) \mid \eta(x))$ and note that

$$\mu(\xi(x) \mid \eta(x)) = \int \chi_{\xi(x)} \, d\mu_{\eta(x)} = E(\chi_{\xi(x)} \mid \eta)(x),$$

where E is the conditional expectation operator from Corollary 3.2.10 (see also Remark A.2.17 for notation). Thus, for $x \in C \in \xi$ we have

$$\mu(\xi(x) \mid \eta(x)) = E(\chi_C \mid \eta)(x)$$

and hence

$$I[\xi \mid \eta] = -\sum_{C \in \xi} \chi_C \log E(\chi_C \mid \eta).$$

[6] Or rather, $\mathcal{P}(\mathcal{T})$ as in Proposition A.1.11.

This allows us to write

$$\begin{aligned} H(\xi \mid \eta) &= -\int \sum_{C\in\xi} \chi_C \log E(\chi_C \mid \eta) = -\sum_{C\in\xi} \int \chi_C \log E(\chi_C \mid \eta) \\ &= -\sum_{C\in\xi} \int E(\chi_C \log E(\chi_C \mid \eta) \mid \eta) = -\sum_{C\in\xi} \int E(\chi_C \mid \eta) \log E(\chi_C \mid \eta). \end{aligned}$$

The last step used Proposition 3.2.11.

Thus, we have now observed that it suffices to show

$$-\sum_{C\in\xi} E(\chi_C \mid \eta_n) \log(E(\chi_C \mid \eta_n)) \xrightarrow[n\to\infty]{L^1} -\sum_{C\in\xi} E(\chi_C \mid \eta) \log(E(\chi_C \mid \eta)). \tag{A.2.14}$$

To show this, we establish

$$-\sum_{C\in\xi} E(\chi_C \mid \eta_n) \xrightarrow[n\to\infty]{L^2} -\sum_{C\in\xi} E(\chi_C \mid \eta). \tag{A.2.15}$$

This implies that (A.2.15) holds for convergence in measure and hence that (A.2.14) holds for convergence in measure. Since the functions in question are bounded by $e \operatorname{card} \xi < \infty$, we obtain (A.2.14) by the Dominated-Convergence Theorem.

To prove (A.2.15) let us note that for $D \in \eta$ one can take $D_n \in \eta_N$ such that $d_\mu(D, D_n) \to 0$ as $n \to \infty$, where $d(A,B) := d_\mu(A,B) := \mu(A \vartriangle B)$ as in (3.4.5). Then $\chi_{D_n} \in L^2(X, \mathcal{A}(\eta_n), \mu)$, and since $E(\chi_D \mid \eta_n)$ is the orthogonal projection of χ_D to $L^2(X, \mathcal{A}(\eta_n), \mu)$ (Example 3.2.14), we use Proposition 3.2.13 to obtain

$$\|E(\chi_D \mid \eta_n) - \chi_D\|_2^2 \le \|\chi_{D_n} - \chi_D\|_2^2 = \mu(D_n \vartriangle D) \to 0.$$

Now, $h := E(\chi_C \mid \eta)$ can be L^2-approximated by linear combinations of χ_D with $D \in \mathcal{A}(\eta)$, so the preceding implies that

$$\|E(h \mid \eta_n) - h\|_2^2 \to 0.$$

Since $E(h \mid \eta_n) = E(\chi_C \mid \eta_n)$ (Proposition 3.2.11), we obtain (A.2.15). □

Remark A.2.19. Using approximations by finite partitions one can show that Theorem A.2.18 holds with the assumption that ξ has finite entropy instead of the assumption that ξ is finite.

For a measure space $(X, \mathcal{A}, \mu)$ and $m \in \mathbb{N}$ consider the space $\mathcal{P}_m$ of all equivalence classes mod 0 of partitions of X into at most m measurable sets. By adding null sets if necessary, we may assume that every partition in $\mathcal{P}_m$ has exactly m elements. For

$\xi, \eta \in \mathcal{P}_m$ consider now the set of bijections σ between the elements of ξ and η and set

$$\mathcal{D}(\xi,\eta) := \min_{\sigma} \sum_{C \in \xi} \mu(C \,\triangle\, \sigma(C)) = \min_{\sigma} \sum_{C \in \xi} d_\mu(C, \sigma(C)). \tag{A.2.16}$$

Obviously $\mathcal{D}$ is a metric. We need the fact that convergence in this metric guarantees convergence in the Rokhlin metric.

Proposition A.2.20. *For $\epsilon > 0$ there is a $\delta > 0$ such that $\mathcal{D}(\xi,\eta) < \delta \Rightarrow d_R(\xi,\eta) < \epsilon$.*

Remark A.2.21. In fact, the metrics $\mathcal{D}$ and d_R are equivalent on the space $\mathcal{P}_m$.

Proof. By symmetry it suffices to estimate $H(\eta \mid \xi)$. If $\mathcal{D}(\xi,\eta) = \delta$ write $\xi = (A_1, \dots, A_m)$, $\eta = (B_1, \dots, B_m)$ in such a way that $\sum_{i=1}^m \mu(A_i \,\triangle\, B_i) = \delta$. For $i \in \{1, \dots, m\}$ such that $\mu(A_i) > 0$ let $\alpha_i := \mu(A_i \setminus B_i)/\mu(A_i)$. Then the contribution of A_i to the expression for $H(\eta \mid \xi)$ in Definition A.2.2 is

$$\begin{aligned}
&-\mu(B_i \cap A_i) \log \frac{\mu(B_i \cap A_i)}{\mu(A_i)} - \sum_{j \neq i} \mu(B_j \cap A_i) \log \frac{\mu(B_j \cap A_i)}{\mu(A_i)} \\
&\leq \mu(A_i)[-(1-\alpha_i)\log(1-\alpha_i) - \alpha_i \log \alpha_i + \alpha_i \log(m-1)] \\
&= \mu(A_i)\Big[(1-\alpha_i)\log\frac{1}{1-\alpha_i} + \alpha_i \log \frac{m-1}{\alpha_i}\Big] \leq \mu(A_i)\log m.
\end{aligned}$$

Here the first inequality follows from Proposition A.2.6(1) by considering the measure induced on $A_i \setminus B_i = \bigcup_{j \neq i}(A_i \cap B_j)$ and estimating the entropy of η with respect to that measure. The last inequality uses convexity of $-\log x$. Thus

$$\begin{aligned}
H(\eta \mid \xi) \leq &\sum_{\mu(A_i) \geq \sqrt{\delta}} \mu(A_i)[-(1-\alpha_i)\log(1-\alpha_i) - \alpha_i \log \alpha_i + \alpha_i \log(m-1)] \\
&+ \sum_{\mu(A_i) < \sqrt{\delta}} \mu(A_i) \log m.
\end{aligned}$$

The second term does not exceed $m \log m \sqrt{\delta}$. To estimate the first, note that

$$\alpha_i \mu(A_i) = \mu(A_i \setminus B_i) = \sum_{j \neq i} \mu(B_j \cap A_i) \leq \sum_{j=1}^{m} \mu(A_j \,\triangle\, B_j) = \delta,$$

so for $\mu(A_i) \geq \sqrt{\delta}$ we get $\alpha_i \leq \sqrt{\delta}$. Now $\varphi(x) := -x \log x - (1-x)\log(1-x)$ is increasing on $(0, 1/2)$, so for $\delta < 1/4$ the first sum is dominated by $\varphi(\sqrt{\delta}) + \sqrt{\delta}\log(m-1)$ and hence $H(\eta \mid \xi) \leq \varphi(\sqrt{\delta}) + \sqrt{\delta}(m \log m + \log(m-1))$. Since $\varphi(x) \xrightarrow[x \to 0]{} 0$, the statement follows. □

A.2.b Entropy of a measure-preserving transformation.

Definition A.2.22. For a measurable partition ξ and a measure-preserving transformation f we define the *joint partition* as follows. For $I \subset \mathbb{R}$ set

$$\xi_I^f := \bigvee_{i \in I \cap \mathbb{Z}} f^i(\xi)$$

and

$$\xi_{-n}^f := \xi_{[-n,0)}^f, \quad \xi_-^f := \xi_{(-\infty,0)}^f, \quad \xi_n^f := \xi_{[0,n)}^f, \quad \xi_+^f := \xi_{[0,\infty)}^f, \quad \xi^f := \xi_{\mathbb{Z}}^f.$$

From now on, unless stated otherwise, we assume that all partitions are finite or countable measurable partitions with finite entropy.

Proposition A.2.23. $\lim_{n\to\infty} \frac{1}{n} H(\xi_{-n}^f)$ *exists (and equals* $\inf_{n\in\mathbb{N}} H(\xi_{-n}^f)/n$*).*

Proof. $H(\xi_{-n-m}^f) \le H(\xi_{-n}^f) + H(\xi_{-m}^f)$ by (A.2.7) and (A.2.3), so the statement follows by the Bowen–Fekete Lemma (Lemma 4.2.7). □

Definition A.2.24. $h(f,\xi) := h_\mu(f,\xi) := \lim_{n\to\infty} H(\xi_{-n}^f)/n = \inf_{n\in\mathbb{N}} H_\mu(\xi_{-n}^f)/n$ is the *measure-theoretic entropy* of the transformation f *relative* to the partition ξ.

The following proposition gives an alternative proof of existence of the limit $h(f,\xi)$ as well as another expression for it.

Proposition A.2.25. $H(\xi \mid \xi_{-n}^f) \searrow h(f,\xi)$ *as* $n \to \infty$.

Proof. We first note that

$$H(\xi_{-n}^f) = \sum_{k=0}^{n-1} H(\xi \mid \xi_{-k}^f).$$

For $n = 1$ this is clear since $H(\xi_{-n}^f) = H(\xi_{(-n,0]}^f)$, and using (A.2.7) we have

$$H(\xi_{-n-1}^f) = H(\xi_{[-n,0]}^f) = H(\xi \vee \xi_{-n}^f) = H(\xi_{-n}^f) + H(\xi \mid \xi_{-n}^f).$$

By the invariance property (A.2.3), this implies the claim.

Since the partition ξ_{-k}^f in the "denominator" is refined as k increases, by Proposition A.2.6(2) the sequence $b_n := H(\xi \mid \xi_{-n}^f)$ of summands is nonincreasing and hence convergent. Thus

$$\lim_{n\to\infty} b_n = \lim_{n\to\infty} \frac{1}{n} \sum_{k=0}^{n-1} b_k = \lim_{n\to\infty} \frac{1}{n} H(\xi_{-n}^f) = h_\mu(f,\xi). \qquad \square$$

Corollary A.2.26. *If $\xi \in \mathcal{P}_H$ then $h(f,\xi) = H(\xi \mid \xi_-^f)$.*

Proof. Combine Proposition A.2.25 with Theorem A.2.18. □

Definition A.2.27. The *entropy* of f with respect to μ (or the entropy of μ) is

$$h(f) := h_\mu(f) := \sup\left\{h_\mu(f,\xi) \mid \xi \in \mathcal{P}_H\right\}.$$

Obviously entropy is invariant under measure-theoretic isomorphism. We will see soon that this definition is more constructive than it seems; in many cases $h_\mu(f) = h_\mu(f,\xi)$ for an appropriately chosen ξ. (See, for example, Theorem A.3.7.)

Recalling the definition of the partition entropy through the information function (A.2.1)–(A.2.2) we can interpret the entropy $h_\mu(f,\xi)$ as the average amount of information provided by the knowledge of the "present state" in addition to the knowledge of an arbitrarily long past. Thus, a system with zero entropy can be viewed as strongly deterministic in the sense that an approximate knowledge of the entire past (that is, the past itinerary with respect to a finite partition) precisely determines the future itinerary.

A.3 Properties of entropy

A.3.a Properties of entropy with respect to a partition. The following proposition summarizes basic properties of the entropy $h(f,\xi)$ as a function of the partition ξ. It prepares the way for subsequent criteria which allow one to calculate the transformation entropy $h(f)$.

Proposition A.3.1. *Let $f\colon (X,\mu) \to (X,\mu)$ be a measure-preserving transformation of a probability space and $\eta,\xi \in \mathcal{P}_H$. Then*

(1) $0 \le \overline{\lim}_{n\to\infty} -(1/n)\log(\sup_{C\in\xi_{-n}^f} \mu(C)) \le h(f,\xi) \le H(\xi)$;

(2) $h(f,\xi \vee \eta) \le h(f,\xi) + h(f,\eta)$;

(3) $h(f,\eta) \le h(f,\xi) + H(\eta \mid \xi)$ *and, in particular, if $\xi \le \eta$ then* $h(f,\xi) \le h(f,\eta)$;

(4) $|h(f,\xi) - h(f,\eta)| \le H(\xi \mid \eta) + H(\eta \mid \xi)$ (*the* Rokhlin inequality);

(5) $h(f,f^{-1}(\xi)) = h(f,\xi) = h(f,f(\xi))$;

(6) $h(f,\xi) = h(f,\xi_{-k}^f) = h(f,\xi_{[-k,k]}^f)$ *for $k \in \mathbb{N}$*;

(7) *if λ is another probability measure and $p \in [0,1]$ then*

$$ph_\mu(f,\xi) + (1-p)h_\lambda(f,\xi) = h_{p\mu+(1-p)\lambda}(f,\xi).$$

Remark A.3.2. Property (4) means that $h(f, \cdot)$ is a Lipschitz function with Lipschitz constant 1 on $(\mathcal{P}_H, d_R)$ (see (A.2.9)).

Proof. The middle inequality in (1) follows directly from Proposition A.2.6(1) and the right inequality follows from Propositions A.2.25 and A.2.6(2).

(2) Since $(\xi \vee \eta)^f_{-n} = \xi^f_{-n} \vee \eta^f_{-n}$, this statement follows from (A.2.7) which is a particular case of Proposition A.2.6(3).

(3) By (A.2.7) $H(\xi^f_{-n}) \le H(\xi^f_{-n} \vee \eta^f_{-n}) = H(\eta^f_{-n}) + H(\xi^f_{-n} \mid \eta^f_{-n})$ and by using Proposition A.2.6(3) inductively, we obtain

$$\begin{aligned}H(\xi^f_{-n} \mid \eta^f_{-n}) &= H(f^{-1}(\xi) \mid \eta^f_{-n}) + H(f^{-1}(\xi^f_{1-n}) \mid f^{-1}(\xi) \vee \eta^f_{-n})\\ &\le H(f^{-1}(\xi) \mid f^{-1}(\eta)) + H(f^{-1}(\xi^f_{1-n}) \mid \eta^f_{-n})\\ &\le H(f^{-1}(\xi) \mid f^{-1}(\eta)) + H(f^{-2}(\xi) \mid f^{-2}(\eta)) + H(f^{-2}(\xi^f_{2-n}) \mid \eta^f_{-n})\\ &\le \cdots \le n\,H(\xi \mid \eta).\end{aligned}$$

Property (4) follows directly from (3).

Property (5) follows from the invariance property (A.2.3) since

$$H((f^{-1}(\xi))^f_{-n}) = H(f^{-1}(\xi^f_{-n})) = H(\xi^f_{-n}) = H(f(\xi^f_{-n})) = H((f(\xi))^f_{-n}).$$

(6) $(\xi^f_{-k})^f_{-n} = \xi^f_{[-n-k,-2]}$ and hence

$$\begin{aligned}h(f, \xi^f_{-k}) = h\Big(f, \xi^f_{[-n-k,-2]}\Big) &= \lim_{n\to\infty} \frac{1}{n} H(\xi^f_{1-n-k})\\ &= \lim_{n\to\infty} \frac{1}{n+k-1} H(\xi^f_{1-n-k}) = h(f, \xi).\end{aligned}$$

The argument for $f(\xi)$ is similar.

Property (7) follows directly from Proposition A.2.6(6). □

A.3.b The generator theorem. We can now formulate some criteria for calculating the entropy of a transformation.

Definition A.3.3. A family $\Xi \subset \mathcal{P}_H$ is said to be *sufficient* with respect to the measure-preserving transformation f if partitions subordinate to partitions of the form $\xi^f_{[-k,k]}$ ($\xi \in \Xi$, $k \in \mathbb{N}$) form a dense subset in $(\mathcal{P}_H, d_R)$ (see (A.2.9)).

Remark A.3.4. Proposition A.2.20 allows us to replace the Rokhlin metric by the metric $\mathcal{D}$ from (A.2.16) in this definition. In the case of a nonatomic Borel measure on a compact metric space a more obvious condition that guarantees sufficiency of a family $\Xi = \{\xi_n\}_{n\in\mathbb{N}}$ is $\operatorname{diam}(\xi_n) \to 0$, where $\operatorname{diam}(\xi) := \sup_{C\in\xi}(\operatorname{diam}(C))$.

Theorem A.3.5 (Kolmogorov–Sinai). $h_\mu(f) = \sup_{\xi\in\Xi} h_\mu(f,\xi)$ *for any sufficient family* Ξ *of partitions.*

Proof. For $\eta \in \mathcal{P}_H$ and $\epsilon > 0$ find $\xi \in \Xi$ and $k \in \mathbb{N}$ such that

$$d_R(\eta,\zeta) = H(\eta \mid \zeta) + H(\zeta \mid \eta) < \epsilon$$

for some partition $\zeta \le \xi^f_{[-k,k]}$. Using consecutively Proposition A.3.1(4), (3), and (6), we obtain

$$h_\mu(f,\eta) \le h_\mu(f,\zeta) + \epsilon \le h_\mu(f,\xi^f_{[-k,k]}) + \epsilon = h_\mu(f,\xi) + \epsilon.$$

Since ϵ is arbitrary, the statement follows. □

Definition A.3.6. A partition ξ is said to be a *generator* for f if $\Xi = \{\xi\}$ is a sufficient family.

The following corollary is the best-known and simplest-sounding criterion for calculating entropy.

Theorem A.3.7 (Kolmogorov–Sinai). *If* ξ *is a generator* with finite entropy *for* f *then* $h_\mu(f) = h_\mu(f,\xi)$.

Remark A.3.8. By Corollary A.2.26, this can be restated as saying that for a generator we have $h_\mu(f) = H(\xi \mid \xi^f_-)$. This was Kolmogorov's original definition of entropy.

We call a partition ξ a *one-sided generator* or *strong generator* if partitions subordinate to partitions of the form $\xi^f_{[1-k,0]}$ ($k \in \mathbb{N}$) are dense in the metric d_R. Clearly, a one-sided generator is a generator.

Proposition A.3.9. *If an invertible measure-preserving transformation possesses a one-sided generator* with finite entropy *then* $h_\mu(f) = 0$.

Proof. If ξ is a one-sided generator, then $\xi^f_{(-\infty,0]}$ is the point partition and hence

$$0 = H(f(\xi) \mid \xi^f_{(-\infty,0]}) = H(f(\xi) \mid (f(\xi))^f_-) = h(f, f(\xi))$$

by Corollary A.2.26. Since f is invertible, this implies $h_\mu(f,\xi) = 0$ by Proposition A.3.1(5). The claim then follows by Theorem A.3.7 because ξ is a one-sided generator and hence a generator. □

Note that the existence of countable sufficient families is ensured by separability of d_R (see (A.2.9), Remark A.2.21). This leads to a slight refinement of Theorem A.3.5 and a general existence theorem for generators. Suppose $\{\zeta_n\}_{n\in\mathbb{N}} \subset \mathcal{P}_H$ is a countable dense family of partitions and define $\xi_n := \bigvee_{i\le n} \zeta_i \in \mathcal{P}_H$. Then this defines an increasing sufficient family, and Theorem A.3.5 becomes the following result:

Proposition A.3.10. *With these choices,* $h_\mu(f) = \lim_{n\to\infty} h_\mu(f,\xi_n)$.

Proposition A.3.11. *An ergodic aperiodic transformation has a one-sided generator.*[7]

Proof [112, Proposition 9.5]. If $\{\zeta_i\}_{i\in\mathbb{N}} \subset \mathcal{P}_H$ is a countable dense family of partitions, A_i are sets of positive measure, N_i are such that $\mu\big(\bigcup_{0\le j\le N_i} f^{-j}(A_i)\big) > 1-2^{-i}$, then a minimal partition ξ that refines all the $\eta_i := (\zeta_i)_{-N_i}^0 \cap A_i$ is a generator because up to Rokhlin distance at most 2^{-i} it is contained in

$$\underbrace{\zeta_i \cap \bigcup_{0\le j\le N_i} f^{-j}(A_i)}_{\zeta_i\cap f^{-j}(A_i)\subset f^{-j}(\eta_i)} \subset (\eta_i)_0^{N_i} \subset (\xi)_0^{N_i}. \qquad \square$$

Remark A.3.12. Ergodicity is assumed here for (significant) convenience, but is not required for the conclusion. Proposition A.3.9 indicates that we should not expect a general existence result for one-sided finite-entropy generators; indeed, by Theorem A.3.7, this can only be the case for systems with finite entropy. In that case, there is even a finite generator for ergodic systems (Krieger's Theorem). Refinements such as this are at the heart of further results such as the Jewett–Krieger Theorem (Remark 3.3.35). At the same time, this also implies that when constructed in this generality, generators cannot be expected to encode any geometric information about a dynamical system.

This is a natural moment to make a connection with topological dynamics.

Proposition A.3.13. *If f is an expansive homeomorphism and $\epsilon > 0$ an expansivity constant, then for any invariant Borel probability measure a partition with diameter less than ϵ whose boundary is a null set is a generator.*

Proof. Expansivity ensures that the partition refines to the point partition under iteration of f. □

Corollary A.3.14. *If f is an expansive homeomorphism, then $\mu \mapsto h_\mu(f)$ is upper semicontinuous on $\mathfrak{M}(f)$ with the weak* topology, that is, if $\mu_n \xrightarrow[n\to\infty]{\text{weakly}} \mu$ in $\mathfrak{M}(f)$, then* $\lim_{n\to\infty} h_{\mu_n}(f) \le h_\mu(f)$.

Proof. If $\mu_n \xrightarrow[n\to\infty]{\text{weakly}} \mu$ let ξ be a finite partition with $\operatorname{diam}\xi < \epsilon$ and $\mu\,\partial\xi = 0$. Then $h_{\mu_n}(f) = h_{\mu_n}(f,\xi) \le \frac{H_{\mu_n}(\xi_{-k}^f)}{k} \xrightarrow[n\to\infty]{} \frac{H_\mu(\xi_{-k}^f)}{k} \xrightarrow[k\to\infty]{} h_\mu(f)$ (Definition A.2.24). □

[7] We do not claim finiteness of its entropy!

A.3.c Basic properties of entropy. The following proposition is a counterpart for measure-preserving transformations to Propositions 4.2.11 and 4.2.12.

Proposition A.3.15. (1) *If* $g\colon (Y,\nu) \to (Y,\nu)$ *is a factor (see Definition* 3.1.20) *of* $f\colon (X,\mu) \to (X,\mu)$ *then* $h_\nu(g) \le h_\mu(f)$.

(2) *If* A *is invariant under* f *and* $\mu(A) > 0$ *then*

$$h_\mu(f) = \mu(A)h_{\mu_A}(f) + \mu(X \setminus A)h_{\mu_{X\setminus A}}(f).$$

(3) *If* μ, λ *are two invariant probability measures for* f *then for any* $p \in [0,1]$,

$$ph_\mu(f) + (1-p)h_\lambda(f) = h_{p\mu+(1-p)\lambda}(f).$$

(4) $h_\mu(f^k) = kh_\mu(f)$ *for any* $k \in \mathbb{N}$, *and* $h_\mu(f^{-1}) = h_\mu(f)$, *so* $h_\mu(f^k) = |k|h_\mu(f)$ *for any* $k \in \mathbb{Z}$.

(5) $h_{\mu\times\lambda}(f \times g) = h_\mu(f) + h_\lambda(g)$.

Proof. (1) For any measurable partition η of Y, the preimage

$$\pi^{-1}(\eta) = \{\pi^{-1}D \mid D \in \eta\}$$

under the factor map π is a measurable partition of X and by definition $H_\mu(\pi^{-1}\eta) = H_\nu(\eta)$ and $h_\mu(f,\pi^{-1}\eta) = h_\nu(g,\eta)$. Thus

$$\begin{aligned} h_\mu(f) = \sup\{h_\mu(f,\xi) \mid H_\mu(\xi) < \infty\} &\ge \sup\{h_\mu(f,\pi^{-1}(\eta)) \mid H_\mu(\pi^{-1}(\eta)) < \infty\} \\ &= \sup\{h_\nu(g,\eta) \mid H_\nu(\eta) < \infty\} = h_\nu(g). \end{aligned}$$

(2) Let ξ be a measurable partition of X, $H_\mu(\xi) < \infty$, and $\zeta = \{A, X\setminus A\}$. By replacing ξ by $\xi \vee \zeta$ if necessary, we may assume that $\xi \ge \zeta$. Then

$$\begin{aligned} H_\mu(\xi^f_{-n}) = \mu(A)H_{\mu_A}(\xi^f_{-n}) &+ \mu(X \setminus A)H_{\mu_{X\setminus A}}(\xi^f_{-n}) \\ &- \mu(A)\log\mu(A) - \mu(X\setminus A)\log\mu(X\setminus A) \end{aligned}$$

by the definition of the conditional measures μ_A and $\mu_{X\setminus A}$, since A is f-invariant. The two last terms are independent of n and vanish in the limit.

(3) Proposition A.3.1(7) implies $h_{p\mu+(1-p)\lambda}(f) \le ph_\mu(f) + (1-p)h_\lambda(f)$.

On the other hand, given $C < ph_\mu(f) + (1-p)h_\lambda(f)$ take $C_1 < h_\mu(f)$ and $C_2 < h_\lambda(f)$ such that $pC_1 + (1-p)C_2 > C$ and partitions ξ_1 and ξ_2 such that

$h_\mu(f,\xi_1) > C_1$ and $h_\lambda(f,\xi_2) > C_2$. Then Proposition A.3.1(7) with $\xi := \xi_1 \vee \xi_2$ implies that

$$\begin{aligned} h_{p\mu+(1-p)\lambda}(f,\xi) &= ph_\mu(f,\xi) + (1-p)h_\lambda(f,\xi) \\ &\geq ph_\mu(f,\xi_1) + (1-p)h_\lambda(f,\xi_2) > pC_1 + (1-p)C_2 > C. \end{aligned}$$

Since $C < ph_\mu(f) + (1-p)h_\lambda(f)$ was arbitrary, this implies

$$h_{p\mu+(1-p)\lambda}(f) \geq ph_\mu(f) + (1-p)h_\lambda(f).$$

(4) If $k \in \mathbb{N}$ then $\frac{1}{n}H_\mu\big(\bigvee_{j=0}^{n-1} f^{-kj}\big(\bigvee_{i=0}^{k-1} f^{-i}(\xi)\big)\big) = \frac{k}{nk}H_\mu\big(\bigvee_{i=0}^{nk-1} f^{-i}(\xi)\big)$ and

$$h_\mu\Big(f^k, \bigvee_{i=0}^{k-1} f^{-i}(\xi)\Big) = k\, h_\mu(f,\xi).$$

Furthermore,

$$h_\mu(f,\xi) = h_\mu(f^{-1},\xi)$$

since $\xi_{-n}^f = f^{-n+1}(\xi_{-n}^{f^{-1}})$.

(5) Let ξ, η be measurable partitions of X and Y, correspondingly, and $\mathcal{N}_X = \{X\}$ and $\mathcal{N}_Y = \{Y\}$ the trivial partitions. Then $\xi \times \eta = (\xi \times \mathcal{N}_Y) \vee (\mathcal{N}_X \times \eta)$, where $\xi \times \mathcal{N}_Y$ and $\mathcal{N}_X \times \eta$ are independent as partitions of $X \times Y$.

By Corollary A.2.7,

$$H_{\mu\times\lambda}(\xi \times \eta) = H_{\mu\times\lambda}(\xi \times \mathcal{N}_Y) + H_{\mu\times\lambda}(\mathcal{N}_X \times \eta) = H_\mu(\xi) + H_\lambda(\eta).$$

Since $(\xi \times \eta)_{-n}^{f\times g} = \xi_{-n}^f \times \eta_{-n}^g$, this implies $h_{\mu\times\lambda}(f \times g, \xi \times \eta) = h_\mu(f,\xi) + h_\lambda(g,\eta)$ and hence $h_{\mu\times\lambda}(f \times g) \leq h_\mu(f) + h_\lambda(g)$. But the family of partitions of $X \times Y$ of the form $\xi \times \eta$ where $H_\mu(\xi) < \infty$ and $H_\lambda(\eta) < \infty$ is sufficient with respect to any measure-preserving transformation of $X \times Y$. Hence $h_{\mu\times\lambda}(f \times g) = h_\mu(f) + h_\lambda(g)$ by Theorem A.3.5. □

A.3.d Ergodic decomposition of entropy. If $\mu \perp \nu$ in Theorem 4.1.3, the statement can be read as one about an invariant partition of the space into two pieces. That kind of statement holds in much greater generality:

Theorem A.3.16. *If η is a measurable f-invariant partition by f-invariant sets, then $h(f) = \int_{X/\eta} h(f\restriction_B)\, d\mu_\eta(B)$.*

The proof of this result requires the development of additional properties of entropy for which it is essential to use infinite partitions. The title of the present section reflects an application of Theorem A.3.16:

Corollary A.3.17. *If η is the ergodic decomposition of (f,μ) (Theorem 3.3.37), then*

$$h(f) = \int_{X/\eta} h(f\restriction_B)\, d\mu_\eta(B).$$

Remark A.3.18. Despite this "linearity," the behavior of measure-theoretic entropy as a function of the measure is rather subtle because it is often not continuous (with respect to the weak topology). The coexistence of this "linearity" with discontinuity is related to the fact that even on the set of ergodic measures entropy is not continuous; for example, a weak limit of periodic δ-measures may have positive entropy. This is, in fact, exactly how we obtain measures with large entropy (Proposition 4.3.14).

We now develop the needed further properties of the entropy with respect to a partition.

Lemma A.3.19. *If $\xi \in \mathcal{P}_H$, η a measurable partition, and $\xi \le \eta$ or $\eta \le \xi$ then*

$$\frac{1}{n} H(\xi_n^f \mid \eta_-^f) \to H(\xi \mid \xi_-^f).$$

Remark A.3.20. By Corollary A.2.26 one can restate this as $\frac{1}{n} H(\xi_n^f \mid \eta_-^f) \to h(f,\xi)$.

Proof. Case 1*: $\eta \le \xi$.* This does not use $H(\xi) < \infty$. Since $f^{-n}(\eta_-^f \vee \xi_{n-1}^f) \nearrow \xi_-^f$, we have $H(\xi \mid f^{-n}(\eta_-^f \vee \xi_{n-1}^f)) \to H(\xi \mid \xi_-^f)$ by Theorem A.2.18. Also, Proposition A.2.6(3) gives

$$\begin{aligned} H(\xi_n^f \mid \eta_-^f) &= H(\xi \mid \eta_-^f) + H(f(\xi) \mid \xi \vee \eta_-^f) + \cdots + H(f^n(\xi) \mid \xi_{n-1}^f \vee \eta_-^f) \\ &= H(\xi \mid \eta_-^f) + H(\xi \mid f^{-1}(\xi \vee \eta_-^f)) + \cdots + H(\xi \mid f^{-n}(\xi_{n-1}^f \vee \eta_-^f)), \end{aligned}$$

which implies the claim.

Case 2*: $\xi \le \eta$.* On one hand $H(\xi_n^f \mid \eta_-^f)/n \le H(\xi_n^f \mid \xi_-^f)/n \to H(\xi \mid \xi_-^f)$ from the previous case. On the other hand, (A.2.8) gives

$$H(\xi_n^f \mid \eta_-^f) = H(\eta_n^f \mid \eta_-^f) - H(\eta_n^f \mid \xi_n^f \vee \eta_-^f) \ge H(\eta \mid \eta_-^f) - H(\eta_n^f \mid \xi_n^f \vee \xi_-^f),$$

so, using Case 1 with ξ and η interchanged,

$$\begin{aligned} \lim_{n\to\infty} \frac{1}{n} H(\xi_n^f \mid \eta_-^f) &\ge H(\eta \mid \eta_-^f) - \lim_{n\to\infty} \frac{1}{n} H(\eta_n^f \mid \xi_n^f \vee \xi_-^f) \\ &= \lim_{n\to\infty} \Big(\frac{1}{n} H(\eta_n^f \mid \xi_-^f) - \frac{1}{n} H(\eta_n^f \mid \xi_n^f \vee \xi_-^f)\Big) \\ &= \lim_{n\to\infty} \frac{1}{n} H(\xi_n^f \mid \xi_-^f) = H(\xi \mid \xi_-^f), \end{aligned}$$

where the penultimate step again used Proposition A.2.6(3). □

If ν is a partition, then one would expect $\nu^f_{(-\infty,-n]}$ to essentially "disappear" as $n \to -\infty$. The following statement is a way of making this precise in a specific context.

Lemma A.3.21. *If $\xi, \eta \in \mathcal{P}_H$ are such that $\xi \le \eta$ and ν is a measurable partition, then $H(\xi \mid \eta^f_- \vee f^{-n}(\nu^f_-)) \to H(\xi \mid \eta^f_-)$.*

Proof. We first treat the case $\xi = \eta$. By Lemma A.3.19 applied to $\eta \le \eta \vee \nu$ and Proposition A.2.6(3) we have

$$\begin{aligned}
H(\eta \mid \eta^f_-) &= \lim_{n\to\infty} \frac{1}{n} H(\eta^f_n \mid \eta^f_- \vee \nu^f_-) \\
&= \lim_{n\to\infty} \frac{1}{n}\Big(H(\eta \mid \eta^f_- \vee \nu^f_-) + \underbrace{H(f(\eta) \mid f(\eta^f_-) \vee \nu^f_-)}_{=H(\eta\mid\eta^f_-\vee f^{-1}(\nu^f_-))} + \cdots + \underbrace{H(f^n(\eta) \mid f^n(\eta^f_-) \vee \nu^f_-)}_{=H(\eta\mid\eta^f_-\vee f^{-n}(\nu^f_-))}\Big) \\
&= \lim_{n\to\infty} H(\eta \mid \eta^f_- \vee f^{-n}(\nu^f_-)).
\end{aligned}$$

For arbitrary $\xi \le \eta$, this and (A.2.8) (twice) now give

$$\begin{aligned}
\lim_{n\to\infty} H(\xi \mid \eta^f_- \vee f^{-n}(\nu^f_-)) &= \lim_{n\to\infty} \Big(H(\eta \mid \eta^f_- \vee f^{-n}(\nu^f_-)) - H(\eta \mid \xi \vee \eta^f_- \vee f^{-n}(\nu^f_-))\Big) \\
&= H(\eta \mid \eta^f_-) - \underbrace{\lim_{n\to\infty} H(\eta \mid \xi \vee \eta^f_- \vee f^{-n}(\nu^f_-))}_{\le H(\eta\mid\xi\vee\eta^f_-)} \\
&= H(\xi \mid \eta^f_-) \\
&\ge H(\xi \mid \eta^f_- \vee f^{-n}(\nu^f_-)).
\end{aligned}$$

□

There is also a formula for computing the entropy with respect to the join of two partitions.

Proposition A.3.22. *If $\xi, \eta \in \mathcal{P}_H$ then $h(f, \xi \vee \eta) = h(f, \eta) + H(\xi \mid \eta^f \vee \xi^f_-)$.*

Proof. By Lemma A.3.19,

$$\begin{aligned}
\frac{1}{n} H(\xi^f_n \vee \eta^f_n \mid \xi^f_- \vee \eta^f_-) &= \frac{1}{n} H(\eta^f_n \mid \xi^f_- \vee \eta^f_-) + \frac{1}{n} \underbrace{H(\xi^f_n \mid \xi^f_- \vee \eta^f_- \vee \eta^f_n)}_{=\sum_{i=0}^n H(\xi\mid\xi^f_-\vee\eta^f_-\vee\eta^f_i)} \\
&\xrightarrow[n\to\infty]{} H(\eta, \eta^f_-) + H(\xi \mid \xi^f_- \vee \eta^f).
\end{aligned}$$

Now use Corollary A.2.26. □

Corollary A.3.23. *If $\eta \in \mathcal{P}_H$ and ξ is fixed by f (that is, consists of f-invariant sets), then $h(f, \xi \vee \eta) = h(f, \eta)$.*

Proof. For $\xi \in \mathcal{P}_H$ this is Proposition A.3.22 (because the last term there vanishes). To reduce to this case take $\xi_n \nearrow \xi$ with finite entropy (and necessarily fixed); the claim then holds for the ξ_n and this gives Corollary A.3.23 by Theorem A.2.18 (Remark A.2.19). □

Corollary A.3.23 lends itself to a convex decomposition of entropy as follows.

Proposition A.3.24. *If $\xi \in \mathcal{P}_H$ and η is a partition fixed by f, then with the notation of* (A.2.6) *and Definition* A.1.6 *we have*

$$h(f,\xi) = \int_{X/\eta} h(f\restriction_B, \xi_B)\, d\mu_\eta(B).$$

Proof. Corollary A.2.26 and (A.2.5) give

$$h(f\restriction_B, \xi_B) = H(\xi_B \mid (\xi_B)_-^{f\restriction_B}) = -\int_B \log \mu(\xi_B(x) \mid (\xi_B)_-^{f\restriction_B}(x))\, d\mu_B$$

so

$$\begin{aligned}\int_{X/\eta} h(f\restriction_B, \xi_B)\, d\mu_\eta(B) &= -\int_{X/\eta}\int_B \log \mu(\xi_B(x) \mid (\xi_B)_-^{f\restriction_B}(x))\, d\mu_B\, d\mu_\eta(B)\\ &= \int_X \log \mu(\xi(x) \mid (\eta \vee \xi_-^f)(x))\, d\mu = H(\xi \mid \eta \vee \xi_-^f),\end{aligned}$$

while Corollaries A.3.23 and A.2.26 give

$$\begin{aligned}h(f,\xi) &= h(f,\xi\vee\eta)\\ &= H(\xi\vee\eta \mid (\xi\vee\eta)_-^f)\\ &= H(\xi\vee\eta \mid \eta\vee\xi_-^f)\\ &= H(\xi \mid \eta\vee\xi_-^f) + H(\eta \mid \xi\vee\eta\vee\xi_-^f)\\ &= H(\xi \mid \eta\vee\xi_-^f)\end{aligned}$$

by Proposition A.2.6(3). □

Proof of Theorem A.3.16. Let $\{\zeta_n\}_{n\in\mathbb{N}} \subset \mathcal{P}_H$ be a countable dense family of partitions and define $\xi_n := \bigvee_{i\le n}\zeta_i \in \mathcal{P}_H$. Then on one hand, Proposition A.3.24 implies

$$h(f,\xi_n) = \int_{X/\eta} h(f\restriction_B, (\xi_n)_B)\, d\mu_\eta(B).$$

On the other hand, Proposition A.3.10 implies

$$h(f,\xi_n) \nearrow h(f) \quad \text{and} \quad h(f\restriction_B, (\xi_n)_B) \nearrow h(f\restriction_B).$$

The claim then follows by the Monotone Convergence Theorem. □

A.3.e The Shannon–Macmillan–Breiman Theorem. Definition A.2.24 suggests that the average measure of elements of ξ^f_{-n} should be about $e^{-nh(f,\xi)}$. For ergodic transformations this turns out to be true in a much stronger sense.

Theorem A.3.25 (Shannon–Macmillan–Breiman). *If $f\colon X \to X$ is a μ-preserving ergodic transformation and ξ a measurable partition with finite entropy, then*

$$-\frac{1}{n}\log\mu(\xi^f_{-n}(x)) \xrightarrow[n\to\infty]{\text{a.e. and in } L^1} h_\mu(f,\xi),$$

where ξ^f_{-n} is as in Definition A.2.22.

Remark A.3.26. To keep the proof simple, we will assume that the partition is finite.

Lemma A.3.27. *There is an h such that $-\lim_{n\to\infty}\frac{1}{n}\log\mu(\xi^f_{-n}(x)) = h$ a.e.*

Proof. We write

$$I_n(x) := I[\xi^f_{-n}](x) = -\log\mu(\xi^f_{-n}(x)).$$

Since

$$f^{-1}(\xi^f_{1-n}(f(x))) = \Big(\bigvee_{i=2}^{n} f^{-i}(\xi)\Big)(x) \supset \xi^f_{-n}(x)$$

and f is measure preserving, we have $I_{n-1}(f(x)) \le I_n(x)$. Thus

$$\varliminf_{n\to\infty}\frac{1}{n}I_n(f(x)) \le \varliminf_{n\to\infty}\frac{1}{n}I_n(x).$$

Corollary 3.3.23 implies that there is an $h \in \mathbb{R}$ such that

$$\varliminf_{n\to\infty}\frac{1}{n}I_n(x) \overset{\text{ae}}{=} h. \tag{A.3.1}$$

To establish

$$\lim_{n\to\infty}\frac{1}{n}I_n(x) \overset{\text{ae}}{=} h, \tag{A.3.2}$$

we show that $\varlimsup_{n\to\infty}\frac{1}{n}I_n(x) \le h$ a.e. Fix $\epsilon > 0$ and $L > 3$. If

$$\alpha_n := \big\{x \in X \mid \tfrac{1}{n}I_n(x) \le h + \epsilon\big\},$$

then (A.3.1) implies that $\mu\big(\bigcup_{n\ge L}\alpha_n\big) = 1$, so there is an $M \ge L$ such that

$$A := \bigcup_{L\le n\le M}\alpha_n$$

satisfies $\mu(A) > 1 - \epsilon$. The definitions of A and α_n yield

$$\forall\, x \in A\, \exists\, q \in \mathbb{N}: \quad L \le q \le M \text{ and } \mu(\xi^f_{-n}(x)) \ge e^{-q(h+\epsilon)}. \tag{A.3.3}$$

The Birkhoff Ergodic Theorem shows that $\frac{1}{n}\sum_{i=0}^{n-1} \chi_A \circ F^i \xrightarrow[n\to\infty]{} \mu(A) > 1 - \epsilon$ a.e., hence in measure, so for $\delta > 0$ there is a $B \subset X$ with $\mu(B) > 1 - \delta$ and an $N \in \mathbb{N}$ with

$$\forall\, x \in B, n > N: \quad \frac{1}{n}\operatorname{card}\{i \mid 0 \le i < n \mid f^i(x) \notin A\} < 2\epsilon. \tag{A.3.4}$$

Claim A.3.28. *If L is large enough, then $-M_n := \operatorname{card} \xi^f_{-n}\restriction_B \le e^{n(h+2\epsilon(1+\log\operatorname{card}\xi))}$ for all large $n \in \mathbb{N}$.*

Alternatively, M_n can be described as the number of elements of ξ^f_{-n} that intersect B in a set of positive measure. We prove this claim below.

The claim implies that the "bad" set

$$\{x \in B \mid \mu(\xi^f_{-n}(x)) < e^{-n(h+2\epsilon(2+\log\operatorname{card}\xi))}\}$$

has measure less than $M_n \cdot e^{-n(h+2\epsilon(2+\log\operatorname{card}\xi))} \le e^{-2n\epsilon}$, which is summable. By the Borel–Cantelli Lemma (Theorem A.3.29) almost every x is in these bad sets for at most finitely many n, so

$$\overline{\lim_{n\to\infty}} \frac{1}{n} I_n(x) \le h + 2\epsilon(2 + \log\operatorname{card}\xi)$$

a.e. in B. Since δ is arbitrary, this holds a.e. on X, and as $\epsilon \to 0$, (A.3.2) follows. □

Here is the measure-theory result we invoked:

Theorem A.3.29 (Borel–Cantelli). *If $\sum_{n\in\mathbb{N}} \mu(A_n) < \infty$ then $\mu(\bigcap_{n\in\mathbb{N}} \bigcup_{i\ge n} A_i) = 0$, that is, almost every point lies in only finitely many A_n.*

Proof. $\mu(\bigcap_{n\in\mathbb{N}} \bigcup_{i\ge n} A_i) \le \mu(\bigcup_{i\ge n} A_i) \le \sum_{i\ge n} \mu(A_i) \to 0$. □

Proof of Claim A.3.28. Take $C \in \xi^f_{-n}$ with $\mu(C \cap B) > 0$. Then (A.3.3) and (A.3.4) imply that for $x \in C \cap B$ there are pairwise disjoint intervals $[m_k, n_k] \subset [1, n] \subset \mathbb{N}$ such that

(1) $L \le n_k - m_k \le M$,

(2) $\sum_k (n_k - m_k) \ge (1 + 2\epsilon)n$,

(3) $f^{m_k}(x) \in A$ and $\mu(\xi^f_{n_k - m_k}(f^{m_k}(x))) \ge e^{(m_k - n_k)(h+\epsilon)}$.

To see this, take $m_1 := \min\{i \in [1,n] \mid f^i(x) \in A\}$, then $n_1 := m_1 + q$, where q is as in (A.3.3) for $f^{m_1}(x)$. Then take $m_2 := \min\{i \in (n_1,n] \mid f^i(x) \in A\}$ and so on.

To see how many different such C there can be note that each such C is determined by a sequence of choices of $C_i \in \xi$ for $i \in [1,n]$. In brief, we have

$$\begin{aligned}\#\text{choices of } C = {} & \#\text{choices of } \{[m_k,n_k]\} \times \#\text{choices on } [m_k,n_k]\\ & \times \#\text{choices off } [m_k,n_k].\end{aligned}$$

For some i there are $\operatorname{card}\xi$ choices, but (3) provides much better control for the collective choice corresponding to $[m_k,n_k]$. Thus, for a given choice of $\{(m_k,n_k)\}$ this allows at most

$$(\operatorname{card}\xi)^{2\epsilon n}\cdot e^{\sum_k (n_k-m_k)(h+\epsilon)} \le e^{n(h+\epsilon+2\epsilon\log\operatorname{card}\xi)}$$

different such C.

On the other hand, the number of choices of $\{(m_k,n_k)\}$ can be bounded by noting that $1 \le k \le K := \lfloor n/L \rfloor$; so we are choosing two subsets of $[1,n]$ (the m_k and the n_k) of cardinality at most K. A (generous) upper bound for the possibilities is given by

$$\Big[\sum_{i\le K}\binom{n}{i}\Big]^2 \le \Big[K\binom{n}{K}\Big]^2.$$

We bound $K\binom{n}{K}$ using $\binom{n}{K} = \frac{n!}{K!(n-K)!}$, the Stirling formula [313]

$$n! = \sqrt{2\pi n}\Big(\frac{n}{e}\Big)^n e^{\zeta_n} \quad\text{with}\quad \frac{1}{12n+1} < \zeta_n < \frac{1}{12n}, \tag{A.3.5}$$

and writing $K = \ell n$,

$$\begin{aligned}K\binom{n}{K} &= \ell n\frac{n!}{(\ell n)!((1-\ell)n)!}\\ &= \frac{\ell n}{\sqrt{2\pi}}\sqrt{\frac{n}{\ell n(1-\ell)n}}\frac{n^n}{(\ell n)^{\ell n}((1-\ell)n)^{(1-\ell)n}}\underbrace{e^{\zeta_n-\zeta_{\ell n}-\zeta_{(1-\ell)n}}}_{<1 \text{ since } a,b\in\mathbb{N}\Rightarrow\frac{1}{a+b}-\frac{1}{a+1}-\frac{1}{b+1}<0}\\ &< \sqrt{\frac{\ell n}{2\pi}}\underbrace{\sqrt{\frac{\ell}{1-\ell}}}_{<1 \text{ for } L>1}\big(\ell^\ell(1-\ell)^{1-\ell}\big)^{-n} < \sqrt{\frac{K}{2\pi}}e^{-[\ell\log\ell+(1-\ell)\log(1-\ell)]n}.\end{aligned}$$

Take L so large that ℓ is small enough for $-[\ell\log\ell + (1-\ell)\log(1-\ell)] \le \epsilon/4$,[8] then take n large enough for $\sqrt{\frac{K}{2\pi}} < \sqrt{\frac{n}{2\pi}} < e^{n\epsilon/4}$ to obtain $K\binom{n}{K} \le e^{n\epsilon/2}$, hence

$$M_n \le e^{n(h+\epsilon+2\epsilon\log\operatorname{card}\xi)}\Big[K\binom{n}{K}\Big]^2 \le e^{n(h+\epsilon+2\epsilon\log\operatorname{card}\xi)}e^{\epsilon n}.$$

□

[8] It is interesting to note that $\ell\log\ell + (1-\ell)\log(1-\ell) = -H(\{[0,\ell],[1-\ell,1]\})$.

Proof of the Shannon–Macmillan–Breiman Theorem. To prove convergence in mean, $-\frac{1}{n}\log\mu(\xi_{-n}^{f}(x)) \xrightarrow[n\to\infty]{L^1} h$, we use uniform integrability.

Definition A.3.30. We say that $\mathcal{F} \subset L^1$ is *uniformly integrable* if for $\epsilon > 0$ there exists a $\delta > 0$ such that $f \in \mathcal{F}$ and $\mu(A) < \delta$ imply $\int |f|\, d\mu < \epsilon$.

Theorem A.3.31 (Vitali). *If $f_n \to f$ in measure and $\{f_n\}_{n\in\mathbb{N}}$ is uniformly integrable, then $f_n \to f$ in L^1.*

Proof. $0 \le g_n := |f_n - f| \to 0$ in measure and is uniformly integrable. Given $\epsilon > 0$ take $\delta > 0$ such that $\int_A g_n < \epsilon/2$ for all n whenever $\mu(A) < \delta$ and $N \in \mathbb{N}$ such that $g_n(x) < \epsilon/2$ for all $n \ge N$ and x outside a set A of measure less than δ. Then $\int g_n = \int_A g_n + \int_{X\setminus A} g_n < \epsilon/2 + \epsilon/2$ for all $n \ge N$. □

Given $\epsilon > 0$, take $\delta \in (0, 1/e)$ such that $-\delta[\log\delta + \operatorname{card}\xi] < \epsilon$. Then

$$\begin{aligned}
\mu(A) < \delta \Rightarrow \int_A \frac{1}{n} I_n\, d\mu &= -\frac{1}{n} \sum_{C\in\xi_{-n}^{f}} \mu(A\cap C) \underbrace{\log\mu(C)}_{\le \log\mu(A\cap C)} \\
&\le -\frac{\mu(A)}{n} \sum_{C\in\xi_{-n}^{f}} \mu(C \mid A) \log(\mu(A)\cdot\mu(C\mid A)) \\
&= \mu(A)\Big[\underbrace{\frac{1}{n} H(\xi_{-n}^{f} \mid A)}_{\le \frac{1}{n}\log\operatorname{card}\xi_{-n}^{f} \le \log\operatorname{card}\xi} - \frac{1}{n} \underbrace{\sum_{C\in\xi_{-n}^{f}} \mu(C\mid A)\log\mu(A)}_{=\log\mu(A)\le -1 \text{ since } \mu(A)\le 1/e}\Big] \\
&\le -\mu(A)[\log\mu(A) + \log\operatorname{card}\xi] < \epsilon,
\end{aligned}$$

which is uniform integrability. Therefore (A.3.2) and Theorem A.3.31 imply

$$\frac{1}{n} I_n(x) \xrightarrow[n\to\infty]{L^1} h = \lim_{n\to\infty} \frac{I_n}{n} = \int \lim_{n\to\infty} \frac{I_n}{n} = \lim_{n\to\infty} \int \frac{I_n}{n} = \lim_{n\to\infty} \frac{1}{n} H(\xi_{-n}^{f}) = h(f,\xi).$$

□

A.3.f Skew products. We now describe a class of examples to demonstrate that the addition of subexponential complexity does not increase the entropy.

Proposition A.3.32. *Consider a probability space (Y, μ), an invertible measure-preserving transformation $f: Y \to Y$, and the transformation*

$$S: X := Y \times S^1 \to X, \quad (y, s) \mapsto (f(y), R_{\phi(y)}(s)),$$

where $\phi: Y \to S^1$ is measurable, R_α is the rotation (as in Example A.1.8), and S^1 carries Lebesgue measure m. Then $h(S) = h(f)$.

Proof. Since f is a factor of S, we have $h(f) \leq h(S)$. The main point is the reverse inequality.

We start with some choices and observations.

- Let $\{B_n\}_{n \in \mathbb{N}}$ be a basis for Y (Remark A.1.2),
- $\alpha_m := \bigvee_{n=0}^{m} \{B_n, Y \smallsetminus B_n\}$,
- $\xi_m := \bigvee_{n=0}^{m} \{B_n \times S^1, X \smallsetminus (B_n \times S^1)\} \nearrow \xi := \{I_y\}_{y \in Y}$, where $I_y := \{y\} \times S^1$,
- $\beta_m := \{[\frac{i}{m}, \frac{i+1}{m})\}_{i=0}^{m-1}$,
- $\zeta_m := \{Y \times [\frac{i}{m}, \frac{i+1}{m})\}_{i=0}^{m-1}$.

By Theorem A.2.18 we have

$$H((\zeta_r)_n^S \mid \xi_m) \xrightarrow[m\to\infty]{} H((\zeta_r)_n^S \mid \xi) \leq \log rn.$$

The latter inequality is due to the fact that no element of ξ is divided into more than rn pieces by $(\zeta_r)_n^S$. This is the main point: n shows up inside the logarithm.

Thus, we can fix $\epsilon > 0$, and for $r \in \mathbb{N}$ choose $n_r \in \mathbb{N}$ such that $\frac{\log rn_r}{n_r} < \frac{\epsilon}{2}$, and

$$m_r := 1 + \max\left\{i \in \mathbb{N} \mid H((\zeta_r)_n^S \mid \xi_i) \geq \log rn + \tfrac{\epsilon}{2}\right\} < \infty.$$

Setting $M_0 := 1$ and $M_r := \max\{M_{r-1}, m_r, r\}$ for $r \in \mathbb{N}$ gives

$$\begin{aligned}
h(S) &\xleftarrow[r\to\infty]{} h(S, \xi_{M_r} \vee \zeta_r) \xleftarrow[k\to\infty]{} \frac{1}{kn_r} \underbrace{H((\xi_{M_r} \vee \zeta_r)_{kn_r}^S)}_{=H((\xi_{M_r})_{kn_r}^S \vee (\zeta_r)_{kn_r}^S) = [H((\xi_{M_r})_{kn_r}^S) + H((\zeta_r)_{kn_r}^S \mid (\xi_{M_r})_{kn_r}^S)]} \\
&\leq \frac{1}{kn_r}\Big[H((\alpha_{M_r})_{kn_r}^f) + \underbrace{\sum_{i=0}^{k-1} H(S^{in_r}(\zeta_r)_{n_r}^S \mid (\xi_{M_r})_{kn_r}^S)}_{\leq kH((\zeta_r)_{n_r}^S \mid \xi_{M_r}) \leq k(\log(rn_r) + \frac{\epsilon}{2})}\Big] \\
&\leq \frac{1}{kn_r} H((\alpha_{M_r})_{kn_r}^f) + \epsilon \xrightarrow[k\to\infty]{} h(f, \alpha_{M_r}) + \epsilon \xrightarrow[r\to\infty]{} h(f) + \epsilon \xrightarrow[\epsilon\to 0]{} h(f).
\end{aligned}$$

□

A.3.g Induced maps.

Definition A.3.33 (First-return map, section). Let (X, μ) be a measure space, $A \subset X$ measurable with $\mu(A) > 0$, $f\colon X \to X$ a measure-preserving transformation, and

μ_A the conditional measure on A (see (3.3.3)). For $x \in X$ let $n_A(x) := \min\{n \in \mathbb{N} \mid f^n(x) \in A\}$. Then the μ_A-preserving transformation

$$f_A\colon A \to A, \quad f_A(x) := f^{n_A(x)}(x) \tag{A.3.6}$$

is called the *first-return map* induced by f on the set A, and A is called a *section* for f if

$$A_f := \bigcup_{n=0}^{\infty} f^{-n} A \overset{\text{ae}}{=} X.$$

That f_A preserves μ_A can be seen by considering one level set of n_A at a time.

As one would expect, the average return time is $1/\mu(A)$:

Proposition A.3.34. *Let (X, μ) be a measure space, $A \subset X$ measurable, $\mu(A) > 0$, $f\colon X \to X$ measure preserving. Then $\int_A n_A \, d\mu = \mu(\bigcup_{n=0}^{\infty} f^n(A)) > 0$.*

Proof. If $A_j := n_A^{-1}(\{j\})$ then the $f^i(A_j)$ for $0 \le i < j \in \mathbb{N}$ are pairwise disjoint, so

$$\begin{aligned}\mu\Big(\bigcup_{n=0}^{\infty} f^n(A)\Big) = \mu\Big(\bigcup_{0 \le i < n \in \mathbb{N}} f^i(A_n)\Big) &= \sum_{0 \le i < n \in \mathbb{N}} \mu(f^i(A_n)) \\ &= \sum_{n \in \mathbb{N}} n\mu(A_n) = \int_A n_A \, d\mu.\end{aligned}$$ □

Lemma A.3.35 (Kac). *If μ is ergodic and $\mu(A) > 0$, then $\int n_A \, d\mu_A = 1/\mu(A)$.*

Proof. $1 = \mu(\bigcup_{n=0}^{\infty} f^n(A)) = \int_A n_A \, d\mu = \mu(A) \int n_A \, d\mu_A$. □

The Poincaré Recurrence Theorem guarantees that for any set of positive measure the map induced by a measure-preserving transformation is defined almost everywhere.

We next show how the entropy of an induced map is related to that of the map from which it arises. One would expect it to scale by the average return time. This is indeed the case. Specifically, we prove the following result:

Theorem A.3.36 (Abramov). *If $f\colon X \to X$ is an invertible μ-preserving map and $A \subset X$ is a section (Definition A.3.33), then $h_\mu(f) = h_{\mu_A}(f_A)\mu(A)$.*

Remark A.3.37. Note that by the Kac Lemma (Lemma A.3.35), this does amount to scaling by the return time and that for ergodic f any set A of positive measure is a section.

Proof (Neveu). The central concern of the proof is to understand the relationship between partitions of X and partitions of A. We pay special attention to their joint partitions (Definition A.2.22) because we will use that by Corollary A.2.26 we have

$$h_\mu(f,\xi) = H_\mu(\xi \mid \xi^f_-)$$

and a corresponding statement for $h_\mu(f_A)$.

A partition ξ of X naturally defines a partition

$$A \cap \xi := \{A \cap C \mid C \in \xi\}$$

whenever $A \subset X$. For a collection A_i of disjoint sets in X and for partitions ξ_i, this gives a partition $\bigcup_i (A_i \cap \xi_i)$ of $\bigcup_i A_i$. The reader is encouraged to verify that if in addition one has a disjoint collection B_j and partitions η_j, then

$$\Big(\bigcup_i (A_i \cap \xi_i)\Big) \vee \Big(\bigcup_j (B_j \cap \eta_j)\Big) = \bigcup_{ij} \big((A_i \cap B_j) \cap (\xi_i \vee \eta_j)\big). \tag{A.3.7}$$

When $\{B_j\}_j = \{X\}$, this reduces to $\big(\bigcup_i (A_i \cap \xi_i)\big) \vee \eta = \bigcup_i A_i \cap (\xi_i \vee \eta)$. We decompose the set A in the theorem according to the return times for f^{-1} to A:

$$A_n := A \cap f(X \smallsetminus A) \cap \cdots \cap f^{n-1}(X \smallsetminus A) \cap f^n(A) \tag{A.3.8}$$

for $n \in \mathbb{N}$. Given a partition ξ of X, the partition

$$\xi_A := \bigcup_{n \in \mathbb{N}} A_n \cap \xi^f_n,$$

of A (where ξ^f_n is as in Definition A.2.22) is convenient for studying the return map. Inductively, we can express $(\xi_A)^{f_A}_{-l}$ in terms of ξ^f_{-n} as follows:

Claim A.3.38. $(\xi_A)^{f_A}_{-l} = \bigcup_{n_1,\dots,n_l \in \mathbb{N}} f^{-n_1}(A_{n_1}) \cap \cdots \cap f^{-n_1-\cdots-n_l}(A_{n_l}) \cap \xi^f_{-n_1-\cdots-n_l}$.

Proof. For $l = 1$ this follows from

$$\begin{aligned}(\xi_A)^{f_A}_{-1} = f_A^{-1}(\xi_A) = f_A^{-1}\Big(\bigcup_{n\in\mathbb{N}} A_n \cap \xi^f_n\Big) &= \bigcup_{n\in\mathbb{N}} f_A^{-1}(A_n \cap \xi^f_n)\\ &= \bigcup_{n\in\mathbb{N}} f^{-n}(A_n \cap \xi^f_n) = \bigcup_{n\in\mathbb{N}} f^{-n}(A_n) \cap \xi^f_{-n}.\end{aligned}$$

Using (A.3.7), the step to $l+1$ is accomplished as follows:

$$\begin{aligned}(\xi_A)^{f_A}_{-l-1} &= f_A^{-1}\Big(\underbrace{\xi_A \vee ((\xi_A)^{f_A}_{-l})}_{=\bigcup_{n\in\mathbb{N}}(A_n\cap\xi^f_n)\vee((\xi_A)^{f_A}_{-l})}\Big)\\ &= \bigcup_{n\in\mathbb{N}} f_A^{-1}\Big((A_n \cap \xi^f_n) \vee \big((\xi_A)^{f_A}_{-l}\big)\Big)\end{aligned}$$

$$
\begin{aligned}
&= \bigcup_{n\in\mathbb{N}} f^{-n}\Big((A_n \cap \xi_n^f) \vee ((\xi_A)_{-l}^{f_A})\Big)\\
&= \bigcup_{n\in\mathbb{N}} \Big(\underbrace{f^{-n}(A_n) \cap f^{-n}(\xi_n^f) \vee f^{-n}\big((\xi_A)_{-l}^{f_A}\big)}_{=(f^{-n}(A_n)\cap\xi_{-n}^f)\vee f^{-n}(\bigcup_{n_1,\dots,n_l\in\mathbb{N}} f^{-n_1}(A_{n_1})\cap\dots\cap f^{-n_1-\dots-n_l}(A_{n_l})\cap\xi_{-n_1-\dots-n_l}^f)}\Big)\\
&= \bigcup_{n,n_1,\dots,n_l\in\mathbb{N}} \Big(f^{-n}(A_n) \cap \dots \cap f^{-n-n_1-n_2-\dots-n_l}(A_{n_l})\Big)\\
&\qquad\qquad \cap \Big(\xi_{-n}^f \vee f^{-n}(\xi_{-n_1-\dots-n_l}^f)\Big)\\
&= \bigcup_{n_1,\dots,n_{l+1}\in\mathbb{N}} f^{-n_1}(A_{n_1}) \cap \dots \cap f^{-n_1-\dots-n_{l+1}}(A_{n_{l+1}}) \cap \xi_{-n_1-\dots-n_{l+1}}^f. \qquad \square
\end{aligned}
$$

The purpose of Claim A.3.38 is to ascertain the following:

Claim A.3.39. *If* $\{f^{-1}(A), X \smallsetminus f^{-1}(A)\} \le \xi$, *then (with the notation of Definition* A.2.22) $(\xi_A)_-^{f_A} = A \cap \xi_-^f$ *and* $(\xi_A)^{f_A} = A \cap \xi^f$.

Proof. The hypothesis means that $f^{-1}(A)$ (and hence $X \smallsetminus f^{-1}(A)$) is subordinate to ξ, that is, a union of elements of ξ. Using (A.3.8), this implies that

$$f^{-n_1}(A_{n_1}) \cap \dots \cap f^{-n_1-\dots-n_l}(A_{n_l})$$

is subordinate to $A \cap \xi_{-n_1-\dots-n_l}^f \le A \cap \xi_-^f$. Thus, Claim A.3.38 implies that

$$(\xi_A)_{-l}^{f_A} \le A \cap \xi_-^f.$$

At the same time, Claim A.3.38 shows that every element of $(\xi_A)_{-l}^{f_A}$ lies in A and in an element of ξ_{-l}^f (since $n_1 + \dots + n_l \ge l$), so

$$A \cap \xi_{-l}^f \le (\xi_A)_{-l}^{f_A} \le A \cap \xi_-^f.$$

Since $\xi_{-l}^f \nearrow \xi_-^f$ as $l \to \infty$, this implies $(\xi_A)_-^{f_A} = A \cap \xi_-^f$.

Analogously, $(\xi_A)_+^{f_A} = A \cap \xi_+^f$, and together these give $(\xi_A)^{f_A} = A \cap \xi^f$. $\square$

Corollary A.2.26 now implies $h_\mu(f,\xi) = H_\mu(\xi \mid \xi_-^f) = \int_X I[\xi \mid \xi_-^f]\, d\mu$ and

$$
\begin{aligned}
\mu(A) h_{\mu_A}(f_A, \xi_A) &= \mu(A) H_{\mu_A}(\xi_A \mid (\xi_A)_-^{f_A}) = \int_A I[\xi_A \mid (\xi_A)_-^{f_A})]\, d\mu\\
\text{[Claim A.3.39]} \quad &= \int_A I[\xi_A \mid \xi_-^f]\, d\mu, = \sum_{n\in\mathbb{N}} \int_{A_n} I[\xi_n^f \mid \xi_-^f]\, d\mu,
\end{aligned}
$$

where I is the conditional information function from (A.2.4), and we used the definition of ξ_A. We thus need to show the following result:

Claim A.3.40.

$$\sum_{n\in\mathbb{N}}\int_{A_n} I[\xi_n^f \mid \xi_-^f]\,d\mu = \int_X I[\xi \mid \xi_-^f]\,d\mu. \tag{A.3.9}$$

Proof. Proposition A.2.6(3) says that $I[\xi_0 \vee \xi_1 \mid \eta] = I[\xi_0 \mid \eta] + I[\xi_1 \mid \eta \vee \xi_0]$, and recursively, this implies $I[\bigvee_{i=0}^{n-1} \xi_i \mid \eta] = \sum_{i=0}^{n-1} I[\xi_i \mid \eta \vee \bigvee_{j=0}^{i-1} \xi_j]$: For $n = 1$ this is clear, and

$$\begin{aligned} I[\textstyle\bigvee_{i=0}^{n} \xi_i \mid \eta] &= I[\textstyle\bigvee_{i=0}^{n-1} \xi_i \mid \eta] + I[\xi_n \mid \eta \vee \bigvee_{i=0}^{n-1} \xi_i] \\ &= \sum_{i=0}^{n-1} I[\xi_i \mid \eta \vee \textstyle\bigvee_{j=0}^{i-1} \xi_j] + I[\xi_n \mid \eta \vee \bigvee_{i=0}^{n-1} \xi_i] \\ &= \sum_{i=0}^{n} I[\xi_i \mid \eta \vee \textstyle\bigvee_{j=0}^{i-1} \xi_j]. \end{aligned}$$

Thus, on the left-hand side of (A.3.9) we can write

$$I[\xi_n^f \mid \xi_-^f] = \sum_{i=0}^{n-1} I[f^i(\xi) \mid \xi_-^f \vee \textstyle\bigvee_{j=0}^{i-1} \xi_j] = \displaystyle\sum_{i=0}^{n-1} I[f^i(\xi) \mid f^i(\xi_-^f)].$$

The conditional information function is "stationary" in that

$$I[f(\xi) \mid f(\eta)] = I[\xi \mid \eta] \circ f.$$

Therefore,

$$\begin{aligned} \sum_{n\in\mathbb{N}}\int_{A_n} I[\xi_n^f \mid \xi_-^f]\,d\mu &= \sum_{n\in\mathbb{N}}\int_{A_n} \sum_{i=0}^{n-1} \underbrace{I[f^i(\xi) \mid f^i(\xi_-^f)]}_{=I[\xi \mid \xi_-^f]\circ f^i}\,d\mu \\ &= \sum_{n\in\mathbb{N}}\sum_{i=0}^{n-1}\int_{f^{-i}(A_n)} I[\xi \mid \xi_-^f]\,d\mu. \end{aligned}$$

Since A is a section, $X = \bigcup_{n\in\mathbb{N}} \bigcup_{i=0}^{n-1} f^{-i}(A_n)$, disjointly by definition of A_n. Thus,

$$\sum_{n\in\mathbb{N}}\sum_{i=0}^{n-1}\int_{f^{-i}(A_n)} I[\xi \mid \xi_-^f]\,d\mu = \int_X I[\xi \mid \xi_-^f]\,d\mu. \qquad \square$$

As noted, (A.3.9) implies $\mu(A)h_{\mu_A}(f_A, \xi_A) = h_\mu(f, \xi)$. To obtain Theorem A.3.36, we pass to suprema over partitions.

Since we used Claim A.3.39, and hence $\{f^{-1}(A), X \setminus f^{-1}(A)\} \leq \xi$, note that for any partition ζ we can choose $\xi := \{f^{-1}(A), X \setminus f^{-1}(A)\} \vee \zeta$, so

$$h_\mu(f) = \sup_{\zeta \in \mathcal{P}_H} h_\mu(f, \zeta) = \sup_{\substack{\{f^{-1}(A), X \setminus f^{-1}(A)\} \\ \leq \xi \in \mathcal{P}_H}} h_\mu(f, \xi),$$

and likewise for the left-hand side:

$$\sup_{\zeta \in \mathcal{P}_H} h_{\mu_A}(f_A, \zeta_A) = \sup_{\substack{\{f^{-1}(A), X \setminus f^{-1}(A)\} \\ \leq \xi \in \mathcal{P}_H}} h_{\mu_A}(f_A, \xi_A).$$

To establish Theorem A.3.36 it thus suffices, by Theorem A.3.5, to check that the partitions ζ_A for $\zeta \in \mathcal{P}_H$ form a sufficient family. This is clear because if $\xi \in \mathcal{P}_H$ and η is any finite-entropy partition of A, then $\zeta := \xi \vee (\eta \cup \{X \setminus A\}) \geq \eta$. □

B

Hyperbolic maps and invariant manifolds

The main purpose of this appendix is to prove the Stable Manifold Theorem (Theorem B.5.2). This is done here because it is a discrete-time statement even though its application to time-t maps of flows gives the counterpart for continuous time. Another reason is that the length of the argument here would disrupt the flow of ideas in the main body of the book, and we have here the space to develop the ideas of the proof to further applications.

This main objective is preceded by work that is on one hand preliminary to it and on the other hand of importance in its own right. This is the Banach Contraction Mapping Principle on one hand, and the study of hyperbolic linear maps in Banach spaces and perturbations of them. The principal application of proof ideas from the Stable Manifold Theorem is the Inclination Lemma.

Section B.7 is important beyond the purpose of this book. It builds on the Stable Manifold Theorem by establishing absolute continuity of the resulting foliations, an essential ingredient for smooth ergodic theory. This result as well is proved for discrete time and applies to time-t maps of flows. The importance beyond this book lies in the fact that we reproduce here a result much more general than needed, which establishes absolute continuity for partially hyperbolic dynamical systems in a rather broad sense. The proof is due to Abdenur and Viana and has not been published elsewhere.

B.1 The Contraction Mapping Principle

The Banach Contraction Principle is a fixed-point theorem that is particularly important in hyperbolic dynamics, and we present it here in order to highlight also the dependence of the fixed point on parameters, which is less commonly examined than its mere existence. It is also interesting that this theorem is established by studying a simple dynamical system, the contraction for which it is named.

A map $f\colon X \to X$ is said to be *contracting* if there exists $\lambda < 1$ such that for any $x, y \in X$,

$$d(f(x), f(y)) \leq \lambda d(x, y). \tag{B.1.1}$$

These maps exhibit both stability of equilibria in the sense of ordinary differential equations and in the sense of persistence under perturbation of the dynamical system.

All orbits tend to a fixed point, and changing the contracting map slightly does not move the fixed point much. Some pertinent observations are cast in terms of a regularity notion that extends the one of Lipschitz continuity in a natural way (see also Definitions 1.9.6, 6.3.1, 9.5.1).

Definition B.1.1 (Lipschitz and Hölder regularity). Let (X, d), (Y, d) be metric spaces. A map $f\colon X \to Y$ is said to be *Lipschitz (continuous)* if there exists $C > 0$ such that $d(x, y) < \epsilon$ implies $d(f(x), f(y)) \le C(d(x, y))$, in which case f is said to be C-Lipschitz, and the *Lipschitz constant* $L(f)$ (or $\mathrm{Lip}(f)$) of f is defined by

$$L(f) := \sup_{x \ne y} \frac{d(f(x), f(y))}{d(x, y)}.$$

We say that f is *bi-Lipschitz* if it is Lipschitz and has a Lipschitz inverse.

A map $f\colon X \to Y$ is said to be *Hölder continuous* with exponent α, or *α-Hölder*, if there exist $C, \epsilon > 0$ such that $d(x, y) < \epsilon$ implies $d(f(x), f(y)) \le C(d(x, y))^\alpha$. A Hölder-continuous map with Hölder-continuous inverse is said to be bi-Hölder.

Remark B.1.2. This notion is both natural and useful in the context of hyperbolic dynamical systems because it corresponds to saying that if $d(x, y)$ tends to 0 exponentially (as a function of some parameter) then so does $d(f(x), f(y))$.

Proposition B.1.3 (Contraction Mapping Principle). *Let X be a complete metric space and $f\colon X \to X$ a contracting map. Then f has a unique fixed point ϕ, and under the action of iterates of f all points converge exponentially to ϕ.*

Indeed, the error at any step can be estimated in terms of the size of the step:

$$d(x, \phi) \le \frac{1}{1-\lambda} d(x, f(x)). \tag{B.1.2}$$

Suppose X, Y are metric spaces, X complete, $f\colon X \times Y \to X$, $\lambda \in (0, 1)$ such that $d(f_y(x), f_y(x')) \le \lambda d(x, x')$ for all $x, x' \in X, y \in Y$. Denote the fixed point of f_y by ϕ_y. Then

(1) $d(\phi_y, \phi_{y'}) \le \frac{1}{1-\lambda} d(f_{y'}(\phi_{y'}), f_y(\phi_{y'}))$;

(2) *if f is continuous then so is $y \mapsto \phi_y$;*

(3) *if $\alpha \in (0, 1]$ and $y \mapsto f_y$ is α-Hölder-continuous,*[1] *then so is $y \mapsto \phi_y$;*

(4) *if X, Y are open subsets of Banach spaces and f is C^r, then so is $y \mapsto \phi_y$, with derivative*

$$(1 - D^Y f\big|_{(y,\phi_y)})^{-1} \circ D^X f\big|_{(y,\phi_y)},$$

where the superscript denotes the differential in the respective space;

[1] Uniformly in x, that is, $\exists C \in \mathbb{R}$ such that $d(f_y(x), f_{y'}(x)) \le C d(y, y')^\alpha$ for all $x \in X$, $y, y' \in Y$.

(5) *if* $\lambda \in (0,1)$ *and*

$$d(f_y(x), f_{y'}(x')) \leq \lambda \max\{d(x,x'), d(y,y')\}$$

for all $x, x' \in X$, $y, y' \in Y$, *then* $d(\phi_y, \phi_{y'}) \leq \lambda d(y,y')$.

Proof. $\{f^n(x)\}_{n\in\mathbb{N}}$ is a Cauchy sequence because if $m \geq n$ then

$$d(f^m(x), f^n(x)) \leq \sum_{k=0}^{m-n-1} \underbrace{d(f^{n+k+1}(x), f^{n+k}(x))}_{\leq \lambda^{n+k} d(f(x),x)} \leq \frac{\lambda^n}{1-\lambda} d(f(x),x) \xrightarrow[n\to\infty]{} 0. \quad \text{(B.1.3)}$$

Then $\phi := \lim_{n\to\infty} f^n(x) = \lim_{n\to\infty} f^{n+1}(x) = \lim_{n\to\infty} f(f^n(x)) = f(\lim_{n\to\infty} f^n(x)) = f(\phi)$ exists since X is complete. Inequality (B.1.1) implies uniqueness,[2] and $m \to \infty$ in (B.1.3) gives

$$d(f^n(x), \phi) \leq \frac{\lambda^n}{1-\lambda} d(f(x), x).$$

This proves exponential convergence and for $n = 0$ gives (B.1.2).

(1) Apply (B.1.2) with $x = \phi_{y'} = f_{y'}(\phi_{y'})$.

(2) and (3) follow from (1), and (4) from the Implicit Function Theorem.

(5): Take $x = \phi_y = f_y(\phi_y)$ and $x' = \phi_{y'} = f_{y'}(\phi_{y'})$ in the assumption and note that the maximum on the right-hand side must be $d(y, y')$. □

Remark B.1.4. This in particular implies continuous dependence of the fixed point on the contraction when one makes C^1-perturbations.

The robustness of the asymptotic behavior of contractions in Proposition B.1.3 has a counterpart for hyperbolic maps, even when they are perturbed so as to be nonlinear.

Theorem B.1.5 (Hyperbolic Fixed-Point Theorem). *If* $A\colon E \to E$ *is a bounded linear map of a Banach space* E *and* $\operatorname{Id} - A$ *is invertible, then a continuous map* $F\colon E \to E$ *has a unique fixed point* ϕ *if* $\lambda := L(F - A)\|(\operatorname{Id} - A)^{-1}\| < 1$. *Furthermore,* ϕ *depends continuously on* F, *and* $\|\phi\| \leq \frac{1}{1-\lambda}\|F(0)\|$.

Remark B.1.6. $(\operatorname{Id} - A)^{-1}$ is bounded by the Open Mapping Theorem.

Proof. ϕ is a solution of $(F - A)(x) = x - A(x) = (\operatorname{Id} - A)x$, hence a fixed point of the λ-contraction $(F - A)(\operatorname{Id} - A)^{-1}$. Apply (B.1.2) with $x = 0$. □

This is analogous to the persistence of the fixed point of a contraction under perturbations, but a hyperbolic fixed point is harder to find. Proposition B.4.7 shows that although the fixed point of a contraction is the limit of the forward orbit of any initial condition, this fails for hyperbolic maps except with a lucky starting point.

[2] $f(x) = x$ in (B.1.1) $\Rightarrow y = x$ or $y \neq f(y)$.

B.2 Generalized eigenspaces

Definition B.2.1. A vector $v \in \mathbb{C}^n$ (or $\mathbb{R}^n$) is a *generalized eigenvector* of degree p for A if for some $\lambda \in \mathbb{C}$ we have $(A - \lambda I)^p v = 0$ and $(A - \lambda I)^{p-1} v \neq 0$. The space of these is the *generalized eigenspace* or *root space* for λ.

Note that generalized eigenvectors of degree 1 are just eigenvectors. Also,

$$(A - \lambda I)^p v = (A - \lambda I)(A - \lambda I)^{p-1} v.$$

So λ is an eigenvalue of A and $(A - \lambda I)^{p-1} v$ is an eigenvector associated with λ.

Proposition B.2.2. *If v is a generalized eigenvector of degree p for eigenvalue λ, then $v, (A - \lambda I)v, \dots, (A - \lambda I)^{p-1} v$ are linearly independent.*

Proof. If $c_1 v + c_2(A - \lambda I)v + \dots + c_{p-1}(A - \lambda I)^{p-1} v = 0$ and $c_i = 0$ for $0 \le i \le k \le p-1$, then

$$c_k (A - \lambda I)^k v = \sum_{j=k+1}^{p-1} c_j (A - \lambda I)^j v.$$

Applying $(A - \lambda I)^{p-k-1}$ to each side gives

$$c_k \underbrace{(A - \lambda I)^{p-1} v}_{\neq 0} = \sum_{j=k+1}^{p-1} c_j (A - \lambda I)^{p-k-1+j} v = 0,$$

since $p - k - 1 + j \ge p$ for $j > k$. Thus, $c_1 = \dots = c_{p-1} = 0$. □

For $A \in \mathcal{M}_n(\mathbb{R})$, let $\mathcal{N}(A) = \ker A$ be the nullspace (or kernel) of A. If λ is an eigenvalue, then

$$\{0\} \subset \mathcal{N}(A - \lambda I) \subseteq \mathcal{N}(A - \lambda I)^2 \subseteq \cdots,$$

while $\dim \mathcal{N}(A - \lambda I)^k \le n$ for k. So there is a smallest $k =: r(\lambda)$ at which the nullspace stabilizes ($\mathcal{N}(A - \lambda I)^k = \mathcal{N}(A - \lambda I)^{k+1}$), and we call $M(\lambda) := \mathcal{N}(A - \lambda I)^{r(\lambda)}$ the *generalized eigenspace* of A for λ.

Lemma B.2.3. *The following hold.*

(1) $\mathcal{N}(A - \lambda I)^k = M(\lambda)$ *for all* $k \ge r(\lambda)$.

(2) $M(\lambda) = \{v \mid (A - \lambda I)^k v = 0$ *for some* $k \ge 1\}$.

(3) $r(\lambda)$ *is the maximal degree of the generalized eigenvectors for* λ.

(4) $\dim M(\lambda) \ge r(\lambda)$.

Proof. (1) Claim: $\mathcal{N}(A-\lambda I)^k = \mathcal{N}(A-\lambda I)^{k+1} \Rightarrow \mathcal{N}(A-\lambda I)^{k+1} \supseteq \mathcal{N}(A-\lambda I)^{k+2}$: If $v \in \mathcal{N}(A-\lambda I)^{k+2}$, then $(A-\lambda I)^{k+1}(A-\lambda I)v = 0$, so $(A-\lambda I)v \in \mathcal{N}(A-\lambda I)^{k+1} = \mathcal{N}(T-\lambda I)^k$, and $0 = (A-\lambda I)^k(A-\lambda I)v = (A-\lambda I)^{k+1}v$, so $v \in \mathcal{N}(A-\lambda I)^{k+1}$.

Property (2) is a consequence of (1).

Property (3) follows from the definition of degree.

(4) Proposition B.2.2 gives $r(\lambda)$ linearly independent vectors in $M(\lambda)$. □

Now let $R(\lambda)$ be the range of $(A-\lambda I)^{r(\lambda)}$. Then $\dim M(\lambda) + \dim R(\lambda) = n$.

Proposition B.2.4. *If λ is an eigenvalue of A, then $M(\lambda)$ and $R(\lambda)$ are A-invariant subspaces and $\mathbb{C}^n = M(\lambda) \oplus R(\lambda)$.*

Proof. We will use that $A(A-\lambda I)^k = (A-\lambda I)^k A$. If $v \in M(\lambda)$, then

$$(A-\lambda I)^{r(\lambda)}Av = A(A-\lambda I)^{r(\lambda)}v = A0 = 0,$$

so $Av \in M(\lambda)$, and $M(\lambda)$ is A-invariant.

If $w \in R(\lambda)$, then $w = (A-\lambda I)^{r(\lambda)}v$ for some v, so

$$A(w) = A(A-\lambda I)^{r(\lambda)}v = (A-\lambda I)^{r(\lambda)}Av \in R(\lambda).$$

Since $\dim M(\lambda) + \dim R(\lambda) = n$, we show $M(\lambda) \cap R(\lambda) = \{0\}$ to complete the proof.

If $0 \neq v \in R(\lambda) \cap M(\lambda)$, then there is a $u \in \mathbb{C}^n$ with $0 \neq v = (A-\lambda I)^{r(\lambda)}u$ and $(A-\lambda I)^{r(\lambda)}v = 0$. Thus, u is a generalized eigenvector of degree greater than $r(\lambda)$, contrary to Lemma B.2.3. □

The eigenvalues of $A \in \mathcal{M}_n(\mathbb{R})$ are the roots of $\det(A-\lambda I) = (-1)^n(\lambda-\lambda_1)^{m_1}\cdots(\lambda-\lambda_p)^{m_p}$, where $\sum_{i=1}^p m_i = n$. The m_i are called the algebraic multiplicities.

Theorem B.2.5. *Let $\lambda_1,\dots,\lambda_p$ be eigenvalues of A with algebraic multiplicity $m_1,\dots,m_p$. Then $\dim M(\lambda_j) = m_j$ for $1 \le j \le p$ and $\mathbb{C}^n = M(\lambda_1) \oplus \cdots \oplus M(\lambda_p)$.*

Proof. Let λ_j be an eigenvalue of A. Since $\mathbb{C}^n = M(\lambda_j) \oplus R(\lambda_j)$ by Proposition B.2.4, A can be represented in a basis as

$$T = \begin{pmatrix} A_1 & 0 \\ 0 & A_2 \end{pmatrix},$$

where A_1 is a $\dim(M(\lambda_j))$ square block. Then

$$\det(A_1-\lambda I)\cdot\det(A_2-\lambda I) = \det(A-\lambda I) = (-1)^n\prod_{k=1}^p(\lambda-\lambda_k)^{m_k}.$$

Also, $(\lambda-\lambda_j)$ does not divide $\det(A_2-\lambda I)$ since λ_j is not an eigenvalue of A_2.

Then $(\lambda - \lambda_j)^{m_j}$ divides $\det(A_1 - \lambda I)$. If there is a $k \neq j$ such that $(\lambda - \lambda_k)$ divides $\det(A_1 - \lambda I)$, then $M(\lambda_j)$ has an eigenvector v for some λ_k which is a generalized eigenvector for λ_j, so

$$0 = (A - \lambda_j I)^p v = (A - \lambda_j I)^{p-1} \underbrace{(A - \lambda_j I)v}_{=(\lambda_k - \lambda_j)v} = (\lambda_k - \lambda_j)(A - \lambda_j I)^{p-1} v \neq 0.$$

Thus $(\lambda - \lambda_k)$ does not divide $\det(A_1 - \lambda I)$ for $k \neq j$, and $\det(A_1 - \lambda I) = (-1)^n(\lambda - \lambda_j)^{m_j}$, and $\dim(M(\lambda_j)) = m_j$.

To prove $\mathbb{C}^n = M(\lambda_1) \oplus \cdots \oplus M(\lambda_p)$, suppose to the contrary that $M(\lambda_j) \cap M(\lambda_k) \neq \{0\}$. Since

$$(A - \lambda_k I)(A - \lambda_j I) = (A - \lambda_j)(A - \lambda_k I),$$

$M(\lambda_j)$ is $(A - \lambda_k I)$-invariant and $M(\lambda_k)$ is $(A - \lambda_j I)$-invariant.

Now suppose there exists a nonzero vector $v \in M(\lambda_j) \cap M(\lambda_k)$ where v is of degree q in λ_k. Then $(A - \lambda_k I)^{q-1} v$ is an eigenvector for λ_k and in $M(\lambda_j) \cap M(\lambda_k)$. From previous arguments this is a contradiction.

Suppose there exists $v_1 + \cdots + v_p = 0$ such that $v_j \in M(\lambda_j)$ for $1 \leq j \leq p$. Let

$$S_j = (A - \lambda_1 I)^{r(\lambda_1)} \cdots (A - \lambda_{j-1} I)^{r(\lambda_{j-1})} (A - \lambda_{j+1} I)^{r(\lambda_{j+1})} \cdots (A - \lambda_p I)^{r(\lambda_p)}.$$

Then $S_j(v_1 + \cdots + v_p) = 0$ if and only if $v_j = 0$. So $v_1 = \cdots = v_p = 0$ and

$$\mathbb{C}^n = M(\lambda_1) \oplus \cdots \oplus M(\lambda_p). \qquad \square$$

The *geometric multiplicity* of λ_j is $\dim \ker(A - \lambda_j I)$. From the previous results the geometric multiplicity is at most the algebraic multiplicity.

B.3 The spectrum of a linear map

If a linear transformation of a finite-dimensional vector space has no eigenvalues on the unit circle, then the space is the direct sum of an expanding subspace (the sum of the generalized eigenspaces for eigenvalues outside the unit circle) and a contracting subspace (the sum of the generalized eigenspaces for eigenvalues inside the unit circle). The purpose of this subsection and the next is to prove the same for transformations of Banach spaces (Theorem B.4.2).

This involves interesting functional analysis that a dynamicist may not otherwise encounter frequently, but the reader may also take the conclusion of Theorem B.4.2 as a definition of hyperbolicity and skip ahead to Section B.5.

We now look at a similarly general context that combines contraction and expansion. Here a linear structure helps separate the two, so the natural generality in which this is effective is a Banach space.

It is convenient to consider Banach spaces over the complex numbers. The results we obtain in this context can be used for real Banach spaces E by passing to the complexification $E_{\mathbb{C}}$ (that is, the space $E \otimes \mathbb{C}$ obtained by allowing complex scalars) and then suitably restricting attention to the real part.

We denote by $B(z,r)$ the ball of radius r around z in $\mathbb{C}$, and by $S(z,r)$ its boundary.

Definition B.3.1. Let E be a Banach space and $A\colon E \to E$ be a bounded linear map, that is, the norm $\|A\| := \sup_{\|v\|=1} \|Av\|$ of A is finite. The *resolvent set* $R(A)$ of A is the set of $\lambda \in \mathbb{C}$ for which $\lambda \operatorname{Id} - A$ has bounded inverse $R_A(\lambda)$, called the *resolvent* of A. We call $\operatorname{sp} A := \mathbb{C} \setminus R(A)$ the *spectrum* of A. The *spectral radius* $r(A)$ of A is defined by $r(A) := \sup\{|\lambda| \mid \lambda \in \operatorname{sp} A\}$.

The *point spectrum* consists of the eigenvalues of A (and $\ker(A - \lambda \operatorname{Id})$ is the corresponding eigenspace).

The *continuous spectrum* is $\big\{\lambda \in \operatorname{sp} A \mid A - \lambda \operatorname{Id} \text{ injective}, \overline{(A - \lambda \operatorname{Id})(E)} = E\big\}$.

The *residual spectrum* is $\big\{\lambda \in \operatorname{sp} A \mid A - \lambda \operatorname{Id} \text{ is injective \& } \overline{(A - \lambda \operatorname{Id})(E)} \neq E\big\}$.

Remark B.3.2. If E is finite-dimensional, then $\operatorname{sp} A$ is the set of eigenvalues: those λ for which $A - \lambda \operatorname{Id}$ is not injective—which is also the set of λ for which $A - \lambda \operatorname{Id}$ is not surjective. In this case the spectral radius is therefore the largest modulus of an eigenvalue. Invertibility is the only issue in this context because all linear maps between finite-dimensional spaces are bounded. By the Open Mapping Theorem a bounded linear bijection between Banach spaces has bounded inverse, so

$$\operatorname{sp} A = \{\lambda \in \mathbb{C} \mid \lambda \operatorname{Id} - A \text{ is not injective}\} \cup \{\lambda \in \mathbb{C} \mid \lambda \operatorname{Id} - A \text{ is not surjective}\}.$$

Accordingly, the three items in Definition B.3.1 are a decomposition of $\operatorname{sp} A$.

Lemma B.3.3. $r(A) \le \|A\| : \ |\lambda| > \|A\| \Rightarrow \lambda \notin \operatorname{sp} A, R_A(\lambda) = \sum_{i=0}^{\infty} \frac{A^i}{\lambda^{i+1}}$ *(Laurent series)*.

Proof. $(\lambda \operatorname{Id} - A) \sum_{i=0}^{n-1} \frac{A^i}{\lambda^{i+1}} = \sum_{i=0}^{n-1} \frac{A^i}{\lambda^i} - \frac{A^{i+1}}{\lambda^{i+1}} = \operatorname{Id} - \frac{A^n}{\lambda^n} \xrightarrow[n\to\infty]{} \operatorname{Id}$. □

The spectral radius provides an asymptotically sharp bound:

Proposition B.3.4 (Gelfand spectral radius formula). $r(A) = \lim_{n\to\infty} \|A^n\|^{1/n}$.

Proof. Since $a_n := \log \|A^n\|$ is subadditive, the limit exists by Lemma 4.2.7. By Lemma B.3.3 the domain of convergence of the Laurent series $\sum_{i=0}^{\infty} A^i/\lambda^{i+1}$ of $R_A(\cdot)$ is $\{|\lambda| > r(A)\}$ while by the root test it is $\{|\lambda| > \lim_{n\to\infty} \|A^n\|^{1/n}\}$. □

Lemma B.3.5. *If A is a bounded linear operator, then $R(A)$ is the natural domain of analyticity of $R_A(\cdot)$. Thus, $R(A)$ is open, and* $\operatorname{sp} A$ *is compact by Lemma* B.3.3.

Proof. We show analyticity on $R(A)$ and that $d(\lambda, \operatorname{sp} A) \ge \|(R_A(\lambda)\|^{-1}$ on $R(A)$; this implies openness and $\|R_A(\lambda)\| \xrightarrow[d(\lambda,\operatorname{sp} A)\to 0]{} \infty$, hence the claim.

If $\lambda \in R(A)$ and $|\mu| < \|R_A(\lambda)\|^{-1}$, then $\|\mu R_A(\lambda)\| < 1$, so $T(\mu) := \sum_{i=0}^{\infty} \mu^i (R_A(\lambda))^{i+1}$ converges,[3] and

$$\begin{aligned} ((\lambda-\mu)\operatorname{Id} - A)T(\mu) &= (\lambda \operatorname{Id} - A)T(\mu) - \mu T(\mu) \\ &= \sum_{i=0}^{\infty} (\mu R_A(\lambda))^i - (\mu R_A(\lambda))^{i+1} = \operatorname{Id}, \end{aligned}$$

so $\lambda - \mu \in R(A)$ and $R_A(\lambda - \mu) = T(\mu)$ is analytic at $\mu = 0$. □

Remark B.3.6 (Resolvent equation). For $\mu, \lambda \in R(A)$, multiplying

$$(\mu \operatorname{Id} - A)(\lambda \operatorname{Id} - A)[R_A(\lambda) - R_A(\mu)] = (\mu \operatorname{Id} - A) - (\lambda \operatorname{Id} - A) = (\mu - \lambda)\operatorname{Id}$$

by $R_A(\lambda)R_A(\mu)$ gives the *resolvent equation*

$$R_A(\lambda) - R_A(\mu) = (\mu - \lambda)R_A(\lambda)R_A(\mu). \tag{B.3.1}$$

Proposition B.3.7. $\operatorname{sp} A \ne \varnothing$ *unless* $E = \{0\}$.

Proof. If $\operatorname{sp} A = \varnothing$, then R_A is entire. It is bounded on $\overline{B(0, 2\|A\|)}$ by compactness, and $\|R_A(\lambda)\| \le \|A\|^{-1}$ for $|\lambda| > 2\|A\|$ because

$$\|(\lambda \operatorname{Id} - A)v\| = \|\lambda(\operatorname{Id} - \tfrac{A}{\lambda})v\| \ge 2\|A\| \cdot \tfrac{1}{2}\|v\|.$$

Being bounded and entire, R_A is constant by the Liouville Theorem, which implies that $\operatorname{Id} = 0$, hence $E = \{0\}$. □

The Liouville Theorem applies to this situation since for a bounded linear functional $f \in E^*$, $f \circ R_A$ is an entire bounded scalar function and hence constant.

If A is diagonal, then clearly $\|A\| = r(A)$. The following counterpart to Proposition 5.1.5 is useful for understanding the dynamics of linear maps even if they cannot be diagonalized.

Proposition B.3.8. *For every $\delta > 0$ there exists an equivalent norm on E with respect to which $\|A\| < r(A) + \delta$. This is called an* adapted *or* Lyapunov norm.

[3] This is the Neumann series for the inverse of $(\lambda - \mu)\operatorname{Id} - A = (\lambda \operatorname{Id} - A) - \mu \operatorname{Id}$.

Proof. Take n such that $\|A^n\| < (r(A)+\delta)^n$ and $|v| := \sum_{i=0}^{n-1} \|A^i v\|(r(A)+\delta)^{-i}$. Then

$$\frac{|Av|}{|v|} = \frac{\sum_{i=1}^{n} \|A^i v\|(r(A)+\delta)^{1-i}}{\sum_{i=0}^{n-1} \|A^i v\|(r(A)+\delta)^{-i}} = (r(A)+\delta)\Big[\underbrace{1 + \frac{\|A^n v\|(r(A)+\delta)^{-n} - \|v\|}{\sum_{i=0}^{n-1} \|A^i v\|(r(A)+\delta)^{-i}}}_{<1}\Big]. \qquad \square$$

Remark B.3.9. One can conclude from this that for any equivalent norm and for every $\epsilon > 0$ there exists C_ϵ such that $\|A^n v\| \le C_\epsilon (r(A)+\epsilon)^n \|v\|$ for any $v \in \mathbb{R}^n$.

Corollary B.3.10. *If* $\operatorname{sp}(A) \subset B(0,1)$, *then there is an equivalent norm on E such that A is a contraction with respect to the metric generated by that norm.*

Proof. Apply Proposition B.3.8 with $0 < \delta < 1 - r(A)$ (> 0 by compactness). □

The concept of exponential convergence does not depend on a particular choice of an equivalent norm. Thus Proposition B.1.3 and Corollary B.3.10 imply the following result:

Corollary B.3.11. *If* $\operatorname{sp}(A) \subset B(0,1)$, *then the positive iterates of every point converge exponentially to the origin. If in addition A is an invertible map, then negative iterates of every point go to infinity exponentially.*

B.4 Hyperbolic linear maps

Next we consider maps with both contraction and expansion.

Definition B.4.1. A bounded linear map A of a Banach space E is said to be *hyperbolic* if $\operatorname{sp} A \cap S(0,1) = \varnothing$. It is said to be *$(\ell^-,\ell^+)$-hyperbolic* if $0 < \ell^- < 1 < \ell^+$ and $\operatorname{sp} A \cap \{z \in \mathbb{C} \mid \ell^- \le |z| \le \ell^+\} = \varnothing$.

Theorem B.4.2. *If E is a Banach space, $A\colon E \to E$ continuous linear, $\gamma := S(0,r) \subset R(A)$, then there are $0 < \ell^- < r < \ell^+$ such that $\{z \in \mathbb{C} \mid \ell^- \le |z| \le \ell^+\} \subset R(A)$, $\lambda(A) := r(A^-) < \ell^-$, and $\mu(A) := 1/r(A^{-1}\restriction_{E^+}) > \ell^+$ (notation as in (2) below), that is, $\|(A^-)^n\| \in O(\lambda^n)$ and $\|(A^+)^{-n}\| \in O(\mu^{-n})$ (see Remark 3.2.18). In particular, if A is hyperbolic ($r = 1$), then there are $0 < \ell^- < 1 < \ell^+$ such that A is (ℓ^-,ℓ^+)-hyperbolic.*

If $\gamma \subset \mathbb{C}$ is a smooth curve bounding a topological disk D and $\operatorname{sp} A \cap \gamma = \varnothing$, *then there are linear subspaces E^- and E^+ of E such that*

(1) $E = E^- \oplus E^+$,

(2) $AE^- \subset E^-$ *(with equality if* $0 \notin \operatorname{sp} A$), $AE^+ = E^+$: *we write* $A^\pm := A\restriction_{E^\pm}$,

(3) $\operatorname{sp} A^- = \operatorname{sp}^- A := \operatorname{sp} A \cap D$, $\operatorname{sp} A^+ = \operatorname{sp}^+ A := \operatorname{sp} A \smallsetminus D$.

Remark B.4.3. If $\ell^- < 1 < \ell^+$, then these conditions in turn imply that A is hyperbolic, so this is a characterization of hyperbolicity.

If $E^\pm$ are both nontrivial, then the spectrum is contained in two annuli. This result readily generalizes to larger numbers of annuli; for instance, if $0 < r_1 < r_2$ and $\operatorname{sp} A \cap S(0, r_i) = \varnothing$, then $\operatorname{sp} A$ lies in the union of three annuli; the corresponding subspaces are $E^-_{r_1}$, $E^+_{r_1} \cap E^-_{r_2}$, and $E^+_{r_2}$. Linear maps for which all three subspaces in this decomposition are nontrivial are said to be *partially hyperbolic* if $r_1 < 1 < r_2$.

As in Corollary B.3.10, there is an *adapted norm* (or *Lyapunov norm*) associated with such (ℓ^-, ℓ^+), that is, a norm $|\cdot|$ equivalent to the given one with

$$\|A^-\| \le \ell^-, \quad \|(A^+)^{-1}\| \le 1/\ell^+, \quad \text{and } |v^- + v^+| = \max(|v^-|, |v^+|) \text{ for } v^\pm \in E^\pm.$$

(Take Lyapunov norms $|\cdot|$ for $A^\pm$ and $|v^- + v^+| := \max(|v^-|, |v^+|)$ for $v^\pm \in E^\pm$.)

Definition B.4.4. If $\ell^- < 1 < \ell^+$, then E^- is called the *contracting* subspace and E^+ the *expanding* subspace.

Remark B.4.5. The expanding subspace is not characterized by the fact that vectors in it expand under iterates of the map—all vectors outside the contracting subspace are expanded by a sufficiently large iterate of the map. The characterization of E^+ is given by the description of Remark B.4.3, namely that preimages contract.

Proof of Theorem B.4.2. Compactness of $\operatorname{sp} A$ implies the first assertions and the existence of a smooth Jordan curve γ' with γ inside it and $\operatorname{sp} A \setminus D$ outside it.

Claim B.4.6. $\pi^- := \frac{1}{2\pi i}\int_\gamma R_A(\lambda)\,d\lambda = \frac{1}{2\pi i}\int_{\gamma'} R_A(\lambda)\,d\lambda$ *is a projection.*

Proof. $\frac{1}{2\pi i}\int_c \frac{1}{\mu-\lambda}\,d\mu = \begin{cases} 1 & \text{if } \lambda \text{ is inside } c, \\ 0 & \text{if } \lambda \text{ is outside } c, \end{cases}$ for $c \in \{\gamma, \gamma'\}$, so

$$\begin{aligned}
\pi^-\pi^- &= \frac{1}{2\pi i}\int_\gamma R_A(\lambda)\,d\lambda \cdot \frac{1}{2\pi i}\int_{\gamma'} R_A(\mu)\,d\mu = \Big(\frac{1}{2\pi i}\Big)^2 \int_\gamma\int_{\gamma'} \underbrace{R_A(\lambda)R_A(\mu)}_{=\frac{R_A(\lambda)-R_A(\mu)}{\mu-\lambda} \text{ by (B.3.1)}}\,d\mu\,d\lambda \\
&= \Big(\frac{1}{2\pi i}\Big)^2 \Big[\int_\gamma R_A(\lambda) \underbrace{\int_{\gamma'} \frac{1}{\mu-\lambda}\,d\mu}_{=2\pi i \text{ since } \lambda\in\gamma \text{ inside } \gamma'}\,d\lambda - \int_{\gamma'} R_A(\mu) \underbrace{\int_\gamma \frac{1}{\mu-\lambda}\,d\lambda}_{=0 \text{ since } \mu\in\gamma' \text{ outside } \gamma}\,d\mu\Big] \\
&= \frac{1}{2\pi i}\int_\gamma R_A(\lambda)\,d\lambda = \pi^-.
\end{aligned}$$

□

(1) $\pi^+ := \operatorname{Id} - \pi^-$ is then also a projection; take $E^\pm := \pi^\pm(E)$.

(2) $A(E^\pm) = A(\pi^\pm(E)) = \pi^\pm(A(E)) \subset \pi^\pm(E) = E^\pm$ because A commutes with $R_A(\cdot)$ and hence with $\pi^\pm$. Then $AE^+ = E^+$ because below we show that $0 \notin \operatorname{sp} A^+$.

(3) $E = E^- \oplus E^+$ and $A(E^\pm) \subset E^\pm$ give $\operatorname{sp} A = \operatorname{sp} A^- \oplus A^+ = \operatorname{sp} A^- \cup \operatorname{sp} A^+$, so we show $\operatorname{sp} A^- \subset D$ and $\operatorname{sp} A^+ \cap D = \varnothing$:

$$\underbrace{(\lambda \operatorname{Id} - A)}_{=(\mu \operatorname{Id} - A) + (\lambda - \mu)\operatorname{Id}} \frac{1}{2\pi i} \int_\gamma \frac{1}{\lambda - \mu} R_A(\mu)\, d\mu = \frac{1}{2\pi i} \int_\gamma R_A(\mu) - \frac{\operatorname{Id}}{\mu - \lambda}\, d\mu$$

$$= \begin{cases} \pi^- & \text{if } \lambda \notin D \cup \gamma, \\ \pi^- - \operatorname{Id} = -\pi^+ & \text{if } \lambda \in D. \end{cases}$$

If $\lambda \notin D \cup \gamma$, restrict to E^- to see that $\lambda \operatorname{Id} - A^-$ is invertible, so $\lambda \notin \operatorname{sp} A^-$, and $\operatorname{sp} A^- \subset D$. If $\lambda \in D$, restrict to E^+ to get $\operatorname{sp} A^+ \cap D = \varnothing$, hence (3). □

We now describe the asymptotics of iterates of a hyperbolic linear map.

Proposition B.4.7. *If E is a Banach space, $A\colon E \to E$ hyperbolic linear, then*

(1) *for every $v \in E^-$, the positive iterates $A^n v$ converge to the origin with exponential speed as $n \to \infty$ and if A is invertible then the negative iterates $A^n v$ go to infinity with exponential speed as $n \to -\infty$;*

(2) *for every $v \in E^+$ the positive iterates of v go to infinity exponentially and if A is invertible then the negative iterates converge exponentially to the origin;*

(3) *for every $v \in E \setminus (E^- \cup E^+)$ the iterates $A^n v$ go to infinity exponentially as $n \to \infty$ and if A is invertible also as $n \to -\infty$.*

Proof. This is mainly a restatement of Theorem B.4.2 and Remark B.4.3. If $v \in \mathbb{R}^n \setminus (E^- \cup E^+)$ write $v = v^- + v^+$ where $v^- \in E^- \setminus \{0\}$, $v^+ \in E^+ \setminus \{0\}$ to get

$$\|A^n v\| = \|A^n(v^- + v^+)\| \geq \|A^n v^+\| - \|A^n v^-\| \geq \lambda^n c \|v^+\| - \lambda^{-n} c' \|v^-\| \geq \lambda^n c'',$$

for large positive n, where $\lambda > 1$ and $c, c', c'' > 0$ do not depend on n.

The argument for negative iterates is the same with v^+ and v^- exchanged. □

With the present notation one can recast Theorem B.1.5 as follows.

Theorem B.4.8 (Hyperbolic Fixed-Point Theorem II). *If A is a (λ, μ)-hyperbolic bounded linear map of a Banach space and $F\colon E \to E$ is such that $\ell := L(F - A) < \epsilon := \min(1 - \lambda, 1 - \mu^{-1})$ (see Definition* B.1.1*), then F has a unique fixed point $\phi \in E$, and $|\phi| < |F(0)|/(\epsilon - \ell)$, where $|\cdot|$ is an adapted norm. Also, ϕ depends continuously on F.*

This version is more explicit about the closeness assumption in terms of known parameters, but it uses hyperbolicity rather than just $1 \in R(A)$.

Proof. Write $E = E^- \times E^+$, $\pi^\pm\colon E \to E^\pm$, $x \mapsto x^\pm$ for the projections, $F^\pm := \pi^\pm \circ F$ and show that $\overline{F}(x) := \big(F^-(x), x^+ + (A^+)^{-1}(x^+ - F^+(x))\big)$ is a $(1+\ell-\epsilon)$-contraction. □

Remark B.4.9. The generality of the present context is motivated by its utility when applied in auxiliary spaces, and the Hyperbolic Fixed-Point Theorem (Theorem B.1.5) can be used to prove a variety of results in hyperbolic dynamical systems, including some of our main theorems such as structural stability [206, Theorem A]. We immediately show one instance of this: Theorem B.1.5 can be greatly amplified by applying the very same result in a suitable infinite-dimensional space to show that the dynamics of the almost-linear map f in Theorem B.1.5 does not only match that of the linear map in that there is a unique fixed point, but that the entire orbit structure of f is the same as that of A.

Theorem B.4.10. *Let A be a (λ, μ)-hyperbolic bounded linear map of a Banach space and f_1, f_2 Lipschitz-continuous maps with $\Delta f_i := f_i - A$ bounded and*

$$\ell := \max L(\Delta f_i) < \epsilon := \min(1-\lambda, 1-\mu^{-1}, \|A^{-1}\|^{-1}). \tag{B.4.1}$$

Then there is a unique continuous map $h = h_{f_1,f_2}\colon E \to E$ such that $f_1 \circ h = h \circ f_2$ and $\Delta h := h - \mathrm{Id} \in \mathcal{E} := C_b(E,E)$ (bounded continuous maps with the sup norm).

Proof. The f_i are invertible: $f_i(x) = y \Leftrightarrow x = A^{-1}(y - \Delta f_i(x))$, and the right-hand side is an $\ell\|A^{-1}\|$-contraction, so there is a unique such x.

We can thus rewrite the desired conclusion as $f_1 \circ h \circ f_2^{-1} = h$ or

$$(A + \Delta f_1) \circ (\mathrm{Id} + \Delta h) \circ f_2^{-1} = \mathrm{Id} + \Delta h$$

or

$$\mathcal{F}(\Delta h) := \underbrace{A \circ \Delta h \circ f_2^{-1}}_{=:\mathcal{A}(\Delta h) \in \mathcal{E}} + \underbrace{\Delta f_1 \circ (\mathrm{Id} + \Delta h) \circ f_2^{-1} + A \circ f_2^{-1} - \mathrm{Id}}_{=:\Delta\mathcal{F}(\Delta h) \in \mathcal{E}} = \Delta h \in \mathcal{E},$$

a fixed-point problem for $\mathcal{F} = \mathcal{A} + \Delta\mathcal{F}$. We have that $\mathcal{A}$ is hyperbolic: $\mathcal{E} = \mathcal{E}^- \oplus \mathcal{E}^+$, where $\mathcal{E}^\pm := C_b(E, E^\pm) = \mathcal{A}(\mathcal{E}^\pm)$, $\|\mathcal{A}^-\| \le \lambda$, and $\|(\mathcal{A}^+)^{-1}\| \le 1/\mu$. Since $L(\Delta\mathcal{F}) \le L(\Delta f_1) < \epsilon$, Theorem B.1.5 provides the desired unique fixed point $\Delta h \in \mathcal{E}$, and $h := \mathrm{Id} + \Delta h$ is the required continuous map. □

This does not quite produce what we promised; for the orbit structures of the maps to be the same, h must be a homeomorphism. This is an easy consequence.

Corollary B.4.11 (Hartman–Grobman). *Let A be a (λ, μ)-hyperbolic bounded linear map of a Banach space, $f\colon E \to E$ Lipschitz with $\Delta f := f - A$ bounded, ϵ as in* (B.4.1), *and $\ell := L(\Delta f) < \epsilon$. Then there is a unique homeomorphism $h\colon E \to E$ depending continuously on f with $h - \mathrm{Id}$ bounded and $h \circ A = f \circ h$.*

Proof. The map h in Theorem B.4.10 is a homeomorphism because $f_1 \circ h_{f_1,f_2} = h_{f_1,f_2} \circ f_2$ and (by symmetry) $f_2 \circ h_{f_2,f_1} = h_{f_2,f_1} \circ f_1$, hence

$$\begin{aligned} f_2 \circ [h_{f_2,f_1} \circ h_{f_1,f_2}] &= h_{f_2,f_1} \circ f_1 \circ h_{f_1,f_2} = [h_{f_2,f_1} \circ h_{f_1,f_2}] \circ f_2, \\ f_1 \circ [h_{f_1,f_2} \circ h_{f_2,f_1}] &= h_{f_1,f_2} \circ f_2 \circ h_{f_2,f_1} = [h_{f_1,f_2} \circ h_{f_2,f_1}] \circ f_1, \end{aligned}$$

so uniqueness in Theorem B.4.10 gives $h_{f_2,f_1} \circ h_{f_1,f_2} = \mathrm{Id} = h_{f_1,f_2} \circ h_{f_2,f_1}$. □

We now describe a localization procedure that connects the global picture in a linear space (such as in Corollary B.4.11) with local analysis on a manifold.

On a smooth compact manifold M we can choose a Riemannian metric, and then there is an open set $B \subset TM$ such that $0 \in B_x := B \cap T_xM$ and $\exp_x \colon B_x \to M$ is an embedding of B_x with $\exp_x(0) = x$.

Theorem B.4.12. *If f is a C^1-diffeomorphism of M with a compact invariant set Λ, take $\epsilon_0 > 0$ and a C^1-neighborhood U of f such that $g(\exp_x(v)) \in \exp_{f(x)}(B_{f(x)})$ for $g \in U$, $x \in \Lambda$, $\|v\| \le 2\epsilon_0$. If $\rho \colon \mathbb{R} \to [0,1]$ is smooth, $\rho([0,1]) = \{1\}$, $\rho([2,\infty)) = \{0\}$, and $\epsilon < \epsilon_0$ and U are sufficiently small, then the* localization

$$G_x(v) := D_xf(v) + \rho(\|v\|/\epsilon)\big(\exp_{f(x)}^{-1} \circ g \circ \exp_x(v) - D_xf(v)\big)$$

of $g \in U$ is arbitrarily uniformly C^1-close to D_xf.

Proof. Near $v = 0$ this is the choice of ϵ, U; for $\|v\| \ge 2\epsilon$ we have equality. □

Remark B.4.13. The point is that the continuous map $G \colon T_\Lambda M \to T_\Lambda M := TM\restriction_\Lambda$ defined by $G\restriction_{T_xM} = G_x$ fibers over f, that is, $G_x(T_xM) \subset T_{f(x)}M$, and satisfies

$$\begin{aligned} G(v) &= D_xf(v) && \text{when } \|v\| \ge 2\epsilon, \\ \exp_{f(x)} G(v) &= g(\exp_x(v)) && \text{when } \|v\| \le \epsilon. \end{aligned}$$

Corollary B.4.11 immediately translates to the following.

Theorem B.4.14 (Hartman–Grobman Theorem). *Let M be a smooth manifold, $U \subset M$ open, $f \colon U \to M$ continuously differentiable, and $p \in U$ a hyperbolic fixed point of f, that is, the differential $D_pf \colon T_pM \to T_pM$ at p is a hyperbolic linear map. Then there exist neighborhoods U_1, U_2 of p and V_1, V_2 of $0 \in T_pM$, as well as a homeomorphism $h \colon U_1 \cup U_2 \to V_1 \cup V_2$ such that $f = h^{-1} \circ Df_p \circ h$ on U_1, that is, the following diagram commutes*:

$$\begin{array}{ccc} U_1 & \xrightarrow{\;f\;} & U_2 \\ {\scriptstyle h}\big\downarrow & & \big\downarrow{\scriptstyle h} \\ V_1 & \xrightarrow{\;D_pf\;} & V_2. \end{array}$$

B.5 Admissible manifolds: The Hadamard method

We prove the existence of unstable manifolds by the Hadamard graph transform method. It obtains unstable manifolds as limits of manifolds of an approximately right kind. More specifically, Hadamard's approach is to consider graphs over the unstable subspace and apply the dynamics to these in order to discern successive improvement that leads to an application of the Contraction Mapping Principle. Figure B.5.2 shows this very idea iconically: The unstable stretch and stable contraction combine to make such graphs "nicer," and they do so in a way that in a suitable norm defines a contraction with a rate that is determined by the contraction and expansion rates. Hadamard's original paper made a point of explaining the core idea well rather than being as strong as possible, and it still makes good reading today [175]. The framework in which we present it has the advantage of producing a result that is more general in ways that are essential for some applications. Specifically, admissible (rather than stable or unstable) manifolds are an important product of these arguments.

Recall that for a linear map $A\colon \mathbb{R}^n \to \mathbb{R}^n$ the set of all eigenvalues of A is denoted by $\mathrm{sp}(A)$ (Definition B.3.1). If A is hyperbolic we define the slowest contraction and expansion rates of A by

$$\lambda(A) := r(A\restriction_{E^-}) = \sup\{|\chi| \mid \chi \in \mathrm{sp}(A),\ |\chi| < 1\},$$
$$\mu(A) := 1/r(A^{-1}\restriction_{E^+}) = \inf\{|\chi| \mid \chi \in \mathrm{sp}(A),\ |\chi| > 1\},$$

where the subspaces E^+ and E^- are as in Definition B.4.4 and Theorem B.4.2.

By Proposition B.3.8 for any $\delta > 0$ one can introduce a norm in $\mathbb{R}^n$ such that $\|A\restriction_{E^-}\| < \lambda(A) + \delta$ and $\|A^{-1}\restriction_{E^+}\| < \mu^{-1}(A) + \delta$.

Now we proceed to the local analysis near a general (nonperiodic) orbit. The differentials of the iterates f^k, $k \in \mathbb{Z}$ along such an orbit cannot be reduced to the iterates of a single linear map but should be viewed as products of different linear maps. Thus, we cannot talk about eigenvalues anymore, but rather should define hyperbolicity in terms of expansion and contraction of tangent vectors. We also generalize the situation somewhat by allowing a more general kind of exponential splitting for linear maps into "fast-expanding" or "fast-contracting" directions and the rest. As in the case of a single point one can choose appropriate coordinate systems centered at the points of the reference orbit and express both the nonlinear map and its differential in those coordinates.

Definition B.5.1. Let $\lambda < \mu$. A sequence of invertible linear maps $L_m\colon \mathbb{R}^n \to \mathbb{R}^n$, $m \in \mathbb{Z}$ is said to admit a (λ, μ)-*splitting* if there exist decompositions $\mathbb{R}^n = E_m^\mu \oplus E_m^\lambda$ such that $L_m E_m^i = E_{m+1}^i$ for $i = \lambda, \mu$ and

$$\|L_m\restriction_{E_m^\lambda}\| \le \lambda, \quad \|L_m^{-1}\restriction_{E_{m+1}^\mu}\| \le \mu^{-1}.$$

We say that $\{L_m\}_{m\in\mathbb{Z}}$ *admits an exponential splitting* or *is partially hyperbolic in the broad sense* if it admits a (λ,μ)-splitting for some λ, μ and $\lambda < 1$, $\dim E_m^\lambda \geq 1$ or $\mu > 1$, $\dim E_m^\mu \geq 1$. We say that $\{L_m\}_{m\in\mathbb{Z}}$ is *hyperbolic* (or *uniformly hyperbolic*) if it admits a (λ,μ)-splitting for some $\lambda < 1 < \mu$. In this case we set $E_m^- := E_m^\lambda$ and $E_m^+ := E_m^\mu$.

By viewing $\mathbb{R}^n$ as a canonical product $\mathbb{R}^k \times \mathbb{R}^{n-k}$ and making a sequence of orthogonal coordinate changes in $\mathbb{R}^n$ one can assume in the previous definition that $E_m^\mu = \mathbb{R}^k \times \{0\}$, $E_m^\lambda = \{0\} \times \mathbb{R}^{n-k}$ for some k, $0 \leq k \leq n$, and all m.

Thus we have reduced the problem of the local behavior of the iterates of a diffeomorphism near a reference orbit to the study of a sequence of local diffeomorphisms $f_m\colon U_m \longrightarrow \mathbb{R}^n$, where each U_m is a neighborhood of the origin in $\mathbb{R}^n$ containing a ball of some fixed radius, fixing the origin and such that the sequence of linear maps at the origin $(Df_m)_0$, $m \in \mathbb{Z}$ admits an exponential splitting. Although we are interested only in points whose successive images stay in the neighborhoods, it is convenient to artificially extend our maps from somewhat smaller neighborhoods to the whole space $\mathbb{R}^n$ using Theorem B.4.12.

Here is the Stable–Unstable Manifold Theorem in the desired generality.

Theorem B.5.2 (Hadamard–Perron Theorem). *Let $\lambda < \mu$, $r \geq 1$, and for each $m \in \mathbb{Z}$ let $f_m\colon \mathbb{R}^n \longrightarrow \mathbb{R}^n$ be a (surjective) C^r-diffeomorphism such that for $(x,y) \in \mathbb{R}^k \oplus \mathbb{R}^{n-k}$,*

$$f_m(x,y) = (A_m x + \alpha_m(x,y),\ B_m y + \beta_m(x,y))$$

for some linear maps $A_m\colon \mathbb{R}^k \longrightarrow \mathbb{R}^k$ and $B_m\colon \mathbb{R}^{n-k} \longrightarrow \mathbb{R}^{n-k}$ with $\|A_m^{-1}\| \leq \mu^{-1}$, $\|B_m\| \leq \lambda$ and $\alpha_m(0) = 0$, $\beta_m(0) = 0$.

Then for $0 < \gamma < \min\left(1, \sqrt{\mu/\lambda} - 1\right)$ and

$$0 < \delta < \min\left(\frac{\mu-\lambda}{\gamma+2+1/\gamma}, \frac{\mu-(1+\gamma)^2\lambda}{(1+\gamma)(\gamma^2+2\gamma+2)}\right),$$

if $\|\alpha_m\|_{C^1} < \delta$ and $\|\beta_m\|_{C^1} < \delta$ for all $m \in \mathbb{Z}$ then there is

(1) *a unique family $\{W_m^+\}_{m\in\mathbb{Z}}$ of k-dimensional C^1-manifolds*

$$W_m^+ = \{(x, \varphi_m^+(x)) \mid x \in \mathbb{R}^k\} = \operatorname{graph}\varphi_m^+$$

and

(2) *a unique family $\{W_m^-\}_{m\in\mathbb{Z}}$ of $(n-k)$-dimensional C^1-manifolds*

$$W_m^- = \{(\varphi_m^-(y), y) \mid y \in \mathbb{R}^{n-k}\} = \operatorname{graph}\varphi_m^-,$$

where $\varphi_m^+ \colon \mathbb{R}^k \to \mathbb{R}^{n-k}$, $\varphi_m^- \colon \mathbb{R}^{n-k} \to \mathbb{R}^k$, $\sup_{m\in\mathbb{Z}} \|D\varphi_m^\pm\| < \gamma$, *and the following properties hold*:

(i) $f_m(W_m^-) = W_{m+1}^-$, $f_m(W_m^+) = W_{m+1}^+$.

(ii) $\|f_m(z)\| < \lambda'\|z\|$ *for* $z \in W_m^-$, *and* $\|f_{m-1}^{-1}(z)\| < (\mu')^{-1}\|z\|$ *for* $z \in W_m^+$, *where* $\lambda' := (1+\gamma)(\lambda + \delta(1+\gamma)) < \frac{\mu}{1+\gamma} - \delta =: \mu'$.

(iii) *Let* $\lambda' < \nu < \mu'$. *If* $\|f_{m+L-1} \circ \cdots \circ f_m(z)\| < C\nu^L\|z\|$ *for all* $L \geq 0$ *and some* $C > 0$ *then* $z \in W_m^-$.

Similarly, if $\|f_{m-L}^{-1} \circ \cdots \circ f_{m-1}^{-1}(z)\| \leq C\nu^{-L}\|z\|$ *for all* $L \geq 0$ *and some* $C > 0$ *then* $z \in W_m^+$.

Finally, if $\mu \geq 1$, *then the families* $\{W_m^+\}_{m\in\mathbb{Z}}$ *consist of* C^r*-manifolds, and if* $\lambda \leq 1$, *then* $\{W_m^-\}_{m\in\mathbb{Z}}$ *consist of* C^r*-manifolds.*

As we intimated before, little of the proof uses the assumption that $\gamma < 1$, giving results about “fast-unstable” manifolds.

The “stationary” special case without dependence of our data on m (corresponding to iterates of a single locally defined map f) may be good to keep in mind on the first reading of the arguments, and it gives the following theorem:

Theorem B.5.3. *Let* p *be a hyperbolic fixed point of a local* C^r*-diffeomorphism* $f \colon U \to M$, $r \geq 1$. *Then there exist* C^r*-embedded disks* W_p^+, $W_p^- \subset U$ *such that*

- $T_pW_p^\pm = E^\pm(Df_p)$,
- $f(W_p^-) \subset W_p^-$, *and*
- $f^{-1}(W_p^+) \subset W_p^+$.

There also exist $\lambda < 1 < \mu$ *such that* $\operatorname{sp} Df_p \cap \{z \in \mathbb{C} \mid \lambda \leq |z| \leq \mu\} = \varnothing$ *and* $C(\delta)$ *such that if* $y \in W_p^-$, $z \in W_p^+$, $m \geq 0$, *then*

$$d(f^m(y), p) < C(\delta)\left(\lambda(Df_p) + \delta\right)^m d(y, p),$$
$$d(f^{-m}(z), p) < C(\delta)\left(\mu^{-1}(Df_p) + \delta\right)^m d(z, p).$$

Furthermore, there exists $\delta_0 > 0$ *such that*

$$\text{if } d(f^m(y), p) \leq \delta_0 \text{ for } m \geq 0 \text{ then } y \in W_p^-,$$
$$\text{if } d(f^m(z), p) \leq \delta_0 \text{ for } m \leq 0 \text{ then } z \in W_p^+.$$

In fact, there exist a neighborhood $O \subset U$ *of* p *and* C^r *coordinates* $\psi \colon O \to \mathbb{R}^n$ *such that* $\psi(W_p^+ \cap O) \subset \mathbb{R}^k \oplus \{0\}$ *and* $\psi(W_p^- \cap O) \subset \{0\} \oplus \mathbb{R}^{(n-k)}$ (adapted coordinates).

Remark B.5.4. The disks W_p^+ and W_p^- are not uniquely defined, but their *germs* are: for any two disks satisfying the assertion of this theorem for W_p^+, their intersection contains a neighborhood of p in each of them. In other words, they are open subsets of a common larger submanifold. The same property holds for W_p^-.

Remark B.5.5 (Application by localization). The Hadamard method plays a central role in the theory of hyperbolic dynamical systems and applications of Theorem B.5.2 require *localization* of the context in which it is stated. Fix $r > 0$ and let

$$D_r = \{(x,y) \in \mathbb{R}^k \oplus \mathbb{R}^{n-k} \mid \|x\| \le r, \|y\| \le r\}, \quad W_{m,r}^{\pm} = W_m^{\pm} \cap D_r.$$

If $\lambda' < 1$ (this is true if $\lambda < 1$ and γ and δ are sufficiently small), then by (ii), $f_m(W_{m,r}^-) \subset W_{m+1,r}^-$ and W_m^- is contracted under the action of f_m. Thus in this case $W_{m,r}^-$ is determined by the action of f_m on D_r only. Similar comments apply to W_m^+ if $\mu' > 1$. Thus in these situations we obtain meaningful objects from local data.

If one tries to apply Theorem B.5.2 via local charts and the previous extension procedure then one obtains meaningful objects (independent of the extensions and determined by local data) *only* in the two cases of the preceding paragraph ($\lambda' < 1$ for W^- or $\mu' > 1$ for W^+).

In particular, in the hyperbolic case for sufficiently small γ, δ we have $\lambda' < 1 < \mu'$ and both W_m^+ and $W_{m,r}^+$ are determined locally. In this case W_m^- and W_m^+ are usually called the *stable manifolds* and the *unstable manifolds* at the origin, correspondingly. Furthermore, we can put $\nu = 1$ in (iii). That shows that stable and unstable manifolds are defined purely topologically, namely,

$$W_m^- = \{z \in \mathbb{R}^n \mid \|f_{m+L-1} \circ \cdots \circ f_m z\| \xrightarrow[L\to\infty]{} 0\},$$
$$W_m^+ = \{z \in \mathbb{R}^n \mid \|f_{m-L}^{-1} \circ \cdots \circ f_{m-1}^{-1} z\| \xrightarrow[L\to\infty]{} 0\}.$$

In the course of the proof we show that the sequence of differentials $(Df_m)_0$, $m \in \mathbb{Z}$ admits a (λ', μ')-splitting. This immediately implies that $T_0 W_m^{\pm} = E_m^{\pm}$.

By considering successive images $p_m = f_{m-1} \circ \cdots \circ f_0(p)$ for $m \ge 0$ and $p_m = f_m^{-1} \circ f_{m+1}^{-1} \circ \cdots \circ f_{-1}^{-1}(p)$ for $m < 0$ of any point $p \in \mathbb{R}^n$ and translating the coordinate systems so that they become centered at p_m, we obtain maps

$$f_m^p(z) = f_m(z + p_m) - p_{m+1}$$

satisfying the hypotheses of the theorem. Thus we can construct manifolds $W_{m,p}^+$ and $W_{m,p}^-$ passing through p and satisfying appropriately modified assertions. In particular, (ii) and (iii) imply that if $W_{m,p}^+ \cap W_{m,q}^+ \ne \varnothing$ then $W_{m,p}^+ = W_{m,q}^+$ (similarly for $W_{m,p}^-$). Furthermore, $f_m(W_{m,p}^{\pm}) = W_{m+1,f_m p}^{\pm}$. Thus, $\mathbb{R}^n$ splits in two ways into invariant families of manifolds. Naturally, the fields of tangent planes to those manifolds are invariant under the differentials Df_m.

The proof of the Hadamard–Perron Theorem (Theorem B.5.2) consists of five steps:

Step 1. Construction of invariant cone families.

Step 2. Construction of invariant sequences of plane fields inside the invariant cone families. Here we also explore other implications of the existence of invariant cones which are used later on a number of occasions.

Step 3. Construction of invariant Lipschitz graphs via an application of the Contraction Mapping Principle to an appropriate operator (graph transform).

Step 4. Verification of differentiability.

Step 5. C^r-smoothness in the hyperbolic case.

In order to distinguish tangent vectors from points in the Euclidean space we usually denote by $(x, y) \in \mathbb{R}^k \oplus \mathbb{R}^{n-k}$ a point in $\mathbb{R}^n$ and by $(u, v) \in \mathbb{R}^k \oplus \mathbb{R}^{n-k} \cong T_{(x,y)}\mathbb{R}^n$ a tangent vector at (x, y).

The remainder of this section carries out this proof in these five steps.

B.5.a Invariant cones. We begin by defining cone fields.

Definition B.5.6. If a normed vector bundle E over a metric space Λ decomposes into $E^1 \oplus E^2$, then the *standard horizontal γ-cone* field is defined by

$$H_p^\gamma := \{u + v \in E_p^1 \oplus E_p^2 \mid \|v\| \le \gamma\|u\|\}.$$

The *standard vertical γ-cone* is

$$V_p^\gamma := \{u + v \in E_p^1 \oplus E_p^2 \mid \|u\| \le \gamma\|v\|\}.$$

By a *cone field* we mean a map that associates to every point $p \in \mathbb{R}^n$ a cone K_p in $T_p\mathbb{R}^n$. These cone fields are said to be bounded if there is a constant c such that

$$\|u + v\|/c \le \|u\| + \|v\| \le c\|u + v\|$$

for all $p \in \Lambda$, $u \in E_p^1$, $v \in E_p^2$. For a given cone K, the *dual cone* K^* is the closure of the complement of K.

If Λ is an invariant set for a diffeomorphism $f\colon M \to M$, then f naturally acts on cone fields on $E := T_\Lambda M$ by

$$(f_*K)_p := Df_{f^{-1}(p)}(K_{f^{-1}(p)}).$$

We say that a cone family K is (strictly) *invariant* if

$$(f_*K)_p \subset \operatorname{Int} K_p \cup \{0\};$$

we write

$$f_*K \Subset K.$$

Let us look at some examples to clarify the picture involved here. In dimension $n = 2$ all cones look alike. A horizontal cone $|x_2| \leq \gamma|x_1|$ is shaded on the left of Figure B.5.1. Its dual cone is a vertical cone given by $|x_1| \leq |x_2|/\gamma$. In dimension $n = 3$

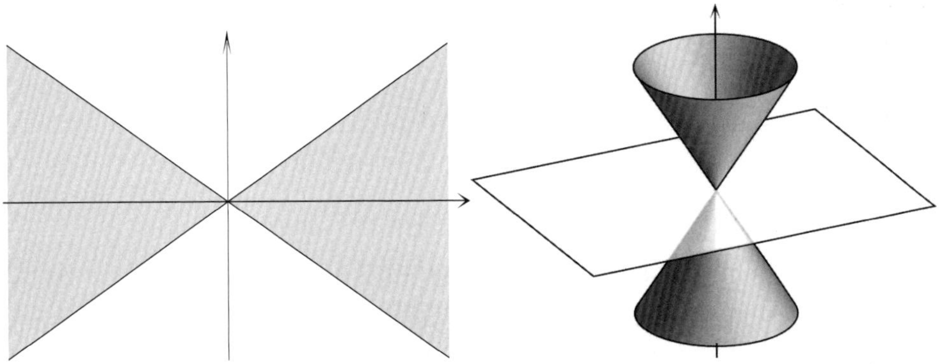

Figure B.5.1. A horizontal cone and a vertical cone. [Reprinted from [213] (© Cambridge University Press, all rights reserved) with permission.]

the following is obviously a cone: let $u = x_1$, $v = (x_2, x_3)$, and $\sqrt{x_2^2 + x_3^2} \leq \gamma|x_1|$; and so is its dual cone, described by $u = (x_2, x_3)$, $v = x_1$, and $|x_1| \leq \sqrt{x_2^2 + x_3^2}/\gamma$. This cone does not look like those designed to hold ice cream.

By a *cone family* we mean a sequence of cone fields (Definition B.5.6). A sequence $f = \{f_m\}_{m\in\mathbb{Z}}$ of diffeomorphisms acts on cone families by

$$(f_*K)_{p,m} = (Df_{m-1})_{f_{m-1}^{-1}(p)}(K_{f_{m-1}^{-1}(p),m-1}).$$

We say that a cone family K is (strictly) *invariant* if

$$(f_*K)_{p,m} \subset \operatorname{Int} K_{p,m} \cup \{0\}.$$

We consider the action of a sequence $f = \{f_m\}_{m\in\mathbb{Z}}$ satisfying the hypotheses of Theorem B.5.2 on the standard horizontal and vertical cone families which assign to $p \in \mathbb{R}^n$ the cones H_p^γ and V_p^γ for all m, correspondingly.

Lemma B.5.7. *If* $\delta < \frac{\mu-\lambda}{2+\gamma+1/\gamma}$ *then*

$$(Df_m)_p(H_p^\gamma) \subset \operatorname{Int} H_{f_m(p)}^\gamma \quad \textit{and} \quad (Df_m)_p^{-1}(V_{f_m(p)}^\gamma) \subset \operatorname{Int} V_p^\gamma.$$

Proof. If $(u,v) \in H_p^\gamma$, that is, $\|v\| \le \gamma\|u\|$, and

$$(u',v') = (Df_m)_p(u,v) = (A_m u + (D\alpha_m)_p(u,v), B_m v + (D\beta_m)_p(u,v))$$

then

$$\|v'\| = \|B_m v + (D\beta_m)_p(u,v)\| \le \|B_m v\| + \|(D\beta_m)_p(u,v)\| < \lambda\|v\| + \delta\|(u,v)\|.$$

We also have

$$\|u'\| = \|A_m u + (D\alpha_m)_p(u,v)\| \ge \|A_m u\| - \|(D\alpha_m)_p(u,v)\| > \mu\|u\| - \delta\|(u,v)\|$$

and since $\|(u,v)\| \le \|u\| + \|v\| \le (1+\gamma)\|u\|$ we get

$$\|u'\| > (\mu - \delta(1+\gamma))\,\|u\|. \tag{B.5.1}$$

Now $\delta < \frac{\mu-\lambda}{2+\gamma+1/\gamma}$, so $\delta(1+\gamma)^2 < \gamma(\mu-\lambda)$, whence

$$\lambda\gamma + \delta(1+\gamma) < \gamma(\mu - \delta(1+\gamma)) \tag{B.5.2}$$

and $\|v'\| < \lambda\|v\| + \delta\|(u,v)\| \le (\lambda\gamma + \delta(1+\gamma))\|u\| < \gamma(\mu-\delta(1+\gamma))\|u\| < \gamma\|u'\|$.

Invariance of vertical γ-cones means that $V_{f_m(p)}^\gamma \subset \operatorname{Int}(Df_m)_p V_p^\gamma$ or, equivalently, $(Df_m)_p(H_p^{1/\gamma}) \subset \operatorname{Int}(H_{f_m(p)}^{1/\gamma})$. But this follows from the observation that the preceding estimates still hold when γ is replaced by $1/\gamma$. □

Let $\tilde{V}^\gamma = f_* V^\gamma$. We now show that vectors in horizontal cones expand and those in vertical cones contract.

Lemma B.5.8.

$$\|(Df_m)_p(u,v)\| > \left(\frac{\mu}{1+\gamma} - \delta\right)\|(u,v)\| \qquad \textit{for } (u,v) \in H_p^\gamma$$

and

$$\|(Df_m)_p(u,v)\| < (1+\gamma)(\lambda+\delta)\|(u,v)\| \quad \textit{for } (u,v) \in \tilde{V}_p^\gamma.$$

Proof. With the above notation, (B.5.1) implies

$$\|(u',v')\| \ge \|u'\| > (\mu - \delta(1+\gamma))\,\|u\| \ge \frac{\mu - \delta(1+\gamma)}{1+\gamma}\|(u,v)\|$$

for $(u,v) \in H_p^\gamma$. If $(u,v) \in \tilde{V}_p^\gamma$, then $(u',v') \in V_{f_m(p)}^\gamma$, and (B.5.1) yields

$$\underbrace{\|(u',v')\|}_{\le\|u'\|+\|v'\|} \le (1+\gamma)\|v'\| < (1+\gamma)[\lambda\|v\| + \delta\|(u,v)\|] \le (1+\gamma)(\lambda+\delta)\|(u,v)\|. \quad \square$$

B.5.b Invariant sequences of plane fields. Now we explore the relation between the existence of an invariant sequence of cones and the exponential splitting for a sequence of linear maps. The conclusions of Lemmas B.5.7 and B.5.8 applied along each orbit make the results of this step applicable to our setting.

Proposition B.5.9. *Let $\lambda' < \mu'$ and $L_m\colon \mathbb{R}^k \times \mathbb{R}^{n-k} \to \mathbb{R}^k \times \mathbb{R}^{n-k}$ a sequence of invertible linear maps such that there is an $\epsilon > 0$ such that for all $m \in \mathbb{Z}$ and $n \in \mathbb{N}$ there are $\gamma_m, \gamma'_m > 0$ for which*

(1) $L_m H^{\gamma_m} \subset \operatorname{Int} H^{\gamma_{m+1}}$;

(2) $L_m^{-1} V^{\gamma'_{m+1}} \subset \operatorname{Int} V^{\gamma'_m}$;

(3) $\|L_{m-1} \circ \cdots \circ L_{m-n}(u,v)\| > \epsilon \mu'^n \|(u,v)\|$ *for* $(u,v) \in H^{\gamma_{m-n}}$;

(4) $\|L_{m-1} \circ \cdots \circ L_{m-n}(u,v)\| < \epsilon^{-1} \lambda'^n \|(u,v)\|$ *for* $(u,v) \in L_{m-n}^{-1} \circ \cdots \circ L_{m-1}^{-1} V^{\gamma'_m}$.

Then

$$E_m^{\mu'} := \bigcap_{i=0}^{\infty} L_{m-1} \circ L_{m-2} \circ \cdots \circ L_{m-i} H^{\gamma_{m-i}}$$

is a k-dimensional subspace inside H^{γ_m} and

$$E_m^{\lambda'} := \bigcap_{i=0}^{\infty} L_m^{-1} \circ L_{m+1}^{-1} \circ \cdots \circ L_{m+i}^{-1} V^{\gamma'_{m+i+1}}$$

is an $(n-k)$-dimensional subspace inside $V^{\gamma'_m}$.

Proof. Since $\mathbb{R}^k \times \{0\} \subset H^{\gamma}$ for all γ, condition (1) implies that

$$\begin{aligned} S_j &:= L_{m-1} \circ L_{m-2} \circ \cdots \circ L_{m-j}(\mathbb{R}^k \times \{0\}) \\ &\subset L_{m-1} \circ L_{m-2} \circ \cdots \circ L_{m-j} H^{\gamma_{m-j}} =: T_j. \end{aligned}$$

For each S_j take an ordered orthonormal basis and consider a subsequence such that the sequences of basis elements all converge. Since the intersection of T_j with the unit sphere is compact it contains the basis consisting of the limits of the basis elements. By the same token any sequence of vectors defined by a fixed set of coefficients converges to a vector in T_j. Hence the span S of the limiting basis belongs to all T_j and thus to the intersection. We need to show that $S = E_m^{\mu'}$.

If $(u,v) \in E_m^{\mu'}$ then, since $S \subset H^{\gamma_m}$ is transverse to $\{0\} \times \mathbb{R}^{n-k}$, we can write $(u,v) = (u,v') + (0,v'')$ with $(u,v') \in S$.

If we let

$$(u_j, v_j) := L_{m-j}^{-1} \circ \cdots \circ L_{m-1}^{-1}(u, v),$$
$$(u_j', v_j') := L_{m-j}^{-1} \circ \cdots \circ L_{m-1}^{-1}(u, v'),$$
$$(u_j'', v_j'') := L_{m-j}^{-1} \circ \cdots \circ L_{m-1}^{-1}(0, v''),$$

then $(u,v) \in E_m^{\mu'}$ implies that $(u_j, v_j) \in H^{\gamma_{m-j}}$ and by (3), $\|(u_j, v_j)\| \le \epsilon^{-1}(\mu')^{-j}\|(u,v)\|$. By the same token $\|(u_j', v_j')\| \le \epsilon^{-1}(\mu')^{-j}\|(u,v')\|$. Thus since $(u_j'', v_j'') \in V^{\gamma'_{m-j}}$ we have by (4) that

$$\begin{aligned}\|v''\| < \epsilon^{-1}(\lambda')^j\|(u_j'', v_j'')\| &\le \epsilon^{-1}(\lambda')^j\left(\|(u_j, v_j)\| + \|(u_j', v_j')\|\right)\\ &\le \epsilon^{-2}\left(\frac{\lambda'}{\mu'}\right)^j\left(\|(u,v)\| + \|(u,v')\|\right)\end{aligned}$$

for all $j \in \mathbb{N}$, whence $v'' = 0$ and $(u, v) \in S$.

The argument for $E_m^{\lambda'}$ is similar, using the family $\{L_m^{-1}\}$ instead of L_m. □

Remark B.5.10. Note that $E_m^{\mu'}$ and $E_m^{\lambda'}$ are unique invariant sequences of subspaces inside the cones H_m^{γ} and $V_m^{\gamma'}$, respectively.

Corollary B.5.11. *If under the assumptions of Proposition* B.5.9 *we have $\lambda' < 1 < \mu'$ then $\{L_m\}$ is a hyperbolic family of linear maps which admits a (λ', μ')-splitting.*

Corollary B.5.12. *If $\gamma < \sqrt{(\mu/\lambda) - 1}$ and*

$$0 < \delta < \min\left(\frac{\mu - \lambda}{\gamma + \frac{1}{\gamma} + 2}, \frac{\mu - (1+\gamma)^2\lambda}{(2+\gamma)(1+\gamma)}\right) \tag{B.5.3}$$

then

$$(E_p^{\mu})_m = \bigcap_{i=0}^{\infty}((f_*)^i H^{\gamma})_{p,m} = \bigcap_{i=0}^{\infty}(f_*(f_*(\dots f_*(H^{\gamma})\dots)))_{p,m}$$

is a k-dimensional subspace inside H_p^{γ},

$$(Df_m)_p(E_p^{\mu})_m = \left(E_{f_m(p)}^{\mu}\right)_{m+1},$$

and

$$\|(Df_m)_p\xi\| \ge \left(\frac{\mu}{1+\gamma} - \delta\right)\|\xi\|$$

for every $\xi \in (E_p^{\mu})_m$.

Similarly $(E_p^\lambda)_m = \bigcap_{i=0}^{\infty}((f_*^{-1})^i V^\gamma)_{p,m}$ *is an* $(n-k)$*-dimensional subspace in* V_p^γ,

$$(Df_m)_p(E_p^\lambda)_m = \left(E_{f_m(p)}^\lambda\right)_{m+1},$$

and $\|(Df_m)_p\xi\| \le (1+\gamma)(\lambda+\delta)\|\xi\|$ *for every* $\xi \in (E_p^\lambda)_m$.

Proof. By Lemmas B.5.7 and B.5.8 and (B.5.3) we can apply Proposition B.5.9 with $\lambda' = (1+\gamma)(\lambda+\delta)$ and $\mu' = (\mu/(1+\gamma)-\delta)$ since under our assumptions $\lambda' < \mu'$ along each orbit of the sequence $\{f_m\}$. □

Lemma B.5.13. *For* $m \in \mathbb{Z}$ *the subspaces* $(E_p^\mu)_m$ *and* $(E_p^\lambda)_m$ *are continuous in* p.

Proof. The vectors $v \in (E_p^\lambda)_m$ are characterized by the inequalities

$$\|(Df_{m+j})(Df_{m+j-1})\cdots(Df_m)_p v\| \le (\lambda')^{j+1}\|v\| \qquad (j \in \mathbb{N}). \tag{B.5.4}$$

For a sequence $p_l \to p$ take orthonormal bases $\xi_1^l, \dots \xi_k^l$ of $(E_{p_l}^\lambda)_m$ and assume without loss of generality that $\lim_{l\to\infty}\xi_i^l = \xi_i$ $(i = 1,\dots,k)$. Since for any fixed i the vectors ξ_i^l satisfy (B.5.4) for all l we conclude by continuity of all Df_m that ξ_i satisfies (B.5.4) and hence $\xi_i \in (E_p^\lambda)_m$. Since $\dim(E_p^\lambda)_m$ does not depend on p this implies that $\lim_{l\to\infty}(E_{p_l}^\lambda)_m = (E_p^\lambda)_m$. □

Here, $(E_p^\mu)_m$ and $(E_p^\lambda)_m$ $(m \in \mathbb{Z})$ are the invariant sequences of plane fields mentioned in the description of the proof.

B.5.c Invariant Lipschitz graphs. To get invariant graphs, that is, a family $\{\varphi_m^+ : \mathbb{R}^k \to \mathbb{R}^{n-k}\}_{m\in\mathbb{Z}}$ of Lipschitz functions such that $f_m(\operatorname{graph}\varphi_m^+) = \operatorname{graph}\varphi_{m+1}^+$ and $\varphi_m^+(0) = 0$, let $C_\gamma(\mathbb{R}^k)$ be the set of functions $\varphi\colon \mathbb{R}^k \to \mathbb{R}^{n-k}$ that are Lipschitz continuous with Lipschitz constant γ. Let $C_\gamma^0(\mathbb{R}^k)$ be the space of $\varphi \in C_\gamma(\mathbb{R}^k)$ such that $\varphi(0) = 0$. The following lemma can be viewed as a nonlinear counterpart of Lemma B.5.7 and shows that the maps f_m act on the spaces $C_\gamma(\mathbb{R}^k)$ and $C_\gamma^0(\mathbb{R}^k)$:

Lemma B.5.14. *If* (B.5.3) *holds and* $\varphi \in C_\gamma(\mathbb{R}^k)$ *then* $f_m(\operatorname{graph}\varphi) = \operatorname{graph}\psi$ *for some* $\psi \in C_\gamma(\mathbb{R}^k)$. *The same holds for* $C_\gamma^0(\mathbb{R}^k)$.

Proof. The map $G_\varphi^m\colon \mathbb{R}^k \to \mathbb{R}^k$ given by

$$G_\varphi^m(x) = A_m x + \alpha_m(x, \varphi(x)) \tag{B.5.5}$$

represents the x-coordinate of f_m acting on graph φ. To show that $f_m(\operatorname{graph}\varphi)$ is a

Figure B.5.2. The graph transform.

graph, we need to prove that G_φ^m is a bijection. Thus for $x_0 \in \mathbb{R}^k$ we need to find a unique $x \in \mathbb{R}^k$ such that $x_0 = G_\varphi^m(x)$ or equivalently

$$x = F(x) := A_m^{-1} x_0 - A_m^{-1} (\alpha_m(x, \varphi(x))) .$$

Now, $F : \mathbb{R}^k \to \mathbb{R}^k$ is a contracting map since

$$\begin{aligned}\|F(x_1) - F(x_2)\| &= \|A_m^{-1}(\alpha_m(x_1, \varphi(x_1)) - \alpha_m(x_2, \varphi(x_2)))\| \\ &\leq \mu^{-1} \|\alpha_m\|_{C^1} \cdot (1 + \gamma)\|x_1 - x_2\| < \delta\mu^{-1}(1 + \gamma)\|x_1 - x_2\|\end{aligned}$$

and $\delta\mu^{-1}(1 + \gamma) < 1$ by the second inequality in (B.5.3). Thus by the Contraction Mapping Principle (Proposition B.1.3) equation F has a unique fixed point, that is, $f_m(\operatorname{graph} \varphi) = \operatorname{graph} \psi$.

Next we show that ψ is γ-Lipschitz continuous. Suppose $\psi(x_1') = y_1'$ and $\psi(x_2') = y_2'$ and take $(x_1, y_1), (x_2, y_2) \in \operatorname{graph} \varphi$ such that for $i = 1, 2$,

$$(x_i', y_i') = f_m(x_i, y_i) = (A_m x_i + \alpha_m(x_i, \varphi(x_i)),\ B_m \varphi(x_i) + \beta_m(x_i, \varphi(x_i))) .$$

Then

$$\begin{aligned}\|y_2' - y_1'\| &= \|B_m(\varphi(x_2) - \varphi(x_1)) + \beta_m(x_2, \varphi(x_2)) - \beta_m(x_1, \varphi(x_1))\| \\ &< \lambda\gamma\|x_2 - x_1\| + \delta(1 + \gamma)\|x_2 - x_1\| \\ &= (\lambda\gamma + \delta(1 + \gamma))\|x_2 - x_1\|\end{aligned} \tag{B.5.6}$$

and

$$\begin{aligned}\|x_2' - x_1'\| &= \|A_m(x_2 - x_1) + \alpha_m(x_2, \varphi(x_2)) - \alpha_m(x_1, \varphi(x_1))\| \\ &> \mu\|x_2 - x_1\| - \delta(1 + \gamma)\|x_2 - x_1\| \\ &= (\mu - \delta(1 + \gamma))\|x_2 - x_1\| .\end{aligned} \tag{B.5.7}$$

Consequently, $\|y_2' - y_1'\| \leq \frac{\lambda\gamma + \delta(1+\gamma)}{\mu - \delta(1+\gamma)} \|x_2' - x_1'\| =: \gamma' \|x_2' - x_1'\|$. But a straightforward calculation shows that the first condition in (B.5.3) is equivalent to $\gamma' < \gamma$. This shows that f_m acts on $C_\gamma(\mathbb{R}^k)$. The same holds for $C_\gamma^0(\mathbb{R}^k)$ since $f_m(0) = 0$. □

Since we eventually want to apply the Contraction Mapping Principle, we introduce a metric on the space $C^0_\gamma(\mathbb{R}^k)$ and show that the action of f_m is a contraction.

Since $\varphi, \psi \in C^0_\gamma(\mathbb{R}^k)$ are Lipschitz continuous with $\varphi(0) = \psi(0) = 0$,

$$d(\varphi, \psi) := \sup_{x \in \mathbb{R}^k \setminus \{0\}} \frac{\|\varphi(x) - \psi(x)\|}{\|x\|}$$

is a well-defined metric. It is easy to check that it is complete.

The next lemma shows that the action of f_m on $C^0_\gamma(\mathbb{R}^k)$ given by

$$f_m(\operatorname{graph} \varphi) = \operatorname{graph}((f_m)_* \varphi)$$

is a contracting map.

Lemma B.5.15. *$d((f_m)_*\varphi, (f_m)_*\psi) \le \frac{\lambda+\delta(1+\gamma)}{\mu-\delta(1+\gamma)} d(\varphi,\psi)$ for $\varphi, \psi \in C^0_\gamma(\mathbb{R}^k)$.*

Proof. Let $\varphi' = (f_m)_*\varphi$ and $\psi' = (f_m)_*\psi$. Using the map G^m_φ defined by (B.5.5) and the fact that $\psi' \in C^0_\gamma(\mathbb{R}^k)$ we have

$$\begin{aligned}
&\|\varphi'(G^m_\varphi(x)) - \psi'(G^m_\varphi(x))\| \\
&\quad \le \|\varphi'(G^m_\varphi(x)) - \psi'(G^m_\psi(x))\| + \|\psi'(G^m_\psi(x)) - \psi'(G^m_\varphi(x))\| \\
&\quad \le \|(B_m(\varphi(x)) + \beta_m(x,\varphi(x))) - (B_m(\psi(x)) + \beta_m(x,\psi(x)))\| \\
&\qquad + \gamma\|G^m_\psi(x) - G^m_\varphi(x)\| \\
&\quad \le \|B_m(\varphi(x) - \psi(x))\| + \|\beta_m(x,\varphi(x)) - \beta_m(x,\psi(x))\| \\
&\qquad + \gamma\|\alpha_m(x,\psi(x)) - \alpha_m(x,\varphi(x))\| \\
&\quad < \lambda\|\varphi(x) - \psi(x)\| + \delta\|\varphi(x) - \psi(x)\| + \gamma\delta\|\varphi(x) - \psi(x)\| \\
&\quad = (\lambda + \delta(1+\gamma))\|\varphi(x) - \psi(x)\|.
\end{aligned}$$

On the other hand,

$$\begin{aligned}
\|G^m_\varphi(x)\| &= \|A_m x + \alpha_m(x,\varphi(x))\| \ge \|A_m x\| - \|\alpha_m(x,\varphi(x))\| \\
&\ge \mu\|x\| - \delta(1+\gamma)\|x\| = (\mu - \delta(1+\gamma))\|x\|.
\end{aligned}$$

Consequently,

$$\begin{aligned}
\frac{\|(f_m)_*\varphi(G^m_\varphi(x)) - (f_m)_*\psi(G^m_\varphi(x))\|}{\|G^m_\varphi(x)\|} &\le \frac{\lambda+\delta(1+\gamma)}{\mu-\delta(1+\gamma)} \cdot \frac{\|\varphi(x)-\psi(x)\|}{\|x\|} \\
&\le \frac{\lambda+\delta(1+\gamma)}{\mu-\delta(1+\gamma)} \cdot d(\varphi,\psi).
\end{aligned}$$

□

Here $\gamma < 1 \Rightarrow \frac{\lambda+\delta(1+\gamma)}{\mu-\delta(1+\gamma)} = \gamma^{-1}\frac{\lambda\gamma+\delta\gamma(1+\gamma)}{\mu-\delta(1+\gamma)} \leq \gamma^{-1}\frac{\lambda\gamma+\delta(1+\gamma)}{\mu-\delta(1+\gamma)} < 1$ by (B.5.3) (see (B.5.2)).

We now denote by C^0_γ the space of families $\{\varphi_m\}_{m\in\mathbb{Z}}$ of functions in $C^0_\gamma(\mathbb{R}^k)$. The action of $f = \{f_m\}_{m\in\mathbb{Z}}$ on the space C^0_γ given by

$$f_m(\operatorname{graph}\varphi_m) = \operatorname{graph}((f_*\varphi)_{m+1})$$

is called the *graph transform*. Lemma B.5.15 shows that the graph transform is a contraction with respect to the metric

$$d(\{\varphi_m\}_{m\in\mathbb{Z}}, \{\psi_m\}_{m\in\mathbb{Z}}) := \sup_{m\in\mathbb{Z}} d(\varphi_m, \psi_m).$$

Since C^0_γ is complete with this metric, the Contraction Mapping Principle (Proposition B.1.3) yields a unique fixed point for this action of f, hence an invariant family $\{\varphi^+_m\}$ of graphs, as claimed.

Remark B.5.16. If $\lambda < 1$ one can show that $\|\varphi^+_m\|_{C^0} < \delta/(1-\lambda)$ by considering only $\varphi \in C^0_\gamma(\mathbb{R}^k)$ bounded by $\delta/(1-\lambda)$ and showing invariance of this condition under f_*. In this case the first estimate in the proof of Lemma B.5.15 also shows that the graph transform is a contraction with respect to the C^0-topology.

To construct the functions φ^-_m one argues along the same lines. Using the estimates obtained in this step, with γ replaced by $1/\gamma$, one shows that the maps Df^{-1}_m act on families of γ-Lipschitz functions $\varphi\colon \mathbb{R}^{n-k} \to \mathbb{R}^k$ vanishing at the origin, and are contracting.

At this point it is natural to prove (ii) since we use the estimates (B.5.6) and (B.5.7). First replace (x_1, y_1) by $(0,0)$ and (x_2, y_2) by $(x, \varphi^+_m x)$ in (B.5.7). Then

$$\begin{aligned}\|f_m(x, \varphi^+_m(x))\| &\geq \|A_m x + \alpha_m(x, \varphi^+_m(x))\| \\ &> (\mu - \delta(1+\gamma))\|x\| \geq \frac{\mu - \delta(1+\gamma)}{1+\gamma}\|(x, \varphi^+_m(x))\|.\end{aligned}$$

On the other hand, applying (B.5.6) to $(0,0)$ and $(\varphi^-_m(y), y)$ and using the fact that φ^-_m are γ-Lipschitz yields

$$\begin{aligned}\|f_m(\varphi^-_m(y), y)\| &\leq (1+\gamma)\|B_m(y) + \beta_m(\varphi^-_m(y), y)\| \\ &< (1+\gamma)(\lambda\|y\| + \delta(1+\gamma)\|y\|) \\ &= (1+\gamma)(\lambda + \delta(1+\gamma))\|(\varphi^-_m(y), y)\|.\end{aligned}$$

B.5.d Differentiability. To prove that the invariant family of functions obtained in the previous step consists of continuously differentiable functions, we introduce the notion of a tangent set for a graph. The results of step 2, the existence of a unique invariant family of continuous plane fields, then imply that the tangent set of each of these graphs is a continuous plane field. But this, by definition, implies that the graphs are graphs of C^1-functions.

Definition B.5.17. Let $\varphi \in C^0_\gamma(\mathbb{R}^k)$, $x \in \mathbb{R}^k$,

$$\Delta_y \varphi := \frac{(y, \varphi(y)) - (x, \varphi(x))}{\|(y, \varphi(y)) - (x, \varphi(x))\|} \quad \text{for } y \neq x,$$

$t_x\varphi := \{v \in T_x\mathbb{R}^n \mid \exists \{x_n\}_{n \in \mathbb{N}} \text{ such that } \lim_{n\to\infty} x_n = x \text{ and } \lim_{n\to\infty} \Delta_{x_n}\varphi = v\}$. Then $\tau_x\varphi := \bigcup_{v \in t_x\varphi} \mathbb{R}v$, where $\mathbb{R}v := \{av \mid a \in \mathbb{R}\}$ is the line containing v, is called the *tangent set* of φ at x, and the (disjoint) union $\tau\varphi := \bigcup_{x \in \mathbb{R}^k} \tau_x\varphi$, the *tangent set* of φ.

Note that since for every $v \in \mathbb{R}^k$ one can choose $y = x + tv$ in the definition, $\tau_x\varphi$ projects onto $\mathbb{R}^k$.

As an example consider $\varphi(x) = x\sin(1/x) \in C^0_\gamma(\mathbb{R})$ for which $\tau_0\varphi = \{(x, y) \in \mathbb{R}^2 \mid |y| \leq |x|\} = H^1_0$. Indeed, for $\varphi \in C^0_\gamma(\mathbb{R}^k)$ and $x \in \mathbb{R}^k$ we always have $\tau_x\varphi \subset H^\gamma_x$, since φ has Lipschitz constant γ. Another important observation is that $\varphi \in C^0_\gamma(\mathbb{R}^k)$ is differentiable at x if and only if $\tau_x\varphi$ is a k-dimensional plane.

We can now show that the invariant family $\varphi^+ = \{\varphi^+_m\}_{m\in\mathbb{Z}}$ obtained in step 3 consists of C^1-functions. Associated with φ^+ is the family $\tau\varphi^+ := \{\tau\varphi^+_m\}_{m\in\mathbb{Z}}$ of tangent sets for the functions φ^+_m, $m \in \mathbb{Z}$. Since φ^+ is an invariant family of functions for $f = \{f_m\}_{m\in\mathbb{Z}}$, the associated family $\tau\varphi^+$ of tangent sets is invariant under the action of the differentials Df_m. In step 2 we showed that any such invariant family inside the γ-cones is contained in the unique invariant family E^+_m of continuous plane fields obtained there. Since every tangent set $\tau_p\varphi^+_m$ projects onto $\mathbb{R}^k$, we conclude that $\tau_p\varphi^+_m = (E^+_p)_m$, that is, the φ^+_m are C^1-functions.

Smoothness of φ^-_m is proved likewise. This ends the proof of (i).

It remains to prove (iii). We remarked after the formulation of the theorem that we can construct the manifolds $(W^-_m)_p$ and $(W^+_m)_p$ for any point $p = (x, y)$. We still have $(W^+_m)_p = \operatorname{graph}(\varphi^+_m)_p$ and $(W^-_m)_p = \operatorname{graph}(\varphi^-_m)_p$ for some γ-Lipschitz functions $(\varphi^+_m)_p\colon \mathbb{R}^k \to \mathbb{R}^{n-k}$ and $(\varphi^-_m)_p\colon \mathbb{R}^{n-k} \to \mathbb{R}^k$ and properties analogous to (i) and (ii).

Lemma B.5.18. *For $p, q \in \mathbb{R}^n$ the intersection $(W^+_m)_p \cap (W^-_m)_q$ is a point.*

Proof. If $z = (x, y) \in (W^+_m)_p \cap (W^-_m)_q$ then $x = (\varphi^-_m)_q(y)$ and $y = (\varphi^+_m)_p(x)$ and hence $x = (\varphi^-_m)_q \circ (\varphi^+_m)_p(x)$. This in turn implies again that $\big(x, (\varphi^+_m)_p(x)\big) \in (W^+_m)_p \cap (W^-_m)_q$. But since we can assume $\gamma < 1$ the map $(\varphi^-_m)_q \circ (\varphi^+_m)_p\colon \mathbb{R}^k \to \mathbb{R}^k$ is a contraction and hence has a unique fixed point. □

Now assume $p \notin (W_m^-)_0$. By Lemma B.5.18 there is a unique $q \in (W_m^-)_0 \cap (W_m^+)_p$. Using (ii) for $(W_m^-)_0$ and $(W_m^+)_p$ we see that

$$\begin{aligned}
&\|f_{m+L-1} \circ \cdots \circ f_m(p)\| \\
&\quad \geq \|f_{m+L-1} \circ \cdots \circ f_m(p) - f_{m+L-1} \circ \cdots \circ f_m(q)\| - \|f_{m+L-1} \circ \cdots \circ f_m(q)\| \\
&\quad \geq (\mu')^L \|p-q\| - (\lambda')^L \|q\| = (\mu')^L \Big(\|p-q\| - \Big(\frac{\lambda'}{\mu'}\Big)^L \|q\|\Big).
\end{aligned}$$

Whenever $\lambda' < \nu < \mu'$ and $C \in \mathbb{R}$ this quantity exceeds $C \cdot \nu^L \|p\|$ for sufficiently large $L \in \mathbb{N}$.

Together with a parallel argument for $(W_m^+)_0$ this proves (iii) and thus also the uniqueness of W_m^+ and W_m^-.

This finishes the proof of the general part of the Hadamard–Perron Theorem.

B.5.e Higher smoothness. To complete the proof of Theorem B.5.2 we now prove that if $\mu \geq 1$ in Theorem B.5.2 then $\{W_m^+\}_{m \in \mathbb{Z}}$ consists of manifolds as smooth as the diffeomorphism. Let Df_m have block form $\begin{pmatrix} A_m^{uu} & A_m^{su} \\ A_m^{us} & A_m^{ss} \end{pmatrix}$ with A_m^{uu} a $(k \times k)$-matrix with $\|(A_m^{uu})^{-1}\| \leq 1/(\mu - \delta)$, A_m^{ss} an $((n-k) \times (n-k))$-matrix with $\|A_m^{ss}\| \leq \lambda + \delta$, and $\|A_m^{su}\| < \delta$, $\|A_m^{us}\| < \delta$. By the preceding steps, notably Lemma B.5.14, we can obtain W_m^+ by taking smooth functions $\varphi_m^0 \in C_\gamma^0(\mathbb{R}^k)$ (such as $\varphi_m^0 = 0$), applying the graph transform repeatedly to obtain families $\{\varphi_m^i\}$ for $i \in \mathbb{N}$, and taking the limit as $i \to \infty$. We plan to show inductively that the $(r+1)$st derivative of φ_m^i converges as $i \to \infty$, so long as f is C^{r+1}. To that end we note that $D\varphi_m^i$ is the graph of a linear map E_m^i from $\mathbb{R}^k$ to $\mathbb{R}^{n-k}$, or, equivalently, the image of the map $\begin{pmatrix} I \\ E_m^i \end{pmatrix} : \mathbb{R}^k \to \mathbb{R}^n$. Note that the image of $D\varphi_m^i$ under Df_m is the image of the linear map

$$\begin{pmatrix} A_m^{uu} & A_m^{su} \\ A_m^{us} & A_m^{ss} \end{pmatrix} \begin{pmatrix} I \\ E_m^i \end{pmatrix} = \begin{pmatrix} A_m^{uu} + A_m^{su} E_m^i \\ A_m^{us} + A_m^{ss} E_m^i \end{pmatrix}.$$

If, referring to (B.5.5), we let $g_m^i := (G^{m-1}_{\varphi_{m-1}^i})^{-1}$ then this has to coincide with the image of $\begin{pmatrix} I \\ E_{m+1}^{i+1} \circ (g_{m+1}^i)^{-1} \end{pmatrix}$ which is the same as that of

$$\begin{pmatrix} A_m^{uu} + A_m^{su} E_m^i \\ (E_{m+1}^{i+1} \circ (g_{m+1}^i)^{-1})(A_m^{uu} + A_m^{su} E_m^i) \end{pmatrix},$$

so

$$(E_{m+1}^{i+1} \circ (g_{m+1}^i)^{-1})(A_m^{uu} + A_m^{su} E_m^i) = A_m^{us} + A_m^{ss} E_m^i.$$

Composing with g^i_{m+1} and differentiating r times we get

$$D^r E^{i+1}_{m+1}(\alpha^u_{m+1,i+1})^{-1} + E^{i+1}_{m+1}(A^{su}_m \circ g^i_{m+1})(D^r E^i_m \circ g^i_{m+1})(Dg^i_{m+1})^{\otimes r}$$
$$= (A^{ss}_m \circ g^i_{m+1})(D^r E^i_m \circ g^i_{m+1})(Dg^i_{m+1})^{\otimes r} + \zeta_{m+1,i+1}(\alpha^u_{m+1,i+1})^{-1},$$

where $\zeta_{m+1,i+1}$ is a polynomial in lower derivatives of E^{i+1}_{m+1} and E^i_m and

$$\alpha^u_{m+1,i+1} := [(A^{uu}_m \circ g^i_{m+1}) + (A^{su}_m \circ g^i_{m+1})(E^i_m \circ g^i_{m+1})]^{-1}.$$

Letting $\alpha^s_{m,i} := (A^{ss}_{m-1} \circ g^{i-1}_m) - E^i_m(A^{su}_{m-1} \circ g^{i-1}_m)$ this yields

$$\begin{aligned} D^r E^i_m &= \alpha^s_{m,i}(D^r E^{i-1}_{m-1} \circ g^{i-1}_m)(Dg^{i-1}_m)^{\otimes r}\alpha^u_{m,i} + \zeta_{m,i} \\ &= \alpha^s_{m,i}(\alpha^s_{m-1,i-1} \circ g^{i-1}_m) \\ &\quad \times (D^r E^{i-2}_{m-2} \circ g^{i-2}_{m-1} \circ g^{i-1}_m)(Dg^{i-2}_{m-1})^{\otimes r}(\alpha^u_{m-1,i-1} \circ g^{i-1}_m)(Dg^{i-1}_m)^{\otimes r}\alpha^u_{m,i} \\ &\quad + \alpha^s_{m,i}(\zeta_{m-1,i-1} \circ g^{i-1}_m)(Dg^{i-1}_m)^{\otimes r}\alpha^u_{m,i} + \zeta_{m,i} \\ &= \cdots . \end{aligned}$$

Applying this inductively we obtain an expression for $D^r E^i_m$ with a leading term involving $D^r E^0_{m-i}$ between i-fold products

$$\alpha^s_{m,i}(\alpha^s_{m-1,i-1} \circ g^{i-1}_m)(\alpha^s_{m-2,i-2} \circ g^{i-2}_{m-1} \circ g^{i-1}_m) \cdots$$

of terms $\alpha^s_{m-l,i-l}$ and

$$\cdots (Dg^{i-3}_{m-2})^{\otimes r}(\alpha^u_{m-2,i-2} \circ g^{i-2}_{m-1} \circ g^{i-1}_m)(Dg^{i-2}_{m-1})^{\otimes r}(\alpha^u_{m-1,i-1} \circ g^{i-1}_m)(Dg^{i-1}_m)^{\otimes r}\alpha^u_{m,i}$$

of $\alpha^u_{m-l,i-l}$ and i occurrences of $(Dg^{i-l-1}_{m-l})^{\otimes r}$. This term goes to 0 uniformly as $i \to \infty$: $\|D^r E^0_{m-i}\|$ is uniformly bounded by choice of φ^0_{m-i} and $\|\alpha^s_{m-l,i-l}\|\|\alpha^u_{m-l,i-l}\| < 1$ uniformly by taking small δ. Finally, the assumption $\mu \geq 1$ of this step ensures that the factors $(Dg^{i-l-1}_{m-l})^{\otimes r}$ cause no exponential growth.

The jth of the remaining i summands in the expression for $D^r E^i_m$ similarly consists of $\zeta_{m-j-1,i-j-1}$ between j-fold products of terms $\alpha^s_{m-l,i-l}$ and $\alpha^u_{m-l,i-l}$ as well as j occurrences of $(Dg^{i-l-1}_{m-l})^{\otimes r}$. As before, these terms tend to 0 uniformly as $j \to \infty$ given uniform control of $\zeta_{m-j-1,i-j-1}$. These, however, involve only lower derivatives of the E^k_l, which are uniformly bounded by induction assumption, as well as derivatives up to order r of coefficients of Df, which are bounded because $f \in C^{r+1}$. Consequently, these remaining terms give partial sums of an exponentially convergent series. We already know that lower-order derivatives of E^i_m converge as $i \to \infty$ and thus conclude that the limit of E^i_m is C^r, as desired. □

Note that $(W^+_m)_p$ and $(W^-_m)_p$ for $p \in \mathbb{R}^n$ depend continuously on p: characterization (iii) of Theorem B.5.2 yields the following proposition:

Proposition B.5.19. *If $p_l \to p \in \mathbb{R}^n$ as $l \to \infty$ and $y_l \in (W_m^+)_{p_l}$ for all $l \in \mathbb{N}$ and $y_l \to y \in \mathbb{R}^n$ as $l \to \infty$ then $y \in (W_m^+)_p$.*

Proof. Fix $L \in \mathbb{N}$. Then Theorem B.5.2(ii) implies for $\nu < \mu'$ that

$$\|f_{m-L}^{-1} \circ \cdots \circ f_{m-1}^{-1}(y_l) - f_{m-L}^{-1} \circ \cdots \circ f_{m-1}^{-1}(p_l)\| \le \nu^{-L}\|y_l - p_l\|$$

for all $l \in \mathbb{N}$. By continuity of the f_m this implies

$$\|f_{m-L}^{-1} \circ \cdots \circ f_{m-1}^{-1}(y) - f_{m-L}^{-1} \circ \cdots \circ f_{m-1}^{-1}(p)\| \le \nu^{-L}\|y - p\|$$

and since L was arbitrary the claim follows by (iii). □

Since on any fixed compact set the assumption that y_l converges is redundant (by passing to a subsequence) this means that $(W_m^+)_{p_l} \to (W_m^+)_p$ when $p_l \to p$. Convergence here is in the pointwise sense of the proposition. Since we know that E_m^+ is continuous, we have continuity of W_m^+ together with its tangent spaces. A similar statement holds for W_m^-.

Another pertinent remark is that we obtain in fact continuous dependence of W^+ and W^- on the family f_m of maps we consider. Since the main ingredient of the proof of the Hadamard–Perron Theorem (Theorem B.5.2) was obtaining the invariant manifolds and their tangent bundles as fixed points of a contraction operator associated with the family f_m, we may use Proposition B.1.3 to infer that the invariant manifolds depend continuously on the diffeomorphisms with respect to the C^1-topology.

Proposition B.5.20. *The invariant manifolds (with the C^1-topology) obtained in the Hadamard–Perron Theorem* (Theorem B.5.2) *depend continuously on the family f_m if we use the C^1-topology ($\{f_m\}_{m\in\mathbb{N}}, \{g_m\}_{m\in\mathbb{N}}$ are C^1-close if $\sup_m d_{C^1}(f_m, g_m)$ is small).*

Remark B.5.21. In the hyperbolic case one can use the C^r-topology for invariant manifolds, and one does indeed obtain continuous dependence on the family f_m in the C^r-topology.

Corollary B.5.22. *If $p_l \to p \in \mathbb{R}^n$ as $l \to \infty$ and $q \in \mathbb{R}^n$ then the sequence y_l given by $(W_m^+)_{p_l} \cap (W_m^-)_q = \{y_l\}$ converges to y given by $\{y\} = (W_m^+)_p \cap (W_m^-)_q$.*

This follows from Proposition B.5.20 and Lemma B.5.18 since the y_l are contained in a compact set because the $(W_m^+)_{p_l}$ are Lipschitz graphs.

Since the construction and characterization of $(W_m^+)_p$ and $(W_m^+)_p$ other than $(W_m^+)_0$ and $(W_m^-)_0$ depend on the behavior of points whose orbits do not stay in a neighborhood of the origin, they depend on the extension chosen in the Extension Theorem (Theorem B.4.12) and do not represent meaningful objects associated with neighborhoods of a reference orbit on a manifold.

B.6 The Inclination Lemma and homoclinic tangles

The graph-transform method also yields the *Inclination Lemma*, that successive images of a disk transverse to the stable manifold of a hyperbolic fixed point accumulate (in the C^1-topology) on the unstable manifold of the point.

Theorem B.6.1 (Inclination Lemma). *Suppose p is a hyperbolic fixed point of a diffeomorphism f and $\mathcal{D}$ is a disk that transversely intersects $W^s(p)$ (and hence has the same dimension as $W^u(p)$). Then the $f^n(\mathcal{D})$ accumulate on $W^u(p)$ in the C^1-topology as $n \to +\infty$. Specifically, for any disk Δ in $W^u(p)$ and any $\epsilon > 0$ there is an $n \in \mathbb{N}$ and a $\mathcal{D}' \subset \mathcal{D}$ such that $d_{C^1}(f^n(\mathcal{D}'), \Delta) < \epsilon$.*

Proof. In order to apply Proposition B.6.2 below, choose adapted coordinates at p. After possibly conjugating these by f^k for some $k \in \mathbb{N}$ these will contain Δ. Now replace $\mathcal{D}$ by $f^m(\mathcal{D})$ for $m \in \mathbb{N}$ such that (using that $\mathcal{D}$ is C^1 and after possibly shrinking $\mathcal{D}$) $\mathcal{D}$ is in the adapted coordinate system and the hypotheses of Proposition B.6.2 hold. The conclusion of Proposition B.6.2 and the Lipschitz convergence of the graph transform then imply the claim because the Lipschitz topology on C^1 submanifolds induces the C^1-topology. □

To state the main lemma it is convenient to use adapted coordinates as in Theorem B.5.3 and to let $\pi_1 : \mathbb{R}^k \oplus \mathbb{R}^{n-k} \to \mathbb{R}^k$ be the projection to the first coordinate.

Proposition B.6.2. *Under the hypotheses of Theorem* B.5.3 *consider C^r adapted coordinates on a neighborhood O of a hyperbolic fixed point p of $f : U \to M$. Given $\epsilon, K, \eta > 0$ there exists an $N \in \mathbb{N}$ such that if $\mathcal{D}$ is a C^1-disk containing $q \in W_p^- \cap O$ with all tangent spaces in horizontal K-cones and such that $\pi_1(\mathcal{D})$ contains an η-ball around $0 \in \mathbb{R}^k \oplus \{0\}$ and $n \geq N$ then $\pi_1(f^n(\mathcal{D})) = W_p^+ \cap O$ and $T_z f^n(\mathcal{D})$ is contained in a horizontal ϵ-cone for every $z \in f^n(\mathcal{D})$.*

Proof. Since $\mathbb{R}^k \oplus \{0\}$ and $\{0\} \oplus \mathbb{R}^{n-k}$ are f-invariant and f is C^1, the differential of f at points $(x, y) \in \mathbb{R}^k \oplus \mathbb{R}^{n-k}$ takes the form

$$Df_{(x,y)} = \begin{pmatrix} A_z^{uu} & A_z^{su} \\ A_z^{us} & A_z^{ss} \end{pmatrix},$$

where

$$\begin{aligned} A_z^{uu} &\in M_{k,k}, & \|(A_z^{uu})^{-1}\| &\leq \frac{1}{\mu - \delta}, \\ A_z^{ss} &\in M_{n-k,n-k}, & \|A_z^{ss}\| &< \lambda + \delta, \\ A_z^{us} &\in M_{n-k,k}, & \|A_z^{us}\| &\in o(\|y\|), \\ A_z^{su} &\in M_{k,n-k}, & A_z^{su} &\in o(\|x\|). \end{aligned}$$

Here $\lambda < 1 < \mu$ are the contraction and expansion rates as before, and we used the notation from Remark 3.2.18. Then δ can be taken arbitrarily small by possibly shrinking the size of the neighborhood O (and replacing $\mathcal{D}$ by its image under an iterate f^n, so that $\mathcal{D}$ intersects the local stable leaf of p in a point in O). Similarly to the proof of smoothness of stable and unstable manifolds it is convenient now to consider planes in horizontal γ-cones as graphs of linear maps whose operator norm (denoted by $\|\cdot\|$) is bounded by γ. After possibly shrinking $\mathcal{D}$ we assume that $\mathcal{D} \cap (\{0\} \oplus \mathbb{R}^{n-k}) = \{z\}$ is a single point. Then our first step consists of showing that $T_{z_n} f^n(\mathcal{D})$ is contained in a horizontal $\epsilon/2$-cone for some $n \in \mathbb{N}$, where $z_i = f^i(z)$. To that end consider a linear map $E_z\colon \mathbb{R}^k \to \mathbb{R}^{n-k}$ with $\|E\| \le K$. Its graph is parametrized as the image of the linear map

$$\begin{pmatrix} I \\ E_z \end{pmatrix} : \mathbb{R}^k \to \mathbb{R}^k \oplus \mathbb{R}^{n-k},$$

where $I\colon \mathbb{R}^k \to \mathbb{R}^k$ is the identity. The image of the graph under Df_z is then the image of the linear map

$$Df_z \circ \begin{pmatrix} I \\ E_z \end{pmatrix} : \mathbb{R}^k \to \mathbb{R}^k \oplus \mathbb{R}^{n-k}.$$

In our coordinates this composition is obtained via the matrix product

$$\begin{pmatrix} A_z^{uu} & A_z^{su} \\ A_z^{us} & A_z^{ss} \end{pmatrix} \begin{pmatrix} I \\ E_z \end{pmatrix} = \begin{pmatrix} A_z^{uu} + A_z^{su} E_z \\ A_z^{us} + A_z^{ss} E_z \end{pmatrix} \tag{B.6.1}$$

with $A_z^{su} = 0$ in this case. Note that $A_z^{uu}\colon \mathbb{R}^k \to \mathbb{R}^k$ is nonsingular, so the image of $\begin{pmatrix} A_z^{uu} \\ A_z^{us} + A_z^{ss} E_z \end{pmatrix}$ is that of $\begin{pmatrix} A_z^{uu} \\ A_z^{us} + A_z^{ss} E_z \end{pmatrix} \circ (A_z^{uu})^{-1} = \begin{pmatrix} I \\ A_z^{us} A_z^{uu-1} + A_z^{ss} E_z A_z^{uu-1} \end{pmatrix}$. In other words, $Df_z(T_z\mathcal{D})$ is the graph of the linear map

$$E_{z_1} = A_z^{us}(A_z^{uu})^{-1} + A_z^{ss} E_z (A_z^{uu})^{-1}.$$

Note that $\|E_{z_1}\| \le \frac{\|A_{z_0}^{us}\|}{\mu-\delta} + \frac{\lambda+\delta}{\mu-\delta}\|E_{z_0}\|$ and inductively

$$\|E_{z_n}\| \le \sum_{i=0}^{n-1} \frac{(\lambda+\delta)^{n-i-1}}{(\mu-\delta)^{n-i}} \|A_{z_i}^{us}\| + \frac{(\lambda+\delta)^n}{(\mu-\delta)^n}\|E_{z_0}\|.$$

Since $\|A_{z_i}^{us}\| \in o(\|y_i\|)$, where $z_i = (x_i, y_i)$, there exists $N \in \mathbb{N}$ such that for $n > N$ both $\sum_{i=N}^{n-1} \frac{(\lambda+\delta)^{n-i-1}}{(\mu-\delta)^{n-i}} \|A_{z_i}^{us}\| < \epsilon/6$ and $\frac{(\lambda+\delta)^n}{(\mu-\delta)^n}\|E_{z_0}\| < \epsilon/6$. If furthermore $N' \in \mathbb{N}$ is such that $\sum_{i=0}^{N-1} \frac{(\lambda+\delta)^{N'+N-i-1}}{(\mu-\delta)^{N'+N-i}} \|A_{z_i}^{us}\| < \epsilon/6$ then for $n \ge N + N' =: N_0$ we have $\|E_{z_n}\| < \epsilon/2$.

After possibly increasing N_0 we may assume that $\|A^{us}_{(x,y)}\| < (1-\lambda-\delta)\epsilon$ whenever $\|(x,y)\| \leq \|z_{N_O}\|$. Consequently, by possibly shrinking $\mathcal{D}$, we may assume that all tangent planes to $f^{N_0}(\mathcal{D})$ lie in horizontal ϵ-cones and that $\|A^{us}_z\| < (1-\lambda-\delta)\epsilon$ for $z \in \bigcup_{i \geq N_0} f^i(\mathcal{D})$. If ϵ is sufficiently small then $\|(A^{uu}_z + A^{su}_z E)^{-1}\| < 1$ whenever $\|E\| < \epsilon$. With this choice of parameters the action of f preserves horizontal ϵ-cones because if $\|E_{z_i}\| < \epsilon$ then (B.6.1) gives

$$\begin{aligned}\|E_{z_{i+1}}\| &\leq \|(A^{us}_{z_i} + A^{ss}_{z_i} E_{z_i})(A^{uu}_{z_i} + A^{su}_{z_i} E_{z_i})^{-1}\| \\ &\leq \|A^{us}_{z_i} + A^{ss}_{z_i} E_{z_i}\| \, \|(A^{uu}_{z_i} + A^{su}_{z_i} E_{z_i})^{-1}\| \\ &< \|A^{us}_{z_i}\| + \|A^{ss}_{z_i} E_{z_i}\| < (1-\lambda-\delta)\epsilon + (\lambda+\delta)\epsilon = \epsilon.\end{aligned}$$

Finally, note that the proof of Lemma B.5.14 shows that $f^n(\mathcal{D})$ covers $W^u_{\mathrm{loc}}(p)$ under the projection π_1 whenever n is sufficiently large. □

Remark B.6.3. It is useful to note (for example, by setting $\mu-\delta = 1$ in the calculations) that expansion in the unstable direction is not used in the proof of Proposition B.6.2; it is needed in Theorem B.6.1 to assert that arbitrarily large disks are limits of $\mathcal{D}$ under the dynamics. This allows us to invoke Proposition B.6.2 for time-t maps of flows by including the flow direction with the unstable one.

We next study horseshoes and a generic mechanism that gives rise to them.

By a rectangle in $\mathbb{R}^n$ we mean a set of the form $\Delta = D_1 \times D_2 \subset \mathbb{R}^k \oplus \mathbb{R}^l = \mathbb{R}^n$, where D_1 and D_2 are disks. We denote by $\pi_1 \colon \mathbb{R}^n \to \mathbb{R}^k$ and $\pi_2 \colon \mathbb{R}^n \to \mathbb{R}^l$ the canonical projections. As in Section B.5 we refer to the $\mathbb{R}^k$-direction as "horizontal" and the $\mathbb{R}^l$-direction as "vertical."

Definition B.6.4. Suppose $\Delta \subset U \subset \mathbb{R}^n$ is a rectangle and $f \colon U \to \mathbb{R}^n$ a diffeomorphism. A connected component $C' = fC$ of $\Delta \cap f(\Delta)$ is said to be *full* (for f) if

(1) $\pi_2(C) = D_2$, and

(2) for any $z \in C$, $\pi_1\restriction_{f(C\cap(D_1\times\pi_2(z)))}$ is a bijection onto D_1.

Geometrically, condition (2) means that the image of every horizontal fiber in C meets Δ and "traverses" Δ completely.

Definition B.6.5. If $U \subset \mathbb{R}^n$ is open then a rectangle $\Delta = D_1 \times D_2 \subset U \subset \mathbb{R}^k \oplus \mathbb{R}^l = \mathbb{R}^n$ is called a *horseshoe* for a diffeomorphism $f \colon U \to \mathbb{R}^n$ if $\Delta \cap f(\Delta)$ contains at least two full components Δ_0 and Δ_1 such that for $\Delta' = \Delta_0 \cup \Delta_1$,

(1) $\pi_2(\Delta') \subset \operatorname{int} D_2$, $\pi_1(f^{-1}(\Delta')) \subset \operatorname{int} D_1$,

(2) $D(f\restriction_{f^{-1}(\Delta')})$ preserves and expands a horizontal cone family on $f^{-1}(\Delta')$,

(3) $D(f^{-1}\restriction_{\Delta'})$ preserves and expands a vertical cone family on Δ'.

Conditions (2) and (3) imply by (the discrete-time version of) Proposition 5.1.7 that $\Lambda := \bigcap_{n\in\mathbb{Z}} f^{-n}(\Delta')$ is a hyperbolic set for f with "almost horizontal" expanding and "almost vertical" contracting directions.

We can now establish a connection between transverse homoclinic points and the existence of horseshoes. Figures 6.5.1 and 6.5.2 illustrate this.

Theorem B.6.6 (Birkhoff–Smale: homoclinic tangles produce horseshoes). *Let M be a smooth manifold, $U \subset M$ open, $f\colon U \to M$ an embedding, and $p \in U$ a hyperbolic fixed point with a transverse homoclinic point q. Then in an arbitrarily small neighborhood of p there exists a horseshoe for some iterate of f. Furthermore, the hyperbolic invariant set in this horseshoe contains an iterate of q.*

Proof. Via adapted coordinates on a neighborhood $\mathcal{O}$ we may assume that the hyperbolic fixed point is at the origin and that $W^u_{\mathrm{loc}}(0) := \mathbb{C}(W^u(0)\cap\mathcal{O},0) \subset \mathbb{R}^k \oplus \{0\}$ (as in Definition 1.6.13) and $W^s_{\mathrm{loc}}(0) := \mathbb{C}(W^s(0)\cap\mathcal{O},0) \subset \{0\}\oplus\mathbb{R}^l$, where $\mathbb{R}^n = \mathbb{R}^k\oplus\mathbb{R}^l$. Let D_1, D_2 be small disks around 0 in $W^u_{\mathrm{loc}}(0)$ and $W^u_{\mathrm{loc}}(0)$, respectively, and $B := D_1 \times D_2$.

Take N_0 minimal such that $q' := f^{-N_0}(q) \in \operatorname{Int} D_1$; since q' is transverse homoclinic we can take $\delta > 0$ sufficiently small so that $D_1 \times \{x\}$ is transverse to $W^s_{\mathrm{loc}}(q') := \mathbb{C}(W^s(p)\cap\Delta, q') \subset W^s(p)$ for $x \in \delta D_2 := \{\delta z \mid z \in D_2\}$, where $\Delta := D_1 \times \delta D_2$. By the Inclination Lemma (Theorem B.6.1) we can choose $\delta > 0$ and $N_1 \in \mathbb{N}$ such that if $\mathcal{D}_z := \mathbb{C}\big(f^{N_1}(D_1\times\{z\})\cap B, f^{N_1}(D_1\times\{z\}\cap W^s_{\mathrm{loc}}(q'))\big)$ for $z \in \delta D_2$, then $T_x\mathcal{D}_z$ is in a horizontal ϵ-cone for $x \in \mathcal{D}_z$, and $\pi_1\mathcal{D}_z = D_1$.

This shows that $\Delta_1 := \bigcup_{z\in\delta D_2}\mathcal{D}_z$ is a full component (Definition B.6.5) of $\Delta \cap f^{N_1}(\Delta)$. We have in fact shown that in a natural sense this component can be taken arbitrarily close to horizontal. Together with $\Delta_0 := \mathbb{C}\left(\Delta\cap f^{N_1}(\Delta), 0\right)$ which is obviously a full component, we thus have verified (1) of Definition B.6.5. It remains to prove the required hyperbolicity. Conditions (2) and (3) of Definition B.6.5 are easy to check for points $x \in f^{-N_1}(\Delta_0)$ since $f^i(x) \in \Delta$ for $i = 1, 2, \dots, N_1$. Consider $f^{-N_1}(\Delta_1)$. Since $f^{N_1}(q')$ is a transverse homoclinic point we can use the decomposition $\mathbb{R}^n = \mathbb{R}^k \oplus \mathbb{R}^l$ to write $Df^{N_1}(q') = \left(\begin{smallmatrix} E & F \\ G & H\end{smallmatrix}\right)$ with E nonsingular. The same holds for all $x \in f^{-N_1}(\Delta_1)$ by our choice of δ. If these differentials do not satisfy (2) and (3), replace q' by $q'' = f^{-m}(q')$ and N_1 by $N_2 = N_1 + m + n$ for some $n, m \in \mathbb{N}$ to be specified later. Then

$$Df^{N_2}(q'') = \begin{pmatrix} A_n & B_n \\ C_n & D_n \end{pmatrix}\begin{pmatrix} E & F \\ G & H \end{pmatrix}\begin{pmatrix} A'_m & B'_m \\ C'_m & D'_m \end{pmatrix}.$$

Since E is nonsingular there exists a $\gamma_0 \in \mathbb{R}$ such that the horizontal γ_0-cone is mapped into the horizontal γ_1-cone with $\gamma_1 < \infty$ by $\left(\begin{smallmatrix} E & F \\ G & H \end{smallmatrix}\right)$. For $\gamma \in \mathbb{R}_+$ take $m \in \mathbb{N}$ such that the horizontal γ-cone is mapped into the horizontal γ_0-cone by $\left(\begin{smallmatrix} A'_m & B'_m \\ C'_m & D'_m \end{smallmatrix}\right)$ and $n \in \mathbb{N}$ such that the horizontal γ_1-cone is mapped into the horizontal γ-cone by $\left(\begin{smallmatrix} A_n & B_n \\ C_n & D_n \end{smallmatrix}\right)$. Thus $Df^{N_2}(q'')$ preserves horizontal γ-cones. Enlarging n, m further, if necessary, shows that $Df^{N_2}(q'')$ expands vectors in γ-cones. Since these estimates can be made uniformly on $f^{-N_2}(\Delta_1)$ and even better estimates hold on $f^{-N_2}(\Delta_0)$, we obtain (2) and (3) of Definition B.6.5. □

B.7 Absolute continuity

The central argument with which the ergodic theory of hyperbolic dynamical systems started is the Hopf argument, and this argument relies on using the Fubini Theorem, that is, absolute continuity of the invariant foliations. This section establishes absolute continuity of these foliations in a discrete-time setting that is general enough to apply directly to time-1 maps of flows. It is, in fact, more general than that, covering partially hyperbolic diffeomorphisms (Definition 5.6.2) and hence also time-1 maps of partially hyperbolic flows. For our purpose, viewing time-1 maps of Anosov flows as partially hyperbolic diffeomorphisms with 1-dimensional center direction shows that the proof of ergodicity also establishes ergodicity of an invariant volume for an Anosov flow. We note that elsewhere, we bypassed the issue of absolute continuity by establishing a weaker property (the Volume Lemma, Proposition 7.4.3) and using the theory of equilibrium states.

B.7.a The Katok example. Before embarking on the proof, we present here the *original* Katok example of a foliation that is not absolutely continuous, as it was recorded and passed on by Keith Burns [180, p. 31].

Let A be the hyperbolic automorphism of the torus $\mathbb{T}^2$ defined by the matrix

$$\begin{pmatrix} 2 & 1 \\ 1 & 1 \end{pmatrix}.$$

There is a family $\{f_t \mid t \in [0,1]\}$ of diffeomorphisms preserving the area m and satisfying the following conditions:

(1) f_t is a small perturbation of A for every $t \in [0,1]$;

(2) f_t depends smoothly on t;

(3) $l'(t) \neq 0$, where $l(t)$ is the larger eigenvalue of the derivative of f_t at its fixed point.

The diffeomorphisms f_t are all Anosov, conjugate to A, and ergodic with respect to m. For any s and t in $[0,1]$, the maps f_s and f_t are conjugate via a unique homeomorphism h_{st} close to the identity, that is, $f_t = h_{st} \circ f_s \circ h_{st}^{-1}$. The homeomorphism h_{st} is Hölder continuous. Let m_{st} be the pushforward of m by h_{st}. Then m_{st} is an ergodic invariant measure for f_t. Using the condition on $l(t)$ and the following lemma, we see that $m \neq m_{st}$ unless $s = t$.

Lemma B.7.1 (de la Llave [246]). *Suppose $f, g\colon \mathbb{T}^2 \to \mathbb{T}^2$ are smooth area-preserving Anosov diffeomorphisms that are conjugate via an area-preserving homeomorphism h. Let p be a periodic point for f with least period k. Then $Df^k(p)$ and $Dg^k(h(p))$ have the same eigenvalues up to sign.*

Proof. Let λ and λ' be the eigenvalues of $Df^k(p)$ and $Dg^k(h(p))$, respectively, inside the unit circle. Since f and g preserve area, the other eigenvalues of $Df^k(p)$ and $Dg^k(h(p))$ are $1/\lambda$ and $1/\lambda'$ respectively. Choose $x \in W^u_{\text{loc}}(p; f) \smallsetminus \{p\}$ and $y \in W^s_{\text{loc}}(p; f) \smallsetminus \{p\}$. Let R_n be the smallest "rectangle" with boundary in $W^s_{\text{loc}}(p; f)$, $W^s_{\text{loc}}(f^{-kn}(x); f)$, $W^u_{\text{loc}}(p; f)$, and $W^u_{\text{loc}}(f^{kn}(y); f)$, and R'_n the smallest "rectangle" with boundary in $W^s_{\text{loc}}(h(p); g)$, $W^s_{\text{loc}}(h(f^{-kn}(x)); g)$, $W^u_{\text{loc}}(h(p); g)$, and $W^u_{\text{loc}}(h(f^{kn}(y)); g)$. Then

$$\lim_{n\to\infty} \frac{\text{area}(R_{n+1})}{\text{area}(R_n)} = \lambda^{2k} \quad \text{and} \quad \lim_{n\to\infty} \frac{\text{area}(R'_{n+1})}{\text{area}(R_n)} = \lambda'^{2k}.$$

On the other hand, the conjugacy h takes R_n to R'_n for any n. Since h is area preserving, it follows that $\lambda = \pm\lambda'$. □

A point is *generic* with respect to an invariant measure if the forward and backward Birkhoff averages of any continuous function are defined at the point and are equal to its integral with respect to the measure. If x is generic for f_s with respect to m, then $h_{st}(x)$ is generic for f_t with respect to m_{st} and hence is not generic for f_t with respect to m, unless $s = t$. (To see this, note that the Birkhoff averages of a continuous function φ along the f_t-orbit of $h_{st}(x)$ are the same as the Birkhoff averages of $\varphi \circ h_{st}$ along the f_s orbit of x.)

Now consider the diffeomorphism $F\colon \mathbb{T}^2 \times [0,1] \to \mathbb{T}^2 \times [0,1]$ given by $F(x,t) = (f_t(x), t)$. We have just observed that for any $x \in \mathbb{T}^2$ the set $H(x) = \{(h_{0t}(x), t) \mid t \in [0,1]\}$ contains at most one element of the set $\mathcal{G}$ of points $(y,t) \in \mathbb{T}^2 \times [0,1]$ such that y is generic for f_t with respect to m.

Now, F is a small perturbation of $A \times \text{Id}_{[0,1]}$ and thus partially hyperbolic. It follows from Theorem B.7.2 that F has a center foliation whose leaves are small perturbations of the intervals $\{x\} \times [0,1]$ for $x \in \mathbb{T}^2$:

Theorem B.7.2 (Hirsch–Pugh–Shub [194, Theorem 7.5]). *Assume that the central distribution E^c for f is integrable, that the corresponding foliation W^c is smooth, and that g is a C^q-diffeomorphism sufficiently close to f in the C^1-topology. Then g is partially hyperbolic with integrable central distribution E_g^c.*

Since F maps the tori $\mathbb{T}^2 \times \{t\}$ into themselves, it is easily seen that the leaves of W_F^c are ℓ-normally hyperbolic for any ℓ, and hence are C^∞ by Theorem B.7.3. On the other hand, for each $x \in \mathbb{T}^2$, the leaf of W_F^c that passes through $(x, 0) \in \mathbb{T}^2 \times [0,1]$ is $H(x)$.

Theorem B.7.3. *Let $f\colon M \to M$ be a partially hyperbolic embedding with integrable central distribution. Then $W^c(x) \in C^\ell$ for every $x \in M$ and ℓ such that the leaves of W_F^c are ℓ-normally hyperbolic.*

The set $\mathcal{G}$ of generic points for F has full measure with respect to m in each torus $\mathbb{T}^2 \times \{t\}$ and hence has full Lebesgue measure in $\mathbb{T}^2 \times [0,1]$, but, as observed above, it intersects each center leaf in at most one point.

To construct an analogous example on $\mathbb{T}^2 \times S^1$ use two periodic points simultaneously instead of the one fixed point. The example here is constructed in such a way that $l(t) = l(s) \Rightarrow t = s$. For a continuous parametrization using $t \in S^1$ this won't work, but starting from the map A^2 instead, which has several fixed points, we use perturbations for which the largest eigenvalues $l_1(t)$ and $l_2(t)$ at two fixed points $x_1(t)$ and $x_2(t)$ satisfy $l_1(t) = l_1(s)$ and $l_2(t) = l_2(s) \Rightarrow t = s \pmod 1$. For example, make $l_1'(t) > 0$ on $(0, 1/2)$, $l_1'(t) < 0$ on $(1/2, 1)$ and $l_2'(t) = 0$ on $(0, 1/2)$, $l_1'(t) > 0$ on $(1/2, 3/4)$, $l_2'(t) < 0$ on $(3/4, 1)$.

B.7.b The absolute continuity theorem. Now we get to work on the proof of absolute continuity. With the conorm from Definition 5.6.1, we make the following definition.

Definition B.7.4. An embedding f is said to be *relatively partially hyperbolic* on Λ if there exists a Riemannian metric called a *Lyapunov metric* in an open neighborhood U of Λ for which there are continuous *functions*

$$0 < \lambda < \zeta \leq \xi < \lambda \quad \text{with } \lambda < 1 < \mu$$

in Λ and a pairwise orthogonal invariant splitting (5.6.2) such that if $x \in \Lambda$ then

$$\begin{aligned} \|d_x f \restriction E^s(x)\| \leq \lambda < \zeta \leq \llbracket d_x f \restriction E^c(x) \rrbracket \\ \leq \|d_x f \restriction E^c(x)\| \leq \xi < \mu \leq \llbracket d_x f \restriction E^u(x) \rrbracket. \end{aligned}$$

We introduce useful terminology and conventions for our local analysis. For $x \in M$ and $n \in \mathbb{Z}$ let $x_n := f^n(x)$, and for $\varphi\colon M \to (0,\infty)$ and $n \in \mathbb{N}$ define $\varphi_0 := 1$,

$$\varphi_n := \varphi \cdot \varphi \circ f \cdot \varphi \circ f^2 \cdots \varphi \circ f^{n-1}$$

and

$$\varphi_{-n} := [\varphi \circ f^{-n} \cdot \varphi \circ f^{1-n} \cdots \varphi \circ f^{-1}]^{-1}.$$

We furthermore choose $R > 0$ small enough that every Riemannian R-ball $B(p,R)$ lies in a foliation box (Definition 7.1.15) and if $q, q' \in B(p,R)$ and $q \in W^s_{\text{loc}}(q')$ then $d(f(q), f(q')) \le \lambda(p) d(q,q')$ (where λ is as in Definition B.7.4). Inductively, this implies that if $q_j, q'_j \in B(p_j, R)$ for $j = 0, \dots, n-1$ and $q \in W^s_{\text{loc}}(q')$ then $d(q_n, q'_n) \le \lambda_n d(q,q')$ where $\lambda_n = \prod_{i=0}^{n-1} \lambda(p_i)$.

Definition 7.1.17 describes the property we will verify for the stable and unstable foliations of a partially hyperbolic dynamical system.

Theorem B.7.5 (Transverse absolute continuity). *The stable and unstable foliations of a $C^{1+\alpha}$ partially hyperbolic diffeomorphism are transversely absolutely continuous with bounded Jacobians (Definition* 7.1.17).

With Proposition 7.1.20 this implies the following theorem:

Theorem B.7.6 (Absolute continuity). *The stable and unstable foliations of a $C^{1+\alpha}$ partially hyperbolic diffeomorphism are absolutely continuous with bounded Jacobians (Definition* 7.1.18).

These results cover our needs because we apply them to time-t maps of hyperbolic flows, where one can take $\zeta = \xi = 1$ in Definition B.7.4. Indeed, these statements go far beyond the applications in this book. They, and their proofs, are included here in this form to put in the published record the argument by Abdenur and Viana with which one obtains this important ingredient for the study of partially hyperbolic diffeomorphisms [288].

Let us explain the strategy before embarking on the proof of Theorem B.7.5. It is remarkably simple even though the full proof is rather long. To begin with, it suffices to prove this for stable holonomies, passing to the inverse then gives absolute continuity of the unstable foliation.

First, consider what happens to the volume of a large disk mapped from one transversal to another along *short* stable leaves: Any difference between the image of the disk and a disk in the second transversal resides in a small neighborhood of the boundary, so the error in volume is a small percentage. If we are concerned about small sets being mapped by the holonomy among not-so-short stable leaves, we push the whole picture forward by the dynamics to make the sets large and the stable leaves short. This reduces the situation to the previous one, except that we have to explain that these images are sufficiently "disk-like" and that the distortions of the set under the dynamics applied to one of the transversals versus the other one are not so large as to make the reduction useless. This uses distortion control along exponentially close orbits in ways we have seen before. The proof we give here is classical, but in the details follows an unpublished manuscript of Abdenur and Viana.

Proof of Theorem B.7.5 (Abdenur–Viana). Referring to Definition 7.1.17 with $\mathcal{F}$ the stable foliation, we will show that

$$m_{\tau_1}(A)/C \le m_{\tau_2}(h_{\tau_1,\tau_2}(A)) \le C m_{\tau_1}(A) \tag{B.7.1}$$

whenever A is a *disk* in τ_1. To see that this implies the same conclusion for any measurable set $E \subset \tau_1$, cover it by disks A_i with $\sum m_{\tau_1}(A_i) \le m_{\tau_1}(E) + \epsilon$ to get

$$m_{\tau_2}(h_{\tau_1,\tau_2}(E)) \le \sum m_{\tau_2}(h_{\tau_1,\tau_2}(A_i)) \le C \sum m_{\tau_1}(A_i) \le C(m_{\tau_1}(E) + \epsilon),$$

where ϵ can be arbitrarily small. The reverse inequality follows from the same argument applied to h_{τ_2,τ_1}. Indeed, this symmetry shows that we only need to establish $m_{\tau_2}(h_{\tau_1,\tau_2}(A)) \le C m_{\tau_1}(A)$ for disks.

Partial hyperbolicity provides $m \in \mathbb{N}$ and $\theta \in (0,1)$ (a measure of the gap between stable and center-unstable behavior) for which

$$\alpha_m(x) := \|Df^m\restriction_{E^s(x)}\| < \theta^2 \min(1, \beta_m(x)) \quad \text{for all } x \in M, \tag{B.7.2}$$

where $\beta_m(x) := \|Df^{-m}\restriction_{E^{cu}(f^m(x))}\|^{-1}$ (in the hyperbolic case we take $\beta_m(x) \equiv 1$). Using a Lyapunov metric or by passing to f^m we may (and will) take $m = 1$ here.

It is useful to get a quantitative measure of closeness of $f^n(\tau_1)$ and E^{cu} that improves exponentially in n. (This is reminiscent of the Inclination Lemma, Theorem B.6.1.)

Lemma B.7.7. *Writing* $d(E_1, E_2) := \max_{v\in E_2,\ \|v\|=1} d(v, E_1)$ *there is a* $K_1 > 0$ *such that* $d(T_{f^n(x)} f^n(\tau_1), E^{cu}(f^n(x))) \le K_1\theta^{2n}$ *for* $n \in \mathbb{N}$ *and* $x \in \tau_1$, *and similarly for* τ_2.

Proof. For a vector $v_n \in T_{f^n(x)} f^n(\tau_1)$ write $v_n = Df^n(x)v$ with $v \in T_x\tau_1$ and decompose $v = v^s + v^{cu}$ with $v^i \in E^i(x)$. Transversality of τ_1 and E^s gives a $K_1 > 0$ that depends only on τ_1 such that $\|v^s\| \le K_1\|v^{cu}\|$. Now, for $i \le n$, (B.7.2) gives

$$\left\| \frac{Df^i(x)v}{\|Df^i(x)v^{cu}\|} - \frac{Df^i(x)v^{cu}}{\|Df^i(x)v^{cu}\|} \right\| = \frac{\|Df^i(x)v^s\|}{\|Df^i(x)v^{cu}\|} \le \theta^{2i}\frac{\|v^s\|}{\|v^{cu}\|} \le K_1\theta^{2i}. \tag{B.7.3}$$

□

We now refine the explanation of our proof strategy a little. While we will indeed apply the partially hyperbolic diffeomorphism f repeatedly to A and $h_{\tau_1,\tau_2}(A)$, the resulting sets are highly distorted, and instead of trying to control their sizes directly, we will instead cover $f^n(A)$ with disks $B_n(x) := B(r(n,x), f^n(x))$ of radius $r(n,x)$ chosen large relative to the distance between $f^n(\tau_1)$ and $f^n(\tau_2)$ but small with respect to the "thinnest" direction of $f^n(A)$. Actually, more to the point, $r(n,x)$ will be chosen small enough for the Jacobian of f^n to be close enough to constant on $f^{-n}(B_n(x))$ (Lemma B.7.14) and to also agree across the holonomy gap at time n (Proposition B.7.15). This amounts to choosing it in the gap between the contracting rates in the stable direction and the rates in the center direction as follows.

Recalling (B.7.2), we note that by continuity of α_1, β_1 in $\alpha_1(x) < \theta^2 \min(1, \beta_1(x))$ we can choose $\delta > 0$ such that $a < \theta^2 \min(1, b)$, where $a(x) := \sup\{\alpha_1(y) \mid d(x,y) < \delta\}$ and $b(x) := \sup\{\beta_1(y) \mid d(x,y) < \delta\}$.

Now,

$$\mu(n,x) := \prod_{i=0}^{n-1} a(f^i(x))$$

is an upper bound for the stable contraction along *any* orbit segment that stays within δ of that of x for the first n steps, and

$$\sigma(n,x) := \prod_{i=0}^{n-1} b(f^i(x))$$

is a corresponding lower bound for center-unstable behavior; the preceding estimates imply

$$\mu(n,x) < \theta^{2n} \min(1, \sigma(n,x))$$

for all x and $n \in \mathbb{N}$. With $c := a/\theta$, the desired radius of balls in $f^n(\tau_1)$ is

$$r(n,x) := \prod_{i=0}^{n-1} c(f^i(x)).$$

As advertised, it satisfies

$$\mu(n,x) \le \theta^n r(n,x) \le \theta^{2n} \min(1, \sigma(n,x)). \tag{B.7.4}$$

Having chosen these radii, we now show that it is not only the last points of an orbit segment that are exponentially close, but the whole segment.

Lemma B.7.8. *If $f^n(\xi) \in B_n(x)$ then $d(f^i(x), f^i(\xi)) \le 3\theta^n$ when $-\frac{\log 2 + \log K_1}{2\log\theta} \le p \le i \le n$.*

Proof. We prove this by induction downward from $i = n$, in which case the conclusion is the definition of $B_n(x)$. More precisely, there is a piecewise smooth curve γ_n in $f^n(\tau_1)$ from $f^n(x)$ to $f^n(\xi)$ with length less than $r(n, x)$, and we show that the length of $\gamma_i := f^{i-n}(\gamma_n)$ is less than $3\theta^n$ for $i \le n$.

Decomposing the tangent vector as $\dot\gamma_i = \dot\gamma_i^s + \dot\gamma_i^{cu} \in E^s \oplus E^{cu}$, Lemma B.7.7 (or (B.7.3)) gives $\|\dot\gamma_i^s\|/\|\dot\gamma_i^{cu}\| \le K_1\theta^{2i}$, so for $i \ge p \in \mathbb{N}$ such that $K_1\theta^{2p} \le 1/2$ we obtain

$$\|\dot\gamma_i\| \le \frac{3}{2}\|\dot\gamma_i^{cu}\| \quad \text{and} \quad \|\dot\gamma_i\| \ge \frac{1}{2}\|\dot\gamma_i^{cu}\|.$$

For purposes of induction suppose now that the claim is known for $i + 1, \dots, n$. To show that the length of $\gamma_i := f^{i-n}(\gamma_n)$ is less than $3\theta^n$ note first that by assumption it is bounded above by

$$\|Df^{-1}\|\ell(\gamma_{i+1}) \le \|Df^{-1}\|3\theta^n,$$

and assume n has been chosen large enough for the right-hand side to be less than δ. This implies that γ_j lies in a δ-ball around $f^j(x)$ for $i \le j \le n$, and we can use the definition of σ:

$$\frac{2}{3}\|\dot\gamma_i\| \le \|\dot\gamma_i^{cu}\| = \|dF^{i-n}\dot\gamma_n^{cu}\| \le \frac{\|\dot\gamma_n^{cu}\|}{\sigma(n-i, f^i(x))} \le 2\frac{\|\dot\gamma_n\|}{\sigma(n-i, f^i(x))},$$

so

$$\ell(\gamma_i) \le \frac{3}{2}\cdot 2\frac{\ell(\gamma_n)}{\sigma(n-i, f^i(x))} \le 3\frac{r(n,x)}{\sigma(n-i, f^i(x))} = 3\frac{r(i,x)r(n-i, f^i(x))}{\sigma(n-i, f^i(x))}.$$

Inequality (B.7.4) now implies the claim: $r(i,x)r(n-i, f^i(x)) \le \theta^i\theta^{n-i}\sigma(n-i, f^i(x))$. □

Having studied the dynamics on transversals, we now start to look at the way the two transversals become closer under repeated application of f. The first statement sounds obvious, but takes some care to establish.

Lemma B.7.9. *There is a $K_2 > 0$ such that $d_s(y, h_n(y)) \le K_2\mu(n, x)$ for $n \in \mathbb{N}$ and $y \in B_n(x)$, where*

$$h_n := h_{f^n(\tau_1), f^n(\tau_2)}$$

and d_s denotes distance within a stable leaf.

Proof. Let $C_1 \ge \sup\{d_s(\xi, h_0(\xi)) \mid \xi \in \tau_1,\ h_0(\xi) \in \tau_2\}$ and write $\xi = f^{-n}(y)$. The choice of θ then implies

$$d_s(f^i(\xi), f^i(h_0(\xi))) \le C_1 \sup\|Df\restriction_{E^s}\| \le C_1\theta^{2i}$$

for all $i \in \mathbb{N}$. While this is an exponential estimate, the point for now is merely that for $p \in \mathbb{N}$ such that $C_1\theta^{2i} < \delta/2$ this implies

$$d(f^i(\xi), f^i(h_0(\xi))) \le d_s(f^i(\xi), f^i(h_0(\xi))) \le C_1 \sup \|Df\restriction_{E^s}\| \le \delta/2$$

for all $i \ge p$. At the same time, taking p as in Lemma B.7.8 gives $d(f^i(x), f^i(\xi)) \le 3\theta^n$ for $p \le i \le n$, where the right-hand side is less than $\delta/2$ if n is chosen large enough. Combining these, we find that $p \le i \le n$ implies

$$d(f^i(x), f^i(h_0(\xi))) < \delta \quad \textit{and} \quad d(f^i(x), f^i(\xi)) < \delta.$$

This allows us to bring in the definition of μ:

$$\begin{aligned} d_s(y, h_n(y)) &\le d(f^n(\xi), f^n(h_0(\xi))) \le \mu(n-p, f^p(x)) d_s(f^p(\xi), f^p(h_0(\xi))) \\ &= \frac{\mu(n,x)}{\mu(p,x)} d_s(f^p(\xi), f^p(h_0(\xi))) < \frac{\delta}{\inf_x \mu(p,x)} \mu(n,x). \end{aligned}$$ □

These preparations will let us show that $B_n(x)$ and $h_n(B_n(x))$ are graphs of maps from E^{cu} to E^s and that these two maps are C^1-close exponentially in n. This will make it possible to compare volumes and is the content of the next two lemmas.

Lemma B.7.10. *There is a disk $D_1 \subset E^{cu}(f^n(x))$ and a C^1-map $g_1 \colon D_1 \to E^s(f^n(x))$ such that $B_n(x) = \operatorname{graph}(g_1)$, and likewise for $h_n(B_n(x))$.*

Proof. Consider the balls $B^{cu}(y,\rho)$ and $B^s(y,\rho)$ around 0 in $E^{cu}(y)$ and $E^s(y)$, respectively, and choose ρ such that the exponential map $\exp_y \colon T_yM \to M$ is an embedding of $B^{TM}(y,\rho) := B^{cu}(y,\rho) \times B^s(y,\rho)$. Then $B_n(f^{-n}(y)) \subset \exp_y(B^{TM}(y,\rho))$ for large enough n, so we can consider $B_n(x)$ as a subset of $T_{f^n(x)}M$. By Lemma B.7.7, $f^n(\tau_1)$, hence $B_n(x)$, is nearly tangent to E^{cu}, hence transverse to E^s. So each $z \in B_n(x)$ corresponds to a unique $(z^{cu}, z^s) \in E^{cu}(f^n(x)) \times E^s(f^n(x))$ with z^{cu}-values in a disk $D_1 \subset E^{cu}(f^n(x)) \ni 0$ and defines a map $g \colon D_1 \to E^s(f^n(x))$, $z^{cu} \mapsto z^s$. Smoothness of the leaves of the foliations implies that g_1 is C^1. □

Lemma B.7.11. *There are $K_3 > 0$, $\alpha \in (0,1)$ with $\|Dg_1\| \le K_3\theta^{\alpha n}$. Likewise for g_2.*

Proof. Write $E^{cu}(y) = \operatorname{graph}(\xi_y)$ for y near $f^n(x)$ with $\xi_y \colon E^{cu}(f^n(x)) \to E^s(f^n(x))$ satisfying $\|\xi_y\| \le C(df^n(x), y)^\alpha$ by Hölder continuity of E^{cu} (Theorem 5.1.16). Since $d_{f^n(\tau_1)}(f^n(x), y) \le r(n,x) \le \theta^n$, allowing for slight distortion under exp gives

$$d(f^n(x), y) \le 2d_{f^n(\tau_1)}(f^n(x), y) \le 2\theta^n,$$

hence $\|\xi_y\| \le C2^\alpha\theta^{\alpha n}$.

Meanwhile, Lemma B.7.7 gives $d(T_yB_n(x), E^{cu}(y)) \le K_1\theta^{2n}$, which implies

$$T_yB_n(x) = \operatorname{graph}(\zeta_y), \quad \zeta_y : E^{cu}(f^n(y)) \to E^s(f^n(y))$$

for large enough n, with $\|\zeta_y - \xi_y\| \le Ld(T_yB_n(x), E^{cu}(y))$ for some L, so

$$\|\zeta_y\| \le \|\xi_y\| + \|\zeta_y - \xi_y\| \le C2^\alpha\theta^{\alpha n} + LK_1\theta^{2n} \le (C2^\alpha + LK_1)\theta^{\alpha n}. \qquad \square$$

We have now achieved in precise terms the first step of the proof strategy: in forward time these balls are close to each other in the C^1-topology, indeed exponentially so. As a result of this, any difference in volume amounts to an exponentially small percentage error concentrated around the edges:

Lemma B.7.12. *For some $\alpha' \in (0, \alpha)$ and $K_4 > 0$, D_1 contains the ball around zero of radius $r(n,x)(1 - K_4\theta^{\alpha' n})$ and is contained in the ball around zero of radius $r(n,x)(1 + K_4\theta^{\alpha' n})$, and likewise for D_2.*

Proof. This follows from the small-angle property $d(T_yB_n(x), E^{cu}(f^n(x))) \le K_1\theta^{2n}$ we just proved for $y \in B_n(x)$ because D_1 is the projection along E^s of $B_n(x)$ to $E^{cu}(f^n(x))$, hence coincides with the $r(n,x)$-ball around 0 up to a factor $e^{\theta^{\alpha' n}}$. $\square$

This allows us to conclude, as planned, that the volume of $B_n(x)$ is changed arbitrarily little under the holonomy for large enough n.

Lemma B.7.13.

$$\sup_{x\in\tau_1}\left|\frac{m_{f^n(\tau_1)}(B_n(x))}{m_{f^n(\tau_2)}(h_n(B_n(x)))} - 1\right| \xrightarrow[n\to\infty]{} 0.$$

Proof. By Lemma B.7.12, D_1 and D_2 contain the projection to $E^{cu}(f^n(x))$ of the ball $A(n,x)$ in $f^n(\tau_1)$ of radius $r(n,x)(1 - K_4\theta^{\alpha' n})$ around 0 and lie in the projection of the ball $C(n,x)$ of radius $r(n,x)(1 + K_4\theta^{\alpha' n})$. Lemma B.7.11 implies

$$\max_{x\in\tau_1}\left|\frac{m_{f^n(\tau_1)}(A(n,x))}{m_{f^n(\tau_1)}(C(n,x))} - 1\right| \xrightarrow[n\to\infty]{} 0 \tag{B.7.5}$$

and

$$\max_{x\in\tau_1}\left|\frac{m_{f^n(\tau_2)}(h_n(A(n,x)))}{m_{f^n(\tau_2)}(h_n(C(n,x)))} - 1\right| \xrightarrow[n\to\infty]{} 0.$$

Writing $P(u, g_k(u)) := u$ for $k = 1, 2$ and large enough $n \in \mathbb{N}$ gives

$$P(A(n,x)) \subset P(h_n(B_n(x))) \subset P(C(n,x)) \tag{B.7.6}$$

while for $k = 1, 2$ and a disk $D \subset f^n(\tau_k)$ the definition of $m_{f^n(\tau_k)}$ gives

$$m_{f^n(\tau_k)}(D) = \iint_{P(D)} \sqrt{\det g^k_{ij}}\, du_1 \ldots du_{\dim E^{cu}}.$$

The coefficients $g_{ij}^k = \delta_{ij} + \frac{\partial g_k}{\partial u_i}\frac{\partial g_k}{\partial u_j}$ of the inner product only involve first derivatives of g_k, so Lemma B.7.11 and (B.7.6) imply that for $\epsilon > 0$ there is an $n \in \mathbb{N}$ with

$$\begin{aligned} m_{f^n(\tau_1)}(A(n,x)) &= \iint_{P(A(n,x))} \sqrt{\det g_{ij}^k}\, du_1 \dots du_{\dim E^{cu}} \\ &\le \iint_{P(h_n(B_n(x)))} \sqrt{\det g_{ij}^k}\, du_1 \dots du_{\dim E^{cu}} + \epsilon m(P(h_n(B_n(x)))) \\ &= (1+\epsilon) m(P(h_n(B_n(x)))), \end{aligned}$$

where m is the standard measure in $T_{f^n(x)}B_n(x)$. Now (B.7.5) lets us replace the left-hand side by $m_{f^n(\tau_1)}(B_n(x))$ and on the right-hand side $m(P(h_n(B_n(x))))$ is arbitrarily close to $m_{f^n(\tau_2)}(h_n(B_n(x)))$.

Similar arguments bound $m_{f^n(\tau_2)}(h_n(B_n(x)))$ in terms of $m_{f^n(\tau_1)}(C(n,x))$. □

As promised in the proof outline, we have now shown that the volume of $B_n(x)$ is essentially preserved by the holonomy. From here it is downhill. We first show that pulling this result back by Df^{-n} involves on either side a distortion that is essentially constant, and the next step is to check that these constants are close enough to each other.

Lemma B.7.14.

$$\sup_{\substack{n\in\mathbb{N}\\ z_1,z_2\in B_n(x)}} \left| \log\det Df^{-n}(z_1)\restriction_{T_{z_1}B_n(x)} - \log\det Df^{-n}(z_2)\restriction_{T_{z_2}B_n(x)} \right| < \infty,$$

and likewise for $\log\det Df^{-n}(z_k)\restriction_{T_{z_k}h_n(B_n(x))}$.

Proof. Note that $|\log\det Df^{-1}(z)\restriction_{A_1} - \log\det Df^{-1}(z)\restriction_{A_2}| \le C_1 d(A_1, A_2)$ for nearby $(\dim E^{cu})$-dimensional subspaces A_k and that Hölder continuity of E^{cu} implies the same for $\log\det Df^{-1}\restriction_{E^{cu}}$. For $z_1, z_2 \in f^{-n(B_n(x))}$, Lemma B.7.7 thus implies

$$\begin{aligned} &\left|\log\frac{\det Df^{-n}(z_1)\restriction_{T_{z_1}B_n(x)}}{\det Df^{-n}(z_2)\restriction_{T_{z_2}B_n(x)}}\right| \\ &\quad\le \underbrace{\left|\log\frac{\det Df^{-n}(z_1)\restriction_{T_{z_1}B_n(x)}}{\det Df^{-n}(z_1)\restriction_{E^{cu}(z_1)}}\right|}_{\le C_1\sum_{i=0}^{n-1} d(T_{f^{-i}(z_1)}f^{-i}(B_n(x)),E^{cu}(f^{-i}(z_1)))} \\ &\qquad + \underbrace{\left|\log\frac{\det Df^{-n}(z_1)\restriction_{E^{cu}(z_1)}}{\det Df^{-n}(z_2)\restriction_{E^{cu}(z_2)}}\right|}_{\le C_2\sum_{i=0}^{n-1} d(f^{-i}(z_1),f^{-i}(z_2))^\alpha} + \underbrace{\left|\log\frac{\det Df^{-n}(z_2)\restriction_{E^{cu}(z_2)}}{\det Df^{-n}(z_2)\restriction_{T_{z_2}B_n(x)}}\right|}_{\le C_1\sum_{i=0}^{n-1} d(T_{f^{-i}(z_2)}f^{-i}(B_n(x)),E^{cu}(f^{-i}(z_2)))} \end{aligned}$$

$$\leq (C_2 + 2C_1K_1) \sum_{i=0}^{n-1} \theta^{\alpha(n-i)} d(z_1, z_2) < \frac{C_2 + 2C_1K_1}{1-\theta^\alpha} d(z_1, z_2),$$

and likewise for $\log \det Df^{-n}(z_k)\restriction_{T_{z_k} h_n(B_n(x))}$. □

A similar argument now establishes our advertised goal:

Proposition B.7.15.

$$\sup_n \log \frac{m_{\mathcal{T}_1}(f^{-n}(B_n(x)))}{m_{\mathcal{T}_2}(h_0(f^{-n}(B_n(x))))} < \infty.$$

Proof. We write the numerator and denominator as

$$m_{\mathcal{T}_1}(f^{-n}(B_n(x))) = \int_{B_n(x)} |\det Df^{-n}(z)\restriction_{T_z B_n(x)}|\, dm_{f^m(\mathcal{T}_1)}(z),$$

$$m_{\mathcal{T}_2}(h_0(f^{-n}(B_n(x)))) = \int_{h_n(B_n(x))} |\det Df^{-n}(z)\restriction_{T_{h_n(z)} h_n(B_n(x))}|\, dm_{f^m(\mathcal{T}_2)}(z)$$

and note that the integrals are over sets of comparable measures $m_{f^m(\mathcal{T}_1)}(B_n(x))$ and $m_{f^m(\mathcal{T}_2)}(h_n(B_n(x)))$, so it suffices to show that the ratio of the integrands is bounded. To that end note first that by the last lemma, we can replace the integrands by their value at any z_1 and $h_n(z_1)$, respectively, up to a uniformly bounded factor. This leaves us to bound

$$\begin{aligned}
\left|\log \frac{\det Df^{-n}(z_1)\restriction_{T_{z_1} B_n(x)}}{\det Df^{-n}(z_1)\restriction_{T_{h_n(z_1)} h_n(B_n(x))}}\right| &\leq \underbrace{\left|\log \frac{\det Df^{-n}(z_1)\restriction_{T_{z_1} B_n(x)}}{\det Df^{-n}(z_1)\restriction_{E^{cu}(z_1)}}\right|}_{\leq C_1 \sum_{i=0}^{n-1} d(T_{f^{-i}(z_1)} f^{-i}(B_n(x)), E^{cu}(f^{-i}(z_1)))} \\
&+ \underbrace{\left|\log \frac{\det Df^{-n}(z_1)\restriction_{E^{cu}(z_1)}}{\det Df^{-n}(h_n(z_1))\restriction_{E^{cu}(h_n(z_1))}}\right|}_{\leq C_2 \sum_{i=0}^{n-1} d(f^{-i}(z_1), f^{-i}(h_n(z_1)))^\alpha} \\
&+ \underbrace{\left|\log \frac{\det Df^{-n}(h_n(z_1))\restriction_{E^{cu}(h_n(z_1))}}{\det Df^{-n}(z_1)\restriction_{T_{h_n(z_1)} h_n(B_n(x))}}\right|}_{\leq C_1 \sum_{i=0}^{n-1} d(T_{f^{-i}(h_n(z_1))} f^{-i}(h_n(B_n(x))), E^{cu}(f^{-i}(h_n(z_1))))} \\
&\leq (C_2 + 2C_1K_1) \sum_{i=0}^{n-1} \theta^{\alpha(n-i)} d(z_1, h_n(z_1)) < \frac{C_2 + 2C_1K_1}{1-\theta^\alpha} < \infty.
\end{aligned}$$

This uniformly controls the distortion effected by h_0 on $f^{-n}(B_n(x))$. □

Proposition B.7.15 says that pullbacks of suitably chosen balls behave well under the holonomy. As promised, a simple covering argument now establishes (B.7.1) and hence completes the proof of Theorem B.7.5.

With A as in (B.7.1) choose $\epsilon > 0$ such that $A_\epsilon := \{y \in A \mid d_{\tau_1}(y, \tau_1 \setminus A) > \epsilon\}$ satisfies

$$\left|\frac{m_{\tau_1}(A_\epsilon)}{m_{\tau_1}(A)} - 1\right| < \frac{1}{2} \quad \text{and} \quad \left|\frac{m_{\tau_1} h_0((A_\epsilon))}{m_{\tau_1} h_0((A))} - 1\right| < \frac{1}{2}.$$

The cover $\mathcal{B}_n := \{B_n(x) \mid x \in A_\epsilon\}$ of A_ϵ satisfies $\bigcup \mathcal{B}_n \subset f^n(A)$ for large n since the radius of $B_n(x)$ is $r(n,x) \le \theta^n \sigma(n,x) < \text{const.}\, \theta^n d(\partial f^n(A), \partial f^n(A_\epsilon))$ by (B.7.4). The Besicovich Covering Theorem provides a countable subcover $C_n \subset \mathcal{B}_n$ of A_ϵ such that no point of A_ϵ is contained in more than ℓ elements of C_n, where ℓ depends only on the dimension of A_ϵ (this is where we use that A is a disk). Thus

$$\frac{1}{2\ell} m_{\tau_1}(A) \le \frac{1}{\ell} m_{\tau_1}(A_\epsilon) \le \sum_{B \in C_n} m_{\tau_1}(f^{-n}(B)) \le \ell m_{\tau_1}(A),$$

and likewise,

$$\frac{1}{2\ell} m_{\tau_2}(h_0(A)) \le \frac{1}{\ell} m_{\tau_2}(h_0(A_\epsilon)) \le \sum_{B \in C_n} m_{\tau_2}(h_0(f^{-n}(B))) \le \ell m_{\tau_2}(h_0(A)),$$

so Theorem B.7.5 follows by applying Proposition B.7.15 to each $B \in C_n$. □

Hints and answers to the exercises

Exercise 1.1. Show and then use that $\mathbb{T} := \{t \mid \varphi^t(x) = x\} \ni 0$ is closed (continuity), $\mathbb{T} + \mathbb{T} = \mathbb{T}$, and (1) $\Leftrightarrow \mathbb{T} = \{0\}$, (2) $\Leftrightarrow \mathbb{T} \neq \{0\}$ and 0 is not an accumulation point of $\mathbb{T}$, (3) $\Leftrightarrow \mathbb{T} = \mathbb{R}$.

Exercise 1.3. $\frac{dH}{dt} = \frac{dH}{x}\frac{dx}{dt} + \frac{dH}{dv}\frac{dv}{dt} = g(x) \cdot v + v \cdot (-g(x)) \equiv 0.$

Exercise 1.13. [168, Theorem X.1.3].

Exercise 1.15. Use Corollary 1.4.24 [168, Corollary X.1.2].

Exercise 1.17. $y \in W^s(\{x\}) \Leftrightarrow \varnothing \neq \omega(y) \subset \{x\} \Leftrightarrow \omega(y) = \{x\} \Leftrightarrow \varphi^t(y) \to x \Leftrightarrow y \in W^s(x)$.

Exercise 1.20. Since this implies (2), it suffices to show that this follows from (4), to which end one can adapt the proof of (4) $\Rightarrow$ (1).

Exercise 1.21. As usual, reflexivity (take $h = \mathrm{Id}$) and symmetry (replace h by h^{-1}) are easy. Transitivity: compose two conjugacies and check that this is as required.

Exercise 1.23. See [268, p. 59].

Exercise 1.24. Same steps as for Exercise 1.21.

Exercise 1.25. Use a circle around the attracting fixed point as a fundamental domain to define the conjugacy analogously to the linear case for all orbits that tend to this equilibrium; extend by continuity to the orbits ending on the saddle.

Exercise 1.26. A fixed point is a constant sequence, and sequences asymptotic to it are those which are eventually (on the left or the right, respectively) constant.

Exercise 1.27. A periodic point is a periodic sequence, so like in Exercise 1.26, sequences asymptotic to it are those which are eventually (on the left or the right, respectively) periodic.

Exercise 1.35. All other points have empty first prolongational limit sets.

Exercise 1.37. Example 1.4.16 is a counterexample.

Exercise 1.41. This can be shown from the definitions or from Theorem 1.5.44 (L induces a continuous injection into $\mathbb{R}$, which is a homeomorphism onto its image by invariance of domain).

Exercise 1.45. Yes: compose $x \mapsto x + \mathfrak{c}(x)$ from Exercise 1.43 with its inverse.

Exercise 1.50. Suppose the latter is not the case and take $f_n \longrightarrow \mathrm{Id}$ commuting with Φ, so $f_n = \varphi^{t_n}$ with $|t_n| \geq \epsilon$. If t is an accumulation point of the t_n then write $t = \lim_k t_{n_k}$ to get $\varphi^t = \lim_k \varphi^{t_k} = \lim_k f_{n_k} = \mathrm{Id}$ so Φ is periodic, contrary to being separating. Thus, the t_n have no accumulation point, and $t_n \longrightarrow \infty$.

Exercise 2.2. This need not be computational, but these computations suffice:

$$
\begin{aligned}
&E \wedge dE(P,V,Q) = \cos^2\theta(\cos^2\theta + \sin^2\theta) + \sin^2\theta(\cos^2\theta + \sin^2\theta) \equiv 1,\\
&E(P) = 1, \quad E(Q) = 0, \quad E(V) = 0,\\
&Z \in \{Q,V\} \Rightarrow dE(P,Z) = \underbrace{\mathcal{L}_P \overbrace{E(Z)}^{\equiv 0}}_{=0} - \underbrace{\mathcal{L}_Z \overbrace{E(P)}^{\equiv 1}}_{=0} - \underbrace{E(\overbrace{[P,Z]}^{\in -\{V,Q\}})}_{=0} = 0,\\
&\zeta^\pm := Q \pm V \Rightarrow [P,\zeta^\pm] = -\cos^2\theta[H,X] \mp \cos\theta[V,X]\\
&\qquad\qquad\qquad\qquad\qquad\quad - \sin^2\theta[H,X] \pm \sin\theta[H,V]\\
&\qquad\qquad\qquad\qquad\qquad = \mp\zeta^\pm.
\end{aligned}
$$

Exercise 2.9. $\theta = (1/2)\sum_{i=1}^n (p_i\, dq_i - q_i\, dp_i)$, $v = -(-q,p)$.

Exercise 2.13. Decompose into blocks $\begin{pmatrix} \lambda & 0 \\ 0 & 1/\lambda \end{pmatrix}$ for real λ, rotations R_α for $\lambda = e^{i\alpha}$, and $\begin{pmatrix} \rho R_\alpha & 0 \\ 0 & \rho^{-1} R_{-\alpha} \end{pmatrix}$ for $\lambda = \rho e^{i\alpha}$.

Exercise 2.14. ω^n is a volume and exterior multiplication on forms induces a multiplicative structure on cohomology, hence the second cohomology of ω is nonzero.

Exercise 2.15. Use the previous exercise.

Exercise 2.16. Use the Moser "homotopy trick" in the proof of the Darboux Theorem (Theorem 2.6.11).

Exercise 2.17. Use rotational symmetry (this is an instance of the Noether Theorem). The integral obtained is angular momentum. Independence can be seen by studying how the integral depends on momenta.

Exercise 2.19. To show that $\omega(v,w)$ depends only on the projection of v and w, use that the projections are along the flow, hence invariant, and $\omega(X, X_H) = 0$ for every $X \in TM_c$.

Exercise 2.20. For $n = 2$ a geodesic is an oriented great circle and hence identified with an oriented plane which in turn is defined by a unit vector (a positive normal). The space of these is S^2. By rotational symmetry the volume is the standard one.

Alternatively take a single great circle together with the unit tangent vectors pointing into one complementary hemisphere as a transversal and compactify by the two tangent directions to again get a sphere.

Exercise 3.1. Although length is preserved in Example 1.1.5, there is no invariant Borel probability measure (any open interval has infinitely many disjoint images of equal measure, which must be 0 for finite total measure). Example 1.1.7 has only the point mass at 0; by similar reasons no open interval in $\mathbb{R} \smallsetminus \{0\}$ has positive measure. As noted, Example 1.3.7 is conjugate to Example 1.1.5 hence has no invariant Borel probability measure. From before, no invariant Borel probability measure has positive measure in the interior of Example 1.3.8, which leaves the Dirac masses at the ends and their convex combinations. Likewise, Example 1.3.13 has only the Dirac mass at the fixed point. More generally, *any* probability measure on the fixed-point set in Example 1.3.15 is an invariant probability measure, and these are it. By like arguments, only the circle of fixed points in Example 1.4.16 supports invariant Borel probability measures, and any Borel probability measure on this circle is an invariant Borel probability measure. In Figure 1.4.1 again only the two fixed points support invariant probability measures, so $\mathfrak{M}(\Phi)$ consists of convex combinations of two Dirac masses, that is, an interval, and likewise for Figure 1.5.4 but with three points (so $\mathfrak{M}(\Phi)$ is now a triangle). In Figure 1.5.11 all interior points are wandering, so the invariant Borel probability measures are supported on the boundary, and any Borel probability measure on the boundary is invariant. Figure 1.1.4 is the most complex. All points are fixed or periodic orbits and hence support a Dirac measure. Standard area is also invariant (this is the Hamiltonian nature of the pendulum), as is area multiplied by any constant of motion as a density, and we can expect a multitude of other invariant measures. However, the aforementioned Dirac measures are the only *ergodic* ones, so $\mathfrak{M}(\Phi)$ is their closed convex hull.

Exercise 3.3. See Exercise 3.2

Exercise 3.9. See Remark 3.3.38.

Exercise 3.10. Scale the generating vector field by a function $f\colon \mathbb{T}^2 \to [0,\infty)$ with $f^{-1}(\{0\}) = \{0\}$ and show that the existence of a nonatomic invariant measure contradicts unique ergodicity of the original flow.

Exercise 3.11. To check Proposition 1.6.9(3), suppose $\varnothing \neq U, V \subset \operatorname{supp}\mu$ are open, hence have positive measure. By ergodicity, $\varphi^{\mathbb{R}}(U)$ has full measure, hence intersects V.

Exercise 3.14. Such functions are constant on $[0,1/2]$ or its complement.

Exercise 3.15. $f \mapsto \int_{[0,1/2]} f \chi_{[0,1/2]} + \int_{(1/2,1]} \chi_{[1/2,1]}$.

Exercise 3.17. Use the argument for (4) $\Rightarrow$ (1) in Proposition 1.6.9: if $U_1, U_2, \dots$ is a base for the topology of $\operatorname{supp}\mu$, then $E_i := \{x \in \operatorname{supp}\mu \mid \varphi^{\mathbb{R}}(x) \cap U_i \neq \varnothing\}$ has full measure by the Birkhoff Ergodic Theorem applied to χ_{U_i} and so then has the desired set $\bigcap_{i\in\mathbb{N}} E_i$.

Exercise 3.18. Example 3.3.15 provides the essential insight since a circle rotation is a factor of a suspension: the exceptional set is $\mathbb{Q}$. The quickest proof is to invoke Remark 3.3.5 and Example 3.3.15; the most satisfying one would be to consider the action of the time-p/q map on $X \times ([0, 1/2q) + \mathbb{Z}/q)$, say.

Exercise 3.26. By Proposition 3.7.4(3) it suffices to show density. To that end show that for the Koopman operator of an ergodic transformation T on a nonatomic space every point of the unit circle is an approximate eigenvalue as follows: if $|\lambda| = 1$, then there are $f_n \in L^2$ with $\|f_n\| = 1$ and $\|U_T f_n - \lambda f_n\| \xrightarrow[n\to\infty]{} 0$.

Exercise 3.27. Otherwise there are $t_i \to \infty$ with $\varphi^{-t_i}(A)$ pairwise essentially disjoint.

Exercise 3.28. f can be taken to be continuous or a characteristic function [229].

Exercise 4.2. The topological entropy is 0 in these cases, so the essential information is from Exercise 3.4.

Exercise 4.4. The lazy solution is to recognize that $\left(\begin{smallmatrix}5&3\\3&2\end{smallmatrix}\right) = \left(\begin{smallmatrix}2&1\\1&1\end{smallmatrix}\right)^2$ and double the entropy in Proposition 4.2.17. Alternatively, follow the steps in the proof of Proposition 4.2.17 for periodic orbit growth and use the equality of these two (Remark 4.2.25). Finally, one can produce a coding for this suspension from that used for Proposition 4.2.17 to compute entropy from the symbolic model as in the proof of Proposition 4.2.17.

Exercise 4.5. Proceed as in Exercise 4.4; here the eigenvalues are $\frac{5\pm\sqrt{21}}{2}$.

Exercise 4.6. Proceed as in Exercise 4.4; here the eigenvalues are $-0.692\dots$, $-0.357\dots, 4.049\dots$.

Exercise 4.7. By Proposition 4.2.15 it is the entropy of the base (Example 4.1.11).

Exercise 4.8. To apply Theorem 4.1.4 compute $\int_{\mathbb{T}^2} r$ and obtain the entropy of the base from Example 4.1.11.

Exercise 4.9. The topological entropy of the map and its suspension is $\log\frac{3+\sqrt{5}}{2} > \log\frac{3+1}{2} = \log 2$ (Proposition 4.2.17), the entropy of the full shift on two symbols and hence an upper bound for that of any subshift. Thus, F_A is not a factor of any shift on two symbols (Proposition 4.2.11).

Exercise 4.10. This is what the previous solution shows.

Exercise 4.11. Yes, combining Δ_3 and Δ_4 in Figure 1.9.3 into a single rectangle gives such a partition. (It is less clear how to find a partition by three pieces, even though entropy considerations are no obstruction.)

Exercise 4.12. Entropy considerations are no obstruction, and the following works. (Regrettably, this partition does not work for Example 1.5.26.)

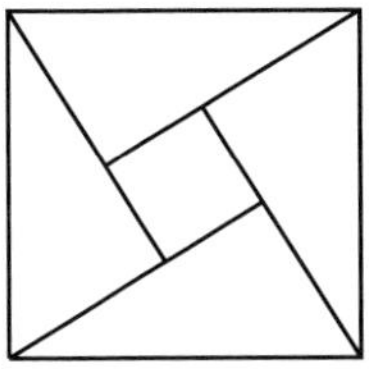

Exercise 4.19. See Remark 4.3.4.

Exercise 4.23. This is a suitable Bernoulli measure.

Exercise 4.24. This will be a Markov measure. For a potential that depends on more than the first two coordinates, one would get a Markov process with "longer memory." Indeed, every Hölder potential is a uniform limit of locally constant potentials, and the corresponding equilibrium states are weak* limits of Markov measures.

Exercise 4.25. Expansivity is invoked solely to obtain (4.3.9), which follows from kinematic expansivity.

Exercise 5.1. Check that S is sufficiently large if

$$\int_S^{S+t} \underline{\lambda}^{-2s} \left(\|D\varphi^s v\|_{\varphi^s(x)}\right)^2 ds < \int_0^t \underline{\lambda}^{-2s} \left(\|D\varphi^s v\|_{\varphi^s(x)}\right)^2 ds$$

for all $t > 0$. To see that there is such an S use that there are $c, \overline{\lambda} > 0$ such that $\|D\varphi^s v\|_{\varphi^s(x)} \geq c\overline{\lambda}^s \|x\|_x$ for all $s \geq 0$.

Exercise 5.4. The rectangle Δ is an isolating neighborhood.

Exercise 5.6. Use the Hartman–Grobman Theorem.

Exercise 5.13. Calculate the volume of a sphere by integrating the volume element generated by orthonormal Jacobi fields.

Exercise 5.14. Use the previous exercise.

Exercise 5.15. Show that $\exp_x \colon T_x\widetilde{M} \to \widetilde{M}$ is a diffeomorphism.

Exercise 5.16. Use the previous exercise.

Exercise 5.17. Use the hexagonal plane tiling to define a torus and take as scatterer a sufficiently large disk as shown here.

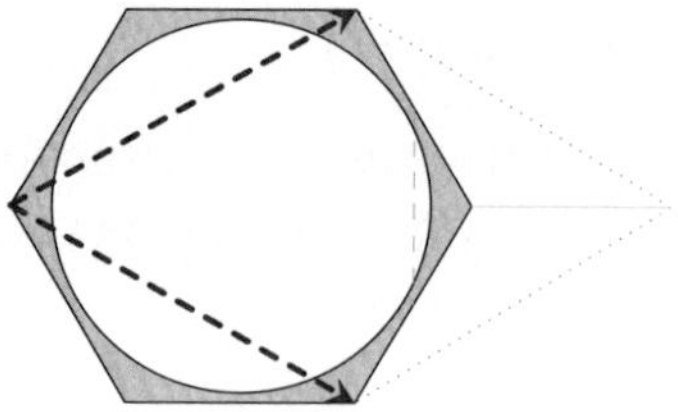

(This was described to us by Yves Coudène; the picture includes the two translation vectors that generate the deck transformation group.) Specify how large is sufficient.

Exercise 6.3. Apply Exercise 5.4 and Theorem 6.2.7 or check directly (noting that here and generally for suspensions $t(x, y) \equiv 0$ in Proposition 6.2.2).

Exercise 6.4. Apply Exercise 5.5 and Theorem 6.2.7 or check directly.

Exercise 6.5. For instance, let $\epsilon, \delta > 0$ be such that for any points δ-close in $\Lambda = \overline{\mathrm{Per}(\Phi)}$, the bracket $x = [p, q]$ is defined and unique, where p and q are periodic points that are δ-close. If $x \notin \Lambda$, then let $\Lambda' = \Lambda \cup \mathcal{O}(x)$. This is a new hyperbolic set and we can form a periodic pseudo-orbit using segments of the orbits of x, p, q, and a transitive orbit. So there is a periodic point arbitrarily close to x (Anosov Closing Lemma), a contradiction. Therefore, $x \in \Lambda$. Thus Λ has a local product structure.

Exercise 6.6. This means that the geodesic flow has periodic points with incommensurate periods, which follows from Theorem 6.2.11 and Remark 2.4.6.

Exercise 6.7. This is essentially a restatement of Theorem 6.2.11(3) (which holds by Remark 2.4.6): each neighborhood of v in $S\Sigma$ contains vectors that generate closed geodesics of incommensurate lengths, so at least one of those lengths is incommensurate with that of γ_v. This is the choice of v_i

Exercise 6.8. Apply Proposition 6.2.17 to the geodesic flow.

Exercise 6.10. Print and cut out the template (making sure it is square). In the following illustrations (from http://www.qland.de/origami/tetra/index.html) the back side of the paper is shown in gray; dotted lines show the current fold and solid lines show previous folds. "Fold" means a "valley fold" as opposed to a "mountain fold"; the fold line is deeper than the adjacent paper.

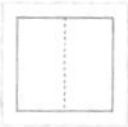

Fold indicated vertical, unfold.

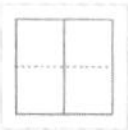

Fold indicated horizontal, unfold.

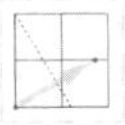

Place the lower left corner on the right half of the horizontal fold, so that the fold goes exactly through the upper left corner, unfold.

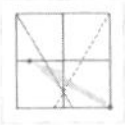

Same procedure with the lower right corner—to left half of the horizontal line.

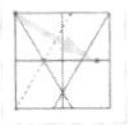

Same procedure with the upper left corner—to right half of the horizontal line.

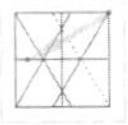

Same procedure with the upper right corner—to right half of the horizontal line.

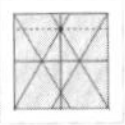

Turn the paper over, so that you see the back side, fold through the point of intersection of the oblique lines on the upper part of the paper, parallel to the edge, unfold.

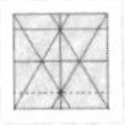

Same on bottom part of the paper, do not unfold.

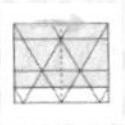

Fold indicated vertical, turn the part on the left of the middle line over the right part.

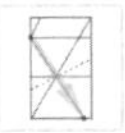

Make an auxiliary fold now: as shown on the illustration, place the upper marked point on the lower point and fix the resulting fold, unfold and turn the part above this fold to the back side of the paper, use the same fold to do this, unfold.

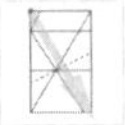

Spread the two layers of paper and bring the marked point down *between* the layers using the fold from the last step.

Turn down the triangle above the dotted line, the front layer to the front, the back layer to the back.

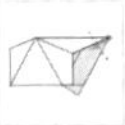

Turn the small triangle on the right side of the dotted line to the left, the front layer to the front, the back layer to the back.

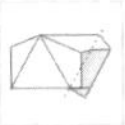

Auxiliary fold: turn the part to the right of the dotted line to the left, fold both layers, unfold, fold again along the same fold to the back, unfold.

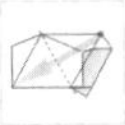

Pull the part on the right of the dotted line over the left part by spreading the two layers, front layer to the front, back layer to the back, fold along the dotted line.

Turn the small triangle to the right of the dotted line to the left, unfold, turn this triangle between the two layers; pleat the whole paper.

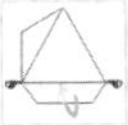

Slightly press the paper at the marked points so that both layers spread and carefully insert the trapezoid on the bottom between the layers; pleat the paper again.

Hold the paper at the marked points and blow sharply between the layers; the tetrahedron should unfold.

Exercise 7.1. Expansivity is inherited by closed invariant subsets by definition; show that specification is inherited by factors.

Exercise 7.2. Implement the argument outlined in Remark 7.3.3.

Exercise 7.5. Consider functions of one coordinate only.

Exercise 7.7. Use Proposition 7.3.15 to show that $h_{\mu_\varphi}(\Phi) = 0$ implies that μ_φ is atomic.

Exercise 7.8. This is Theorem 7.4.14. OR, the equation in Theorem 7.4.10 with vol on the right-hand side holds vol-a.e. by the Birkhoff Ergodic Theorem (Theorem 3.2.15) and for μ_{SRB} by Theorem 7.4.10, so these invariant Borel probability measures coincide (and the attractor is M). Alternatively, use Theorem 7.4.13.

Exercise 7.10. Expansivity enters solely through Proposition 4.2.18 and Lemma 4.3.20, which hold for kinematic-expansive flows (Exercise 4.25).

Exercise 7.11. Use Exercises 7.11 and 4.25.

Bibliography

[1] J. Aczél, *Lectures on functional equations and their applications*, Mathematics in Science and Engineering, vol. 19, Academic Press, New York-London, 1966. Translated by Scripta Technica. Supplemented by the author. Edited by Hansjorg Oser. Zbl 0139.09301 MR 0208210

[2] Ilesanmi Adeboye, Harrison Bray, and David Constantine, *Entropy rigidity and Hilbert volume*, Discrete Contin. Dyn. Syst. **39** (2019), no. 4, 1731–1744. Zbl 07042655 MR 2276493

[3] Jiweon Ahn, Manseob Lee, and Jumi Oh, *Measure expansivity for C^1-conservative systems*, Chaos Solitons Fractals **81** (2015), part A, 400–405. Zbl 1355.37002 MR 3426052

[4] Jiweon Ahn, Manseob Lee, and Jumi Oh, *Corrigendum to: Measure expansivity for C^1-conservative systems [Chaos, Solitons & Fractals, 81PA (2015) 400–405; Zbl 1355.37002 MR 3426052]*, Chaos Solitons Fractals **82** (2016), 155. MR 3433573

[5] Ethan Akin, Mike Hurley, and Judy A. Kennedy, *Dynamics of topologically generic homeomorphisms*, Mem. Amer. Math. Soc. **164** (2003), no. 783, viii+130. Zbl 1022.37010 MR 1980335

[6] Warren Ambrose, *Representation of ergodic flows*, Ann. of Math. (2) **42** (1941), 723–739. Zbl 0025.26901 MR 0004730

[7] Warren Ambrose and Shizuo Kakutani, *Structure and continuity of measurable flows*, Duke Math. J. **9** (1942), 25–42. Zbl 0063.00065 MR 0005800

[8] Nalini Anantharaman, *Precise counting results for closed orbits of Anosov flows*, Ann. Sci. Éc. Norm. Sup. (4) **33** (2000), no. 1, 33–56. Zbl 0992.37026 MR 1743718

[9] Alexander Andronov and Lev Pontrjagin, *Systèmes grossiers*, Comptes Rendus (Doklady) de l'Académie des Sciences de l'URSS **14** (1937), no. 5, 247–250. Zbl 0016.11301

[10] Dmitry V. Anosov, *Geodesic flows on closed Riemann manifolds with negative curvature*, Proceedings of the Steklov Institute of Mathematics, No. 90 (1967). Translated from the Russian by S. Feder, American Mathematical Society, Providence, RI, 1969. Zbl 0176.19101 MR 0242194

[11] Dmitry V. Anosov, *About one class of invariant sets of smooth dynamical systems*, Proceedings 5th International Conference on nonlinear oscillations, vol. 2, Kiev, 1970, pp. 39–45. Zbl 0243.34085

[12] Dmitry V. Anosov and Yakov G. Sinaĭ, *Certain smooth ergodic systems*, Uspehi Mat. Nauk **22** (1967), no. 5 (137), 107–172. Zbl 0177.42002 MR 0224771

[13] N. Aoki and K. Hiraide, *Topological theory of dynamical systems*, North-Holland Mathematical Library, vol. 52, North-Holland, Amsterdam, 1994, Recent advances. Zbl 0798.54047 MR 1289410

[14] Marie-Claude Arnaud, *Le "closing lemma" en topologie C^1*, Mém. Soc. Math. Fr. (N.S.) (1998), no. 74, vi+120. Zbl 0920.58039 MR 1662930

[15] Vladimir I. Arnold, *Small denominators and problems of stability of motion in classical and celestial mechanics*, Uspehi Mat. Nauk **18** (1963), no. 6 (114), 91–192. Zbl 0135.42701 MR 0170705

[16] Vladimir I. Arnold, *Mathematical methods of classical mechanics*, second ed., Graduate Texts in Mathematics, vol. 60, Springer, New York, 1989. Translated from the Russian by K. Vogtmann and A. Weinstein. Zbl 0692.70003 MR 997295

[17] Masayuki Asaoka, *On invariant volumes of codimension-one Anosov flows and the Verjovsky conjecture*, Invent. Math. **174** (2008), no. 2, 435–462. Zbl 1154.37013

[18] Masayuki Asaoka and Kei Irie, *A C^∞ closing lemma for Hamiltonian diffeomorphisms of closed surfaces*, Geom. Funct. Anal. **26** (2016), no. 5, 1245–1254. Zbl 1408.37092 MR 3568031

[19] Joseph Auslander, *Generalized recurrence in dynamical systems*, Contributions to Differential Equations **3** (1964), 65–74. Zbl 0152.21503 MR 0162238

[20] Artur Avila, Marcelo Viana, and Amie Wilkinson, *Absolute continuity, Lyapunov exponents and rigidity I: geodesic flows*, J. Eur. Math. Soc. (JEMS) **17** (2015), no. 6, 1435–1462. Zbl 1352.37084 MR 3353805

[21] Viviane Baladi and Mark Demers, *On the measure of maximal entropy for finite horizon Sinai billiard maps*, 2018, arXiv:1807.02330.

[22] Viviane Baladi, Mark F. Demers, and Carlangelo Liverani, *Exponential decay of correlations for finite horizon Sinai billiard flows*, Invent. Math. **211** (2018), no. 1, 39–177. Zbl 1382.37037 MR 3742756

[23] Werner Ballmann, *Lectures on spaces of nonpositive curvature*, DMV Seminar, vol. 25, Birkhäuser, Basel, 1995. With an appendix by Misha Brin. Zbl 0834.53003 MR 1377265

[24] J. Banks, J. Brooks, G. Cairns, G. Davis, and P. Stacey, *On Devaney's definition of chaos*, Amer. Math. Monthly **99** (1992), no. 4, 332–334. Zbl 0758.58019 MR 1157223

[25] Thierry Barbot, *Caractérisation des flots d'Anosov en dimension 3 par leurs feuilletages faibles*, Ergodic Theory Dynam. Systems **15** (1995), no. 2, 247–270. Zbl 0826.58025

[26] Thierry Barbot, *Flots d'Anosov sur les variétés graphées au sens de Waldhausen*, Ann. Inst. Fourier (Grenoble) **46** (1996), no. 5, 1451–1517. Zbl 0861.58028 MR 1427133

[27] Thierry Barbot, *Generalizations of the Bonatti-Langevin example of Anosov flow and their classification up to topological equivalence*, Comm. Anal. Geom. **6** (1998), no. 4, 749–798. Zbl 0916.58033 MR 1652255

[28] Thierry Barbot, *Plane affine geometry and Anosov flows*, Ann. Sci. Éc. Norm. Supér. (4) **34** (2001), no. 6, 871–889. Zbl 1098.37513

[29] Thierry Barbot and Carlos Maquera, *Transitivity of codimension-one Anosov actions of $\mathbb{R}^k$ on closed manifolds*, Ergodic Theory Dynam. Systems **31** (2011), no. 1, 1–22. Zbl 1213.37049 MR 2755918

[30] Thierry Barbot and Carlos Maquera, *Nil-Anosov actions*, Math. Z. **287** (2017), no. 3-4, 1279–1305. Zbl 1380.37060 MR 3719536

[31] Luis Barreira and Yakov Pesin, *Lectures on Lyapunov exponents and smooth ergodic theory*, Smooth ergodic theory and its applications (Seattle, WA, 1999), Proc. Sympos. Pure Math., vol. 69, American Mathematical Society, Providence, RI, 2001, Appendix A by M. Brin and Appendix B by D. Dolgopyat, H. Hu and Pesin, pp. 3–106. Zbl 0996.37001 MR 1858534

[32] Luis Barreira and Yakov Pesin, *Nonuniform hyperbolicity*, Encyclopedia of Mathematics and Its Applications, vol. 115, Cambridge University Press, Cambridge, 2007, Dynamics of systems with nonzero Lyapunov exponents. Zbl 1144.37002 MR 2348606

[33] Luis Barreira and Claudia Valls, *Hölder Grobman-Hartman linearization*, Discrete Contin. Dyn. Syst. **18** (2007), no. 1, 187–197. Zbl 1120.37013 MR 2276493

[34] June Barrow-Green, *Poincaré and the three body problem*, History of Mathematics, vol. 11, American Mathematical Society, Providence, RI; London Mathematical Society, London, 1997. Zbl 0877.01022 MR 1415387

[35] Thomas Barthelmé, Christian Bonatti, Andrey Gogolev, and Federico Rodriguez Hertz, *Anomalous Anosov flows revisited*, 2017, arXiv:1712.07755.

[36] Thomas Barthelmé and Sergio R. Fenley, *Counting periodic orbits of Anosov flows in free homotopy classes*, Comment. Math. Helv. **92** (2017), no. 4, 641–714. Zbl 1386.37028 MR 3718484

[37] Thomas Barthelmé and Andrey Gogolev, *A note on self orbit equivalences of Anosov flows and bundles with fiberwise Anosov flows*, 2017, to appear in Mathematical Research Letters; arXiv:1702.01178.

[38] Edward Belbruno, *Fly me to the moon, an insider's guide to the new science of space travel*, Princeton University Press, Princeton, NJ, 2007. With a foreword by Neil deGrasse Tyson. Zbl 1146.70001 MR 2391999

[39] Yves Benoist, *Convexes divisibles*, C. R. Acad. Sci. Paris Sér. I Math. **332** (2001), no. 5, 387–390. Zbl 1010.37014 MR 1826621

[40] Yves Benoist, *Convexes divisibles. II*, Duke Math. J. **120** (2003), no. 1, 97–120. Zbl 1037.22022 MR 2010735

[41] Yves Benoist, *Convexes divisibles. I*, Algebraic groups and arithmetic, Tata Inst. Fund. Res., Mumbai, 2004, pp. 339–374. Zbl 1084.37026 MR 2094116

[42] Yves Benoist, *Convexes divisibles. III*, Ann. Sci. Éc. Norm. Sup. (4) **38** (2005), no. 5, 793–832. Zbl 1085.22006 MR 2195260

[43] Yves Benoist, *Convexes divisibles. IV. Structure du bord en dimension 3*, Invent. Math. **164** (2006), no. 2, 249–278. Zbl 1107.22006 MR 2218481

[44] Yves Benoist, *Convexes hyperboliques et quasiisométries*, Geom. Dedicata **122** (2006), 109–134. Zbl 1122.20020 MR 2295544

[45] Yves Benoist, *A survey on divisible convex sets*, Geometry, analysis and topology of discrete groups, Adv. Lect. Math. (ALM), vol. 6, Int. Press, Somerville, MA, 2008, pp. 1–18. Zbl 1154.22016 MR 2464391

[46] Yves Benoist, Patrick Foulon, and François Labourie, *Flots d'Anosov à distributions de Liapounov différentiables. I*, Ann. Inst. H. Poincaré Phys. Théor. **53** (1990), no. 4, 395–412, Hyperbolic behaviour of dynamical systems (Paris, 1990). Zbl 0723.58040 MR 1096099

[47] Gerard Besson, Gilles Courtois, and Sylvestre Gallot, *Entropies et rigidités des espaces localement symétriques de courbure strictement négative*, Geom. Funct. Anal. **5** (1995), no. 5, 731–799. Zbl 0851.53032 MR 1354289

[48] Gérard Besson, Gilles Courtois, and Sylvestre Gallot, *Minimal entropy and Mostow's rigidity theorems*, Ergodic Theory Dynam. Systems **16** (1996), no. 4, 623–649. Zbl 0887.58030 MR 1406425

[49] George D. Birkhoff, *On the periodic motions of dynamical systems*, Acta Math. **50** (1927), no. 1, 359–379. Zbl 53.0733.03 MR 1555257

[50] George D. Birkhoff, *Dynamical systems*. With an addendum by Jürgen Moser. American Mathematical Society Colloquium Publications, vol. IX, American Mathematical Society, Providence, RI, 1966. Zbl 0171.05402 MR 0209095

[51] George D. Birkhoff, *Nouvelles recherches sur les systèmes dynamique*, Mem. Pontif. Acad. Sci. Novi Lyncaei, III, Ser. 1, 1934, pp. 85–216 (Reprinted in [52, 530–661]). Zbl 0016.23401

[52] George David Birkhoff, *Collected mathematical papers (in three volumes). Vol. II: Dynamics (continued), physical theories*, Editorial committee: D. V. Widder (Chairman), C. R. Adams, R. E. Langer, Marston Morse, and M. H. Stone, Dover Publications, New York, 1968, vol. 2. Zbl 0225.01009 MR 0235972

[53] Jeff Boland and Florence Newberger, *Minimal entropy rigidity for Finsler manifolds of negative flag curvature*, Ergodic Theory Dynam. Systems **21** (2001), no. 1, 13–23. Zbl 0992.53027 MR 1826659

[54] Jeffrey Boland, *On rigidity properties of contact time changes of locally symmetric geodesic flows*, Discrete Contin. Dynam. Systems **6** (2000), no. 3, 645–650. Zbl 1009.37021 MR 1757392

[55] Christian Bonatti, Lorenzo J. Díaz, and Marcelo Viana, *Dynamics beyond uniform hyperbolicity*, Encyclopaedia of Mathematical Sciences, vol. 102, Springer, Berlin, 2005, A global geometric and probabilistic perspective, Mathematical Physics, III. Zbl 1060.37020 MR 2105774

[56] Christian Bonatti and Rémi Langevin, *Un exemple de flot d'Anosov transitif transverse à un tore et non conjugué à une suspension*, Ergodic Theory Dynam. Systems **14** (1994), no. 4, 633–643. Zbl 0826.58026 MR 1304136

[57] Rufus Bowen, *Entropy-expansive maps*, Trans. Amer. Math. Soc. **164** (1972), 323–331. Zbl 0229.28011 MR 0285689

[58] Rufus Bowen, *The equidistribution of closed geodesics*, Amer. J. Math. **94** (1972), 413–423. Zbl 0249.53033 MR 0315742

[59] Rufus Bowen, *Periodic orbits for hyperbolic flows*, Amer. J. Math. **94** (1972), 1–30. Zbl 0254.58005

[60] Rufus Bowen, *Some systems with unique equilibrium states*, Math. Systems Theory **8** (1974/75), no. 3, 193–202. Zbl 0299.54031 MR 0399413

[61] Rufus Bowen, *Mixing Anosov flows*, Topology **15** (1976), no. 1, 77–79. Zbl 0321.58018

[62] Rufus Bowen and Brian Marcus, *Unique ergodicity for horocycle foliations*, Israel J. Math. **26** (1977), no. 1, 43–67. Zbl 0346.58009 MR 0451307

[63] Rufus Bowen and David Ruelle, *The ergodic theory of Axiom A flows*, Invent. Math. **29** (1975), no. 3, 181–202. Zbl 0311.58010 MR 0380889

[64] Rufus Bowen and Peter Walters, *Expansive one-parameter flows*, J. Differential Equations **12** (1972), 180–193. Zbl 0242.54041

[65] Harrison Bray, *Nonuniform hyperbolicity in Hilbert geometries*, ProQuest LLC, Ann Arbor, MI, 2016, Ph.D. Thesis – Tufts University. MR 3527292

[66] M. I. Brin, *Topological transitivity of a certain class of dynamical systems, and flows of frames on manifolds of negative curvature*, Funktsional. Anal. i Priložen. **9** (1975), no. 1, 9–19. Zbl 0357.58011 MR 0370660

[67] M. Brin and M. Gromov, *On the ergodicity of frame flows*, Invent. Math. **60** (1980), no. 1, 1–7. Zbl 0445.58023 MR 582702

[68] M. Brin and H. Karcher, *Frame flows on manifolds with pinched negative curvature*, Compositio Math. **52** (1984), no. 3, 275–297. Zbl 0561.58039 MR 756723

[69] M. I. Brin and Ja. B. Pesin, *Partially hyperbolic dynamical systems*, Izv. Akad. Nauk SSSR Ser. Mat. **38** (1974), 170–212. Zbl 0304.58017 MR 0343316

[70] Michael Brin and Garrett Stuck, *Introduction to dynamical systems*, Cambridge University Press, Cambridge, 2015. Corrected paper back edition of the 2002 original [MR 1963683]. Zbl 1319.37001 MR 3558919

[71] H. W. Broer, B. Hasselblatt, and F. Takens (eds.), *Handbook of dynamical systems. Vol. 3*, Elsevier/North-Holland, Amsterdam, 2010. Zbl 1216.37002 MR 3292649

[72] Idel U. Bronstein and Alexander Ya. Kopanskiĭ, *Smooth invariant manifolds and normal forms*, World Scientific Series on Nonlinear Science. Series A: Monographs and Treatises, vol. 7, World Scientific, River Edge, NJ, 1994. Zbl 0974.34001 MR 1337026

[73] D. S. Broomhead and Eugene Gutkin, *The dynamics of billiards with no-slip collisions*, Phys. D **67** (1993), no. 1-3, 188–197. Zbl 0785.70013 MR 1234441

[74] Marco Brunella, *Separating the basic sets of a nontransitive Anosov flow*, Bull. London Math. Soc. **25** (1993), no. 5, 487–490. Zbl 0790.58028 MR 1233413

[75] Marc Burger, Alessandra Iozzi, François Labourie, and Anna Wienhard, *Maximal representations of surface groups: symplectic Anosov structures*, Pure Appl. Math. Q. **1** (2005), no. 3, Special Issue: In memory of Armand Borel. Part 2, 543–590. Zbl 1157.53025 MR 2201327

[76] K. Burns, V. Climenhaga, T. Fisher, and D. J. Thompson, *Unique equilibrium states for geodesic flows in nonpositive curvature*, Geom. Funct. Anal. **28** (2018), no. 5, 1209–1259. Zbl 1401.37038 MR 3856792

[77] K. Burns, D. Dolgopyat, and Ya. Pesin, *Partial hyperbolicity, Lyapunov exponents and stable ergodicity*, J. Statist. Phys. **108** (2002), no. 5-6, 927–942, Dedicated to David Ruelle and Yasha Sinai on the occasion of their 65th birthdays. Zbl 1124.37308 MR 1933439

[78] Keith Burns and Mark Pollicott, *Stable ergodicity and frame flows*, Geom. Dedicata **98** (2003), 189–210. Zbl 1027.37019 MR 1988429

[79] Keith Burns, Charles Pugh, Michael Shub, and Amie Wilkinson, *Recent results about stable ergodicity*, Smooth ergodic theory and its applications (Seattle, WA, 1999), Proc. Sympos. Pure Math., vol. 69, American Mathematical Society, Providence, RI, 2001, pp. 327–366. Zbl 1012.37019 MR 1858538

[80] Keith Burns, Charles Pugh, and Amie Wilkinson, *Stable ergodicity and Anosov flows*, Topology **39** (2000), no. 1, 149–159. Zbl 0930.37008 MR 1710997

[81] Keith Burns and Amie Wilkinson, *Stable ergodicity of skew products*, Ann. Sci. Éc. Norm. Sup. (4) **32** (1999), no. 6, 859–889. Zbl 0942.37015 MR 1717580

[82] Keith Burns and Amie Wilkinson, *Dynamical coherence and center bunching*, Discrete Contin. Dyn. Syst. **22** (2008), no. 1-2, 89–100. Zbl 1154.37332 MR 2410949

[83] Keith Burns and Amie Wilkinson, *On the ergodicity of partially hyperbolic systems*, Ann. of Math. (2) **171** (2010), no. 1, 451–489. Zbl 1196.37057 MR 2630044

[84] Clark Butler, *Characterizing symmetric spaces by their Lyapunov spectra*, 2017, arXiv:1709.08066.

[85] Clark Butler, *Rigidity of equality of Lyapunov exponents for geodesic flows*, J. Differential Geom. **109** (2018), no. 1, 39–79. Zbl 1391.37032 MR 3798715

[86] Oliver Butterley and Carlangelo Liverani, *Smooth Anosov flows: correlation spectra and stability*, J. Mod. Dyn. **1** (2007), no. 2, 301–322. Zbl 1144.37011 MR 2285731

[87] Mary L. Cartwright, *Forced oscillations in nonlinear systems*, Contributions to the Theory of Nonlinear Oscillations, Annals of Mathematics Studies, no. 20, Princeton University Press, Princeton, NJ, 1950, pp. 149–241. Zbl 0039.09901 MR 0035355

[88] Mary L. Cartwright and John E. Littlewood, *On non-linear differential equations of the second order. I. The equation* $\ddot{y} - k(1 - y^2)y + y = b\lambda k \cos(\lambda t + a), k$ *large*, J. London Math. Soc. **20** (1945), 180–189. Zbl 0061.18903 MR 0016789

[89] Nikolai I. Chernov, *On statistical properties of chaotic dynamical systems*, Sinaĭ's Moscow Seminar on Dynamical Systems, Amer. Math. Soc. Transl. Ser. 2, vol. 171, American Mathematical Society, Providence, RI, 1996, pp. 57–71. Zbl 0846.58042 MR 1359093

[90] Nikolai I. Chernov, *Markov approximations and decay of correlations for Anosov flows*, Ann. of Math. (2) **147** (1998), no. 2, 269–324. Zbl 0911.58028 MR 1626741

[91] Nikolai I. Chernov and Cymra Haskell, *Nonuniformly hyperbolic K-systems are Bernoulli*, Ergodic Theory Dynam. Systems **16** (1996), no. 1, 19–44. Zbl 0853.58081 MR 1375125

[92] Richard C. Churchill, John Franke, and James Selgrade, *A geometric criterion for hyperbolicity of flows*, Proc. Amer. Math. Soc. **62** (1976), no. 1, 137–143 (1977). Zbl 0316.58015 MR 0428358

[93] Vaughn Climenhaga, Gerhard Knieper, and Khadim War, *Uniqueness of the measure of maximal entropy for geodesic flows on certain manifolds without conjugate points*, 2019, arXiv:1903.09831.

[94] Vaughn Climenhaga and Daniel J. Thompson, *Unique equilibrium states for flows and homeomorphisms with non-uniform structure*, Adv. Math. **303** (2016), 745–799. Zbl 1366.37084 MR 3552538

[95] Pierre Collet, Henri Epstein, and Giovanni Gallavotti, *Perturbations of geodesic flows on surfaces of constant negative curvature and their mixing properties*, Comm. Math. Phys. **95** (1984), no. 1, 61–112. Zbl 0585.58022 MR 757055

[96] G. Contreras, R. Iturriaga, G. P. Paternain, and M. Paternain, *Lagrangian graphs, minimizing measures and Mañé's critical values*, Geom. Funct. Anal. **8** (1998), no. 5, 788–809. Zbl 0920.58015 MR 1650090

[97] Israel P. Cornfeld, Sergey V. Fomin, and Yakov G. Sinaĭ, *Ergodic theory*, Grundlehren der Mathematischen Wissenschaften [Fundamental Principles of Mathematical Sciences], vol. 245, Springer, New York, 1982. Translated from the Russian by A. B. Sosinskiĭ. Zbl 0493.28007 MR 832433

[98] Yves Coudène, *Ergodic theory and dynamical systems*, Universitext, Springer, London; EDP Sciences [Les Ulis], 2016. Translated from the 2013 French original [MR 3184308] by Reinie Erné. Zbl 1368.37001 MR 3586310

[99] Ethan M. Coven and Zbigniew H. Nitecki, *On the genesis of symbolic dynamics as we know it*, Colloq. Math. **110** (2008), no. 2, 227–242. Zbl 1142.37013 MR 2353908

[100] David Cowan, *A billiard model for a gas of particles with rotation*, Discrete Contin. Dyn. Syst. **22** (2008), no. 1-2, 101–109. Zbl 1153.37364 MR 2410950

[101] David Cowan, *Rigid particle systems and their billiard models*, Discrete Contin. Dyn. Syst. **22** (2008), no. 1-2, 111–130. Zbl 1153.37365 MR 2410951

[102] C. Cox and R. Feres, *No-slip billiards in dimension two*, Dynamical systems, ergodic theory, and probability: in memory of Kolya Chernov, Contemp. Math., vol. 698, American Mathematical Society, Providence, RI, 2017, pp. 91–110. Zbl 06951245 MR 3716087

[103] Jane Cronin, *Ordinary differential equations*, third ed., Pure and Applied Mathematics (Boca Raton), vol. 292, Chapman & Hall/CRC, Boca Raton, FL, 2008, Introduction and qualitative theory. Zbl 1151.34001 MR 2372271

[104] Sylvain Crovisier, *Une remarque sur les ensembles hyperboliques localement maximaux*, C. R. Math. Acad. Sci. Paris **334** (2002), no. 5, 401–404. Zbl 1042.37022 MR 1892942

[105] Nurlan S. Dairbekov and Gabriel P. Paternain, *Longitudinal KAM-cocycles and action spectra of magnetic flows*, Math. Res. Lett. **12** (2005), no. 5-6, 719–729. Zbl 1089.37041 MR 2189233

[106] N. S. Dairbekov and G. P. Paternain, *On the cohomological equation of magnetic flows*, Mat. Contemp. **34** (2008), 155–193. Zbl 1196.37079 MR 2588611

[107] Nurlan S. Dairbekov and Gabriel P. Paternain, *Rigidity properties of Anosov optical hypersurfaces*, Ergodic Theory Dynam. Systems **28** (2008), no. 3, 707–737. Zbl 1144.53097 MR 2422013

[108] Alan Dankner, *On Smale's Axiom* A *dynamical systems*, Ann. of Math. (2) **107** (1978), no. 3, 517–553. Zbl 0366.58009 MR 0488161

[109] David DeLatte, *Nonstationary normal forms and cocycle invariants*, Random Comput. Dynam. **1** (1992/93), no. 2, 229–259. Zbl 0778.58058 MR 1186375

[110] David DeLatte, *On normal forms in Hamiltonian dynamics, a new approach to some convergence questions*, Ergodic Theory Dynam. Systems **15** (1995), no. 1, 49–66. Zbl 0820.58052 MR 1314968

[111] Manfred Denker and Ernst Eberlein, *Ergodic flows are strictly ergodic*, Advances in Math. **13** (1974), 437–473. Zbl 0283.28012 MR 0352403

[112] Manfred Denker, Christian Grillenberger, and Karl Sigmund, *Ergodic theory on compact spaces*, Lecture Notes in Mathematics, vol. 527, Springer, Berlin-New York, 1976. Zbl 0328.28008

[113] Robert L. Devaney, *A first course in chaotic dynamical systems*, Addison-Wesley Studies in Nonlinearity, Addison-Wesley, Advanced Book Program, Reading, MA, 1992. Theory and experiment. With a separately available computer disk. Zbl 0768.58001 MR 1202237

[114] Robert L. Devaney, *An introduction to chaotic dynamical systems*, Studies in Nonlinearity, Westview Press, Boulder, CO, 2003. Reprint of the second (1989) edition. Zbl 0695.58002 MR 1979140

[115] Claus I. Doering, *Persistently transitive vector fields on three-dimensional manifolds*, Dynamical systems and bifurcation theory (Rio de Janeiro, 1985), Pitman Res. Notes Math. Ser., vol. 160, Longman Sci. Tech., Harlow, 1987, pp. 59–89. Zbl 0631.58016 MR 907891

[116] Dmitry Dolgopyat, *On decay of correlations in Anosov flows*, Ann. of Math. (2) **147** (1998), no. 2, 357–390. Zbl 0911.58029 MR 1626749

[117] Dmitry Dolgopyat, *Prevalence of rapid mixing in hyperbolic flows*, Ergodic Theory Dynam. Systems **18** (1998), no. 5, 1097–1114. Zbl 0918.58058 MR 1653299

[118] Dmitry Dolgopyat, *Prevalence of rapid mixing. II. Topological prevalence*, Ergodic Theory Dynam. Systems **20** (2000), no. 4, 1045–1059. Zbl 0965.37032 MR 1779392

[119] Dmitry Dolgopyat and Mark Pollicott, *Addendum to: "Periodic orbits and dynamical spectra" [Ergodic Theory Dynam. Systems* **18** *(1998), no. 2, 255–292; Zbl 0915.58088 MR 1619556 (2000h:37031)] by V. Baladi*, Ergodic Theory Dynam. Systems **18** (1998), no. 2, 293–301. Zbl 0922.58065 MR 1619557

[120] Dmitry Dolgopyat and Amie Wilkinson, *Stable accessibility is C^1 dense*, Astérisque (2003), no. 287, xvii, 33–60. Geometric methods in dynamics. II. Zbl 1213.37053 MR 2039999

[121] Victor Donnay and Daniel Visscher, *A new proof of the existence of embedded surfaces with Anosov geodesic flow*, Regul. Chaotic Dyn. **23** (2018), no. 6, 685–694. Zbl 1410.37032

[122] Victor J. Donnay and Charles C. Pugh, *Anosov geodesic flows for embedded surfaces*, Astérisque (2003), no. 287, xviii, 61–69. Geometric methods in dynamics. II. Zbl 1054.37009 MR 2040000

[123] Pierre Duhem, *The aim and structure of physical theory*, Princeton Science Library, Princeton University Press, Princeton, NJ, 1991. With a foreword by Louis de Broglie. Translated from the second French edition by Philip P. Wiener. Reprint of the 1954 English translation. With an introduction by Jules Vuillemin. Zbl 0058.17503 MR 1145487

[124] Henry A. Dye, *On groups of measure preserving transformations. I*, Amer. J. Math. **81** (1959), 119–159. Zbl 0087.11501 MR 0131516

[125] Henry. A. Dye, *On groups of measure preserving transformations. II*, Amer. J. Math. **85** (1963), 551–576. Zbl 0191.42803 MR 0158048

[126] Patrick Eberlein, *When is a geodesic flow of Anosov type? I,II*, J. Differential Geometry **8** (1973), 437–463; **8** (1973), 565–577. Zbl 0285.58008 Zbl 0295.58009 MR 0380891

[127] Paul Ehrenfest and Tatiana Ehrenfest, *Begriffliche Grundlagen der statistischen Auffassung in der Mechanik*, In Encyklopaedie der Mathematischen Wissenschaften, vol. 4 (1912), 1–190. Zbl 43.0763.01

[128] Yong Fang, *A dynamical-geometric characterization of the geodesic flows of negatively curved locally symmetric spaces*, Ergodic Theory Dynam. Systems **35** (2015), no. 7, 2094–2113. Zbl 1352.37094 MR 3394109

[129] Albert Fathi and Pierre Pageault, *Aubry-Mather theory for homeomorphisms*, Ergodic Theory Dynam. Systems **35** (2015), no. 4, 1187–1207. Zbl 1343.37012 MR 3345168

[130] Albert Fathi and Pierre Pageault, *Smoothing Lyapunov functions*, Trans. Amer. Math. Soc. **371** (2019), no. 3, 1677–1700. Zbl 1410.37025 MR 3894031

[131] Jacob Feldman and Donald Ornstein, *Semirigidity of horocycle flows over compact surfaces of variable negative curvature*, Ergodic Theory Dynam. Systems **7** (1987), no. 1, 49–72. Zbl 0633.58024 MR 886370

[132] Sérgio R. Fenley, *Anosov flows in 3-manifolds*, Ann. of Math. (2) **139** (1994), no. 1, 79–115. Zbl 0796.58039 MR 1259365

[133] Bernold Fiedler (ed.), *Handbook of dynamical systems. Vol. 2*, North-Holland, Amsterdam, 2002. Zbl 0982.37002 MR 1900651

[134] Todd Fisher, *Hyperbolic sets that are not locally maximal*, Ergodic Theory Dynam. Systems **26** (2006), no. 5, 1491–1509. Zbl 1122.37022 MR 2266370

[135] Livio Flaminio, *Local entropy rigidity for hyperbolic manifolds*, Comm. Anal. Geom. **3** (1995), no. 3-4, 555–596. Zbl 0852.58068 MR 1371210

[136] Patrick Foulon, *Entropy rigidity of Anosov flows in dimension three*, Ergodic Theory Dynam. Systems **21** (2001), no. 4, 1101–1112. Zbl 1055.37031

[137] Patrick Foulon and Boris Hasselblatt, *Contact Anosov flows on hyperbolic 3-manifolds*, Geom. Topol. **17** (2013), no. 2, 1225–1252. Zbl 1277.37057 MR 3070525

[138] Patrick Foulon and François Labourie, *Sur les variétés compactes asymptotiquement harmoniques*, Invent. Math. **109** (1992), no. 1, 97–111. Zbl 0767.53030 MR 1168367

[139] John E. Franke and James F. Selgrade, *Abstract ω-limit sets, chain recurrent sets, and basic sets for flows*, Proc. Amer. Math. Soc. **60** (1976), 309–316 (1977). Zbl 0334.58010 MR 0423423

[140] John Franks, *Anosov diffeomorphisms on tori*, Trans. Amer. Math. Soc. **145** (1969), 117–124. Zbl 0191.21604 MR 0253352

[141] John Franks, *Anosov diffeomorphisms*, Global Analysis (Proc. Sympos. Pure Math., vol. XIV, Berkeley, Calif., 1968), American Mathematical Society, Providence, RI, 1970, pp. 61–93. Zbl 0207.54304 MR 0271990

[142] John Franks, *A new proof of the Brouwer plane translation theorem*, Ergodic Theory Dynam. Systems **12** (1992), no. 2, 217–226. Zbl 0767.58025 MR 1176619

[143] John Franks and Bob Williams, *Anomalous Anosov flows*, Global theory of dynamical systems (Proc. Internat. Conf., Northwestern University, Evanston, Ill., 1979), Lecture Notes in Mathematics, vol. 819, Springer, Berlin, 1980, pp. 158–174. Zbl 0463.58021 MR 591182

[144] David Fried, *Transitive Anosov flows and pseudo-Anosov maps*, Topology **22** (1983), no. 3, 299–303. Zbl 0516.58035 MR 710103

[145] Hans-Otto Georgii, *Gibbs measures and phase transitions*, second ed., De Gruyter Studies in Mathematics, vol. 9, Walter de Gruyter, Berlin, 2011. Zbl 1225.60001 MR 2807681

[146] Marlies Gerber, Boris Hasselblatt, and Daniel Keesing, *The Riccati equation: pinching of forcing and solutions*, Experiment. Math. **12** (2003), no. 2, 129–134. Zbl 1059.34004 MR 2016702

[147] Étienne Ghys, *Flots d'Anosov sur les* 3*-variétés fibrées en cercles*, Ergodic Theory Dynam. Systems **4** (1984), no. 1, 67–80. Zbl 0527.58030 MR 758894

[148] Etienne Ghys, *Flots d'Anosov dont les feuilletages stables sont différentiables*, Ann. Sci. Éc. Norm. Sup. (4) **20** (1987), no. 2, 251–270. Zbl 0663.58025

[149] Étienne Ghys, *Déformations de flots d'Anosov et de groupes fuchsiens*, Ann. Inst. Fourier (Grenoble) **42** (1992), no. 1-2, 209–247. Zbl 0759.58036 MR 1162561

[150] Étienne Ghys, *Rigidité différentiable des groupes fuchsiens*, Inst. Hautes Études Sci. Publ. Math. (1993), no. 78, 163–185 (1994). Zbl 0812.58066 MR 1259430

[151] Paolo Giulietti, Carlangelo Liverani, and Mark Pollicott, *Anosov flows and dynamical zeta functions*, Ann. of Math. (2) **178** (2013), no. 2, 687–773. Zbl 1418.37042 MR 3071508

[152] James Gleick, *Chaos, making a new science*, Penguin Books, New York, 1987. Zbl 0706.58002 MR 1010647

[153] Sue Goodman, *Dehn surgery on Anosov flows*, Geometric dynamics (Rio de Janeiro, 1981), Lecture Notes in Mathematics, vol. 1007, Springer, Berlin, 1983, pp. 300–307. Zbl 0532.58021 MR 1691596

[154] Geoffrey R. Goodson, *Chaotic dynamics*, Cambridge Mathematical Textbooks, Cambridge University Press, Cambridge, 2017. Fractals, tilings, and substitutions. Zbl 1387.37001 MR 3617651

[155] Anna Grant, *Surfaces of negative curvature and permanent regional transitivity*, Duke Math. J. **5** (1939), no. 2, 207–229. Zbl 0021.23603 MR 1546119

[156] Matthew Grayson, Bruce Kitchens, and George Zettler, *Visualizing toral automorphisms*, Math. Intelligencer **15** (1993), no. 1, 63–66. Zbl 0772.57035

[157] Matthew Grayson, Charles Pugh, and Michael Shub, *Stably ergodic diffeomorphisms*, Ann. of Math. (2) **140** (1994), no. 2, 295–329. Zbl 0824.58032 MR 1298715

[158] Judy Green and Jeanne LaDuke, *Pioneering women in American mathematics*, History of Mathematics, vol. 34, American Mathematical Society, Providence, RI; London Mathematical Society, London, 2009. The pre-1940 PhD's. Zbl 1180.01002 MR 2464022

[159] Stéphane Grognet, *Flots magnétiques en courbure négative*, Ergodic Theory Dynam. Systems **19** (1999), no. 2, 413–436. Zbl 0935.53037 MR 1685401

[160] Jorge Groisman and Zbigniew Nitecki, *Foliations and conjugacy: Anosov structures in the plane*, Ergodic Theory Dynam. Systems **35** (2015), no. 4, 1229–1242. Zbl 1355.37049 MR 3345170

[161] Jacob Gross, *What is... Riemannian holonomy*, Notices Amer. Math. Soc. **65** (2018), no. 7, 795–796. Zbl 1398.53002

[162] A. A. Gura, *The horocycle flow on a surface of negative curvature is separating*, Mat. Zametki **36** (1984), no. 2, 279–284. Zbl 0563.58024 MR 759440

[163] Martin Guterman and Zbigniew Nitecki, *Differential equations with linear algebra*, third ed., The Saunders Series, Saunders College Publishing (CBS College Publishing), Philadelphia, 1986. Zbl 0661.34001

[164] Misha Guysinsky, *Smoothness of holonomy maps derived from unstable foliation*, Smooth ergodic theory and its applications (Seattle, WA, 1999), Proc. Sympos. Pure Math., vol. 69, American Mathematical Society, Providence, RI, 2001, pp. 785–790. Zbl 1013.37038 MR 1858554

[165] Misha Guysinsky, Boris Hasselblatt, and Victoria Rayskin, *Differentiability of the Hartman-Grobman linearization*, Discrete Contin. Dyn. Syst. **9** (2003), no. 4, 979–984. Zbl 1024.37022 MR 1975364

[166] Jaques Hadamard, *Les surfaces à courbures opposées et leurs lignes géodesiques*, J. Math. Pures Appl. (5) **4** (1898), 195–216. Zbl 29.0522.01

[167] Jaques Hadamard, *Sur l'itération et les solutions asymptotiques des équations différentielles*, Bull. Soc. Math. France **29** (1901), 224–228. Zbl 32.0314.01

[168] Jack K. Hale, *Ordinary differential equations*, second ed., Robert E. Krieger, Huntington, NY, 1980. Zbl 0433.34003 MR 587488

[169] Ursula Hamenstädt, *Invariant two-forms for geodesic flows*, Math. Ann. **301** (1995), no. 4, 677–698. Zbl 0821.58033 MR 1326763

[170] Michael Handel and William P. Thurston, *Anosov flows on new three manifolds*, Invent. Math. **59** (1980), no. 2, 95–103. Zbl 0435.58019 MR 577356

[171] Philip Hartman, *On local homeomorphisms of Euclidean spaces*, Bol. Soc. Mat. Mexicana (2) **5** (1960), 220–241. Zbl 0127.30202 MR 0141856

[172] Boris Hasselblatt, *Horospheric foliations and relative pinching*, J. Differential Geom. **39** (1994), no. 1, 57–63. Zbl 0795.53026 MR 1258914

[173] Boris Hasselblatt, *Periodic bunching and invariant foliations*, Math. Res. Lett. **1** (1994), no. 5, 597–600. Zbl 0845.58043 MR 1295553

[174] Boris Hasselblatt, *Introduction to hyperbolic dynamics and ergodic theory*, Ergodic theory and negative curvature, Lecture Notes in Mathematics, vol. 2164, Springer, Cham, 2017, http://www.springer.com/cda/content/document/cda_downloaddocument/9783319430584-c1.pdf?SGWID=0-0-45-1628926-p180166381, pp. 1–124. Zbl 1394.37001 MR 3588132

[175] Boris Hasselblatt, *On iteration and asymptotic solutions of differential equations by Jacques Hadamard*, Ergodic theory and negative curvature, Lecture Notes in Mathematics, vol. 2164, Springer, Cham, 2017. Translation of [167], pp. 125–128. MR 3588133

[176] B. Hasselblatt and A. Katok (eds.), *Handbook of dynamical systems. Vol. 1A*, North-Holland, Amsterdam, 2002. Zbl 1013.00016 MR 1928517

[177] Boris Hasselblatt and Anatole B. Katok, *A first course in dynamics*, Cambridge University Press, New York, 2003. With a panorama of recent developments. Zbl 1027.37001 MR 1995704

[178] B. Hasselblatt and A. Katok (eds.), *Handbook of dynamical systems. Vol. 1B*, Elsevier, Amsterdam, 2006. Zbl 1081.00006 MR 2184980

[179] Boris Hasselblatt, Zbigniew Nitecki, and James Propp, *Topological entropy for nonuniformly continuous maps*, Discrete Contin. Dyn. Syst. **22** (2008), no. 1-2, 201–213. Zbl 1153.37319 MR 2410955

[180] Boris Hasselblatt and Yakov Pesin, *Partially hyperbolic dynamical systems*, Handbook of dynamical systems. Vol. 1B, Elsevier, Amsterdam, 2006, pp. 1–55. Zbl 1130.37355 MR 2186241

[181] Boris Hasselblatt, Yakov Pesin, and Jörg Schmeling, *Pointwise hyperbolicity implies uniform hyperbolicity*, Discrete Contin. Dyn. Syst. **34** (2014), no. 7, 2819–2827. Zbl 1358.37062 MR 3177662

[182] Boris Hasselblatt and Jörg Schmeling, *Dimension product structure of hyperbolic sets*, Modern dynamical systems and applications, Cambridge University Press, Cambridge, 2004, pp. 331–345. Zbl 1147.37310 MR 2093308

[183] Boris Hasselblatt and Jorg Schmeling, *Dimension product structure of hyperbolic sets*, Electron. Res. Announc. Amer. Math. Soc. **10** (2004), 88–96. Zbl 1068.37016 MR 2084468

[184] Boris Hasselblatt and Amie Wilkinson, *Prevalence of non-Lipschitz Anosov foliations*, Electron. Res. Announc. Amer. Math. Soc. **3** (1997), 93–98. Zbl 0887.58043 MR 1465582

[185] Boris Hasselblatt and Amie Wilkinson, *Prevalence of non-Lipschitz Anosov foliations*, Ergodic Theory Dynam. Systems **19** (1999), no. 3, 643–656. Zbl 1069.37031 MR 1695913

[186] Allen Hatcher, *Algebraic topology*, Cambridge University Press, Cambridge, 2002. Zbl 1044.55001 MR 1867354

[187] Shuhei Hayashi, *Connecting invariant manifolds and the solution of the C^1 stability and Ω-stability conjectures for flows*, Ann. of Math. (2) **145** (1997), no. 1, 81–137. Zbl 0871.58067 MR 1432037

[188] Shuhei Hayashi, *Correction to: "Connecting invariant manifolds and the solution of the C^1 stability and Ω-stability conjectures for flows" [Ann. of Math. (2)* **145** *(1997), no. 1, 81–137; Zbl 0871.58067 MR 1432037 (98b:58096)]*, Ann. of Math. (2) **150** (1999), no. 1, 353–356. Zbl 1157.37317 MR 1715329

[189] Shuhei Hayashi, *Stability of dynamical systems [translation of Sūgaku* **50** *(1998), no. 2, 149–162; MR 1648432 (99j:58115)]*, Sugaku Expositions **14** (2001), no. 1, 15–29, Sugaku Expositions. Zbl 0988.37022 MR 1834910

[190] Nicolai T. A. Haydn, *Canonical product structure of equilibrium states*, Random Comput. Dynam. **2** (1994), no. 1, 79–96. Zbl 0810.58030 MR 1265227

[191] Gustav A. Hedlund, *Fuchsian groups and transitive horocycles*, Duke Math. J. **2** (1936), no. 3, 530–542. Zbl 0015.10201 MR 1545946

[192] Gustav A. Hedlund, *Fuchsian groups and mixtures*, Ann. of Math. (2) **40** (1939), no. 2, 370–383. Zbl 0020.40302 MR 1503464

[193] Federico Rodriguez Hertz, *Global rigidity of certain abelian actions by toral automorphisms*, J. Mod. Dyn. **1** (2007), no. 3, 425–442. Zbl 1130.37013 MR 2318497

[194] Morris W. Hirsch, Charles C. Pugh, and Michael Shub, *Invariant manifolds*, Lecture Notes in Mathematics, vol. 583, Springer, Berlin-New York, 1977. Zbl 0355.58009 MR 0501173

[195] Ale Jan Homburg, *Atomic disintegrations for partially hyperbolic diffeomorphisms*, Proc. Amer. Math. Soc. **145** (2017), no. 7, 2981–2996. Zbl 1375.37069 MR 3637946

[196] Eberhard Hopf, *Fuchsian groups and ergodic theory*, Trans. Amer. Math. Soc. **39** (1936), no. 2, 299–314. Zbl 0014.08303 MR 1501848

[197] Eberhard Hopf, *Statistik der geodätischen Linien in Mannigfaltigkeiten negativer Krümmung*, Ber. Verh. Sächs. Akad. Wiss. Leipzig **91** (1939), 261–304. Zbl 0024.08003

[198] Eberhard Hopf, *Statistik der Lösungen geodätischer Probleme vom unstabilen Typus. II*, Math. Ann. **117** (1940), 590–608. Zbl 0023.26801 MR 0002722

[199] Tim J. Hunt and Robert S. MacKay, *Anosov parameter values for the triple linkage and a physical system with a uniformly chaotic attractor*, Nonlinearity **16** (2003), no. 4, 1499–1510. Zbl 1040.37072 MR 1986308

[200] Steven Hurder and Anatole B. Katok, *Differentiability, rigidity and Godbillon-Vey classes for Anosov flows*, Inst. Hautes Études Sci. Publ. Math. (1990), no. 72, 5–61 (1991). Zbl 0725.58034 MR 1087392

[201] Kei Irie, *Dense existence of periodic Reeb orbits and ECH spectral invariants*, J. Mod. Dyn. **9** (2015), 357–363. Zbl 1353.37125 MR 3436746

[202] Konrad Jacobs, *Lipschitz functions and the prevalence of strict ergodicity for continuous-time flows*, Contributions to Ergodic Theory and Probability (Proc. Conf., Ohio State Univ., Columbus, Ohio, 1970), Springer, Berlin, 1970, pp. 87–124. Zbl 0201.38302 MR 0274709

[203] Jean-Lin Journé, *A regularity lemma for functions of several variables*, Rev. Mat. Iberoam. **4** (1988), no. 2, 187–193. Zbl 0699.58008

[204] Steven Kalikow and Randall McCutcheon, *An outline of ergodic theory*, Cambridge Studies in Advanced Mathematics, vol. 122, Cambridge University Press, Cambridge, 2010. Zbl 1192.37004 MR 2650005

[205] Boris Kalinin, *Livšic theorem for matrix cocycles*, Ann. of Math. (2) **173** (2011), no. 2, 1025–1042. Zbl 1238.37008 MR 2776369

[206] Kazuhisa Kato and Akihiko Morimoto, *Topological stability of Anosov flows and their centralizers*, Topology **12** (1973), 255–273. Zbl 0261.58004 MR 0326779

[207] Anatole B. Katok, *Dynamical systems with hyperbolic structure*, Ninth Mathematical Summer School (Kaciveli, 1971) (Russian), Izdanie Inst. Mat. Akad. Nauk Ukrain. SSR, Kiev, 1972, Three papers on smooth dynamical systems, pp. 125–211. MR 0377991

[208] Anatole B. Katok, *Time change, monotone equivalence, and standard dynamical systems*, Dokl. Akad. Nauk SSSR **223** (1975), no. 4, 789–792. Zbl 0326.28025 MR 0412383

[209] Anatole B. Katok, *Monotone equivalence in ergodic theory*, Izv. Akad. Nauk SSSR Ser. Mat. **41** (1977), no. 1, 104–157, 231. Zbl 0379.28008 MR 0442195

[210] Anatole B. Katok, *Lyapunov exponents, entropy and periodic orbits for diffeomorphisms*, Inst. Hautes Études Sci. Publ. Math. (1980), no. 51, 137–173. Zbl 0445.58015 MR 573822

[211] Anatole B. Katok, *Entropy and closed geodesics*, Ergodic Theory Dynam. Systems **2** (1982), no. 3-4, 339–365 (1983). Zbl 0525.58027 MR 0721728

[212] Anatole B. Katok, *Four applications of conformal equivalence to geometry and dynamics*, Ergodic Theory Dynam. Systems **8*** (1988), no. Charles Conley Memorial Issue, 139–152. Zbl 0668.58042 MR 0967635

[213] Anatole B. Katok and Boris Hasselblatt, *Introduction to the modern theory of dynamical systems*, Encyclopedia of Mathematics and Its Applications, vol. 54, Cambridge University Press, Cambridge, 1995. With a supplementary chapter by Katok and Leonardo Mendoza. Zbl 0878.58020 MR 1326374

[214] A. Katok, G. Knieper, M. Pollicott, and H. Weiss, *Differentiability and analyticity of topological entropy for Anosov and geodesic flows*, Invent. Math. **98** (1989), no. 3, 581–597. Zbl 0702.58053 MR 1022308

[215] A. Katok, G. Knieper, M. Pollicott, and H. Weiss, *Differentiability of entropy for Anosov and geodesic flows*, Bull. Amer. Math. Soc. (N.S.) **22** (1990), no. 2, 285–293. Zbl 0717.58045 MR 1013257

[216] Anatole Katok, Gerhard Knieper, and Howard Weiss, *Formulas for the derivative and critical points of topological entropy for Anosov and geodesic flows*, Comm. Math. Phys. **138** (1991), no. 1, 19–31. Zbl 0749.58041 MR 1108034

[217] A. Katok and A. Kononenko, *Cocycles' stability for partially hyperbolic systems*, Math. Res. Lett. **3** (1996), no. 2, 191–210. Zbl 0853.58082 MR 1386840

[218] Anatole B. Katok and Ralf J. Spatzier, *First cohomology of Anosov actions of higher rank abelian groups and applications to rigidity*, Inst. Hautes Études Sci. Publ. Math. (1994), no. 79, 131–156. Zbl 0819.58027 MR 1307298

[219] Anatole B. Katok and Ralf J. Spatzier, *Differential rigidity of Anosov actions of higher rank abelian groups and algebraic lattice actions*, Trudy Matematicheskogo Instituta Imeni V. A. Steklova. Rossii skaya Akademiya Nauk **216** (1997), no. Din. Sist. i Smezhnye Vopr., 292–319. Zbl 0938.37010 MR 1632177

[220] Anatole B. Katok, Jean-Marie Strelcyn, François Ledrappier, and Feliks Przytycki, *Invariant manifolds, entropy and billiards; smooth maps with singularities*, Lecture Notes in Mathematics, vol. 1222, Springer, Berlin, 1986. Zbl 0658.58001 MR 872698

[221] Gerhard Keller, *Equilibrium states in ergodic theory*, London Mathematical Society Student Texts, vol. 42, Cambridge University Press, Cambridge, 1998. Zbl 0896.28006 MR 1618769

[222] Wilhelm Klingenberg, *Riemannian manifolds with geodesic flow of Anosov type*, Ann. of Math. (2) **99** (1974), 1–13. Zbl 0272.53025 MR 0377980

[223] Gerhard Knieper, *The uniqueness of the measure of maximal entropy for geodesic flows on rank* 1 *manifolds*, Ann. of Math. (2) **148** (1998), no. 1, 291–314. Zbl 0946.53045 MR 1652924

[224] Gerhard Knieper, *Hyperbolic dynamics and Riemannian geometry*, Handbook of dynamical systems, vol. 1A, North-Holland, Amsterdam, 2002, pp. 453–545. Zbl 1049.37020 MR 1928523

[225] Paul Koebe, *Riemannsche Mannigfaltigkeiten und nicht euklidische Raumformen (i)*, Sitzungsberichte der Preussischen Akademie der Wissenschaften (1927), 164–196. Zbl 53.0548.04

[226] Mickaël Kourganoff, *Anosov geodesic flows, billiards and linkages*, Comm. Math. Phys. **344** (2016), no. 3, 831–856. Zbl 1364.37079 MR 3508162

[227] Mickaël Kourganoff, *Embedded surfaces with Anosov geodesic flows, approximating spherical billiards*, 2016, arXiv:1612.05430.

[228] Mickaël Kourganoff, *Uniform hyperbolicity in nonflat billiards*, Discrete Contin. Dyn. Syst. **38** (2018), no. 3, 1145–1160. Zbl 1394.37050 MR 3808990

[229] Ulrich Krengel, *On the speed of convergence in the ergodic theorem*, Monatsh. Math. **86** (1978/79), no. 1, 3–6. Zbl 0352.28008 MR 510630

[230] Isabel S. Labouriau and Alexandre A. P. Rodrigues, *On Takens' last problem: tangencies and time averages near heteroclinic networks*, Nonlinearity **30** (2017), no. 5, 1876–1910. Zbl 1370.34073 MR 3639293

[231] Pierre-Simon de Laplace, *Philosophical essay on probabilities*, Sources in the History of Mathematics and Physical Sciences, vol. 13, Springer, New York, 1995. Translated from the fifth (1825) French edition, and with notes and a preface by Andrew I. Dale. Zbl 0810.01015 MR 1325241

[232] Joel L. Lebowitz and Oliver Penrose, *Modern ergodic theory*, Physics Today **26** (1973), no. 2, 23–29 (English).

[233] François Ledrappier, *Mesures d'equilibre d'éntropie complètement positive*, Astérisque **50** (1977), 251–272. Zbl 0376.28015

[234] François Ledrappier, *Harmonic measures and Bowen-Margulis measures*, Israel J. Math. **71** (1990), no. 3, 275–287. Zbl 0728.53029 MR 1088820

[235] François Ledrappier, Yuri Lima, and Omri Sarig, *Ergodic properties of equilibrium measures for smooth three dimensional flows*, Comment. Math. Helv. **91** (2016), no. 1, 65–106. Zbl 1366.37087 MR 3471937

[236] F. Ledrappier and L.-S. Young, *The metric entropy of diffeomorphisms*, Bull. Amer. Math. Soc. (N.S.) **11** (1984), no. 2, 343–346. Zbl 0558.58007 MR 752794

[237] F. Ledrappier and L.-S. Young, *The metric entropy of diffeomorphisms. I. Characterization of measures satisfying Pesin's entropy formula*, Ann. of Math. (2) **122** (1985), no. 3, 509–539. Zbl 0605.58028 MR 819556

[238] F. Ledrappier and L.-S. Young, *The metric entropy of diffeomorphisms. II. Relations between entropy, exponents and dimension*, Ann. of Math. (2) **122** (1985), no. 3, 540–574. Zbl 1371.37012 MR 819557

[239] Manseob Lee and Jumi Oh, *Measure expansive flows for the generic view point*, J. Difference Equ. Appl. **22** (2016), no. 7, 1005–1018. MR 3567278

[240] Seunghee Lee and Junmi Park, *Expansive homoclinic classes of generic C^1-vector fields*, Acta Math. Sin. (Engl. Ser.) **32** (2016), no. 12, 1451–1458. Zbl 1379.37051 MR 3568075

[241] Norman Levinson, *A second order differential equation with singular solutions*, Ann. of Math. (2) **50** (1949), 127–153. Zbl 0041.42311 MR 0030079

[242] Douglas Lind and Brian Marcus, *An introduction to symbolic dynamics and coding*, Cambridge University Press, Cambridge, 1995. Zbl 1106.37301 MR 1369092

[243] John E. Littlewood, *On non-linear differential equations of the second order. IV. The general equation $\ddot{y}+kf(y)\dot{y}+g(y) = bkp(\phi)$, $\phi = t+\alpha$*, Acta Math. **98** (1957), 1–110. Zbl 0081.08401 MR 0090732

[244] Carlangelo Liverani, *On contact Anosov flows*, Ann. of Math. (2) **159** (2004), no. 3, 1275–1312. Zbl 1067.37031 MR 2113022

[245] Carlangelo Liverani, *On the work and vision of Dmitry Dolgopyat*, J. Mod. Dyn. **4** (2010), no. 2, 211–225. Zbl 1209.37041 MR 2672294

[246] Rafael de la Llave, *Smooth conjugacy and S-R-B measures for uniformly and non-uniformly hyperbolic systems*, Comm. Math. Phys. **150** (1992), no. 2, 289–320. Zbl 0770.58029 MR 1194019

[247] Rafael de la Llave, *Analytic regularity of solutions of Livsic's cohomology equation and some applications to analytic conjugacy of hyperbolic dynamical systems*, Ergodic Theory Dynam. Systems **17** (1997), no. 3, 649–662. Zbl 0883.58024 MR 1452186

[248] Rafael de la Llave, José M. Marco, and Roberto Moriyón, *Canonical perturbation theory of Anosov systems and regularity results for the Livšic cohomology equation*, Ann. of Math. (2) **123** (1986), no. 3, 537–611. Zbl 0603.58016 MR 840722

[249] Frank Löbell, *Uber die geodätischen Linien der Clifford–Kleinschen Flächen*, Mathematische Zeitschrift **30** (1929), 572–607. Zbl 55.0959.01

[250] Ricardo Mañé, *Expansive diffeomorphisms*, Dynamical systems—Warwick 1974 (Proc. Sympos. Appl. Topology and Dynamical Systems, University Warwick, Coventry, 1973/1974; presented to E. C. Zeeman on his fiftieth birthday), Lecture Notes in Mathematics, vol. 468, Springer, Berlin, 1975, pp. 162–174. Zbl 0309.58016 MR 0650658

[251] Ricardo Mañé, *Ergodic theory and differentiable dynamics*, Ergebnisse der Mathematik und ihrer Grenzgebiete (3) [Results in Mathematics and Related Areas (3)], vol. 8, Springer, Berlin, 1987. Translated from the Portuguese by Silvio Levy. Zbl 0616.28007 MR 889254

[252] Anthony Manning, *There are no new Anosov diffeomorphisms on tori*, Amer. J. Math. **96** (1974), 422–429. Zbl 0242.58003 MR 0358865

[253] Brian Marcus, *Reparameterizations of uniquely ergodic flows*, J. Differential Equations **22** (1976), no. 1, 227–235. Zbl 0295.28022 MR 0422578

[254] Brian Marcus, *Ergodic properties of horocycle flows for surfaces of negative curvature*, Ann. of Math. (2) **105** (1977), no. 1, 81–105. Zbl 0322.28012 MR 0458496

[255] Brian Marcus, *The horocycle flow is mixing of all degrees*, Invent. Math. **46** (1978), no. 3, 201–209. Zbl 0395.28012 MR 0488168

[256] Grigoriy A. Margulis, *On some aspects of the theory of Anosov systems*, Springer Monographs in Mathematics, Springer, Berlin, 2004. With a survey by Richard Sharp: Periodic orbits of hyperbolic flows. Translated from the Russian by Valentina Vladimirovna Szulikowska. Zbl 1140.37010 MR 2035655

[257] Mario Martelli, *Introduction to discrete dynamical systems and chaos*, Wiley-Interscience Series in Discrete Mathematics and Optimization, Wiley-Interscience, New York, 1999. Zbl 1127.37300 MR 1819750

[258] Shigenori Matsumoto, *Codimension one Anosov flows*, Lecture Notes Series, vol. 27, Seoul National University, Research Institute of Mathematics, Global Analysis Research Center, Seoul, 1995. Zbl 0827.58044 MR 1330920

[259] John Milnor, *Fubini foiled: Katok's paradoxical example in measure theory*, Math. Intelligencer **19** (1997), no. 2, 30–32. Zbl 0883.28004 MR 1457445

[260] Calvin C. Moore, *Exponential decay of correlation coefficients for geodesic flows*, Group representations, ergodic theory, operator algebras, and mathematical physics (Berkeley, Calif., 1984), Math. Sci. Res. Inst. Publ., vol. 6, Springer, New York, 1987, pp. 163–181. Zbl 0625.58023 MR 880376

[261] Kazumine Moriyasu, Kazuhiro Sakai, and Wenxiang Sun, *C^1-stably expansive flows*, J. Differential Equations **213** (2005), no. 2, 352–367. Zbl 1066.37009 MR 2142370

[262] Jürgen Moser, *The analytic invariants of an area-preserving mapping near a hyperbolic fixed point*, Comm. Pure Appl. Math. **9** (1956), 673–692. Zbl 0072.40801 MR 0086981

[263] Jürgen Moser, *Stable and random motions in dynamical systems*, Princeton Landmarks in Mathematics, Princeton University Press, Princeton, NJ, 2001. With special emphasis on celestial mechanics. Reprint of the 1973 original. With a foreword by Philip J. Holmes. Zbl 0991.70002 MR 1829194

[264] Jürgen Moser and Eduard J. Zehnder, *Notes on dynamical systems*, Courant Lecture Notes in Mathematics, vol. 12, New York University, Courant Institute of Mathematical Sciences, New York; American Mathematical Society, Providence, RI, 2005. Zbl 1087.37001 MR 2189486

[265] J. von Neumann, *Zur Operatorenmethode in der klassischen Mechanik*, Ann. of Math. (2) **33** (1932), no. 3, 587–642. Zbl 58.1270.04 MR 1503078

[266] J. von Neumann, *Zusätze zur Arbeit "Zur Operatorenmethode... "*, Ann. of Math. (2) **33** (1932), no. 4, 789–791. Zbl 0005.31504 MR 1503096

[267] Sheldon E. Newhouse, *On codimension one Anosov diffeomorphisms*, Amer. J. Math. **92** (1970), 761–770. Zbl 0204.56901 MR 0277004

[268] Zbigniew Nitecki, *Differentiable dynamics. An introduction to the orbit structure of diffeomorphisms*, The M.I.T. Press, Cambridge, Mass.-London, 1971. Zbl 0246.58012 MR 0649788

[269] Davi Obata, *Symmetries of vector fields: the diffeomorphism centralizer*, 2019, arXiv:1903.05883.

[270] Donald Ornstein and Benjamin Weiss, *On the Bernoulli nature of systems with some hyperbolic structure*, Ergodic Theory Dynam. Systems **18** (1998), no. 2, 441–456. Zbl 0915.58076 MR 1619567

[271] Donald S. Ornstein, *Imbedding Bernoulli shifts in flows*, Contributions to Ergodic Theory and Probability (Proc. Conf., Ohio State University, Columbus, Ohio, 1970), Springer, Berlin, 1970, pp. 178–218. Zbl 0227.28013 MR 0272985

[272] Donald S. Ornstein, *Ergodic theory, randomness, and dynamical systems*, Yale University Press, New Haven, Conn.-London, 1974, James K. Whittemore Lectures in Mathematics given at Yale University, Yale Mathematical Monographs, No. 5. Zbl 0296.28016 MR 0447525

[273] Valery I. Oseledec, *A multiplicative ergodic theorem. Characteristic Ljapunov, exponents of dynamical systems*, Trudy Moskov. Mat. Obšč. **19** (1968), 179–210. Zbl 0236.93034 MR 0240280

[274] Jacob Palis Jr. and Welington de Melo, *Geometric theory of dynamical systems*, Springer, New York-Berlin, 1982. An introduction. Translated from the Portuguese by A. K. Manning. Zbl 0491.58001 MR 669541

[275] J. Palis and M. Viana, *On the continuity of Hausdorff dimension and limit capacity for horseshoes*, Dynamical systems, Valparaiso 1986, Lecture Notes in Mathematics, vol. 1331, Springer, Berlin, 1988, pp. 150–160. Zbl 0661.58023 MR 961098

[276] Carlos Frederico Borges Palmeira, *Open manifolds foliated by planes*, Ann. Math. (2) **107** (1978), no. 1, 109–131. Zbl 0382.57010 MR 0501018

[277] William Parry, *Bowen's equidistribution theory and the Dirichlet density theorem*, Ergodic Theory Dynam. Systems **4** (1984), no. 1, 117–134. Zbl 0567.58014 MR 758898

[278] William Parry and Mark Pollicott, *An analogue of the prime number theorem for closed orbits of Axiom A flows*, Ann. of Math. (2) **118** (1983), no. 3, 573–591. Zbl 0537.58038 MR 727704

[279] William Parry and Mark Pollicott, *Zeta functions and the periodic orbit structure of hyperbolic dynamics*, Astérisque (1990), no. 187-188, 268. Zbl 0726.58003 MR 1085356

[280] William Parry and Klaus Schmidt, *Natural coefficients and invariants for Markov-shifts*, Invent. Math. **76** (1984), no. 1, 15–32. Zbl 0563.28008 MR 739621

[281] Gabriel P. Paternain, *Geodesic flows*, Progress in Mathematics, vol. 180, Birkhäuser, Boston, MA, 1999. Zbl 0930.53001 MR 1712465

[282] Miguel Paternain, *Expansive flows and the fundamental group*, Bol. Soc. Brasil. Mat. (N.S.) **24** (1993), no. 2, 179–199. Zbl 0796.58040 MR 1254982

[283] Frédéric Paulin, Mark Pollicott, and Barbara Schapira, *Equilibrium states in negative curvature*, Astérisque (2015), no. 373, viii+281. Zbl 1347.37001 MR 3444431

[284] Mauricio M. Peixoto, *Structural stability on two-dimensional manifolds*, Topology **1** (1962), 101–120. Zbl 0107.07103 MR 0142859

[285] Oskar Perron, *Über Stabilität und asymptotisches Verhalten der Integrale von Differential-gleichungssystemen*, Math. Z. **29** (1929), no. 1, 129–160. Zbl 54.0456.04 MR 1544998

[286] Yakov B. Pesin, *Families of invariant manifolds that correspond to nonzero characteristic exponents*, Izv. Akad. Nauk SSSR Ser. Mat. **40** (1976), no. 6, 1332–1379, 1440. Zbl 0372.58009 MR 0458490

[287] Yakov B. Pesin, *Dimension theory in dynamical systems*, Chicago Lectures in Mathematics, University of Chicago Press, Chicago, IL, 1997. Contemporary views and applications. Zbl 0895.58033 MR 1489237

[288] Yakov B. Pesin, *Lectures on partial hyperbolicity and stable ergodicity*, Zurich Lectures in Advanced Mathematics, European Mathematical Society (EMS), Zürich, 2004. Zbl 1098.37024 MR 2068774

[289] Karl Petersen, *Ergodic theory*, Cambridge Studies in Advanced Mathematics, vol. 2, Cambridge University Press, Cambridge, 1989. Corrected reprint of the 1983 original. Zbl 0676.28008 MR 1073173

[290] Sergei Yu. Pilyugin and Sergey Tikhomirov, *Lipschitz shadowing implies structural stability*, Nonlinearity **23** (2010), no. 10, 2509–2515. Zbl 1206.37012 MR 2683779

[291] Michel Plancherel, *Beweis der Unmöglichkeit ergodischer mechanischer Systeme*, Ann. Phys. (4) **42** (1913), 1061–1063. Zbl 44.1049.04

[292] Joseph F. Plante, *Anosov flows*, Amer. J. Math. **94** (1972), 729–754. Zbl 0257.58007 MR 0377930

[293] J. F. Plante, *Anosov flows, transversely affine foliations, and a conjecture of Verjovsky*, J. London Math. Soc. (2) **23** (1981), no. 2, 359–362. Zbl 0465.58020 MR 609116

[294] R. V. Plykin, *Sources and sinks of* A-*diffeomorphisms of surfaces*, Mat. Sb. (N.S.) **94(136)** (1974), 243–264, 336. Zbl 0324.58013 MR 0356137

[295] Henri Poincaré, *Sur le problème des trois corps et les équations de la dynamique*, Acta mathematica **13** (1890), 1–270. Zbl 22.0907.01

[296] Henri Poincaré, *Sur la théorie cinétique des gaz*, Rev. Gén. Sci. Pures Appl. **5** (1894), 513–521; C. R. Acad. Sci., Paris **116** (1893), 1165–1166. Zbl 25.1819.02

[297] Mark Pollicott, *On the rate of mixing of Axiom A flows*, Invent. Math. **81** (1985), no. 3, 413–426. Zbl 0591.58025 MR 807065

[298] Mark Pollicott, *Symbolic dynamics for Smale flows*, Amer. J. Math. **109** (1987), no. 1, 183–200. Zbl 0628.58042 MR 878205

[299] Mark Pollicott, *Exponential mixing for the geodesic flow on hyperbolic three-manifolds*, J. Statist. Phys. **67** (1992), no. 3-4, 667–673. Zbl 0892.58060 MR 1171148

[300] Mark Pollicott, *Periodic orbits and zeta functions*, Handbook of dynamical systems, vol. 1A, North-Holland, Amsterdam, 2002, pp. 409–452. Zbl 1050.37009 MR 1928522

[301] Mark Pollicott and Michiko Yuri, *Dynamical systems and ergodic theory*, London Mathematical Society Student Texts, vol. 40, Cambridge University Press, Cambridge, 1998. Zbl 0897.28009 MR 1627681

[302] Charles Pugh and Michael Shub, *The Ω-stability theorem for flows*, Invent. Math. **11** (1970), 150–158. Zbl 0212.29102 MR 0287579

[303] Charles Pugh and Michael Shub, *Ergodicity of Anosov actions*, Invent. Math. **15** (1972), 1–23. Zbl 0236.58007 MR 0295390

[304] Charles Pugh and Michael Shub, *Stably ergodic dynamical systems and partial hyperbolicity*, J. Complexity **13** (1997), no. 1, 125–179. Zbl 0883.58025 MR 1449765

[305] Charles Pugh and Michael Shub, *Stable ergodicity and julienne quasi-conformality*, J. Eur. Math. Soc. (JEMS) **2** (2000), no. 1, 1–52. Zbl 0964.37017 MR 1750453

[306] Charles Pugh, Michael Shub, and Alexander Starkov, *Corrigendum to: "Stable ergodicity and julienne quasi-conformality" [J. Eur. Math. Soc. (JEMS)* **2** *(2000), no. 1, 1–52; Zbl 0964.37017 MR 1750453]*, J. Eur. Math. Soc. (JEMS) **6** (2004), no. 1, 149–151. MR 2041009

[307] Charles C. Pugh, *The Closing Lemma and structural stability*, Bull. Amer. Math. Soc. **70** (1964), 584–587. Zbl 0125.32301 MR 0163038

[308] Charles C. Pugh, *The closing lemma*, Amer. J. Math. **89** (1967), 956–1009. Zbl 0167.21803 MR 0226669

[309] Charles C. Pugh, *An improved closing lemma and a general density theorem*, Amer. J. Math. **89** (1967), 1010–1021. Zbl 0167.21804 MR 0226670

[310] Marina Ratner, *Anosov flows with Gibbs measures are also Bernoullian*, Israel J. Math. **17** (1974), 380–391. Zbl 0304.28011 MR 0374387

[311] Marina Ratner, *The rate of mixing for geodesic and horocycle flows*, Ergodic Theory Dynam. Systems **7** (1987), no. 2, 267–288. Zbl 0623.22008 MR 896798

[312] Victoria Rayskin, *α-Hölder linearization*, J. Differential Equations **147** (1998), no. 2, 271–284. Zbl 0941.37008 MR 1634012

[313] Herbert Robbins, *A remark on Stirling's formula*, Amer. Math. Monthly **62** (1955), 26–29. Zbl 0068.05404 MR 0069328

[314] Clark Robinson, *Dynamical systems*, second ed., Studies in Advanced Mathematics, CRC Press, Boca Raton, FL, 1999. Stability, symbolic dynamics, and chaos. Zbl 0914.58021 MR 1792240

[315] R. Clark Robinson, *An introduction to dynamical systems—continuous and discrete*, second ed., Pure and Applied Undergraduate Texts, vol. 19, American Mathematical Society, Providence, RI, 2012. Zbl 1277.37001 MR 3012659

[316] Arthur Rosenthal, *Beweis der Unmöglichkeit ergodischer Gassysteme*, Ann. Phys. (4) **42** (1913), 796–806. Zbl 44.1049.03

[317] Daniel J. Rudolph, *Fundamentals of measurable dynamics*, Oxford Science Publications, The Clarendon Press, Oxford University Press, New York, 1990. Ergodic theory on Lebesgue spaces. Zbl 0718.28008 MR 1086631

[318] David Ruelle, *Flots qui ne mélangent pas exponentiellement*, C. R. Acad. Sci. Paris Sér. I Math. **296** (1983), no. 4, 191–193. Zbl 0531.58040 MR 692974

[319] David Ruelle, *Historical behaviour in smooth dynamical systems*, Global analysis of dynamical systems—Festschrift dedicated to Floris Takens for his 60th birthday (Henk W. Broer, Bernd Krauskopf, and Gert Vegter, eds.), Institute of Physics Publishing, Bristol, 2001, pp. 63–66. Zbl 1067.37501 MR 1858471

[320] David Ruelle, *Differentiation of SRB states for hyperbolic flows*, Ergodic Theory Dynam. Systems **28** (2008), no. 2, 613–631. Zbl 1165.37011 MR 2408395

[321] David Ruelle and Amie Wilkinson, *Absolutely singular dynamical foliations*, Comm. Math. Phys. **219** (2001), no. 3, 481–487. Zbl 1031.37029 MR 1838747

[322] Robert J. Sacker and George R. Sell, *Existence of dichotomies and invariant splittings for linear differential systems. III*, J. Differential Equations **22** (1976), no. 2, 497–522. Zbl 0338.58016 MR 0440621

[323] Barbara Schapira, *Dynamics of geodesic and horocyclic flows*, Ergodic theory and negative curvature, Lecture Notes in Mathematics, vol. 2164, Springer, Cham, 2017, pp. 129–155. Zbl 1394.37002 MR 3588134

[324] J. Schmeling, *Hölder continuity of the holonomy maps for hyperbolic basic sets. II*, Math. Nachr. **170** (1994), 211–225. Zbl 0819.58029 MR 1302376

[325] Jörg Schmeling and Rainer Siegmund-Schultze, *Hölder continuity of the holonomy maps for hyperbolic basic sets. I*, Ergodic theory and related topics, III (Güstrow, 1990), Lecture Notes in Mathematics, vol. 1514, Springer, Berlin, 1992, pp. 174–191. Zbl 0766.58043 MR 1179182

[326] Peter Scott, *The geometries of* 3*-manifolds*, Bull. London Math. Soc. **15** (1983), no. 5, 401–487. Zbl 0561.57001 MR 705527

[327] Wladimir P. Seidel, *On a metric property of Fuchsian groups*, Proceedings of the National Academy of Sciences **21** (1935), 475–478. Zbl 0012.15501

[328] Laura Senos, *Generic Bowen-expansive flows*, Bull. Braz. Math. Soc. (N.S.) **43** (2012), no. 1, 59–71. Zbl 1320.37011 MR 2909923

[329] Michael Shub, *Global stability of dynamical systems*, Springer, New York, 1987. With the collaboration of Albert Fathi and Rémi Langevin. Translated from the French by Joseph Christy. Zbl 0606.58003 MR 869255

[330] Nándor Simányi, *Conditional proof of the Boltzmann-Sinai ergodic hypothesis*, Invent. Math. **177** (2009), no. 2, 381–413. Zbl 1178.37038 MR 2511746

[331] Nándor Simányi, *Singularities and non-hyperbolic manifolds do not coincide*, Nonlinearity **26** (2013), no. 6, 1703–1717. Zbl 1314.37023 MR 3065929

[332] Nandor Simanyi, *Further developments of Sinai's ideas: The Boltzmann-Sinai Hypothesis*, The Abel Prize 2013–2017 (Helge Holden and Ragni Piene, eds.), Springer, Heidelberg, 2019, pp. 287–298.

[333] Slobodan Simić, *Codimension one Anosov flows and a conjecture of Verjovsky*, Ergodic Theory Dynam. Systems **17** (1997), no. 5, 1211–1231. Zbl 0903.58026 MR 1477039

[334] Ya. G. Sinaĭ, *Topics in ergodic theory*, Princeton Mathematical Series, vol. 44, Princeton University Press, Princeton, NJ, 1994. Zbl 0805.58005 MR 1258087

[335] Yakov G. Sinaĭ, *Dynamical systems with elastic reflections. Ergodic properties of dispersing billiards*, Uspehi Mat. Nauk **25** (1970), no. 2 (152), 141–192. Zbl 0252.58005 MR 0274721

[336] Stephen Smale, *On dynamical systems*, Bol. Soc. Mat. Mexicana (2) **5** (1960), 195–198. Zbl 0121.17906 MR 0141855

[337] Stephen Smale, *A structurally stable differentiable homeomorphism with an infinite number of periodic points*, Qualitative methods in the theory of non-linear vibrations (Proc. Internat. Sympos. Non-linear Vibrations, vol. II, 1961), Izdat. Akad. Nauk Ukrain. SSR, Kiev, 1963, pp. 365–366. Zbl 0125.11601 MR 0160220

[338] Stephen Smale, *Diffeomorphisms with many periodic points*, Differential and Combinatorial Topology (A Symposium in Honor of Marston Morse), Princeton University Press, Princeton, NJ, 1965, pp. 63–80. Zbl 0142.41103 MR 0182020

[339] Stephen Smale, *Differentiable dynamical systems*, Bull. Amer. Math. Soc. **73** (1967), 747–817. Zbl 0202.55202 MR 0228014

[340] Wenxiang Sun and Edson Vargas, *Entropy of flows, revisited*, Bol. Soc. Brasil. Mat. (N.S.) **30** (1999), no. 3, 315–333. Zbl 0954.37005 MR 1726916

[341] Per Tomter, *Anosov flows on infra-homogeneous spaces*, Global Analysis (Proc. Sympos. Pure Math., vol. XIV, Berkeley, Calif., 1968), American Mathematical Society, Providence, RI, 1970, pp. 299–327. Zbl 0207.54502 MR 0279831

[342] Per Tomter, *On the classification of Anosov flows*, Topology **14** (1975), 179–189. Zbl 0365.58013 MR 0377992

[343] Masato Tsujii, *Exponential mixing for generic volume-preserving Anosov flows in dimension three*, J. Math. Soc. Japan **70** (2018), no. 2, 757–821. Zbl 1393.37028 MR 3787739

[344] Paulo Varandas, *A version of Kac's lemma on first return times for suspension flows*, Stoch. Dyn. **16** (2016), no. 2, 1660002, 12. Zbl 1354.37026 MR 3470551

[345] Alberto Verjovsky, *Codimension one Anosov flows*, Bol. Soc. Mat. Mexicana (2) **19** (1974), no. 2, 49–77. Zbl 0323.58014 MR 0431281

[346] Peter Walters, *Invariant measures and equilibrium states for some mappings which expand distances*, Trans. Amer. Math. Soc. **236** (1978), 121–153. Zbl 0375.28009 MR 0466493

[347] Peter Walters, *On the pseudo-orbit tracing property and its relationship to stability*, The structure of attractors in dynamical systems (Proc. Conf., North Dakota State University, Fargo, N.D., 1977), Lecture Notes in Mathematics, vol. 668, Springer, Berlin, 1978, pp. 231–244. Zbl 0403.58019 MR 518563

[348] Peter Walters, *An introduction to ergodic theory*, Graduate Texts in Mathematics, vol. 79, Springer, New York-Berlin, 1982. Zbl 0475.28009 MR 648108

[349] Lan Wen, *On the C^1 stability conjecture for flows*, J. Differential Equations **129** (1996), no. 2, 334–357. Zbl 0866.58050 MR 1404387

[350] Warren White, *An Anosov translation*, Dynamical systems (Proc. Sympos., Univ. Bahia, Salvador, 1971), Academic Press, New York, 1973, pp. 667–670. Zbl 0285.58009 MR 0356134

[351] Isaac Wilhelm, *Celestial chaos: The new logics of theory-testing in orbital dynamics*, Studies in History and Philosophy of Science Part B: Studies in History and Philosophy of Modern Physics **65** (2019), 97–102. Zbl 1409.70016

[352] Amie Wilkinson, *Conservative partially hyperbolic dynamics*, Proceedings of the International Congress of Mathematicians. Volume III, Hindustan Book Agency, New Delhi, 2010, pp. 1816–1836. Zbl 1246.37054 MR 2827868

[353] Jean-Christophe Yoccoz, *Introduction to hyperbolic dynamics*, Real and complex dynamical systems (Hillerød, 1993), NATO Adv. Sci. Inst. Ser. C Math. Phys. Sci., vol. 464, Kluwer Acad. Publ., Dordrecht, 1995, pp. 265–291. Zbl 0834.54023 MR 1351526

[354] Cheng Bo Yue, *On the Sullivan conjecture*, Random Comput. Dynam. **1** (1992/93), no. 2, 131–145. Zbl 0782.58014 MR 1186371

[355] Chengbo Yue, *Smooth rigidity at infinity of the universal cover of closed manifolds of negative curvature*, Complex analysis and its applications, OCAMI Stud., vol. 2, Osaka Munic. Univ. Press, Osaka, 2007, pp. 107–110. Zbl 1141.37014 MR 2405703

[356] Wenmeng Zhang, Kening Lu, and Weinian Zhang, *Differentiability of the conjugacy in the Hartman-Grobman theorem*, Trans. Amer. Math. Soc. **369** (2017), no. 7, 4995–5030. Zbl 1362.37053 MR 3632558

[357] Wenmeng Zhang and Weinian Zhang, *α-Hölder linearization of hyperbolic diffeomorphisms with resonance*, Ergodic Theory Dynam. Systems **36** (2016), no. 1, 310–334. Zbl 1354.37030 MR 3436764

[358] Joseph D. Zund, *George David Birkhoff and John von Neumann: a question of priority and the ergodic theorems, 1931–1932*, Historia Math. **29** (2002), no. 2, 138–156. Zbl 1003.01011 MR 1896971

Index of persons

Index

Index of theorems